T0388342

# TRENDS IN MARITIME TECHNOLOGY AND ENGINEERING

Trends in Maritime Technology and Engineering comprises the papers presented at the 6th International Conference on Maritime Technology and Engineering (MARTECH 2022) that was held in Lisbon, Portugal, from 24-26 May 2022. The Conference has evolved from the series of biennial national conferences in Portugal, which have become an international event, and which reflect the internationalization of the maritime sector and its activities. MARTECH 2022 is the sixth of this new series of biennial conferences.

The book covers all aspects of maritime activity, including in Volume 1: Structures, Hydrodynamics, Machinery, Control and Design. In Volume 2: Maritime Transportation and Ports, Maritime Traffic, Safety, Environmental Conditions, Renewable Energy, Oil & Gas, and Fisheries and Aquaculture.

# Proceedings in Marine Technology and Ocean Engineering



**ABOUT THE SERIES**

The 'Proceedings in Marine Technology and Ocean Engineering' series is devoted to the publication of proceedings of peer-reviewed international conferences dealing with various aspects of 'Marine Technology and Ocean Engineering'. The Series includes the proceedings of the following conferences: the International Maritime Association of the Mediterranean (IMAM) Conferences, the Marine Structures (MARSTRUCT) Conferences, the Renewable Energies Offshore (RENEW) Conferences and the Maritime Technology (MARTECH) Conferences. The 'Marine Technology and Ocean Engineering' series is also open to new conferences that cover topics on the sustainable exploration and exploitation of marine resources in various fields, such as maritime transport and ports, usage of the ocean including coastal areas, nautical activities, the exploration and exploitation of mineral resources, the protection of the marine environment and its resources, and risk analysis, safety and reliability. The aim of the series is to stimulate advanced education and training through the wide dissemination of the results of scientific research.

**BOOKS IN THE SERIES**

Volume 1: Advances in Renewable Energies Offshore, 2019, C. Guedes Soares (Ed.).
Volume 2: Trends in the Analysis and Design of Marine Structures, 2019, J. Parunov and C. Guedes Soares (Eds.).
Volume 3: Sustainable Development and Innovations in Marine Technologies, 2020, P. Georgiev and C. Guedes Soares (Eds.).
Volume 4: Developments in the Collision and Grounding of Ships and Offshore Structures, 2020, C. Guedes Soares (Ed.).
Volume 5: Developments in Renewable Energies Offshore, 2021, C. Guedes Soares (Ed.).
Volume 6: Developments in Maritime Technology and Engineering, 2021, C. Guedes Soares and T.A. Santos (Eds.)
Volume 7: Developments in the Analysis and Design of Marine Structures, 2022, J. Amdahl and C. Guedes Soares (Eds.).
Volume 8: Trends in Maritime Technology and Engineering, 2022, C. Guedes Soares and T.A. Santos (Eds.)

Proceedings in Marine Technology and Ocean Engineering (Print): ISSN: 2638-647X
Proceedings in Marine Technology and Ocean Engineering (Online): eISSN: 2638-6461

PROCEEDINGS OF THE 6TH INTERNATIONAL CONFERENCE ON MARITIME TECHNOLOGY AND ENGINEERING (MARTECH 2022), LISBON, PORTUGAL, 24–26 MAY 2022

# Trends in Maritime Technology and Engineering

Volume 2

*Editors*

C. Guedes Soares

*Centre for Marine Technology and Ocean Engineering (CENTEC), Técnico Lisboa, Universidade de Lisboa, Portugal*

T.A. Santos

*Ordem dos Engenheiros, Portugal*

CRC Press is an imprint of the
Taylor & Francis Group, an **informa** business
A BALKEMA BOOK

*CRC Press/Balkema is an imprint of the Taylor & Francis Group, an informa business*



*Typeset by Integra Software Services Pvt. Ltd., Pondicherry, India*



*Library of Congress Cataloging-in-Publication Data*

A catalog record has been requested for this book

Published by: CRC Press/Balkema
Schipholweg 107C, 2316 XC Leiden, The Netherlands
e-mail: Pub.NL@taylorandfrancis.com
www.routledge.com – www.taylorandfrancis.com

ISBN: 978-1-032-33572-8 (Set Hbk + Multimedia)
ISBN: 978-1-032-33573-5 (Set Pbk)
ISBN: 978-1-003-32025-8 (eBook)
DOI: 10.1201/9781003320258

Volume 1:
ISBN: 978-1-032-33574-2 (Hbk)
ISBN: 978-1-032-33577-3 (Pbk)
ISBN: 978-1-003-32027-2 (eBook)
DOI: 10.1201/9781003320272

Volume 2:
ISBN: 978-1-032-33583-4 (Hbk)
ISBN: 978-1-032-33578-0 (Pbk)
ISBN: 978-1-003-32028-9 (eBook)
DOI: 10.1201/9781003320289

*Trends in Maritime Technology and Engineering – Guedes Soares & Santos (eds)*


# Table of contents

## *Maritime traffic*



## *Safety*

*Environmental conditions*



*Renewable energy*

*Oil & gas*

 *ISBN 978-1-032-33583-4*

# Preface

Since 1987, the Naval Architecture and Marine Engineering branch of the Portuguese Association of Engineers (Ordem dos Engenheiros) and the Centre for Marine Technology and Ocean Engineering (CENTEC) of the Instituto Superior Técnico (IST), University of Lisbon, (formerly Technical University of Lisbon) have been organizing national conferences on Naval Architecture and Marine Engineering. Initially, they were organised annually and later became biannual events.

These meetings had the objective of bringing together Portuguese professionals giving them an opportunity to present and discuss the ongoing technical activities. The meetings have been typically attended by 150 to 200 participants and the number of papers presented at each meeting was in the order of 30 in the beginning and 50 at later events.

At the same time as the conferences have become more mature, the international contacts have also increased and the industry became more international in such a way that the fact that the conference was in Portuguese started to hinder its further development with wider participation. Therefore, a decision was made to experiment with having also papers in English, mixed with the usual papers in Portuguese. This was first implemented in the First International Conference of Maritime Technology and Engineering (MARTECH 2011), but, subsequently, four more MARTECH conferences have been organized, namely in 2014, 2016, 2018 and 2020.

For the Sixth International Conference of Maritime Technology and Engineering (MARTECH 2022), a total of around 200 abstracts have been received and 136 papers were finally accepted.

The Scientific Committee had a major role in the review process of the papers although several other anonymous reviewers have also contributed and deserve our thanks for the detailed comments provided to the authors allowing them to improve their papers. The participation is coming from research and industry from almost every continent, which is also a demonstration of the wide geographical reach of the conference.

The contents of the present books are organized in the main subject areas corresponding to the sessions in the Conference and within each group the papers are listed by the alphabetic order of the authors.

We want to thank all contributors for their efforts and the sponsoring Institutions for their support, hoping that this Conference will be continued and improved in the future.

*C. Guedes Soares & T.A. Santos*

*Trends in Maritime Technology and Engineering – Guedes Soares & Santos (eds)*
*© 2022 copyright the Editor(s), ISBN 978-1-032-33583-4*

# Organisation

## Conference Chairs

Carlos Guedes Soares, *Técnico Lisboa, Universidade de Lisboa, Portugal*
Dina Dimas, *Ordem dos Engenheiros, Portugal*

## Organizing Committee

Tiago A. Santos, *Ordem dos Engenheiros & Técnico Lisboa, Universidade de Lisboa, Portugal*
Ângelo Teixeira, *Técnico Lisboa, Universidade de Lisboa, Portugal*
José Manuel Cruz, *Ordem dos Engenheiros, Portugal*
Manuel Ventura, *Técnico Lisboa, Universidade de Lisboa, Portugal*
Francisco C. Salvado, *Ordem dos Engenheiros, Portugal*
José Gordo, *Técnico Lisboa, Universidade de Lisboa, Portugal*
Pedro Pereira da Silva, *Ordem dos Engenheiros, Portugal*
José Gaspar, *Técnico Lisboa, Universidade de Lisboa, Portugal*
Raúl Caria, *Ordem dos Engenheiros, Portugal*

## Scientific Committee

Ahmed Dursun Alkan, *Yildiz Technology University, Turkey*
Jorgen Amdahl, *NTNU, Norway*
Ermina Begovic, *UNINA, Italy*
Kostas Belibassakis, *NTUA, Greece*
Elzbieta Bitner-Gregersen, *DNV, Norway*
Rui Carlos Botter, *University of São Paulo, Brazil*
Gabriele Bulian, *University of Trieste, Italy*
Nian Zhong Chen, *Tianjin University, China*
Ranadev Datta, *IIT Kharagpur, India*
Nastia Degiuli, *University of Zagreb, Croatia*
Vicente Díaz Casas, *Universidad A Coruña, Spain*
Leonard Domnisoru, *Univ. Dunarea de Jos Galati, Romania*
Soren Ehlers, *Technische Universität Hamburg, Germany*
Saad Eldeen, *Port Said University, Egypt*
Bettar Ould el Moctar, *Univ. of Duisburg-Essen, Germany*
Segen F. Estefen, *UFRJ, Brazil*
Selma Ergin, *Istanbul Technical University, Turkey*

Massimo Figari, *University of Genova, Italy*
Conceição J. Fortes, *LNEC, Portugal*
Yordan Garbatov, *Técnico Lisboa, Portugal*
Sergio Garcia, *University of Cantabria, Spain*
Henrique Gaspar, *NTNU, Norway*
Peter Georgiev, *TU Varna, Bulgaria*
Hercules Haralambides, *Erasmus Univ. Rotterdam, The Netherlands*
Spyros Hirdaris, *Aalto University, Finland*
Zhiqiang Hu, *Newcastle University, UK*
Dominic Hudson, *University of Southampton, UK*
Xiaoli Jiang, *TUDelft, The Netherlands*
Debabrata Karmakar, *National Inst. Techn. Karnataka, India*
Faisal Khan, *Texas A&M University, USA*
Pentti Kujala, *Aalto University, Finland*
Iraklis Lazakis, *Univ. of Strathclyde, UK*
Xavier Martínez-Garcia, *Uni. Politécnica de Catalunya, Spain*
Alba Martínez-López, *Univ. Las Palmas Gran Canária, Spain*
Jakub Montewka, *Gdynia Maritime University, Poland*
Muk Chen Ong, *University of Stavanger, Norway*
Joško Parunov, *University of Zagreb, Croatia*
Apostolos Papanikolaou, *NTUA, Greece*
L. Prasad Perera, *Arctic University of Norway, Norway*
Jasna Prpić-Oršić, *University of Rijeka, Croatia*
Harilaos N. Psaraftis, *DTU, Denmark*
Franco Quaranta, *University of Naples, Italy*
Suresh Rajendran, *IIT Madras, India*
Jonas W. Ringsberg, *Chalmers Univ. of Technology, Sweden*
Cesare Rizzo, *University of Genova, Italy*
Liliana Rusu, *University Dunarea de Jos Galati, Romania*
Xinghua Shi, *Jiangsu University of Science & Technology, China*
António Souto-Iglesias, *Univ. Politécnica de Madrid, Spain*
Nicholas Tsouvalis, *NTUA, Greece*
Michele Viviani, *University of Genova, Italy*
Deyu Wang, *Shanghai Jiao Tong University, China*
Pieter van Gelder, *TU Delft, The Netherlands*
Xinping Yan, *Wuhan University of Technology, China*
Zaili Yang, *Liverpool John Moores Univ., UK*
Xueqian Zhou, *Harbin Engineering University, China*

## Technical Programme & Conference Secretariat

Maria de Fátima Pina, *Técnico Lisboa, Universidade de Lisboa, Portugal*
Sandra Ponce, *Técnico Lisboa, Universidade de Lisboa, Portugal*
Sónia Vicente, *Técnico Lisboa, Universidade de Lisboa, Portugal*

*Trends in Maritime Technology and Engineering – Guedes Soares & Santos (eds)*


# Sponsors

Sponsored by

Supported by

*Maritime transportation & ports*

*Trends in Maritime Technology and Engineering – Guedes Soares & Santos (eds)*


# The effects of operational and environmental conditions in cruise ship emissions in port areas

H. Abreu, V. Cardoso & T.A. Santos
*Centre for Marine Technology and Ocean Engineering (CENTEC), Instituto Superior Técnico, Universidade de Lisboa, Lisbon, Portugal*

ABSTRACT: Cruise ships produce significant levels of emissions, both greenhouse gases and air pollution, primarily as a consequence of propulsion power needs, but also due to the significant installed power of auxiliary engines. These emissions, particularly those occurring inside the area of major cruise ports, have caused significant concern among local populations and attracted the attention and study of environmental organizations worldwide. However, a number of technical parameters related to the ship's operational patterns, such as ship speed, electric power demand and local environmental conditions, namely tidal current speed, have influence on the levels of cruise ship emissions in port areas. This paper presents a methodology for calculating ship emissions based on the details of the ship's route within the port area, describes its implementation in a computational tool and applies it to estimate emissions for a typical cruise ship call within the port of Lisbon, considering the effects of the mentioned operational parameters and environmental conditions. The results allow conclusions to be taken regarding the variability of total emissions under different conditions.

## 1 INTRODUCTION

The shipping industry is currently under significant pressure to increase its sustainability, in line with global trends applicable to all sectors of human activity. Sustainability encompasses a wide range of dimensions, but air emissions arising from exhaust gases are currently the main focus of interest worldwide. Air emissions include two broad categories: greenhouse gases and air pollution gases. The first include mainly, and among others, carbon dioxide ($CO_2$) and methane ($CH_4$), gases known to play a major role in global warming, thus leading to climate change. Air pollution gases include primarily sulphur dioxides ($SO_x$), nitrogen dioxides ($NO_x$) and particulate matter (PM) of various dimensions. These chemical substances are known to have a significant impact on human health. Therefore, while greenhouse gases have a "global" effect, air pollution acts at a more "local" level, raising concern among impacted populations.

The main focus of air emissions studies has been on cargo ships, as shown in the multiple studies carried out by the International Maritime Organization (IMO). Indeed, between 2000 and 2020, four separate studies were carried out by IMO, the latest being IMO (2020). At European level, the European Commission is promoting a strong push towards greening many aspects of European economic and social activities, including transport and mobility, having published in 2019 the broad strategy known as the *European Green Deal*, see European Commission (2019). In parallel, the European Maritime Safety Agency (EMSA) monitors all aspects of shipping and ports sustainability, including air emissions from ships, see EMSA (2021). In 2019, most of the port calls in the EU were made by Ro-pax ships (41 %) and passenger ships (18 %). The first comprises ferries used in short and medium distance routes between European ports, while the second is cruise ships and other smaller passenger ships.

As regards cruise ships, traffic is focused on a few ports such as Barcelona, Civitavecchia, Naples, Dubrovnik, Piraeus, Palma de Maiorca, and others, that receive typically hundreds of calls per year. It is not surprising, therefore, that such cities arise as heavily impacted by air pollution in studies focused specifically on cruise ships, such as that of Transport & Environment (2019). The list in this study includes also in a position of some relevance two Portuguese ports: Funchal (Madeira) and Lisbon. In addition, a significant number of scientific studies have also been published aiming to evaluate the total emissions from ships in general, and in some cases specifically from cruise ships, receiving a common denomination of "emissions inventories". The impact of ship traffic on the air quality of the cities to which the ports belong is of general concern and it is one of the motivations of the present study. Studies on this matter have been performed for the ports of Naples - Toscano *et al.* (2021) - and Venice - Contini *et al.* (2011).

DOI: 10.1201/9781003320289-1

This paper contributes to the literature by providing an innovative analysis of the impact of uncertainties in the ship's technical characteristics (rated engine's power) and in the ship's operational patterns in the calculated levels of cruise ship emissions in port areas, thus influencing emissions inventories. The ship's operational patterns are taken in this paper to include the ship speed and electric power demand, while the local environmental conditions include tidal current speed and temperature conditions (impacting on electrical power demand). These parameters have been chosen as they are well known to have major impacts in terms of power demand onboard cruise ships.

Accordingly, calculated emissions from cruise ships need to be taken with care as actual values may vary significantly with the above parameters. A case study will be presented containing the results of calculations carried out for a cruise ship call in the port of Lisbon. Numerical results are obtained with a computational tool that calculates air emissions from ships considering the specific details of a given call in port.

This paper, apart from this introduction, is organized in the following manner. Section 2 presents a literature review, while section 3 details the numerical model. Section 4 presents the details of the case study, including the definition of a number of scenarios under which emissions are evaluated. Finally, section 5 summarizes the conclusions of the paper.

## 2 LITERATURE REVIEW

Cruise shipping has become very significative in two of the most important Portuguese ports: Lisbon and Funchal. In Lisbon, 2018 was one of the best years for this activity with 339 calls, in which 237 of those are ships in transit, 83 turnarounds, and 19 interporting, as may be seen in Figure 1. These calls resulted from 123 different ships. In the previous year, 2017, the port registered 330 calls, again by 123 different ships, according with the cruise activity reports of the Lisbon Port Authority, APL (2017, 2018). Interestingly, the study on ship emissions carried out by Transport & Environment (2019) reports, for the year of 2017, the number of cruise ships in Lisbon to be 115 and the total port call time to be 7,953 hours, resulting in little more than 69 hours per ship (assuming that each ship calls about 3 times in Lisbon per year, the average port call time would be around 23 hours). However, APL (2019) reports that the average port call time was only 16h39m, a value in line with the common operational pattern of cruise ships: arrival early in the morning and departure in the evening. Therefore, it is not entirely clear what is the source and precise meaning of the numbers in the ship emissions study mentioned above.

Considering the importance of the activity of cruise ships in Lisbon, Santos *et al.* (2021a) have studied this topic for the Atlantic coast of the Iberian Peninsula. In order to support such study a database of cruise ship technical characteristics has been developed that includes over 200 ships of various sizes, enabling Santos *et al.* (2021b) to proceed with a characterization of the cruise ship fleet calling in the port of Lisbon, based on the identification of the ships provided in APL (2018). Such information could in the future support a detailed evaluation of cruise ship emissions in this port.

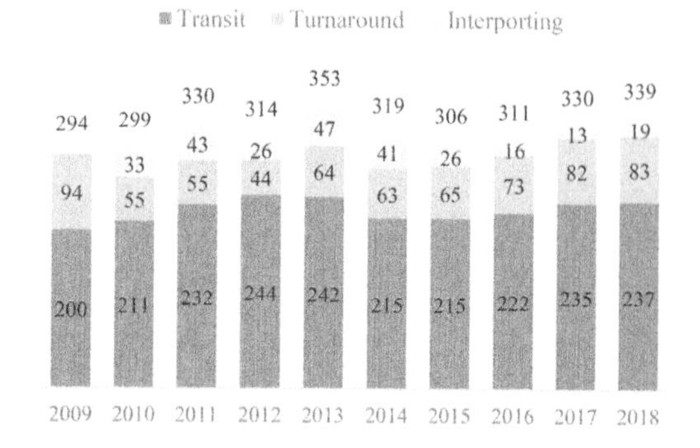


Figure 1. Cruise ship calls in the port of Lisbon (2009-2018) (Source: APL (2018)).

The literature on ship emissions inventories is quite rich and even applications to cruise ships in ports are common, with Trozzi *et al.* (1995) being one of the first examples. Some of the earlier studies include Howitt *et al.* (2010), who focused on the New Zealand case, and Hulskotte *et al.* (2010), analyzing the emissions occurring at berth, a topic already empirically examined by Cooper (2003). The impact of emissions in British Columbia was studied by Poplawski *et al.* (2011). The number of calls in European ports has determined also some interest in Europe on this topic, with studies for Piraeus (Greece) carried out by Tzannatos (2010a,b). Castells Sanabra *et al.* (2014) evaluated emissions in Spanish ports arising from cruise ships, assuming average values for main and auxiliary engine power, as necessary. Maragkogianni and Papaefthimiou (2015) extended air emissions studies to a number of significant cruise ports in Greece. More recently, Tichavska and Tovar (2015a,b) studied the air emissions and their impact in the major Spanish cruise port of Las Palmas.

Many of these studies are based on the bottom-up approach that uses AIS data as a basis for determining emissions. AIS data must be complete, particularly with no geographical gaps and with ship draught indication. The effects of waves, wind and currents are also significant but generally cannot be accounted for accurately due to lack of information. A considerable volume of technical information about the ships needs to be available namely main and auxiliary engine powers, but ship databases are often incomplete. Even if these details are known, there remains the issue of the main and auxiliary engine power usage onboard vessels in the various phases of operation: transit (cruising), maneuvering and hoteling (at berth). These and still other uncertainties affect the accuracy of emissions estimates, with further details being given in Tichavska and Tovar (2015b) and Psaraftis and Kontovas (2021).

Considering these uncertainties, the interest that cruise activity presents for the port of Lisbon and the mentioned inaccuracies noted in some studies, it was decided to undertake numerical calculations of ship emissions for one port call, while addressing some of the most relevant uncertainties mentioned above, as applicable to the specific conditions of the port of Lisbon. The objective is to evaluate the effects of such uncertainties in the calculated air emissions at the simple level of one single cruise ship port call. The calculations use freely available data on the ship's trajectory within the port area. The technical characteristics of the ship under study are entirely known and the electrical load balance is also available from Cuenca (2018). Emissions are evaluated using a numerical model developed by Ramalho and Santos (2021a,b) for application to intermodal transport chains, which was adapted to the specific case of cruise ship calls in port areas.

## 3 NUMERICAL MODEL FOR SHIP EMISSIONS

The total emissions of $SO_x$, $NO_x$, PM and $CO_2$ in the port area are given by the sum of emissions in each link that composes the trajectory of the ship in the port area and the emissions at berth. The basic procedure for doing this consists in computing the fuel consumption by using link characteristics, such as distance, speed and ship specifications (installed power, service speed). In the end, the mass of pollutants emitted is obtained by applying emission coefficients as will be described in this section.

For the vessel's main engine, the fuel consumption $FC_{ME}$ is calculated for each link $k$ part of path $j$ and pair origin/destination (o/d) $i$. The numerical tool may assess multiple paths and pairs o/d if needed. Fuel consumption depends on the engine power demand $P_{EF}$, in kW, on the specific fuel consumption of the main engine $SFC_{ME}$, in g/kWh, on the vessel's speed in the link, $LSpeed_{ijk}$, in km/h and on the link's length $LDist_{ijk}$ in km:

$$FC_{ME,ijk} = SFC_{ME} \times P_{EF} \times \frac{LDist_{ijk}}{LSpeed_{ijk}} \quad (1)$$

Additionally, for a cruise ship, the number of consumables does not change significantly along the trajectory within the port, meaning that the displacement is approximately constant in all links and the engine power demand may be simplified to a cubic relation in terms of the speed.

The total power consumed onboard, however, is not only composed by the power required to sail the vessel, but also by the power consumed by onboard facilities, normally called hotel power or hotel load. In this ship, this power comes from a Power Take-Off (shaft alternator). The power taken out of the PTO is considered constant for a maneuvering profile and linearly dependent on the vessel's speed in the transit condition:

$$FC_{PTO,ijk} = SFC_{ME} \times SssPTO \times LoadPTO \times \frac{LDist_{ijk}}{LSpeed_{ijk}} \quad (2)$$

in which $SssPTO$ is the total PTO power installed onboard and $LoadPTO$ is the usage percentage, function or not of the link speed.

For navigating condition, the PTO percentage is based on a linear regression using two known hotel power conditions, taken from Cuenca (2018), as shown in Table 1. The data belongs to the vessel considered in Cuenca (2018), which is the same taken for this study. Its characteristics are further described in Table 2.

Table 1. Hotel power and power take-off utilization.

| Speed [knots] | Hotel power [kW] | PTO utilization [%] |
|---|---|---|
| 11 | 1,303 | 0.33 |
| 15 | 1,391 | 0.35 |

Each link that composes the path of the ship in the port area has to be classified as regards which operating condition the ship is in. The adopted criterium for classifying links was the percentage of use of the installed power, called load power $Load_{ij}$. When the main engine power computed by IMO formulation is lower than 25% of the installed power, the vessel is in maneuvering profile and the hotel and thrusters' power is equal to 2,875 kW, as mentioned in Cuenca (2018).

In addition to the main engine fuel consumption, auxiliary engines can be activated to supply the required power or work as spare power sources onboard. Similarly, the fuel consumption of the auxiliary engines can be calculated by the equation below, depending on the time and engine power demand at sea in kW. By assumption, auxiliary engines' specific fuel consumption is not dependent on their load, and so $SFC_{AE}$ is assumed constant, but just a percentage $LoadAE$ of the total auxiliary engine power $SssPAE$ is used at sea operating activity.

$$FC_{AE,ijk} = SFC_{AE} \times LoadAE \times SssPAE_{ij} \times \frac{LDist_{ijk}}{LSpeed_{ijk}} \quad (3)$$

In a similar way to the previously used percentages, *LoadAE* is dependent on which operating condition the vessel is considered to be in the link. In this paper, these values assume two different sets, one related to the real use of the total power installed as described in Cuenca (2018) and another based on regressions of similar vessels.

To conclude, the fuel consumption at berth considers the use of main and auxiliary installed power during the time at berth:

$$FC_{Port} = (SFC_{AE} \times LoadAE_{port} \times SssPAE_{ij} + SFC_{ME} \times LoadME_{port} \times SssPME_{ij}) \times TPort_{ij} \quad (4)$$

The amount of emissions per pollutant in each link $k$ of path $j$ and the emissions at berth are obtained by multiplying the fuel consumption by the emission factors in g/g. Emissions per passenger are computed by dividing the total emissions by the vessel's passenger capacity.

Differently from the other pollutants, the emission method for computing $NO_x$ and PM emissions does not use direct coefficients of mass of pollutant per mass of fuel consumption. Thus, the conversion of $NO_x$ and PM energy-based emission factors in g/kW to fuel-based emission factors in g/g applies the $SFC_{base}$ as:

$$EF_f = \frac{EF_e}{SFC_{base}} \quad (5)$$

On the other hand, $PM_{10}$ energy-based emission factors $EF_e$ are also indirect functions of the engine load and fuel sulphur content and $NO_X$ factor depends on the engine Tier and rotation speed. However, above 10% MCR these factors can be considered constant. When the engine load is below 10% MCR, the emission factors increase due to lower combustion efficiency. Adjustment factors that represent this change are included.

## 4 CASE STUDY

### 4.1 *Cruise ship call*

The objective of this case study is to illustrate how the numerical calculation method described above may be used to estimate cruise ship emissions (or, in fact, emissions of any ship type) during a typical call in a port. Let us take as an example a port call carried out in September 2021 in the port of Lisbon, shown in Figure 2. The emissions are to be estimated for the complete trajectory of the ship throughout the estuary of river Tagus, from the moment the ship enters the area of jurisdiction of the port authority (which starts at the mouth of the river) to the moment it leaves this area at the same point. The ship entered the port at approximately 06:29 local time, thus in the early hours of the day (as usual for these ships) and departed the following day at 00:15 local time, which is later than usual in this port, since most cruise ships depart in the evening just before night falls, as shown in Santos *et al.* (2021). The ship's trajectory through the port and its speed profile was taken from website VesselFinder, which provides AIS data.

The information in this site does not have a high degree of definition (short time lapses), as provided directly by the AIS system, but is still detailed enough to define accurately the navigational details

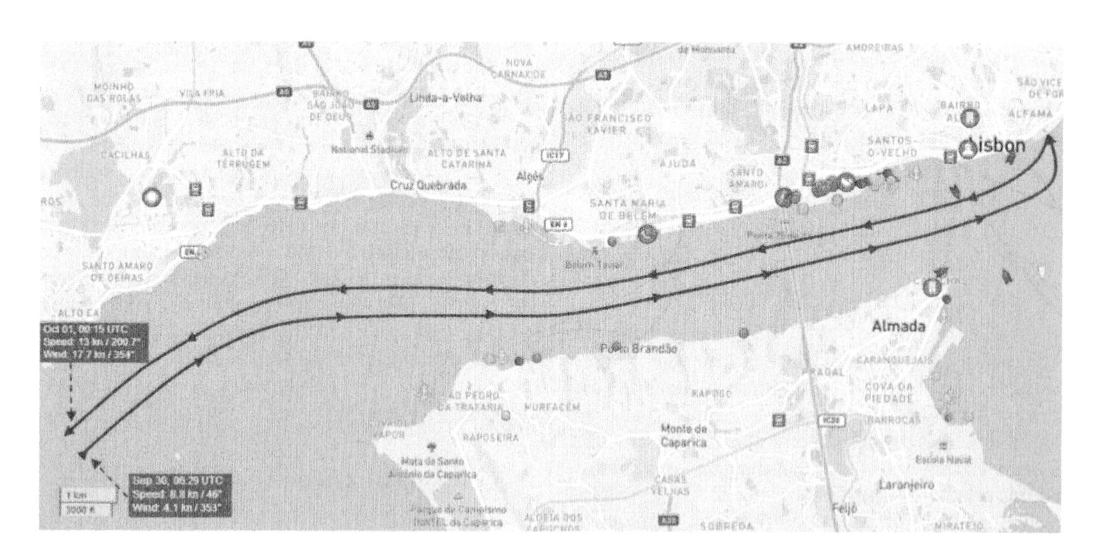


Figure 2. Cruise ship call in the port of Lisbon (Source: own elaboration based on data from VesselFinder website).

during the port call. It should be pointed out that is not the purpose of this paper to consider in detail all the technical aspects connected with the extraction and analysis of large volumes of AIS data, but merely to use a sufficient amount of such data to characterize a call of a ship in a given port.

In the case of this call, a total of 25 points were available, fully characterizing the path within river Tagus of the cruise ship from the moment of arrival at the mouth of the river to the moment of departure. The total path measures about 35.5km, equivalent to 16.19 nautical miles (nm) and this implies that, on average, there are 0.77 nm between every two consecutive waypoints. Most of these waypoints are provided with a time interval of 5 minutes. This degree of precision is considered sufficient to estimate emissions throughout this port call and assess the likely variations in emissions under different assumptions.

Figure 3 shows the speed profile during the port call, being possible to see an initial phase at approximately 6 knots, probably corresponding to a slow down while waiting for pilot arrival, followed by navigation at 14 knots up the river to the cruise terminal. Speed is then progressively reduced while approaching the berth and during maneuvering. Upon departure, the speed is quickly increased to 13 knots and is kept at approximately that level until leaving port, except for a waypoint, which presents a speed slightly below 7 knots, corresponding to a decrease of speed as needed for disembarking the pilot.

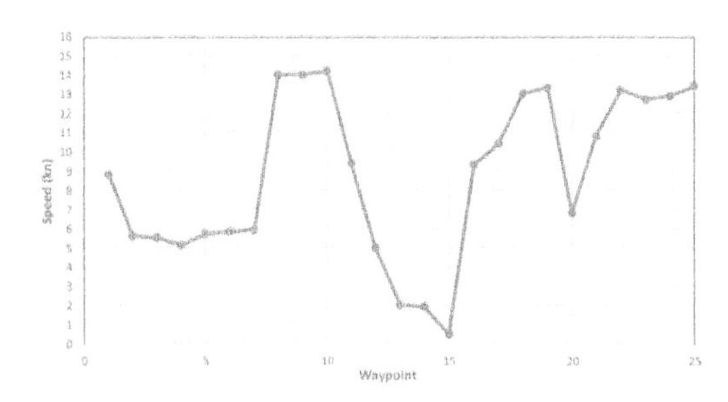


Figure 3. Speed profile throughout the port call.

Summarizing, the ship arrived at the Lisbon Cruise Terminal at about 07:55 and finally left the same terminal at 23:10 of the same day. It took about 1:25 hours to reach the cruise terminal in the morning and only about 45 minutes at night to reach the port entrance. The weather conditions were good, with wave height less than 1 meter and light wind ranging from 4.1 knots in the evening (entry in port) to 6.6 knots at night (departure from port). These conditions are not believed to affect significantly the propulsion power needs.

The cruise ship carrying out this particular port call is relatively small, belonging to the so-called expedition/niche market segment, and has a capacity for 200 passengers. Table 2 summarizes the technical characteristics of the ship, which is diesel-mechanical and fitted with two shafts and propellers.

Table 2. Cruise ship technical characteristics.

| | |
|---|---|
| Length over all (m) | 126.0 |
| Length between perpendiculars (m) | 113.2 |
| Breadth (m) | 19.0 |
| Depth (m) | 7.0 |
| Summer draught (m) | 4.75 |
| Gross tonnage | 9,934 |
| Main engines power (kW) | 2x2,665 |
| Service speed (kn) | 16 |
| Auxiliary engines power (kW) | 1x1,900 |

The ship's machinery consists of two main engines with a rated power of 2665 kW and two diesel generators with a rated power of 950 kW. There are also two power take offs rated at 2000 kW each. Both main engines and generators operate on marine diesel oil (MDO). Table 3 shows the electric load balance in various conditions, including transit, maneuvering, berth and berth in hot conditions. The last condition has been estimated from the at berth condition, but with two chiller units working at 90% of average power rather than just one chiller unit at 60%, as is the case in the at berth condition. This condition is meant to reproduce conditions on a typical hot day in Lisbon and does not even resort to the third chiller unit with which this ship is equipped. It is a conservative condition from the point of view of power demand in port and thus leads to higher emissions. Table 3 also shows the load in the main engines and generators (percentage for port and starboard units) in various conditions.

Table 3. Propulsion power and electric load in different conditions of operation.

| | Transit 5-11 kn | Transit 11-15 kn | Maneuv. | Berth | Berth Hot |
|---|---|---|---|---|---|
| Propulsion (kW) | 1500 | 3500 | 1100 | 0 | 0 |
| Hotel+thrust. (kW) | 1303 | 1391 | 2875 | 864 | 1194 |
| Main Engines (%) | 82-29 | 96-96 | 79-79 | 0-0 | 0-0 |
| Diesel Gens (%) | 0-0 | 0-0 | 0-0 | 91-0 | 63-63 |

### 4.2 *Scenario definition*

Based on this information, a number of scenarios will be analyzed that consider different ship speed, power demand for hotel loads and local environmental conditions (tidal current). Table 4 indicates the scenarios under consideration, the first one being the actual situation experienced in this port call, while the remaining scenarios correspond to variations where certain critical operational and environmental conditions are changed. Scenario 1 corresponds to the existing ship call with the speed profile given in Figure 3, hotel loads as per Table 3 and assuming that the ship was at berth with the hotel load of 864 kW throughout the day.

Table 4. Summary of scenarios for cruise ship call.

| | |
|---|---|
| Scenario 1 | Original port call characteristics |
| Scenario 2A,2B | Estimated ship characteristics |
| Scenario 3 | Estimated electrical load |
| Scenario 4 | Speed limited to 10 knots |
| Scenario 5 | Hotel power increased to 1,194 kW |
| Scenario 6 | With 3 knots tidal current |

Scenario 2 is based on scenario 1 but assuming that the technical characteristics of the ship were not known, so that an estimate of these would be necessary. For this purpose, the database of cruise ships previously used in Santos *et al.* (2021a, b) is used to evaluate the main and auxiliary engine powers of a cruise ship of this size (gross tonnage). This statistical approach is frequently used in shipping emission inventories when the precise technical details of the ships are not known. The mentioned database allows the conclusion that propulsion power for ships of approximately 10,000 GT (generally diesel-mechanic ships) could be anywhere between 6,000 kW and 10,000 kW, depending on the design service speed and age of the ship. The same database shows that auxiliary power for these ships could range between 2,000 kW and 3,000 kW. Based on these values, in this scenario 2, it is considered that the ship may have, respectively, 6,000 kW and 2,000kW of main and auxiliary power (Scenario 2A) or 10,000 kW and 3,000 kW (Scenario 2B).

Scenario 3 considers that the electric load in the various operating conditions is not known, due to the unavailability of electric load balance, and therefore the standard percentages found in the literature for the power demand in typical conditions (sailing, maneuvering, berthing) are used rather than precise values from the electric load balance.

Scenario 4 corresponds to the same ship trajectory in port but with the ship's speed restricted to 10 knots in all those waypoints where it was exceeded. It is worth pointing out that the port of Lisbon authority, in its safety regulations, see APL (2021), does not specify maximum limits to ship speed in the river. It is thus an option of the ship's captain (upon the proposal of the pilot) to sail at more or less speed, depending on specific weather conditions and operational needs.

Scenario 5 corresponds to a situation as in scenario 1 but in which the hotel power demand while at berth is increased to 1,194 kW, representing the conditions of a hot summer day. No increase in power demand was considered while sailing in the river as this is carried out in the early morning and late-night hours, when temperatures are lower.

Scenario 6 corresponds to a situation again similar to Scenario 1 but in which the ship, when sailing up and down the river, is assisted by tidal current, thus representing an optimistic (but realistic) situation. In river Tagus, tidal currents may be as large as 4.85 knots in certain seasons in the port access channel, see Santos (2017). However, it was decided to consider in the calculations a tidal current at a moderate level of 3 knots, considered representative of the overall average across the path of the ship in normal tides.

In all scenarios, the effects of fouling and weather conditions are neglected as the ship is new (little additional drag from fouling) and weather effects (additional resistance) within the estuary are negligible.

### 4.3 *Results for scenarios 1, 2A, 2B and 3*

In scenarios 2A, 2B and 3, the technical characteristics of the cruise ship are estimated. For scenario 2, the installed power of the vessel and the engine loads are unknown. The installed powers considered for the main and auxiliary engines are, respectively, 6,000kW and 2,000 kW for scenario 2A, and 10,000kW and 3,000kW for scenario 2B. For scenario 3, the main and auxiliary powers are known but the load factors are unknown.

For all these scenarios, the load factors were estimated based on the electrical load balance of two passenger vessels and two Ro-Pax vessels as well as information reported in Cali Air Resources Board (2011). Table 5 shows the loads in the main and auxiliary engines considered in scenarios 2A, 2B, and 3 while in transit, at berth (port) and maneuvering.

Table 5. Estimated load factors of main and auxiliary engines used in scenarios 2A, 2B and 3.

| Load factors | Transit | Port | Maneuvering |
|---|---|---|---|
| Main Engine | 0.80 | 0.20 | 0.20 |
| Auxiliary Engine | 0.45 | 0.39 | 0.79 |

Figure 4 presents the total power comparison between Scenarios 2 and the real condition (Scenario 1) in which the line represents the speed in each link and the bars the total power used in that link. The links cover the transit of the ship when entering and leaving the port. The maneuver line indicates from which speed the transit condition starts to be considered. Given the significantly higher power considered in Scenario 2B, the effective power for each link has distant values from the ones in Scenario 1. Scenario 2A, on the other hand, shows closer values to Scenario 1, given the closer installed power estimation to the real case. In this case, the difference in effective power comes mainly from the estimated loads of main and auxiliary engines.

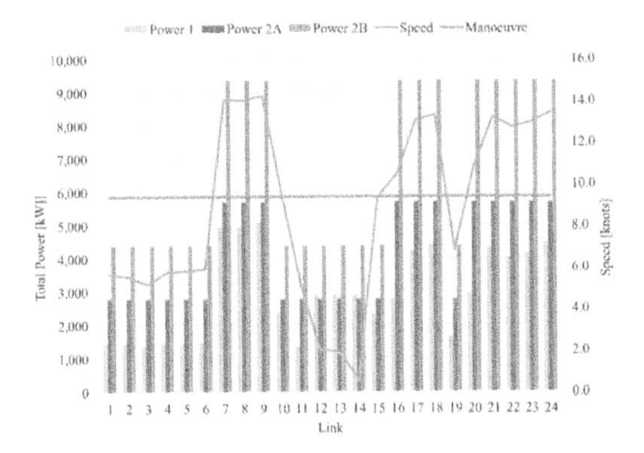


Figure 4. Power and speed for Scenarios 1, 2A and 2B.

For these three scenarios, Figure 5 presents the translation of the power results into emissions along the complete route into the port, including also emissions at berth. The difference in emissions is justified by the variation of loads between the original electrical load and the one assumed for these cases. For the sake of example, in Scenario 1, the auxiliary engines do not operate in transit, while in Scenarios 2 and 3 they operate with a 45% load. For maneuvering profiles, auxiliary engines are not used in Scenario 1, but they operate with 79% load in Scenarios 2 and 3. Since in Scenario 3 the value of the installed power is known and used as input, the emissions results are closer to Scenario 1, but still more than the double.

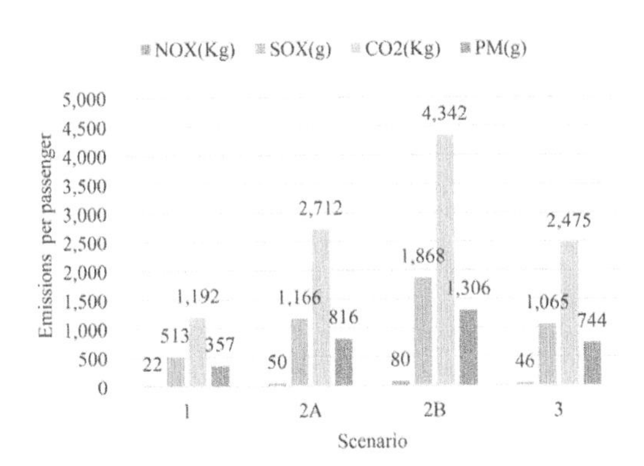


Figure 5. Air pollutants emissions for Scenario 2A, 2B and 3.

## 4.4 *Results for scenario 4*

Scenario 4 simulates the case in which the vessel speed would be limited to 10 knots inside the port area. Since the power is proportional to the cube of the speed, the power required to displace the ship is lower for those links where the speed is limited. The same type of effect is considered for the hotel power consumption, but with a linear dependence of speed. Figure 6 presents the total power compared with the basic condition (Scenario 1) in which the lines represent the speed in each link and the bar the total power used in this link. Since no difference is considered for the time at berth or total power installed (main or auxiliary engines) compared to the basic condition, the power consumption berthed is the same from Scenario 1.

The comparison between the emission levels outside berth on Scenario 1 and the case in which the vessel speed is limited to 10 knots can be seen in Figure 7. All emissions for the limited speed case were decreased at least 10% and the highest reduction was noticed in $SO_x$ of about 12% emission. Differently than the case with favorable current (see below), the range of main engine loads are similar to Scenario 1 and the energy-based pollutants – PM and $NO_x$ - have been reduced too.

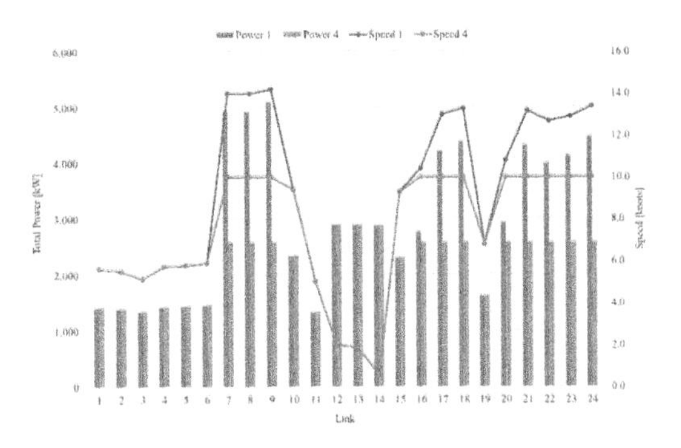

Figure 6. Power and speed for Scenario 4.

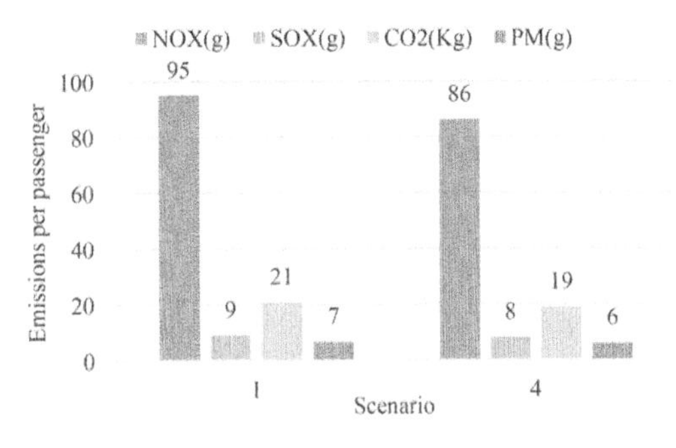


Figure 7. Air pollutants emissions outside berth for Scenario 4.

## 4.5 *Results for scenario 5*

Scenario 5 simulates a day in which more power is required to maintain the cruise ship facilities (air conditioning) running in better (hotter) climate conditions for passengers and crew onboard. Thus, no changes are made for the transit and maneuvering profile when compared to the original case, but the power consumed at berth is increased to 1,194 kW, while in the basis condition it was 864kW. The 38% rise in power is comparable to the power increase that one of the passenger vessels mentioned in the report of California Air Resources Board (2011) presents for hoteling in summer conditions.

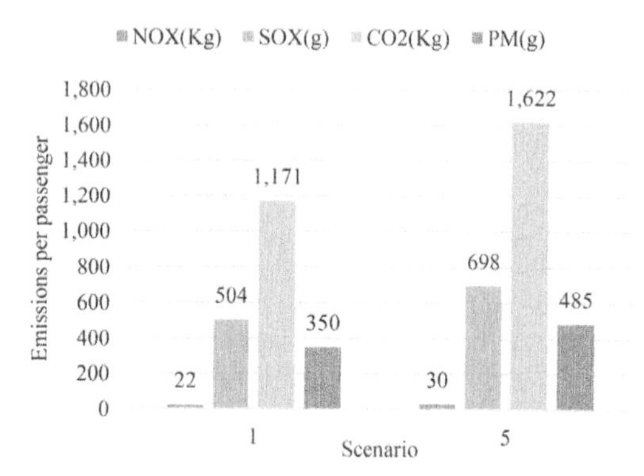


Figure 8. Air pollutants emissions at berth for Scenario 5.

Figure 8 shows the emissions at berth for Scenarios 1 and 5. Notably, there is an increment of 38% for all pollutants considered. Since all power used when the vessel is anchored comes from the auxiliary engine, the difference in the emissions is directly related to the increment of 38% of the power used, explaining the constant difference from all pollutants.

## 4.6 *Results for scenario 6*

In Scenario 6 a favorable current induces a speed reduction of 3 knots, leading to a significant reduction in the required power for propulsion. Figure 9 presents the power consumption for the basic condition (Scenario 1) and the current condition (Scenario 6). In addition, it is presented the speed registered in each link for both scenarios. For links 12, 13 and 14, however, the speed was considered the same since it was already lower than 3 knots in the original scenario.

It is possible to notice that the main engine's effective power is significantly reduced for those links where the speed is different (Figure 9). The most significant reduction occurs in those links

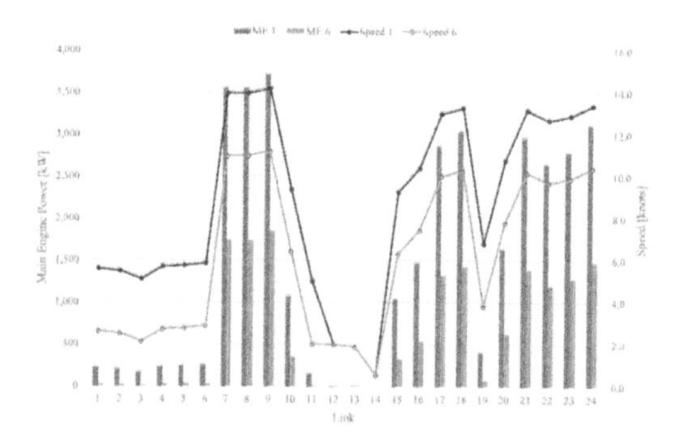

Figure 9. Power and speed for scenario 6.

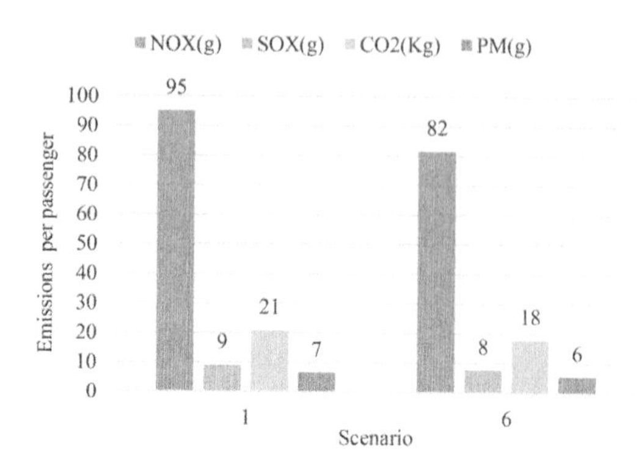


Figure 10. Air pollutants emissions outside berth for Scenario 6.

where the speed has been decreased from 13-14 knots to 10-11 knots, leading to a decrease in propulsive power consumption of 1884 kW, or 51%.

Figure 10 shows that, in a favorable current condition, the emissions of all pollutants were reduced at least by 35% while the vessel is sailing in the channel access. In general, the reduction percentage was close for all pollutants, but $NO_x$ emission has presented the highest reduction with 35.9% decrement, followed by 35.1%, 35%, and 34.9% for PM, $CO_2$, and $SO_X$, respectively.

## 4.7 *Summary of results*

Figure 11 summarizes the results obtained for the various scenarios as regards total emissions (transit, maneuvering, and at berth) for the various pollutants. Most of these emissions occur at berth.

It may be seen that there is a wide variation of emissions (for all pollutants) between the various scenarios. When the ship's technical characteristics (main and auxiliary engines rated power) are not known estimates of air pollutant emissions rely on estimates of these characteristics and cruise ships vary significantly in such technical characteristics. In scenarios 2A and 2B the two possible extremes for main and auxiliary engine power, as taken from a ship database, were considered and this is shown to significantly affect the calculated emissions. Results in scenario 2A are closer to the basis condition (for which all data about the ship is known) but still much higher.

In scenario 3 the hotel load is calculated using the standard load factors in the literature and values are generally twice the value of the basis scenario since more power is being consumed at berth. Scenario 5 corresponds to a hotter day in port, with more hotel load due to air conditioning system power demand, and emissions are for that reason higher than in the basis scenario.

Finally, in Scenarios 4 and 6, the effects of a decreased speed in transit and a certain degree of current assistance in transit are shown to have little effect on the total emissions due to the fact the emissions mainly occur at berth.

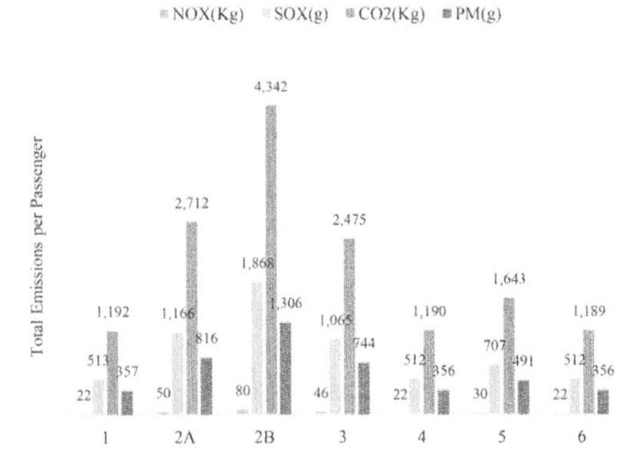


Figure 11. Air pollutants emissions summary for the six scenarios.

## 5 CONCLUSIONS

Considering the results presented in the previous sections, it can be seen that there is a wide variation of emissions (for all pollutants) between the various scenarios, but most of the emissions consistently occur during the period the ship is berthed. For the scenarios in which the ship's technical characteristics (main and auxiliary engines rated power) are unknown, the estimates of air pollutant emissions rely on estimates of these characteristics. These estimates are found to significantly affect the total calculated emissions.

The environmental conditions (hot day) have also been found to significantly affect the total air pollutant emissions, and this is especially important considering that most emissions occur at berth. Thus, any uncertainty related to the real hotel load while at berth significantly affects the accuracy of calculated emissions.

The conditions that consider a 10 knots speed limit (Scenario 4) and a favorable current of 3 knots (Scenario 6) do show benefits on the total emissions. However, as the time and power spent at berth are the same as for the basic condition (Scenario 1), the benefit is not that significant considering the dominance of emissions at berth.

It is also possible to state that, due to relatively high time spent at berth, variations on the vessel's operating conditions (speed, tide assistance) would only show proportionally relevant benefits if they are related to a reduction of berth time or berth hotel load. This conclusion is not exclusively relevant for the port of Lisbon, as it can be extrapolated to any port subjected to cruise ship calls with significant time at berth.

In any case, the results of this case study indicate that port emission inventories are significantly affected by several uncertainties related to technical and environmental parameters, leading potentially to significantly different calculated emissions. In particular, it is quite frequent that ship databases do not include all existing cruise ships and that, even the ship is included, technical characteristics related to the propulsion and auxiliary power generation machinery are not complete or accurate. Furthermore, even the operational procedures used onboard by the crew may also influence emissions.

These conclusions are relevant for most cruise ports and advises a word of caution when considering numbers provided for cruise ship emissions in port, especially if the methodologies, assumptions and available ship technical information used to generate such numbers are not clearly outlined.

Finally, the project of the Lisbon Port Authority to install in the cruise terminal electrical facilities that would be able to meet the energy demands of cruise ships (shore supply) appears to be an appropriate one. Such facilities have typically a cleaner power source (lower emissions) and can provide most – if not all – electrical needs of the ships, which would significantly reduce the environmental footprint of berthed vessels. This finding is also applicable to most cruise ports.

## ACKNOWLEDGEMENTS

This research was funded by Fundação para a Ciência e a Tecnologia (FCT) under grant number PTDC/ECI-TRA/28754/2017 and is in line with the Strategic Research Plan of the Centre for Marine Technology and Ocean Engineering (CENTEC), which is financed by FCT under contract UIDB/ UIDP/ 00134/2020. The authors would like to thank the contribution of Pedro Martins in the gathering of numerical data supporting this study.

## REFERENCES

APL (2017, 2018, 2019). Cruise ship traffic – activity report (in Portuguese), Lisbon Port Authority, Lisbon, Portugal.

APL (2021). Port safety regulations. Lisbon Port Authority, Lisbon, Portugal.

California Air Resources Board (2011). Emissions Estimation Methodology for Ocean-Going Vessels, Planning and Technical Support Division Report, Los Angeles, USA.

Castells Sanabra, M., Usabiaga Santamaría, J.J., Martínez De Osés, F.X., (2014). Manoeuvring and hoteling external costs: enough for alternative energy sources? Maritime Policy & Management, Vol. 41, pp. 42–60.

Contini, D., Gambaro, A., Belosi, F., De Pieri, S., Cairns, W. R. L., Donateo, A., … & Citron, M. (2011). The direct influence of ship traffic on atmospheric PM2. 5, PM10 and PAH in Venice. Journal of Environmental Management, 92(9), 2119–2129.

Cooper, D., (2003). Exhaust emissions from ships at berth. Atmospheric Environment, Vol. 37, pp. 3817–3830.

Cuenca (2018). Study of the energy consumption in cruise ships. Analysis of the power management and fuel consumption, MSc thesis. University of Sevilla, Spain.

EMSA (2021). European Maritime Transport Environmental Report 2021, European Maritime Safety Agency, Lisbon, Portugal.

European Commission (2019). The European Green Deal, Communication from the European Commission to the European Parliament, European Council, the European Economic and Social Committee and the Committee of the Regions, Brussels, Belgium.

Howitt, O.J.A., Revol, V.G., Smith, I.J., Rodger, C.J. (2010). Carbon emissions from international cruise ship passengers' travel to and from New Zealand. Energy Policy, Vol. 38(5), pp. 2552–2560.

Hulskotte, J.H.J., Denier van der Gon, H.A.C. (2010). Fuel consumption and associated emissions from seagoing ships at berth derived from an on-board survey. Atmospheric Environment, Vol. 44, pp. 1229–1266.

IMO (2020). Fourth GHG Study, International Maritime Organization, London, United Kingdom.

Maragkogianni, A., Papaefthimiou, S. (2015). Evaluating the social cost of cruise ships air emissions in major ports of Greece. Transportation Research, Part D, Vol. 36, pp. 10–17.

Poplawski, K., Setton, E., McEwen, B., Hrebenyk, D., Graham, M., Keller, P. (2011). Impact of cruise ship emissions in Victoria, BC, Canada. Atmospheric Environment, Vol. 45, pp. 824–833.

Psaraftis, H.N., Kontovas, C.A. (2021). Decarbonization of Maritime Transport: Is There Light at the End of the Tunnel? Sustainability, 13, 237.

Ramalho, M.M., Santos, T.A. (2021a). Numerical Modeling of Air Pollutants and Greenhouse Gases Emissions in Intermodal Transport Chains. Journal of Marine Science and Engineering, 9, 679.

Ramalho, M.M., Santos, T.A. (2021b). The Impact of the Internalization of External Costs in the Competitiveness of Short Sea Shipping. Journal of Marine Science and Engineering, 9, 959.

Santos, R.G. (2017). Dinâmica num sector distal do Estuário do Tejo com base em dados oceanográficos, sedimentológicos e de nanoplâncton calcário (in Portuguese). MSc thesis, Universidade de Lisboa, Portugal.

Santos, T.A., Martins, P.A., Guedes Soares, C. (2021a). Cruise shipping on the Atlantic coast of the Iberian Peninsula. Maritime Policy & Management, Vol 48(1), pp. 129–145.

Santos, T.A., Martins, P., Guedes Soares, C. (2021b). Characterization of the cruise ship fleet calling in the port of Lisbon, in *Developments in Maritime Technology and Engineering*, Edited by C. Guedes Soares, T.A. Santos, pp. 91–100, CRC Press, London.

Tichavska, M., Tovar, M. (2015a). Environmental cost and eco-efficiency from vessel emissions in Las Palmas Port. Transportation Research, Part E, Vol. 83, pp. 126–140.

Tichavska, M., Tovar, M. (2015b). Port-city exhaust emission model: an application to cruise and ferry operations in Las Palmas Port. Transportation Research, Part A, Vol. 78, pp. 347–360.

Toscano, D., Murena, F., Quaranta, F., & Mocerino, L. (2021). Assessment of the impact of ship emissions on air quality based on a complete annual emission inventory using AIS data for the port of Naples. Ocean Engineering, 232, 109166.

Transport & Environment (2019). One Corporation to Pollute Them All - Luxury cruise air emissions in Europe. European Federation for Transport and Environment AISBL, Brussels, Belgium.

Trozzi, C., Vaccaro, R., Nicolo, L. (1995). Air pollutants emissions estimate from maritime traffic in the Italian ports of Venice and Piombino. The Science of the total Environment, Vol. 169, pp. 257–263.

Tzannatos, E. (2010a). Ship emissions and their externalities for the port of Piraeus – Greece. Atmospheric Environment, Vol. 44, pp. 400–407.

Tzannatos, E. (2010b). Cost assessment of ship emission reduction methods at berth: the case of the Port of Piraeus, Greece. Maritime Policy & Management, Vol. 37, pp. 427–445.

*Trends in Maritime Technology and Engineering – Guedes Soares & Santos (eds)*


# Improving the environmental performance of shipping and maritime transport - Highlights of the maritime emissions workshop

E. Altarriba, S. Rahiala, T. Tanhuanpää & M. Piispa
*South-Eastern Finland University of Applied Sciences, Kotka, Finland*

ABSTRACT: The European Commission's legislative proposal entitled "Fit for 55" sets out guidelines for the maritime sector in the Member States. In Finland, foreign trade is exceptionally dependent on smooth maritime transport: A requirement to stay on schedule with low-emissions approach, including competitive transport pricing and periodically challenging ice conditions makes for a challenging combination. These challenges are being studied in project called the "MEPTEK". During this project, a maritime emissions workshop was arranged on 6 October 2021. The attendees were mainly participants from different authorities, research institutions, and representatives of the commercial maritime sector. The workshop highlighted the environmental impact of scrubbers, emission regulations and monitoring methods, EU maritime proposals on emission trading and fuels, winter shipping system, and emission managements proposals from the perspective of authorities, researchers, and shipowners. The objective of this paper is to summarize the highlights of the workshop, from a scientific, legal, and practical perspective.

## 1 INTRODUCTION

### 1.1 *Aim and scope*

The aim of this paper is to summarize the hi ghlights of the maritime emissions workshop arranged online on 6 October 2021. The participants were mainly researchers, authorities, and maritime operators. In the context of major ongoing changes, such as green transition, forums for cooperation and discussion are important in order to reflect on the progress, consensus and perspectives of other parties. This reduces confrontation and promotes understanding of common goals.

### 1.2 *Background*

At present, reducing emissions caused by maritime sector is topical. In order to slow down climate change, the production of greenhouse gas (GHG) emissions must be reduced (IMO, 2020). The maritime sector is a relatively small segment in terms of total GHG emissions (approximately 3%), but achieving carbon neutrality in this sector is challenging. However, there are no simple ways to achieve global carbon neutrality, and therefore solutions need to be widely sought. In the maritime sector, replacing fossil oil-based fuels with an energy-efficient and cost-effective alternative is not an easy process (Wahlström et al., 2006). In addition, other harmful emission components, such as $NO_X$, $SO_2$, black carbon or microparticles also require attention (Bouman et al., 2017), making emission reduction a complex and multidimensional process.

In July 2021, the European Commission published a proposal called the Fit for 55-action program with ambitious targets to promote carbon neutrality in the European Union (EU, 2021). The new emission reduction requirements for the maritime sector are part of this proposed regulation package. These include the gradual extension of the emission trading system (ETS) to the maritime sector, minimum levels of fuel taxation, and the targeted reduction of the carbon intensity of fuels. The new regulations apply to vessels of more than 5000 GT in commercial marine traffic.

However, the consequences of these proposed practices vary from region to region. The reorganization of logistic chains, with the aim of circumventing emissions trading obligations by exploiting a geographically suitable third country's ports are one of the issues raised that could lead to carbon leakage (EU, 2021). In Finland, the biggest challenges are meeting the requirements for winter seaworthiness, including the large share of RORO/ROPAX-shipping (Jalkanen & Johansson, 2019). The consequences are emphasized, as maritime transport plays a significant important role in Finland's foreign trade (Finnish customs, 2020).

### 1.3 *Study structure*

This paper proceeds as follows: Section 2 briefly presents Finland's marine and shipping environment, as

DOI: 10.1201/9781003320289-2

shipping plays a key role in Finland's foreign trade, where winter conditions and archipelago coastline are special features of this region. Section 3 refers to the topics presented and addressed in the workshop. Section 4 discusses the highlights of the workshop and the lessons learned from the expert presentations.

## 2 THE FINNISH SEAFARING ENVIRONMENT

Finland is one of the most dependent countries on maritime transport in the world. According to customs statistics (see Figure 1), the maritime sector accounts for 91.1-92.0% of total exports, or 41.9-46.7 million tons of cargo. The emphasis on imports is not so large, but the dominance of maritime transport is very clear: The share varies between 76.6-78.6% or 42.0-47.1 million tons (Finnish customs, 2020).

Enabling winter sea traffic is a prerequisite for Finland's foreign trade (Matala & Sahari, 2017). Winter shipping refers to an entity in which the performance of vessels in ice is one important aspect that complements e.g., adequate icebreaker capacity, VTS control and fairway design. Port operations also need to adapt to prevailing winter conditions.

Prevailing ice conditions vary not only with the frost season, but also with wind conditions (FMI, 2021): A flat-fixed ice field poses challenges for ships with a lower ice rating, but these conditions are still quite predictable. In the Gulf of Bothnia, ice thickness usually varies from 40 to 80 cm, whereas the thickness of the ice in the Gulf of Finland is often less than 40 cm.

The ice field moving with the wind brings great challenges for ships (Lensu & Goerlandt, 2019). Ice embarkments formed in front of the coastal archipelago or in the collision of ice fields are breakable just by powerful icebreakers. The Gulf of Finland and the Kvarken includes challenging areas due to packed ice. Compression on the ice field quickly closes the fairways opened by the icebreakers and hangs ships in the ice. In Figure 2, the ice area during a normal winter 2020-2021 is illustrated in gray (FMI, 2021). At that time, the voyages through the ice are almost 200 NM from the edge of the ice to the farthest ports in the area. During severe ice winters, such as 2010-2011, the ice field coverage continues until Gotland. However, cargo does not wait, so shipping must be able to operate even in these difficult conditions.

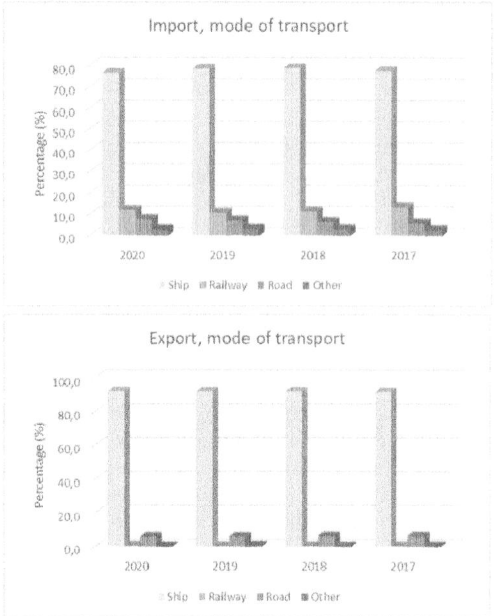

Figure 1. Export and import in Finland.

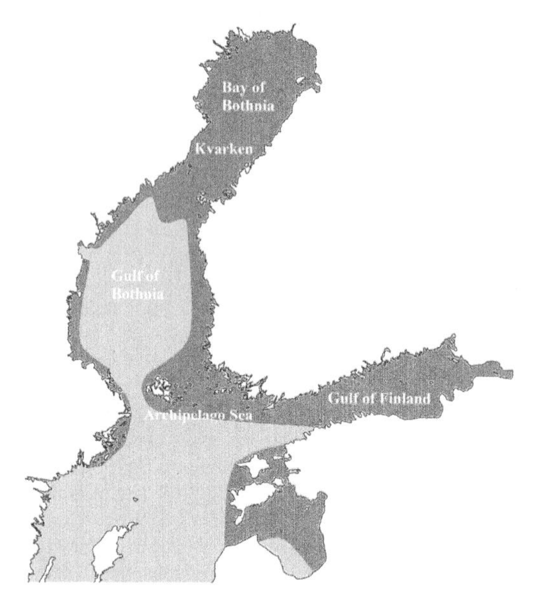


Figure 2. Ice coverage in the Baltic Sea (2020-2021).

Ice conditions vary by sea area: Freezing begins in late autumn in the Bay of Bothnia and gradually extends south. During normal winters, the Gulf of Finland also freezes, but the Baltic Sea south of usually remains an ice-free area. However, exceptions can occur during severe ice winters (Räisänen, 2017).

The Archipelago Sea is located between Finland and Sweden, including the autonomous state of Åland. On a global scale, the area forms a special marine environment (Andersson & Eklund, 1999). Maritime traffic in the area is lively: Several shipping companies operate RORO/ROPAX type vessels through the area, connecting Finnish and Swedish logistic chains. On winding fairways, cruising speeds often remain low, main engines are operated at varying part loads and auxiliary engines are kept running for emergencies. A similar operating profile carries on as the voyage continues in the Stockholm archipelago.

In addition to the Archipelago Sea, RORO/ROPAX-type vessels are common on many lines in the Baltic Sea region (Jalkanen & Johansson, 2019; Raza et al., 2019). On relatively short voyages, these ships can be loaded quickly and those provides good services for passengers during the voyages. The flexibility of cargo arrangements is also high. However, the fuel consumption of this type of vessels is quite high due to the high ice-rating (typically 1A Super) and the schedules adjusted to the daily rhythms.

## 3 THE WORKSHOP

The workshop was organized as an online event with five expert presentations, followed by discussion sessions. What was special about the workshop was the time set aside for discussion, which was considerably longer than a normal conference schedule. Participants also actively participated in the discussions and left the Chair only to facilitate the discussion, which made the workshop highly interactive. This was certainly due to the interdisciplinary nature of the workshop, topicality of the theme, and activity of the authorities and other actors working in the maritime sector. The presentations covered the following topics: Impact of sulfur scrubbers, emission regulations and control, proposed extension of the emission trading system (ETS), and the applicability of low-carbon fuels in the maritime sector, winter shipping and its characteristics in terms of environmental legislation, and finally emission regulations from the point of view of shipowners.

### 3.1 *Sulfur scrubbers*

The Baltic Sea was included in the SECA (sulfur emission control area) area in 2015 with a 0.1% limit. At present, experience has been gained over several years, so comprehensive assessment of the real impact can be done (Jonson et al., 2019; Prause & Olaniyi, 2019). With the entry into force of the new regulations, sulfur scrubbers were installed in some ships, but a large part of the fleet instead switched to low-sulfur fuels. The installation and maintenance costs of open-loop scrubbers are the most affordable, but their washing water is discharged into the sea. The closed-loop system is based on sodium hydroxide and fresh water washing liquid. In this system, washing water remains in the loop. The most common technology is hybrid models, where the user can choose the operating principle. This is affected not only by regulations (e.g., restricted zones in Germany where the discharge of washing waters is prohibited), but also by the environmental conditions in the Baltic Sea: The open-loop system requires salt water in order to function well, and the salinity level of brackish water in the Baltic Sea is low in the Gulf of Bothnia and in parts of the Gulf of Finland.

Sewage from closed-loop systems must be discharged to ports. The reception of wastewater is included in the legal obligations of ports, but the price level can be determined independently. Therefore, the operation of an open-loop systems is more affordable for shipowners, as the volumes of wastewater are high in any case.

However, the real environmental impact of scrubber wastewaters is not well known (Endres et al., 2018; Ytreberg et al., 2019). This is also the basis for Germany's decision to ban any wastewater discharges. The impact is less focused on sulfur compounds, but other chemicals such as heavy metals, PAHs and organic hydrocarbons, whose impact on the environment is still poorly understood.

### 3.2 *Existing and proposed regulations*

Annex VI of the MARPOL Convention defines the framework for air emissions (IMO, 2021). The SECA restrictions can be fulfilled by using low-sulfur or alternative fuels (e.g., LNG), emissions reduction techniques (such as scrubbers), or, in some cases, battery technology. The restrictions are monitored in several ways. Controlling fuel distribution is an effective preventive action. On port entrance fairways, sniffer units can be applied to remotely monitor passing ships. Where appropriate, targeted inspections shall be carried out, including port state control or separate on-site-inspections. The inspections are focused on documentary checks (e.g., international air pollution prevention certificate, IAPP; or engine international air pollution prevention certificate, EIAPP) and, if necessary, the fuel can be sampled by a rapid analysis tester. If the sulfur content of the sample exceeds the limits, laboratory tests are performed.

The NECA (nitrogen emission control area) restrictions significantly reduce the eutrophication load in the Baltic Sea (Jutterström et al., 2021), which is a major problem in the shallow sea, where the water changes slowly. From 2021, TIER III-requirements are applied to installed engines, which significantly tightens NOx regulations. In practice, TIER III requires catalytic converters (SCR), an exhaust gas recirculation system (EGR), hybrid solutions or alternative fuel solutions such as LNG.

Regulatory oversight of the NECA restrictions is challenging. The NOx limits are measured at a load of 75% according to a weighted average set by the IMO. In addition, the emission levels are proportional to the engine power (g/kWh). As a result, regulatory control focuses on documentary control (e.g., IAPP & EIAPP + NOx technical file) and confiding that the manufacturer's declaration of conformity is also valid in practice. However, actual emission levels may differ (Herdzik, 2011), especially when operating with specific operating profiles in an archipelago area or in icy conditions. The technical condition of the systems also affects their efficiency. In future, sniffer units will also be applied for the rough monitoring of NOx levels, but due to previously mentioned reasons, these are likely to remain indicative methods only. If the automatic reporting of ship voyage data is regulated in the future, the possibilities for monitoring NOx emissions will be significantly improved.

Reducing $CO_2$ emissions requires a reduction in fuel consumption. Attention to this is paid at the IMO level (energy efficiency design index EEDI, ship energy efficiency management plan SEEMP, energy efficiency existing ship index EEXI, carbon intensity indicator CII), but also by shipowners for whom fuel is a significant cost item (Mocerino & Rizzuto, 2019a; Mocerino & Rizzuto, 2019b). As a result, market-based reduction methods often come up in discussions alongside regulatory actions. The IMO aims to reduce greenhouse gas emissions from maritime transport by 50% from 2008 levels by 2050 (IMO, 2020). Under pressure to tighten its aims, this strategy will be updated until 2023. However, the European Union is setting more ambitious goals, as Fit for 55-proposal includes the extension of the emission trading system (ETS) to the maritime sector. The proposal is currently being prepared by the Parliament (EU, 2003; EU, 2021). The aim is to design a more efficient emission trading system by giving up free allowances and enabling the auctioning of all allowances and tightening total emission level. For maritime transport, the ETS is proposed to be based on the MRV database (EU, 2015) and is intended to be an open system, so the shipowner can buy (or sell) allowances from other sectors in the ETS. As a result, emission reduction investments in energy (or other industrial) sector affect the total price of allowances. In the maritime sector, the ETS has been proposed to start in stages over the period 2023-2026, but this schedule is so ambitious that delays are possible. Free allocation of allowances or other regulatory simplifications for winter shipping (or in other regional special treatments) are not included in the emission trading system (EU, 2021). In addition, a 50% share of emissions trading is proposed for voyages to ports in third countries ("MEXTRA50").

Reducing the greenhouse gas intensity of marine fuels is also part of the Fit for 55-package (EU, 2021). This initiative, called the FuelEUMaritime, is proposed to come into force in 2025 (-2%) from the 2020 level, after which the restrictions will be tightened over five-year periods and will reach -75% in 2050. This obligation explicitly obliges the use of alternative fuels and does not consider other emission reduction methods, such as increasing energy efficiency etc. The obligation applies to the same fleet as the ETS, but the emissions coverage is wider, including $CO_2$, $CH_4$ and $N_2O$. The FuelEU-Maritime also considers life-cycle greenhouse gas emissions of the fuels, contrary to the ETS (EU, 2021). In addition, from 2030 onwards, container and passenger ships will have to use shore power (Coppola et al., 2016) or zero-emission technology in ports.

### 3.3 *Practical operating environment*

Winter shipping conditions vary considerably in the Baltic Sea region. The variability in icy winter conditions is also large, but long-term trends have been observed since the $17^{th}$ century. In principle, global warming reduces the mass of ice in the sea but does not necessarily alleviate conditions; Winters have varying periods of heat and cold, and increasing windy conditions, including winter storms, further facilitate the movement of the ice formed. The proliferation of energy-efficient EEDI-ships does not make things easier. In open-water optimized hulls often run into problems under icy conditions and cannot be reach a high ice-class rating (Bergström & Kujala, 2020). These types of vessels are also not the best ones to be towed by an icebreaker. However, as the economic activity of the State cannot be stopped for winter, significant investment will be needed in icebreaking equipment. This issue is also a safety risk. When low-powered vessels sail in succession, keeping only short safety margins on the fairway paths opened by a breaker, the stalling of one vessel always increases a risk of collision.

There is a general will to reduce emissions from shipping, but there is still a lack of consensus on the right approach. This is complicated by the fact that shipping is international, but mankind is divided into national states. However, the implementation of the new regulations will require major investments in fleets by shipowners, and a long-lived fleet does not regenerate in an instant. The lifespan of the vessels is about 30-40 years and investment decisions range from tens of millions to hundreds of millions of euros, depending on the type of vessel. Decision-making in the IMO is constantly progressing from the shipowner's perspective. At the same time, regional decision-making (e.g., the EU) and other local constraints create a situation where anticipating regulatory development is challenging: The need for predictable and harmonized regulation is to be expected, especially when deciding on investments.

The development of fossil fuel-independent shipping is challenging, but successes have been achieved: For example, in the IMO's fourth GHG study, maritime transport accounts for around 3% of total greenhouse gas emissions, but due to improved energy efficiency, the increase in emissions no longer follows the increase of transport volumes (IMO, 2020). LNG is likely to be seen as a transitional fuel choice, but the transitional target is still open. Ammonia, LBG (liquefied biogas) or synthetic fuels are possible options. However, production volumes do not meet the needs of maritime transport at all, and modification of the distribution infrastructure (including end-user needs) takes time.

The production of synthetic fuels (or green ammonia) requires a lot of carbon-free electricity to be climate-friendly option (Wilson & Styring, 2017). The well-to-wake life cycle of the fuel is essential to consider. On short voyages, such as ferry lines, electric hybrid vessels are likely to become more common.

## 4 DISCUSSION OF LESSONS LEARNT

Achieving a sustainable transition is an important aim during the green transition (EU, 2019). In the worst case, solving one problem brings several other problems for example, open-loop sulfur scrubbers or previously applied TBT-anti-fouling coatings. In addition, importance of advanced port services plays a key role especially in the Baltic Sea region, where distances are relatively short. Digitalization offers many new tools that can be a threat or a solution (de Anders Gonzalez et al., 2021).

The pressure to develop the environmental friendliness of maritime transport is currently high in Finland (Zandersen et al., 2019), including expectations of politicians, authorities, customers, shipowners, and stakeholders. However, as a clear technological solution is missing, investment will be done knowing that these are transitional solutions. This also makes some operators cautious about investing, which can even slow down the green transition. This has also been stated by others, such as Sala et al. (2015) who highlighted the importance of stakeholder involvement in sustainability assessment: "the current research challenges call for co-production of knowledge through the collaboration and participation of different stakeholders". They further demand concrete contribution and involvement of stakeholders in all the steps of the sustainable development process.

In the workshop, the partial overlap of emission regulation provoked much debate: What will happen to the EU ETS emission trading system if the IMO introduces a market-based emission reduction mechanism? However, in a survey conducted in the workshop, participants estimated that ETS would have a greater impact on strategies and investments than IMO's decisions: 39% considered the ETS to be a good tool for reducing $CO_2$ emissions.

Carbon leakage risk also came to the fore: The Fit for 55-package has discussed this issue, but mainly from the perspective of voyages to third countries. In the Baltic Sea region, short sea shipping is often operated by RORO/ROPAX ships of more than 5000 GT, and if transport tariffs increase significantly, cargo transport may shift to road traffic. A separate emission trading system planned for road transport can compensate for this. However, the challenging anticipation of emissions trading allowance prices makes it difficult to predict this issue. The size limit of 5000 GT in the EU regulations was also discussed with carbon leakage risk. The vessels below 5000 GT constitute a minor proportion of $CO_2$ emissions (EU, 2021). However, some participants were concerned about the future incentive to build vessels, which would be barely below the proposed size limit. This would increase the total number of operating vessels, resulting in increased emissions. In winter conditions, this phenomenon would also lead to an increased need for icebreaking services. Some attendants pointed out that a smaller size limit would increase the administrative burden of the shipowners, an argument by which the EU justifies the 5000 GT boundary (EU, 2021). Furthermore, the proposed limit is in line with the MRV-regulation.

The ETS is likely to reduce emissions significantly more in sectors other than shipping. However, reducing emissions from shipping is expensive compared to many other industries (Solakivi et al., 2020). The utilization of auction proceeds is also the subject of discussion: Directing the revenue to the EU budget will make the auction a tax-like payment. The mandatory use of auctioning revenues for climate actions in Member States (including the reduction of transport emissions) and increasing funding for innovation and modernization of emission trading (and their extension to the maritime sector) probably will increase the system's acceptance. Research and development are also seen as very necessary, but in the end, ships applying polluting technology can only be replaced with cleaner ones through direct investments.

Shipowners are not alone in the ambiguity of the technological aspect. The workshop participants (35 persons) were allowed to respond to a questionnaire that mapped their personal views on alternative solutions (9 answers). The survey presented the options for technical solutions and their likelihood after 20 years. The options were "probable", "possible", "improbable" or "I can't say". The results are shown in Figure 3.

In this survey, responses confirmed the foggy outlook. The strongest probable or possible level is achieved with options for which technical capability already exists (e.g., batteries/hybrid solutions, LNG, schedule/route optimization, shore power and treatment systems such as scrubbers, EGR, SCR or DWI). On the other hand, some respondents do not see treatment systems as a solution. This may be due to the assumption that the propulsion system will be completely changed. The respondents are quite contradictory about ammonia or methanol, including biomethanol. The outlook for LNG also varies surprisingly (as for LBG), given the high expectations in recent years. Perhaps a little surprisingly, fuel cells (Coppola et al., 2020) were viewed positively, although little was mentioned in the discussions. There is also no strong potential seen in carbon capture systems, perhaps in part because of the novelty of this technological innovation and the difficulty of integrating it into the ship. Synthetic fuels seem to set expectations.

The utilization of wind-based systems did not receive attention in the workshop. In any case, these

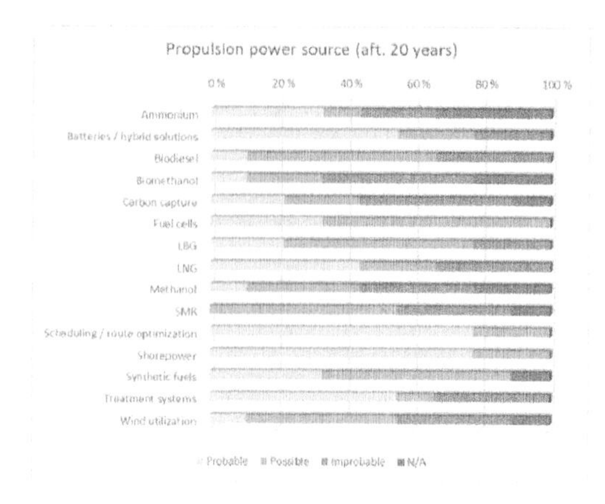


Figure 3. Propulsion power source after 20 years.

are auxiliary systems whose main function is to reduce fuel consumption. In Finland, the rotor sail system was tested on the LNG ROPAX ferry operating from Turku to Stockholm (years 2019-2021) but after the pilot period, the equipment was dismantled: According to the shipowner, the efficiency was not sufficient considering circumstances of the archipelago sailing and winter operation (Turun sanomat, 2021). On short routes in the Baltic Sea, the possibility of influencing the routes according to weather conditions is very limited.

Nuclear power is currently a highly politicized energy form. Often people group themselves as supporters or opponents of nuclear power. However, similar groupings are not as visible when discussing other forms of marine energy sources. Much research is currently being carried out to develop SMR (small modular reactor) type reactors for markets (Söderholm, 2013). If these reactor types reach commercially competitive levels, it may be possible to spread them to power source technology for maritime transport. Nuclear power could be an energy source of the future, which is not talked about presently.

The workshop committed different stakeholders (researchers, authorities, ship owners) and provided access to transparent information and a forum for common discussion of topical issues. According to Sala et al. (2015), in decision making, the common setting of sustainability objectives with stakeholders in mid-term planning at local scale is essential. The workshop highlighted the special requirements of Finland in the emission regulation development when the decision making is still in the process. Further, Sala et al. (2015) require the integration of stakeholders' requirements in technology sustainability assessment, which was accomplished in the workshop with different surveys and discussions, and revealed stakeholders' opinion related to technical options.

## 5 CONCLUSIONS

Summary of the main highlights of the workshop (in alphabetical order):

- Reducing $CO_2$ emissions, including preventing carbon leakage, has a key role to play in slowing down climate change. The climate benefits of emission reduction actions will be lost if emission sources are simply relocated.
- Geopolitically harmonized regulations will facilitate the political adoption of new legislation when investments in emission reduction technologies are appropriate over a wider geographical area.
- Predictability of regulations makes investing in green technology more economically secure. If investments become technologically obsolete rapidly as regulation changes, the situation will be frustrating for all parties. This often leads to resistance to any change.
- Investing auctioning revenues in processes that really contribute to the green transition will make the ETS politically more acceptable among shipowners. Otherwise, it is easily perceived as a tax-like payment.
- Investing in research and development: Further clarification is needed on unclear options and for the appropriate development of legislation. On the other hand; in addition to know-how, commercially available technologies at a competitive cost level are needed too.
- The importance of well-to-wake-analysis to find the most cost-effective way to reduce emissions: The ETS supports to reach this goal, but individual solutions are not enough to solve complex problems such as climate change.
- Winter shipping is not an option, but a condition set by circumstances. However, the challenge is to identify which concrete legislative actions would be a solution, considering the objectives of harmonization of international regulations.

## ACKNOWLEDGEMENTS

Thanks are given to the European Regional Development Fund for funding the MEPTEK (Benchmarking study of marine emission reduction technologies) research project and for making the workshop possible as a part of the project. Special thanks for the experts who presented at the workshop (Annukka Lehikoinen, Kotka Maritime Research Centre; Jukka-Pekka Jalkanen, Finnish Meteorological Institute; Ville-Veikko Intovuori, Finnish Transport and Communications Agency Traficom; Niina Honkasalo, Ministry of Transport and Communications; Jarkko Toivola, Finnish Transport Infrastructure Agency; Mats Björkendahl, Finnish Shipowners' Association). Thanks for participants and other project partners.

## REFERENCES


de Anders Gonzales, O., Koivisto, H., Mustonen, J.M. & Keinänen-Toivola, M.M. 2021. Digitalization in just-in-time approach as a sustainable solution for maritime logistics in the Baltic Sea region. *Sustainability* 13: article 1173.

Andersson, K. & Eklund, E. 1999. Tradition and innovation in coastal Finland: The transformation of the Archipelago Sea region. *Sociologia ruralis* 39(3): 377–393.

Bergström, M. & Kujala, P. 2020. Simulation-based assessment of the operational performance of the Finnish-Swedish winter navigation system. *Applied sciences* 10: article 6747.

Bouman, E.A., Lindstad, E., Rialland, A.I. & Strömman, A.H. 2017. State-of-the-art technologies, measures and potential for reducing GHG emissions from shipping – A review. *Transportation research part D* 52: 408–421.

Coppola, T., Fantauzzi, M., Miranda, S. & Quaranta, F. 2016. Cost/benefit analysis of alternative systems for feeding electric energy to ships in port from ashore. *2016 International annual conference (AEIT), IEEE 2016.*

Coppola, T., Micoli, L. & Turco, M., 2020. State of the art of high-temperature fuel cells in maritime applications. *2020 International symposium on power electronics, electrical drives, automation and motion (SPEEDAM), IEEE 2020.*

Endres, S., Maes, F., Hopkins, F., Houghton, K., Mårtensson, E.M., Oeffner, J., Quack, B., Singh, P. & Turner, D. 2018. A new perspective at the ship-air-sea-interface: The environmental impacts of exhaust gas scrubber discharge. *Frontiers in marine science* 5: article 139.

European commission. 2003. Directive 2003/87/EC of the European parliament and of the council of 13 October 2003 establishing a scheme for greenhouse gas emissions allowance trading within the Community and amending Council Directive 96/61/EC

European commission. 2015. Regulation (EU) 2015/757 of the European parliament and of the council of 29 April 2015 on the monitoring, reporting and verification of carbon dioxide emission from maritime transport, and amending Directive 2009/16/EC.

European commission. 2019. White paper 2011, roadmap to a single European transport area – Towards a competitive and resource efficient transport system. Modified: 3.10.2019.

European commission. 2021. Proposal for a directive of the European parliament and of the council amending Directive 2003/87/EC establishing a system for greenhouse gas emission allowance trading within the Union, decision (EU) 2015/1814 concerning the establishment and operation of a market stability reserve for the Union greenhouse gas emission trading scheme and regulation (EU) 2015/757.

Finnish customs. 2020. *International trade transports*. Helsinki: Finnish customs statistics.

Finnish meteorological institute. 2021. Ice season in the Baltic Sea. Website [referred 11 Nov 2021]: https://en.ilmatieteenlaitos.fi/ice-season-in-the-baltic-sea

Herdzik, J. 2011. Emissions from marine engines versus IMO certification and requirements of TIER III. *Journal of KONES powertrain and transport* 18(2): 161–167.

International maritime organization. 2020. *Fourth IMO GHG study 2020*. London: IMO.

International maritime organization. 2021. *MARPOL Convention 73/78, Annex VI*. London: IMO.

Jalkanen, J-P. & Johansson, L. 2019. Emissions from Baltic Sea shipping in 2006 – 2018. *Maritime Working Group Lisbon, Portugal, 23-26 September 2019.*

Jonson, J.E., Gauss, M., Jalkanen, J-P. & Johansson, L. 2019. Effects of strengthening the Baltic Sea ECA regulations. *Atmospheric chemistry and physics* 19: 13469–13487.

Jutterström, S., Moldan, F., Moldanová, J., Karl, M., Matthias, V. & Posch, M. 2021. The impact of nitrogen and sulfur emissions from shipping on the exceedance of critical loads in the Baltic Sea region. *Atmospheric chemistry and physics* 21(20): 15827–15845.

Lensu, M. & Goerlandt, F. 2019. Big maritime data for the Baltic Sea with a focus on the winter navigation system. *Marine policy* 104: 53–65.

Matala, S. & Sahari, A. 2017. Small nation, big ships winter navigation and technological nationalism in a peripheral country, 1878-1978. *History and technology* 33(2): 220–248.

Mocerino, L. & Rizzuto, E., 2019a. Preliminary approach to the application of the environmental ship index. In Georgiev & Soares (eds.), *Sustainable development and innovations in marine technologies*. London: Taylor & Francis group.

Mocerino, L. & Rizzuto, E., 2019b. Computer model application to the evaluation of energy efficient measures for cruise ships. In Georgiev & Soares (eds.), *Sustainable development and innovations in marine technologies*. London: Taylor & Francis group.

Prause, G. & Olaniyi, E.O. 2019. A Compliance cost analysis of the SECA regulation in the Baltic Sea. *Entrepreneurship and sustainability issues* 6(4): 1907–1921.

Raza, Z., Woxenius, J. & Finnsgård, C. 2019. Slow steaming as part of SECA compliance strategies among RORO and ROPAX shipping companies. *Sustainability*: 11 (5):article 1435.

Räisänen, J.A. 2017. Future climate change in the Baltic Sea region and environmental impacts. In Storch, (ed.), *Oxford research encyclopedias: Climate science*. Oxford: Oxford university press.

Sala, S., Ciuffo, B. & Nijkamp, P. 2015. A systemic framework for sustainability assesment. *Ecological Economics* 119: 314–325.

Solakivi, T., Jalkanen, J-P., Perrels, A., Kiiski, T. & Ojala, L. 2020. *Merenkulun päästökaupan vaikutukset*. Helsinki: Valtioneuvoston selvitys 2020:1.

Söderholm, K. 2013. *Licensing model development for small modular reactors (SMRs) – focusing on the Finnish regulatory framework*. Lappeenranta: Lappeenranta university of technology, dissertation.

Turun Sanomat. 2021. *Viking Line luopuu roottoripurjeesta Viking Grace-aluksella*. Published in newspaper in 23 April 2021.

Ytreberg, E., Hassellöv, I-M., Nylund, A.T., Hedblom, M., Al-Handal, A.Y. & Wulff, A. 2019. Effects of scrubber washwater discharge on microplankton in the Baltic Sea. *Marine pollution bulletin:* 145: 316–324.

Wahlström, J., Karvosenoja, N. & Porvari, P. 2006. *Ship emissions and technical emission reduction potential in the Northern Baltic Sea*. Helsinki: Reports of Finnish environment institute 8/2006.

Wilson, I.A.G. & Styring, P. 2017. Why synthetic fuels are necessary in future energy systems. *Frontiers in energy research* 5: article 19.

Zandersen, M., Hyytiäinen, K., Meier, H.E.M., Tomczak, M. T., Bauer, B., Haapasaari, P.E., Olensen, J.E., Gustafsson, B.G., Refsgaard, J.C., Fridell, E., Pihlainen, S., le Tissier, M.D.A., Kosenius, A-K. & van Vuuren, D.P. 2019. Shared socio-economic pathways extended for the Baltic Sea: exploring long-term environmental problems. *Regional environmental change* 19: 1073–1086.

*Trends in Maritime Technology and Engineering – Guedes Soares & Santos (eds)*


# Risk assessment and dynamic simulation of concentralized distribution logistics of the cruise construction build-in material

Z.M. Cui
*School of Transportation and Logistics Engineering, Wuhan University of Technology, Wuhan, China*

H.Y. Wang
*School of Transportation and Logistics Engineering, Wuhan University of Technology, Wuhan, China*
*National Engineering Research Center for Water Transport Safety, Wuhan University of Technology, Wuhan, China*

J. Xu
*Shanghai Waigaoqiao Shipbuilding Co., Ltd., Shanghai, China*

ABSTRACT: Concentralized distribution logistics of the cruise construction build-in material (CDL-CCBIM) is a complex project involving multi-participation and multi-factor coordination. It is affected by many uncertain factors such as internal link, external environment and cooperative relationship. To improve the risk management level of CDL-CCBIM can help to guarantee the progress and quality requirements of cruise construction. In this paper, A risk assessment index system based on a work breakdown structure-risk breakdown structure (WBS-RBS) matrix was established to deal with the complex risks of concentralized disruption logistics system. A system dynamic (SD) model was constructed to simulate changes in system risk. Based on sensitivity analysis, critical risk factors were determined. Subsequently, analyzed the risk level trend with time and impact of dynamic changes of each risk factor on the risk of concentralized distribution logistics. The results show that the risk level of the CDL-CCBIM is on the rise when no measures are taken. Over time, the risk is transmitted in the direction of increasing energy due to the dynamic evolution among influencing factors, and the plan change has the greatest impact on the risk of CDL-CCBIM.

## 1 INTRODUCTION

Cruise, known as "maritime mobile city", integrating navigation, sightseeing, tourism, leisure and entertainment functions, has been known and loved by the public in recent years. It has become one of the fastest growing tourism formats (Wondirad A. 2019). According to Cruise Lines International Association, demand for cruise in China was 2.1% in 2018, second only to the United States. China dominates the passenger share of the Asian cruise market, accounting for more than 70% of the Asian cruise passengers in 2018 (Li et al. 2021). The market is expected to reach 10 million by 2026. However, China's cruise construction is still in the exploratory stage. In the case of the imbalance between supply and demand of international cruise construction, it is timely to develop domestic cruise based on the huge demand of China's cruise market. Compared with ordinary merchant ships, the core value of cruise is more embodied in soft indicators, such as leisure and comfort experience, service assurance level, health, and safety assurance, etc.(Pan 2021), which are related to the build-in material. It can be seen that the quality of build-in engineering is very critical to the cruise construction. Build-in engineering is the core and most competitive part of cruise construction. The value of material involved is more than 50% of the total ship, and the construction cost accounts for 70%. Different from integration of hull and outfitting, cruise build-in engineering is an extremely complex project, involving more kinds of material, more extensive participants, and more difficult coordination. It is understood that China's first domestic large cruise has more than 25 million parts, of which the build-in engineering accounts for nearly 40% of the cruise. Most of them are provided by foreign suppliers, which are of high value, easy to damage, and high cost of secondary supply. Strict quality and progress requirements during the cruise construction have put forward higher requirements for build-in construction. For example, the progress of each public area is related to the progress of the whole ship and other public areas. Therefore, this also puts forward higher requirements for the progress, quality and coordination of the CDL-CCBIM.

DOI: 10.1201/9781003320289-3

The risk of CDL-CCBIM mainly comes from internal links, external environment and cooperation between participants. If the risks that may arise in the process cannot be effectively identified, assessed, predicted and controlled, it will lead to the interruption or destruction of the logistics distribution logistics process. Further, due to the interaction of risk factors, some risk may transfer to other links, resulting in plan management disorder and this may seriously affect the cruise construction. It is necessary to analyze the risk factors, risk mechanism and its effect on the system risk level from the perspective of the system. Through risk assessment and risk evolution situation, the risk of CDL-CCBIM can be managed and controlled in a targeted manner at the right time.

At present, there are few researches in the field of logistics distribution logistics of cruise construction material. However, in order to ensure that the construction of the cruise is carried out as planned, the risk of logistics distribution logistics cannot be ignored. In this paper, we focus on the analysis of concentralized distribution logistics of cruise construction build-in material process. Firstly, we identify the risk factors that affect the efficiency of distribution logistics from two aspects of workflow and risk link. Then, the causal relationship between risk factors is analyzed qualitatively and quantitatively. Finally, a SD model is constructed to simulate the risk level of CDL-CCBIM and compare the risk consequences caused by different threats. The purpose of this paper is to provide a decision-making reference for the risk management, minimize cruise construction delays caused by disruptions in concentralized distribution logistics of build-in material.

The remainder of the paper is as follows. Section 2 gives a literature review about risk management in shipbuilding logistics and application of system dynamics in risk management. Section 3 describes the background and the method used in this article. Then, section 4 takes a case simulation, and the sensitivity analysis is carried out. According to the results, section 5 shows relevant discussions. Finally, section 6 gives the conclusions of this paper.

## 2 LITERATURE REVIEW

At present, the research on cruise construction logistics is mostly concentrated in the fields of material distribution path optimization, material ordering strategy selection, inventory management and control. Only a few studies from the perspective of material concentralized distribution logistics risk. In order to expand the research basis, we can learn from the research on risk management in shipbuilding logistics and other complex systems.

### 2.1 *Risk management in shipbuilding logistics*

The shipbuilding industry must cope with high complexity of the shipbuilding process because of the large number of complex parts and interim products in ships (Sender et al. 2020). To ensure an efficient and competitive shipbuilding process, it is necessary to speed up the shipbuilding logistics process and avoid disrupting. Shipbuilding logistics is not only the procurement and transportation of materials, it focuses on production logistics and supply logistics (Sender et al. 2020). The complex system with not only internal dependencies but also external influences from natural and industrial environments.

Chen (2011) used supply-chain Operations Reference-Model (SCOR) to analyze the causes of risk in shipbuilding supply chain from four stages: planning, procurement, construction, and ship delivery. The risk evaluation system of shipbuilding supply chain was constructed from the aspects of external environment, production plan, procurement, construction, delivery, and cooperation. The evaluation model was established by fuzzy comprehensive evaluation theory, and relevant control measures were put forward. Starting from the idea of sustainability, Ozturkoglu et al. (2019) identified the risk factors related to human, economy, and environment in ship construction recycling logistics. Analyzed the causal relationship and importance ranking among the prevention factors by using FUZZY-DEMATEL method, and put forward comprehensive management suggestions. Crispim et al. (2020) combined Delphi method and Bayesian network model, proposed a risk assessment framework for military shipbuilding projects. Through probability analysis, it was concluded that production, contract, demand, and plan were the key control objects in shipbuilding projects. Wang et al. (2019) combined with the concentralized distribution logistics process of the cruise construction outfitting material, identified six risk factors including supply, warehousing, outsourcing, transportation, external environment, and cooperation. And evaluated each factor based on fuzzy mathematics theory.

Furthermore, Yue et al. (2008) put forward that the shipbuilding supply chain risk refers to the negative impact of uncertain factors on the members of the supply chain or the damage to the operating environment, which leads to failure in achieving the target plan. The author also states that potential risks in the shipbuilding supply chain would disrupt the cooperation between nodes. In order to explain the risks more clearly, he divides the risks into external environmental risks such as poor transportation environment and international politics, together with operational risks including information interruption, supply risk, organizational risk, information risk, etc. Zhu et al.(2021) integrated the flexibility theory into the shipbuilding material supply chain and explored the flexibility of supply, logistics, organization and quality in combination with the needs of shipbuilding enterprises. Combining QFD, fuzzy theory and DEMATEL, Design a more flexible supply chain for shipbuilding materials. Diaz et al. (2020) proposed a simulation framework for shipbuilding supply chain risk management based on the real-time data of Industry 4.0 and a simulation model that supports risk management and decision-making. Mello & Strandhagen (2011) propose that effective supply chain management can be used to mitigate ship supply chain risks, such as improving relationships with suppliers and using appropriate information tools.

### 2.2 *System dynamics used in risk management*

SD is a computer simulation technique proposed by Dr. Forrester of the Massachusetts Institute of Technology (MIT) to study the dynamics of the system. Compared with other methods, SD is more suitable for studying the correlation feedback of risk factors in complex systems. It can examine dynamic behaviors of complex systems from a macro and holistic-thinking perspective (Ding et al. 2018). Baliwangi et al. (2007) established an improved SD risk model for risk management simulation and discussed the advantages of system behavior in risk management. At present, system dynamics is widely used in (Yue et al. 2008)n risk management in various fields, such as supply chain management (Song et al. 2021, Chen et al. 2020, Peng et al. 2014), logistics management (Kwesi-Buor et al. 2019), and project management (Liu et al. 2021, Yu et al. 2019).

Li et al. (2018) analyzed the relationship between the risks of prefabricated housing assembly schedule and simulated the influence of different risk factors on the assembly schedule by using system dynamics method. Han et al. (2021) through the system dynamics method, analyzed the impact of dynamic changes of each risk factor on the risk accident of the integrated pipe corridor, and realized the evaluation and analysis of the accident risk of the integrated pipe corridor. Chen et al. (2021) built a system dynamics model to reveal the complex relationship among risk factors of the urban metro system and simulate the vulnerability of system. Zarghami & Dumrak (2021) used SD model as a potent tool for analyzing the future evolution of social vulnerability to natural disasters by creating qualitative and quantitative causal models. Su et al. (2020) adopted the system dynamics method to establish a model for studying urban traffic congestion system, added some practical policies to the model for policy analysis.

### 2.3 *Summary*

The above literature review reveals that the shipbuilding logistics and shipbuilding supply chain are faced with diversified risk. It is necessary to take effective measures to assess and mitigate the consequences of risks. However, there are a few researches on the risk of concentralized distribution logistics of cruise construction material. Meanwhile, most of the existing literature focuses on analyzing and assessing risks, and there is a lack of consideration of the interaction between factors. This paper attempts to make a comprehensive analysis of the concentralized distribution logistics system of the cruise construction build-in material. Then, identify the risk factors and analyze the relationship between them. Finally, through system dynamics theory to determine the critical risk factors and analyze the evolution trend of risk level.

## 3 MATERIALS AND METHODS

### 3.1 *Background*

Cruise construction build-in material refer to cabin rooms, public areas, entertainment, and leisure equipment, etc., mainly including top and bottom slots, ceiling thin walls, door/window buckets, sanitary units, air conditioning system, furniture, and shipowner supplies. It contains a very wide range of parts and components, and is the largest and most extensive component in the cruise package. Take China's cruise under construction as an example, the quantity related to build-in material includes 38,000 square meters of public area and 2826 cabin rooms. In addition, the build-in material has a higher requirement for transportation and storage due to its flammable properties and decorative properties.

In the construction process of China's first cruise, the build-in engineering borrowed advanced management experience from Europe and adopted EPC (Engineering Procurement Construction) mode. "Design - Purchase – Build" general contract is carried out, and the main contractor management department of the shipyard is responsible for planning, supervision, and coordination. According to the cruise construction planning and the material arrival planning, the subcontractors or suppliers will deliver the material to shipyard. The distribution center is responsible for the storage of material. The production management department shall apply for material to the distribution center according to the actual requirements of the installation site. Then, the distribution center shall pick material according to the pallet distribution planning and deliver them to the production site for installation by subcontractors on schedule. Concentralized distribution logistics system of the cruise construction build-in material is shown in Figure 1.

### 3.2 *Risk identification of concentralized distribution logistics*

The study used the Work Breakdown Structure-Risk Breakdown Structure (WBS-RBS) method to ensure the accuracy, reliability and effectiveness of the risk identification. The WBS-RBS method was first proposed and applied by PMI (Ragele C et al.2005). WBS is a systematic decomposition of the complex work to object blocks. RBS decomposes the risk factors that may cause complex work disruption.

It starts from higher levels and going down to finer level risks. The main risk identification steps of CDL-CCBIM based on WBS-RBS are as follows:

Step 1: Determining the objectives of risk identification. The objectives and the scope of risk identification should be determined according to the risk assessment requirements.

Step 2: Decomposition of concentralized distribution logistics work. The concentralized distribution logistics process of cruise construction build-in material can be decomposed step by step according to the order of the project and sub-projects. The project is decomposed into work units suitable for risk identification.

Step 3: Decomposition of concentralized distribution logistics risk. The risk sources are decomposed by considering the project characteristics and expert investigation. Similarly, the decomposed risk units must be suitable for risk identification.

Step 4: Developing WBS – RBS risk identification matrix. Based on the analysis of WBS and RBS, the risk matrix is formed by coupling.

Step 5: Preparing risk identification list. Eliminate risk factors that have none or little relation to the CDL-CCBIM according to expert advice. Prepare a list of risks for the CDL-CCBIM by combining their characteristics and requirements.

3.2.1 *WBS decomposition of the CDL-CCBIM*

According to the actual survey results of cruise construction company, the concentralized distribution logistics process of cruise construction build-in material can be divided into Arrival of materials (W1), Storge(W2), Matching Pallets (W3) and Distribution (W4). The WBS is shown in Figure 2.

3.2.2 *RBS decomposition of CDL-CCBIM*

The risk sources of the CDL-CCBIM are divided into five risk factors: Human (R1), Equipment (R2), Material (R3), Environment (R4), and Management (R5). The RBS is shown in Figure 3.

3.2.3 *Developing WBS – RBS risk identification matrix*

The matrix is based on a case studies and experts' opinions, it is developed by using WBS process as a row vector and RBS basic risk source as a column vector as shown in Table 1. In this table, 0 stands for no risks generated from coupling and 1 stands for risks can be generated from the coupling. Also, in different positions, 1 stands for different risk incidents.

3.2.4 *Establish a risk assessment index system of the CDL-CCBIM.*

The risk of CDL-CCBIM can be summarized as several uncertain factors that affect the logistics progress and efficiency, lead to the interruption of concentralized distribution logistics, and eventually fail to meet the build-in material demand plan. It is a multi-level and multi-dimensional system problem. One risk factor may cause multiple risk events to occur, and the occurrence of one risk event may also be caused by multiple risk factors. Based on WBS-RBS risk identification matrix above, combined with experts filtering and supplementation. The environmental system, resource system and management system are divided, regard these three subsystem risks as the primary risk indicators.

(1) Environmental risk. The impact of the environment on the concentralized logistics distribution system is complex, including the natural environment and the social environment, affecting the arrival schedule and delivery efficiency of material.
(2) Resource risk. Resources are the core of the concentralized distribution logistics system, which refers to the human, material, equipment involved in the process.
(3) Management risk. Management system refers to the process management, information sharing, collaboration involved in completing CDL-CCBIM, which is the key points and difficulties of the logistics process.

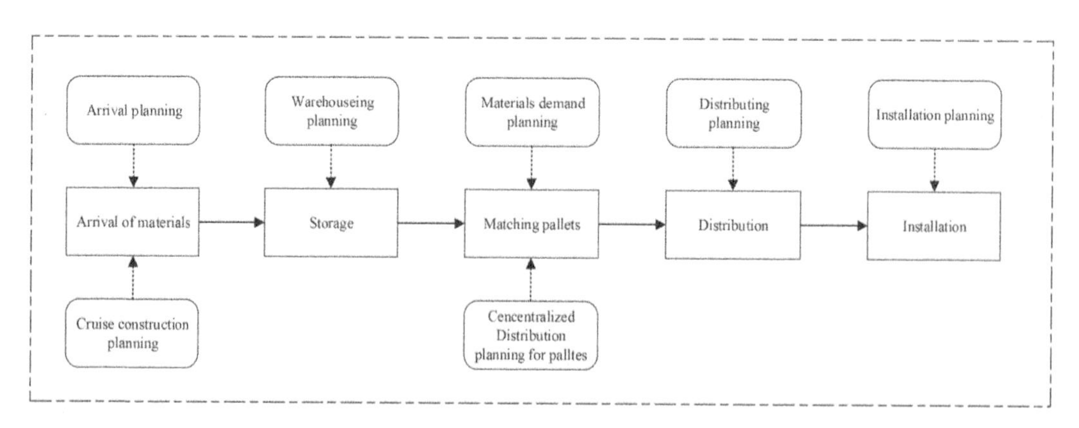


Figure 1. Concentralized distribution logistics system of cruise construction build-in material.

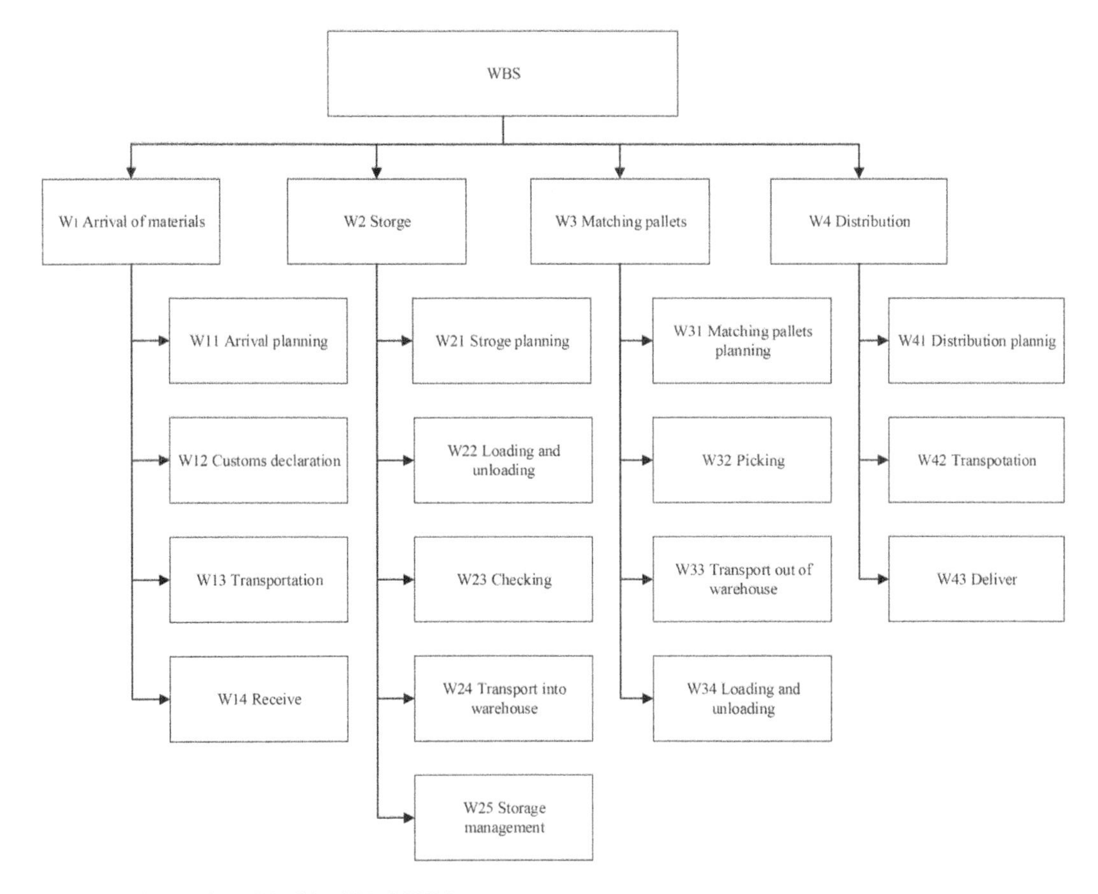


Figure 2. Diagram for WBS of the CDL-CCBIM.

The risk indicators system of CDL-CCBIM is constructed, as shown in Table 2.

### 3.3 *System dynamics model construction*

#### 3.3.1 *Model development*

SD is a combination of the qualitative method and quantitative method. To establish the system dynamics model, we must first determine the causal circuit diagram of each factor. This article specifically constructs the causal circuit diagram for the comprehensive risk of CDL-CCBIM, the primary risk, secondary risk, and the third-level risk. Through the investigation and analysis of cruise construction company, combined with experts' advice. The risks are connected one by one to determine the casual loop between the third-level risks. Eventually, risk propagation path of CDL-CCBIM is obtained. The final causal loop diagram is shown in Figure 4.

The behavior of system dynamics analysis is based on the interaction of the internal elements of the system, and assumes that the changes outside the system do not have an essential impact on the system behavior. This article only considers the impact of the three subsystems of environment, resource, management, and their interactions on the risk level of CDL-CCBIM, and does not consider the effects of other factors such as economy and technology that cause system risk. Based on the overall analysis of the causal relationship and feedback mechanism of the influencing factors between and within the subsystems, the system dynamics software Vensim_PLE is used to draw the stock and flow diagram. As can be seen from Figure 5, it is used to analyze the cumulative effect of risk factors on the system and the rate of change.

#### 3.3.2 *Weight calculation for risk factors*

This paper uses the analytic hierarchy process (AHP) based on triangular fuzzy numbers to determine the index weights. AHP was first proposed in the mid-1970s by Thomas Saaty (Moumeni et al. 2021). In general, AHP follows four basic steps: modeling, scoring, prioritizing, and synthesizing. Using triangular fuzzy numbers to replace the precise scales in traditional analytic hierarchy process can weaken the uncertainty and ambiguity of the

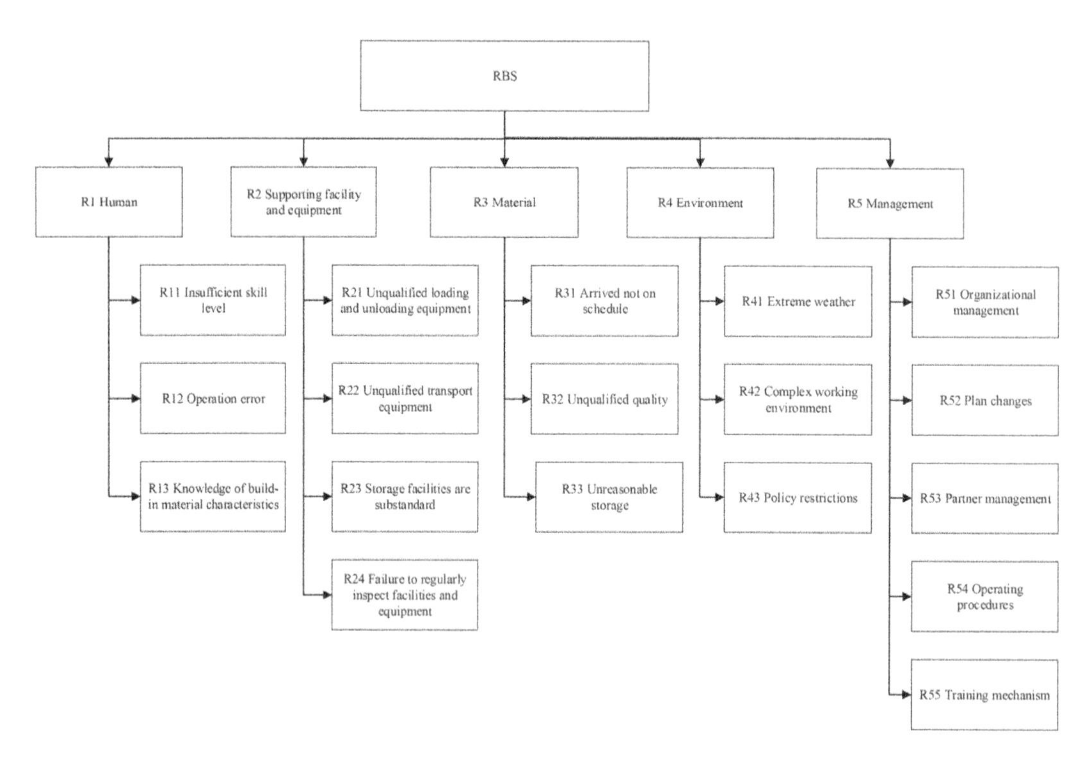


Figure 3. Diagram for RBS of the CDL-CCBIM.

Table 1. WBS-RBS risk identification matrix.

| | | W | | | | | | | | | | | | | | | |
|---|---|---|---|---|---|---|---|---|---|---|---|---|---|---|---|---|---|
| | | W1 | | | | W2 | | | | | W3 | | | | W4 | | |
| | | W11 | W12 | W13 | W14 | W21 | W22 | W23 | W24 | W25 | W31 | W32 | W33 | W34 | W41 | W42 | W43 |
| R | R1 | R11 0 | 0 | 1 | 0 | 0 | 1 | 1 | 1 | 1 | 0 | 1 | 1 | 1 | 0 | 1 | 0 |
| | | R12 0 | 1 | 1 | 0 | 0 | 1 | 1 | 1 | 1 | 0 | 1 | 1 | 1 | 0 | 1 | 0 |
| | | R13 0 | 1 | 1 | 0 | 0 | 1 | 1 | 1 | 1 | 0 | 1 | 1 | 1 | 0 | 1 | 0 |
| | R2 | R21 0 | 0 | 0 | 0 | 0 | 1 | 0 | 0 | 0 | 0 | 0 | 0 | 1 | 0 | 0 | 0 |
| | | R22 0 | 0 | 1 | 0 | 0 | 0 | 0 | 1 | 0 | 0 | 0 | 1 | 0 | 0 | 1 | 0 |
| | | R23 0 | 0 | 0 | 0 | 0 | 0 | 0 | 0 | 1 | 0 | 1 | 0 | 0 | 0 | 0 | 0 |
| | | R24 0 | 0 | 1 | 0 | 0 | 1 | 0 | 1 | 1 | 0 | 1 | 1 | 1 | 0 | 1 | 0 |
| | R3 | R31 1 | 1 | 1 | 0 | 0 | 0 | 0 | 0 | 0 | 0 | 0 | 0 | 0 | 0 | 0 | 0 |
| | | R32 0 | 0 | 1 | 0 | 0 | 1 | 0 | 1 | 1 | 0 | 1 | 1 | 1 | 0 | 1 | 0 |
| | | R33 0 | 0 | 0 | 0 | 1 | 0 | 0 | 0 | 1 | 0 | 0 | 0 | 0 | 0 | 0 | 0 |
| | R4 | R41 0 | 0 | 0 | 1 | 0 | 0 | 0 | 0 | 0 | 0 | 0 | 0 | 0 | 0 | 1 | 0 |
| | | R42 0 | 0 | 1 | 1 | 0 | 1 | 1 | 1 | 1 | 0 | 1 | 1 | 1 | 0 | 1 | 1 |
| | | R43 0 | 1 | 0 | 0 | 0 | 0 | 0 | 0 | 0 | 0 | 0 | 0 | 0 | 0 | 0 | 0 |
| | R5 | R51 1 | 1 | 1 | 1 | 1 | 1 | 1 | 1 | 1 | 1 | 1 | 1 | 1 | 1 | 1 | 1 |
| | | R52 1 | 0 | 0 | 0 | 1 | 0 | 0 | 0 | 0 | 1 | 0 | 0 | 0 | 1 | 0 | 0 |
| | | R53 1 | 1 | 0 | 0 | 0 | 0 | 0 | 0 | 0 | 0 | 0 | 0 | 0 | 0 | 0 | 0 |
| | | R54 0 | 0 | 1 | 0 | 0 | 1 | 1 | 1 | 1 | 0 | 1 | 1 | 1 | 0 | 1 | 1 |
| | | R55 0 | 0 | 1 | 0 | 0 | 1 | 1 | 1 | 1 | 0 | 1 | 1 | 1 | 0 | 1 | 0 |

Table 2. A list of risk factors in CDL-CCBIM.

| Primary risk | Secondary risk | Third-level risk |
|---|---|---|
| Environmental risk A | Natural risk $A_1$ | Bad weather $A_{11}$ |
| | | Emergency $A_{12}$ |
| | Social risk $A_2$ | Changes in policy and regulation $A_{21}$ |
| | | Changes in ship market environment $A_{22}$ |
| Resource risk B | Human risk $B_1$ | Insufficient skills of employee $B_{11}$ |
| | | Operating error $B_{12}$ |
| | | Insufficient employee knowledge training of build-in material $B_{21}$ |
| | Material risk $B_2$ | Material arrival delay $B_{22}$ |
| | | Material arrival planning does not match demand planning $B_{23}$ |
| | | Material quality problems $B_{24}$ |
| | | Low rates of pallets matching $B_{25}$ |
| | Supporting facility and equipment risk $B_3$ | Insufficient storage space $B_{31}$ |
| | | Poor storage environment $B_{32}$ |
| | | Lack of operating equipment $B_{33}$ |
| | | Equipment failure $B_{34}$ |
| Manage risks C | Information sharing risk $C_1$ | Internal information is not shared $C_{11}$ |
| | | Information is not shared between partners $C_{12}$ |
| | Collaboration risk $C_2$ | Insufficient internal cooperation $C_{21}$ |
| | | Insufficient cooperation with partners $C_{22}$ |
| | Manage process risk $C_3$ | Plan change $C_{31}$ |
| | | Difficulty in resource coordination $C_{32}$ |
| | | Irregular operation management $C_{33}$ |
| | | Training mechanism problems $C_{34}$ |
| | | Insufficient investment $C_{35}$ |
| | | Partner problems $C_{36}$ |

experts' personal preferences, and make the evaluation results more objective and accurate. The calculation steps are as follows:

(1) Through expert interviews, a triangular fuzzy matrix $Q = (q_{ij})_{n\times n}$ is constructed according to the questionnaire. The triangular fuzzy number $q_{ij} = (l_{ij}, m_{ij}, u_{ij})$.Where $l$ is the lower bound of the fuzzy set, $m$ is the median value, and $u$ is the upper bound.
(2) Carry out the consistency test. The consistency test of the median matrix is carried out, and the largest eigenvalue $\lambda_{\max}$ of the median matrix $M = (m_{ij})_{n\times n}$ is calculated according to the traditional AHP.
(3) Constructing the Fuzzy Judgment Factor Matrix $E = (e_{ij})_{n\times n}$.

$$e_{ij} = \begin{cases} 1, i = j \\ 1 - s_{ij}, i \neq j \end{cases} \tag{1}$$

$$s_{ij} = \frac{u_{ij} - l_{ij}}{2m_{ij}} \tag{2}$$

Where $s_{ij}$ is the standard deviation rate, reflecting the degree of ambiguity of expert judgement. The larger $s_{ij}$ is, the greater the ambiguity is, and the less reliable the evaluation result is.

(4) Calculate adjustment matrix O. Where $O = M \times E$.
(5) The matrix O is converted to the judgment matrix P with a diagonal of 1. According to the traditional AHP, MATLAB is used to calculate the weight of indicators, and the consistency test is carried out.

Invite 7 industry experts (2 are university professors engaged in logistics risk management, 5 from a cruise construction company, respectively responsible for cruise build-in engineering management and concentralized distribution logistics management) to fill out the questionnaire.

Taking the weight calculation of the environmental system, resource system, and management system on the risk of concentralized distribution logistics of cruise construction build-in material as an example, the judgment matrix P is obtained. The index weight calculation and inspection results are shown in Table 3.

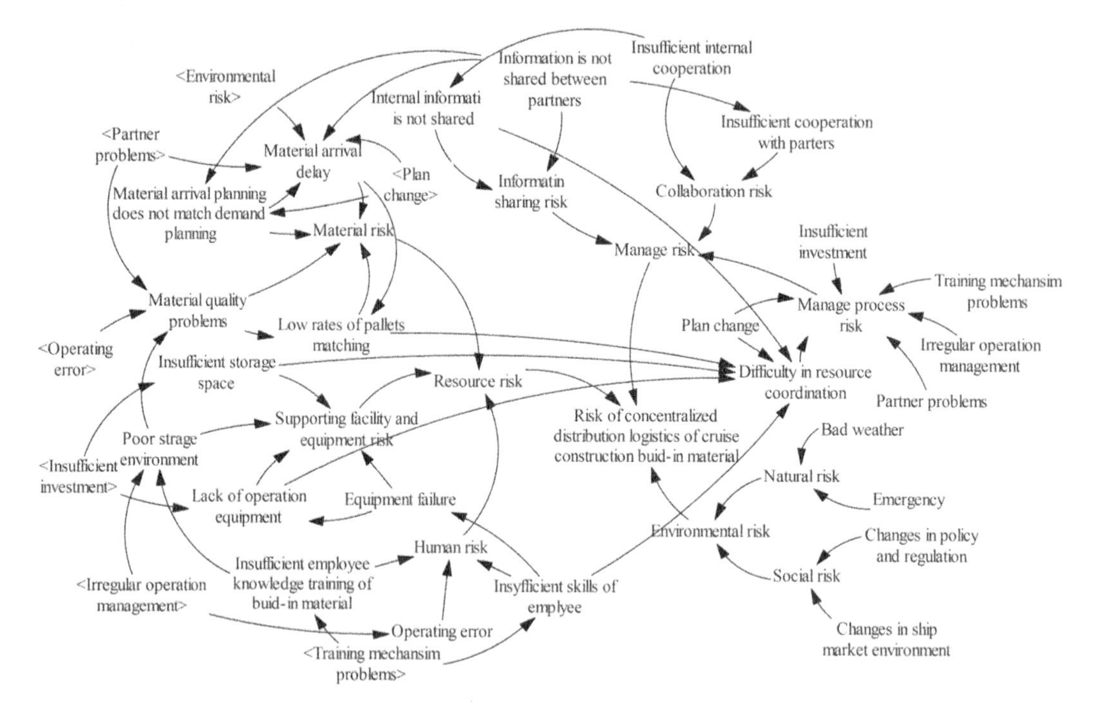


Figure 4. Causal circuit diagram.

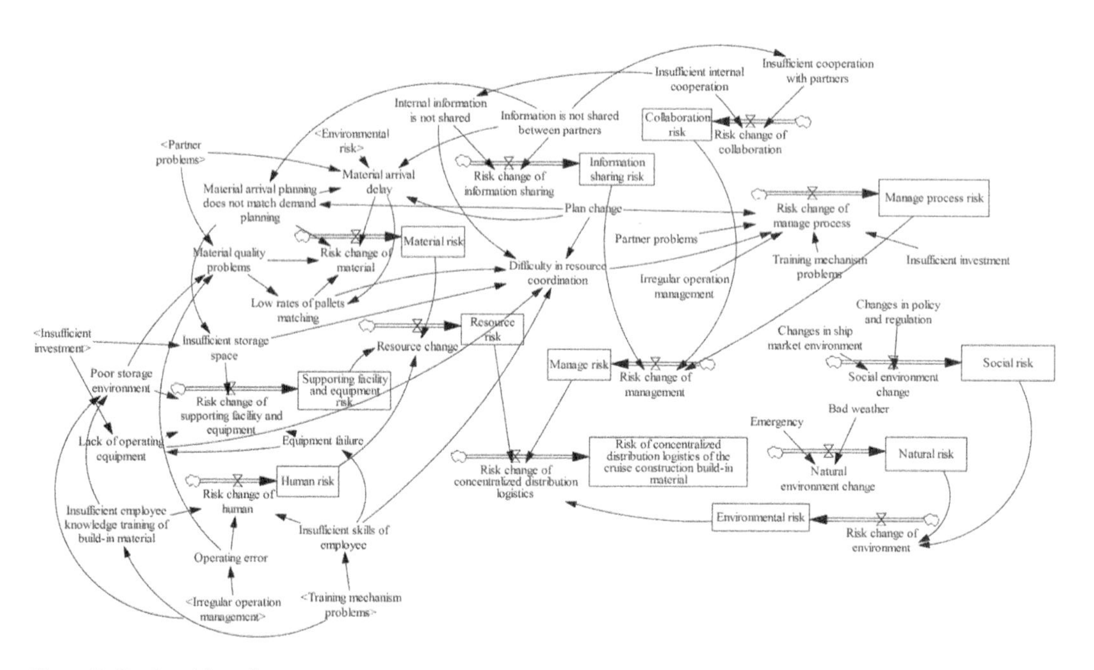


Figure 5. Stock and flow diagram.

Table 3. Weights of each subsystem of RCDL-CCBIM.

| Index | A | B | C | Weight |
|---|---|---|---|---|
| A | 1.000 | 0.174 | 0.134 | 0.068 |
| B | 5.642 | 1.000 | 0.738 | 0.382 |
| C | 8.208 | 1.515 | 1.000 | 0.550 |

$\lambda_{max} = 3.065, CI = 0.033, CR = 0.056 < 0.100$
Pass the consistency test

According to the same method, the weight value of other factors in the risk index system for the CDL-CCBIM can be calculated (Table 4).

3.3.3 The main system dynamics equations in the model.

The variables in the model are set according to the results of the questionnaire and interview with experts to obtain objective and reasonable formulas. The main logic equations in the mode are shown as follows, where the INTEG function is the integral of the variable over a period. The parameters in the formula are obtained from Table 5.

(1) Risk of CDL-CCBIM=INTEG (risk change of concentralized distribution logistics, initial value)
(2) A= INTEG (risk change of environment, initial value)
(3) B= INTEG (risk change of resource, initial value)
(4) C= INTEG (risk change of management, initial value)
(5) Risk change of concentralized distribution logistics=0.068* A+0.382* B+0.550* C
(6) Risk change of environment=0.466* $*A_1$ $+0.534*A_2$
(7) Natural environment change=0.321*$A_{11}$ +0.679*$A_{12}$
(8) Social environment change=0.500* $A_{21}$ +0.500* $A_{22}$
(9) Risk change of resource= 0.183*$B_1$ +0.503*$B_2$ +0.314*$B_3$

Table 4. Weights of risk factors.

| Primary risk | Secondary risk | Three-level risk | Weight |
|---|---|---|---|
| Environmental risk A | Natural risk $A_1$ | Bad weather $A_{11}$ | 0.010 |
| | | Emergency $A_{12}$ | 0.022 |
| | Social risk $A_2$ | Changes in policy and regulation $A_{21}$ | 0.018 |
| | | Changes in ship market environment $A_{22}$ | 0.018 |
| Resource risk B | Human risk $B_1$ | Insufficient skills of employee $B_{11}$ | 0.013 |
| | | Operating error $B_{12}$ | 0.004 |
| | | Insufficient employee knowledge training of build-in material $B_{13}$ | 0.022 |
| | Material risk $B_2$ | Material arrival delay $B_{21}$ | 0.098 |
| | | Material arrival planning does not match demand planning $B_{22}$ | 0.080 |
| | | Material quality problems $B_{23}$ | 0.014 |
| | | Low rates of pallets matching $B_{24}$ | 0.038 |
| | Supporting facility and equipment risk $B_3$ | Insufficient storage space $B_{31}$ | 0.055 |
| | | Poor storage environment $B_{32}$ | 0.027 |
| | | Lack of operating equipment $B_{33}$ | 0.022 |
| | | Equipment failure $B_{34}$ | 0.008 |
| Manage risks C | Information sharing risk $C_1$ | Internal information is not shared $C_{11}$ | 0.065 |
| | | Information is not shared between partners $C_{12}$ | 0.094 |
| | Collaboration risk $C_2$ | Insufficient internal cooperation $C_{21}$ | 0.030 |
| | | Insufficient cooperation with partners $C_{22}$ | 0.027 |
| | Manage process risk $C_3$ | Plan change C31 | 0.105 |
| | | Difficulty in resource coordination $C_{32}$ | 0.079 |
| | | Irregular operation management $C_{33}$ | 0.049 |
| | | Training mechanism problems $C_{34}$ | 0.007 |
| | | Insufficient investment $C_{35}$ | 0.025 |
| | | Partner problems $C_{36}$ | 0.069 |

(10) Risk change of human= $0.334*B_{11}$ $+0.092*B_{12}$ $+0.574*B_{13}$
(11) Risk change of material= $0.424*B_{21}$ $+0.347*B_{22}$ $+0.063*B_{23}$ $+0.166*B_{24}$
(12) Risk change of supporting facility and equipment risk= $0.387*B_{31}+0.242*B_{32}$ $+0.300*B_{33}$ $+0.071*B_{34}$
(13) Risk change of management= $0.289*C_1$ $+0.103*C_2$ $+0.608*C_3$
(14) Risk change of information sharing= $0.409*C_{11}$ $+0.591*C_{12}$
(15) Risk change of collaboration= $0.534*C_{21}$ $+0.466*C_{22}$
(16) Risk change of manage process=$0.314*C_{31}$ $+0.236*C_{32}+0.207*C_{33}+0.147*C_{34}$ $+0.074*C_{35}$ $+0.022*C_{36}$

## 4 EXPERIMENT AND RESULTS ANALYSIS

The initial time of the simulation is 0, the end time is 36, and the step is 1 month.

### 4.1 *Analysis of initial risk level simulation results*

Verifying the validity of the model is the first step of system simulation. Select the primary and secondary indicators of the concentralized distribution logistics risk, combined with the Vensim_PLE software, and simulate whether the trend of the level value is in line with the actual situation. As can be seen from Figure 6, except for the small change in the risk level of the environmental system, the risk level of the management system and resource system presents an overall upward trend. The upward rate accelerates in the middle and late construction period. This is because in the middle and late stage of cruise construction, the hull construction is basically completed, and the build-in engineering becomes the main work. The build-in material arrives centrally, and the operation pressure of concentralized distribution logistics increases. The risk level of the three subsystems in the first 18 months is management system, resource system and environment system in descending order. Later, due to the exposure of management risk, multi-dimensional risk factors act together, interact with each other, and gradually accumulate impacts on material, human, supporting facility and equipment. The risk level of the resource system increases significantly in the later period and is higher than the risk level of the management system.

In Figure 7 and Figure 8, the risk level of human, material, supporting facility and equipment, information sharing, collaboration, and management process is basically consistent with the risk level of resource system and management system. With the advance of build-in engineering of cruise construction, the risk level of CDL-CCBIM is on the rise due to the complex interaction of various factors. The simulation results are roughly the same as those in the actual process of cruise ship construction, which verifies that the model is reliable and reasonable (Zheng et al. 2021).

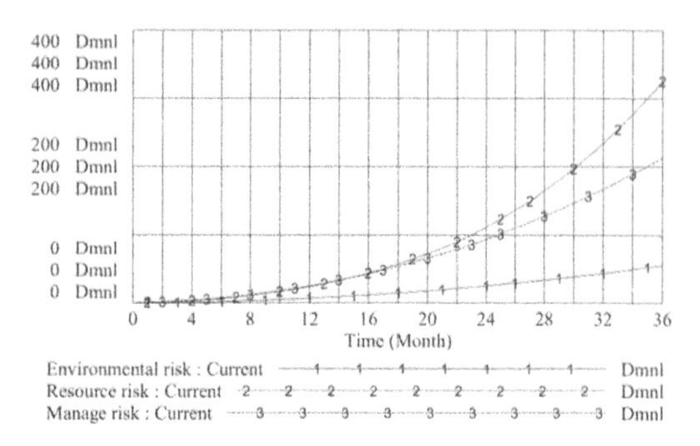


Figure 6. The risk level of the primary indicators of the concentralized distribution logistics.

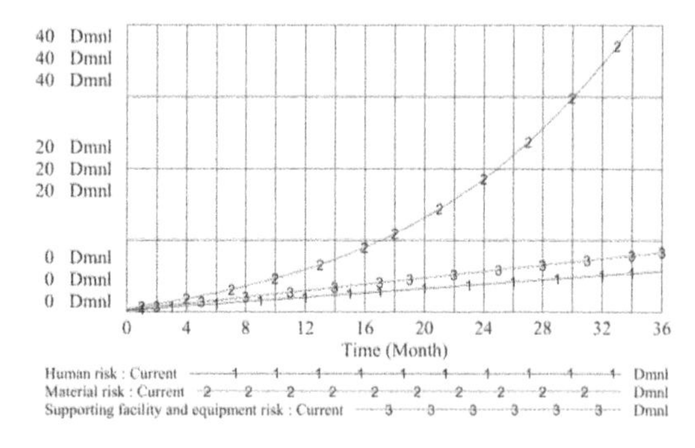


Figure 7. The risk level of secondary indicators of resource subsystem.

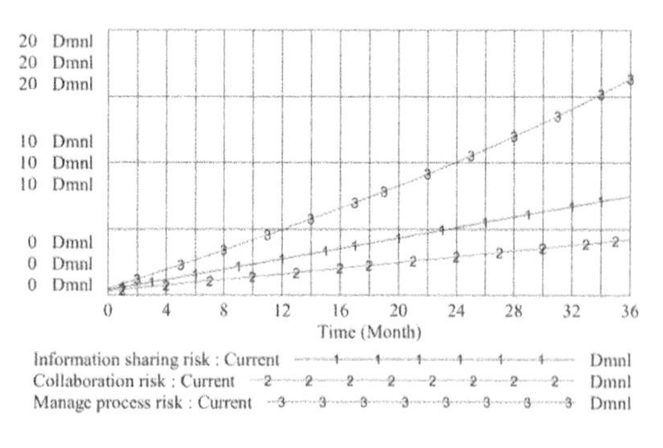


Figure 8. The risk level of secondary indicators of management subsystem.

Based on the validity verification of the model, the trend of risk level of CDL-CCBIM is simulated without taking any risk prevention measures. As can be seen from Figure 9, the risk level of CDL-CCBIM

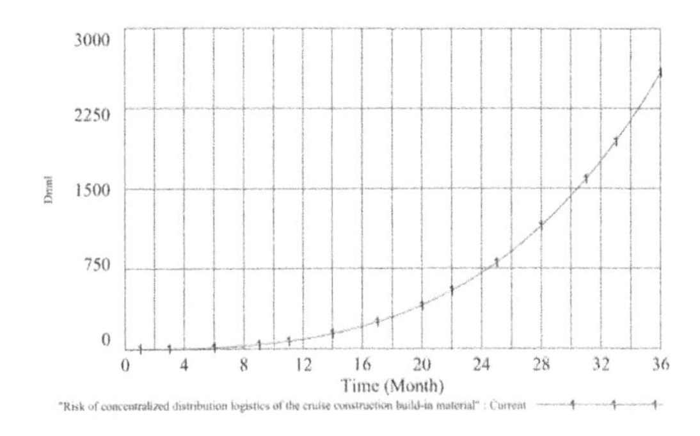


Figure 9. The risk level of the concentralized distribution logistics of the cruise construction build-in material over time.

has continued to rise, and finally reached the highest risk level of 2584.25. It is because the losses caused by risk are cumulative and irreversible. If risk is not controlled, it is very likely that the concentralized distribution logistics system of build-in material will be paralyzed, the progress of the build-in engineering will be interrupted, and the efficiency of cruise construction will be affected. At the same time, the risk level increased slowly from 0 to 12 months, and then the increase rate of risk level value accelerated significantly. It is because the build-in engineering is in the middle and later stages of the cruise construction cycle. The build-in material arrives in large quantities in the middle and late stages, and the concentralized distribution logistics of material is carried out on a large scale. As time goes on, under the interactive influence of various factors, the overall risk level of system increases at a faster rate. The risk level of CDL-CCBIM with time is shown in Table 5.

## 4.2 *Sensitivity analysis of system simulation*

Sensitivity refers to the magnitude to which a change of a single factor affects the system results. For the SD model, the sensitivity analysis is usually realized by changing the values of constants and then comparing the results. It is worth noting that a more sensitive variable leads to a higher variation in the system output. Several critical variables or parameters are selected for sensitivity analysis. For example, the initial value of the insufficient investment risk in the management process, partners problems risk, irregular operation management risk, training mechanism problem risk, and plan change risk was reduced by 50% to simulate the model. Comparing the data under different scenarios with the initial current, the change trend of the risk level of CDL-CCBIM is shown in Figure 10. In order to clearly compare the risk level values in each period, the Table Time is used for analysis, and the data change values are shown in Table 6.

The influence degree of the change of key risk factors on the risk level of CDL-CCBIM is compared and analyzed. Taking the initial current as the calculation standard, the results are as follows: changing the risk of insufficient investment leads to a change of 1.908% in system risk, changing the risk of partners problem leads to a change of 1.915% in system risk, changing the risk of irregular operation management leads to a change of 4.033% in system risk, changing the risk of training mechanism problem

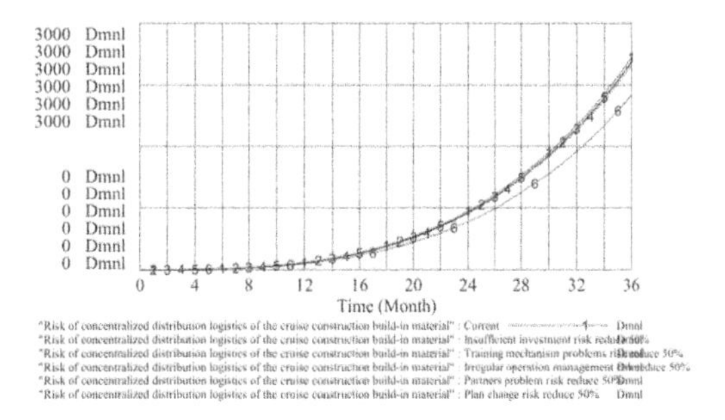


Figure 10. Simulation of system risk level under different adjustment schemes.

Table 5. The risk level of the concentralized distribution logistics of the cruise construction build-in material with time.

| Time (Month) | Concentralized distribution logistics risk level | Time (Month) | Concentralized distribution logistics risk level | Time (Month) | Concentralized distribution logistics risk level |
|---|---|---|---|---|---|
| 0 | 0 | 13 | 113.544 | 25 | 805.228 |
| 1 | 0.419618 | 14 | 140.819 | 26 | 909.979 |
| 2 | 1.2083 | 15 | 172.323 | 27 | 1024.27 |
| 3 | 2.6365 | 16 | 208.412 | 28 | 1148.67 |
| 4 | 4.97574 | 17 | 249.455 | 29 | 1283.77 |
| 5 | 8.49961 | 18 | 295.832 | 30 | 1430.18 |
| 6 | 13.4846 | 19 | 347.94 | 31 | 1588.56 |
| 7 | 20.2109 | 20 | 406.192 | 32 | 1759.57 |
| 8 | 28.9647 | 21 | 471.016 | 33 | 1943.9 |
| 9 | 40.0336 | 22 | 542.858 | 34 | 2142.28 |
| 10 | 53.7176 | 23 | 622.182 | 35 | 2355.47 |
| 11 | 70.3206 | 24 | 709.471 | 36 | 2584.25 |
| 12 | 90.1552 | | | | |

Table 6. Some synthetical risk values of the risk system with time.

| Time (Month) | Current | Insufficient investment risk reduce 50% | Partners problem risk reduce 50% | Irregular operation management risk reduce 50% | Training mechanism problems risk reduce 50% | Plan change risk reduce 50% |
|---|---|---|---|---|---|---|
| 0 | 0 | 0 | 0 | 0 | 0 | 0 |
| 6 | 13.48 | 13.35 | 13.35 | 13.20 | 13.32 | 12.21 |
| 12 | 90.16 | 88.66 | 88.66 | 87.07 | 88.30 | 76.11 |
| 18 | 295.83 | 290.30 | 290.28 | 284.38 | 288.95 | 243.74 |
| 24 | 709.47 | 695.75 | 695.71 | 681.07 | 692.40 | 580.27 |
| 30 | 1430.18 | 1402.66 | 1402.58 | 1373.21 | 1395.94 | 1171.01 |
| 36 | 2584.25 | 2535.85 | 2535.70 | 2484.06 | 2524.03 | 2128.47 |

leads to a change of 2.386% in system risk,2.386% and changing the risk of plan change leads to a change of 21.414% in system risk. According to this, the degree of influence on the risk level of CDL-CCBIM are as follows: plan change, irregular operation management, training mechanism problem, partners problems and insufficient investment.

## 5 DISCUSSION

Without considering the risk prevention strategy, this paper simulates the trend of the risk level of CDL-CCBIM. Results show that the interactions between risks can cause rise in concentralized distribution logistics risk of cruise construction build-in material throughout whole phase. This interaction between risks is called risk coupling. In the field of risk management, risk coupling refers to the degree of interdependence and interaction among different risk factors in the process of complex system activities (Xue et al. 2013). In the risk management of CDL-CCBIM, it is necessary to clarify the path of risk propagation and determine the interaction between risks. The management of key risks and their front-end risk portfolio can block the path of risk propagation from the root cause. This is also the work to be carried out in the next stage of this paper, to explore the impact of different combinations of prevention strategies on the risk level.

Cruise construction is a complex system engineering. Planning management is the key link, involving production planning, purchase planning, material arrival planning, storage planning, distribution planning, installation planning and delivery planning (Zheng et al. 2021). Therefore, the resources coordination and planning control of cruise construction have a strong complexity. Once a certain link occurs different degrees of disjointed planning problems, will lead to production delays. As can be seen from the section 4, the plan change has the greatest impact on the risk of CDL-CCBIM. Material demand planning and arrival planning will seriously affect storage planning, pallets match planning and distribution planning. In turn, distribution planning also affects production planning. The risk propagation path and consequence of the interruption of concentralized distribution logistics in cruise construction caused by plan change deserve more attention. Improve the flexibility of the plan and establish a rapid response to the disruption of logistics distribution system caused by the plan change. Strengthen the coordination and supervision of the main contractor and subcontractor of the build-in project in material distribution information sharing.

## 6 CONCLUSIONS

Because of the complexity, multi-participation of CDL-CCBIM, there are various risk factors that affect sustainable production. The particularity of build-in engineering becomes the most complicated part of cruise construction, how to conduct effective risk management and how to predict and respond to risks in advance will become a necessary research work. In this paper, a system dynamics model is constructed to evaluate and simulate the risk of CDL-CCBIM. The main contribution of this paper are as follows:

(1) A new perspective of risk management of CDL-CCBIM is put forward. Based on analyzing the investigation results, a clear and structured risk assessment index system is constructed. Combined with SD theory, the causal interaction between system structure and risk factors was clarified.
(2) The SD model can be constructed to quantitatively simulate the risk level of CDL-CCBIM in different periods and provide risk control parameters. The simulation results show that when no preventive strategy is adopted, the risk level shows an upward trend, and the highest risk value reaches 2584.25. With the interaction of various risk factors, the rising rate accelerates in the middle and late period, which affects the construction progress of build-in engineering.

(3) By changing the risk values of key factors, and comparing the overall risk level of the system. The plan change risk has the greatest impact on CDL-CCBIM. Managers should attach importance to the ability to cope with and balance the plan changes for the life cycle of cruise construction.

However, As the build-in engineering of the first cruise in China has just started, the identification of the risk factors of the concentralized distribution logistics of build-in material in this paper is inevitable to omit. And the model constructed in this article is based on experts' inquiry. In the next step, risk thresholds can be set to timely control system risk, and the implementation effects of different prevention strategies can be studied.

## REFERENCES


Baliwangi, L., Arima, H., Artana, K.B. and Ishida, K. (2007) Risk modification through system dynamics simulation. Proceedings of the IASTED International Conference on Modelling and Simulation, 350–354.

Chen B. (2011) Research on risk management of shipbuilding supply chain based on SCOR. JiangSu University of Science and Technology

Chen S., M. Zhang, Y. Ding, et al. (2020) Resilience of China's oil import system under external shocks: A system dynamics simulation analysis. *Energy Policy*, 146, 111795.

Diaz R., K. Smith, B. Acero, et al. 2020. Developing an Artificial Intelligence Framework to Assess Shipbuilding and Repair Sub-Tier Supply Chains Risk. In *2nd International Conference on Industry 4.0 and Smart Manufacturing (ISM)*, 996–1002. Electr Network.

Ding Z., M. Zhu, V. W. Y. Tam, et al. (2018) A system dynamics-based environmental benefit assessment model of construction waste reduction management at the design and construction stages. *Journal of Cleaner Production*, 176, 676–692.

Kwesi-Buor J., D. A. Menachof & R. Talas (2019) Scenario analysis and disaster preparedness for port and maritime logistics risk management. *Accident Analysis & Prevention*, 123, 433–447.

Li H., S. Meng & H. Tong (2021) How to control cruise ship disease risk? Inspiration from the research literature. *Marine Policy*, 132, 104652.

Liu A., K. Chen, X. Huang, et al. (2021) Dynamic risk assessment model of buried gas pipelines based on system dynamics. *Reliability Engineering & System Safety*, 208, 107326.

Mello M. H. & J. O. Strandhagen (2011) Supply chain management in the shipbuilding industry: challenges and perspectives. *Proceedings of the Institution of Mechanical Engineers Part M-Journal of Engineering for the Maritime Environment*, 225, 261–270.

Moumeni M., R. Nozaem & M. Dehbozorgi (2021) Quantitative assessment of the relative tectonic activity using the analytical hierarchy process in the northwestern margin of the Lut Block, Central Iran. *Journal of Asian Earth Sciences*, 206, 104607.

Pan Y. (2021) Alternative ideas for localization of cruise ship's build-in materials. *China Ports*, 10, 31–33.

Peng M., Y. Peng & H. Chen (2014) Post-seismic supply chain risk management: A system dynamics disruption analysis approach for inventory and logistics planning. *Computers & Operations Research*, 42, 14–24.

Ragele C, Hillson D, Grimaldi S.2005. Understanding project risk exposure using the two-dimensional risk breakdown matrix. Pennsylvania: Project Management Institute.

Sender J., S. Klink & W. Flügge (2020) Method for integrated logistics planning in shipbuilding. *Procedia CIRP*, 88, 122–126.

Song S., J. C. L. Goh & H. T. W. Tan (2021) Is food security an illusion for cities? A system dynamics approach to assess disturbance in the urban food supply chain during pandemics. *Agricultural Systems*, 189, 103045.

Wondirad A. (2019) Retracing the past, comprehending the present and contemplating the future of cruise tourism through a meta-analysis of journal publications. *Marine Policy*, 108.

Xue Y., Y. Liu & T. Zhang (2013) Research on formation mechanism of coupled disaster risk. *Nature Reviews Disease Primers*, 22, 44–50.

Yu K., Q. Cao, C. Xie, et al. (2019) Analysis of intervention strategies for coal miners' unsafe behaviors based on analytic network process and system dynamics. *Safety Science*, 118, 145–157.

Yue W., Q. Zhang & Ieee. 2008. Research on the Shipbuilding Supply Chain Risk Control. In *IEEE International Conference on Automation and Logistics*, 2205-+. Qingdao, PEOPLES R CHINA.

Zheng Y., J. Ke & H. Wang (2021) Risk Propagation of Concentralized Distribution Logistics Plan Change in Cruise Construction. *Processes*, 9.

Zhu J., H. Wang & J. Xu (2021) Fuzzy DEMATEL-QFD for Designing Supply Chain of Shipbuilding Materials Based on Flexible Strategies. *Journal of Marine Science and Engineering*, 9.

*Trends in Maritime Technology and Engineering – Guedes Soares & Santos (eds)*
 *ISBN 978-1-032-33583-4*

# Short sea shipping gas emissions and dispersion

Y. Garbatov
*Centre for Marine Technology and Ocean Engineering (CENTEC), Instituto Superior Técnico, Universidade de Lisboa, Lisboa, Portugal*

P. Georgiev
*Technical University of Varna, Varna, Bulgaria*

ABSTRACT: This work aims to analyse the gas emissions of a multipurpose fleet operating in the Black Sea. The vessels were designed to satisfy all naval architectural characteristics employing the lifecycle cost assessment and accounting for the energy efficiency. The gas emissions of NOx, SO2 and PM of the fleet of five vessels during the voyage, port operation and queuing in the seaport in front of the port of Varna are evaluated. The geographical coordinates of queuing locations, number of queuing ships that are seen as a source of pollution, wind speed and direction, global solar radiation, geographical receptor coordinates are used to execute the Gaussian dispersion model over a distance coinciding with seaports where the ships are queuing. The ground-level concentrations on preselected receptors in Varna are estimated. The analysis concluded that the high levels of air emissions indicate tendencies for possible violation of the maximum allowable concentrations.

## 1 INTRODUCTION

Despite an already developed network of short sea shipping in Europe, there is a need to employ new technologies to comply with the latest environmental legislation (Santos & Guedes Soares, 2017, 2019, ESSN, 2020).

The gas emissions like nitrogen oxides, sulphur dioxide and particulate matter have a significant impact (Sorte et al., 2020) by violating the maximum allowable concentrations that reflect the air quality in the populated harbour areas (Lonati et al., 2010) and human health brook (Lu et al., 2006). Additionally, more pollution from ships can be seen originating from the oil spill, ballast water, solid waste etc., (Eyring et al., 2005). There is a significant impact of gas emissions on air quality during ship navigation, queuing, and port operations (Toscano & Murena, 2019). Approximately 70% of the ship emissions are estimated to occur within 400 km from onshore (Eyring et al., 2010).

The particular matters and nitrogen oxides originating from ships were examined by (Cohan et al., 2011), identifying that the affected area was within 2–6 km of the port. The ships operating in port showed that the relatively high sulphur content (up to 3.5%) in marine fuels led to elevated sulphur dioxide emissions (Davies, 2014).

The ship emission of nitrogen oxides, sulphur dioxide and particulate matter can be modelled by employing different dispersion models considering wind speed and direction, global solar radiation over a distance. For this purpose, Gaussian and Eulerian dispersion models have been developed and widely used (Gibson et al., 2013, Milazzo et al., 2017).

A stochastic model for predicting the dispersion of the pollution concentration has been developed by Garbatov & Georgiev (2022), where the uncertainties from the input data and mathematical model were accounted for in the exceedance of the threshold limits demonstrating its importance for a regulatory program evaluation.

The study presented here is based on the well-known methodologies for ship emission estimation and impact valuation. The analysis brings more inside the problem of controlling the ship's environmental performance when navigating in the open sea, queuing in front of the port or during port operations.

However, the transport demand continues growing, and the emissions need to be reduced in Europe to comply with the level stipulated by (Parlament, 2016, EU, 2021). In this respect, ship emissions mitigation and control are of significant importance. The objectives here are to quantify the emissions of a new design fleet of multipurpose ships operating in the Black Sea basin and their impact on air quality during the voyage, queuing, and port operation that can be further used in optimising the implementation of alternative ship power supply and to introduce more stringent maritime policies aimed at reducing emissions to air from ships.

DOI: 10.1201/9781003320289-4

The study includes several sections covering the introduction, emission estimation of fleet operating between the ports in the Black Sea basin, followed by the use of the Gaussian Dispersion Model for defining the gas concentration reaching the port of Varna and their impact on preselected populated locations and finally some conclusions are derived.

## 2 EMISSION ESTIMATION

A fleet of five multipurpose ships is analysed concerning the generated gas emissions. The fleet is used to transport cargoes between Varna, Odesa, Novorossiysk, Poti and Istanbul (Georgiev & Garbatov, 2021), see Figure 1. The characteristics of the ships range as $L \in [130,150]$ *m*, $B \in [24.53,28.3]$ *m*, $D \in [11.3,13.04]$ *m*, $C_b = 0.63$, *m*, $SFC_{PE} \in [167,182]$ *g/kWh*, $MCA \in [1320,1540]$ *kW*, $SFC_{AE} = 189 g/kWh$, $v_s = 16$ *knots*. The distance and transported cargo, *nm/tonnes* are Varna - Poti 613/ 2,121,435, Varna - Istanbul 149 /1,123,242, Varna - Novorossiysk 440/579,172 and Varna - Odessa 244 /569,358.

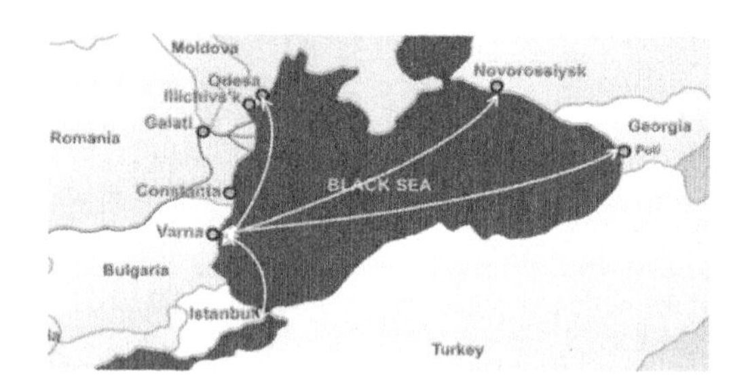


Figure 1. Black Sea routes.

The implementation of environmental requirements for existing ships, the Energy Efficiency Design Index, *EEDI* reference, is defined based as suggested in (MEPC, 2011) and its required value as stipulated in (MEPC, 2012), which is adjusted by a reduction factor relative to *EEDI* baseline, as a function of the propulsion and auxiliary engine power, deadweight and service speed. The energy efficiency design index, *EEDI*, calculated for the analysed ships is given in Figure 2.

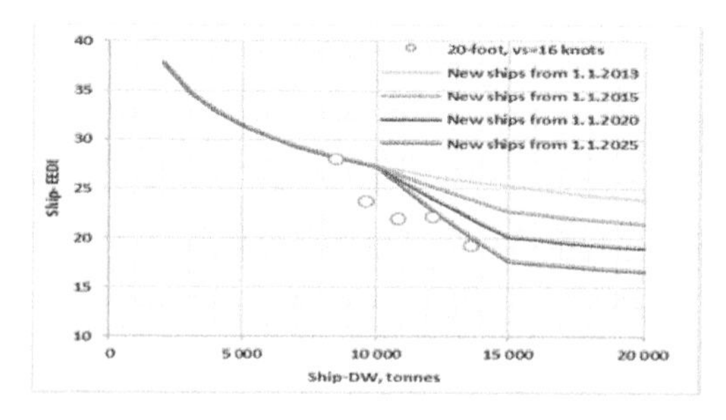


Figure 2. Energy efficiency design index.

The gas emissions are estimated following the methodology for quantifying ship emissions as stipulated by the (EMEP/EEA, 2019) air pollutant emission inventory guidebook using the bottom-up approach.

Estimating the gas ship emissions, the ship's operational activities are navigation, port, and queuing. The last one is the ship's time when it is anchored, waiting for the next voyage or cargo loading and unloading operation. The total ship emissions are the sum of the emissions related to all operational activities during one voyage. Special attention is paid to ship queuing and the consequences of emission pollution.

The time the ship spent in port, $T_{port}$ in one voyage is defined assuming that is based on the crane capacity, $C_{crane}$ =360 *TEU*-containers per day, the number of cranes used for loading and unloading in the port and can be estimated as:

$$N_{cranes} = 0.0187 L_{oa} + 0.3572 \tag{1}$$

leading to:

$$T_{port} = \frac{2 N_{ports} N_{TEU}}{3 C_{crane} N_{cranes}}, \text{ day} \tag{2}$$

where $N_{ports}$ is the number of port visits during one voyage and $N_{TEU}$ is the number of containers loaded and unloaded during one voyage.

The queuing time, $T_{queuing}$ in one voyage is assumed as:

$$T_{queuing} = \frac{3\left(2 N_{ports} - 1\right)}{24}, \text{ day} \tag{3}$$

The navigation time, $T_{navigation}$ in one voyage is defined as:

$$T_{navigation} = \frac{D_{round}}{24 v_s}, \text{ day} \tag{4}$$

where $D_{round}$ is the round distance and $v_s$ is the ship speed.

The voyage time, $T_{voyage}$ is estimated as:

$$T_{voyage} = T_{navigation} + T_{queuing} + T_{port} \tag{5}$$

The gas emissions are defined as the installed propulsion and auxiliary engine power and used fuel type. It has been assumed that the propulsion and the auxiliary engines use Marine Diesel Oil, *MDO*.

The engine load factors are defined as the actual power output relative to the maximum power. The engine time factors are estimated as the actual time spent in specific operational activity relative to the total duration. The factors assumed for calculating the gas pollutants, uses $LF_{PE,i}$, which is the propulsion engine load factor, $LF_{AE,i}$ is the auxiliary engine load factor, and $TF_{PE,i}$ is the propulsion engine time of operation during the different activities in the voyage, where $LF_{PE,nav}$=80%, $LF_{PE,port} = 20\%$, $LF_{PE,queuing} = 20\%$, $TF_{PE,port} = 5\%$, $TF_{PE,queing} = 100\%$, $LF_{AE,nav} = 30\%$, $LF_{AE,port} = 40\%$, $LF_{AE,queuing} = 50\%$.

The emissions depend on several factors, including the propulsion and auxiliary engine and fuel types. The ships analysed here are assumed to be Medium-Speed Diesel, *MSD*, and the fuel type is *MDO*, and the emission factors (EU, 2021) used in the analysis are $EF_{PE,\ NOx} = 10.6\ g/kWh$, $EF_{PE,\ SO2} = 6.8g/kWh$ $EF_{PE,\ PM10} = 1.2g/kWh$, $EF_{AE,\ NOx} = 13.9\text{g/kWh}$, $EF_{AE,\ SO2} = 6.5g/kWh$, $EF_{AE,\ PM10} = 0.4\text{g/kWh}$.

The gas emissions, $E_i$ are estimated by multiplying the time spent in different activities, $T_i$ with the sum of the installed propulsion, $P_{PE}$ and auxiliary, $P_{AE}$ engine power, load factors and the operational and emission factors for propulsion, $EF_{PE,i}$ and auxiliary, $EF_{AE,i}$ engines. The emissions for the $i^{th}$ operating mode (Ramalho & Santos, 2021) are calculated as follows:

$$E_i = \left[\left(P_{PE}LF_{PE,i}EF_{PE,i}TF_{PE,i}\right) + \left(P_{AE}LF_{AE,i}EF_{AE,i}TF_{AE,i}\right)\right]T_i \tag{6}$$

The estimation for the fuel consumption and gas emissions of the fleet ships voyaging between Varna – Odesa – Novorossiysk – Poti - Istanbul is shown in Figure 3 and Figure 4. It is also assumed that the more significant number of arrivals in the port will produce more emissions, and with this respect, five ships are considered queuing simultaneously. The ship emissions Varna – Poti - Varna and different emissions for different routes are presented in Figure 5.

Some segments of maritime transport operate today with ships of very high speeds, which requires a high-power machine, with high fuel consumption and large production quantities of polluting gases. A comparison is made between three fleets of the same ships operating between the same ports with different service speeds of 12, 16 and 18 knots. The estimated gas emissions for the three fleets are shown in Figure 6.

It appears that the decrease in the fleet speed generates a significant reduction in the power of ships, therefore lower fuel consumption and also less removal of gas emissions from propulsion and auxiliary engines. However, the slowdown in ship operation needs to increase the number of ships. Even with an increase in the number of ships, the fuel consumption, and emissions of gases from the propulsion and auxiliary engines with lower speeds are lower than the fleets with higher speeds.

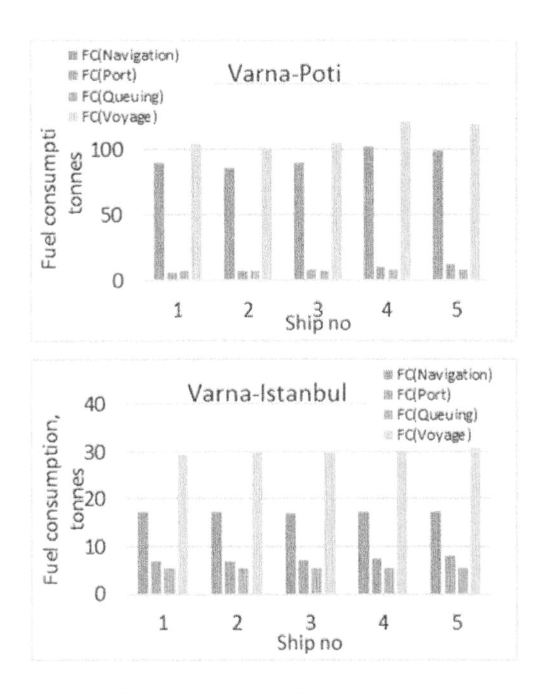


Figure 3. Fuel consumption, Varna – Poti - Varna (up), Varna – Istanbul - Varna (down).

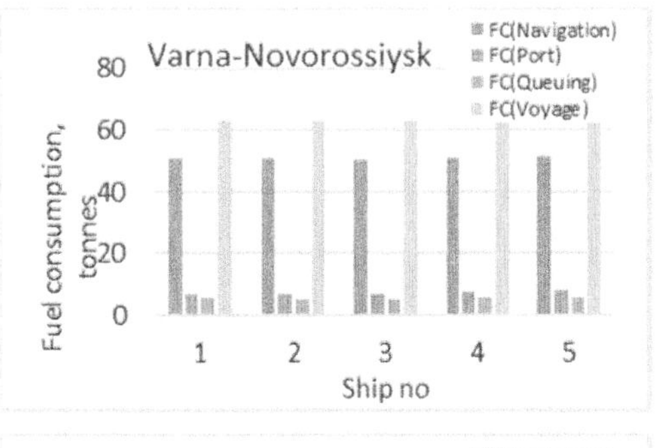


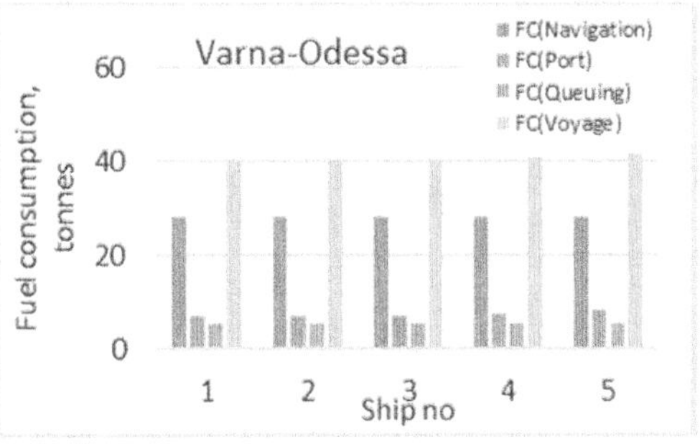


Figure 4. Fuel consumption, Varna – Poti - Varna (up), Varna – Istanbul - Varna (down).

The total damage cost as a result of pollution generated by the fleet transporting goods in different routes, $D_j$ is estimated as:

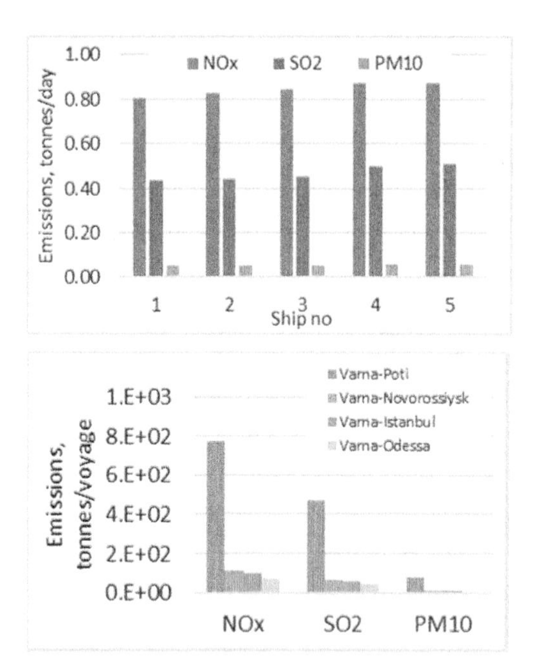


Figure 5. Ship emissions Varna – Poti - Varna (up) and for different routes (down).

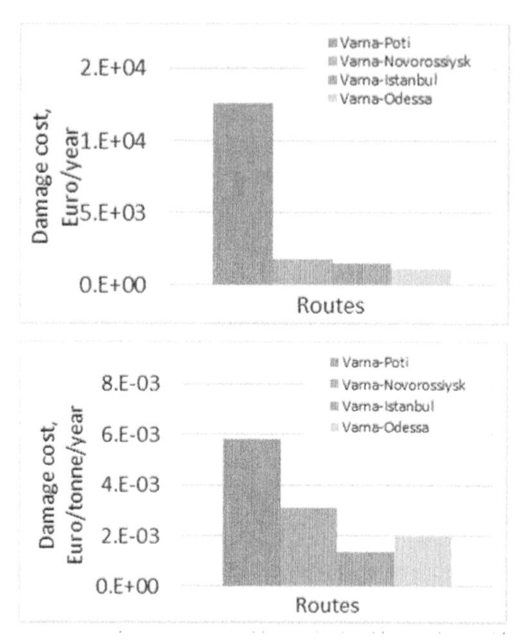


Figure 7. Feet damage cost, *€/year* (up), *€/tonne/year* (down) through different routes.

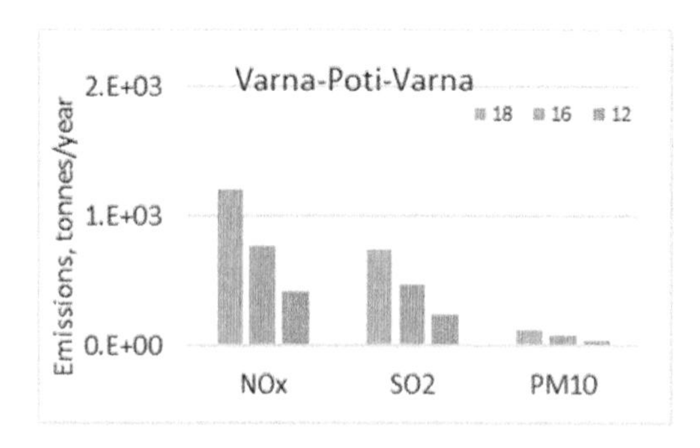


Figure 6. Emissions for different speeds.

$$D_j = d_{j,\,SO_2} E_{j,\,SO_2} + d_{j,\,NO_x} E_{j,NO_x} + d_{j,PM_{10}} E_{j,\,PM_{10}} \tag{7}$$

where $E_{j,\,k}$ is the total emissions of pollutant $k$ in the route $j$, $d_{j,\,k}$ is the damage cost of pollutant $k$ in route $j$. The average damage costs as defined by the national agencies for maritime emissions related to the health effect, crop loss, biodiversity loss, material damage etc., the average damage cost for $NO_x$, $SO_2$ and $PM_{10}$ is 7.8*€/kg*, 11.1*€/kg* and 17.1*€/kg* (van Essen et al., 2019).

The damage cost produced by the fleet circulating in different routes is estimated as $D = \sum_j D_j$. The estimate made with Eqn 7 is used to compare the damage cost between various scenarios without considering the complex atmospheric interaction and relationships between the emissions. Figure 7 shows the feet pollution damage cost in different routes in *€/year* (up) and *€/tonne/year* (down). Figure 7 (up) shows that the longer distance routes contribute more to the pollution damage. However, when the pollution damage is normalised by the cargo transported, the most extended distance dominates the pollution damage. Some of the shortest distance is not that critical, as, for example, the route Varna-Istanbul-Varna moves to the fourth position due to its cargo volume.

## 3 GAUSSIAN PLUME DISPERSION

The Gaussian Dispersion Model, *GDM* (Pasquill, 1961, Henderson-Sellers & Allen, 1985), is widely used in predicting the maximum ground-level concentration from a continuous point-source plume at a given distance downwind. The *GDM* provides a solution to the dispersion of pollutants under specific constraints. The model predicts the concentrations at any given location, for any set of contaminants, the nature of surface downwind from the stack, meteorological conditions, and any period. It is assumed that particulate matter less than $20\mu m$ behaves as gas and can be analysed using the *GDM*. However, the currently available *GDM* is not fully capable of their predictions. The application of the *GDM* is presented in Figure 8.

It is also assumed that the effluents leave the stack with sufficient momentum and buoyancy. The hot gases continue to rise, and the plume is deflected along its longitudinal axis, proportional to the wind speed.

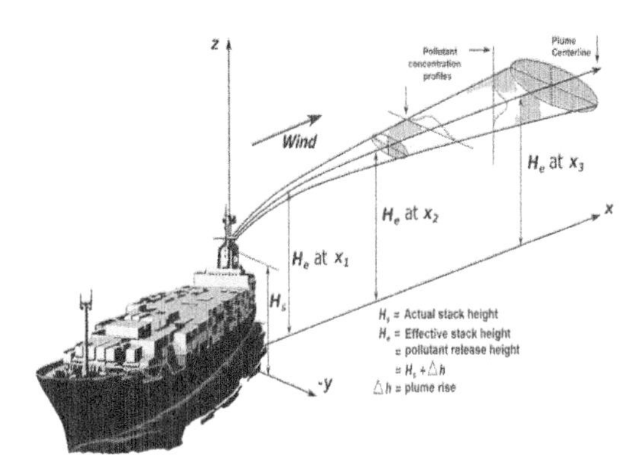


Figure 8. Gaussian Dispersion Model (Garbatov & Georgiev, 2022).

The three-dimensional *GDM* for estimating the concentration of emissions downwind from a pollution source, accounting for ground reflection, is defined as (Sutton, 1932):

$$C(x,y,z) = \frac{Q}{2\pi\sigma_y(x)\sigma_z(x)u} exp\left(-\frac{y^2}{2\sigma(x)_y^2}\right) \left\{exp\left[-\frac{(z+H)^2}{2\sigma(x)_z^2}\right] + exp\left[-\frac{(z-H)^2}{2\sigma(x)_y^2}\right]\right. \quad (8)$$

where the mass-flow rate of emissions is given as $Q$, $u$ is the wind speed, $H$ is the effective height of the plume, considering the stack, $h_s$ and plume rise $\Delta h$. The first term represents the gas concentration of the centreline of the plume. The second term shows the concentrations moving in the lateral direction, and the third term shows how the concentrations move in the vertical direction.

The concentration at the ground level $z = 0$, accounting for a no ground reflection, used to analyse particles, nitric acid vapour propagation is defined as:

$$C(x,y,0) = \frac{Q}{2\pi\sigma_y(x)\sigma_z(x)u} exp\left(\frac{-y^2}{2\sigma(x)_y^2}\right) exp\left(-\frac{H^2}{2\sigma(x)_z^2}\right) \quad (9)$$

The plume can expand outward from the centreline in the horizontal, $y$ and vertical, $z$ planes. The sigma terms give the extent of the plane expansion, $\sigma_y(x)$ and $\sigma_z(x)$. They are dependent on the stability of the idealised atmospheric conditions and the arbitrary distance $x$ from the source to the receptor.

The idealised *GDM* conditions are subjected to a point source continuous mass-flow rate of emissions, $Q$, as Pasquill (1961) defined, where a stability category is assumed to characterise a stable atmosphere. A horizontal wind velocity is constant. The wind is also considered to be in the direction of the receptor, where the vertical and crosswind dispersion terms, $\sigma y(x)$ and $\sigma z(x)$, according to the Gaussian distribution, are constants to a given distance downwind. The dispersion of the emissions assumes an expanding cone-shaped plume implicitly requiring homogeneous turbulence throughout the $x$, $y$, and $z$ -planes of the plume.

The estimation of the *GDM*, where $C(x,y,z)$ is for one-hour average concentrations. There is no deposition, washout, chemical conversion, or absorption of pollutants by the ground or other physical bodies, and no chemical conversion of the contaminants like a secondary formation of *PM* or ozone formation.

The dispersion of the pollutants is a function of the stability of the environment into which they are released. In an unstable atmosphere, the pollutants tend to rise unhindered unless acted upon by other forces, such as the incoming air mass. In a stable atmosphere, the contaminants remain at or near the height of the release point and are assumed to travel downwind until they reach the ground.

Pasquill (1961) introduced six categories, *A, B, C, D, E*, and *F*, to define the strength of incoming sunlight (insolation) and cloud cover to represent atmospheric stability. *A* is the most unstable environment, and *F* is the most stable. The stability classes *A, B*, and *C*, reflect unstable atmospheric conditions; category *D* reflects neutral stability characteristics; and categories *E* and *F* reflect stable conditions, where F reflects the most extreme harsh conditions and is rarely used in research.

The mean wind speed of 2*m/s* to3 *m/s* is related to the *E* stability class. Because pollutant concentrations increase with lower wind speeds, the lower of the range was selected for this analysis to reflect the most conservative case. Thus, a mean wind speed of 2*m/s* is applied to this analysis.

As the pollutants travel within the plume in the downwind direction and throughout the distance to $x$, the plume expands to some size in $y$, and $z$ -directions. The degree to which the plume extends with the length in vertical and horizontal directions is defined by the dispersion coefficients $\sigma_y(x)$ and $\sigma_z(x)$ representing the atmospheric stability and turbulence category in the downwind distance, $x$, from the air pollution source.

The vertical and crosswind dispersion occurs according to the Gaussian distribution. It is constants to a given downwind distance, $x$, where the expansion of the plume assumes an expanding conical plume implicitly requiring homogeneous turbulence throughout the plume. The concentrations are symmetric about the $y$, and $z$ -axes. Wind speed varies with height, and it is measured at the plume height.

The source and wind coordinate are needed in defining the dispersion of the concentration. The source geographical location for any arbitrary i$^{th}$ source is determined as $x_{S,i}$, $y_{S,i}$, $z_{S,i}$, the source coordinate system is defined as $x_i$, $y_i$, $z_i$, the wind coordinate system is $x_W$, $y_W$, $z_W$, where the $x_W$ axis is parallel to the direction of the wind velocity, u. The receptor locations are identified as $x_{R,j}$, $y_{R,j}$, $z_{R,j}$ for any arbitrary receptor j. The vertical coordinates are the same for the wind, source, and geographical location, $z = z_W = z_{S,i}$. The transformation from the geographical location to the source coordinate system is given as:

$$\begin{aligned} x_{i,j} &= \left(x_{S,i} - x_{R,j}\right) sin\theta + \left(y_{R,i} - y_{S,i}\right) cos\theta \\ y_{i,j} &= \left(x_{S,i} - x_{R,j}\right) cos\theta + \left(y_{R,i} - y_{S,i}\right) sin\theta \end{aligned} \tag{10}$$

where $\theta$ represents the angle of the wind direction, accounting for the wind direction, the concentration can be calculated as:

$$C_{i,j}\left(x_{i,j}, y_{i,j}, 0\right) = \frac{Q}{2\pi\sigma_y\left(x_{i,j}\right)\sigma_z\left(x_{i,j}\right)u} exp\left(-\frac{y^2}{2\sigma\left(x_{i,j}\right)_y^2}\right) exp\left(-\frac{H^2}{2\sigma\left(x_{i,j}\right)_z^2}\right) \tag{11}$$

The ground reflection increases the concentration at the level of $z$ =0, but it does not continue indefinitely. It is expected that the diffusion decreases the concentration in the $y$, and $z$ -directions. The concentration estimates for different sampling times can be calculated for other periods.

The *GDM* is a good numerical model for estimating the gas concentration on a scale of several kilometres. For finer scales needs to account for the dispersion very close to the source. For uniformly wind concentration of well-defined sources seems to be correct for up to several tens of kilometres. For longer distances, the pollution dispersion model needs to be coupled with the weather prediction code.

The concentration at any receptor is defined as a sum of the source pollutions are reaching the location of the receptor, $x_{R,j}, y_{R,j}$ estimated as:

$$C_j\left(x_{R,j}, y_{R,j}, 0\right) = \sum_i C_{i,j}\left(x_{i,j}, y_{i,j}, 0\right) \tag{12}$$

Several steps in analysing the gas emissions of nitrogen oxides, sulphur dioxide and particulate matter of a multipurpose ship fleet of five vessels during the queuing in the seaport in front of the port of Varna are employed. The first step is related to identifying the geographical coordinates of the seaport (queuing locations) and queuing ships are seen as a source of pollution, next is to identify the weather conditions, the number of queuing ships and the type of pollutant (nitrogen oxides, sulphur dioxide and particulate matter), wind speed and direction, global solar radiation, geographical receptor coordinates and finally using of the Gaussian dispersion model to identify the pollution concentration at any receptor.

## 4 POLLUTION CONCENTRATION IMPACT DUE TO QUEUING SHIPS

The Gaussian dispersion model is employed here to estimate the background concentration of $NO_x$, $SO_2$ and $PM_{10}$ emitted from a multipurpose ship fleet composed of five ships queuing in the front of the port of Varna. Two queuing places are analysed: Seaport *A* and *B*. Blue and yellow rectangles mark the queuing ports on the map shown in Figure 9. Dots mark the locations of ships for Seaport *A* and *B*, respectively. The ships are identified as the source of pollution. Ten points located on the onshore line of the city of Varna are defined as receptors marked by triangles on the map in Figure 9. The wind is 2 *m/s*, and the direction of the wind is shown in Figure 9 with an arrow, and the stack height for all ships is assumed 12 *m*.

The emissions of $NO_x$, $SO_2$ and $PM_{10}$ generated from the fleet of multipurpose ships are given in Table 1.

Using *GDM* and accounting for the point sources of the gas emission and assuming that the stationary conditions exist in the short run, the pollution concentration reaching the receptors are estimated. Then the sum of the contamination from all sources on

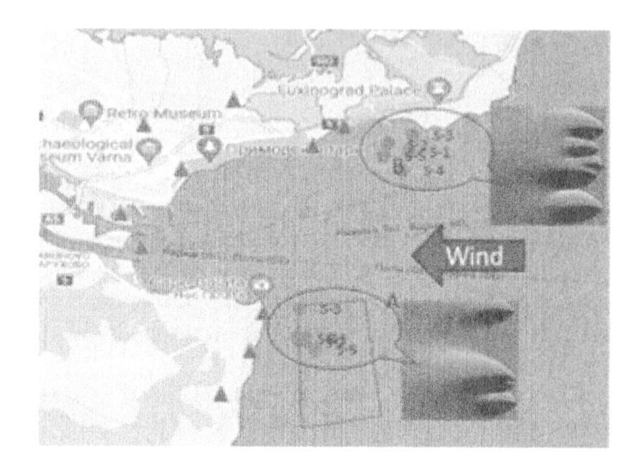


Figure 9. Sources (ships) and receptors.

Table 1. Emission of $NO_x$, $SO_2$ and $PM_{10}$.

| Ship n° | $NO_x$, kg/day | $SO_2$, kg/day | $PM_{10}$, kg/day |
|---|---|---|---|
| 1 | 805 | 435 | 52 |
| 2 | 824 | 444 | 52 |
| 3 | 846 | 454 | 53 |
| 4 | 924 | 500 | 60 |
| 5 | 947 | 511 | 60 |

each receptor gives the total concentration reaching any receptor.

The straightforward application of the *GDM* gives the pollution dispersion as a function of the $x$ and $y$-directions, as can be seen in Figure 10 and Figure 11.

To mitigate the pollution level from different ships, an optimisation procedure is employed, where the gas emission from each ship, represented as a source of pollution, is reduced to meet the receptor Maximum Allowable Concentration standards by minimising the air pollution cost expenses, cleaning only the excessive emission. The cost of cleaning is considered here as 0.5 kg/€. The Maximum Allowable Concentration, *MAC,* is considered as $MAC_{NOx}$ =30 $\mu g/m^3$, $MAC_{SO2}$=20 $\mu g/m^3$ and $MAC_{PM10}$=20 $\mu g/m^3$.

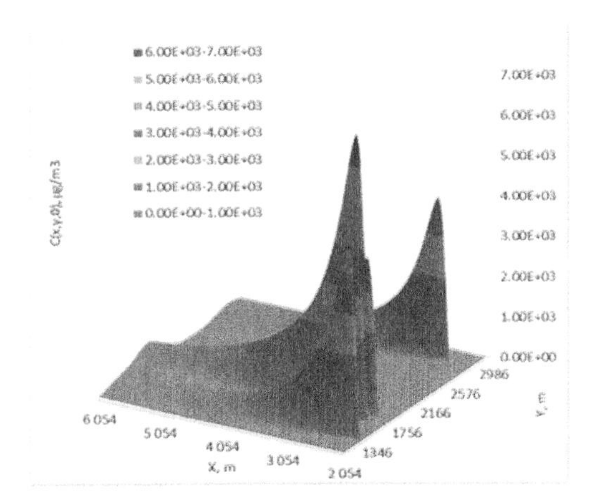

Figure 10. $NO_x$ dispersion, Seaport *A*.

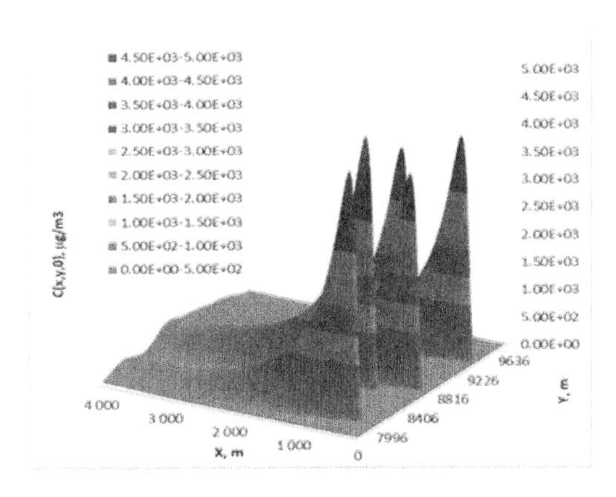

Figure 11. $NO_x$ dispersion, Seaport *B*.

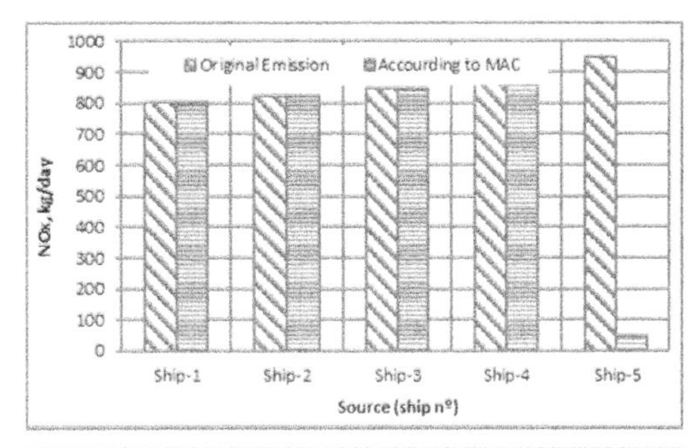


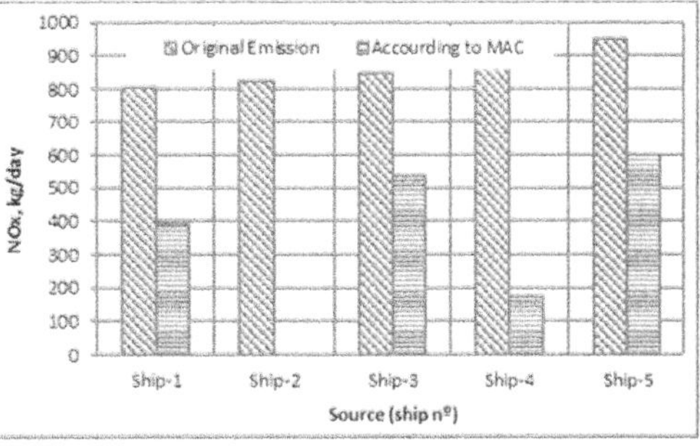


Figure 12. Emissions concentration, Seaport *A* (up) *B* (down), $NO_x$.

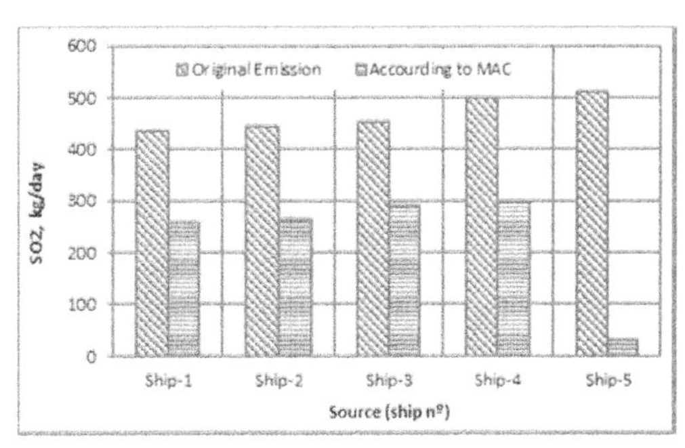


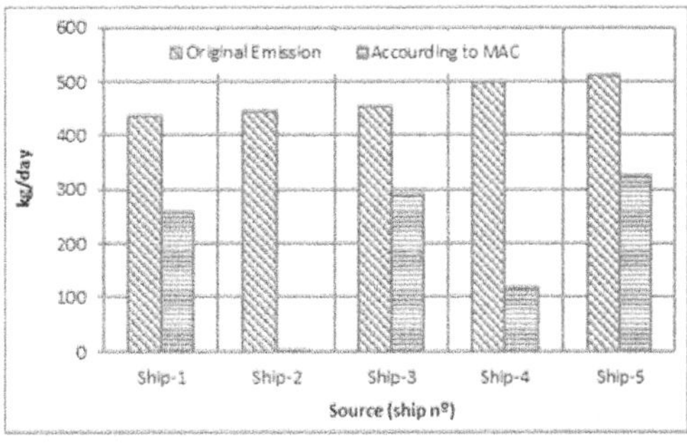


Figure 13. Emissions concentration, Seaport *A* (up) and Seaport *B* (down), $SO_2$.

Emissions for $NO_x$, $SO_2$ and $PM_{10}$ originating from different ships, waiting in Seaport *A* and *B* and their mitigated values to satisfy the *MAC* for any individual ship are shown in Figure 12 to Figure 14.

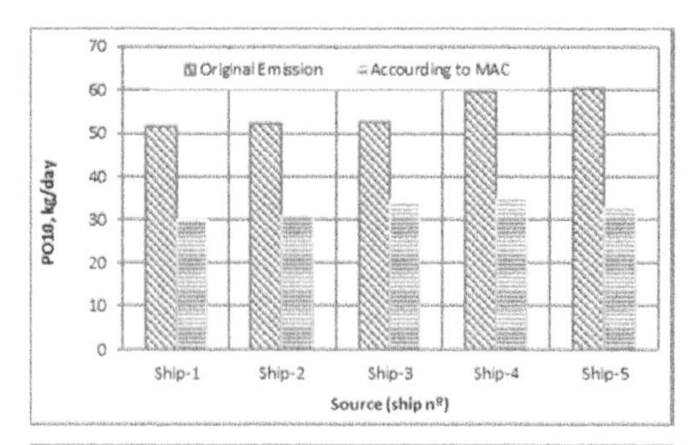


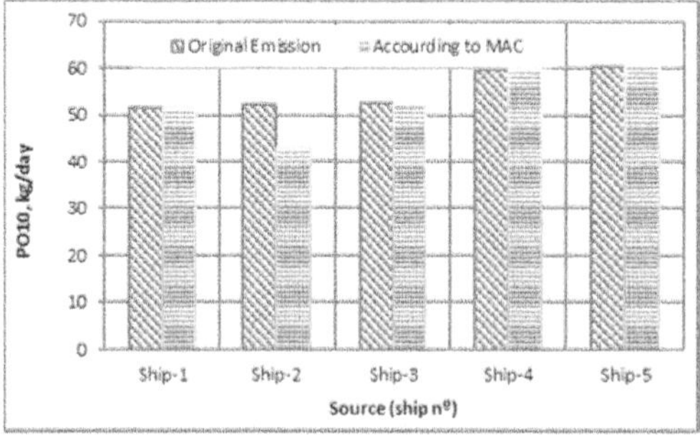


Figure 14. Emissions concentration, Seaport *A* (up) and *B* (down), $PM_{10}$.

## 5 CONCLUSIONS

The study analysed the emissions generated by a multipurpose ship fleet designed to operate in the Black Sea basin. Numerical models have been employed in estimating the pollution caused during the navigation, queuing, and port operations aimed at promoting the integration of maritime transportation in eco-friendly transport, contributing to the development of adequate transport policies in mitigating the gases emitted from ships and implementing the new technological solution in the marine transport sector.

Four different routes, $NO_x$, $SO_2$ and $PM_{10}$ emissions were analysed for any segment of the voyage operation and each route. The damage cost produced by the fleet circulating in different routes showed that the longer distance routes contribute more to the pollution damage. However, when the pollution damage is normalised by the cargo transported, the most extended distance dominates the pollution damage. Some of the shortest distance is not that critical.

The present study also analysed the effect of emitted gases on the multipurpose ship fleet during the queuing and their dispersion and local impact. Considering the meteorological parameters of the seaports and their locations, the Gaussian dispersion model has been employed, estimating the ground-level concentrations on preselected receptors on the coast of Varna, indicating violation of the maximum allowable concentrations in the specific condition in the present study where the analysis was made.

The present study can be extended considering the real-time ship voyaging and dynamic change of the meteorological conditions in examining the air quality and mitigating the gases produced by ships and can be further used in optimising the implementation of alternative ship power supply and to introduce more stringent maritime policies aimed at reducing emissions to air from ships.

## ACKNOWLEDGEMENTS

This work was performed within the Research Plan of the Technical University of Varna, which is financed by the State Budget under the contract NP8/2021.

## REFERENCES

Cohan, A., Wu, J. & Dabdub, D. 2011. High-resolution pollutant transport in the San Pedro Bay of California. *Atmos. Pollut. Res.*, 2, 237–246.

Davies, M. 2014. Ship emissions from Australian ports. *WIT Trans. Ecol Environ.*, 183, 327–337.

EMEP/EEA 2019. Technical guidance to prepare national emission inventories.

ESSN. 2020. *MSC* [Online]. Available: https://www.msc.com/che/our-services/trade-services/European-short-sea-network.

EU. 2021. *Quantification of Emissions from Ships Associated with Ship Movements between Ports in the European Community* [Online]. European Commission: Directorate-General for Mobility and Transport. Available: https://ec.europa.eu/environment/air/pdf/chapter1_ship_emissions.pdf.

Eyring, V., Isaksen, I. S. A., Berntsen, T., Collins, W. J., Corbett, J. J., Endresen, Ø., Grainger, R. G., Moldanova, J., Schlager, H. & Stevenson, D. S. 2010. Transport impacts on atmosphere and climate. *Shipping. Atmos. Environ*, 44, 4735–4771.

Eyring, V., Kohler, H., van Aardenne, J. & Lauer, A. 2005. Emissions from international shipping: 1. The last 50 years. *Journal of Geophysical Research*, 110, D 17305.

Garbatov, Y. & Georgiev, P. 2022. Stochastic air quality dispersion model for defining queuing ships seaport location. *Journal of Marine Science and Engineering*, 10, 140.

Georgiev, P. & Garbatov, Y. 2021. Multipurpose vessel fleet for short black sea shipping through multimodal transport corridors. *Brodogradnja*, 72, 79–101.

Gibson, M. D., Kundu, S. & Satish, M. 2013. Dispersion model evaluation of PM 2. 5, NOx and SO2 from point and major line sources in Nova Scotia, Canada using AERMOD Gaussian plume air dispersion model. *Atmospheric Pollution Research*, 4, 157–167.

Henderson-Sellers, B. & Allen, S. E. 1985. Verification of the plume rise/dispersion model U.S.P.R.: plume rise for single stack emissions. *Ecological Modeling*, 30, 209–227.

Lonati, G., Cernuschi, S. & Sidi, S. 2010. Air quality impact assessment of at-berth ship emissions: Case-study for the project of a new freight port. *Sci Total Environ*, 409, 192–200.

Lu, G., Brook, J. R., Alfarra, M. R., Anlauf, K., Leaitch, W. R., Sharma, S., Wang, D., Worsnop, D. R. & Phinney, L. 2006. Identification and characterization of inland ship plumes over Vancouver. *Atmos. Environ.*, 40, 2767–2782.

MEPC 2011. Amendments to the annex of the protocol of 1997 to amend the International Convention for the prevention of pollution from ships, 1973, as modified by the protocol of 1978 relating to MEPC 203 (62). *The Marine Environment Protection Committee (MEPC)*.

MEPC 2012. Guidelines for Calculation of Reference Lines for use with the Energy Efficiency Design Index (EEDI) MEPC 215 (63). *Marine Environment Protection Agency (MEPC)*.

Milazzo, M. F., Ancione, G. & Lisi, R. 2017. Emissions of volatile organic compounds during the ship-loading of petroleum products: Dispersion modelling and environmental concerns. *J Environ Manage*, 204, 637–650.

Parlament, E. 2016. Council of the European Union. Directive (EU) 2016/2284 of the European Parliament and of the Council of 14 December 2016 on the reduction of national emissions of certain atmospheric pollutants, amending Directive 2003/35/EC and repealing Directive 2001/81/EC (Text with EEA relevance). *J. Eur. Union*, 36, 1–31.

Pasquill, F. 1961. The Estimation of the Dispersion of Windborne Material. *Meteorological Magazine*, 90, 33–49.

Ramalho, M. & Santos, T. A. 2021. Numerical Modeling of Air Pollutants and Greenhouse Gases Emissions in Intermodal Transport Chains. *Journal of Marine Science and Engineering*, 9.

Santos, T. A. & Guedes Soares, C. 2017. Development dynamics of the Portuguese range as a multi-port gateway system, *Journal of Transport Geography*, 60, 178–188.

Santos, T. A. & Guedes Soares, C. 2019. Methodology for container terminal potential hinterland characterization in a multi-port system subject to a regionalization process. *Journal of Transport Geography*, 75, 132–146.

Sorte, S., Rodrigues, V., Borrego, C. & Monteiro, A. 2020. Impact of harbour activities on local air quality: A review. *Environ. Pollut.*, 257, 113542.

Sutton, O. G., 1932, A theory of eddy diffusion in the atmosphere, Proc. Roy. Soc., London, 143–165.

Toscano, D. & Murena, F. 2019. Atmospheric ship emissions in ports: A review. Correlation with data of ship traffic. *Atmos. Environ.*, 4, 100050.

van Essen, H., van Wijngaarden, L., Schroten, A., Sutter, D., Bieler, C., Maffii, S., Brambilla, M., Fiorello, D., Fermi, F., Parolin, R. & Beyrouty, K. 2019. Handbook on the external costs of transport. *In:* European Commission, D.-G. F. M. a. T. (ed.).

*Trends in Maritime Technology and Engineering – Guedes Soares & Santos (eds)*


# Evaluation of decarbonization strategies for existing ships

C. Karatug & Y. Arslanoglu
*Maritime Faculty, Istanbul Technical University, Tuzla, Istanbul, Turkey*

C. Guedes Soares
*Centre for Marine Technology and Ocean Engineering (CENTEC), Instituto Superior Técnico, Universidade de Lisboa, Lisbon, Portugal*

ABSTRACT: Various approaches such as usage of alternative fuels, implementation of slow steaming, adaptation of carbon capture system, are evaluated to determine if they meet the decarbonization aim introduced by International Maritime Organization. These methods can be categorized under two headings; utilization and adaptation of alternative marine fuels and alternative energy sources and application of operational efficiency increasing methods. These approaches allow maritime companies to increase the energy efficiency of marine vessels, reduce fuel consumption and the amount of emissions emitted to the atmosphere. To achieve the decarbonization aim, these strategies could be used in a singular way or in a hybrid design according to its suitability, availability, and meet desired performance. The advantageous and disadvantageous sides of each method were discussed, in detail. As a result of the study, significant findings have been presented to maritime companies, stakeholders, and researchers.

## 1 INTRODUCTION

Shipping, which is the basis of international commerce, carries the vast majority of cargo around the world. Between 2008 and 2018, the global commerce fleet increased by approximately 4.6 percent each year, reaching 117,000 ships (EMSA, 2019). On the other hand, shipping makes a significant contribution to air pollution and climate change owing to its large volume (Uyanık et al., 2020). The amount of fuel consumption was projected to be between 250 and 350 million tonnes between 2007 and 2012, with fossil fuels accounting for the bulk (IMO, 2014).

Based on a recent greenhouse gas (GHG) study conducted by the International Maritime Organization (IMO), ship-emitted GHG emissions increased by 9.6% between 2012 and 2018. In 2018, it was stated that maritime transportation was responsible for 2.89 per cent of global anthropogenic emissions (IMO, 2020). The spatial GHG spread of the emissions emitted to the atmosphere by ships is illustrated in Figure 1. On the subject of emission releases, container ships, bulk carriers, and oil tankers have a huge portion among all kinds of marine vessels (Balcombe et al., 2019).

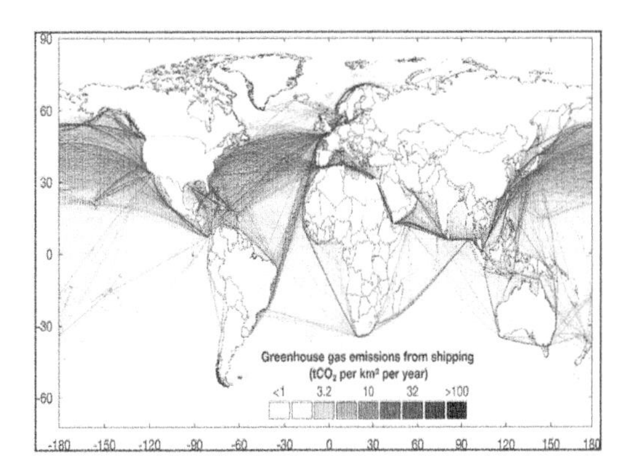


Figure 1. Global spread of ship-related GHG emissions (Balcombe et al., 2019).

Through Annex VI of the International Convention for the Prevention of Pollution from Ships (MARPOL), the IMO aimed to minimize pollutants released to the atmosphere by ships as well as enhance ship energy efficiency. In this sense, The Energy Efficiency Design Index (EEDI) was established to decrease $CO_2$ emissions from newly built ships by improving technical efficiency (IMO, 2012b).

DOI: 10.1201/9781003320289-5

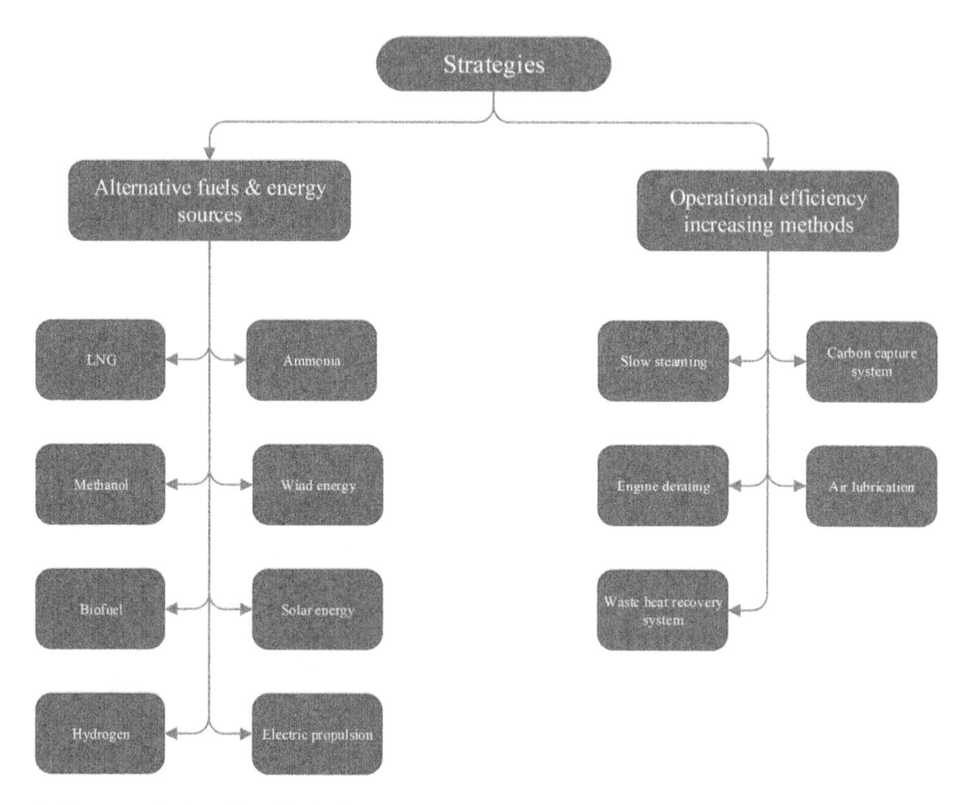


Figure 2. The categorization of decarbonization strategies.

Besides, MARPOL introduced the Ship Energy Efficiency Management Plan (SEEMP) to improve fuel efficiency for both new and existing ships through operational improvements (IMO, 2012a). In this regard, optimization of the engine speed, the fuel consumption of vessels, and emission released by ships are significant. For instance, an optimization procedure has been developed to minimize fuel consumption (Tadros et al., 2019) and the amount of emission released (Tadros et al., 2020). In (Mocerino et al., 2021), an emission model on the Ricardo Wave simulation software has been developed and the results have been validated by real ship data. Similarly, a model where a ship's speed and amount of fuel consumption prediction is carried out has been demonstrated (Moreira et al., 2021).

In addition to efficiency indexes, the IMO has determined the Initial Strategy of maritime transportation for reducing ship-based GHG emissions (IMO, 2018). It is aimed with the strategy that the net global ship-based GHG emissions decrease by 50 per cent in 2050, as well as reducing carbon intensity by at least 40 per cent in 2030 and 70 per cent in 2050, compared to 2008. The strategies to reaching these specific targets are not completely clear, but several approaches are investigating by researchers and maritime companies to meet the decarbonization aim. The strategies could be categorized under two headings as shown in Figure 2.

In this paper, information about the stated methods in Figure 2 has been briefly presented. Then, their advantageous specifications and barriers of application have been discussed in detail. According to the findings, pathways to meet the decarbonization aim were proposed to researchers, maritime companies, and stakeholders.

The rest of the paper is organized as follows. Section 2 gives brief information about the approaches. The comparison was realized in Section 3. Finally, in Section 4, significant findings were presented and effective pathways for achieving IMO's target were proposed.

## 2 STRATEGIES FOR ACHIEVING DECARBONIZATION STRATEGY

### 2.1 *Alternative fuels & energy sources*

#### 2.1.1 *Liquefied Natural Gas (LNG)*

LNG is a fuel that is increasingly being utilized as the main source of energy for marine vessels. It is also able to use with other fuels in dual-fuel engines (Bilgili, 2021). Due to recent international restrictions, developing technology, and maritime field economics, LNG is becoming a more popular marine fuel. In the case of comparing with HFO, LNG emits 25% less $CO_2$ (Iannaccone et al., 2020), generates less NOx and PM after combustion, and does not produce SOx (Kim et al., 2020). In addition, LNG is a cost-effective option compared to other alternative marine fuels.

2.1.2 *Methanol*
Methanol is suitable for utilization to reduce emissions to meet the IMO 2020 limitations (Zincir et al., 2019). Methanol, often known as $CH_3OH$, is a basic oxygenated hydrocarbon that is one of the top five most traded compounds in the world (Verhelst et al., 2019). It has a sulfur-free structure but at the same time, it is a corrosive liquid fuel. It may combust easily with $CO_2$ and $H_2O$, producing no SOx and just a few amounts of NOx and PM.

It is possible to generate methanol from either coal or natural gas. Because of its toxicity nature, it is a risky alternative maritime fuel with a low flash-point. It is a particularly flammable gas since it has a calorific value of 20,000 MJ/t (Bilgili, 2021). Methanol has been used as a source of energy in several major naval operations and commercial ventures (Liu et al., 2019). However, due to the low flash point, it does not comply with the requirement of SOLAS which is get enforced by IMO, but this might be remedied by using double-wall construction (Ammar, 2019).

2.1.3 *Biofuels*
Biofuels have traditionally been derived from plant-based sugars and oils such as palm, soybean, and rapeseed (Hsieh & Felby, 2017), and they can be available in the forms of solid, gas, or liquid. Marine biofuels may be generated in a method that works with existing marine engines, pipelines, and bunker infrastructure, resulting in minimal adaptation costs (Florentinus et al., 2012). Depending on the property, nature, and processing techniques of the bio feedstock, biofuels are expected to reduce $CO_2$ emissions by 25% to 100% (ITF/ OECD, 2018). Thus, they ensure assurance on decreasing $CO_2$ emissions as well as meet IMO's decarbonization strategy. However, a low amount of sulfur emission could be released as a result of the usage. On the other hand, precautions should be handled while choosing biofuels for ship usage, as some of them tend to oxidize and decay over a period of time.

First-generation conventional biofuels include straight vegetable oil (SVO), hydrotreated vegetable oil (HVO), fatty acid methyl ester (FAME), and bioethanol. Advanced biofuels that are more appropriate for maritime usage contain Fischer-Tropsch diesel (FT-Diesel), pyrolysis oil, lignocellulosic ethanol (LC Ethanol), bio-methanol, dimethyl-ether (made from bio-methanol), and bio-LNG. It is determined that conventional biofuels release more GHG than advanced biofuel (Balcombe et al., 2019).

Even biofuels consider as an alternative fuel solution, there are some concerns about their availability, sustainability, and production and supply maturity. There are limitations on biofuels availability such as dense agricultural inputs, production space requirements, and trade-offs between food crop and fuel crop generation. Aquatic organisms such as algae which are the production feedstock of advanced biofuels offer better sustainability to biofuels such as bioethanol, biodiesel, and biohydrogen than first-generation biofuels (Juneja et al., 2013).

2.1.4 *Hydrogen*
Hydrogen fuel cells might transform the chemical energy of the fuel into DC power (Barbir, 2013). Fuel cell technology, as compared to internal combustion engines, is a greatly promising technology (Yan et al., 2020) because of its better energy efficiency and less environmental impact (Bicer & Dincer, 2018).

Fuel cells are classified into five varieties based on the type of electrolyte used: Proton exchange membrane fuel cell (PEMFC), alkaline fuel cell (AFC), phosphoric acid fuel cell (PAFC), molten carbonate fuel cell (MCFC) and solid oxide fuel cell (SOFC) (Inal et al., 2022; Inal & Deniz, 2020).

A pilot project, known as "Viking Lady", has made the first effort at using fuel cells in maritime transportation. On the project, LNG was employed as the main power source of propulsion in a diesel engine, with hydrogen or methanol-fueled fuel cell acting as a backup. Consequently, the amount of SOx, NOx, and $CO_2$ emitted to the atmosphere was reduced by 100, 85, and 20 percent, respectively (Pospiech, 2012).

2.1.5 *Ammonia*
Ammonia ($NH_3$) is an alternative marine fuel that is being extensively researched for potential global use. It's a carbon-free alternative fuel that's employed in sectors including healthcare, polymers, textiles, cosmetics, nutrition, and electronics (Hansson et al., 2020). It could be used at gas turbines, diesel engines, and fuel cells (Kim et al., 2020). While ammonia includes one nitrogen atom and three hydrogen atoms, it is also a sulfur-free molecule as well as being carbon-free. Therefore, CO, $CO_2$, or SOx were not released into the atmosphere as a result of ammonia's combustion. 10 bar pressure at ambient temperature or atmospheric pressure at 33°C is enough for the liquefication of ammonia. However, in comparison to other alternative fuels, the ignition of ammonia is difficult and is harmful to both human health and the environment. Additionally, it is corrosive to the fuel system and its related components (Zincir, 2020).

2.1.6 *Wind energy*
Wind energy could be obtained from different strategies that are Flettner rotors, kites, wind turbines, and wing sails.

Even when the ship is cruising faster than the wind, Flettner rotors, one of the strategies, could generate power, but it should be underlined that wind direction has major importance during the conversion process. studies related to Flettner rotors show that this approach could reduce fuel consumption by up to 20 percent (Talluri et al., 2016). The E-Ship1 pilot project, which was finished in 2010, carried out a novel rotor application in maritime. The ship has 3.5 MW x 2 diesel engines and 4 x Flettner rotors, and the energy provided by the hybrid system enables the ship to navigate at 17.5 knots (Lu & Ringsberg, 2020).

In the different approaches, smooth sails and kites have a greatly easier application due to wind force being used directly. The method has advantageous on no requirement of additional space and applicability on any type of ship. However, its performance is directly dependent on the wind direction and speed. The kite system could ensure energy contribution up to 50 percent based on the weather and sea conditions (Leloup et al., 2016).

#### 2.1.7 *Solar energy*

Solar power could be generated from the solar photovoltaic panels that are more suitable for marine application by converting the energy of solar beams to electrical energy. It is a green energy source, limitless, and zero-emission energy-producing method. Nowadays, even though there are some samples where the ship propulsion system is driven by solar energy as the main power, it is generally used as a part of a hybrid power unit (Tang et al., 2018). Solar energy is an alternative power source that could be reasonably beneficial when used as an auxiliary energy source to the ship power unit. In the maritime field, Auriga Leader, Solar Sailor, and Emerald Ace are three sample ships that have been used as a model for the use of solar power on ships (Karatuğ & Durmuşoğlu, 2020).

#### 2.1.8 *Electric propulsion*

An electric propeller system is an approach that could be designed on the basis of different kinds of energy sources such as solar systems, wind assistance, flywheels, battery units, diesel generators as standalone or in a hybrid way. One or more electric motors for each propeller assign during the utilization. The approach has noticeable dynamics for some conditions such as stopping, manoeuvrability, starting, etc. Less fuel consumption is ensured by the adaptation of this approach. Depending on the system design and components, this approach has less vibration compared to conventional diesel systems. Thus, more comfortable working and living places are provided for seafarers on board (Sulligoi et al., 2016). Also, lower NOx emission emits to the atmosphere through an electrical propulsion system than a mechanical propulsion system since desired power generates by more than one engine system. And this strategy provides fewer maintenance loads. On the contrary, electrical losses from the convertors and electric motors increase. Also, the risk of the ship's black-out is getting high due to possible sharp changes in voltage and frequency of the propulsion unit. This leads to lower system reliability (Geertsma et al., 2017).

### 2.2 *Vessel efficiency enhancement methods*

#### 2.2.1 *Slow steaming*

The benefits of slow steaming could vary depending to ship types, routes, and sizes(Faber et al., 2012). For instance, full speed could be defined as 23 and 25 knots for container ships while slow steaming and super slow are considered as 20–22 knots and 17–19 knots, respectively (Maloni et al., 2013). Because of the slow steaming, navigation time and duration of the port might be extended. However, it enables to decrease fuel consumption and corresponded amount of emissions by improving energy efficiency at the operational level (Lee et al., 2015; Maloni et al., 2013). IMO was specified in the third GHG study that big-sized commercial marine vessels such as container ships, oil tankers, and bulk carriers reduced fuel consumption by 30 percent by slow steaming (IMO, 2014).

#### 2.2.2 *Engine derating*

Propeller determination, main engine decision, and intended operational speed matching must be done appropriately during ship design. This would allow for the most efficient operating of the ships (Tadros et al., 2021). According to the vessel's management, if the vessel is demanded to run at a reduced speed continuously, the ship's speed might be optimized to the desired level to minimize fuel consumption and emissions. This process is called the derating of the engine. The strategy could be made in two ways permanent and temporary. 10-12 per cent of decrease of fuel consumption is ensured by the method as well as enhancing energy efficiency level of the ship (MAN, 2019).

#### 2.2.3 *Waste heat recovery system*

Approximately half of the entire fuel energy provided to a marine vessel's diesel power unit is lost to the environment (MAN Diesel&Turbo, 2012). A waste heat recovery system (WHRS) creates electricity by retrieving heat energy that would otherwise be lost to the environment during thermal processes while consuming no additional fuel.

The gain from the fuel based on using the WHRS is estimated to be between 4 and 16 per cent (Senary et al., 2016; Singh & Pedersen, 2016). Steam Rankine Cycle, Organic Rankine Cycle (ORC), and Kalina Cycle are some examples of the use of WHRS in a marine application. The system's period of redemption might be lower than three years but it should be underlined that it could be varies based on the savings from the fuel consumption (Theotokatos & Livanos, 2012).

### 2.2.4 *Carbon capture system*

The maritime industry has recently familiar with exhaust after-treatment technologies due to meeting IMO restrictions on NOx and SOx with selective catalytic reduction and scrubber technologies, respectively. In the subject of the decarbonization target introduced by the IMO, the carbon capture method is an encouraging strategy to remove $CO_2$. It is possible to decrease $CO_2$ in the exhaust gas at about a rate of 85-90 per cent by filtering within Calix's RECAST system (Sweeney, 2018). It was stated in the feasibility study conducted by Det Norske Veritas (DNV) and Process Systems Enterprise Ltd. (PSE) that it is possible to decrease $CO_2$ emission via the carbon capture method by approximately up to 65 per cent (DNV & PSE Ltd., 2016).. In a carried out case study conducted on a 3800 TEU container ship, it is presented that the carbon capture method is effective to meet IMO 2050 aim (Oh et al., 2022). Different $CO_2$ elimination approaches such as carbon capture system, LNG usage, speed reduction have been separately evaluated and compared for use in different ship types. It is determined that speed reduction is a cost-beneficial way for ships with low freight values while carbon capture system offers a significant solution for high-speed crafts (Güler & Ergin, 2021).

### 2.2.5 *Air lubrication*

The friction force of the water has a negative impact on the hull of the ship. The effect of the force enhances with the rising of the ship's speed. To decrease and minimize the specified resistance, the air lubrication method is a very promising strategy that put an air layer between ship hull and seawater. Also, it could be used on any kind of marine vessel. There are various strategies for the application such as bubble drag reduction, transitional air layer reduction, air layer drag reduction, partial cavity drag reduction, and multi-wave partial cavity drag reduction (Mäkiharju et al., 2012). Some parameters should be considered during conducting the EEDI calculation of the ship based on the air lubrication strategy such as hull surface with covering air bubbles, the thickness of air layer, decrease the ratio of frictional resistance, the possible difference of the propulsion efficiency because of the system adaptation, and difference of the resistance based on a piece of extra equipment (IMO, 2013). It is stated that with the air lubrication method, increasing energy efficiency and corresponding reducing amount of emissions could be achieved by 15% for oil tankers, 7.5% for containers (Wartsila, 2008).

## 3 ADVANTAGEOUS SIDES AND BARRIERS OF THE STRATEGIES

Each method has advantages and disadvantages in terms of implementation. In the case of the adaptation, the characteristics of strategies should be carefully analyzed and the feasibility of the application should be measured with great attention. The advantageous sides and the constraints corresponding to each approach have been attempted to be presented in detail in this section. In this sense, significant indicators and parameters have been presented in Table 1, comparatively.

Table 1. Advantageous sides and barriers of stated approaches.

| Method | Advantageous sides | Disadvantageous sides |
|---|---|---|
| LNG as an alternative fuel | • It is influential to reduce $CO_2$ by approx. 20 percent.<br>• About LNG as an alternative marine fuel, there is some regulation.<br>• It is cheaper compared to fossil fuel.<br>• It is not corrosive and toxic for human health.<br>• The infrastructure for the bunkering process is developing. | • Methane slip is a significant concern for LNG usage.<br>• It is still working on the preventing boil-off issue.<br>• Compared to fossil fuel, it has less energy capacity.<br>• It is not enough to meet IMO's decarbonization aim in a singular way. |
| Methanol as an alternative fuel | • It is effective to reduce $CO_2$ by approx. 25 percent.<br>• It is approved by IMO for usage as an alternative marine fuel.<br>• Compared to other alternative marine fuels, there is a need for small changes on the existing system for usage and storage.<br>• Existing infrastructure could be converted easily for supplying ports and ships. | • Energy capacity of methanol is less than fossil fuel.<br>• SCR system could be needed for complying with the IMO Tier III emission level.<br>• It is not enough to meet IMO's decarbonization aim in a singular way.<br>• Compared to other alternative marine fuels, it is more flammable. |

(*Continued*)

Table 1. (Cont.)

| Method | Advantageous sides | Disadvantageous sides |
|---|---|---|
| | • In the last GHG study conducted by IMO, it is presented as the 4th most common marine fuel. | • It is poisonous for human health, and corrosive for some materials. |
| Biofuels as an alternative fuel | • Biofuels might provide the decrease of air pollution caused by ships at a vast rate.<br>• Some of the biofuels could be utilized as an alternative marine fuel with small changes on engines, bunkering infrastructure, fuel supply arrangements, and storage.<br>• Since biofuels have biodegradable features, compared to fuel oil they have a lower risk in the case of oil spill accidents. | • Biofuels require a pilot fuel system for combustion since they cannot self-ignite because of the feature of low cetane number.<br>• The availability and sustainability of these fuels are weak. Also, usability and system infrastructure still need development.<br>• Regulation related to usage of biofuels as a alternative marine fuel is unclear. |
| Hydrogen as an alternative fuel | • The noise and vibration level of fuel cell utilization is lower.<br>• Compared to conventional engines, electric motors that contained in system design and transmit power to propeller have higher efficiency.<br>• In the world fleet, some marine vessels run with hydrogen in the waters. | • The main concerns for the hydrogen are its availability and volumetric energy capacity.<br>• During the utilization on board, a large fuel storage capacity is required.<br>• System infrastructure and design should be handled, carefully.<br>• On the IGF Code, carry of the liquid hydrogen is not specified.<br>• The initial capital cost is expensive.<br>• It is not enough to meet IMO's decarbonization aim in a singular way. |
| Ammonia as an alternative fuel | • The carbonization target could be achieved via ammonia usage.<br>• It can use on the duel-fuel newly-developed engines.<br>• There are some ways to produce from fossil fuel such as renewable energy source, carbon capture method. | • It requires a high amount of pilot oil to ignite.<br>• Compared to fuel oil, its energy density is less.<br>• The bunkering infrastructure needs to evolve.<br>• Its investment and operational cost are high.<br>• It is corrosive for some materials, and toxic and dangerous for crew onboard.<br>• On the IGF Code, it is not specified.<br>• The application of the SCR system could be needed to eliminate NOx emission after combustion. |
| Wind Energy & Solar Energy | • They are green, limitless, and environmentally friendly alternative power sources.<br>• They could be used in a standalone system as well as in a hybrid way. | • Generation of the power dependent on the environmental conditions such as wind force, wind direction, sunny or cloudy day, etc.<br>• They are not enough to meet power demand of the marine vessel, individually<br>• Initial capital and operational expenses are high. |
| Electric propulsion | • It is an effective strategy to meet decarbonization strategy as well as zero-emission ship target. | • System results such as environmental impact, economic indicators vary based on the including power sources and system arrangements. |

(*Continued*)

Table 1. (Cont.)

| Method | Advantageous sides | Disadvantageous sides |
|---|---|---|
| | • The efficiency of electric motors is higher compared to conventional engines.<br>• The system design allows all kinds of power sources to be designed in a hybrid way within the framework of different arrangements. | |
| Slow steaming | • Slow steaming is highly beneficial strategy to decrease fuel consumption, operational costs, and emissions.<br>• By adapting slow steaming, $CO_2$ emission might be reduced approx. 25-30 percent. | • Based on the difference in the operational status of engine and ship, the risk of corrosion and fouling rises. |
| Engine derating | • With engine derating, fuel consumption might be reduced approx. 10-12 percent.<br>• According to decrease in number of fuel consumption, emissions released to the atmosphere might be prevented. | • The investment cost is high.<br>• To optimize turbocharger capacity, calculations should be realized in detail.<br>• Since changes are revealed as a result of the difference in power forces, mechanical tests should be made again, carefully. |
| Waste heat recovery systems | • The savings from the ship's fuel consumption could be achieved by approx. 4-16 percent.<br>• Based on the reduction of fuel consumption, emissions and operational expenses decrease.<br>• Its capital cost is a decently high but acceptable level for ship lifespan. | • Thermal calculations should be realized in detail with high attention. Otherwise, system efficiency decreases.<br>• Specifications of the liquid used in the thermal processes and its convenience should be analyzed, carefully. |
| Carbon capture system | • It is a highly beneficial and effective strategy to meet IMO's decarbonization aim.<br>• It provides an edge over the future subjects for maritime such as carbon credit costs and the Carbon Intensity Indicator (CII). | • The investment cost of the system is high.<br>• It needs enough places to store $CO_2$ in liquid form.<br>• Extra power generation and fuel consumption are demanded as a result of system application. |
| Air Lubrication | • The release of the emission could be decreased approx. 7.5-15 percent.<br>• Energy efficiency level of the ship is improved.<br>• The method allows more flexible cruise and maneuverability based on the removing surface resistance.<br>• In the world fleet, there are ships equipped with air lubrication system. | • The investment cost of the system is relatively high.<br>• Areas are required for the component of the new system design.<br>• Efficiency of the system could be differ based on the sea status. |
| Sources | (ABS, 2020a, 2020b,2021; Alvela et al., 2018; Ampah et al., 2021; Andrews & Shabani, 2012; Balcombe et al., 2018, 2019; Cheliotis et al., 2021; Chu Van et al., 2019; Florentinus et al., 2012; Gilbert et al., 2018; Hansson et al., 2020; Hsieh & Felby, 2017; IMO, 2014, 2020; Mallouppas & Yfantis, 2021; MAN, 2019; MAN Energy Solutions, 2020; Natural Resources Canada, 2013; Salarkia & Golabi, 2021; Senary et al., 2016; Valera, H., & Agarwal, 2019; Wan et al., 2015; Xing et al., 2021; Zhou & Wang, 2014) | |

## 4 CONCLUSIONS

One of the significant issues of marine transportation is the reduction of carbon emissions generated by ship exhaust gas, as well as the decrease of SOx and NOx emissions. In this regard, the maritime industry's path has been established by the IMO's Initial GHG Strategy, which was presented in 2018. There are various viable techniques that marine companies might implement to attain the stated purpose, which can be classified under two main headings: *(i)* Utilization and adaptation of alternative marine fuels & alternative energy sources, *(ii)* Application of operational efficiency increasing methods.

In this paper, brief information about methods such as ammonia, methanol, biofuels usage as an alternative fuel, solar and wind power as an alternative energy source, and implementation of slow steaming, carbon capture as operational beneficial methods, etc. have been presented. Besides, their advantageous and disadvantageous sides have been examined, comparatively. In line with conducted study, the main findings of the paper are as follows:

1. In the literature, there are lots of papers that studied the applicability of different types of alternative marine fuels. Among these fuels, the usage of LNG and methanol is not enough to meet the decarbonization target of IMO in a singular way. On the other hand, ammonia and biofuels are more suitable alternative fuels for the stated aim.
2. It is observed that the energy density of alternative marine fuels is less than fossil fuels. Therefore, they require to pilot an oil system to ignite.
3. Currently, there are no particular regulations regarding the use of most alternative maritime fuels. Furthermore, these systems' maturity, availability, sustainability, and bunkering infrastructure are considerably weak.
4. Alternative power sources such as solar and wind are insufficient to satisfy the IMO's carbon reduction goal, but an electric propulsion system design based on a hybrid of some of these methods might be a suitable solution. However, capital and operational expenses are unclear since they vary depending on system design and components.
5. Slow steaming is an effective strategy, particularly for cargo ships. Even though it does not fulfill the carbon reduction aim on its own, it provides benefits when combined with other ship design strategies, such as the use of alternative marine fuel in the main engine and adaptation of alternative energy sources.
6. Engine derating provides some benefits to shipowners about energy efficiency and reduction fuel consumption and emission, but it should be made with caution. The crucial calculations and the ship's futuristic management plan during the conversion should be considered since the system's recovery is hard in the future.
7. The waste heat recovery system is a very effective strategy because it allows savings from existing waste energy. The payback period of the system is acceptable considering a ship's lifespan. The system ensures a decrease in the amount of fuel consumption and emission as well as improving energy efficiency.
8. The carbon capture technology has a high initial investment cost, and experts are now studying its applicability in the marine industry. It's a really promising technique for reducing carbon emissions. One of the key challenges is the system's storage capacity related to CO2. It could be stated that this approach is one of the most promising strategies for maritime companies during the removal of carbon emissions.

## REFERENCES


ADDIN Mendeley Bibliography CSL_BIBLIOGRAPHY ABS. (2020a). Ammonia As Marine Fuel. In *NH3 Fuel Conference* (Issue October).

ABS. (2021). *Methanol as marine fuel* (Issue February).

Alvela, M., Carroll, D. C., & Marten, I. (2018). *Global gas report 2018*. 2–5.

Ammar, N. R. (2019). An environmental and economic analysis of methanol fuel for a cellular container ship. *Transportation Research Part D: Transport and Environment, 69*, 66–76.

Ampah, J. D., Yusuf, A. A., Afrane, S., Jin, C., & Liu, H. (2021). Reviewing two decades of cleaner alternative marine fuels: Towards IMO's decarbonization of the maritime transport sector. *Journal of Cleaner Production, 320*(May), 128871.

Andrews, J., & Shabani, B. (2012). Where does Hydrogen Fit in a Sustainable Energy Economy? *Procedia Engineering, 49*, 15–25.

Balcombe, P., Brierley, J., Lewis, C., Skatvedt, L., Speirs, J., Hawkes, A., & Staffell, I. (2019). How to decarbonise international shipping: Options for fuels, technologies and policies. *Energy Conversion and Management, 182*(January), 72–88.

Balcombe, P., Speirs, J., Johnson, E., Martin, J., Brandon, N., & Hawkes, A. (2018). The carbon credentials of hydrogen gas networks and supply chains. *Renewable and Sustainable Energy Reviews, 91*, 1077–1088.

Barbir, F. (2013). *PEM fuel cells: theory and practice*. https://books.google.com/books?hl=tr&id=elO7n4Z6uLoC&oi=fnd&pg=PP1&ots=Lwal0DEHNZ&sig=U0AT2Ky3KALPOsbHjg8OoPo2wC0

Bicer, Y., & Dincer, I. (2018). Environmental impact categories of hydrogen and ammonia driven transoceanic maritime vehicles: A comparative evaluation. *International Journal of Hydrogen Energy, 43*(9), 4583–4596.

Bilgili, L. (2021). Comparative assessment of alternative marine fuels in life cycle perspective. *Renewable and Sustainable Energy Reviews, 144*, 110985.

Cheliotis, M., Boulougouris, E., Trivyza, N. L., Theotokatos, G., Livanos, G., Mantalos, G., Stubos, A., Stamatakis, E., & Venetsanos, A. (2021). Review on the safe use of ammonia fuel cells in the maritime industry. *Energies, 14*(11), 1–20.

Chu Van, T., Ramirez, J., Rainey, T., Ristovski, Z., & Brown, R. J. (2019). Global impacts of recent IMO regulations on marine fuel oil refining processes and ship emissions. *Transportation Research Part D: Transport and Environment, 70*, 123–134.

DNV, & PSE Ltd. (2016). *DNV and PSE report on ship carbon capture & storage*. https://www.psenterprise.com/news/news-press-releases-dnv-pse-ccs-report.

EMSA. (2019). *The World Merchant Fleet: Statistics from Equasis*. http://www.emsa.europa.eu/equasis-statistics/items.html?cid=95&id=472

Faber, J., Nelissen, D., Hon, G., Wang, H., & Tsimplis, M. (2012). *Regulated Slow Steaming in Maritime Transport. An Assessment of Options, Costs and Benefits*.

Florentinus, A., Hamelinck, C., Van den Bos, A., Winkel, R., & Cuijpers, M. (2012). *Potential of biofuels for shipping. Final Report*.

Geertsma, R. D., Negenborn, R. R., Visser, K., & Hopman, J. J. (2017). Design and control of hybrid power and propulsion systems for smart ships: A review of developments. *Applied Energy, 194*, 30–54.

Gilbert, P., Walsh, C., Traut, M., Kesieme, U., Pazouki, K., & Murphy, A. (2018). Assessment of full life-cycle air emissions of alternative shipping fuels. *Journal of Cleaner Production*, *172*, 855–866.

Güler, E., & Ergin, S. (2021). An investigation on the solvent based carbon capture and storage system by process modeling and comparisons with another carbon control methods for different ships. *International Journal of Greenhouse Gas Control*, *110*, 103438.

Hansson, J., Brynolf, S., Fridell, E., & Lehtveer, M. (2020). The Potential Role of Ammonia as Marine Fuel—Based on Energy Systems Modeling and Multi-Criteria Decision Analysis. *Sustainability 2020, Vol. 12, Page 3265*, *12*(8), 3265.

Hsieh, C.-W. C., & Felby, C. (2017). *Biofuels for the marine shipping sector Biofuels for the marine shipping sector An overview and analysis of sector infrastructure, fuel technologies and regulations*.

Iannaccone, T., Landucci, G., Tugnoli, A., Salzano, E., & Cozzani, V. (2020). Sustainability of cruise ship fuel systems: Comparison among LNG and diesel technologies. *Journal of Cleaner Production*, *260*, 121069.

IMO. (2012a). *Guidelines For The Development of A Ship Energy Efficiency Management Plan (SEEMP). MEPC.282(70)*.

IMO. (2012b). *Guidelines on The Method of Calculation of The Attained Energy Efficiency Design Index (EEDI) For New Ships. MEPC.308(73)*.

IMO. (2013). *MEPC.1/Circ.815 17. Guidance on Treatment of Innovative Energy Efficiency Technologies for Calculation and Verification of the Attained EEDI*.

IMO. (2014). *Third IMO GHG study executive summary*.

IMO. (2018). *Initial IMO strategy on reduction of GHG emissions from ships*.

IMO. (2020). *Fourth IMO GHG study executive summary*.

Inal, O. B., Charpentier, J. F., & Deniz, C. (2022). Hybrid power and propulsion systems for ships: Current status and future challenges. *Renewable and Sustainable Energy Reviews*, *156*, 111965.

Inal, O. B., & Deniz, C. (2020). Assessment of fuel cell types for ships: Based on multi-criteria decision analysis. *Journal of Cleaner Production*, *265*, 121734.

ITF/ OECD. (2018). Decarbonising Maritime Transport. Pathways to zero-carbon shipping by 2035. In *International Transport Forum*. https://www.itf-oecd.org/sites/default/files/docs/decarbonising-maritime-transport.pdf

Juneja, A., Ceballos, R. M., & Murthy, G. S. (2013). Effects of Environmental Factors and Nutrient Availability on the Biochemical Composition of Algae for Biofuels Production: A Review. *Energies 2013, Vol. 6, Pages 4607-4638*, *6*(9), 4607–4638.

Karatuğ, Ç., & Durmuşoğlu, Y. (2020). Design of a solar photovoltaic system for a Ro-Ro ship and estimation of performance analysis: A case study. *Solar Energy*, *207* (April), 1259–1268.

Kim, H., Koo, K. Y., & Joung, T.-H. (2020). A study on the necessity of integrated evaluation of alternative marine fuels, *4*(2), 26–31.

Lee, C. Y., Lee, H. L., & Zhang, J. (2015). The impact of slow ocean steaming on delivery reliability and fuel consumption. *Transportation Research Part E: Logistics and Transportation Review*, *76*, 176–190.

Leloup, R., Roncin, K., Behrel, M., Bles, G., Leroux, J. B., Jochum, C., & Parlier, Y. (2016). A continuous and analytical modeling for kites as auxiliary propulsion devoted to merchant ships, including fuel saving estimation. *Renewable Energy*, *86*, 483–496.

Liu, M., Li, C., Koh, E. K., Ang, Z., & Lam, J. S. L. (2019). Is methanol a future marine fuel for shipping? In *Journal of Physics: Conference Series*. IOP Publishing.

Lu, R., & Ringsberg, J. W. (2020). Ship energy performance study of three wind-assisted ship propulsion technologies including a parametric study of the Flettner rotor technology. *Ships and Offshore Structures*, *15*(3), 249–258.

Mäkiharju, S. A., Perlin, M., & Ceccio, S. L. (2012). On the energy economics of air lubrication drag reduction. *International Journal of Naval Architecture and Ocean Engineering*, *4*(4), 412–422.

Mallouppas, G., & Yfantis, E. A. (2021). *Decarbonization in Shipping Industry : A Review of Research, Technology Development, and Innovation Proposals*.

Maloni, M., Paul, J. A., & Gligor, D. M. (2013). Slow steaming impacts on ocean carriers and shippers. *Maritime Economics and Logistics*, *15*(2), 151–171.

MAN. (2019). *Propulsion of 2,200-3,000 teu container vessels Container Feeder Modern two-stroke engine technology for a modern vessel type MAN Energy Solutions Propulsion of 2,200-3,000 teu container vessels 2 Future in the making 3*.

MAN Diesel&Turbo. (2012). *Thermo efficiency system for reduction of fuel consumption and CO2*.

MAN Energy Solutions. (2020). Managing methane slip. In *Man*.

Mocerino, L., Guedes Soares, C., Rizzuto, E., Balsamo, F., & Quaranta, F. (2021). Validation of an Emission Model for a Marine Diesel Engine with Data from Sea Operations. *Journal of Marine Science and Application*, *20*(3), 534–545.

Moreira, L., Vettor, R., & Guedes Soares, C. (2021). Neural network approach for predicting ship speed and fuel consumption. *Journal of Marine Science and Engineering*, *9*(2), 1–14.

Natural Resources Canada. (2013). Liquefied Natural Gas: Properties and Reliability. *Government of Canada, c*, 1–3.

Oh, J., Anantharaman, R., Zahid, U., Lee, P. S., & Lim, Y. (2022). Process design of onboard membrane carbon capture and liquefaction systems for LNG-fueled ships. *Separation and Purification Technology*, *282*, 120052.

Pospiech, P. (2012). *World's first fuel-cell ship FCS Alsterwasser proves its reliability*.

Salarkia, M. R., & Golabi, S. (2021). *Liquefied Natural Gas (LNG): Alternative Marine Fuel Restriction and Regulation Considerations, Environmental and Economic Assessment*. *10*(4), 44–59.

Senary, K., Tawfik, A., Hegazy, E., & Ali, A. (2016). Development of a waste heat recovery system onboard LNG carrier to meet IMO regulations. *Alexandria Engineering Journal*, *55*(3), 1951–1960.

Singh, D. V., & Pedersen, E. (2016). A review of waste heat recovery technologies for maritime applications. *Energy Conversion and Management*, *111*, 315–328.

Sulligoi, G., Vicenzutti, A., & Menis, R. (2016). All-electric ship design: From electrical propulsion to integrated electrical and electronic power systems. *IEEE Transactions on Transportation Electrification*, *2*(4), 507–521.

Sweeney, B. (2018). Shipping CO2 emissions can be eliminated by Calix RECAST system. *Energy Systems Conference*.

Tadros, M., Ventura, M., & Guedes Soares, C. (2019). Optimization procedure to minimize fuel consumption of a four-stroke marine turbocharged diesel engine. *Energy*, *168*, 897–908.

Tadros, M., Ventura, M., & Guedes Soares, C. (2020). Optimization of the Performance of Marine Diesel Engines to Minimize the Formation of SOx Emissions. *Journal of Marine Science and Application*, *19*(3), 473–484.

Tadros, M., Vettor, R., Ventura, M., & Guedes Soares, C. (2021). Coupled engine-propeller selection procedure to minimize fuel consumption at a specified speed. *Journal of Marine Science and Engineering*, *9*(1), 1–13.

Talluri, L., Nalianda, D. K., Kyprianidis, K. G., Nikolaidis, T., & Pilidis, P. (2016). Techno-economic and environmental assessment of wind-assisted marine propulsion systems. *Ocean Engineering*, *121*, 301–311.

Tang, R., Wu, Z., & Li, X. (2018). Optimal operation of photovoltaic/battery/diesel/cold-ironing hybrid energy system for maritime application. *Energy*, *162*(2018), 697–714.

Theotokatos, G., & Livanos, G. (2012). Techno-economical analysis of single pressure exhaust gas waste heat recovery systems in marine propulsion plants, *227*(2), 83–97.

Uyanık, T., Karatuğ, Ç., & Arslanoğlu, Y. (2020). Machine learning approach to ship fuel consumption: A case of container vessel. *Transportation Research Part D: Transport and Environment*, *84*, 102389.

Valera, H., & Agarwal, A. K. (2019). Methanol as an Alternative Fuel for Diesel Engines. *Methanol and the Alternate Fuel Economy*, 9–33.

Verhelst, S., Turner, J. W., Sileghem, L., & Vancoillie, J. (2019). Methanol as a fuel for internal combustion engines. *Progress in Energy and Combustion Science*, *70*, 43–88.

Wan, C., Yan, X., Zhang, D., Shi, J., Fu, S., & Ng, A. K. Y. (2015). Emerging LNG-fueled ships in the Chinese shipping industry: a hybrid analysis on its prospects. *WMU Journal of Maritime Affairs*, *14*(1), 43–59.

Wartsila. (2008). *Boosting energy efficiency*. https://home.hvl.no/ansatte/jjo/ftp/hydrodynamikk/Notater/WartsilaShipEfficiencySept08.pdf

Xing, H., Stuart, C., Spence, S., & Chen, H. (2021). Alternative fuel options for low carbon maritime transportation: Pathways to 2050. *Journal of Cleaner Production*, *297*, 126651.

Yan, H., Wang, G., Lu, Z., Tan, P., Kwan, T. H., Xu, H., Chen, B., Ni, M., & Wu, Z. (2020). Techno-economic evaluation and technology roadmap of the MWe-scale SOFC-PEMFC hybrid fuel cell system for clean power generation. *Journal of Cleaner Production*, *255*, 120225.

Zhou, P., & Wang, H. (2014). Carbon capture and storage —Solidification and storage of carbon dioxide captured on ships. *Ocean Engineering*, *91*, 172–180.

Zincir, B. (2020). A short review of ammonia as an alternative marine fuel for decarbonised maritime transportation. *International Conference on Energy, Environment and Storage of Energy*, 373–380.

Zincir, B., Deniz, C., & Tunér, M. (2019). Investigation of environmental, operational and economic performance of methanol partially premixed combustion at slow speed operation of a marine engine. *Journal of Cleaner Production*, *235*, 1006–1019.

*Trends in Maritime Technology and Engineering – Guedes Soares & Santos (eds)*


# Design exploration of autonomous container transshipment for interterminal transport based on current technology

K. Kruimer, J. Jovanova & D. Schott
*Delft University of Technology, Delft, The Netherlands*

ABSTRACT: In recent years the operations of container terminals have become more and more autonomous with the introduction of various types of autonomous terminal equipment, such as Automated Guided Vehicles. However, the transshipment of containers from ship to shore is still not yet fully automated. As designs for fully automated container barges are currently in development, a next step can be taken into exploration of designing autonomous transshipment machines for these new barges, which will open the way for container terminals to become integrated and efficient multimodal hubs in the future. This paper presents exploratory designs of an autonomous device for the ship to shore container transshipment, specifically to be used for (autonomous) container barges. The machine should be deployable in any port, with little to no need of fixed infrastructures, such as tracks. Exploration of transshipment concepts fit for use in a fully automated environment, as well as evaluation of initial design solutions in their capacity to improve the productivity of these transshipment concepts is showed in this work.

## 1 INTRODUCTION

### 1.1 *The need for autonomous container transshipment*

Currently the waiting times for barges to be loaded and unloaded in the ports of Rotterdam and Antwerp are between 12 and 120 hours (Contargo, 2018). One of the reasons for this is the fact that container barges share their quay space and transshipment equipment with seagoing vessels and feeder ships. Currently, the productivity of container terminals will have strong peaks throughout a week: Whenever a large vessel arrives and is unloaded, many different barges and feeders will call for service at the same time. When this happens, barges will have the lowest priority getting serviced, as they are the smallest vessels. Also, the Ship To Shore (STS) cranes that are used for loading and unloading the barges are optimized for the larger seagoing vessels. This means the use of the same STS cranes for barges is inefficient, as their height and outreach are up to six times larger than they need to be for handling barges.

To solve the problem of high waiting times in the near future, one way is to apply logistic optimization of call times and service windows throughout a week in order to decrease the peak loads of terminals. However, when looking at long term development of container terminals and seeing the ports of the future as 'Smart Green Ports as Integrated Efficient multimodal hubs' (Magpie, 2022), another solution could be to come up with a design of a fully automated container transshipment device, suited specifically for the use on automated barges, or perhaps even smaller automated container vessels. This way, the inefficient use of large STS cranes on barges will also no longer be necessary.

When designing a new type of specialized equipment for the loading and unloading of container barges, a logical step is to design for autonomous use from the start, given the development in autonomous technology. In recent years the operations of container terminals have become more and more autonomous due to the introduction of various types of autonomous terminal equipment, such as Automated Guided Vehicles (AGV's). These autonomous vehicles have brought many gains in the efficiency and flexibility of container terminals (Schulze & Wülner, 2006). However, the transshipment of containers from ship to shore is still performed by machines that are partially humanly operated. This is caused by current legislation that prohibits the use of fully autonomous equipment in areas where humans are present. As ships currently still use human operators, such as lashers, in their loading and unloading operations, fully autonomous container transshipment is not yet a possibility; humans will have to operate the cranes when the spreader reaches the deck of a ship.

When humans are no longer needed on board of vessels during loading and unloading operations, only then can fully autonomous container transshipment become a reality. Currently, few companies are working on designs for fully automated container barges. Therefore, a logical next step in the development of

DOI: 10.1201/9781003320289-6

fully automated container ports, is to design an autonomous container transshipment device to handle the automated barges.

A second use for these small autonomous container transshipment devices could be to create a more efficient way to interterminally transport containers within the same port. Currently, when interterminal transport is performed by land, the train or truck routes are relatively long, as they have to go all the way around the bay. This becomes visible in Figure 1. When small autonomous ships, loaded by small autonomous transshipment devices are used instead, the bay can simply be crossed directly, creating smaller transport routes. This way of transport is not performed at this moment, but could be an implementation when looking at the ports of the future as smart, green, efficient, multimodal hubs (Magpie, 2022).

During this exploratory design phase there will be a special focus on the port of Rotterdam, since it has been the biggest port in Europe in terms of container throughput for many years (Notteboom, 2010) and contains some of the most modern deep sea container terminals in the world, located at the Maasvlakte. Not only does this mean that a good design solution would have most impact in this port, it is also one of the better examples of a 'port of the future' in terms of currently existing ports. However, for a design solution to be a good one, it should also be easily implementable in other ports of the world. For this reason, we explore designs without the need of any specialized quay infrastructure, such as train tracks or rail. This ensures that the design will be a flexible one, as it makes the solution easy to be implemented on almost any existing terminal, regardless of the type of existing infrastructure.

## 2 PROBLEM ANALYSIS

### 2.1 *Design method*

Our aim is to explore conceptual designs of autonomous transshipment equipment to load and unload containers from autonomous barges and AGV's, that is efficient, sustainable, ready to be implemented in the coming years and viable for use in the Port of Rotterdam.

In this case, the autonomy of the concept itself is not yet implemented in the design exploration. Our assumption is that the technology to make container transshipment devices work autonomously is largely already developed. The aim is to create a conceptual mechanical design that suits the ports of the future best, given the fact that they will function autonomously.

We will use a design method that divides the design problem into design phases and is based on the model of Pahl en Beitz (Roozenburg & Eekels, 1998). The different phases are shown below:

1. Orient and Analyze
   a. Analyze design problem
   b. Determine main and sub functions
   c. Create list of requirements
2. Explore and Experiment
   a. Determine (partial) solutions
   b. Explore different design directions
3. Develop Alternatives and Concept Choice
   a. Create complete concepts
   b. Estimate exptected the performances
   c. Compare and choose concepts
4. Optimize and Detail
5. Create and Test
6. Analyse and Iterate

In this exploratory conceptual design phase, the main focus is to identify different (innovative) solutions and to estimate their performance and feasibility. For this reason, we put a special emphasis on the phases: 2b. '*Explore different design directions*', 3a. '*Create complete concepts*' and 3b.'*Estimate the expected performances*'.

The explored design directions are used to create subsolutions to improve a single functionality of an existing crane. These subsolutions can be incorporated in the transhipment exploratory concepts. Whereas the design of the transshipment device concepts is meant to explore a fully functional system, on which the innovative subsolutions could be implemented, but not as a requirement.

### 2.2 *Current situation: Container freight in the port of Rotterdam*

The current system of container shipping in Rotterdam knows a handful of key hubs and nodes in the form of container terminals. The biggest of these is the Maasvlakte, containing five deep sea container terminals (Port of Rotterdam, 2021). At the Maasvlakte all containers from seagoing vessels are received, transshipped, temporarily stored and transported into the hinterland, and vice versa. As the 2020 annual throughput of containers in Rotterdam was 14.3 million TEU and currently 38% of containers between the Maasvlakte and hinterland are being transported by inland shipping, this makes it one of the busiest parts of the port of Rotterdam, with a surplus of seagoing vessels, feeder ships and barges either being loaded or unloaded, or waiting to be loaded or unloaded. As stated in Section 1.1, the current waiting time to be loaded or unloaded for container barges, having the lowest priority of all ship types, is up to 120 hours.

As mentioned, the Maasvlakte contains five modern deep sea container terminals.

1. Hutchison Ports ECT Euromax
2. Hutchison Ports ECT Delta
3. Hutchison Ports Delta II
4. APM Terminals Maasvlakte II
5. Rotterdam World Gateway

Figure 1. The deep sea terminals of the Maasvlakte.

The abovementioned terminals function in a similar fashion (Gharehgozli, 2019): The terminals are designed around the loading and unloading of the largest category of seagoing vessels. This is done by using large STS cranes, mostly of the 'Super-post -Panamax' size. Feeder size container vessels and container barges will mostly have a specially designated space on quay for loading and unloading. Here, they will usually be serviced by Panamax size STS cranes, which are still oversized for the use on these types of vessels. Also, it is not uncommon for barges and feeders to be serviced by the larger STS cranes used for seagoing vessels, when space is available.

The remainder of the journey of a container is as follows: The cranes transship the container to and from the ships onto AGVs. The AGVs are used to transport the containers to and from stacking cranes, which autonomously stack the containers onto large stacks that will contain hundreds up to thousands of containers. From here the containers wait to be loaded onto another ship, barge, train or truck.

Next to the five deep sea terminals, there are five inland shipping container terminals part of the port of Rotterdam as well. One of the terminals is located at the Maasvlakte, the others are spread throughout the port of Rotterdam and are mostly used as hubs to get containers to and from the Maasvlakte. The operations in these hubs is very similar to the operations in the large deep sea terminals. However, as the terminals are used just for barges, the terminals are smaller, the capacity is lower, and the used equipment is smaller. Most hubs make use of Panamax size STS cranes, which are able to stack containers underneath their bridge. These cranes can then be used to load the containers onto other vehicles at a later point.

### 2.3 *Container barge sizes*

Container barges are built in a variety of sizes. The most used sizes are listed below. These are not the only container barge types in existence. However, other sizes are uncommon.

1. Small container barge
   a. Length: 63 m
   b. Width: 7 m - 2 containers
   c. Capacity: 34 TEU
2. Standard container barge
   a. Length: 110 m
   b. Width: 11.40 m - 4 containers
   c. Capacity: 200 TEU
3. Large container barge (Jowi-class)
   a. Length: 135 m
   b. Width: 14 or 17m – 5 or 6 containers
   c. Capacity: 420 or 500 TEU

We assume that these barges will be smaller than the Jowi-class container barges, the maximum barge size for which the transshipment apparatus will be designed is the standard container barge size, containing container stacks of four containers wide and four containers high.

### 2.4 *Functions*

The main function of the to be designed apparatus is: To tranship containers from a container barge to an AGV and vice versa.

The sub functions the apparatus needs to fulfill in order to complete is main function are: To pick and release a container, to lift and lower a container, to horizontally transport a container, to support the weight of the loaded container, the moment of the loaded container and the own weight of the apparatus, and to maneuver itself on quay.

### 2.5 *List of requirements*

The list of requirements for the conceptual design of the container shipment apparatus will be subdivided into two different categories. The first category is the 'functional demands'. This list is comprised of all hard demands that at least need to be met in order to create a functional design. Secondly, there will be a list of 'performance criteria'. Once all functional demands are met, the performance criteria will be used to measure the performance of a design or concept to evaluate its functionalities and compare these with other designs or concepts.

The functional demands of the apparatus are the following:

1. The apparatus can pick and release a 40 ft container.
2. The apparatus can fully load and unload a container barge. I.e. it can reach any container stacked on board of a container barge without aid from other terminal equipment.

3. The apparatus can lift and lower the weight of a 40 ft container plus its maximum weight capacity.
4. The apparatus can horizontally transport a fully loaded 40 ft container between an AGV located on the quay and a container barge.
5. The minimum outreach for the apparatus will be large enough to service a container barge with stacks of up to 4 containers wide.
6. The apparatus can place or retrieve a 40 ft container on top of an AGV.
7. The apparatus can place or retrieve a 40 ft container loaded at any place on a container barge with maximum stack dimensions of 4 containers wide and 4 containers high.
8. The apparatus will be able to transport itself to any location on quay where a barge is docked or a barge will dock.
9. The apparatus will function fully autonomously.

The list of performance criteria of the apparatus is the following:

1. The apparatus can be implemented into existing container terminals, while changing the terminals' infrastructure as little as possible.
2. The apparatus will disrupt or negatively influence the operations of existing terminal equipment as little as possible.
3. The average transshipment time of a container from barge to AGV and vice versa will be as low as possible.
4. The installation and setup time of the apparatus when deployed to a new barge docked somewhere on the quay is as low as possible.
5. The speed the apparatus can transport itself with on the quay between calls is as high possible.
6. The average energy consumption for transferring a single container from barge to AGV and vice versa is as low as possible.
7. The average energy consumption for the apparatus to transport itself on quay between calls is as low as possible.
8. The average local CO2 emission of the apparatus in operation is as low as possible.
9. The wind speeds in which the apparatus can still continue operations will be as high as possible.
10. The continuous operation of the apparatus is as inexpensive as possible.
11. The manufacturing of the apparatus is as inexpensive as possible.

## 3 EXPLORE DESIGN DIRECTIONS

In order to create a working concept design, the next step in the design phases is to explore partial solutions for the sub functions of the machine. When these partial solutions are combined, in theory, a concept is created that will fulfill the main function. These partial solutions are determined structurally in the form of a morphological chart. However, since not all solutions in a morphological chart are useful or innovative, the chart itself is left out of this paper. Instead, the most innovative or otherwise promising solutions are highlighted and more thoroughly discussed in this chapter.

To assess the usefulness of these subsolutions, a cycle time analysis is made for small STS cranes, based on the half cycle time estimation of STS cranes (Zrnic & Bosnjak, 2005). For a generic small STS crane used for barges, a half cycle time is estimated to be a total of roughly 34 seconds. This estimation includes similar travel speeds for the hoists and trolleys of the small STS crane as for commonly used bigger cranes, but over shorter distances. Also, the dwell times are estimated to be lower as rope lengths are much shorter. It should be stated that actual cycle times are dependent on many factors, such as crane type and weather. The times stated in this paper are merely assumptions.

For each subsolution an estimation will be made in terms of the possible improvement in cycle time it could bring.

### 3.1 *Magnet pick up for empty containers*

The industry standard for picking and releasing containers is the use of a spreader with automatic twist locks. In this case, the corner castings of a container are inserted by a steel pin. When this pin is then twisted by 90°, it will be locked inside the corner casting by its own shape, forming a rigid connection between container and spreader. To insert the steel pin inside the corner casting, the pin needs to be lowered inside a 50 mm by 100 mm window, considering the shape of both the pin and the hole in the corner casting (ISO 1161, 2016). This is a relatively precise job, as the spreader will often be suspended by cables at a height of over 10 meters. When looking at cycle time analysis of a crane using a twist lock spreader and a newly proposed type of spreader in need of less accuracy when attaching, the usefulness of such an innovative idea can be assessed.

A new method to pick containers could be the use electrically powered magnets. Looking at magnets that are currently used for lifting in industry, an average surface area needed to be able to lift 1 ton, is roughly 450 cm$^2$, or a surface of roughly 20 cm to 20 cm (Mitari Hijstechniek, 2021). This means that a fully loaded 40 ft container, weighing roughly 40 tons, cannot be suspended by magnets on the containers corner castings. In fact, as the surface of a single corner castings is roughly 216 cm$^2$, even the 4-ton empty weight of a 40 ft container cannot be suspended on the corner castings. This is a problem, as a standard container is not ISO-tested to be lifted at any place other than its corner castings. However, the empty weight of a container is not nearly as high as its fully loaded weight. Therefore, it might be possible to lift empty containers using magnets attached elsewhere on the roof of container.

By using this assumption, three different concepts for a magnet pick up to be used on empty containers are suggested. Concept 1 and concept 2 make use of the flat section between the corner castings on the short side of a container for attaching. However, the gain in placement range for these concepts is either so small or even negative, that they will bring no improvement. Concept 3 (Figure 2) will attach onto the middle section of the roof of a container. As this section is mostly ribbed instead of flat, concept 3 will need a much larger electromagnetic surface to rigidly connect to the roof.

As the placement range of magnet concept 3 is much larger than that of regular twist lock spreaders, it is assumed that the aligning and picking time of a small STS crane could be halved, resulting in an improvement of roughly 5 seconds.

Table 1. Placement range of different grabbing mechanisms for container spreaders.

| Concept | X-range [mm] | Y-range [mm] |
|---|---|---|
| Twist lock spreader | 50.5 | 98 |
| Magnet concept 1 | 300 | 45 |
| Magnet concept 2 | 300 | 70 |
| Magnet concept 3 | 800 | 800 |

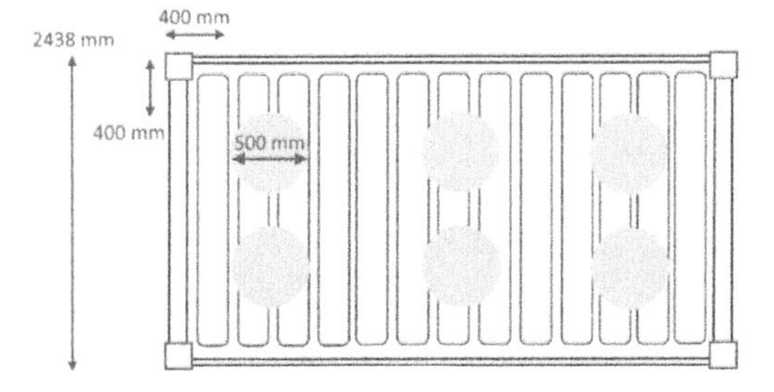


Figure 2. Magnet grab concept 3.

## 3.2 *Scissor mechanism for trolley cables*

Another concept that might have the potential to increase the productivity of a crane is a variation on the so called 'Variable level lifting platform for a cargo container crane' (Hasegawa, 1996). This device, shown in Figure 3, is comprised of a scissor mechanism that is placed over the cables between a trolley and spreader. It will effectively lower the rope distance between the trolley and the spreader to the distance between the spreader and the underside of the adjustable platform.

Hasegawa proposes this device to be used for large STS cranes with long rope lengths, in the case that they are continuously picking containers stacked relatively low. The platform will be lowered to a set distance and will remain at this distance for lifting and lowering operations.

For the use on barges, this will be less effective, as the distance between the trolley and spreader is much lower and more often used in its entirety. Instead, for the use specialized STS cranes for barges, it might be best to permanently connect the scissor device to the spreader and trolley. Meaning, the scissor mechanism will move up and down with the spreader on every container lift. When using the device in this way, the vertical load of the spreader and a container is still carried by the trolley cables. However, the swaying motions in X- and Y-directions are now countered by the scissor mechanism, which only has one degree of freedom: translation on the Z-axis.

As the payload sway for a small STS crane can be reduced by using this device, the assumption is made that both the aligning time for picking a container and for releasing a container can be reduced by roughly 20%, resulting in a cycle time improvement of roughly 4 seconds.

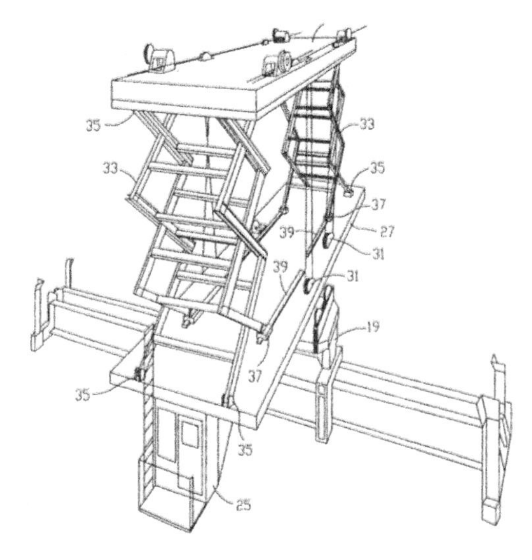


Figure 3. Sway reduction mechanism (Hasegawa, 1996).

## 3.3 *Motion compensated boom*

At this point, cranes are being operated using complex control systems to keep the payload steady and in an optimal trajectory. These control systems are mostly based on decreasing the payload swing due to accelerations and decelerations (Hoang et all, 2014) and (Nidhil Wilfred et all, 2014). However, wind forces on the payload will always be present during operations, causing disturbances and sway.

To counter any disturbances in a cranes' X-direction, a control system using the acceleration and deceleration of the trolley can be implemented, much similar to the current control systems used for overhead cranes. To counter disturbances in a cranes' Y-direction, there currently exists no viable solution. A solution could be to implement a motion compensation system in this direction, using a sliding boom.

When the sway of a payload can actively be countered, the assumption is made that the aligning times for both picking and releasing containers can be reduced by roughly 30%, resulting in a cycle time improvement of roughly 6 seconds.

## 4 INITIAL CONCEPTS

The exploration in partial solutions for the sub functions of the machine can then be integrated in different configurations of initial concepts. The three proposed subsolutions for increasing crane productivity are made for use on an overhead STS crane type, making use of a trolley. The solutions can be used on these types of cranes, but will not alter the original conceptual design of the crane.

Next to these innovative solutions, three new conceptual designs for an autonomous container transshipment are made, specialized for the use on barges. For all these concepts, dimensions and weights are based on stability and tire load. All devices are stable in a statical analysis excluding wind, swinging and lifting, and the maximum reaction forces are not exceeding the tire limit.

### 4.1 *Concept 1*

The first concept, shown in Figure 7 and characteristics in Table 2, is based on a straddle carrier. It is a widened straddle carrier with a boom connected to its topside and a trolley with spreader driving along the boom. On the landside of the boom, a counterweight is connected to ensure the stability of the concept.

The process of this concept is as follows:

- A container located on a barge is picked by a regular spreader.
- A trolley lifts the container and spreader and moves them to shore.
- The container and spreader pass through the portal.
- The container is dropped on to an AGV parked inside the portal of the device.
- This process is repeated for the entire row.
- The device moves along to the next row.

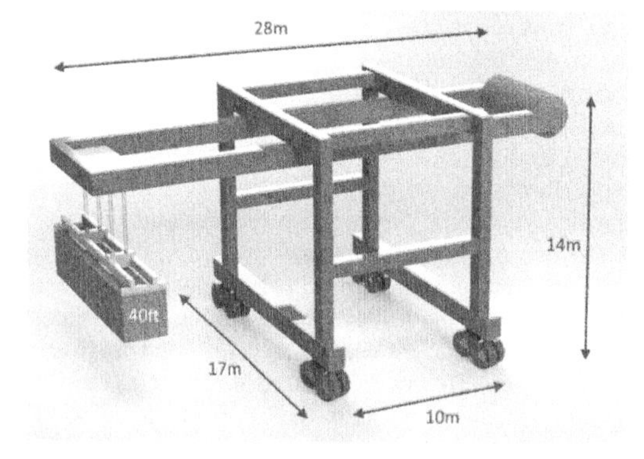


Figure 4. Concept 1.

Table 2. Characteristics of concept 1.

| | |
|---|---|
| SWL | 40 ton |
| Service weight | 120 ~140 ton |
| Counterweight | 15 ton |
| Outreach | 11 m |
| Lifting height | 10 m |
| Tire size | 18.00R33 |
| Tire amount | 16 |
| Max tire load | 16 ton |

### 4.2 *Concept 2*

The second concept, shown in Figure 5 and characteristics in Table 3, has a lifting mechanism based on that of an excavator. This concept is based on the 'Mantsinen 200R' (Mantsinen Group, 2008), which is a multifunctional heavy lifting machine, capable of being deployed in many different environments, as it is supported by caterpillar tracks. For this conceptual design the choice was made to use regular tires instead of caterpillar tracks. Its SWL is dependent on the distance of the payload to the device, as in this concept the moment of the payload needs to be countered as well, instead of just the weight of the payload when using a trolley.

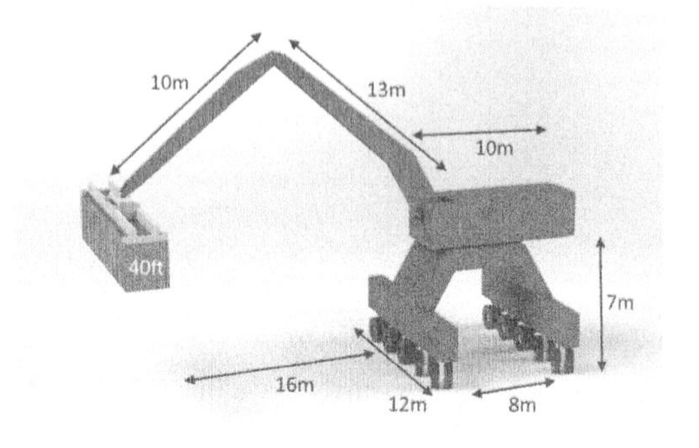


Figure 5. Concept 2.

The process of this concept is as follows:

- A container located on a barge is picked by a regular spreader.
- Hydraulic cylinders on the boom and stick lift the container and transport it to shore.
- The container is dropped on to an AGV parked between the device and the waterside.
- This process is repeated for the entire row.
- The device moves along to the next row.

Table 3. Characteristics of concept 2.

| | |
|---|---|
| SWL | 40 ton at 16m |
| Service weight | 200 ~ 240 ton |
| Outreach | 23 m |
| Lifting height | 10 m |
| Tire size | 18.00R33 |
| Tire amount | 20 |
| Max tire load | 18 ton |

### 4.3 *Concept 3a*

The third concept, shown in Figure 6 and characteristics in Table 4, is based on a generic construction crane. It uses a trolley to transport the container to shore, just as concept 1. However, the structure much narrower, with a smaller boom and supported by a single pillar, instead of on its four corners.

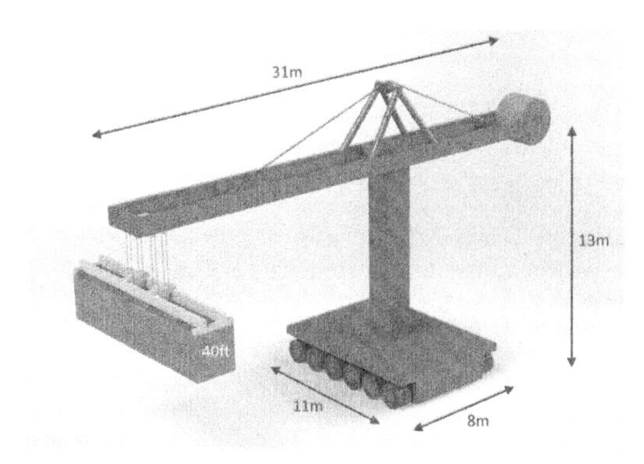


Figure 6. Concept 3a.

The process of this concept is as follows:

- A container located on a barge is picked by a regular spreader.
- A trolley lifts the spreader and container and moves them to shore
- The container is dropped on to an AGV parked between the device and the waterside.
- This process is repeated for the entire row.
- The device moves along to the next row.

Table 4. Characteristics of concept 3a.

| | |
|---|---|
| SWL | 40 ton |
| Service weight | 160 ~200 ton |
| Counter weight | 20 ton |
| Outreach | 15 m |
| Lifting height | 13 m |
| Tire size | 18.00R33 |
| Tire amount | 20 |
| Max tire load | 18 ton |

### 4.4 *Concept 3b*

As both concept 1 and concept 3a have a relatively long boom, these concepts become less mobile, since large structures will have more trouble maneuvering themselves on quay. A solution for this problem might be to make the boom of these concepts deployable. Comparing both concepts, it becomes clear that creating a deployable version is best done for concept 3a. This is because concept 3a has a larger boom, creating more space for a deployable mechanism.

There are multiple ways to make a boom deployable. A telescoping mechanism is not useful, as it requires the boom to incrementally decrease its height and width. This is problem, since the trolley needs a straight set of rails to roll over, which is impossible when the width of the boom becomes smaller over its length. Another way to make the boom deployable is by dividing it into sections that can rotate compared to each other.

Firstly, it is a possibility to rotate the boom upwards. However, this is not practical, as this will heighten the center of gravity of the crane and create a larger moment arm for wind forces, making the crane less stable and less mobile.

Secondly, the boom can be rotated sideways. This is unpractical, because the crane then either becomes unsymmetrical, or the boom needs to be split in two along its width, creating a solution with much more moving parts.

The final option is to rotate the boom downwards. This means the center of gravity will drop lower to the ground, creating a more stable structure to maneuver around. The downside of this is that the downward hinge will make the boom lose all of its stiffness in negative Y-direction, meaning it can no longer hold a payload. This effect must be countered by creating a mechanism that locks the different booms rigidly together when in deployed configuration.

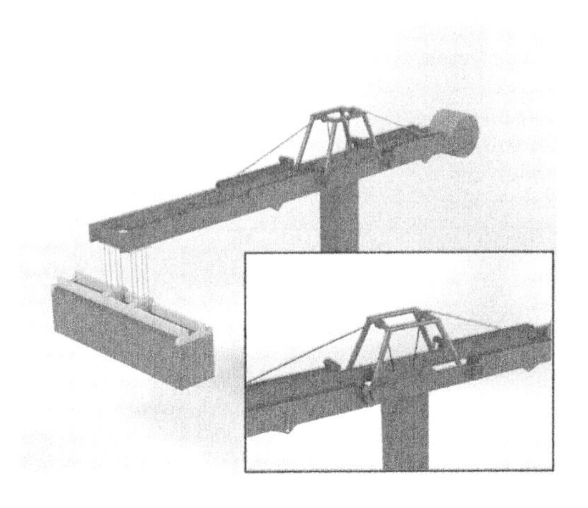

Figure 7. Concept 3b Deployable boom in deployed state.

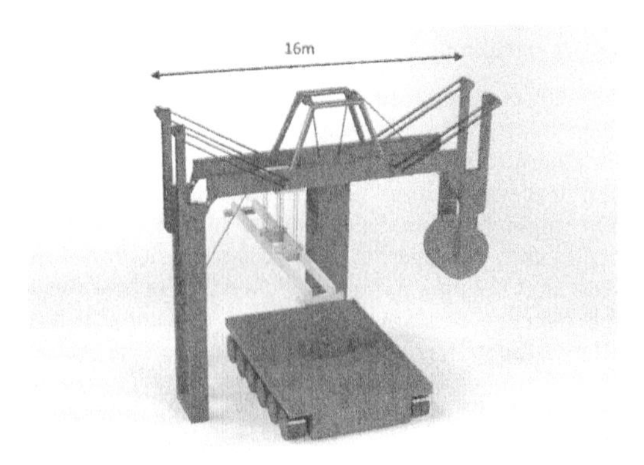


Figure 8. Concept 3b deployable boom in retracted state.

The process of loading and unloading barges is exactly similar for concept 3b as for concept 3a. The process of deploying the boom of concept 3b from a retracted position is as follows:

- The forward and rearward boom sections of the crane are in retracted position, supported by the hinge and hanging downwards.
- Electric motors will serve as a winch to tighten a rope and pully system, this will rotate both boom sections upwards.
- Once both forward and rearward boom sections are horizontal, a mechanical pin system will lock both sections rigidly to the middle section of the boom.
- The electric motors are disengaged and the mechanical lock will hold all boom sections in place.
- When retracting the boom sections, the abovementioned process is followed in reverse.

Table 5. Characteristics of concept 3b.

| | |
|---|---|
| SWL | 40 ton |
| Service weight | 180 ~220 ton |
| Counter weight | 20 ton |
| Outreach | 15 m |
| Lifting height | 13 m |
| Tire size | 18.00R33 |
| Tire amount | 20 |
| Max tire load | 19 ton |

## 5 EXPLORATORY CONCEPT COMPARISON

To compare the four proposed concepts in terms of their usefulness in different situations or terminals, the performance criteria of Section 2.5 are summarized into a more generic set of criteria. This is done because the performances of the concepts can at this point only be evaluated by looking at their mechanical design, and no actual details like cost estimations or energy consumptions are known. The criteria are:

- Productivity: Concepts using a simpler or more efficient method of transshipment will likely have a higher productivity.
- Complexity: A complex solution with many different components will result in larger capital costs as well as operational costs of a machine.
- Mobility: A longer, wider or taller machine is less maneuverable, making it less mobile.
- Weather Resistant (WR): Light, slender and less stable concepts will be less resistant to work in extreme environments.

Table 6. Concept comparison.

| Concept | Productivity | Complexity | Mobility | WR |
|---|---|---|---|---|
| 1 | + | ++ | ± | + |
| 2 | - | - | + | ++ |
| 3a | + | + | - | - |
| 3b | + | – | + | - |

### 5.1 *Discussion*

Concept 1 and 3 make use of a trolley system to move containers to and from shore. This is relatively fast transshipment method, enabling the concept to have a high productivity. Secondly, the square frame composed of steel beams is both simple and stable, as it is supported at four corners. This makes it a light machine with a low level of complexity and relatively high weather resistance. However, concept 1 has a large, undeployable boom, making it less mobile.

Concept 2 needs a large amount of torque and power to be able to lift a payload at distance. This is caused by the fact that the excavator arms need to counter the moment created by the payload at distance as well as the weight of the payload. To make this work, at least five large hydraulic pumps need to be used (Mantsinen Group, 2008). This makes for a slow and complex machine. However, as the excavator arms are deployable and able to withstand wind loads better than trolley cables, concept 2 will have a high mobility and weather resistance.

Concept 3a will be very similar to concept 1 when looking at its operation, allowing for an equal productivity. On the other hand, the large single pillar support as well as the long boom make it a concept that is more complex, less mobile and less weather resistant compared to concept 1.

Concept 3b tackles the mobility issues of concept 3a, but this will cause large increase in complexity, because of the added electric drives, pulley systems and the locking mechanism and reinforcement for the boom. As the operations of concept 3a and concept 3b are completely similar, their productivity and weather resistance is equal.

## 6 CONCLUSIONS

When looking at the ports of the future as smart, green, efficient, multimodal hubs that function fully autonomously, the following can be concluded about the conceptual design of autonomous container transshipment devices to handle autonomous container barges:

The new concepts show promise, but the eventual choice of a concept will be dependent on many factors, including the location of the port and the location of the terminal within the port.

At terminals, where large volumes of containers are transshipped daily, the best solution would be to make use of a large fleet of cranes similar to concept 1. It is a cheap and productive option, that is relatively weather resistant. Also, making use of a large fleet of cranes reduces the problem of the lower mobility of the concept, as travel distances for individual cranes become shorter in a large fleet.

In places where weather and wind play a small role in the daily operations of a terminal, concept 3a and 3b could also be implemented.

At terminals located in extreme environments and heavily suffering from bad weather, concept 2 could be a solution for a container transshipment device.

Considering the proposed subsolutions, the following can be stated: Firstly, the only way a magnet pick up can improve productivity of a crane, is by lifting a container by its roof with an oversized magnet. This will only work for empty container. Secondly, a scissor mechanism and a motion compensated boom have a similar potential to improve the productivity of an STS crane. However, the scissor mechanism is a much cheaper addition to a crane, compared to the complexity and added systems an adjustable boom will bring.

As these design are only exploratory and conceptual, more design and evaluation work should be done to verify the feasibility of the concepts and to introduce autonomy into the concepts, as this has not yet been done. Also, more research should be done to conclude whether the added functionality making a deployable crane will outweigh the increased complexity and costs that this will add.

## ACKNOWLEDGEMENTS

This project has received funding from the European Union's Horizon 2020 research (CEF 2014-2020) and innovation programme under Grant Agreement 101036594.

## REFERENCES

Contargo (2018), Waiting times in seaports of rotterdam and Antwerp .Available:https://www.contargo.net/en/newsarchive/2018-03-06_wartzeiten_in_dern_se.

Gharehgozli, A. (2019): Container terminal layout design: transition and future. California State University. Maritime Economics & Logistics.

Hasegawa, S. (1996). Variable level lifting platform for a cargo container crane (U.S. Patent No. 5,538,382). U.S. Patent and Trademark Office. https://rb.gy/ik0fb0

Hoang, Q. H. et all (2014). Trajectory planning for overhead crane by trolley acceleration shaping. Hanoi University of Science and Technology: Springer

ISO 1161:2016 (2016): Series 1 freight containers – Corner and intermediate fittings – Specifications. International Organization for Standardization.

MAGPIE (2022): Smart Green Ports as Integrated Efficient multimodal Hubs: https://www.magpie-ports.eu/

Mantsinen Group Ltd Oy (2010): Mantsinen 200 R and 200 ER Hybrilift. Ylamylly, Finland

Mitari Hijstechniek BV. (2021): PLM-A600 Lifting Magnet.

Nidhil Wilfred, K. J. et all (2014) Container Crane Control using Sliding Mode Control. VIT University: International Journal of Engineering Research & Technology.

Notteboom, T. E. (2010). Concentration and the formation of multi-port gateway regions in the European container port system: An update. Journal of Transport Geography, 18(4),567–583

Port of Rotterdam (2021): Hub network with dedicated intra port barge shuttles. Rotterdam, the Netherlands.

Roozenburg, N. F. M., & Eekels, J. (1998). Productontwerpen, structuur en methoden (Tweede Druk). Den Haag: Lemma.

Schulze, L & Wüllner, A (2006): The Approach of Automated Guided Vehicle Systems. University of Hannover: IEEE

Zrnic, N. & Petkovic, Z. & Bosjnak, S. (2005). Automation of Ship-To-Shore Container Cranes: A Review of State-of-the-Art. University of Belgrade, Faculty of Mechanical Engineering: FME Transactions

*Trends in Maritime Technology and Engineering – Guedes Soares & Santos (eds)*


# Environmental assessment of the scrubbers' use in regular traffic between the Canary Islands and the Iberian Peninsula

A. Martínez-López
*Department of Mechanical Engineering, University of Las Palmas de Gran Canaria, Spain*

Á. Marrero
*Department of Naval and Industrial Engineering, University of A Coruña, Spain*
*Center for Innovation in Transport (CENIT), CIMNE-UPC, Barcelona, Spain*

Y. Martín-Cruz
*Department of Process Engineering, University of Las Palmas de Gran Canaria, Spain*

M. Míguez González
*Integrated Group for Engineering Research, University of A Coruña, Spain*

ABSTRACT: The exhaust gas cleaning systems (scrubbers) have proven to be an effective mitigation system for pollutant air emissions. Aside from meeting the current emissions normative (Annex VI of MARPOL 73/78), the scrubbers have resulted in broadly fulfilling the IMO 2015 Guidelines for Exhaust Gas Cleaning Systems (resolution MEPC.259(68)) related to discharge. However, in recent years' ports have increased restrictions on the use of open-loop scrubbers, as a number of studies have identified deficiencies in the current regulations, especially regarding the limitations of metal concentrations in wash water discharge. This paper quantitatively assesses the overall environmental consequences of open and closed-loop scrubbers' operation by considering not only air emission reductions, but also the ecotoxicity and eutrophication impact of wash water discharge. Therefore, a pollution cost model is introduced where these concepts are jointly evaluated to calculate the environmental performance of scrubbers. In order to test the model's effectiveness and to provide quantitative information, the paper analyzes results from an application case for the regular container fleet between the Canary Islands and the Iberian Peninsula.

## 1 INTRODUCTION

The scrubber, or exhaust gas cleaning system (EGCS) is a mitigation system able to meet efficiently with the Global Sulphur Cap requirements (they entered into force in January 2020, where the permitted Sulphur content was reduced from 3.5% to 0.5%, by keeping 0.1% in Emission Control Areas- Annex VI of MARPOL).

EGCS involves spraying vessel engines' exhaust gases with seawater (open-loop system) or with freshwater (closed-loop system) to remove $SO_X$ and particulate matter. In the open-loop system, the alkalinity of the seawater is used to clean the gases, and the wash water is discharged to the sea (as it is single-use); whereas in the closed-loop system a base to the freshwater (sodium hydroxide) must be added to be effective. However, in the latter case the wash water is recirculated (multiple-use) and only a small amount (the bleed-off) is discharged into the sea. Consequently, the effluent's 'volume overboard' for closed-loop scrubbers is very small in comparison to open-loop scrubbers ($0.3m^3$/MWh versus $45m^3$/MWh, Ytreberg et al., 2021). Finally, hybrid scrubbers enable operators to shift between the open- and closed-loop systems.

Despite of the initial advantage shown by the operative of closed-loop scrubbers, currently 81.29% of scrubbers are open-loop and only 16.72% are hybrid (DNV statistics, 2021). This fact, along with the significant increase in the use of scrubbers in recent years (4,584 vessels in 2021 versus 740 in 2018 according to DNV statistics, 2021), has led to a rising concern about the environmental consequences of scrubbers' discharge. Proof of this is the review of the current regulation (2015 IMO Guidelines for Exhaust Gas Cleaning Systems -Resolution MEPC 259(68)-) through 2020 draft IMO Guidelines for Exhaust Gas Cleaning Systems (2020 EGCS Guidelines) adopted by the Sub-Committee on

DOI: 10.1201/9781003320289-7

Pollution Prevention and Response (PPR 7), in February 2020.

Whereas 2015 EGCS Guidelines only established limits for PH, PAHphe, turbidity and nitrates in the scrubber's wash water, Appendix 3 of 2020 EGCS Guidelines recommends, but does not enforce to, monitor relevant data on discharge water for future assessments of the EGCS technology: 16 Polycyclic Aromatic Hydrocarbons (PAH) and 9 metals. All of these are recognized as contaminants (Directive 2013/39/EU) that have a significant impact on marine ecotoxicity (Ytreberg, et al., 2021; Hermansson, et al., 2021, Faber et al., 2021).

As a consequence of this, a deregulation perception exists, and it has driven to unilateral actions from the countries (Germany, Belgium, Singapore, among others) to ban open-loop scrubber use in their waters.

Even though increasing attention has been paid in recent years to the impact of discharges from scrubbers, very few studies have jointly analyzed the environmental advantage given by their pollutant emissions' reduction versus the disadvantages from wash water discharges. Additionally, most of them are restricted to North Sea and Baltic Sea context. Among them, two are specially significantly due to the relevance of their findings: Hermansson et al., 2021 concluded that the scrubbers not only transfer the environmental load from the atmosphere to the marine environment but also introduce new contaminants like Cr and alter the atmospheric dispersion of air pollutants (Hermansson et al., 2021). Likewise, Ytreberg et al., 2021 determined that there are no environmental advantages for scrubbers in economic terms, and even the phasing out of HFO use is suggested with or without scrubbers by Hermansson et al., 2021.

This paper attempts to broaden knowledge about the environmental performance of scrubbers beyond the meeting of the regulation. To this aim, the paper offers an environmental assessment method for the scrubbers, where the emissions' reduction provided by the EGCS is quantified in monetary terms versus the increase of the ecotoxicity and eutrophication value provided to the marine environment. To offer quantitative information in monetary terms, the method is applied to assess the environmental convenience of scrubbers' use in regular traffic between the Canary Islands and the Iberian Peninsula.

## 2 METHODOLOGY

The assessment method attempts to provide a quantification of the pollutant impact from several options which enable to meet the Global Sulphur Cap normative. As a first approach, the method assumes the following environmental impacts: climate change, air quality, ecotoxicity of the marine environment and marine eutrophication (based only on the Nitrogen concentrations). Regarding the former two (*CEM*), the following pollutants were considered ($U=\{1,..,u\}$): $SO_X$ (acidifying substances), $NO_x$ (ozone precursors), $PM_{2.5}$, $PM_{10}$ (particulate matter mass), and the greenhouse gases $CO_2$. In parallel, the contaminants ($P=\{1,..,p\}$): the PAHs and the metals collected in Appendix 3 of the 2020 EGCS guidelines were taken into account to evaluate ecotoxicity (*EME*). Finally, the effects on marine eutrophication due to the Nitrogen in the scrubbers' discharge are also evaluated (*EUT*). The Pollutant Impact per trip (PI €/trip, see equation 1) is the result of the addition of the previous pollutant categories (*CEM, EME and EUT*), assuming the three possible navigation stages ($S=\{1,..,s\}$): open sea shipping, manoeuvring (port pilot, tug service, and mooring time), and "at berth" (loading/unloading operations).

$$\mathrm{PI} = \sum_{s=1}^{3} CEMs + \sum_{s=1}^{3} EMEs + \sum_{s=1}^{3} EUTs; \forall \mathrm{s} \in \mathrm{S} \tag{1}$$

Thus, the PI offers a decision-making tool based on environmental criteria for compliance options.

The assessment of the pollutant air emissions is based on a modification of the model published by Martínez-López et al., 2018. In this case, LNG is not considered as an alternative solution in a first approach - the $CH_4$ pollutant has been replaced by $PM_{10}$ - since this latter pollutant has proved to have a significant impact on human health.

$$\mathrm{CEM}_1 = \sum_{u=1}^{5} (EG_{1u} \times CF_{1u} \times TVB_1); \forall \mathrm{u} \in \mathrm{U} \tag{2}$$

$$\mathrm{CEM}_2 = 1/2 \times \sum_{f=1}^{2} CEM_{2f}; \forall \mathrm{f} \in \mathrm{F} \tag{3}$$

$$\mathrm{CEM}_{2\mathrm{f}} = \sum_{u=1}^{5} (EG_{2u} \times CF_{2ufv} \times TVB_2); \forall \mathrm{f} \in \mathrm{F} \wedge \forall \mathrm{v} \in \mathrm{V} \tag{4}$$

$$\mathrm{CEM}_3 = 1/2 \times \sum_{f=1}^{2} CEM_{3f}; \forall \mathrm{f} \in \mathrm{F} \tag{5}$$

$$\mathrm{CEM}_{3\mathrm{f}} = \sum_{u=1}^{5} (EG_{3u} \times CF_{3ufv} \times TVB_3); \forall \mathrm{f} \in \mathrm{F} \wedge \forall \mathrm{v} \in \mathrm{V} \tag{6}$$

Equations 2–6 show the environmental impact of the pollutant air emissions by considering the emission factors per pollutant and navigation stage ($EG_{su}$; $\forall s \in S \wedge \forall u \in U$), and the unitary costs for them ($CF_{sufv}$; $\forall s \in S \wedge \forall u \in U \wedge \forall f \in F \wedge \forall v \in V$). The emission factors are taken from the calculation tools developed by the Technical University of Denmark (Kristensen & Bingham, 2020; Kristensen & Psaraftis, 2016).

Since the emission factors for particulate matters are aggregated in these calculation tools, to meet the $PM_{2.5}$ and $PM_{10}$ emissions, the relationship between them - as published in the 'EMEP/EEA air pollutant emission inventory guidebook, 2019' - for several fuels was also considered.

According to the emission factor units (kg/h or kg/nautical mile), the maritime distance per trip (*DM*) of the time invested at every navigation stage ($TVB_s$; $\forall s \in S$) were also considered (see equations 2, 4 and 6). Finally, the unitary costs (€/kg pollutant) involve the marginal social cost pricing principle for air pollution (external cost categories: health effects, crop loss, biodiversity loss, material damage). For this, the costs take into consideration the geographical location of the emission sources (-country-F $=\{1 ..., f\}$) and their population density (V $=\{1,..,v\}$). Moreover, for $CO_2$ the climate change avoidance costs (medium values) are assumed. Successive updates of these costs are published by the European Commission in the Handbook on the External Costs of Transport (last updated in 2019; Van Essen et al., 2019).

## 2.1 *Scrubbers' water discharge assessment*

The quantification in monetary terms of the ecotoxicity for scrubbers' discharges (*EME* in €/trip) in the marine environment follows the method introduced by Ytreberg et al. (2021) but has been adapted to the navigation stages in the Atlantic Ocean (*EMEs*, $\forall s \in S$ -see equation 7-).

$$EMEs = (ETP_{marine})_p \times Vs \times \Delta\rho_p \times EPE; \quad \forall s \in S \wedge \forall p \in P \tag{7}$$

Where:

$(ETP_{marine})_p$ (kg1,4DCB-eq/kg pollutant): ecotoxicological midpoint characterization factor to the ocean water for every contaminant (P=$\{1,..,p\}$).

$V_s$(l/trip): Discharge volume of scrubbers for every navigation stage (S=$\{1,..,s\}$).

$\Delta\rho_p$ (kg pollutant/l): Increase in concentrations for every contaminant (P=$\{1,..,p\}$) regarding their base concentrations in the pristine sea water.

EPE(€/kg1,4DCB-eq): Monetary value for marine ecotoxicity.

This approach is based on the use of ReCiPe characterization factors (Huijbregts, 2017a) to quantify the accumulative toxicity from different contaminants to several receiving compartments. The harmonized factors ($ETP_{marine}$), expressed as 1,4-dicholorobenzene equivalents (1,4-DCB-eq) for substances discharged to the marine environment, can be taken from Huijbregts et al. (2017b). Table 1 collects those contaminants included in Appendix 3 of the 2020 EGCS guidelines (P=$\{1,..,p\}$;16 Polycyclic Aromatic Hydrocarbons (PAH) and 9 metals) along with their characterization factors (hierarchist perspective).

As can be seen in Table 1, even though all metals have characterization factors, the factors for six of the 16 Polycyclic aromatic hydrocarbons are unavailable.

The Marine eutrophication ($EUT_s$ $\forall s \in S$ in €/trip) by Nitrogen is calculated through expression 8.

$$EUTs = Vs \times \Delta\rho_N \times EPF; \forall s \in S \tag{8}$$

Where, aside from the discharge volumes of scrubbers for each navigation stage (Vs, $\forall s \in S$), the increase in Nitrogen concentration ($\Delta\rho_N$ in kg/l) in pristine seawater is considered along with monetary value for the marine eutrophication (EPF in €/Kg N).

Even though, several methods, based on Willingness to pay (WTP) of societies for damage cost, exist for monetizing the eutrophication and ecotoxicity impact on the marine environment, only Ecovalue (Ahlroth, and Finnveden, 2011) and Environmental Prices (based on ReCiPe methodology, De Bruyn et al., 2018), methods take into account uniquely marine eutrophication (Arendt et al., 2020). In turn, Ecovalue method only assesses environmental impacts in Sweden.

Table 1. Characterization factors for contaminants of scrubbers' wash water.

| CAS Registry Number | Polycyclic Aromatic Hydrocarbons (PAH): 16 EPA PAHs | $ETP_{marine}$ (kg1,4DCB-eq/kg poll.) Hierarchist perspective |
|---|---|---|
| 83329 | Acenaphthene | 1.17E+01 |
| 602879 | Acenaphthylene | #N/D |
| 120127 | Anthracene | 3.06E+02 |
| 56553 | Benzo-a-anthracene | 1.70E+03 |
| 50328 | Benzo-a-pyrene | 2.92E+02 |
| 205992 | Benzo-b-fluoranthene | #N/D |
| 191242 | Benzo-g,h,i-perylene | #N/D |
| 207089 | Benzo-k-fluoranthene | #N/D |
| 218019 | Chrysene | #N/D |
| 53703 | Dibenzo-a,h-anthracene | 1.55E+01 |
| 206440 | Fluoranthene | 3.85E+02 |
| 86737 | Fluorene | 3.12E+00 |
| 193395 | Indeno-1,2,3-pyrene | #N/D |
| 91203 | Naphthalene | 2.13E+00 |
| 85018 | Phenanthrene | 4.73E+01 |
| 129000 | Pyrene | 1.30E+03 |
| | **Metals** | |
| 7440439 | Cd | 1.96E+02 |
| 7440508 | Cu | 1.57E+03 |
| 7440020 | Ni | 3.21E+02 |
| 7439921 | Pb | 9.53E+00 |
| 7440666 | Zn | 3.42E+02 |
| 7440382 | As | 2.12E+02 |

(*Continued*)

Table 1. (Cont.)

| CAS Registry Number | Polycyclic Aromatic Hydrocarbons (PAH):16 EPA PAHs | $ETP_{marine}$ (kg1,4DCB-eq/kg poll.) Hierarchist perspective |
|---|---|---|
| 7440473 | Cr | 3.22E+02 |
| 7440622 | V | 4.55E+02 |
| 7782492 | Se | 1.06E+02 |
| 7439976 | Hg* | 7.09E+02 |

* *Due to the high toxicity of mercury, this substance was included despite it was not collected in the EGCS guidelines 2020*

## 3 APLLICATION CASE

The previously introduced method will be applied to a container fleet (see Table 2) with linear traffic between the Canary Islands and the Iberian Peninsula. Thus, through this application case, the environmental consequences of compliance options with the Global Sulphur Cap will be assessed.

In order to simplify the evaluation, a regular route between Las Palmas and Cadiz port will be assumed as a base route (*DM*=687 n.m.). Likewise, the current operative pattern has been taken per trip for the calculations: 37.13 hours for free sailing ($TVB_1$); 0.5 hours/port for maneuvering time ($TVB_2$) and six hours for "at berth" time ($TVB_3$).

Table 2. Main features for a container vessel in the application case.

| Features | |
|---|---|
| Lt (m) | 148.00 |
| Lpp (m) | 137.82 |
| B(m) | 20.50 |
| D (m) | 11.17 |
| T (m) | 8.20 |
| Service speed (kn) | 18.50 |
| Main engine (BHP kW) | 8,300 |
| TEUs | 869 |
| TEUs (reefer) | 234 |
| Auxiliary engines (kWe) | 3X590 |
| PTO (kw) | 1800 |
| Bow thruster (kW) | 880 |
| Lightweight (t) | 4,666.21 |

### 3.1 *Assumptions and constraints for operativity*

The scrubber configuration assumption was based on just one system for the main engine (MAN B&W G50ME-C9.6-LPSCR, (MAN B&W two-stroke propulsion marine engine). This was mainly due to the generating sets size (MAN 5L23/30DF): 590 kWe at 750 rpm (four-stroke marine engines) and the electric load balance (see Table 3).

Table 3. Electricity generating capacity planning.

| Navigation stage | %BHP main engine (kW) | Required electrical power (kW) | Capacity planning* |
|---|---|---|---|
| Free Sailing | 78.00% | 1,570 | PTO 87.22% |
| Maneuvering | 24.31% | 2,400 | PTO81% +2XMMAA80% |
| "at berth" | 0.00% | 1,470 | 3xMMAA 83% |

* *MMAA=Auxiliary engines (generating sets)*

According to this, the generating sets will predominantly be operative in port, where 0.1%S fuel is required in EU ports or equivalent emissions through mitigation systems- Directive 2005/33/EC; amending Directive 1999/32/EC- and due to most EU countries have forbidden open-loop scrubbers (Spain among them) and to the generating sets' technical characteristics (medium speed engines), 0.1% MGO or LNG fuels must be used to maintain engine performance. Consequently, the exhaust cleaning system will address emissions from the main engine (see Table 4).

Two consequences of the scrubbers' settings have been evaluated in terms of efficiency: the increase in the required propulsion power and the additional electrical power needed to operate the EGCS. To evaluate both aspects, the technical features of scrubbers given by MAN (2020) and Alfa Laval (2021) were assumed (see Table 4).

Table 4. Scrubbers' setting implications for practical case.

| Increase | Open-loop | Closed-loop |
|---|---|---|
| Lightweight (t) | 18 | 21 |
| Additional back pressure in main engine (ΔSFOC g/kWh) | +1 | +1 |
| Pumps' EGCS (kW) | 49 | 49 |
| Water treatment System (kW) | | 40 |
| Δ Required power (main engine) kW | +0.6% | +0.6% |
| Δ Required electrical power (kW) | 49 | 89 |

### 3.2 *Scenarios of analysis*

Three alternative abatement solutions are evaluated to fulfil the Global Sulphur Cap: HFO operating with an open-loop scrubber; MGO and HFO operating with a closed-loop scrubber (see Table 5). In turn, the following sulphur contents will be assumed for free sailing stage: 0.5%S for MGO and 1%, 2% and 3.5%S for HFO. Therefore, whereas for MGO only one scenario exists, for the use of HFO with scrubber, three different scenarios are considered (see Table 5).

Table 5. Assumptions for the scenarios of analysis.

| Abatement solutions | Scenarios | Navigation stages | Fuel for main and auxiliary engines |
|---|---|---|---|
| **Scrubber (open-loop)** | Scenario 1: 3.5%S HFO | Free sailing | 3.5%S HFO |
| | | Maneuvering | 0.1%S MGO |
| | | "at berth" | 0.1%S MGO |
| | Scenario 2: 2%S HFO | Free sailing | 2%S HFO |
| | | Maneuvering | 0.1%S MGO |
| | | "at berth" | 0.1%S MGO |
| | Scenario 3: 1%S HFO | Free sailing | 1%S HFO |
| | | Maneuvering | 0.1%S MGO |
| | | "at berth" | 0.1%S MGO |
| **MGO** | 0.5% S | Free sailing | 0.5%S MGO |
| | | Maneuvering | 0.1%S MGO |
| | | "at berth" | 0.1%S MGO |
| **Scrubber (closed-loop)** | Scenario 1: 3.5%S HFO | Free sailing | 3.5%S HFO |
| | | Maneuvering | 3.5%HFO+ 0.1%S MGO |
| | | "at berth" | 0.1%S MGO |
| | Scenario 2: 2%S HFO | Free sailing | 2%S HFO |
| | | Maneuvering | 2%HFO+ 0.1%S MGO |
| | | "at berth" | 0.1%S MGO |
| | Scenario 3: 1%S HFO | Free sailing | 1%S HFO |
| | | Maneuvering | 1%HFO+ 0.1%S MGO |
| | | "at berth" | 0.1%S MGO |

In all cases, the auxiliary engines operate with 0.1%S MGO (see Table 3), since this is required to operate in EU ports (Directive 2005/33/EC). For the same reason when the main engine is operating in the maneuvering stage (see Table 3), for all alternatives different from closed-loop scrubber use, 0.1%S MGO will be assumed as fuel. In port operations, only the closed-loop scrubber is permitted (maneuvring). Thus, in this case, the closed-loop scrubbers operate with the same fuel as that for free navigation (HFO for the different scenarios, see Table 3 and Table 5).

### 3.3 *Unitary costs for air pollutants and wash water contaminants*

The unitary costs for air emissions ($CF_{sufv}$; $\forall s \in S \wedge \forall u \in U \wedge \forall f \in F \wedge \forall v \in V$; see equations 2-6) published by the European Environmental Agency in the Handbook on the External Costs of Transport (Van Essen et al., 2019) were used for the analysis. Thus, the climate change avoidance cost was taken for $CO_2$ emissions, whereas in the case of the other air pollutants, the national average of damage costs for transport emissions were considered (see Table 6).

To this aim the published values have been updated to January 2021 by considering the CPI of Spain (CPI=6.2%-2016-2021), from the National Statistics Institute of Spain (2021). Moreover, due to the Cadiz and Las Palmas hinterlands having over 0.5 million inhabitants, both ports have been considered metropolitan zones ($V = \{1,..,v\}$). This involves higher values for the unitary costs of $PM_{2.5}$ emissions.

Table 6. Unitary costs for air pollutants $CF_{sufv}$.

| Pollutant ($u$) | $CF_{1u}$ (€/kg)* | $CF_{2u11}$(€/kg)** $CF_{3u11}$(€/kg) |
|---|---|---|
| $SO_X$ | 3.72 | 7.22 |
| $NO_x$ | 9.03 | 4.03 |
| $PM_{2.5}$ | 7.64 | 369.58 |
| $PM_{10}$ | 4.35 | 12.64 |
| $CO_2$ | 0.11 | 0.11 |

* *$CF_{1u}$ involves Atlantic Ocean values*
** *$CF_{2ufv}$= $CF_{3ufv}$= $CF_{2u11}$=$CF_{3u11}$, the unitary costs for maneuvering and "at berth" stages are related to Spain and metropolitan hinterlands*

Likewise, as a first approach to the quantification of the scrubbers' discharges impact, the environmental price method was chosen due to the following reasons: the method includes prices for both impacts evaluated in this analysis: ecotoxicity and eutrophication. Moreover, this method provides a specific monetization for marine eutrophication and finally, the method is suitable to be applied in the locations of this framework. Environmental prices (De Bruyn et al., 2018) for marine ecotoxicity (*EPE*), were estimated according to the Life Cycle Assessment using the ReCiPe methodology (under the hierarchist perspective). Thus, Ecotoxicity price for average EU28 emissions has been updated (from 2015 to 2021) by taking CPI (CPI: 7.7% Eurostat, 2021): *EPE*=0.007959 €/kg1,4 DCB-eq (see equation 7). The same reference and update was applied to the environmental price for marine eutrophication *EPF*=3.35 €/kg N (see equation 8).

### 3.4 *Characterization of the discharge water for scrubbers.*

Concentrations ($\rho_p$; $\forall p \in P$) for the selection of metals and PAHs in the scrubbers' discharges must be met to quantify in monetary terms their impact on ecotoxicity (*EMEs* $\forall s \in S$, see equation 8). To this aim the average concentrations (µg/l for every pollutant) given by Hermansson et al. (2021) have been assumed for open- and closed-loop modes (see Table 7). The authors show a processed database of samples of scrubbers' discharge for 41 vessels by operating in open- and closed-loop modes and sailing for different seas. The final samples taken by the authors (see Table 7) involve a wide %MCR range and %S contents for the fuel from 0.7 to 3.2%. Despite the dispersion of these samples, the average concentrations are in line with others, such as those obtained by Ytreberg et al. (2021) (based on the database of Jalkanen et al., 2020). For that reason, these average values have been assumed for the analysis. Moreover, due to the scale of the contaminants' concentrations in pristine seawater (non-contaminated seawater pattern-see Table 7), the increase in contaminants regarding the pristine seawater will be taken as equivalent to the overall discharged contaminants ($\Delta\rho_p = \rho_p$; $\forall p \in P$, see equation 7).

The pristine seawater pattern was obtained from previous researches of chemical analysis of the North-East Central Atlantic Ocean, being this zone the closest one to the study case.

With the aim of determining the volume of wash water discharge for each and every navigation stage ($V_s; \forall s \in S$), a relationship between discharge flow rates (m3/MW.h) and the power developed by the engine (%MCR) at %S fuel, was obtained from the database published by Hermansson et al. (2021). In such a way, more accurate discharge rates than those provided by the scrubber manufacturers (45 m3/MW.h for open-loop systems and for closed mode system 0.1-0.3 m3/MW.h) allow us to obtain more realistic results. Likewise, real measures of total Nitrogen of the samples for different %S and %MCR were taken from Hermansson et al.'s (2021) database. These N concentrations were compared with open and closed mode results provided by the samples of Kjølholt et al., (2012) and proved to be very close to each other's: $\rho_N$: 0.20-0.60 mg/l for open mode and $\rho_N$: 24-120 mg/l for closed mode.

Table 7. Average concentrations for pollutants of scrubbers' discharges $\rho_p$.

| Polycyclic Aromatic Hydrocarbons (PAH) | Open loop scrubber discharge (µg/l) | Closed loop scrubber discharge (µg/l) | Pristine * Sewater (µg/l) (max) |
|---|---|---|---|
| Acenaphthene | 0.19 | 0.47 | 0.0001 [1] |
| Acenaphthylene | 0.12 | 0.09 | 0.0002 [2] |
| Anthracene | 0.08 | 1.55 | <0.0001 [1] |
| Benzo-a-anthracene | 0.12 | 0.3 | 0.0001 [2] |
| Benzo-a-pyrene | 0.05 | 0.06 | <0.0001 [1] |
| Benzo-b-fluoranthene | 0.04 | 0.14 | <0.0001 [1] |
| Benzo-g,h,i-perylene | 0.02 | 0.07 | <0.0001 [1] |
| Benzo-k-fluoranthene | 0.01 | 0.02 | <0.0001 [1] |
| Chrysene | 0.19 | 0.5 | 0.0004 [2] |
| Dibenzo-a, h-anthracene | 0.03 | 0.03 | 0.001 [9] |
| Fluoranthene | 0.16 | 0.63 | <0.0001 [1] |
| Fluorene | 0.46 | 1.32 | 0.0005 [2] |
| Indeno-1,2,3-pyrene | 0.07 | 0.04 | 0.001 [9] |
| Naphthalene | 2.81 | 2.08 | <0.0001 [2] |
| Phenanthrene | 1.51 | 5 | 0.0006 [1] |
| Pyrene | 0.31 | 0.76 | 0.0002 [1] |
| **Metals** | | | |
| Cd | 0.8 | 0.55 | 0.07 [10] |
| Cu | 36 | 480 | 0.10 [4] |
| Ni | 48 | 2700 | 0.26 [3] [4] |
| Pb | 8.8 | 7.7 | 0.004 [3] |
| Zn | 110 | 370 | 0.80 [4] |
| As | 6.8 | 22 | 1.18 [5] [6] |
| Cr | 15 | 1300 | 0.16 [7] |
| V | 170 | 9100 | 4.9 [4] |
| Se | #N/D | #N/D | 0.05 [6] |
| Hg | 0.09 | 0.07 | 0.0004 [3] [8] |

(Continued)

Table 7. (Cont.)

[1] Nizzetto et al., 2008;
[2] Vecchiato et al., 2018;
[3] Pohe et al., 2011;
[4] Prego et al., 2013;
[5] Wurl et al., 2015;
[6] Cutter and Cutter et al., 1995;
[7] Sirinawin et al., 2000;
[8] Mason et al., 2012;
[9] Law et al., 1997;
[10] Bruland, 1980.

## 4 RESULTS FOR APPLICATION CASE

The calculation of the emission factors ($EG_{su}$; $\forall s \in S \wedge \forall u \in U$ see equations 2-6) has assumed a minimum reduction of 55% for particulate matter ($PM_{2.5}$ and $PM_{10}$) and 98% of $SO_X$ emissions for scrubbers' use. The maneuvering and "at berth" stage's emission factors are the same for all scenarios (%S fuels) with open-loop scrubbers (in all cases 0.1%S is required for operation) and the emission factors in these stages are also very close to MGO 0.5%S solution for the same reason. Therefore, significant differences are only found in free sailing in terms of air emissions between open scrubber and MGO alternatives. Contrary to this, the main differences in air emissions between open- and closed- loop scrubbers are obtained during the maneuvering stage, where the closed-loop scrubbers operate (through PTO; see Table 3) along with the auxiliary engines to supply electric energy. This can be evaluated in monetary terms (see Table 8) ($CEM; \forall s \in S$).

The impact of the scrubbers' discharge has been less significant than expected regarding air emissions.

Assuming the discharge patterns of Table 6, total 0.194 kg 1,4 DCB-eq/m$^3$ were obtained for open loop and 6.31 kg 1,4 DCB-eq/m$^3$ for closed-loop. Despite the higher impact on the ecotoxicity of closed- loop scrubbers' discharge, their absolute impact in this regard (*EME*, see Table 8) is much less significant due to their low discharge rate (Vs; $\forall s \in S$, see Table 8). Contrary to this is the impact on eutrophication (*EUT*), where high Nitrogen concentrations for closed-loop mode ($\rho_N$: 24-120 mg/l) versus open-loop ($\rho_N$: 0.20-0.60 mg/l) have led to similar results for both options (see Table 8).

Paying attention to the total pollutant impact per trip (PI €/trip), the options for scenario 3 (HFO 1%S) have proved to be the most sustainable, regardless of the operation mode (open- or closed-loop) of the scrubbers. Indeed, the air emissions component (CEM) is the most influential, as this is several orders of magnitude higher than the components related to the scrubbers' discharge components (EME and EUT see Table 8).

Table 8. The pollutant Impact per trip and their components for the different scenarios.

| Abatement solutions | Scenarios | Navigation stages | CEMs (€/trip) | CEM (€/trip) | Vs(m3) | EMEs (€/trip) | EUTs (€/trip) | EME (€/trip) | EUT (€/trip) | PI (€/trip) |
|---|---|---|---|---|---|---|---|---|---|---|
| **Scrubber (open-loop)** | Scenario 1: 3.5%S for HFO | Free sailing ($CEM_1$) | 19,783.77 | 21,814.36 | 18,115.72 | 28.02 | 33.98 | 28.02 | 33.98 | 21,876.36 |
| | | Maneuvering ($CEM_2$) | 536.99 | | 0.00 | 0.00 | 0.00 | | | |
| | | "at berth" ($CEM_3$) | 1,493.60 | | 0.00 | 0.00 | 0.00 | | | |
| | Scenario 2: 2%S for HFO | Free sailing ($CEM_1$) | 18,959.82 | 20,990.41 | 13,702.50 | 21.19 | 25.70 | 21.19 | 25.70 | 21,037.30 |
| | | Maneuvering ($CEM_2$) | 536.99 | | 0.00 | 0.00 | 0.00 | | | |
| | | "at berth" ($CEM_3$) | 1,493.60 | | 0.00 | 0.00 | 0.00 | | | |
| | Scenario 3: 1%S for HFO | Free sailing ($CEM_1$) | 18,627.12 | 20,657.71 | 12,133.45 | 18.77 | 14.63 | 18.77 | 14.63 | 20,691.11 |
| | | Maneuvering ($CEM_2$) | 536.99 | | 0.00 | 0.00 | 0.00 | | | |
| | | "at berth" ($CEM_3$) | 1,493.60 | | 0.00 | 0.00 | 0.00 | | | |
| MGO 0.5% S | | Free sailing ($CEM_1$) | 19,433.36 | 21,465.28 | 0.00 | 0.00 | 0.00 | 0.00 | 0.00 | 21,465.28 |
| | | Maneuvering ($CEM_2$) | 537.04 | | 0.00 | 0.00 | 0.00 | | | |
| | | "at berth" ($CEM_3$) | 1,494.89 | | 0.00 | 0.00 | 0.00 | | | |
| **Scrubber (closed-loop)** | Scenario 1: 3.5%S for HFO | Free sailing ($CEM_1$) | 19,965.13 | 22,196.56 | 60.09 | 3.02 | 24.15 | 3.08 | 24.67 | 22,224.32 |
| | | Maneuvering ($CEM_2$) | 736.54 | | 1.30 | 0.07 | 0.52 | | | |
| | | "at berth" ($CEM_3$) | 1,494.89 | | 0.00 | 0.00 | 0.00 | | | |
| | Scenario 2: 2%S for HFO | Free sailing ($CEM_1$) | 19,086.77 | 21,097.64 | 57.98 | 2.91 | 23.31 | 2.98 | 23.83 | 21,124.45 |
| | | Maneuvering ($CEM_2$) | 515.98 | | 1.30 | 0.07 | 0.52 | | | |
| | | "at berth" ($CEM_3$) | 1,494.89 | | 0.00 | 0.00 | 0.00 | | | |
| | Scenario 3: 1%S for HFO | Free sailing ($CEM_1$) | 18,739.87 | 20,661.68 | 57.98 | 2.91 | 16.70 | 2.97 | 17.04 | 20,681.70 |
| | | Maneuvering ($CEM_2$) | 426.92 | | 1.17 | 0.06 | 0.34 | | | |
| | | "at berth" ($CEM_3$) | 1,494.89 | | 0.00 | 0.00 | 0.00 | | | |

## 5 CONCLUSIONS AND DISCUSSION

The assessment method introduced in this paper attempts to surpass the classical environmental evaluation of air emissions by including increases in the marine eutrophication and ecotoxicity due to scrubbers' discharges. The method is applied to a regular shipping line for a container vessel operating between the Canary Islands and the Iberian Peninsula.

Despite of the fact that previous research based on Baltic and North Sea has warned about the significance of the impact of open-loop scrubbers' wash water on the marine environment, however the results achieved in this paper show low relevance against air pollutant emissions (0.10%-0.3% contribution from scrubbers' discharges to total pollutant impact of the vessel per trip) by assuming The values of the inputs considered in this paper are in line with those as sumed by previous studies. However, we have detected significant differences among the unitary costs used to evaluate in monetary terms the impact of scrubbers' effluents.Thus, whereas this paper assumes for 2021 environmental prices:0.0079€/kg 1,4 DCB-eq and 3.349€/kg $N_{tot}$, authors such as Ytreberg et al., take for 2010 prices 0.64-1.13€/kg 1,4 DCB-eq and a range of 52-60 €/kg $N_{tot}$.

Likewise, is notwithstanding the difference of values for monetizing the eutrophication and marine ecotoxicity provided by Ecovalue and Environmental Prices (2019 values, respectively):1.03€/kg 1,4 DCB-eq versus 0.0074€/kg 1,4 DCB-eq and 7.729 €/kg $N_{tot}$ versus 3.325€/kg $N_{tot}$.To this regard, it is necessary to bear in mind the influence of the societies' localization (Willingness to pay) on the values: Ecovalue is referred to Sweden and Environmental prices to the EU28 average. This reality emphasizes the need for further research into monetary values of the marine environmental with greater accuracy levels.

In parallel, results achieved also show that the additional energy consumed by main engines for scrubber use (free sailing) is not as relevant as expected: close to 1% for open-loop mode and 1.3% for closed-loop mode. Moreover, the results achieved in the application case suggests a further research line focused on the impact of the closed-loop scrubbers on marine eutrophication.

## ACKNOWLEDGEMENTS

This work has been supported by 'Canary smart specialization strategy programme RIS-3, 2020′ co-funded by European Regional Development Fund (2014-2020). (REPORT project, grant agreement no. ProID2020010056).

## REFERENCES

Ahlroth, S.; Finnveden, G. (2011) ''Ecovalue08—A new valuation set for environmental systems analysis tools". Journal of Cleaner Production. 2011, 19 (17-18) pp: 1994–2003.

Arendt R, Bachmann TM, Motoshita M, Bach V, Finkbeiner M. (2020)"Comparison of Different Monetization Methods in LCA: A Review". *Sustainability*.2020;12 (24):10493. https://doi.org/10.3390/su122410493

Alfa Laval (2021) Alfa Laval PureScrub H2O. https://www.alfalaval.com/globalassets/documents/products/process-solutions/scrubber-solutions/purescrub-h2o/purescrub-h2o_product-leaflet.pdf (On-line Access 17/11/21)

Bruland, K. W. (1980). ''Oceanographic distributions of cadmium, zinc, nickel and copper in the North Pacific". *Earth and Planetary Science Letters*. Vol 47 (2): 176–198.

Cutter G. A.; Cutter, L. S. (1995). ''Behavior of dissolved antimony, arsenic and selenium in the Atlantic Ocean". *Marine Chemistry*. 49. 4. 295–306. https://doi.org/10.1016/0304-4203(95)00019-N

DNV-GL statistics, (2021). Scrubber statistics - DNV-GL. https://afi.dnvgl.com/Statistics?repId=2 (On-line Access 17/11/21)

De Bruyn, S.; Bijleveld, M.; de Graa_, L.; Schep, E.; Schroten, A.; Vergeer, R.; Ahdour, S. (2018) *Environmental Prices Handbook*; CE Delft: Delft, The Netherlands, 2018.

EMEP/EEA air pollutant emission inventory guidebook (2019). 1.A.1 Energy industries. https://www.eea.europa.eu/publications/emep-eea-guidebook-2019/part-b-sectoral-guidance-chapters/1-energy/1-a-combustion/1-a-1-energy-industries/view (on-line access 23/ 11/ 2021)

EUROSTAT (2021) ttps://ec.europa.eu/eurostat/databrowser/view/PRC_IPC_G20__custom_1687250/default/table?lang=en (On-line Access 17/11/21)

Faber, M.; Peijnenbur, W.; Smit, C.E. (2021). Environmental risks of scrubber discharges for seawater and sediment. Preliminary risk assessment for metals and polycyclic aromatic hydrocarbons. National Institute for Public Health and the Environment. RIVM letter repor 2021-0048.

Finnveden, G.; Eldh, P.; Johansson, J. (2006) ''Weighting in LCA based on ecotaxes: Development of a mid-point method and experiences from case studies". The International Journal of Life Cycle Assessment 11 (S1):81–88.

Hermansson A. L., Hassellov I.M., Moldanova J., Ytreberg E. (2021). ''Comparing emissions of polyaromatic hydrocarbons and metals from marine fuels and scrubbers" *Transportation Research Part D* 97 (2021) 102912.

Huijbregts, M., Steinmann, Z., Elshout P., Stam G., Verones F.,Vieira M., Hollander A., Zijp M., van Zelm R. (2017a). *ReCiPe 2016: a harmonized life cycle impact assessment method at midpoint and endpoint level* Report I: characterization, RIVM Report 2016-0104a. National Institute for Public Health and the Environment (Netherlands).

Huijbregts M., Steinmann Z. Elshout P., Stam G., Verones F., Vieira M., Zijp M., Hollander A., Van Zelm R. (2017b) ''ReCiPe2016: a harmonised life cycle

impact assessment method at midpoint and endpoint level" *Int J Life Cycle Assess* (2017) 22:138–147
Kjølholt J., Aakre S., Jürgensen C., Lauridsen J. (2012). Assessment of possible impacts of scrubber water discharges on the
Marine environment. Environmental Project No. 1431, 2012. Danish Environmental Protection Agency.
Kristensen H.O and Bingham H., (2020). Project no. 2016-108: ''Update of decision support system for calculation of exhaust gas emissions". Report no.06, HOK Marineconsult ApS and Technical University of Denmark, May 2020.
Kristensen, H.O., Psaraftis, H., (2016). Project no. 2014-122: ''Mitigating and reversing the side-effects of environmental legislation on Ro-Ro shipping in Northern Europe", Work Package 2.3, Report no. 07 HOK Marineconsult ApS and Technical University of Denmark, August 2016.
Law, R.J., Dawes, V.J., Woodhead, R.J., Matthiessen, P., 1997. ''Polycyclic aromatic hydrocarbons (PAH) in seawater around England and Wales". *Mar. Pollut. Bull.* 34, 306–322.
MAN (2020) *Emission Project Guide MAN energy solutions, 2020.*https://man-es.com/applications/projectguides/2stroke/content/special_pg/PG_7020-0145.pdf (On-line Access 17/11/21)
Martínez-López A, Caamaño Sobrino P, Chica González M, Trujillo L (2018). ''Choice of propulsion plants for container vessels operating under Short Sea Shipping conditions in the European Union: An assessment focused on the environmental impact on the intermodal chains". *Proceedings of the Institution of Mechanical Engineers, Part M: Journal of Engineering for the Maritime Environment.* 2019;233(2):653–669. doi:10.1177/1475090218797179
Mason, R.P.; Choi, A.L.; Fitzgerald, W. F.; Hammerschmidt, C. R.; Lamborg, C.H.; Soerensen, A. L.; Sunderland, E.M. (2012). ''Mercury biogeochemical cycling in the ocean and policy implications". *Environmental Research.* 119. 101–117. https://doi.org/10.1016/j.envres.2012.03.013.
National Statistics Institute of Spain (2021). https://www.ine.es/dyngs/INEbase/en/operacion.htm?c=Estadistica_C&cid=1254736176802&menu=ultiDatos&idp=1254735976607 (On-line Access 17/11/21)
Nizzetto, L.; Lohmann, R.; Gioia, R.; Jahnke, A.; Temme, C.; Dachs, J.; Herckes, P.; Di Guardo, A.; Jones, K.C. (2008). ''PAHs in Air and Seawater along a North–South Atlantic Transect: Trends, Processes and Possible Sources". *Environmental Science & Technology.* 42, 5, 1580–1585. https://doi.org/10.1021/es0717414
Pohl, C.; Croot, P-L.; Hennings, U.; Daberkow, T.; Budeus, G.; Rutgers, M.; Loeff, V.D. (2011). ''Synoptic transects on the distribution of trace elements (Hg, Pb, Cd, Cu, Ni, Zn, Co, Mn, Fe and Al) in surface waters of the Northern- and Southern East Atlantic". *Journal of Marine Systems.* 84. 1-2. 28–41. https://doi.org/10.1016/j.jmarsys.2010.08.003.
Prego, R.; Santos-Echeandia, J.; Bernández, P.; Cobelo-García, A.; Varela, M. (2013). ''Trace metals in the NE Atlantic coastal zone of Finisterre (Iberian Peninsula): terrestrial and marine sources and rates of sedimentation". *Journal of Marine Systems.* 126. 69–81. http://dx.doi.org/oi: 10.1016/j.jmarsys.2012.05.008
Sirinawin, W.; Turner, D. R.; Westerlund, S. (2000). ''Chromium (VI) distributions in the Arctic and the Atlantic Oceans and a reassessment of the oceanic Cr cycle". *Marine Chemistry.* 71. 3-4. 265–262. https://doi.org/10.1016/S0304-4203(00)00055-4.
Van Essen H., Van Wijngaarden L., Schroten A., Sutter D., Bieler C., Maffii S., Brambilla M., Fiorello D., Fermi F., Parolin R, and El Beyrouty K., (2019). ''Handbook on the external costs of transport. Version 2019." European Commission – DG Mobility and Transport. Unit A3 — Economic analysis and better regulation, January 2019.
Vecchiato, M.; Turetta, C.; Patti, B.; Barbante, C.; Piazza, R.; Bonato, T.; Gambaro, A. (2018). ''Distribution of fragances and PAHs in the surface seawater of the Sicily Channel, Central Mediterranean". *Science of The Total Environment.* 634. 983–989. https://doi.org/10.1016/j.scitotenv.2018.04.080
Wurl, O.; Shelley, R. U.; Landing W. M.; Cutter, G. A. (2015). ''Biogeochemistry of dissolved arsenic in the temperate to tropical North Atlantic Ocean". *Deep Sea Research Part II: Topical Studies in Oceanography.* 116. 240–250. https://doi.org/10.1016/j.dsr2.2014.11.008.
Ytreberg E., Astrom S., Fridell E. (2021) ''Valuating environmental impacts from ship emissions – The marine perspective"*Journal of Environmental Management*, 282 (2021) 111958.

*Trends in Maritime Technology and Engineering – Guedes Soares & Santos (eds)*


# Optimization model for integrated port terminal management

F.G.G. Pereira, J.P.G. Cruz & R.C. Botter
*Department of Naval and Oceanic Engineering, University of Sao Paulo, Sao Paulo, Brazil*

L.T. Robles
*Federal University of Maranhão, Maranhão, Brazil*

ABSTRACT: Iron ore is the main Brazilian exported commodity, the mineral and steel industries are searching for efficient production, cost reductions and higher profitability. This paper presents an optimization model to support the storage stockyard management, which integrate the train discharging and ships loading operations, reducing operating times, maximizing the stockyards occupation, and increasing the volume shipped in a period. An operational investigation was carried out to identify the main rules and restrictions of a terminal. Linear Programming (LP) techniques were applied to integrate a strategic and a tactical model. Furthermore, perform the scheduling of trains and ship berthing to minimize non-operational time, providing information to ensure stockpiles while the ship is loaded, consequently minimizing total ship service time. The results are satisfactory, however, for large instances and high levels of operations detail, a combination of heuristic methods is recommended to optimize processing time, a common difficulty for the topic.

## 1 INTRODUCTION

The integrated management of operations in port terminals specialized in exporting iron ore involves complex processes and operational routines, which consider the reception of trains from mine to port, handling and storage of products at the terminal and loading of ships, ensuring the integrity of facilities and quality of the cargo formed for each ship. These processes must be managed and optimized seeking the rational use of resources and the reduction of operating costs involved. The challenge of managing the mass balance of an ore port terminal is a daily activity for its workers. It is in this challenge that this research seeks to support port managers in decision-making, at strategic and managerial levels, so that they can prepare their short, medium and long-term planning, and manage their daily activities. The integration of planning and scheduling decisions is critical to achieving efficient and reliable port terminal operation, as well as good inventory management. Menezes, Mateus & Ravetti. (2016a, 2016b and 2016c) developed a generic mathematical model for the integration of planning and scheduling decisions at a solid bulk port terminal, iron ore exporter. The problem was to define the quantity and destination of each order entering or leaving the terminal, establishing a set of viable routes for storing and shipping the products on time, minimizing operating costs. Boland, et al. (2011) discuss the use of an optimization model in operations involving stockyard s in a coal export supply chain, with customized cargo assembly, in which stocks are assembled from coal delivered by trains originating in mines. Barros et al. (2011) used a linear programming model for the transport and berth allocation problem in ports, which is associated with the management of stock levels, by an algorithm based on Simulated Annealing, which employs problem-specific heuristics. Pratap et al. (2015) studied the operations of a coal importing port to develop a decision support model that deals with port dynamics and helps in better decision-making in different scenarios, taken by port authorities. Operational rules are incorporated into the model-to-model port operations in its various processes, based on mixed integer scheduling and heuristic methods. Irfanbabu, et al. (2015) employed two Greedy construction algorithms based on heuristics to improve the throughput capacity of the port terminal, minimizing the delay of ships at the port terminal. In the management of stockyards integrated with the railway unloading and ship loading processes, it is essential to know the arrival and departure schedule of cargo, as well as the availability of storage stockyards and equipment. The stockyard equipment, stacker and reclaimers are programmed, according to the arrangement of areas available for stockpiling and the materials stored in the stockyards. The section 2 presents challenges faced in developing an optimization model for integrated port processes. Section 3 presents the problem of integrated planning of port operations, which

DOI: 10.1201/9781003320289-8

motivated this research. The train and ship integrated scheduling model is described in section 4, computational experiment and analysis of results in section 5, finally, the conclusion in section 6.

## 2 INTEGRATED PORT OPTIMIZATION CHALLENGES

Solutions to problems in planning the allocation of piles and sequencing of port processes is not trivial and requires detailed research to identify existing alternatives, to propose and build models with greater effectiveness and potential for use. Given the degree of complexity, it can be classified as an NP-hard problem (Marcos Junior et al, 2018). For Tang, Sun, & Liu (2015), solving the entire master problem in an optimal way using hierarchy of sub-problems can generate significant results, however, very time-consuming. To overcome this challenge, the author suggests the use of the Benders decomposition algorithm, applied to help break the unfeasibility and further accelerate the convergence of results. For high levels of detail in the operations of a solid bulk terminal, Boland et al. (2012) states that an integrated problem is too complex to use mathematical programming, in terms of computational performance, due to the excessive number of variables and constraints. Therefore, they developed constructive heuristics to solve the problem. The contribution of this research is in the development of a model that applies Linear Programming (LP), which can solve real problems and assist in strategic decisions. The model optimizes the scheduling and sequencing of the queue of ships and trains together, considering the stock flow in the stockyard and moving loads as one of the model's inputs, thus minimizing demurrage costs.

## 3 PROBLEM OF PORT OPERATIONAL INTEGRATED PLANNING

The products arrive by train to the unloading station. Car dumpers unload iron ore in belt conveyors, which, through programmed routes, transport it for stockpiling in stockpiles or for loading directly onboard ships. Ore piles can remain in the stockyard s until the scheduled ship and berth facilities are available to begin loading. There is only one berth for ships to be loaded. The scheduling of entry and departure of ships must respect the queue for use of the navigation channel. The carrying capacity of trains at the mines, in terms of the number of trains, and the volume that can be transported per day on the railway is limited. Spaces in stockyard s or pads have specific capacities. Given the existing restrictions, it is estimated to verify whether different combinations of allocation of lots, train specific product, in the stockpiles improve or worsen unloading in car dumpers and loading of ships.

### 3.1 *Railway unloading process*

Each specific train that arrives at the terminal waits in line for its opportunity to unload at any of the existing car dumpers. Upon arrival, there is a line that can access any of the turners. When a car dumper is available and is scheduled to unload, the train leaves the queue and does not return. The sequencing of trains starts at the railway stockyard. When a given turner is about to finish unloading a train, another one is cadenced and sent to the terminal. The next train can position itself in the sequence of another train, as per the schedule. Trains must arrive at the terminal to guarantee the mass balance between the cargo to be unloaded and the cargo to be loaded. In the case of unloading trains directly for loading the ship, it is evaluated whether the product is the same as the ship and whether equipment is available, with the possibility that the operation is carried out. Once the train and location in the stockyard are selected, the dumper is considered occupied and unloading is carried out in two stages: the first goes from the moment the train leaves the railway stockyard until its positioning in the car dumper, maneuvering and circulation time not considered in the model; and the second goes from the beginning of the wagon unloading operation to the end of the last wagon characterized as the total unloading time of the train.

### 3.2 *Planning and allocation of stockpiles*

The stockyard is set in terms of its occupation both in terms of capacity and type of product. The same area can be used by more than one product, however, never simultaneously. The proposed optimization model considers a stockpile standard, product type and storage capacity to test product allocation combinations in the stockyards. A stockpile cannot be manipulated by 2 equipment at the same time. There can be a combination of simultaneous recovery with up to two reclaimers, if there are pads available, and the charging limit capacity is respected. Onboard train operation does not occur concurrently with recovery with stockyard equipment. The proposed model does not consider the set of system routes. In this case, the product allocation configurations in the stockyards are analyzed, according to the empty spaces, to minimize the distances between the piles formed with the same type of product and the transfer of reclaimers during the loading of a certain ship.

### 3.3 *Ship scheduling*

For a ship to be berthed at the terminal, a berth must be available for loading, otherwise the ship waits in line at the anchorage area. The loading schedule must ensure that a minimum stock of products is available in the stockyards or on a route during ship loading. The ships sequencing can be changed, depending on cargo availability and demurrage projection. Only when the ship finishes berthing, its waiting time in queue is ended. When a ship is berthed and released, it waits a pre-operation time before loading. This time is necessary to complete the system configurations, regularize documenttation and to wait for the material to travel the route from the stockyard to the pier. Loading starts when the product starts to be loaded on the ship. The stockyard management looks for the pile with the product and the recovery equipment, which uses a designated route for the shiploader, always seeking to maximize its use. During the shipping operation, a quantity of product can be shipped from the car dumper. In this case, train on board or direct boarding is configured. Once loading is complete, the shiploader is released, and the ship waits for a post-operation period to start the unberthing maneuver operation. Only after the unberthing maneuver is completed, the ship is considered finished, and the berth is released to the next ship.

### 3.4 *Boarding schedule*

Each ship has, depending on its capacity, a distribution of the number of holds and available cargo, which may comprise more than one product. The model is simplified and does not consider loading plans. In practice, a shipping line is assigned, respecting the maximum capacity of the system. The model uses the concept of commercial fee, which already considers the time penalized by process stoppages during the service of the ship. If the ship has docked and its cargo is incomplete in the stockyard, a routine situation that can be configured, it can wait for direct to the ship. When fully recovering a stockpile, the space, that is, the pad is immediately freed for the allocation of another product to be stockpiled. The terminal seeks the optimized number of pads needed to unload a certain amount of product, specific to form a ship's cargo, in order to reduce train unloading and ship loading times, maximize stockyard occupancy, and reduce costs operational.

## 4 OPTIMIZATION MODEL FOR INTEGRATED PORT TERMINAL MANAGEMENT

The proposed optimization model was developed using Linear Programming (PL) techniques to solve real problems. The model for strategic decisions optimizes the scheduling and sequencing of the queue of ships and trains together, considering the variation of stock in the stockyards and the ship loading flow as one of the model's input data, thus minimizing cost with demurrage. Ship and train sequencing data is used to designate origin and destination, that is, each pad to a ship and each train load to a pad or direct to board. In other words, the model must select, for an origin represented by the turner-product pair, the best possible destination, that is, which stockyard and pad are available for the operation, and which would be the best choice at that moment to unload and stockpile the loading a train in a queue. The model was implemented in the Python programming language, using the Solver Gurobi libraries to perform the optimization with the Branch and Bound method. The model deals with problems at the tactical and strategic level. Incorporating tactical-level decision will help give operational-level decision makers more flexibility and predictability (Kim, Koo, & Park, 2009).

### 4.1 *Train and ship integrated scheduling model*

The model optimizes the scheduling and sequencing of the queue of ships and trains together, considering stock, movement of products and stockyard equipment as one of the model's input data, thus minimizing cost with demurrage, charge payable to the ship owner in respect of failure to load or unload the ship within the time agreed, indicated for strategic decisions. This model aims to optimize schedules with a horizon of up to 30 days of operation.

#### 4.1.1 *Index*

The index represents the queue of ships and trains, the amount of available car dumper, the working day and cargo handling:

I. i: ship i = 1, 2, …, N;
II. n: ship queue position n = 1st, 2nd, …, N;
III. m: train queue position;
IV. v: Rotary car dumper;
V. d: working day;
VI. t: train;
VII. p: product.

#### 4.1.2 *Parameters*

The parameters consider operational data such as product handled, train and ship times, as well as available inventory:

I. B: Infinitely large number;
II. N: Number of ships;
III. V: Number of dumpers available for unloading trains;
IV. M: Number of trains served per turner;
V. T: Number of trains served per day;
VI. D: Number of days in the planning horizon;
VII. Q: Quantity of products;
VIII. Cdemurragei: Demurrage cost for ship i;
IX. Cdespatchi: Ship i dispatch award;
X. TTi: Total loading time of ship i;
XI. ETAi: Date of availability of vessel i to start loading;
XII. Turntimei: Administrative time consumed by vessel i;
XIII. TDi: Time available for ship i to complete loading without being penalized with demurrage;

XIV. TTtd: Time spent unloading train t on day d;
XV. ETAt: Availability date of the train t to start unloading;
XVI. Deadline: Deadline for train t on day d to finish unloading;
XVII. Cargopi: Loading of product p on ship i;
XVIII. Cargopt: Loading of product p on train t;
XIX. Stockypilep: Product stocked available in the stockyard at the beginning of planning;
XX. TDesat: Average time for the ship to unberth from the terminal from the end of loading.

4.1.3 *Decision variables related to ships and trains*

These variables represent the sequencing and costs or rewards of ships:

I. TDemurragei: Used demurrage time on ship i;
II. TDespatchi: Despach time used on ship i;
III. DICn: Start date loading position n of the queue of ships;
IV. DTCn: Loading end date position n of the queue of ships;
V. DICi: Start date loading vessel i;
VI. DTCi: End date loading vessel i;
VII. FG1_DTCin: Backlash variable;
VIII. FG2_DTCin: Backlash variable;
IX. Yin: Binary variable equal to 1 if ship i occupies position n of the ship's queue, 0 otherwise.

The decision variables related to the train's queue are:

I. Ymvdt: Binary variable equal to 1 if train t occupies position m in the train queue of turner v on day d;
II. DIDesc_mvd: Train unloading start date at position m of the dumper queue v on day d;
III. DTDesc_mvd: Train unloading end date at position m of the dumper queue v on d day.

4.1.4 *Decision variables to the integration of trains, ships, available stockpile and shipments*

These variables are used for model integration and linearization:

I. EDPIpn: Stock available in the product stockyard p at the beginning of ship loading in position n of the queue of ships;
II. Cargo_pn: Load of product p on the ship in position n of the ship queue;
III. Qntpn: Quantity of product p unloaded by the trains during the loading execution of the ship allocated in position n;
IV. Cargopmvd: Loading of product p on the ship from position m of dumper v on d day;
V. Bin_TDmvdn: Binary equal to 1 if the product of the train unloaded in the m position of the dumper v on d day is available for the ship from the n position until the end of the ship loading, 0 otherwise;
VI. FG_Sup_TDmvdn: Positive slack variable, indicates that the train finished unloading before the end of ship n;
VII. FG_Inf_TDmvdn: Negative slack variable, indicates that the train has finished unloading after the end of ship n;
VIII. CargoDpmvdn: Cargo of the product p of the train unloaded in position m of the car dumper v on day d available for ship from position n until the end of loading the ship;
IX. FG_I_CargoDpmvdn: Backlash variable for calculating the CargoDpmvdn;
X. FG_S_CargoDpmvdn: Backlash variable for calculating the CargoDpmvdn.

4.1.5 *The objective function*

The model performs the sequencing of trains and ships to generate the first schedule of operations, thus obtaining the sequences and the respective start and end dates of ship loading and trains unloading as parameters used in the stockyards management. The objective function seeks to reduce the delay between serving ships and, consequently, the cost of demurrage.

$$\min \mathrm{TDemurrage}_i * \mathrm{Cdemurrage}_i - \mathrm{TDespatch}_i * \mathrm{Cdespatch}_i \tag{1}$$

I. Functions on ships:

$$DTC_n - DIC_n = \sum_{i=1}^{N} \mathrm{Y}_{in} * \mathrm{TT}_i \quad \forall n \in \{1, N\} \tag{2}$$

$$DIC_n \geq DTC_{n-1} \quad \forall n \in \{2, N\} \tag{3}$$

$$\sum_{i=1}^{N} Y_{in} = 1 \quad \forall n \in \{1, N\} \tag{4}$$

$$\sum_{n=1}^{N} Y_{in} = 1 \quad \forall i \in \{1, N\} \tag{5}$$

$$DIC_n \geq \sum_{i=1}^{N} Y_{in} * ETAi \quad \forall n \in \{1, N\} \tag{6}$$

$$(ETA_i + Turntime_i + TD_i) + \mathrm{Tdemurrage}_i = DTC_i + \mathrm{TDespatch\_i} \quad \forall i \in \{1, N\} \tag{7}$$

$$DTC_i - DIC_i = TT_i \quad \forall i \in \{1, N\} \tag{8}$$

$$DTC_i - DTC_n + \mathrm{FG1DTC_{in}} - \mathrm{FG2DTC_{in}} = 0 \quad \forall n, i \in \{1, N\} \tag{9}$$

$$(\mathrm{FG1DTC_{in}} + \mathrm{FG2DTC_{in}}) \leq \mathrm{B}(1 - \mathrm{Y_{in}}) \quad \forall n, i \in \{1, N\} \tag{10}$$

$$YCi, n = (Bin - 0 \text{ ou } 1) \quad \forall n, i \in \{1, N\} \quad (11)$$

$$FG1_{DTC_{in}}, FG2_{DTC_{in}}, DTC_i, DIC_i, DTC_n, DIC_n, Tdemurrage_i, TDespatch_i \geq 0 \quad \forall n, i \in \{1, N\} \quad (12)$$

The constraint (2) ensures that the difference between the beginning and the end of the ship loading in the N of the queue must be the estimated total time of loading. A (3) maintains the stream of the start time and end of loading the ship in the N position in the queue. A (4) and (5) guarantee the flow of the vessel queue. A (6) ensures that the start date of loading a ship is only after its ETAI. A (7) is a linearized function for calculating time used in Dispatch and Demurrage. A (8), (8) and (10), based on the DTCN loading date of the ship in the N ship position, define the end date of the DTCI loading of the ship I occupy the N of the queue. Already the restrictions (11) and (12) are restrictions that ensures the variable type.

II. Functions on trains:

$$DTDesc_{mvd} - DIDesc_{mvd} = \sum_{t=1}^{T} TT_{td} * Y_{tmvd} \quad \forall m \in \{1, M\}, \forall v \in \{1, V\}, \forall d \in \{1, D\} \quad (13)$$

$$DIDesc_{mvd} + \left(1 - \sum_{t=1}^{T} Y_{tmvd}\right) \geq DTDesc_{(m-1)vd}; \quad \forall m > 1, \forall m \in \{2, M\}, \forall v \in \{1, V\}, \forall d \in \{1, D\} \quad (14)$$

$$\sum_{m=1}^{M} \sum_{v=1}^{V} Y_{tmvd} = 1, \quad \forall t\{1, T\}, \forall d \in \{1, D\} \quad (15)$$

$$\sum_{t=1}^{T} T_{tmvd} = 1, \forall m = 1, \ldots, M; \quad \forall v = 1, \ldots, V; \forall d = 1, \ldots, D \quad (16)$$

$$DIDesc_{mvd} \geq \sum_{t=1}^{T} Y_{tmvd} * ETA_{td}, \forall m = 1, \ldots, M; \quad \forall v = 1, \ldots, V; \forall d = 1, \ldots, D \quad (17)$$

$$DTDesc_{mvd} \leq \sum_{t=1}^{T} Y_{tmvd} * DataDispT_{td}, \quad \forall m = 1, \ldots, M; \forall v = 1, \ldots, V; \forall d = 1, \ldots, D \quad (18)$$

$$\sum_{t=1}^{T} Y_{tmvd} \leq \sum_{t=1}^{T} Y_{t(m-1)vd}, \quad \forall m \in \{2, M\}, \forall v \in \{1, V\}, \forall d \in \{1, D\} \quad (19)$$

$$Ytmvd = (Bin - 0 \text{ ou } 1); DTDesc_{mvd}, DIDesc_{mvd} \geq 0, \forall m\ 1, \ldots, M; \forall v = 1, \ldots, V; \forall d = 1, \ldots, D; \forall t = 1, \ldots, T \quad (20)$$

The restriction (13) ensures that the difference between the start date and end of the unloading of the train in the m of the car dumper v on day d is equal to the total time spent on the unloading of the train t on day d allocated in this position. A (14) ensures the flow of time. A (15) and (16) guarantee the flow of the queue. A (17) allows the start of train unloading only after its ETAt and (18) obliges the unloading of the train occurs within the deadline, (19) ensures row flow in the turners and (20) represents type of variables. Integration factors between the two rows, ships and trains, are their respective loads and inventory availability, being considered inventory in the patio and inventory in the transfer, in the programmed trains, so that a scheduled train to finalize the unloading during loading A ship can have their stock allocated directly on the ship without going through the stockyards.

III. Train and ships scheduling functions:

$$EDPI_{pn} + Qnt_{pn} - CargoProg_{pn} >= 0, \quad \forall p \in \{1, P\}; \forall n \in \{1, N\} \quad (21)$$

$$CargoProg_{pn} = \sum_{i=1}^{N} Y_{in} * CargoNavio_{pi}, \quad \forall p \in \{1, P\}; \forall n \in \{1, N\} \quad (22)$$

$$CargoProgTrain_{pmvd} = \sum_{t=1}^{T} Y_{tmvd} * CargoTrain_{ptd}, \quad \forall p \in \{1, P\}; \forall m \in \{1, M\}; \forall v \in \{1, V\}; \forall d \in \{1, D\} \quad (23)$$

$$EDPI_{pn} = Stockypile_p \quad \forall p \in \{1, P\}; n = 1 \quad (24)$$

$$EDPI_{pn} = EDPI_{p(n-1)} + Qnt_{p(n-1)} - CargoProg_{p(n-1)}, \forall p \in \{1, P\}; \forall n \in \{2, N\} \quad (25)$$

$$Qnt_{pn} = \sum_{m=1}^{M} \sum_{v=1}^{V} \sum_{d=1}^{D} CargoD_{pmvdn}, \quad \forall p \in \{1, P\}; n = 1 \quad (26)$$

$$Qnt_{pn} = \left(\sum_{m=1}^{M}\sum_{v=1}^{V}\sum_{d=1}^{D} CargoD_{pmvdn} - \sum_{k=1}^{n-1} Qnt_{p[k]}\right), \forall p \in \{1,P\}; \forall n \in \{2,N\} \tag{27}$$

$$\begin{aligned} &DTC_n - (TDesat) - DTDesc_{mvd} \\ &- FGSupTD_{mvdn} + FGInfTD_{mvdn} = 0, \\ &\forall n \in \{1,N\}; \forall m \in \{1,M\}; \\ &\forall v \in \{1,V\}; \forall d \in \{1,D\} \end{aligned} \tag{28}$$

$$\begin{aligned} &FGSupTD_{mvdn} \leq B * BinTD_{mvdn}, \\ &\forall n \in \{1,N\}; \forall m \in \{1,M\}; \\ &\forall v \in \{1,V\}; \forall d \in \{1,D\} \end{aligned} \tag{29}$$

$$\begin{aligned} &CargoD_{pmvdn} - CargoProgTrain_{pmvd} \\ &+ BinTD_{mvdn} - FGSCargoD_{pmvdn} \\ &+ FGICargoD_{pmvdn} = 1, \forall n \in \{1,N\}; \\ &\forall p \in \{1,P\}; \forall m \in \{1,M\}; \\ &\forall v \in \{1,V\}; \forall d \in \{1,D\} \end{aligned} \tag{30}$$

$$\begin{aligned} &FGSCargoD_{pmvdn} \leq B * (1 - BinTD_{mvdn}), \\ &\forall n \in \{1,N\}; \forall p \in \{1,P\}; \forall m \in \{1,M\}; \\ &\forall v \in \{1,V\}; \forall d \in \{1,D\} \end{aligned} \tag{31}$$

$$\begin{aligned} &FGICargoD_{pmvdn} \leq B * (1 - BinTD_{mvdn}), \\ &\forall n \in \{1,N\}; \forall p \in \{1,P\}; \forall m \in \{1,M\}; \\ &\forall v \in \{1,V\}; \forall d \in \{1,D\} \end{aligned} \tag{32}$$

$$\begin{aligned} &EDPI_{pn}, CargaProg_{pn}, Qnt_{pn}, CargaProgTrem_{pmvd}, \\ &FGSupTD_{mvdn} \geq 0; FGInfTD_{mvdn}, CargaD_{pmvdn}, \\ &FolgaICargaD_{pmvdn}, FolgaSCargaD_{pmvdn} \geq 0; \\ &BinTD_{mvdn} = (\text{Bin} - 0 \text{ ou } 1) \end{aligned} \tag{33}$$

The stock cohesion in the patio with the quantity loaded in ships is guaranteed by restriction (21). A (22) Calculates the product load p to be charged on the n. A (23) Calculates the product load p unloads in the m position, from the car dumper v, on day d. A (24) ensures that the inventory variable available in the patio at the beginning of loading the first ship is equivalent to the initial stock in the patio. A (25) estimates the stock in the patio based on the scheduled loading and product entries up to the time of loading the ship. Restrictions (26) and (27) estimate the amount of products in the patio until the time of loading the ship. Restrictions (28) and (29) confer the charging dates of ships and unloading the trains to verify that the loads of the trains will be available on the patio during ship loading. Restrictions (30), (31) and (32) are disjunctive constraints that indicate whether bin_tdmvdn binary is equal to 1, train loading in the MVD indexes is available for the vessel in the index position n. Finally, the restriction (33) indicates the type of data of each variable.

## 5 COMPUTATIONAL EXPERIMENTS AND ANALYSIS OF RESULTS

In a real situation, the model can help in the construction of the daily production planning of a port terminal, which seeks to meet the monthly, quarterly and annual plans. Experiments and tests for validation of the model and identification of the solution limitations were performed. The instances used to analyze the approach, the model implementation procedure and the results of the experiments, which were satisfactory. A reduced instance was used to exemplify the use of the model.

### 5.1 *Instance's construction*

For numerical experimentation, the model was implemented in the Python programming language, using the Solver Gurobi libraries to perform the optimization with Solver's native Branch and Bound methods. The instance created for model evaluation uses part of actual data of train and ship cargo size, estimated arrival dates (ETA) and expected dates to process loading and unloading. An environment with 2 turners operating in 3 stockyards each with resources of respectively 6, 10 and 12 tons per pad

Table 1. Stockyard at the beginning of planning.

| | Pad 1 | Pad 2 | Pad 3 | Pad 4 | Pad 5 | Pad 6 | Pad 7 | Pad 8 | Pad 9 | Pad 10 |
|---|---|---|---|---|---|---|---|---|---|---|
| Stockyard 1 | 6000 | 6000 | 3000 | 0 | 6000 | 6000 | 576 | 6000 | 6000 | 6000 |
| Stockyard 2 | 10000 | 10000 | 10000 | 0 | 10000 | 10000 | 0 | 10000 | 10000 | 10000 |
| Stockyard 3 | 12000 | 12000 | 12000 | 12000 | 12000 | 0 | 12000 | 12000 | 0 | 5900 |

P001 P002

Table 2. Ship data used in the instance.

| Ship | Product | Cargo | Turn time | Daily on Board | Demur. Rate | Disp. Rate | ETA_i | DT_i (days) | TT_I (days) |
|---|---|---|---|---|---|---|---|---|---|
| N001 | P002 | 89.300 | 08:00:00 | 80.000 | 50 | 25 | 43927,125 | 1,12 | 1,04 |
| N002 | P002 | 97.600 | 08:00:00 | 80.000 | 50 | 25 | 43928,5 | 1,22 | 1,11 |
| N003 | P001 | 100.900 | 08:00:00 | 80.000 | 50 | 25 | 43929,04167 | 1,26 | 1,13 |

Table 3. Train data used in the instance.

| Train | Product | Cargo | ETA_td | Day (d) | DataDispT_td | TT_td |
|---|---|---|---|---|---|---|
| T001 | P001 | 14201 | 5/4/20 9:36 | 1 | 06/04/2020 | 0,12 |
| T002 | P002 | 13828 | 5/4/20 14:24 | 1 | 06/04/2020 | 0,12 |
| T003 | P001 | 12258 | 5/4/20 14:24 | 1 | 06/04/2020 | 0,10 |
| T004 | P001 | 14393 | 5/4/20 16:48 | 1 | 06/04/2020 | 0,12 |
| T005 | P002 | 12027 | 6/4/20 4:48 | 2 | 07/04/2020 | 0,10 |
| T006 | P002 | 13221 | 6/4/20 12:00 | 2 | 07/04/2020 | 0,11 |
| T007 | P001 | 13654 | 6/4/20 14:24 | 2 | 07/04/2020 | 0,11 |
| T008 | P002 | 14070 | 6/4/20 16:48 | 2 | 07/04/2020 | 0,12 |
| T009 | P001 | 12228 | 7/4/20 4:48 | 3 | 08/04/2020 | 0,10 |
| T010 | P002 | 12819 | 7/4/20 9:36 | 3 | 08/04/2020 | 0,11 |
| T011 | P001 | 14458 | 7/4/20 20:09 | 3 | 08/04/2020 | 0,12 |
| T012 | P002 | 12359 | 7/4/20 20:09 | 3 | 08/04/2020 | 0,10 |

was considered. A period in which 3 ships were loaded with two different products was also adopted for experimentation. During this period, about 100 trains with different types of products were unloaded, however, in the experiments, only a few trains that transport the two products corresponding to the ship's cargoes were considered. As for the stockyard map at the beginning of planning, a reduced example was created, with 10 pads each stockyard, loaded as in Table 1. Parameters N, V, M and P used are 3 ships, 2 car dumpers, considering that each turner can serve up to 2 trains per day and 2 products. In addition, they were considered D = 3 days, arriving T = 4 trains per day and each ship leading on average TDesat = 3.5 to ship unberthing. In the stockyards, as shown Table 1, it has stockyardp equal to 100,900 kg of the product P001 and 114,576 kg of the product P002, with TL= 3 stockyards, TB = 10 pads per patio, K_L = 6, 10 and 12 thousand tons. The Flbr parameter considers that the recovery 1 cannot act in any stockyard of the patio 3 and the recovering 2 cannot be present in any stockyard of the patio 1. The length of the pad used is equivalent to 10 m each. The instance with the data of the ships used is described in Table 2 and the trains in Table 3.

The stockyard equipment, stacker and reclaimers used in the stockyard operations, and the stockpiling according to the arrangement of areas available are managed at an operational level and are not modeled in this case, providing an opportunity for the evolution of this research.

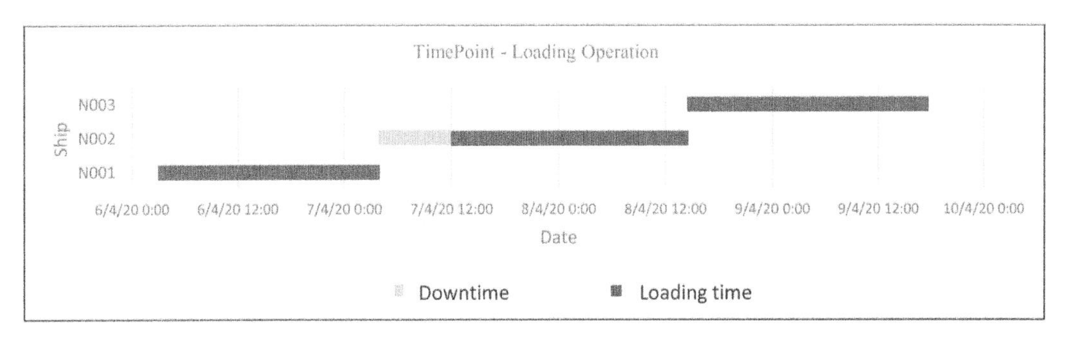


Figure 1. Ship sequence Gantt.

Table 4. Ship schedule consolidation.

| Ship | Products | Start Loading | End Loading | Demurrage/ Despatch | Pads Reclaimed | Cargo on Board |
|---|---|---|---|---|---|---|
| N001 | P002 | 06/04/2020 11:08 | 07/04/2020 12:00 | 0 | - | Train 5; Train 6; Train 7; Train 8; Train 12. |
| N002 | P002 | 07/04/2020 12:00 | 08/04/2020 14:31 | 0 | Stockyard 1 Pad 6 - 5999; Stockyard 2 Pad 6 - 8992; Stockyard 2 Pad 7 - 10000; Stockyard 2 Pad 3 - 2992; Stockyard 3 Pad 3 - 11999; Stockyard 2 Pad 4 - 9999; Stockyard 3 Pad 4 – 1. | |
| N003 | P001 | 08/04/2020 14:31 | 09/04/2020 17:42 | 0 | Stockyard 1 Pad 7 - 6000; Stockyard 2 Pad 7 - 5000; Stockyard 1 Pad 8 - 5999; Stockyard 2 Pad 8 - 8997; Stockyard 1 Pad 9 - 6000; Stockyard 2 Pad 9 - 10000; Stockyard 2 Pad 1 - 9999; Stockyard 3 Pad 1 - 11999; Stockyard 2 Pad 2 - 9998; Stockyard 3 Pad 2 – 11249. | - |

### 5.2 *Model running*

The execution of the model occurred where the train sequencing and vessels model is first carried out to generate the first production schedule and, having the sequences and the respective dates of start and end of the shipments of the ships and discharges of the trains as parameters used for the yard decisions. It was considered that every train that completely performed its discharge operation during loading a ship with the same product should be sent directly on-board Ship and, if more than one train is in this situation at the same time, the largest load should go straight on-board, and the others must be temporarily allocated into the patio and then be sent on board.

### 5.3 *Experiment results*

The solution of the models shows that the sequence N001 must be followed first, followed by N002 and N003. The start and end dates for loading ships, as well as the period in which the port would have been idle are shown in Graph 1. The consolidation of results is described in Table 4, where the column "Cargo on Board" represents the train loads that must go directly on board or temporarily to the stockyard for each ship, the column "Recovered Pads" indicates the pads that will be recovered for load each ship, where the code "P1 Pad 6 – 5999" indicates that 5999 tons must be recovered in Stockyard 1 at Pad 6.

## 6 CONCLUSIONS

Problems in planning stockyards s for storage of solid bulk and sequencing of port processes are not trivial. Despite solving a real problem, the model proved to be complex to be implemented only with Linear Programming (LP). For large instances and high levels of detail of operations, a combination of heuristic methods is recommended to optimize processing time and obtain satisfactory results, a difficulty also observed by other authors. This search can be extended in some directions. The current model does not consider the possibility of route conflict between two stockyards, sequence of movements of the stockyard machines, and operational level failures and problems not known in advance. However, it provides tactical and strategic information for decision-making at operational levels. In practice, the model can be used and updated with new data daily, in order to incorporate the dynamics that occurred in the operational programming. Incorporating tactical-level decision will help give operational-level decision makers more flexibility and predictability. The use of traditional approaches and optimization tools is difficult and, therefore, new approaches such as evolutionary algorithms and constructive heuristics to solve the problem are increasingly used. Regarding heuristics, most of the cited authors claim that the method is more efficient in producing a viable solution when compared to a commercial solver, obtaining. Furthermore, modifications and adaptations according to the different rules of the ports can be easily implemented.

It is common for companies in the mining and steel industry to seek solutions that allow the integration of the entire logistics chain and optimization of strategic, tactical and operational planning. Incorporating operational conditions such as process failures, corrective and preventive maintenance, product quality and outages due to external causes is complex and requires the supply of information in real time. A dynamic optimization model can be obtained from the discussed approach. It is hoped that this research can guide the future development of tools for the planning and programming of the bulk mineral production system, in order to optimize processes from beneficiation in mines, purchase of iron ore, rail transport to delivery of products to customers in different markets. The industry is increasingly interested in solutions and technologies that can improve overall efficiency, reduce demurrage costs, improve use of its assets, better manage its inventories and obtain integrated planning from independent areas.

## REFERENCES


Barros, V. H., Costa, T. S., Oliveira, A. C., & Lorena, L. A. (2011). Model and heuristic for berth allocation in tidal bulk ports with stock level constraints (Vols. Computers & Industrial Engineering 60 (2011) 606–613). Elsevier Ltd. doi:10.1016/j.cie.2010.12.018.

Boland, N., Gulczynski, D., Jackson, M., Savelsbergh, M., & Tam, M. (2011). Improved stockyard management strategies for coal. Perth, Australia: 19th International Congress on Modelling and Simulation.

Irfanbabu, S. A., Pratap, S., Lahotia, G., Fernandes, K. J., Tiwaria, M. K., Mountc, M., & Xiong, Y. (2015). Minimizing delay of ships in bulk terminals by simultaneous ship scheduling, stockyard planning and train scheduling (Vols. Maritime Economics & Logistics (2015) 17, 464–492). (M. Publishers, Ed.), doi:10.1057/mel.2014.20.

Kim, B.-I., Koo, J., & Park, B. S. (2009). A raw material storage stockyard allocation problem for a large-scale steelwork. The International Journal of Advanced Manufacturing. Int J Adv Manuf Technol (2009) 41:880–884.

Marcos Junior, W. J., Lopes, Á. D., Salles, J. L., & Rocha, H. R. (2018). Modelo matemático para otimização na alocação pilhas em um pátio de estocagem de portos e carregamento de navios (Vol. VIII Congresso Brasileiro de Engenharia de Produção). Ponta Grossa: ABEPRO.

Menezes, G. C., Mateus, G. R., & Ravetti, M. G. (2016a). Heuristic methods to solve a production planning and scheduling problem in bulk ports. doi:https://www.researchgate.net/publication/311202729.

Menezes, G. C., Mateus, G. R., & Ravetti, M. G. (2016b). A hierarchical approach to solve a production planning and scheduling problem in bulk cargo terminal (Vols. Computers & Industrial Engineering 97 (2016) 1–14). Elsevier Ltd.

Menezes, G. C., Mateus, G. R., & Ravetti, M. G. (2016c). A branch and price algorithm to solve the integrated production planning and scheduling in bulk ports. (E. B. V., Ed.) Belo Horizonte, Brazil: European Journal of Operational Research 258 (2017) 926–937. doi:http://dx.doi.org/10.1016/j.ejor.2016.08.073.

Pratap, S., Daultani, Y., Tiwari, M. K., & Mahanty, B. (2015). Rule based optimization for a bulk handling port operations. New York: Springer Science+Business Media New York 2015. doi:http://dx.doi.org/10.1007/s10845-015-1108-7.

Tang, L., Sun, D., & Liu, J. (2015). Integrated Storage Space Allocation and Ship Scheduling Problem in Bulk Cargo Terminals. Leicestershire, England: Loughborough University Institutional Repository.

*Trends in Maritime Technology and Engineering – Guedes Soares & Santos (eds)*
 *ISBN 978-1-032-33583-4*

# The East-European maritime ports hinterland and the influence of road transport network

F. Rusca, A. Rusca, E. Rosca, C. Oprea, O. Dinu & A. Ilie
*Faculty of Transports, University Politehnica of Bucharest, Bucharest, Romania*

ABSTRACT: The importance of maritime ports in logistics chains relates to the size and shape of ports hinterland. The development of new methods and techniques in the field of transportation such as intermodal approach, multimodal approach, containerized transport has led to changes in the structure and shape of the territory served by maritime ports. In case of East-European maritime ports some regional characteristics have an influence over the shape size of ports hinterlands. A study case made for a Romanian maritime port is developed in the paper. The main cargo flows, wheat, which transit Constanta maritime port have origin in Romanian territory served by road network. Taking in consideration the legislation limitation an analysis is made for the hinterland of the maritime port. The "time travel" border calculated using macrosimulation model help us to understand the hinterland shape. The results obtained show the coverage percentage from wheat production inside the hinterland with "travel time" border set at one day.

## 1 INTRODUCTION

### 1.1 *The main aspects about maritime ports hinterland*

The ports represent the connection between the land transport network and the maritime routes. The area covered by a maritime port can have different shapes being influenced by a number of aspects such as: land connections, location of production areas, geographical landforms or national borders.

In the case of East-European maritime ports the historic events, like the war against Ottoman Empire did not encourage a good development for maritime ports. Particularizing for Romanian maritime ports, analyzed in this paper the land near Black Sea was integrated in the national administration only in the last 100 years. This aspect has led to an atypical development of the hinterland of these ports.

The connection with the territory is limited on the road network at two routes that involve crossing the Danube River.

Using a macroscopic simulation model in which the main road arteries are inserted, a shape of the port hinterland dependent on a temporal attribute can be identified.

### 1.2 *The particularities of East-European maritime ports-case of Romanian ports*

The main East-European countries with maritime ports are Romania and Bulgaria with ports at Black Sea. The connections with these ones are made using land network (road or railway) and in some cases using river connection (Danube-Black Sea Channel).

In both cases the territory covered by the maritime ports is described by long distances, mountain areas and large rivers to cross.

In these two countries the land network has a territorial development using a Hub and Spoke typology with the capital city, Bucharest and Sofia in the center of the structure. The main port at Black Sea is Constanta Port located on Romanian shore with big values for wheat transit flows. The main destinations for these ones are situated in Egypt, United Arab Emirates and Qatar. The wheat production zones are located mainly in the southern part of Romania but can be found also in the northern and western parts of Romanian territory (Figure 1).

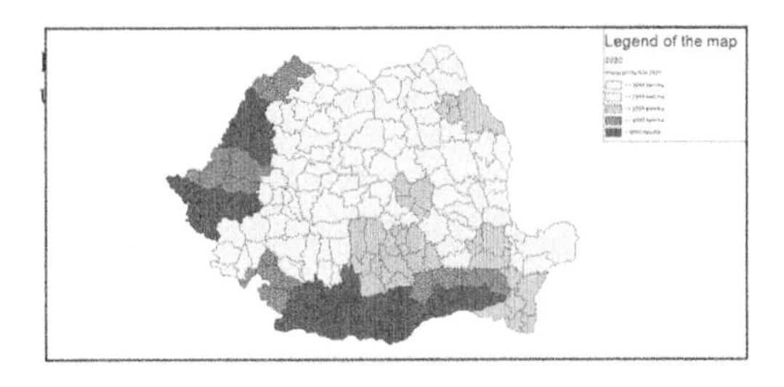


Figure 1. Wheat production share in Romania (source: created using data from Romanian Statistics National Institute/).

DOI: 10.1201/9781003320289-9

Because the Romanian maritime ports are located in the South-East part of the country the production zones from West or North are connected with these ones using land network for large distances.

## 2 LITERATURE REVIEW

The subject of ports hinterlands can be found in a large set of scientific papers and studies. Most of them are dedicated to Western European, American and Asian ports. We started our research with two reviewed articles (Lee et al., 2008, Sdoukopoulos & Boile, 2020). The selection of literature is separated on evolutive stages of port hinterland over the time and the typology of the territory served by maritime ports. One of the most important landmarks is Taaffe model for colonial hinterland development presented in Taaffe et al. (1963). This important research is preceded by studies of demarcation approaches of hinterland made in an early period starting in 1908 (Chisholm, 1908). Furthermore, the influence of commodity types in setting the shape of hinterland is observed by Sargen (1938) and Morgan (1948).

The quantitative analyses are made for the case of American ports (Mayer, 1957) and for British seaports (Bird, 1963). The concepts presented in these studies are useful to understand the basic aspects used in defining the shape of maritime port hinterland and to find also the similarities in the development of Eastern European ports situated on the shore of the Black Sea.

The end of the twentieth century and current period was marked by the introduction in the studies of concepts such as hub ports, containerization, intermodality. The influence of intermodal transport by reducing the transport cost is evaluated by Hayuth (1981). The new definition of hinterland as '*spatial systems in which port users i.e. shippers, carriers and forwarders interact via cargo flows and information exchanges*' is given by van Klink (1998). The inner patterns of port cities in emerging countries and shape of hinterland were described by Gleave (1997). The land network through the development of transport corridors has an influence over the shape of area served by maritime ports and over services like short sea shipping (Santos et al. 2021, Santos et al. 2022). The methodology used to assess the economic significance and geographic reach of port investments was developed in 2018 by Santos et al 2018.

## 3 THE HINTERLAND ASPECTS AND EAST EUROPEAN PARTICULARITIES

The word "hinterland" appears for the first time in English introduced by a geographer, G. Chisholm in 1908 and symbolizes "the land behind". He uses this concept to make a reference to area served by a port. First ports at Black Sea, in the Eastern part of Europe, appear in Antiquity when Greek colonists founded city-states in this part of the world. The land allocated to these ones was around the port and served the cities in providing the necessary products for daily living. The period faced with the invasion of the migrating peoples and the Ottoman Empire led to their degradation, being abandoned by the inhabitants. Gaining independence and restoring control of the riparian states over the coast overlapped with the industrial age. The ports in this region, since then, have been developed based on modern principles adapted to a large volume of industrial goods. Our study shows the main models used in literature to evaluate the hinterland of the maritime ports and search the connection between these ones and the specific situation of maritime ports at Black Sea.

According to Lee et al. (2008), there are two main patterns: the port development models in Western countries and the port development models in emerging countries. First of them are separated in two categories one for the case of ports from North America and one for the ports from Western Europe. Regarding to North American continent, Eastern and Western coasts are characterized by a large concentration of maritime ports and markets. Logistic chains with large capacity ensure the connections between ports. In Europe the inland markets is more important and ports are developed as an extension for large cities located at maximum distances of 100 km (Figure 2).

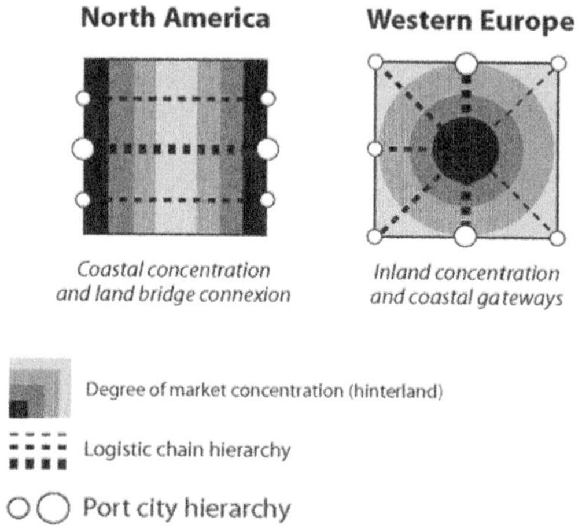

Figure 2. Western countries pattern port development (source Lee et al. 2008).

The characteristics of the European Maritime Traffic are described in Chirosca & Rusu (2020). Each European Sea were studied for evaluating the main types of area-specific ships and the density of maritime traffic.

The description of port development models starts with Taaffe model. Considering a new territory, the evolution of the land network is closely related to the evolution of the ports (Figure 3). Thus, Taaffe et al (1963) defines 6 stages of expansion of the land network in a new territory, served in a first stage only by ports:

- first stage is characterized by scattered ports with no connection to land
- in the second stage some penetration lines are developed, and a port concentration is visible
- development of feeders is part of the third stage
- in the fourth stage begin to appear interconnection inside of the land
- in the fifth stage a complete interconnection is finished
- in the sixth stage main transport corridors attract important transport flows

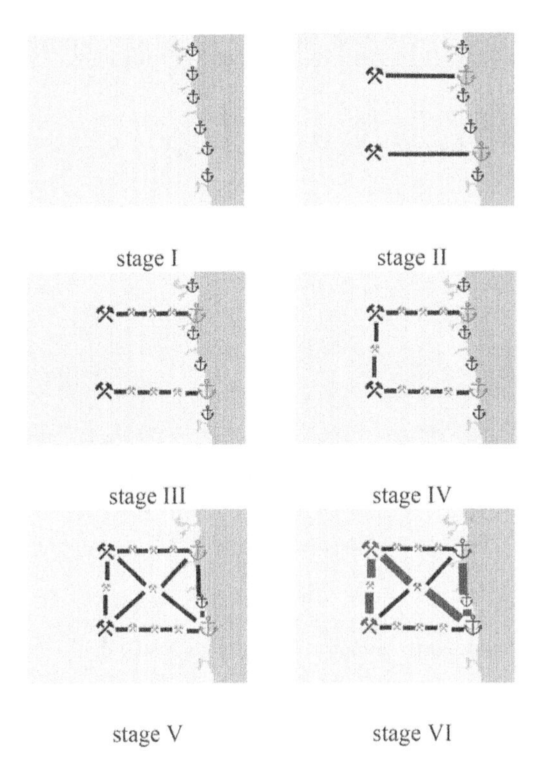


Figure 3. Ports development in new territory (source Taaffe et al. 1963).

In the case of Romanian ports located at Black Sea the development model can be associated with Western Europe ones where the main market is located inland but because the coast administration is active since 1878 some aspects from Taaffe model can be considered. The next five stages can be identified in the port-hinterland system development:

- Romanian coast administration begins with a strategic plan for development of Constanta port with a rail bridge: 1878-1966
- Road network is developed by building two bridges over the Danube River to connect the maritime coast with the inland cities: 1966-1987
- Secondary ports are built in Northern and Southern part of Constanta County
- Highway which connects the maritime shore with the Romanian Capital city is succeeded to increase the capacity of land transport network
- The third bridge is being built over the Danube River (the third longest bridge in Europe)-in progress

The evolution of the spatial system around maritime ports in which the actors involved (carriers, shippers and forwarders) interact through cargo flow and also through data exchanges, can be described in four stages of development (van Klink& et al., 1998):

- port city: where the maritime port is developed as a trade center dedicated to local industries. Spatial functions are reduced and connection with land transport network has a low influence over port activities
- port area: where the maritime port evolves to an industrial center. Spatial functions are developed with existing connections between industrial factories and land supply points
- port region: where the maritime port is the main hub in the region, with connection at transport corridors from land network.
- port network: all maritime ports from the region are integrated in a logistic system with delimited hinterlands covered by dedicated land transport networks, sharing capacity for cargo flows and sharing real-time data about cargo flows. The spatial functions are developed using a complex structure.

The biggest and most important Romanian port, Constanta Port, has, as we already presented, a long history. His role as port city has begun in Hellenic Period when Greek navigators colonized the coast. After First World War, the port and land around it evolves to an industrial center. At this moment, we can identify a role of port region. An interesting aspect is the fact that during the communist period the port was closer to a network structure with other ports in the region than it is now.

The complexity of the hinterland can be evaluated using a four-layer approach. In this direction there are defined the location layer, the infrastructure layer, the transport layer and finally the logistical layer. These are ranked starting from location layer, which is the most important in research and stops at the end at logistical layer, which reflects the most evaluated attributes of the hinterland. The infrastructure layer contains the description of different nodes and links within the cargo transport system. The transport layer contains information about transport, handling process, storage, consolidation process, etc. (Figure 4).

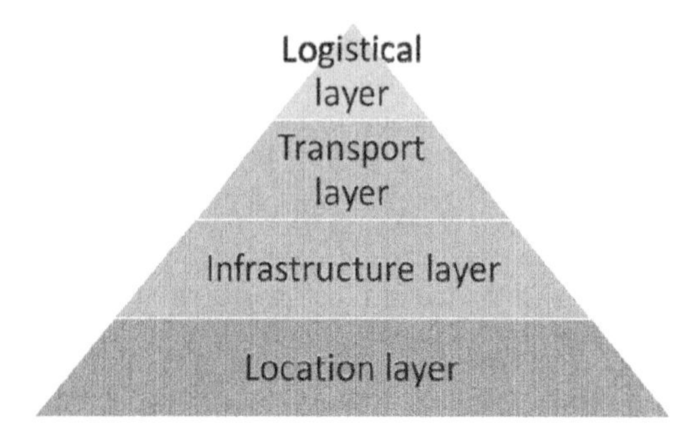


Figure 4. A multi-layer approach (after Notteboom& Rodrigue. 2007).

The development of new methods and techniques in the field of transportation such as intermodal approach, a multimodal approach, containerized transport have led to changes in the structure and shape of the territory served by maritime ports. Temporal parameters become important in the market-share process of competing maritime ports (van Klink& et al., 1998). The increase of the accessibility along transport corridors leads to an expansion of the port hinterland along these ones. Adjacent freight flows converge at the transport corridor where these ones are aggregated into compact traffic units. The advantages of this type of transport system increase the competitiveness of the port. According to these principles, four types of hinterlands of maritime ports are identified in relation to the transport characteristics of goods (Ferrari et al., 2011):

- agricultural hinterland with products presenting bulk characteristics, low mass weight, without overlapping
- solid mineral hinterland with products presenting also bulk characteristics and overlapping, but with a big value for mass weight
- chemical products hinterland with bulk characteristics, sometime the goods can be overlapped, but with specified special conditions at transport (dangerous goods)
- manufactured products hinterland presenting a high overlapping percent and a big value for mass weight.

The main type of hinterland which can be identified for Constanta port is agricultural. The Romanian country has important production zones for wheat and other cereals. The large distances and limitation induced by the capacity of land network influence the shape of this type of hinterland. In the next chapter, using a software tool, for transport network modelling, we present the shape of hinterland taking in consideration the attributes of Romanian national road network, the national legislation, and the location of the main production zones.

## 4 THE CASE OF CONSTANTA PORT WHEAT HINTERLAND

The name of the port and city comes from the Byzantine Emperor Constantine. The home of the exiled Roman poet Ovidius, the archeological artifacts, founded in port area, are dated I Century BC when the port was named Tomis. The port was making the connection between Greek navigators and the native populations. The current configuration has been designed since 1857. Anghel Saligny (1854-1925), the most famous engineer in Romania, had a special contribution. Wheat silos built under his guidance are still used today. The cereals products are the main cargo flow transiting through Constanta maritime port (over 21 million tons in 2019 according to port administration). Other important flows are oil products, mineral products, natural and chemical fertilizers, and other products. The maps with cargo operators is presented in Figure 5.

Figure 5. The cargo operators in Constanza port (blue-general cargo, green-cereals, orange-oil products, brown-ore scrap and yellow-containers terminals).

The connection with production areas, presented in Figure 1, is made using land network (road and rail). The capacity limitation and a high value of passenger traffic flow in the holiday summer period of railway network, that connects the Black Sea Romanian shore with inland zones increase the importance of road network in serving Constanta Port Hinterland. To evaluate the shape of it we use a macrosimulation model developed in PTV Visum software. To calculate the travel time an initial allocation is made using Equilibrium Assignment Algorithm (Figure 6). The procedure is in accordance with Wardrop's first principle: "Every individual road user chooses his route in such a way that his trip takes the same time on all alternative routes and that switching route would only increase personal journey time." (***PTV Visum Manual.2021).

The map and traffic flows loaded on links of the road network (Figure 7) are obtained in the previous research of the authors (Rusca et al, 2015). The updating of the model is made for the state of the road network in 2021.

To evaluate the shape of hinterland we consider the European and Romanian legislation. According with these rules an employee can drive up with some minor exceptions to a maximum limit of 9 hours per day. After 4.5 hours driving period, the driver must take an uninterrupted break of at least 45 minutes.

Using a time length of 4.5, 6 and 9 hours a shape of hinterland located at one day travel time can be obtained using macrosimulation model (Figure 8, Figure 9).

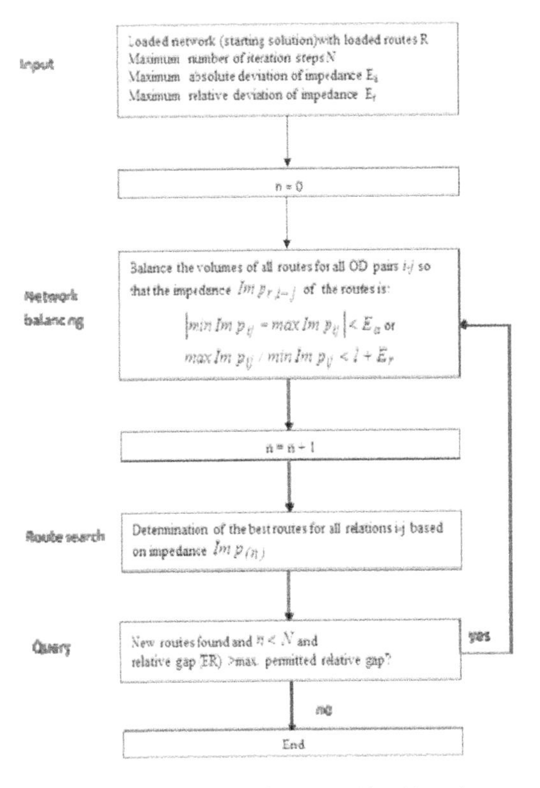


Figure 6. Equilibrium Assignment Algorithm (source ***PTV Visum Manual.2021).

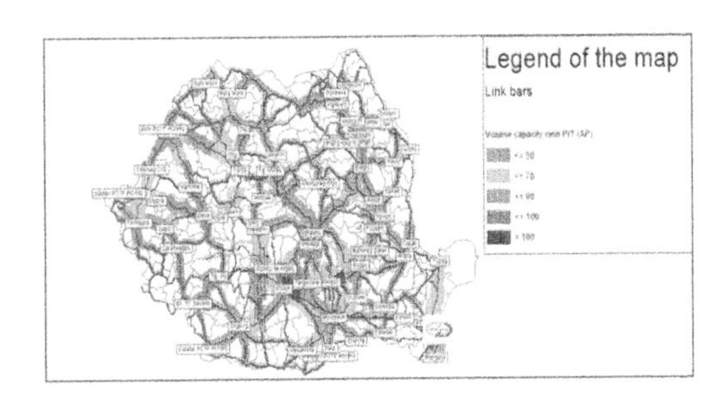


Figure 7. The road traffic flows used in the model (source: VISUM macrosimulation model).

The values of wheat production areas separated in relation with travel length to Constanta Port are shown in Figure 10. According to this 29% of production volume is in area covered by time isochrone of 4 hours and 30 minutes. For the case of isochrones of 6 hours and 9 hours the production quantity covered decrease at 27% and 21% from the total quantity of national production. A quantity of 23% from total wheat production in Romania is located at more one-day travel time to Constanta Port. This distance, counted in travel time, can reduce the interest for a logistic chain with a maritime component starting from Constanta Port with an influence over the shape of Constanta maritime port hinterland.

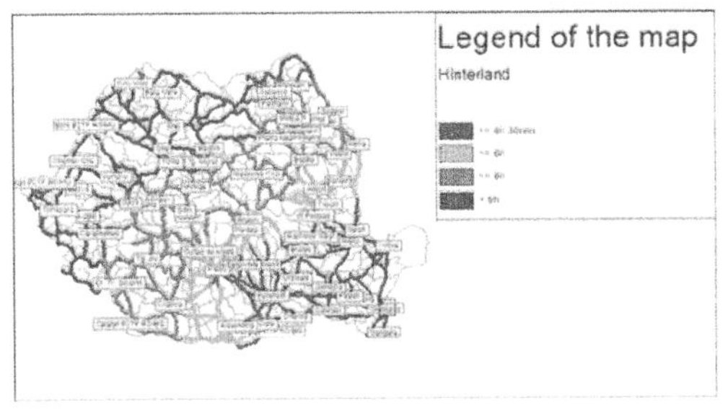


Figure 8. The isochrones map on road network for the case of Constanta maritime port (source: VISUM macrosimulation model).

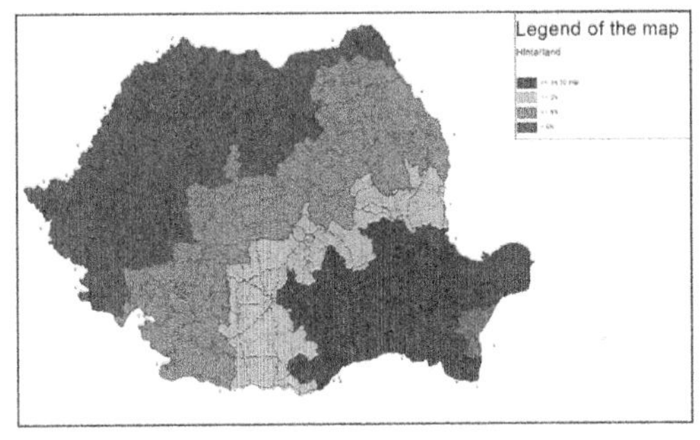


Figure 9. The time limit of wheat hinterland for the case of Constanta maritime port (source: VISUM macrosimulation model).

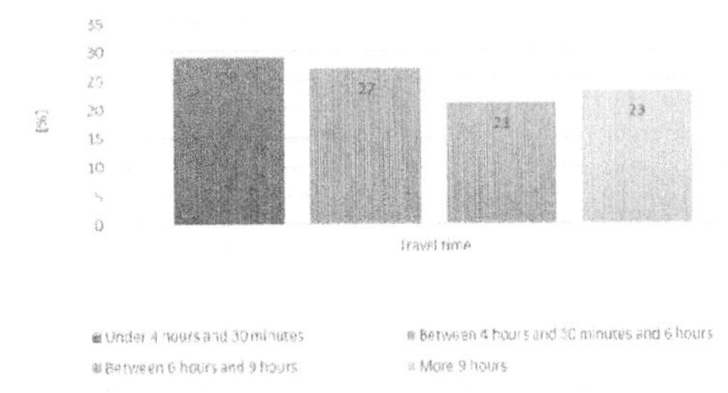


Figure 10. The wheat production areas separated by distances counted in travel time to Constanta Port.

The economic implications for the port future development can be summarized as follows:

- the land transports network development plan must be used as an influencing factor in feasibility studies on the development of port facilities (the "9 hours' time travel" border can be extended to the West)
- the main production areas are in South and East, covered by a hinterland with a "9 hours' time travel" border but some other production areas from West (23% from total production) are not covered. The port administration must promote to the national authorities the development of the transport infrastructure to these areas

The road network allows to have a hinterland who cover the most important production area of wheat. The main transport corridor connects all parts of Romania, but the limitation of road capacity and traffic flows conduce to an increased travel time. Using a macrosimulation model developed in PTV Visum and taking in consideration the European legislation and Romanian legislation can be calculated the "travel time" border of Constanta maritime port. The results obtained show a coverage rate of 77% from national wheat production inside the hinterland with "travel time" border set at one day.

## 5 CONCLUSIONS

The shape of maritime port hinterlands depends on a set of factors like land network development, the type of cargo, the historic development and other. A research study in this direction must take in consideration the local aspects with influence over the activity of logistic chain in the area of maritime ports. In our paper we try to expose the specific case of Constanta maritime port hinterland for wheat cargo. A "travel time" border calculated in accordance with European and Romanian legislation was used to estimate the shape of hinterland. Using the production value of wheat is calculated the percentage of national map covered by hinterland estimated with the method previous presented. The results are useful in elaborating the strategic map for land network development to increase the economic impact of wheat production through export to other countries using maritime transport.

Our future research will be made to evaluate maritime port hinterlands through syncromodal logistics concepts which are taking in consideration the dynamicity of logistics chains at the confluence between road network, rail network and inland waterway network.

## REFERENCES


***. 2021. PTV VISUM Manual, PTV Group, 2021: 403

Bird, J.H. 1963. The Major Seaports of the United Kingdom. Hutchinson & Co., London.

Chirosca, A.M. & Rusu, L. 2020. Sea state characteristics and the maritime traffic in the European seas. *International Multidisciplinary Scientific GeoConference: SGEM. Sofia*. 20(3.1).

Chisholm, G. 1908. Handbook of Commercial Geography. Longmans, Green and Co., London.

Ferrari, C. & Parola, F.& Gattorna, E. 2011. Measuring the quality of port hinterland accessibility: The Ligurian case. *Transport Policy* 18:382–391.

Gleave, M. 1997. Port activities and the spatial structure of cities: the case of freetown, Sierra Leone. *Journal of Transport Geography*. 5(4):257–275.

Hayuth, Y. 1988. Rationalization and deconcentration of the U.S. container port system. *The Professional Geographer* 40(3): 279–288. https://europa.eu/youreurope/business/human-resources/transport-sector-workers/road-transportation-workers/index_en.htm

Lee, S-W., Song, D-W., Ducruet, C. 2008. A tale of Asia's world ports: The spatial evolution in global hub port cities. *Geoforum* 39(1):372–385.

Mayer, H.M., 1973. Some geographic aspects of technological change in maritime transportation. *Economic Geography*. 49(2):145–155.

Morgan, F.W., 1948. The pre-war hinterlands of the German North Sea ports. *Transactions and Papers* 14: 45–55.

Notteboom, T. & Rodrigue, J.-P. 2007. Re-assessing port-hinterland relationships in the context of global commodity chains In: *Ports, Cities and Global Supply Chains*. Ed. Wang, J., Notteboom, T., Olivier, D., Slack, B. Ashgate. Aldershot.

Rusca, F., Rosca, E., Rusca, A., Roșca, M. & Burciu, S. 2015. The influence of transport network vulnerability for maritime ports. *MS&E*. 95(1).

Santos, A.M.P., Salvador, R., Dias J.C.Q. & Guedes Soares, C. 2018. Assessment of port economic impacts on regional econ-omy with a case study on the Port of Lisbon, *Mari-time Policy & Management*, 45 (5):684–698

Santos, T.A., Escabelado, J., Martins, P. & Guedes Soares, C. 2021. Short sea shipping routes hinterland delimitation in the European Atlantic Area. *Developments in Maritime Technology and Engineering*.

Santos, T.A., Fonseca, M., Martins, P. & Guedes Soares, C. 2022. Integrating Short Sea Shipping with Trans-European Transport Networks. *Journal of Marine Science and Engineering*. 10(2).

Sargent, A.J. 1938. Seaports and Hinterlands. Adam and Charles Black. London.

Sdoukopoulos, E. & Boile, M. 2020. Port-hinterland concept evolution: A critical review. *Journal of Transport Geography*, 86.

Taaffe, E.J. Morrill, R.L. & Gould, P.R. 1963. Transport expansion in underdeveloped countries: a comparative analysis. *Geographical Review* 53 (4):503–529.

van Klink, H.A. 1998. The port network as a new stage in port development: the case of Rotterdam. *Environment and Planning A: Economy and Space* 30(1):143–160.

*Trends in Maritime Technology and Engineering – Guedes Soares & Santos (eds)*
*© 2022 copyright the Author(s), ISBN 978-1-032-33583-4*

# Emerging ICT in port operations: Case studies

F. Russo & G. Musolino
*Dipartimento di Ingegneria dell'Informazione, delle Infrastrutture e dell'Energia Sostenibile (DIIES), Università Mediterranea di Reggio Calabria, Italy*

ABSTRACT: Geographical location, infrastructures, and services are traditionally the pillars of a port. In the last years, emerging Information and Communication Technologies (ICTs) were identified as a new pillar. Today new benefits and challenges are connected with the introduction of shared emerging ICTs among decision-makers inside ports. The crucial issue concerns the fact that several decision-makers could share a decision about a single port operation. Therefore, the effectiveness and efficiency of ports depend on how the interactions between the decision-makers are solved. The paper describes Transportation System Models to explain how ports could benefit, in terms of costs reduction, of the potentialities offered by emerging ICTs in solving interactions among decision-makers. Moreover, the paper presents a critical analysis of some representative case studies of the presence of Port Community Systems and emerging ICTs inside ports, in order to aggregate the observed trends in some macro-classes.

## 1 INTRODUCTION

Geographical location, infrastructures, and services were traditionally the pillars of a port in terms of its capacity to compete with other ports (UNCTAD, 1994; 1999; Russo & Musolino, 2020). The first ports were built in the proximity of the cities, and they were the gate for the exchange of freight among cities. The port-city model worked for centuries and these ports are classified as first-generation ports. In middle of XX century, some new ports were built close to industrial areas, mainly finalized to oil and chemical, ensuring transport and commercial services to industrial plants. These ports belong to the class of second-generation ports. In the last decade of the XX century, third-generation ports emerged, mainly due to the worldwide large-scale containerization, becoming a generator of value added inside the international value chain rather than only a centre of cost. At the beginning of XXI century, growing ships' dimensions, alliances in the shipping market, and shared hinterland access determined pressure on individual closer ports to cooperate, leaving the traditional attitude to compete (Martín-Alcalde et al., 2016; Russo & Musolino, 2021a). The four-generation ports were born.

In the last years, emerging ICTs were identified as a new pillar of port competitiveness (Carlan et al., 2016; Russo & Musolino, 2021b). Ports entered in a new generation: the fifth. Ports' issues were initially solved with stand-alone ICT solutions adopted by each decision-maker, which generated efficiencies in the three main port flows: cargo, information, and financial.

Today new benefits and challenges are connected with the introduction of shared emerging ICTs among decision-makers inside ports. The crucial issue concerns the fact that several decision-makers could share a decision about a single port operation in real time. Therefore, the effectiveness and efficiency of ports depend on how the interactions between the decision-makers are solved.

The paper is articulated into two main parts.

The first one presents the theoretical approaches (Transport System Models, TSMs) to explain how ports could benefit, in terms of costs reduction, of the potentialities supplied by emerging ICTs in solving interactions among decision-makers.

The second one presents a critical analysis of some representative case studies of the presence of ICT inside ports, in order to aggregate the observed trends in some macro-classes. The macro-classes are related to the presence of different evolutions of Port Community Systems (PCSs) and, in particular, to the diffusion of emerging technologies present in the market as: Internet of Things (IoT), Big Data (BD), BlockChain (BC), Digital Twin (DT), Artificial Intelligence (AI).

The main results of the study and its novelty concern two elements. On the one hand, the extension of TSMs to port systems, highlighting the problem of a non-single decision-maker (two or more) in some port operations. On the other hand, the possibility of aggregating the real observed cases in some representative macro-cases of presence of ICT inside the ports, in order to be able to study the resulting systems with TSM methods.

DOI: 10.1201/9781003320289-10

The work represents a step in analyzing the capacity of Transport System Models (TSMs) in evaluating the role of emerging ICTs in the modification of port operation generalized cost, considering the different decision-makers. In this way the work is useful both to port planners and technicians of port manufacturing equipment companies, because it allows to identify the advantages obtained with the introduction of PCS in increasingly advanced versions

The remaining part of the paper is articulated in four section. The first section synthetically describes the role of TSMs for the estimation of the cost for a unit of load (e.g. a container traveling along a path) travelling inside a port taking into consideration the presence of emerging technologies. In the second section a review of case studies of existing PCSs in relation to their evolution (waves). The third section describe the contribution of emerging technologies in the evolution of PCSs, with a focus on IoT. Finally, the conclusions and the research perspective are reported in the last sections.

## 2 TRANSPORT SYSTEM MODELS (TSM)

Port operations are associated with cargo movements and with information and financial transactions, in which different decision-makers are involved. Today emerging technologies play an important role in the modification of port operations cost. A current research line concerns the building of theoretical models using the formal equations of TSMs for the estimation of the cost for a unit of load, (e.g., a container traveling along a path, composed of a sequence of port operations inside a port taking into consideration the presence of emerging technologies).

The issue that arises inside the ports concerns the presence of more than one decision-maker in managing port operations, especially the ones involving information and financial exchanges. It is often more time consuming to solve administrative and financial problems than to execute the physical movement of goods. Times connected to administrative and financial operations may be cut down today by the emerging technologies, allowing different interacting decision-makers to optimize the solution of problems.

The theoretical model is based on TSMs equations (Cascetta, 2013). Few works in the current literature address the issue of estimating the increasing utilities of several decision-makers according to shared decisions by using TSMs (see Russo & Musolino, 2021b, and the references included). TSMs simulate a transport system, in which transport supply and travel demand interact.

TSMs are composed by three main models. The transport supply model estimates the users' utilities (or costs) deriving from the use of transport infrastructures and services. The common approach used is the topological model, given by a network model, with links, nodes and cost functions (e.g. time-flow relationship). The travel demand model simulates user choices based on the performance of infrastructure and services. Travel demand models can be behavioural or non-behavioural. In the behavioural approach (Ben-Akiva & Lerman, 1984), demand models can be stochastic or deterministic according to whether the (dis) utility associated with each user's choice is a random variable; or a deterministic variable. The supply-demand interaction model simulates the interaction between the user's choices and the performance of the infrastructures and the services. The model uses the topological-behavioural paradigm. The demand-supply interaction models can be classified into (Cascetta, 2013; Cantarella et al., 2019): static and dynamic, according to whether they allow to simulate a transport system in stationary or evolving conditions due to travel demand peaks or temporary changes in network capacity. Static models can be divided into: free-flowing models, such as 'network loading', and models based on an equilibrium approach, such as 'user equilibrium'. A detailed description of static models is reported in Wardrop (1952), Sheffi (1985), Cantarella & Cascetta (1995), and Cascetta (2013).

The three TSM models must be reformulated for the specific port condition as follows, in order to provide a solution of the issue presented above.

The *transport supply model* must simulate the performances (costs) of:

- the port operations, subdivided into physical movements, financial and information transactions;
- the emerging technologies, which allow decisions to be shared.

The *travel demand model* must simulate the behaviour of the decision-makers (or stakeholders), such as:

- public authorities, port companies, terminal companies, port agencies and services, land transport operators, on the land side;
- ship owners, shipping lines, customers, on the sea side.

The *supply-demand interaction model* identifies the links between port operations and decision-makers. The model highlights the role of emerging ICTs in cost functions and, therefore, allows to evaluate different ICT scenarios with different levels of shared decision among decision-makers.

As far as concerns the emerging technologies, it is worth noting that each port decision-maker has developed its own stand-alone ICT system. The more proprietary these ICT systems are, the less they can be shared with other decision makers. The proprietary approach allows full integration of emerging technologies within the company, ensuring high security in data control but makes it difficult (or impossible) to interact with other ICT systems. At the same time, the different flows of goods, information, and monetary value are governed by different decision-makers, which are often in opposite roles.

## 3 CASE STUDY: PORT COMMUNITY SYSTEMS

The European Port Community System Association (Morton, 2016) defines a Port Community System (PCS) as "a neutral and open electronic platform enabling intelligent and secure exchange of information between public and private stakeholders in order to improve the efficiency and competitive position of the sea and airports' communities". It "optimises, manages and automates smooth port and logistics processes through a single submission of data and by connecting transport and logistics chains".

Literature reviews on PCS are presented in several papers (see, among the others, Carlan et al., 2016; Mendes Costante et al., 2019; Moros-Daza et al., 2020, and the reported references). Carlan et al. (2016) focused on the associated costs and benefits of implementation of PCS. They classified existing studies in qualitative vs quantitative. The majority of the studies adopted a qualitative (descriptive) approach. Few studies had quantitative approaches, in defining indicators to measure the strategic and operational efficiency of PCS or in using quantitative data to assess the features implemented by PCSs. Moreover, the studies are mainly oriented in identifying and describing the benefits of PCS, rather than their costs, and mainly the utility generated. Ultimately, there are still difficulties in finding common ground to assess the impact of PCS on port competitiveness, due to the great variety of the existing PCS' s architecture and purpose. Moros-Daza et al. (2020) introduced the concept of PCS waves, in describing groups of PCS sharing a common deployment time in terms of normative technical and geo-economic settings. They identified four PCS waves from 1982 (the year of the first PCS) to today. Each wave is determined based on technological, political and legal, environmental, and economical factors, with some of its defining milestones marked over the time-line. A sketch of the normative and technological elements of each wave is reported in Table 1. A common economic element among the countries adopting the PCS platforms during the different waves, is their open economy towards external trade (trade index over 40% of their GDP).

Some Italian ports developed their own PCSs. The port of Ravenna, La Spezia and Genova introduced a second wave PCS platforms in 2005; and, the port of Trieste the third wave PCS platform Sinfomar, in 2014. Today an ongoing project is active in the port of Livorno (Pagano et al., 2016).

According to Italian legislation, the digital link between ports and the supply chain is in charge of the Italian National Logistics Platform (NLP). The National Strategic Plan for Ports and Logistics (Italian Ministry of Infrastructures and Transport, 2015) specified that the integration between ports and the supply chain "must take place through a cooperative modular architecture that allows the integration of information and of services related to road transport

Table 1. Waves in the development of PCS.

| Wave | Period | Normative/ Concepts | Technology |
|---|---|---|---|
| I | 1982-2000 | EDI agreement (paperless procedures) | Application Server Provider (ASP): distributed systems with an architectural middle-tier client-server |
| II | 2001-2011 | Paperless Trading Initiatives, Vessel Traffic Monitoring System Regulation for shared IT | Service-Oriented Architectures (SOA): from multi-tier distributed computer systems to decentralized web-based distributed systems |
| III | 2002-2017 | eCustoms Law and Law for paperless customs | from Web-Services to Online-Services on the web and mobile applications from automated procedures to smart procedures |
| IV | 2008- | blockchain, industry 5.0, artificial intelligence, virtual reality | cloud-services micro-data centers |

(source: elaboration from Moros-Daza et al., 2020)

[..] and node management. The national strategy assumed that in some ports the PCS were already well-established and that the need was rather to making homogeneous systems that were "born different". The NLP was therefore a facilitator of this process, without replacing the single existing PCS.

This process advances with two successive legislative interventions. The first is the memorandum of understanding of 2017 between Italian Ministry of Transport and UIRNET platform, which envisages the creation of a single PCS and encourages the transition from a federative to a centralized approach at national level. The second one of 2017 changed the framework defined in the PSNPL of 2015 by establishing the National Logistics Monitoring System (NLP, named SinaMoLo). SinaMoLo is also powered by some national "thematic" DSSs already operative (see Table 2). The PCS and the NLP operate in two distinct areas.

In 2018 a directive Italian Ministry of Transport, "Guidelines for homogenizing and organizing the PCSs through the NLP", imposes the migration of the existing PCSs besides the individual Port Authorities in the private cloud of the NLP developed by the Italian Ministry of Transport. The above operation was at the center of a legal dispute until an agreement was reached, according to which Logistica Digitale should manage the NPL until 2022. The

Italian Government, however, took over the direct management of the NPL in 2021, transferring "everything that has been built or is in progress in the implementation of every agreement" to the operational company of the Italian Ministry of Transport, Rete Autostrade Mediterranee (RAM).

Table 2. National Decision Support Systems (DSS) and port functions in Italy.

| | FUNCTIONS(*) | | | | |
|---|---|---|---|---|---|
| DSS(**) | T | L | N | C | D |
| SIMPT | X | X | | | |
| PIL | X | X | | | |
| PIC | X | | | | |
| PMIS | | | X | | |
| AIDA | | | | X | |
| SISTRI | | | | | X |

(*) T=Transportation, L=Logistics, N=Navigation, C=Custom, D=Dangerous.
(**) AIDA, Automazione Integrata Dogane Accise, SISTRI, Sistema Informativo Tracciabilità Rifiuti, SIMPT, Sistema Informativo per il Monitoraggio e la Pianificazione dei Trasporti, PMIS, Port Management Information System, PIL, Piattaforma Integrata della Logistica, PIC, Piattaforma Integrata della Circolazione.

A great number of PCSs have been established at international level. Table 3 presents the name of the PCS in the main ports (Notteboom et al., 2022).

TradeXchange of port of Singapore is a single electronic window for workflow entries, submissions and inquiries relating to ports, airports, maritime authorities, customs and agencies. PortBase in port of Rotterdam was obtained from the merger between Rotterdam's Port infolink and Amsterdam's PortNET. PortBase is an online platform for communication among freight forwarders, agents and terminals, customs. APCS in Antwerp port is a communication platform between all the stakeholders belonging to the defined categories of functions: transport/logistics, navigation, custom and dangerous goods. Los Angeles Port Optimizer is a cloud-based platform connecting the port and the stakeholders to optimize the cargo flows. DAKOSY in Hamburg port is a web-based freight tracking and order entry system for shippers and freight forwarders. FIRST in port of New York is an Internet-based, real-time network that integrates sources of freight location and status into a single web portal to allow port users to access cargo and port information. Finally, the valenciaportpcs.net of Valencia port is an online platform linking freight forwarders, customs, terminal information systems and gate management systems.

Table 3. PCS in a selection of ports.

| Port | Ranking | MTEUs(*) | PCS Name |
|---|---|---|---|
| Singapore | 2 | 36,60 | TradeXchange |
| Rotterdam | 11 | 14,51 | PortBase |
| Antwerp | 13 | 11,10 | APCS(**) |
| Los Angeles | 17 | 9,50 | Port Optimizer |
| Hamburg | 19 | 8,70 | DAKOSY |
| New York | 23 | 7,20 | FIRST(***) |
| Valencia | 27 | 5,30 | valenciaportpcs.net |

(*) Year 2018, (**) Antwerp Port Community System, (***) Freight Information Real-time System for Transport.

## 4 EMERGING TECHNOLOGIES BEYOND THE PCS

As seen in the previous paragraph, the PCS with its various articulations of functions and actors, represents the reference for a fifth-generation port. It was also highlighted that in the context of the PCS there have been and continue to progress in the system that have been defined as "waves" to mark the difference with the "generations" that identify the ports.

The PCS of the fourth wave is characterized by blockchain, industry 5.0, artificial intelligence, virtual reality (Moros-Daza et al. 2020, Bavassano et al., 2020).

The field of emerging technologies in ports has widened even more with the growth of quality and reliability of these technologies.

It is therefore necessary to consider, in addition to those mentioned above:

- the increasing massive presence of Internet of Things (IoT);
- the significant and directly uncontrollable increase in stored information that increasingly requires the use of Big Data (BD);
- the need to evaluate in advance all processes by making a Digital Twin (DT).

An evolution of PCS is being developed which, in addition to the technologies of the fourth, also contains IoT, BD, DT. There are no, or in any case no identifiable, ports that have internalized all the technologies referred to inside the PCS. It is therefore useful to refer to the most advanced uses of these technologies.

The scientific literature itself has not been able, some time, to internalize the most advanced uses. For this reason, reference will be made in this section also to the technical news from where it is possible to make a survey of what is happening. The following part of the sections presents a focus on the Internet of Things.

An element that is changing the structure of the PCS is the IoT. The possibility of directly interconnecting the objects (containers or other unit of loads) inside the port, without human intervention, makes it possible to deeply modify the cost function in all its terms.

The cost modification includes multiple aspects ranging from the macro ones related to the real-time identification of all container positions, to the modification of the internal contents both in terms of only rearranging the quantities and types of the products contained, and in terms of modification of nature same of the products, modifying the added value.

The continuous increase in the use of IoT is in turn connected with the subsequent generations of IoT available on the market.

According to Atzori et al. (2017), three evolutionary phases can be identified in the IoT. In their original note, the phases are indicated as generations, to avoid misunderstandings with the generations of the ports, the term "phase" is used in this paper. Each phase can be characterized by three elements: key enabling technologies; main reference architectural solutions; functions related to the evolution of stakeholders and objects. Table 4 shows a summary of the three-phase schematization.

It can be useful to recall some relevant papers regarding IoT implementations. The first important use of the IoT in ports concerns the improvement of the performance of operations related to transport, logistics and security. The work of Abderrahmen et al. (2017) analyzes the case study of the port of Le Havre. The work of Metin et al. (2018) extends the possibility given by IoT by enabling an artificial intelligence system and generating a DT. This development is implemented in the port of Rotterdam with the aim of satisfying the needs of port users with a higher level of efficiency, transparency and value.

The use of IoT has expanded over time, and has extended to the possibility of supporting a method for assessing the environmental impact of the port, in real time (Široka et al. 2021). This use of IoT has been tested in the ports of Bordeaux (France), Monfalcone (Italy), Thessaloniki (Greece).

Finally, it can be recalled the work of Bracke et al. (2021) in which together with the real-time measurements of environmental conditions and energy consumption, the consolidated function of remote assistance in operations is carried out. To pursue these objectives, a platform is developed that has three roles: to provide uniform access to data from a high number of devices and protocols; protect access to data; adapt to the growing volumes of data it has to acquire and process. The experimentation refers to mid-sized ports, as the ports of the Port System Authority of the Northern Tyrrhenian Sea (Italy), for the Autoridad Portuaria de Baleares (Spain), and for Piraeus (Greece).

Table 4. Phases in the development of IoT.

| Generation | Enabling Technology | Functions |
|---|---|---|
| I (the tagged things) | RFID technology<br>WSN, Telemetry, and SCADA<br>EPCglobal Network (RFIDs in IoT) | cost-effective way to tag objects and give them an identity |
| II (full interconnection of things and the (social) web of things) | IP network<br>Web of Things technologies (HTTP-based services)<br>Social Networks Services (SNS) platforms | integrate (moving) objects into the internetworking community<br>sensor nodes become devices of the WWW<br>exploit social networking in the IoT |
| III (social objects and cloud computing) | Semantic Sensor Network Ontology (SSN) and Information-Centric Network (ICN) paradigm<br>Integration of the IoT into cloud computing<br>Evolved RFID in the IoT | Social IoT solutions (objects participate in communities of objects, create groups of interest, and take collaborative actions) |

(elaboration from Atzori et al., 2017)

As mentioned above, alongside the scientific literature, which can naturally report the results obtained after a few years, there is an important technical literature published by the manufacturers. In this context it is useful to recall TIC 4.0 initiative that has been endorsed by the Federation of European Private Port Companies and Terminals (FEPORT) and of the Port Equipment Manufacturers Association (PEMA). The mission of TIC 4.0 is to promote, define and adopt standards that will enable cargo handling industry to embrace the 4th industrial revolution. It can be cited the iTerminals 4.0 project, has been funded in 2018 by the EU's Connecting Europe Program. (TIC, 2018). The project is coordinated by the Valenciaport Foundation. In particular, new IoT solutions and pilot projects are developed in container terminals located in eight main ports of the trans-European transport network: Antwerp, Dunkerque, Genoa, Malta, Montoir, Rouen, Sines, Thessaloniki.

## 5 CONCLUSIONS

The paper starts from the theoretical models, based on TSMs equations, to address the issue of estimating the increasing utilities of several decision-makers according to shared decisions. The background theory of the TSMs is the topological-behavioural equilibrium paradigm, which is today commonly shared in the scientific literature. In the authors' knowledge, the general framework of TSMs only recently started to be used to explain how ports could benefit, in terms of costs reduction, of the potentialities offered by emerging ICTs (Russo and Musolino, 2021b).

A relevant section of the paper is dedicated to a critical analysis of some representative case studies of the presence of ICT inside ports, in order to aggregate the observed trends in some macro-classes. The macro-classes are related to the presence of different evolutions of Port Community Systems (PCSs) and, in particular, to the diffusion of one emerging technology present in the market, as the Internet of Things.

The work aims to extend the TSMs from the consolidated field of passengers' mobility and freight distribution on land transport networks to the field of maritime ports.

Future work concerns the analysis of the impacts of other classes of emerging technologies on decision-makers, such as blockchain, big data, artificial intelligence and digital twin and the specification of the TSM models that incorporate the impacts of the emerging technologies on the port operations, the stakeholders and the port functions.

## REFERENCES


Abderrahmen, B., Duvallet, C. & Sadeg B. 2017. The Internet of Things for smart ports: Application to the port of Le Havre. *Proceedings of IPaSPort, May 3-4, 2017.* Normandie University, Le Havre, France.

Atzori L., Iera A. & G. Morabito. 2017. Understanding the Internet of Things: definition, potentials, and societal role of a fast evolving paradigm. *Ad Hoc Networks*, Volume 56, Pages 122–140.

Bavassano G., Ferrari C., Tei A. 2020. Blockchain: How shipping industry is dealing with the ultimate technological leap. *Research in Transportation Business & Management*, Volume 34, 100428.

Ben-Akiva M. & S.R. Lerman. 1984. D*iscrete choice analysis. Theory and application to travel demand*, MIT Press.

Bracke, V., Sebrechts, M., Moons, B., Hoebeke, J., De Turck, F., Volckaert, B. 2021. Design and evaluation of a scalable Internet of Things backend for smart ports. *Software: Practice and Experience* 51(7),1557–1579.

Cantarella G.E., Cascetta E. 1995. Dynamic Processes and Equilibrium in Transportation Networks, *Transportation Science*, vol. 29, 305–329.

Cantarella G.E., Watling D., de Luca S. & R. Di Pace. 2019. *Dynamics and Stochasticity in Transportation* Systems, *Tools for Transportation Network Modelling*, 1st Edition. Elsevier.

Carlan, V., Coppens, F., Sys, C., Vanelslander, T., Van Gastel, G. 2020. Blockchain Technology as Key Contributor to the Integration of Maritime Supply Chain? *Maritime Supply Chains*, 229–259.

Carlan, V., Sys, C., Vanelslander, T. 2016. How Port Community Systems Can Contribute to Port Competitiveness: Developing a Cost-Benefit Framework. *Research in Transportation Business Management* 19, 51–64.

Cascetta E. 2013. *Transportation systems engineering: theory and methods*. Volume 49. Springer Science & Business Media.

Italian Ministry of Infrastructures and Transport. 2015. *Piano Strategico Nazionale della Portualità e della Logistica*, Report Finale.

Martín-Alcalde, E., Saurí, S. & Ng, A. K. Y. 2016. Port-Focal Logistics and the Evolution of Port Regions in a Globalized World, in: Tae-Woo Lee and K. Cullinane (eds.) *Dynamic Shipping and Port Development in the Globalized Economy Volume 2: Emerging Trends in Ports*, P. Mcmillan, Palmgrave, pp 102–127.

Mendes Constante J., Lucenti K. & S. Deambros. 2019. *International case studies and good practices for implementing Port Community Systems*. Inter-American Development Bank. Technical Note n. IDB-TN-1641.

Metin O., Jaber M., Imran M.A. 2018. Energy-aware smart connectivity for IoT networks: Enabling smart ports. *Wireless Communications and Mobile Computing* Volume 2018, Article ID 5379326.

Moros-Daza, A., Amaya-Mier, R., Paternina-Arboleda, C. 2020. Port Community Systems: A Structured Literature Review. *Transportation Research Part A*: Policy and Practice, Volume 133, 27–46.

Morton R. 2016. *Port Community Systems: critical tools of Trade Facilitation* IPCSA.

Notteboom T., Pallis A. & J.P. Rodrigue. 2022. *Port Economics, Management and Policy*. 1st edition. Routledge, Taylor & Francis, London, UK.

Pagano P., Falcitelli M., Ferrini S., Papucci F., De Bari F. & Querci A. 2016. Complex Infrastructures: The Benefit of ITS Services in Seaports. In *Intelligent Transportation Systems: From Good Practices to Standards", 1st Edition*, CRC Press.

Russo, F., Musolino, G. 2020. Quantitative Characteristics for Port Generations: the Italian Case Study. *International Journal Transport Development and Integration* 4(2).

Russo, F., Musolino, G. 2021a. Case Studies and Theoretical Approaches in Port Competition and Cooperation. *Lecture Notes in Computer Science (including subseries Lecture Notes in Artificial Intelligence and Lecture Notes in Bioinformatics)*, 12958 LNCS, pp. 198–212.

Russo F., Musolino, G. 2021b. The Role of Emerging ICT in the Ports: Increasing Utilities According to Shared Decisions. *Frontiers in Future Transportation*, 2:722812.

Santos T.A., Guedes Soares, C. 2017. Development dynamics of the Portuguese range as a multi-port gateway system, *Journal of Transport Geography* 60, 178–188.

Santos, T.A., Guedes Soares, C. 2020. Short sea shipping in the age of information and Communications technology. In (eds.) Santos, T.A., Guedes Soares, C., *Short Sea Shipping in the Age of Sustainable Development and Information Technology*. Routledge, Taylor & Francis Group, London, UK, pp. 277–296,

Sheffi Y. 1985. Urban Transportation Networks. Prentice-Hall, Englewood Cliffs, New Jersery.

Široka M., Piličić S., Milošević T., Lacalle I., Traven L. 2021. A novel approach for assessing the ports' environmental impacts in real time: the IoT based port environmental index, *Ecological Indicators*, Volume 120, 106949.

UNCTAD 1994. *Port Marketing and the challenge of the Third Generation Port*. Geneva, Switzerland: Trade and Development Board Committee on Shipping ad-hoc Inter-government Group of Port Experts.

UNCTAD 1999. *Fourth-generation Port: Technical Note*. Ports Newsletter N. 19, Prepared by UNCTAD Secretariat. Geneva, Switzerland.

Wardrop J.G. 1952. Some theoretical aspects of road traffic research, *In: Proceedings of Institution of Civil Engineers, Part II*, 325–378.

TIC 2018. *https://iterminalsproject.eu/about/* (last view February 2022).

*Trends in Maritime Technology and Engineering – Guedes Soares & Santos (eds)*


# Geographical scope of competitiveness of short sea shipping and freight railways in the Atlantic Corridor

T.A. Santos, M.Â. Fonseca, P. Martins & C. Guedes Soares
*Centre for Marine Technology and Ocean Engineering (CENTEC), Instituto Superior Técnico, Universidade de Lisboa, Lisbon, Portugal*

ABSTRACT: Short sea shipping competes with other components of the Trans-European Transport Networks such as road and, in particular, rail freight corridors, in the transportation of freight along the coastlines of the European Union. This paper aims at contributing to the identification of the geographical scope of competitiveness of short sea shipping when competing with rail freight corridors, considering also that both need to be integrated in door-to-door intermodal transport chains. The importance of value of time for freight is also assessed as this factor assumes significant relevance when pure road haulage is also a competitor. A model of transport networks is used to support a numerical model of transport cost, transit time and generalized transport cost, with calculations being carried out for multiple pairs origin/destination on a door-to-door basis. These models are applied in a case study relating to the corridor from Northern Portugal to Northern Europe (Atlantic Corridor). Conclusions are drawn regarding the most suitable transport chains for different destinations and values of time.

## 1 INTRODUCTION

The need to shift long distance freight transportation from the road to other more environmentally friendly modes of transportation is on top of the agenda worldwide and the European Commission (2019) has made it a main priority of its transport policy. The European Green Deal in fact prescribes that a substantial part of inland freight carried today by road should be shifted to rail and inland waterways. Furthermore, from 2011, a target of shifting 30% of road freight over 300 km to other transport modes by 2030, and more than 50% by 2050 had already been adopted, see European Commission (2011). This 2011 White Paper indicated that this shift should be facilitated by efficient and green freight corridors and the development of appropriate infrastructure.

The White Paper also recognized that it is highly likely that freight shipments over short and medium distances (below about 300 km) will rely, in the foreseeable future, on road haulage. In fact, more than three quarters of road shipments correspond to distances below 150 km, so road haulage will still play a very significant role in the future, and most certainly in the distribution of freight from seaports and inter-modal terminals to final destinations. Short sea shipping (SSS), inland waterways (IWW) and railways should therefore take the role of carrying freight across long distances, with pre-carriage and on-carriage remaining in the hands of road haulage.

These broad EU transport policies are all the more relevant for member states located further away from Central Europe but having a large part of its foreign trade (goods) concentrated on a few major trading partners in the Union, mainly concentrated in Northwestern and Central Europe. That is the case of Portugal, whose main trading partners, in addition to Spain, are France, Germany, Italy and the United Kingdom. Furthermore, ports in the Northern range such as Rotterdam and Antwerp are frequently used as transshipment hubs feeding other smaller ports in Northern Europe. Accordingly, it is not surprising that the topic of developing infrastructure allowing improved freight transportation to Northern and Central Europe is very much present in Portugal.

One example of this interest is a study by the Portuguese Industry Confederation, CIP (2015), that includes useful notes related to the problem of the competitiveness of freight transportation between Portugal and Northern Europe. This organization reports that 70% of Portuguese foreign trade occurs with the European Union, CIP (2016), and out of this trade, about two thirds of the physical goods are carried by road with the remaining part being carried by sea in containers. Rail freight transportation along this axis is almost non-existent, although Rail Freight Corridor 4 (RFC 4), see RailNetEurope (2020) and Atlantic Corridor (2019), could be used. It is important to clarify here that in this paper we will consider that the expression Atlantic Corridor encompasses not only RFC 4 (also designated as

DOI: 10.1201/9781003320289-11

Atlantic Corridor) but also other modes of transportation (most notably SSS, but also road haulage) suitable to carry freight between the broad area of the Western Iberian Peninsula and continental Northwestern Europe.

Taking in consideration the persistent use of the road mode of transportation even for long distances, countries bearing the consequences of the intense road traffic are increasingly monitoring heavy goods vehicle (HGV) traffic in critical points of the road network such as the Pyrenees crossings and the Spanish-Portuguese borders, see MTMAU (2020) and METD (2019), with clear indications that road freight traffic is again increasing since at least 2014. Consequently, many EU countries are increasingly establishing legal and administrative barriers to control HGV traffic in their roads. Alternative options to road haulage in this corridor could be an increased utilization of SSS and rail freight corridors (RFC), but just-in-time logistics often makes it difficult to use such modes of transportation. In fact, the findings of an inquiry to Portuguese stakeholders, namely industrial companies and sectorial organizations, regarding the suitability of different transport modes for carrying goods to Northern and Central Europe, CIP (2016), allows the conclusion that SSS is not considered to be adapted to modern logistic needs, but rail could be used with advantage in this corridor.

This paper seeks to add on the existing body of knowledge by providing quantitative insight on these difficult problems, especially for peripheral countries, which have been largely discussed up until now on a predominantly qualitative basis. The approach in this paper is mainly quantitative, through the use of a transport network model comprising road, rail and maritime sub-models, all interconnected at relevant nodes (representing seaport terminals or intermodal terminals). Transport cost and time are then computed over door-to-door transport chains, using a numerical model that includes also time and cost involved in cargo handling operations in terminals.

This approach has been presented in Santos *et al.* (2021) and applied in that paper to the competition between SSS (container ship or roll-on/roll-off (Ro-Ro) ship based) and road haulage, the two modes currently used in this corridor. The main finding is that the geographical scope of NUTS 2 regions attracted to SSS intermodal solutions is much wider when using Ro-Ro ships than when using container ships, due to the effect of larger transit times in the latter case.

Considering these findings, this paper seeks to extend the scope of the analysis to the case in which a direct railway service using RFC 4 is available between Portugal and central Germany (Mannheim). At this stage, no such direct railway service exists along this rail freight corridor, although it has existed for a brief period of time in the past and is planned to be revived in the near future. The effects of a full range of different values of time (VoTs) are also evaluated in terms of its effect in the geographical scope of SSS competitiveness when comparing with road haulage and a rail service based on RFC 4.

The paper is organized in the following manner. Section 2 presents a literature review on this theme focusing on multimodal competition in this corridor and on the concept of value of time. Section 3 presents a numerical model for assessing multimodal transport chains. Section 4 presents the case study to be carried out in this paper, while section 5 details the numerical results dealing with freight transport be-tween Portugal and Northern Europe, allowing conclusions to be drawn in section 6.

## 2 LITERATURE REVIEW

### 2.1 *Multimodal corridors studies*

The Atlantic Corridor is one example out of many major corridors for freight transportation identified in the EU. The SuperGreen project, for example, identified as many as 60 corridors and finally concentrated the analysis on 9 corridors with high "greening potential", Psaraftis & Panagakos (2012). These corridors have been studied by different authors in recent years focusing on environmental aspects, mainly greenhouse gas emissions, but also on external costs of transportation, see Psaraftis & Panagakos (2012) and Zis *et al.* (2019). Other authors have considered primarily the economic side of the competitiveness of SSS against road haulage, in specific corridors, for example in the Baltic Sea, Ng (2009), in the Western Mediterranean Sea, Morales-Fusco *et al.* (2012), or around the Italian peninsula, Lupi *et al.* (2017). The Atlantic Corridor has also been considered, both from the economic and environmental points of view in Santos *et al.* (2020), Santos & Guedes Soares (2019) and Ramalho & Santos (2021), and, in the case of these works, including not only the SSS leg but the entire door-to-door transport chain.

Most of these studies consider mainly transport parameters such as transit time and cost, greenhouse gases emissions and external costs. Besides these main factors, there are others with impact in the utilization of SSS in multimodal logistics supply chains, such as service frequency, quality, reliability and commercial attitude, see Paixão-Casaca & Marlow (2009). In addition to general studies (EU wide) on these factors, there is at least one such study dedicated to the Atlantic Corridor by CIP (2016), that presents the results of questionnaires sent to Portuguese industrial companies (taking the role of shippers) relative to the performance of SSS, rail and road haulage in this corridor. Table 1 summarizes the views of the respondents regarding three modes of transportation and five main decision parameters: suitability for door-to-door, transit time, transport cost, frequency and reliability. Most stakeholders showed similar views but their answers indicate that their experience with SSS is mainly related to containerized cargo, rather than Ro-Ro cargo.

Analysis of the answers indicates that many respondents mention that SSS is more suitable for coastal areas than for origins and destinations within the continental mass of Northern Europe, as might be expected. SSS in considered unable to deliver adequate solutions for door-to-door transport mainly due to low speed. Furthermore, SSS needs to rely on road haulage (or eventually freight train plus road) to perform the pre-carriage and on-carriage to the final destination, involving cargo handling, delays and extra costs. However, overall, the respondents indicate that SSS is cheaper. In turn, discontinuities in the transport chain and the slowness of the ship cause longer transit times on the door-to-door transport and this is seen as a substantial problem by approximately two-thirds of respondents, especially for destinations away from the coastal area. Furthermore, delays in seaports and intermodal terminals may lead to extra cargo storage costs.

Table 1. Summary of respondent views on transport mode performance (authors elaboration based on CIP (2016)).

| | Short Sea Shipping | Road | Rail |
|---|---|---|---|
| Door-to-door | No. Requires road haulage and cargo handling | Yes | No. Requires road haulage and cargo handling |
| Transit time | Long | Short | Long |
| Transport cost | Low. Increases because of road haulage | High | Average. Increases because of road haulage |
| Frequency | Typically low. May be increased if there is sufficient demand | As required by logistic needs | Typically low. May be increased if there is sufficient demand |
| Reliability | Low. Cargo handling between modes and ship operation lead to delays | High. | Low. Cargo handling between modes and train slot availability lead to delays |

The final conclusion in CIP (2016) is that EU transport policies focused on reducing long distance freight transportation, without a suitable alternative in the Atlantic Corridor, are likely to affect very negatively the competitiveness of the Portuguese economy. In particular, SSS is deemed to be not suitable as a replacement for road haulage but railways could form a competitive alternative for regions in Central Europe located far from coastal

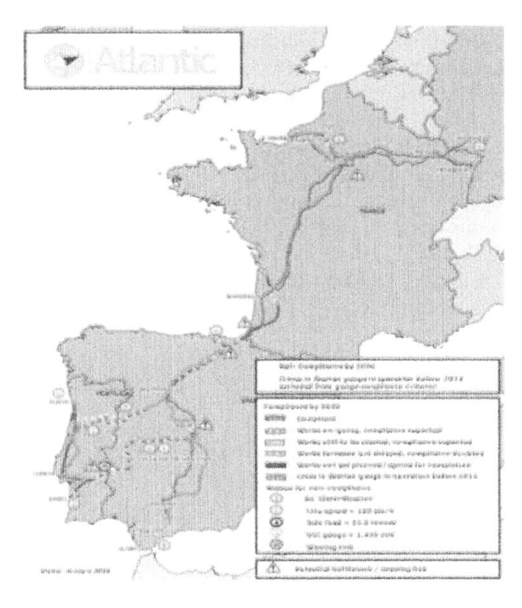


Figure 1 . Status of compliance of Rail Freight Corridor 4 with EU standards (Source: Atlantic Corridor (2020)).

ports. However, RFC 4, which stretches from Portuguese ports (through Spain and France) to Central Germany (Mannheim) as shown in Figure 1, still uses Iberian gauge tracks up to the French border, is partly not electrified in Portugal and Spain and still is not fitted with the European Railway Traffic Management System (ERTMS). These shortcomings imply that it is currently not the desired competitive alternative for freight transportation to Central Europe.

### 2.2 *Value of time for cargo*

Considering the importance attributed to the time factor by shippers, consequence of the extensive application of just-in-time logistic strategies in EU industry and distribution, it is appropriate to resort in transport analyses to the use of the concept of generalized transportation cost (GTC) that depends on cost and time values for transportation, but also on the value of time for the cargo. However, the VoT variable is a difficult parameter to estimate in a precise manner, since it depends on many factors. The main factors influencing the VoT relate to cargo details: (a) value; (b) nature; and (c) urgency. Accordingly, higher cargo VoT corresponds to high value cargo, more perishable or urgent cargos (or a combination of all), and the opposite is true for low VoT. Other parameters, related to the above, such as the country (or the region) where the transport operation takes place or even the transportation mode, can also influence VoT.

There is an extensive body of literature that presents different concepts of VoT, such as those reported in Wang *et al.* (2015), different models for the quantification of VoT, Zis & Psaraftis (2017), as well as different actual values for VoT, as reported in Zamparini & Reggiani (2007) and Feo-Valero *et al.* (2011). The disparity of VoT values observed in the literature confirms the difficulty in quantifying VoT and the heterogeneity of values reported. In this regard, Figure 2 compiles a significant number of VoTs, with a variation between 2 and 47 €/hour.trailer, with an average value of €20.8 €/hour.trailer.

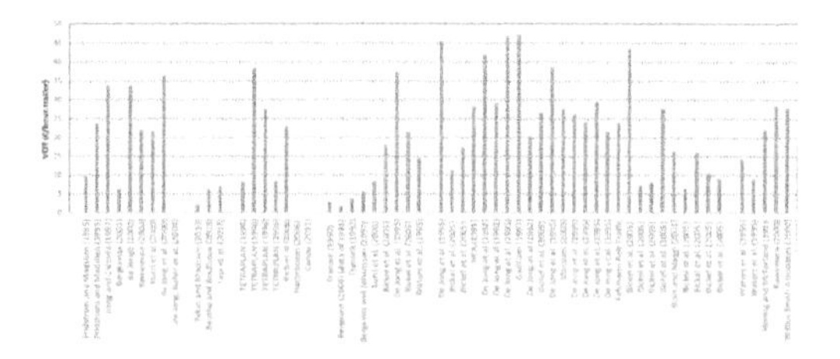

Figure 2. Values of Time (VoT) compiled from various references in Feo-Valero *et al.* (2011).

These values provide a range of VoT that will be used in the evaluation of the preferences of NUTS 2 regions for transport chains, when considering the generalized transportation cost as the decision parameter. The values depend on the characteristics of the cargo and, in order to narrow down the range of relevant VoTs for the Portuguese case, a brief analysis of Portuguese exports was carried out. Using the database from the Kennedy School of Management (2018) focusing on Portugal, it is possible to identify the set of the 20 most exported products (services are excluded). The analysis of these products indicates a wide variety of products: (a) low-value goods, as paper for graphic purposes or articles from natural cork; (b) middle-value goods, as parts of motors vehicles, wines and olive oil; and, (c) high-value goods, as leather footwear and medicines. However, textiles, footwear, food and beverages, electrical apparatus and automobile parts appear to be the most significant products and most of these fall on the lower to mid-range. It is thus likely that for most Portuguese cargos carried to other EU countries by road (or even by sea in containers) the VoT lies in the low to middle-range, that is from about 6 to 20 €/hour.trailer, unless the urgency of delivery or high value of cargo imposes higher values. However, in this paper VoT will be considered to range from €2/hour.trailer to €47/hour.trailer in order to illustrate the significant changes in transport chain preferences across the geographical area under study.

## 3 NUMERICAL MODEL

### 3.1 *Transport chain evaluation*

The numerical model used in this paper has been presented and discussed in references Santos *et al.* (2020, 2021) and Santos & Guedes Soares (2019), and it is thus presented below in a summary manner. The model allows the calculation of transportation time and cost and the generalized transportation cost using a model of the transport networks available in a given geographical area under study. Several alternative transport chains (designated as paths, *p*) between the same pair of origin and destination (O/D) may be set up (and grouped in sets, *k*) and multiple pairs O/D may be evaluated in a single run.

The model calculates the transport time over different modes of transportation, namely road, rail, inland waterways and maritime (using Ro-Ro or container ships), denoted as, respectively, $T_{RD}$, $T_{RL}$, $T_{IW}$, $T_{RR}$ and $T_{CC}$. The transportation time taken along path *p* of set *k* is then given by:

$$T_{TRkp} = T_{RD_{kp}} + T_{RL_{kp}} + T_{IW_{kp}} + T_{RR_{kp}} + T_{CC_{kp}} \quad (1)$$

The time taken in each mode is calculated over a network composed of nodes and links, where each link is characterized by a distance and typical average speed (depending on type of road, for example). Nodes, when they are of seaport terminal type or intermodal terminal type, are characterized by typical average dwell time and thus contribute to transportation time.

Similarly, the model also calculates the total transportation cost in a given path. This is again composed of the sum of the costs associated with each mode of transportation:

$$C_{TRkp} = C_{RD_{kp}} + C_{RL_{kp}} + C_{IW_{kp}} + C_{RR_{kp}} + C_{CC_{kp}} \quad (2)$$

This cost also includes those costs associated with cargo transfers (both handling and storage) between transport modes occurring in particular nodes of the network (seaport terminals or intermodal terminals).

Finally, the total transportation time and cost on a given path, *p*, are combined to produce a generalized transportation cost, given by:

$$GTC_{kp} = C_{kp} + VoT.T_{kp} \quad (3)$$

where *VoT* represents the value of time for the cargo in monetary units per hour and trailer. Finally, it should be pointed out that this numerical model is sensitive to many parameters involved in intermodal transport operations, as listed in Table 2.

Table 2. Transport operations parameters covered in the numerical model.

| Transport parameter | Transport modes and TEN-T components |
|---|---|
| 1 | Road haulage tariffs and road tolls (if not included in road haulage cost) |
| 2 | Rail tariffs (if not included in rail freight rates) |
| 3 | IWW freight rates |
| 4 | SSS freight rates (Ro-Ro) |
| 5 | SSS freight rates (container ships) |
| 6 | Bunker surcharges or similar (through "toll" mechanism) |
| 7 | Canal fees (through "toll" mechanism) |
| 8 | Distance between origin and destination (split per mode) |
| 9 | Road speed, type, congestion |
| 10 | Road haulage driver's resting times (not using two drivers) |
| 11 | Rail speed |
| 12 | IWW speed |
| 13 | SSS speed |
| 14 | Ship service frequency (through adjusting port time) |
| 15 | Cargo handling charges |
| 16 | Storage costs (beyond free time) |
| 17 | Efficiency of seaport terminals (through port time) |
| 18 | Efficiency of intermodal terminals (through terminal time) |
| 19 | Value of time for cargo |
| 20 | Chokepoints (borders, locks, canals) |

### 3.2 *Computational tool*

The numerical model above has been implemented in a dedicated computational tool (Intermodal Analyst). Figure 3 shows the general functioning of this tool, with a number of data files on the left, which are read by the tool according with the specification created by the user in the "Log File". The tool then produces numerical results that are stored in output files, as shown on the right. The necessary data include the network database specification, the cost functions database (covering road, rail, IWW and SSS modes), the transport chain (specification of chains to be evaluated) database and the cargo amounts (per pair origin/destination) database. At least one transport chain needs to be specified and, if there are several alternative transport chains for a single pair origin/destination, these should be listed together in a set of paths. If statistical data on volumes of cargo (intermodal units) is available for each pair origin/destination, this information is included in the cargo database. The main output of the tool consists of time and cost involved in each transport chain (these might be unimodal or multimodal), split per transport mode, including also time spent and costs incurred in seaport terminals and intermodal terminals.

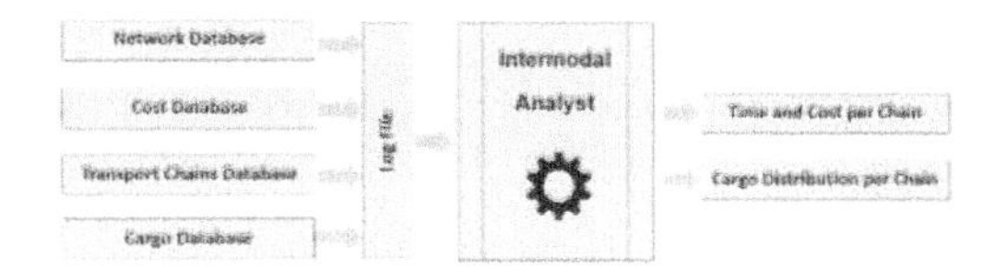


Figure 3 . Flowchart of computational tool.

This computational tool runs in a personal computer and the computational time is generally very low and proportional to the number of transport chains under evaluation. The numerical results regarding transportation cost, transportation time and generalized transportation cost have been thoroughly analysed and are displayed in the next sections using maps built in a geographic information system (QGIS 3.14), using a uniform distribution and classification system.

## 4 CASE STUDY DEFINITION

### 4.1 *Transport chains*

In order to carry out the case study a number of transport chains has been specified, using the transport network outlined in Figure 4. This figure shows a maritime route between Leixões and Rotterdam (identified in Santos *et al.* (2022) as very competitive), which may be used by a Ro-Ro ship or a container ship, railway lines forming part of RFC 1 (Rhine-Alpine Corridor) and RFC 4 (Atlantic Corridor), inland waterway between Rotterdam and Duisburg (Rhine river) and road network (shown only as an example for one city, Wurzburg). The network model actually comprises a full definition of the core and comprehensive road network (Portugal, Spain, France, Germany, Benelux) as defined in the TEN-T network definition, European Parliament and Council (2013).

Figure 4 also shows the locations of the various seaport and intermodal terminals involved in the various transport chains and the approximate path of the roads, railway lines and maritime routes. A number of alternative transport chains have been specified for the numerical study (most comprise also, in addition to the transport modes indicated below, road haulage in Portugal and Northern Europe, respectively pre-carriage and on-carriage). Table 3 shows the alternative transport chains specified for this case study (most comprise also, in addition to the transport modes indicated below, road haulage in Portugal and Northern Europe, respectively pre-carriage and on-carriage).

The different transport chains run between an origin in Portugal (Porto, main city of this country's northernmost NUTS 2 region) and NUTS 2 regions covering Germany, Benelux and northern half of France. These NUTS 2 regions were chosen as these

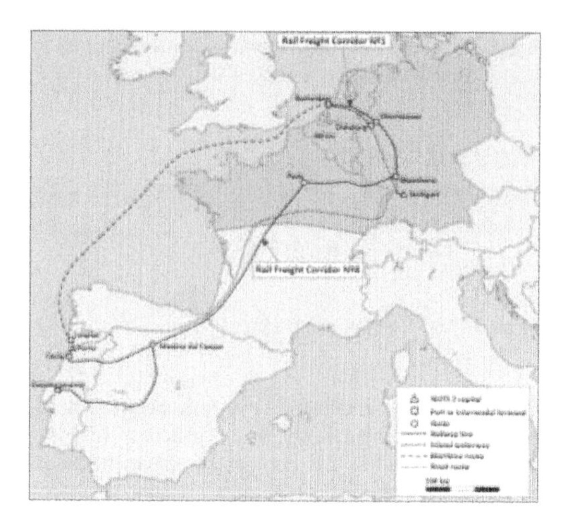

Figure 4. Intermodal routes supporting transport chains between Porto and Stuttgart.

Table 3. Definition of the alternative transport chains in the case study.

| Transport Chain | Transport modes and TEN-T components |
|---|---|
| TC1 | Road haulage Portugal-Northern Europe |
| TC2 | SSS (Ro-Ro ship) Leixões-Rotterdam |
| TC3 | SSS (Container ship) Leixões-Rotterdam |
| TC4 | SSS (Ro-Ro ship) Leixões-Rotterdam + RFC 1 (to Oberhausen) |
| TC5 | SSS (Ro-Ro ship) Leixões-Rotterdam + IWW (Rhine to Duisburg) |
| TC6 | RFC 4 (Entroncamento-Mannheim) |
| TC7 | RFC 4 (Cacia-Mannheim) |
| TC8 | SSS (Ro-Ro ship) Leixões-Rotterdam + RFC 1 (to Mannheim) |

are the ones easily reachable from the port of Rotterdam. Regions in the southern half of France are not considered to be susceptible to gains in competitiveness or attraction, when considering the use of the port of Rotterdam, for freight originating in Porto region in Portugal. Porto was chosen as the common origin for transport chains as its NUTS 2 in Portugal comprises a large part of the manufacturing industry in the country (mostly small and medium enterprises) and its trade is mainly focused on the EU.

## 4.2 *Transport parameters specification*

Table 4 shows the load/unloading costs and dwell times in seaport and intermodal terminals. Costs applicable to seaport terminals have been taken from port tariffs, APL (2020). Costs applicable to intermodal terminals have been taken from Infraestruturas de Portugal (2017) and when not available, taken in line with the values shown in Black *et al.* (2003) and Martins (2015). A comprehensive review of cost parameters in intermodal operations can be found in Santos & Guedes Soares (2017). Regarding the dwell time in terminals, it may be seen that the container terminal is specified to have the largest dwell time (48 hours), while the Ro-Ro terminal has a dwell time of only 6 hours. These values vary significantly between terminals and countries, and the values taken here are below those given in PIANC (2014), but this is found to be feasible within the scope of short sea shipping. Regarding dwell times in rail and IWW terminals, the full transfer of cargo (needed because of different rail gauges) between trains (with 32 car-go units) may take 8 hours, Martins (2013). Therefore, it seems reasonable to assume that in Germany an average dwell time of cargo units of 6 hours is feasible and in Portugal twice as much (12 hours). For inland waterways terminals, the average dwell time was taken as slightly larger than in rail terminals, and it has been assumed at 9 hours.

Table 4. Costs and average dwell times in seaport and intermodal terminals.

| Terminal | Cost Unload (€) | Cost Load (€) | Dwell time (Hours) | Free Time (Hours) | Time Cost (€/day) |
|---|---|---|---|---|---|
| Rail (Cacia) | 12.5 | 12.5 | 12.0 | 96.0 | 10.0 |
| Rail (Entroncamento) | 12.5 | 12.5 | 12.0 | 96.0 | 10.0 |
| Rail (Oberhausen) | 25.0 | 25.0 | 6.0 | 48.0 | 2.0 |
| Rail (Mannheim) | 25.0 | 25.0 | 12.0 | 48.0 | 2.0 |
| Container (Leixões) | 0.0 | 142.7 | 48.0 | 120.0 | 1.8 |
| Container (Rotterdam) | 25.0 | 120.0 | 48.0 | 96.0 | 2.0 |
| Ro-Ro (Leixões) | 25.0 | 25.0 | 6.0 | 48.0 | 2.0 |
| Ro-Ro (Rotterdam) | 50.0 | 50.0 | 6.0 | 96.0 | 2.0 |
| IWW (Duisburg) | 25.0 | 25.0 | 9.0 | 96.0 | 2.0 |

It has been assumed that road haulage would conform with standard rest times, see European Parliament and Council (2006), including a night time rest of 12 hours, which exceeds slightly EU regulations. It is assumed that road haulage is using only one driver. Furthermore, regarding the utilization of SSS (Ro-Ro or container ships), it is assumed that unaccompanied transport is used and road haulage is arranged for carriage to the port of loading and from the port of unloading towards the final destination.

Regarding costs of transport modes, the cost functions shown in Figure 5 have been considered. It should be noted that pre and on carriage represent 25-40% of total costs of intermodal transport (thus impacting heavily on its competitiveness), see Hitjens *et al.* (2020). Its cost per tonne-km may rise to several times that of rail for short distances and, therefore, intermodal terminals need to be located near shippers/receivers. The specific costs used in this model have been aligned with these qualitative observations. Furthermore, Martins (2015) indicates that for international freight transport across large distances, values per km vary between 0.7 and 1.2 €. Further evidence on these values may be found in CEGE (2014) and Reis (2014).

Figure 5 shows that at short-medium distances (up to 500 km) Ro-Ro or container ships present the highest specific costs, followed by road haulage, while rail and inland water-ways present the lowest costs. All transport modes increase in specific costs especially for short distances, but it should be noted that rail, IWW and SSS are relatively less used for these shorter distances. Regarding the high distance range (above 750 km), it may be seen that road haulage is the most expensive mode, followed by rail and inland waterways. Maritime modes present the lowest specific costs for distances above 1500 km. It is important to note that all these values are external to the numerical tool and may be replaced or updated according with specific circumstances.

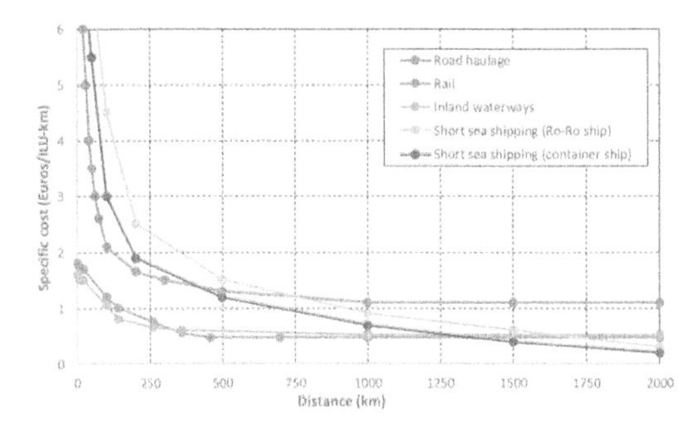


Figure 5. Transport modes specific costs.

In the case of maritime transportation, the values for this specific route (Leixões to Rotterdam), have been corrected according with information from shipping companies, to 1050 €/FEU for the Ro-Ro ship and 740 €/FEU for the containership. Regarding rail cost, it is indicated in Martins (2013) that the cost per unit would be 1281 € (0.47 €/km) for long-distance. For inland waterways it is known by direct information that the cost of shipping a FEU from Rotterdam to Duisburg would be about 120 € over a distance of 230 km, implying a specific cost of 0.52 €/km. These costs for trains and barges relate to the case in which these vehicles travel fully loaded.

## 5 NUMERICAL RESULTS

### 5.1 *Transport cost using different transport chains*

Figure 6 shows the total transport cost (door-to-door) from Porto to different NUTS 2 regions, using a Ro-Ro ship coupled to road haulage (TC2). The Ro-Ro ship is used to reach Rotterdam and as the road distance from this port increases, so do transport costs. A similar pattern could be observed for the other transport chains, but now with transport cost increasing progressively from Rotterdam, Oberhausen, Duisburg or Mannheim, depending on which intermodal terminal is being used.

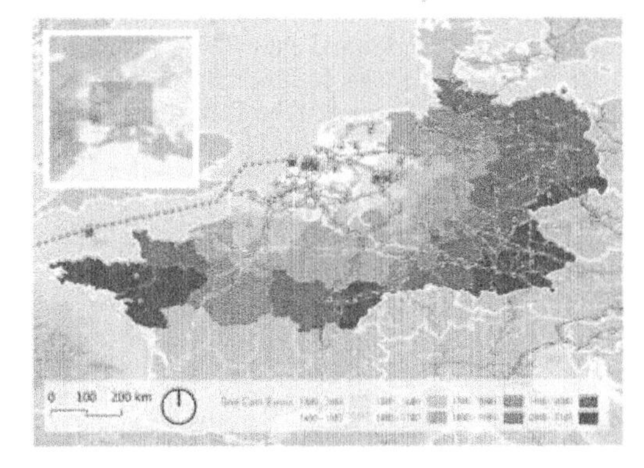


Figure 6. Total transport cost (door to door, in euros) - transport Chain TC2 SSS (Ro-Ro ship) Leixões-Rotterdam.

Although the increase in the total cost of transportation must be read continuously throughout the territory, significant differences in total costs are observed between NUTS 2 regions. Table 5

Table 5. Transport chains, minimum value, maximum value and difference for total transport cost (€).

| Transport Chain | Minimum Value (€) | Maximum Value (€) | Range (€) |
|---|---|---|---|
| TC1: Road haulage Portugal-Northern Europe | 1500 | 2707 | 1207 |
| TC2: SSS (Ro-Ro ship) Leixões-Rotterdam | 1380 | 2165 | 785 |
| TC3: SSS (Container ship) Leixões-Rotterdam | 1230 | 2021 | 791 |
| TC4: SSS (Ro-Ro ship) Leixões-Rotterdam + RFC N°1 (to Oberhausen) | 1612 | 2399 | 787 |
| TC5: SSS (Ro-Ro ship) Leixões-Rotterdam + IWW (Rhine to Duisburg) | 1596 | 2379 | 783 |
| TC6: Rail Freight Corridor N.° 4 (Entroncamento-Mannheim) | 1973 | 2745 | 772 |
| TC7: Rail Freight Corridor N.° 4 (Cacia-Mannheim) | 1503 | 2275 | 772 |
| TC8: SSS (Ro-Ro ship) Leixões-Rotterdam + RFC N.° 1 (to Mannheim) | 1710 | 2481 | 771 |

summarizes minimum, maximum and range of differences in transportation cost to the different NUTS 2 regions. The range of differences is significantly larger for TC1 as it results from wider differences in road distance, as compared with the other chains.

These results allow a comparative analysis of transport costs, measured by differences between different options. The first such analysis concerns the competitiveness of TC1 in the studied territory. This transport chain, as shown in Figure 7, when compared to the TC2 (SSS), shows significantly greater competitiveness (lower cost) mainly in French territory.

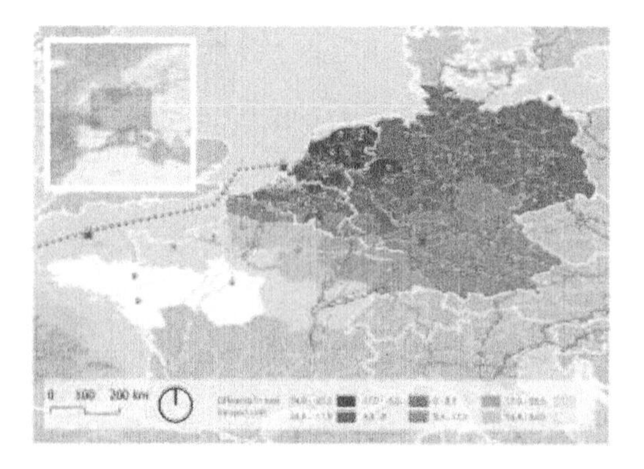

Figure 7. Difference in total transport cost (in percentage) between SSS and road transportation from Portugal (TC2 and TC1).

Figure 8 shows the most cost-effective transport option for each NUTS 2 region in these conditions. It should be noted that TC4 (Ro-Ro/Rail to Oberhausen), TC6 (Rail from Entroncamento) and TC8 (Ro-Ro/Rail to Mannheim) do not appear in Figure 8 as they are not competitive transport options for any NUTS 2 region, when compared to the other chains. Also, TC1 (pure road haulage) has been excluded due

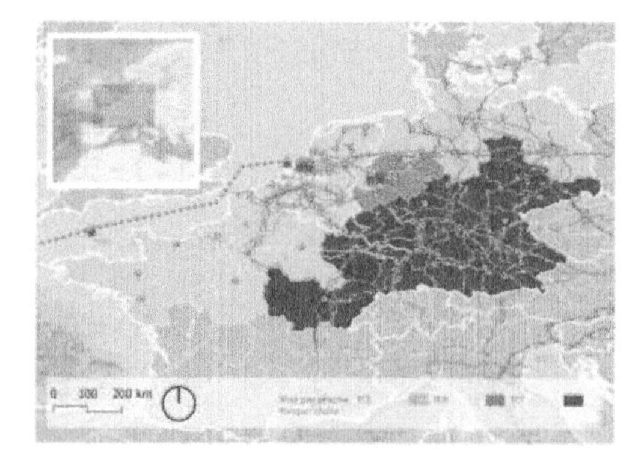

Figure 8. Most cost-effective transport chains for Northern European NUTS2 (among TC2, TC4, TC5, TC6, TC7 and TC8).

to the costs presented, but also due to its limited competitiveness in French territory, as well as TC3, due to the high transport time shown in Table 6 below.

Table 6. Transport chains, minimum value, maximum value and difference for total transport time (hours).

| Transport Chain | Minimum Value (h) | Maximum Value (h) | Range (h) |
|---|---|---|---|
| TC1: Road haulage Portugal-Northern Europe | 44,7 | 73,3 | 28,6 |
| TC2: SSS (Ro-Ro ship) Leixões-Rotterdam | 75,2 | 98,5 | 23,3 |
| TC3: SSS (Container ship) Leixões-Rotterdam | 159,1 | 182,3 | 23,3 |
| TC4: SSS (Ro-Ro ship) Leixões-Rotterdam + RFC Nº1 (to Oberhausen) | 84,8 | 108,1 | 23,3 |
| TC5: SSS (Ro-Ro ship) Leixões-Rotterdam + IWW (Rhine to Duisburg) | 104,8 | 128,2 | 23,3 |
| TC6: Rail Freight Corridor N.º 4 (Entroncamento-Mannheim) | 104,4 | 127,6 | 23,2 |
| TC7: Rail Freight Corridor N.º 4 (Cacia-Mannheim) | 85,9 | 109,1 | 23,2 |
| TC8: SSS (Ro-Ro ship) Leixões-Rotterdam + RFC N.º 1 (to Mannheim) | 100 | 123,3 | 23,2 |

It may be seen that TC2 and TC7 determine the boundary between the NUTS 2 regions for which SSS or rail is most competitive. In fact, on TC2 (and partly on TC5) the most competitive regions in terms of transport costs are found along the Atlantic coast. At a distance greater than approximately 300km from the coast, TC7 is the most competitive in terms of transport cost, thus being consistent with EU policy objectives. It is also useful to show the cost differences implied by the most competitive transport chains (all including SSS to Rotterdam) in comparison with railway direct from Portugal (TC7). The aim is to assess how cost competitive are these chains compared with TC7. Figure 9 shows such differences, with negative values indicating that chains based on SSS offer cheaper options than TC7 and positive values indicating costlier options. One can note that for regions around Rotterdam the savings are about € 500 to € 600. In general, for coastal NUTS, SSS based transport chains are cost competitive. Only for those NUTS 2 regions for which TC7 was the most cost-effective option, the additional cost is positive and comprised between € 100 and € 207. These values may be considered as acceptable for most shippers as they correspond to about 5-10% of transport costs. This implies that transport chains based on SSS are not significantly less competitive (in terms of cost) than railway directly from Portugal.

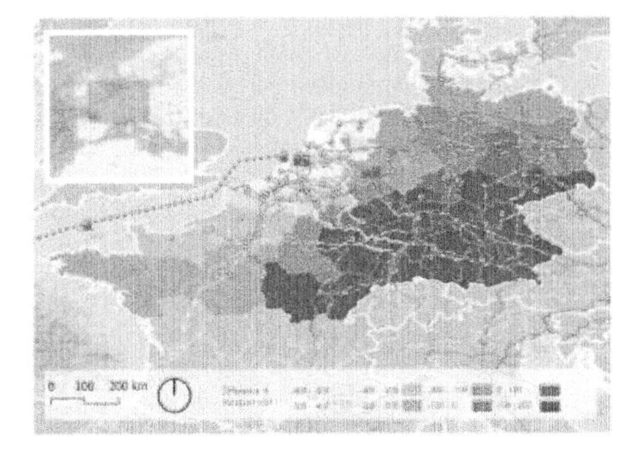

Figure 9. Difference in transport cost for Northern European NUTS2 regions (between TC2, TC5 and TC7).

## 5.2 *Transport time using different transport chains*

The results of an analysis similar to that of the previous section, but now focused on transport time are now presented. Figure 10 shows the transport time using when using TC7.

If the different chains are examined for transport time, the results are shown in Table 6, which indicates the maximum and minimum values of transport time as well as their difference. TC1 is the most effective chain, with the maximum transport time value (73.3 hours) below the minimum transport time values for the remaining transport chains. In the same context, the least effective transport chain is TC3 (based on container ships), considering that the mini-mum transport time value (159.1 hours) is higher than the maximum transport time value of the remaining transport chains.

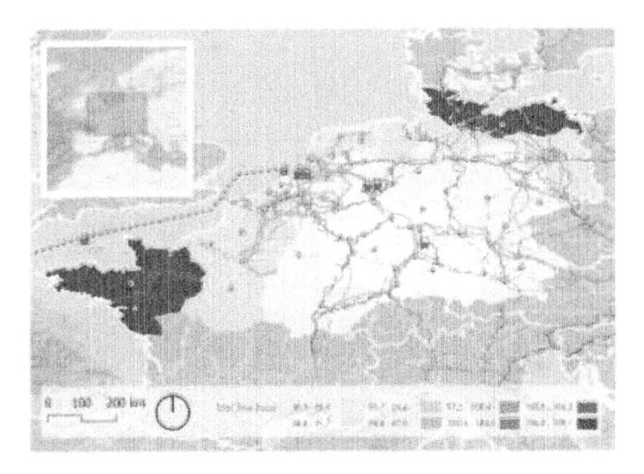

Figure 10. Total transport time (door to door, in hours), transport chain TC7 Rail Freight Corridor 4 (Cacia-Mannheim).

When comparing the transport time effectiveness of the proposed transport chains, removing only TC1 and TC3 because these chains are, respectively, the fastest and the slowest, TC2 stand out as the most effective, losing this condition only for TC4 in Dresden and for TC7 in Besançon (France) and between Munich and Chemnitz (Germany), as shown in Figure 11. It is worth reminding that TC2 is the Ro-Ro ship to Rotterdam and distribution of cargo units using road haulage. This implies that the use of RFC 4 (TC7) is rarely more time competitive than TC2.

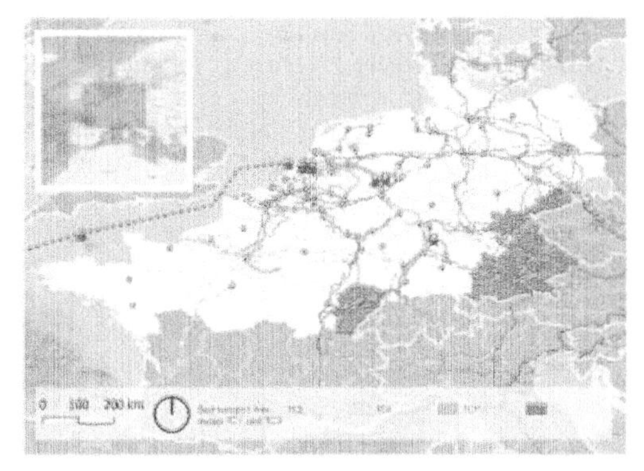

Figure 11. Most time effective transport chains for Northern European NUTS 2 regions (among TC2, TC4, TC5, TC6, TC7 and TC8).

Figure 12 shows the differences in transit time between TC2, TC4 and TC7 (the most effective chains according with Figure 13) and TC1. As may be seen, all these chains involve longer transit times ranging from 10.2 hours to 51.8 hours. However, throughout the territories of Netherlands and Germany, TC1 offers shorter transit times by only

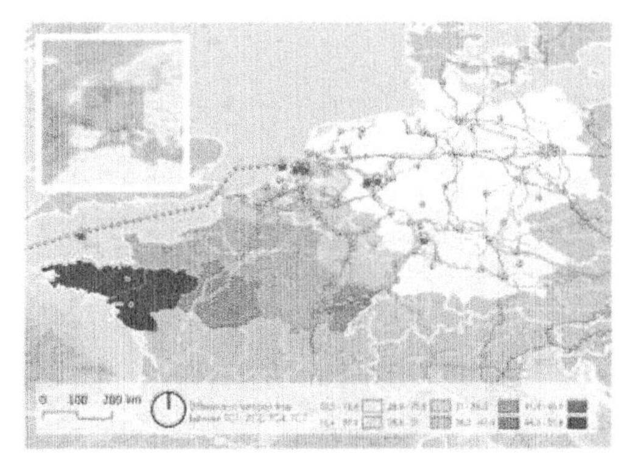

Figure 12. Difference in transport time for Northern European NUTS 2 regions (TC2, TC4 and TC7 compared with TC1).

between 10.2 and 20.6 hours, that is between half a day and a full day more of transport time. This means that alternative transport chains to TC1 (fully road) imply only moderately higher transport times, that are not critical for not urgent cargos. However, for just-in-time cargos this certainly explains the fact that road haulage is preferred. In fact, the situation only becomes very disadvantageous for France, with transport times larger by one and a half days or more compared to road haulage direct from Portugal.

### 5.3 *Generalized cost of transportation using different transport chains*

In the previous sections the effect of transportation and time were analysed separately. This section aims at putting together these two factors through the use of the concept of generalized cost of transportation. As this concept involves a valuation of time for cargo and this process produces very different values, depending on the type of cargo and the specific circumstances under which the transportation is carried out, a set of different VoTs, as identified in section 2.2, will be used: € 2.0, € 6.82, € 20.80 and € 47.0. Table 7 summarizes the minimum and maximum values of Generalized Transport Cost for the 8 transport chains and 4 different values of VoT.

The values of VoT lead to reconfigurations of the competitiveness of different transport chains, when analysed using the GTC. Figure 13 provides an example of GTC (based on VoT of € 20.80) for the various NUTS 2 regions when considering transport chain TC1, which is based on full road haulage. As expected, the GTC increases with distance from Portugal. The pattern of variations in GTC is slightly different for some areas than that shown for transport cost, but in general differences are not very significant.

Figures 14, 15 and 16 represent the most competitive transport chains for each NUTS 2 region, based on GTC calculated with VoT of, respectively, € 2.00, € 6.82, € 20.80 and € 47.00. Considering Figure 16a (GTC calculated with a VoT of € 2.00), it may be con-firmed that TC1 is the most competitive chain for most of France. A significant difference comparing with Figure 8 is the loss of competitiveness of TC5 and the inclusion of TC4 (rail using RFC 1), even considering the small value of time (€2.00).

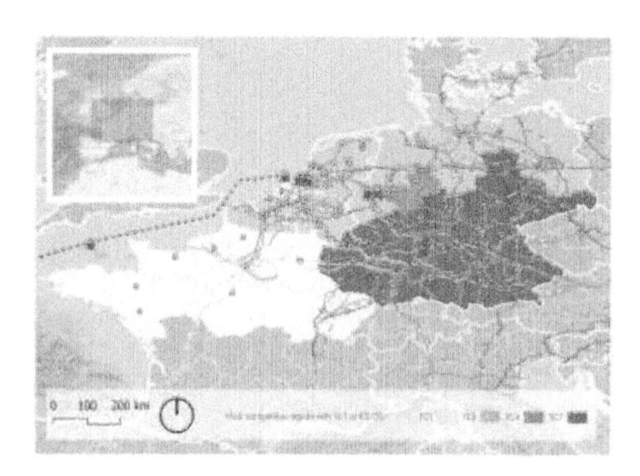

Figure 13. Transport chain TC1 generalized transport cost based on value of time (VoT) at €20,80.

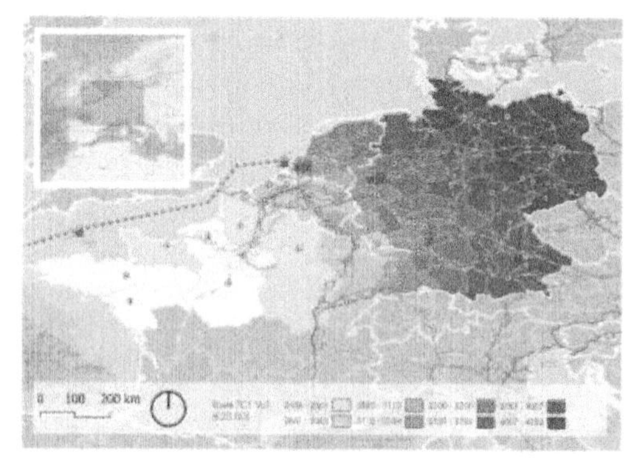

Figure 14. Most effective transport chains for Northern European NUTS 2 regions (among TC1, TC2, TC4, TC5, TC6, TC7 and TC8) considering GTC based on VoT of €2.00.

Table 7. Generalized Transport Cost chains, based on VoT of €2,00, €6,82, €20,80 and €47,00, with minimum and maximum values (€).

| | VoT at €2,00 | | VoT at €6,82 | | VoT at €20,80 | | VoT at €47,00 | |
|---|---|---|---|---|---|---|---|---|
| TC | Min Val | Max Val | Min Val | Max Val | Min Val | Max Val | Min Val | Max Val |
| TC1 | 1596 | 2853 | 1811 | 3207 | 2436 | 4232 | 3607 | 6153 |
| TC2 | 1533 | 2362 | 1895 | 2836 | 2947 | 4213 | 4917 | 6793 |
| TC3 | 1553 | 2386 | 2319 | 3264 | 4543 | 5814 | 8711 | 10591 |
| TC4 | 1782 | 2615 | 2190 | 3136 | 3376 | 4647 | 5597 | 7480 |
| TC5 | 1806 | 2635 | 2311 | 3253 | 3777 | 5045 | 6523 | 8403 |
| TC6 | 2182 | 3000 | 2685 | 3615 | 4145 | 5399 | 6880 | 8743 |
| TC7 | 1675 | 2493 | 2089 | 3019 | 3291 | 4545 | 5542 | 7404 |
| TC8 | 1910 | 2728 | 2392 | 3322 | 3791 | 5045 | 6412 | 8274 |

Changes in territorial competitiveness of the different transport chains become more evident when the VoT increases still further. Figure 15 shows the results for a VoT of €20,80, indicating that TC4 chain loses its place as a competitive option for Ruhr region in Germany, with TC2 chain becoming there the most competitive. This transport chain also penetrates further into regions where, previously, TC7 was the most competitive chain. However, the major difference is the high competitiveness of TC1 across most of Belgium, Luxembourg and western Germany. The reason for this is that TC1 (pure road haulage from Portugal), is very time competitive and, for medium range values of VoT, it becomes the most competitive option for more regions.

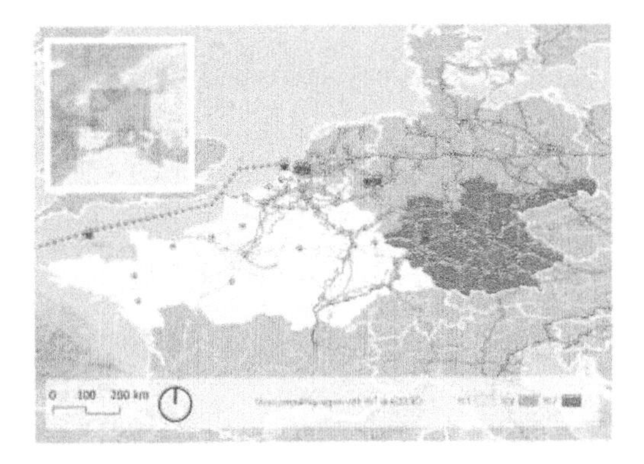

Figure 15. Most effective transport chains for Northern European NUTS 2 regions (among TC1, TC2, TC4, TC5, TC6, TC7 and TC8) considering GTC based on VoT of €20.80.

A very significant increase in VoT to €47.00, see Figure 16, shows the most profound changes in the competitiveness of transport chains. Road haulage (TC1) completely overcomes TC7 (rail to Mannheim) and also reduces significantly the regions where TC2 is most competitive. This chain remains competitive across northern Germany. The reason for this increase in competitiveness of TC1 is the very high value of time, representative of either high-value cargos or very urgent cargos, leading to large competitiveness gains for road haulage (TC1).

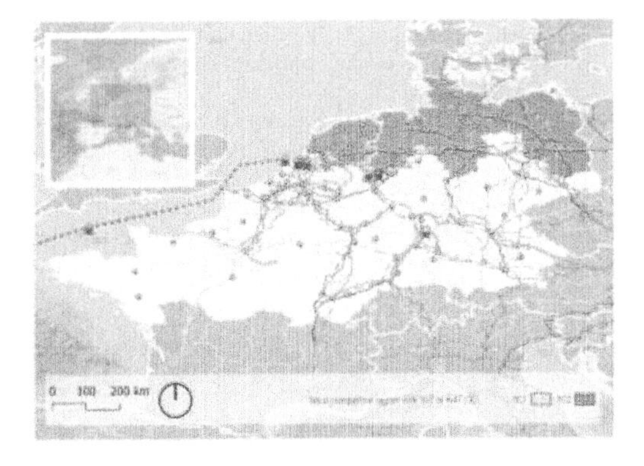

Figure 16. Most effective transport chains for Northern European NUTS 2 regions (among TC1, TC2, TC4, TC5, TC6, TC7 and TC8) considering GTC based on VoT of €47.0.

## 6 CONCLUSIONS

This paper presents a numerical model for evaluating the competitiveness of intermodal chains. This model is used in combination with a model of transport networks in western Europe. These numerical models have been applied to assess the relative competitiveness of short sea shipping (SSS), rail freight transportation and road haulage in the Atlantic Corridor. The analysis if focused on a particular SSS route between Portugal (port of Leixões) and the Netherlands (port of Rotterdam), but this route is deemed to be representative of the situation in the Atlantic Corridor.

The results presented in this paper for the port of Rotterdam confirm that SSS services are very competitive up to approximately 300 km from the Atlantic coastline. As it is EU policy the promotion of modal shift away from road for distances above 300 km, this finding is in line with such policies. For larger distances from Rotterdam (regions deep in the European landmass), SSS intermodal solutions comprising also rail or IWW, may be used with cost savings in comparison with road transportation. However, these options are slightly less competitive than a direct railway service from Portugal to Mannheim (using RFC 4). Regarding the transport time, the results confirm that pure road haulage from Portugal remains a highly time effective solution. However, the Ro-Ro ship solution coupled with distribution from Rotterdam using road haulage is also highly time effective. In general, rail freight transportation (RFC 4) directly from Portugal does not take less time than SSS (using Ro-Ro ships.

Finally, considering the generalized transport costs, it may be seen that, regarding SSS solutions based on Ro-Ro ships, for low to medium values of time (VoT), these solutions remain competitive for most NUTS in Germany and Benelux. Even for high VoT, these transport chains are competitive for northern Germany. This is in stark contrast with the transport chain based on SSS with container ships, which is not competitive for any NUTS for medium and high VoT. The conclusion is that Ro-Ro ships offer clear advantages in terms of transport time and this translates into competitive advantages from the point of view of generalized transport cost.

Although RFC 4 is competitive for NUTS 2 deep in the continental mass, it is necessary to point out that railway lines in this corridor still need substantial investments in both Spain and Portugal. Considering that intermodal alternatives based on the port of Rotterdam are nearly as competitive even for regions deep in the Europe, SSS arises as a competitive solution for freight transportation in the Atlantic Corridor. EU public policies should re-assess cost-benefit analyses of the investments in these rail corridors to peripheral countries, duly considering the option of supporting SSS integrated with already well developed rail corridors in Northern Europe to reach regions far from the coast, thus promoting in the near term the modal shift away from road.

## ACKNOWLEDGEMENTS

This research was funded by Fundação para a Ciência e a Tecnologia (FCT) under grant number PTDC/ECI-TRA/28754/2017 and is in line with the Strategic Research Plan of the Centre for Marine Technology and Ocean Engineering (CENTEC), which is financed by FCT under contract UIDB/UIDP/00134/2020. The authors would like to thank the European Commission, DG MOVE, TENtec Information System, for the permission to use the representation of the TEN-T network in the maps presented in this paper.

## REFERENCES

APL (2020). Container terminal tariffs. APL, Lisbon, Portugal.

Atlantic Corridor (2019). Corridor Information Document. Atlantic Corridor, Bordeaux, France. (https://www.atlantic-corridor.eu/, accessed on 29/ 06/2021).

Atlantic Corridor (2020). Fourth Work Plan of the European Coordinator Carlo Secchi, May, DG Mobility and Transport, Brussels, Belgium (https://ec.europa.eu/transport/sites/default/files/atlworkplanivweb.pdf, accessed on 08/ 07/2021).

Black, I., Seaton, R., Ricci, A., Enei, R. (2003). Final Report: Actions to Promote Intermodal Transport. Final Report of EU Project RECORDIT (Real Cost Reduction of Door-To-Door Intermodal Transport).

CEGE (2014). Container terminals in the region of Lisbon-Setúbal – a Comparative Analysis. Technical study, ISEG – School of Economics and Management, Lisbon, Portugal.

CIP (2015). Logistics in Portugal (in Portuguese). CIP, Lisbon, Portugal.

CIP (2016). Survey on international goods transportation (in Portuguese), Report Ref-106, CIP, Lisbon, Portugal.

European Commission (2011). White Paper Roadmap to a Single European Transport Area – Towards a competitive and resource efficient transport system, COM (2011) 0144, Brussels, Belgium.

European Commission (2019). The European Green Deal, COM (2019) 640 final, Brussels, Belgium.

European Parliament and Council (2006). Harmonisation of certain social legislation relating to road transport and amending Council Regulations (EEC) No 3821/85 and (EC) No 2135/98 and repealing Council Regulation (EEC) No 3820/85. Regulation (EC) No 561/2006, Strasbourg, France.

European Parliament and Council (2013). Regulation (EU) No 1315/2013 on Union guidelines for the development of the trans-European transport network and repealing Decision No 661/2010/EU. Strasbourg, France.

Feo-Valero, M., García-Menéndez, L, Garrido-Hidalgo, R. (2011). Valuing Freight Transport Time using Transport De-mand Modelling: A Bibliographical Review, Transport Reviews, 31:5, 625–651. http://dx.doi.org/10.1080/01441647.2011.564330

Hintjens, J., van Hassel, E., Vanelslander, T., Van de Voorde, E. (2020). Port Cooperation and Bundling: A Way to Reduce the External Costs of Hinterland Transport. Sustainability, 12, 9983. http://dx.doi.org/10.3390/su12239983

Infraestruturas de Portugal (2017) Regulation on access and tariffs of the rail freight terminals of Leixões and Bobadela (in Portuguese), Lisbon, Portugal.

Kennedy School of Government (2018), Atlas of Economic Complexity. (https://atlas.cid.harvard.edu/countries/179, accessed on 30/ 11/2021).

Lupi, M., Farina, A., Orsi, D., Pratelli, A. (2017). The capability of Motorways of the Sea of being competitive against road transport. The case of the Italian mainland and Sicily. Journal of Transport Geography, 58, 9–21.

Martins, P. (2015). Road freight transportation: the dilemma railway versus road in Portugal (in Portuguese), MSc thesis, Faculty of Economics, University of Porto, Portugal.

Martins, S. (2013). Economic evaluation of the project "Rail freight corridor Lisbon-Germany" developed by DB Schenker (in Portuguese), MSc thesis, Faculty of Sciences, University of Lisbon, Portugal.

METD (2019). Portuguese-Spanish Observatory, 9th Report (in Portuguese), November, GEE (METD), Lisbon, Portugal.

Morales-Fusco, P., Saurí, S., Lago, A. (2012). Potential freight distribution improvements using motorways of the sea. Journal of Transport Geography, vol. 24, pp. 1–11.

MTMAU (2020). Spanish-French Observatory of traffic in the Pirineus, Documento nº9 (in Spanish), December, Publication Center (MTMAU), Madrid, Spain.

Ng, A. K.Y. (2009). Competitiveness of short sea shipping and the role of port: the case of North Europe, Maritime Policy & Management, Vol. 36, 4, pp. 337–352.

Paixão Casaca, A.C., Marlow, P.B., 2009. Logistics strategies for short sea shipping operating as part of multimodal transport chains. Maritime Policy & Management, Vol. 36 1, pp. 1–19.

PIANC (2014). Masterplans for the development of existing ports, Report n° 158, Brussels, Belgium.

Psaraftis, H.N., Panagakos, G. (2012). Green corridors in European surface freight logistics and the SuperGreen project. Procedia - Social and Behavioral Sciences 48, pp. 1723–1732.

RailNetEurope (2020), Rail Freight Corridors Map 2020, RNE RailNetEurope, Vienna, Austria. (https://rne.eu/rail-freight-corridors/, accessed on 29/ 06/2021).

Ramalho, M.M., Santos, T.A. (2021). Numerical modelling of air pollutants and greenhouse gases emissions in intermodal transport chains, Journal of Marine Science and Engineering, 9(6), 679.

Reis, V. (2014). Analysis of mode choice variables in short-distance intermodal freight transport using an agent-based model. Transportation Research Part A, 61, 100–120.

Santos, T.A., Fonseca, M.Â., Guedes Soares, C. (2022). Integrating short sea shipping with Trans-European Transport Networks. (forthcoming).

Santos, T.A., Guedes Soares, C. (2017). Modeling of Transportation Demand in Short Sea Shipping, Maritime Economics and Logistics, 19:4, pp 695–722.

Santos, T.A., Guedes Soares, C. (2019). Short sea shipping and the promotion of multimodality in the European Atlantic Area. Proc. of the 27th Annual Conference of the International Association of Maritime Economists, Athens, Greece.

Santos, T.A., Guedes Soares, C. (2020). Assessment of transportation demand on alternative short sea shipping services considering external costs, in Integration of the Maritime Supply Chain: evolving from collaboration processed to maritime supply chain network, Vanelslander, T., Sys, C. (Eds.), Elsevier, pp. 13–45.

Santos, T.A., Ramalho, M., Guedes Soares, C. (2020). Sustainability in short sea shipping–based intermodal transport chains, in Short Sea Shipping in the Age of Sustainable Development and Information Technology, Santos, T.A., Guedes Soares, C. (Eds.), pp. 89–115, Routledge, Taylor & Francis Group, London, UK.

Wang, S., Qu, X., Yang, Y. (2015). Estimation of the perceived value of transit time for containerized cargoes. Transportation Research Part A, 78, 298–308.

Zamparini, L., Reggiani, A. (2007). Freight Transport and the Value of Travel Time Savings: A Meta-analysis of Em-pirical Studies, Transport Reviews, 27:5, 621–636.

Zis, T.P.V., Psaraftis, H.N. (2017). The implications of the new Sulphur limits on the European Ro-Ro sector. Transportation Research Part D, Vol. 52, pp. 185–201.

Zis, T.P.V., Psaraftis, H.N., Panagakos, G., Kronbak, J. (2019). Policy measures to avert possible modal shifts caused by sulphur regulation in the European Ro-Ro sector. Transportation Research PartD, Vol 70, pp. 1–17.

Zis, T.P.V., Psaraftis, H.N., Tillig, F., Ringsberg, J.W. (2020). Decarbonizing maritime transport: a Ro-Pax case study. Research in Transportation Business & Management, Vol. 37, 100565.

*Maritime traffic*

*Trends in Maritime Technology and Engineering – Guedes Soares & Santos (eds)*
 *ISBN 978-1-032-33583-4*

# A framework for characterizing the marine traffic of the continental coast of Portugal using historical AIS data

B. Lee & P. Silveira
*Centre for Marine Technology and Ocean Engineering (CENTEC), Instituto Superior Técnico, Universidade de Lisboa, Portuga*
*Escola Superior Náutica Infante D. Henrique (ENIDH), Paço de Arcos, Portugal*

H. Loureiro & A.P. Teixeira
*Centre for Marine Technology and Ocean Engineering (CENTEC), Instituto Superior Técnico, Universidade de Lisboa, Lisbon, Portugal*

ABSTRACT: The paper proposes a framework to thoroughly analyze, compile and characterize the maritime traffic patterns as to further improve the ability to monitor maritime traffic, plan voyages, assess risks, and prevent accidents. The framework decodes historical Automatic Identification System messages and processes the data by an unsupervised machine-learning density-based spatial clustering algorithm combined with the Douglas-Peucker algorithm to compress the ship trajectories. Then, probabilistic methods are used to yield a set of relevant statistics that adequately characterize the routes and the vessels' expected behavior. The framework is implemented in a software tool and tested on a historical AIS dataset to characterize the maritime traffic along the Portuguese continental coast.

## 1 INTRODUCTION

The study of the characteristics of the local maritime traffic is of great importance, not only to maintain the operational efficiency, but also to ensure the safety at sea. Learning its traffic patterns provides support to maritime surveillance as it gives information on the usual trajectories of ships along the main traffic routes (Rong *et al.* 2020). Ship motion patterns can be extracted by sorting ship trajectories following the same route using statistical analysis techniques (Etienne *et al.* 2012), unsupervised learning methods (Vespe *et al.* 2012), or clustering methods (de Vries & Someren, 2012; Pallotta *et al.*, 2013). After extracted, each motion pattern can be used to characterize and predict the maritime traffic on a macro scale (Rong *et al.* 2022) and its risks (Silveira *et al.* 2013).

The maritime traffic information is provided by Automatic Identification System (AIS) messages that include dynamic information such as the geographic position of the ship, its speed, heading and navigational status. They also include the ship's static information such as its Maritime Mobile Service Identity (MMSI), IMO number and its characteristics (IMO, 2002). Currently, ships are required to have on board AIS transponders that automatically broadcast messages to other ships nearby, AIS base stations and satellites, to keep track of the ship.

In general, the transmission time interval of an AIS position report is around 2 to 13 seconds for vessels underway and 3 minutes for anchored vessels (ITU, 2014). A trajectory can be seen as a discrete sample of coordinate points broadcasted by the AIS transmitter. If the sampling rate is higher, the more accurate the trajectory can be described, but at the same time it will decrease the computational efficiency of further processes due to the massive amount of data. Nevertheless, the communication occurs more frequently when compared to the number of the actual changes of navigational state of the ship (Zhao *et al.*, 2018). Therefore, there is an overwhelming amount of AIS data just for one single voyage, being most of them redundant. Since there is no human or machine capable to deal with that amount of data, it is fundamental to compress the trajectories by eliminating these redundant data before proceeding with any kind of analysis (Zhang *et al.* 2016).

Douglas & Peucker (1973) developed an algorithm to reduce the number of trajectory points by using line segments to approximate the original trajectory using a single threshold parameter. The approach has proven to be one of the most accurate line simplification algorithms, suitable for simplifying AIS trajectories (Zhang *et al.* 2016).

DOI: 10.1201/9781003320289-12

Although several compression methods have been developed, some problems remain, mainly when defining the value of the threshold for the compression, since it depends on the area of navigation, vessels' sizes, type of trajectories, among others (Zhao *et al.* 2018). Defining the DP threshold holds a significant problem for most studies since none of them reaches a consensus on its optimal value. If this value is very high, the sampling rate is low and the trajectory is over compressed, which means important key points are lost. On the other hand, if the value of the parameter is very low, the sampling rate is high, and the trajectory is under compressed. So, attaining a good balance between compressing the redundant information and maintaining the feature points of the trajectory is in fact very difficult. To define the DP threshold, Zhang *et al.* (2016) evaluated first the ship domain size and concluded that 0.8 times the ship's length can be chosen as the maximum value for the DP threshold. This has been used by Zhang *et al.* (2018) who combined the Douglas-Peucker (DP) algorithm with the Sliding Window (SW) algorithm and the Density-Based Spatial Clustering of Applications with Noise (DBSCAN) to identify turning nodes, to then compute the optimal path from the starting to the ending turning point. Rong *et al.* (2020) used the DP algorithm with the threshold set at 500 m, to isolate key features of a certain trajectory along the Portuguese continental coast, and then characterized the route and detected anomalies in the traffic. Others such as Liu *et al.* (2019) developed an adaptive threshold DP algorithm based on the channel characteristics and the distances and characteristics of all feature points, which complemented with the SW algorithm and the DBSCAN clustering method, could compress the AIS trajectories while maintaining the main geometrical features of the Yangtze River inland waterway.

This study aims at characterizing the maritime traffic along the Portuguese continental coast and at assessing probabilistically its main features. The AIS messages are first decoded and grouped according to the pair starting - ending locations. Then, cleaning and compression algorithms are implemented and used to filter and eliminate some position errors and compress and simplify the trajectories, so that statistical analyses can be performed to describe the maritime traffic.

This paper is organized in 4 main sections. Section 2 describes the methodology of the present study, explaining in detail each algorithm developed. Section 3 presents the results of a specific case as an example of application of the algorithms. Lastly, section 4 presents the conclusions and possible future developments of this study.

## 2 METHODOLOGY

The goal of this paper is to propose an approach to characterize and analyze probabilistically the maritime traffic. The steps of the proposed approach are detailed in the following sections and represented in Figure 1.

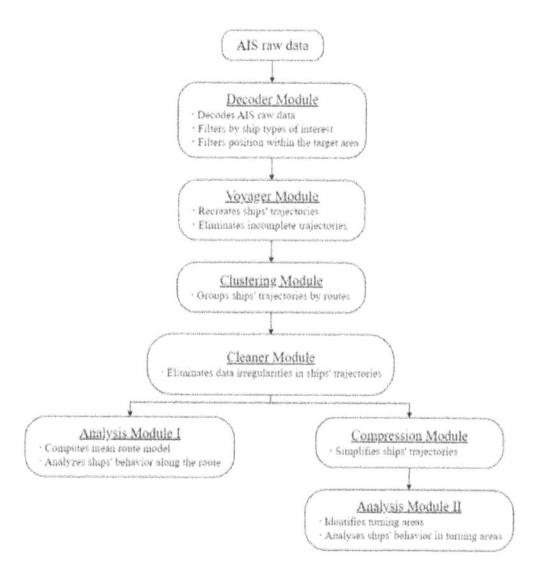


Figure 1. Flowchart of the maritime traffic characterization algorithm.

### 2.1 *Nomenclature*

For the purposes of this paper the following definitions apply.

- A datapoint is a single AIS message of types 1, 2, or 3 along with its contents, namely longitude, latitude, and SOG (Speed Over Ground).
- A voyage is a set of consecutive datapoints, pertaining to the same vessel, i.e., same MMSI, corresponding to that vessel moving between an origin and a destination, which usually fall either on ports or on entries/exits of the geographical area.
- A route is a set of voyages, not limited by MMSI, with matching origin and destination points.
- A waypoint is a datapoint where the voyage's trajectory shifts (e.g., directional change).
- A simplified voyage is the set of waypoints of a voyage representing the voyage compressed.
- A mean route is the interpolation of the geographic coordinates of a route.

### 2.2 *Decoding*

The AIS is a statutory required system for vessels above 20 meters, representing the totality of vessels of interest for this study, and thus AIS data is strictly formatted by types, with each containing a clear set of information. Since this study deals with shipping routes and associated motion patterns of cargo vessels, most AIS data messages can be swiftly dismissed – leaving only four types of messages of interest: types 1, 2, and 3, containing motion data of the vessels, and type 5 that describe static data of a vessel.

These messages are decoded with an AIS decoder module, inputting raw AIS data files, which are files containing all received messages in a single day. The decoder module processes these files in three steps: it first checks if the message is type 1, 2, 3, or 5, if so, decodes it, and then stores it in a separate file – either in a static data file (for type 5 messages) or in the motion data file for the MMSI referenced in the message.

### 2.3 *Voyages extraction*

The Voyager Module extracts voyages from the MMSI data files outputted by the Decoder module. Here a voyage is a set of datapoints (each obtained from an AIS message type 1,2, or 3) corresponding to a period of continuous motion by one vessel. As such, the Voyager module starts with a new empty voyage, reads the datapoints (one at a time) in the MMSI data file, and adds it to the voyage, until the voyage ends.

The module finishes a voyage if with the next datapoint the voyage has a period at least one hour of datapoints with low or no speed (i.e. SOG under two knots) or if the next datapoint is more than sixty minutes later than the current one. The first condition aims to have voyages finishing at ports and the latter prevents voyages from having unacceptable time gaps, which is an issue in AIS messages transmitted far from the AIS based receiver stations. Once a voyage is closed, if there are still datapoints left (as long as their SOG is over two knots), the Voyager module starts a new voyage and keeps adding datapoints until the end conditions are achieved. The outcome of this module are files, each containing the datapoints of a single voyage, for a single vessel.

### 2.4 *Clustering*

In this study shipping routes are synonymous to voyage clusters and the Clustering Module identifies them. For every voyage there is an origin cluster and a destination cluster and if two voyages share both, then they belong to the same route.

The origin/destination clusters are found using a hybrid solution, i.e. port locations are defined in advance while the entry/exit points are clustered by the DBSCAN algorithm.

### 2.5 *Cleaning*

There are several cleaning tasks throughout the process, and they perform different functions according to the step where they are applied. There is a balance to be attained between allowing erroneous datapoints and deleting valid voyage information. Cleaning is performed as needed, and it is thus, discussed in terms of the specific issues it seeks to correct.

The Decoder Module suffers perhaps the most extensive cleaning of all. The first issue is geographic related. If a vessel is far from the coast, the AIS data tend to be erratic and imprecise with gaps in the messages. The second issue concerns the flow to and from Spanish waters, which is covered in this analysis. To fix both issues, the study is limited to a geographic area, with longitudes between $-11.4°$ and $-6.4°$ and latitudes between $35°$ and $42°$.

Another issue solved by cleaning is the removal of non-cargo vessels. The MMSI data file is purged of any messages relating to MMSI not corresponding to vessel types between 60 to 68, 70 to 74, or 80 to 84 (AIS codes for passenger, cargo and tanker vessels, respectively).

An important issue is related to the data itself. Messages type 1,2, and 3 are reliable but can exhibit some errors. The most common is impossible data, i.e. speeds above 50 knots or incompatible headings. These are usually a small fraction of the messages and are summarily deleted. Cleaning is applied also to delete small and not complete voyages.

There are also data irregularities in some trajectories such as datapoints with position information inconsistent with the previous datapoint given the course and speed registered. In this context, two consecutive positions reports are consistent if the distance d between positions is less or equal to $d_{max}$, with $d_{max} = 3 \times SOG_i \times \Delta t + 0.01$ if $SOG_i < 3$ and $d_{max} = 2 \times SOG_i \times \Delta t$ otherwise, being $SOG_i$ the SOG in knots of a previous report and $\Delta t$ the time variation in hours between them, which cannot be greater than 5 minutes. Also, they are consistent if the difference between the COG computed using the information from the previous report and the actual value for the COG registered in the next datapoint is not greater than 10 degrees and, if the SOG does not suffer a large variation for a small $\Delta t$. These threshold parameters are totally empiric and have been successfully tested with the dataset used, although further investigation may result in some adjustments on the values. To clean those irregularities, a Cleaner Module is developed to:

- Eliminate datapoints with COG greater or equal to 360.0 or with SOG greater or equal to 102.3, because it means there is no information available for those parameters and higher values are not allowed.
- Split the trajectory into consistent sequences;
- Evaluate those sequences by checking against each other for overlapping. If they do not overlap,

they are put together and then the set of non-overlapping consistent sequences that has a maximum time span is kept, while the other sequences that are not on the set are eliminated.

### 2.6 *Voyages compression*

After grouping the voyages by shipping routes and filtering any position errors, each voyage is simplified by the Compression Module using the DP algorithm followed by the SW algorithm.

In this study a DP threshold parameter of $0.8 \times L$, where $L$ is the ship's length in nautical miles is adopted. Also, a course variation of more than 5 degrees is considered as a waypoint applied to the SW algorithm.

Figure 2 illustrates the main steps of the DP algorithm. It starts by connecting the starting point $p_0$ and the ending point $p_{11}$ to generate an approximated line segment, represented by the long dash line. Then, considering the Euclidean distance of the remaining points from the original trajectory, represented by the dotted solid line, to the approximated line segment, the algorithm ignores the ones that are smaller than the pre-defined compression threshold parameter $\tau$ and it considers and adds the furthest one to the approximated segment line, like shown in Figure 2a. After this first step, the approximated trajectory is now composed by two-line segments, $p_0p_7$ and $p_7p_1$, and are treated independently. For each line segment the algorithm repeats the process and adds the furthest point to the approximated trajectory, like shown in Figure 2b. This is an iterative process that repeats until every point of the original trajectory is within the pre-defined threshold, as illustrated in Figure 2c.

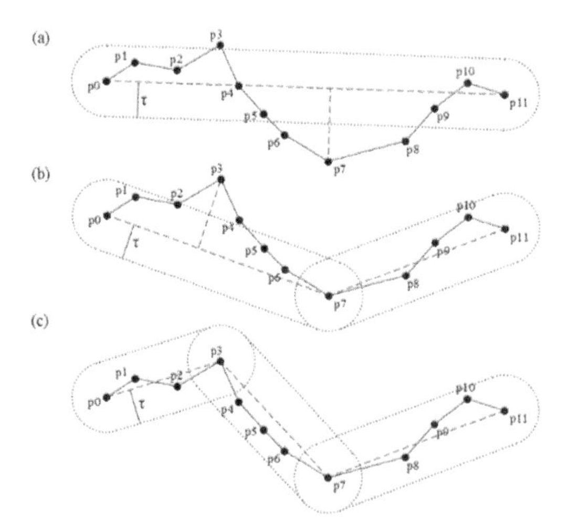


Figure 2. Illustration of the main steps of the Douglas-Peucker algorithm, adapted from Zhang et al. (2016).

### 2.7 *Statistical analysis*

Having obtained the corresponding simplified voyages, a route can be characterized by the following elements:

- the mean route and its boundary;
- lateral distance distribution (LD) along the route;
- distributions of SOG along the route;
- probability intervals of LD and SOG along the route (90% probability intervals);
- the turning areas;
- vessel behavior in the turning areas.

The route is divided into gates that could be evenly spaced (one gate every 10% of the route length, for example) or can be tactically placed at points of interest, for instance, before or after a Traffic Separation Schemes (TSS).

The mean route is determined by a cubic spline of the mean trajectory points along a route. These mean points are simply the mean of the geographical coordinates of all datapoints of a route within a certain gate.

The lateral distance datapoints represent the geographical distance between them and the mean route. Both the lateral distance and SOG distributions along the route are calculated at these gates.

At each gate, the Kernel Density Estimation (KDE) and Gaussian Distributions (GD) are adopted to describe the uncertainty on the lateral distance and SOG datapoints. For each distribution, the 90% probability intervals are calculated from the corresponding 5% and 95% percentiles of the variables.

The route boundary is then defined by a geographical area polygon corresponding the previously calculated 90% probability intervals.

The turning areas are locations where major trajectory directional changes occur, and they can be easily identified once all the trajectory voyages belonging to the same route are plotted. These areas are typically found when approaching to land and moving away from it, when preparing to enter and exit the TSS, and when inside them.

Both the route sections and the turning areas analyses use the KDE, defined in Eq.(1), to better represent the distribution of the boundaries distance and SOG of the AIS data.

$$\hat{f}(x) = \frac{1}{nh} \sum_{i=1}^{n} \phi\left(\frac{x - x_i}{h}\right) \tag{1}$$

where $n$ is the number of points of the sample and $h$ the bandwidth of the kernel function.

## 3 RESULTS

This section presents the results of applying the methodology to a route connecting the ports of Lisbon and Leixões. The raw AIS data are from the period between the 9th of June and the 9th of July of 2008. The route model consisting of the mean route and its boundary are presented first, following by the modeling of the SOG along the route, and finally the ship's behavior at the turning areas. The route is divided into eleven evenly distributed gates, as shown in Figure 3.

### 3.1 *Route model*

This route has been taken by 19 distinct vessels in 29 voyages, mostly cargo vessels. On average there are 7391 datapoints throughout this route and the compression algorithms reduce them to 33 points without losing its initial shape, approximately 0.44% of the initial sample size. Figure 3 shows the compressed version of each trajectory, together with the mean route extracted from the original trajectories and Figure 4 shows the route boundaries.

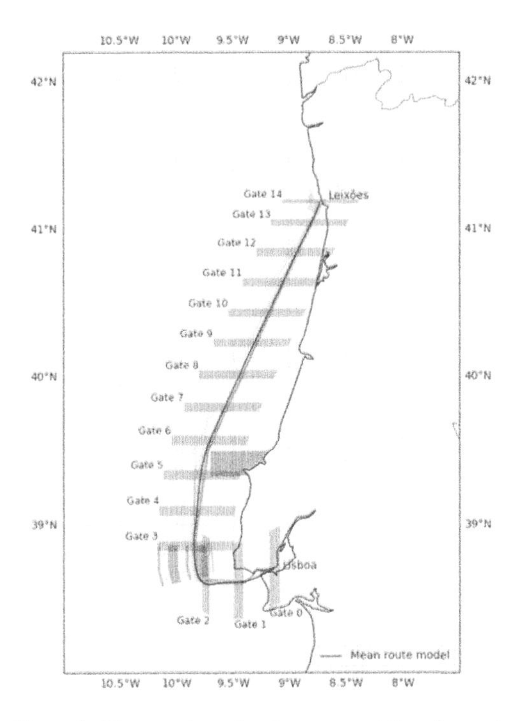


Figure 3. Mean route and gates representation for route Lisbon-Leixões.

### 3.2 *Lateral distribution along the route*

At each gate, the GD and KDE distributions are used to describe the lateral uncertainty of the ship trajectories. Figure 5 shows the two lateral density functions at gate zero, located at 1% of the route length. Although KDE is more flexible, the Gaussian distribution has little deviation from the KDE and seems to be a good model for describing the ship lateral distance.

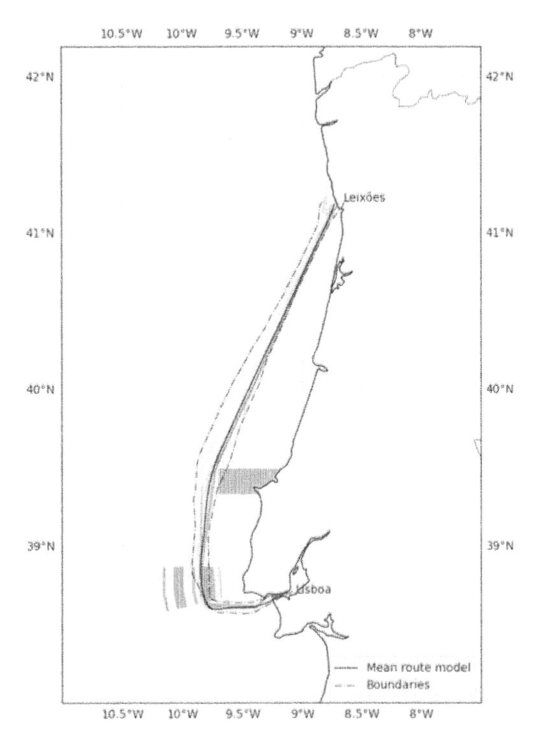


Figure 4. Mean route and boundary representation for route Lisbon-Leixões.

Figure 6 and Figure 7 show, respectively, the 95% and 5% percentiles of the ship lateral distance at the route gates calculated from the GD and KDE models. It can be seen that the KDE distribution is shifted and tend to predict higher upper bounds (95% percentiles) and lower bounds (5%) than those derived by the Gaussian distribution. However, both models result in similar widths for the route boundary.

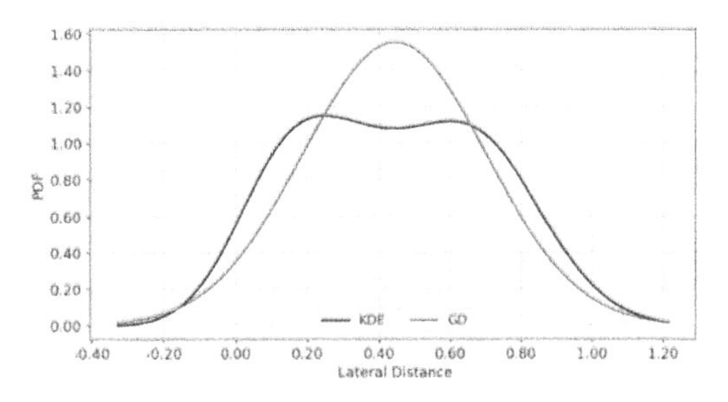


Figure 5. Lateral distributions at gate 0 of route Lisbon-Leixões.

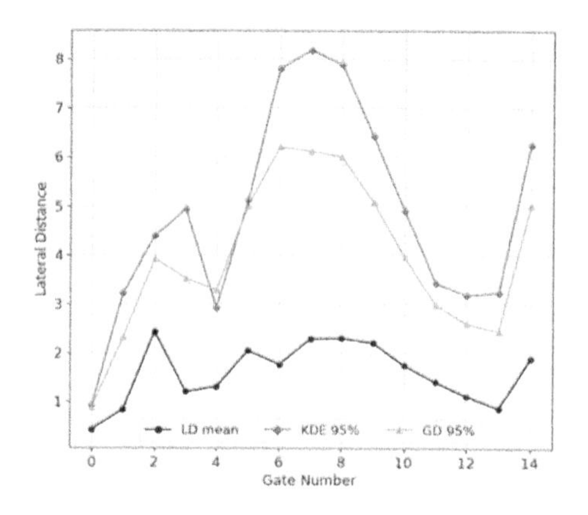


Figure 6. LD's 95% percentiles from the GD and KDE models.

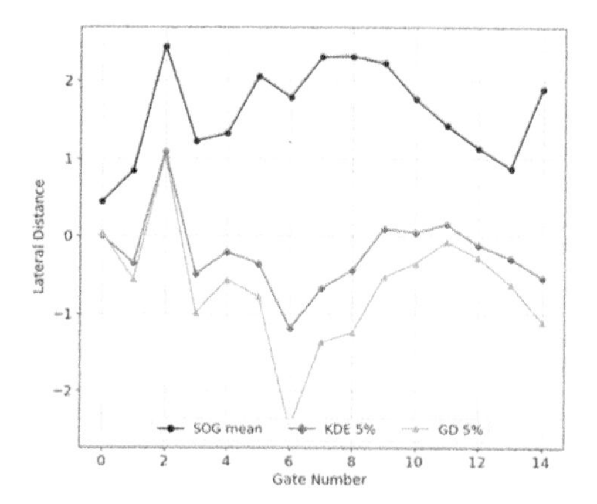


Figure 7. LD's 5% percentiles from the GD and KDE models.

### 3.3 *Speed of ground along the route*

The same analysis is conducted for the speed of ground of the ships at the route gates. Figure 8 and Figure 9 show both models fitted to the ships' SOG at route gates 0 and 4, respectively. Figure 10 and Figure 11 show the 95% and 5% percentiles of the ships' SOG along the route gates. The figures show that both models predict similar SOG upper bounds (95% percentiles), but the KDE model tends to estimate lower 5% percentiles, reflecting its better capability for representing the lower tail of the SOG distributions, as illustrated in Figure 8 and Figure 9. Moreover, the SOG shows to be steady throughout the route.

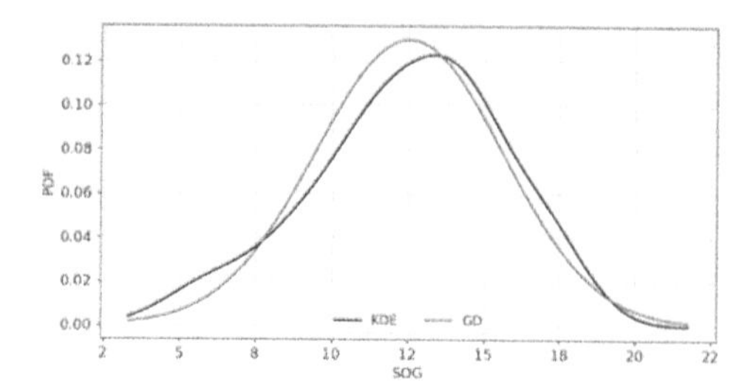


Figure 8. SOG distributions at gate 0 of route Lisbon-Leixões.

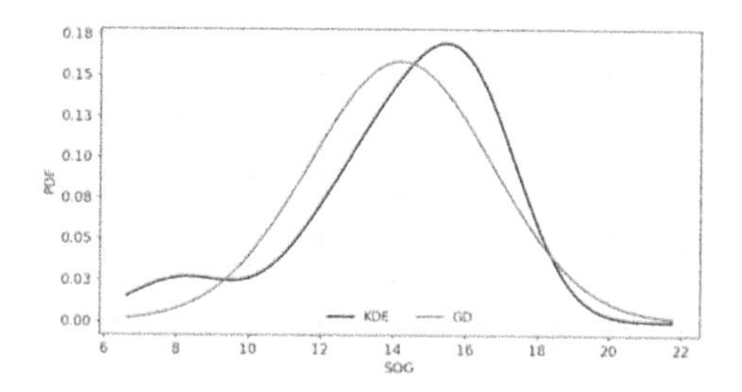


Figure 9. SOG distributions at gate 4 of route Lisbon-Leixões.

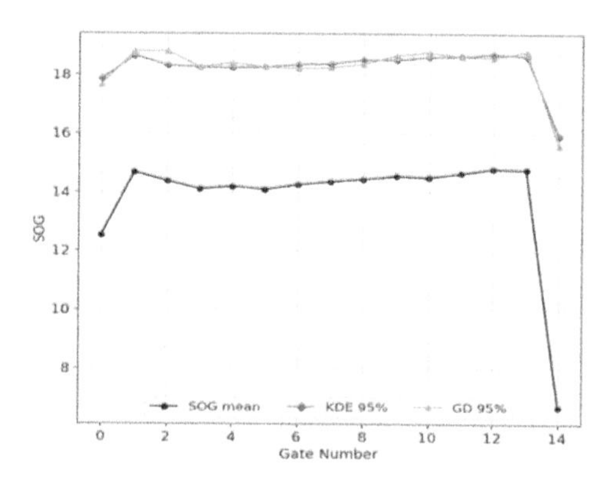


Figure 10. SOG's 95% percentiles from the GD and KDE models.

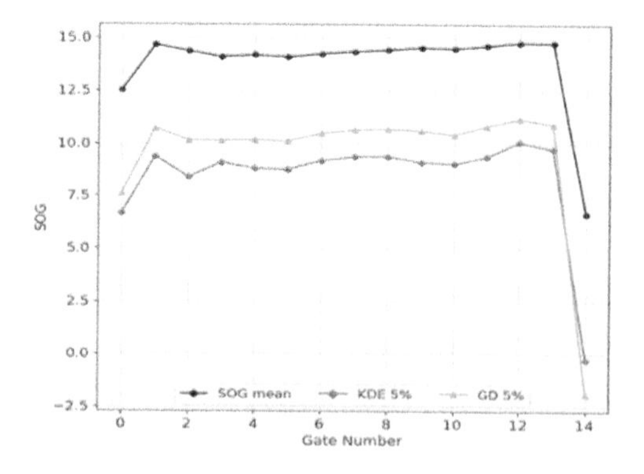


Figure 11. SOG's 5% percentiles from the GD and KDE models.

### 3.4 *Analysis of the turning areas*

Regarding the turning areas, as it was mentioned before, they can be easily identified when analysing Figure 3. It is possible to identified six turning areas, numbered from A to F, as represented in Figure 12, where most of the vessels tend to change their course.

To analyze the spatial distribution of the waypoints in each turning area, a 2D KDE method is applied to those areas. Each area is discretized into a cell grid and the spatial density distribution for each cell is computed using the 2D KDE method. In Figure 13 is represented the spatial density distribution for the six turning areas, that characterize the regions where most course changes tend to happen.

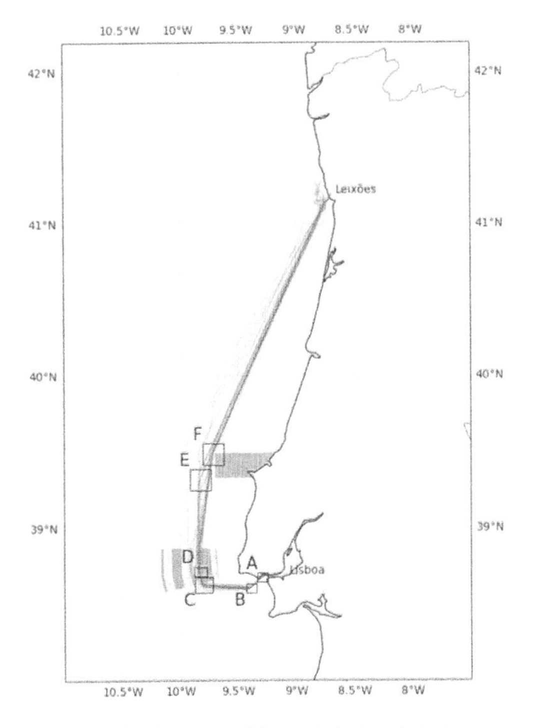


Figure 12. Turning areas of the route Lisbon-Leixões.

Another parameter of interest for caracterizing the ship's behavior in those turning areas is the SOG. For each turning area a speed distribution is computed using the KDE method, as shown in Figure 14.

Turning area C is the area that registered more waypoints followed by turning areas A and B. Starting from turning area A, a mean SOG value of 14.75 knots is registered with a maximum value of 20.8 knots resulting from the currents' velocity in the river channel. From turning area A to turning area B, the SOG distribution curve in Figure 14 shifts to the right side as expected since this area is where the ships tend to increase their velocity when leaving the port of Lisbon. Consequently, the SOG mean value in turning area B also increases, registering a value of 15.01 knots. Passing turning area B the SOG values decrease, registering a mean value of 14.17 knots in turning area C and the distribution curve

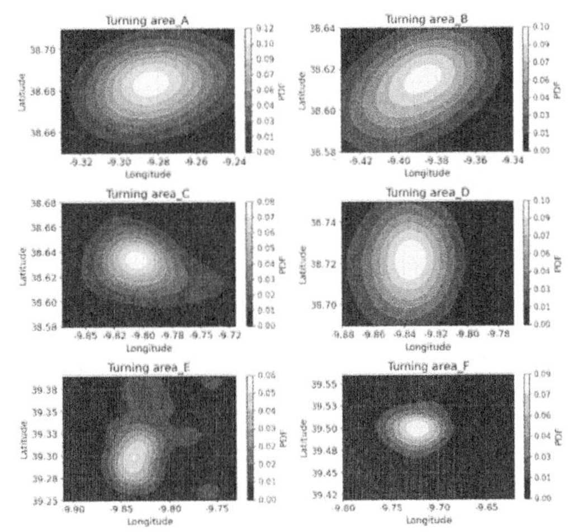


Figure 13. Spatial density distributions for the main turning areas.

shifts to the left. Turning area C is located at the TSS entrance and is where major course changes occur. All these and the values for the remaining areas can be found in Table 1, where a statistical summary of the each distribution can be found.

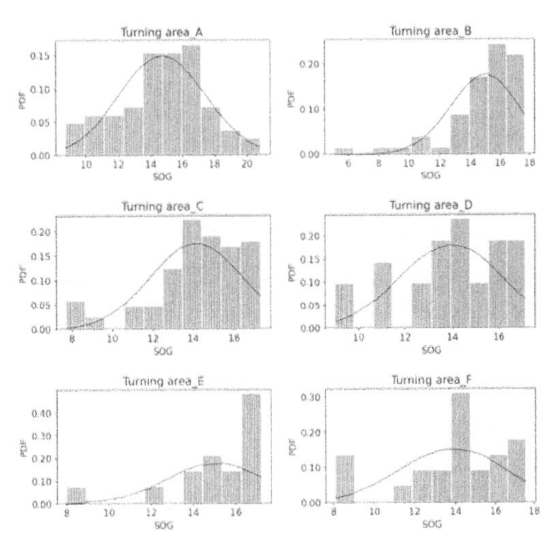


Figure 14. Ship speed distribution in the turning areas.

Table 1. Ship speed statistical summary of each turning area.

| Turning Area | Nº data points | Min | Max | Mean | Std |
|---|---|---|---|---|---|
| A | 71 | 8,70 | 20,80 | 14,75 | 2,68 |
| B | 66 | 5,10 | 17,70 | 15,01 | 2,28 |
| C | 94 | 7,70 | 17,30 | 14,17 | 2,27 |
| D | 26 | 9,00 | 17,20 | 13,98 | 2,22 |
| E | 16 | 8,00 | 17,20 | 15,11 | 2,29 |
| F | 24 | 8,10 | 17,60 | 13,94 | 2,65 |

## 4 CONCLUSIONS

This paper presents a methodology to characterize the maritime traffic along the Portuguese coast. To accomplish this, several algorithms are developed to decode AIS raw data, construct ships' trajectories, group them in routes, filter erroneous datapoints, simplify those trajectories and analyze them. A specific route is adopted to test and show the applicability of the algorithms developed.

In general, the trajectories are well reconstructed and grouped. The empirical threshold values applied to clean the erroneous datapoints as well as the DP and SW thresholds used for voyages compression work relatively well in this case study. The route model developed provides a good description of the geographical distribution of the route's voyages.

Lateral and SOG distributions at gates along the route are stablished, which have the potential to be used for ship abnormal behavior detection. The SOG shows to be steady throughout the route, but the lateral distributions tend to be wider in the middle gates of the route.

When analyzing the spatial distribution of the waypoints in each turning area, a specific region is identified as the area where most ships tend to change course. Relatively to the SOG in those areas, the mean value for each turning area do not vary significantly. It is important to notice that the dataset of route Lisbon-Leixões is relatively small, hence the values obtained in the turning areas may not be representative, requiring confirmation from a lager sample.

The of algorithms developed in this paper are able to characterize individual maritime routes, as shown by their application to the Lisbon to Leixões route, and can be generalized to any number of routes, making them a suitable tool to fully characterize the maritime traffic off the continental coast of Portugal.

## ACKNOWLEDGEMENTS

Paper contributes to the project "Monitoring and Surveillance of Maritime Traffic off the continental coast of Portugal (MoniTraffic)", which has been co-funded by "Fundo Azul" programme of the Portuguese Directorate-General for Maritime Policy, under contract no. FA-04-2017-005.

This work also contributes to the Strategic Research Plan of the Centre for Marine Technology and Ocean Engineering (CENTEC), which is financed by the Portuguese Foundation for Science and Technology under contract UIDB/UIDP/00134/2020.

## REFERENCES

de Vries, G. K. D., & Someren, M. van. (2012). Machine learning for vessel trajectories using compression, alignments and domain knowledge. *Expert Systems with Applications*, 39, 13426–13439.

Douglas, H.D. and Peucker, K.T., 1973. Algorithms for the Reduction of the Number of Points Required to Represent a Digitized Line or its Caricature. *The International Journal for Geographic Information and Geovisualization*, 10 (2), 112–122.

Etienne, L., Devogele, T., and Bouju, A., 2010. Spatio-Temporal Trajectory Analysis of Mobile Objects Following the Same Itinerary. *The International Archives of the Photogrammetry, Remote Sensing and Spatial Information Sciences*, 38 (II), 86–91

IMO. (2002). *Guidelines for the Onboard Operational Use of Shipborne Automatic Identification Systems (AIS)*, International Maritime Organization (IMO).

ITU, 2014. Technical characteristics for an automatic identification system using time division multiple access in the VHF maritime mobile frequency band, M.1371-5, International Telecommunication Union (ITU).

Liu, J., Li, H., Yang, Z., Wu, K., Liu, Y., and Liu, R.W., 2019. Adaptive Douglas-Peucker Algorithm with Automatic Thresholding for AIS-Based Vessel Trajectory Compression. *IEEE Access*, 7, 150677–150692.

Pallotta, G., Vespe, M., and Bryan, K., 2013. Vessel pattern knowledge discovery from AIS data: A framework for anomaly detection and route prediction. *Entropy*, 15 (6), 2218–2245

Rong, H., Teixeira, A. P., & Guedes Soares, C. (2020). Data mining approach to shipping route characterization and anomaly detection based on AIS data. *Ocean Engineering*, 198, 106936.

Rong, H., Teixeira, A.P., and Guedes Soares, C., 2022. Maritime traffic probabilistic prediction based on ship motion pattern extraction. *Reliability Engineering and System Safety*, 217, 108061.

Silveira, P., Teixeira, A.P., and Guedes Soares, C., 2013. Use of AIS Data to Characterise Marine Traffic Patterns and Ship Collision Risk off the Coast of Portugal. *Journal of Navigation*, 66 (06), 879–898

Vespe, M., Visentini, I., Bryan, K., and Braca, P., 2012. Unsupervised learning of maritime traffic patterns for anomaly detection. *9th IET Data Fusion & Target Tracking Conference (DF&TT 2012): Algorithms & Applications*, (August 2015), 14–14.

Zhang, S.K., Liu, Z.J., Cai, Y., Wu, Z.L., and Shi, G.Y., 2016. AIS Trajectories Simplification and Threshold Determination. *Journal of Navigation*, 69 (4), 729–744.

Zhang, S., Shi, G., Liu, Z., Zhao, Z., and Wu, Z., 2018. Data-driven based automatic maritime routing from massive AIS trajectories in the face of disparity. *Ocean Engineering*, 155 (1550), 240–250.

Zhao, L., Shi, G., & Yang, J. 2018. Ship Trajectories Pre-processing Based on AIS Data. *The Navigational Journal*, 71(05), 1–21.

# Preliminary analysis of the fishing activity in Portugal

E. Lotovskyi & A.P. Teixeira
*Centre for Marine Technology and Ocean Engineering (CENTEC), Instituto Superior Técnico, Universidade de Lisboa, Lisbon, Portugal*

P. Silveira
*Centre for Marine Technology and Ocean Engineering (CENTEC), Instituto Superior Técnico, Universidade de Lisboa, Lisbon, Portugal*
*Escola Superior Náutica Infante D. Henrique (ENIDH), Paço de Arcos, Portugal*

E. Torrão
*Escola Superior Náutica Infante D. Henrique (ENIDH), Paço de Arcos, Portugal*

ABSTRACT: Despite more than 3000 national fishing vessels land daily in Portuguese waters, the fishing activity and the spatial-temporal patterns of this fleet are still poorly characterised and documented. This poses some difficulties to a rational approach for maritime operation planning, control, monitoring, safety and security of the fishing activity. This paper conducts a preliminary analysis of the fishing activity in Portugal. First, the most used fishing gears and a portrait of the Portuguese licensed fishing fleet are presented. Current monitoring systems used to support control and surveillance tasks are described and their limitations discussed. The trajectories of fishing vessels in a selected area are then analysed using Automatic Identification System data. The vessels' trajectories are grouped by practiced fishing gear and a preliminary characterization of the phases and of the motion patterns per group is conducted.

## 1 INTRODUCTION

Portugal has a long tradition of fishery activity at sea representing the third highest seafood consumption per capita in Europe (FAO, 2020). In recent years, Portugal produced 190 thousand tons of fisheries and aquaculture (FAO, 2020) and 264 million euros of gross value added (INE, 2018). Thus, the monitoring of fishing vessels is crucial for economic impacts assessment (Campos et al., 2021), for fishing effort estimation (Zhao et al., 2021; Russo et al., 2020; Campos et al., 2018), for demographic analysis of the species population (Pilar-Fonseca et al., 2014; Gerritsen et al., 2012), for fishing activity vigilance (Rowlands et al., 2019; Ferreira et al., 2017), and for monitoring of marine environment (Liu & Tsai, 2011, Franklin 2008).

The correct monitoring requires knowledge on the different operational phases of a fishing voyage (Lopes et al., 2018). Unfortunately, there is a lack of information in the literature about fishing vessel methods and strategies, that makes it difficult to assess accurate fishing patterns and in estimating fishing effort (Campos et al., 2018).

In order to track and monitor fishing vessel in activity, European Union (EU) implemented the Vessel Monitoring System (VMS) (Ferreira et al., 2017). VMS is a satellite-based monitoring system, which at regular time intervals collects information for fisheries authorities, such as: date and time, geographical coordinates (i.e. latitude and longitude), Speed Over Ground (SOG), and Course Over Ground (COG). VMS is mandatory for all vessels with length overall (LOA) equal or above 12 meters, however, due to existing derogations, the system is available for ships with 15 m or above (Lopes et al., 2018). In Portugal, fishing vessels transmit VMS messages typically every two hours by mobile satellite communications (EC, 2009b). The discrepancy between temporal scale of fishing activity and VMS time stamp may cause discrepancies in operational phases identification within fishing trip, leading to errors in fishing pattern characterization, as reported by Katara & Silva (2017) and Lopes et al. (2018) for seine and trawl fisheries, respectively.

Automatic Identification System (AIS) is a shoreand satellite-based automatic tracking system that can be used in fishing pattern characterization. AIS broadcasts every few seconds via a VHF transceiver 4 types of information (IMO, 1998): static data (e.g. IMO number, name, size and type of the vessel), dynamic data (e.g. geographical position, SOG, COG, rate of turn (ROT), heading), voyage

DOI: 10.1201/9781003320289-13

related data (e.g. draught, cargo, destiny), short safety-related messages. Initially, AIS was implemented as navigational aid to avoid vessel collisions only for large ships and all passenger vessels. Today, any EU flagged fishing vessel with LOA of more than 15 m must have an AIS transponder onboard, too (EC, 2011).

Comparing AIS with VMS, the first represents some weaknesses. Unlike VMS, AIS system can be, intentionally or not, turned off (Serra-Sogas et al., 2021). In some areas, AIS has low spatial coverage of both satellite and shore based AIS receivers (Shepperson et al., 2018). In addition, AIS messages are send with irregular intervals (Campos et al., 2018). Notwithstanding the shortcomings, the use of AIS in scientific and management applications is increasing due to its high temporal resolution and ease of data access (Russo et al., 2020).

AIS and VMS do not provide the information about the vessel status (underway/fishing phase), thus the data must be analysed and processed to obtain operational fishing trajectory patterns. During the voyage, a fishing vessel may sail, search, fish or just drift. To identify the operational phase, the sequential change in vessel speed is assessed. Bastardie et al. (2010) use the segmented regression of speed to determine automatically break points of the different fishing phases. However, the methodology is not suited to gillnet and seiners. Hintzen et al. (2012) developed an open-source software that uses logbook and VMS data as input variables to distinguish fishing from other activities, but a more generic method across fisheries is missing. Campos et al. (2021) and Russo et al. (2020) used pre-defined speed ranges to characterize the fishing activity patterns. However, because the speed ranges indicate different values for the same fishing phases, establishing reliable spatial-temporal patterns of fishing activity is challenging with these speeds.

The main objective of this paper is to provide a preliminary analysis of the fishing activity in Portuguese waters. For this purpose, a portrait of the Portuguese licensed fishing fleet is presented first. The most used fishing gears and their operational phases are described. Fishing vessels are analysed per fleet segment and not as individual part; thus, the study is divided in two assessments: trawl fishing and purse seine fishing. Finally, a preliminary characterization of the phases and of the motion patterns per group at a selected area using AIS data from the Portuguese Vessel Traffic Service control centre (CCTMC) is conducted.

## 2 PORTUGUESE FISHING FLEET PORTRAIT

On the 9th of July, 2021, the Portuguese fishing fleet consisted of 3146 licensed vessels distributed through the Continent (32 ports), the Autonomous Regions of the Azores (4 ports), and Madeira (1 port) (DGRM, 2021). In mainland Portugal, the largest distribution of ships by port is concentrated in the Centre of the country, where 30.3% of the total fleet is distributed across 5 ports (see Figure 1). In turn, Algarve is the area of mainland Portugal with the largest number of ports where fishing vessels are registered. The smallest number of fishing vessels are in Madeira, only 0.2% of the total fleet.

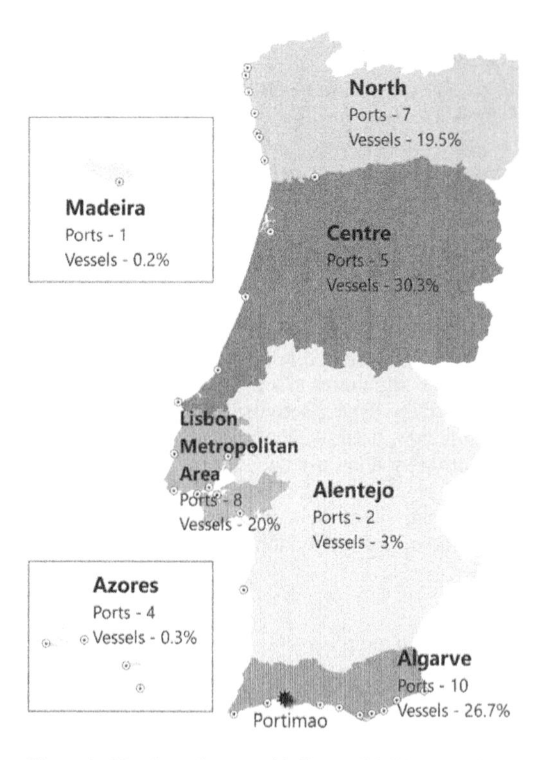


Figure 1. Number of ports with licensed fishing vessels by regions of Portugal.

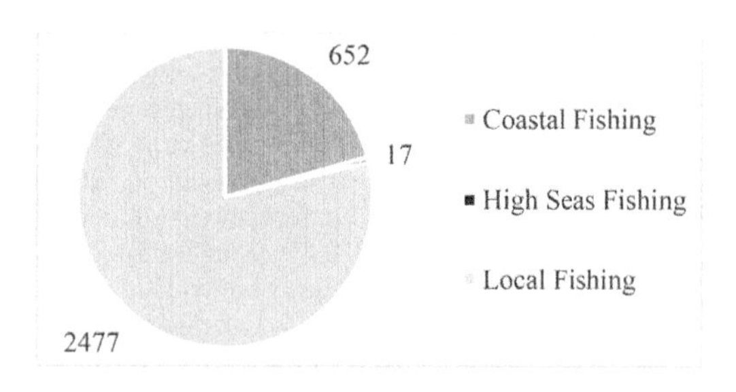


Figure 2. Number of licensed fishing vessels per category.

The Portuguese fishing fleet is divided into three categories (DGRM, 2021):

- Local fishing – small vessels that operate in oceanic and inland waters. LOA is less than 9 m and maximum propulsion power is 75 kW (i.e. 100 hp).

- Coastal fishing –vessels with a propulsion power of 26 kW or more (i.e. 35 hp). LOA ranged between 9 m and less than 35 m.
- High Seas fishing – vessels operating beyond 12 nm having a tonnage capacity of more than 100 Gross Tonnage (GT) and a minimum autonomy of 15 days.

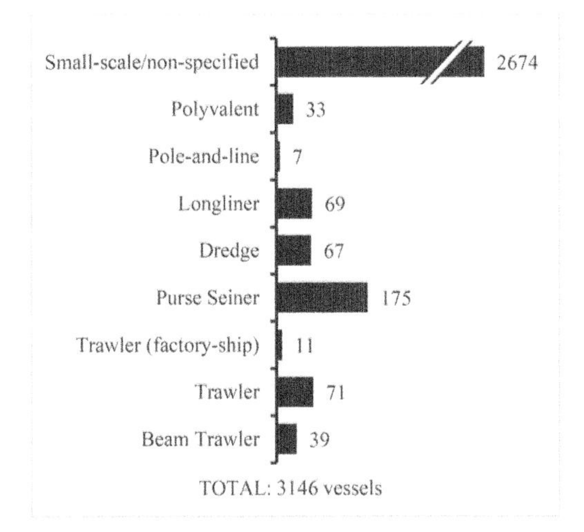


Figure 3. Number of fishing vessels by fishing gear.

A major part of the licensed fishing vessels belongs to Local fishing category representing almost 79% of the fleet, while Coastal fishing has 20%, and High Seas fishing has only 17 vessels (see Figure 2). The average age of the registered fleet is around 33 years and, in terms of licensed fleet, around 24 years. The Portuguese fishing fleet is characterized by a high proportion of small-scale fishing vessels, with over 90% of them with LOA less than 12 m and a low GT. Figure 3 shows the number of licensed vessels by fishing gear.

## 3 FISHING ACTIVITY DESCRIPTION

To define operational patterns, the fishing vessels are analysed per fleet segment. Due to variability of fishing effort and of stock availability, the same spatial-temporal factors may occur inside a given fleet segment (Pilar-Fonseca et al., 2014). Thus, the study is divided in two assessments: trawl fishing and purse seine fishing.

### 3.1 *Trawl fishing*

Trawling is the fishing method in which a cone-shaped body of netting (i.e. trawl) is towed behind one or two vessels (i.e. trawler) to catch the organisms that are in the path of the net (He et al., 2021).

Depending on the net position in the water column, two types of trawling are distinguished: bottom trawls and midwater trawls. In the bottom trawls, the net is towed on the seabed and designed to catch the demersal species (e.g. crustacean, flatfish). In the midwater trawls, the net is towed in mid- and surface water to catch pelagic or semi-demersal fish.

In Portugal, trawlers operate along the entire mainland coast with some restrictions (DGRM, 2021). Except for some small vessels that can operate inside the 6 nm limit, as long as outside the limit between Cabo Raso and Cabo Espichel, trawling is only authorized beyond 6 nm from the coast. Trawlers typically operate between the 6nm distance limit and the 200 m bathymetric curve, while bottom trawlers mostly operate up to 1000 m depths.

A trawl fishing voyage is composed by the following sequence:

- Underway – navigation from port to the fishing grounds.
- Searching – fish aggregation seeking and evaluation of its catchability.
- Shooting the net in water column either with bottom contact or in midwater.
- Trawling haul – towing of the net towards the school of fish or crustaceans.
- Hauling back the trawl.
- Either searching for a new fish school or proceeding back to port, not necessary the same as the initial one.

### 3.2 *Purse seine fishing*

Purse seine fishing is a method that employs a long wall of netting (i.e. seine) that is deployed vertically in the water to encircle shoaling pelagic species near the surface (He et al., 2021). Top edge of the net (i.e. headrope) is made of numerous floats; while the bottom edge (i.e. footrope) has weights to increase the sinking velocity and purse rings through which a purse line passes, allowing the net to be closed like a purse and retaining the fish collected. This fishing is operated by a single vessel (i.e. seiner or mother ship) with or without an auxiliary skiff.

In Portugal, purse seine is authorized out of 0.25 nm away from the coast line and, between 0.25 nm and 1 nm, only at depths greater than 20 m (DGRM, 2021).

The following operations constitute a purse seine fishing voyage:

- Underway – navigation from port to the fishing grounds.
- Searching – fish aggregation seeking and evaluation of its catchability.
- Encircling of a target catch. In this study, encircling is realised by mother ship, while a support skiff is launched to the water with first end of the seine net prepared for shooting.
- Hauling back the purse seine.
- Underway – seiner sails back to any port.

## 4 METHODOLOGY

AIS messages of type 1, 3 and 5 were used to assess fishing vessel operational patterns. AIS message type 5 is Class A ship static and voyage related data that is broadcasted every 6 min (ITU, 2014) and provides the information regarding MMSI number and ship type. AIS messages type 1 and type 3 are position reports that broadcast the information (e.g. MMSI number, SOG, COG, geographical coordinates) every 2 to 10 s while underway, and every 3 min while at anchor. Besides, these messages contain the information regarding navigation status (e.g. underway using engine, at anchor, engaged in fishing), ROT, and heading, too. However, due to the class of equipment, sensor availability and quality of user input, these data are not always accurate, as in the case of navigation status, or even absent, as the case of ROT and heading.

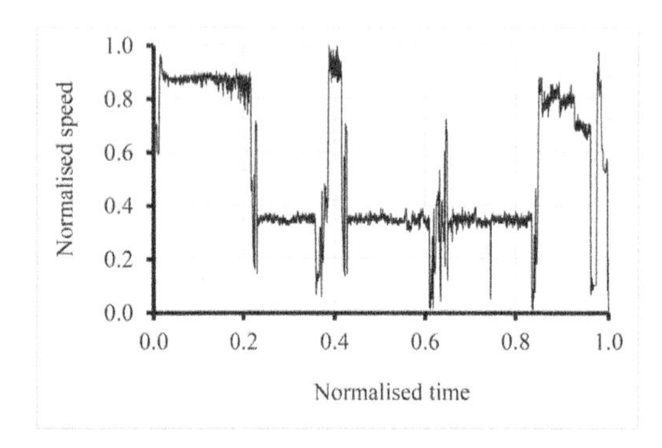


Figure 4. Example of trawling speed extracted from AIS data.

The data were collected between 07.05.2019 and 19.08.2019, a period of 104 days. All AIS messages were decoded and inserted to an open-source relational database MySQL (MySQL, 2021), where all data were processed and analysed.

Operational patterns of fishing vessels are analysed in a selected area. In this study, only the vessels operating close to Portimão were considered (see Figure 1). For this purpose, first, the geographic square window of Portimão was defined. Then, all fishing vessels that cross this window at least once were registered. As result, 26 ships were obtained: 12 trawlers, 12 seiners, and 2 polyvalent fishing vessels. Small fishing vessels with LOA less than 15 m are not included, because AIS is not required for this type of vessels (EC, 2011). Please note, the use of DGRM ´s database (DGRM, 2021), where the port of registry of each ship is shown, is not recommended to identify the vessels from Portimão, as the fishing may occur far from the port of registry (e.g. longliners).

The final criterion for defining the study sample is that the voyage's departure and arrival ports must both be in Portimão. As a result, 5 trawlers and 5 purse seiners are obtained. The LOA classification of trawlers is 24-40 m and of seiners is 18-24 m (EC, 2009a). Due to duration of trawl fishing being much longer that of purse seine fishing, the analysis considers 1 voyage per trawler and 2 voyages per seiner. All voyages are equally distributed through the available time frame of AIS data.

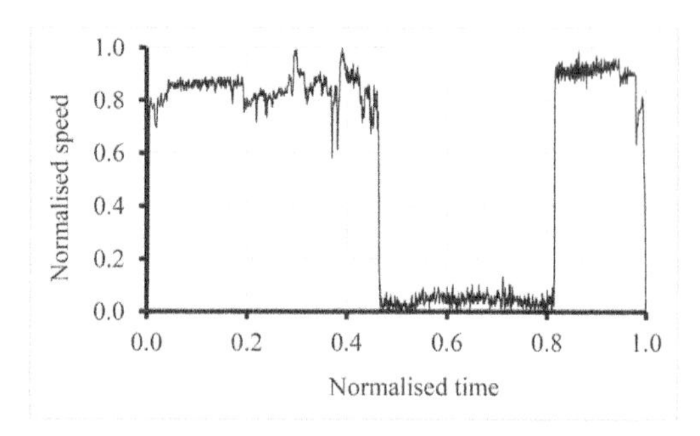


Figure 5. Example of purse seining speed extracted from AIS data.

The speed versus time plot of each voyage was retrieved in order to assess the operational patterns.

Figure 4 and Figure 5 represent examples for trawling and purse seining speed plots extracted from AIS data, respectively, where the speed is normalised by the maximum speed, and the time by the total voyage time.

To separate the spatial fishing activity of the individual voyage on distinct operational phases, the speed behaviour and geographic location are analysed. Rijnsdorp et al. (1998), Eastwood et al. (2007), Pilar-Fonseca et al. (2014) concluded that high speed corresponds to navigation to or between fishing grounds, while medium/low speed means the fishing operation. Besides, Deng et al. (2005) proposed to analyse the spatial relationships between successive position records. Through the open-source Geographic Information System QGIS (QGIS, 2021), it is possible to view the fishing trajectory point by point. The variation in the distance between two successive points on the trajectory allows to separate the speeds into more than two levels (i.e. navigation or fishing), thus allowing to distinguish all the fishing phases.

Speed as a function of time is measured to analyse the behaviour of fishing operational patterns. In each voyage, both variables speed and time are normalised using Equation 1,

$$x' = \frac{x - \min(x)}{\max(x) - \min(x)} \tag{1}$$

where $x =$ original value (i.e. speed or time); $\max(x) =$ maximum value of $x$; $\min(x) =$ minimum value of $x$; $x'$ = normalised value (i.e. normalised speed or normalised time).

It is possible to determine the average normalised speed in each fishing phase with the respective normalised time by converting every speed of the voyage into normalised values with Equation 2. Equation 3 is used to calculate the standard deviation (SD or $\sigma$), which measures the dispersion of data in comparison to the determined normalised average. Equation 4 uses the coefficient of variation (COV) to express the precision of the determined amount of dispersion.

$$\bar{x} = \frac{\sum_{i=1}^{N} x_i}{N} \quad (2)$$

$$\sigma = \sqrt{\frac{\sum_{i=1}^{N} (x_i - \bar{x})^2}{N}} \quad (3)$$

$$COV = \frac{\sigma}{\bar{x}} \quad (4)$$

where $x_i$ = observed values of the sample items; $N$ = size of the sample; $\bar{x}$ = mean value of the observations.

## 5 DISCUSSION AND RESULTS

### 5.1 *Trawl fishing*

Figure 6 Shows the average normalised speed as a function of the normalised time of the trawl fishing voyage. The result is obtained by analysing each speed plot of the trawling voyage and calculating the average values in each fishing phase. As can be seen, the underway phase occurs twice throughout the voyage. At the beginning of the trip, the underway phase corresponds to the departure from Portimão and arrival to fishing ground, while at the end of the plot, the underway phase corresponds to the opposite route.

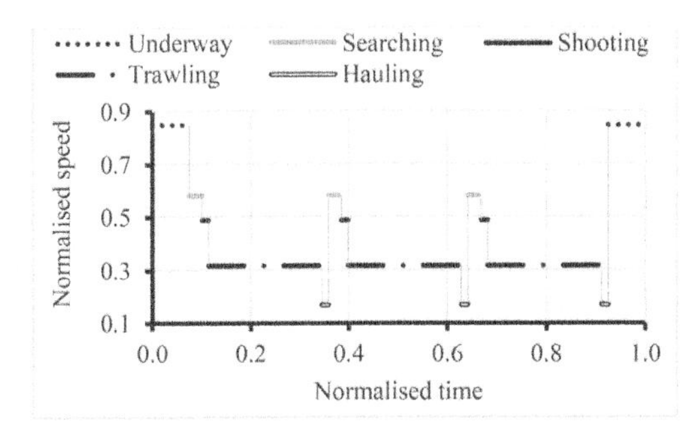


Figure 6. Average normalised speed as a function of the normalised time of the trawl fishing voyage.

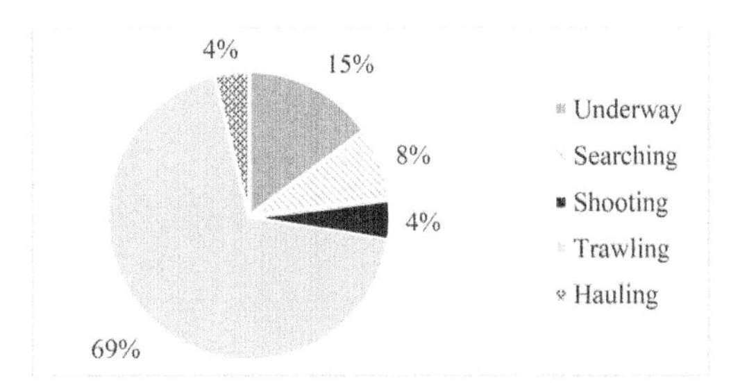


Figure 7. Percentage of time for each trawl fishing phase.

Please note, since the departure and arrival ports are the same, the total underway time has been split into two equal parts. In trawling, the shooting of the net occurs several times per voyage: between 3 to 10 times. In order to simplify the visual understanding of the results, only 3 sequences of searching-shooting-trawling-hauling are presented, although a vessel may have more fishing cycles.

#### 5.1.1 *Assessment of fishing operation time by phase*

From Figure 6, the time of each fishing phase is determined and summarised in Figure 7. As can be seen, the longest fishing phase is trawling representing 69% of the total voyage. Besides, trawler spent 8% of the time searching for fish aggregation that is half of the underway time. Thus, trawl vessels are sailing around 23% of the time at high speed.

Table 1 is obtained from the analysis of each fishing phases in absolute hours. The COV is low in underway phase, but SD is high in all other phases. This is because each analysed voyage has a different number of hauls. Thus, Table 2 presents the normalised time of searching-shooting-trawling-hauling cycle per haul. The COV for searching is high, too, while in other phases it significantly decreases. The duration of searching phase varies significantly because of the numerous unforeseen factors, such as, weather conditions, current, location, fishing aggregation and its catchability level. Thus, the COV of searching is 64%.

Table 1. Total duration in hours of each trawl fishing phase: (1) Underway; (2) Searching; (3) Shooting; (4) Trawling; (5) Hauling; (6) Total.

| | (1) | (2) | (3) | (4) | (5) | (6) |
|---|---|---|---|---|---|---|
| Mean | 4.41 | 2.62 | 1.77 | 26.40 | 1.64 | 36.83 |
| SD | 0.38 | 1.21 | 0.87 | 11.82 | 0.98 | 14.60 |
| COV | 0.09 | 0.46 | 0.50 | 0.45 | 0.60 | 0.40 |

#### 5.1.2 *Assessment of fishing operation speed by phase*

Figure 8 Shows the normalised speed per each fishing phase. In underway, shooting, and trawling haul the normalised speed is 0.85, 0.49, and 0.32, respectively, with low SD. This allows to state that the normalised

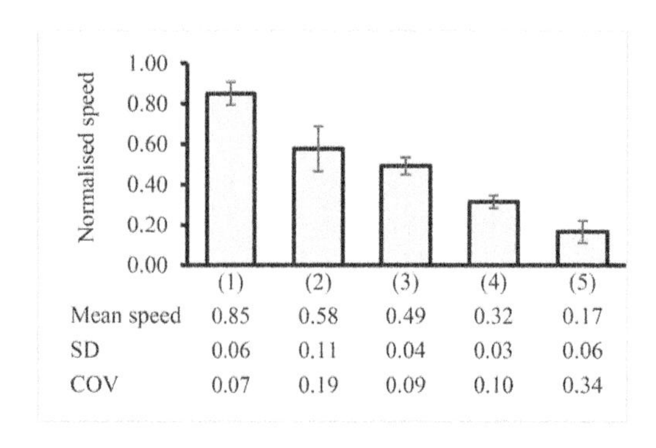


Figure 8. Normalised speed per each trawl fishing phase: (1) Underway; (2) Searching; (3) Shooting; (4) Trawling; (5) Hauling.

Table 2. Normalised duration of phases (in %).

| | Searching | Shooting | Trawling | Hauling |
|---|---|---|---|---|
| Mean | 0.10 | 0.05 | 0.80 | 0.05 |
| SD | 0.07 | 0.01 | 0.05 | 0.01 |
| COV | 0.64 | 0.28 | 0.06 | 0.26 |

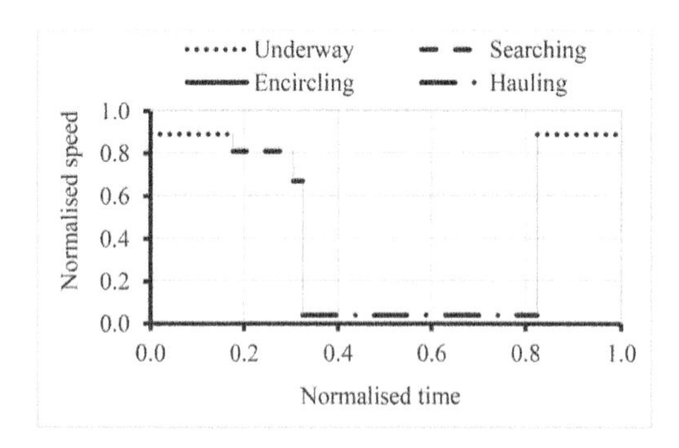


Figure 9. Average normalised speed as a function of the normalised time of the purse seine fishing voyage.

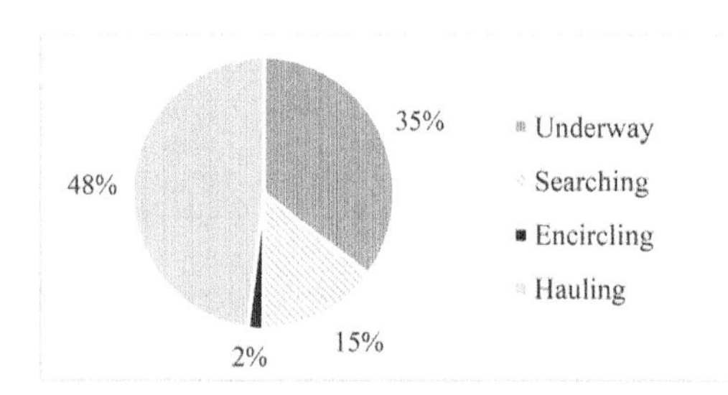


Figure 10. Percentage of time for each purse seine fishing phase.

speed is a good variable to distinguish different phases of fishing. Hauling has a high COV (i.e. 34%) as it presents the average normalised velocity close to zero (i.e. 0.17). In searching, COV of 19% is expected due to a significant variation in speed, while the vessel is exploring potential fishing grounds.

## 5.2 *Purse seine fishing*

Following the same procedure as in trawl fishing, Figure 9 shows the average normalised speed as a function of the normalised time of the purse seine fishing voyage. In this case, the underway phase occurs twice throughout the voyage, too. However, unlike in trawl fishing, in seiners the fishing catch occurs once per voyage, so in Figure 9 a single sequence of searching-encircling -hauling is presented.

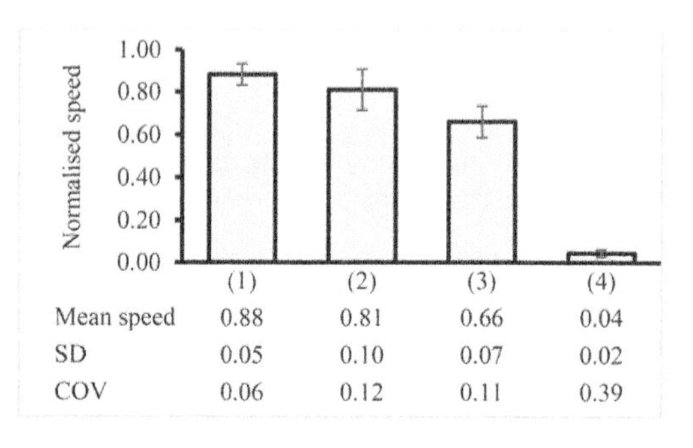


Figure 11. Normalised speed per each purse seine fishing phase: (1) Underway; (2) Searching; (3) Encircling; (4) Hauling.

Table 3. Duration in hours of each purse seine fishing phase: (1) Underway; (2) Searching; (3) Encircling; (4) Hauling; (5) Total.

| | (1) | (2) | (3) | (4) | (5) |
|---|---|---|---|---|---|
| Mean | 1.70 | 0.81 | 0.09 | 2.22 | 4.82 |
| SD | 0.56 | 0.66 | 0.01 | 0.51 | 1.24 |
| COV | 0.33 | 0.82 | 0.15 | 0.23 | 0.26 |

### 5.2.1 *Assessment of fishing operation time by phase*

From Figure 9, the time of each fishing phase is determined and summarised in Figure 10. It can be seen that the longest fishing phase is hauling representing 48% of the total voyage. The second longest phase is underway with 35%. Although seiners can operate close to the coast and the total fishing voyage is short, underway proportion represents a third of the complete voyage.

Table 3 is obtained from the analysis of each fishing phases in absolute hours. A high coefficient of variation can be observed in underway and searching phases. This is because distinguishing between underway and searching at the data collection level is not very precise. Two procedures were used to separate these two phases: speed behaviour and geographic location assessment. Amplitude of speed variation in searching is usually greater than in underway phase. However, it is not always feasible to clearly distinguish the point at which one phase transits to the next. As a result, spatial distribution of vessels' position records is used. By analysing the geographic locations of seiners, it is possible to identify the vessel's usual fishing area, so when a ship enters it, its status changes from underway to searching. However, because the typical fishing area is so varied, it is difficult to identify the exact location of transition between phases. Therefore, underway has a high COV (i.e. 33%), which in fact is too high since all ships leave and return to the same port.

Searching is very difficult to forecast due to weather conditions, current, and dimension of shoaling species, resulting in a COV of 82%. While encircling and hauling represent similar behaviour in all analysed voyages having low COV. Overall fishing time has COV of 26%, that may be lower if searching phase estimation was more precise.

#### 5.2.2 *Assessment of fishing operation speed by phase*

Figure 11 Shows the normalised speed per each fishing phase. In underway, searching, and encircling the normalised speed is 0.88, 0.81, and 0.66, respectively, with low SD. This allows to state that the normalised speed is a good variable to distinguish different phases of fishing. Hauling has a high COV (i.e. 39%) as it presents the average normalised velocity close to zero (i.e. 0.04).

## 6 CONCLUSIONS

Monitoring of fishing vessels is crucial for economic impacts and fishing effort estimations, for demographic analysis of the species population and for marine surveillance. The objective of this paper is to provide a preliminary analysis of the fishing activity in Portugal and of the vessels' trajectories using historical AIS data.

A portrait of the Portuguese licensed fishing fleet shows that Algarve has the largest number of ports where fishing vessels are registered. As follows, Portimão is selected as a case study area.

AIS does not provide the information about the vessels operational phase status. Therefore, the sequential change in vessel speed is assessed, accompanied by geographic coordinates of fishing trajectory. Variation in the distance between two successive points on the trajectory makes possible to separate the speeds into more than two levels (i.e. navigation or fishing), thus allowing to distinguish all the fishing phases.

The vessels' trajectories are grouped by practiced fishing gear (i.e. trawl fishing and purse seine fishing) and a preliminary characterization of the phases and of the motion patterns per group is conducted.

In trawl fishing the longest operational phase is trawling haul that represents 69% of the total voyage. The duration of the searching phase is very difficult to forecast, due to numerous unforeseen factors, such as, weather conditions, current, location, fishing aggregation and its catchability level. Trawl vessels are sailing at high speed around 23% of the time. Assessment of operation speed by phase provided results with low coefficient of variation. Only in searching the normalised speed has a high standard deviation caused by a significant variation in speed to explore the location for potential fishing ground.

In purse seine fishing the longest fishing phase is hauling representing 48% of the total voyage. Although seiners can operate close to the coast and the total fishing voyage is short, underway proportion represents a third of the complete voyage. At the data collection level, distinguishing between underway and searching is not very precise. Amplitude of speed variation in searching is usually greater than in underway phase. However, it is not always feasible to identify. By analysing the geographic locations of seiners, it is possible to identify the vessels' usual fishing ground. However, this area is so varied, that it becomes difficult to identify the exact location of passage between phases. As in trawl fishing, the duration of searching phase is very difficult to forecast. Assessment of fishing operation speed by phase provided results with low coefficient of variation. This allows to state that the normalised speed is a good variable to distinguish different phases of fishing.

## ACKNOWLEDGEMENTS

Paper contributes to the project "Monitoring and Surveillance of Maritime Traffic off the continental coast of Portugal (MoniTraffic)", which has been co-funded by "Fundo Azul" programme of the Portuguese Directorate-General for Maritime Policy, under contract no. FA-04-2017-005. This work also contributes to the Strategic Research Plan of the Centre for Marine Technology and Ocean Engineering (CENTEC), which is financed by the Portuguese Foundation for Science and Technology under contract UIDB/UIDP/00134/2020.

## REFERENCES

Bastardie, F., Nielsen, J. R., Ulrich, C., Egekvist, J., & Degel, H. (2010). Detailed mapping of fishing effort and landings by coupling fishing logbooks with satellite-recorded vessel geo-location. *Fisheries Research*, *106*(1), 41–53.

Campos, A., Henriques, V., Erzini, K., & Castro, M. (2021). Deep-sea trawling off the Portuguese continental coast—Spatial patterns, target species and impact of

a prospective EU-level ban. *Marine Policy, 128* (February), 104466.
Campos, A., Lopes, P., Fonseca, P., Araújo, G., & Figueiredo, I. (2018). Fishing patterns for a portuguese longliner fishing at the gorringe seamount—a first analysis based on ais data and onboard observations. *Maritime Transportation and Harvesting of Sea Resources, 2* (February), 1241–1244.
Deng, R., Dichmont, C., Milton, D., Haywood, M., Vance, D., Hall, N., & Die, D. (2005). Can vessel monitoring system data also be used to study trawling intensity and population depletion? The example of Australia's northern prawn fishery. *Canadian Journal of Fisheries and Aquatic Sciences, 62*(3), 611–622.
DGRM. (2021). *Official site of DGRM*, https://www.dgrm.mm.gov.pt/en/web/guest/frota. Data Accessed: 2021-11-01.
Eastwood, P. D., Mills, C. M., Aldridge, J. N., Houghton, C. A., & Rogers, S. I. (2007). Human activities in UK offshore waters: An assessment of direct, physical pressure on the seabed. *ICES Journal of Marine Science, 64*(3), 453–463.
EC. (2009a). *2010/93/: Commission Decision of 18 December 2009 adopting a multiannual Community programme for the collection, management and use of data in the fisheries sector for the period 2011-2013 (notified under document C(2009) 10121).*
EC. (2009b). *Council Regulation (EC) No 1224/2009 of 20 November 2009 establishing a Union control system for ensuring compliance with the rules of the common fisheries policy.*
EC. (2011). *Commission Directive 2011/15/EU of 23 February 2011 amending Directive 2002/59/EC of the European Parliament and of the Council establishing a Community vessel traffic monitoring and information system Text with EEA relevance.*
FAO. (2020). *FAO Yearbook. Fishery and Aquaculture Statistics 2018/FAO annuaire. Statistiques des pêches et de l'aquaculture 2018/FAO anuario.* Estadísticas de pesca y acuicultura 2018. FAO.
Ferreira, J. C., Lage, S., Pinto, I., & Antunes, N. (2017). Fishing monitor system data: A naïve bayes approach. *Advances in Intelligent Systems and Computing, 557*, 582–591.
Franklin, E. C. (2008). An assessment of vessel traffic patterns in the Northwestern Hawaiian Islands between 1994 and 2004. *Marine Pollution Bulletin, 56*(1), 150–153.
Gerritsen, H. D., Lordan, C., Minto, C., & Kraak, S. B. M. (2012). Spatial patterns in the retained catch composition of Irish demersal otter trawlers: High-resolution fisheries data as a management tool. *Fisheries Research, 129–130*, 127–136.
He, P., Chopin, F., Suuronen, P., Ferro, R. S. T., & Lansley, J. (2021). Classification and illustrated definition of fishing gears. *FAO Fisheries and Aquaculture Technical Paper No. 672.* FAO.
Hintzen, N. T., Bastardie, F., Beare, D., Piet, G. J., Ulrich, C., Deporte, N., Egekvist, J., & Degel, H. (2012). VMStools: Open-source software for the processing, analysis and visualisation of fisheries logbook and VMS data. *Fisheries Research, 115–116*, 31–43.
IMO. (1998). Resolution MSC.74(69) adoption of new and amended performance standards. *MSC 69/22/Add.1, 74* (May), 20.
INE. (2018). *Estatísticas da Pesca 2017.* Instituto Nacional de Estatística. [in Portuguese].
ITU. (2014). *Recommendation ITU-R M.1371-5. Technical characteristics for an automatic identification system using time division multiple access in the VHF maritime mobile frequency band.* ITU.
Katara, I., & Silva, A. (2017). Mismatch between VMS data temporal resolution and fishing activity time scales. *Fisheries Research, 188*, 1–5.
Liu, T. K., & Tsai, T. K. (2011). Vessel traffic patterns in the Port of Kaohsiung and the management implications for preventing the introduction of non-indigenous aquatic species. *Marine Pollution Bulletin, 62*(3), 602–608.
Lopes, P., Campos, A., Fonseca, P., Parente, J., & Antunes, N. (2018). The impact of different vms data acquisition rates on the estimation of fishing effort—an example for portuguese coastal trawlers. *Maritime Transportation and Harvesting of Sea Resources, 2* (February), 1263–1266.
MySQL. (2021). *Official site of MySQL*, https://www.mysql.com/. Data Accessed: 2021- 11-01.
Pilar-Fonseca, T., Campos, A., Pereira, J., Moreno, A., Lourenço, S., & Afonso-Dias, M. (2014). Integration of fishery-dependent data sources in support of octopus spatial management. *Marine Policy, 45*, 69–75.
QGIS. (2021). *Official site of QGIS*, https://www.qgis.org/en/site/. Data Accessed: 2021- 11-01.
Rijnsdorp, A. D., Buys, A. M., Storbeck, F., & Visser, E. G. (1998). Micro-scale distribution of beam trawl effort in the southern North Sea between 1993 and 1996 in relation to the trawling frequency of the sea bed and the impact on benthic organisms. *ICES Journal of Marine Science, 55*(3), 403–419.
Rowlands, G., Brown, J., Soule, B., Boluda, P. T., & Rogers, A. D. (2019). Satellite surveillance of fishing vessel activity in the Ascension Island Exclusive Economic Zone and Marine Protected Area. *Marine Policy, 101* (April 2018), 39–50.
Russo, E., Monti, M. A., Mangano, M. C., Raffaetà, A., Sarà, G., Silvestri, C., & Pranovi, F. (2020). Temporal and spatial patterns of trawl fishing activities in the Adriatic Sea (Central Mediterranean Sea, GSA17). *Ocean and Coastal Management, 192* (May).
Serra-Sogas, N., O'Hara, P. D., Pearce, K., Smallshaw, L., & Canessa, R. (2021). Using aerial surveys to fill gaps in AIS vessel traffic data to inform threat assessments, vessel management and planning. *Marine Policy, 133* (December 2020), 104765.
Shepperson, J. L., Hintzen, N. T., Szostek, C. L., Bell, E., Murray, L. G., & Kaiser, M. J. (2018). A comparison of VMS and AIS data: The effect of data coverage and vessel position recording frequency on estimates of fishing footprints. *ICES Journal of Marine Science, 75*(3), 988–998.
Zhao, Z., Hong, F., Huang, H., Liu, C., Feng, Y., & Guo, Z. (2021). Short-term prediction of fishing effort distributions by discovering fishing chronology among trawlers based on VMS dataset. *Expert Systems with Applications, 184*, 115512.

*Trends in Maritime Technology and Engineering – Guedes Soares & Santos (eds)*


# Low-cost tracking system to infer fishing activity from small scale fisheries in Scotland

A. Mujal-Colilles
*Scottish Oceans Institute, University of St. Andrews, Scotland, UK*
*Department of Nautical Sciences and Engineering, UPC-BarcelonaTECH, Catalunya, Spain. Serra Húnter Fellow*

T. Mendo, R. Swift & M. James
*Scottish Oceans Institute, University of St. Andrews, Scotland, UK*

S. Crowe & P. McCann
*IT Research Support, University of St. Andrews, Scotland, UK*

ABSTRACT: Large Scale Fisheries (LSF), with Length Over All (LOA) larger than 12 to 15 m are generally subject to national and international regulations that require they carry some form of tracking system that is capable of reporting their position with a prescribed frequency and accuracy. However, Small Scale Fisheries (i.e. LOA < 15 m) although representing ~90% of the world's fishers they are generally not subject to such regulation and thus important data related to the location, intensity and type of fishing activity for the majority of SSF is unknown. LSF are tracked using an Automatic Identification System (AIS) and/or Vessel Monitoring System (VMS). The need to develop tracking systems suitable for use in SSF is increasing as regulators and many of those involved in this fishing sector recognize the potential advantages of collecting this data. Funded by Scottish Government, this research represents a structured analysis of track data from a trial of a low cost tracking system being conducted in the Outer Hebrides, Scotland. Approximately ~40 creel fishing vessels (LOA < 12 m) have been fitted with the tracking system since November 2020.

## 1 INTRODUCTION

Marine ecosystems are impacted by fishing and other anthropogenic pressures including climate change (Sumaila et al., 2011). Fishing in some countries remains politically iconic even when its economic value is marginal. However, for many countries fishing provides significant socio-economic benefits to fragile coastal communities and may represent a major source of nutrition. According to the Food and Agriculture Organization of the United Nations around 90% of the world fishers are defined as Small Scale Fishers (SSF), most of them based in Asia and Africa (FAO, 2021). However, SSF in western countries can also represent a significant percentage of registered fishing vessels. In Europe, ~80% of fishing vessels are regarded as SSF (EC, 2021b) and in Scotland, this is ~ 75% of the fleet equating to ~1,400 vessels. There is little information on the fishing effort within SSF even within developed countries. Fundamental to understanding the scale and impact of SFF requires data on the temporal and spatial distribution of fishing effort (Mendo et al., 2019a; Mendo et al., 2019b; Parnell et al., 2010). Whilst vessel track data are collected for many larger scale fisheries using a variety of methods including Vessel Monitoring Systems (VMS), Automatic Identification Systems (AIS) and Remote Electronic Monitoring (REM) systems (James et al., 2018; Shepperson et al., 2018), SSF are not usually tracked.

Although there is no current legal requirement for 12-15m and under vessels to carry tracking equipment, many European countries are considering the introduction of this technology (EC, 2021a). In addition, some sectors of the SSF are, of their own volition, starting to collect tempo-spatial data on their fisheries to ensure that they can provide evidence of their fishing activities (Mendo, et al. 2019a). The collection of vessel track and catch data are required as the foundation for more sustainable fisheries management and are increasingly needed in a marine spatial planning context to inform decision makers with respect to competing demands on the allocation of marine resource space.

The objective of this contribution is to show the implementation of a low-cost tracking system in fishing vessels using creel (pots and traps) in the Outer Hebrides, Scotland. This low-cost system is designed in order to gather information for further

DOI: 10.1201/9781003320289-14

implementation of creel limitations. Therefore, it is also an opportunity to obtain fishing behaviour information using machine learning techniques from low-cost tracking systems.

## 2 METHODOLOGY

### 2.1 *Area of interest*

The Outer Hebrides or Western Isles is formed by a group of more than 70 islands located at the North West of Scotland, UK, as shown in Figure 1. The main islands from North to South are Lewis and Harris, North Uist, Benbecula, South Uist and Barra and concentrate around 95% of the total population of the area. The maximum distance North-South is ~215 km and East-West is ~50 km.

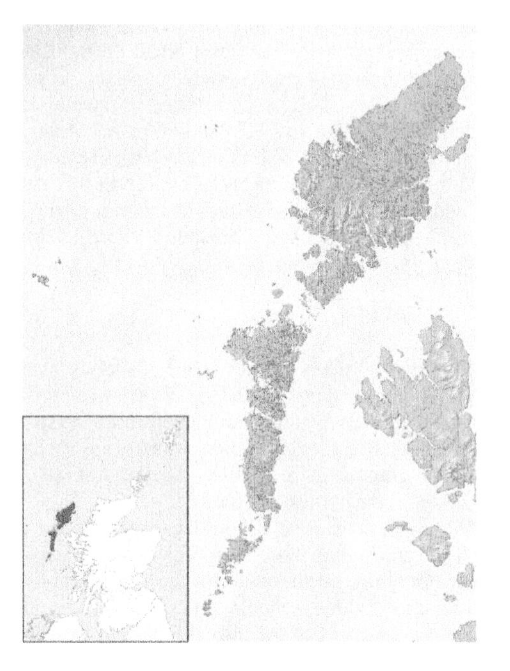

Figure 1. Location of the Outer Hebrides in a Scotland map. Source: Wikipedia.

According to (ICES, 2019), in 2019 there were a total of 366 vessels fishing in the area of the Western Isles, 57% of which were vessels with a Length Over All (LOA) <12 m. Around one third of the total landed weight was attributed to shellfish species which include *edible/brown crab*, *velvet crab*, *Norway lobster* (Nephrops), and *European lobster* (Lobster). These are species targeted with creels (formally designated as *"pots and traps")*. The total number of vessels dedicated to creeling activity in the Western Islands, with a LOA< 12 m, is ~ 143.

### 2.2 *Tracking system*

A total of 37 vessels were equipped with FMB204 ©Teltonika devices by the end of October 2020 with reliable data from the majority of them recorded since 1$^{st}$ November 2020. These are vessels fishing mainly with creels operating in around 70% of the coast of the Western Isles (to the south of the islands of Lewis and Harris).

These FMB204 trackers are principally used in the road transport sector but this model of tracker is robust, IP67 rated with high gain internal GNSS and GSM antennas together with an integrated accelerometer and on-board memory. Operating on voltages ranging from 6-30 V the trackers fitted to the vessels in this trial are supplied with power directly from the vessel, but a solar powered version of the systems has also been developed by the University of St Andrews. An internal battery allows the tracker to continue to operate for up to four days without external power. The trackers are fully configurable using preparatory software provided by Teltonika.

The original tracker configuration was set to record (poll) GNSS data every 60 seconds or with changes in Course Over Ground (COG) of >=1 degree, changes in speed more than 1m/s or a minimum distance of 10 m travelled. During periods of activity lower than these thresholds the units were configured to report position every hour.

The devices stop recording data when the internal voltage is < 3V and start recording when the internal voltage is > 3.5V. This is set up in order to stop recording when vessels are moored during more than 6 days and may affect the initial points of the following trip since the device needs a minimum voltage to start transmitting.

Data is transmitted via GSM using a multinetwork data SIM card and will transmit on 2G, 3G and 4G networks. Data transfer rates are up to 240 kbps depending on the signal. The device can store ~ 128 Mb of data in its internal memory which can contain around 6 months of data transferred in a few minutes. The open source Traccar advanced programming interface (API) (Traccar, 2021) is used to deliver the data to an open source PostgreSQL database running on an MS Azure cloud server. The tracker configuration defines a "trip" as vessel data captured after 00:00:01 each day.

With this configuration the distribution of GNSS polling frequency varied from 1 second to more than 10 hours, with most of the data recorded at intervals of less than 10 seconds (Figure 2). Points with a sampling frequency larger than 1 hour are attributed to periods of low activity, where the vessel does not have activity or the internal voltage falls below 3V.

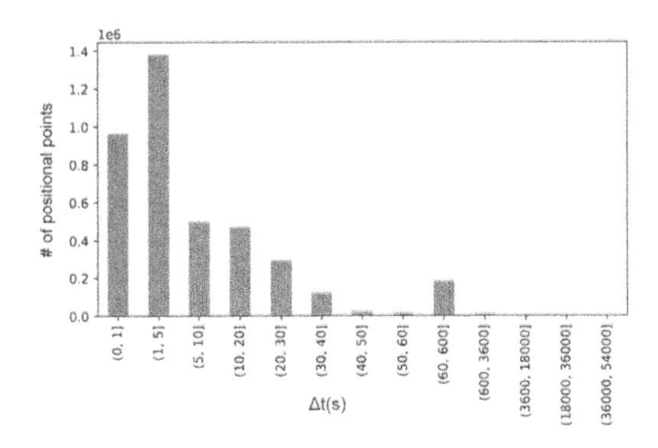


Figure 2. Sampling frequency distribution (in seconds).

## 3 PRE-PROCESS

Raw data (4,234,543 points) was cleaned as follows:

1. Latitude-longitude points outside a reference frame
2. Duplicate points. In latitude-longitude and trip ID.
3. Single-point trips. Remove trips with only one point.
4. Filter points inland with a buffer of 10 m around the coast, assuming the standard error of good GNSS signal.
5. Remove points yielding unrealistic velocities. Points with speed > 25 knots.

The total number of points removed from the data set in each step are shown in Table 1, where the process ends with 3,950,314 positional points.

Table 1. Percentage of data removed after every step.

| Pre-Process step | % of removed points from origin |
|---|---|
| Latitude-longitude | 2.7 |
| Duplicates | 0.51 |
| Single-point trips | 0.11 |
| Inland | 2.7 |
| Speed > 25knots | 0.8 |
| Single-point trips | 0.001 |
| Total points removed | 6.7 (3950314 points) |

Figure 3 shows most trips have < 100 points/trip. However, since the post-process of the data requires interpolation, only single-point trips were removed initially. This step was repeated after filtering the data with respect to speed.

The process of filtering points that occur inland was achieved using a buffer from a shapefile of the Outer Hebrides (source: https://marine.gov.scot/maps/ shapefile). It is important to note that these particular fisheries do not always use a harbour or port and vessels

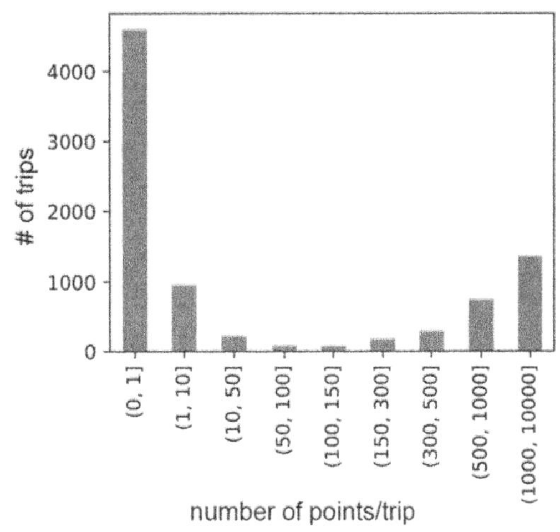


Figure 3. Distribution of the number of position points per trip.

may be anchored in sheltered bays. Therefore, the use of a buffer from the coastline also removes some anchorages. Five different buffer distances were tested to assess the influence on the number of position points/trip. Figure 4 shows a distribution of the remaining points/trip when spatial filters, with different distances from the coastline, were applied. Removing position points in-land has a larger effect on trips having less than 10 points/trip. The spatial filtering process also fails to exclude some single point trips. However, increasing the buffer distance failed to increase the removal of single-point trips significantly and thus a buffer of 10 m was used taking into account a GNSS error of ~10m and retaining fishing activity that takes place close to shore.

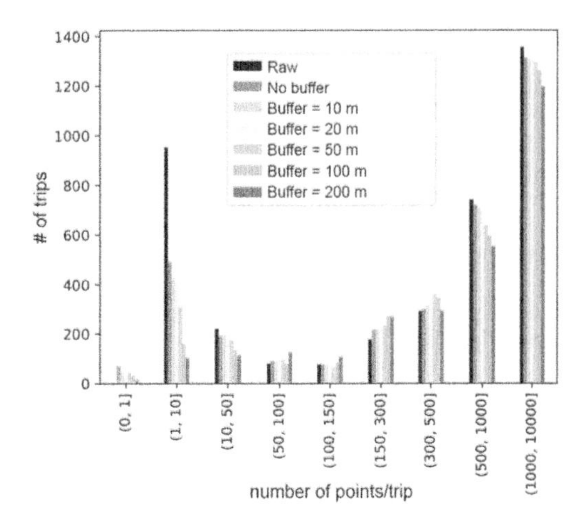


Figure 4. Distribution of the trips containing N points/trip after removing the points within a buffer of M meters (with M = 10,20,50,100 and 200 m).

Positional records representing unrealistic velocities (>25 knots) were removed recursively since velocities are inferred from geo-spatial data instead of using the velocity yielded by the tracker. An analysis carried out comparing inferred velocity versus tracker measured velocity indicated that later are being overestimated. Moreover, velocities are given in km/h with a very low precision resulting in steps of 0.3 knots when converting units. The speed filter removes 0.8% of the data, as detailed in Table 1.

As stated previously, single-point trips remaining after the spatial and speed filtering are again removed and represent around 0.001%, see Table 1.

## 4 RESULTS AND DISCUSSION

### 4.1 *Coverage*

The rationale for using a low cost, robust GNSS tracking system which transmits data via GSM technology to a dedicated secure server is two fold: fishers are reluctant to share their vessel track data openly – particularly with other fishers. The data store and forward capability of the tracking device together with the operating range of the vessels in this fishery means that it is usually possible for vessels to be in range of at least a 2G network signal. Previous research using AIS for SFF data collection in Scotland (James et al., 2018) revealed some areas where AIS reception was poor or absent close to land; data quality issues with respect to the polled frequency of the data provided by commercial AIS data providers and; some concerns from fishers that AIS positional data can be seen by anyone with an AIS receiver.

Whilst near real time position reporting via GSM is possible, there is no practical requirement from a fisheries management perspective for data to be transmitted with such frequency. In addition, the data store and forward capability of the Teltonika device ensures that GNSS data continues to be collected and is transmitted automatically as soon as a GSM signal is available. Predicted mobile phone network coverage for the Western Isles (see https://www.signalchecker.co.uk/) suggests reasonably comprehensive coverage for the EE and Vodaphone. The distribution of track data shown in Figure 5 suggests that for practical purposes it is possible to secure GSM transmitted vessel track data within the area of the trial, with the caveat that GNSS coverage is not universal particularly in areas close to the coast (< 100m) where some data may be lost.

### 4.2 *Data quality*

Some indicators were used to validate the quality of the data. First, the sampling frequency distribution shown in Figure 2 was examined for a specific bias towards the identity of the vessels after preprocessing the data in Figure 6. Whilst this relationship has not yet been tested statistically the

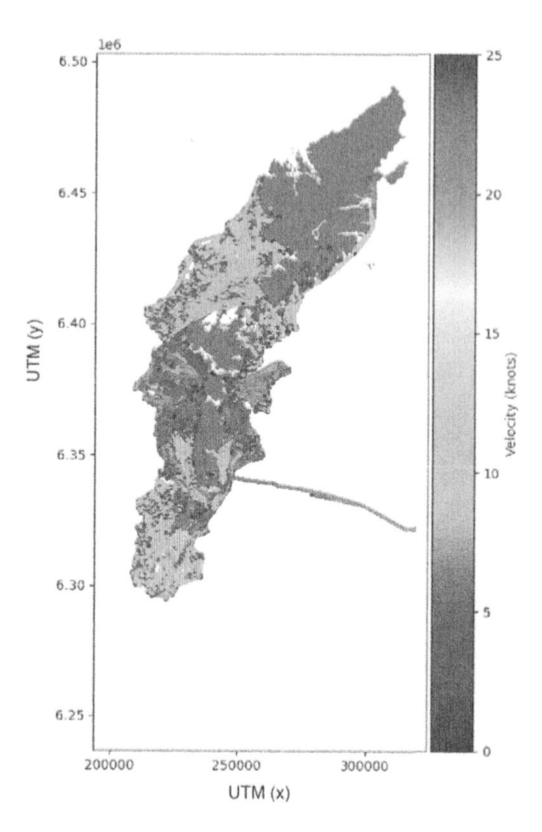


Figure 5. Geospatial raw data in UTM coordinates. Colors represent the instant velocity.

scatterplot strongly suggests that there is no tracking device or vessel related bias in reporting frequency. Moreover, points with shorter distances and larger $\Delta t$'s were associated with periods of vessel inactivity and the device's internal battery was too low (<3V) to record data. Figure 6 also indicates that the spatial filtering using a buffer of 10m from the coastline did not remove all the position points from port or anchorage areas.

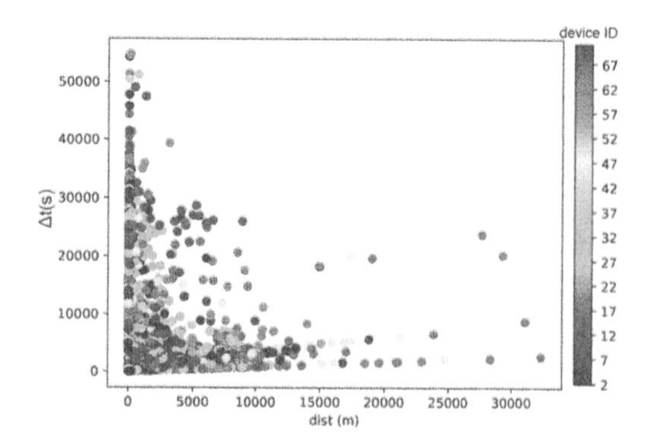


Figure 6. Scatterplot of the sampling frequency vs the spatial distance between consecutive points coloured by device id.

To confirm the utility of trip definition (24 hours – midnight to midnight) configured by default in the tracker, the hourly distribution of the recorded points was analysed. Figure 7, shows that most of the activity (99.7%) is carried out during daylight hours confirming that a trip-day definition is useful in this particular SSF because there is little overnight activity. This is also consistent with the original configuration of the trackers where the sampling frequency is set to 1 hour if the vessels are not active.

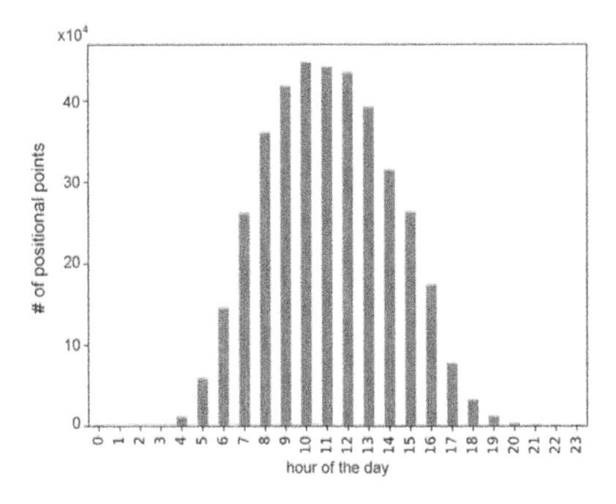


Figure 7. Hourly distribution of recorded points.

A more detailed analysis on the trip duration both in time and space is shown in Figure 8. Two analyses were carried out to check the distribution of the trips recorded between November 2020 and October 2021, both included: a) the distribution of trips and b) the distribution of points. Blank spots in Figure8a show areas with less than 0.1% of trips whereas in Figure 8b white areas represent no data (e.g. there are no points in trips with a total length between 5 and 10 km and lasting 10 to 12 hours). In Figure 8a, trips lasting 18-24 hours are trips where the spatial filtering does not remove the anchorage or ports areas and points within the resting time are still accounted for and they represent 7% of the total trips. However, 4% of trips correspond to a total fishing trip length of < 1h, these are vessels that are inactive for prolonged periods of days or weeks. Figure 8a, shows that 25% of the trips are 5 to 10h in duration with a total distance between 10 to 50km. About 13% of trips are > 75 km and represent movement between ports or anchorages. This activity takes place for operational reasons including seasonal changes in the location of fishing grounds.

If we look at the distribution of points, Figure 8b, results are not essentially different from Figure 8a, but they help to confirm the results previously

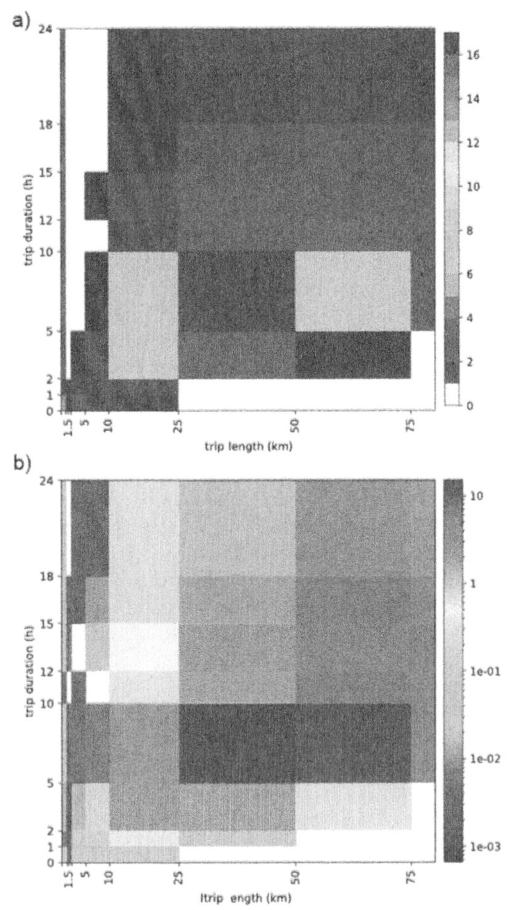


Figure 8. 2D distribution of pre-processed trips using the total trip duration in hours and the total trip length in km with respect to a) percentage of trips or b) percentage of points in logarithmic scale.

obtained. Trips no longer than 1 km account for 17% of all data points and represents a combination of continued positional reporting by the tracker whilst the vessel is at anchor or in port or when the internal battery power of the device is < 3V. More than 60% of points relate to trips > 10 km and > 5 h. Therefore, short trips are not common in this particular SSF, where < 1% of the trips are between 1 and 2 km.

## 5 CONCLUSIONS

This research provides an analysis of the quality of SFF track data derived from a low cost tracking device designed principally for road transport use. Knowledge of the limitations of such devices is important as this will help to inform decisions with respect to the use of such technology in relation to fisheries management and compliance. The devices tested in this research do not have external aerials for the reception of GNSS signal or the reception and transmission of GSM data. The ability of the devices to store and forward GNSS data via GSM means that the acquisition of near real time positional data may be compromised where the GSM signal is insufficient. However, for this fishery, near real time reporting is not important. Securing GNSS data of sufficient resolution to detect fishing activity is the primary goal. The analysis performed in this research has been designed to identify data gaps and inform the configuration of the tracking device to optimize data acquisition whilst avoiding the collection and transmission of irrelevant data.

Our results suggest that the geographic coverage of the tracking system with respect to the GSM signal is not a barrier to the transmission of data in the area of the trial. Whilst the reporting of GNSS positions is reasonably acceptable, there are small discontinuities in GNSS data within 50 m of the coast. We are now investigating the spatial and temporal nature of these discontinuities in GNSS data to assess their significance.

Previous research (Mendo et al., 2019a,b) has demonstrated that positional data collected every 60 seconds is sufficient to infer where vessels using pots and creels are hauling their fishing gear (Mendo et al, 2019). The results of the research presented here, have been used to refine the configuration of the tracking device to amend thresholds for course over ground, speed and minimum distance variables.

This research represents the first structured analysis of the GNSS data being collected as part of a scaled trial of a low cost tracking system in SFF. The results are helping to establish protocols for the pre-processing of the raw data, identifying data gaps and providing rationales for defining the thresholds we are using to optimize data acquisition.

## ACKNOWLEDGEMENTS

This work was funded by the Scottish Government through Marine Scotland. Authors would also like to thank the Western Isles Fishermen's Association. AMC acknowledges the support from the Serra Húnter program of the Generalitat de Catalunya.

## REFERENCES

EC. (2021a). *Inspections, monitoring and surveillance.* https://ec.europa.eu/oceans-and-fisheries/fisheries/rules/enforcing-rules/inspections-monitoring-and-surveillance_en

EC. (2021b). *Small-scale fisheries.* https://ec.europa.eu/oceans-and-fisheries/fisheries/rules/small-scale-fisheries_en

FAO. (2021). *Fisheries and Aquaculture - Fisheries and Aquaculture - Small-scale fisheries around the world.* https://www.fao.org/fishery/en/ssf/world/en

ICES. (2019). *International council for the Exploration of the Sea.*

James, K. M., Campbell, N., Vidarsson, J. R., Vilas, C., Plet-Hansen, K. S., Borges, L., González, Ó., van Helmond, A. T. M., Pérez-Martín, R. I., Antelo, L. T., Pérez-Bouzada, J., & Ulrich, C. (2018). Tools and technologies for the monitoring, control and surveillance of unwanted catches. In *The European Landing Obligation: Reducing Discards in Complex*, Multi-Species *and Multi-Jurisdictional Fisheries* (pp. 363–382). Springer International Publishing. https://doi.org/10.1007/978-3-030-03308-8_18

James, M., Mendo, T., Jones, E. L., Orr, K., McKnight, A., & Thompson, J. (2018). AIS data to inform small scale fisheries management and marine spatial planning. *Marine Policy*, *91*, 113–121. https://doi.org/10.1016/j.marpol.2018.02.012

Mendo, T., Smout, S., Photopoulou, T., & James, M. (2019). Identifying fishing grounds from vessel tracks: model-based inference for small scale fisheries. *Royal Society Open Science*, *6*(10), 191161. https://doi.org/10.1098/rsos.191161

Mendo, T., Smout, S., Russo, T., D'Andrea, L., James, M., & Maravelias, C. (2019). Effect of temporal and spatial resolution on identification of fishing activities in small-scale fisheries using pots and traps. *ICES Journal of Marine Science*, *76*(6), 1601–1609. https://doi.org/10.1093/icesjms/fsz073

Parnell, P. E., Dayton, P. K., Fisher, R. A., Loarie, C. C., & Darrow, R. D. (2010). Spatial patterns of fishing effort off San Diego: Implications for zonal management and ecosystem function. *Ecological Applications*, *20*(8), 2203–2222. https://doi.org/10.1890/09-1543.1

Shepperson, J. L., Hintzen, N. T., Szostek, C. L., Bell, E., Murray, L. G., & Kaiser, M. J. (2018). A comparison of VMS and AIS data: the effect of data coverage and vessel position recording frequency on estimates of fishing footprints. *ICES Journal of Marine Science*, *75*(3), 988–998. https://doi.org/10.1093/icesjms/fsx230

Sumaila, U. R., Cheung, W. W. L., Lam, V. W. Y., Pauly, D., & Herrick, S. (2011). Climate change impacts on the biophysics and economics of world fisheries. In *Nature Climate Change* (Vol. 1, Issue 9, pp. 449–456). Nature Publishing Group. https://doi.org/10.1038/nclimate1301

Traccar. (2021). *Traccar.* https://www.traccar.org/

*Trends in Maritime Technology and Engineering – Guedes Soares & Santos (eds)*


# Identification of ship trajectories when approaching and berthing in Sines port based on AIS data

H. Rong, A.P. Teixeira & C. Guedes Soares
*Centre for Marine Technology and Ocean Engineering (CENTEC), Instituto Superior Técnico, Universidade de Lisboa, Lisbon, Portugal*

ABSTRACT: A data mining approach is adopted for identifying ship behavior in the approaches and inside the port area using the Automatic Identification System (AIS) data, covering different ship types, sizes and final terminals within the port. The approach consists of two steps: (1) clustering ship trajectories in the port area and identifying the characteristics of the clusters; (2) determining the waypoints along the ship routes by which ships approaching to the terminals based on the analysis of ship trajectories within the clusters. The clustering results present both the ship behavior patterns and the area where significant changes in the behavior patterns for ship path, which are the dominant behavior attributes for ships in ports. The proposed method is applied to the ship trajectories entering the Sines Port, and the results of this study demonstrate that AIS data can be used to provide a systematic understanding of ship behavior even inside a port area.

## 1 INTRODUCTION

In the maritime domain, the need to ensure the safety of navigation has led to the implementation of the Automatic Identification System (AIS), enforced by the International Maritime Organization (IMO) for ships over 300 tons and all passenger ships. The system enables automatic transmission of ship movement data (e.g. position, speed, Course Over Ground, report time) and static data (e.g. Maritime Mobile Service Identity (MMSI), ship type and dimensions) between ships and shore-based stations. In recent years, large datasets of AIS data have become available for ship trajectory data mining in various locations (Tu et al., 2018).

Due to the high sampling rate and wide coverage of AIS data, it has gradually become the main data source for ship behaviour recognition (Mestl et al., 2016; Pallotta et al., 2013; Rong et al., 2022a), maritime traffic characterization (Zhang et al., 2021; Varlamis et al., 2020; Zhang et al., 2019) and ship collision risk assessment (Cai et al, 2021; Silveira et al., 2021; Teixeira & Guedes Soares, 2018; Zhang et al., 2016).

Gao & Shi (2020) presented a method to recognize ship-handling behaviour patterns when ships conduct ship-collision avoidance research and developed routing plans. Vitali et al. (2020) have analysed AIS data to estimate the speed loss of container ships in different weather conditions. A data mining approach was presented by (Rong et al., 2020) for characterization of the maritime traffic off the coast of Portugal in a probabilistic manner. In addition, the relationship between ship collision hotspots and the local traffic characteristics was studied by Rong et al., (2021). Rawson & Brito (2021) investigated the practical applications of domain analysis in waterway risk assessment.

Previous research work on extracting maritime traffic from AIS data have proved that valuable knowledge about ship behaviour can be extracted through the analysis of historical data. Ships following the same motion pattern behave similarly and move along the optimized ship route, therefore, the ship trajectories provided by AIS data can be grouped according to the similar behaviour of ships, and the traffic group thus provides an overview of the general motion pattern of ships.

Murray et al. (2020) presented a ship trajectory prediction method that utilizes a cluster of historical ship trajectories to facilitate the prediction of a selected vessel. To enhance the understanding of ship behaviour in the port of Rotterdam, Zhou et al. (2019) performed a clustering analysis on inbound and outbound ship trajectories of Rotterdam Port. Xiao et al. (2017) applied lattice-based DBSCAN algorithm to extract ship motion pattern, and by incorporating both the ship motion pattern and ship behaviour knowledge, a ship traffic prediction method is proposed that capable of accurately predicting maritime traffic 5, 30 and 60 minutes ahead. Rong et al. (2022b) proposed a maritime traffic probabilistic prediction method based on ship motion pattern extraction, considering ship destination prediction and ship trajectory prediction within a specific route.

DOI: 10.1201/9781003320289-15

The maritime traffic off the Portuguese coast has been characterized by Silveira et al. (2013) and the traffic in a more restricted area like the Azores archipelago was studied by Rong et al. (2021), and even the more restricted area of the Tagus estuary was studied (Rong et al. 2015) and routes were identified and used for track planning of autonomous vessels (Xu et al. 2019). The present work studies an even more localized area, to understand the dynamic behaviour of ships and the traffic flow when entering a port and berthing. An unsupervised data mining approach is used to analyse ship trajectories and extract the ship motion patterns entering a port area from historical AIS data.

## 2 METHODOLOGY

As shown in Figure 1, the framework of ship behaviour characterization in port areas consists of three phases which are data pre-processing, ship motion pattern extraction and waypoint identification. To extract the ship motion patterns in an unsupervised way, a clustering-based method will be presented to group ship trajectories according to trajectory similarity and ship static characteristics including ship types, ship sizes. The ship motion pattern extraction starts at terminal areas determination based on Density-based spatial clustering of applications with noise (DBSCAN) algorithm. Then, ship trajectories are grouped considering different ship types, ship sizes and terminals.

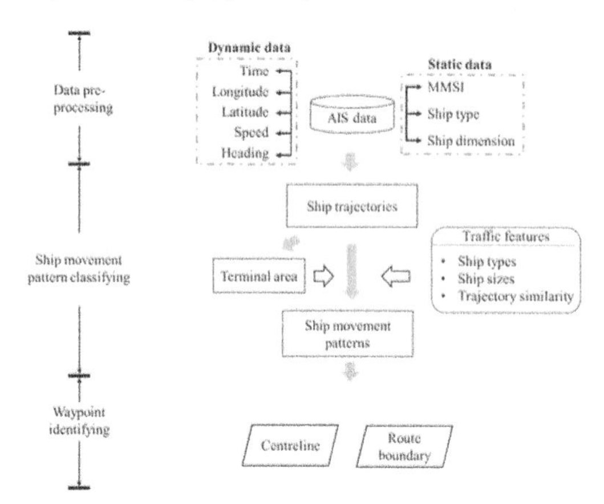


Figure 1. Framework for ship behaviour characterization in port area.

In terms of identifying waypoints, the direction change of individual ship trajectories is analysed. In addition, the waypoints along the ship routes approaching port terminals are identified considering the clustering of kinetic characteristics of the ships.

### 2.1 *Terminal area identification*

The terminal refers to a specific location at a port where a vessel can be moored, usually for loading and unloading, and where the ship speed equals zero. Terminals are designated by the management of a facility like port authority or harbourmaster. Therefore, the terminal areas can be identified by clustering ship static positions (speed equals zero).

In this study, the static points in ship trajectories are first extracted and then the DBSCAN algorithm is performed to cluster the static points to form the terminal areas. The DBSCAN algorithm is a density-based clustering algorithm that was presented by Ester et al. (1996). The reason of applying DBSCAN algorithm in this study is that given a set of points in an area, it groups together points that are close to each other (points with enough neighbours) and mark outlier points located in low-density regions.

In the clustering procedure, the points are classified as core points, density-reachable points and outlier points. DBSCAN algorithm requires two parameters: $\epsilon$ and *minPts*. The parameter $\varepsilon$ specifies the radius of a neighbourhood with respect to the point $p$. The parameter *minPts* is the minimum number of points required to form a dense region within the $\epsilon$ radius. DBSCAN algorithm starts at looking for core points, i.e., points $p \in \mathcal{D}$ such that $|N_\varepsilon(p)| \geq minPts$, where $N_\epsilon(p)$ is the $\epsilon$-neighborhood of $p$. Once a core point is found, such neighbourhood points are density-reachable from $p_i$ and belong to the same cluster. Then, the points that are not density-reachable from other points are considered as outlier points.

As shown in Figure 2, the radius of the large solid and dash-line circles in the figure is $\epsilon$. The red point is detected as core points because its neighbourhood area contains more than 3 points (given $minPts = 3$). In the meanwhile, all the neighbour points (represented as yellow points in Figure 2) are marked as reachable points and grouped in the same cluster. Blue points are neither core points nor reachable points but are reachable from a reachable point, and thus belong to the same cluster. Hollow circles are outlier points that are neither a core point nor density reachable.

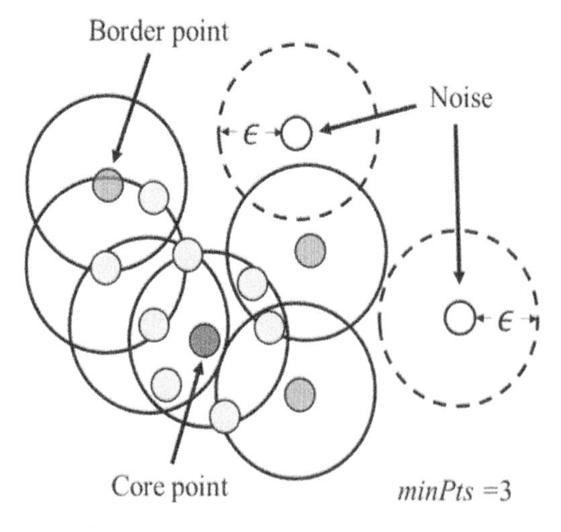


Figure 2. Illustration of the DBSCAN algorithm.

### 2.2 *Trajectory similarity*

As ships entering or leaving a port exhibit well defined motion patterns. Extracting the ship motion patterns help in understand the specific manoeuvres of ships entering or leaving the port. Therefore, a clustering procedure is performed to group similar trajectories based on DBSCAN algorithm, and the distance between ship trajectories is measured by Hausdorff distance.

The Hausdorff distance measures how far two subsets are from each other. In this context, the Hausdorff distance is adopted as an objective measure of each trajectory similarity. This technique is well suited for clustering ship trajectories into different groups according to their similarity (Rong et al 2015).

Let $traj_a = \{p_a^1, \dots p_a^i, \dots, p_a^{N_a}\}$ and $traj_b = \{p_b^1, \dots p_b^j, \dots, p_b^{N_b}\}$ be two time-series of ship trajectories, the Hausdorff distance $d_H(traj_a, traj_b)$ is defined as:

$$d_H(traj_a, traj_b) = max\left\{\max_{p_a^i}\min_{p_b^j} d\left(p_a^i, p_b^j\right), \max_{p_b^j}\min_{p_a^i} d\left(p_a^i, p_b^j\right)\right\} \quad (1)$$

where $d\left(p_a^i, p_b^j\right)$ represents the distance between data point $p_a^i$ in ship trajectory $traj_a$ and data point $p_b^j$ in ship trajectory $traj_a$, respectively. In this study, the Haversine formula is applied to calculate the distance between latitude $(lat)$ and longitude $(lng)$ coordinates.

### 2.3 *Characterization of traffic flow*

After identifying the motion patterns of ships approaching the different terminals, the ship behaviours in each motion pattern is characterized, as shown in Figure 1. In particular, the waypoints that ships follow to approach the terminal areas and the geographical areas where speed changes are frequently observed are studied. In addition, the traffic flow is also characterized in terms of ship type, speed and length distributions.

## 3 CASE STUDY

The Port of Sines, due to its geophysical characteristics, is one of the main ports of Portugal in terms of throughput of containers, natural gas, coal, oil and its derivatives (Santos & Guedes Soares, 2017). The AIS data employed in this paper are from 1st October to 31st December 2015, and the study area is bounded by parallels 37.85°N and 38°N, and by meridians 8.82°W and 9°W. The collected AIS data in the study area contain 169652 messages and 753 ship trajectories, as shown in Figure 3. It is observed that the AIS data within the port of Sines generally shows a certain geographic clustering of the ship trajectories as individual ships follow a similar path to the terminals. These clusters of ship trajectories to the terminals show the median path and the path dispersion in the area, which can provide useful statistics about the preferred manoeuvres of ships.

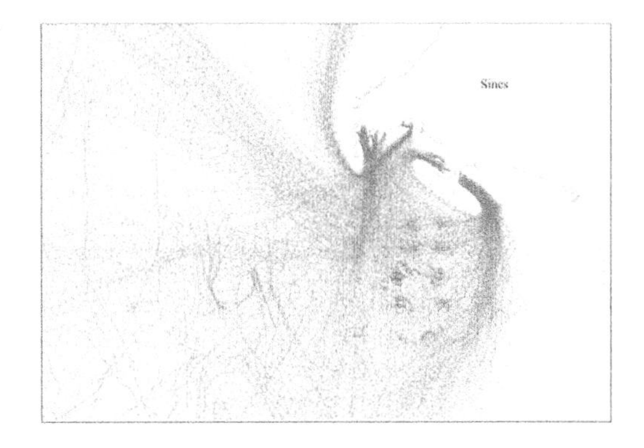


Figure 3. Three months AIS data of ships entering Port of Sines.

Figure 4 Shows the ship length distribution for different ship types. According to Figure 1, container ships account for the largest proportion (46.5%), followed by tankers (27.1%), and then other (17.9%) and fishing boat (8.5%). In terms of container ships, there are 96.6% container ships that exceed 100 meters and the length beyond 200 meters account for 67.1%. When investigating the composition of tankers, 63.2% of tankers have a medium length (100-200 meters) and 15.7% of tankers have a large length (>200 meters). The length of all fishing boat and other ships is less than 100 meters. The ship length distribution for different ship types reveals that the port of Sines mainly handles different types of container ships and tankers.

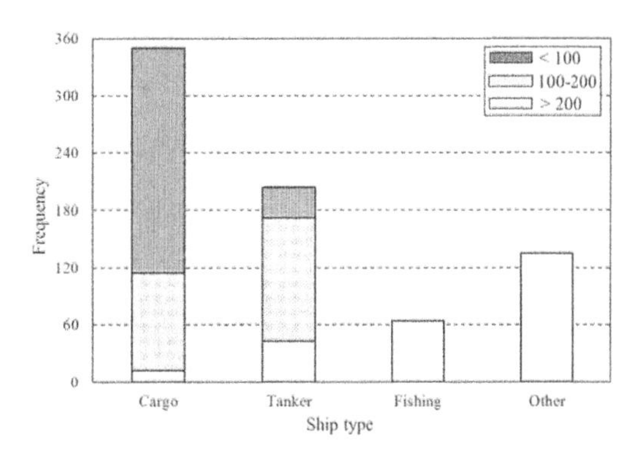


Figure 4. Ship length distribution divided by ship type.

As shown in Figure 5a, the ship static positions represented by grey circles are obviously clustered at the terminal areas and anchoring area. As already mentioned, the ship static positions are clustered by the DBSCAN algorithm. Compared with centroid-based clustering, the DBSCAN algorithm can determine arbitrarily shaped clusters and does not require the number of clusters as *priori*. However, there is no general way of choosing its parameters *minPts* and the radius $\epsilon$. The setting of *minPts* requires some knowledge on the data density on the study area. In the present case, the algorithm has shown to be robust with respect to *minPts,* as similar clustering structures are obtained for a wide range of *minPts* values (between 5 to 8). The parameter $\epsilon$ is set to 600 meters and *minPts* is set to 8. It should be mentioned that by setting a smaller value different clustering structures can be observed. By applying the DBSCAN algorithm to ship static positions, there are 5 clusters identified that are represented by a triangle in colours and outlier points are represented by black points. The detected areas (represented by the grey polygons in Figure 5b) are formed by the outline of clustered static points.

Figure 6 shows the trajectories of container ships, tankers and fishing vessels. It is found that all fishing vessels' destination is Terminal 2. The fishing harbour is located inner Port of Sines and is formed by an inner basin sheltered by a breakwater that offers good protection for docking and anchoring. Regarding container ships, all the container ships are berthing at Terminal 3 and Terminal 4. Moreover, most container ships berthing at the Terminal 4, that accounts for 90.3% (as shown in Figure 7 by green trajectories). Most of tankers berth at Terminal 1 and the other tankers berth at Terminal 3 and Terminal 4, as shown in Figure 6b.

In addition to the terminal areas detected from the clustering of static positions, the frequent ship motion patterns can be mined from the historical ship trajectories. In order to capture the behaviour of ships that enter or leave the port, individual ship trajectories are grouped according to the terminal areas that ship trajectories approach or leave from. A similarity-based clustering algorithm is applied to filter outlier trajectories from each main route based on the Hausdorff distance. This technique is well suited for clustering ship trajectories into different groups according to their similarity (Rong et al 2015).

The traffic groups generally show a certain geographic clustering of the ship trajectories as individual ships tend to follow a similar path. In this study, 12 traffic groups are detected based on the clustering of ship trajectories, as shown in Figure 8. The traffic groups consist of ship motion patterns from the open sea area to the terminal areas 1 to 4, and show the median path and the spread of traffic in the area which can provide useful statistics about the traffic situation.

The ship type distributions of each traffic group are illustrated in Table 1. The number of ship trajectories in each traffic group reveals that different ship types follow the corresponding traffic route to the terminals. For example, traffic group No. 1 represents the traffic of tankers entering and leaving Terminal 1 to the north-west corner of the area; traffic group No. 2 shows the traffic between Terminal 1 and Terminal 4; traffic group No. 4 represents the traffic of fishing vessels from the mid-point of the northern border of the area to Terminal 2, and traffic group No. 12 is the container traffic from mid-point of the southern border of the area to Terminal 5.

As an example, the traffic group No.7 that enters Terminal 2 is illustrated in Figure 9a. The ships in the traffic group are initially move southward and then turn port-side with destination to Terminal 2. This traffic group contains 112 ship trajectories and totally

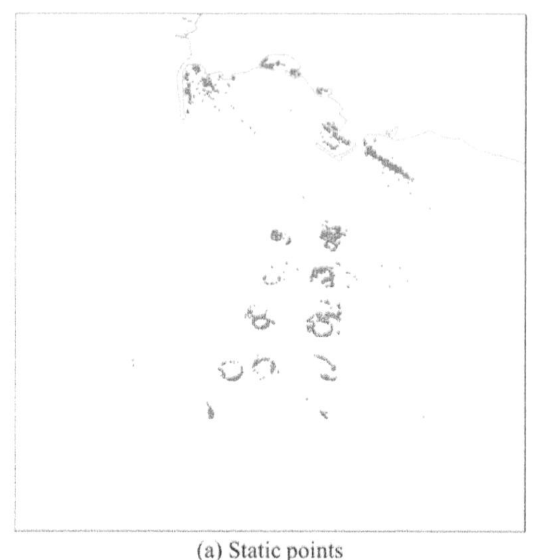

(a) Static points

(b) Clustering of static points

Figure 5. Clustering of static points based on DBSCAN algorithm.

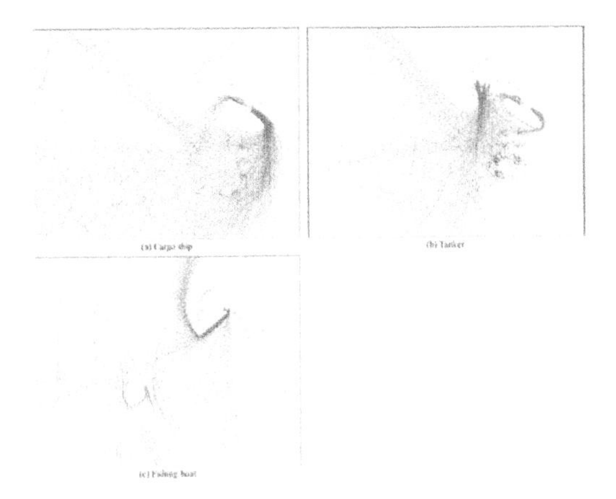

Figure 6. Ship trajectories for different ship types.

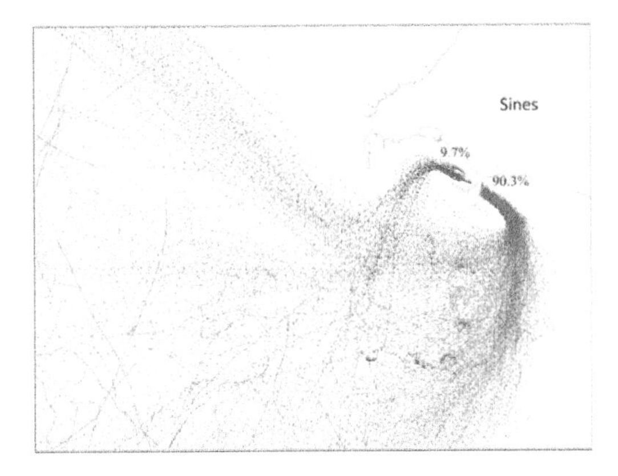


Figure 7. Ship trajectories of container ships berthing at different terminals.

6810 data points. Knowing where the ships within the traffic group change their heading is important in the behaviour analysis of the motion pattern.

Therefore, the turning points of ship trajectories are detected by applying Douglas-Peucker algorithm (Douglas & Peucker, 1973). The DP algorithm has been widely adopted in ship trajectory compression and is considered as one of the most effective trajectory simplification methods to compress ship trajectories. As shown in Figure 9b, turning points (represented by the yellow points) are detected over the traffic group. Topographically, a turning section (represented by the grey polygon in Figure 9b) refers to an area where turning points are frequently observed. In this study, the turning section is shaped via a spatial distribution given by the coordinates of the turning points detected from ship trajectories within the traffic group.

With the classification of ship trajectories into traffic groups, ship trajectories can be further analysed to determine the path by which ships approach the terminals. As the traffic group provides an

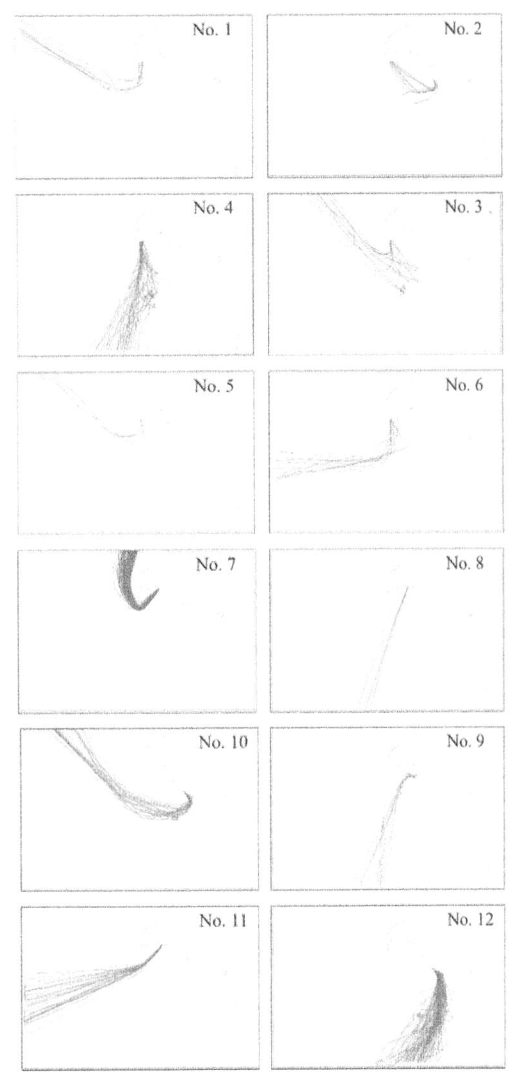


Figure 8. Traffic groups detected from ship trajectories.

overview of the general motion pattern that can be used to infer the behaviour of ships following it, the corresponding specific manoeuvres can be characterized by the ship trajectories within the traffic group. Figure 10 illustrates the centreline and boundary of the traffic group estimated by statistical analysis of ship trajectories within the traffic group. It can be seen that the centreline expresses the motion tendency of the traffic group and the boundaries well exhibit the spatial extension of the ship trajectories following the centreline. Figure 11 demonstrates the route centrelines of all traffic groups.

Three ship trajectories are selected to illustrate ships entering or leaving terminals and in port behaviour. The ship attributes are listed in Table 2.

Table 1. Ship type distribution of traffic groups.

| | Ship type | | |
|---|---|---|---|
| Traffic group | Container | Tanker | Fishing |
| 1 | 0 | 12 | 0 |
| 2 | 0 | 17 | 11 |
| 3 | 0 | 11 | 0 |
| 4 | 0 | 48 | 0 |
| 5 | 0 | 6 | 0 |
| 6 | 0 | 15 | 0 |
| 7 | 0 | 0 | 112 |
| 8 | 0 | 0 | 7 |
| 9 | 0 | 0 | 23 |
| 10 | 9 | 4 | 0 |
| 11 | 37 | 0 | 0 |
| 12 | 113 | 0 | 0 |

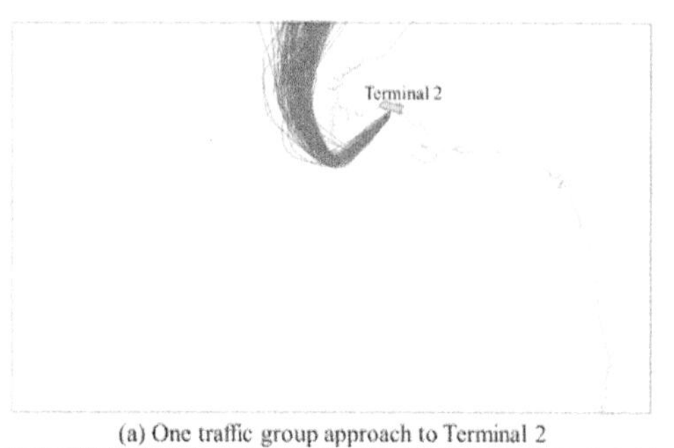


(a) One traffic group approach to Terminal 2

(b) Turning points and turning section detected from ship trajectories

Figure 9. Turning points and turning sections detected from one traffic group.

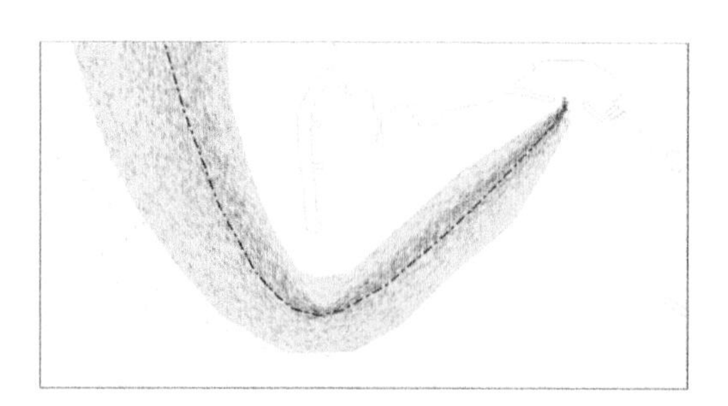

Figure 10. Centreline and boundary of the traffic group.

Figure 11. Route centreline of the traffic groups.

Table 2. Ship attributes of three ship trajectories.

| MMSI | Ship type | Length (m) | Breadth (m) |
|---|---|---|---|
| 256XXX000 | Tanker | 184 | 27 |
| 253XXX000 | Container | 277 | 40 |
| 224XXX000 | Chemical tanker | 97 | 14 |

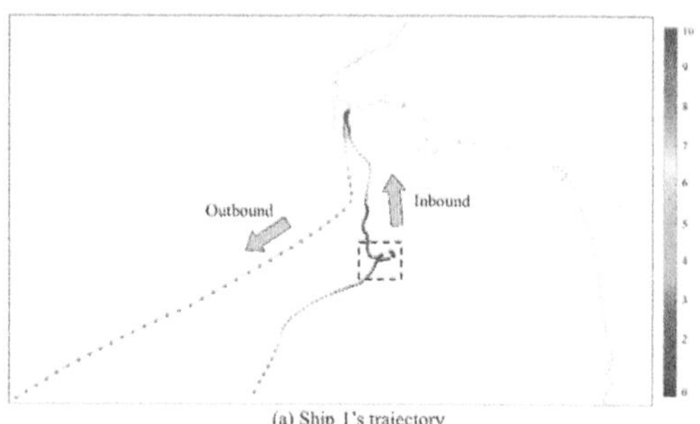


(a) Ship 1's trajectory

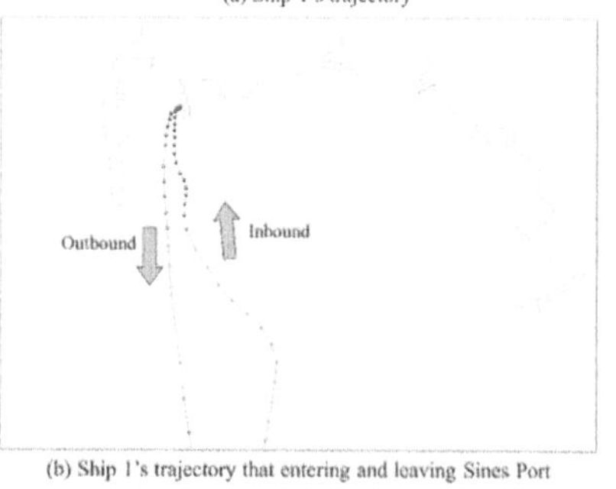


(b) Ship 1's trajectory that entering and leaving Sines Port

Figure 12. Ship 1's trajectory.

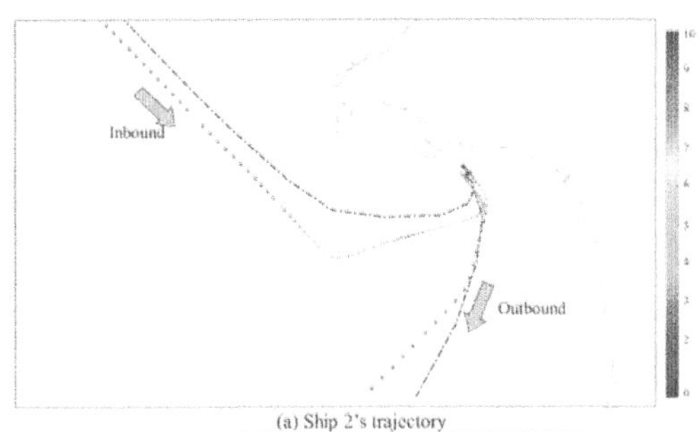


(a) Ship 2's trajectory

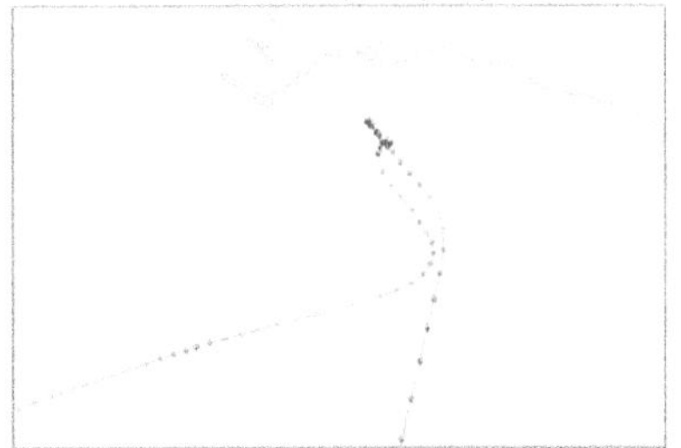

(b) Ship 2's trajectory that entering and leaving Sines Port

Figure 13. Ship 2's trajectory.

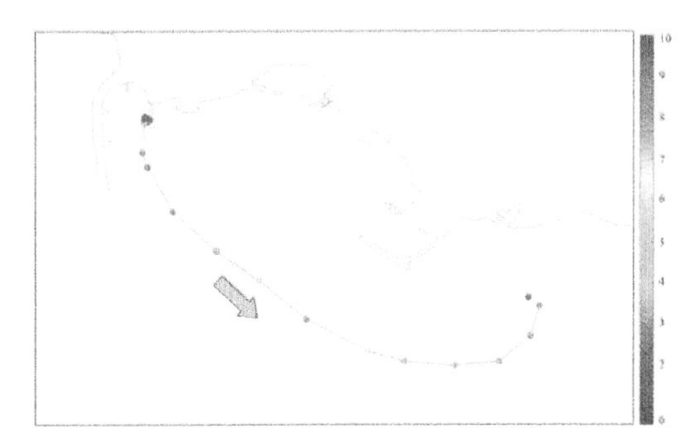

Figure 14. Ship 3's trajectory.

Figure 12a shows ship1's trajectory entering and leaving the port of Sines. In the figure, the colours closer to red indicate higher ship speed and colours closer to blue indicate lower ship speed. It is seen that ship 1 initially decreases the speed until it reaches the anchoring area. Then, after the anchorage, ship 1 starts entering the port. As shown in Figure 12b, ship 1 gradually decreases its speed from 6-knot to 1-knot, whereas, ship 1 is constantly correcting the course in order to berth at Terminal 1. After the berthing, ship 1 turns portside slightly and increases the speed to 10-knot for leaving the port.

Figure 13a shows a container ship entering and leaving the port of Sines. The dashed line in Figure 13a represents the centreline of traffic group No. 11 and No.12. According to Figure 13a, ship 2's trajectory follows well the motion patterns. When ship 2 approaches the port, its trajectory is parallel to the centreline. Then, ship 2 turns portside and starts to decrease speed at the same time. As shown in Figure 12b, once ship 2 reaches Sines Port, it makes a sharp turn and continues decreasing the speed until it reaches Terminal 4. After the berthing, ship 2 turns portside slightly and increases the speed to 10 knots following the centreline of traffic group No. 12 for leaving the port.

Figure 14 shows the trajectory of a chemical tanker moving from Terminal 1 to Terminal 4. It is seen that ship 3 increases the speed to near 8 knot and then turns portside and slightly decreases the speed to 3 knot until ship 3 reaches Terminal 4.

## 4 CONCLUSIONS

This study adopts a data mining approach for characterization of the ship behaviour in a port area in terms of ship type, length and final terminals. The ship trajectories are divided into traffic groups according to their similarity, and the centreline and boundary of each traffic group are determined based on the analysis of the ship trajectories within the traffic groups.

A statistical analysis of the traffic flow entering and leaving the port of Sines shows that the traffic mainly consists of container ships, tankers and fishing boats. In addition, the ship length distributions for different ship types show that the port of Sines handles container ships and tankers of different sizes. By clustering static points based on the DBSCAN algorithm, four terminal areas and one anchorage area are determined.

Then, ship trajectories are divided into traffic groups according to their destination and trajectory similarity. There are 12 traffic groups that represent the typical ship motion patterns in the port area. It is found that different traffic routes are used by different ship types. The ship trajectories within each traffic group are further analysed to estimate the route centreline and boundary which represent the ship motion tendency and spatial extension of the traffic group, respectively. At last, three ships are selected to illustrate their trajectories entering and leaving terminals and the in-port behaviour.

## ACKNOWLEDGEMENTS

The paper contributes to the project "Safety System for Manoeuvring and Moored Ships in Ports (BlueSafePort)", which has been co-funded by "Fundo Azul" programme of the Portuguese Directorate-General for Maritime Policy, under contract no. FA-04-2017-016. This work also contributes to the Strategic Research Plan of the Centre for Marine Technology and Ocean Engineering (CENTEC), which is financed by the Portuguese Foundation for Science and Technology under contract UIDB/UIDP/00134/2020.

## REFERENCES


Cai, M. Y.; Zhang, J. F.; Zhang, D.; Yuan, X. L., and Guedes Soares, C. 2021. Collision risk analysis on ferry ship in Jiangsu Section of the Yangtze River based on AIS data. Reliability Engineering and System Safety.; 215: 107901.

Douglas, H.D., Peucker, K.T., 1973. Algorithms for the Reduction of the Number of Points Required to Represent a Digitized Line or its Caricature. Int. J. Geogr. Inf. Geovisualization 10, 112–122. https://doi.org/10.1002/9780470669488.ch3.

Gao, M., Shi, G., 2020. Ship-handling behavior pattern recognition using AIS sub-trajectory clustering analysis based on the T-SNE and spectral clustering algorithms. Ocean Eng. 205, 106919. https://doi.org/10.1016/j.oceaneng.2020.106919.

Mestl, T., Tallakstad, K.T., Castberg, R., 2016. Identifying and Analyzing Safety Critical Maneuvers from High Resolution AIS Data. TransNav 10, 69–77. https://doi.org/10.12716/1001.10.01.07.

Murray, B., Perera, L.P., 2020. A dual linear autoencoder approach for vessel trajectory prediction using historical AIS data. Ocean Eng. 209, 107478. https://doi.org/10.1016/j.oceaneng.2020.107478.

Pallotta, G., Vespe, M., Bryan, K., 2013. Vessel pattern knowledge discovery from AIS data: A framework for anomaly detection and route prediction. Entropy 15, 2218–2245. https://doi.org/10.3390/e15062218.

Rawson, A., Brito, M., 2021. A critique of the use of domain analysis for spatial collision risk assessment. Ocean Eng. 219, 108259. https://doi.org/10.1016/j.oceaneng.2020. 108259.

Rong, H.; Teixeira, A. P., and Guedes Soares, C. 2015. Simulation and analysis of maritime traffic in the Tagus River Estuary using AIS data. Guedes Soares, C. & Santos T.A. (Eds.). Maritime Technology and Engineering. London, UK: Taylor & Francis Group; pp. 185–194.

Rong, H.; Teixeira, A. P., and Guedes Soares, C. Spatial-temporal analysis of ship traffic in Azores based on AIS data. Guedes Soares, C. & Santos T. A., (Eds.). Developments in Maritime Technology and Engineering. London, UK: Taylor and Francis; 2021; pp. Vol 1, pp. 185–192.

Rong, H., Teixeira, A.P., Guedes Soares, C., 2022a. Ship collision avoidance behaviour recognition and analysis based on AIS data. Ocean Eng. 245, 110479. https://doi.org/10.1016/j.oceaneng.2021.110479.

Rong, H., Teixeira, A.P., Guedes Soares, C., 2022b. Maritime traffic probabilistic prediction based on ship motion pattern extraction. Reliab. Eng. Syst. Saf. 217, 108061. https://doi.org/10.1016/j.ress.2021.108061.

Rong, H., Teixeira, A.P., Guedes Soares, C., 2021. Spatial correlation analysis of near ship collision hotspots with local maritime traffic characteristics. Reliab. Eng. Syst. Saf. 209, 107463. https://doi.org/10.1016/j.ress.2021.107463.

Rong, H., Teixeira, A.P., Guedes Soares, C., 2020. Data mining approach to shipping route characterization and anomaly detection based on AIS data. Ocean Eng. 198, 106936. https://doi.org/10.1016/j.oceaneng.2020.106936.

Santos, T. A. and Guedes Soares, C. 2017. Development dynamics of the Portuguese range as a multi-port gateway system. Journal of Transport Geography. 60:178–188.

Silveira, P. A. M.; Teixeira, A. P., and Guedes Soares, C. 2013. Use of AIS Data to Characterise Marine Traffic Patterns and Ship Collision Risk off the Coast of Portugal. Journal of Navigation. 66(6):879–898.

Silveira, P., Teixeira, A.P., Figueira, J.R., Guedes Soares, C., 2021. A multicriteria outranking approach for ship collision risk assessment. Reliab. Eng. Syst. Saf. 214, 107789. https://doi.org/10.1016/j.ress.2021.107789.

Teixeira, A.P., Guedes Soares, C., 2018. Risk of maritime traffic in coastal waters, in: Proceedings 37th International Conference on Ocean, Offshore and Arctic Engineering (OMAE2018). Paper No: OMAE2018-77312, V11AT12A025. https://doi.org/10.1115/OMAE2018-77312.

Tu, E., Zhang, G., Rachmawati, L., Rajabally, E., Huang, G. Bin, 2018. Exploiting AIS Data for Intelligent Maritime Navigation: A Comprehensive Survey from Data to Methodology. IEEE Trans. Intell. Transp. Syst. 19, 1559–1582. https://doi.org/10.1109/TITS.2017.2724551.

Varlamis, I., Kontopoulos, I., Tserpes, K., Etemad, M., Soares, A., Matwin, S., 2020. Building navigation networks from multi-vessel trajectory data. Geoinformatica. https://doi.org/10.1007/s10707-020-00421-y.

Vitali, N.; Prpic-Oršic, J., and Guedes Soares, C. 2020. Coupling voyage and weather data to estimate speed loss of container ships in realistic conditions. Ocean Engineering. 210: 106758.

Xiao, Z., Ponnambalam, L., Fu, X., Zhang, W., 2017. Maritime Traffic Probabilistic Forecasting Based on Vessels' Waterway Patterns and Motion Behaviors. IEEE Trans. Intell. Transp. Syst. 18, 3122–3134. https://doi.org/10.1109/TITS.2017.2681810.

Zhang, J.F., Teixeira, A.P., Guedes Soares, C., Yan, X., Liu, K., 2016. Maritime Transportation Risk Assessment of Tianjin Port with Bayesian Belief Networks. Risk Anal. 36, 1171–1187. https://doi.org/10.1111/risa.12519.

Zhang, J. F.; Wan, C. P.; He, A.; Zhang, D., and Guedes Soares, C. 2021. A two-stage black-spot identification model for inland waterway transportation. Reliability Engineering and System Safety. 213:107677.

Zhang, L., Meng, Q., Fwa, T.F., 2019. Big AIS data based spatial-temporal analyses of ship traffic in Singapore port waters. Transp. Res. Part E Logist. Transp. Rev. 129, 287–304. https://doi.org/10.1016/j.tre.2017.07.011.

Zhou, Y., Daamen, W., Vellinga, T., Hoogendoorn, S.P., 2019. Ship classification based on ship behavior clustering from AIS data. Ocean Eng. 175, 176–187. https://doi.org/10.1016/j.oceaneng.2019.02.005.

*Trends in Maritime Technology and Engineering – Guedes Soares & Santos (eds)*
 *ISBN 978-1-032-33583-4*

# Ship abnormal behaviour detection off the continental coast of Portugal

H. Rong, A.P. Teixeira & C. Guedes Soares
*Centre for Marine Technology and Ocean Engineering (CENTEC), Instituto Superior Técnico, Universidade de Lisboa, Lisbon, Portugal*

ABSTRACT: In alternative to rule-based positional anomalies that do not explicitly rely on historical maritime traffic information, a data-driven method for ship anomaly detection is proposed. The approach is derived from a maritime traffic normalcy model that is constructed based on historical ship trajectories provided by Automatic Identification System data. Positional and motion abnormal behaviours are formulated probabilistically based on statistical models of the lateral distribution of the ships' trajectories and their motion characteristics such as speed and course changes along the routes. The approach is applied to historical maritime traffic data off the continental coast of Portugal.

## 1 INTRODUCTION

Maritime transportation contributes to 90% of the international trade by volume and, therefore, safety and security issues have always been a major concern in the maritime sector (Teixeira and Guedes Soares 2018).

The Automatic Identification System (AIS), enforced by the International Maritime Organization (IMO) for ships over 300 tons and all passenger ships, has introduced a new data source for maritime traffic studies. AIS allows automatic transmission of spatial-temporal data (e.g. position, speed, Course Over Ground) every 2-10s and static data (e.g. Maritime Mobile Service Identity (MMSI), ship type and size) between ships and shore-based stations. Large datasets of historical ship trajectories provided by AIS have been used in many maritime traffic studied, such as for maritime traffic characterization (Wu *et al.,* 2016, Kang *et al.,* 2018; Silveira *et al.,* 2013); for collision risk assessment (Mou *et al.,* 2010; Qu *et al.* 2011; Goerlandt *et al.,* 2012; Dinis *et al.,* 2020; Silveira *et al.* 2021), for ship behaviour analysis (Du *et al.* 2020; Gao and Shi 2020; Rong *et al.* 2022a); for ship trajectory prediction (e.g. Rong *et al.,* 2019; Rong *et al.,* 2022b); for anomaly detection (e.g. Riveiro *et al.,* 2018; Rong *et al.,* 2020a); among others. The AIS data have been also used to develop models to assess the static (Dinis *et al.,* 2020) and dynamic (Yu *et al.* 2021) risk characteristics of ships and of the maritime traffic (e.g. Rong *et al.,* 2020b; Rong *et al.,* 2021)

Tu *et al.* (2018) provided a comprehensive perspective on the use of AIS data for smart navigation in which maritime anomaly detection, ship route estimation, collision prediction and path planning are reviewed. Pallotta *et al.* (2013) presented a machine learning approach to extract knowledge of the maritime motion patterns, and applied it for ship route prediction. A data mining approach was presented by Rong *et al.* (2020a) for characterization of the maritime traffic off the coast of Portugal in a probabilistic manner. In addition, the relationship between ship collision hotspots and the local traffic characteristics was studied by Rong *et al.* (2021).

The surveillance of the maritime traffic is of great importance for maritime safety and security, and the detection of ship abnormal behaviour is one of the critical tasks of maritime traffic surveillance.

In the maritime domain, ship abnormal behaviour such as an unexpected stop, deviation from a regular ship route, traffic direction violations among others, may indicate threats and dangerous situations.

Riveiro *et al.* (2018) presented a comprehensive literature review on maritime abnormal behaviour detection. Basically, the methods and techniques for maritime abnormal behaviour detection can be categorized, based on the data processing strategy, into rule-based approaches, data-driven approaches and hybrid approaches. A rule-based approach seeks for predefined patterns, areas or dynamic characteristics on the ship trajectories or specific trajectory evolutions. The European Maritime Safety Agency (EMSA 2018) has developed the Automated Behaviour Monitoring (ABM) system to support maritime surveillance. The system includes a set of predefined rule-based scenarios in which ships arriving at specific areas, following unusual routes and located in specific areas can be automatically detected. Roy (2010) implemented a rule-based automated reasoning tool in an expert system for supporting maritime abnormal behaviour detection with semantic information. The rule-based approach defines the

DOI: 10.1201/9781003320289-16

abnormal behaviour in a supervised way. However, as the abnormal behaviours occur rarely and are different from one case to another, the application of rule-based approach in maritime surveillance is limited.

Data-driven approaches assume that most of the historical ship trajectories reflect the "normal" behaviour and aim at establishing a maritime traffic normalcy model based on historical ship trajectories. Ristic *et al.* (2008) extracted traffic motion patterns in ports and waterways using AIS data and trained an adaptive Kernel Density Estimation (KDE) method as motion anomaly detector. Rong *et al.* (2019) used ship trajectories within a motion pattern to develop a probabilistic trajectory prediction method based on Gaussian Process models and applied the prediction model for real-time ship abnormal behaviour detection. Vespe *et al.* (2012) proposed an efficient representation and consistent knowledge of ship behaviour based on AIS data and from which low-likelihood ship behaviour can be detected. Mascaro *et al.* (2014) applied a dynamic Bayesian Network model to detect ship trajectory deviations from normal patterns.

The data-driven approaches construct traffic normalcy models in an unsupervised way on large scale datasets. Hybrid approached combine data-driven and rule-based approaches in a semi-supervised way by which complex critical events or situations can be detected. Pristrom *et al.* (2016) proposed an analytical model incorporating Bayesian reasoning to estimate the likelihood of a ship being hijacked in the Western Indian or Eastern African region. The proposed model considers the characteristics of the ship, environmental conditions, and the maritime security measures in place in an integrated manner.

Data-driven based approaches are extensively applied for maritime traffic characterization. Basically, ship traffic statistical analyses are performed to provide quantitative representations of the traffic features. The derived traffic features can be used for constructing a traffic normalcy model (or distribution) to help distinguish the normal and abnormal ship behaviours. In this study, an unsupervised method for extracting and characterizing ship routes and detecting different abnormal behaviours is presented.

## 2 METHODOLOGY

In the context of maritime traffic, ship abnormal behaviour, such as deviation from a standard ship route, unexpected low speed, motion direction anomaly, among others, may indicate threats and dangers related to, for example, ship-ship collision, hijacking or piracy. The detection of such events is of importance for implementing proactive measures. However, due to information overload and fatigue, it is difficult for humans to track all ship trajectories in real time and detect the abnormal behaviour. Therefore, there is a need for automated detection of abnormal trajectory patterns. In this study, a data-driven approach for learning a maritime traffic normalcy model and detecting ship abnormal behaviour is presented.

The approach consists of building a maritime traffic normalcy model based on the analysis of historical ship trajectories and to apply the normalcy model for ship abnormal behaviour detection. As shown in the flowchart of Figure 1, the proposed approach for ship abnormal behaviour detection includes three main steps: 1) Grouping the ship trajectories according to corresponding motion patterns; 2) Constructing a normalcy model for ship routes based on analysis of ship trajectory clusters in terms of lateral distribution along the ship route, speed distribution and Course Over Ground (COG) distribution; 3) Using the normalcy model for ship abnormal behaviour detection. It should be mentioned that in this study the ship route is assumed as a set of straight legs connecting turning sections. The ship route is characterized probabilistically along the ship route in terms of lateral distribution, direction and speed profiles, which allow for the characterization of the typical behaviour of ships navigating along a specific route.

### 2.1 *Ship route extraction*

Grouping and analysing the ship trajectories following a particular ship route is important for characterization of the ship behaviour. In this study, ship route extraction refers to grouping ship trajectories that start at the same origin, follow the same ship route and end at the same destination. The first step for ship route extraction is the identification of origin-destination areas in the study area, and then grouping the ship trajectories that belong to the same itinerary. More detailed information on ship route extraction can be found in the study by Rong *et al.* (2020b).

### 2.2 *Ship route characterization*

The characterisation of all abnormal behaviours or situations a priori is usually a difficult task, since the behaviour sought is relatively rare and the set of examples of such unusual ship behaviours is limited. In contrast to the scarcity of ship abnormal behaviour, there is a large amount of data corresponding to normal routine behaviour of ships, which enables the maritime traffic normalcy model to be built.

Before detecting the ship abnormal behaviour, the ship normal behaviour in specific ship routes needs to be characterized, so as to generate normal behaviour estimations against which monitored behaviour can be compared. The lateral distribution of ship trajectory along the ship route, and the speed and COG profiles are considered for building the ship route normalcy model. In this study, the normalcy model can be estimated by a series of Gaussian distributions. This probabilistic normalcy representation supports anomaly detection of ship behaviour that deviate from the standard behaviour. More specifically, the lateral distribution of the ship trajectory enables to determine the route boundary and detect the off-route behaviour, whereas the speed and COG profiles enable to determine ship speeds or directions not compatible with the ship route.

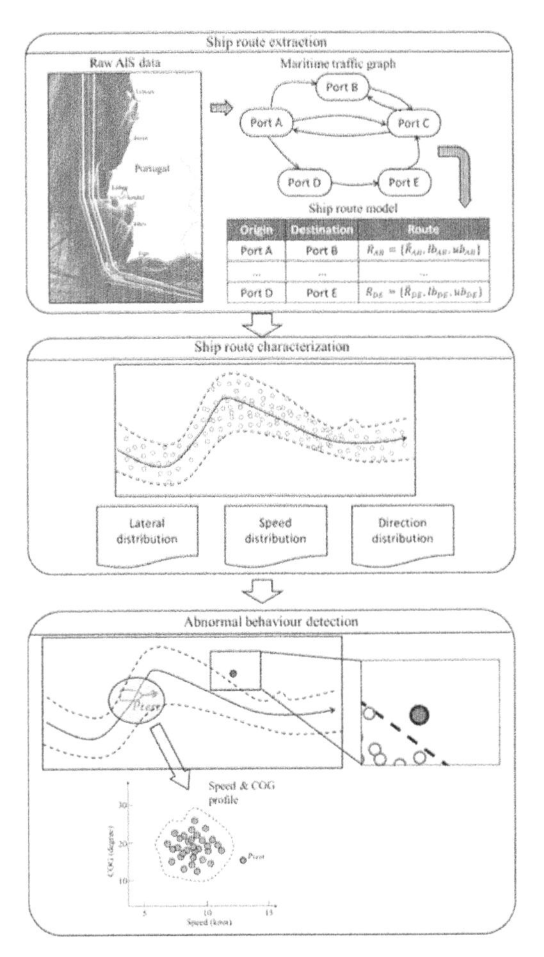


Figure 1. Flowchart of the maritime traffic characterization and abnormal behaviour detection approach.

## 2.3 *Abnormal behaviour detection*

Rong *et al.* (2020a) have shown that valuable knowledge regarding the ship behaviour along a specific ship route can be derived from the analysis of ship trajectories. In this study, three types of ship abnormal behaviour are addressed which are off-route behaviour, unexpected speed and ship direction not compatible with the route, as shown in Figure 2bcd.

In practice, classifying ship position into a specific route is essential for tracking ships. The task of route classification consists of determining whether the ship position is located inside the route boundary. As shown in Figure 2a, the ship is first assigned to the route based on its positional information only. Once the ship is associated to a specific route, then, off-route behaviour is triggered when a ship is found travelling outside the route boundary by calculating the probability under the Gaussian distribution function at the ship lateral position, as shown in Figure 2b. In addition, compatibility tests between the ship kinematic status and the speed and direction profiles along the route can be performed. An unexpected speed (speed is too high or too low) and a ship direction not compatible with the route can be determined, as shown in Figure 2cd, respectively.

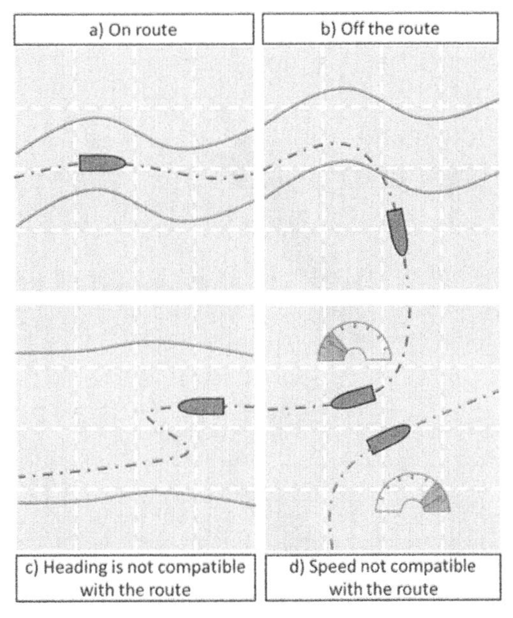


Figure 2. Abnormal behaviour types considered in this study.

# 3 CASE STUDY

The maritime traffic off the continental coast of Portugal is formed by ship routes connecting the northern Europe and the Mediterranean Sea and by ships bound to and leaving from national ports. Two Traffic Separation Schemes (TSSs) located off Cape Roca and off Cape St. Vincent organize the maritime traffic off the coast of Portugal. The available AIS historical dataset refers to the period from 1st October to 31th December, 2015, and the study area in this study is bounded by parallels 36°N and 42°N, and by meridians 7°W and 11°W.

## 3.1 *Ship route extraction*

By applying the Order Points to Identify the Clustering Structure (OPTICS) algorithm to static positions and enter/exit points, the origin-destination areas are determined. Then, individual ship trajectories are grouped, and the ship route boundaries are estimated by the analysis of the ship trajectories within the traffic groups (Rong *et al.,* 2022b). As shown in Figure 3, there are 68 ship routes identified off the coast of Portugal. Each ship route is defined by the route centreline and route boundaries. The route centreline is represented by a polyline estimated by statistical analysis of aligned ship trajectories within the route. The route boundaries express the spatial variability of the ship trajectories around the route centreline, which is probabilistically described by the lateral distribution of ship positions.

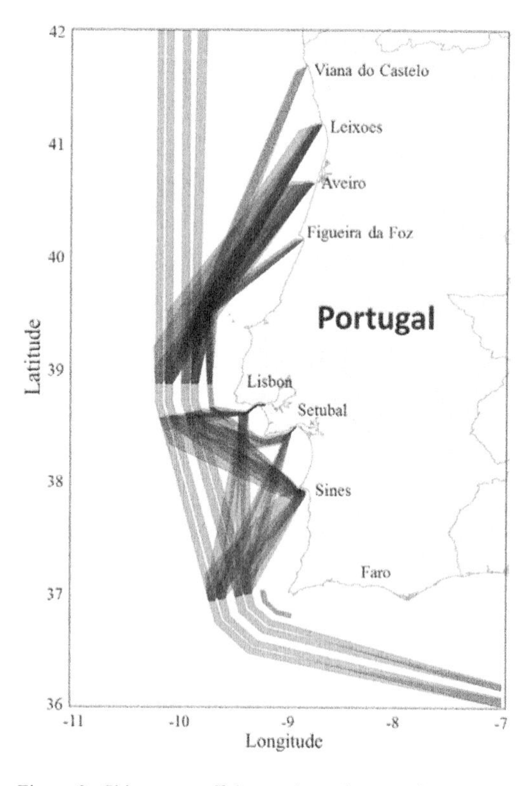


Figure 3. Ship routes off the continental coast of Portugal.

An example of a ship route connecting the ports of Leixões and Lisbon is shown in Figure 4. The ships from the port of Leixões initially move southward and then turn eastward to approach the port of Lisbon. There are 49 ship trajectories, containing 1714 data points, detected from historical AIS data that follow the ship route, as shown in Figure 4a. The ship route, including the route centreline (represented by a dash line) and route boundary, estimated from the ship trajectories are shown in Figure 4b. According to Figure 4c, the ship positions (represented by grey arrows) are well located within the route boundaries and the route centreline represents the tendency of the ships' motion along the ship route. The grey polygon in Figure 4b represents the spread of the ship route, and ships traveling outside of the polygon are considered as off-route behaviour.

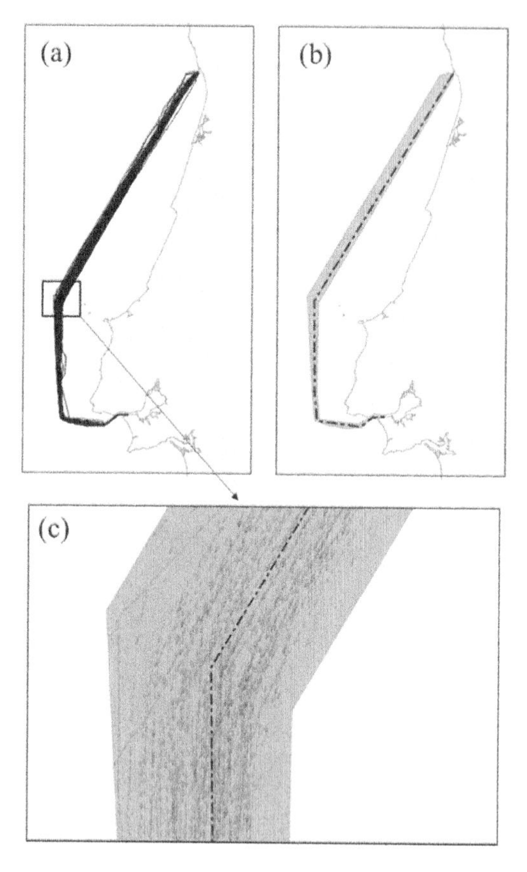


Figure 4. Example of a ship route from the port of Leixões to the port of Lisbon: (a) Ship trajectories; (b) Ship route model; (c) Mean route and route boundary.

### 3.2 *Ship route characterization*

To better capture the pattern of the ships' behaviour along the ship route, the ship route is divided into a set of straight legs connecting turning sections (Rong *et al.,* 2020a). The turning section detection is facilitated by clustering the turning points of ship trajectories that are detected based on a ship trajectory compression method. As shown in Figure 5, four turning point clusters, represented by red, green, yellow and blue circles, are detected based on the DBSCAN algorithm. The corresponding turning sections are formed by the convex polygon of turning points clusters.

Figure 6 illustrates the results of the ship route segmentation. The ship route is divided by four turning sections, and there are three main straight sections, as shown in Figure 6. In Section 1, ships move southwest and then turn south towards the TSS off Cape

Roca, and finally ships turn east towards the port of Lisbon. The average speed of the straight sections indicate that ship gradually decelerate from port of Leixões to port of Lisbon.

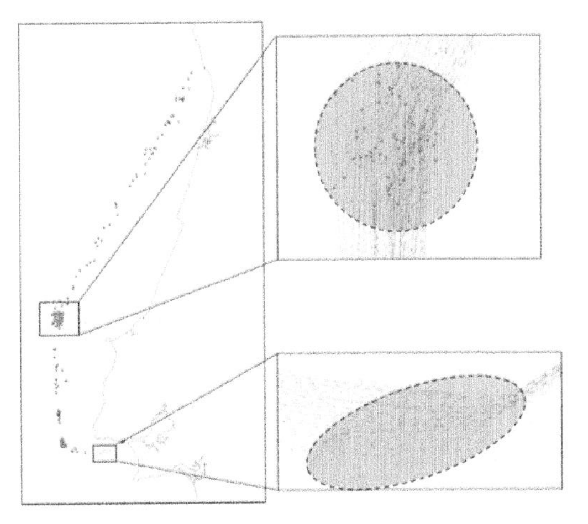

Figure 5. Turning sections determined by clustering trajectory turning points.

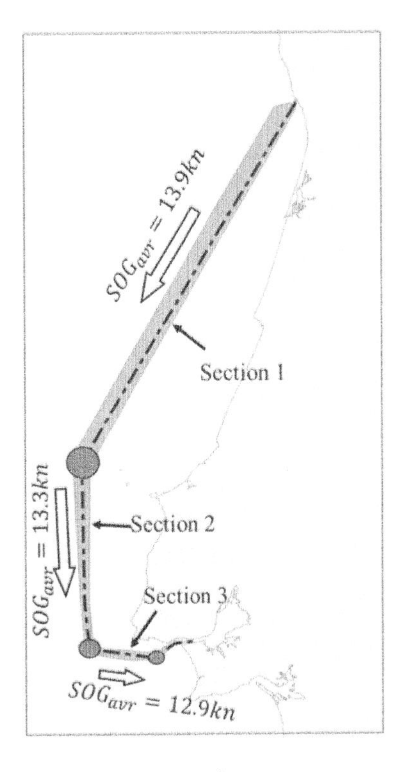


Figure 6. Ship route segmentation.

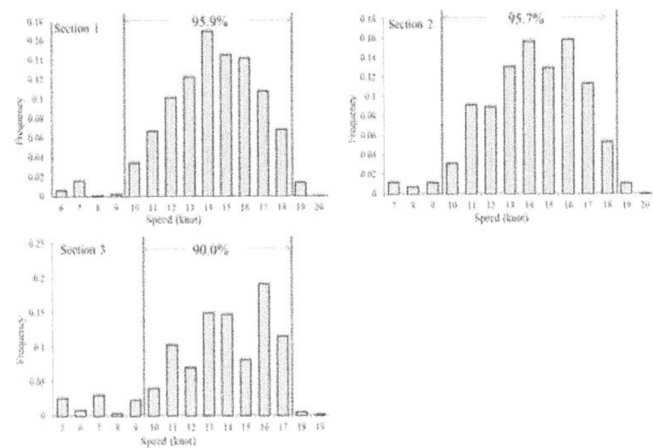


Figure 7. Speed distributions along the three route sections.

Figure 7 and Figure 8 illustrate the speed and COG distributions of ships in the three sections. In terms of speed distributions, it is seen that most ships in the route sail at 10~18 knot (95.9% in Section 1, 95.7% in Section 2 and 90.0% in Section 3, respectively). Figure 8 shows the COG distribution of along the three sections. Significant differences between the COG distributions can be observed. The COG of most ships is between 200~212 degrees, 170~186 degrees and 94~108 degrees (93.8% in Section 1, 96.6% Section 2 and 82.7% Section 3, respectively), which is in line with the trend of the ship route.

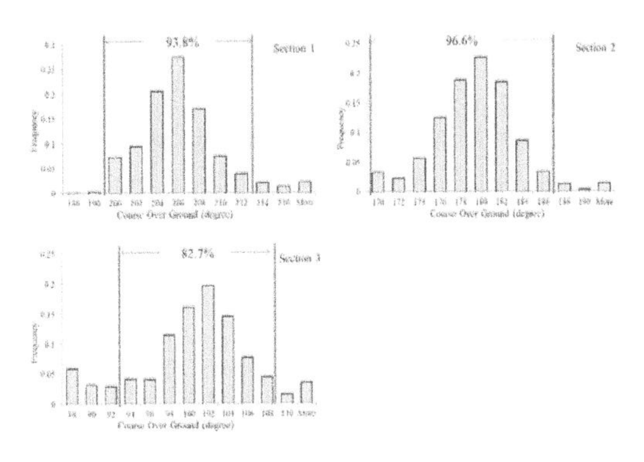


Figure 8. COG distributions along the three route sections.

The lateral distribution reflects the spatial extension of the ship routes. Figure 9 shows the lateral distributions of the ship trajectories in the three sections. On the *y-axis*, the zero value corresponds to the position of the estimated route centreline. A positive value for *y* refers to ships to the starboard side of the route centreline. The lateral distribution along the ship route in Section 1 and Section 2 are calculated from the available AIS data at every 10 nautical miles (NM), and in Section 3 are calculated at every 2 NM. The grey box indicates the 95% confidence interval estimated by the

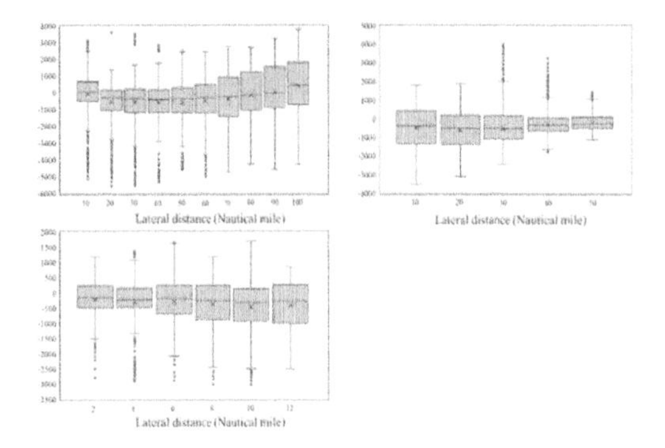


Figure 9. Lateral distributions along the three route sections.

series of route widths in each section. The comparison of the lateral distributions of the three sections shows that the deviation of lateral distribution in Section 1 and Section 2 are larger than in Section 3, which means that ship trajectories in Section 3 are more concentrated on the route centreline. It is found that the route width changes along the ship route (Figure 6). For example, at the south part of Section 2 the route is narrower than in other locations of the Section 2. According to Figure 9a, the route width estimated by AIS data varies considerably. The reason is that the ships in Section 1 sail at an open sea area and, consequently, the variability of the ship trajectories along the route is higher. The route width distribution of Section 2 (Figure 9b) indicates that the route becomes narrower from north to south. The reason is that the maritime traffic at the end of Section 2 converges by the TSS off Cape Roca.

### 3.3 *Abnormal behaviour detection*

In this study, the abnormal behaviour is considered as a deviation from the normality, and the normalcy model is built from statistical analysis of ship trajectory data, in terms of lateral, speed and COG distributions. Figures 10-12 show three types of ship abnormal behaviours detected based on the method proposed in this study. The red markers highlight the abnormal behaviour, and the blue markers indicate the normal behaviour.

Ship off-route behaviour may be caused by several reasons: the ship may deviate from the route to avoid other ships or turn towards the destination (or next waypoint) too soon or too late. As shown in Figure 10, the ship is assigned to the route from the port of Leixões to the port of Lisbon. Based on the assumption that the ship lateral distribution is given by a normal distribution along the ship route, the probability that the ship belongs to the route can be calculated by the ship lateral position. Then, the off-route behaviour is detected as shown in Figure 10a. Figure 10b shows the ship lateral position (represented by a dash line) and the ship in-route probability (represented by a solid

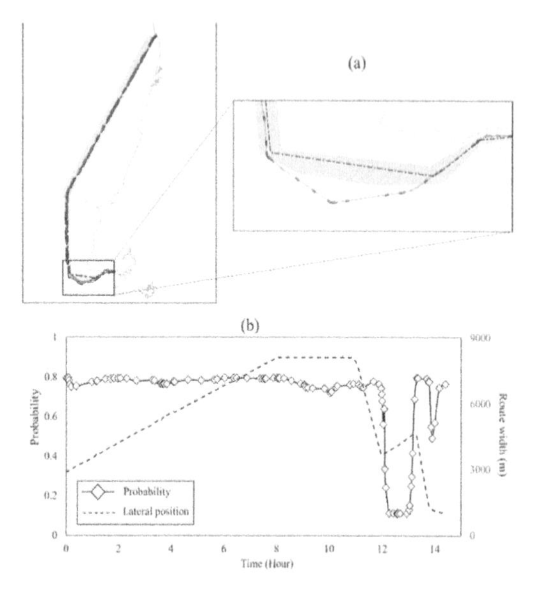


Figure 10. Detection of a ship off-route behaviour: (a) Off-route trajectory; (b) Sequential estimation of ship in-route probability and lateral position.

line and a hollow rectangle). It is seen that the ship in-route probability decreases dramatically when the ship trajectory is outside the ship route.

Ships navigating from one port to another typically prefer to use an effective path and exhibit similar speed and direction. Therefore, ship abnormal behaviour can be detected if the ship speed or direction considerably deviate from the normalcy speed and direction route models. As shown in Figure 11a, a ship initially moves eastward with a speed in accordance with the ship route. Its positions are shown by blue markers that represent a normal ship

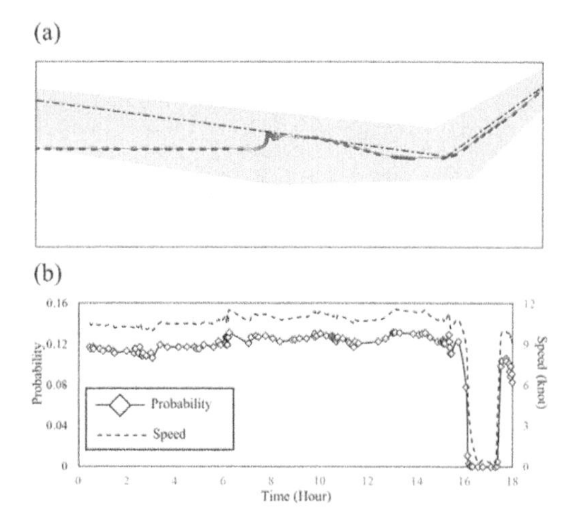


Figure 11. Detection of a ship speed abnormal behaviour: (a) Speed abnormal behaviour; (b) Sequential estimation of ship speed probability and ship speed.

behaviour. Then, the ship speed reduces dramatically from 11 knot to 1 knot, and the speed probability decreases to 0.01, which results in the detection of a speed abnormal behaviour, as shown in Figure 11b.

Figure 12 shows the detection of a ship COG not compatible with the ship route. In Figure 12a, the ship position and motion, represented by blue markers, are compatible with the ship route. Then, the ship starts heading southward and makes a U-turn. Even though, the positional features are still compatible with the routes, the probability decays dramatically meaning that the ship's COG is incompatible with the normal movement of this route. The abnormal behaviour is highlighted by red markers and after the double U-turns the ship changes again into the normal motion flow of the ship route, represented by the position blue markers. The abnormal behaviour detection results suggest that the method can detect deviations from the expected and regular motion in each route.

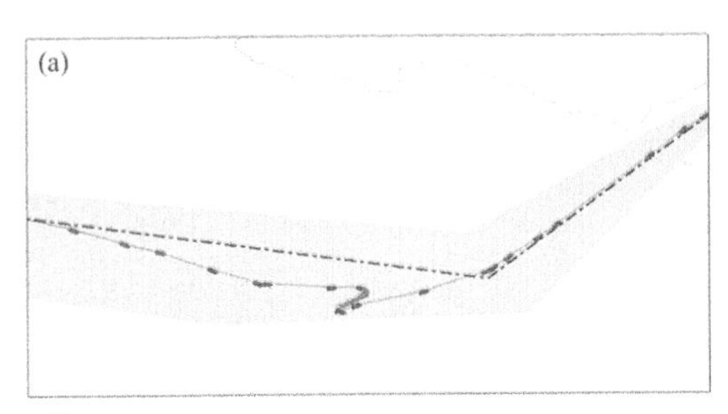


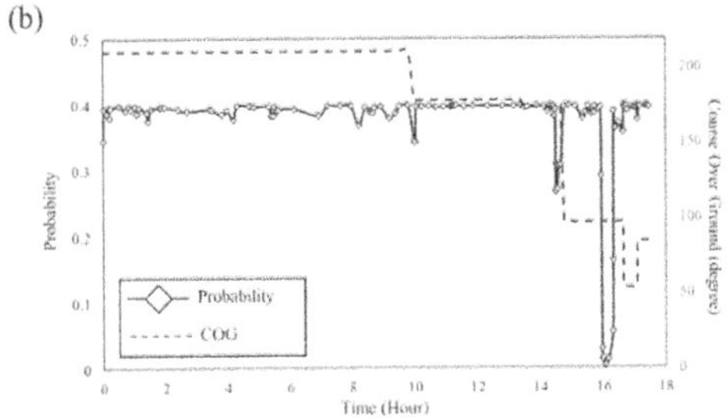


Figure 12. Detection of a ship direction abnormal behaviour: (a) Ship direction abnormal behaviour; (b) Sequential estimation of ship direction probability and ship COG.

## 4 CONCLUSIONS

The proposed ship abnormal behaviour detection method provides a straightforward way to detect different types of ship abnormal behaviours based on maritime traffic normalcy models constructed for specific ship routes. The ship route is extracted in an unsupervised way from historical AIS data, and the characterization of the ship routes is based on statistical models of lateral position, speed and Course Over Ground of ship trajectories within the ship routes. The off-route behaviour, unexpected speed and heading not compatible with the route can be effectively detected in a probabilistic manner.

The maritime traffic off the continental coast of Portugal is crossed by a ship traffic network formed by 68 ship routes derived from historical AIS data. The ship route connecting the ports of Leixões and Lisbon is selected to demonstrate the proposed approach. The ship route is divided into a set of turning sections and straight sections. A statistical analysis of the ship route at each straight section shows that the average speed of the ships decreases gradually along the ship route. In addition, the ship speed in the three sections is mainly in the 10-18 knot range and the COG distributions reflect the ship route tendency. The lateral distribution of the ship trajectories along the ship route provides an indication on the spatial extension of the route, which can be used for ship off-route behaviour detection.

In this study, three ship trajectories are used for demonstrating the abnormal behaviour detection method based on the constructed ship route normalcy model. The abnormal behaviour detection results suggest that the proposed approach can detect ship off-route behaviours, unexpected speeds and headings not compatible with the route.

## ACKNOWLEDGEMENTS

The paper contributes to the project "Monitoring and Surveillance of Maritime Traffic off the continental coast of Portugal (MoniTraffic)", which has been co-funded by "Fundo Azul" programme of the Portuguese Directorate-General for Maritime Policy, under contract no. FA-04-2017-005 and to the project "Integrated System for Traffic Monitoring and Maritime Risk Assessment (MoniRisk)", which has been co-funded by the European Regional Development Fund (Fundo Europeu de Desenvolvimento Regional (FEDER) and by the Portuguese Foundation for Science and Technology (Fundação para a Ciência e a Tecnologia – FCT) under contract no. 028746. This work also contributes to the Strategic Research Plan of the Centre for Marine Technology and Ocean Engineering (CENTEC), which is financed by the Portuguese Foundation for Science and Technology under contract UIDB/UIDP/00134/2020.

## REFERENCES

Dinis, D., Teixeira, A.P., and Guedes Soares, C., 2020. Probabilistic approach for characterising the static risk of ships using Bayesian networks. *Reliability Engineering & System Safety*, 203 (June), 107073.

Du, L., Goerlandt, F., Valdez Banda, O.A., Huang, Y., Wen, Y., and Kujala, P., 2020. Improving stand-on ship's situational awareness by estimating the intention

of the give-way ship. *Ocean Engineering*, 201 (December 2019), 107110.
EMSA, 2018. *Automated Behaviour Monitoring*. European Maritime Safety Agency. European Union (EU).
Gao, M. and Shi, G., 2020. Ship-handling behavior pattern recognition using AIS sub-trajectory clustering analysis based on the T-SNE and spectral clustering algorithms. *Ocean Engineering*, 205 (January), 106919.
Goerlandt, F., Montewka, J., Lammi, H., and Kujala, P., 2012. Analysis of near collisions in the Gulf of Finland. *In*: C. Bérenguer, A. Grall, and C. Guedes Soares, eds. *Advances in Safety, Reliability and Risk Management*. Taylor & Francis Group, London, 2880–2886.
Kang, L., Meng, Q., and Liu, Q., 2018. Fundamental diagram of ship traffic in the Singapore Strait. *Ocean Engineering*, 147 (October 2017), 340–354.
Mascaro, S., Nicholson, A., and Korb, K., 2014. Anomaly detection in vessel tracks using Bayesian networks. *International Journal of Approximate Reasoning*, 55 (1 PART 1), 84–98.
Mou, J.M., Tak, C. van der, and Ligteringen, H., 2010. Study on collision avoidance in busy waterways by using AIS data. *Ocean Engineering*, 37 (5–6), 483–490.
Pallotta, G., Vespe, M., and Bryan, K., 2013. Vessel pattern knowledge discovery from AIS data: A framework for anomaly detection and route prediction. *Entropy*, 15 (6), 2218–2245.
Pristrom, S., Yang, Z., Wang, J., and Yan, X., 2016. A novel flexible model for piracy and robbery assessment of merchant ship operations. *Reliability Engineering and System Safety*, 155, 196–211.
Qu, X., Meng, Q., and Suyi, L., 2011. Ship collision risk assessment for the Singapore Strait. *Accident Analysis and Prevention*, 43 (6), 2030–2036.
Ristic, B., La Scala, B., Morelande, M., and Gordon, N., 2008. Statistical analysis of motion patterns in AIS data: Anomaly detection and motion prediction. *Proceedings of the 11th International Conference on Information Fusion, FUSION 2008*, 1–7.
Riveiro, M., Pallotta, G., and Vespe, M., 2018. Maritime anomaly detection: A review. *Wiley Interdisciplinary Reviews: Data Mining and Knowledge Discovery*, 8 (5), 1–19.
Rong, H., Teixeira, A.P., and Guedes Soares, C., 2019. Ship trajectory uncertainty prediction based on a Gaussian Process model. *Ocean Engineering*, 182 (December 2018), 499–511.
Rong, H., Teixeira, A.P., and Guedes Soares, C., 2020a. Data mining approach to shipping route characterization and anomaly detection based on AIS data. *Ocean Engineering*, 198 (August 2019), 106936.
Rong, H., Teixeira, A.P., and Guedes Soares, C., 2020b. Risk of Ship Near Collision Scenarios off the Coast of Portugal. *In*: *Proceedings of the 29th European Safety and Reliability Conference, ESREL 2019*. 3660–3666.
Rong, H., Teixeira, A.P., and Guedes Soares, C., 2021. Spatial correlation analysis of near ship collision hotspots with local maritime traffic characteristics. *Reliability Engineering and System Safety*, 209 (August 2020), 107463.
Rong, H., Teixeira, A.P., and Guedes Soares, C., 2022a. Ship collision avoidance behaviour recognition and analysis based on AIS data. *Ocean Engineering*, 110479.
Rong, H., Teixeira, A.P., and Guedes Soares, C., 2022b. Maritime traffic probabilistic prediction based on ship motion pattern extraction. *Reliability Engineering and System Safety*, 217, 108061.
Roy, J., 2010. Rule-based expert system for maritime anomaly detection. *In*: *SPIE Defense, Security, and Sensing*. Orlando, Florida, United States.
Silveira, P., Teixeira, A.P., Figueira, J.R., and Guedes Soares, C., 2021. A multicriteria outranking approach for ship collision risk assessment. *Reliability Engineering and System Safety*, 214 (December 2020), 107789.
Silveira, P., Teixeira, A.P., and Guedes Soares, C., 2013. Use of AIS Data to Characterise Marine Traffic Patterns and Ship Collision Risk off the Coast of Portugal. *Journal of Navigation*, 66 (06), 879–898.
Teixeira, A.P. and Guedes Soares, C., 2018. Risk of maritime traffic in coastal waters. *In*: *Proceedings of the International Conference on Offshore Mechanics and Arctic Engineering (OMAE)*. June 17-22, 2018, Madrid, Spain, 1–10.
Tu, E., Zhang, G., Rachmawati, L., Rajabally, E., and Huang, G. Bin, 2018. Exploiting AIS Data for Intelligent Maritime Navigation: A Comprehensive Survey from Data to Methodology. *IEEE Transactions on Intelligent Transportation Systems*, 19 (5), 1559–1582.
Vespe, M., Visentini, I., Bryan, K., and Braca, P., 2012. Unsupervised learning of maritime traffic patterns for anomaly detection. *9th IET Data Fusion & Target Tracking Conference (DF&TT 2012): Algorithms & Applications*, (November), 14–14.
Wu, X., Mehta, A.L., Zaloom, V.A., and Craig, B.N., 2016. Analysis of waterway transportation in Southeast Texas waterway based on AIS data. *Ocean Engineering*, 121, 196–209.
Yu, Q., Teixeira, A.P., Liu, K., Rong, H., and Guedes Soares, C., 2021. An integrated dynamic ship risk model based on Bayesian Networks and Evidential Reasoning. *Reliability Engineering & System Safety*, 107993.

*Trends in Maritime Technology and Engineering – Guedes Soares & Santos (eds)*



# Characterisation of ship routes off the continental coast of Portugal using the Dijkstra algorithm

P. Silveira
*Centre for Marine Technology and Ocean Engineering (CENTEC), Instituto Superior Técnico, Universidade de Lisboa, Lisbon, Portugal*
*Escola Superior Náutica Infante D. Henrique (ENIDH), Paço de Arcos, Portugal*

A.P. Teixeira & C. Guedes Soares
*Centre for Marine Technology and Ocean Engineering (CENTEC), Instituto Superior Técnico, Universidade de Lisboa, Lisbon, Portugal*

ABSTRACT: The paper uses the well-known Dijkstra's shortest path algorithm to systematically characterize the maritime traffic of the continental coast of Portugal. The approach consists of using historical Automatic Identification System data to build a graph in which the nodes are cells of a grid covering the geographical area being studied and the weights of directional edges are inversely related to ship movements between cells. This route identification method is applied to characterize the maritime traffic pattern off the continental Portuguese coast, as well as between two other boundaries identified as potentially interesting through observation of traffic density maps. A similar approach is used to obtain the speed profiles along routes that are adequate to determine the global speed trends.

## 1 INTRODUCTION

The need to improve the safety of navigation has led to the implementation of coastal and port Vessel Traffic Services (VTS) that rely on several systems to provide monitoring, control, and information services. Among these systems, the Automatic Identification System (AIS) has become an important source of information for studying maritime traffic and associated risks (Silveira et al., 2013), as reviewed by Tu et al. (2018).

The ships' position, course, heading, speed, dimensions, type, draught, destination and other important data are broadcasted at regular intervals by AIS transceivers and can be stored for later analysis, making AIS data a valuable resource to researchers.

The general concept of information sharing and particularly route exchanging to improve maritime situational awareness and the safety of navigation has been explored, namely in the scope of several EU research projects such as the EfficienSea and the MONALISA projects. The idea is that a limited number of waypoints ahead of the present position are broadcasted using AIS to mitigate misunderstandings and hopefully aid in collision avoidance decisions (Porathe et al., 2014). It is possible to incorporate pre-shared routes in decision aiding systems such as the Marine Traffic Alert and Collision Avoidance System (MTCAS) (Denker and Hahn, 2016), inspired by the Traffic Collision Avoidance System (TCAS) implementation of the Airborne Collision Avoidance System (ACAS).

A great amount of research work exists on the topic of route or trajectory identification and its applications to AIS data, resulting in a variety of methods for route extraction, prediction and anomaly detection (Etienne et al., 2010; Vespe et al., 2012; Pallotta et al., 2013; Rong et al., 2019, 2020). The knowledge of route crossing areas, lateral distribution of traffic, ship dimensions and ship speeds is required to apply probabilistic models to estimate collision frequency, such as the one proposed by Pedersen (1995) and applied by Kujala et al. (2009). Safe path identification is an important step in the development of certified safe routes, which have the potential to become one of the key enablers for future autonomous vessel operations and may facilitate route exchange mechanisms, thus playing a role in future methods to reduce collision risk.

In this paper, AIS data is used to characterize maritime traffic off the coast of Portugal, by identifying the most used routes between important ports and area boundaries using the well-known Dijkstra shortest path algorithm. Issues related with the decoding and pre-processing of data are briefly described and traffic statistics are presented, along with a traffic density map extracted from data. The same algorithm is used to obtain a speed profile of ships using one of the identified routes.

DOI: 10.1201/9781003320289-17

## 2 AIS DATA DECODING AND PRE-PROCESSING

AIS messages are transmitted in an encoded format and must be decoded to be interpreted. The integrity of all AIS messages can be verified using a checksum that is found at the end of each message. However, it is possible to find messages with correct checksums but wrong bit lengths for the specified message type (Last et al., 2014). In addition, a number of other errors may be found in AIS data, resulting from poor human input, incorrect installation or faulty sensors. Harati-Mokhtari et al. (2007) found errors in various message fields, such as Maritime Mobile Service Identity (MMSI) number, vessel type, ship's name and call sign, navigational status, length and beam, position, draught, destination and Estimated Time of Arrival (ETA), heading, Course Over Ground (COG) and Speed Over Ground (SOG).

Tu et al. (2018) performed a data quality assessment on the completeness and resolution of data samples retrieved from several AIS data providers. This assessment was focused on four aspects: position precision, time stamp resolution (interval between two consecutive AIS messages), data completeness and erroneous/corrupted entries. These authors found that longitude and latitude are rarely invalid, but heading, status and Rate of Turn (ROT) are quite often invalid. As for Course Over Ground (COG) and Speed Over Ground (SOG) data entries, these occasionally have wrong values. SOG entries, in particular, may contain isolated extremely large or negative values, due to noise contamination during measure or transmission period.

Decoding AIS messages is a straightforward process, although decoders must be prepared for possible deviations from the standard. All messages used in the scope of this paper were decoded using purposely developed Python scripts. There are 27 different message types defined in (ITU, 2014). Message types 1, 2, 3 and 5 are used to study traffic and extract routes throughout this paper. Message types 1, 2 and 3 are class A position reports, while message type 5 contains class A static and voyage related data. The AIS data used in the scope of this work was recorded by Centro de Controlo de Tráfego Marítimo do Continente (CCTMC), the Portuguese coastal VTS centre, between 05-05-2012 and 21-06-2013. Position reports from outside a geographical area bounded by meridians 007.4° W and 010.24° W and by parallels 36.08° N and 41.86° N were ignored.

## 3 MARITIME TRAFFIC STATISTICS AND VISUALIZATION

For traffic characterization, Table 1 shows the traffic statistics obtained by sampling active ships from data at 30 minutes intervals. Cargo ships correspond to AIS ship type codes 70 through 79, tankers correspond to AIS ship type codes 80 through 89, passenger ships correspond to AIS ship type codes 60 through 69, fishing vessels correspond to AIS type code 30 and all other AIS ship types are included in the class *Other*. In SOG statistics, values over 35 knots, which amounted to 0.001% of all records, were disregarded. In length statistics, values of length equal to zero were also disregarded, amounting to 3.301% of all records. In draught statistics, values of draught equal to zero were equally disregarded, amounting to 10.35% of all records. Messages with correct checksum were not deleted, even if one or more of the fields was disregarded when producing statistics for a specific variable. The draught field is expected to be changed by the operator, so it seems to be more error-prone than the length field that is set during the installation of the AIS equipment, which in turn has more errors than the SOG field, that is automatically updated with information from a GNSS.

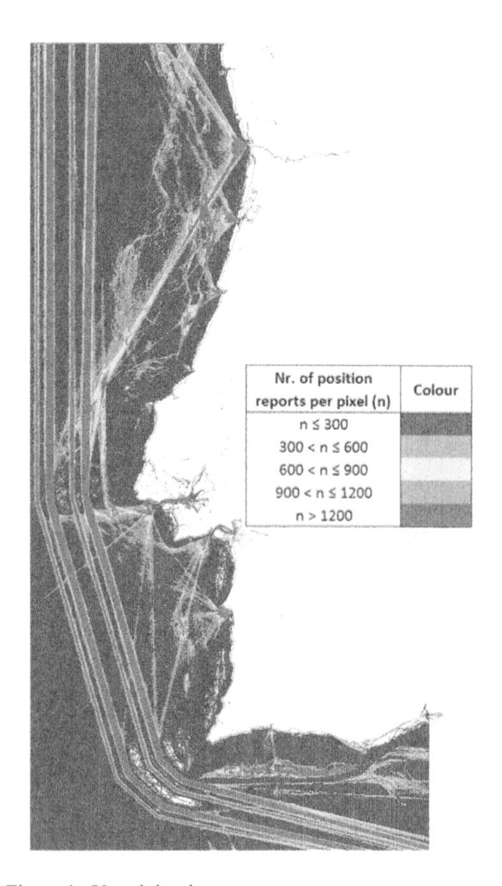


Figure 1. Vessel density map.

Visualizing shipping density is of great value for improving maritime situational awareness (Wu et al., 2017). Since the advent of AIS, it has been possible to gather a great amount of information on maritime traffic, creating the need to learn how to best visualize and use such information. Visual representations of AIS information can help in obtaining useful knowledge from data (Chen et al., 2016).

Table 1. Traffic statistics.

| Ship type | | Nav. status | Mean | Stand. Dev. |
|---|---|---|---|---|
| All | Number of ships | Underway | 113.37 | 13.22 |
| | | Other | 94.96 | 15.29 |
| | Length (m) | Underway | 141.05 | 80.78 |
| | | Other | 85.33 | 63.39 |
| | SOG (knots) | Underway | 10.41 | 5.23 |
| | | Other | 1.29 | 3.09 |
| | Draught (m) | Underway | 7.04 | 2.92 |
| | | Other | 5.81 | 4.18 |
| Cargo | Number of ships | Underway | 94.27 | 12.09 |
| | | Other | 46.98 | 9.11 |
| | Length (m) | Underway | 153.32 | 75.94 |
| | | Other | 129.57 | 53.21 |
| | SOG (knots) | Underway | 11.36 | 4.34 |
| | | Other | 0.34 | 1.72 |
| | Draught (m) | Underway | 7.40 | 2.84 |
| | | Other | 6.56 | 3.28 |
| Tanker | Number of ships | Underway | 31.67 | 5.36 |
| | | Other | 15.29 | 3.89 |
| | Length (m) | Underway | 164.89 | 62.82 |
| | | Other | 144.36 | 57.67 |
| | SOG (knots) | Underway | 11.42 | 3.26 |
| | | Other | 0.43 | 1.86 |
| | Draught (m) | Underway | 8.58 | 2.96 |
| | | Other | 6.82 | 2.26 |
| Passenger | Number of ships | Underway | 4.97 | 2.17 |
| | | Other | 9.85 | 4.08 |
| | Length (m) | Underway | 114.33 | 97.22 |
| | | Other | 57.15 | 49.05 |
| | SOG (knots) | Underway | 8.17 | 8.15 |
| | | Other | 2.08 | 5.23 |
| | Draught (m) | Underway | 4.79 | 2.83 |
| | | Other | 6.83 | 8.85 |
| Fishing | Number of ships | Underway | 2.43 | 1.25 |
| | | Other | 19.29 | 6.37 |
| | Length (m) | Underway | 34.95 | 20.90 |
| | | Other | 26.15 | 5.64 |
| | SOG (knots) | Underway | 3.49 | 3.72 |
| | | Other | 2.98 | 3.12 |
| | Draught (m) | Underway | 4.31 | 1.11 |
| | | Other | 3.65 | 0.75 |
| Other | Number of ships | Underway | 11.89 | 3.78 |
| | | Other | 18.84 | 4.74 |
| | Length (m) | Underway | 68.80 | 62.30 |
| | | Other | 46.04 | 36.88 |
| | SOG (knots) | Underway | 5.11 | 5.88 |
| | | Other | 1.51 | 3.20 |
| | Draught (m) | Underway | 4.91 | 2.41 |
| | | Other | 4.01 | 2.51 |

Wu et al. (2017) define vessel density, traffic density and AIS receiving frequency, proposing the equations to compute each one of these values using a grid-based method. Vessel density considers the time spent by each ship in a given grid cell, while traffic density is related to the number of grid cell crossings during a given period. These researchers demonstrate vessel density, traffic density and AIS frequency maps, and compare traffic density maps of raw and pre-processed data, using a proposed pre-processing algorithm that uses empiric thresholds.

Figure 1 shows a vessel density map of AIS position reports, obtained after pre-processing the data to remove errors and inconsistencies. This bitmap image is obtained using a grid-based method, where each image pixel value is related to the number of AIS position reports originating from the corresponding grid element.

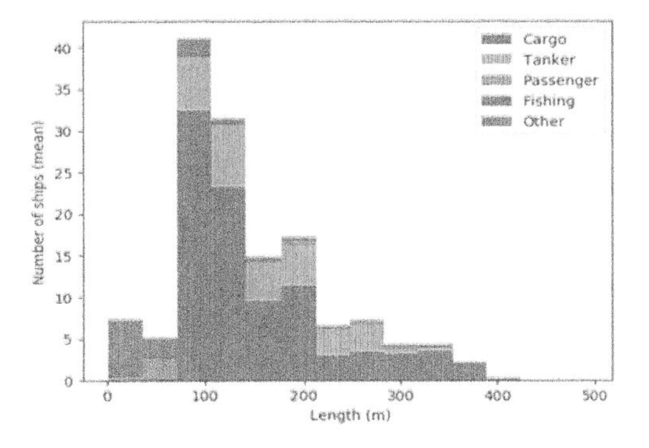


Figure 2. Histogram of ship lengths.

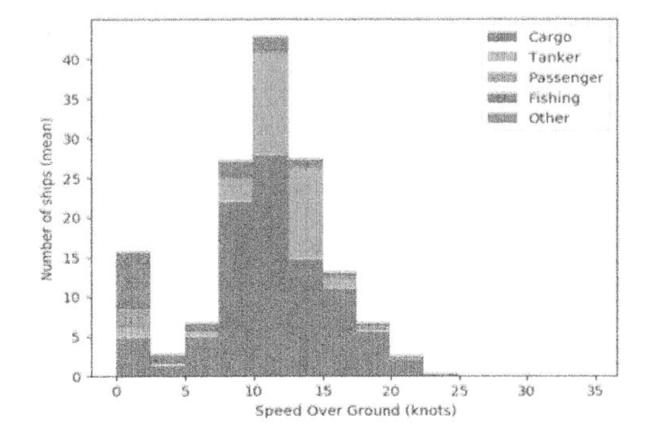


Figure 3. Histogram of SOG.

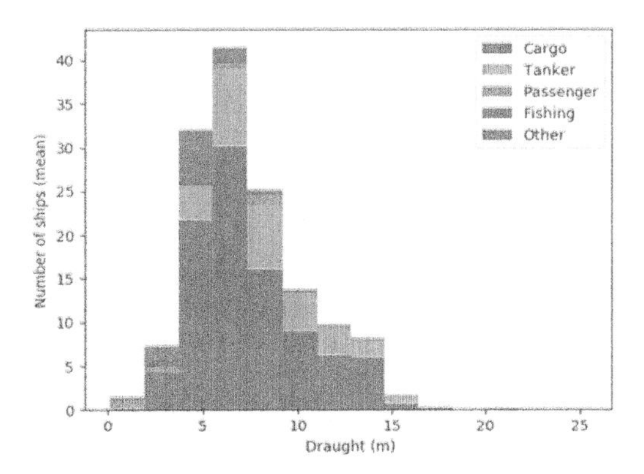


Figure 4. Histogram of ship draughts.

Figure 2 to Figure 4 show histograms of length, SOG and draught of ships with navigational status "under way using engine", respectively, based on the same samples. Fishing vessels, vessels in the class *Other* and small passenger ships (ferries) are more common in the lower lengths, while cargo ships and tankers are the majority for lengths over eighty meters. Since the continental coast of Portugal is exposed to ocean swells and weather systems, ship speeds can vary considerably depending on the direction of winds, waves and the ships' intended courses. However, differences between vessel types are observable in Figure 4. As stated in the previous paragraph, since the draught field is updated by the operator in the AIS equipment, it is expected to be less reliable.

## 4 ROUTEING SYSTEMS OFF THE PORTUGUESE COAST

Maritime traffic patterns are related to the measures implemented to organize traffic. In fact, according to the (IMO, 2020), the number of collisions and groundings has often been dramatically reduced by Traffic Separation Schemes (TSS) and other ship routeing measures established in most of the major congested shipping areas around the world. Since the AIS data used in this paper is sourced from the Portuguese coastal VTS, routeing systems and special areas implemented in this area are briefly described in this section.

There are two Traffic Separation Schemes (TSS) and one area to be avoided (ATBA) off the Portuguese continental coast. These marine routeing systems, shown in Figure 5 and Figure 6, contribute to the safety of life at sea, protection of the marine environment and to the safety of navigation in convergence or high shipping density areas. The TSSs off the Portuguese continental coast are named "Off Cape Roca" and "Off Cape S. Vicente". Both have 5 traffic lanes each (Figure 5 and Figure 6):

- A southbound lane for ships carrying dangerous or pollutant cargoes in bulk (A).
- A southbound lane for ships not carrying dangerous or pollutant cargoes in bulk (B).
- A northbound lane for ships carrying dangerous or pollutant cargoes in bulk (C).
- A northbound lane for ships not carrying dangerous or pollutant cargoes in bulk (D).
- A lane closer to the coast for ships that fulfil specific requirements regarding the ports of origin and destination, as well as the type of cargoes carried. In the "Off Cape Roca" TSS this is a 2-way traffic lane. In the "Off Cape S. Vicente" TSS this is a southbound traffic lane (E).

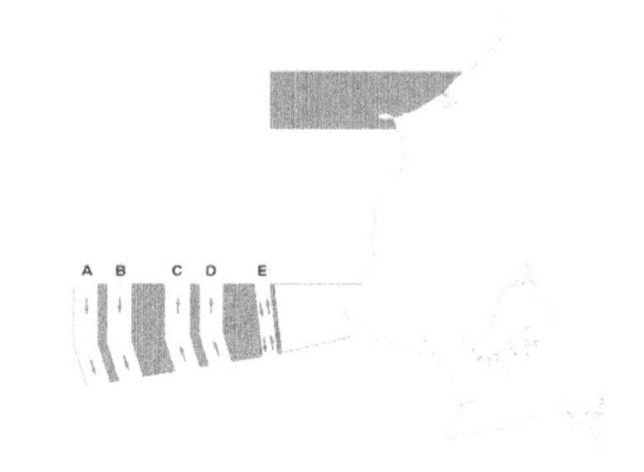


Figure 5. Off Cape Roca TSS and Berlengas ATBA.

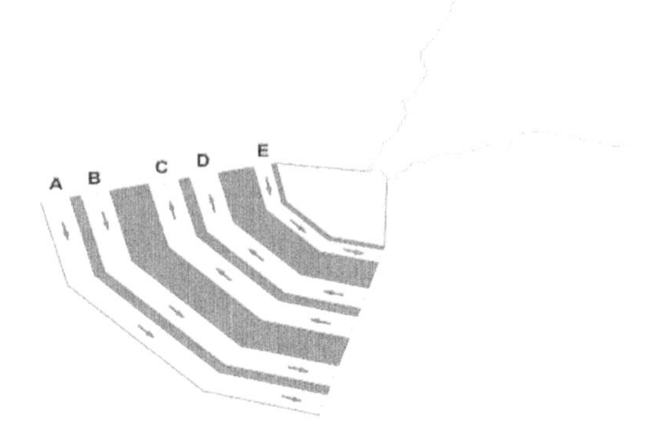


Figure 6. Off Cape S. Vicente TSS.

As for the area to be avoided protecting the environmentally sensitive Berlengas archipelago, it is not to be crossed by vessels over 300 GT or vessels carrying dangerous or pollutant cargoes, unless duly authorized by competent authorities.

## 5 USING DIJKSTRA'S ALGORITHM TO IDENTIFY POTENTIAL SAFE ROUTES

A great amount of research exists on the topic of route or trajectory identification and its applications to AIS data (Pallotta et al., 2013; Vespe et al., 2012; Etienne et al. 2010).

In this section, the well-known Dijkstra's shortest path algorithm (Dijkstra, 1959) is used for establishing routes that have potential to be deemed safe, by extracting from historical AIS data the most used path between an origin and a destination. The Dijkstra algorithm, as well as several improved versions of the original method, have been widely used in different contexts such as for ship route planning, weather routing, logistics management, and many other network optimization problems that can be formulated as shortest path problem (Takashima et al., 2009; Mannarini et al., 2013; Neumann, 2016; Vettor and Guedes Soares, 2016).

This approach consists of using AIS data to build a graph in which the nodes are cells of a grid covering the geographical area being studied and the weights of directional edges are inversely related to ship movements between cells (Silveira et al., 2019). Based on this graph, the Dijkstra algorithm is used to identify the most used route by ships between two locations. A grid with square cells is defined, covering the reference area, and for which the cell size matches a required resolution. Once the grid is defined, a weighted directed edge graph that represents the ship traffic is created. This graph may be created using all the AIS messages or from selected specific ship types and/or length intervals. Each grid cell is potentially a graph node. AIS messages are processed in the order they were received and for each position report (AIS message types 1, 2 and 3) the ship position is matched to a grid cell. When a ship changes cell, the following process takes place:

1. if the origin or destination cells are not yet graph nodes, these nodes are added to the graph;
2. if there is no edge connecting the origin and destination nodes, an edge is created with weight equal to one, otherwise, the edge weight is incremented by one.

When all AIS messages are processed the graph creation process ends. At this stage, the weight of each edge of the graph represents the number of times a transition between the origin and destination cells occurred.

The Dijkstra (1959) algorithm finds the shortest path between nodes in a directed graph, where weights represent distances between nodes (or vertices). The graph weights must be non-negative. Since in this case the intention is to find the "most used" route, the edge weights are inverted and all weights remain positive to apply the algorithm. This transformation is performed in the following way:

$$w_{ij} = w_{max} - w_{ij} + 1 \tag{1}$$

where $w_{ij}$ is the weight of the directed edge between node $i$ and node $j$ and $w_{max}$ is the maximum edge weight in the graph. Once the edge weights are recalculated, the Dijkstra algorithm is used to find the most used trajectory between an origin node and any other node in the graph. However, it may be more interesting to find the best route between a group of origin cells (origin area) and a group of destination cells (destination area), because one cell may represent a very small geographical area, depending on the grid resolution.

Regarding the grouping of destination cells and since the Dijkstra algorithm solution includes the minimal cost path from an origin cell to any other cell in the graph, it is only necessary to choose the best solution from the set of destination cells. As for the grouping of origin cells, an extra step is added, which consists of joining all origin cells that are connected through edges to non-origin cells, updating the edge weights as cells are joined, until only one origin cell remains. The path with the lowest cost is found by applying the Dijkstra algorithm on this updated graph.

The choice of the length of each grid cell not only has impacts on path resolution, but may also bias the resulting route towards the path chosen by slower ships. Class A AIS transmits position reports every 2 to 10 seconds, depending on ship speed and rate of turn. If the grid cell dimensions are small enough for faster ships to skip one or more cells in transitions between cells, the paths taken by slower ships end up contributing to more edge weights, therefore biasing the results. To avoid this, the length $L_C$ of cell sides in nautical miles (nm) should be chosen according to the following rule:

$$L_C\ (nm) \geq \frac{V_f(kts) \times RI_{max}(s)}{3600\ (s)} \tag{2}$$

where $V_f$ is the estimated or observed speed of the faster ship and $RI_{max}$ is the maximum reporting interval. $L_C$ is equivalent to the distance covered by the faster ship during the largest position reporting interval. Another problem related to the choice of $L_C$ may occur when this value is too small when the amount of AIS data is also small that, to the limit, may result in all edge weights taking the value 1. In this case, assuming ships have equal speeds, the algorithm will choose the longest path, for which the sum of inverted edge weights is smaller.

The route identification process previously described is applied to identify routes between the main continental ports of Portugal and also the north and east boundaries of the main shipping routes that use the two Traffic Separation Schemes off the continental Portuguese coast, as well as 2 other boundaries identified as potentially interesting from observing Figure 1. The control areas are shown in Figure 7 and Figure 8, along with the TSS limits.

Table 2. Number of voyages between control areas.

| From\To | A | B | C | D | E | F | G | H | I | J | K | L | M |
|---|---|---|---|---|---|---|---|---|---|---|---|---|---|
| A | -- | 103 | 1297 | 387 | 252 | 788 | 312 | 434 | 4 | 0 | 10229 | 39 | 273 |
| B | 161 | -- | 12 | 5 | 2 | 2 | 16 | 0 | 0 | 0 | 41 | 0 | 1 |
| C | 1219 | 18 | -- | 59 | 32 | 445 | 67 | 246 | 3 | 0 | 496 | 3 | 23 |
| D | 518 | 13 | 61 | -- | 68 | 27 | 41 | 13 | 3 | 1 | 150 | 1 | 28 |
| E | 253 | 9 | 62 | 71 | -- | 11 | 11 | 2 | 1 | 0 | 104 | 0 | 13 |
| F | 670 | 12 | 429 | 51 | 38 | -- | 98 | 184 | 23 | 0 | 538 | 187 | 76 |
| G | 374 | 19 | 98 | 37 | 17 | 85 | -- | 47 | 5 | 0 | 358 | 14 | 41 |
| H | 450 | 0 | 194 | 39 | 7 | 156 | 44 | -- | 10 | 2 | 386 | 59 | 56 |
| I | 8 | 0 | 1 | 4 | 0 | 26 | 7 | 10 | -- | 13 | 96 | 0 | 0 |
| J | 3 | 0 | 0 | 1 | 0 | 0 | 0 | 0 | 12 | -- | 284 | 0 | 5 |
| K | 11737 | 24 | 396 | 230 | 59 | 653 | 423 | 609 | 98 | 265 | -- | 21 | 25 |
| L | 26 | 1 | 12 | 1 | 2 | 142 | 8 | 41 | 0 | 0 | 35 | -- | 11 |
| M | 442 | 8 | 71 | 28 | 54 | 39 | 104 | 16 | 2 | 4 | 28 | 9 | -- |

Paths between origin and destination shown in Figure 7
Paths between origin and destination shown in Figure 8
Paths between origin and destination, Figure 7 and Figure 8

A maximum time interval threshold of 10 minutes between consecutive position reports within a voyage is used, to ensure that the vessels were continuously tracked along the voyage. The number of identified voyages between the defined control areas is shown in Table 2 for all ship types and ship lengths. The result of the application of this route identification method is shown in Figure 7 and Figure 8. The method is applied to build graphs for cargo ships (ship types 70-79) and tankers (ship types 80-89) for the origins and destinations in Table 2. Highlighted cells in the table correspond to paths between the respective origin and destination areas shown in the previously mentioned Figures. When creating the graphs, a lower limit of 50 meters for ship length was defined, and the chosen cell side length was 0.1 nautical miles, which is adequate for ship speeds up to 30 knots and maximum reporting intervals of 10 seconds, even allowing for slightly faster ships or longer reporting intervals (up to 36 knots or 12 seconds).

Table 3 and Table 4 show the mean value of the weight of edges connecting the nodes that are in the most used path, before applying the weight inversion procedure, for cargo ships and tankers, respectively. These values provide some insight on the validity of the method for each application because a value close to one indicates that the path is taken by only one vessel, therefore losing its meaning as the most used path.

Table 3. Mean value of the edge weights before inversion, most used path, cargo ships.

| From\To | A | B | C | D | E | F | G | H | I | J | K | L | M |
|---|---|---|---|---|---|---|---|---|---|---|---|---|---|
| **A** | – | 2.2 | 12.4 | 7.6 | 5.1 | 9.2 | 10.0 | 9.2 | – | – | 387.0 | 1.5 | 15.4 |
| **B** | 4.6 | – | 1.4 | 1.3 | 1.0 | 1.0 | 2.4 | – | – | – | 2.7 | – | 1.0 |
| **C** | 9.6 | 1.3 | – | 3.5 | 1.6 | 14.2 | 3.9 | 2.5 | – | – | 13.9 | 1.0 | 2.0 |
| **D** | 12.8 | 1.3 | 2.8 | – | 1.8 | 1.1 | 3.2 | 1.0 | – | 1.0 | 5.6 | 1.0 | 2.0 |
| **E** | 10.9 | 1.0 | 7.7 | 1.6 | – | 1.1 | 1.4 | 1.0 | – | – | 4.7 | – | 1.8 |
| **F** | 2.6 | 1.4 | 11.7 | 2.6 | 3.4 | – | 5.3 | 3.7 | – | – | 11.9 | 10.1 | 4.2 |
| **G** | 15.8 | 2.3 | 6.1 | 2.7 | 1.2 | 4.4 | – | 1.4 | – | – | 11.1 | 1.4 | 3.1 |
| **H** | 10.7 | – | 2.8 | 1.0 | 1.5 | 1.4 | 1.5 | – | – | – | 2.3 | 2.3 | 3.3 |
| **I** | – | – | – | – | – | – | – | – | – | – | – | – | – |
| **J** | 1.0 | – | – | 1.0 | – | – | – | – | – | – | 1.2 | – | 1.1 |
| **K** | 303.2 | 1.8 | 12.2 | 5.9 | 4.3 | 13.8 | 13.8 | 5.1 | – | 1.5 | – | 1.1 | 1.4 |
| **L** | 2.4 | 1.0 | 1.5 | 1.0 | 1.1 | 7.2 | 1.0 | 1.8 | – | – | 1.1 | – | 1.0 |
| **M** | 23.8 | 1.1 | 6.7 | 2.0 | 3.5 | 2.8 | 7.4 | 1.6 | – | 1.0 | 1.0 | 1.0 | – |

Table 4. Mean value of the edge weights before inversion, most used path, tankers.

| From\To | A | B | C | D | E | F | G | H | I | J | K | L | M |
|---|---|---|---|---|---|---|---|---|---|---|---|---|---|
| **A** | – | – | 5.0 | 3.6 | – | 9.4 | 1.5 | 13.4 | – | – | 171.2 | 6.1 | 3.0 |
| **B** | – | – | – | – | – | – | – | – | – | – | 1.0 | – | – |
| **C** | 4.0 | 1.0 | – | 1.0 | – | 1.5 | 1.2 | 5.7 | – | – | 4.8 | – | 1.0 |
| **D** | 4.4 | – | 2.2 | – | – | 1.0 | – | 2.5 | – | – | 2.4 | – | – |
| **E** | – | – | – | – | – | – | – | – | – | – | – | – | – |
| **F** | 6.1 | – | 2.4 | 1.0 | – | – | 2.4 | 9.5 | – | – | 4.6 | 1.0 | 1.0 |
| **G** | 1.6 | – | 1.6 | 1.0 | – | 2.6 | – | 5.2 | – | – | 1.9 | 1.0 | 1.0 |
| **H** | 22.0 | – | 5.5 | 2.5 | – | 11.1 | 5.0 | – | – | – | 1.2 | 4.2 | 1.0 |
| **I** | – | – | – | 1.0 | – | – | – | – | – | – | – | – | – |
| **J** | – | – | – | – | – | – | – | – | – | – | – | – | – |
| **K** | 388.2 | 1.0 | 3.7 | 2.5 | – | 9.0 | 1.7 | 2.1 | 1.0 | – | – | 1.1 | 1.0 |
| **L** | 1.6 | – | 1.0 | – | – | – | 1.1 | 3.0 | – | – | 1.1 | – | 1.1 |
| **M** | 6.3 | 1.0 | 1.0 | – | – | 1.0 | 1.0 | 1.1 | – | – | 1.0 | 1.0 | – |

Figure 9 Shows one of the identified routes for cargo ships, from the area (C) to the area (H), overlaid on a heatmap of the edge weights for the corresponding graph obtained with QGIS, an open-source geographical information system, which uses a Kernel Density Estimation (KDE) algorithm with a quartic function as a kernel. This is a particularly interesting case, because of the co-existence of a two-way traffic lane, closer to shore, that can be used by vessels that fulfil certain requisites, and of a regular southbound traffic lane for ships not carrying dangerous cargo in bulk. This results in two main routeing options that are observable in the figure. The most used option in practice is naturally the one for which the distance is shorter, and that is also the one returned by the algorithm.

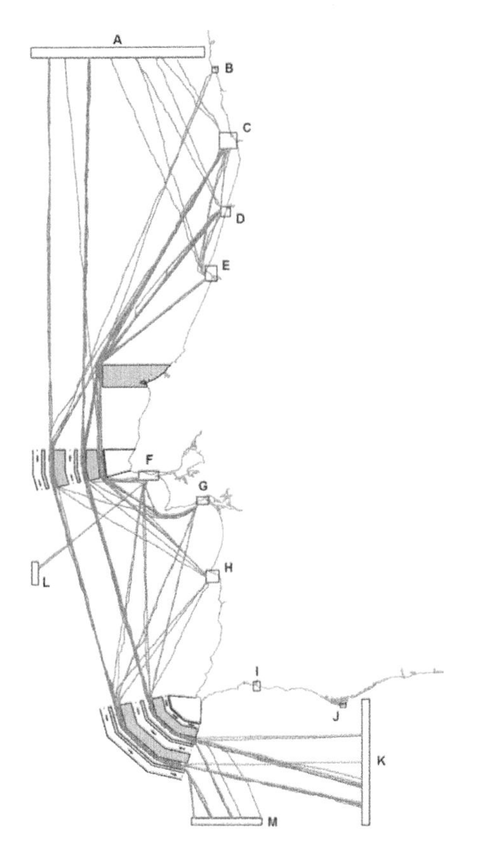


Figure 7. Identified cargo ships' routes.

## 6 SPEED PROFILE OF THE MOST USED PATH

The procedure to create the graph that represents the traffic density, described in the previous section, can be used to create a second graph in which the edge weights are the average speed of ships making the transition between grid cells. The most used path returned by the algorithm is a sequence of the traffic density graph nodes, connected by edges. A speed profile of the most used path is obtained from the weights of the edges connecting the same nodes, but on the average speed graph.

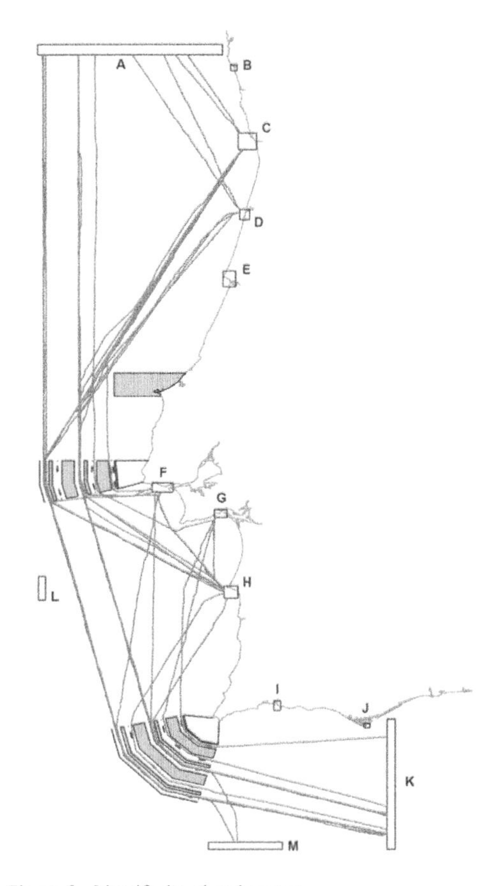


Figure 8. Identified tankers' routes.

Two examples of speed profiles obtained using this method are presented. Figure 10 shows the speed profile for the cargo ship route from area (C) to the area (H), depicted in Figure 9. Figure 11 shows the speed profile for the tanker route between the same origin and destination areas. Due to the large oscillations in average transition speed between nodes, a Savitsky-Golay filter (Savitzky and Golay, 1964) is used to smooth the profiles and show the general speed variation trend along the route. Cargo ship speeds are globally higher on the route between the two ports and an expected decrease in speed is observable at the final part of both routes, when ships are approaching the destination port. A number of factors can contribute to the oscillations observed in the speed profiles, e.g. wind speed and direction, wave height and direction, ship dimensions and loading condition. Thus, additional environmental and

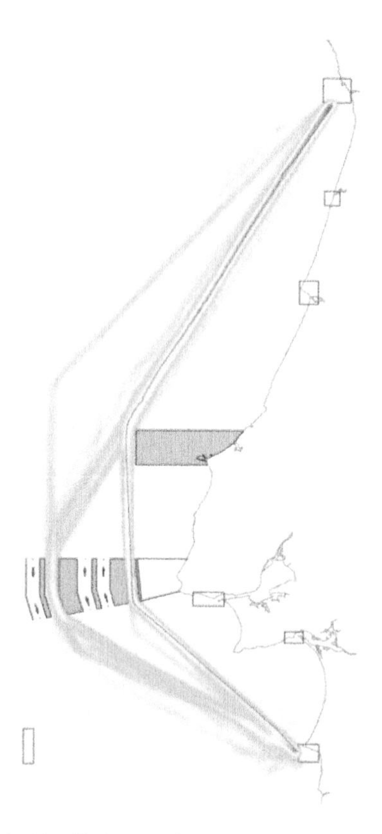

Figure 9. Identified cargo ship route from area (C) to area (H), overlaid on a heatmap of edge weights of the corresponding traffic density graph.

ship related data are required to better characterize the speeds practiced by ships using a specific route. However, the proposed method is satisfactory when analysing general trends in speed along a route.

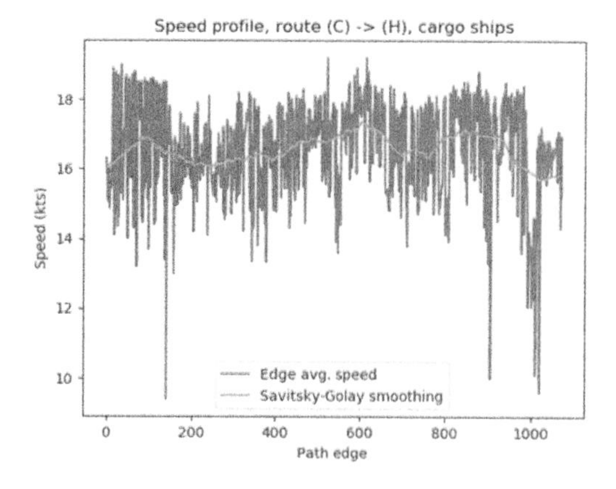


Figure 10. Speed profile, route (C) → (H), cargo ships.

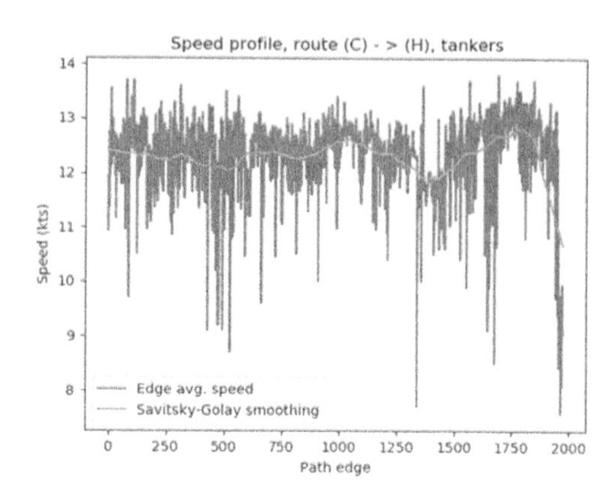


Figure 11. Speed profile, route (C) → (H), tankers.

## 7 CONCLUSIONS

Traffic statistics and visualization tools provide data and good insights on patterns and traffic flows. Such information is used in this paper to define the control areas that contain the origin and destination of the most relevant routes. A Dijkstra algorithm-based approach is used to identify the most used path between the aforementioned origin and destination areas. The approach matches nodes in the graph required to apply the Dijkstra algorithm to cells on a grid covering a geographical area of interest, using inverted number of transitions between cells as the graph edges' weights. The maximum speed of the ships limits the choice of the dimension of the grid cells, since smaller cell sizes may result in the skipping of some transitions. As an example, the length of each cell side must be at least 154 metres in areas where the maximum ship speed is 30 knots, for 10-second reporting intervals. This is acceptable for some applications, such as coastal route identification, but may be an issue where greater refinement is required. It is possible to overcome this limitation by interpolating ship positions between reports, which has not been done in the present application. While this approach is certainly effective when the quantity of data is appropriate, as shown in Figure 9, the same cannot be stated when all the edge weights are close to 1, since this means there is no obvious most used path. Grid resolution can be decreased to compensate, but at the cost of less accuracy in the resulting most used path. The proposed graph-based method to obtain the speed profiles of the identified routes is adequate to determine the global speed trends, although its use as a route characterization tool may be improved by additional data, particularly weather and ship-related data.

## ACKNOWLEDGEMENTS


Paper contributes to the project "Monitoring and Surveillance of Maritime Traffic off the continental coast of Portugal (MoniTraffic)", which has been co-funded by "Fundo Azul" programme of the Portuguese Directorate-General for Maritime Policy, under contract no. FA-04-2017-005. This work also contributes to the Strategic Research Plan of the Centre for Marine Technology and Ocean Engineering (CENTEC), which is financed by the Portuguese Foundation for Science and Technology under contract UIDB/UIDP/00134/2020.

## REFERENCES

Chen C, Wu Q, Zhou Y, et al (2016) Information visualization of AIS data. In: 2016 International Conference on Logistics, Informatics and Service Sciences (LISS). 24-27 July, 2016, Sydney, Australia, pp 1–8.

Denker C, Hahn A (2016) MTCAS - An Assistance System for Maritime Collision Avoidance. In: Proceedings of the 12th International Symposium on Integrated Ship's Information Systems & Marine Traffic Engineering Conference ISIS-MTE. 31 August-02 September, 2016, Hamburg, Germany.

Dijkstra EW (1959) A Note on Two Problems in Connexion with Graphs. Numerische Mathematik 1:269–271. http://dx.doi.org/10.1007/BF01386390.

Etienne L, Devogele T, Bouju A (2010) Spatio-Temporal Trajectory Analysis of Mobile Objects Following the Same Itinerary. The International Archives of the Photogrammetry, Remote Sensing and Spatial Information Sciences 38:86–91. http://dx.doi.org/10.1007/s11263-012-0594-8.

Harati-Mokhtari A, Wall A, Brooks P, Wang J (2007) Automatic identification system (AIS): data reliability and human error implications. Journal of Navigation 60:373–389. http://dx.doi.org/10.1017/S0373463307004298.

IMO (2020) IMO - Ships' routeing. http://www.imo.org/en/OurWork/Safety/Navigation/Pages/ShipsRouteing.aspx. Accessed 23 Sep 2020.

ITU (2014) Recommendation ITU-R M.1371-5: Technical characteristics for an automatic identification system using time division multiple access in the VHF maritime mobile frequency band. International Telecommunications Union.

Kujala P, Hänninen M, Arola T, Ylitalo J (2009) Analysis of the marine traffic safety in the Gulf of Finland. Reliability Engineering and System Safety 94:1349–1357. http://dx.doi.org/10.1016/j.ress.2009.02.028.

Last P, Bahlke C, Hering-Bertram M, Linsen L (2014) Comprehensive Analysis of Automatic Identification System (AIS) Data in Regard to Vessel Movement Prediction. Journal of Navigation 67:791–809. http://dx.doi.org/10.1017/S0373463314000253.

Mannarini G, Coppini G, Oddo P, Pinardi N (2013) A Prototype of Ship Routing Decision Support System for an Operational Oceanographic Service. TransNav, the International Journal on Marine Navigation and Safety of Sea Transportation 7: 53–59. http://dx.doi.org/10.12716/1001.07.01.06.

Neumann T (2016) Vessels Route Planning Problem with Uncertain Data. TransNav, the International Journal on Marine Navigation and Safety of Sea Transportation 10:459–464. http://dx.doi.org/10.12716/1001.10.03.11.

Pallotta G, Vespe M, Bryan K (2013) Vessel pattern knowledge discovery from AIS data: A framework for anomaly detection and route prediction. Entropy 15:2218–2245. http://dx.doi.org/10.3390/e15062218.

Pedersen PT (1995) Collision and Grounding Mechanics. In: Proceedings of the WEMT'95, Ship Safety and Protection of the Environment. 17-19 May, 1995, Kopenhagen, Denmark, pp 125–157.

Porathe T, Borup O, Jeong JS, et al (2014) Ship traffic management route exchange: acceptance in Korea and Sweden, a cross cultural study. In: Proceedings of the International Symposium Information on Ships. 4-5 September, Hamburg, Germany, pp 64–79.

Rong H, Teixeira AP, Guedes Soares C (2020) Data mining approach to shipping route characterization and anomaly detection based on AIS data. Ocean Engineering 198:106936. http://dx.doi.org/10.1016/j.oceaneng.2020.106936.

Rong H, Teixeira AP, Guedes Soares C (2019) Ship trajectory uncertainty prediction based on a Gaussian Process model. Ocean Engineering 182:499–511. http://dx.doi.org/10.1016/j.oceaneng.2019.04.024.

Savitzky A, Golay MJE (1964) Smoothing and Differentiation of Data by Simplified Least Squares Procedures. Analytical Chemistry 36:1627–1639. http://dx.doi.org/10.1021/ac60214a047.

Silveira P, Teixeira AP, Guedes Soares C (2013) Use of AIS data to characterise marine traffic patterns and ship collision risk off the coast of Portugal. Journal of Navigation 66:879–898. http://dx.doi.org/10.1017/S0373463313000519.

Silveira P, Teixeira AP, Guedes Soares C (2019) AIS based shipping routes using the Dijkstra algorithm. TransNav, the International Journal on Marine Navigation and Safety of Sea Transportation 13: 565–571. http://dx.doi.org/10.12716/1001.13.03.11.

Takashima K, Mezaoui B, Shoji R (2009) On the Fuel Saving Operation for Coastal Merchant Ships using Weather Routing. The International Journal on Marine Navigation and Safety of Sea Transportation 3:401–406.

Tu E, Zhang G, Rachmawati L, et al (2018) Exploiting AIS Data for Intelligent Maritime Navigation: A Comprehensive Survey from Data to Methodology. IEEE Transactions on Intelligent Transportation Systems 19:1559–1582. http://dx.doi.org/10.1109/TITS.2017.2724551.

Vespe M, Visentini I, Bryan K, Braca P (2012) Unsupervised learning of maritime traffic patterns for anomaly detection. In: 9th IET Data Fusion & Target Tracking Conference (DF&TT 2012): Algorithms & Applications. 16-17 May, 2012, London, UK.

Vettor R, Guedes Soares C (2016) Development of a ship weather routing system. Ocean Engineering 123:1–14. http://dx.doi.org/10.1016/j.oceaneng.2016.06.035.

Wu L, Xu Y, Wang Q, et al (2017) Mapping Global Shipping Density from AIS Data. Journal of Navigation 70:67–81. http://dx.doi.org/10.1017/S0373463316000345.

*Trends in Maritime Technology and Engineering – Guedes Soares & Santos (eds)*
 *ISBN 978-1-032-33583-4*

# A dynamic Rapid-exploring Random Tree algorithm for collision avoidance for multi-ship encounter situations under COLREGs

H. Zhang & J.F. Zhang
*Intelligent Transportation System Research Center, Wuhan University of Technology, Wuhan, China*
*National Engineering Research Center for Water Transport Safety (WTS Center), Wuhan University of Technology, Wuhan, China*
*Inland Port and Shipping Industry Research Co. Ltd. of Guangdong Province, Guangdong, China*

T. Shi
*School of Mechanical and Electronic Engineering, Wuhan University of Technology, Wuhan, China*

C. Guedes Soares
*Centre for Marine Technology and Ocean Engineering (CENTEC), Instituto Superior Técnico, Universidade de Lisboa, Lisbon, Portugal*

ABSTRACT: The Intelligent Ship (IS) should have the ability to find a safe and economical path within a limited time under complex environment to avoid static and dynamic obstacles, while following the requirements of International Regulations for Preventing Collisions at Sea (COLREGs). In this paper, a dynamic Rapid-exploring Random Tree (RRT) algorithm is proposed under multi-ship encounter situations in open waters. First, the Quaternion Ship Domain (QSD) model is introduced to reflect collision risk between ships. The collision detection procedure between two moving ships under dynamic situations is constructed. Then, a tree-grow strategy is formulated to find the feasible paths for own ship from start to destination. A random and uniform sampling method within an ellipse is formulated to perform path re-planning. The turning angle between consecutive segments of the path tree is also restricted according to ship maneuverability. A multi-ship encounter situation is formulated to evaluate its feasibility and efficiency.

## 1 INTRODUCTION

Intelligent Ships (IS) technologies have been receiving increasing attentions in recent years. According to its degree of autonomy, IS can be divided into four levels, i.e. fully controlled by onboard seafarers, remote control with seafarers on board, remote control without seafarers on board and fully autonomy. As it is widely accepted by authorities and academic community that the main causes of maritime accidents (especially collisions) are human errors or human failures (Fan et al., 2018; 2020; Wu et al, 2022), IS technologies have advantages in assisting making decisions or even eliminating human involvement (Huang et al., 2020).

Many research projects, including the Maritime Unmanned Navigation through Intelligence in Networks (MUNIN), Advanced Autonomous Waterborne Applications (AAWA), and AUTOSHIP, reveal a strong demand for an effective collision avoidance decision-making system capable of coping with complex navigation environment (Wang et al., 2022), as a prerequisite for IS.

IS technologies mainly include navigation situation awareness (Chen et al., 2020), risk assessment (Qu et al., 2011), decision making and operations (Zhang et al., 2015). Among the above, the performance of collision avoidance decision-making system is one of the kernel components in dealing with complex navigation situations (e.g. multi-ship encounter situations, multiple static obstacles). The existing researches in collision avoidance decision-making are in general divided into two categories, which are real-time/iterative decision-making, and path planning. The main idea of the former is to divide the decision procedure into a series of stages, i.e. Observation, Inference, Prediction and Decision (OIPD) (Wang et al., 2020). The ships make sequential decisions in the present cycle that is determined by the combination of the decisions in the previous cycle and the intent prediction of the target ships.

The main challenges for such models include that the intent prediction of target ships has large uncertainty, and the calculation efficiency should be guaranteed in order to make real-time decisions. On the

DOI: 10.1201/9781003320289-18

contrary, path planning is to find a safe path that is constructed by series of waypoints from a macro perspective and the ships would avoid all the static and dynamic obstacles by following the planned path. In general, there should be some restrictions on the turning angles between waypoints because the ships have limited maneuverability, so any too sharp turnings are neither practicable nor economical.

Path planning algorithms was originally applied to the ships' long-range navigation, like weather routing (Gkerekos & Lazakis, 2020). It was later extended to solving the ship collision avoidance problems (Li et al. 2021). It should be noted that the two types of collision avoidance decision making mode do not have strict bound and they can be used to solve the same problem.

In this study, a path planning algorithm under multi-ship encounter situations is formulated by considering the International Regulations for Preventing Collisions at Sea (COLREGs). The original RRT algorithm has been modified by adding some functions (e.g. dynamic obstacles collision check function, navigation safety restrictions) in order to build the dynamic RRT algorithm that can generate a safe and feasible path for vessels sailing in open waters.

The rest of the paper is organized as follows: A literature review on ship collision avoidance algorithms is discussed in Section 2. In section 3, the dynamic path planning methodology is formulated, including dynamic collision check, Quaternion Ship Domain model and dynamic path planning algorithms. A comprehensive simulation on the path planning under multi-ship encounter situations is performed and discussed in section 4. Finally, conclusions are summarized in section 5.

## 2 LITERATURE REVIEW

### 2.1 *Collision risk assessment*

With the continuous development of IS technologies and applications, the researches on collision avoidance decision-making have been attracting increasing attentions in recent years. The researches mainly include encounter situation awareness and collision risk evaluation (Yoshida et al., 2020), anti-collision procedure or algorithms formulation (Zhang et al., 2015).

Ship Domain (SD) model was one of the earliest and most widely used collision risk quantification model. SD is an area around a ship where any violation of static or dynamic obstacles are not allowed in order to maintain navigation safety. The size of SD is determined by many factors, including ship characteristics (i.e. ship type, size, maneuverability), navigation waters (i.e. open waters, port waters, restricted waters), and environments (i.e. visibility, wind, current). In general, SD models can be divided into crisp and fuzzy versions. The crisp version assumes that the collision risk is 1 when the domain violation occurs and 0 otherwise. On the contrary, the SDs are treated as fuzzy boundaries and the collision risk ranges between 0 and 1 and its value is determined by the degree of violation. Besides SD model, collision risk can also be measured by synthesizing different Risk influencing Factors (RIFs) like Time to Closest Point of Approach (TCPA), Distance to CPA (DCPA), relative velocity, bearing angle, among others (Cai et al., 2021). They can be combined using algorithms like fuzzy logic, Analytic Hierarchy Process (AHP), Evidence Reasoning (ER) and so on. Moreover, the Velocity Obstacle (VO) model (Chen et al., 2018) is another type of collision risk evaluation approach from geographical perspective between encounter ships using the idea of velocity field.

Besides collision risk quantification, its spatial evolution under dynamic environment is also significant for collision avoidance decision-making. The reason is that different intentions of target ships may sometime result in substantial different strategies by the own ship. Wang et al. (2020) proposed a sequential collision risk evaluation and inference system under uncertainties. They proposed a four-stage procedure called observation-inference-prediction-decision (OIPD). The decision from the former cycle will have influence the intention prediction in the next circulation and the algorithm is performed in an iterative mode. The advantage is that it can adapt to the mutations in the risk evolution. However, the reliability of the approach should be further investigated due to its probabilistic nature.

### 2.2 *Collision avoidance decision-making*

The researches mainly include ship collision avoidance decision-making and ship path planning. The former focused on the operations of ships from a micro perspective under different encounter situations in different types of waters, whereas the latter tries to find a safe path for a ship from the start to destination in a macro perspective. However, they do not have clear boundaries and path planning can also be applied to multi-ship encounter situations.

In general, collision avoidance procedures are performed in distributive mode (Perera et al. 2011; Zhang et al., 2015; Hu et al., 2020), which means that each ship should make decisions from its own perspective while considering the COLREGs requirements, the intention of target ships as well as uncertainties. Zhang et al. (2015) proposed a distributive and close-loop anti-collision decision support formulation considering COLREGs. The model can find a safe path for ships when all the ships are complying with COLREGs, and even some of the ships violate the regulations. The algorithm found the safe path with a series of waypoints in the first step, and then follow the path by considering ship maneuverability. Hu et al. (2020) proposed a similar decision-making procedure by involving a Collision Risk Index (CRI) based on fuzzy logic algorithm.

The decision on course alteration as well as changing speed are considered when the CRI reaches a threshold. However, the maneuverability was not considered in the study.

There are also many path planning algorithms that have been successfully introduced into the maritime transportation field. For instance, Xie et al. (2019) proposed a multi-direction A* algorithm for path planning of working ships navigating within wind farm waters. Artificial Potential Field (APF) model was first formulated by considering that obstacles have repulsion and the destination has gravitation to the ships. Then the penalty function was constructed to stimulate ships navigating along the shortest path. There are also many types of deterministic and heuristic path planning approaches, including the Dijkstra algorithm, Ant Colony Optimization (ACO), Genetic Algorithm (GA), as well as some advanced machine/deep learning approach like Long Short Term Memory (LSTM) and Deep Reinforce Learning (Sawada et al., 2021). Most of them works well in certain circumstance. But their expansibility to various application scenarios should be further investigated.

Rapid-exploring Random Tree (RRT) is another heuristic path planning algorithm that has been introduced into ship collision avoidance. The main idea of RRT is performing random searching with fixed step in the searching space, until reaching the destination. The searching procedure usually makes a trade-off between efficiency and randomness. RRT was first introduced to robot path planning and works well in many applications (Li et al., 2020). In recent years, some researches extended its application to ship path planning. Chiang & Tapia (2018) proposed a COLREG-RRT algorithm for surface vehicle navigation. The path planning was realized by constructing virtual obstacles in the tree growth procedure and a scenario containing 20 obstacle ships and many static obstacles was simulated.

The results indicate that the model has high success rate and has good adaptation to the environment change. However, it still cannot guarantee that the ship can find a safe and feasible path. Enevoldsen et al. (2021) proposed an informed RRT* algorithm, in which a feasible path was formulated using the traditional RRT algorithm and the planned path can be shortened with iterative update. The model also takes into consideration of ship maneuverability restrictions, so that the path has good smoothness. Generally speaking, the RRT algorithms for ship path planning still have some issues to be further studies. For example, the ships should make collision detection under dynamic situations, rather than only in static environment. The ships also need to consider the COLREGs requirement as well as the uncertainties of target ship dynamics. Once one of the encounter ships has found a safe path, how other ships adjust and coordinate with the new situation? These problems should be considered in a comprehensive way.

## 3 COLLISION AVOIDANCE PATH PLANNING MODEL

### 3.1 *The RRT algorithm*

The main idea of RRT algorithm is exploring a safe path by growing a tree from the start point with fixed steps, until reaching the destination. As present in Figure 1, in each step, the algorithm would randomly sample a point ($X_{rand}$) from the solution space with a large probability, or select the destination with a small probability. By doing this, the algorithm can make a trade-off between randomness and goal-orientation. Then it will be connected with the nearest node in the existing tree ($X_{nearest}$) and found the new node ($X_{new}$) with the fixed distance. If the path from $X_{nearest}$ to $X_{new}$ is collision free, the new node will be added to the tree. Otherwise, it will sample a new node in the same way. The procedure will continue until reaching the destiny and a safe path can be found. The pseudo-code of RRT algorithm is present in Table 1.

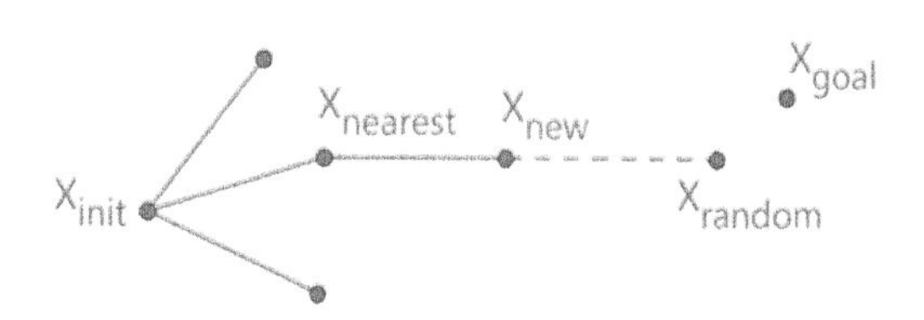


Figure 1. How RRT algorithm expanding spanning tree.

RRT algorithm has advantage in finding a safe path with satisfactory time limit even in some very complex environment. However, there are two issues that need to be considered when applying to ship collision avoidance. The first one is the collision detection. RRT algorithm was initially used in static environment and the collision detection is straightforward. However, both own ship and target ships are moving continuously and the collision detection should consider collision risk between them. The second issue is that ship maneuverability should be taken into consideration, which means that the turning angle between nodes should not be too large. Moreover, the path should be as short as possible by smoothing and re-planning.

### 3.2 *Dynamic collision check method*

In order to deal with the first problem, a dynamic collision check method is proposed in this section by involving Ship Domain model. It is realized by predicting the dynamics of target ships and making domain violation accordingly.

Table 1. Pseudo-code of RRT algorithm.

| Algorithm 1 RRT | |
|---|---|
| Input: $P_{start}$, $P_{goal}$, $O_{sta}$, $O_{dyn}$, $X_{movelength}$ | |
| Output: TailNode | |
| 1: | $\tau \leftarrow \{P_{start}\}$ |
| 2: | for t = 1, 2, ..., n do |
| 3: | $X_{rand} \leftarrow$ Rand(0, 1); |
| 4: | if $X_{rand} > 0.05$ then |
| 5: | $P_{samp} \leftarrow$ FreeSpaceSampling(); |
| 6: | else |
| 7: | $P_{samp} \leftarrow P_{goal}$; |
| 8: | end if |
| 9: | $P_{nearest} \leftarrow$ GetNearest($\tau$, $P_{samp}$); |
| 10: | $P_{new} \leftarrow$ NewNode($P_{samp}$, $P_{nearest}$, $X_{movelength}$); |
| 11: | if CoChek($P_{nearest}$, $P_{new}$, $O_{sta}$, $O_{dyn}$) then |
| 12: | SetParent($P_{nearest}$, $P_{new}$) |
| 13: | $\tau \cup^{\{P_{new}\}}$ |
| 14: | else |
| 15: | Continue; |
| 16: | end if |
| 17: | if InGoalRegion($P_{new}$) then |
| 18: | if CoChek($P_{nearest}$, $P_{new}$, $O_{sta}$, $O_{dyn}$) then |
| 19: | SetParent($P_{new}$, $P_{goal}$); |
| 20: | Break; |
| 21: | else |
| 22: | Continue; |
| 23: | end if |
| 24: | end if |
| 25: | end for |
| 26: | return $P_{goal}$; |

### 3.2.1 *Ship Domain model*

Ship Domain is an area around the ship where any violations of static or dynamic obstacles are not allowed, otherwise the collision risk are not acceptable or cannot be neglected. The size and shape of Ship Domain are influenced by many factors, including traffic density, visibility, ship size and speed, and maneuverability. There are many types of Ship Domain model, either crisp or fuzzy versions. In this study, the Quaternion Ship Domain (QSD) model proposed by Wang (2013) is introduced, as shown in Figure 2. The model can be applied to open waters and it takes COLREGs into consideration with different safety radius in different bearing angles. The parameters of quaternion dynamic ship domain as follows:

$$\begin{cases} R_{fore} = \left(1 + 1.34\sqrt{k_{AD}^2 + (k_{DT}/2)^2}\right)L \\ R_{aft} = \left(1 + 0.67\sqrt{k_{AD}^2 + (k_{DT}/2)^2}\right)L \\ R_{starb} = (0.2 + k_{AD})L \\ R_{port} = (0.2 + 0.75k_{AD})L \end{cases} \quad (1)$$

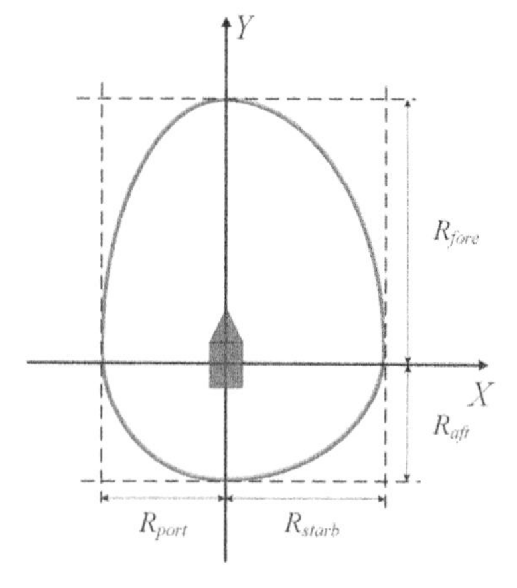


Figure 2. Quaternion dynamic ship domain.

The parameter $v$ is vessel speed. The parameter $L$ is vessel length and other parameters are defined as follows:

$$\begin{cases} k_{AD} = 10^{0.3591lgv+0.0952} \\ k_{DT} = 10^{0.5441lgv-0.0795} \end{cases} \quad (2)$$

### 3.2.2 *Dynamic collision check*

Based on the QSD model, the dynamic collision check method can be formulated. It is the premise of the sampling method of RRT algorithm for ship collision avoidance. As can be seen from Figure 3, one more parameter of $t$ is added to the vertices of the tree, which means the time of reaching the correspondent vertices. Assume the new sampling node is node $V_{n+1}$, the time interval of sailing from node $V_n$ to node $V_{n+1}$ can be decomposed into many pieces and they are considered as many consecutive moments. The positions, speeds and courses of own ship and target ship can be predicted accordingly and collision check for each moment can be performed accordingly.

If any domain violation occurs, i.e. the interval node $V_n'$ falls within the domain area of target ship, the path from $V_n$ to $V_{n+1}$ is supposed to be obstructed. If there is no ship domain violation in all the moments, the new path is considered to be collision-free and can be added to the tree. It should be noted that in the above analysis, the target ship is assumed to keep its course and speed. However, the target ship may plan a new path beforehand and its trajectory prediction should be on the basis of its new path.

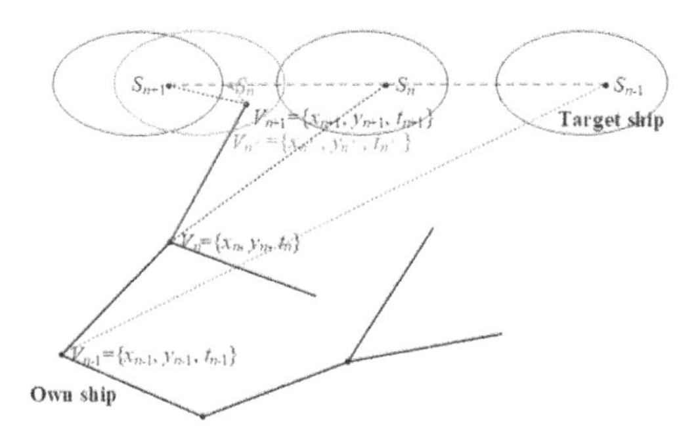


Figure 3. An illustrative demonstration of dynamic collision check.

By adding the dynamic collision check method to the original version of RRT algorithm, it can generate a path that can not only avoid statics obstacles but also can avoid other sailing ships in dynamic environment. In other words, because of the probability space of RRT algorithm is complete, the algorithm can find the safe path if it exists. The pseudo-code of dynamic collision check algorithm is present in Table 2.

### 3.3 *Path optimization with restrictions*

Now we have a path from start to destination that is generated by the dynamic RRT algorithm. However, the path still needs to be optimized by considering path smoothness and COLREGs compliance. On the one hand, RRT algorithm is in nature non-asymptotically optimality and the generated path can be optimized. On the other hand, the ship sometimes cannot follow the path because of too large turning angle to be operated with limited maneuverability.

With respect to the above problems, we should find a way to optimize the path and give the new path some restrictions once it is generated. The idea of Informed RRT* is introduced, in which the path can be optimized by randomly sampling within an ellipse around the start and destination. One issue that need to be resolved is sampling within the eclipse randomly and uniformly. In order to realize this, the ellipse is decomposed with a series of triangles. We can sample within the triangle randomly and uniformly. The probability of the sample falling within any triangle is in proportion to its area. By doing so the samples can be evenly distributed within the ellipse. The generated path will be optimized iteratively within the new sampling space and the path can be shortened accordingly while guarantee safety.

Table 2. Pseudo-code of dynamic collision check algorithm.

| Algorithm 2 CollisionCheck(CoChek) | |
|---|---|
| Input: $P_{start}$, $P_{end}$, $O_{sta}$, $O_{dyn}$ | |
| Output: Bool | |
| 1: | $t \leftarrow$ GetSailTime($P_{start}$, $P_{end}$); |
| 2: | $T \leftarrow 0$; |
| 3: | $t_0 \leftarrow$ GetStartTime($P_{start}$); |
| 4: | $n \leftarrow len(O_{dyn})$ |
| 5: | While $T <= t$ do |
| 6: | $P_{pos} \leftarrow$ GetSelfPosition($t_0 + T$); |
| 7: | for $i = 1, 2, \ldots, n$ do |
| 8: | $P_{dyn} \leftarrow$ GetDynPosition($O_{dyn}[i]$); |
| 9: | $C_{dyn} \leftarrow$ GetDynCourse($O_{dyn}[i]$); |
| 10: | $S_{dyn} \leftarrow$ GetDynSpeed($O_{dyn}[i]$); |
| 11: | $D_{dyn} \leftarrow$ GetDynDomain($P_{dyn}$, $C_{dyn}$, $S_{dyn}$); |
| 12: | if $P_{pos}$within($D_{dyn}$) then |
| 13: | return False; |
| 14: | end if |
| 15: | end for |
| 16: | $T \leftarrow T + 1$; |
| 17: | end while |
| 18: | return True; |

Another issue that should be considered is COLREGs compliance for the path planning. According to COLREGs, when two power-driven vessels are meeting with collision risk, the ships who has a target ship on its starboard is the give-way ship and should try its best to cross from the stern of the target ship. This requirement implies that the give-way ship is highly recommended to turn starboard rather than port to keep clearance. In order to follow such regulation in the path planning algorithm, it is assumed that the path optimization is performed by sampling only in the half ellipse on the right side along the course. The results of the sampling method are present in Figure 4.

Moreover, it should be noted that there is another restriction on the maximum turning angle of the ships. In dealing with this, a maximum turning angle is assigned for the tree growing. Only the nodes whose turning angle is smaller than the maximum turning angle are treated as valid. Otherwise, resampling is performed to search new available nodes. It should be noted that by doing so, the sampling efficiency will be reduced to some degree because invalid samples would sometimes appear in the algorithm. The pseudo-code of ellipse sampling is presented in Tables 3 and 4, respectively.

By combining the ellipse sampling method and the informed RRT* algorithm, the path optimization can be realized. The pseudo-code of path

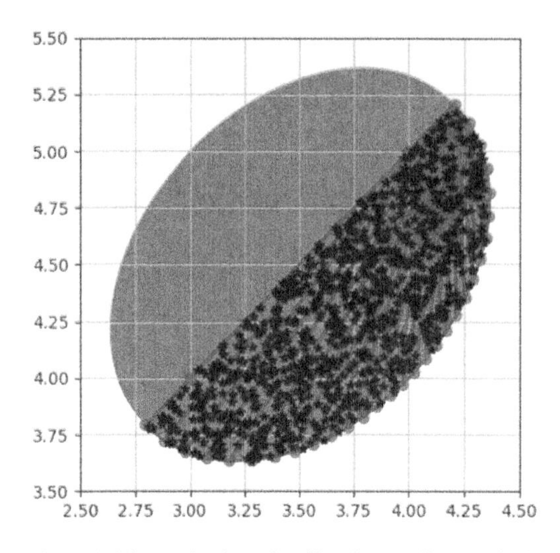


Figure 4. The randomly and uniformly sampling results in half ellipse area.

Table 3. Pseudo-code of EllipseSampling.

Algorithm 3 EllipseSampling

Input: $P_{start}$, $P_{goal}$, $L_{path}$
Output: Point

| | |
|---|---|
| 1: | $L_{major} \leftarrow L_{path}$; |
| 2: | $D \leftarrow \text{GetDistance}(P_{start}, P_{goal})$; |
| 3: | $L_{minor} \leftarrow \text{GetMinorAxis}(L_{major}, D)$; |
| 4: | $E_{space} \leftarrow \text{GetEllipse}(P_{start}, P_{goal}, L_{major}, L_{minor})$; |
| 5: | for $i = 1, 2, \ldots, n$ do |
| 6: | $P_{new} \leftarrow \text{GetPointOnEllipse}(E_{space}, i)$; |
| 7: | $S_{ellipsepoints} \cup^{\{P_{new}\}}$; |
| 8: | end for |
| 9: | $S_{triangles} \leftarrow \text{GetTriangles}(S_{ellipsepoints})$; |
| 10: | $X_{rand} \leftarrow \text{Rand}(0, 1)$; |
| 11: | $T \leftarrow \text{ChooseTriangle}(S_{triangles}, X_{rand})$; |
| 12: | $P_{samp} \leftarrow \text{TriangleSampling}(T)$; |
| 13: | return $P_{samp}$; |

optimization is present in Table 4. The main idea of path optimization is demonstrated in Figure 5. Once the initial path of $Path_0$ has been obtained, the algorithm will search within the eclipse of $E_0$ for a better path. By doing so, the path length will be shortened. The procedure will be repeated for many times, until a satisfactory path is obtained.

### 3.4 *Dynamic RRT algorithm*

Now the dynamic RRT collision avoidance algorithm starts to be constructed. The procedure used RRT algorithm with dynamic collision checking method to generate a feasible path first. Then path optimization is used with restrictions to optimize the initial path and make it shorter and smoother. After that, the algorithm will generate a better path.

Table 4. Pseudo-code of path optimization.

Algorithm 4 PathOptimizing

Input: $P_{start}$, $P_{goal}$, $O_{sta}$, $O_{dyn}$, $L_{path}$, $X_{movelength}$
Output: TailNode

| | |
|---|---|
| 1: | $\tau \leftarrow \{P_{start}\}$; |
| 2: | for $i = 1, 2, \ldots, n$ do |
| 3: | $E_{space} \leftarrow \text{EllipseBuilding}(P_{start}, P_{end}, L_{path})$; |
| 4: | $E_{halfspace} \leftarrow \text{GetHalfEllipse}(E_{space})$; |
| 5: | $P_{samp} \leftarrow \text{EllipseSampling}(E_{halfspace})$ |
| 6: | $S_{points} \leftarrow \text{GetProperPoints}(\tau)$; |
| 7: | $S_{points} \cup^{\{P_{start}\}}$; |
| 8: | $P_{nearest} \leftarrow \text{GetNearest}(S_{points})$; |
| 9: | $P_{new} \leftarrow \text{GenNode}(P_{samp}, P_{nearest}, X_{movelength})$; |
| 10: | if $\text{CoChek}(P_{nearest}, P_{new}, O_{sta}, O_{dyn})$ then |
| 11: | $\text{SetParent}(P_{nearest}, P_{new})$; |
| 12: | $\tau \cup^{\{P_{new}\}}$; |
| 13: | else |
| 14: | Continue; |
| 15: | end if |
| 16: | if $\text{InGoalRegion}(P_{new})$ then |
| 17: | if $\text{CoChek}(P_{nearest}, P_{new}, O_{sta}, O_{dyn})$ then |
| 18: | $\text{SetParent}(P_{nearest}, P_{new})$; |
| 19: | Break; |
| 20: | else |
| 21: | Continue; |
| 22: | end if |
| 23: | end if |
| 24: | end for |
| 25: | return $P_{goal}$ |

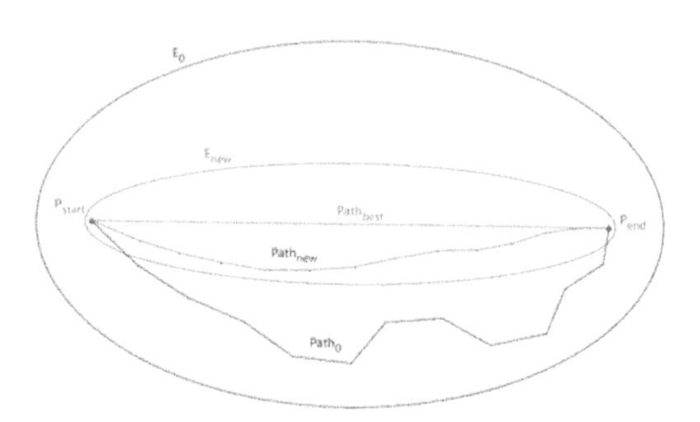


Figure 5. An illustrative demonstration of path optimization.

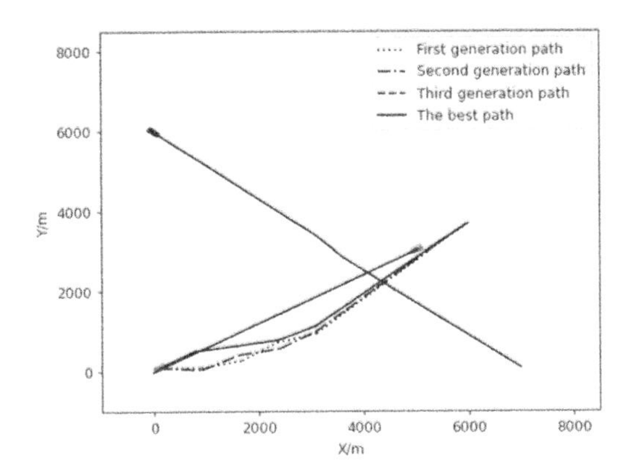


Figure 6. Path optimization result.

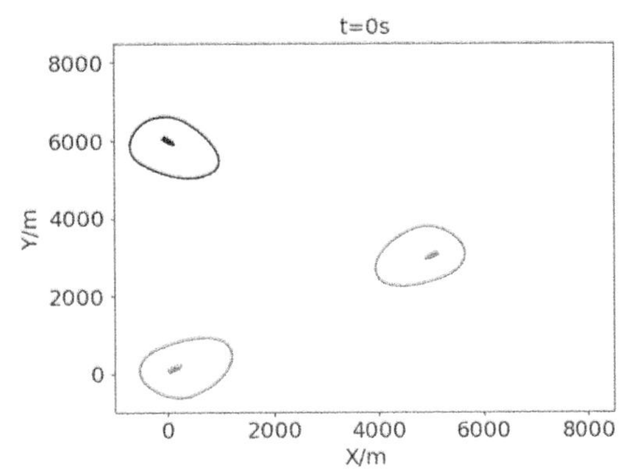


Figure 7. Initial multi-ship encounter situation.

Table 5. Pseudo-code of dynamic RRT algorithm.

```
Algorithm 5 Dynamic RRT

Input: P_start, P_goal, O_sta, O_dyn, X_movelength
Output: TailNode
 1:   P_rrt ← RRT(P_start, P_goal, O_sta, O_dyn, X_movelength);
 2:   L_best ← GetPathLength(P_rrt)
 3:   for i = 1, 2, ..., n do
 4:   P_new ← PathOptimizing(P_start, P_goal, O_sta,
 5:   O_dyn, L_best, X_movelength);
 6:   L_new ← GetPathLength(P_new);
 7:   if L_new < L_best then
 8:   L_best ← L_new;
 9:   P_best ← P_new;
10:   end if
11:   end for
12:   return P_best;
```

Table 6. Parameters of the encounter ships.

| Ships | Start point | Destiny point | Velocity ($kn$) | Length ($m$) | Course (°) |
|---|---|---|---|---|---|
| S1 | (100, 100) | (6000, 3700) | 18.0 | 50 | 58.61 |
| S2 | (5038, 3027) | (0, 0) | 16.0 | 50 | 239.00 |
| S3 | (0, 6000) | (7000, 100) | 16.5 | 50 | 130.13 |

## 4 SIMULATION AND ANALYSIS

### 4.1 *Encounter situation setup*

In order to evaluate the performance of the proposed path planning algorithm, simulations are performed in this section. An encounter situation including three ships are formulated, as shown in Figure 7. The detailed parameters of the three ships are present in Table 6. Both head-on and crossing situations can be found between the three ships and they have collision risk between each other and collision avoidance decision-making are necessary.

### 4.2 *Results analysis*

According to COLREGs, all the ships should be treated as give-way vessels. S1 should give way to S2, S2 should give way to S3 and S3 should give way to S1. By using the dynamic path planning algorithm considering COLREGs, the planned paths of the three ships are presented in Figure 8. It should be noted that the order of the ships' decision makings has influence on the final paths. In the first simulation, S1 made the path planning first, followed by S3 and S2. After S1 has planned its path, S3 will make dynamic collision check based on the new path of S1. The same is for S2. It can be seen from Figure 8 that all the ships keep clearance of each other. S2 did not take any turning actions. The reason is that S1 and S3 have already considered to give way to S2 and no more actions are necessary.

According to Figure 9, the first encounter between S1 and S2 occurred between 260$s$ and 340$s$. During the time, DCPA is larger than 1000$m$ and TCPA gradually reduced to be negative. The distance between these two ships is smaller than 1000$m$. But there is no ship domain violation. Another encounter occurred around 480s between S1 and S3. Because S1 has already made collision avoidance action in the first encounter, S3 only need a little adjustment of course to avoid collision to S1. At that time TCPA also gradually reduced to negative. In summary, the decision made by S1 is the most complex with many turning points. The main reason is that S1 made decisions in the first order and has to deal with complex encounter situation.

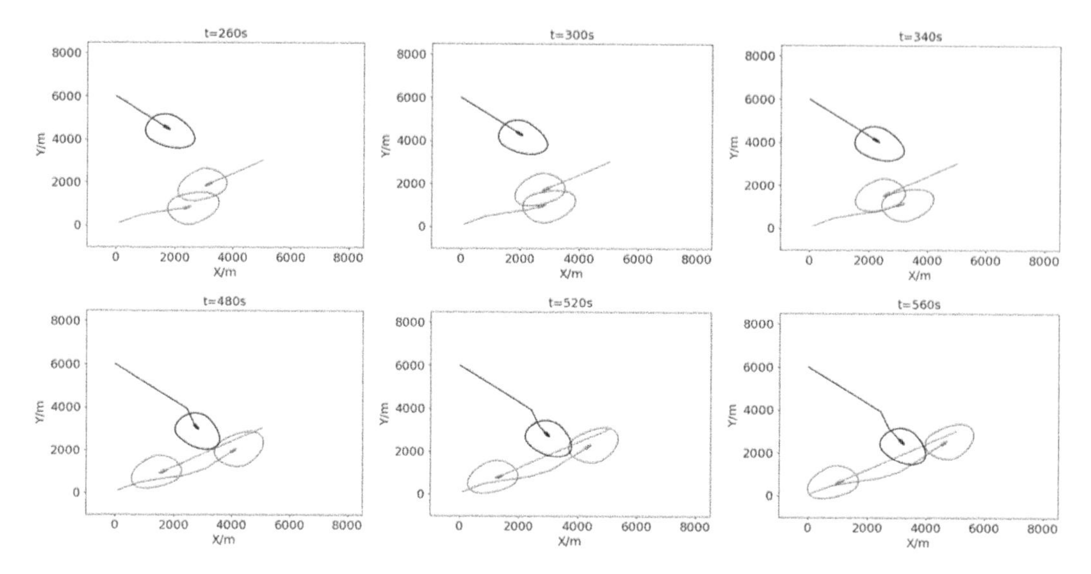


Figure 8. Ship sailing paths with ship domains in the first simulation.

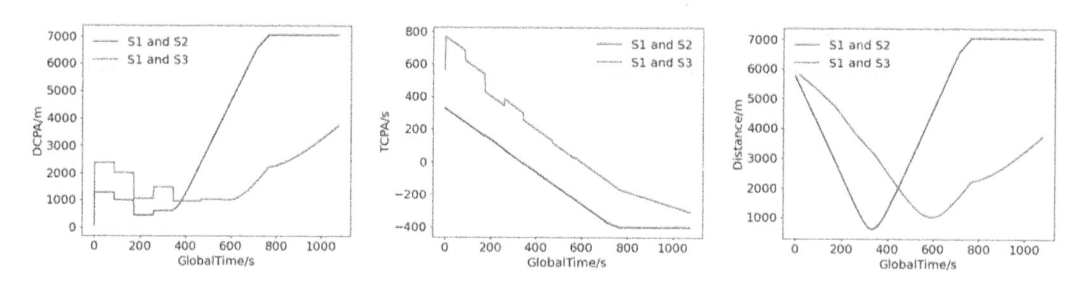


Figure 9. DCPA, TCPA and relative distance between pairwise ships in the first simulation.

This reflects that the order of path planning for the ships has influence on the final results.

The DCPA, TCPA and relative distances between pairwise ships are present in Figure 9. During the whole process, there are two encounter situations for S1, with S2 for the first time, and S3 for the second time. However, S3 should give way to S1 in the second encounter.

### 4.3 *Comparative simulation and analysis*

In order to further evaluate the influence of the decision-making sequence on the results. One more simulation is performed. In this simulation, S3 made the decision first, followed by S1 and S2. The paths generated by the algorithm is present in 10, and the DCPA, TCPA relative distances between pairwise ships are present in Figure 11. According to Figure 10, S1 take a more apparent action to avoid collision compared with the first simulation. In the meantime, S1 has more turning points than the first simulation. S2 did not take any actions, which is the same with the first simulation.

Comparing Figure 8 and Figure 10, it can be seen that during the encounter of S1 and S2, the distance between them is too close. Besides that, the planned paths for S1 and S3 are less complex than the first simulation. In term of the path length, the total length of the three ships' trajectories is 22137.51$m$, and in the second simulation it is 22179.41$m$. They do not have much distinction. But the turning operations are quite different in the two simulations. In summary, the coordination among the encounter ships under complex situations is important in simplifying the collision avoidance path planning.

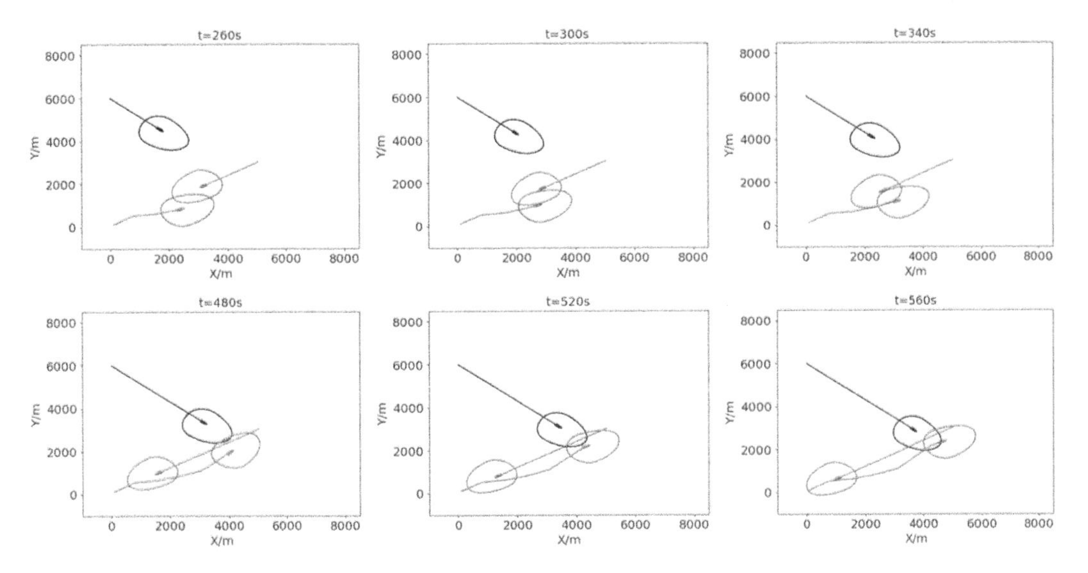


Figure 10. Ship sailing tracks with ship domains in the second simulation.

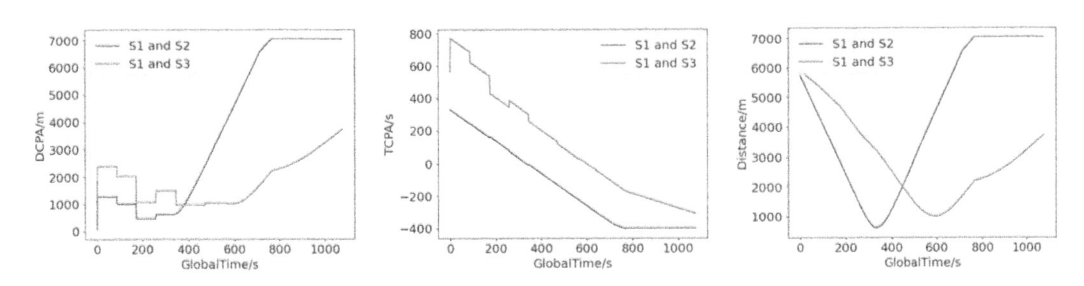


Figure 11. DCPA, TCPA and distance between pairwise ships in the second simulation.

## 5 CONCLUSIONS

In this paper, a path planning methodology is proposed for collision avoidance decision making by introducing RRT algorithm. The original version of RRT algorithm is modified from two perspectives. The first is dynamic collision detection approach been proposed under moving own ship and target ships environment. The second is optimizing the path by re-planning while considering the requirements of COLREGs. The simulation results indicate that the proposed method can find a collision-free path with satisfactory path length under multi-ship encounter situations. The results also indicate that the order of decision making among the ships has influence on the final trajectories. Sometimes the planned trajectory is complex with multiple turning points. Further researches are necessary on the cooperation and coordination mechanism among the encounter ships.

## ACKNOWLEDGEMENTS

This work was supported by National Natural Science Foundation of China (52071247, 5192010-5014), Hubei Provincial Natural Science Foundation of China (2019CFA039, 2021EHB007), Technical Innovation Project of Jiangsu Province (2020Z02), Innovation and entrepreneurship team import project of Shaoguan city (201208176230693), European Union's Horizon 2020 research and innovation programme under the Marie Skłodowska-Curie SAFEMODE (814961) and ENHANCE (823904).

## REFERENCES

Chen, Z., Chen, D., Zhang, Y., Cheng, X., Zhang, M., Wu, C. 2020. Deep learning for autonomous ship-oriented small ship detection. Safety Science. 13: 104812.

Chen, P., Huang, Y., Mou, J., et al. 2018. Ship collision candidate detection method: a velocity obstacle approach. Ocean Engineering. 170: 186–198.

Chiang, H.T.L., Tapia, L. 2018. COLREG-RRT: an RRT-based COLREGS-compliant motion planner for surface vehicle navigation. IEEE Robotics and Automation Letters, 3(3): 2024–2031.

Cai M., Zhang J., Zhang D., Yuan X., Guedes Soares, C. 2021. Collision risk analysis on ferry ships in Jiangsu section of the Yangtze River based on AIS data. Reliability Engineering & System Safety. 215(2): 107901.

Enevoldsen, T.T., Reinartz, C., Galeazzi, R. 2021. COLREGs-Informed RRT* for Collision Avoidance of Marine Crafts. arXiv preprint arXiv:2103: 14426.

Fan, S., Zhang, J., Blanco-Davis, E., Yang, Z., Yan, X. 2020. Maritime accident prevention strategy formulation from a human factor perspective using Bayesian Networks and TOPSIS. Ocean Engineering. 210: 107544.

Fan, S., Zhang, J., Blanco-Davis, E., Yang, Z., Wang, J., Yan, X. 2018. Effects of seafarers' emotion on human performance using bridge simulation. Ocean Engineering. 170: 111–119.

Gkerekos, C., Lazakis, I. 2020. A novel, data-driven heuristic framework for vessel weather routing. Ocean Engineering. 197: 106887.

Huang, Y., Chen, L., Chen, P., Negenborn, R. R., Van, G. P. H. A. J. M. 2020. Ship collision avoidance methods: State-of-the-art. Safety science. 121: 451–473.

Hu, Y., Zhang, A., Tian, W., Zhang, J., Hou, Z. 2020. Multi-ship collision avoidance decision-making based on collision risk index. Journal of Marine Science and Engineering. 8: 640.

Li, L., Wu, D., Huang, Y., Yuan, Z. M. 2021. A path planning strategy unified with a COLREGS collision avoidance function based on deep reinforcement learning and artificial potential field. Applied Ocean Research. 113: 102759.

Li, Y., Wei, W., Gao, Y., Wang, D., Fan, Z. 2020. PQ-RRT*: An improved path planning algorithm for mobile robots. Expert Systems with Applications. 152: 113425.

Perera, L. P.; Carvalho, J. P., and Guedes Soares, C. Fuzzy-logic based decision making system for collision avoidance of ocean navigation under critical collision conditions. Journal of Marine Science and Technology. 2011; 16(1):84–99.

Qu, X., Meng, Q., Suyi, L. 2011. Ship collision risk assessment for the Singapore Strait. Accident Analysis & Prevention. 43(6): 2030–2036.

Sawada, R., Sato K., Majima T. 2021. Automatic ship collision avoidance using deep reinforcement learning with LSTM in continuous action spaces. Journal of Marine Science and Technology. 26(2): 509–524.

Tam, C., Bucknall, R., Greig, A. 2009. Review of collision avoidance and path planning methods for ships in close range encounters. The Journal of Navigation. 62(3): 455–476.

Wang, C., Wang, N., Xie, G., Su, S. F. 2022. Survey on Collision-Avoidance Navigation of Maritime Autonomous Surface Ships. Offshore Robotics. 1–33.

Wang, N. 2013. A novel analytical framework for dynamic quaternion ship domains. Journal of Navigation. 66(2): 265–281.

Wang, T., Wu, Q., Zhang, J., Wu, B., Wang, Y. 2020. Autonomous decision-making scheme for multi-ship collision avoidance with iterative observation and inference. Ocean Engineering. 197: 106873.

Wu, B.; Yip, T. L.; Yan, X. P., and Guedes Soares, C. 2022; Review of techniques and challenges of human and organizational factors analysis in maritime transportation. Reliability Engineering and System Safety. 219:108249.

Xie, L., Xue, S., Zhang, J., Zhang, M., Tian, W., Haugen, S. 2019. A path planning approach based on multi-direction A* algorithm for ships navigating within wind farm waters. Ocean Engineering. 184: 311–322.

Yoshida, M., Shimizu, E., Sugomori, M., Umeda, A. 2020. Regulatory requirements on the competence of remote operator in maritime autonomous surface ship: situation awareness, ship sense and goal-based gap analysis. Applied Sciences. 10(23): 8751.

Zhang, J., Zhang, D., Yan, X., Haugen, S., Guedes Soares, C. 2015. A distributed anti-collision decision support formulation in multi-ship encounter situations under COLREGs. Ocean Engineering. 105: 336–348.

*Safety*

*Trends in Maritime Technology and Engineering – Guedes Soares & Santos (eds)*


# A review of failure causes and critical factors of maritime LNG leaks

M. Abdelmalek & C. Guedes Soares
*Centre for Marine Technology and Ocean Engineering (CENTEC), Instituto Superior Tecnico, Universidade de Lisboa, Lisbon, Portugal*

ABSTRACT: A review of the various failure causes of LNG leaks onboard the maritime LNG units, other than external accidental causes, is presented. The risk of LNG leaks is a primary safety challenge against the expansion in the maritime LNG segment, in particular, the LNG fuelled ship class. Therefore, the identification of the various causes of leaks and their critical development factors is the starting point for managing the risk of accidental releases onboard this segment of ships. The effect of the various technical, human, organisational, and environmental operational risk factors on developing these failure causes is discussed. Afterwards, the critical factors of realising LNG leaks onboard each facility type are highlighted. Further discussion regarding the shortage in number of competent crew who will be needed to fulfil the future workforce demand onboard LNG fuelled ships is also given.

## 1 INTRODUCTION

The number of maritime LNG units has increased significantly in the last decade. As of the end of 2020, there are around 844 ships in operation are carrying LNG onboard as a cargo or as a ship fuel, and in addition, there are 427 ships on order (Songhurst 2019, GIIGNL 2021, MOL 2021, Nerheim *et al.* 2021). Further expansion is expected in the marine LNG segment due to the expected rapid growth in the worldwide LNG demand (GIIGNL 2021), and the increasing attractiveness of LNG as a ship fuel in light of the new International Maritime Organisation (IMO) environmental requirements for reducing exhaust emissions from ships (Hansson *et al.* 2019, Ampah *et al.* 2021, Wang *et al.* 2021a).

Indeed, the maritime LNG segment has a good reputation in terms of operational safety since the launch of the first large scale LNGC in 1964 (Pelto *et al.*1982). Only few events of unignited accidental leaks of LNG ships have been reported in the literature e.g. (Ostvik *etl al.* 2005, Paltrinieri *et al.* 2015). However, more efforts are needed nowadays for ensuring a leak free maritime LNG segment in order to avoid the potential severe consequences of LNG leaks (Pitblado *et al.* 2004, Alderman 2005).

The following section of this research presents a background knowledge on the various LNG ship types and systems, in addition to, the various consequences of LNG leaks. Afterwards, section three presents the various failure causes (FCs) of leaks of LNG systems. The presented FCs in section three are the ones that occur due to internal failures of the containment systems. However, leaks that occur due to external events such as fire and explosion, or collision and grounding are not covered in this research. Furthermore, a criticality identification of the operational factors that affect the development of the various FCs on each facility type is conducted in this section. Moreover, in section 4, discussions of the related operational safety challenges to LNG leakage failures are presented with concentration on the effect of lack of competent crew of the LNG fuelled vessels. Lastly, section 5 presents a summary of the work performed in this research.

## 2 BACKGROUND KNOWLEDGE

### 2.1 *Marine LNG facilities and systems*

As of the end of 2020, there are approximately 844 marine LNG units in operation and 427 units on order as indicated in Table 1. In this research, marine LNG units are classified as transporting units and in-situ units. The first category represents LNG carriers (LNGCs), LNG bunkering vessels (LNGBVs) and LNG fuelled-ships (LNGFSs). In addition, the second category represents the floating LNG units (FLNGs) which are subdivided into floating production, storage and offloading vessels (LNG FPSOs) and floating storage and regasification units (FSRUs) (Aronsson 2012).

Marine LNG systems involve four systems which are: i) cargo containment systems (CCS), ii) transfer systems, iii) regasification systems, and iv) liquefaction systems. In the vast majority of LNG ships, LNG is stored in membrane tanks or in IMO independent tanks of type A (prismatic), B (spherical) or

DOI: 10.1201/9781003320289-19

Table 1. Number of operational and on order LNG marine units.

| Categories | Transporting | | | In-situ | |
|---|---|---|---|---|---|
| Ship type | LNGCs | LNGBVs | LNGFSs | FSRUs | LNG FPSOs |
| In operation | 576 | 23 | 198 | 43 | 4 |
| On order | 119 | 21 | 277 | 7 | 3 |
| Sources | (Songhurst 2019, GIIGNL 2021, MOL 2021, Nerheim *et al.* 2021) | | | | |

C (pressurized). LNG is typically stored below -162°C, and slightly above atmospheric pressure (0.25 – 0.7 barg), or higher than 2 barg in case of type C tanks (SIGTTO 2000).

A traditional LNG transfer system consists of LNG pumps, or pressure build up units (PBUs) onboard some of the LNGFSs, LNG lines and their fittings (e.g. valves and expansion bellows), and LNG receiving or discharge manifolds (SIGTTO 2000, Chorowski *et al.* 2015). Furthermore, LNG regasification systems are used onboard LNGFSs (Chorowski *et al.* 2015) and FSRUs (Martins & Schleder 2012). In regasification systems, LNG is pumped, evaporated, and then compressed. Propane or water-glycol intermediate fluid vaporisers (IFVs) are the widely used LNG vaporisers in the marine regasification systems (Egashira 2013).

However, in some of the FSRU designs, IFVs are used to heat LNG from -162°C to -10°C before using seawater open rack vaporisers (ORVs) to avoid seawater freezing (Martins & Schleder 2012). Afterwards, natural gas is compressed under pressure range from 46 to 130 bar in case of FSRU (Strande & Johnsson 2013). In the case of LNGFSs, natural gas is compressed to 250 – 300 bar in the high pressure fuel gas supply systems (FGSSs) (Chorowski *et al.* 2015).

Moreover, in liquefaction systems of the LNG FPSOs, natural gas is compressed at least to 60 bar before liquefaction, such as the case in the coil wound heat exchangers (CWHEs) (Pek & van der Velde n.d., Bukowski *et al.* 2013). However, an LNG FPSO can liquefy up to 3.6 million tonnes per annum (MMTPA) of LNG as of the case in the Shell Prelude (Yussoff 2019). On the other hand, capacities of the boil-off gas liquefaction systems of LNGCs can range between 3-7 tonne per hour (t/h), and the range of the systems' pressure falls between 4.5 to 45 bar (Gomez *et al.* 2014, Vorkapić *et al.* 2016).

### 2.2 *Consequences of LNG leaks*

LNG is odourless, nontoxic, noncorrosive, and lighter than water (Alderman 2005). In addition, LNG boils at -162°C at atmospheric pressure, and its vapour has a flammability range between 5% to 15% (Nerheim *et al.* 2021). The minimum ignition energy (MIE) of a flammable mixture of LNG vapour and air at 20°C and atmospheric pressure is 0.28 mJ (Vandebroek & Berghmans 2012). However, whether they are ignited or not, LNG leaks can cause significant consequences to human, environment and assets.

Direct contact between human skin and LNG leaks can cause significant cryogenic burns and can also be lethal (Alderman 2005, Ostvik *et al.* 2005). Personnel who are engulfed in LNG vapour can suffer damage to lung, hypothermia, and/or Oxygen deficiency asphyxiation (Foss 2012, SEPS 2018). Furthermore, LNG leaks will result in brittle fractures to ship's hull and decks which are made of carbon steel alloys (Nam *et al.* 2021). This event has been observed in several historical LNG releases (Ostvik *et al.* 2005).

Another consequence of the non-ignited LNG leaks is rapid phase transition (RPT), which is a physical explosion that occurs when spill of LNG contacts a warm surface (e.g. seawater) that can result in a rapid evaporation of LNG (Alderman 2005). The probability of RPT occurrence is positively correlated with the propane and ethane fractions of the spilled LNG (Melhem *et al.* 2006). The resulting pressure waves from RPT are able to cause damages to the surrounding objects without further effects on objects that located at longer distances (Pitblado *et al.* 2004). Further, Large releases of LNG are significant to the environment due to the high 20-year, and 100-year global warming potentials (GWP) of methane, which are 72, and 25, respectively (ICCT 2017).

On the other hand, ignition of LNG leaks can result in several scenarios of fire and explosion. Immediate ignition of LNG can lead to pool fires (Betteridge *et al.* 2014), or jet fires (Zhang *et al.* 2019). Pool fires can lead to severe damage to ship structure if it is sustained for long period due to the high burning temperature of LNG which is approximately 1500° C (Petti *et al.* 2013). On the other hand, jet fires result in higher heat flux than pool fires, and can lead to severe damages rapidly if impinged on adjacent equipment (Roberts *et al.* 2001). Furthermore, LNG containment systems (e.g. pressure vessel) that engulfed in pool fires, or impinged by jet fires can lead to the occurrence of boiling liquid expanding vapour explosions (BLEVE) (Pitblado 2007). However, cold BLEVEs can also occur to LNG containment systems (e.g. pipelines or pressure vessels) if the insulation failed and the expanded LNG was not relieved properly through the pressure relief system (PRS) (Ustolin *et al.* 2020).

Delayed ignition of LNG can lead to flash fires or vapour cloud explosions (VCE) (Cracknell & Carsley 1997, Atkinson *et al.* 2017). Flash fires occur when the flame propagation speed of the vapour cloud is subsonic (deflagration), while VCE is resulting from a supersonic flame propagation (detonation) (Oran *et al.* 2020). Flash fires occur in open air or non-congested areas, while VCEs require congested or confined area to develop the turbulent effect of detonation development. However, a flash fire develops rapidly as a fireball (in less than 3 seconds) radiating heat corresponding to the amount of the burned LNG vapour without causing significant pressure waves (Pitblado *et al.* 2004, Margolin 2013). The resulting heat flux from flash fires can cause severe injuries to humans present in their harmful radiation zone (Sintef 2003). While VCEs can result in significant damages due to their consequent pressure waves.

## 3 LNG LEAKS OF MARINE LNG SYSTEMS

### 3.1 *Leakage failure causes of marine LNG systems*

Loss of containment of marine LNG systems can occur due to several FCs. In this research, several hazard identification and risk assessment reports, and specialised books in maritime LNG operations have been reviewed for identifying the most common driving causes of leaks from the maritime LNG systems (e.g. Pelto *et al.* 1982, SIGTTO 2000, Dogliani 2002, Ostvik *et al.* 2005). Based on the reviewed literature it has been found that the vast majority of LNG leakage FCs can be categorised under the following groups:

i) Time-dependent structural failures,
ii) Operational error structural failures,
iii) Operational error only,
iv) Mechanical equipment failure, and
v) Accidental events.

Accidental events leakage FCs represents the resulting LNG leaks from accidental loads such as collision, grounding, fire and explosion loads, dropped objects, and extreme weather conditions. However, LNG leaks' accidental FCs are out of the scope of this study, so that no further information are presented regarding them.

#### 3.1.1 *Time-dependent structural failure causes*

This group of FCs of the maritime LNG systems represents structural damages that develop over the time while LNG systems are functioning under the safe operating conditions. In detail, structural failures in this case are developing due to the exposure to dynamic (cyclic) elastic loads. The most affected system by these loads is the CCS. However, CCS of LNG are subjected to various sources of dynamic loads during operation which are wave induced loads, cyclic thermal stresses, and pressure variation (Huther *et al.* 1981, DNV 2017a, Park *et al.* 2021).

The continuous exposure of LNG CCSs and the other LNG systems onboard the various facilities to these cyclic loads contribute to the propagation of ductile fatigue cracks at welds and structural elements of theses equipment, although the imposed loads are less than the yield strength of their structures. Fatigue S-N curves are describing the relation between stress levels, and the number of load cycles of each stress level that will cause structural damage (DNV 2017a).

Moreover, vibrations that result from ship's propulsion system, and machinery can contribute to the development of fatigue cracks, as well as, causing fractures to equipment structures in case of resonance (Bass *et al.* 1976). However, marine LNG systems are designed to sustain the variable operational loads for more than 20 years of life span (DNV 2017a, Youn & Kim 2019). Adding to that, LNG containment systems are thoroughly inspected every 5 years during ship's docking (Youn & Kim 2019), so that all the detected structural defects to the CCS and the other equipment will be repaired before floating the vessel (DNV 2017b). However, ship survey can also be conducted in-situ for LNG floating units i.e. FLNGs (DNV 2015).

Furthermore, the risk of internal corrosion failure of LNG containment systems is deemed to be negligible due to the non-corrosive nature of LNG, and also due to the high corrosion resistance of the manufacturing materials of these systems such as Fe-36NI (Invar) and 304L stainless steel (INCO n.d. LR 2011, Wang *et al.* 2020). In addition, insulation of the weather exposed LNG equipment (e.g. cargo pipes), and the manufacturing materials of these equipment (e.g. stainless steel) are primary factors of avoiding the risk of external corrosion structural failure of these equipment (Tasneef 2016, JM 2020). On the other hand, the risk of LNG leak due to internal corrosion and erosion should be considered for the LNG vaporisers that are using seawater as a LNG heating medium (Egashira 2013).

#### 3.1.2 *Operational structural failure causes*

Operational structural failures can occur due to critical cargo loading conditions, or erroneous operation (direct operator error). The related FCs to the critical cargo loading conditions are affecting the structural safety of LNG CCSs. The underlying FCs of this type are sloshing loads (LR 2012), and rollover phenomenon (Kulitsa & Wood 2018, Zakaria *et al.* 2019). On the other hand, operational structural FCs of the erroneous operation are rapid cooling, over pressurisation, and under pressurisation of the various marine LNG systems.

##### 3.1.2.1 Sloshing

According to DNV (2016), "*sloshing is a violent resonant motion of the free surface of a liquid cargo inside a moving tank, e.g. standing waves in partially filled cargo tanks onboard an LNGC*". In detail, the natural period of LNG inside a partially filled tank has a high chance to coincide with the natural period of the wave induced ship motion (Masanobu *et al.* 2015). The realisation of this condition will result in significant pressure loads on the internal longitudinal and transverse structures of the tank causing a potential damage to the tank's internal structures such as membrane and insulation materials. And in some conditions damage can be extended to the inner hull of the ship (LR 2012).

Membrane tanks are the most vulnerable type to sloshing loads due to the presence of vertical walls in the longitudinal and transverse directions. Therefore, restrictions to the filling levels of membrane tanks are applied to avoid the risk of sloshing. LR (2012) stated that, sloshing critical region falls in a filling ranges from 10% to 70% of a membrane tank height, see Figure 1. As indicated in Figure 1, the filling range from 10% to 40% is the most sloshing critical region, and this range is classified as a critical filling zone of the membrane LNGCs during the cargo loading operations from Shell Prelude LNG FPSO (Shell 2019).Other than that, sloshing loads are not significant to type B spherical tanks due to the high strength of their structures, as well as, due to the spherical shape of the tank (van Twillert 2015).

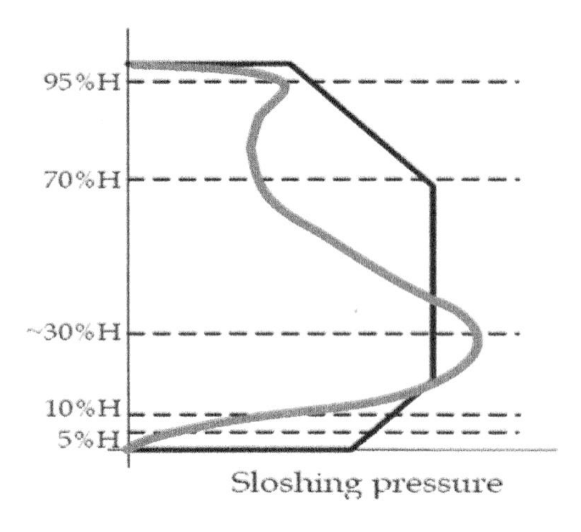


Figure 1. Typical variation of sloshing pressure with fill height (LR 2012).

3.1.2.2 Rollover phenomenon

Rollover is a rapid propagation of LNG vapours inside a storage tank due to LNG stratification (Kulitsa & Wood 2018). Stratification exists inside a tank due to the difference between the densities of the stratified LNG layers. The primary cause of LNG stratification is the mix between different sources of LNG inside the same tank, for example, adding a new cargo to a remaining heal in an LNG tank. Also, auto stratification of the same type of LNG that contains nitrogen can occur due to weathering i.e. storing LNG inside a tank for long period of time (GIIGNL 2012).

Assuming that a tank has two stratified LNG layers as shown in Figure 2, and the leaking heat through insulation is warming up both LNG layers. The upper (lighter) LNG saturated layer will release the added heat through boil-off and will continue to be denser as it produces BOG. On the other hand, the added heat to the lower (denser) LNG layer will not fully convicted through the upper layer, because the upper layer has a higher latent heat vaporisation value (Wlodek 2019). The resulting delay in heat convection from the lower to the upper LNG layers will create a superheated LNG region at the interface between both layers (Zakaria *et al.* 2019). The higher the density difference between both layers, the more the superheating of LNG at the interface will be (Wang *et al.* 2021b).

The superheating of the lower layer will continually decreasing the density of that layer. When the densities of both layers equate, the superheated layer will start to move upward to release the developed superheat causing a rapid LNG vapour evolution (10 – 30 times of the regular BOG rate) (GIIGNL 2012, Wang *et al.* 2021b). The rapid development of LNG vapours will increase the pressure inside the tank, and consequently excessive vapours will be released to the atmosphere through the venting system. The vented vapour from the vent mast will fall down, and once the vapour temperature reaches -110° C, it will be lighter than air, and then it will disperse creating a large hazardous flammable cloud (SIGTTO 2012). Rollover can also cause significant pressure rise which can affect the structural integrity of the tank if the amount of the evaporated LNG is considerably higher than the capacity of the PRS (SIGTTO 2012).

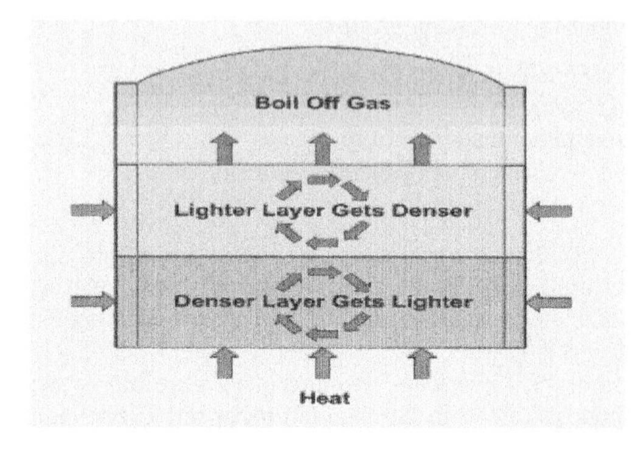


Figure 2. LNG convection cells under stratification (GIIGNL 2012).

3.1.2.3 Rapid cooling of LNG equipment

LNG equipment are designed to operate safely under cryogenic temperature (-161° C) given that the cool down process to that temperature is carried out according to the instructions of the equipment manufacturer. The cooling down instructions of LNG equipment are developed to ensure that the exerted transient thermal stresses on LNG equipment during the cooling down process are not exceeding the strength of the equipment's material. For instance, the cooling down rate of a warm LNG tank should not exceed -10° C per hour for avoiding the excessive transient thermal loads (SIGTTO 2000).

The occurrence of rapid cooling will not allow a gradual heat transfer across the entire thickness of the metal. And then fractures will nucleate and propagate at the exposed surface due to the resulted excessive transient tensile stresses, although the temperature of the exposed surface does not exceed the safe operating limit of the material (Lu & Fleck 1998). Furthermore, exposure of a pressurised equipment to rapid cooling is more dangerous because the pressure induced tensile stresses will increase the occurrence potential and the severity of fractures. This condition is known in nuclear industry as Pressurised Thermal Shock (PRT) (Araneo & D'Auria 2012).

3.1.2.4 Over pressurisation and under pressurisation of LNG equipment

Accidental over pressurisation of LNG equipment in the presence of failure or reduced functionality of the PRS is a sufficient condition to cause structural failure leak if the applied pressure loads have exceeded the structural strength of these equipment. The primary contributing causes of over pressurisation are misoperation or failure of manual and automatic valves, or failure of monitoring and alarm systems (Pelto *et al.* 1982, Kim *et al.* 2005). For instance, over pressure inside LNG CCSs can occur due to instrumentation error or erroneous isolation of vapour return line during cargo loading. Moreover, failure of drying (inerting) LNG systems prior to cargo operations can lead to ice and hydrate formation, and accordingly causing blockage and overpressure of the transfer system which can lead to structural failure leaks (SIGTTO 2000).

Conversely, the continuous build-up of negative pressure inside cargo tanks can also lead to structural collapse (implosion) to some of the LNG tank types (DNV 2017c, p.135), given that vacuum breakers had failed to operate (Kim *et al.* 2005). The developing scenarios of under pressurisation are similar to those of the over pressurisation events. However, under pressurisation can be realised if vapour return line was isolated while discharging LNG from the tank, as well as, due to thermal oxidation or vapour re-liquefaction (Pelto *et al.* 1982, DNV 2017c, p.135).

3.1.3 *Operational error only failure causes*

LNG leaks can also occur without suffering any structural failures to equipment. FCs of this type occurs due to direct human errors or failure of the instrumentation and alarm systems. The underlying FCs of this group are similar to those of the over pressurisation, but without any failures to PRS. Overfilling of CCSs during cargo loading is a well representative example of this group of LNG leakage FCs. Ostvik *et al.* (2005), stated that three overfilling events had occurred in the maritime LNG industry. Furthermore, rollover can also be classified under this group given that excessive vapours have been released without posing any structural defect to the tank.

Other than that, failure of maintenance operations of LNG equipment can also lead to accidental releases of LNG. Accidental release in this case can occur during the performance or after the completion of maintenance operations. In the earlier case, leaks occur due to errors in purging or isolation of equipment, while in the latter, leaks occur due to improper reassembly of equipment or because of the use of wrong spare parts (Sklet 2006). Moreover, the use of wrong or defected gaskets during external LNG transfer operation (e.g. LNG fuelling operation) is one of the possible operational error leak FCs.

3.1.4 *Mechanical equipment failure leaks*

Mechanical equipment failure leaks are resulting from wear and tear of sealing elements of mechanical equipment such as gaskets and seals of flanges and valves (LR 2011). Therefore, maritime LNG facilities that involve large amount of continuously operating mechanical equipment have more frequency of occurrence of LNG leaks. This is particularly one of the major sources of leakage risk of FLNGs which are continuously handling large quantities of LNG. However, the higher amount and larger sizes of equipment onboard LNG FPSOs make them the more susceptible to this type of FCs than FSRUs (Dogliani 2002).

## 3.2 *Critical operational leaks' development factors*

LNG leakage FCs are mostly realised due to erroneous human interventions, lack of detecting failure signs, and deviations from the safe operating conditions. These direct and indirect human errors occur due to the lack of situational awareness of human elements (Di Mattia 2013). On the other hand, the resulting LNG leaks from mechanical failures due to poor equipment design and quality have not occurred in the modern history of the marine LNG systems. However, prevention of accidental maritime leaks require efficient management of the various operational technical, human, organisational and environmental factors that, if poorly managed/addressed, can contribute in realising the various LNG leakage FCs (Xuan *et al.* 2019).

The presented critical operational factors (COFs) in Table 2 are used in the extended criticality assessment of the various operational technical, human, organisational, and environmental aspects which their statuses have direct effect on realising LNG leakage FCs. The identification of the presented COFs in Table 2 is performed based on the presented selection criteria in Abdelmalek & Guedes Soares (2021). However, in the literature COFs are also called risk influencing factors (RIFs) (Sklet 2005), and the influencing factors of human performance only are called performance shaping factors (PSF) (Di Mattia 2013).

Generally, the presented COFs in Table 2 are the factors that with the highest importance in relation to the development of LNG leakage failures onboard all the facility types. But some of these COFs are specifically more critical than others to each facility type. The selection of the highly critical COFs to each facility type is identified based on two different aspects.

Firstly, COFs that are properly addressed during the design phase risk studies but their performance is subjected to high fluctuation during operations are deemed to be highly critical COFs. For example, workload and work complexity are of higher importance to the risk of leak on FLNGs than on LNGCs. And secondly, COFs with high degree of uncertainty at the current state of knowledge should be well monitored and well managed during the operations phase. For instance, competence of the available seafarers who will be needed to fill the future workforce demand are more challenging to LNGFSs segment than to FLNGs.

Table 2. Generic COFs of LNG leakage FCs.

| COFs group | Generic COFs |
|---|---|
| Technical | Equipment design, system complexity, human-machine interface (HMI), and technical condition. |
| Human | Competence, workload and stress, supervision, communications, safety awareness, and quality of maintenance and inspection. |
| Organisational | Control of work system (CoW) quality, procedures quality, process safety culture, shore supportability, and training quality. |
| Environmental | Working environment, ergonomics, weather condition (wave height and wind speed), and visibility. |

3.2.1 *Critical COFs of the in-situ LNG marine units*

The presence of continuously operating equipment with considerable quantities of LNG under high pressure (e.g. liquefaction system of LNG FPSOs) onboard this class of facilities poses significant risk of operational structural failures (erroneous operation), operational error only, and mechanical equipment failure FCs. Examples of these FCs are maintenance failure, failure of equipment start-up, and leakage of valves' seals. Moreover, the resulting human errors from the lack of situational awareness are deemed to be the critical contributing elements to realising the operational error leakage FCs (Di Mattia 2013).

In detail, the resulting human errors from lack of competence and lack of the supporting organisational factors have low likelihood of occurrence onboard FLNGs. Because companies operating these vessels have sufficient operational experience in the LNG segment, and also have strong operational safety standards and auditing systems. Adding to that the stable expansion in the number of FLNGs allows sufficient time for generating crews with adequate competence to fill the future workforce demand. However, dynamic COFs such as workload and stress, communications (e.g. quality of handover), and weather conditions are more critical to FLNGs due to the complexity of their systems and operational tasks. Other than that, COFs to mechanical equipment failures, such as technical condition and quality of maintenance and inspection tasks, are more significant to LNG FPSOs, which according to (Dogliani 2002) are maintenance intensive facilities.

Furthermore, FSRUs are the most class of LNG marine units that subjected to high chance of mixing significant quantities of different sources of LNG in the same tank, which is the primary cause of occurrence of rapid rollovers (Kulitsa & Wood 2018). Rollover can be avoided by using the appropriate filling methodology of the added cargo (i.e. bottom filling for lighter LNG and vice versa) or by the early detection of the triggering signs to rollover (GIIGNL 2012, Kulitsa & Wood 2018).

The avoidance of rollover occurrence relies significantly on the competence of the gas engineer, the technical condition of the tank alarm and instrumentation systems, and the interaction between both of them i.e. human machine interface (HMI). These factors are also critical for preventing the risk of overfilling while receiving cargo from LNGCs, due to the frequent cargo receiving operations of FSRUs. Other than that, the risk of sloshing to the weather exposed membrane FLNGs is well managed by using specially reinforced tank designs which allow the safe operation at any filling levels (GTT 2020). And also adaptive operation planning can be used to minimise the time of operation while tank levels are in the barred zone (Rokstad 2010).

3.2.2 *Critical COFs of the transporting marine LNG units*

3.2.2.1 Critical COFs to LNGCs

Cargo transfer systems on board LNGCs are usually operating during cargo operations unless internal transfers are required. However, Ostvik *et al.* (2005) revealed that the majority of the historical LNG leaks from LNGCs had occurred during cargo operations. In detail, mechanical valve failures and overfilling were the most contributing FCs of LNG leaks during cargo operations. Also, (Paltrinieri *et al.* 2015), represented that high wind speed had resulted in two different events of sheared loading arm during cargo operations. Therefore, technical condition of cargo systems including their alarm and instrumentation systems, as well as, weather condition, and HMI are deemed to be highly critical to leakage risk of LNGCs.

Other than that, the expected future expansion in the worldwide demand of LNG (from 351 MMTPA to approx. 600 MMTPA in 2025) will lead to significant increment in the worldwide fleet of LNGCs (GIIGNL 2020). This rapid increment introduces an explicit operational safety challenge due to the lack of competent workforce (Odedra 2013). Because the lack of position specific experience is one of the primary barriers to building the required level of situational awareness to perform critical tasks safely (Di Mattia 2013). Therefore, lack of competence is classified as a highly critical factor to the operational error leakage FCs due to the associated uncertainty of this factor in light of the current state of knowledge.

Fortunately, no LNG external leakage due to structural failures of tanks of LNGCs had been reported in the literature. Sloshing risk in LNGCs is also deemed to be well managed as far as ships are loaded outside of the sloshing barred zone (10% to 70%) (LR 2012). Only some sloshing damages to the primary barrier of membrane tanks in the upper chamfers region had been observed, although the involved ships in these incidents were sailing while carrying LNG outside of the barred zone (LR 2012)

Other than that, LNGCs might be faced with the risk of sloshing under special situations. For example, an LNGC could be enforced to sail with partial load in case of disconnecting from a weather exposed terminal due to an emergency situation. In that case, cargo consolidation plan should be implemented as soon as possible for avoiding the risk of sloshing onboard the ship. For instance, cargo consolidation should be done in less than 8 hours at any LNGC that prematurely disconnecting from Shell Prelude FPSO (Shell 2019). Other than that, the risk of fatigue failures to the CCS is sensitive to various COFs such as weather conditions, and quality of and adherence to the cargo filling procedures and checklists.

#### 3.2.2.2 Critical COFs to LNGBVs

Critical COFs to leakage FCs of LNGBVs are similar to those of the LNGCs. This is because LNGBVs and LNGCs are, to a large extent, equipped with similar cargo storage and transfer systems. However, in the meantime, there are only 23 LNGBVs in operation (GIIGNL 2012), and most of the bunkering operations nowadays are carried out in non-weather exposed areas i.e. inside the port area, so that sloshing and excessive ship motion fatigue risks are not highly critical to LNGBVs under the current conditions. From another perspective stringent LNG bunkering procedures will be required for avoiding the risk of accidental releases inside the port area during bunkering operation see, for example (SSPA 2017).

Another challenge for the potential future growth of the number of LNGBVs in performing the bunkering operations at weather exposed areas in which an LNGBV can sail with a partially filled tanks to serve more than one LNGFS or while returning back to the port. For this reason, appropriate selection of the CCS design of each LNGBV should be determined based on the specific operational conditions of each ship to avoid the risk of sloshing, see for example (Masanobu *et al.* 2015). The potential mixing of different sources of LNG in the same tank to fulfil the bunkering demand should be considered as one of the future operational safety concerns of LNGBVs. Therefore, in this special case, critical COFs to rollover risk of FSRUs will also be applicable to LNGBVs.

#### 3.2.2.3 Critical COFs to LNGFSs

SEA-LNG (2021) stated that, around 30% of the gross tonnage on order are LNGFSs. The confidence of relying on LNG as a marine fuel is originated from the technical maturity, and the safe operation records of the maritime FGSSs. One of the main factors of achieving a safe performance in this class of ships is that most of the LNGFSs are Norwegian flag ships and/or owned by top class shipping companies (e.g. CMA CGM), where strong regulations and safety culture are implemented, and competent crewmembers are present (Rodrigues 2013, Laribi & Guy 2020).

The lack of competent crew with specific experience in operating FGSSs, and the current level of process safety culture of the majority of seafarers are important operational safety challenges to the rapid expansion of the LNGFSs number. Especially for the small-scale shipping companies, and ships that are planned to operate in developing countries' regions. Therefore, it is foreseeable that special considerations should be given to the related COFs to human knowledge (e.g. competence, safety culture, and training quality) in order to control the risk of the resulting LNG leaks from the related FCs to human errors. Also, the availability of technical remote support with the FGSSs' manufacturers, and online support of ship management are important factors to reduce the related uncertainties to the critical operational decisions in the right time.

## 4 DISCUSSION

### 4.1 *Operational safety challenges of leakage failure in the maritime LNG segment*

The exerted dynamic loads on marine LNG units such as wave motion, and cyclic pressure and temperature loads (Huther *et al.* 1981, DNV 2017a, Park *et al.* 2021), are primary causes of failure of CCSs of these units. Therefore, operational instructions of the manufacturers of the CCSs must be followed for the sake of avoiding the resulting structural damages from excessive fatigue loads. Sloshing is the most dangerous FC to LNG CCSs, in particular, the membrane tank type. The risk of sloshing to membrane CCSs can be avoided by ensuring that partial loading within the barred zone (10% to 70%) is avoided (LR 2012).

The adherence to the safe cargo loading levels is significant to the membrane type LNGCs and LNGBVs which are subjected to sail in rough weather conditions. However, operating with partially filled tanks cannot be avoided onboard FLNGs and LNGFSs. Therefore, special tank designs, operational constraints, and cargo planning solutions can be used to avoid the risk of sloshing on those classes of ships (Rokstad *et al.* 2010, Masanobu *et al.* 2015, GTT 2020).

Furthermore, rollover can lead to massive release of LNG vapours, and under certain conditions structural damages to CCSs can occur (GIIGNL 2012, Zakaria *et al.* 2019). Therefore, triggering events to rollover must be strictly avoided. The primary cause of rollover is the mix of different cargo sources with different densities in the same tank. However, this condition can result in rollover within 12 hours (Kulitsa & Wood 2018). Therefore, corrective actions must be implemented as soon as possible upon the detection of any triggering signs of rollover. FSRU ships are the most susceptible class to suffering a rollover event. However, the detection of the rollover's triggering signs, and the efficiency of the required corrective actions are sensitive to the performance of human elements, the condition of the alarm and instrumentation systems, and HMI.

Moreover, the large amount of the continuously operating mechanical equipment onboard FLNGs makes failure to seals and gaskets one of the significant leakage FCs in this segment of LNG units. The quality of the performed maintenance and inspection activities, in addition to, automated failure detectability are crucial for avoiding the occurrence of these FCs onboard FLNGs. Other than that, performance of human elements, and quality of the organisational aspects require proper management in operations phase of LNG units for the sake of avoiding the resulting leaks from operational errors.

The identification of the critical COFs to the risk of operational error leakage FCs require extended understanding of the operational safety challenges of each segment of the marine LNG units. The highly critical COFs to the risk of operational error leakage FCs onboard FLNGs are considered to be the related factors to complexity, and work load. The related factors to competence and organisational standards are deemed to be less challenging to FLNGs due to the availability of the experienced crew and the availability of strong organisational factors of the operators of these units.

In contrast, LNG systems and the pace of workload onboard the transporting LNG marine units are less complex than those of the FLNGs. However, this class of LNG ships is faced with the lack of competent seafarers who will be needed to fulfil the future workforce demand in this shipping segment (Odedra 2013, Rodrigues 2013). On the other hand, the expansion of using LNG as an alternative ship fuel confronts an additional safety challenge which is the lack of the operational experience of the non-LNG shipping companies who will need to use LNG FGSSs onboard their vessels.

In Table 3, a summary of the highly critical COFs of each LNG facility type is presented. However, the criticality assessment of these factors does not lessen the importance of the remaining COFs, but it is highlighting where more efforts need to be spent to ensure the safe operation of each facility type.

### 4.2 *Operational safety barriers of LNGFSs fleet expansion*

Lack of number of seafarers who are competent to operate LNGFSs safely is the primary operational safety barrier against the expansion of this segment of ships (Rodrigues 2013). The primary area of concern in this regard is the lack of knowledge about LNG FGSSs, and the associated hazards of LNG leaks, and lack of individual experience in operating LNGFSs. In addition, marine LNG FGSSs are considered as a small-scale oil and gas process segment,

Table 3. Highly critical COFs of each facility type.

| | Marine LNG facility typesx | | | | |
|---|---|---|---|---|---|
| Critical COFs | LNGCs | LNGBVs | LNGFSs | LNG FPSO | FSRU |
| Technical | Technical condition of cargo system | Design of CCS, and technical condition | Technical condition of the exposed bunkering equipment. | System complexity, technical condition and HMI | System complexity, technical condition and HMI |
| Human | Competence | Competence | Competence and safety awareness, supervision. | Workload and stress, and communication | Workload and stress, and communication, and competence |
| Organisational | Quality of cargo filling and consolidation procedures | Quality of cargo filling procedures | Training quality, process safety culture, and shore supportability | CoW system integrity. | CoW system integrity. |
| Environmental | Weather condition | Weather condition | Ergonomics | Weather condition and working Env. | Weather condition and working Env. |

and accordingly marine crew who are needed to operate them must have adequate level of process safety awareness.

Therefore, the IMO has stated the minimum level of training requirements and competency level of each rank onboard ships that operate with low flash point fuels in the resolution MSC.397(95) (IMO 2015). Further training requirements can also be applied based on the regional or flag state requirements (Rodrigues 2013).

However, training can only help unexperienced seafarers to perceive the operational aspects of FGSSs, and LNG hazards. But specific operational experience of LNG FGSSs is needed for developing the required level of situational awareness for avoiding the resulting leaks from human errors during the various critical LNG operations (for instance, bunkering operations) (Di Mattia 2013). Other than that, implementation of organisational elements such as CoW system, and monitoring of operational safety performance are paramount for ensuring that operations are carried out safely, as well as, for capturing and learning from all the operational unsafe acts and conditions.

Therefore, shipping companies who are intending to operate LNGFSs have to ensure the compliance with the minimum training requirements of the IMO MSC.397(95) (Kamb 2020), as well as, the continual monitoring and improvement of the occupational and process safety awareness of their crewmembers. In addition, company specific training campaigns, the use of simulator assisted training, and periodical crew assessments through competency assurance programs are reliable solutions for reducing the lack of experience gap, and hence assuring the safe operation of this class of vessels (Odedra 2013, MOL 2015).

## 5 CONCLUSIONS

The maritime LNG segment has been witnessing a significant expansion and evolution since the onset of the last decade. The introduction of new LNG floating facility types, and the evolving increase in the number of LNGCs, and LNGFSs have introduced more risk to the maritime industry. The primary risk of this segment of ships is the resulting consequences from LNG leaks (Pitblado *et al.* 2004, Alderman 2005). However, prevention of risk of leaks onboard these facilities require sufficient identification and control of the resulting FCs of those leaks.

Moreover, direct and indirect human errors are judged, in various literature, to be the primary factors of LNG leaks. Therefore, critical COFs which have the most effect on realising these human errors must be clearly identified, monitored, and managed during the operational phase of these vessels. Other than that, the existing shortage in number of competent crewmembers who are needed to fill the future workforce demand in the maritime LNG segment can be resolved by intensive training of crewmembers and the continual improvement and monitoring of their performance during the operation phase of the ship.


## ACKNOWLEDGEMENTS

This work contributes to the Strategic Research Plan of the Centre for Marine Technology and Ocean Engineering (CENTEC), which is financed by the Portuguese Foundation for Science and Technology (Fundação para a Ciência e Tecnologia - FCT) under contract UIDB/UIDP/00134/2020.


## REFERENCES


Abdelmalek, M. & Guedes Soares, C. 2021. 'Performance-based leading risk indicators of safety barriers on board liquefied natural gas carriers'. Developments in Maritime Technology and Engineering. Guedes Soares, C. & Santos T. A (Eds). Taylor and Francis. London, UK. Vol 1, pp. 211–220.

Alderman, J. A. 2005. 'Introduction to LNG safety', in *Process Safety Progress*, pp. 144–151.

Ampah, J. D., Yusuf, A. A,. Afrane, S., Jin, C. & Liu, H. 2021. 'Reviewing two decades of cleaner alternative marine fuels: Towards IMO's decarbonization of the maritime transport sector', *Journal of Cleaner Production* Elsevier Ltd, 320, p. 128871.

Araneo, D. A. & D'Auria, F. 2012. 'Methodology for pressurized thermal shock analysis in nuclear ower plant', *Intech*, (Applied fracture mechanics), pp. 264–279.

Aronsson, E. 2012. *FLNG compared to LNG carriers 'Requirements and recommendations of LNG production facilities and re-gas units'*, Master Thesis. *Chalmers University of Technology, Gothenburg, Sweden.*

Atkinson, G., Gowpe E., Halliday, J. & Painter, D. 2017. 'A review of very large vapour cloud explosions: Cloud formation and explosion severity', *J Loss Prevent Procees Indus*. 48, pp. 367–375.

Bass, R. L., Hokanson, J. C. & Cox, P. A. 1976. *A study to obtain verification of Liquid natural gas (LNG) tank loading driteria*. Washington DC, USA.

Betteridge, S. Hoyes, J. R. Grant, S. E. & Ivings, M. J. 2014. 'Consequence modelling of large LNG pool fires on water', in *Institution of Chemical Engineers Symposium Series*. Edinburgh, UK, pp. 1–11.

Bukowski, J. D., Liu, Y. N., Pillarella, M. R., Boccella, S. J. & Kennington, W. A. 2013. 'Natural gas liquefaction technology for floating LNG facilities', in *IGRC*. Seuol, S. Korea pp. 1342–1353.

Chorowski, M., Duda, P., Polinski, J & Skrzypacz, J. 2015. 'LNG systems for natural gas propelled ships', *IOP Conference Series: Materials Science and Engineering*, 101(1). Tuscon, AZ, USA.

Cracknell, R. F. & Carsley, A. J. 1997. 'Cloud fires - a methodology for hazard consequence modelling', *Institution of Chemical Engineers Symposium Series*, 1997(141), pp. 139–150.

Di Mattia, D. G. 2013. 'Evaluation and mitigation of human error during LNG tanker offloading, storage and revaporization through enhanced team situational analysis', in *IGT International Liquefied Natural Gas Conference Proceedings*, pp. 1524–1554.

DNV. 2017a. *RULES FOR CLASSIFICATION Ships Edition January 2018 Amended July 2018 Part 5 Ship types Chapter 7 Liquefied gas tankers*. Hovik, Norway.

DNV. 2017b. *Strength Analysis of Liquefied Gas Carriers with Independent Type A Prismatic Tanks*. Hovik, Norway.

DNV. 2015. *Rules for Classification of Floating LNG/LPG Production, Storage and Loading Units*. Hovik, Norway.

DNV. 2016. *Sloshing analysis of LNG membrane tanks*. Hovik, Norway.

DNV. 2017. *RULES FOR CLASSIFICATION Ships Part 7 Fleet in service Chapter 1 Survey requirements for fleet in service*. Hovik, Norway.

Dogliani, M. 2002. 'Safety assessment of LNG Offshore Storage and Regasification Unit', in *Gastech Conference Proceedings*. Doha, Qatar.

Egashira, S. 2013. 'LNG vaporizer for LNG re-gasification terminal', in *Kobelco Technology review*, pp. 64–69.

Foss, M. M. 2012. *LNG Safety and Security Aspects*. Texas, USA

GIIGNL. 2012. *Rollover in LNG storage tanks, GIIGNL. Paris, France*.

GIIGNL. 2021. *GIIGNL Annual Report 2021, GIIGNL*. Paris, France.

Gomez, J. R., Gomez, M. R., Garcia, R. F. & Catoria, A. M. 2014. 'On board LNG reliquefaction technology: A comparative study', *Polish Maritime Research*, 21(1), pp. 77–88.

GTT. 2020. 'GTT membranes safeguard LNG on Prelude, Coral South'. Offshore Magazine. p 1–5.

Hansson, J. Mansson, S., Brynolf, S. & Grahn, M. 2019. 'Alternative marine fuels: Prospects based on multi-criteria decision analysis involving Swedish stakeholders', *Biomass and Bioenergy*. pp. 159–173.

Huther, M., Benoit, F. & Poudret, J. 1981. 'Fatigue Analysis Method for LNG Membrane Tank Details', in *Extreme Loads Response Symposium*. VA, USA.

ICCT. 2017. *Greenhouse Gas Emissions From Global Shipping, 2013-2015., ICCT*. Washigton DC. USA

IMO. 2015. 'Resolution MSC.397(95) (adopted on 11 June 2015) amendments to part A OF The STCW code, 397 (June).

INCO. n.d. 36% Nickel-Iron alloy for low temperature services. New York, USA.

JM. 2020. 4 key characteristics for LNG pipe insulation. Available at: https://www.jm.com/en/blog/2020/abril/4-key-characteristics-for-lng-pipe-insulation/ (Accessed: 16 Feb 2022).

Kamb, R. 2020. *LNG fueled vessel crew training and competence*. USA.

Kim, H., Koh, J. -S., Kim, Y. & Theofanous, G. 2005. 'Risk assessment of membrane type LNG storage tanks in Korea-based on fault tree analysis', *Korean Jour Chem Eng*, 22(1), pp. 1–8.

Kulitsa, M. & Wood, D. A. 2018. 'LNG rollover challenges and their mitigation on Floating Storage and Regasification Units: New perspectives in assessing rollover consequences', *J Loss Prevent Procees Indus*. 54, pp. 352–372.

Laribi, S. & Guy, E. 2020. 'Promoting LNG as a marine fuel in norway: Reflections on the role of global regulations on local transition niches', *Sustainability*, 12(22), pp. 1–17.

LR. 2011. Peters Shipyard I-tanker (Hazard identification study): Liquefied natural gas powered inlan waterways chemical tanker. Lloyds Register, Rotterdam, The Ntherlands.

LR. 2012. *Guidance on the operation of membrane LNG ships to reduce the risk of damage due to sloshing*. Lloyds Register, London, UK.

Lu, T. J. & Fleck, N. A. 1998. 'The thermal shock resistance of solids', *Acta Materialia*, 46(13), pp. 4755–4768.

Margolin, S. 2013. *Flash Fire (duration and heat flux), Westex*. Texas, USA.

Martins, M. R. & Schleder, A. M. 2012. 'Reliability analysis of the regasification system on board of a FSRU using Bayesian networks', *Natural Gas - Extraction to End Use*, pp. 143–158.

Masanobu, T., Hiroki, K. & Kazuo, W. 2015. 'Intrinsically safe cryogenic cargo containment system of IHI-SPB LNG tank', *IHI ER*, 47(2), pp. 27–33.

Melhem, G. A., Ozog, H. & Saraf, S. 2006. *Understand LNG rapid phase transition (RPT)*. ioMosaic, New Hampshire, USA

MOL. 2015. *Investing in Human Capital to implement 'STEER FOR 20201'*

MOL (2021) *Current status and future prospects of LNG fuel for ships*. Available at: https://www.mol-service.com/news/ng-as-ships-fuel/ (Accessed: 24 Oct. 2021).

Nam, W., Mokhtari, M. & Amdahl, J. 2021. 'Thermal analysis of marine structural steel EH36 subject to non-spreading cryogenic spills. Part I: experimental study', *Ships and Offshore Structures*. Taylor & Francis, 0(0), pp. 1–9.

Nerheim, A. R., Æsøy, V. & Holmeset, F. T. 2021. 'Hydrogen as a maritime fuel–can experiences with LNG be transferred to hydrogen systems?', *J Mar Sci Eng*, 9(743).

Odedra, A. 2013. 'The importance of shipping to a global LNG business', *IGT International Liquefied Natural Gas Conference Proceedings*, 2, pp. 1392–1402.

Oran, E. S., Chamberlain, G. & Pekalski, A. 2020. 'Mechanisms and occurrence of detonations in vapor cloud explosions', *Prog in Energy Combu Sci*. 77, p. 100804.

Ostvik, I., Vanem, E. & Castello, F. 2005. *HAZID for LNG tankers*. SAFEDOR-D-4.3.1-2005-11-29 rev.3.

Paltrinieri, N., Tugnoli, A. & Cozzani, V. 2015. 'Hazard identification for innovative LNG regasification technologies', *Rel Eng Sys Saf*, 137, pp. 18–28.

Park, Y. Il, Cho, J. S. & Kim, J. H. 2021. 'Structural integrity assessment of independent type-C cylindrical tanks using finite element analysis: Comparative study using stainless steel and aluminum alloy', *Metals*, 11(10).

Pek, B. & van der Velde, H. n.d. *A high capacity floating LNG design*. Shell, The Netherlands.

Pelto, P. J., Baker, E. G., Holter, G. M. & Power, T. B. (1982) *An Overview Study of LNG Release Prevention and Control Systems*. Washington, USA.

Petti, J. P., Lopez, C., Figueroa, V., Kalan, R. J., Wellman, G., Dempsey, J., Villa, D. & Hightower, M. 2013. *LNG vessel cascading damage structural and thermal analyses, SNL*. New Mexico, USA.

Pitblado, R. 2007. 'Potential for BLEVE associated with marine LNG vessel fires', *J Haz Mat*, 140(3), pp. 527–534.

Pitblado, R. M., Baik, J., Hughes, G. J., Ferro, C. & Shaw, S. J. 2004. 'Consequences of LNG Marine Incidents', in *CCPS Conference Orlando June 29-July 1*, pp. 1–20.

Roberts, T. Bucklan, I., Beckett, H., Hare, J. & Royle, M. 2001. 'Consequences of jet-fire interaction with vessels containing pressurised, reactive chemicals', *IChemE Symposium Series No.148*, (No.148), pp. 147–166.

Rodrigues, A. P. 2013. *The training of officers and crew of LNG-fuelled vessels: a case study of Norway*. Master Thesis. Chalmers University, Guthenburg, Sweden.

Rokstad, E., Erikstad, S. O. & Fagerholt, K. 2010. 'A decision support model for minimising sloshing risk in LNG discharge operations', in *Ship Technology Research*. Taylor & Francis, pp. 154–161.

SEA-LNG. 2021. *LNG-Fuelled vessels approaching 30% of orders*. Available at: https://sea-lng.org/2021/09/lng-fuelled-vessels-approaching-30-of-orders/ (Accessed: 6 Jan. 2022).

SEPS. 2018. *Cryogenic Liquid/Solid Safety*. Glasgow, UK.

Shell. 2019. *Guidance on conduction sloshing risk assessment for operators of membrane LNG carriers*. Rev. 2, Shell Australia.

SIGTTO. 2000. *Liquefied gas handling principles on ships and in terminals*. Third edit. London, England: Witherby & Co Ltd.

SIGTTO. 2012. *Guidance for the Prevention of Rollover in LNG Ships*. London, UK.

Sintef. 2003. *Handbook for fire calculationns and fire risk assessment in the process industry*. Trondheim, Norway.

Sklet, S. 2005. *Safety Barriers on Oil and Gas Platforms, Science And Technology*. PhD Thesis. NTNU, Trondheim, Norway.

Sklet, S. 2006. 'Hydrocarbon releases on oil and gas production platforms: Release scenarios and safety barriers', *J Loss Prevent Process Indus*, 19(5), pp. 481–493.

Songhurst, B. 2019. *Floating LNG Update – Liquefaction and Import Terminals*. Oxford, UK.

SSPA. 2017. 'Safety manual on LNG bunkering procedures for the Port of Helsinki', Helsinki, Finland.

Strande, R. & Johnsson, T. 2013. *Completing the LNG value chain, Wärtsilä Technical Journal*.

Tasneef. 2016. Guide for the design and operation of liquefied natural gas (LNG) carriers. Abu Dhabi, UAE.

Van Twillert, M. J. 2015. *The effect of sloshing in partially filled spherical LNG tanks on ship motions*. Master thesis. Delft University of Technology, Netherlands.

Ustolin, F., Paltrinieri, N. & Landucci, G. 2020. 'An innovative and comprehensive approach for the consequence analysis of liquid hydrogen vessel explosions', *J Loss Prevent Procees Indus*, 68, 104323.

Vandebroek, L. & Berghmans, J. 2012. 'Safety aspects of the use of LNG for marine propulsion', *Procedia Engineering*, 45, pp. 21–26.

Vorkapić, A., Kralj, P. & Bernečić, D. 2016. 'Ship systems for natural gas liquefaction', *Sci J Mar Res*, 30, pp. 105–112.

Wang, Q., Shen, J., Hu, S., Zhao, G., & Zhou, J. 2020 'Microstructure and mechanical properties of Fe-36Ni and 304l dissimilar alloy lap joints by pulsed gas tungesten arc weding'. Material 13(8): 4016.

Wang, Y., Wright, L. & Zhang, P. 2021a. 'Economic Feasibility of LNG Fuel for Ocean-going Ships: A Case Study of Container Vessels', *Maritime Technology and Research*, 3(2), pp. 202–222.

Wang, Z. Han, F., Liu, Y. & Li, w. 2021b. 'Evolution process of liquefied natural gas from stratification to rollover in tanks of coastal engineering with the influence of baffle structure', *J Mar Sci Eng*, 9(95), pp. 1–16.

Wlodek, T. 2019. 'Analysis of boil-off rate problem in Liquefied Natural Gas (LNG) receiving terminals.', in *IOP Conference Series: Earth and Environmental Science*. 2014. Krakow, Poland.

Xuan, S. Hu, S., Li, Z., HU, Q. & Xi, Y. 2019. 'Dynamics simulation for process risk evolution mode on fueling of LNG-fueled vessel', J. Mar Sci Eng. (7) 299. pp. 1–21.

Youn, I.-H. & Kim, S.-C. 2019. 'Preventive maintenance topic models for LNG containment systems of LNG marine carriers using dock specifications', *Applied Sciences (Switzerland)*, 9(6).

Yussoff, I. I. 2019. 'Petronas floating LNG - Completing the LNG value chain through technology', *EGYPS*. Cairo, Egypt.

Zakaria, Z., Baslasl, M. S. O., Samsuri, A. Ismail, I., Supee, A. & Haladin, N. B. 2019. 'Rollover Phenomenon in Liquefied Natural Gas Storage Tank', *J Fail. Anal. and Preven*. Springer US, 19(5), pp. 1439–1447.

Zhang, Q. xi, Liang, D. & Wen, J. 2019. 'Experimental study of flashing LNG jet fires following horizontal releases', *J Loss Prevent Procees Indus*, 57, pp. 245–253.

*Trends in Maritime Technology and Engineering – Guedes Soares & Santos (eds)*


# Investigating ship system performance degradation and failure criticality using FMECA and Artificial Neural Networks

A.A. Daya & I. Lazakis
*Department of Naval Architecture Ocean and Marine Engineering, University of Strathclyde, Glasgow, UK*

ABSTRACT: The goal of all maintenance methods is to eliminate failures or reduce their occurrence. Extended downtime on key ships systems such as power generation plants can lead to undesirable consequences beyond economic and operational losses, especially considering naval vessels. One solution to overcome this challenge is through a system-specific analysis that identifies the most critical component and possible causes of delays be it technical or logistics. In this regard, this paper presents a methodology using FMECA approach that adopts the risk priority number differently to identify Mission Critical Components. This was supported with ANN classification using unsupervised learning to identify patterns in the data that signifies the onset of performance degradation and potential failures onboard an OPV. The study has identified some critical components and failure patterns that contribute to extended downtime based on survey and machinery maintenance reports. Recommendations were provided on preventing/mitigating the failures and how to prioritise existing ship systems maintenance.

## 1 INTRODUCTION

Ships of any type are structures that are operated through a network of systems, the majority of which operate interdependently for their correct functioning. The systems onboard ships are connected to enable economic and efficient operations and maintenance of all equipment/systems. However, a challenge lies in failure of equipment that provides utility to other ship systems. Therefore maintenance efforts are directed to ensure these failures do not occur and if they do the impact can be managed efficiently (Lazakis & Ölçer, 2016). More so, when considered against the cost of routine maintenance which accounts for more than 14 % of ships operating cost and increases as the ship age (Stopford, 2010). Therefore, when this cost is considered against the impact of unscheduled maintenance and the associated operational delays, it becomes a major concern for ship operators. Maintenance uncertainties on ships that belong to law enforcement agencies such as the navy and coast guard ships may not be measured in economic terms but could be the difference between life and death (Goossens & Basten, 2015). It then becomes necessary that ships maintenance adopts a flexible maintenance approach that ensures an efficient and cost-effective ships operational availability (Cheliotis et al., 2019). In this regard other maintenance styles were introduced to overcome some of the challenges when using traditional maintenance (Gits, 1992). Similarly (Shafiee, 2015) provides a review on maintenance selection strategies which highlighted the dynamics involve in maintenance selection, especially in a complex environment such as ships.

Planned Maintenance System (PMS) has remained the mainstay of ships maintenance for both civil and defence sectors (Lazakis et al., 2018, New, 2012). Increasing number of research conducted on ship maintenance has shown that the prepared maintenance onboard ships is preventive maintenance system followed by predictive maintenance system (Lazakis et al., 2018,Velasco-Gallego & Lazakis, 2020). Nonetheless, risk and criticality approach to ship system maintenance are increasingly used to overcome critical component failures or emergency failure events underway especially with advent of unmanned ships(Eriksen et.al 2021) Therefore, based on the foregoing, there is the interest to adopt more efficient maintenance systems, but the challenge is how these technologies can be incorporated by organisations and ship owners/operators. The challenges due to cost of technology can be attributed to installation of new sensors, system upgrades and the cost of training to match new technologies (Mihanović et al 2016). Notwithstanding the need to improve the flexibility of onboard maintenance and the current regulations towards reducing emissions and the strategy by some Original Equipment Manufacturers (OEM) to adopt remanufacturing which offers some discount for operators participating in

DOI: 10.1201/9781003320289-20

the scheme would help change the dynamics of maintenance to be more efficient (International Resource Panel, 2017,IACS Rec, 2018).Moreover, it has been established that the cost of maintenance increases while the equipment ages which can be controlled with more optimised maintenance strategy (Lazakis et al., 2019).

In this regard, this paper presents a methodology based on the combination of FMECA and ANN employing engine sensor records, maintenance, and repair data reports for a set of diesel generators. Therefore, the research provides a comprehensive criticality approach to fault and component reliability that can be replicated onboard or shore maintenance organisations to provide an efficient maintenance practice.

The paper is presented in 5 sections as follows: Section 1 provides an introduction to topic in hand; section 2 highlights a critical review of reliability analysis tools. Section 3 presents the paper methodology while section 4 discusses the paper results as applied on a case study navy patrol vessel. Finally, the conclusion and feature work are presented in section 5.

## 2 CRITICAL REVIEW ON RELIABILITY ANALYSIS TOOLS

Reliability analysis has historically aided maintenance planning since the advent of organised maintenance approaches that evolved from breakdown of simple machines to condition monitoring based predictive analysis (Fred & Geitner, 2006). In this regard, several maintenance methods were advanced to address peculiar problems faced by organisations (Martin et al., 2017). In early maintenance planning, simple methods that can be used to calculate equipment reliability or availability such as mean time to failure (MTTF), mean time between failure (MTBF) failure rates($\lambda$) were used (Lazakis & Ölçer, 2016, Palmer, 2010 and Dhillon, 2006). However, system complexity and the increasing use of electrical components such diodes, valve and software made necessary to adopt other means for system reliability analysis to account for the nature of static failures which are not related to wear and tear. Accordingly, the progressive advance in maintenance has been made possible through the use of reliability analysis tools in various industries such as aerospace and defence (NASA, 2002, Hillier et al, 2003, Kimera & Nangolo, 2020). On the other hand, maintenance within the shipping industry is increasingly getting scrutinised due to regulations including climate change concerns; all this push for the adoption of advanced technologies would require ship operators to adopt addit ional reliability measures (IACS Rec, 2018, ISO 17359:2018, 2018). Likewise Classification Societies require ships to have a standard maintenance documentation and strategy prior to certifying Class qualification (ClassNK, 2017, ABS, 2016).

Research in various fields of maintenance has shown that reliability analysis tools play a vital role in maintenance plaining (Chemweno et al., 2018, Kabir, 2017). The emergence of Reliability Centred Maintenance has brought to the fore the relevance of risk and criticality in equipment maintenance, which focuses maintenance on safety of operations and system reliability (Mokashi et al, 2002, NASA, 2002). In this regard, authors have provided in depth research on the application of reliability analysis tools in the various industries. A criticality-based maintenance for coal power plant used FMECA to drive Risk Priority Number (RPN) aimed at identifying critical components in the plant in order to avoid unplanned shutdown was presented (Melani et al., 2018). System reliability analysis using tools such as FTA and FMECA have found wide application in the nuclear industry especially in the energy sector (Volkanovski et al., 2009). A great deal of research has been made in the maritime sector on the use of reliability tools to improve safety, risk reduction and achieving reliability for ships and offshore wind turbines(Lazakis et al., 2016, 2018, Leimeister & Kolios, 2018).

In the shipping industry and extensive use of Fault Tree Analysis (FTA), Reliability Block Diagrams, (RBD), Event Tree Analysis (ETA), Failure Mode Effect Analysis (FMEA) and other variants of these tools were used to ensure the emplacement of robust maintenance regimes. Therefore, the adoption of method that combines 2 or more reliability tools in order to overcome deficiencies or take advantage of the other tool as shown in (Lazakis et al., 2016, Raptodimos & Lazakis, 2017, Emovon et al., 2018) is increasingly adopted to improve robustness in analysis. Identifying the critical component in system reliability is a dominant area in maintenance research for instance Melani *et al.*(2018) used FMECA for critical component identification in coal fire power plant, while Cheliotis et al (2022) presented a combination of machine learning fault mapping and Bayesian Networks in order to assess increase ship machinery reliability. A combination of reliability tools and ANN was used to develop a predictive condition monitoring by (Lazakis et al., 2018) which shows the competitive flexibility that can be driven to the use of reliability tools and numerical methods in system reliability analysis. The criticality of system, component or event in FMEA is derived by the use of Risk Priority Number (RPN) (Cicek & Celik, 2013, Sharma & Rai, 2021, Fred & Geitner, 2006, Rausand et al., 2021). Reliability analysis tools examine risk of failure by considering quantitative and qualitative aspects. These tools can be grouped into those that are deductive and or inductive based as shown in Table 1.

Table 1. Deductive and Inductive Reliability analysis tools (adapted from (System Reliability Theory, 2021).

| Model/Method | Deductive (Backward) | Inductive (forward) |
|---|---|---|
| FMECA | Yes | Yes |
| Fault Tree Analysis | Yes | No |
| Event Tree Analysis | No | Yes |
| Reliability Block Diagram | Yes | No |
| Bayesian Networks | Yes | Yes |

### 2.1 *Failure Mode Effects and Criticality Analysis (FMECA)*

FMECA is a reliability evaluation technique to determine the effect of system and equipment failures. FMECA is composed of 2 analyses, FMEA and Criticality Analysis (CA). The FMEA is focused on how equipment and system have failed or may fail to perform their function and the effects of these failures, to identify any required corrective actions for implementation to eliminate or minimize the likelihood of severity of failure. The UK MoD defines Criticality assessment as a means of establishing the risk to platform and personnel arising from the occurrence of a failure mode.

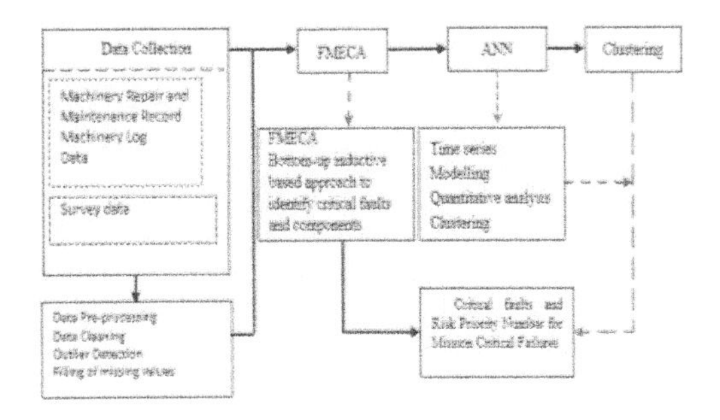


Figure 1. Methodology developed.

It is based on a combination of the worst case consequences of the event coupled with the probability of its occurrence and detection(MoD UK, 2000). Therefore, the Criticality of a failure is determined through a ranking method to obtain RPN which is a product of the risk factors Detectability, Severity and Likelihood.

Stamatis (2003)presented a system of ranking of risk factors to obtain the RPN using numerical scale alongside discrete linguistic terms to reduce ambiguity in linear scale ranging between 1-5 or 1-10. The scale is the same irrespective of which one is used however for uniformity, is best not to combine the scales i.e. 1-5 or 1-10 in one study or analysis. Moreover, FME(C)A is a team activity, it is important that everyone understand what is required as regards the inputs to provide. On the other hand, a system to account for different experience level in a team is equally important, as skills, competence and knowledge usually comes with training, age in service, and exposure to certain responsibilities. Therefore, Liu *et al.*, (2016) presented a system of RPN ranking integrating linear scale with linguistics meaning to address the subjectiveness of FMEA team members. Similarly (Sharma and Rai, 2021,Ilbahar *et al.*, 2018)provided a methodology of evaluating FMECA using linear scale RPN and provides weights to account for importance for individual inputs. Linear approach alongside discreate linguistic terms for conducting FMECA are widely used especially in risk analysis in various field covering humanities to engineering to help improve the clarity and reduce subjectiveness (Tan *et al.*, 2011, Liu *et al.*, 2016, Mutlu & Altuntas, 2019). Accordingly, the flexibility with which FMECA can be interpreted has made it widely accepted across many disciplines from engineering, humanities to medical sciences.

### 2.2 *Artificial Neural Network*

ANN are widely used for statistical analysis and data modelling commonly applied as alternatives to standard nonlinear regression or cluster analysis. Hence, their extensive use in classification, forecasting such as diagnosis, signal processing, speech and image recognition (Gurney, 1997, Mandic & Chambers, 2001). ANN can be defined as interconnected assembly of simple processing elements (units or nodes) whose functionality is loosely based on the animal brain neural. The networks have a processing ability stored in the interunit connection strengths, weights, obtained by a process of adaptation to, or learning from, a set of training patterns (Gurney, 1997). The computational models or nodes are connected through weights that are adapted during use or training to improve performance (Mandic & Chambers, 2001). Therefore, the ability of the ANNs to learn, identify patterns and predict them has made their application in the field maintenance very widely, (Raptodimos & Lazakis, 2018, Vanem & Brandsæter, 2019, Stetco et al., 2019, Lugosch et al., 2020). The process involves the basic node which provides a linear combination of $N$ weights $w_1 \ldots\ldots, w_N$ and $N$ inputs $x_1, \ldots, x_N$ and passes the result through a nonlinearity $\Phi$ as show in equation 1.

$$y = \sum_{i=1}^{N} w_i * x_i + w_0 \tag{1}$$

#### 2.2.1 *ANN self organising maps*

ANN Self Organising Maps (SOM) have been applied in the field of maintenance for machinery health analysis and prediction of machinery condition by various authors. As an unsupervised learning method ANN SOM are effective for data analysis and clustering as presented in (Yu et al., 2015) and used for identifying nonlinear latent features from high dimensional data. Therefore, riding on the existing success and procedures in the use of ANN for machinery data analysis, this research will employ ANN for fault classification and detection, fault/condition prediction and machinery remaining useful life analysis (Wu et al., 2007). ANN approach for fault detection was applied with FTA to identify critical component of a diesel generators in a research presented by (Raptodimos & Lazakis, 2017). In some cases, machinery fault data are recorded without identifying the fault signals, therefore this requires data clustering. Clustering is a form of unclassified machine learning which is applied for machinery diagnostics (Gkerekos et al., 2019). The advantages of using clustering models help identify possible clusters as well as the most influential clusters in the data. SOM consists of competitive layer which can classify a dataset of vectors with any number of dimensions as the number neurons in the layer and are good for dimensionality reduction as presented in (Raptodimos & Lazakis, 2018, Ponmalai & Kamath, 2019).

## 3 METHODOLOGY

The methodology provides a holistic maintenance strategy to cover the entire ship system in a manner that enables flexibility in assigning component maintenance priority or scheduling. The combination of systems onboard ships makes it unsuitable to have a single approach to maintenance. This is more so, when additional consideration is given to ship operators in developing countries where access to technology and original equipment manufacturers is limited and, in some cases, restricted; leading to extended downtime for some critical on-board equipment usually ignored in most cases. In this regard, the present methodology provides an efficient approach to component/equipment failure and degradation analysis. This is because the nature of failure and equipment performance degradation varies a lot from component, equipment, or sub-system as such the need to consider multiple analysis tools to enable a more efficient and flexible methodology. Figure 1 shows the overall methodology of the research. In this paper, Failure Mode Effect and Criticality analysis is used to identify Mission Critical component to ship operations.

### 3.1 *FMECA*

FMECA has been widely adopted in many fields beyond engineering to analyse what can wrong, how it could go, why it goes wrong, and how it can be corrected or addressed. The Criticality Analysis (CA) provides a means of identifying the events, occurrence or components that need more attention to avoid more serious or catastrophic situations. FMECA is a bottom-up approach which provides a systematic methodology to gain deep insight on failures and their course on an equipment or system. Therefore, measuring criticality in FMECA helps to explicitly bring out the most critical component failure which can assist in maintenance actions and planning.

Table 2. Sample FMECA table.

| Sub system | Component | Function | Description of Failure | | | Effects of Failure | |
|---|---|---|---|---|---|---|---|
| | | | Mode | Causes | Detection | Local | Global |
| | | | | | | | |
| | | | | | | | |

| Effects of Failure | | Safeguards | | Criticality | Severity | Likelihood | RPN |
|---|---|---|---|---|---|---|---|
| Influence | TTR | Prevention | Mitigation | 1-10 | 1-10 | 1-10 | CxSxL |
| | | | | | | | |

Therefore, the criticality ranking based on risk use a combination of the consequence (severity) of the failure and the anticipated likelihood of the consequence occurring (ABS, 2015). Criticality analysis will highlight failure modes with probability of occurrence and severity of consequence, allowing corrective actions to be implemented where they produce greatest impact. Given the overall lack of reliability data for many marine systems and components, performing an assessment on qualitative level based on experience and knowledge of the system is sometimes the only means by which to achieve a meaningful criticality assessment. Accordingly, research presents the combination of FMECA and ANN for the investigation of mission critical components of a marine diesel generator. For the FMECA a survey was conducted based on presented in the form of common failures that contributes to about 40 per cent in DGs non availability. Therefore, respondents were asked to rank faults/failures based on 3 criteria on a linear scale from one (1) to ten (10). These criteria were Criticality, Severity and Likelihood as define below:

Criticality: Criticality determines the immediate impact of failure event to the equipment availability and functions. Therefore, a failure mode due to which the ship will not achieve one or more of the mission's targets and /or the safety of whole vessel is at risk until the failure is rectified (NASA, 2008).

Severity: Severity assesses how the failure impacts on the operational availability of the equipment or system regarding normal operation and the duration it takes to be repaired or restored to normal operational levels. Severity is described as the worst potential consequence of the failure determined by the degree of injury, property damage or system damage that could occur (System Reliability Theory, 2021).

Likelihood: This refers to the failure rate of the component including possibility and frequency of the fault occurring over a certain time frame (MIL-STD 1629A, 1980).

### 3.1.1 *Estimated RPN*

The RPN was calculated from the obtained population mean by multiplying each criterion based on the assigned weights according to the seniority of the respondents as a percentage of the original value using equation 2 and 3. The linear values used for the criteria was between | 1 – 10 | therefore was 0 ≤RPN ≥ 300. In this regard to obtain the Mission Criticality the RPN was normalised to ≤100 using the min-max normaliser equation 4.

$$Weighted\ average\ w = \frac{\sum_{i=1}^{n} w_i X_i}{\sum_{i=1}^{n} w_i} \quad (2)$$

$$RPN = \sum\nolimits_{i=\leq 1} Cw_i \times Sw_i \times Lw_i \quad (3)$$

$$RPN_{norm} = \frac{X - min(X)}{max(X) - min(X)} = Mission\ Criticality \quad (4)$$

## 3.2 *Artificial neural network clustering*

In general, there are 2 types of machine learning namely supervised and unsupervised learning (Mahantesh Nadakatti, 2008, Cipollini *et al.*, 2018). The supervised machine learning is used to train a labelled model using a labelled that, that is the features to be looked out are already know there the input data (Gkerekos et al., 2017). On the other hand, unsupervised learning deals with unlabelled data which means the algorithm will identify the unique features in the data and partition it accordingly (Coraddu *et al.*, 2016, Vanem & Brandsæter, 2019). Unsupervised learning is useful for exploring data in order to understand the natural patten of the data especially when there is no specific information about significant incidents in the data that can easily point to some fault indicators (Cipollini et al., 2018). The data collected included hourly machinery log of the DGs hence with no indication of failure or maintenance periods. Therefore, the best method to get the information was to conduct cluster analysis, consequently ANN SOM was used for dimensionality reduction and clustering in the collected data from the case study ship.

Implementing SOM requires the initial training which composes of three phases namely, competition, cooperation and adaption (Kohonen, 2013). The neurons are trained during the competition by competing with each other, whereby the neuron having weight vector closest to the input signal vector is declared as the winner neuron or the Best Matching Unit (BMU). The process can be demonstrated thus; taken the input signal vector to be represented by $I = [I_1, I_2, I_3 \ldots I_n]^T$ and the weight vector is represented by $W = [W_1, W_2, W_3 \ldots W_n]^T$. The difference between the weight vector and input signal vector is computed as the Euclidean Distance (E) between them given by equation 5.

$$E = \| I - W \| \sqrt{\sum_{i=1}^{n} (I_i - W_i)^2} \quad (5)$$

The above equation determines the neuron with the smallest E, which is also the BMU. This is followed by the cooperation phase where the direct neighbourhood neurons of the BMU are identified. The third phase is the adaptation process which neurons are selectively tuned to adopt a specific pattern on the lattice that corresponds to a specific feature of the input vector. The tuning function is written as:

$$W(t+1) = W(t) + \alpha(t)\theta(t)[I(t) - W(t)] \quad (6)$$

Where $\alpha(t)$ is the tuning rate and $\theta(t)$ is the exponential neighbour function; $\alpha(t)$ decrease exponentially with further iteration hence refining the training process, this can be represented in the following equation.

$$\alpha(t) = \alpha_0 e^{\left(-\frac{t}{\lambda}\right)} \quad (7)$$

where $\alpha_0$ is the initial learning rate and $\lambda$ is the time constant given by.

$$\lambda = \frac{N}{\sigma_0} \quad (8)$$

In the above equation N is the total number of training samples and $\sigma_0$ is the radius of the map. The radius is calculated as the Euclidean distance between the coordinates of the outmost neuron and the centre neuron

$$\sigma_0 = \| T_{outmost} - T_{centre} \| \quad (9)$$

In equation 9, $\mathbf{T}_{outmost}$ and $\mathbf{T}_{centre}$ stands for the coordinate of the outmost and central neurons respectively. The overall process is an iterative one to identify the closest neuron to the BMU, thereby fitting the data to required cluster using θ(t) equation 10.

$$\theta(t) = \left( \frac{\| T_j - T_{BMU} \|^2}{2\sigma(t)^2} \right) \quad (10)$$

$$\sigma(t) = \sigma_0 exp\left(\frac{-t}{\lambda}\right) \quad (11)$$

In equation 10 $T_j = [t_j^1 t_j^2]$ which denotes the coordinates of each neuron in a 2D map, $T_{BMU}$ is the coordinate of the best matching unit and σ(t) is the radius of the neighbourhood as shown in equation 11. Therefore, the neurons will keep on updating to get the BMU; In this respect, SOM uses unsupervised learning to produce a map of the input thus providing a good solution for interpreting highly dimensional data making it a good candidate in machinery fault diagnosis.

## 4 RESULTS AND DISCUSSIONS

A case study to demonstrate the proposed approach was conducted on an OPV power generation plant consisting of 4 diesel generators. All generators are rated at load speed of 1800rpm and maximum output 400KW with an overload capacity of 110% for 1 hr in 12hrs. The DGs are 4-stroke, 12-cylinder, V-type, direct injection, sea water intercooled turbocharge and water cooled. All DGs are capable of independent operation or in parallel and can as well change over automatically in case of failure. Overall machinery health monitoring is achieved via as set of sensors capable of shutting the DG or setting up an alarm at certain threshold as shown in Table 3.

Table 3. DG Operating parameters.

| Parameter | Abbreviation | Operating Ranges | | Alarm |
|---|---|---|---|---|
| | | Min | Max | |
| Lubricating Oil Pressure | LoP | 0.4 Mpa | 0.55 Mpa | >0.6 |
| Cooling Fresh Water Temperature | FWT(A/B) | 75 $^{O}$C | 80 $^{O}$C | >85 $^{O}$C |
| Lubricating Oil Temperature | LoT | 30 $^{O}$C | 110 $^{O}$C | >120 $^{O}$C |
| Fresh water pressure | FWP | 0.02 Mpa | 0.25Mpa | >0.3 |
| Exhaust Gas temperature | EGT(A/B) | 220 $^{O}$C | 400 $^{O}$C | >520 |
| Engine Speed | RPM | 1789 RPM | 1850 RPM | 2052 RPM |
| Power Out Put | KW | 0 | 440KVA | 440Kva |
| Generator running hours | HRS | ≥ 2000hours | | |

A survey was conducted to get the opinion of operators and administrators in an organisation with a fleet strength of more about 40 ships of various sizes mainly used for security patrols. The survey consisted of about 20 questions on various types of faults and failure conditions covering DG system including the alternator. Overall, there were 22 respondents of mixed experience and qualification, mainly consisting of 2 specializations; Marine Engineering and Weapon Electrical Engineer. The approach is adopted in order to account for expert knowledge, organisational peculiarities, and challenges to do with access to original equipment manufacturers representatives.

The experience level of respondents varies between 4 to 28 years of service considering position occupied. In this regard appropriate statistical models were used to gain insight to the data. The survey was conducted mainly to quantify the 3 criteria needed for the calculating the RPN which are Criticality (C), Severity (S) and Likelihood (L) (MIL-STD 1629A, 1980). Thereafter, the sample and population of mean of the seven groups were taken using equation 12.

$$population\ mean\ \mu = \frac{\sum x}{N} \quad (12)$$

Where μ is the population mean, x = data values, N = number of samples

Table 4 shows the ranks and weights applied to the groups. Therefore, to accommodate other operational peculiarities mission critical will be used to replace criticality for evaluating the critical components as compared to the traditional way of measuring criticality by evaluating detectability, severity, and likelihood. In the context of the research criticality is looking at the immediate impact of the failure event on the equipment or platform availability and readiness.

A summary of the FMECA result is presented Table 5 Presented are 22 out of about 81 failures analysed due to paper space limitations. The presented failure represents top 50 per cent of the analysed failures on the diesel generator, this ranking provides very important information on how the failure add to the Mission Criticality of the component.

An important case is the cylinder head bolts which has a very short TTR however have very high CA and RPN scores, this underscores the importance of FMECA in maintenance analysis. A further look at the table indicates that failures with longer TTR had higher RPN than those with lower score, again this can provide some guidance to operators on the need to look at skills, type of spare parts holding onboard and procurement process. Therefore, the CA score and the RPN provide a strong base for the next aspect of the analysis which is geared towards identifying features in the machinery health parameters.

The next aspect of the case study is the clustering analysis of the unlabelled DG machinery health data with ANN SOM, a summary of the data is presented in Table 6. The analysis provides, insight on the main groups and the health parameters that can be used for further diagnostics analysis. The data range used for the SOM analysis is presented in Table 7; as can be seen there are 2 abnormal ranges to account for Operator and OEM limits. This disparity was obtained from the operator and shows while the DGs are all rated at 450KW maximum output they were unable to sustain loads beyond 250KW, about 60 percent of the rated output.

Table 4. Rank and Weights method.

| Ranks | Exp.(years) | No. | W* | Positions | W* | Total A | Total B | Applied weight (%) |
|---|---|---|---|---|---|---|---|---|
| Slt | 0–5 | 2 | 50 | WKO/WKD | Non | 50 | Non | 0.5 |
| Lt | 5–11 | 2 | 60 | WKDWEO/MEO | 10 | 60 | 70 | 0.6 |
| Lt Cdr | 11–15 | 4 | 60 | WEO/MEO | 10 | 60 | 70 | 0.7 |
| Cdr | 15–20 | 5 | 65 | FSWEO/FSMEO | 15 | 65 | 80 | 0.8 |
| Capt | 20–24 | 3 | 70 | FSMO/FSG CMDR | 20 | 70 | 90 | 0.9 |
| Cdre | 24–28 | 2 | 80 | FSMO/FSG CMDR | 20 | 80 | 100 | 1 |
| R/Adm | 28–35 | 2 | 100 | FSG CMDR | Non | 100 | 100 | 1 |

* W= Weights

Therefore, clustering will provide a hint on how these operating parameters are related to one another and possibly a clue to fault development. Normal data cleaning and filling of missing data was conducted in order to improve model quality. The initial training was conducted with 6 inputs as shown in Table 7. The data consist of 1800 timestamp data points out partitioned in to 70/30 for training and validation, training was completed after 200 epochs, due to the size of the data. Consequently, the generated topology presented 5 distinct clusters which are good representation of the input data. A critical look at the Table 5 it can be concluded that the DG has never exceeded 60 per cent of its rated output.

The weight input in Figure 2a shows an 8-by-8 two-dimensional map of 100 neurons during the training. The colour variation in the map topology indicates the strength of connection between the neurons; lighter colours indicate short and strong connections while darker colours indicate distant and weak connections. Similarity, the difference in pattern colours indicates how correlated the data cluster are to one another. Accordingly, the cluster weights of Exhaust Gas Temperature B-bank (EGTB) and Fresh Water Temperature B-bank (FWTB) showed a strong correlation which is not present in Fresh Water Temperature A-bank (FWTA) and FWTB. There is also a strong correlation between Power Output in Kilo Watts (KW) and Exhaust Gas Temperature A-bank (EGTA) and to an extent Lubricating Oil Temperature (LoT), therefore the 3 parameters provide a good set of health indicators for further analysis. On the other hand, the slight disparity between both EGT- A/B and FWT - A/B could be an indication of a more serious problem that operators may need to further investigate. Moreover, the fact the DGs are not able to generate beyond 60 per cent of the rated capacity could be due to over rating or dynamic in balance in the DGs. Overall, the SOM neighbourhood weight distances in Figure 2b, shows relatively high dimensional data with about 5 clusters as can be seen within the lower left centre having distinct clusters compare to the lower right end. In contrast the upper right is equally a different concentration of clusters, hence an indication of varying health parameters in the data that points to normal and abnormal data conditions.

Table 5. Mission Criticality and RPN values.

| Sub system | Component | Failure Mode | Time to repair | C | S | L | RPN | N. RPN |
|---|---|---|---|---|---|---|---|---|
| **Cylinder Block** | Crankcase | 1.Cracking | 1-3months | 7 | 6 | 3 | 224 | 65 |
| | Cylinder liner | 2. Cracks<br>3. Scuffing<br>4. Seizure | 1wk-3months (spare parts availability) | 7 | 7 | 4 | 196 | 56 |
| | Cylinder head bolts | 5. Loose<br>6. Not tight | 1-3hrs | 7 | 6 | 8 | 336 | 100 |
| | Top Cylinder gasket | 7. Burnt<br>8. Material Failure | 10-24hrs | 6 | 5 | 5 | 207 | 60 |
| | Cylinder head O-ring | 9.Deformation | 2 wk-2 months | 7 | 6 | 5 | 207 | 60 |
| **Power Take Off** | Crank Shaft | 10. Surface roughness<br>11. Mis alignment | 1 month | 7 | 7 | 3 | 200 | 58 |
| | Journal Bearing | 12.Friction and seizure | 6hrs-2 days<br>1-2 months (OEM to supply spares) | 7 | 7 | 3 | 222 | 64 |
| **Cooling System** | Heat Exchanger Tubes | 13. Scale build-up<br>14. Leakages | 30min-6hrs | 5 | 5 | 5 | 221 | 64 |
| | FW circulation pump | 15. No water supply<br>16. Drop in pressure | 2hrs-4weeks | 6 | 6 | 4 | 180 | 51 |
| | SW pump assembly | 17. No SW supply<br>18.Drop in pressure | 2-4hrs | 6 | 5 | 4 | 179 | 51 |
| | Fuel Quality | 19. Loss of power<br>20. Erratic operation<br>21.Filter blockage<br>22. Sludge accumulation in tanks | 1-2weeks | 6 | 6 | 7 | 196 | 56 |

Table 6. Data Summary.

| | RPM | LoP | FWTA | FWTB | LoT | FWP | EGTA | EGTB | HRS | KW |
|---|---|---|---|---|---|---|---|---|---|---|
| count | 150 | 150 | 150 | 150 | 150 | 150 | 150 | 150 | 150 | 150 |
| mean | 1800 | 0.5 | 66.1 | 68.8 | 84.4 | 0.08 | 334.7 | 317.6 | 2527 | 128 |
| std | 2.9 | 0.1 | 3.4 | 3.8 | 4.7 | 0.01 | 39.3 | 38.9 | 2703 | 34.7 |
| min | 1783 | 0.33 | 40.7 | 42.7 | 41.6 | 0.05 | 161.2 | 146.9 | 523 | 65 |
| 25% | 1799 | 0.38 | 65.2 | 67.7 | 82.4 | 0.07 | 310.2 | 287.5 | 603.3 | 100 |
| 50% | 1800 | 0.56 | 66.2 | 68.8 | 84.6 | 0.07 | 339.5 | 325.8 | 636.5 | 130 |
| 75% | 1801 | 0.57 | 67.4 | 70.3 | 86.4 | 0.08 | 352 | 337.5 | 6341 | 140 |
| max | 1812 | 0.86 | 74.1 | 77.1 | 94 | 0.12 | 426.8 | 408.1 | 6379 | 240 |

Table 7. DG health parameter range.

| DG health parameter | Normal range | Abnormal range | | Alarm |
|---|---|---|---|---|
| | | Operator | OEM | |
| Freshwater Temperature A/ B-Bank | 76-82 | 85 C | 90 C | 90-92 C |
| Exhaust gas Temperatures A/ B-Bank | 250-520 | 480 C | 500 C | 520 C |
| Lub Oil Temperature | 40-95 | 90 | 110 | 113 |
| Lub oil Pressures | 0.45-0.6 | 0.8 | 0.1 | 0.12 |
| Engine power output (kilowatt) | 100-350KW | 240KW | 400KW | 440KW |

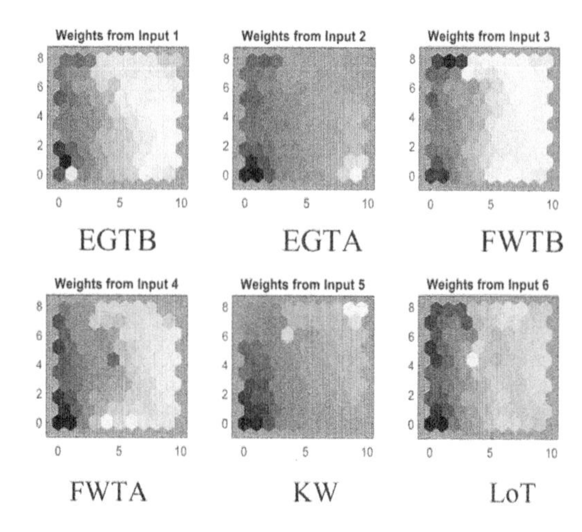


Figure 2a. Fisrt training set data.

It further points to the variation presented in Table 7, in that the OEM values may prove too high above the normal operating range, but could be confused by different operators. Nonetheless, the fact that DG has not been operated beyond 60 percent rated output means that at 50 % output which is about 200KW, the DGs health parameters are already at maximum load capacity. In this regard the darker clusters which indicates weak connections are indication of abnormal readings within the data. Hence this could the reason why, the SOM Neighbour weight distance were heterogeneous and provides a useful indication on degradation in DG performance.

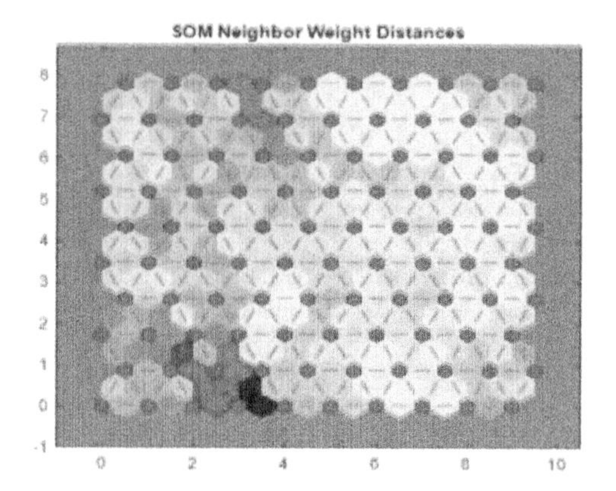


Figure 2b. SOM Neighbour Weight Distance.

## 5 CONCLUSIONS AND FUTURE WORK

In this paper a methodology combining modified approach of using FMECA RPN and ANN SOM to identify mission critical failure event and related as well as data clustering for anomaly dictation was presented using of Maintenance repair and overhaul data from a set of DGs onboard an OPV. A survey to get expert opinion was conducted to produce the criticality, severity and likelihood scores therefore giving a strong credibility to obtained scores. In this regard a number of mission critical components and failure events were identified. A further analysis to investigate DGs health parameters based on hourly log recordings was conducted using ANN SOM. The clustering analysis was conducted using 6 out of 10 parameters collected from the DGs. The SOM neighbourhood weight distance indicates a heterogenous data distribution showing about 5 distinct clusters that can be useful for future degradation and diagnostics analysis. Therefore, in this research, a method of using expert knowledge was implemented to conduct FMECA that was used to identify mission critical components on a DG. The cluster analysis with ANN SOM provided key insight into machinery health parameters that revealed some sharp contrasts between the exhaust gas temperature of A and B banks. Conversely a strong correlation in the features of KW, LoT and EGTA was identified which can be used for further analysis of the DG health parameters. Accordingly, it can be gleaned from the FMECA results that the mission critical components are those influence by high temperature conditions.

In this regard future research will focus on identifying clusters and clearly mapping out the data point they represent. The outcome of mission criticality analysis will be used as input for a maintenance decision making analysis that can help prioritise spare parts holding onboard and shape future procurement choices. FMECA and ANN-SOM has proven to be a good combination of tools for reliability and machinery health analysis in the context of this research and engineering field in general. Therefore, future research direction will take a look into ANN-SOM application for performance degradation and fault classification, while the FMECA analysis will be used for decisions support analysis using another tool. Moving forward this research has provided the data input for implementing machinery health condition monitoring and reliability analysis platform for onboard system and equipment with a focus on failure events and component criticality.

## REFERENCES


ABS. (2015). *Guidance Notes on Failure Mode and Effects Analysis (FMEA) for Classification* (Vol. 2015, Issue MAY).

ABS. (2016). *Equipment Condition Monitoring Techniques. April*, 101.

Chemweno, P., Pintelon, L., Muchiri, P. N., & Van Horenbeek, A. (2018). Risk assessment methodologies in maintenance decision making: A review of dependability modelling approaches. *Reliability Engineering and System Safety*, *173*(January), 64–77. https://doi.org/10.1016/j.ress.2018.01.011.

Cicek, K., & Celik, M. (2013). Application of failure modes and effects analysis to main engine crankcase explosion failure on-board ship. *Safety Science*, *51*(1), 6–10. https://doi.org/10.1016/j.ssci.2012.06.003.

Cipollini, F., Oneto, L., Coraddu, A., Murphy, A. J., & Anguita, D. (2018). Condition-based maintenance of naval propulsion systems: Data analysis with minimal feedback. *Reliability Engineering and System Safety*, *177*, 12–23. https://doi.org/10.1016/j.ress.2018.04.015.

ClassNK. (2017). Good Maintenance On Board Ships. *Hull Construction and Piping On Deck, February.*

Coraddu, A., Oneto, L., Ghio, A., Savio, S., Anguita, D., & Figari, M. (2016). Machine learning approaches for improving condition-based maintenance of naval propulsion plants. *Proceedings of the Institution of Mechanical Engineers Part M: Journal of Engineering for the Maritime Environment*, *230*(1), 136–153. https://doi.org/10.1177/1475090214540874.

Dhillon, B. S. (2006). Maintainability Maintenance and Reliability for Engineers. In *CRC*. https://doi.org/10.5772/16225.

Emovon, I., Norman, R. A., & Murphy, A. J. (2018). Hybrid MCDM based methodology for selecting the optimum maintenance strategy for ship machinery systems. *Journal of Intelligent Manufacturing*, *29*(3), 519–531. https://doi.org/10.1007/s10845-015-1133-6..

Eriksen, S., Utne, I. B., & Lützen, M. (2021). An RCM approach for assessing reliability challenges and maintenance needs of unmanned cargo ships. *Reliability Engineering and System Safety*, *210* (July 2020), 107550. https://doi.org/10.1016/j.ress.2021.107550..

Fred, H. P. B., & K. Geitner. (2006). Maximizing Machinery Uptime. In *Practical Machinery Management for Process Plants* (Vol. 5). https://www.sciencedirect.com/science/article/pii/S1874694206800180.

Gits, C. W. (1992). Desing of Maintenance Concepts. In *University of Eindhoven* (Vol. 91, Issue 5). https://doi.org/10.1017/CBO9781107415324.004.

Gkerekos, C., Lazakis, I., & Theotokatos, G. (2017). Ship machinery condition monitoring using performance data through supervised learning. *RINA, Royal Institution of Naval Architects - Smart Ship Technology 2017.*

Gkerekos, Christos, Lazakis, I., & Theotokatos, G. (2019). Machine learning models for predicting ship main engine Fuel Oil Consumption: A comparative study. *Ocean Engineering*, *188*(June), 106282. https://doi.org/10.1016/j.oceaneng.2019.106282.

Goossens, A. J. M., & Basten, R. J. I. (2015). Exploring maintenance policy selection using the Analytic Hierarchy Process; An application for naval ships. *Reliability Engineering and System Safety*, *142*, 31–41. https://doi.org/10.1016/j.ress.2015.04.014.

Gurney K. (1997). Introduction to neural networks. In *The Lancet* (Vol. 346, Issue 8982). https://doi.org/10.1016/S0140-6736(95)91746-2.

Hillier, F. S., Price, C. C., & Austin, S. F. (2003). *Reliability: A Practitiner's Guide.*

IACS Rec. (2018). *A GUIDE TO MANAGING MAINTENANCE IN ACCODANCE WITH THE REQUIREMENTS OF THE ISM CODE. 74*, 1–9.

Ilbahar, E., Karaşan, A., Cebi, S., & Kahraman, C. (2018). A novel approach to risk assessment for occupational health and safety using Pythagorean fuzzy AHP & fuzzy inference system. *Safety Science*, *103*(July2017), 124–136. https://doi.org/10.1016/j.ssci.2017.10.025.

International Resource Panel. (2017). *Promoting remanufacturing, refurbishment, repair, and direct reuse.*

ISO 17359:2018.(2018). Condition Monitorinf and Diagnostics of Machines. In *Condition Monitoring and Diagnostics of Machines* (Vol. 01, p. 36).

Kabir, S. (2017). An overview of fault tree analysis and its application in model based dependability analysis. *Expert Systems with Applications*, *77*, 114–135. https://doi.org/10.1016/j.eswa.2017.01.058.

Kimera, D., & Nangolo, F. N. (2020). Maintenance practices and parameters for marine mechanical systems: a review. *Journal of Quality in Maintenance Engineering*, *26*(3), 459–488. https://doi.org/10.1108/JQME-03-2019-0026.

Kohonen, T. (2013). Essentials of the self-organizing map. *Neural Networks*, *37*, 52–65. https:/doi.org/10.1016/j.neunet.2012.09.018.

Lazakis, I., Dikis, K., Michala, A. L., & Theotokatos, G. (2016). Advanced Ship Systems Condition Monitoring for Enhanced Inspection, Maintenance and Decision Making in Ship Operations. *Transportation Research Procedia*, *14*(0), 1679–1688. https://doi.org/10.1016/j.trpro.2016.05.133.

Lazakis, I., Gkerekos, C., & Theotokatos, G. (2019). Investigating an SVM-driven, one-class approach to estimating ship systems condition. *Ships and Offshore Structures*, *14*(5), 432–441. https://doi.org/10.1080/17445302.2018.1500189.

Lazakis, I., & Ölçer, A. (2016). Selection of the best maintenance approach in the maritime industry under fuzzy multiple attributive group decision-making environment. *Proceedings of the Institution of Mechanical Engineers Part M: Journal of Engineering for the Maritime Environment*, *230*(2), 297–309. https://doi.org/10.1177/1475090215569819.

Lazakis, I., Raptodimos, Y., & Varelas, T. (2018). Predicting ship machinery system condition through analytical reliability tools and artificial neural networks. *Ocean Engineering*, *152*(February2017), 404–415. https://doi.org/10.1016/j.oceaneng.2017.11.017.

Leimeister, M., & Kolios, A. (2018). A review of reliability-based methods for risk analysis and their application in the offshore wind industry. *Renewable and Sustainable Energy Reviews*, *91*(April), 1065–1076. https://doi.org/10.1016/j.rser.2018.04.004.

Liu, H.-C., You, J.-X., Chen, S., & Chen, Y.-Z. (2016). An integrated failure mode and effect analysis approach for accurate risk assessment under uncertainty An integrated failure mode and effect analysis approach for accurate risk assessment under uncertainty. *IIE Transaction*. https://doi.org/10.1080/0740817X.2016.1172742.

Liu, H. C., You, J. X., Chen, S., & Chen, Y. Z. (2016). An integrated failure mode and effect analysis approach for accurate risk assessment under uncertainty. *IIE Transactions (Institute of Industrial Engineers)*, *48*(11), 1027–1042. https://doi.org/10.1080/0740817X.2016.1172742.

Lugosch, L., Nowrouzezahrai, D., & Meyer, B. H. (2020). *Surprisal-Triggered Conditional Computation with Neural Networks*. 1–16. https://arxiv.org/abs/2006.01659.

Mahantesh Nadakatti, A. R. A. S. K. (2008). Artificial intelligence-based condition monitoring for plant maintenance .pdf. *Artificial Intelligence-Based Condition Monitoring Mahantesh*, *28*(2), 10.

Mandic, D. P., & Chambers, J. A. (2001). *Recurrent Neural Networks for Prediction Authored (Hardback); 0-470-84535-X (Electronic) RECURRENT NEURAL NETWORKS FOR PREDICTION* (Vol. 4).

Martin, B., McMahon, M., Riposo, J., Kallimani, J., Bohman, A., Ramos, A., & Schendt, A. (2017). A Strategic Assessment of the Future of U.S. Navy Ship Maintenance: Challenges and Opportunities. In *A Strategic Assessment of the Future of U.S. Navy Ship Maintenance: Challenges and Opportunities*. https://doi.org/10.7249/rr1951.

Melani, A. H. A., Murad, C. A., Caminada Netto, A., Souza, G. F. M. de, & Nabeta, S. I. (2018). Criticality-based maintenance of a coal-fired power plant. *Energy*, *147*, 767–781. https://doi.org/10.1016/j.energy.2018.01.048.

Mihanović, L., Ristov, P., & Belamarić, G. (2016). Use of new information technologies in the maintenance of ship systems. *Pomorstvo*, *30*(1), 38–44. https://doi.org/10.31217/p.30.1.5.

MIL-STD 1629A. (1980). *MIL-STD-1629A* (Vol. 12, Issue 6). Departmen of Defense, Defense Standards. https://doi.org/10.1016/0010-4485(80)90109-8.

MoD UK. (2000). Defence Standard 02-45 (NES 45). *Ministry of Defence, Defence Standard 02-45 (NES 45)*, *45* (2), 178.

Mokashi, A. J., Wang, J., & Vermar, A. K. (2002). A study of reliability-centred maintenance in maritime operations. *Marine Policy*, *26*(5), 325–335. https://doi.org/10.1016/S0308-597X(02)00014-3.

Mutlu, N. G., & Altuntas, S. (2019). Risk analysis for occupational safety and health in the textile industry: Integration of FMEA, FTA, and BIFPET methods. *International Journal of Industrial Ergonomics*, *72*(May), 222–240. https://doi.org/10.1016/j.ergon.2019.05.013.

NASA. (2002). *Nasa - Fault - Tree - Handbook*.

NASA. (2008). RCM GUIDE RELIABILITY-CENTERED MAINTENANCE GUIDE For Facilities and Collateral Equipment. *Nasa, September*, 472.

New, C. (2012). *RCM in the Royal Navy - Developing a Risk Based Policy for Intergrating safety and Mainteance Management. January 2012*, 25–26.

Palmer, R. D. (2010). *Maintenance Planining and Scheduling Handbook*.

Ponmalai, R., & Kamath, C. (2019). *Self-Organizing Maps and Their Applications to Data Analysis*.

Raptodimos, Y., & Lazakis, I. (2017). Fault tree analysis and artificial neural network modelling for establishing a predictive ship machinery maintenance methodology. *RINA, Royal Institution of Naval Architects - Smart Ship Technology 2017*.

Raptodimos, Yiannis, & Lazakis, I. (2018). Using artificial neural network-self-organising map for data clustering of marine engine condition monitoring applications. *Ships and Offshore Structures*, *13*(6), 649–656. https://doi.org/10.1080/17445302.2018.1443694.

Rausand, M., Barros, A., & Høyland, A. (2021). *System reliability theory: models, statistical methods, and applications*.

Shafiee, M. (2015). *Maintenance strategy selection problem: an MCDM overview*. 23. https://doi.org/10.1108/JQME-04-2016-0014.

Sharma, G., & Rai, R. N. (2021). Modified failure modes and effects analysis model for critical and complex repairable systems. In *Safety and Reliability Modeling and its Applications* (First Edit). Elsevier Ltd. https://doi.org/10.1016/b978-0-12-823323-8.00016-7.

Stamatis, D. H. (2003). Failure Mode and Effect Analysis: FMEA From Theory to Execution. In *Technometrics* (Vol. 38, Issue 1). https://doi.org/10.1080/00401706.1996.10484424.

Stetco, A., Dinmohammadi, F., Zhao, X., Robu, V., Flynn, D., Barnes, M., Keane, J., & Nenadic, G. (2019). Machine learning methods for wind turbine condition monitoring: A review. *Renewable Energy*, *133*, 620–635. https://doi.org/10.1016/j.renene.2018.10.047.

Stopford, M. (2010). Maritime Economics. In *Choice Reviews Online* (Vol. 47, Issue 07). https://doi.org/10.5860/choice.47-3934.

*System Reliability Theory*. (2021). CRC Press.

Tan, Z., Li, J., Wu, Z., Zheng, J., & He, W. (2011). An evaluation of maintenance strategy using risk based inspection. *Safety Science*, *49*(6), 852–860. https://doi.org/10.1016/j.ssci.2011.01.015.

Vanem, E., & Brandsæter, A. (2019). Unsupervised anomaly detection based on clustering methods and sensor data on a marine diesel engine. In *Journal of Marine Engineering and Technology*. https://doi.org/10.1080/20464177.2019.1633223.

Volkanovski, A., Čepin, M., & Mavko, B. (2009). Application of the fault tree analysis for assessment of power system reliability. *Reliability Engineering and System Safety*, *94*(6), 1116–1127. https://doi.org/10.1016/j.ress.2009.01.004.

Wu, S., Gebraeel, N., Lawley, M. a, & Yih, Y. (2007). A Neural Network Integrated Decision Support System for Condition-Based Optimal Predictive Maintenance Policy. *Ieee Transactions on Systems, Man, and Cybernetics*, *37*(2), 226–236.

Yu, H., Khan, F., & Garaniya, V. (2015). Risk-based fault detection using Self-Organizing Map. *Reliability Engineering and System Safety*, *139*, 82–96. https://doi.org/10.1016/j.ress.2015.02.011.

*Trends in Maritime Technology and Engineering – Guedes Soares & Santos (eds)*
*© 2022 copyright the Author(s), ISBN 978-1-032-33583-4*

# A neurophysiological data driven framework for assessing mental workload of seafarers

S. Fan & Z. Yang
*Liverpool Logistics, Offshore and Marine (LOOM) Research Institute, Liverpool John Moores University, Liverpool, UK*

ABSTRACT: Neurophysiological analysis has been used to tackle human factors in such sectors as health and aviation. However, there is little relevant research in the maritime industry in which human errors significantly contribute to accident occurrence. This paper pioneers a conceptual framework for assessing the mental workload of seafarers in maritime operations. It will enable the maritime mental workload assessment in a quantitative manner and hence can be used to test, verify and train the seafarers' safety behaviours. A case study on ship collision avoidance is conducted to demonstrate the feasibility of the framework using ship bridge simulation. The new framework using neurophysiological data can effectively evaluate the contribution of mental workload to human errors and operation risks and it opens a new maritime human reliability analysis stream. The findings can also provide valuable insights for evaluating seafarers' behaviours in remote control centre within the context of autonomous ships.

## 1 INTRODUCTION

Over 80% maritime accidents are directly and indirectly caused by human errors (Chang et al., 2021). Human factors contribute to errors and mistakes which cause severe accidents. Control of human errors often acts as an interruption to terminate the chain of system failures to prevent the occurrence of accidents. It is therefore essential to analyse and tackle human factors in maritime operations for safety at sea. The statistical analysis from the accident reports within 2012 to 2017 by the Marine Accident Investigation Branch (MAIB) and the Transportation Safety Board of Canada (TSB) (Fan et al., 2020a) reveals that 30.77% % of maritime accidents are associated with "poor communication and coordination", 32.69% "under ineffective supervision and supports of the bridge team", 37.50% "without clear order", 32.69% "seafarers' unfamiliarity with equipment, insufficient capability, or ill-preparedness" and 15.38% "with poor lookout". Obviously, the non-technical skills of seafarers on board significantly influence maritime safety, and investigations on human and organisational factors have been on the top of the research agenda in maritime safety (Kim, 2020, Fan et al., 2020b).

Among non-technical skills, seafarers' individual factors are as critical as corporative behaviours and appropriate competencies in terms of the impact on maritime safety. There is evidence showing 21.63% marine accidents with operators in a low state of alertness, 16.35% under distracted condition or insufficient attention of operators, 13.46% in fatigue, 4.81% with cognitively overload, 1.92% with a low level of arousal, panic, anger, or unhappiness (Fan et al., 2020a). In addition, referring to accident investigation reports, it is difficult to identify and measure such individual factors through the current available techniques/methods in maritime accident investigations. Previous studies on human factors (Akhtar and Utne, 2014, Akhtar and Utne, 2015, Besikci et al., 2016) reveal that there is a positive correlation between individual human factors and behaviours. To evaluate the human factors, it is important to measure them in order to define and quantify the associated risks and hence provide a comprehensive perspective for human reliability in maritime safety operations.

The significant individual factors are mental workload, fatigue, inattention, and distraction. It is also evident that mental workload affects other factors (Lim et al., 2018, Wu et al., 2017). With the development of non-invasive technology application in the transport field, functional near-infrared spectroscopy (fNIRS) becomes an effective mean to quantify neurophysiological factors in the process of human operations. However, use of neurophysiological techniques in general and fNIRS in particular in maritime human reliability analysis is fragmented and scanty (Fan et al., 2017). There is no robust methodology to support the relevant study and experiments and allow their advantages to be fully explored in maritime safety. The need of developing the new

DOI: 10.1201/9781003320289-21

methodology becomes more significance as far as the arguable growing mental workload for the operators working in the remote control center of autonomous ships is concerned.

This paper aims to develop a neurophysiological data-driven framework for assessing mental workload of seafarers. It serves for the marine industry to evaluate the mental workload for training or human reliability assurance purpose. It can also provide a novel research direction for maritime safety researchers to explore human and organisational factors. In addition, the framework is illustrated on the basis of the real case study investigating seafarers' mental workload in a ship simulator inFan et al. (2021). The structure of the paper is illustrated as follows. Section 2 describes the human factors influencing maritime accidents. Section 3 draws out the methods of the individual human factors in maritime sector. Next, the date-driven framework is proposed in Section 4. Then, the case study of applying the framework to analyse the officers' mental workload in ship collision avoidance via ship bridge simulation is conducted in Section 5. It is followed by the conclusion in Section 6.

## 2 HUMAN FACTORS INFLUENCING MARITIME ACCIDENTS

It is critical to develop a data-driven mental workload assessment framework for ship safety, as it relates to manoeuvring behaviours and the competence of seafarers. The criticality is evident by both maritime accident reports and lessons learned from historical failure data of maritime accident investigation authorities/organisations (e.g., IMO and MAIB). There were a significant number of maritime accidents associated with individual human factors. The examples are shown in Table 1.

The above four individual human factors represent the high risks in the process of maritime operations. Faced with the harsh working environment onboard, seafarers often take the irregular rest patterns within the existence of time-zone crossings, noise, heat, cold, vibration and motion of vessels. The individual human factors which are affected by these circumstances could result in the reduction of seafarers' performance on duty. In addition, it takes a challenge for them to maintain good performance for multi-tasks to a satisfactory level, including navigation, cargo handling, watchkeeping, communication, emergency response, paper charts, maintenance, administration, and human resources management. Given such a complication, it is necessary and significant to develop a new and robust methodology for assessing mental workload, for which the current existing human reliability analysis methods fail.

Table 1. Individual human factors in maritime accidents.

| Human factor | Consequences |
|---|---|
| Mental workload* | A collision between the stern trawler Karen and a dived Royal Navy submarine in the Irish Sea (MAIB 20/2016).<br>The bulk carrier Heloise collided with the tug Ocean Georgie Bain while transiting the St. Lawrence River (TSB M13L0123). |
| Fatigue** | The bulk carrier Muros ran aground on Haisborough Sand on the east coast of the United Kingdom (MAIB 22/2017).<br>The fishing vessel Louisa foundered with the loss of three lives while at anchor off the Isle of Mingulay in the Outer Hebrides (MAIB 17/2017). |
| Distracted*** | The high-speed passenger catamaran Typhoon Clipper collided with the workboat Alison adjacent to Tower Millennium Pier, River Thames, London (MAIB 24/2017).<br>The Cyprus registered cargo ship Daroja and the St Kitts and Nevis registered oil bunker barge Erin Wood collided at 4 nautical miles southeast of Peterhead, Scotland (MAIB 27/2016). |
| Situational awareness**** (SA) | The ro-ro passenger ferry Hebridesthe lost control, grounding when approaching Lochmaddy, North Uist, Scotland (MAIB 20/2017).<br>The scallop dredger St Apollo grounded on a rocky shelf at the eastern entrance to the Sound of Mull, Scotland (MAIB 14/2016). |

* It reflects operators' response to the given tasks and explains the mental resources required to complete or deal with such tasks.

** It is associated with long working hours, heavy workload, or under the pressure.

*** It shows the insufficient attention to the core tasks with high priority.

**** Lack of SA is associated with being distracted, which disables individuals or the team to identify the hazards and maneuver properly.

## 3 ANALYSIS METHODS OF INDIVIDUAL HUMAN FACTORS IN MARITIME ACCIDENTS

Seafarers are required to learn multiple skills, which include the identification of malfunctions, workload management, watchkeeping, implementation of the best solution, response to the changes of the information, clear and concise communication of information, concentration management, and ability of handling stress, etc. (O'Connor and Long, 2011). In the process of executing actions on board, individuals' unsafe acts including errors or violations can be explained by examining individual human factors. The investigation on individual factors complements human reliability analysis by measuring human's response to critical situations, which further describes the causes of maritime accidents. In addition, there is a close link between mental workload and tasks. The more sophisticated the tasks, the more mental workload is required to accomplish the tasks. It has been widely applied to evaluate the task performance of operators and the practical capability of the designed system (Ngodang et al., 2012, Dijksterhuis et al., 2011).

Regarding the Maritime Autonomous Surface Ship (MASS) operations, the Human Factors Analysis and Classification System-Maritime Accidents (HFACS–MA) framework was conducted to emphasise the significance of maintaining psychoand physiological conditions for remote operators (Wróbel et al., 2021). It further proves the critical role of individual human factors in maritime operations for autonomous shipping. However, although such factor-related research was conducted in road traffic accidents (Boyle et al., 2008, Rakauskas et al., 2008) and aviation transportation (Ayaz et al., 2012, Gateau et al., 2015), seafarers' individual factor analysis is largely scanty (Lim et al., 2018, Fan et al., 2018), revealing a significant research gap to fulfill.

Mental workload has been described as being responsible for the majority of road traffic accidents (Dijksterhuis et al., 2011). Evidence shows that both high and low levels of mental workload lead to human errors for the driver by inducing insufficient perception and attention. Typically, when the mental workload is beyond the threshold, it leads to worse performance of participants (Molteni et al., 2008). However, in the maritime sector, Lim et al. (2018) suggested that the workload and stress of the majority of trainees were relieved due to shared work and responsibility when the experienced captain was present, while the latter had the highest workload levels. In addition, mental workload is related to the experience of operators. Experienced drivers are supposed to induce less mental workload than novices because of effective automation through practice (Patten et al., 2004). Moreover, mental fatigue was assessed with deep learning algorithms in the maritime domain using neuroimaging and physiological sensors including the ECG, EMG, body temperature sensor, EEG, EOG, and eye tracker (Monteiro et al., 2019), which showed high levels of mental fatigue classification accuracy using convolutional neural networks. From a neuroscience perspective, the increasing mental workload was associated with increasing prefrontal cortex activation (Ayaz et al., 2012), while a low level of mental workload leads to decreases in prefrontal cortex activation (Molteni et al., 2008). There is a demand for mental workload study in maritime operations to measure the brain activity or physiological activities of operators using novel techniques (Fan et al., 2017). The applications and advantages for various typical techniques used for human factors investigation are given in Table 2. It can be seen that individual human factors require neurophysiological technique to objectively indicate the relevant human response and evaluate the performance. Therefore, the second research gap is that rare maritime safety study has focused on neurophysiological methodology to study mental workload of seafarers. Therefore, The ship simulator is

Table 2. Typical neurophysiological technique applications.

| Technique | Application | Advantage |
|---|---|---|
| Eye tracking | It is integrated with simulators, motion capture devices, and augmented reality, to understand operators' gaze patterns in aviation (Martinez-Marquez et al., 2021). | Wearable device applied in various fields. |
| Heart rate | It is usually integrated with brain activity measurement to assess mental workload of operators in shipping (Wu et al., 2017). | Easy to be used and low cost. |
| Electroencephalogram (EEG) | It has the advantage of greater time resolution, so as to monitor emotion, mental workload and stress in maritime operations (Hou et al., 2015). | Sensitive neuroimaging sensors with greater time resolution. |
| functional near-infrared spectroscopy (fNIRS) | It is a common technique applied in real-world scenarios (Christian et al., 2013). It can be used to study the association between haemoglobin levels and white matter (Rozanski et al., 2014), working memory (Fishburn et al., 2014). | Greater spatial resolution and less crosstalk between sites, sensitive to the cognitive load and state. |

widely used for crew training. The IMO utilised the simulation for crews' training based on the simulation training requirements (A-I/6: Training and Assessment) in the Standards of Training, Certification and Watchkeeping Convention (STCW 78-95). The awareness of the significance of human factors among navigation and maritime safety was aroused. However, human performance in maritime operations is related to various elements, such as task demands, prior experience, personality, voyage segment, workload, etcetera (Ngodang et al., 2012). To understand and improve the human performance in maritime operation, the ship simulator is considered as a tool to establish complex scenarios and analyse human performance in this dedicated system. However, most of simulation studies rely on questionnaires or subjective expert evaluation, through which it is impossible to measure mental states of seafarer. Previous research on maritime human reliability analysis does not incorporate neurophysiological analysis into maritime simulation for seafarers' performance assessment. This study pioneers a neurophysiological data-driven framework for assessing mental workload of seafarers and analyses individual human factors in maritime operations.

## 4 A DATA-DRIVEN MENTAL WORKLOAD ASSESSMENT FRAMEWORK

A data-driven mental workload assessment framework is outlined in this section to address the identified research gaps on human factors in maritime operations. It is developed by 1) referring to a large amount experiments in the other industries such as road and air transportation and 2) experience gained from a few successful experiments in maritime operations by the authors (e.g., the case in Section 5). Through neurophysiological measurement, the framework creates a new paradigm on quantitative mental workload assessment for seafarers. It aids the use of neuropsychological data to evaluate human performance through ship simulators. The new framework consists of seven critical mental workload assessment steps, 1) criterion definition, 2) training requirement, 3) recruitment of professional participants at the pre-examination stage; 4) simulation-based experiment, 5) debrief, at the intra-examination; 6) data analysis and 7) dissemination at the post-examination stages. These steps are naturally linked with the critical activities for conducting objective workload assessment using neurophysiological data. Such activities are described in the ensuing sections, while the interconnected flow among the steps is shown in Figure 1.

The pre-examination stage consists of criterion definition, training requirement, and participant recruitment. The defined criteria are configured from

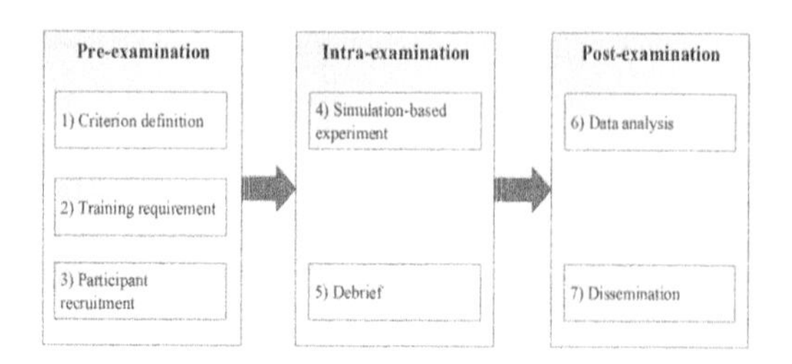


Figure 1. A data driven mental workload assessment framework.

experiment's scenario design, figuring out the input/output data and dependent/independent variables. Due to the nature of neurophysiological science, it is critical to prepare an ethical application and get its approval in the first step, before the commencement of experimental study. Next, it is necessary to train all researchers/personnel involved in the experiments to ensure they can properly use the equipment for data collection. It has to be done before the recruitment of the participants of the experiment. In the third step, the participants will be recruited referring to relevant criteria. The participant needs to be familiar with the experiment environment and specific requirements depending on designed scenarios. All participants will receive a full explanation of the purpose, procedures, risks, and benefits of the experiment. The participant information sheet ethically approved should be informed and signed by any participant before the experiment.

The intra-examination stage is quite diverse and depends on what specific research questions to be solved. Although both ship simulation and real ships can be used to support the experiment, the former is much more economical and safer than the latter. At this stage, the simulation-based experiment is carried out. The neurophysiological data will be collected by different tools (e.g., EEG, fNIRS, heart rate, eye tracking) and questionnaires will be collected for statistics analysis and workload assessment. At the end of the experiment, the debrief is supposed to be taken between the instructor and participants, which mimics the real maritime training process.

The post-examination stage is related to data pre-processing, data analysis, and dissemination of the research findings. Data pre-processing consists of checking discontinuities, spikes and interpolation of neurophysiological data, applying a frequency filter to reduce the noises, and removing motion artefacts. Then, statistical methods and/or other algorithms will be utilised to analyse the purified data after pre-processing. The analysis findings can be disseminated through publications that encourage multi-discipline research, and provide guidelines and recommendations on maritime training and risk management of maritime operations from practical perspectives.

## 5 CASE STUDY

To demonstrate the generic conceptual framework, a case study using a ship bridge simulator to investigate seafarers' mental workload in ship collision avoidance was conducted (Fan et al., 2021). The case aimed at investigating how mental workload influences neurophysiological activation and decision making of experienced and inexperienced deck officers concerning collision avoidance. It was developed with simulated watchkeeping tasks in a ship bridge simulator. The mental workload was induced by a voyage along a North/South axis and participants were required to keep watch over 180° field-of-view of the open sea. This watchkeeping period was terminated when participants spotted a "target" vessel. The decision-making period was from the end of watchkeeping period to the action made for collision avoidance. There were non-distraction and distraction groups. Here, the participants in the distraction group were required to report vessel positions and answering questions at specific points of time, which is the common task requiring temporal mental workload in the real world. During the process, fNIRS data and subjective questionnaires were collected to measure the neurophysiological activation. The time and distance of two ships in encounter were recorded to evaluate the human performance given such situations, when participants made manoeuvres for collision avoidance.

According to the conceptual framework, the seven steps of the mental workload assessment for human factors in maritime operations are demonstrated below.

### 5.1 *Criterion definition*

The criteria are defined based on the research questions and research aims. In order to induce mental workload in ship bridge, there were two groups of participants involving in two levels of tasks requiring high and low mental workload. The high workload situation was developed by more distracted tasks in addition to the task in the low workload situation. In addition, the difference of experienced and inexperienced deck officers was represented in different professional qualifications. Therefore, the study used a mixed design, where two groups of participants were allocated to 1) experienced group and 2) inexperienced group, depending on their professional qualifications. Both groups undertook the same watchkeeping scenarios, which were presented in 1) non-distraction and 2) distraction.

The experimental procedure is in accordance with the principles set out in the Declaration of Helsinki and was reviewed and approved by the ethics committee of the host institution of the experiment. The experimental protocol for the study was approved by the institutional ethics committee prior to data collection. They were provided with written informed consent for participation and were well trained for the study.

### 5.2 *Training requirement*

The investigator received appropriate training on the use of fNIRS equipment before conducting the pilot test and experiments. There were a group of experts involving a senior professor of rich experience of using fNIRS for mental workload assessment, the manufacturer of the fNIRS equipment used in the experiment and two experts of nautical science and ship simulation from the host institution. The possible problems (e.g., noisy data due to head movements) that could occur during the experiments were assumed and control measures were prepared in advance.

### 5.3 *Recruitment of professional participants*

To fulfil the requirement for mixed design, 40 (a 2×2 matrix (workload condition x experience) with 10 people in each "cell" of the design) was the minimum conventional sample size in order to get a sampling distribution that approximates normality (Fan et al., 2021). At last, a total of 41 participants were recruited from the Nautical Institute London Branch and the host institution. Inclusion of participant recruitment is limited to adults (>18 years old), without head injury conditions or suffering from high blood pressure since this may affect the results from fNIRS. In the study, any person suffering from anxiety condition or receiving medication for anxiety condition is excluded. Participants were divided into two groups based on their navigation experience. Twenty experienced seafarers having an average age at 44.60 (SD = 15.47) include master mariner (MM), chief mate (CM), and officer of the watch (OOW). Twenty-one inexperienced seafarers of an average age at 24.76 (SD = 5.25) are AB (Able seaman) and cadets. Raw NASA-TLX data of 41 participants were kept for behaviour performance analysis. However, there was a severe "detector saturation" of data collection for one inexperienced participant. This raw fNIRS data was deleted for not being recorded correctly. Therefore 20 pieces of data for experienced seafarers who had 213.4 months (SD=188.8) experience at sea, and 20 for inexperienced seafarers who had 27.2 months (SD=30.5) experience at sea, were obtained for further analysis, see Table 3 (Fan et al., 2021). All participants were asked to take 10 minutes to attend the training session of the bridge simulator before the experiment. It helped them to be familiar with the equipment and the ship bridge simulator.

### 5.4 *Simulation-based experiments*

The two different conditions for (A) non-distraction group and (B) distraction group can be demonstrated by allocating the reporting missions at a specific time interval, seen in Figure 2.

Table 3. Background data on experienced and inexperienced groups (Fan et al., 2021).

| Group | STCW qualification | Experience at sea (month) |
|---|---|---|
| Experienced | MM, CM, OOW | 213.4 (SD=188.8) |
| Inexperienced | AB, Cadets | 27.2 (SD=30.5) |

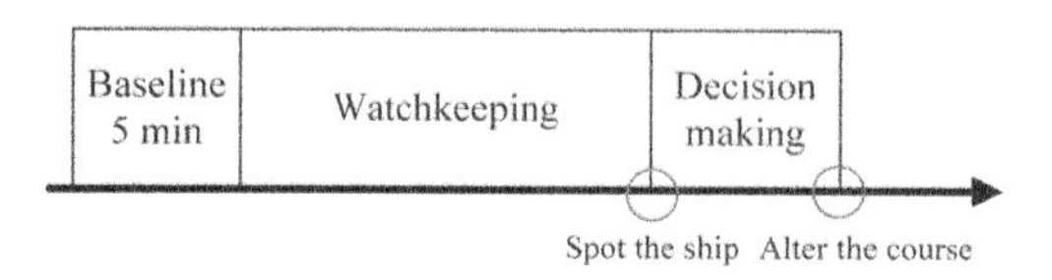


Figure 2. The fNIRS data collection procedure.

The participant wore the fNIRS skullcap containing infrared sensors and detectors, allowing to see the blood volume, oxygenated and deoxygenated blood flow in the prefrontal cortex, indicating what procedures induce the highest mental activity thus showing the change of the mental states of seafarers during navigation scenarios. The recorded measures were: 1) fNIRS data of the frontal cortex, in order to measure levels of the blood volume, oxygenated and deoxygenated blood flow in the prefrontal cortex to indicate the mental activity; 2) the NASA Task Load Index (TLX), to assess subjective levels of perceived workload.

### 5.5 *Debrief*

After the session, the participants received 5 min debrief from instructors to review their actions. They were not given any judgements but the manoeuvring results of altering the course of ship in debrief. This step helped instructors review participants' behaviours and sort out the results of sessions.

### 5.6 *Data analysis*

In pre-processing data stage, raw fNIRS data (15 channels × 2 wavelengths) was pre-processed using nirsLAB software. The Interpolate function was used to fill the data in each channel where there was detector saturation. However, for those channels which lost too much data, this function was not applicable. Then the data quality function was applied to check and identify any 'poor quality' channels in which the signal was too weak. After removing discontinuities and spike artefacts, a filter was applied in order to reduce high-frequency instrument noise and physiological noise such as fast cardiac oscillations. The pre-processed data was imported for haemodynamic states calculation using the modified Beer-Lambert law (Sassaroli and Fantini, 2004). It reveals the changes in oxygenated haemoglobin (HbO), deoxygenated haemoglobin (HbR) and total haemoglobin (Hb).

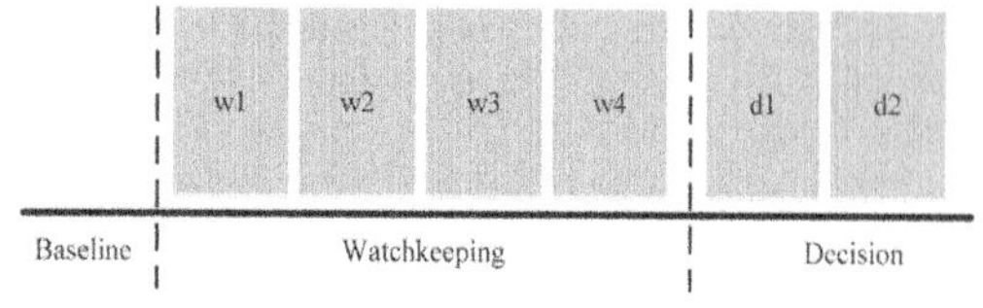


Figure 3. Averaged fNIRS data in the procedure.

For data analysis, a transformation on the data named Correlation-Based Signal Improvement (CBSI) forces HbO and HbR to be negatively correlated and controls for head movement, developed by Cui et al. (2010). As HbR is transformed into the inverse of HbO after this point, only HbO data were used in the subsequent analyses. The analysis was conducted to determine differences between left, central, and right prefrontal cortex activities. In order to create ANOVA models for statistical testing, the channels of HbO were divided into three regions of interest: left, central, and right. According to multiple references including the sample frequency of fNIRS data, the time of watchkeeping and decision phases, and the methodology used in psychophysiological literature, the period of watchkeeping during the task scenario was divided into four periods of equal duration for each participant, and the decision-making phase of the task was divided into two periods of equal duration, shown in Figure 3. These six segments and three regions of interest were analysed in the models.

### 5.7 *Dissemination*

The study was done with simulated watchkeeping tasks in the ship bridge simulator, using fNIRS technology to measure neurophysiological activation. The results show that experienced participants were considered to believe they have better performance than inexperienced people. It also illustrates better performance for experienced seafarers because they made decisions earlier, which leads to successful collision avoidance. In addition, participants under distraction were considered to require more temporal demand than those without distraction. In terms of fNIRS data, it shows significant differences in the right prefrontal cortex of the brain.

It is evident that the HbO of experienced participants was higher at the right lateral region of interest than inexperienced ones, but only during the w4 (watchkeeping) period when the target ship was spotted. Further, the average HbO of the distraction group was significantly higher during the two periods of decision-making that happened after the ship was spotted. The contribution of mental workload to marine risks is then demonstrated on the publication Fan et al. (2021). The findings not only experimentally prove the best practice in nautical science training but also provide quantitative assessment results for both experience and inexperience seafarers, which no doubt will improve human training and reliability in maritime operations towards safety at sea.

The dissemination of the study includes: 1) academic publications and conference presentation on the research, 2) demonstration of the show case to maritime students, 3) proposing a new indicator to assess human factors, especially for maritime training and risk management.

## 6 CONCLUSION

To address maritime human errors, this paper proposes a conceptual mental workload assessment framework for seafarers in maritime operations. It creates a new paradigm on maritime human factor analysis. The proposed framework consists of pre-examination, intra-examination, and post-examination stages, demonstrating all the critical mental workload assessment steps. Each of these steps is linked with the critical activities for conducting objective workload assessment using neurophysiological data. Further, the new framework enables quantitative mental workload assessment for seafarers using neuropsychological data to test the seafarers' behaviour. A case study was conducted to validate the framework using a ship bridge simulator, with 40 qualified professionals recruited. The results show that the mental workload assessment framework can effectively evaluate neurophysiological response of seafarers and consolidate maritime human reliability analysis from a neuroscience perspective. The proposed framework enables mental workload analysis of seafarers using neurophysiological data, filling the blank space for neuroscience applications into maritime transportation. In this way, it helps the maritime organisations understand the workload of seafarers with different qualification levels and recommend guidelines to improve seafarers' certificate training. Furthermore, the development of autonomous ships stimulates the new requirements for operators working at remote control centre. This framework provides insights for evaluating seafarers' behaviours regarding new requirements and scenarios for autonomous ships.


## ACKNOWLEDGEMENTS

This work is financially supported by a European Research Council project (TRUST CoG 2019 864742).


## REFERENCES


Akhtar, M.J. & Utne, I.B. (2014) Human fatigue's effect on the risk of maritime groundings - A Bayesian Network modeling approach. *Safety Science*, 62, 427–440.

Akhtar, M.J. & Utne, I.B. (2015) Common patterns in aggregated accident analysis charts from human fatigue-related groundings and collisions at sea. *Maritime Policy & Management*, 42, 186–206.

Ayaz, H., Shewokis, P.A., Bunce, S., Izzetoglu, K., Willems, B. & Onaral, B. (2012) Optical brain monitoring for operator training and mental workload assessment. *NeuroImage*, 59, 36–47.

Besikci, E.B., Tavacioglu, L. & Arslan, O. (2016) The subjective measurement of seafarers' fatigue levels and mental symptoms. *Maritime Policy & Management*, 43, 329–343.

Boyle, L.N., Tippin, J., Paul, A. & Rizzo, M. (2008) Driver performance in the moments surrounding a microsleep. *Transportation Research Part F: Traffic Psychology and Behaviour*, 11, 126–136.

Chang, C.-H., Kontovas, C., Yu, Q. & Yang, Z. (2021) Risk assessment of the operations of maritime autonomous surface ships. *Reliability Engineering & System Safety*, 207, 107324.

Christian, H., Dominic, H., Ole, F., Johannes, H., Felix, P. & Tanja, S. (2013) Mental workload during n-back task —quantified in the prefrontal cortex using fNIRS. *Frontiers in Human Neuroscience*, 7, 935.

Cui, X., Bray, S. & Reiss, A.L. (2010) Functional near infrared spectroscopy (NIRS) signal improvement based on negative correlation between oxygenated and deoxygenated hemoglobin dynamics. *NeuroImage*, 49, 3039–3046.

Dijksterhuis, C., Brookhuis, K.A. & De Waard, D. (2011) Effects of steering demand on lane keeping behaviour, self-reports, and physiology. A simulator study. *Accident Analysis & Prevention*, 43, 1074–1081.

Fan, S., Blanco-Davis, E., Yang, Z., Zhang, J. & Yan, X. (2020a) Incorporation of human factors into maritime accident analysis using a data-driven Bayesian network. *Reliability Engineering & System Safety*, 203, 107070.

Fan, S., Blanco-Davis, E., Zhang, J., Bury, A., Warren, J., Yang, Z., Yan, X., Wang, J. & Fairclough, S. (2021) The Role of the Prefrontal Cortex and Functional Connectivity during Maritime Operations: An fNIRS study. *Brain and Behavior*, 11, e01910.

Fan, S., Yan, X., Zhang, J. & Wang, J. (2017) A review on human factors in maritime transportation using seafarers' physiological data. *2017 4th International Conference on Transportation Information and Safety (ICTIS)*. IEEE.

Fan, S., Zhang, J., Blanco-Davis, E., Yang, Z., Wang, J. & Yan, X. (2018) Effects of seafarers' emotion on human performance using bridge simulation. *Ocean Engineering*, 170, 111–119.

Fan, S.Q., Zhang, J.F., Blanco-Davis, E., Yang, Z.L. & Yan, X.P. (2020b) Maritime accident prevention strategy formulation from a human factor perspective using

Bayesian Networks and TOPSIS. *Ocean Engineering*, 210, 12.

Fishburn, F.A., Norr, M.E., Medvedev, A.V. & Vaidya, C.J. (2014) Sensitivity of fNIRS to cognitive state and load. *Front Hum Neurosci*, 8, 76.

Gateau, T., Durantin, G., Lancelot, F., Scannella, S. & Dehais, F. (2015) Real-time state estimation in a flight simulator using fNIRS. *Plos One*, 10, e0121279.

Hou, X., Liu, Y., Sourina, O. & Muellerwittig, W. (2015) CogniMeter: EEG-based Emotion, Mental Workload and Stress Visual Monitoring. *International Conference on Cyberworlds*.

Kim, D.-H. (2020) Human factors influencing the ship operator's perceived risk in the last moment of collision encounter. *Reliability Engineering & System Safety*, 203, 107078.

Lim, W.L., Liu, Y.S., Subramaniam, S.C.H., Liew, S. H.P., Krishnan, G., Sourina, O., Konovessis, D., Ang, H.E. & Wang, L.P. (2018) EEG-based mental workload and stress monitoring of crew members in maritime virtual simulator. In Gavrilova, M. L., Tan, C. J. K. & Sourin, A. (Eds.) *Transactions on Computational Science Xxxii: Special Issue on Cybersecurity and Biometrics*. Cham, Springer International Publishing Ag.

Martinez-Marquez, D., Pingali, S., Panuwatwanich, K., Stewart, R.A. & Mohamed, S. (2021) Application of eye tracking technology in aviation, maritime, and construction industries: a systematic review. *Sensors*, 21, 4289.

Molteni, E., Butti, M., Bianchi, A.M. & Reni, G. (2008) Activation of the prefrontal cortex during a visual n-back working memory task with varying memory load: a near infrared spectroscopy study. *International Conference of the IEEE Engineering in Medicine & Biology Society*.

Monteiro, T.G., Zhang, H., Skourup, C. & Tannuri, E.A. (2019) Detecting mental fatigue in vessel pilots using deep learning and physiological sensors. *2019 ieee 15th international conference on control and automation (icca)*. IEEE.

Ngodang, T., Murai, K., Hayashi, Y., Mitomo, N., Yoshimura, K., Hikida, K. & Ieee (2012)A study on navigator's performance in ship bridge simulator using heart rate variability. *Proceedings 2012 Ieee International Conference on Systems, Man, and Cybernetics*.

O'Connor, P. & Long, W.M. (2011) The development of a prototype behavioral marker system for US Navy officers of the deck. *Safety Science*, 49, 1381–1387.

Patten, C.J.D., Kircher, A., Östlund, J. & Nilsson, L. (2004) Using mobile telephones: cognitive workload and attention resource allocation. *Accident Analysis & Prevention*, 36, 341–350.

Rakauskas, M.E., Ward, N.J., Boer, E.R., Bernat, E.M., Cadwallader, M. & Patrick, C.J. (2008) Combined effects of alcohol and distraction on driving performance. *Accident Analysis & Prevention*, 40, 1742–1749.

Rozanski, M., Richter, T.B., Grittner, U., Endres, M., Fiebach, J.B. & Jungehulsing, G.J. (2014) Elevated levels of hemoglobin A1c are associated with cerebral white matter disease in patients with stroke. *Stroke*, 45, 1007–11.

Sassaroli, A. & Fantini, S. (2004) Comment on the modified Beer–Lambert law for scattering media. *Physics in Medicine and Biology*, 49, N255.

Wróbel, K., Gil, M. & Chae, C.-J. (2021) On the influence of human factors on safety of remotely-controlled merchant vessels. *Applied Sciences*, 11, 1145.

Wu, Y., Miwa, T. & Uchida, M. (2017) Using physiological signals to measure operator's mental workload in shipping – an engine room simulator study. *Journal of Marine Engineering & Technology*, 16, 61–69.

*Trends in Maritime Technology and Engineering – Guedes Soares & Santos (eds)*


# Applying the SAFEPORT system in a storm situation

A.H. Gomes, L.V. Pinheiro & C.J.E.M. Fortes
*Hydraulics and Environment Department, National Laboratory of Civil Engineering, Lisbon, Portugal*

J.A. Santos
*Instituto Superior de Engenharia de Lisboa, Instituto Politécnico de Lisboa, Portugal*
*Centre for Marine Technology and Ocean Engineering (CENTEC), Instituto Superior Técnico, Universidade de Lisboa, Lisboa, Portugal*

ABSTRACT: This paper presents an application of the SAFEPORT safety system. It consists of a forecast and early warning system for emergency situations related to navigation, berthing, and mooring in port areas. The prototype of the system is being developed for the Port of Sines. The main objective of this paper is to describe the application of the numerical models in SAFEPORT to simulate the behavior of three different ships moored in three terminals of the Port of Sines, subjected to the sea-wave conditions of the Dora storm. The 3-day advance sea-wave forecasts provided by the WAM model were used. Tide levels and currents were obtained from the XTide model. It was concluded that the SAFEPORT system was able to forecast possible hazards and issue the expected alerts regarding the moored ships' motions and the forces on their mooring lines associated to the Dora storm.

## 1 INTRODUCTION

The significant increase of cargo volume in the Iberian Peninsula as well as the increase in the size of the ships calling at its ports has driven several port-expansion plans. The new ships require deeper waters, which leads them to be berthed in very exposed locations. In addition, the number of major storms that formed in the Atlantic Ocean reached a new maximum in 2020. The harbors of the Portuguese West coast are exposed to such storms, and this is a key factor for the safety of ships moored there.

In this new scenario, to ensure the efficiency of port operations, as well as the required safety levels during a storm situation, it is important to predict, in advance, the sea agitation inside the port and its consequences for ships that will enter the port or are berthed and moored inside. Due to the lack of systems addressing the hazards related to ships maneuvering and moored in ports, the SAFEPORT system, based on the HIDRALERTA system (Fortes *et al.*, 2015, 2020; Poseiro *et al.*, 2017 and Pinheiro *et al.*, 2020), is being developed as part of the BlueSafePort project.

SAFEPORT is a forecast and Early Warning System (EWS) for emergency situations related to navigation, mooring, and berthing in port areas. It provides 72-hour forecasts of the characteristics of sea agitation, its consequences related to ship movements and/or forces in the mooring systems and the associated hazard levels. A prototype of the SAFEPORT system is being developed for the Port of Sines.

This paper presents an application of the SAFEPORT system, more specifically, the application of the numerical models developed to simulate the behavior of three different ships docked and moored at three terminals of the Port of Sines subjected to the sea-wave conditions of storm Dora, which reached mainland Portugal on December 4$^{th}$, 2020.

To execute the numerical models for wave propagation and moored ship behavior, the safety system, herein presented, uses the integrated numerical tool, SWAMS — Simulation of Wave Action on Moored Ships (Pinheiro *et al.* 2013).

After this introduction, section 2 presents the SAFEPORT system in terms of its operation and numerical models. Section 3 describes the case study, the numerical models used, and the results produced by the SAFEPORT system. Finally, section 5 presents the main conclusions provided by this research.

## 2 THE SAFEPORT SYSTEM

The SAFEPORT EWS issues daily forecasts for the next 72 hours, at three-hour intervals, of sea agitation within a port and its consequences on ships (in maneuvering or moored).

### 2.1 *Methodology*

The SAFEPORT safety and alert system, similarly to the HIDRALERTA system (Poseiro, 2019 &

DOI: 10.1201/9781003320289-22

Pinheiro *et al.*, 2020), comprises 4 modules: I - Sea-wave characterization; II - Navigation in Port Areas; III - Monitoring; IV - Risk Assessment.

The first two modules integrate a set of numerical models. Some models are included in the SWAMS tool (Pinheiro *et al.* 2013). Those associated with the behavior of maneuvering ships are not included in this work and are carried out by the Centre for Marine Technology and Ocean Engineering (CENTEC). Numerical simulations run on the Central Node for Grid Computing (NCG) of the Portuguese Infrastructure for Distributed Computing (INCD), a 64-node high performance computing facility.

The third module consists of in situ monitoring required to validate the results produced by the numerical models.

The last module deals with risk assessment for moored ships, which is performed through the definition of hazard levels from 0 to 2 for the moored ships' motions and from 0 to 3 for the forces on their mooring lines.

Concerning the moored ships' motions, 0 corresponds to no danger (green symbol), 1 corresponds to a situation of restricted loading and unloading operations (yellow symbol) and 3 corresponds to the maximum warning level (red symbol). The limits imposed on the ships' motions are the recommended ones in PIANC (1995).

As for the forces on ships' mooring lines, the hazard levels depend on the Maximum Breaking Load (MBL) of the mooring lines (OCIMF, 1992). 0 corresponds to no danger (green symbol), 1 corresponds to 50% of MBL (yellow symbol), 2 corresponds to 80% of MBL (orange symbol) and 3 corresponds to 100% of MBL (red symbol).

Figure 1 shows the symbols used in the system to issue alerts.

Figure 1. Symbols used by the SAFEPORT system to alert ships' motions danger (left) and ships' mooring systems failure danger (right).

## 2.2 *SWAMS numerical software package*

SWAMS (Pinheiro *et al.* 2013), acronym for Simulation of Wave Action on Moored Ships, is an integrated numerical tool capable of simulating the response of a moored ship within a harbor, subjected to the action of sea waves, wind and currents. This tool consists of a graphical user interface and a set of modules that deal with the execution of numerical models for wave propagation and the behavior of moored ships inside harbor basins.

SWAMS consists of 2 modules (Figure 2): the WAVEPROP module for wave propagation and the MOORNAV module for moored ship behavior. The objective of the first module is to determine the sea-wave characteristics in the study area. The second module estimates ship movements and the forces exerted on the mooring system.

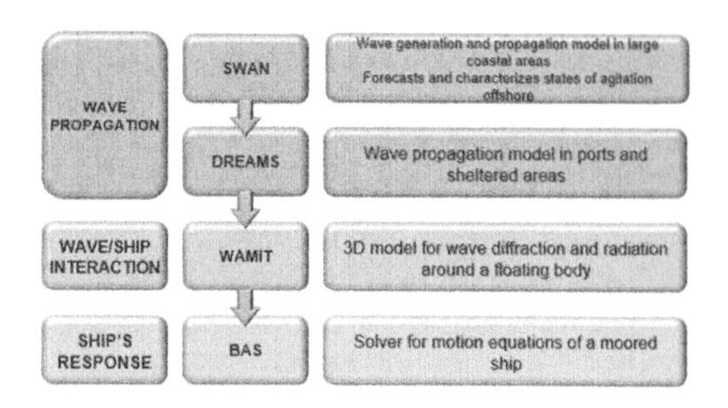


Figure 2. Structure of SWAMS numerical software package used in the case study (adapted from Pinheiro et al., 2013).

The WAVEPROP wave propagation module includes 3 numerical models for wave propagation and a finite element mesh generator, namely:

- SWAN (Booij *et al.*, 1996) is a spectral nonlinear model based on the wave action conservation equation, which simulates the propagation of irregular wave spectrum;
- DREAMS (Fortes, 2002) is a linear finite element model, based on the mild slope equation, to simulate the propagation of monochromatic waves;
- BOUSS-WMH (Pinheiro *et al.*, 2011) is a nonlinear finite element model, based on the extended Boussinesq equations deduced by Nwogu (1993), being able to simulate the propagation of regular and irregular waves;
- GMALHA (Pinheiro *et al.*, 2008) is a triangular finite element mesh generator specially defined to be used by DREAMS and BOUSS-WMH models, being the node density of the meshes variable according to the local wavelength and its construction optimized to reduce computational resources.

The MOORNAV moored ship behavior module includes 2 numerical models (Santos, 1994), namely:

- WAMIT (Korsemeyer *et al.*, 1988) which solves, in the frequency domain, the radiation and diffraction problems of the interaction between a free-floating body and the incident waves;
- BAS (Mynett *et al.*, 1985) that assembles and solves, in the time domain, the equations of motion of a moored ship, considering the time series of forces due to the waves incident on the ship, the ship's impulse response functions and the constitutive relations of the mooring system elements (mooring lines and fenders).

The numerical model SWAN transfers the wave characteristics from the offshore area to the harbor entrance. The DREAMS model and the BOUSS-WHM model, in turn, transfer the wave characteristics from the harbor entrance area to the harbor area, using the harbor mesh generated by GMALHA. The WAMIT numerical model determines the response of the free-floating ship to incident monochromatic waves. Then, with the hydrodynamic information obtained, it is possible to determine the moored ship response through the BAS numerical model.

## 3 PORT OF SINES CASE STUDY

The Port of Sines is a deep-water port located on the west coast of mainland Portugal (Figure 3). The port has 7 terminals, namely: the Liquid Bulk Terminal (TGL), the Liquified Natural Gas Terminal (TGN), the Petrochemical Terminal (TPQ), the Sines Container Terminal or Terminal XXI (TCS), the Sines Multipurpose Terminal (TMS), the Fishing Port and the Sines Marina.

Given its national relevance, its continuous economic growth and constant expansion, the port of Sines has been the subject of several research projects. The prototype of the SAFEPORT system, for example, has been developed and validated for the Port of Sines. For that, SWAMS (Pinheiro *et al.* 2013) was employed to simulate the behavior of three different ships docked and moored at three terminals of the Port of Sines subjected to the sea-wave conditions of storm Dora, namely: an oil tanker at the TGL, a general cargo ship at the TMS, and a container ship at the TCS (Figure 3).

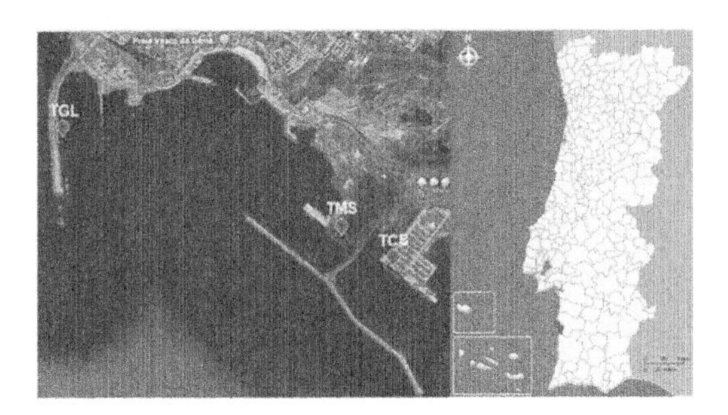


Figure 3. Location of the Port of Sines and the TGL, the TMS, and the TCS.

Storm Dora reached its highest intensity during the afternoon of 4th December 2020, and the early morning of 5th December 2020. According to Civil Protection, there were more than 400 incidents along the Portuguese territory, most of which caused minor material damage. It was a storm characterized by wind gusts of more than 100 km/h (62 mph) in the areas near the coast, snow in the north of the country, rain, and rough sea with records of maximum wave height of 10.3 meters at the wave buoy in front of the Port of Sines.

### 3.1 *Basic Data*

The 3-day advance forecast of the offshore sea agitation was obtained through the European Centre for Medium-Range Weather Forecasts (ECMWF) (Persson, 2001), which uses the WAM model (WAMDI Group, 1988) used by. The forecasts for 4 and 5 December were collected every 3 hours for the following sea wave parameters: significant wave height ($H_s$), peak wave period ($T_p$) and mean wave direction ($\theta_m$) (Figure 4).

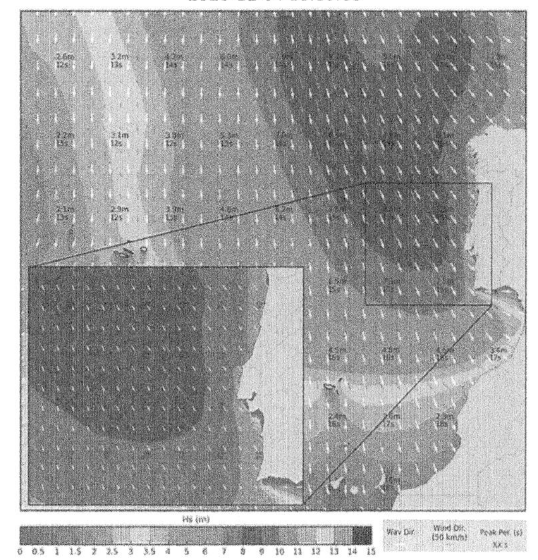


Figure 4. WAM model results offshore the Port of Sines ($H_s$, $T_p$ and $\theta_m$) for 4th December 2020, 6 p.m.

The WAM model predicted that the waves formed due to storm Dora would have a predominant $\theta_m$ of north, with $H_s$ ranging between 3 and 8 m and a maximum $T_p$ of approximately 18 s.

The tide levels were estimated with the XTide model (Flater, 1998). Wind data was obtained from the NAVGEM model (Reynolds et al., 2011). Bathymetry was provided by Sines and Algarve Ports Administration. The general geometric characteristics of the ships are shown in Table 1.

Table 1. General geometric characteristics of the simulated ships.

| | Draft | Beam | Length overall |
|---|---|---|---|
| Ship | m | m | m |
| Oil Tanker | 22.0 | 26.5 | 340 |
| General Cargo | 10.5 | 30.0 | 220 |
| Container | 8.0 | 19.0 | 120 |

### 3.2 *Sea-waves propagation and characterization*

#### 3.2.1 *SWAN numerical model application*

The SWAN numerical model (Booij *et al.*, 1996) was applied to propagate the sea agitation parameters, from offshore to the entrance of the port of Sines. The theoretical JONSWAP spectrum was assumed to represent the real spectrum of the waves approaching the port.

To achieve a better numerical performance, the model domain was discretized into three nested rectangular grids (Figure 5). The physical phenomena accounted by SWAN were diffraction and dissipation by bottom friction. The simulations were performed in the two-dimensional stationary mode. In the third, the smallest, mesh one point was defined, in the vicinity of the port of Sines, where the wave characteristics i.e., $H_s$, $T_p$ and $\theta_m$, were extracted.

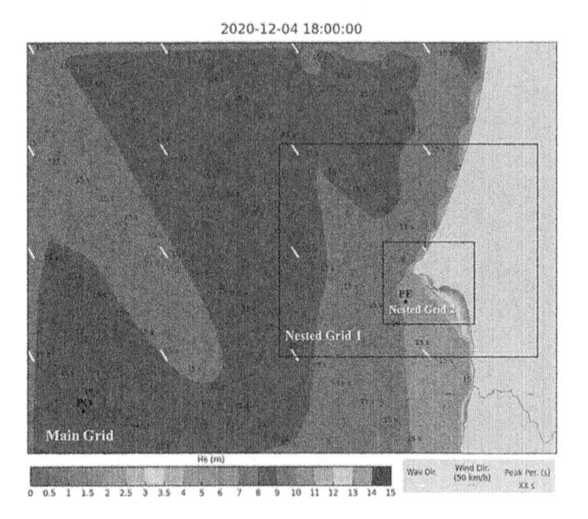


Figure 5. SWAN model results at port of Sines ($H_s$, $T_p$ and $\theta_m$) for 4th December 2020, 6 p.m. PO stands for the North offshore point of the WAM model and PE stands for the transfer point from SWAN to DREAMS.

On the 1st day of the storm, the observed waves in front of the port had their $T_p$ between 15 s and 17 s, and $H_s$ between 3 m and 7.55 m (maximum Hs occurred at 6 p.m.). The most frequent situation corresponds to $H_s$ between 6 m and 6.5 m. As for $\theta_m$, Northwest incidences represent the predominant direction of the waves. For the 2nd day of storm Dora, the sea agitation conditions were less severe. A maximum $H_s$ of 6.94 m was obtained at 12 a.m. with a $T_p$ of 15.09 s. The predominant $\theta_m$ was also northwest.

### 3.3 *DREAMS numerical model application*

The DREAMS model simulates the propagation of monochromatic waves on gently sloping bottoms, considering the combined effects of refraction, diffraction and partial or total reflection of the port area boundaries.

To do the modelling, a finite element mesh generated by GMALHA (Pinheiro *et al.*, 2008) was used to characterize the morphology of the Sines harbor. In addition, a file containing information about the absorption coefficients of each section of the port boundary was assigned to the model. Results (Figure 6 and Figure 7) were extracted at three points near the three terminals where the ships' behavior was simulated.

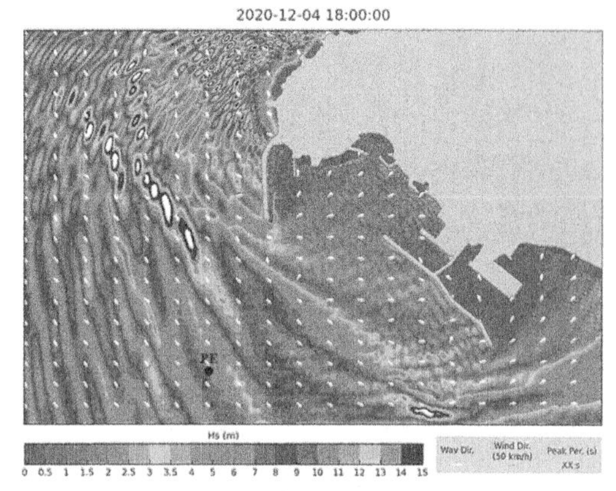


Figure 6. DREAMS model results at port of Sines (significant wave height and mean wave direction) for 4th December 2020, 6 p.m. PE stands for the transfer point from SWAN to DREAMS.

Sea waves approach the TGL with $\theta_m$ from south to southwest and $H_s$ ranging between 0.1 m and 1.2 m. The peak was recorded at 9 p.m., on December - 4th, with $H_s = 1.2$ m, $T_p = 15$ s and $\theta_m = 210°$.

The TMS, in turn, is affected by waves coming from the west with $H_s$ not exceeding 0.7 m. The peak occurred at 6 p.m., on December 4th, with $H_s = 0.7$ m, $T_p = 15$ s and $\theta_m = 280°$.

Finally, at the container terminal, the incident wave action was characterized by south swells with Hs ranging from 0.12 m to 0.84 m. The peak occurred at 3 a.m., with $H_s = 0.8$ m, $T_p = 17$ s and $\theta_m = 174°$. Although wave heights offshore ($H_s \sim 4$ m) were not close to the highest recorded (at 6 p.m. on December 4th), the waves rotating slightly to the west led to the largest impact inside the port.

### 3.4 *Behavior of the moored ships*

#### 3.4.1 *WAMIT numerical model application*

The WAMIT model (Korsemeyer *et al.*, 1988) was applied to calculate the hydrodynamic coefficients of the free ships, i.e., the damping and added-mass coefficients. For this purpose, one needs the geometry of the submerged hulls discretized into rectangular/triangular flat panels and the mass distribution (inertias). of the ships. The submerged hull of the oil tanker was

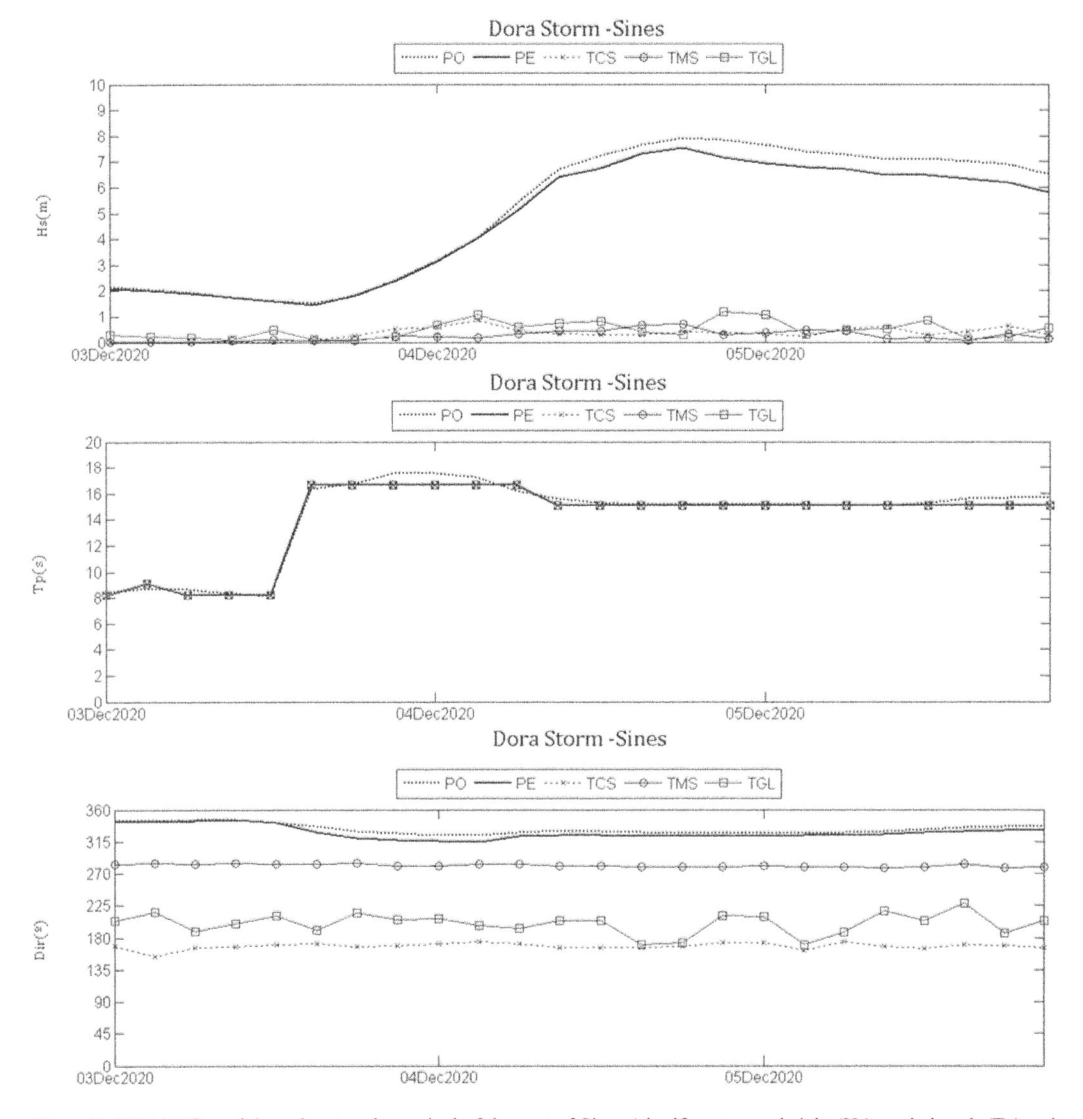


Figure 7. DREAMS model results at each terminal of the port of Sines (significant wave height (Hs), period peak (Tp) and mean wave direction (Dir)). PO stands for the North offshore point of the WAM model and PE stands for the transfer point from SWAN to DREAMS.

discretized with 1004 panels, the general cargo ship with 1992 panels and the container ship with 3464 panels (Figure 8). The tool used was NPP, acronym for Nautical Pre-Processor (Santos, 1994).

The WAMIT simulations were performed for the possible wave directions approaching each terminal and a range of 89 frequencies. Once the damping coefficients of the ships were computed, the impulse response functions (Figure 9) were obtained. The infinite frequency added masses were estimated based on the corresponding frequency domain added mass values for a given frequency and on the corresponding retardation functions.

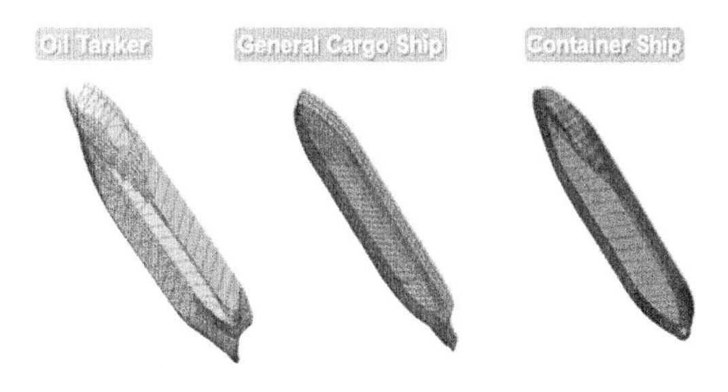


Figure 8. Panel discretization for the 3 ships.

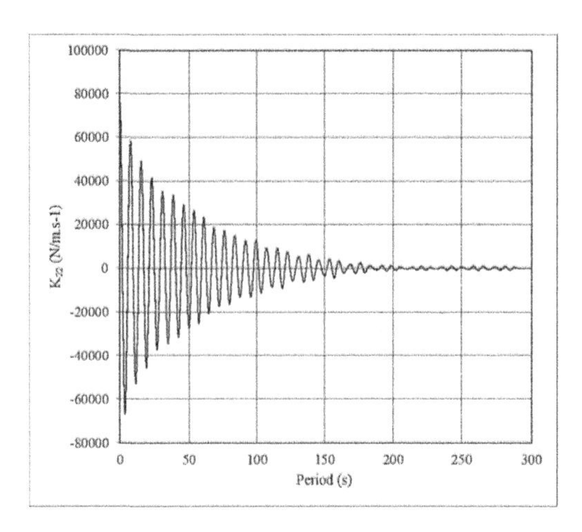


Figure 9. Impulse response (top) and infinite frequency added-mass (bottom) of the general cargo ship for the motion mode j, k = 2, 2.

### 3.4.2 *BAS numerical model application*

The application of the BAS model (Mynett *et al.*, 1985) to the moored ships allows one to obtain the motion of these ships and the forces in their mooring systems. As input parameters, in addition to the sea wave conditions, the BAS model used the information computed by the WAMIT model, the coordinates of the mooring points and the coordinates of the contact points between the ships and the fenders.

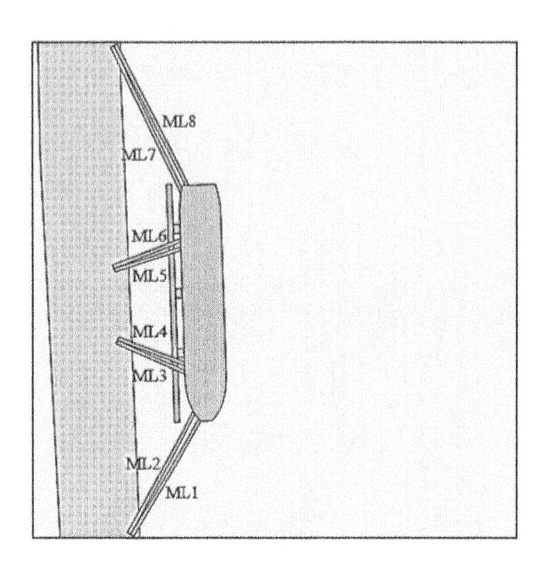


Figure 10. Oil tanker ship mooring system.

For the oil tanker and the general cargo ship, 8 mooring lines grouped in two and 3 fenders were defined (Figure 10 and Figure 11). For the container ship model, a total of 10 mooring lines and 5 fenders were defined (Figure 12).

The constitutive relations for all the mooring lines are linear. For an elongation of 4%, the maximum load on mooring lines of the oil tanker is 2100 kN, on the general cargo ship is 1900 kN and, on the container is 1860 kN.

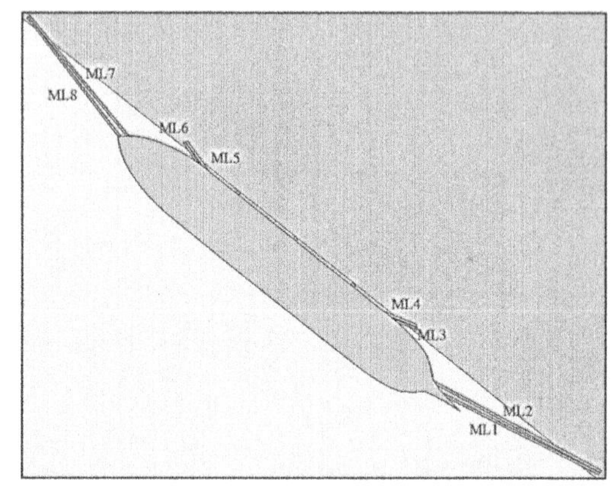


Figure 11. General cargo ship mooring system.

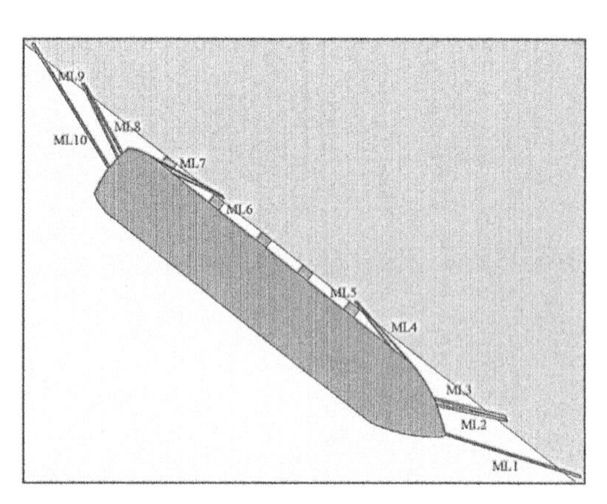


Figure 12. Container ship mooring system.

The same constitutive relations were considered for the oil tanker and the general cargo ships' fenders: non-linear range from 0 kN to 880 kN maximum load, with a maximum deflection of 0.4 m. As for the container ship, its fenders have a linear compression with a maximum force of 8900 kN for a deflection of 1 m.

## 4 RESULTS AND DISCUSSION

Sea wave results provided by the WAM and SWAN models show that at 6 p.m. on December 4th, 2020, corresponds to the most critical time of storm Dora. However, according to the DREAMS model results, within the port, i.e., in the vicinity of the terminals under study, the storm peaks occurred at different times (Figure 7).

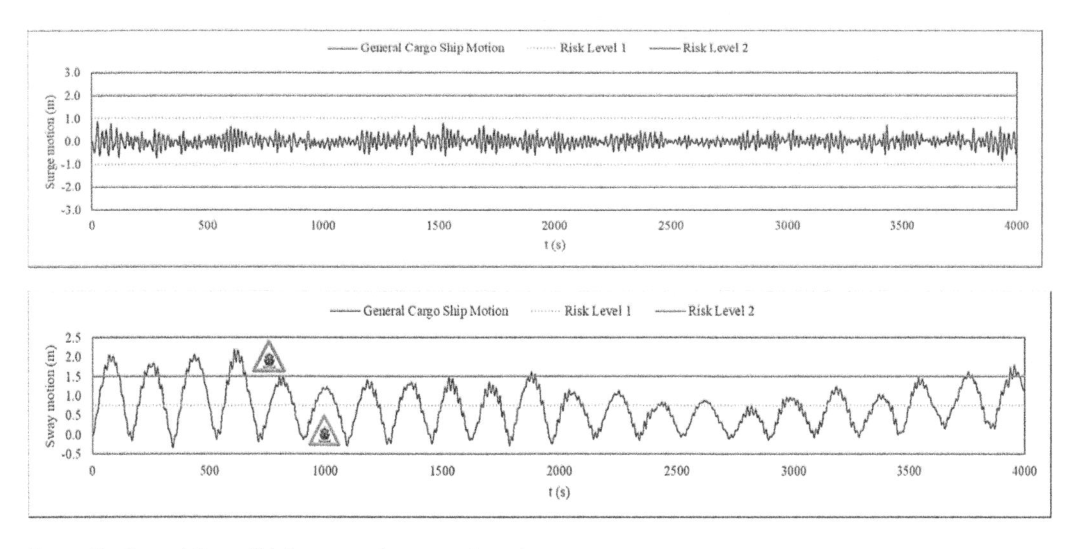


Figure 13. General Cargo Ship's surge and sway motions alerts.

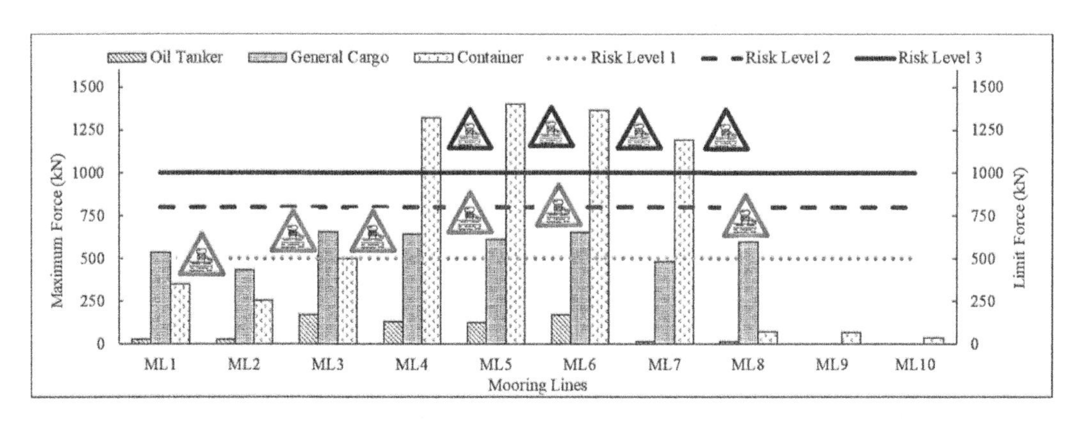


Figure 14. Forces on the ships' mooring lines alerts.

To comply with the objectives of the SAFE-PORT system, the ships' motions and the forces on their mooring lines were analyzed, considering the storm peak recorded in the vicinity of the terminals where they are docked. Thus, Figure 13 presents some results for the general cargo ship (ship motions), at 6 p.m., on December 4$^{th}$, obtained by the SAFEPORT system. Figure 14 displays the forces on the ships' mooring lines at the time when the peak occurred.

During storm Dora, no alerts were issued concerning the oil tanker moored at the liquid bulk terminal. Despite being the most exposed ship, given its large displacement, no excessive motions of the ship were forecast and, consequently, the forces on its mooring lines are not expected to reach the maximum limits.

As for the general cargo ship moored at the multipurpose terminal, yellow and red alerts (danger level 1 and 2, respectively) were issued to sway and heave motions. 6 yellow alerts, i.e., danger level 2, were issued to mooring lines 1, 3, 4, 5, 6 and 8. The mooring lines received a maximum load of more than 500 kN. For the others no alerts were issued. This is an expected result, since among the 3 terminals, the TMS is the most protected from sea waves.

The container ship docked at the container terminal was the most affected ship by the storm Dora. Therefore, the system issued 2 red alerts to surge and sway motions, and 1 yellow alert to heave motions. 4 yellow alerts, i.e., danger level 2, to mooring lines 4, 5, 6 and 7. They were exposed to a maximum load of more than 1000 kN. No alerts were issued for the remaining lines.

## 5 CONCLUSIONS

The SAFEPORT system is intended to predict, alert, and assess the occurrence, 72 hours in advance, of emergency situations associated with the safety of moored ships. To this end, it is essential that the system operates properly in current and storm situations.

This research is part of the SAFEPORT system testing phase. The performance of the system in a storm situation was tested. The developments made to date consist of the application of the prototype developed for the port of Sines in the simulation of the impact of storm Dora on three ships moored at different terminals of the port.

The alerts issued by the system were in accordance with the reported occurrences. It should be noted that these are only preliminary results, as the system is still being validated and tested.

The SAFEPORT system proved to be a useful tool in managing the safety of ships moored in the port of Sines during storm Dora. To ensure that the system provides consistent and reliable information in any situation, further tests and in-depth validation with buoy data are under development.

## ACKNOWLEDGEMENTS

The authors would like to thank the BlueSafePort project (ref: FA_04_2017_016), the National Infrastructure for Distributed Computing (INCD) for granting access to the digital infrastructure to support research and the Sines and Algarve Ports Administration.

## REFERENCES

Booij, N.R., Holthuijsen, L.H. & Ris, R.C. (1996). The SWAN wave model for shallow water. *ICCE'96*: 668-676. Orlando.

Flater, D. (1998). XTide Manual: Harmonic Tide Clock and Tide Predictor. *Technical Report, USA*. URL: https://flaterco.com/xtide

Fortes, C.J.E.M. (2002). *Transformações não lineares de ondas em portos. Análise pelo método dos elementos finitos*. Tese de doutoramento. Lisboa: IST/DEM.

Fortes, C.J.E.M.; Reis, M.T.; Poseiro, P.; Santos, J.A.; Garcia, T.; Capitão, R.; Pinheiro, L.; Reis, R.; Craveiro, J.; Lourenço, I.; Lopes, P.; Rodrigues, A.; Sabino, A.; Araújo, J.P.; Ferreira, J.C.; Silva, S.F.; Raposeiro, P.; Simões, A.; Azevedo, E.B.; Reis, F.V.; Silva, M.C.; Silva, C.P. (2015). Ferramenta de Apoio à Gestão Costeira e Portuária: o Sistema HIDRALERTA. *In Atas do VIII Congresso sobre Planeamento e Gestão das Zonas Costeiras dos Países de Expressão Portuguesa*. Aveiro, outubro 2015.

Korsemeyer, F.T., Lee, C.-H., Newman, J.N. & Sclavounos, P.D. (1988). The analysis of wave effects on tension-leg platforms. Proc. *7th International Conference on Offshore Mechanics and Arctic Engineering*: 1–14. Texas: Houston.

Mynett, A.E., Keunig, P.J. & Vis, F.C. (1985). The dynamic behaviour of moored vessels inside a harbour configuration. *Int. Conf. on Numerical Modelling of Ports and Harbour* 23–25: Cranfield: BHRA. The Fluid Engineering Centre, April 1985. England: Birmingham.

Reynolds, C. A., McLay, J. G., Goerss, J. S., Serra, E. A., Hodyss, D. & Sampson, C. R. (2011). Impact of Resolution and Design on the U.S. *Navy Global Ensemble Perfomance in the Tropics, Monthly Weather Review* 139: 2145–2155.

Nwogu, O. (1993). Alternative form of Boussinesq equations for near-shore wave propagation. *J. Waterway, Port, Coastal, and Ocean Engineering* 119(6): 618–638.

OCIMF (1992) Mooring equipment guidelines. Witherby e Co. Ltd.

Persson, A. (2001). User Guide to ECMWF Forecast Products. Meteorological Bulletin M3.2. *ECMWF*: 115.

PIANC - Permanent International Association of Navigation Congresses (1995). Criteria for movements of moored ships in harbors. Technical report Permanent International Association of Navigation Congresses, PIANC Supp.to bulletin no. 88.

Pinheiro, L. V.; Fortes, C. J.; Fernandes J. L. (2008) Gerador de Malhas de Elementos Finitos para a Simulação Numérica de Propagação de Ondas Marítimas. *Revista Internacional de Métodos Numéricos para Cálculo y Diseño en Ingeniería (RIMNI)* 24(4).

Pinheiro L., Fortes C.J.E.M., Santos J.A., Fernandes L. & Walkley M. (2012). Boussinesq-type Numerical Model for Wave Propagation Near Shore and Wave Penetration in Harbors. *Maritime Engineering and Technology*. Guedes Soares, C., Garbatov, Y., Sutulo, S. & Santos T. A. (Eds.), Taylor & Francis Group, London, UK, pp. 505–512.

Pinheiro, L. V., CJEM, F., Santos, J. A. & Fernandes, J. L. M. (2013). *Numerical software package SWAMS – Simulation of Wave Action on Moored Ships*. Proc. PIANC 3nd Mediterranean Days of Coastal and Port Engineering, 22 a 24 de maio, Marseille, França.

Pinheiro, L., Fortes, C., Reis, M. T., Santos, J. & Soares, C. G. (2020). Risk forecast system for moored ships. *In proceedings of vICCE (virtual International Conference on Coastal Engineering)*. 6–9 October.

Poseiro, P.; Gonçalves, A.B.; Reis, M.T.; Fortes, C.J.E.M. (2017). Early warning systems for coastal risk assessment associated with wave overtopping and flooding. *Journal of Waterway, Port, Coastal, and Ocean Engineering*.

Santos, J.A. (1994). MOORNAV – Numerical model for the behaviour of moored ships. *Final report, Projecto NATO PO-Waves* (3/94-B). Lisbon.

WAMDI Group (1988). The WAM Model - A third generation ocean wave prediction model. *J. Physical Oceanography* (18): 1775–1810.

*Trends in Maritime Technology and Engineering – Guedes Soares & Santos (eds)*


# Simulation of search and rescue operations of the continental coast of Portugal

B. Lee & A.P. Teixeira
*Centre for Marine Technology and Ocean Engineering (CENTEC), Instituto Superior Técnico, Universidade de Lisboa, Lisbon, Portugal*

ABSTRACT: Search and Rescue (SAR) operations include the management and coordination between the decision and the rescue teams, the prediction of the location of the drifting object and the choice of the search pattern to execute the search operation. This paper addresses the two last problems, corresponding to the operation planning phase. A SAR planning tool based on the Monte Carlo Simulation method is developed to provide decision support on where to allocate the search effort and on which search pattern to use for an efficient and successful SAR operation. The tool incorporates a drift model with uncertainty propagation to define the probability density maps of the object location based on the initial position and the effect of winds and currents. Then, different search strategies are simulated to assess their efficiency. Occurrences in Portuguese waters are used to demonstrate the applicability of the developed search planning tool.

## 1 INTRODUCTION

Although Portugal is a relatively small country, it has a vast maritime territory where many vessels navigate every day due to its geographic location. This together with severe weather conditions observed mainly during the winter but also in the summer due to strong winds, increases the probability of maritime accidents. Therefore, a good SAR system is needed to prevent any casualties resulting from these potential accidents and incidents.

SAR decision support systems (DSSs) are used to help the SAR Mission Coordinator (SMC) to better prepare and plan the SAR mission. As examples of SAR DSS current in use are the Search and Rescue Optimal Planning System (SAROPS), developed by the US Coast Guard, which provides capabilities for search theory-based search planning (Kratzke et al., 2010), the CANSARP, developed by the Canadian Coast Guard, and the OVERSEE, developed by the Portuguese Navy together with the Portuguese company *Critical Software*. They estimate the search area and recommend search paths for several search units that maximize the Probability of Detection (POD) (Breivik et al., 2013), and therefore the Probability of Success (POS), using probabilistic models and Monte Carlo Simulation.

A SAR DSS is a search planning tool that simulates search operations to assess their efficiency in achieving the operation success, which incorporates a drift model to define the search area where the target should be found (Allen & Plourde, 1999).

The problem of trajectory prediction of drifting objects under the effect of current and wind has been addressed by several authors (e.g. Allen & Plourde, 1999; Zhang et al., 2017). In particular, Zhang et al. (2017) have proposed a probabilistic model to predict the object's velocity and position at every moment in a sequential way using current and wind velocities near the object derived from spatial correlated velocity fields.

The problem of search planning has been studied by Zhang et al. (2016). Two search patterns have been simulated and their performance assessed, provided that the probability distribution map of the missing object is available.

The present paper develops a SAR planning tool that addresses both the uncertainty on the position of the missing objects and the search planning problems. First, the probability distribution map of the missing object is derived using a drift model and then different search strategies are simulated to assess their efficiency measured in terms of the probability and time to find the missing object.

The paper is organised in five sections. Section 2 introduces the basis of the formulation for predicting the trajectory of a drifting object at sea and the approach adopted to assess the efficiency of a search operation. Section 3 presents the implementation of the developed SAR tool to study the efficiency of maritime search operations and the results of its application are discussed in section 4. Finally, the conclusions are drawn in section 5.

DOI: 10.1201/9781003320289-23

## 2 SAR OPERATIONS PLANNING

To plan a SAR operation, the SMC must follow six specific steps: identify the search target, define the datum, establish the search area, select the appropriate search pattern, determine the desired area coverage and then, develop a practical search plan.

Nowadays there are devices, such as Emergency Position Indicating Radio Beacon (EPIRB) with a GPS incorporated in it, that transmit a distress signal when activated. But sometimes there are cases when this device is lost or not activated due to malfunction during the accident and the people in distress become lost in the middle of the vast sea. In these cases, the planning stage consists in establishing a search area to start the search, and then decide on the best approach to execute it.

### 2.1 *Search area*

Establishing a search area for the Search and Rescue Units (SRUs) locate and rescue the people in distress is the key point for a successful SAR operation. To determine this search area, the SMC must estimate the different trajectories which those people may follow while drifting. The motion of an unpowered floating object is caused by a combination of sea currents, waves, and wind. Since the oceans are in constant motion, the time factor is what determines the survival of the people in distress in every SAR cases. Hence, to guarantee a successful operation, the SMC must define a search area smaller as possible, where the person can be found with a reasonable and predictable level of certainty.

The motion due to the wind and waves, which is defined as leeway, is a crucial factor when analysing the drift of a floating object. According with Fitzgerald et al. (1993) leeway is the motion of a floating object relatively to the surface currents measured between 0.3 meter and 1.0 meter of depth and caused by waves and winds adjusted to a 10 meters height reference, which can be decomposed into downwind and crosswind leeway components.

Following the study developed by Allen and Plourde (1999), the leeway motion is characterized by the leeway angle, $L_\alpha$, the leeway speed, $|L|$, the downwind and crosswind components of leeway, the leeway rate and the relative wind direction (*RWD*). With the leeway angle together with the leeway speed it is possible to define the downwind component of leeway (*DWL*) and the crosswind component of leeway (*CWL*), as shown in Eqs. (1) and (2).

$$|DWL| = |L|\sin(90^o - L_\alpha) \quad (1)$$

$$|CWL| = |L|\cos(90^o - L_\alpha) \quad (2)$$

In terms of vectors, the leeway vector, $L$, can be written as Eq. (3).

$$L = DWL + CWL \quad (3)$$

The motion of a drifting object is the result of several forces acting upon its surface such as, wind, water currents, gravitational and buoyancy force, but is extremely difficult to compute due to the irregular geometry of the real-world objects. It is important to notice that waves are usually left out due to the Stroke drift which is mainly downwind and therefore is difficult to separate it from the leeway drift, and it is considered already included in the empirical leeway coefficients (Breivik & Allen, 2008).

Ni et al. (2010) presented a leeway dynamic model to study the mechanisms behind the leeway concept. This model only considers the current and wind influences, and the long-term drifting, therefore the transient state is neglected. Also, the acceleration can be ignored since the floating object will rapidly reach its steady velocity, according to Breivik & Allen (2008). According to the law of motion the resultant of forces must be equal to zero when the object moves at a steady velocity, meaning that the forces are facing each other. As a result, the equation of motion can be written as Eq. (4).

$$\frac{1}{2}(C_D A\rho)_1|U_w - U_o|(U_w - U_o) = \frac{1}{2}(C_D A\rho)_2 |U_o - U_c|(U_o - U_c) \quad (4)$$

where $U_o$ is the object velocity, $U_w$ is the wind velocity, $U_c$ is the current velocity, $C_D$ is the drag coefficient, $A$ is the cross-sectional area exposed to the fluid, $\rho$ is the density of the fluid, and the subscripts 1 and 2 refers to the environment in which the object is exposed, air and water, respectively.

Assuming Eq. (5), Eq. (4) can be simplified into Eq. (6)

$$\frac{(C_D A\rho)_1}{(C_D A\rho)_2} = \lambda^2 \quad (5)$$

$$\lambda(U_w - U_o) = U_o - U_c \quad (6)$$

Rearranging Eq. (6), the velocity of the floating object is given by Eq. (7)

$$U_o = \frac{1}{1+\lambda}U_c + \frac{\lambda}{1+\lambda}U_w \quad (7)$$

The leeway velocity can be written in function of $U_w$ and $U_c$, as in Eq. (8)

$$U_L = U_o - U_c = \frac{\lambda}{1+\lambda}(U_w - U_c) \quad (8)$$

Considering the factor $\lambda/(1+\lambda)$ as the leeway rate $f$, the leeway velocity can be written as Eq. (9) and consequently, the velocity of the floating object as Eq. (10).

$$U_L = f(U_w - U_c) \tag{9}$$

$$U_o = U_L + f(U_w - U_c) \tag{10}$$

The leeway velocity vector is composed by a downwind velocity component $DWL$ and a crosswind velocity component $CWL$ given by Eqs. (11) and (12) respectively. From those equations the leeway angle $L_a$ is defined as Eq. (13).

$$DWL = f\left[|U_w| - U_c \cdot \frac{U_w}{|U_w|}\right]\frac{U_w}{|U_w|} \tag{11}$$

$$CWL = -fU_c + f\left(U_c \cdot \frac{U_w}{|U_w|}\right)\frac{U_w}{|U_w|} \tag{12}$$

$$\tan L_a = \frac{|CWL|}{|DWL|} = \frac{\left|-U_c + \left(U_c \cdot \frac{U_W}{|U_w|}\right)\frac{U_W}{|U_w|}\right|}{\left||U_w| - U_c \cdot \frac{U_W}{|U_w|}\right|} \tag{13}$$

To estimate the drift trajectory of the object, information on the local wind, surface current and the characteristics of the object needs to be provided (Breivik, 2008). Since those values are not a - hundred percent accurate, it is important to consider the uncertainties associated to those parameters to generate a result with an acceptable level of accuracy. The common way to address the problem related to the uncertainties is using probabilistic formulations. For parameters such as velocity and direction of both wind and current, Last Known Position (LKP) and leeway coefficients of the floating objects, they are characterized by a mean value and a variance. According to Allen (2005), the consideration of the possibility that a floating object could jibe will make the simulation closer to the reality, resulting in an increase of the uncertainties.

The floating objects are characterized by leeway coefficients, obtained through field experiments carried in the past, compiled by Allen and Plourde (1999) into a database and updated by Allen (2005).

### 2.2 *Search strategy*

The search phase is a very complex, demanding, time consuming and expensive task, which involves different parties and could require a large number of limited resources (Frost, 1996). Therefore, it is essential to prepare beforehand to successfully execute the search operation in the most efficiently way. In case of SAR operations, is the SMC who must plan the best course of action to perform the search for the people in distress, after establishing the search area. Each operation plan is elaborated based on search models, which are used to study the efficiency of search operations within the scope of SAR operations.

The Search Theory studies a way to combine and make use of the limited resources more efficiently when locating a missing object at sea, whose location is unknown. Its objective is, mainly, to maximize the probability to detect the object with the resources available but, it can also be to minimize the time spent in finding it (Frost & Stone, 2001).

According to Koopman (1946), a basic problem of optimal search has as objective maximize the probability of success on finding a target over some possible area with a limited amount of resources, and it is normally characterized by: a probability of containment (*POC*), a detection function that relates the search effort density with the probability of detection (*POD*), a constrained amount of search effort, and an optimization criterion for the probability of success (*POS*). The solution of this basic problem of optimal search should provide to the search planner some idea of the quantity of search effort and the places where it should be spent in order to achieve its goals (Frost & Stone, 2001).

The *POS* is determined using Eq. (14)

$$POS = POC \times POD \tag{14}$$

The *POC* measures the possibility of the search target being within the limits of the search area and is related to the search area, which is usually calculated from a Multivariate Gaussian Distribution, normally centered at the origin of the coordinate system, (0,0) (Frost & Stone, 2001). The *POD* is a measure of sensor performance and describes the capability of the SRU in detecting and recognizing the target during the search operation, but it will only detect if the target is inside the designated search area. This probability is obtained by using Eq. (15).

$$POD = 1 - e^{-C} \tag{15}$$

$$with,\ C = \frac{Z}{Total\ area} \tag{16}$$

The parameter $C$ is designated as coverage factor and is a relative measure of how thoroughly an area has been searched. $Z$ is also known as search effort and is calculated using Eq. (17), where $z$ corresponds to the distance covered by the SRU in a straight line, also known as effort, and $W$ corresponds to the sweep width.

$$Z = z \times W \tag{17}$$

$$z = v \times t \tag{18}$$

The amount of effort expended for the search, calculated by Eq. (18), depends on the SRU velocity $v$ and it is limited by its time of endurance. The sweep width characterizes the average ability of the SRU in detecting a target under specific set of conditions such as, the characteristics of the sensor, which can be an electronic device or simply the human eye, the characteristics of the target, and the environmental conditions at the time of the search. Koopman (1946) defines the sweep width as Eq. (19).

$$W = \frac{Numb.\ of\ objects\ detected\ per\ unit\ time}{(Numb.\ of\ objects\ per\ unit\ area) \times v} \tag{19}$$

Another important concept in Search Theory is the lateral range, defined as the perpendicular distance between the search target and the searcher at the closest point of approach.

The sweep over the area to search is done in a methodical manner by following a search pattern, which is the itinerary scheme performed by the SRUs during the search operation to cover the established search area. This pattern is composed by search legs.

There are different types of search patterns, and it is the task of the SMC to choose the appropriate pattern for each search operation after defining the search area. To choose the pattern to use, the characteristics of the detection sensor in use, such as the maximum detection range, which is the maximum distance in which the sensor might detect the target, or the beam sighting distance, which corresponds to the distance in which the sensor can detect the target must be known. By summing the beam sighting distance from both sides of the SRU, the SMC obtain the value of the sweep width. Knowing this value, it allows him to determine the right distance for the track spacing ($S$), which corresponds to the distance between successive search legs in a search pattern and its value should correspond to the value of the sweep width to have a perfect coverage.

The search patterns will have different levels of coverage according to their pattern parameters described above and, consequently different values of *POD*. From all the existing patters, the most common ones for SAR operations at sea are the expansion square search (SS), the sector search (VS), the parallel sweep search (PS) and the creeping line search (CS).

## 3 SAR DECISION SUPPORT TOOL

The algorithm developed within the scope of this study took the work developed by Vettor and Guedes Soares (2015) in consideration, and is designed with the same outline as the DSS for SAR operations presented by Abi-Zeid et al. (2019). The SAR tool algorithm is implemented in *MATLAB*, and it is divided in three different modules: *DRIFT MODULE*, *SEARCH MODULE* and *EFFICIENCY EVALUATION MODULE*, as shown in Figure 1.

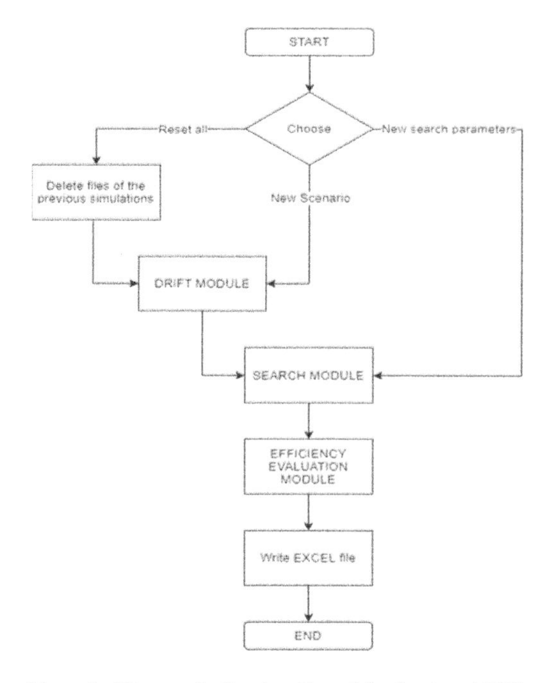


Figure 1. Diagram for the algorithm of the developed SAR tool.

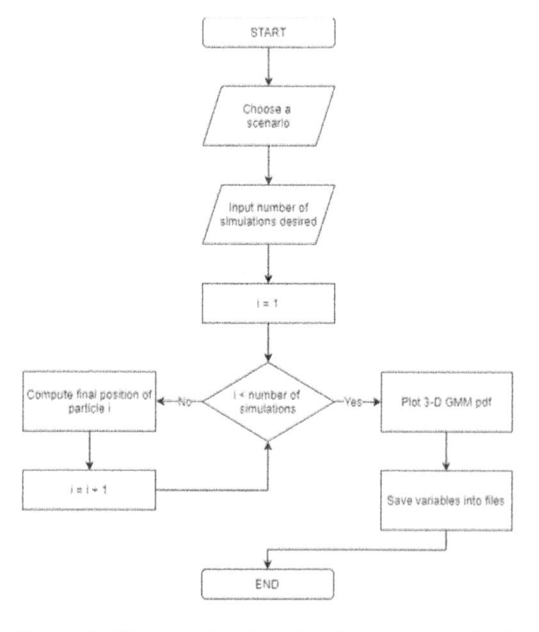


Figure 2. Diagram for the algorithm of the DRIFT MODULE.

### 3.1 *Drift model*

*DRIFT MODULE*, presented in Figure 2, starts by creating a SAR case containing all the information available related to the distress occurrence, the characteristics of the vessel, its LKP and other relevant details, according to the scenario defined by the user. Also, it asks the user to input the number of Monte Carlo Simulations (MCSs) desired. After having all the details for the drift simulation, the algorithm computes the final position of the target for each simulation. To compute the final position of each target, the model considers the uncertainties associated with the drift parameters, described by probability distributions, and uses the coefficients of the drift objects from Allen (2005).

The drift model will register the position of each floating objects randomly generated every hour within a period of time previously stablished. After the drift simulation of every floating object, their final location will be spread all over a certain area where it will be possible to distinguish one or more locations with a higher probability of occurrence, and when moving further away from those locations, the number of occurrences will decrease. Then those results are modelled and plotted into a three-dimensional Gaussian Mixture Model (GMM), and from there, *SEARCH MODULE* starts.

This model also takes into account the orientation of the floating object relatively to the downwind direction, and the possibility of jibing is considered in order to define the direction of the drift, making the simulation closer to the reality.

In the beginning, besides the number of MCSs desired, it also asks the user to choose a predefined scenario. The scenario considers a particular floating object, the time of the initial and final location of the object, the velocity and direction of the wind and current, the geographic coordinates of the initial position, and parameters to estimate the initial position in cartesian coordinates.

### 3.2 *Search model*

After obtaining the probability density map for the location of the missing object, which represents the total size of the search area obtained, the algorithm initiates *SEARCH MODULE*, represented in the diagram in Figure 3, by asking the user to input the values for each search parameters, such as the velocity of the SRU, the sweep width, the track spacing, and the confidence level, and again, the number of simulations desired, since the search model also uses MCS. Then, targets are randomly generated according to the number of simulation desired, the search area is reduced conforming to the confidence level inputted, and the model starts to simulate two different search patterns, according to the search parameters inputted by the SMC. The search patterns in focus are the PS and the SS, represented in Figure 4, and the sweep performed by both patterns follows the steps represented in the diagram from Figure 5.

During the sweep, the SRU moves in small distance increments defined by its velocity ($v$) and a given time step ($T$), inputted by the user, as presented in Equation (20). But, in reality, the SRU moves continuously and scans the search area at the same time, but for implementation purposes, the SRU is defined to scan its surroundings at every time step.

$$SRU\ distance\ increment = v * T \tag{20}$$

To simplify the algorithm, it is assumed that the sensor used is "perfect", meaning that its sweep width will correspond to its maximum detection distance and, it can certainly detect the drifting object within that range. The range is defined by a detection probability function which relates the probability in detecting the object and the distance between the object and the sensor (Zhang et al., 2016), as represented in Eq. (21), where $d$ represents that distance, and $\alpha$ and $\beta$ are model parameters related to search condition factors such as the weather conditions at the time of the search and the detection ability of the SRU.

$$P(d) = \begin{cases} 1 - e^{-\left(\frac{d-\alpha}{\beta}\right)^2}, & \alpha \leq x \\ 0 \quad , & x < \alpha \end{cases} \tag{21}$$

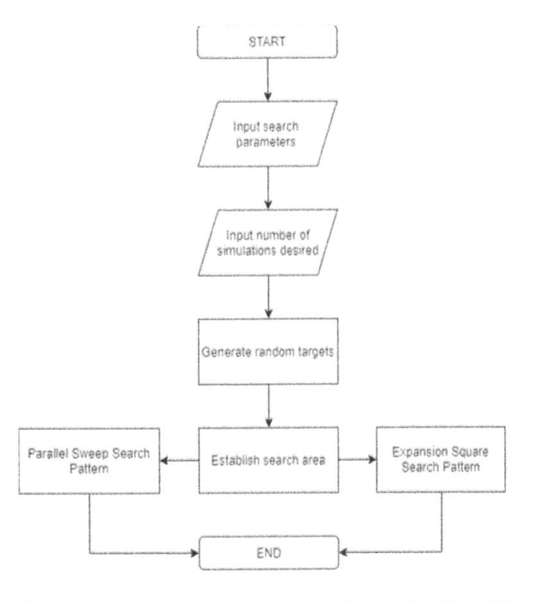


Figure 3. Diagram for the algorithm of *SEARCH MODULE*.

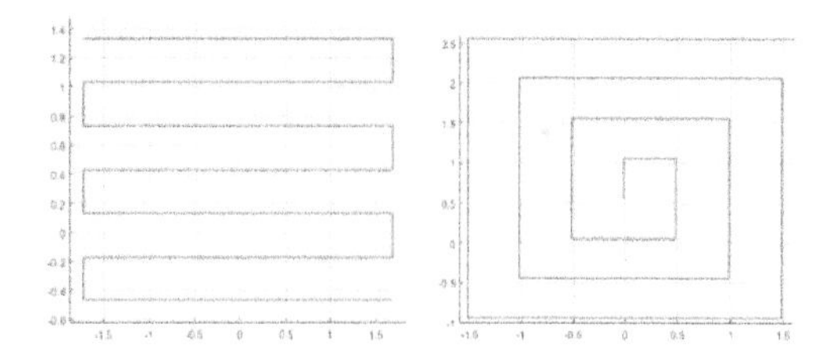

Figure 4. Search model's parallel sweep (PS) pattern (left side) and expansion square (SS) pattern (right side) considered in the search model.

According to the search parameters inputted by the user, the SRU sweeps the search area defined and if it detects the target, the algorithm registers the time to detect it and advances for the search simulation of another target. If not, the search simulation stops when reaching the end of the search area and immediately advances for the next one. This keeps continuing until the algorithm simulates the search for every random target generated.

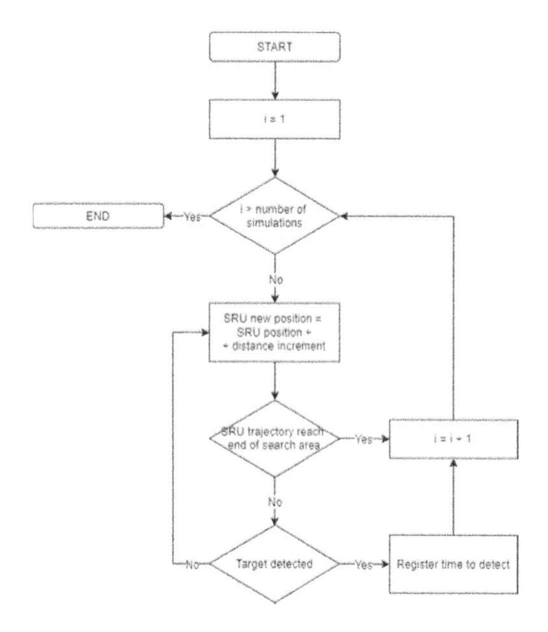


Figure 5. Diagram for both algorithms of the PS and SS patterns.

## 3.3 *Output results*

The SAR tool algorithm ends with the evaluation of the efficiency of the search simulation performed by using the PS pattern and the SS pattern. For each algorithm run, the algorithm checks how many targets are generated inside the defined search area, and how many targets are detected, and then computes the probabilities $P(In)$ and $P(Detect)$. With those both probabilities, the algorithm computes the $P(Detect \cap^{I} n)$, which corresponds to the *POS*, using Eq. (22).

$$P(Detect \cap^{I} n) = P(In) \times P(Detect) \quad (22)$$

Since the detection time is one of the variables to assess the efficiency of a search operation, the algorithm also computes the mean time of detection. Then, the algorithm finalizes by recording those results to facilitate their analysis, together with the search parameters inputted by the user.

## 4 RESULTS ANALYSIS AND DISCUSSION

To study the efficiency of SAR operations, the SAR decision support tool is used to simulate the predetermined scenarios. For each scenario, the search parameters are analysed to understand how they affect the *POS* and the time used to detect the missing target. In the present paper only the results of Scenario 1 are presented in detail.

In Scenario 1, after drifting for 5 hours in Portuguese waters, with winds at 25 knots pointing in the 165deg and currents at 0.1 knots pointing in the 30deg, the possible final location of the fishing vessel is given by the probability density distribution presented in Figure 6, obtained from the drift model for 2000 simulations.

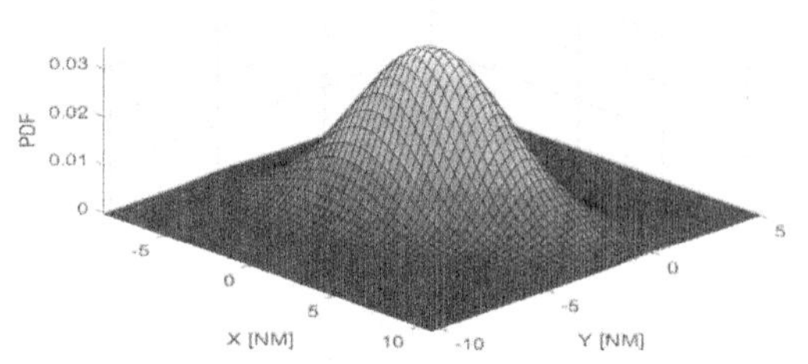


Figure 6. Probability density distribution for the final location of the target in Scenario 1.

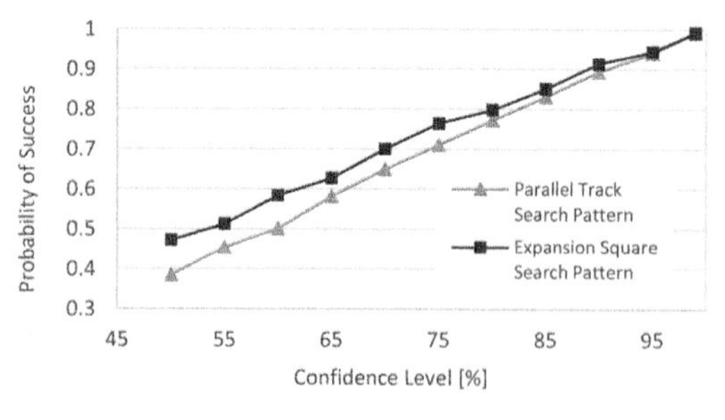


Figure 7. Probability of success of the SAR mission for the two trajectories, using different values of confidence level in Scenario 1.

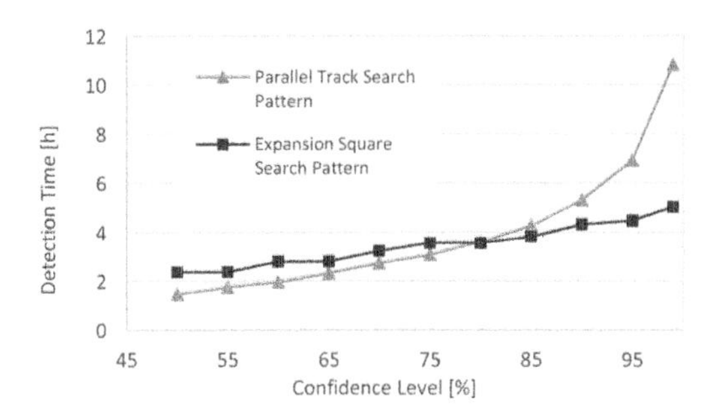


Figure 8. Time to detect the target for the two trajectories, using different values of confidence level in Scenario 1.

Figure 7 and Figure 8 present the evolution of the POS and the time to detect the missing object when varying the confidence level from 50% to 99%, for determining the search area, respectively. As expected, when increasing the confidence level, the POS and the detection time also increases. This increase is due to the size of the search area, when smaller is the confidence level value smaller is the search area. Since the POS is affected by the POC and the POD, lower values of POS mean that either one of those probabilities are low. In this case, it is the *POC* that affect the *POS*, since the number of targets randomly generated inside the designated search area increases with its expansion, see Figure 9.

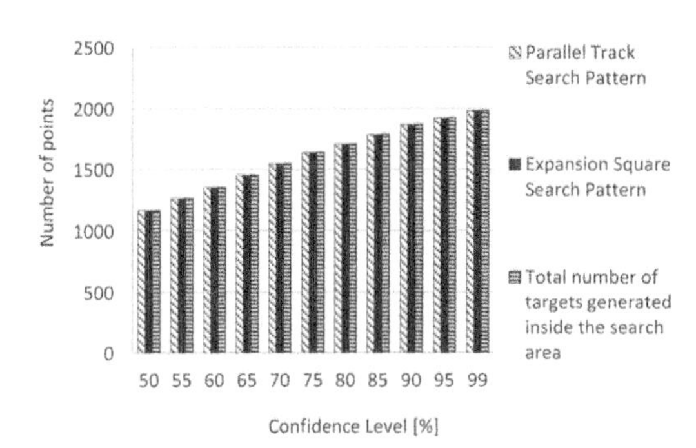


Figure 9. Number of targets detected inside the search area for the two trajectories, using different values of confidence level in Scenario 1.

When analysing the developments of the *POS* and the detection time by varying the track spacing from 250m to 2500m, presented in Figure 10 and Figure 11, it can be observed that the increase of the track spacing causes a decrease of those variables. This variation is mostly because the capability of the SRU in detect also decreases. Since the sweep width is equal to 1000m, the value for track spacing should also be 1000m for a perfect coverage of the search area. Surpassing that value, the number of targets detected start to decrease and consequently the *POS* also decreases. In case of the time to detect, see Figure 11, it decreases since the space between two search legs become larger and therefore, fewer search legs are needed to compose the search trajectory.

In Figure 12 and Figure 13, a variation of the sweep width from 500m to 2000m is considered. As expected, the *POS* increases with the increase of detection range, but when it reaches until a certain point, in this case the 1000m, the *POS* starts to increase very slowly when surpassing the perfect coverage mark, see Figure 12. Also, as expected the detection time decreases since the detection range gets bigger, see Figure 13.

An elasticity analysis of the *POS* and the detection time is performed with the objective of identifying which search parameter has the most influence over them when increasing 10% each one at the time.

When analysing Figure 15 and Figure 14, it is possible to observe that, the confidence level is the search parameter that vary the most for both variables when performing the PS pattern, surpassing the elasticity factor equal to 1. The other parameters have a very little impact or zero impact, except for the track spacing that causes some variation on the detection time, but not significant as the confidence level.

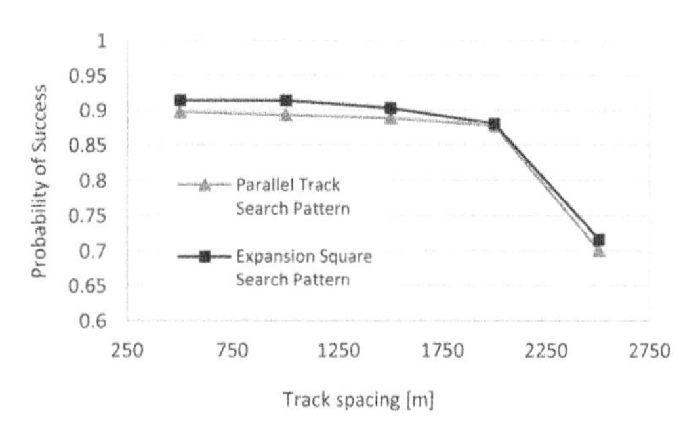


Figure 10. Probability of success of the SAR mission for the two trajectories, using different values of track spacing in Scenario 1.

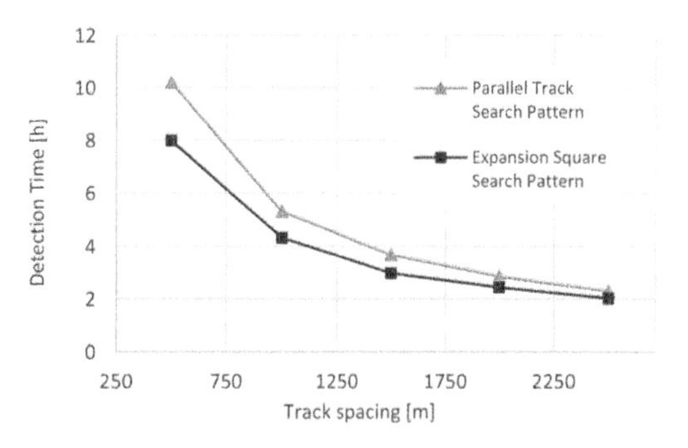


Figure 11. Time to detect the target for the two trajectories, using different values of track spacing in Scenario 1.

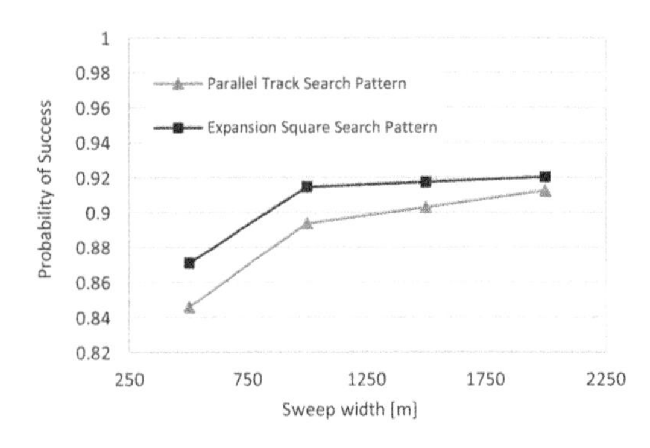


Figure 12. Probability of success of the SAR mission for the two trajectories, using different values of sweep width in Scenario 1.

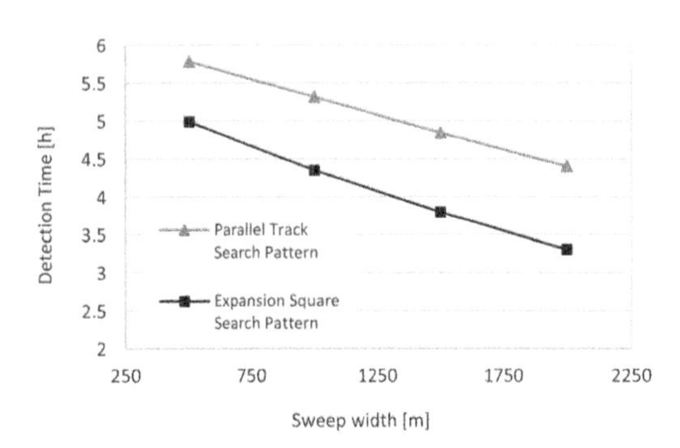


Figure 13. Time to detect the target for the two trajectories, using different values of sweep width in Scenario 1.

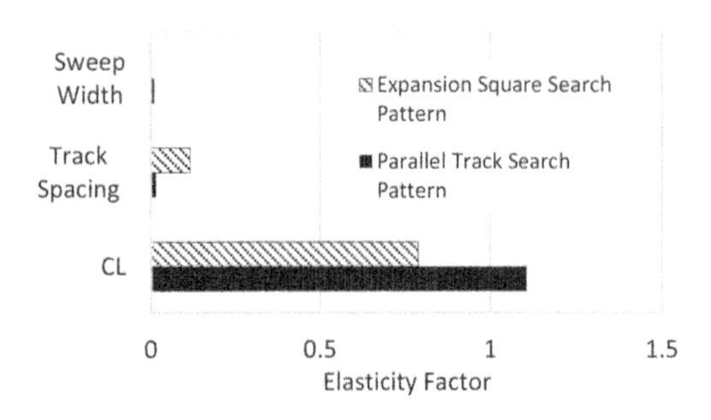


Figure 14. Elasticity analysis of the probability of success in Scenario 1, when increasing each search parameter by 10%.

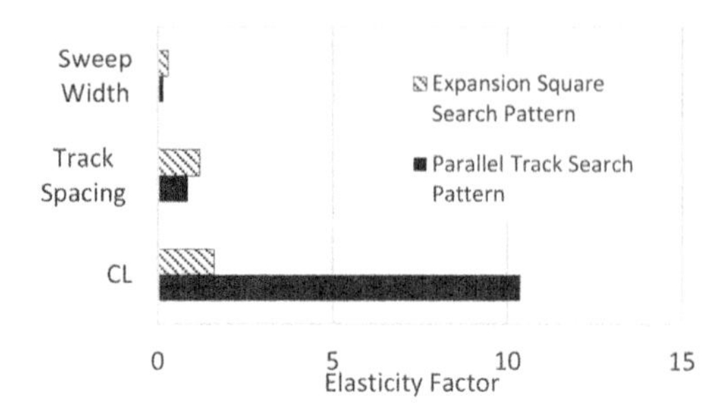


Figure 15. Elasticity analysis of the detection time in Scenario 1, when increasing each search parameter by 10%.

## 5 CONCLUSIONS

In this paper SAR operations are simulated using a developed SAR planning tool. The efficiency of each search operation is measured through the Probability of Success, which represents the capability of the SRU in detect the target, and the time need to detect it.

It is shown that the Probability of Success is affected by the weather conditions at the time of the mission, the size of the search area which depends on the sea state conditions, the characteristics of the target and the characteristics of the SRU, such as the detection range and the track spacing. The detection time is influenced by the location of the target and the SRU detection range. Increasing the track spacing results in the decrease of the Probability of Success at one point and the decrease of the detection time since the SRU covers the search area much faster. The increase of the sweep width leads an increase of the Probability of Success until a certain level because the Probability of Detection will increase slowly after that same level with the increase of search effort which leads to a decrease on the detection time due to the detection range increase. For every simulation, two search patterns are tested, the parallel sweep pattern and the expansion square pattern. When balancing the Probability of Success and the time to detect obtained in every simulation, it is possible to conclude that the expansion square pattern exceeds the parallel sweep pattern in performance. Also, the elasticity analyses performed on the two efficiency measure variables conclude that the parameter that influence the most in both variables is the confidence level.

## ACKNOWLEDGEMENTS

The paper contributes to the project "Monitoring and Surveillance of Maritime Traffic off the continental coast of Portugal (MoniTraffic)", funded by the

"Fundo Azul" programme of the Portuguese Directorate-General for Maritime Policy (Direção-Geral de Política do Mar, DGPM), under the contract FA-04–2017–005.

## REFERENCES

Abi-Zeid, I., Morin, M., & Nilo, O. (2019). Decision support for planning maritime search and rescue operations in Canada. *ICEIS 2019 - Proceedings of the 21st International Conference on Enterprise Information Systems*, *1*, 316–327.

Allen, A. A. (2005). Leeway divergence. *Technical Report No. CG-D-05-05*, U.S. Coast Guard Research and Development Center.

Allen, A. A., & Plourde, J. V. (1999). Review of Leeway: Field Experiments and Implementation. *Technical Report No. CG-D-08-99*, U.S. Coast Guard Research and Development Center.

Breivik, Ø. (2008). *Leeway model documentation v2.5*. Available at: http://www.ifremer.fr/Sar-Drift/Exchange-Repository/Leeway_documentation_v2.5.pdf.

Breivik, Ø., & Allen, A. A. (2008). An operational search and rescue model for the Norwegian Sea and the North Sea. *Journal of Marine Systems*, *69*(1–2), 99–113.

Breivik, Ø., Allen, A. A., Maisondieu, C., & Olagnon, M. (2013). Advances in search and rescue at sea. *Ocean Dynamics*, 63(1), 83–88.

Fitzgerald, B. R., Finlayson, D., Cross, J. F., & Allen, A. A. (1993). Drift of Common Search and Rescue Objects, Phase II. Contract report for Transportation Development Centre, Transport Canada, Montreal.

Frost, J. R. (1996). *The theory of search: A Simplified Explanation*. Annandale, VA, Soza & Co. Ltd. and Office of Search and Rescue, US Coast Guard.

Frost, J. R., & Stone, L. D. (2001). Review of Search Theory: Advances and Applications to Search and Rescue Decision Support. *Technical Report No. CG-D-15-01*, U.S. Coast Guard Research and Development Center.

Koopman, B. O. (1946). Search and Screening. *Operations Evaluation Group Report No. 56*. Operations Evaluation Group. Office of the Chief of Naval Operations. US Navy Department.

Kratzke, T. M., Stone, L. D., & Frost, J. R. (2010). Search and Rescue Optimal Planning System. *13th Conference on Information Fusion*. Edinburgh, UK, 26–29 July 2010.

Ni, Z., Qiu, Z., & Su, T. C. (2010). On predicting boat drift for search and rescue. *Ocean Engineering*, 37, 1169–1179.

Vettor, R., & Guedes Soares, C. (2015). Computational system for planning search and rescue operations at sea. *Procedia Computer Science*, 51, 2848–2853.

Zhang, J.F., Teixeira, A.P., Guedes Soares, C. and Yan, X.P. (2016), Study on path planning strategies for search and rescue, *Maritime Technology and Engineering 3*, Guedes Soares, C. & Santos T. A., (Eds.), Taylor & Francis Group, London, UK, pp. 937–942.

Zhang, J., Teixeira, A. P., Guedes Soares, C., & Yan, X. (2017). Probabilistic modelling of the drifting trajectory of an object under the effect of wind and current for maritime search and rescue. *Ocean Engineering*, 129, 253–264.

*Trends in Maritime Technology and Engineering – Guedes Soares & Santos (eds)*


# Influence of fire-fighting intervention on fire spread characteristics in ship engine room

C.F. Li, J.Y. Mao, Z.X. Kang, S.Z. Zhao & H.L. Ren
*College of Shipbuilding Engineering, Harbin Engineering University, Harbin, China*
*HEU-UL International Joint Laboratory of Naval Architecture and Offshore Technology, Harbin, China*

ABSTRACT: The purpose of this paper is to investigate and evaluate the ship engine room fire spreading characteristics basis on the effect of fire-fighting interventions. The Fire Dynamics Simulator (FDS) was employed to simulate fire scene in engine room and the fire spreading characteristics were obtained by varying ventilation conditions and parameters of water sprinkler. The results show that the pool fire can be extinguished easier with the increasing of spray speed, which is due to the water mist with higher spray speed have larger spray area. Meanwhile, the temperature in engine room would fall down more rapidly with the increasing spray area due to the higher evaporation rate. Besides, the arrangement of ventilation conditions has obvious effect on fire spreading characteristics, which is due to the oxygen exhausted leads to the lack of combustion medium.

## 1 INTRODUCTION

The engine room fire is one of the main types of ship accidents. The engine room has a high probability of fire due to the dense arrangement of oil piping, numerous combustible materials and restricted space and narrow passages. Once a fire occurs, it spreads quickly and is difficult to put out, and the safety of people and property is seriously threatened. The usual design of sprinkler systems is generally based on the requirements of the fire code. The movement of smoke after the action of water sprinklers under different ventilation conditions is rarely taken into account.

Once a fire occurs in a ship, it is necessary to consider the effectiveness of the fire extinguishing medium interacting with the fire according to combustion properties of fire source, and then it should take measures such as launching automated water sprinklers or releasing carbon dioxide extinguishing agent or sealing the cabin vents to ensure the safety of personnel and property. Among them, automated water sprinklers are widely used in various fields due to their many advantages such as high efficiency, stability, low cost, sustainability and environmental-friendliness. At present, the further research on the pool fire in closed space based on fluid dynamics technology and fire source combustion characteristics have been studied by many scholars. The use of water sprinkler systems in enclosed spaces has been studied by many scholars since the 1960s, and the research in this area was mainly experimental. Cooper (1995) was the first to investigate the mechanism of smoke movement under sprinkler action and proposed an empirical formulation of the sprinkler model. Ingason (1992) of the Swedish National Testing and Research Institute conducted a large number of full-scale experiments to study the interaction between sprinkler and smoke exhaust systems under different fire environmental conditions by varying the fire growth and smoke exhaust methods. Then McGrattan et al (1998) of the National Institute of Standards and Technology NIST conducted several full-scale experiments to explore the interaction of water sprinklers, smoke extraction and other measures in similar large warehouses and plants in the event of a fire, and refined its numerical calculation software FDS based on the experimental results. Braun et al (1992) constructed diesel and polyethylene pellet fire experiments at different heat release powers to analyze and evaluate the effectiveness of smoke extraction systems by comparing the FDS numerical simulation results with the data obtained in four large shipboard fire tests. The results show that the environmental parameters of the compartment corridor close to the fire source can be brought to survival conditions by isolating the fire compartment and adjacent space ventilation.

The further research on the fire spread characteristics under the intervention of fire protection system had been studied by Lin et al (2019), Liu et al (2020) and Lee et al (2018).The interaction between fine water mist and fire smoke were studied by Lee (2008) and Fang (2006) who worked in the University of Science and Technology of China. Feng et al (2018) built a full-size naval habitation compartment and corridor test platform to study the fire extinguishing capability

DOI: 10.1201/9781003320289-24

and effectiveness of water sprinkler system in naval habitation compartment fires based on FDS simulation of real place scenario items and water sprinkler system settings. Fu et al (2003) used a large vortex simulation method to simulate the interaction problem between fire and water sprinklers behind the obstacle pool to explore the effectiveness of sprinklers on the control of fire behind the obstacle pool. The above studies used closed space models with simple structures, and also did not consider the influence of each parameter change of fire-fighting equipment on the fire spread characteristics, so they could not effectively study the real fire situation in the cabin area.

In order to investigate the influence of fire-fighting intervention on the spreading characteristics of ship cabin fire, fire simulation in the cabin area under fire-fighting intervention will be carried out based on fire dynamics simulator according to the characteristics of ship cabin arrangement. A theoretical basis for the fire-fighting system arrangement and the development of personnel evacuation plan for cabin fire is provided, by studying the temperature distribution law, smoke spreading characteristics of cabin fire under fire-fighting system intervention and different ventilation conditions by changing the main parameters of fire-fighting equipment and vent size.

## 2 BASIC PRINCIPLE OF FDS

The low Mach compressible flow in FDS can be described by Navier-Stokes equations, which has been widely accepted in the field of fluid dynamics. The N-S equations of compressible flow are as follows:

equation of mass conservation:

$$\frac{\partial \rho}{\partial t}+\frac{\partial \rho u_j}{\partial x_j}=0 \tag{1}$$

momentum-conservatione quation:

$$\frac{\partial \rho}{\partial t}+\frac{\partial \rho u_i u_j}{\partial x_j}=\rho f_{\mathrm{i}}-\frac{\partial p}{\partial x_i}+\frac{\partial t_{ji}}{\partial x_j} \tag{2}$$

energy-conservation equation:

$$\frac{\partial \rho h}{\partial t}+\frac{\partial \rho h u_j}{\partial x_j}=\frac{\partial p}{\partial t}+u_j\frac{\partial p}{\partial x_j}+\phi-\frac{\partial q_j}{\partial x_j}+\dot{q}_m \tag{3}$$

gaseous state equation:

$$pV=nRT \tag{4}$$

The mass conservation equation can also be written in the form of composition equation:

$$\frac{\partial \rho Y_{\mathrm{L}}}{\partial t}+\frac{\partial(\rho Y_L u_j)}{\partial x_j}=\frac{\partial}{\partial x_j}\left(\rho D_L\frac{\partial Y_L}{\partial x_j}\right)+\dot{m}_L \tag{5}$$

where $\rho$ is gas density; $u_j$ is gas density; $f_i$ is volume force; $p$ is pressure; $\tau_{\mathrm{ji}}$is shear stress; $h$ is enthalpy; $\phi$ is the dissipative function; $q_j$ is the heat flux for transmission and radiation; $\dot{q}_m$ is the heat release rate per unit volume of fuel; $n$ is the number of gas molecules per unit volume $R$ is a gas constant; $T$ is the temperature; $Y_L$ is the mass fraction of composition equation $L$; $D_L$ is the diffusion coefficient of composition equation $L$; $\dot{m}_L$ is the formation rate or consumption rate of composition equation L in the unit volume.

## 3 MATHEMATICAL MODEL OF SPRAY SYSTEM

### 3.1 *Triggering formula for water mist spray*

The general water spray system in fire dynamics simulator is triggered by temperature. Heskestad & Bill (1988) gave the trigger formula of the spray model:

$$\frac{\mathrm{d}T_L}{\mathrm{d}t}=\frac{\sqrt{|u|}}{\mathrm{RTI}}(T_G-T_L)-\frac{\mathrm{C}}{\mathrm{RTI}}(T_L-T_m)-\frac{\mathrm{C}_2}{\mathrm{RTI}}\beta|u| \tag{6}$$

where $u$ is the velocity vector of gas; $T_L$ is the junction temperature; $T_G$ is the temperature of the gas near the junction; $T_m$ is the nozzle temperature (set as ambient temperature); $\beta$ is the volume fraction of water in steam; RTI is the sensitivity of the detector; C is the heat transfer loss factor through the equipment; $\mathrm{C}_2$ is an empirical constant.

### 3.2 *Distribution of droplet size in water spray*

When water droplets are ejected from the nozzle, the original droplet size and velocity need to be determined in order to describe the trajectory line of the water droplets. The droplet size and distribution are determined by a random distribution function.

When the particle size of water mist is $d_r \leq d_m$, it can be described by a log-normal distribution, the distribution function is:

$$\mathrm{F}(d_r)=\int_0^{d_r}\frac{1}{\sigma d_r\sqrt{2\pi}}\exp\left[-\frac{\ln{(d_r/d_m)}^2}{2\sigma^2}\right]\mathrm{d}d_r \tag{7}$$

When the particle size of water mist is $d_r > d_m$, it can be described by the Rosin-Rammler distribution, the distribution function is:

$$F(d_r) = 1 - \exp\left[-0.693\left(\frac{d_r}{d_m}\right)^{\gamma}\right] \quad (8)$$

where $d_m$ is the average particle size of water mist; $d_r$ is the particle size of water mist; $\sigma$ and $\gamma$ are measured by experiments.

The average particle size of water mist is a variable determined by the outlet diameter, working pressure and geometric size of the nozzle. Yu (1985) gave a relational expression to determine the average particle size of water mist:

$$\frac{d_m}{D} \propto We^{-1/3} \quad (9)$$

where $D$ is the nozzle outlet diameter; $We$ is the Weber number. In the process of numerical simulation, not all droplets will be simulated, but a part of the sample is extracted to simulate. Generally, 5000 droplets per second per nozzle are simulated. The relationship between flow rate and droplet diameter can be calculated:

$$\dot{m}\delta t = C\sum_{i=1}^{N}\frac{4}{3}\pi\rho_w\left(\frac{d_i}{2}\right)^3 \quad (10)$$

where $\dot{m}$ is the mass flow rate of water leaving the nozzle; $\delta t$ is the time step; $N$ is the number of droplets calculated at each time step; $d_i$ is the particle size of water mist; C is the calculation constant.

## 4 ENGINE ROOM FIRE MODEL

The engine room has a high probability of fire because of the dense arrangement of oil piping, numerous combustible materials and restricted space and narrow passages. Once a fire occurs, it often causes serious consequences. Therefore, this paper will carry out the numerical simulation of the fire in the cabin area under the intervention of water mist fire, and establish the cabin model of the cabin area, including the main engine and related auxiliary equipment. The fire situation under different working conditions is simulated by changing the ventilation conditions and water spray device parameters (spray speed and the particle size of water mist). After that, the fire spread characteristics is obtained.

In order to truly simulate the fire spread, FDS is used to construct the fire scene model of the three-compartment with the engine room as the center. The length, width and height of the model respectively are 16m × 8m × 10m. The design parameters of the compartment are shown in Table 1. In order to analyze the influence of ventilation conditions on the smoke spread in the engine room under fire intervention, thermocouple is arranged at the engine room Z= 4 m, and a cap with free switch is set on the upper part of the engine room. The longitudinal and transverse sections of cabin are shown in Figure 1 and Figure 2. In addition, the bilge curvature is ignored and the double bottom structure are simplified modeled.

Table 1. Cabin design parameters.

| The parameter name | Numerical value |
|---|---|
| Deck materials | Q235 (8mm) |
| The wall material | Q235 (8mm) |
| Atmospheric pressure | 101325Pa (1atm) |
| The outside temperature | 20 °C |
| Humidity | 50% |
| Lower limit of oxygen mass fraction | 15% |

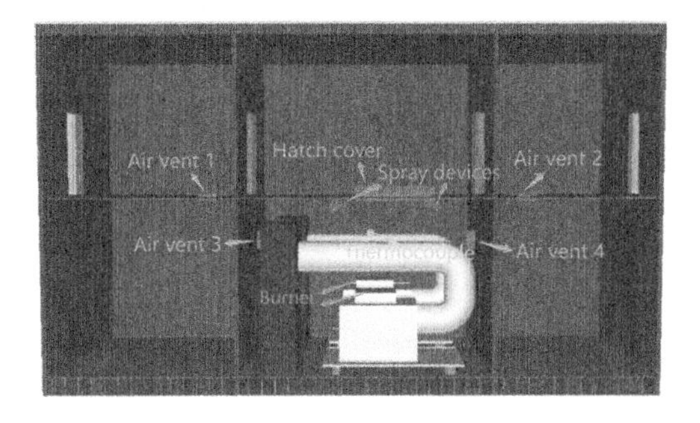


Figure 1. Longitudinal section of engine room cabin model.

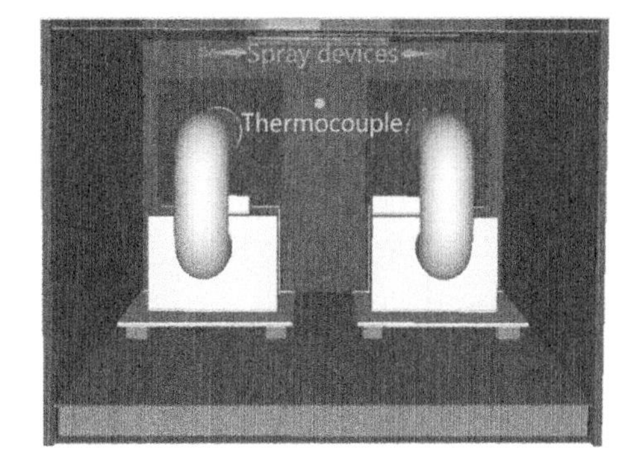


Figure 2. Transverse section of engine room segment model.

In order to study the fire spread characteristics of engine room with different parameters of spray system under different ventilation conditions. The comparative analysis is studied by changing the velocity, the particle size of water mist and the ventilation conditions on fire spread characteristics. The engine room of the ship belongs to the serious danger level II site according to the classification of the fire hazard level of the site in accordance with the design specification of automatic sprinkler system. In order to ensure that the drooping spray equipment reaches the standard fire coating area, the square layout method is shown in Figure 3.

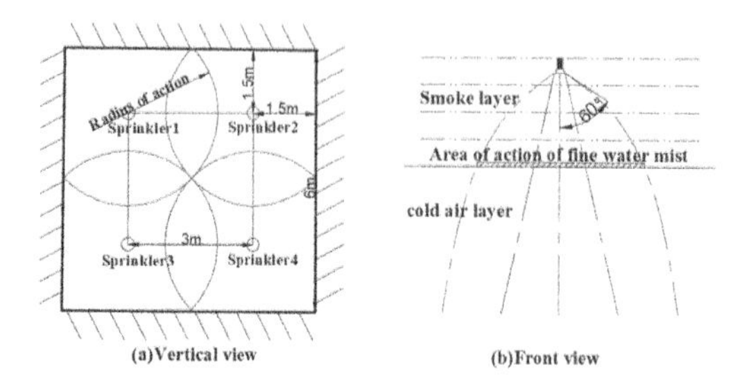


Figure 3. Schematic diagram of water mist interacts with smoke layer.

## 5 NUMERICAL SIMULATION OF THE CABIN FIRE UNDER FIRE INTERVENTION

### 5.1 *Temperature distribution of engine room fire after sealing*

Water mist parameters, which includes the spray flow, the spray speed and the particle size of water mist are important factors affecting the fire extinguishing effect. Firstly, the influence of different spray speeds on the fire extinguishing effect is studied by keeping the particle size of water mist maintains about 500μm and changing the spray speed of water mist under the condition of sealing the hatch cover. The parameter settings of nozzle are listed in Table 2.

Table 2. Design parameters of Nozzles.

| Parameter | value |
|---|---|
| Spray flow | 16 L/min |
| working pressure | 1.5 Mpa |
| Number of droplets per second | 5000 |
| Particle size of fine water mist | 500μm |
| Activation temperature of nozzle | 74°C |
| Minimum spray angle | 0° |
| Maximum spray angle | 60° |

The time history curve of flue gas temperature in engine room under closed hatch cover with different spray speed is shown in Figure 4. It can be found that the temperature decreases rapidly when the spray equipment is involved in t=120 s,, and a short boiling phenomenon occurs in the following few seconds (about 10 seconds), then the smoke layer temperature rises and fluctuate for a period of time (about 50 seconds)again. Subsequently, the temperature of the upper flue gas layer dropped sharply, and its rate gradually slows down. Due to the timely sealing operation by the crew, the fire extinguishing effect is better when the water mist spray speed is 20 m/s and 15 m/s. However, the fire extinguishing effect is invalid when the water mist spray speed is 5 m/s and 10 m/s. It is worth noting that the water mist spray speed of 10m/s is better in the later fire extinguishing stage.

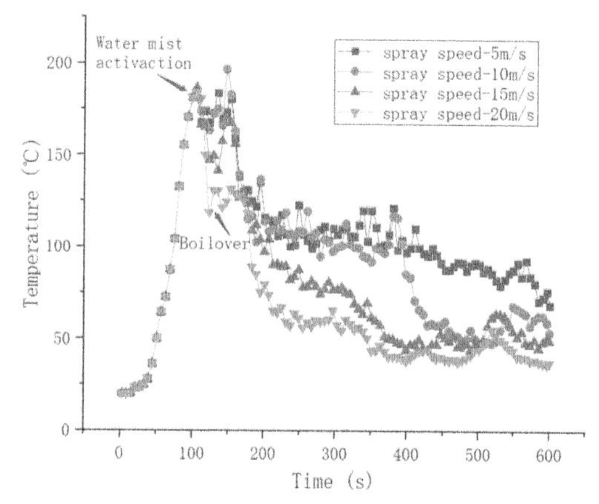


Figure 4. Time history curve of flue gas temperature in engine room under closed hatch cover with different spray speed.

The temperature slice map of engine room under closed hatch cover with different spray speed is shown in Figure5. It can be found that the smoke generated by the oil pool fire before t=150s is gathered in the upper part of the engine room without spreading through the vertical vent between the cabins. Due to the intervention of spraying equipment, the curtain formed by water mist spray can effectively inhibit the spread of smoke from oil pool fire, and the water mist with spraying speed of 20 m/s has obvious inhibitory effect in t=200s. Due to the intensified interaction between water mist and the buoyancy of thermal plume, the flue gas in the engine room first spreads to the adjacent cabin through the vertical vent after a while (about 100 seconds), and then affects the upper cabin through the vertical vent. The water mist with the spraying speed of 20 m/s effectively inhibited the spread of oil pool fire in the engine room in t=600 s. Due to the ventilation conditions of the upper cabin, the temperature maintains about 120 °C. The speed of water mist spray is 5m/s and 10m/s, which cannot effectively cover the smoke spread range. It can be concluded that when the flow rate and the particle size of water mist maintain a constant the fire suppression effect is the best with the water mist spray speed of 20 m/s. In the range of the spray speed from 5m/s to 20m/s, the effect of decreasing the fire temperature is more significant when the spraying speed increases faster. The reason is that the water mist with rising spray speed has more momentum, which is conducive to overcoming the flame thermal buoyancy and penetrating the thermal plume to cool effectively, and the increase of water mist flux per unit area is more likely to destroy the balance of oil mist combustion. However, due to violent gas exchange and air disturbances, smoke spreads through vents to various compartments but most of them are less affected.

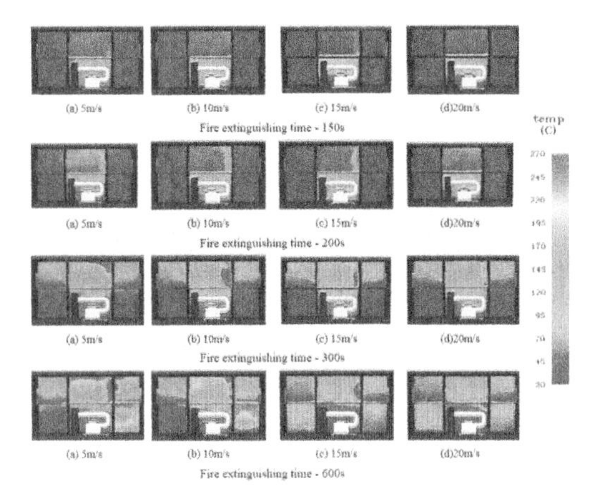

Figure 5. Temperature slice map of engine room under closed hatch cover with different spray speed (color code unit: °C).

## 5.2 *Effect of particle size of water mist after cabin sealing on cabin fire temperature*

The influence of water mist particle size on fire is further analyzed. The cabin fire is simulated when the cabin cover is closed after timely operation, and the optimal spray speed is maintained at 20 m/s. The influence of water mist particle size after cabin closure on cabin fire temperature is studied by changing the particle size of water mist. The parameter settings of nozzle are listed in Table 3.

Table 3. Design parameters of Nozzles.

| Parameter | value |
|---|---|
| Spray flow | 16 L/min |
| working pressure | 1.5 Mpa |
| Number of droplets per second | 5000 |
| Spray speed | 20m/s |
| Activation temperature of nozzle | 74°C |
| Minimum spray angle | 0° |
| Maximum spray angle | 60° |

The time history curve of flue gas temperature in engine room under closed engine room with different particle size of water mist is shown in Figure 6. It can be concluded that when the spray equipment is involved in the engine room fire in t=150s, the temperature of the engine room decreases rapidly after the spray equipment with the involvement of water mist particle size of 500 μm. Meanwhile, the temperature of the engine room maintains about 100°C at t=200 s, and then the temperature decreases slowly. The water mist can effectively inhibit the engine room oil pool fire. The fire extinguishing effect of spray equipment with the particle sizes of 2000μm and 3000μm is the most obvious within 50s after intervention, but the temperature slowed down after t=200s, and the water mist cannot effectively reduce the flue gas temperature in the engine room. It can be found that the water mist releases with the particle size of over 1000μm cannot directly act on the flame to extinguish the fire due to the obstacle effect. The larger particle size of water mist quickly settles and is wasted so that it is unable to participate in the fire extinguishing process leading to the decrease of fire extinguishing efficiency.

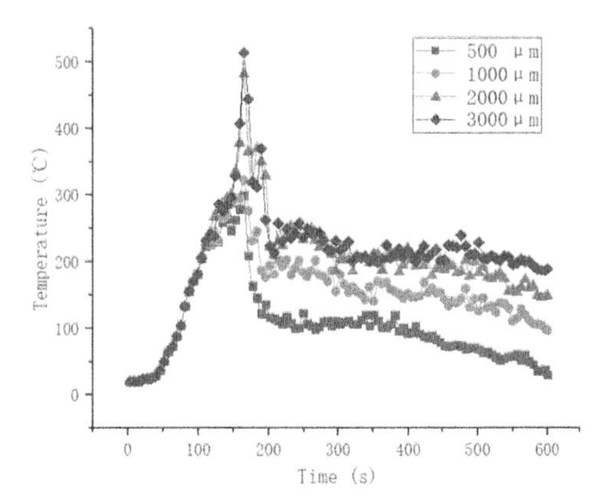


Figure 6. Time history curve of flue gas temperature in engine room under closed engine room with different particle size of water mist.

The Temperature slice map of engine room under closed hatch cover with different particle size of water mist is shown in Figure7. It can be found that due to the temperature of the flue gas layer above the engine room is effectively inhibited and increased in t=150 s, the inhibition effect of water mist particle size of 500μm is the best. The temperature of upper flue gas ascends obviously, and the maximum temperature reaches about 320°C at t=200 s. The interaction between water mist and flue gas in the engine room is intensified at 300 s, and then spreads through the vent. The spread order is engine room→ the right cabin of the engine room → the upper deck cabins. The spray equipment with water mist particle size of 500μm has the lowest temperature in the engine room, but the right cabin of the engine room is greatly affected by smoke at t=600s. The influence of flame radiation is ignored when the spray equipment with water mist particle size of 3000μm is involved, and the upper deck cabin is most affected by the smoke due to the limitation of ventilation conditions.

It can be concluded that the spray devices maintain the same flow rate and spray speed, and then change the particle size of the water mist. When the particle size of the water mist is changed into the smaller, it represents the larger relative surface area. It will make the upward thermal buoyancy driving effect more obvious. Meanwhile, it is easier to disperse and be sucked into the surface of the spray flame, and take more heat to achieve the cooling effect on the flame. At the same time, the smaller particle size of water mist can effectively force the deposition of smoke particles and inhibit the harm of smoke to ship personnel.

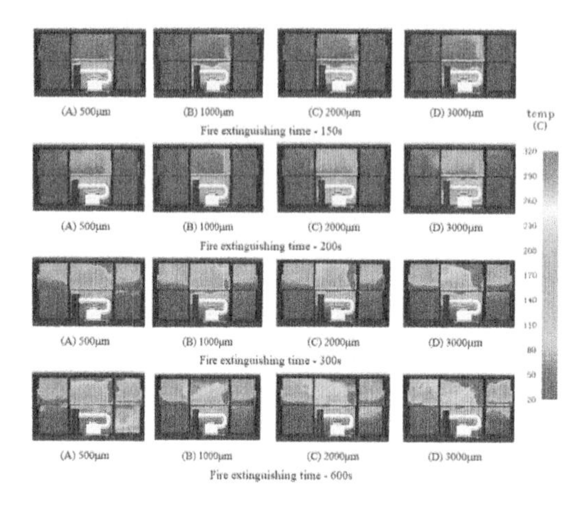


Figure 7. Temperature slice map of engine room under closed hatch cover with different particle size of water mist (color code unit: °C).

In order to further analyze the water mist fire extinguishing process, diagram of flue gas particles and water mist particle size of 500μm is shown in Figure 8. It can be found that in the fire extinguishing process, the water mist particles have good dispersion in the cabin and are greatly affected by the thermal buoyancy of the smoke, which can effectively reduce the temperature of the smoke layer in the upper cabin. A large number of water mist directly settles or accumulates in the lower space of the engine room after t=400s, and the water mist particles overcome the thermal buoyancy effect to break through the smoke layer, thereby reducing the oxygen concentration near the flame, significantly inhibiting the spread of the oil pool fire under the engine room, and achieving the effect of rapid fire suppression.

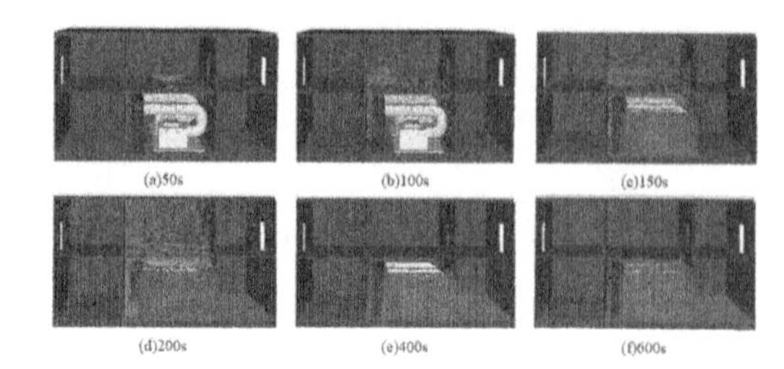


Figure 8. Diagram of flue gas particles and water mist particle size of 500μm.

### 5.3 *Comparative analysis of engine room fire temperature distribution under different ventilation conditions*

Taking into account the influence of ventilation conditions on the efficiency of water mist fire extinguishing, ventilation conditions are set to three most common hatch cover states. (1) The timely hatch cover sealing operation in the event of fire; (2) when the fire occurred, the cabin is not sealed in time, the hatch cover is at 15° (the projected area is used as the area of the enclosed cabin cover); (3) the hatch cover is completely opened, as shown in Figure 9. The spray speed of 20 m/s and the particle size of water mist of 500μm are set as the best water spray extinguishing effect. The parameter settings of nozzle are listed in Table 4.

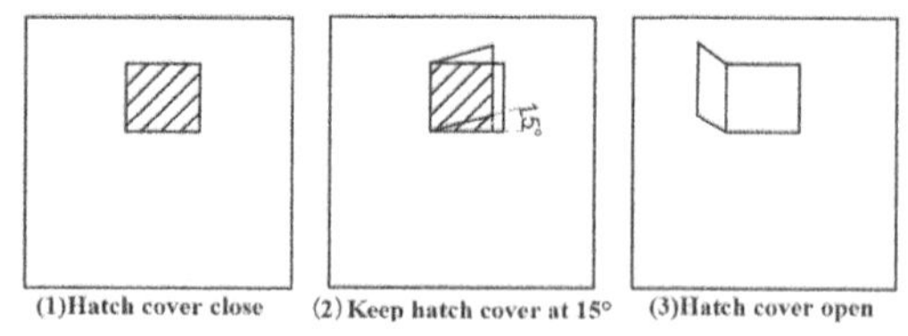


Figure 9. Schematic diagram of hatch cover.

The temperature slices of the engine room under different ventilation conditions is shown in Figure 10. It can be seen that the smoke rapidly spreads to the upper deck cabin when the hatch cover is completely opened in t=120 s. Subsequently, the water mist intervened in the fire process. When the hatch cover is completely opened at t=200 s, the rising process of the flame is in a whirlpool state. However, the large amount of smoke generating under the condition of cap closure is suppressed in the upper space of the cabin due to the limitation of ventilation conditions. When the hatch cover is opened at 15°, part of the oxygen escaped to the upper cabin with the interaction of water mist and flue gas at t=400 s leading to the fire spread. Due to the thermal buoyancy of the flame and air convection, the flue gas gradually diffuses to other cabins at t=600s. The upper deck cabin under the closed hatch cover is basically filled with smoke.

Table 4. Design parameters of Nozzles.

| Parameter | value |
|---|---|
| Spray flow | 16 L/min |
| working pressure | 1.5 Mpa |
| Number of droplets per second | 5000 |
| Spray speed | 20m/s |
| Activation temperature of nozzle | 74°C |
| Minimum spray angle | 0° |
| Maximum spray angle | 60° |
| Particle size of fine water mist | 500μm |

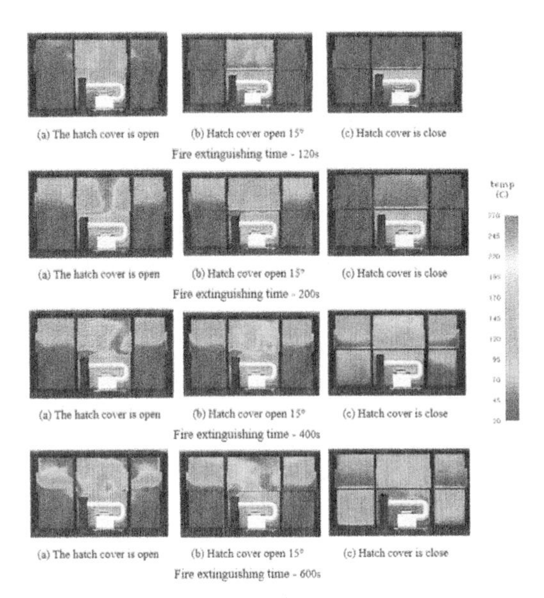


Figure 10. Temperature section under different ventilation conditions.

The diagram of Maximum temperature distribution of engine room under different ventilation conditions is shown in Figure 11. It can be found that when the spray equipment is involved in the pool fire, the flashover phenomenon will occur when the ventilation area reaches a certain value, resulting in a second sharp rise in temperature. It can be concluded that the oxygen asphyxiation has obvious effect on extinguishing oil fires, and it is necessary to seal the hatch cover first when the cabin fire occurs. The closed hatch cover forces the smoke to gather in the cabin, and the water mist can squeeze and isolate the oxygen around the flame. At the same time, the closed hatch cover separates the combustion medium, so the fire extinguishing efficiency of the spray equipment is the most effective. However, due to increased gas exchange and air disturbance, smoke spreads through vents to various compartments but most of them has little impact.

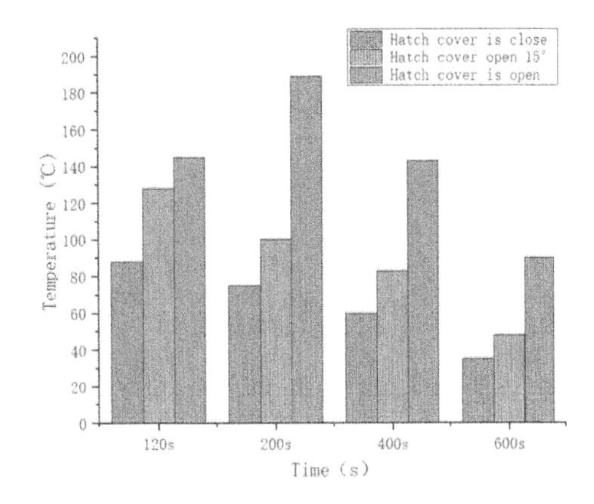


Figure 11. Maximum temperature distribution of engine room under different ventilation conditions.

## 6 CONCLUSIONS

In this paper, the fire extinguishing situation of spray equipment for rapid suppression of pool fire in engine room under different ventilation conditions is studied, and the influence of particle size of water mist on fire extinguishing efficiency is further analyzed. The conclusions are as follows:

(1) Due to the limitation of spray flow on ships, when the spray speed under the same ventilation conditions is faster, the contact area between water mist and thermal plume will be larger, which is more conducive to the absorption of heat and the pool fire is more likely to be extinguished. When the hatch cover is closed, it will be suitable to choose the nozzle with flow rate of 16L/min and spray speed of 20 m/s to extinguish the fire caused by diesel combustion in the marine engine room.

(2) The oxygen suffocation caused by the rapid spraying of a large number of water mist in a closed space has a significant effect on the extinguishing of oil pool fires. When a fire occurs in the engine room, it is necessary to seal the cabin first and use water mist under such a closed cabin space to achieve rapid fire extinguishing. However, when the hatch cover is open and the hatch cover is kept at 15°, the fire extinguishing efficiency of water mist is greatly reduced, and sprinklers with spray speeds of 5 m/s and 10 m/s will be accompanied by boiling phenomenon which leads to the rising of temperature and reducing the fire extinguishing efficiency.

(3) When the spray equipment is involved in the pool fire, the flashover will occur when the ventilation area reaches a certain value, resulting in a second sharp rise in temperature.

(4) Under the premise of constant spray flow rate and spray speed, the water mist of smaller particle size will lead to the larger relative surface area, and the

upward thermal buoyancy driving effect is obvious. It is easier to disperse and be sucked into the surface of the spray flame, then realizing the cooling effect of the flame. When the spray speed is 15m/s, the fire extinguishing efficiency is the highest when the particle size is 500μm. However, when the particle size is greater than or equal to 1000μm, due to the fact that the pool fire is hidden under the oil pipeline, the fire extinguishing effect is obviously deteriorated and even reaches the fire extinguishing failure state due to the decrease of the water mist flux mainly relying on the flame entrainment and dispersion to the flame surface.

## ACKNOWLEDGEMENTS

This research was supported by the Science Fund Project of Heilongjiang Province (LH2020E078) and the National Natural Science Foundation of China (52171305).

## REFERENCES

Braun, E., Lowe, D. L., Jones, W. W., Tatem, P., Carey, R., & Bailey, J. 1992. Comparison of full scale fire tests and a computer fire model of several smoke ejection experiments. *NASA STI/Recon Technical Report N, 93*, 23280.

Cooper, L.Y. 1995. The interaction of an isolated sprinkler spray and a two-layer compartment fire environment. *International Journal of Heat and Mass Transfer*, 38 (4),679–690.

Cooper, L.Y. 2000. Simulating the opening of fusible-link-actuated fire vents. *Fire safety journal*, 34(3),219–255.

Fang, Y.D. 2006. Study on interaction of water mist with fire smoke. University of Science and Technology of China, Hefei. 52–61. (*in Chinese*)

Feng, T.H., Liu, B.Y., Zhang, H.S. 2018. Experimental study on fire in naval cabin under water spray system. *Fire Technology and Product Information*, 31(9): 15–18. (*in Chinese*)

Fu, C.L, Liu, N.S, Lu X.Y. 2003. Numerical Simulation of the Interaction between Pool Fire behind an Obstruction and Water Spray, *Fire Safety Science*, 12(3): 121–129. (*in Chinese*)

Heskestad, G., & Bill Jr, R. G. 1988. Quantification of thermal responsiveness of automatic sprinklers including conduction effects. *Fire Safety Journal, 14*(1-2), 113–125.

Ingason, H. 1992. *Interaction between sprinklers and fire vents. Full scale experiments. Brandforsk project* 406–902.

Lin, Z., Bu, R.W., Zhao, J.M. 2019. Numerical investigation on fire-extinguishing performance using pulsed water mist in open and confined spaces. *Case Studies in Thermal Engineering*, 13: 100402. (*in Chinese*)

Liu, H.R. Wang, C., De cachinho cordeiro, I.M.C. 2020. Critical assessment on operating water droplet sizes for fire sprinkler and water mist systems. *Journal of Building Engineering*, 28: 100999. (*in Chinese*)

Lee, D. H., Paik, J. K., & Seo, J. K. 2018. Efficient water deluge nozzles arrangement on offshore installations for the suppression of pool fires. *Ocean Engineering, 167*, 293–309.

Lee, K.Y. 2008. Studies on stable characters of smoke layer un-der sprinkler spray in fires. University of Science and Technology of China, Hefei.: 40–45. (*in Chinese*)

McGrattan, K. B., Hamins, A., & Stroup, D. W. 1998. *Sprinkler, smoke & heat vent, draft curtain interaction: large scale experiments and model development* (p. 101). NIST.

Yu, H.Z. 1986. Investigation of spray patterns of selected sprinklers with the FMRC drop size measuring system. In First International Symposium on Fire Safety Science (pp. 1165–1176).

*Trends in Maritime Technology and Engineering – Guedes Soares & Santos (eds)*


# The fire risk assessment of ship power system under engine room fire

C.F. Li, H.Y. Zhang, Y. Zhang & J.C. Kang
*College of Shipbuilding Engineering, Harbin Engineering University, Harbin, China*
*HEU-UL International Joint Laboratory of Naval Architecture and Offshore Technology, Harbin, China*

ABSTRACT: This study proposes a risk assessment method for engine room fire safety evaluation based on the Expert Comprehensive Evaluation (ECE) method combined with the Fuzzy Fault Tree Analysis (FFTA). The composition of main engine system in the engine room and the failure logic of each subsystem were analyzed, the fuzzy fault tree of the ship engine room fire was constructed, and the system failure probability and the importance of basic events were calculated, the risk assessment of ship engine room fire was carried out. The results demonstrate that the contact of the cable insulation layer and the combustible materials around the piping system with the high temperature hot wall is the main form of fire, and the fuel oil system of the main engine is the area with high probability of fire. The proposed approach eliminates the impact of incomplete statistical data in the risk assessment process, improving the accuracy and credibility of the results.

## 1 INTRODUCTION

Fire accidents would seriously threaten the safety of ship navigation. The fires in engine room account for up to two-thirds of ship fire accidents, which are particularly difficult to tackle and will directly destroy power system operation. In recent years, ship fire accidents have occurred frequently. In 2017 alone, there were dozens of accidents caused by ship engine room fires. According to IMO (2019), of the 1,400 incidents reported between 2000 and 2017, 270 were fire and explosion-related, or 19.2% of reported incidents, while engine room fires accounted for more than 75% of all ship fires (IMO, 2019). The engine room space of the ship is small, the pipelines are crisscrossed, the fire load is large, the ignition point is hidden, the fire spreads quickly and it is difficult to extinguish. The engine room is the power center of the ship. Once a fire occurs, it is very likely to cause the ship to lose power, or even cause the ship to be damaged and sunk, which seriously threatens the navigation safety of the ship. Therefore, it is of great significance to carry out fire risk assessment of ship power system under the action of ship engine room fire.

In the process of risk assessment, the incompleteness of calculation data is a key factor affecting the accuracy and credibility of assessment results. The expert comprehensive evaluation method is based on the fuzzy set theory, which converts the expert's fuzzy language judgment of events into fuzzy numbers for quantitative calculation, and then obtains the failure probability of the system. Kaptan (2021) adopted the Bow-Tie method based on fuzzy set theory to convert the linguistic variables of experts into quantitative calculation data suitable for the fuzzy logic method, and analyzed the risk of anchor loss in anchorage operations. In terms of fire risk assessment methods in ship engine room: Huang et al (2009) used the fuzzy comprehensive evaluation method to establish a fire risk assessment model for engine room based on fuzzy recognition mode, and took an oil slick recovery ship as an example to assess the fire risk of its engine room. Liu et al (2010) proposed a Bayesian network-based fire risk analysis method for ship engine room, established a Bayesian network model for ship engine room fire, and carried out probabilistic reasoning on risk factors. However, none of the above methods can give quantitative results of system damage. Quantitative evaluation methods commonly used in the field of marine engineering include: Analytic Hierarchy Process (Yang et al, 2017), fault tree method (Pan, 2018), numerical simulation method (Xu, 2016). Sarıalioglu et al (2020) used the HFACS method to classify the contributing factors of engine room fires into a hierarchical structure. The possible accident scenarios and probabilities are calculated using the FFTA method. The cause of the accident was studied, and suggestions for preventing similar accidents in the future were put forward. Among them, the fuzzy fault tree method is based on the fault tree theory in reliability analysis, based on probability theory and Boolean algebra, and uses a logical block diagram to express the relationship between system components. It is an effective tool for quantitative assessment of system fire risk (Lavasani et al, 2015). The fire fault tree model of the ship power system is established by the fuzzy fault tree method, and the fuzzy probability of the system bottom event is obtained by the comprehensive

DOI: 10.1201/9781003320289-25

evaluation method of experts. On the one hand, the ambiguity of evaluation language and cognitive uncertainty in the fire risk assessment of ship power system can be solved. On the other hand, it overcomes the problem of lack of statistical data foundation in fuzzy fault tree operation. Compared with FTA, ETA, BBN, AHP, Delphi and other methods, these are good methods for reliability assessment, but when applied to actual engineering project assessment, they will be limited by factors such as incomplete data, so this is the advantage of the method proposed in this paper, and it is a method worthy of study for fire risk assessment of ship power system.

In order to reasonably evaluate the fire risk of the ship power system under the action of the ship engine room fire, considering the ambiguity and uncertainty of equipment damage in the system, this paper will use the expert comprehensive evaluation combined with the fuzzy fault tree method (ECE-FFTA) method to analyze the ship power system. The failure probability of the system and each subsystem, through the order of the importance of the basic events in the system, to find the weak parts of the system, and to study and establish the fire risk assessment model of the ship power system.

## 2 METHODS

Based on the ECE-FFTA method, this paper proposes a fire risk assessment method for ship power system. This method uses fuzzy language to invite experts to judge the basic events of ship power system, and input the aggregated fuzzy numbers after normalization of opinions into the fault tree, the aggregated fuzzy number of the upper system is solved.

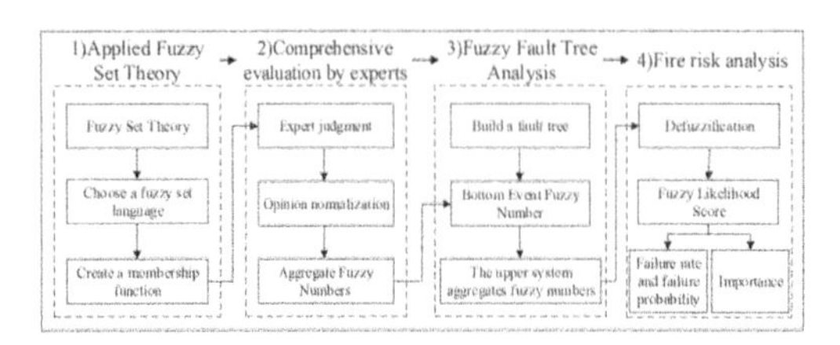


Figure 1. Calculation flow chart based on FFTA-ECE evaluation method.

After defuzzification, the fuzzy probability of the top event and subsystem can be obtained. Finally, the vulnerability of the ship power system can be evaluated and analyzed by the annual fire frequencies and importance. This method can be used to deal with the problems of uncertainty and ambiguity in the research object, and solve the problem of lack of bottom event probability in the fault tree analysis method. The specific flow chart is shown in Figure 1.

### 2.1 *Fuzzy set theory*

The fuzzy judgment language is introduced from the fuzzy set theory as the research object, and the appropriate membership function is established. After the operation and transformation, the research object is quantitatively calculated. A fuzzy set is defined as:

$$A = \{(x, \mu_A(x)) | x \in U, \mu_A \in [0, 1]\} \quad (1)$$

where, $A$ is the fuzzy set belonging to $U$; $\mu_A$ is the membership function of the fuzzy set $A$; $\mu_A(x)$ represents the membership degree of any element $x$ in the fuzzy set.

(1) Establish a membership function

The reasonable selection of the membership function will determine the accuracy of the evaluation results (Tang et al, 2016). The use of triangular and trapezoidal fuzzy numbers to establish membership functions can better express the possibility and degree of certainty of the event. These two classes of fuzzy numbers can more effectively map fuzzy languages into fuzzy membership functions (Figure 2). Its expression is as follows:

Membership function for very low risk states:

$$f_{VL} = \begin{cases} 1 & , \ 1.142E-4 \leq x \leq 1.245E-4 \\ \frac{-x+1.370E-4}{0.125E-4} & , \ 1.245E-4 < x \leq 1.370E-4 \\ 0 & , \ otherwise \end{cases} \quad (2)$$

Membership function for low-risk states:

$$f_L = \begin{cases} \frac{x-1.245E-4}{0.125E-4} & , \ 1.245E-4 \leq x \leq 1.370E-4 \\ \frac{-x+1.522E-4}{0.152E-4} & , \ 1.370E-4 < x \leq 1.522E-4 \\ 0 & , \ otherwise \end{cases} \quad (3)$$

Membership function for middle low risk states:

$$f_{ML} = \begin{cases} \frac{x-1.370E-4}{0.152E-4} & , \ 1.370E-4 \leq x \leq 1.522E-4 \\ 1 & , \ 1.522E-4 < x \leq 1.712E-4 \\ \frac{-x+1.957E-4}{0.245E-4} & , \ 1.712E-4 < x \leq 1.957E-4 \\ 0 & , \ otherwise \end{cases} \quad (4)$$

Membership function for a medium risk state:

$$f_{ML} = \begin{cases} \frac{x-1.712E-4}{0.245E-4} & , \ 1.712E-4 \leq x \leq 1.957E-4 \\ \frac{-x+2.283E-4}{0.326E-4} & , \ 1.957E-4 < x \leq 2.283E-4 \\ 0 & , \ otherwise \end{cases} \quad (5)$$

Membership function for middle high risk states:

$$f_{MH}=\begin{cases}\frac{x-1.957E-4}{0.326E-4}, & 1.957E-4\leq x\leq 2.283E-4\\ 1, & 2.283E-4<x\leq 2.740E-4\\ \frac{-x+3.425E-4}{0.685E-4}, & 2.740E-4<x\leq 3.425E-4\\ 0, & otherwise\end{cases} \quad (6)$$

Membership function for high-risk states:

$$f_{H}=\begin{cases}\frac{x-2.740E-4}{0.685E-4}, & 2.740E-4\leq x\leq 3.425E-4\\ \frac{-x+4.566E-4}{1.141E-4}, & 3.425E-4<x\leq 4.566E-4\\ 0, & otherwise\end{cases} \quad (7)$$

Membership function for very high risk states:

$$f_{VH}=\begin{cases}\frac{x-3.425E-4}{1.141E-4}, & 3.425E-4\leq x\leq 4.566E-4\\ 1, & 4.566E-4\leq x\leq 6.849E-4\\ 0, & otherwise\end{cases} \quad (8)$$

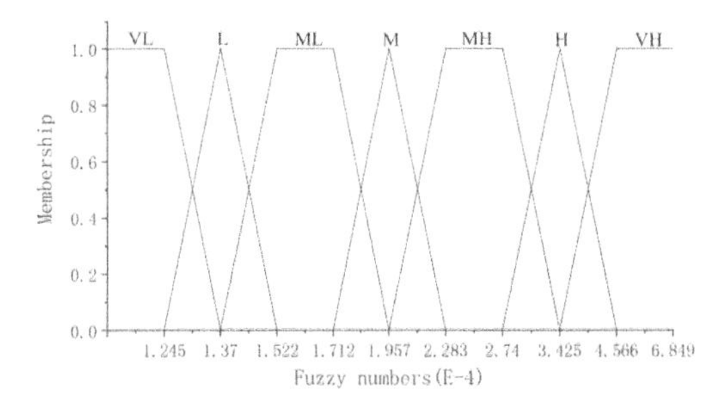


Figure 2. Membership function.

## 2.2 *Expert comprehensive evaluation method*

The comprehensive evaluation method of experts is to invite relevant experts to use fuzzy language to judge the basic events, and convert the fuzzy language into triangular fuzzy numbers or trapezoidal fuzzy numbers for quantitative calculation. In order to accurately obtain the fuzzy probability of the event and reduce the subjective influence of the expert opinion, the weight factor of each expert can be determined according to the professional status, work experience and education level of each expert, and the expert opinions can be normalized. Table 1 shows the scoring rules for the importance of experts. The importance score of experts is the product of three scores of expert title, work experience and education level.

Table 1. Expert importance scoring rules.

| Project | Classification | Score |
|---|---|---|
| Job title | Positive height | 5 |
| | Deputy high | 4 |
| | Intermediate | 3 |
| Work experience | >20years | 5 |
| | 15-20 years | 4 |
| | 10-15 years | 3 |
| | 5-10 years | 2 |
| | <5 years | 1 |
| Education level | PhD | 5 |
| | Master | 4 |
| | Bachelor | 3 |

Table 2 shows the fuzzy numbers corresponding to linguistic variables. The commonly used fuzzy linguistic variables are very low, low, middle low, medium, middle high, high and very high.

Table 2. Fuzzy numbers corresponding to language variables.

| Linguistic variables | Fuzzy numbers |
|---|---|
| VL(Very low) | (1.14E-4,1.14E-4,1.25E-4,1.37E-4) |
| L(Low) | (1.25E-4,1.37E-4,1.37E-4,1.52E-4) |
| ML(Middle low) | (1.37E-4,1.52E-4,1.71E-4,1.96E-4) |
| M(Medium) | (1.71E-4,1.96E-4,1.96E-4,2.28E-4) |
| MH(Middle high) | (1.96E-4,2.28E-4,2.74E-4,3.42E-4) |
| H(High) | (2.74E-4,3.42E-4,3.42E-4,4.57E-4) |
| VH(Very high) | (3.42E-4,4.57E-4,6.85E-4,6.85E-4) |

The specific process of the consensus aggregation method of expert opinions is as follows (Li et al, 2019):

1) When the fuzzy failure probability of the system is calculated by the comprehensive evaluation method of experts, the approval degree of opinions between any two experts is expressed as:

$$S_{ij}=S(A_i,A_j)=1-\frac{1}{4}\sum_{k=1}^{4}\left|a_{ik}-a_{jk}\right| \quad (9)$$

Where, $A_i$ is the opinion of the expert $E_i$, $A_i=(a_{i1},a_{i2},a_{i3},a_{i4})$; $A_j$ is the opinion of the expert $E_j$, $A_j=\left(a_{j1},a_{j2},a_{j3},a_{j4}\right)$.

2) A consensus matrix can be constructed from the approval degrees of opinions between two experts as follows:

$$M=\begin{bmatrix} S_{11} & S_{12} & \cdots & S_{1n}\\ S_{21} & S_{22} & \cdots & S_{2n}\\ \vdots & \vdots & \ddots & \vdots\\ S_{n1} & S_{n2} & \cdots & S_{nn}\end{bmatrix} \quad (10)$$

Obviously, when $i = j$, $S_{ij} = 1$.

3) The average recognition of each expert is:

$$A(E_i) = \frac{1}{n-1}\sum_{j=1, j\neq i}^{n} S_{ij} \tag{11}$$

4) The relative recognition of each expert is:

$$R(E_i) = A(E_i)/\sum_{i=1}^{n} A(E_i) \tag{12}$$

5) The importance of each expert is:

$$IM(E_i) = \frac{score(i)}{\sum_{i=1}^{n} score(i)} \tag{13}$$

Where, $score(i)$ is the importance score of the expert $E_i$.

6) The weight coefficient of experts is:

$$w(E_i) = \alpha IM(E_i) + (1-\alpha)R(E_i) \tag{14}$$

Where, $\alpha$ is the relaxation factor representing the importance of the expert's personal experience and opinion, usually $\alpha$ taken as 0.5.

7) The normalized result of expert opinion is:

$$\lambda_j = \sum_{i=1}^{n} w_i \cdot \lambda_{ij},\ j = 1, 2, \ldots m \tag{15}$$

Where, $\lambda_j$ is the aggregated fuzzy number of basic events $j$; $\lambda_{ij}$ is the fuzzy number corresponding to the opinion of expert $E_i$; $m$ is the number of basic events.

### 2.3 *Fuzzy fault tree analysis*

Input the aggregated fuzzy number of the bottom event after normalization into the fuzzy fault tree, and the aggregated fuzzy number of the top event and the upper system can be obtained by formula (15) (16).

1) "AND" gate:

$$P(E_{upper}) = \prod_{i=1}^{n} p(BE_i) \tag{16}$$

2) "OR" gate:

$$P(E_{upper}) = 1 - \prod_{i=1}^{n}(1 - p(BE_i)) \tag{17}$$

Where, $P(E_{upper})$ denotes the probability of occurrence of upper-level events; $p(BE_i)$ denotes the probability of occurrence of basic events $i$.

For trapezoidal fuzzy numbers $q_1 = (a_1, b_1, c_1, d_1)$ and $q_2 = (a_2, b_2, c_2, d_2)$, the operation rules are:

$$\begin{aligned} q_1 \oplus q_2 &= (a_1 + a_2, b_1 + b_2, c_1 + c_2, d_1 + d_2) \\ q_1 \Theta q_2 &= (a_1 - d_2, b_1 - c_2, c_1 - b_2, d_1 - a_2) \\ q_1 \otimes q_2 &= (a_1 a_2, b_1 b_2, c_1 c_2, d_1 d_2) \end{aligned} \tag{18}$$

According to the trapezoidal fuzzy number, the probability of the upper event in the fuzzy fault tree is calculated, and the algorithm is as follows:

1) "AND" gate:

$$q_{AND} = \prod_{i=1}^{n} q_i = \left(\prod_{i=1}^{n} a_i, \prod_{i=1}^{n} b_i, \prod_{i=1}^{n} c_i, \prod_{i=1}^{n} d_i,\right) \tag{19}$$

2) "OR" gate:

$$\begin{aligned} q_{OR} = 1\Theta \prod_{i=1}^{n}(1\Theta q_i) = \\ \Big[1 - \prod_{i=1}^{n}(1 - a_i), 1 - \prod_{i=1}^{n}(1 - b_i), \\ 1 - \prod_{i=1}^{n}(1 - c_i), 1 - \prod_{i=1}^{n}(1 - d_i),\Big] \end{aligned} \tag{20}$$

### 2.4 *Fire risk analysis*

(1) Defuzzification

To represent the probability of occurrence of the top event, the aggregated fuzzy number of the top event in the fuzzy fault tree should be converted into a clear value. This process, called defuzzification, uses the center of gravity method (COG) to convert the fuzzy number into a fuzzy likelihood score (FPS) (Yu et al, 2019), which is expressed as:

$$X^* = \frac{\int x\mu_A(x)dx}{\int \mu_A(x)dx} \tag{21}$$

For trapezoidal fuzzy numbers, the above formula can be expressed as:

$$X^* = \frac{1}{3}\frac{(d+c)^2 - dc - (a+b)^2 + ab}{d+c-b-a} \quad (22)$$

(2) Importance analysis

Importance is an important indicator to measure the contribution of basic events to the probability of occurrence of top events. By analyzing the importance of each basic event and sorting the importance of system components, the weak links in the system can be determined, and the system can be more targeted equipment for protection (Yu et al, 2019), its expression is:

$$I_{FV}^{i} = \frac{1 - \prod_{j=1}^{N}\left(1 - P_j^i\right)}{p_t} \quad (23)$$

Where, $p_j^i$ represents the probability of occurrence of the $i$ th basic event in the $j$ th smallest cut set; $p_t$ represents the probability of occurrence of the top event.

## 3 RISK ASSESSMENT OF SHIP POWER SYSTEM UNDER ENGINE ROOM FIRE

Taking the Panamax container ship as an example, the damage risk of the ship power system under the action of engine room fire is carried out based on the ECE-FFTA method.

### 3.1 *Construction of fuzzy fault tree for ship power system*

The ship power system is mainly composed of three subsystems: the main engine fuel system, the propulsion motor room, and the gas turbine room. The main engine fuel system provides fuel oil for the main engine of the ship, the propulsion motor room is the space for the power plant of the ship, and the gas turbine room is the space for the mechanical device where the fuel energy is converted into useful work. A fire in any of the subsystems can result in the loss of functionality of the entire ship power system. The fault tree model of each subsystem is established respectively, the fault tree information is shown in the Table 3, and the fault tree is shown in the Figure 3:

Table 3. Failure information of Engine room fire.

| Codes | Description |
|---|---|
| M1 | Main engine fuel system fire |
| M2 | Propulsion motor room fire |
| M3 | Gas turbine cabin fire |

#### 3.1.1 *Main engine fuel system fire*

The function of the main engine fuel system (M1) is to supply quantitative and constant pressure fuel to the main engine injectors, as the power source for the main engine to run. In this system, the fuel supply tank overflows, the valve or flange leaks, the filter leaks and other reasons cause the fuel to be exposed to the environment, contact with high temperature hot walls, open flames and other fire sources are the main factors that cause the main engine fuel system to catch fire. The fire logic gate and time information of the main engine fuel system fire are listed in the Table 4, and the fault tree diagram is drawn in the Figure 4.

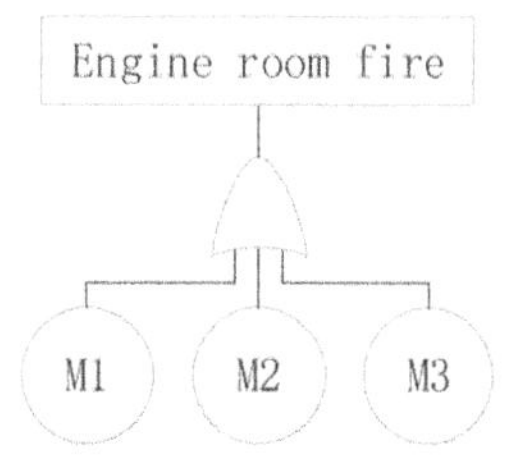


Figure 3. Fault Tree of Engine room fire.

#### 3.1.2 *Propulsion motor room fire*

Propulsion motor room (M2), there are rotating electrical machines that provide propulsion power for the ship, including propulsion generators and propulsion motors. The high-temperature hot surface caused by the long-term operation of the mechanical equipment in the cabin contacted the flammable materials in the cabin to cause fire, and the failure of electrical equipment caused the fire to be the typical cause of fire failure in the cabin. The fire logic gate and time information of the Propulsion motor room fire are listed in the Table 5, and the fault tree diagram is drawn in the Figure 5.

#### 3.1.3 *Gas turbine room fire*

Gas turbine room is an internal combustion power machine that converts the energy of fuel into useful work. The fire caused by the high temperature operation of the machinery and equipment in the cabin and the fire caused by the aging of electrical equipment components are the main causes of fire. The fire logic gate and time information of the Gas turbine room fire are listed in the Table 6, and the fault tree diagram is drawn in the Figure 6.

### 3.2 *Comprehensive assessment of ship power system fire experts*

In order to obtain information on the failure of ship power system equipment, five experts were invited to judge the basic events of the system. The five experts are chief engineers, senior engineers and

Table 4. Logic gates and principal events of Main engine fuel system fire.

| Codes | Description | Codes | Description |
|---|---|---|---|
| G1 | Fire and explosion caused by diesel fuel supply tank generates combustible material | E9 | Electrical failure caused the electrical appliance to overheat |
| G2 | Fire and explosion caused by fuel overflow in diesel supply tank | G6 | Fire and explosion caused by heavy fuel oil tanks produce combustible materials |
| E1 | Diesel supply tank oil overflow | G7 | Fire and explosion caused by oil spill of heavy fuel oil tank |
| G3 | Ignition source | E10 | Heavy fuel oil tank fuel overflow |
| E2 | Insulation failure on the surface of the exhaust pipe of the host | G8 | Fire and explosion caused by volatilization of heavy fuel oil tanks |
| E3 | Welding slag left over during repair welding | E11 | Heavy fuel oil tank volatilization |
| E4 | Insulation failure on the surface of boiler exhaust pipe | G9 | Oil leakage from valve or flange causes fire and explosion |
| G4 | Fire and explosion caused by volatilization of oil and gas in diesel supply tank | E12 | Valve or flange leakage |
| E5 | Diesel supply tank volatilization of oil and gas | E13 | Insulation failure on the surface of the exhaust pipe of the auxiliary machine |
| E6 | Illegal smoking | G10 | Fire and explosion caused by circulating pump produces combustible material |
| E7 | Static electricity produces an open flame | E14 | Circulating pump leaks |
| G5 | Fire and explosion caused by supply pump produces combustible materials | G11 | Fire and explosion caused by oil filter produces combustible materials |
| E8 | Oil supply pump leaks | E15 | Oil filter leaks |

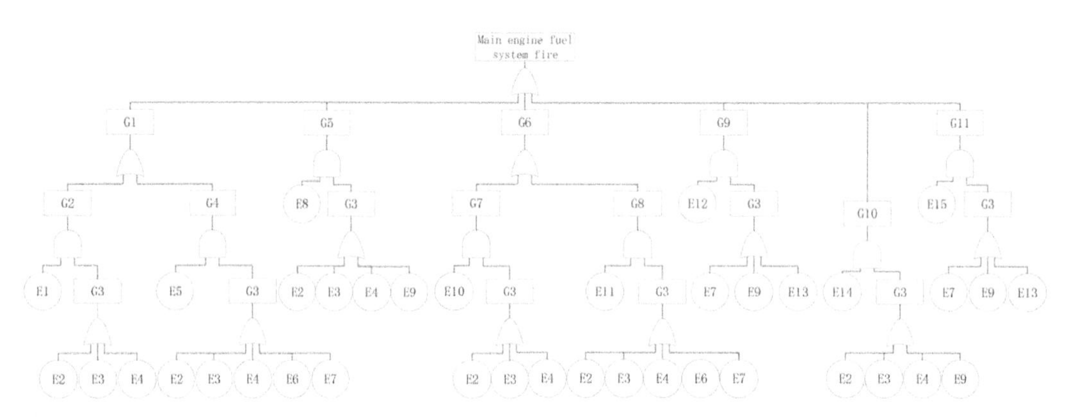


Figure 4. Fault Tree of Main engine fuel system fire.

Table 5. Logic gates and principal events of Propulsion motor room fire.

| Codes | Description | Codes | Description |
|---|---|---|---|
| G12 | Electrical equipment causes fire | E21 | Electrical aging |
| G3 | Ignition source | G15 | Combustible |
| E16 | High temperature live conductor caused by circuit failure | E22 | Cable insulation layer, protective layer |
| G13 | High temperature hot wall | E23 | Surrounding combustibles |
| E17 | Cooling failure in the motor room | G16 | Hot work in the Propulsion motor room caused fire |
| E18 | Excessive temperature in special parts | E24 | Combustible gas accumulates in the cabin |
| G14 | Electric spark | G17 | Open flame |
| E19 | Electrical leakage | E25 | Welding during maintenance in the engine room |
| E20 | Electrical short circuit | E26 | Cutting during maintenance in the engine room |

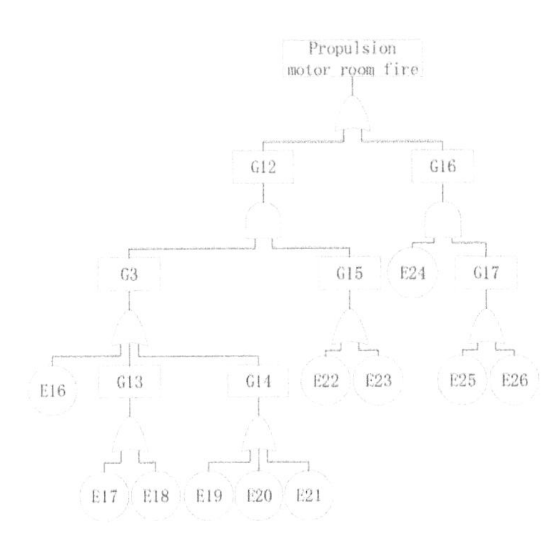


Figure 5. Fault Tree of Propulsion motor room fire.

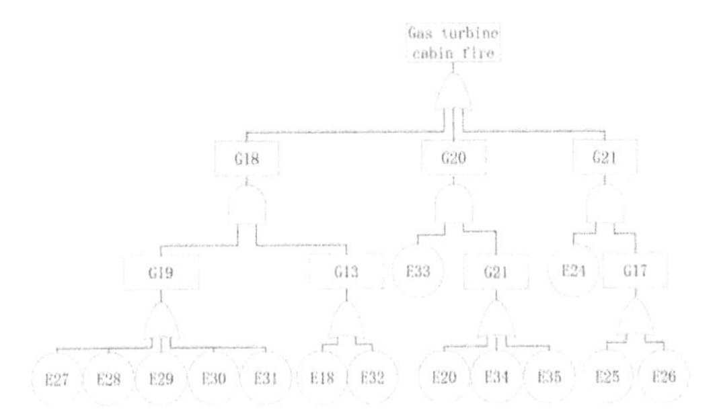


Figure 6. Fault Tree of Gas turbine room fire.

Table 6. Logic gates and principal events of Gas turbine room fire.

| Codes | Description |
|---|---|
| G18 | Fire caused by machinery and equipment |
| G19 | Oil leak |
| E27 | Leakage of oil filter |
| E28 | Leakage of oil heater |
| E29 | Leakage of oil supply pipes and joints |
| E30 | Fuel tank drip |
| E31 | Lubricant leak |
| G13 | High temperature hot wall |
| E18 | Excessive temperature in special parts |
| E32 | Failure of the coolant in the gas tank |
| G20 | Fire caused by electrical equipment |
| E33 | Cable or surrounding combustibles |
| G21 | Getting hot or sparking |
| E20 | Electrical short circuit |
| E34 | The components of the electric control box are aging |
| E35 | Aging of cable insulation material |
| G21 | Hot work in the Gas turbine cabin caused fire |
| E24 | Combustible gas accumulates in the cabin |
| G17 | Open flame |
| E25 | Welding during maintenance in the engine room |
| E26 | Cutting during maintenance in the engine room |

engineers who have long been engaged in ship system design, operation and maintenance in relevant scientific research institutes, as well as ships engaged in ship system design, operation and maintenance. Faculty of Vitality Studies in Colleges and Universities.

According to the importance score rules, the importance scores of each expert are shown in Table 7, and their importance coefficients are 0.328, 0.246, 0.197, 0.131 and 0.098 respectively. The above experts are invited to use fuzzy language to judge the failure of the basic events of the ship power system, and then normalize the judgments of the relevant basic events. The fuzzy numbers of basic events after expert evaluation and normalization are shown in Table 8.

## 3.3 *Fire risk analysis of ship power system*

### 3.3.1 *The annual fire frequencies analysis of ship power system*

According to the quantitative calculation of the fuzzy fault tree, the annual fire frequencies of the ship power system is 5.232E-06 $h^{-1}$, and the calculation results of the annual fire frequencies of each subsystem are shown in Table 9.

The calculation results show that the annual fire frequencies of the main engine fuel system is 2.55705E-6$h^{-1}$. Compared with the other two subsystems, the main engine fuel system has the highest fire frequencies, and its annual fire frequencies is twice that of the other subsystems. Because the main engine fuel system is an important system for supplying fuel to the main engine of the ship, it is one of the most critical systems in the ship power system, and it is the basis for ensuring the normal navigation of the ship. There are many piping systems in the system area, which is prone to fuel leakage problems. The annual fire frequencies of the propulsion motor room is 1.55272E-6$h^{-1}$, and the annual fire frequencies of the gas turbine room is 1.12261E-6$h^{-1}$. The annual fire frequencies of the propulsion motor room fire is slightly higher than that of the gas turbine room fire. Compared with the gas turbine room, the annual fire frequencies of the propulsion motor room is slightly higher. Because the propulsion motor room is mostly electrical equipment. As an important system in the ship power system, the propulsion motor room is responsible for the propulsion of the ship. The electrical equipment in the cabin continues to operate under high load, resulting in a local temperature rise of some equipment in the cabin, and it is easy to cause fire after contact with the surrounding combustibles such as aging components. The risk of fire caused by mechanical equipment in the gas turbine cabin should not be underestimated. The specific annual fire frequencies of the subsystem results is shown in Figure 7.

The ranking of typical failure modes of ship power system fire is shown in Figure 8, in which the fire caused by electrical equipment is the main

Table 7. The scores of expert importance.

| Expert | Job title | Score | Work experience | Score | Education level | Score | Total score |
|---|---|---|---|---|---|---|---|
| 1 | Chief engineer | 5 | 18 | 4 | PhD | 5 | 100 |
| 2 | Professor | 5 | 13 | 3 | PhD | 5 | 75 |
| 3 | Senior engineer | 4 | 12 | 3 | PhD | 5 | 60 |
| 4 | Associate Professor | 4 | 9 | 2 | PhD | 5 | 40 |
| 5 | Engineer | 3 | 5 | 2 | PhD | 5 | 30 |

Table 8. Failure judgment of experts on basic events.

| Numbering | E1 | E2 | E3 | E4 | E5 | Aggregate Fuzzy Numbers |
|---|---|---|---|---|---|---|
| E1 | VH | H | H | H | VH | (3.021E-04,3.895E-04,4.837E-04,5.512E-04) |
| E2 | H | VH | VH | VH | H | (3.139E-04,4.095E-04,5.433E-04,5.908E-04) |
| E3 | H | H | M | H | MH | (2.419E-04,2.960E-04,3.029E-04,3.944E-04) |
| E4 | MH | H | MH | H | M | (2.226E-04,2.675E-04,2.888E-04,3.697E-04) |
| E5 | ML | M | ML | M | L | (1.484E-04,1.669E-04,1.756E-04,2.019E-04) |
| E6 | M | M | ML | L | L | (1.498E-04,1.687E-04,1.725E-04,1.977E-04) |
| E7 | H | M | M | M | ML | (1.931E-04,2.280E-04,2.308E-04,2.837E-04) |
| E8 | H | ML | M | L | H | (1.984E-04,2.367E-04,2.410E-04,3.029E-04) |
| E9 | VH | H | H | MH | VH | (2.892E-04,3.706E-04,4.724E-04,5.321E-04) |
| E10 | H | ML | MH | VH | H | (2.392E-04,2.961E-04,3.472E-04,4.137E-04) |
| E11 | ML | H | L | ML | L | (1.634E-04,1.891E-04,1.973E-04,2.389E-04) |
| E12 | MH | M | M | MH | MH | (1.855E-04,2.145E-04,2.411E-04,2.940E-04) |
| E13 | H | MH | M | M | MH | (2.075E-04,2.464E-04,2.636E-04,3.309E-04) |
| E14 | L | ML | H | ML | L | (1.592E-04,1.835E-04,1.909E-04,2.296E-04) |
| E15 | M | ML | VH | M | H | (2.127E-04,2.597E-04,3.092E-04,3.457E-04) |
| E16 | VH | H | L | L | VH | (2.479E-04,3.149E-04,4.091E-04,4.402E-04) |
| E17 | M | MH | L | L | MH | (1.636E-04,1.864E-04,2.036E-04,2.428E-04) |
| E18 | H | VH | L | MH | H | (2.467E-04,3.081E-04,3.665E-04,4.283E-04) |
| E19 | L | MH | H | H | L | (1.951E-04,2.319E-04,2.422E-04,3.054E-04) |
| E20 | H | MH | MH | VH | H | (2.524E-04,3.130E-04,3.701E-04,4.463E-04) |
| E21 | M | H | H | VH | MH | (2.464E-04,3.055E-04,3.501E-04,4.172E-04) |
| E22 | VH | H | H | MH | MH | (2.674E-04,3.365E-04,4.111E-04,4.810E-04) |
| E23 | H | MH | VH | MH | H | (2.572E-04,3.205E-04,3.836E-04,4.575E-04) |
| E24 | H | L | L | MH | VH | (2.085E-04,2.539E-04,2.955E-04,3.435E-04) |
| E25 | H | H | H | MH | H | (2.611E-04,3.231E-04,3.307E-04,4.380E-04) |
| E26 | M | H | H | L | MH | (2.105E-04,2.525E-04,2.594E-04,3.289E-04) |
| E27 | MH | ML | VH | L | ML | (1.912E-04,2.301E-04,2.945E-04,3.242E-04) |
| E28 | MH | ML | H | L | L | (1.760E-04,2.050E-04,2.214E-04,2.725E-04) |
| E29 | MH | MH | H | L | L | (1.891E-04,2.220E-04,2.444E-04,3.050E-04) |
| E30 | M | M | MH | L | H | (1.837E-04,2.144E-04,2.235E-04,2.722E-04) |
| E31 | M | M | VH | L | L | (1.904E-04,2.292E-04,2.744E-04,2.947E-04) |
| E32 | M | MH | L | L | ML | (1.548E-04,1.751E-04,1.882E-04,2.210E-04) |
| E33 | VH | H | MH | MH | M | (2.482E-04,3.091E-04,3.860E-04,4.412E-04) |
| E34 | M | L | MH | M | MH | (1.694E-04,1.940E-04,2.100E-04,2.507E-04) |
| E35 | MH | MH | H | MH | H | (2.231E-04,2.676E-04,2.976E-04,3.820E-04) |

failure mode, and its annual fire frequencies is the highest. There are two main reasons. One is that the equipment comes into contact with combustibles such as cable insulation and protective layers after high temperature, causing fires, and the other is that electrical equipment is aging and faulty, causing cabin fires. The surface temperature of the equipment in the cabin should be monitored mainly, the inspection of the electrical equipment in the cabin should be strengthened, the equipment failure should be found and eliminated in time, and the aging components should be replaced. Fire and explosion caused by diesel supply tank and heavy fuel oil tank are also an important failure mode with

Table 9. The annual fire frequencies of Subsystem annual fire frequencies.

| Subsystem | Annual fire frequencies ($h^{-1}$) |
|---|---|
| Main engine fuel system fire | 2.55705E-6 |
| Propulsion motor room fire | 1.55272E-6 |
| Gas turbine room fire | 1.12261E-6 |

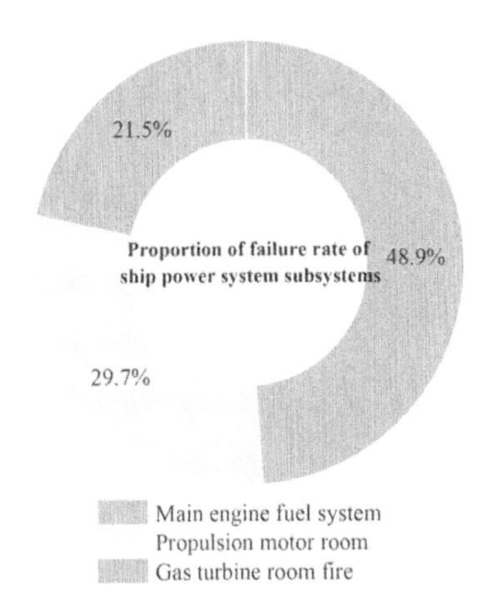


Figure 7. Proportion of annual fire frequencies of ship power system subsystems.

high annual fire frequencies. Insulation protection of high-temperature surfaces such as pipes, and strengthening inspections to prevent fuel leakage accidents, thereby preventing fires. The main failure mode of fire failure caused by mechanical equipment is high. The damaged oil filter and heater should be replaced in time, the fuel supply pipes and joints should be checked to prevent fuel leakage, and the mailbox coolant should be replaced in time to avoid overheating of components, thereby reducing the risk of fire.

### 3.3.2 *Basic event Importance analysis*

In order to clarify the susceptibility of the ship power system to the fire in the ship engine room, find out the weak links in the system, and sort the basic events according to their importance. The higher the importance, the more important the basic events, and the more attention should be paid to them. The calculation results of the importance of basic events are shown in Table 10, and Figure 9 shows the changes in the importance levels of basic events.

From Table 10 and Figure 7, the analysis results of the importance of basic events of ship power system fire, it can be found that the cable insulation layer, protective layer and surrounding combustibles are the key factors leading to ship power system fire. High temperature surfaces and open flames are prone to fire accidents. The inspection of cable

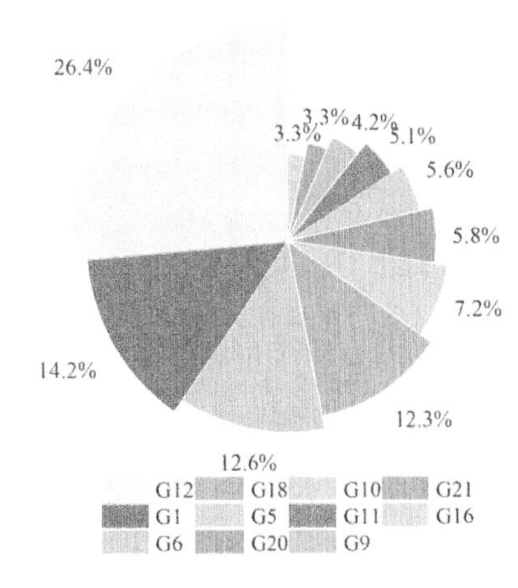


Figure 8. Ranking of Typical Failure Modes of each subsystem.

Table 10. Importance and ranking of basic events.

| Rank | Importance | Numbering | Rank | Importance | Numbering |
|---|---|---|---|---|---|
| 1 | 23.922 | E22 | 19 | 4.904E-02 | E16 |
| 2 | 22.716 | E23 | 20 | 4.873E-02 | E5 |
| 3 | 22.079 | E33 | 21 | 4.100E-02 | E12 |
| 4 | 1.383E-01 | E2 | 22 | 3.991E-02 | E7 |
| 5 | 1.238E-01 | E18 | 23 | 3.771E-02 | E25 |
| 6 | 9.311E-02 | E3 | 24 | 3.420E-02 | E19 |
| 6 | 8.738E-02 | E1 | 25 | 2.923E-02 | E26 |
| 8 | 8.645E-02 | E4 | 26 | 2.786E-02 | E17 |
| 9 | 7.570E-02 | E9 | 27 | 2.702E-02 | E27 |
| 10 | 6.954E-02 | E8 | 28 | 2.603E-02 | E13 |
| 11 | 6.581E-02 | E10 | 29 | 2.565E-02 | E31 |
| 12 | 6.556E-02 | E24 | 30 | 2.514E-02 | E29 |
| 13 | 5.562E-02 | E11 | 31 | 2.337E-02 | E30 |
| 14 | 5.409E-02 | E14 | 32 | 2.289E-02 | E28 |
| 15 | 5.406E-02 | E20 | 33 | 2.060E-02 | E35 |
| 16 | 5.299E-02 | E32 | 34 | 1.446E-02 | E34 |
| 17 | 5.136E-02 | E21 | 35 | 1.227E-02 | E6 |
| 18 | 4.911E-02 | E15 | | | |

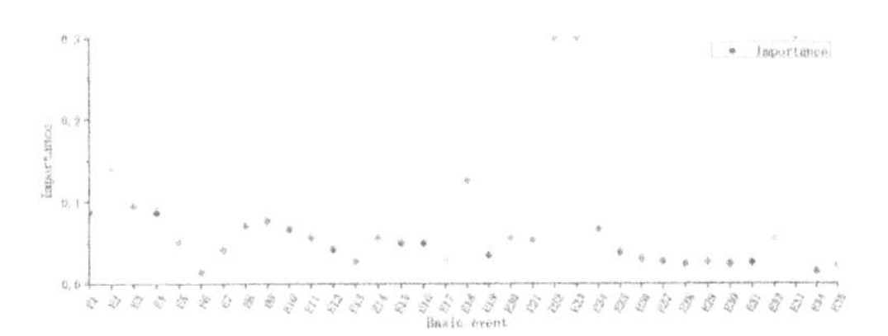

Figure 9. Changes in importance level of basic fire events in ship power system.

insulation and protection layer should be strengthened, aging and damaged cables should be replaced in time, and combustibles in the cabin should be cleaned up in time to avoid the accumulation of combustibles, prevent items in the cabin from igniting, and reduce the risk of fire.

## 4 CONCLUSIONS

Based on ECE-FFTA, this paper establishes a fire risk assessment method for ship power system based on system annual fire frequencies and basic event importance as risk assessment indicators. The fire risk assessment of the ship power system under the action of the ship engine room fire was carried out, and the following conclusions were obtained:

1. The annual fire frequencies of the main engine fuel system is the highest. The main failure mode is the fire of the main engine fuel system caused by the fuel leakage of the diesel supply tank and the heavy fuel oil tank. As an important system to ensure the normal navigation of the ship, the inspection of the parts prone to fuel leakage should be strengthened. Prevent explosion accidents caused by fire and reduce the risk of ship fire.
2. The annual fire frequencies of the propulsion motor room and the gas turbine room is relatively high. The fire caused by electrical equipment and mechanical equipment is the main reason for the fire in these two subsystems. The accumulation of combustibles, these risk factors will increase the risk of fire, and inspections in the area should be strengthened to prevent fires.
3. The ship fire risk assessment method established in this paper has reference significance for the fire protection design of the ship power system and the fire risk assessment of other important systems and compartments.

## ACKNOWLEDGEMENTS

This research was supported by the Science Fund Project of Heilongjiang Province (LH2020E078) and the National Natural Science Foundation of China (52171305, 52101305).

## REFERENCES

Huang Y.S., Chen M.H., Ma Y. 2009. Fire risk evaluation of the engine room. *China Shiprepair*, 22(5),15–18. (*in Chinese*)

IMO, 2019. Global Integrated Shipping Information System (GISIS). International Maritime Organization, United Kingdom.

Kaptan M. 2021. Risk assessment of ship anchorage handling operations using the fuzzy bow-tie method. *Ocean Engineering*, (236),1–9.

Li P.C., Zhou H.J., Zhou G.J. 2019. Fuzzy dynamic fault tree analysis based on expert comprehensive evaluation. *Ship Science and Technology*, 41(10),194–197. (*in Chinese*)

Liu Z.J., Ji Z.S., Lin Y. 2010. Fire Risk Analysis in Ship Engine Room Based on Bayesian Networks. *Shipbuilding of China*, 51(3),199–204. (*in Chinese*)

Lavasani S.M., Ramazali N., Sabzalipour F., Akyuz E. 2015. Utilisation of Fuzzy Fault Tree Analysis (FFTA) for quantified risk analysis of leakage in abandoned oil and natural-gas wells. Ocean Engineering, 108,729–737.

Pan J.W. 2018. Fire and Explosion Risk Analysis of FPSO based on ETA and FTA. Harbin Engineering University, Harbin. (*in Chinese*)

Sarıalioglu S., Ugurlu O., Aydın M., Vardar B., Wang J. 2020. A hybrid model for human-factor analysis of engine-room fires on ships: HFACS-PV&FFTA. *Ocean Engineering*, 217,1–17.

Tang L., Zhang H.P. 2016. Improved Fuzzy Evaluation Based on BP Network Membership Function of the Transformer Condition Assessment. *Computer & Digital Engineering*, 44(3),414–417. (*in Chinese*)

Xu C.Y. 2016. Numerical study on characteristics of ship engine room fires. Harbin Institute of Technology, Harbin. (*in Chinese*)

Yang G.F., Qiu D.F., Pan J.J., Teng Y.N., Yong X.Y. 2017. Corrosion risk assessment and comprehensive evaluation of ship sea water pipe systems. *Chinese Journal of Ship Research*, 12(3),143–148. (*in Chinese*)

Yu J.X., Chen H.C., Yu Y., Yang Z.L. 2019. A weakest t-norm based fuzzy fault tree approach for leakage risk assessment of submarine pipeline. *Journal of Loss Prevention in the Process Industries*, 62,1-12.

Yu W.J., Guo G.P., Wu B. 2019. Risk Prediction of LNG Ship during Loading/Unloading in the Yangtze River Based on Fuzzy Fault Tree. *Journal of Transport Information and Safety*, 37(5),46–52. (*in Chinese*)

*Trends in Maritime Technology and Engineering – Guedes Soares & Santos (eds)*


# Preliminary hazard analysis of vessel maneuvers in access channels to port terminals

M.C. Maturana, D.T.M.P de Abreu & M.R. Martins
*Analysis Evaluation and Risk Management Laboratory – LabRisco, São Paulo, Brazil*

ABSTRACT: This paper presents the general aspects of the Preliminary Hazard Analysis (PHA) and its application in the study of two vessels maneuvers in the evolution basin and in the access channel to a typical terminal in Sepetiba Bay, on the Brazilian coast. This analysis was helped by real-time simulations of selected events (among the most critical, as indicated in the PHA sessions). Overall, the maneuvers analyzed present moderate risk – even though, 14 hazardous events were classified as not tolerable, and five mitigation and contingency measures were indicated aiming at their treatment.

## 1 INTRODUCTION

Preliminary Hazard Analysis (PHA) has its origins in the military sector (Department of Defense, 2000), but is widely recommended for qualitative risk studies in the waterway sector (IMO, 2015) and (PETROBRAS, 2015), allowing to organize and consolidate knowledge of the safety aspects of vessel maneuvers in access channels to port terminals – in addition to allowing for the proposition of mitigation or contingency actions. In general, the PHA focuses on unwanted events, its chance of occurrence and consequences. Thus, the PHA works as a starting point for understanding the risk associated with vessel operations, and for identifying the events that contribute most significantly to the risk.

In addition to presenting the general aspects of the PHA, this paper presents the study of two vessels maneuvers in the evolution basin and in the access channel to a terminal in Sepetiba Bay, on the Brazilian coast. This analysis was helped by real-time simulations of selected events (along with the most critical, as indicated in the PHA sessions). Overall, the maneuvers analyzed present moderate risk – even though, 14 hazardous events were classified as not tolerable, and five mitigation and contingency measures were indicated aiming at their treatment.

Briefly, the next section discusses the general aspects of the technique, and section 3 presents the analysis procedure used in the case study. The results of this methodology in the analysis of maneuvers in the evolution basin and in the access channel to a typical terminal in Sepetiba Bay, on the Brazilian coast, are presented in section 4.

## 2 GENERAL ASPECTS OF THE PRELIMINARY HAZARD ANALYSIS

Overall, PHA focuses on materials and hazardous areas, building on a precursor hazard analysis study. The analysis begins by formulating a list of hazards and hazardous situations, considering the various characteristics of the system under study. As hazardous situations are identified, the potential causes, effects, and possible preventive and corrective measures are also listed. In this process, one or more analysts assess the significance of the hazards and can assign a frequency category and a severity category to each situation. Thus, hazards can be ranked, allowing for the prioritization of safety improvement actions (which emerge from the analysis) with the greatest positive impact on the risk. Next, the situations for which the use of PHA is recommended, the expected results with the application of this technique and the required resources are discussed.

### 2.1 *Application recommendation*

Typically, PHA is recommended for the preliminary phases of system development, when experience provides little or no insight into potential safety issues, and when there is little detailed information about design or operating procedures. However, when there is a need to rank the hazards, and circumstances do not allow the use of a more extensive technique, PHA can also be conveniently applied to the analysis of large projects in operation. In addition, the PHA is recommended when there is a need to select or compare design alternatives, or to improve the safety features of the system - with the

DOI: 10.1201/9781003320289-26

proposition of actions to mitigate the risk associated with the hazard events deemed most relevant.

### 2.2 *Expected results*

The PHA results in a description of the hazards related to a system and provides a qualitative (or semi-quantitative) classification of hazardous situations. This data can be used, for example, to prioritize actions to reduce or eliminate the probability of exposure (e.g., of operators and the environment) to the hazard.

### 2.3 *Required resources*

The use of PHA requires that risk analysts have access to the facility's design criteria (including operating criteria), equipment and material specifications, and other sources of information that may be considered pertinent to the study object. The PHA can be carried out by one or more professionals with knowledge of the system (experts to be consulted by risk analysts) - it should be noted that the PHA requires many judgments from the professionals involved and, thus, the depth and quality of the analysis depends on the experience of risk analysts and experts consulted in the PHA sessions. Overall, considering preparation time, runtime and documentation time, the PHA can be completed in a few weeks of continuous work, even for more complex systems (CCPS, 2008).

## 3 PHA ANALYSIS PROCESS

Once the scope of the analysis has been defined, the PHA consists of the following steps (Hammer, 1972; Greenberg & Cramer, 1991; Stephenson, 1991): a) preparation for the analysis; b) carrying out the analysis, and; c) documentation of results.

### 3.1 *Analysis preparation*

In this step, the PHA requires the review team to gather and become familiar with the information available about the system, as well as any relevant information – e.g., from similar systems or operations, or from different systems that use the same equipment. The analysis team should gather experiences from any possible source – including hazard studies of similar systems, experiences relating to similar systems or operations, and checklists with issues relevant to the specific system or operation.

For the effectiveness of this step, a conceptual description of the system must be made available, detailing the basic design parameters, the reactions and the products involved (e.g., chemicals), as well as the main types of equipment (e.g., pressure vessels, heat exchangers). Operational objectives and basic operating requirements can also help define hazard events and the operating environment.

### 3.2 *Analysis performance*

The main objective of PHA is to identify hazards and incidental situations that may result in unintended consequences, and to identify criteria or design alternatives that help reduce the chance of exposure to identified hazards and/or decrease the consequences of possible accidents – depending on the experience of the team responsible for this judgment. To do so, the team involved should consider:

- Hazardous equipment and materials (e.g., fuels, highly reactive chemicals, toxic substances, explosives, high pressure systems, systems that store energy);
- Interfaces between system equipment and safety-related materials (e.g., interaction between materials, fire ignition and propagation, protection systems);
- Environmental conditions that can influence system equipment and materials (e.g., extreme environmental conditions, vibrations, floods, extreme temperatures, electrical discharges, humidity);
- Operating, testing, maintenance and emergency procedures (focusing on the importance of human error, actions to be taken by the operator, accessibility of equipment, layout, and personal protection);
- Support equipment (e.g., storage, testing equipment, training, utilities);
- Safety-related equipment (e.g., mitigation systems, redundancies, fire suppression, personnel protection equipment).

For each part of the system, analysts identify the hazards and assess the possible causes and effects of potential accidents involving those hazards. Analysts do not usually seek to develop an exhaustive list of causes. Otherwise, the causes are listed in sufficient numbers to judge the credibility of a potential accident – whether it is credible or not.

Thus, the team assesses the effects of each accident – these effects must represent the worst-case impacts associated with the potential accident – and classifies each incidental situation according to frequency and consequence categories, which combine to define the risk categories. The organization employing the PHA should define the categories so that the review group can properly judge the hazards.

Finally, the team lists viable alternatives for the mitigation of hazard events and to limit their potential consequences, with a focus on substantiating the basic risk control strategies to be practiced – thus, the ultimate goal of PHA is beyond full development of incidental scenarios, or even the complete assessment of the risk associated with these scenarios, i.e., the objective is to define the best risk control strategy to be implemented.

### 3.3 *Results documentation*

PHA results are conveniently recorded in a table showing identified hazards (or hazardous situations), causes, potential consequences, hazard categories, and any suggested corrective or mitigation actions. Table 1 presents the format proposed by the MIL-STD-882 standard (Department of Defense, 2000). This table can be changed, for example, to record responsibilities for monitoring and executing key points, or to reflect the current state of implementation of suggested corrective (or improvement) actions.

Table 1. Typical format of a PHA table.

| *AREA:* | | | *DATE OF MEETING:* | |
|---|---|---|---|---|
| *NÚMBER OF CONTROL:* | | | *TEAM MEMBERS:* | |
| *Hazard* | *Cause* | *Effects* | *Hazard category* | *Mitigation actions/Contingency actions* |
| | | | | |

The information that must be included in the completion of each field in Table 1 is as follows:

- *Hazard:* name of the hazard event and brief description that ensures its clear and unambiguous identification;
- *Cause:* description of the means that can lead to the realization of the dangerous event (i.e., an accident) and categorization of the frequency of occurrence of the event;
- *Effects:* description of the possible undesired consequences of the realization of a hazardous event and categorization of damage to property, people and the environment;
- *Hazard category:*results from the combination of frequency and consequence categories. Therefore, it bases the assessment of risk tolerability;
- *Mitigation actions:* list of actions that can be taken to reduce the frequency of occurrence associated with the hazard event;
- *Contingency actions:* list of actions that can be taken to reduce the damage resulting from the occurrence of the accident associated with the hazard event.

Once completed, the PHA table will be a support tool for decision-making with all the risk control and safeguard measures suggested during the analysis. In the end, the total risk of the analyzed operation will be given by the combination of the risk associated with all the potential hazard events considered. Additionally, the hazard events listed can be ordered according to their level of risk, and those with less tolerability can be prioritized in later steps of risk management.

## 4 PHA OF TWO VESSELS MANEUVERS IN SEPETIBA BAY

In this application, a preliminary phase of preparation for the analysis and three PHA sessions were carried out, as indicated in the flowchart in Figure 1.

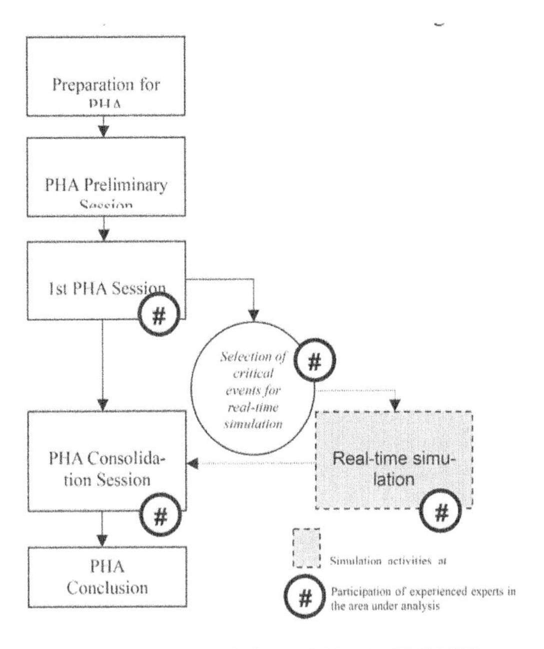


Figure 1. PHA and simulation activities (at TPN-USP).

For all these activities, the support of experts in the maneuvers under study was foreseen – pilots from the local pilotage zone, captains of the vessels involved and other specialists familiar with similar operations and in the channel design. In the activities highlighted by the symbol # in Figure 1, these experts were consulted directly, in meetings dedicated to the PHA. The PHA Consolidation Session was carried out after the real-time simulations in the TPN-USP (Numerical Offshore Tank Laboratory of the University of São Paulo, Tannuri et al. 2014; Queiroz Filho et al. 2014) of the hazard events selected in the PHA Initial Session.

The TPN-USP maintains a maneuvering simulation center for port, rivers and offshore operations. Its core simulator code (named SMH) – Tannuri et al. (2014) and Queiroz Filho et al. (2014) discuss the maneuver models implemented in this code (e.g., hull and propeller hydrodynamic forces, wind and shallow-waters effects, steering gears) – is adequate for low-speed maneuvers and has undergone a rigorous calibration process, based on results from sea-trials and captains and pilots' feedback. More details about the simulator and its reliability can be found in Tannuri and Martins (2018). Nevertheless, considering that simulators are imperfect (since mathematical models are simplifications of reality), in this work the vessel behavior in extreme conditions was evaluated by experts considering the limits imposed by standards and recommendations such as PIANC (2014), USACE (1980), etc. In order to predict the behavior of the ship, therefore, in this work the results of the simulator and the evaluation of the experts who participated in the simulations were considered.

The data collected in the preparation step were reviewed by the participants in the 1st Session and in the PHA Consolidation Session. The results of this step are presented in subsection 4.1. In general, the steps proposed in section 3 of this paper were followed in the PHA sessions, with results presented in subsection 4.2. The selection of events for simulation at TPN-USP are discussed in subsection 4.3. The analysis of the results is presented in section 4.4.

### 4.1 *Preparation for PHA*

The PHA preparation step included the survey and selection of resources to support the PHA sessions, also comprising the familiarization of the risk analysis team with the standard vessels, and with the area of maneuver. At this stage, the following information regarding the maneuvers of interest were gathered and studied:

- Description of the vessels involved in the maneuvers;
- Description of maneuvers to be considered;
- Characterization of the terminal access channel under study – especially in relation to the presence of stones, sandbanks and limits for navigation;
- Descriptions of the areas approaching the terminal, anchoring areas, evolution basin, mooring and terminal systems;
- Description of local traffic;
- Most recent bathymetries available;
- Aids to navegation;
- Description of similar maneuvers already performed;
- Results of simulations of similar maneuvers already performed;
- List of areas potentially affected by accidents in maneuver (e.g., environmental preservation areas, cities);
- List of areas that potentially influence the risk (e.g., buildings, cargo warehouses);
- Survey of environmental conditions (especially in relation to variations in current, wind and visibility in the analyzed stretch).

It is noteworthy that the absence of relevant information or the presence of uncertainties in relation to the information presented led the risk analysis team to adopt a conservative posture in the aspects directly affected.

This information supported the preparation of a preliminary list of hazards and hazard events worked on during the PHA sessions.

### 4.2 *Vessels involved in the analyzed maneuvers*

The standard vessels shown in Table 2 were considered in the PHA – this table also presents the general characteristics of these reference vessels. As for the loading conditions of the standard vessels, for the Suezmax Conventional, three possibilities were considered in each maneuver: a) a fully loaded vessel, b) a half-load vessel, and; c) a vessel in ballast; and, for the Suezmax DP (with dynamic positioning system), two possibilities were considered for each maneuver: a) navigation in the inlet channel: fully loaded and half-loaded vessel; b) navigation in the exit channel: half-load vessel and ballast vessel. The draft for the loaded vessel was presented in Table 2 – in column "T" –, for the other loading conditions the following drafts were considered (for both standard vessels): half-load vessel: 12.5 m; b) for the vessel in ballast: 8.0 m. For both standard vessels, a distance of 2.5 m between the double hull layers (both on the bottom and on the side) and 12 cargo tanks (6 on each side) are considered. As for the help of tugboats, four tugboats were considered, 3 of 60 TBP (Ton Bollard Pull) and 1 of 70 TBP.

Table 2. Standard vessels considered in the PHA.

| *Vessel* | *DWT [t]* | *LOA [m]* | *B [m]* | *T [m]* |
|---|---|---|---|---|
| **Suezmax Conventional** | 160,000 | 278.5 | 48 | 17.2 |
| **Suezmax DP** | 160,000 | 278.5 | 48 | 17.2 |

### 4.3 *Characterization of navigation channels*

The characterization of the navigation channels in the scope of the PHA was part of the preparation for the risk analysis and, among other peculiarities, helped in the identification of hazards and potential aggravating elements of the risk – such as types of vessels that frequent the region (and their respective loads), flow of vessels with passengers and structures that can be hit in cases of collision.

Regarding the traffic of the standard vessels (through the access channel to the terminal under study), the PHA considers the frequency of 28 vessels per month in each direction (entering or leaving terminal) – 14 for each standard vessel. This frequency was explored in the categorization of each hazard event according to its frequency.

A relevant point for defining the scenarios considered in the PHA was the understanding of the tidal cycle in Sepetiba Bay (RJ). Thus, maneuvers performed both with ebb tide current and with flood tide current were considered.

### 4.4 *Maneuver phases*

In this PHA, four maneuver phases were considered: 1) Channel navigation (entering); 2) Turning (entering); 3) Turning (leaving), and; 4) Channel navigation (leaving). In general, the actions of the crew in these maneuver phases involve the decision on how to navigate (e.g., vessel by their own means or with the help of tugboats), control of the vessel (e.g., maintaining the distance from the vessel to the reference isobaths) – with monitoring of navigation (both visually and by instruments), the comparison between the desired state and the actual state (e.g., vessel movement trend), and planning and execution of corrective action (involving the activities of command, execution and verification of command over the vessel) – the monitoring of traffic, and the combination of maneuver with other vessels. A more detailed discussion of the actions of operators in restricted waters can be found in Abreu et al. (2020).

### 4.5 *PHA Sessions*

The identified hazard events were evaluated by experts as to their probability of occurrence and possible impacts on people, on the assets of the organizations involved and on the environment. For this purpose, all the information collected in the preliminary analysis step (see summary in section 4.1) was presented to the experts, and a methodology for classifying these events was defined.

Thus, for the categorization of hazards, their causes and consequences, the risk matrix presented in Table 3 was used as a reference – based on the N-2782 standard (PETROBRAS, 2015). In Table 3, the risk classes are divided into Tolerable (T), Moderate (M) or Not Tolerable (NT). However, given that the measures of severity of consequences can be evaluated according to different dimensions of damage - to people, to the company assets and to the environment (the category "image" was not considered in this analysis) –, different levels risk factors could be associated with each dimension.

Differently from the consequences categories, the frequency categories of the N-2782 standard do not directly adjust to the risk analysis in vessel maneuvers in port accesses. In that standard, the frequency categories are based on similar installations and the concept of the installation's useful life, which can lead to inconsistencies when applied in the study of specific maneuvers.

An alternative for frequency categorization is presented in the guide for Formal Safety Assessment (FSA) (IMO, 2015), which is based on the useful life of vessels or fleet of vessels, but is also limited for application in the case under study – since it would be necessary to establish the fractions of the lifetime of the reference vessels dedicated to maneuvers in the terminal under study. Therefore, in order to adapt the frequency categories to the context of the maneuvers under analysis, the categories presented in Table 4 were used, focused on the expected interval between similar occurrences.

Table 3. Risk matrix.

| Frequency categories \ Consequence categories | Negligible (I) | Marginal (II) | Mean (III) | Critical (IV) | Catastrophic (V) |
|---|---|---|---|---|---|
| Frequent (E) | M | M | NT | NT | NT |
| Likely (D) | T | M | M | NT | NT |
| Unlikely (C) | T | T | M | M | NT |
| Remote (B) | T | T | T | M | M |
| Extremely remote (A) | T | T | T | T | M |

In this work, using Table 4, it was considered that each vessel type presents a specific annual frequency for maneuvers in the terminal under study – see subsection 4.1.

The next items present the results of the 1st PHA Session reviewed in the PHA Consolidation Session. In summary, the identified hazards and the hazard events inferred in the analysis of the maneuvers are presented, in addition to the PHA table.

Table 4. Frequency categories.

| Categories | Description |
|---|---|
| **Extremely remote (A)** | Possible, but no references on similar terminals |
| **Remote (B)** | Likely to occur once every 7,200 similar maneuvers (1/40 years) |
| **Unlikely (C)** | Likely to occur once every 1,800 similar maneuvers (1/10 years) |
| **Likely (D)** | Likely to occur once every 360 similar maneuvers (1/2 years) |
| **Frequent (E)** | Likely to occur once every 90 similar maneuvers (1/6 months) |

### 4.6 *Hazard event identification*

The hazard events were raised from the hazards identified in the different parts that make up the complex in which the navigation channels are located, including vessels in transit, and with reference to the standard vessels – the hazards that (case exposed) may result in direct or indirect consequences for the vessel's maneuver, with damage to life, property or the environment.

The hazard events associated with the maneuvers in the terminal under study (identified in the survey session carried out for the PHA) were the following (considered in each maneuver phase):

- Delay in tugboat response;
- Blackout;
- Adverse current conditions;
- Adverse wind conditions;
- Human error in maintaining control of the vessel;
- Error in communication with tugboats;
- Failure in the steering system (rudder locked in non-neutral position);
- Failure in the steering system (rudder locked in neutral position);
- Propulsion system failure;
- Unavailability of tugboat;
- Interaction with submarine;
- Towing rope breakage.

### 4.7 *PHA table*

The hazard events identified for the maneuvers were included in the PHA table (a part of the table was presented in the Appendix 1), organized according to the maneuver in which they may occur and associated with the frequency, consequence and risk categories. The other fields in this table refer to: description of the hazard event; description of causes and mitigation actions; description of consequences and contingency actions; and classification of risk tolerability (Tolerable - T, Moderate - M or Not Tolerable - NT). It should be noted that, in the PHA table (exemplified in Appendix 1), the events were also distinguished according to the tidal current observed during the maneuver (flood or ebb), with the vessel type involved and its load condition (loaded, half loaded or ballasted). In addition, for hazard events that have more than one cause, combinations of frequencies associated with each were considered.

Overall, 48 hazard events were evaluated in the PHA sessions for standard vessels maneuvering through the channel (from the terminal under study) for each possible combination (tidal current/standard vessel/loading condition). The final assessment of risk tolerability for each hazard event took into account the worst score in each of the three dimensions – environment, property and people – as seen in the PHA table clipping in Appendix 1. From this point of view, considering the Suezmax Conventional vessel, there were 34 events (11.81%) with tolerable final rating (T) and 246 events (85.42%) with moderate rating (M), while 8 (2.78%) events were classified as non-tolerable (NT). For the Suezmax DP vessel, 23 events (11.98%) were classified as tolerable (T), 163 events (84.90%) were classified as moderate (M), and 6 events (3.13%) were classified as non-tolerable (NT). Tables 5 and 6 present the results by maneuver phase for each vessel type (respectively for the Suezmax Conventional and for the Suezmax DP).

In all, 57 events (11.88%) were classified as tolerable (T), 409 events (85.21%) were classified as moderate (M), and 14 events (2.92%) were classified as non-tolerable (NT). The analysis of these results is presented in subsection 4.4. Before this discussion, however, and noting that the results presented in the PHA table were consolidated in the last PHA session (conducted based on the dynamic simulation rounds at the TPN-USP), the next section discusses the criteria used in the event selection for real-time simulation.

Table 5. Number of hazard events by risk category per maneuver for the Suezmax Conventional.

| Suezmax Conventional<br>Maneuver phases | T | M | NT |
|---|---|---|---|
| Channel navigation (entering) | 7 | 61 | 4 |
| Turning (entering) | 11 | 61 | 0 |
| Turning (leaving) | 12 | 60 | 0 |
| Channel navigation (leaving) | 4 | 64 | 4 |

Table 6. Number of hazard events by risk category per maneuver for the Suezmax DP.

| Suezmax DP<br>Maneuver phases | T | M | NT |
|---|---|---|---|
| Channel navigation (entering) | 2 | 42 | 4 |
| Turning (entering) | 7 | 41 | 0 |
| Turning (leaving) | 8 | 40 | 0 |
| Channel navigation (leaving) | 6 | 40 | 2 |

### 4.8 *Criteria for selecting events and simulation at TPN-USP (with examples)*

For the selection of simulated maneuvers in the TPN-USP, the following criteria were used:

- *Risk tolerability*: the maneuver steps associated with events classified as NT (Not Tolerable) in the initial PHA sessions were selected – considering the final classification of the hazard event;
- *Uncertainties in classification*: the maneuver steps related to the events or maneuver conditions that generated doubts among the participants in the initial PHA sessions were selected, and those that made it difficult to assign the frequency and consequence categories – the main doubts and questions among participants in the PHA sessions

focused on the possibility of towline breakage during different maneuvering scenarios;

- *Indications by experts*: hazard events indicated by the experts for simulation were selected in order to validate what was assumed during the PHA (e.g., feasibility of employing some contingency measure) – the experts (who participated in the PHA) had the opportunity to propose simulations to validate the effects of failures considered in the initial PHA sessions; in this sense, simulations of failure in the steering system or in the propulsion system during navigation in the channel were listed for different maneuver scenarios –, and;
- *High severity*: events that, despite not being classified as NT, cause concern because of the high severity attributed to their consequence ("Critical" or "Catastrophic") were selected – in PHA, the events of the turning phase in the evolution basin during the entrance to the terminal were selected.

### 4.9 *Results analysis*

As pointed out in subsection 4.3, a significant part of the hazard events – 11.81% for Suezmax Conventional and 11.98% for Suezmax DP – was in the Tolerable (T) class to risk. For these events, the combination between the frequency of their causes and their potentially harmful consequences do not represent a significant risk to the operation. It is noteworthy, however, that this assessment is based on the current state of knowledge. Therefore, these events cannot be ignored over time – as more information is collected, indicating greater frequencies and/or consequences of the hazard events in question, the risk classification will need to be revised.

In this PHA, most hazard events were judged as Moderate (M) – between 85.42% for the Suezmax Conventional vessels and 84.90% for the Suezmax DP. The classification of an event as M does not translate into its immediate disregard for the treatment of risk. In this sense, the ALARP (As Low As Reasonably Practicable) principle should be considered, i.e., that mitigation and contingency measures are adopted, as long as they are feasible in technical-economic terms.

At the end of the PHA Consolidation Session, 14 hazard events remained classified as non-tolerable risk (NT). These events are associated with the navigation stages of standard vessels in the access channel to the terminal under study, representing 2.92% of the hazard events listed for these vessels.

### 4.10 *Hazard events that most contribute to risk*

Hazard events classified as NT associated with the Suezmax Conventional vessel have a relatively high frequency – "Probable (D)" –, being related to adverse wind conditions during navigation in the channel (both entering and leaving). These events can lead to grounding with little possibility of drilling the double hull (due to the vessel's low speed) and collision with signal buoys, resulting in severe damage to the vessel, loss of signal buoy and light injuries to people – to help understand the severity of these events and the respective mitigation measures (discussed later), Figure 2 illustrates an example of a critical part of the channel, exposing possible obstacles (rocks and shallow depths) for the vessel in case of strong wind. These events were, therefore, associated with the "Critical (IV)" category of consequence severity (due to the possibility of severe damage to the vessel).

The considerations presented in the previous paragraph for the NT events associated with the Suezmax Conventional vessel can be used for the NT events associated with the Suezmax DP – these events have a relatively high frequency – "Probable (D)"– and have critical consequences due to the possibility of severe damage to the vessel.

### 4.11 *Mitigation and contingency measures*

For the hazard events classified as NT in the PHA table – related to adverse wind conditions – it was proposed, as a mitigation measure, to monitor the environmental conditions to support the suspension of the maneuver, i.e., to reduce the chance of exposing the vessel to unfavorable environmental conditions. It was suggested that information should arrive in advance of about 2 hours (indications of adverse wind conditions) at the terminal under study.

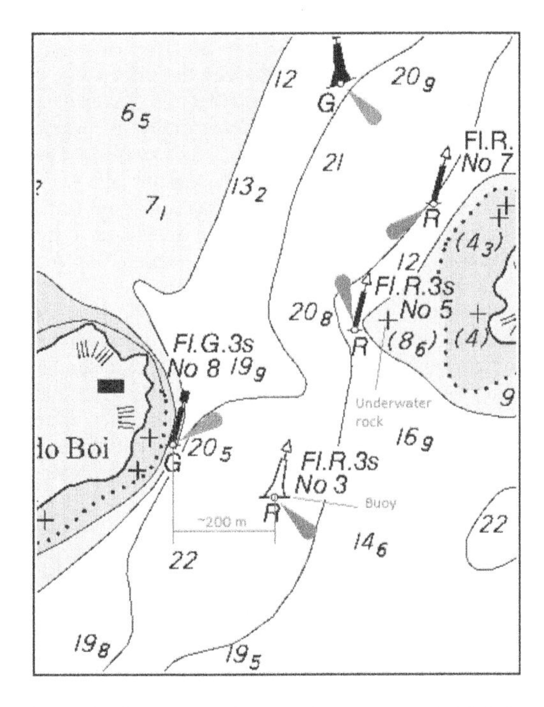


Figure 2. Critical part of navigation in strong wind (DHN, 2022).

Also for NT events, as contingency measures were proposed: a) the establishment of reference speed bands along the channel, limiting the damage associated with the possible grounding of the vessel; b) actions to ensure speed control during the maneuver, so as not to compromise the performance of the tugs in an emergency; c) use of the second pilot, aiming to improve the response to emergencies – both in terms of response time and emergency management; d) rectify the navigation channel, increasing the margin for maneuvers, and reducing the chance of grounding.

It is expected that the adoption of these measures will be enough to reduce the risk in relation to that considered by the experts consulted in the PHA Consolidation Session.

The mitigation and contingency measures proposed in this PHA are considered in addition to those already provided for by law. Thus, to reduce risk, complementary actions must be comprehensive enough to deal with the magnitude of harm predicted in the PHA Table.

## 5 CONCLUSIONS AND FINAL CONSIDERATIONS

The PHA discussed in this paper serves as a starting point for: a) understanding the total risk associated with the operations of Suezmax oil tankers at the terminal at the Sepetiba Bay under study; b) the identification of the events that contribute most significantly to the total risk, and; c) the proposition of mitigation and contingency measures for the identified hazard events.

During the PHA sessions, the hazard events were surveyed and studied and, through the PHA table, it was possible to compare the hazard events related to each type of operation (for different standard vessels, loadings, maneuver phases and tidal current). From the events discussed in the PHA sessions, the maneuver scenarios and the aggravating conditions to be simulated at the TPN-USP were identified. This selection considered participants' uncertainties and events that most contribute to the risk.

Real-time simulations reduced the uncertainties observed in the PHA sessions, and allowed for a more objective survey of mitigation and contingency actions for events considered to be of greater risk. Insights gained from the simulations were explored in the PHA Consolidation Session, reviewing or sustaining the results discussed in the initial sessions.

This PHA presents a general picture of the risk associated with the cases studied, being valid for the conditions exposed throughout this paper. Despite the consistency of these results, it is noteworthy that the study presented does not aim to support by itself a final verdict on the feasibility of different types of operations from the point of view of risk. In-depth analysis of each specific hazard event (e.g., application of quantitative risk analysis techniques such as those discussed in Maturana & Martins, 2019, and in Maturana et al. 2021) can change the event risk classification considered in the PHA and, ultimately, reclassify as tolerable or moderate hazard events preliminarily classified as non-tolerable, or even reveal their intractability – reducing the conservatism adopted during the PHA (Martins & Maturana, 2010, 2013).

This deepening in the understanding of hazard events will also be useful to support the analysis of the feasibility of the proposed mitigation and contingency measures, allowing risk sensitivity studies to these measures and the comparison of their cost-benefit relationships – it is expected that the implementation of the mitigation and contingency measures proposed in the PHA will reduce the risk associated with the operation; the quantification of this reduction, however, was not evaluated in this PHA.

## REFERENCES

Abreu, D.T.M.P., Maturana, M.C., López Droguett, E.A. & Martins, M.R. 2020. Human Reliability Analysis of Ship Maneuvers in Harbor Areas. Journal of Offshore Mechanics and Arctic Engineering 142 (6).

CCPS. 2008. Guidelines for Hazard Evaluation Procedures. Center for Chemical Process Safety. New York: John Wiley & Sons, Inc., 2008.

Department of Defense. 2000. Military Standard System Safety Program Requirements. MIL-STD-882D. Washington, DC: s.n., 2000.

DHN. 2022. Carta Raster do Porto de Itaguaí. Número 1623. Diretoria de Hidrografia da Marinha (DHN), Centro de Hidrografia da Marinha (CHM). Available at: https://www.marinha.mil.br/chm/dados-do-segnav/cartas-raster

Greenberg, H. R. & Cramer, J. J. 1991. Risk Assessment and Risk Management for Chemical Process Industry. New York: Van Nostrand Reinhold, 1991.

Hammer, W. 1972. Handbook of System and Product Safety. New York: Prentice Hall, Inc., 1972.

IMO. 2015. Guidelines for Formal Safety Assessment (FSA) for use in the IMO rule-making process. s.l.: INTERNATIONAL MARITIME ORGANIZATION, 2015.

Martins, M.R. & Maturana, M. C. 2013. Application of Bayesian Belief networks to the human reliability analysis of an oil tanker operation focusing on collision accidents. Reliability Engineering & System Safety 110, 89–109.

Martins, M.R. & Maturana, M.C. 2010. Human error contribution in collision and grounding of oil tankers. Risk Analysis: An International Journal 30 (4), 674–698.

Maturana, M.C. & Martins, M. R. 2019. Technique for Early Consideration of Human Reliability: Applying a Generic Model in an Oil Tanker Operation to Study Scenarios of Collision. Journal of Offshore Mechanics and Arctic Engineering 141 (5).

Maturana, M.C., Martins, M.R. & Melo, P.F.F.F. 2021. Application of a quantitative human performance model to the operational procedure design of a fuel storage pool cooling system. Reliability Engineering & System Safety 216.

PETROBRAS. 2015. N-2782. Técnicas Aplicáveis à Análise de Riscos Industriais. s.l.: CONTEC - Comissão de normatização Técnica, 2015.
PIANC. 2014. Harbor approach Channels. Design Guidelines. Maritime Navigation. Commission, Report No. 121.
Queiroz Filho, A.N., Zimbres, M. Tannuri, E. A. 2014. Development and Validation of a Customizable DP System for a Full Bridge Real Time Simulator. In: ASME 33th International Conference on Ocean, Offshore and Arctic Engineering OMAE 2014, EUA.
Stephenson, J. 1991. System Safety 2000 - A Practical Guide for Planning, Managing, and Conducting System Safety Programs. New York: Van Nostrand Reinhold, 1991.
Tannuri, E.A., Rateiro, F., Fucatu, C. H., Ferreira, M. D., Mastti, I.Q., Nishimoto, K. 2014. Modular Mathematical Model for a Low-Speed Maneuvering Simulator. In: ASME 33th International Conference on Ocean, Offshore and Arctic Engineering OMAE 2014, EUA.
Tannuri, E.A. & Martins, G.H.A. 2018. Application of a maneuvering simulation center and pilots expertise to the design of new ports and terminals and infraestructure optimization in Brazil. PIANC-World Congress Panama City, Panama.
USACE. 1980. Engineering and Design–Layout and Design of Shallow-Drift Waterways. U.S. Army Corps of Engineers, Engineering Manual 1110-2-1611, Washington, DC, December.

# 6. APPENDIX 1

| Phase Description | | | | Channel navigation (entering) | | Phase ID | B | | | | | | | | | | | |
|---|---|---|---|---|---|---|---|---|---|---|---|---|---|---|---|---|---|---|
| | Scenario | | | | | Cause | | | Consequences | | | Consequence categories | | | Risk class | | | |
| Index | Standard vessel | Loading | Tidal | Hazard event | Description of the hazard event | Description | Frequency | Mitigation measures | Description | Damage (environmental, property, people) | Contingency measures | Environment (En) | Property (Pr) | People (Pe) | En | Pr | Pe | FJ |
| B.1.1.1.1 | Suezmax conventional | Fully loaded | Flood | Failure in the steering system (rudder locked in neutral position) | Failure of steering system causes rudder lock at 0°. | • failure of physical components of the system without the possibility of timely maintenance. | Unlikely (C) | | • grounding; • collision with signal buoys. | • negligible environmental damage; • severe damage to the ship; • loss of signal buoy; • minor injuries to people. | • establish reference speed ranges along the channel; • actions to ensure speed control during the maneuver; • employment of the second pilot; • rectify the navigation channel between buoys 10 and 14. | Negligible (I) | Critical (IV) | Marginal (II) | T | M | T | M |
| B.1.1.1.2 | Suezmax conventional | Fully loaded | Flood | Failure in the steering system (rudder locked in non-neutral position) | Failure of the steering system causes the rudder to lock in a non-neutral position. | • failure of physical components of the system without the possibility of timely maintenance. | Unlikely (C) | | • grounding; • collision with signal buoys. | • negligible environmental damage; • severe damage to the ship; • loss of signal buoy; • minor injuries to people. | • establish reference speed ranges along the channel; • actions to ensure speed control during the maneuver; • employment of the second pilot; • rectify the navigation channel | Negligible (I) | Critical (IV) | Marginal (II) | T | M | T | M |

| | | | | | | | | | | | | | | | | | | |
|---|---|---|---|---|---|---|---|---|---|---|---|---|---|---|---|---|---|---|
| | | | | | | | | | | | between buoys 10 and 14. | | | | | | | |
| B.1.1.1.3 | Suezmax conventional | Fully loaded | Flood | Propulsion system failure | Loss of propulsion due to failure of mechanical systems. | ▪ failure of physical components of the propulsion system without the possibility of timely maintenance. | Remote (B) | | ▪ grounding;<br>▪ collision with signal buoys. | ▪ negligible environmental damage;<br>▪ severe damage to the ship;<br>▪ loss of signal buoy;<br>▪ minor injuries to people. | ▪ establish reference speed ranges along the channel;<br>▪ actions to ensure speed control during the maneuver;<br>▪ employment of the second pilot;<br>▪ rectify the navigation channel between buoys 10 and 14. | Negligible (I) | Critical (IV) | Marginal (II) | T | M | T | M |
| B.1.1.1.4 | Suezmax conventional | Fully loaded | Flood | Blackout | Failure in the ship's power generation or distribution systems leaves the ship adrift. | ▪ failure of physical components of generation, distribution systems or their auxiliaries without the possibility of timely maintenance. | Remote (B) | | ▪ grounding;<br>▪ collision with signal buoys. | ▪ negligible environmental damage;<br>▪ severe damage to the ship;<br>▪ loss of signal buoy;<br>▪ minor injuries to people. | ▪ establish reference speed ranges along the channel;<br>▪ actions to ensure speed control during the maneuver;<br>▪ employment of the second pilot;<br>▪ rectify the navigation channel between buoys 10 and 14. | Negligible (I) | Critical (IV) | Marginal (II) | T | M | T | M |
| B.1.1.1.5 | Suezmax conventional | Fully loaded | Flood | Adverse wind conditions | Occurrence of unfavorable wind conditions during the maneuver, both in terms of intensity and direction. | ▪ error in monitoring/predicting environmental conditions;<br>▪ sudden change in | Likely (D) | ▪ improvement of the metaoceanographic monitoring system. | ▪ grounding;<br>▪ collision with signal buoys. | ▪ negligible environmental damage;<br>▪ severe damage to the ship;<br>▪ loss of signal buoy; | ▪ establish reference speed ranges along the channel;<br>▪ actions to ensure speed control | Negligible (I) | Critical (IV) | Marginal (II) | T | NT | M | NT |

(*Continued*)

| | | | | | | | | | | | | | | | | |
|---|---|---|---|---|---|---|---|---|---|---|---|---|---|---|---|---|
| | | | | | | environmental conditions. | | | | • minor injuries to people. | during the maneuver;<br>• employment of the second pilot;<br>• rectify the navigation channel between buoys 10 and 14. | | | | | | 
| B.1.1.1.6 | Suezmax conventional | Fully loaded | Flood | Adverse current conditions | Occurrence of unfavorable current conditions during the maneuver, both in terms of intensity and direction. | • error in monitoring/predicting environmental conditions;<br>• sudden change in environmental conditions. | Likely (D) | • improvement of the metaoceanographic monitoring system. | • grounding;<br>• collision with signal buoys. | • negligible environmental damage;<br>• severe damage to the ship;<br>• loss of signal buoy;<br>• minor injuries to people. | • establish reference speed ranges along the channel;<br>• actions to ensure speed control during the maneuver;<br>• employment of the second pilot;<br>• rectify the navigation channel between buoys 10 and 14. | Negligible (I) | Mean (III) | Marginal (II) | T | M | M | M |
| B.1.1.1.7 | Suezmax conventional | Fully loaded | Flood | Human error in maintaining control of the vessel | Human control error causes a dangerous situation that leads the ship to significantly deviate from its route. | • wrong execution of command;<br>• wrong command order. | Unlikely (C) | • second pilot. | • grounding;<br>• collision with signal buoys. | • negligible environmental damage;<br>• severe damage to the ship;<br>• loss of signal buoy;<br>• minor injuries to people. | • establish reference speed ranges along the channel;<br>• actions to ensure speed control during the maneuver;<br>• employment of the second pilot;<br>• rectify the navigation channel between buoys 10 and 14. | Negligible (I) | Critical (IV) | Marginal (II) | T | M | T | M |

| | | | | | | | | | | | | | | | | | | |
|---|---|---|---|---|---|---|---|---|---|---|---|---|---|---|---|---|---|---|
| B.1.1.1.8 | Suezmax conventional | Fully loaded | Flood | Delay in tug-boat response | Tug response time inadequate to the needs of the maneuver. | • tug commander response latency;<br>• operating conditions unfavorable to the efficient performance of the tug. | Unlikely (C) | | • grounding;<br>• collision with signal buoys. | • negligible environmental damage;<br>• severe damage to the ship;<br>• loss of signal buoy;<br>• minor injuries to people. | • establish reference speed ranges along the channel;<br>• actions to ensure speed control during the maneuver;<br>• employment of the second pilot;<br>• rectify the navigation channel between buoys 10 and 14. | Negligible (I) | Critical (IV) | Marginal (II) | T | M | T | M |
| B.1.1.1.9 | Suezmax conventional | Fully loaded | Flood | Unavailability of tugboat | Mechanical failure leads to the tug's inability to perform its function. | • failure of the tug's propulsion and/or steering systems;<br>• blackout of the tug. | Unlikely (C) | | • grounding;<br>• collision with signal buoys. | • negligible environmental damage;<br>• severe damage to the ship;<br>• loss of signal buoy;<br>• minor injuries to people. | • establish reference speed ranges along the channel;<br>• actions to ensure speed control during the maneuver;<br>• employment of the second pilot;<br>• rectify the navigation channel between buoys 10 and 14. | Negligible (I) | Critical (IV) | Marginal (II)) | T | M | T | M |
| B.1.1.1.10 | Suezmax conventional | Fully loaded | Flood | Towing rope breakage | Tow rope breaks during maneuvering. | • excessive strain on the tow rope;<br>• degraded tow rope. | Unlikely (C) | | • collision with signal buoys;<br>• towing rope hits ship's crew member. | • loss of signal buoy;<br>• crew member serious injury or fatality. | • establish reference speed ranges along the channel;<br>• actions to ensure speed control during the maneuver;<br>• employment of the second pilot; | Negligible (I) | Mean (III) | Critical (IV) | T | M | M | M |

(*Continued*)

| | | | | | | | | | | | | | | | | | | |
|---|---|---|---|---|---|---|---|---|---|---|---|---|---|---|---|---|---|---|
| | | | | | | | | | | | • rectify the navigation channel between buoys 10 and 14. | | | | | | | |
| B.1.1.1.11 | Suezmax conventional | Fully loaded | Flood | Error in communication with tugboats | Misinterpretation of command given to tugboat. | • human error of the pilot or tug commander;<br>• interference with communication equipment. | Likely (D) | | • grounding;<br>• collision with signal buoys. | • negligible environmental damage;<br>• moderate damage to the ship;<br>• loss of signal buoy;<br>• minor injuries to people. | • establish reference speed ranges along the channel;<br>• actions to ensure speed control during the maneuver;<br>• employment of the second pilot;<br>• rectify the navigation channel between buoys 10 and 14. | Negligible (I) | Mean (III) | Marginal (II) | T | M | M | M |
| B.1.1.1.12 | Suezmax conventional | Fully loaded | Flood | Interaction with submarine | Submarine leaving base without prior notice | • communication failure;<br>• lack of procedure. | Unlikely (C) | • establish communication procedure;<br>• submarine establish alternative channel. | • collision with submarine;<br>• collision with signal buoys. | • negligible environmental damage;<br>• light damage to the ship;<br>• moderate damage to the submarine. | | Negligible (I) | Critical (IV) | Marginal (II) | T | M | T | M |

*Trends in Maritime Technology and Engineering – Guedes Soares & Santos (eds)*
 *ISBN 978-1-032-33583-4*

# Risky maritime encounter prediction via ensemble machine learning

M.F. Oruc & Y.C. Altan
*Özyeğin University, Istanbul, Turkey*

ABSTRACT: As the demand for maritime transport is enhancing, ship-ship encounters are increasing. Narrow and congested waterways are primarily risky areas for potential consequences. The purpose of this study is to predict non-accidental risky encounters between two ships without distance between ships as a model variable on narrow and congested waterways via machine learning. A novel framework is developed to model risky encounters as predictable events. Site specific ship domain approach is used as the risk labeling criteria. To overcome imbalance in the nature of non-accidental critical events, ensemble machine learning algorithms are adopted. Historical navigational statistics of different sectors of the waterway are integrated to prediction model. The approach is tested on a historical AIS dataset from the Strait of Istanbul. To evaluate the methodology, accuracy, precision, recall and roc-auc metrics are used. K-fold cross validation and permutation based feature importance tests are performed. Each 4 out of 5 risky encounters are successfully predicted. Developed methodology can be used by vessels as an early potential collision avoidance alert system.

## 1 INTRODUCTION

Maritime traffic safety is being defined as an accelerating concern by researchers and authorities (Montewka et al., 2014). With the increasing trend in international trade, demand for maritime transport is deepening. Larger and faster ships are being built each year to meet the growing demand. The increasing number of vessels is being primarily observed in narrow and congested waterways and ports (Mou et al., 2010). Ship traffic is getting more complex in these areas, and the increasing volume of traffic is resulting in more ship-ship encounters (Altan & Otay, 2017). This trend proposes fundamental risks on ship-ship collisions, as this type of accident is one of the most frequent ones and grounding (Soares and Teixeira, 2001). Collisions have potentially drastic consequences in waterways and ports, where high traffic density and urban habitat are present. To overcome these risks, innovative methodologies are needed (Zhang et al., 2021).

Researchers developed various frameworks to mitigate the risks of ship-ship collisions. Previous works include traffic simulation methodologies (Bergström et al., 2016; Goerlandt and Kujala, 2011; Mazurek et al., 2020), statistical analysis (Bye & Aalberg, 2018; Kum & Sahin, 2015), probabilistic and stochastic modeling (Otay et al., 2003; Zhang et al., 2013), and Bayesian networks (Dinis et al., 2020; Montewka et al., 2010; Zhang et al., 2020). While these methods provide valid analyses to model conditions on the occurrence of collisions, they rely heavily on historical records of accidents and their surrounding conditions. With focusing on past collisions, these studies have been subject to certain limitations. First, ship-ship collisions are infrequent in the historical record, considering the richness of vessel movements (Eleftheria et al., 2016). Also, these records provide a limited set of details on the occurrence conditions (Lappalainen et al., 2011). Other issues related to accident statistics are inaccuracy of data, subjective reporting conditions, and presence of data in specific regions, which is an issue of geographical conditions based bias (Sormunen et al., 2016). On the other hand, wide-scale adoption of the Automatic Identification System (AIS) provided detailed intelligence on vessel movements.

Researchers recently adopted proxy methodologies to better understand ship-ship collisions with the AIS data as a crucial resource (Du et al., 2020). Rather than analyzing the accident itself, emphasis is put on risky encounters or near misses in ship-ship encounters. Examining risky encounters proposes significant advantages when compared to collisions. As they happen more frequently in traffic conditions, AIS is available to set up large databases of

DOI: 10.1201/9781003320289-27

encounters to conduct sound statistical learning methods. With the AIS being transmitted from everywhere in the world, risky encounters enable examination of specific local conditions and geographically unbiased modeling with up to date data. Among techniques to model near misses, ship domain has been an important measure to detect the level of risk associated with each encounter (Szlapczynski & Szlapczynska, 2021).

Researchers use ship domains to assess encounter risk. (Fujii & Tanaka, 1971) have initially introduced ship domain in their leading research. Fujii and Tanaka described the ship domain as the area around the own ship from which other ships must tend to avoid. Empirical, knowledge-based, and analysis-based are the main ship domain methodology classifications. Approaches have increasingly been used in collision risk estimation and non-accidental critical event research. (Weng et al., 2012) estimated vessel collision frequency in the Singapore Strait via ship domain violation principles. (Rong et al., 2015) used ship domain to detect near-collision scenarios in Tagus River. (Watawana & Caldera, 2018) used machine learning to classify non-accidental critical events via ship domain approach. (Goerlandt et al., 2017) have implemented domain-based analyses of near collisions in the Baltic Sea. (van Iperen, 2015) adopted an empirical ship domain to analyze encounter types and respective suitable domain geometries. (Zhang et al., 2016, 2015) proposed the vessel conflict ranking operator (VCRO) to assess conflict risks based on encounter conditions, using the elliptical ship domain. While previous studies have analyzed the conditions that lead to near misses, potential future collision predictions on the individual ship level have not been conducted. Assessment of future risks for vessels during their cruise may help to mitigate risks and help captains make more informed decisions.

Vessel Traffic Service (VTS) and AIS are helpful resources for improving captains' decisions. VTS presents information about vessels in the area and assists navigational decisions through images and relevant maritime information. With the help of improved situational awareness, potential maritime risks are overseen. AIS is also helpful in navigational awareness, while the difference is that captains can track other ships at greater distances. Ships can plan courses before approaching ports and waterways by overcoming possible conflicts through this.

Among AIS and VTS, the versatility and volume of AIS data present opportunities. AIS represents historical records of messages, where detailed information about ships' location, speed, size, heading, and other relevant properties can be found. Regular and orderly transmission of AIS messages helps advanced data manipulation and mining techniques be implemented to analyze deeper than individual message content. With its vast availability and versatility, AIS can be used to conduct machine learning based predictions on potential ship-ship collisions. The most commonly adopted machine learning methodology can be mentioned as clustering in the scope of research on near misses and ship-ship collisions (Rong et al., 2021). Recently, (Zhang et al., 2020)'s work demonstrated an implementation of convolutional neural network (CNN) to classify risky encounters via modelling navigational data as representative images, where expert judgments are benefited. However, the main emphasis has been on identifying the risk associated with different encounter types and quantifying these factors. Predicting risky encounters via machine learning before the event happens is conducted by (Oruc & Altan, 2022), a rather new perspective in the current literature.

This paper introduces a method to predict potential ship-ship collisions via variables calculated through AIS data in narrow and congested waterways. The method initially contains a data mining application to model ship-ship encounters and labeling of risky and non-risky encounters through ship domain as the risk criteria. A machine learning framework is introduced based on the encounter model and risk representation to predict risky encounters without distance between two ships as a predictor variable. Bagging, boosting, and sampling techniques are used to overcome class imbalance present in the nature of risky encounters. The resulting methodology is adopted to a dataset captured from the Strait of Istanbul and presented as a case study.

The remaining part of the paper is as follows: In section 2, problem statement, application area and background on the data is presented. In section 3, methodological framework is provided. Through this section, encounter model, supervised machine learning model with ensemble methods and testing approach are described. Section 4 is formed of results and discussion. In this section, quantitative results of prediction algorithms are presented. A discussion on outcomes is detailed. In conclusions, findings and contributions are summarized.

## 2 PROBLEM STATEMENT AND APPLICATION AREA

### 2.1 *Problem statement*

The main focus of this study is predicting non-accidental critical events between two ships before they happen via non-distance related variables. To conduct prediction, scalable and explainable machine learning methodologies are used. Via excluding distance between ships from predictors, this study is positioned to contribute decision support systems for safe navigation in narrow and congested waterways via early alert mechanisms for potential collision avoidance. Also, it is aimed to provide an approach for imbalanced data focused ensemble machine learning algorithms to be used in determining potentially risky encounters.

### 2.2 *Application area*

The Strait of Istanbul (SOI), also known as the Bosphorus, is chosen as the application area for the developed methodology. The area connects the Black Sea and the Sea of Marmara. It is a part of the Turkish Straits, the only sea passage from the Black Sea to the Aegean Sea and the Mediterranean Sea. Also, the strait is positioned in the middle of Europe and Asia. In addition to its unique location, the Bosphorus is a major sea passage and home to busy transit and local maritime traffic (Akten, 2004). While transit traffic runs through the northbound and southbound lines, local traffic runs through westbound and eastbound lines to connect Europe and Asia. The SOI is one of the most challenging waterways for vessels. The complexity of the strait is sourced by the narrowness of the sea passage, crossing the intersection of local and transit traffic and present currents (Altan & Otay, 2017; Tan & Otay, 1999; Yazici & Otay, 2009). Nearly 40000 transit ships pass through the SOI, each year. With local traffic intersecting the transit traffic, the SOI is unique in the number and variety of ship-ship encounters. From the SOI, AIS data was collected between March 2014 and February 2015. The dataset has been stored in a relational database and managed through Structured Query Language (SQL). Dataset covers the size of 94 Gigabytes (GB). More details on the properties of the source data can be found in (Altan & Otay, 2017).

## 3 METHODOLOGY

In this section, the methodological approach is outlined. Encounter model, encounter variables, machine learning model, and testing process are presented. Firstly, the encounter model is introduced. In the scope of the encounter model, the transformation of the raw AIS dataset to ship-ship interactions is conducted. Also, the labeling process for risky and non-risky encounters is presented. Secondly, recorded variables from the encounter model are presented to be used in the prediction step. In the machine learning model, model building steps and utilized algorithms are demonstrated. Also, the specific approach towards the imbalanced nature of the dataset is detailed. Lastly, the testing process is presented. Through the testing process, the robustness of the model is ensured via presented techniques. Figure 1 provides the main steps the methodology.

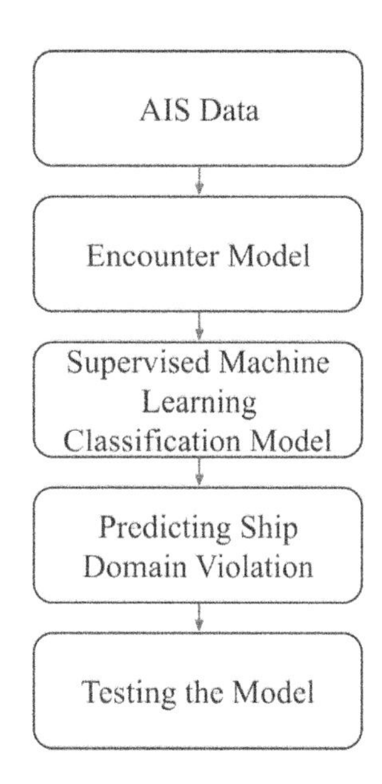


Figure 1. Methodological chart.

### 3.1 *Encounter model*

The raw AIS dataset is transformed to represent each ship-ship encounter in a certain time and geographic interval to analyze ship-ship interactions. The resulting encounter model has been the basis for applied machine learning algorithms to predict near accidental critical events. The ship domain approach has been utilized as the safety criteria between two ships in the model development process. In the resulting ship-ship interaction dataset, each encounter is labeled in binary classification as a risky or non-risky encounter based on ship domain violation occurrence.

A spatial scanning algorithm is introduced to understand vessel interactions. Detected target vessels are recorded around each vessel and these positionings are defined as encounters between ships, based on the closest passage between each vessel. To calculate the distance between each ship for the case of encounter, Equation 2 is utilized. To calculate distance between two vessels, longitude and latitude of each vessel are used. To overcome synchronization issue in AIS messages' exact time, a linear interpolation process is conducted. Below, Equation 2 is presented. Here, (xi,yi) represent the position of the own ship. (xj,yj) represents the position of the target ship for the exact moment.

$$D = \sqrt{(x_i - x_j)^2 + (y_i - y_j)^2} \tag{1}$$

Each encounter situation is examined based on the ship domain from the perspective of the own ship to label as a risky or non-risky encounter. To define the safety measure, approach by (Fujii and Tanaka,

1971) has been adopted. In the adopted ship domain approach, the area that should not be violated by the target ship around the own ship. Ship domain is defined in circular geometry to assure symmetry around vessels, also in line with (Mou et al., 2010; Weng et al., 2012). To calculate violation of ship domain from the perspective of the own ship, Equation 2 is utilized. In this equation, D is distance between two ships, l is the length of the own ship, and r is the ship domain radius for the own ship. Violation indicator is defined as:

$$f(x) = \begin{cases} 0, & D - (l * r) > 0 \\ 1, & D - (l * r) \leq 0 \end{cases} \quad (2)$$

Circular ship domain is calculated via a site-specific ship domain approach (Altan and Meijers, 2021). Based on site-specific conditions, two different circle radii multipliers are defined for ships dependent on their LOA as presented in Table 1. These multipliers are 1.75 for ships shorter than 157 m and 3 for ships longer than 157 m, which is the length threshold for the SOI ship domain approach provided by (Altan and Meijers, 2021).

Table 1. Ship domain calculation based on length threshold.

| Ship Length (m) | Site Specific Ship Domain (m) |
|---|---|
| Length (LOA) ≤ 157 m | 1.75*LOA |
| Length (LOA) > 157 m | 3*LOA |

This approach can be described as an empirical ship domain, based on (Szlapczynski & Szlapczynska, 2021)'s classification. A limitation of the encounter model is on COLREGs not being taken into account, where the passage from each side of the ship is modeled as the same degree of violation or non-violation of ship domain. As a result, each vessel encounter is represented, and binary labeling of risky and non-risky encounters is followed. Table 2 represents the encounter model. A more detailed discussion on the encounter model can be found in (Oruc and Altan, 2022).

## 3.2 *Model variables*

AIS data present a rich source of information on vessels and their static and dynamic properties. As a result of the encounter model, dynamic information of each vessel at the exact moment of encounter is acquired. Also, static information of each vessel is also obtained from the records. These variables present the basis for developing prediction model types of encounters. Along with individual vessel-based information, site-specific statistics based on the location of each encounter are also integrated to the machine learning model. To specify location of each encounter, the SOI is represented in 13 different sectors, as provided in Figure 2. In preparation of sectoral division, navigational pattern changes are benefited, and each sector is represented with minimum of navigational patterns' change.

In the scope of this study, two types of variables are integrated into the prediction model, namely numeric data in continuous form and numeric data in categorical form. Individual vessel-based information is represented in the continuous form. Site-specific statistics for encounter occurrence time and location are represented as categorical data. Utilized continuous data for individual vessels are the own ship's length and the target ship's speed.

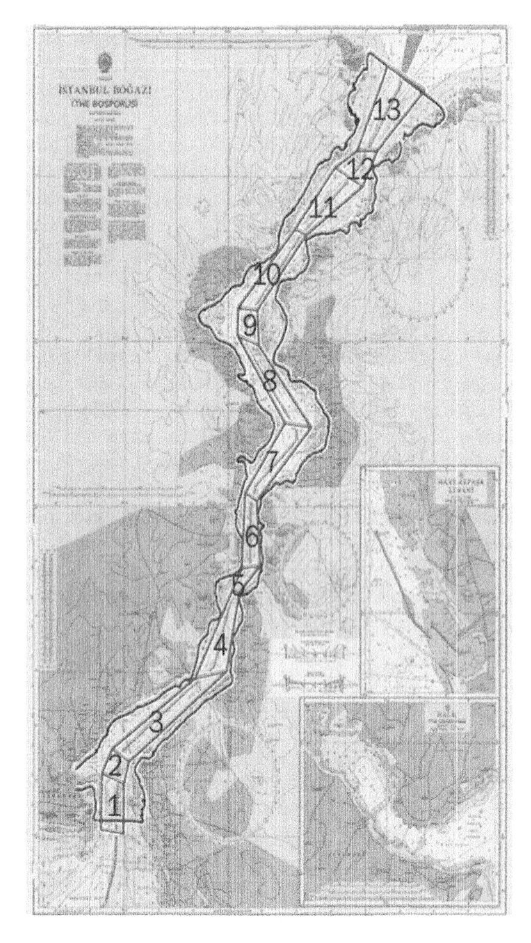


Figure 2. Sectors of the Strait of Istanbul (Altan and Otay, 2017).

Among two types of inputs, categorical data are calculated from the historical AIS dataset and define each information about each encounter's sector along the SOI. These variables are sectoral historical traffic density, sectoral average historical vessel speed for navigational traffic direction, sectoral average heading for each navigational traffic direction, based on the northern hemisphere.

Table 2. Overview of the machine learning model variables.

| Field Explanations | Variables | Values |
|---|---|---|
| Inputs for the Machine Learning Model: Continuous Fields | Own Ship's Length (m) | 157 |
| | Compared Ship's Speed (m/s) | 4.5 |
| | Traffic Density (Ships/hour/nm^2) | 15.9 |
| Inputs for the Machine Learning Model: Categorical Fields Based on Each Encounter's Sector | Northbound Average SOG (knots) | 8.68 |
| | Southbound Average SOG (knots) | 11.12 |
| | Eastbound Average SOG (knots) | 10.76 |
| | Westbound Average SOG (knots) | 11.06 |
| | Northbound Average Heading (0-360 Degree) | 1.87 |
| | Southbound Average Heading (0-360 Degree) | 192.69 |
| | Eastbound Average Heading (0-360 Degree) | 100.11 |
| | Westbound Average Heading (0-360 Degree) | 302.97 |
| Output for the Machine Learning Model | Violation Indicator (0 or 1) (Site-specific Ship Domain) | 1 |

Detailed breakdown of sectoral values can be found in (Altan and Otay, 2017). In Table 2, a representation of the inputs and output of the machine learning model is provided. This table represents a single encounter record between ships, while entire dataset includes 73.543 encounters.

### 3.3 *Machine learning based prediction model*

The former steps are followed to implement supervised machine learning applications to predict risk associated with each encounter. Processing raw AIS data to an encounter model, labelling each encounter based on safety criteria, and preparing model variables laid the foundations for prediction.

Through violation indicator in Equation 2, each ship-ship encounter is labeled to be either risky or non-risky encounter. Since ship domain violation is one of the most prominent and basic safety measures, considering the complexity of machine learning based prediction's introduction, provided safety criteria is selected. Distance between ships is only used in risk labelling step, which is the output of the machine learning model. Among inputs of the prediction model, distance is entirely excluded. Through this approach, it became possible to conduct prediction independent of the distance. In short, considering the distance's direct impact in the output of the machine learning model, exclusion of distance from inputs of the prediction model has been required. This approach of prediction helps to identify key factors of risky encounters except close proximity and provides a framework to alert maritime authorities before a risky encounter happens.

Since safe interaction between ships vastly outnumber risky interactions, encounter dataset for ship-ship interactions can be identified as an imbalanced dataset. Traditional machine learning algorithms have difficulty predicting occasions that are minority in a given dataset, where majority data points largely outnumber one another (Sun et al., 2009). To overcome the challenge presented by imbalance, machine learning algorithms tailored for imbalanced datasets are developed and applied in the scope of this paper.

Several approaches are helpful to deal with class imbalance. Oversampling and undersampling are leading methodologies to overcome the imbalance in the classification algorithms. Oversampling refers to generation of synthetically created data points from minority classes to improve the balance in the data and increase the robustness of the prediction (Quinlan, 1996). Undersampling can be explained as rearranging the data points belonging to the majority class, so that label distribution in a dataset can be close to each other (Liu et al., 2009). Boosting and bagging are also other leading properties of ensemble learning. The bagging technique refers to the sampling of training data to smaller groups, where different algorithms can run on different samples, and each resulting classification can be compared. Boosting is also similar to bagging, where the difference is on the weighted approach (Breiman, 2001). In boosting, each misclassified item class on different training set samples obtains higher weights to be classified correctly in future samples. This way, if a class is being largely misclassified, boosting balances the situation. This paper uses EasyEnsemble (Liu, 2009), RUSBoost (Seiffert et al., 2008) and XGBoost (Chen & Guestrin, 2016) meta-algorithms to predict risky encounters. Since each meta-algorithm requires a base algorithm, logistic regression and decision tree are selected as the base algorithms. Logistic regression conducts linear separations in the feature space. Then, it models the probability of class assignments for each data point. The decision tree lies on the principle of dividing the input space to smaller parts, then being further divided by other features along the path (Alpaydin, 2004). Based on the separations of data based on interval boundaries, each data point gets to be assigned to a particular class. Algorithms are two of the most used ones in data mining applications, due to their explainability and robustness. Python packages of algorithms are used. Applicable versions of algorithms are present in scikit-learn (Pedregosa Fabian et al., 2011), imbalanced-learn (Lemaıtre et al., 2017) and xgboost (Chen &

Guestrin, 2016) packages. Models are coded in Python open source programming language.

### 3.4 *Testing the model*

In the testing process, k-fold cross-validation technique is used to ensure robustness of the model. Also, to demonstrate relative contribution of features to the model, permutation importance test is implemented (Breiman, 2001). With the help of this test, model interpretability is ensured and each variable's contribution to the prediction is visualized. K-fold cross-validation works via dividing data to k randomly generated groups, conducting training on each k-1 part of the data, and testing via the remaining part. This approach helps to reduce the impact of biased randomization of training data; this way, each subsegment of the data is tested. In order to evaluate the results of predictions, confusion matrix based evaluation metrics are used. Confusion matrices are one of the most valuable methodologies to assess outcomes of classification models (James et al., 2000). Table 3 represents a confusion matrix.

The imbalanced nature of the dataset requires a specific approach to evaluate prediction results. Precision, recall and ROC-AUC score are the primary metrics to understand results. The precision score is the rate of true positives with proportion to sum of true positives and false positives. Precision score explains the selectiveness of an algorithm while classifying a data point as positive, which is an instrumental approach in imbalanced datasets (Sun et al., 2009). Recall score is the rate of true positives to the sum of true positives and false negatives. Recall score determines the correctness of classifying minority classes. ROC-AUC score is one of the primary ways to determine the overall quality of a predictor's performance. The score is a representative of the area under the Receiver Operating Characteristics curve, which is a graphical representation of binary classification success (Fawcett, 2006). Equations (3-5) represent the equation for each presented metric based on provided confusion matrix.

Table 3. Confusion matrix.

| | | Condition | |
|---|---|---|---|
| | Total | False | True |
| Prediction | False | True Negative | False Negative |
| | True | False Positive | True Positive |

$$\frac{True\ Positive + True\ Negative}{Total\ Population} = Accuracy \quad (3)$$

$$\frac{True\ Positive}{True\ Positive + False\ Positive} = Precision \quad (4)$$

$$\frac{True\ Positive}{True\ Positive + False\ Negative} = Recall \quad (5)$$

## 4 RESULTS AND DISCUSSION

The developed approach is performed on the AIS dataset collected from the Strait of Istanbul. The ship-ship encounter model is implemented to obtain an encounter dataset based on the raw AIS dataset. Statistical results are integrated into the dataset among individual ship-based model variables as categorical variables. The resulting dataset covers 73.543 ship-ship encounters, where each encounter is labeled as either risky or non-risky encounters in binary order. On the encounter model, machine learning based prediction algorithms are implemented. Decision Tree and Logistic Regression are used as base algorithms for EasyEnsemble and RUSBoost meta-algorithms. Also, XGBoost algorithm's results are also presented with evaluation metrics. 10-Fold cross-validation technique has been applied to ensure the robustness of the model. Each algorithm's evaluation metrics are presented as the average of 10 folds of cross-validation. Table 4 presents the results for each prediction model.

Table 4. Predicting risky encounters: Results for ensemble machine learning methodologies.

| | Mean Score from Test Set in K-Fold (7-Fold) Validation | | | |
|---|---|---|---|---|
| | Precision | Recall | Accuracy | ROC-AUC |
| Easy Ensemble L.R. | 0.6167 | 0.7895 | 0.816 | 0.8296 |
| Easy Ensemble D.T. | 0.6487 | 0.7801 | 0.8222 | 0.8494 |
| RUS Boost L. R. | 0.5163 | 0.7216 | 0.7523 | 0.7681 |
| RUS Boost D. T. | 0.6495 | 0.7778 | 0.8216 | 0.844 |
| XGBoost | 0.7268 | 0.7003 | 0.8079 | 0.8632 |

Results show that risky encounters can be correctly predicted without distance-related variables with certain success. Based on results, EasyEnsemble with Logistic Regression outperformed other algorithms in terms of recall score. XGBoost achieved the highest precision score. Also, the highest ROC-AUC score has been achieved by the same algorithm. The interpretation would be on the selectiveness of XGBoost algorithm to correctly predict the risky encounters, while minimizing the number of false positives. RUSBoost and EasyEnsemble based models result with higher recall scores and less precision scores. These models present higher recall scores than XGBoost, where these algorithms correctly classify a higher number of risky encounters. This variation can be interpreted as XGBoost being in the position of missing more risky encounters, while it also leads to less number of non-risky encounters to be predicted as a risky one. Both model types can be considered successful, even though a discrepancy is present. Since both types are successful in certain ways, the use case of algorithms can determine the most feasible one. XGBoost may be the leading model for the use case where minimization of false positives is prioritized. For the case of achieving the highest rate of predicting risky encounters regardless of the number of false positives, EasyEnsemble and RUSboost based models can represent a more feasible outcome.

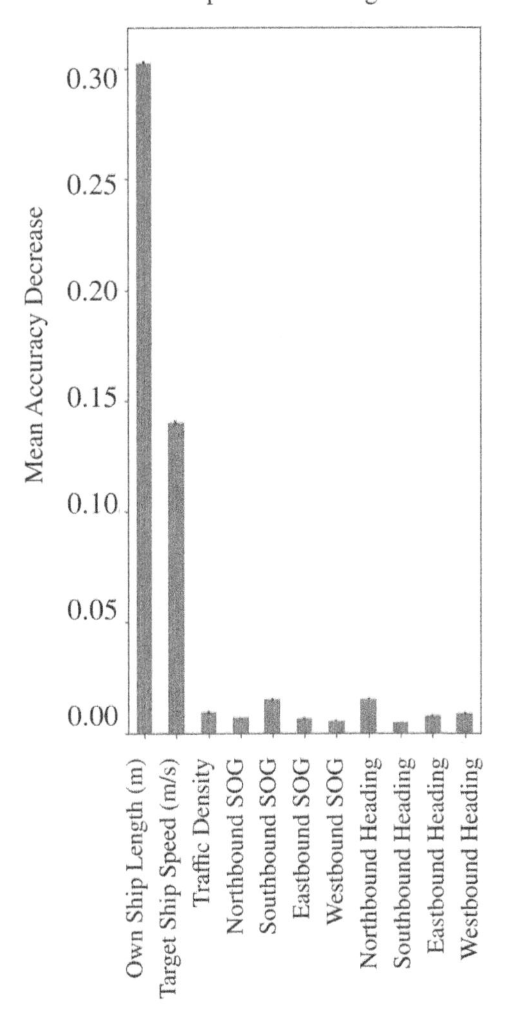


Table 5. Feature importance test results.

In Table 5, results of the feature importance test is presented. Through this test, contribution of each feature to the predictive capability of the model is measured. Results show that the most distinctive features are own ship's length and target ship's speed. Also, average speed for southbound ships and average heading for northbound ships are also other leading contributors. Own ship's length and target ship's speed can be expressed as highly influential variables while predicting risky encounters between two ships, based on the results.

## 5 CONCLUSIONS

In this study, a novel methodology to predict non-accidental critical events via non-distance related variables is presented. Through this process, site specific risk identification approach is used. The main contribution of this research is introduction of risky encounter prediction framework without distance between two ships as a decision factor. Further, ensemble learning methodologies on the imbalanced ship-ship encounter dataset are demonstrated. Developed approach is tested on a dataset captured from the Strait of Istanbul and presented as a case study. Results show that almost each 4 out of 5 risky encounters can be successfully predicted.

After development of the ship-ship encounter model and labelling risky and non-risky encounters via ship domain as the risk criteria, supervised machine learning algorithms are applied. Through this approach, a framework for predicting risky encounters before they obtain presence in close proximity is presented. Specifically, ensemble learning based algorithms with bagging and boosting methodologies are adopted. To ensure robustness, 7-fold cross validation is applied. Also, permutation based variable importance test is presented. EasyEnsemble and XGBoost algorithms provided the most successful results on different evaluation metrics. XGBoost has been the most successful model in terms of precision with 0.7268 and roc-auc score as 0.8632, while EasyEnsemble has achieved the highest recall score with 0.7895.

Since the model development process relies on historical AIS data, presented approach can be applied and tested in other narrow and congested waterways. The study can be further improved via integrating other potential variables, such as weather conditions and ocean currents to predict potential accidents. Also, different risk criteria can be implemented to benchmark with the ship domain approach.

## REFERENCES


Akten, N., 2004. Analysis of Shipping Casualties in the Bosphorus. J. Navig. 57, 345–356. http://dx.doi.org/10.1017/S0373463304002826 .

Alpaydin, E., 2004. Introduction to Machine Learning. The MIT Press.

Altan, Y.C., Meijers, B.M., 2021. Ship Domain Variations in the Strait of Istanbul, in: WCTRS SIGA2 2021 Conference. Antwerp, Belgium.

Altan, Y.C., Otay, E.N., 2017. Maritime Traffic Analysis of the Strait of Istanbul based on AIS data. J. Navig. 70, 1367–1382. http://dx.doi.org/10.1017/S0373463317000431

Bergström, M., Erikstad, S.O., Ehlers, S., 2016. Assessment of the applicability of goal- and risk-based design on Arctic sea transport systems. Ocean Eng. 128, 183–198. http://dx.doi.org/10.1016/J.OCEANENG.2016.10.040

Breiman, L., 2001. Random forests. Mach. Learn. 45, 5–32. http://dx.doi.org/10.1023/A:1010933404324

Bye, R.J., Aalberg, A.L., 2018. Maritime navigation accidents and risk indicators: An exploratory statistical analysis using AIS data and accident reports. Reliab. Eng. Syst. Saf. 176, 174–186. http://dx.doi.org/10.1016/j.ress.2018.03.033

Chen, T., Guestrin, C., 2016. XGBoost: A scalable tree boosting system. Proc. ACM SIGKDD Int. Conf. Knowl. Discov. Data Min. 13-17-August-2016, 785–794. http://dx.doi.org/10.1145/2939672.2939785

Dinis, D., Teixeira, A.P., Guedes Soares, C., 2020. Probabilistic approach for characterising the static risk of ships using Bayesian networks. Reliab. Eng. Syst. Saf. 203, 107073. http://dx.doi.org/10.1016/J.RESS.2020.107073

Du, L., Goerlandt, F., Kujala, P., 2020. Review and analysis of methods for assessing maritime waterway risk based on non-accident critical events detected from AIS data. Reliab. Eng. Syst. Saf. http://dx.doi.org/10.1016/j.ress.2020.106933

Eleftheria, E., Apostolos, P., Markos, V., 2016. Statistical analysis of ship accidents and review of safety level. Saf. Sci. 85, 282–292. http://dx.doi.org/10.1016/J.SSCI.2016.02.001

Fawcett, T., 2006. An introduction to ROC analysis. Pattern Recognit. Lett. 27, 861–874. http://dx.doi.org/10.1016/j.patrec.2005.10.010.

Fujii, Y., Tanaka, K., 1971. Traffic Capacity. J. Navig. 24, 543–552. http://dx.doi.org/10.1017/S0373463300022384

Goerlandt, F., Goite, H., Valdez Banda, O.A., Höglund, A., Ahonen-Rainio, P., Lensu, M., 2017. An analysis of wintertime navigational accidents in the Northern Baltic Sea. Saf. Sci. 92, 66–84. http://dx.doi.org/10.1016/j.ssci.2016.09.011.

Goerlandt, F., Kujala, P., 2011. Traffic simulation based ship collision probability modeling. Reliab. Eng. Syst. Saf. 96, 91–107. http://dx.doi.org/10.1016/J.RESS.2010.09.003

Guedes Soares, C., Teixeira, A.P., 2001. Risk assessment in maritime transportation. Reliab. Eng. Syst. Saf. 74, 299–309. http://dx.doi.org/10.1016/S0951-8320(01)00104-1

James, G., Witten, D., Hastie, T., Tibshirani, R., 2000. An introduction to Statistical Learning, Current medicinal chemistry. http://dx.doi.org/10.1007/978-1-4614-7138-7

Kum, S., Sahin, B., 2015. A root cause analysis for Arctic Marine accidents from 1993 to 2011. Saf. Sci. 74, 206–220. http://dx.doi.org/10.1016/J.SSCI.2014.12.010

Lappalainen, J., Vepsäläinen, A., Salmi, K., Tapaninen, U., 2011. Incident reporting in Finnish shipping companies. WMU J. Marit. Aff. 2011 102 10, 167–181. http://dx.doi.org/10.1007/S13437-011-0011-0

Lemaıtre, G., Nogueira, F., Aridas char, C.K., 2017. Imbalanced-learn: A Python Toolbox to Tackle the Curse of Imbalanced Datasets in Machine Learning. J. Mach. Learn. Res. 18, 1–5. http://dx.doi.org/10.5555/3122009.3122026

Liu, T.Y., 2009. EasyEnsemble and feature selection for imbalance data sets, in: Proceedings - 2009 International Joint Conference on Bioinformatics, Systems Biology and Intelligent Computing, IJCBS 2009. pp. 517–520. http://dx.doi.org/10.1109/IJCBS.2009.22

Liu, X.Y., Wu, J., Zhou, Z.H., 2009. Exploratory undersampling for class-imbalance learning. IEEE Trans. Syst. Man, Cybern. Part B Cybern. 39, 539–550. http://dx.doi.org/10.1109/TSMCB.2008.2007853

Mazurek, J., Montewka, J., Smolarek, L., 2020. A simulation model to support planning of resources to combat oil spills at sea. Dev. Collis. Grounding Ships Offshore Struct. - Proc. 8th Int. Conf. Collis. Grounding Ships Offshore Struct. ICCGS 2019 355–363. http://dx.doi.org/10.1201/9781003002420-44/SIMULATION-MODEL-SUPPORT-PLANNING-RESOURCES-COMBAT-OIL-SPILLS-SEA-MAZUREK-MONTEWKA-SMOLAREK

Montewka, J., Ehlers, S., Goerlandt, F., Hinz, T., Tabri, K., Kujala, P., 2014. A framework for risk assessment for maritime transportation systems—A case study for open sea collisions involving RoPax vessels. Reliab. Eng. Syst. Saf. 124, 142–157. http://dx.doi.org/10.1016/J.RESS.2013.11.014

Montewka, J., Hinz, T., Kujala, P., Matusiak, J., 2010. Probability modelling of vessel collisions. Reliab. Eng. Syst. Saf. 95, 573–589. http://dx.doi.org/10.1016/j.ress.2010.01.009

Mou, J.M., Tak, C. van der, Ligteringen, H., 2010. Study on collision avoidance in busy waterways by using AIS data. Ocean Eng. 37, 483–490. http://dx.doi.org/10.1016/j.oceaneng.2010.01.012

Oruc, M.F., Altan, Y.C., 2022. Predicting the risky encounters without distance knowledge between the ships via machine learning algorithms (In Review). Ocean Eng.

Otay, E.N., Otay, E.N., Özkan, Ş., 2003. Stochastic Prediction of Maritime Accidents in the strait of Istanbul. Proc. 3RD Int. Conf. OIL SPILLS Mediterr. BLACK SEA Reg. 92–104.

Pedregosa Fabian, F., Michel, V., Grisel OLIVIERGRISEL, O., Blondel, M., Prettenhofer, P., Weiss, R., Vanderplas, J., Cournapeau, D., Pedregosa, F., Varoquaux, G., Gramfort, A., Thirion, B., Grisel, O., Dubourg, V., Passos, A., Brucher, M., Perrot andÉdouardand, M., Duchesnay, A., Duchesnay EDOUARDDUCHESNAY, Fré., 2011. Scikit-learn: Machine Learning in Python. J. Mach. Learn. Res. 12, 2825–2830.

Quinlan, J.R., 1996. Bagging, Boosting, and C4.5, cs.ecu.edu.

Rong, H., Teixeira, A., Guedes Soares, C., 2015. Evaluation of near-collisions in the Tagus River Estuary using a marine traffic simulation model. Zesz. Nauk. Akad. Morskiej w Szczecinie nr 43 (115), 68–78.

Rong, H., Teixeira, A.P., Guedes Soares, C., 2021. Spatial correlation analysis of near ship collision hotspots with local maritime traffic characteristics. Reliab. Eng. Syst. Saf. 209, 107463. http://dx.doi.org/10.1016/j.ress.2021.107463

Seiffert, C., Khoshgoftaar, T.M., Van Hulse, J., Napolitano, A., 2008. RUSBoost: Improving classification performance when training data is skewed, in: Proceedings - International Conference on Pattern Recognition. Institute of Electrical and Electronics Engineers Inc. http://dx.doi.org/10.1109/icpr.2008.4761297

Sormunen, O., Hänninen, M., Kujala, P., 2016. Marine traffic, accidents, and underreporting in the Baltic Sea. undefined. http://dx.doi.org/10.17402/134

Sun, Y., Wong, A.K.C., Kamel, M.S., 2009. Classification of imbalanced data: A review. Int. J. Pattern Recognit. Artif. Intell. 23, 687–719. http://dx.doi.org/10.1142/S0218001409007326

Szlapczynski, R., Szlapczynska, J., 2021. A ship domain-based model of collision risk for near-miss detection and Collision Alert Systems. Reliab. Eng. Syst. Saf. 214, 107766. http://dx.doi.org/10.1016/J.RESS.2021.107766

Tan, B., Otay, E., 1999. MODELING AND ANALYSIS OF VESSEL CASUALTIES RESULTING FROM TANKER TRAFFIC THROUGH NARROW WATERWAYS. undefined. http://dx.doi.org/10.1002/(SICI)1520-6750(199912)46:8

van Iperen, E., 2015. Classifying Ship Encounters to Monitor Traffic Safety on the North Sea from AIS Data. TransNav, Int. J. Mar. Navig. Saf. Sea Transp. 9, 51–58. http://dx.doi.org/10.12716/1001.09.01.06

Watawana, T., Caldera, A., 2018. Analyse Near Collision Situations of Ships Using Automatic Identification System Dataset, in: 5th International Conference on Soft Computing and Machine Intelligence, ISCMI 2018. Institute of Electrical and Electronics Engineers Inc., pp. 155–162. http://dx.doi.org/10.1109/ISCMI.2018.8703228

Weng, J., Meng, Q., Qu, X., 2012. Vessel collision frequency estimation in the Singapore Strait. J. Navig. 65, 207–221. http://dx.doi.org/10.1017/S0373463311000683

Yazici, M.A., Otay, E.N., 2009. A Navigation Safety Support Model for the Strait of Istanbul. J. Navig. 62, 609–630. http://dx.doi.org/10.1017/S0373463309990130

Zhang, D., Yan, X.P., Yang, Z.L., Wall, A., Wang, J., 2013. Incorporation of formal safety assessment and Bayesian network in navigational risk estimation of the Yangtze River. Reliab. Eng. Syst. Saf. 118, 93–105. http://dx.doi.org/10.1016/J.RESS.2013.04.006

Zhang, M., Montewka, J., Manderbacka, T., Kujala, P., Hirdaris, S., 2021. A Big Data Analytics Method for the Evaluation of Ship - Ship Collision Risk reflecting Hydrometeorological Conditions. Reliab. Eng. Syst. Saf. 213, 107674. http://dx.doi.org/10.1016/J.RESS.2021.107674

Zhang, M., Zhang, D., Yao, H., Zhang, K., 2020. A probabilistic model of human error assessment for autonomous cargo ships focusing on human–autonomy collaboration. Saf. Sci. 130, 104838. http://dx.doi.org/10.1016/J.SSCI.2020.104838

Zhang, W., Feng, X., Goerlandt, F., Liu, Q., 2020. Towards a Convolutional Neural Network model for classifying regional ship collision risk levels for waterway risk analysis. Reliab. Eng. Syst. Saf. 204, 107127. http://dx.doi.org/10.1016/j.ress.2020.107127

Zhang, W., Goerlandt, F., Kujala, P., Wang, Y., 2016. An advanced method for detecting possible near miss ship collisions from AIS data. Ocean Eng. 124, 141–156. http://dx.doi.org/10.1016/j.oceaneng.2016.07.059

Zhang, W., Goerlandt, F., Montewka, J., Kujala, P., 2015. A method for detecting possible near miss ship collisions from AIS data. Ocean Eng. 107, 60–69. http://dx.doi.org/10.1016/j.oceaneng.2015.07.046

*Trends in Maritime Technology and Engineering – Guedes Soares & Santos (eds)*


# BlueSafePort Project - safety system for maneuvering and moored ships at the Port of Sines

L.V. Pinheiro, C.J.E.M. Fortes & A.H. Gomes
*Hydraulics and Environment Department, National Laboratory of Civil Engineering, Lisbon, Portugal*

J.A. Santos
*Instituto Superior de Engenharia de Lisboa, Instituto Politécnico de Lisboa, Portugal*

C. Guedes Soares
*Center for Marine Technology and Ocean Engineering (CENTEC), Instituto Superior Técnico, Universidade de Lisboa, Lisbon, Portugal*

ABSTRACT: This paper presents a safety system for maneuvering and moored ships at the Port of Sines developed in the scope of the BlueSafePort project. The system is described in terms of its development and architecture, the functioning of its numerical models, the data flow and processing and the risk assessment used to issue alerts. The web platform and the mobile application where the relevant results to assess the referred risks are disseminated are also presented. The system provides 3-day advance forecasts of sea agitation, wind and tide conditions, as well as ship motions, identifying situations that could have serious consequences for ships maneuvering or moored inside the port. The system issues alerts associated with risk levels based on safety and operability criteria for ships' motions and forces in their mooring systems.

## 1 INTRODUCTION

Port terminals downtimes lead to large economic losses and largely affect the port's overall competitiveness. It's essential to reduce the port's vulnerability by increasing its planning capacity and efficient response to emergency situations.

The BlueSafePort project aims at developing a safety system (both on web and mobile platforms) for forecasting and alerting emergency situations related to navigating, docking and moored ships in ports. A similar system was developed to the Praia da Vitória Port, followed by the ports of S. Roque do Pico and Madalena do Pico, in the Azores archipelago, under the HIDRALERTA and ECOMARPORT projects (Fortes *et al.*, 2015, 2020; Poseiro *et al.*, 2017 and Pinheiro *et al.*, 2020). Now it is being developed for the port of Sines, Portugal, adding to it the capability of forecasting risks for maneuvering (entering or leaving the port) and docking ships, and a mobile application.

The system uses available forecasts of regional wind and sea-wave characteristics offshore, together with astronomical tidal data as inputs to a set of numerical models. These numerical models provide accurate estimates of wave characteristics inside the port (including nonlinear wave interactions) and of the ship's response to those waves and wind forcings. The wave effects in terms of excessive vertical movements of a maneuvering ship that enters or leaves a harbor basin or in terms of forces on mooring lines and fenders as well as of motions of a ship moored at a quay, are then compared with pre-set maximum values. Probability assessment of exceedance of those values results in a risk level assessment. Finally, based on the forecasted risk level, emergencies, and situations where the safety of port operations is at risk can be foreseen in advance and corresponding warnings can be issued.

All the information provided by this system is available in a dedicated website and a mobile application. Port stakeholders will benefit from a decision-support tool in order to timely implement mitigation measures to avoid accidents and reduce economic losses. Therefore, this will be a valuable tool for monitoring and optimizing maritime operations in a port environment, such as, for example, route definition, berthing and mooring optimization systems.

The system prototype under development is the Port of Sines, but it will be flexible and scalable so that it can be easily replicated in other ports.

This paper is structured in 6 sections. The first section provides a brief background of the BlueSafePort project. Section 2 presents the safety system in terms

DOI: 10.1201/9781003320289-28

of its architecture and development. The development and adaptation of the numerical models implemented in the system are shown in section 3. Section 4 presents the qualitative risk assessment used as a foundation for issuing alerts to emergency situations related to navigation. Section 5 presents the web and mobile platforms where the results issued by the system will be disseminated. Finally, in the last section, the final considerations of the paper are made.

## 2 THE SAFETY SYSTEM

On a daily basis, the safety system provides, 72 hours in advance and at 3-hour intervals, forecasts and alerts of emergency situations associated with maneuvering and moored ships due to extreme met-ocean conditions. The corresponding risk levels are based on safety and operational criteria for ship movements and for the stresses on the mooring system elements.

The developed system was implemented and will be validated in three terminals of the Port of Sines. This ensures the flexibility and scalability of the proposed system.

### 2.1 *Architecture*

The SAFEPORT system is structured in 4 modules (Figure 1): I - Sea wave characterization; II - Navigation in port areas; III - Monitoring and IV - Risk assessment and forecasting.

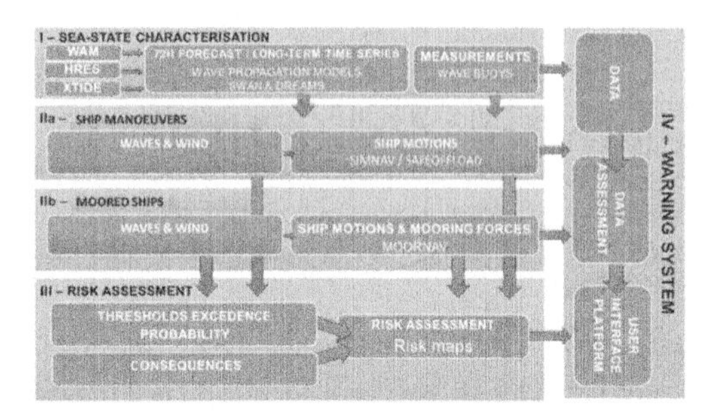


Figure 1. Safety system architecture.

The purpose of module I is the characterization of the sea agitation offshore, at the port entrance and in several points within the port basin, in terms characteristics of the wave states. In addition, this module contains the forecast of other phenomena involved in the propagation of sea waves, namely wind fields, tide levels and currents. This first module includes most of the numerical models implemented in the system, namely, the WAM, NAVGEM, XTide, SWAN and DREAMS models. It also includes sea wave data collected *in situ*, essential in the validation of the sea wave characterization models.

Module II simulates the behavior of the ships, both maneuvering and moored, providing time series of the ship motions as well as of the forces on the mooring system elements (mooring lines and fenders). The results of this module are obtained by employing the WAMIT and BAS models.

Module III deals with monitoring the behavior of the ships in maneuvering and moored, by collecting and processing in situ data, necessary to validate the results produced by the WAMIT and BAS numerical models.

Module IV concerns the qualitative analysis of the risk associated with excessive values of ship motion and forces on the ships' mooring systems. This analysis produces a risk score which is the product of the degree of occurrence, established from ranges for the probability of occurrence of the hazardous event, and the degree of consequence associated with that event.

The safety system issues warnings based on the forecasts waves, the atmospheric conditions, and the risk levels.

### 2.2 *Development*

The main actions considered in the development of the safety system are as follows:

1. Data collection and survey;
2. Installation of two pressure sensors to register sea agitation;
3. Installation of a gyroscope on a moored ship to register the movements;
4. Development and adaptation of the numerical models;
5. Numerical simulations and systematic tests to validate the results with *in situ* measurements;
6. Simulation of ship maneuvers at the entrance and exit of the port based on the information provided by the system;
7. Implementation of the system in the operational prototype (Port of Sines);
8. Feeding of a database with the operational registers in order to learn and adjust the system;
9. Development of the web and mobile applications to disseminate the results.

## 3 THE NUMERICAL MODELS

This section presents the development and adaptation of the numerical models in the operational prototype, the Port of Sines. Procedures were established (script routines), based on python programming, to obtain the forecasts from the European Centre for Medium-Range Weather Forecasts, ECMWF (Persson, 2001) which provides WAM and HIRES model simulations. XTide provides tide levels. Further scripts couple all the numerical models and create outputs in various formats (images, maps, graphs and datafiles).

### 3.1 *Operational prototype*

The system was developed and will be validated for the Port of Sines and easily adaptable to other ports.

The Port of Sines is located at 37° 57′N and 08° 53′ W, and is the only deep-water port in Portugal, with natural depths up to -28 m ZH. Three breakwaters protect the port basin: the so-called west breakwater, with a length of approximately 1600 m, and the function of directly sheltering the petrochemical (TPQ), liquid bulk (TGL) and natural gas (TGN) terminals; the so-called east breakwater, which is 1000 m long and shelters the port facilities of the multipurpose terminal (TMS); and the so-called eastern breakwater, 1500 m long and constantly extended, shelters the container terminal or terminal XXI (TCS), also in constant extension.

For the first application of this system, an oil tanker at TGL, a general cargo ship at TMS and a container ship at TCS were chosen (Figure 2).

Figure 2. Operational prototype, the Port of Sines, and the chosen terminals, TGL, TMS and TCS.

### 3.2 *Sea wave characterization models*

An early warning system is only possible due to the accuracy achieved by met-ocean forecasting models at regional and local level, which can provide, a few days in advance, estimates for the relevant parameters.

WAM is a third-generation ocean wave prediction model (WAMDI Group, 1988), used by the European Centre for Medium-Range Weather Forecasts, ECMWF (Persson, 2001). Its implementation in the safety system enables forecasts, 72 hours in advance (with results every 3 hours), of the significant height ($H_s$), the peak period ($T_p$) and the average direction ($\theta_m$) of the wave states (Figure 3).

Predictions of wind fields and sea level of astronomical tide are obtained from NAVGEM model (Whitcomb, 2012) (Figure 3) and XTide model (Flater, 2016), respectively. These constitute the required boundary forcing data to run the SWAN model.

SWAN (Booij *et al.*, 1999) is a spectral nonlinear model based on the wave action conservation equation, which simulates the propagation of irregular wave spectrum. The SWAN model was used to perform the propagation of the sea agitation from offshore to the coast. For this model a discretization for the directional spectrum as well as bathymetric meshes have been established (Figure 4).

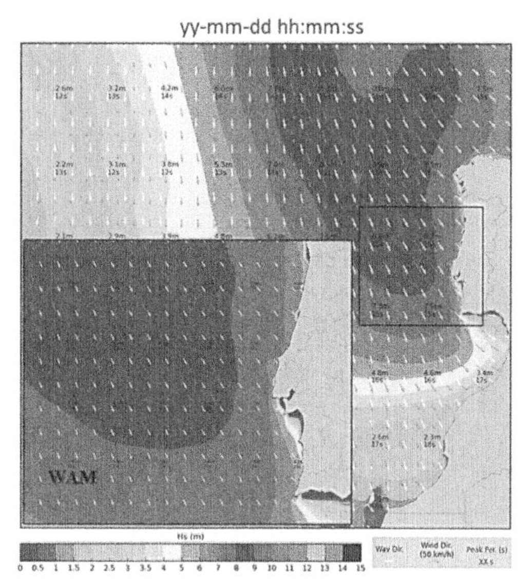


Figure 3. Graphical representation of the WAM model results for waves and HIRES model results for wind field.

The boundary conditions in the larger bathymetric grid are the predictions of wave state characteristics provided by the WAM model, the regional wind fields used in the SWAN model result from the NAVGEM model while the tide levels, considered constant at all points in the SWAN domain, result from the XTide model.

In the system herein developed, the SWAN model operates in stationary mode, having been calibrated all the parameters necessary for this operation. One point was defined, near the Sines buoy, to extract the results of the model. The transfer of the characteristics of the wave states into the port is performed with the DREAMS model (Fortes, 2002). DREAMS is a linear finite element model, based on the mild slope equation, to simulate the propagation of monochromatic waves in port basins. For this model, it was generated, with the GMALHA tool (Pinheiro, 2008), a finite element mesh that characterizes the port area, in terms of its bottom morphology and its solid boundary (Figure 5). The solid boundary was characterized in terms of its structures and their reflection coefficients.

The boundary conditions of the model calculation domain were defined from the $H_s$, the $T_p$ and $\theta_m$

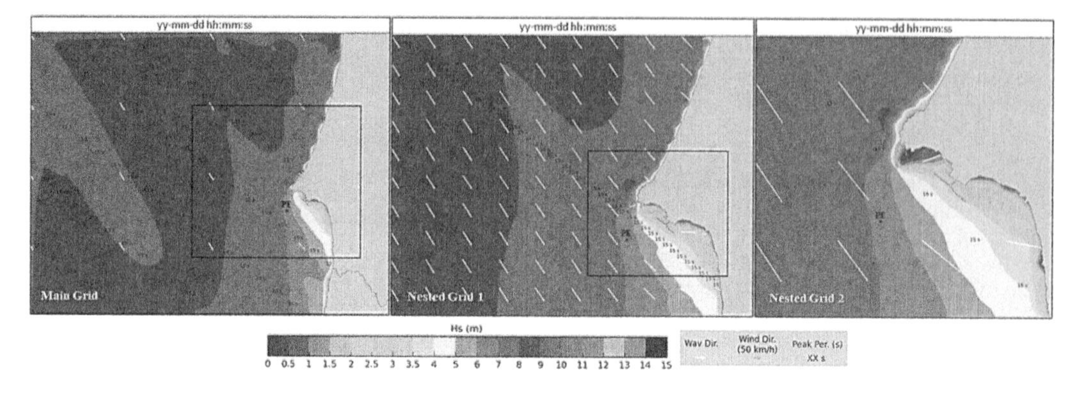


Figure 4. Graphical representation of the SWAN model results at the three nested grids.

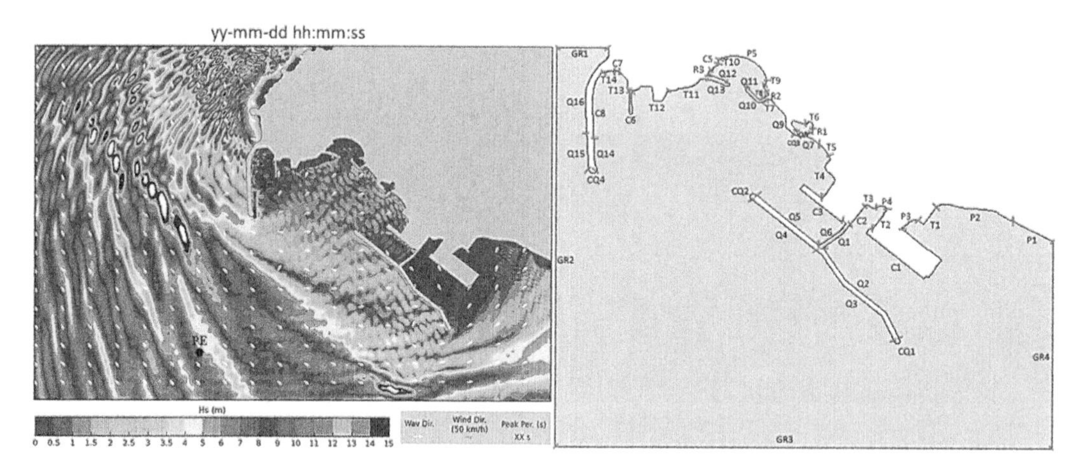


Figure 5. Graphical representation of the DREAMS model results (left) and the contour of the solid boundary (right).

extracted at the point defined in the SWAN model. The DREAMS model provides the characteristics of the sea agitation inside the sheltered area of the port required to assess the behavior of the moored ships.

### 3.3 *Navigation in port area models*

The numerical models of this module deal with the time-domain formulations, relating instantaneous values of forces and movements, to evaluate the behavior of the moored ships.

The WAMIT model (Korsemeyer *et al.*, 1988 and Newman & Sclavounos, 1988) uses a panel method for solving in the frequency domain radiation and diffraction problems of a free-floating body. This model uses the second Green identity to determine the intensity of the source and dipole distributions in the panels of the hull's wetted surface discretization. The forces along each of the six degrees of freedom of the ship motion are determined for regular incident waves that hit the ship. Its application provides the required information to run the BAS model, namely: the wave exciting forces and moments, the impulse response functions obtained from the damping coefficients, and the added-mass coefficients for infinite frequency, obtained from the frequency domain added-mass coefficients and the corresponding impulse response functions.

To produce the expected results, WAMIT was provided with the models of the ships to be simulated, discretized into flat panels. The selected ships represent, in the most comprehensive way possible, the ships that operate in the Port of Sines, more specifically in the bulk liquid (TGL), container (TCS) and multipurpose (TMS) terminals, which will benefit most from the safety system.

All the hydrodynamic coefficients and transfer functions from the WAMIT numerical model, are stored in a database and stored into the system.

The BAS model uses the impulse response, the mass (including added mass) and hydrostatic restoration matrices, together with the time series of the forces exerted by the waves on the ship and the constitutive relations of the mooring system elements (mooring lines and fenders) to set up the equations of motion of the moored ship.

The forces due to mooring lines and fenders can be determined from their constitutive relations. The wave diffraction forces result from a synthetic time series generated from the characteristic values of the wave field obtained with DREAMS model.

Strictly speaking, this is a set of six equations whose solutions are the time series of the ship movements along each of her six degrees of freedom, $X_j(t)$ as well as of the forces in the mooring lines and fenders.Figure 6 shows the results of the safety system in terms of the forces on the mooring system of a generic tanker moored at the TGL.

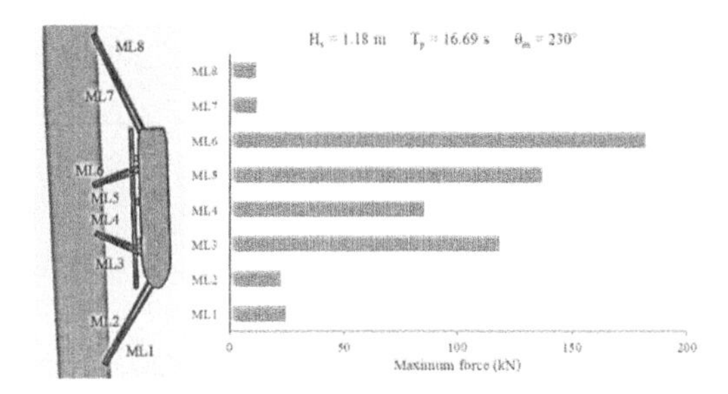


Figure 6. Oil tanker ship moored at TGL. Wave conditions and maximum forces on the mooring lines provided by the safety system on December 4th, 2020, at 9 p.m.

### 3.4 *Numerical models validation*

An important data source that allows a solid validation of the numerical models is the Sines wave buoy, named Sines 1D, installed on the -90 m bathymetric (ZH) in front of the port. The system downloads, daily, the 30-minute measurement values from the wave buoy (significant heights, average directions and peak periods).

For the validation of the SWAN model forcing boundary wave data, requests were submitted to ECMWF to obtain reanalysis data from the ERA4 and ERA5 models corresponding to 40 years of data between 1988 and 2018 of wave and wind regimes (Figure 7). Correlation coefficients were computed for Hs (r=0.949), Tp (r=0.888) and Dirm (r=0.624), and the results showed that the ECMWF WAM model results have an excellent agreement with measurements in terms of Hs and Tp but a very good agreement in terms of wave direction.

For the validation of the SWAN model results, a data extraction point on the Sines 1D buoy location is requested as output of the model. This allows real-time validation of the system's meteo-oceanographic forecasts (Figure 8). Correlation coefficients computed for Hs (r=0.92), Tp (r=0.770) and Dirm (r=0.74), and the results showed that the SWAN model results still have an excellent agreement with measurements in terms of Hs. As for Tp correlation is not as good as ECMWF source data. In terms of wave direction, agreement is slightly better in SWAN model results than the one obtained with the ECMWF source data.

Other sources of model validation include *in situ* data collection of wave characteristics inside the port basin as well as ship motions and mooring forces. To

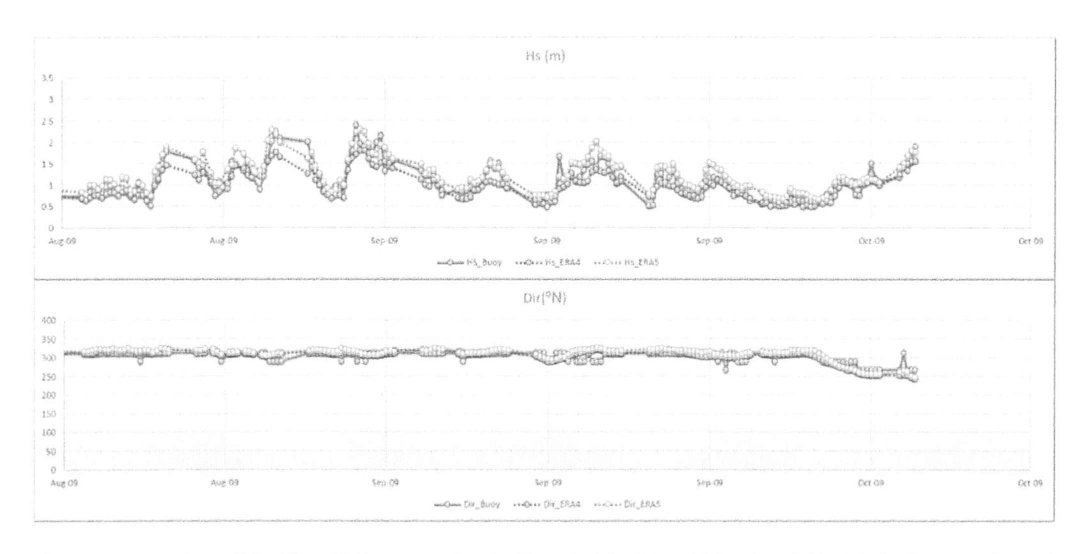


Figure 7. Comparison of the Sines 1D buoy records (significant height, Hs and Direction, Dir) with the forecasts obtained at the closest ECMWF point, using two datasets: Era4 and Era5. Period from August to September 2009.

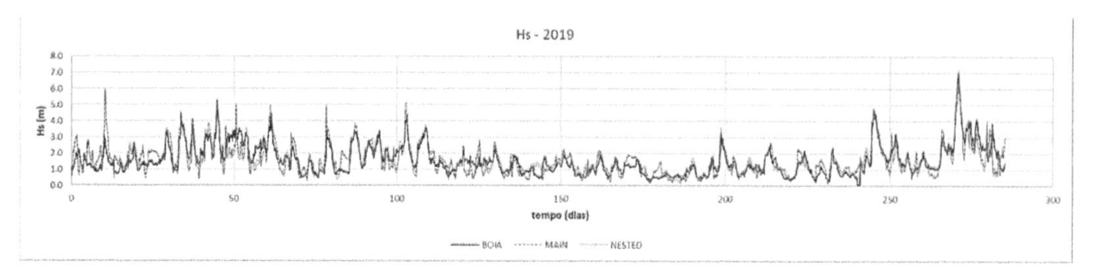


Figure 8. Comparison of the Sines 1D buoy records with the SWAN numerical model results (extracted from the main grid - MAIN and the nested grid 1 - NESTED), in the year 2019.

accomplish this task, it is intended to install pressure sensors near the terminals, gyroscopes on ships and forces sensors on moorings. This data will enable the validation of the sea agitation characteristics inside the Port of Sines, as well as the behavior of the simulated ships.

## 4 RISK ANALYSIS AND WARNING SYSTEM

A risk analysis combines the possible consequences of an event with the probability of its occurrence. The safety system herein presented, issues alerts based on a qualitative risk assessment, associated with the probability of exceeding a pre-established threshold for the amplitude of ships' motions and for the forces on their mooring system (Pinheiro *et al.*, 2018).

Excessive ship motions make loading and unloading operations unviable and can also cause excessive forces on ship's mooring systems. The breakage of a mooring element can lead to considerable damage.

The pre-established thresholds considered are those recommended by organizations concerned with maritime and port activities. Table 1 shows the pre-set limits for ships' motions amplitude (PIANC, 1995 & PIANC, 2012) and Table 2 the ones for the forces on the mooring systems (OCIMF, 1992 & PIANC 1995). Considering the nature of the activities, the characteristics of the vessel and the need to ensure the safety of persons and infrastructure, consequence levels have been attributed to the threshold values. These values can be adjusted to reflect each port administration internal criteria and rules.

Table 1. Consequence levels of exceeding movement amplitude (Pinheiro et al., 2018).

| | | Surge | Sway | Heave | Roll | Pitch/ Yaw |
|---|---|---|---|---|---|---|
| Consequence | Level | m | m | m | ° | ° |
| Insignificant | 0 | 0.1 | 0.10 | 0.1 | 0.5 | 0.1 |
| Mild | 1 | 0.3 | 0.20 | 0.2 | 1.0 | 0.3 |
| Serious | 2 | 0.4 | 0.25 | 0.3 | 1.3 | 0.4 |
| Critical | 3 | 0.5 | 0.30 | 0.4 | 1.5 | 0.5 |

Table 2. Consequence levels of exceeding forces on mooring system (Pinheiro et al., 2018).

| | | Mooring Lines | Fenders |
|---|---|---|---|
| Consequence | Level | KN | KN |
| Insignificant | 0 | 40%MBL | 50%ML |
| Mild | 1 | 50%MBL | 70%ML |
| Serious | 2 | 80%MBL | 90%ML |
| Critical | 3 | MBL | ML |

To associate the consequence levels with the risk levels, 3600 seconds time series are produced for each variable (ship motions and mooring system forces). Each time series undergoes a Fourier transform and a power density spectrum is obtained. From this spectrum, statistical information can be derived from spectral moments (Pinheiro *et al.*, 2018). The statistical distribution is assumed to be a Rayleigh distribution (Longuet-Higgins, 1952).

Thus, relating the probability of exceedance of the Rayleigh distribution, discretized into levels, with the consequence levels, it is possible to obtain risk levels for each variable (Table 3).

Table 3. Risk levels (Pinheiro et al., 2018).

| Exceedance Probability (P) | | Consequence levels Insig. | Mild | Serious | Critical |
|---|---|---|---|---|---|
| Levels | | 0 | 1 | 2 | 3 |
| Rare (P<0.001%) | 0 | 0 | 0 | 0 | 0 |
| Unlikely (P<0.1%) | 1 | 0 | 1 | 2 | 3 |
| Possible (0.1<P<10%) | 2 | 0 | 2 | 4 | 6 |
| Critical (P>10%) | 3 | 0 | 3 | 6 | 9 |

Based on the risk levels, the system issue alerts, through the definition of danger alerts (color coded):

- Green alert: Risk level of 0 or 1. No danger
- Yellow alert: Risk level of 2 or 3. Low danger Loading and unloading operations conditioned.
- Orange alert: Risk level of 4 or 6. Moderate danger Loading and unloading operations cannot be performed.
- Red alert: Risk level of 9. Maximum danger Loading and unloading operations are suspended. Possibility of breakage of mooring system elements and structural damage.

Figure 9 shows the symbols used in the safety system to issue the color alerts.

Figure 9. Symbols used by the SAFEPORT system to alert ships' motions danger (left) and ships' mooring systems failure danger (right).

## 5 DISSEMINATION OF RESULTS

A web platform (HTML page) and a mobile application (for android platforms only) were developed for the dissemination of the results. In these, it is possible to access and view, in graphic and numerical form, the forecasts of sea agitation, wind, currents and occurrence of long waves and monitor possible emergency situations in real time. In addition, it is also possible to visualize the alerts for navigation and access the system database and continuous validations (forecasts *vs* observations).

The development of the web page (Figure 9) for Port of Sines has some particularities inherent to the port in question and with features, such as the simulation of ships in maneuver. The mobile application (Figure 10) has the same functionalities as the web platform, presented in a simpler manner.

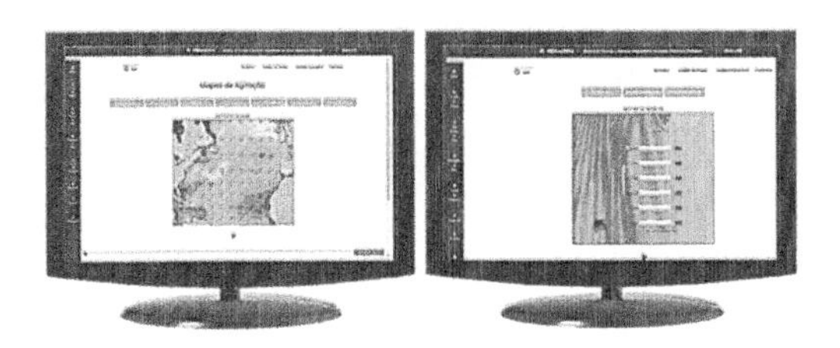

Figure 10. Web tool for the dissemination of results.

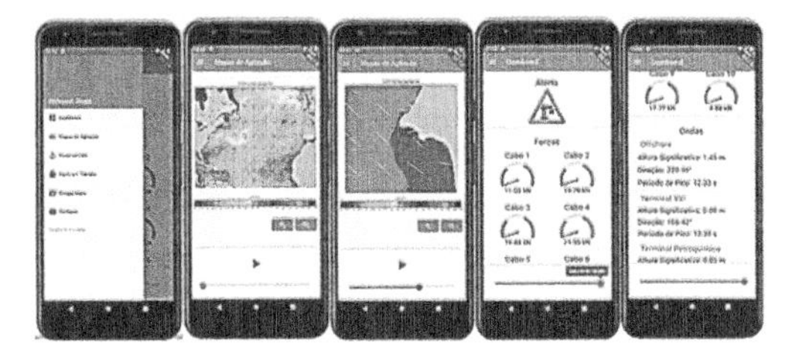

Figure 11. Mobile tool for the dissemination of results.

Furthermore, daily reports in pdf, constructed with the 3-day forecasts of the system's wave and navigation results, are sent by email to the interested parties.

## 6 FINAL CONSIDERATIONS

The ultimate goal of this project is to reduce the vulnerability of ports by increasing their capacity for efficient planning and responses to emergency situations. In addition, this project aims to provide greater access to information and communication by providing specific information on a specific ship and/or port terminal.

The ability to anticipate the effects on ships of potentially hazardous sea agitation and weather conditions enables to take informed decisions on navigation routes and berthing procedures and to increase the safety of ships moored and in maneuvering. Thus, the system will alert the responsible authorities in order for them to take timely mitigation measures to avoid accidents and reduce economic losses.

## ACKNOWLEDGEMENTS

The authors would like to thank the BlueSafePort project (ref: FA_04_2017_016), the National Infrastructure for Distributed Computing (INCD) for granting access to the digital infrastructure to support research and the Sines and Algarve Ports Administration.

## REFERENCES

Booij, N.R., Holthuijsen, L.H. & Ris, R.C. (1996). The SWAN wave model for shallow water. *ICCE'96*: 668–676. Orlando.

Flater, D. (1998). XTide Manual: Harmonic Tide Clock and Tide Predictor. *Technical Report, USA*. URL: https://flaterco.com/xtide

Fortes, C.J.E.M. (2002). *Transformações não lineares de ondas em portos. Análise pelo método dos elementos finitos*. Tese de doutoramento. Lisboa: IST/DEM.

Fortes, C.J.E.M.; Reis, M.T.; Poseiro, P.; Santos, J.A.; Garcia, T.; Capitão, R.; Pinheiro, L.; Reis, R.; Craveiro, J.; Lourenço, I.; Lopes, P.; Rodrigues, A.; Sabino, A.; Araújo, J.P.; Ferreira, J.C.; Silva, S.F.;

Raposeiro, P.; Simões, A.; Azevedo, E.B.; Reis, F.V.; Silva, M.C.; Silva, C.P. (2015). Ferramenta de Apoio à Gestão Costeira e Portuária: o Sistema HIDRALERTA. *In Atas do VIII Congresso sobre Planeamento e Gestão das Zonas Costeiras dos Países de Expressão Portuguesa*. Aveiro, outubro 2015.

Korsemeyer, F.T., Lee, C.-H., Newman, J.N. & Sclavounos, P.D. (1988). The analysis of wave effects on tension-leg platforms. Proc. *7th International Conference on Offshore Mechanics and Arctic Engineering*: 1–14. Texas: Houston.

Longuet-Higgins M.S. (1952). On the statistical distribution of the heights of sea waves. *J. Marine Research*. Vol XI, No. 3.

Mynett, A.E., Keunig, P.J. & Vis, F.C. (1985). The dynamic behaviour of moored vessels inside a harbour configuration. *Int. Conf. on Numerical Modelling of Ports and Harbour* 23-25: Cranfield: BHRA. The Fluid Engineering Centre, April 1985. England: Birmingham.

Newman, J. N. & Sclavounos, P. D. (1988). The Computation of Wave Loads on Large Offshore Structures. *In Boss 88 Conference*: 1–19, Norway.

OCIMF (1992) Mooring equipment guidelines. Witherby e Co. Ltd.

Persson, A. (2001). User Guide to ECMWF Forecast Products. *Meteorological Bulletin M3.2. ECMWF*: 115.

PIANC - Permanent International Association of Navigation Congresses (1995). Criteria for movements of moored ships in harbors. *Technical report Permanent International Association of Navigation Congresses*. PIANC Supp.to bulletin no. 88.

PIANC. (2012). Guidelines for berthing structures related to thrusters. *PIANC*. Brussels, Belgium.

Pinheiro, L., Fortes, C. & Santos, J. (2018). Risk analysis and management of moored ships in ports. *the 37th International Conference on Ocean, Offshore and Arctic Engineering (OMAE2018)*. June 17-22, Madrid, Spain.

Pinheiro, L., Fortes, C., Reis, M. T., Santos, J. & Soares, C. G. (2020). Risk forecast system for moored ships. *In proceedings of vICCE (virtual International Conference on Coastal Engineering)*. 6-9 de Outubro.

Poseiro, P.; Gonçalves, A.B.; Reis, M.T.; Fortes, C.J.E.M. (2017). Early warning systems for coastal risk assessment associated with wave overtopping and flooding. *Journal of Waterway, Port, Coastal, and Ocean Engineering*.

WAMDI Group (1988). The WAM Model - A third generation ocean wave prediction model. *J. Physical Oceanography* (18): 1775–1810.

*Trends in Maritime Technology and Engineering – Guedes Soares & Santos (eds)*
 *ISBN 978-1-032-33583-4*

# Numerical simulations of potential oil spills near Fernando de Noronha archipelago

P.G.S.C. Siqueira, J.A.M. Silva, M.L.B. Gois, H.O. Duarte, M.C. Moura, M.A. Silva & M.C. Araújo
*Federal University of Pernambuco, Recife, Brazil*

ABSTRACT: Oil spills in the ocean are a significant threat that causes catastrophic impacts on coastal countries' marine environment and ecosystems. Therefore, it is essential to assess the risk of potential oil spills and provide orientation on the best strategies to mitigate impacts. This paper focuses on simulating potential oil spills from oil tankers navigating near Fernando de Noronha Archipelago (FNA), considering the oil's physical and chemical transformations and transport in the ocean. Using the Lagrangian model MEDSLIK-II, we simulate various scenarios: the amount of oil spilled, based on the tank capacity of typical oil tankers; the location of the hypothetical spillage, placed in ship routes near the FNA; and the various meteo-oceanographic conditions (e.g. currents and wind velocities). The simulation results are the oil concentrations that can reach the FNA. When integrated into a risk assessment for oil spills, it improves the exposure estimates to the risks in this region.

## 1 INTRODUCTION

'Oil spills in the ocean are a significant threat that has caused catastrophic impacts on coastal countries' marine environment and ecosystems (Chen et al., 2019). The high number of vessels, including oil tankers that circulate the globe and extreme events such as storms and tropical cyclones due to global warming, increase the risk of potential oil spills affecting oceanic islands (Queiroz et al., 2019). Despite the immense efforts of international and national maritime authorities over the years to enhance ship safety, many shipping accidents still occur (Ung, 2019; ITOPF, 2021). For instance, in the 2010s, a total of 63 spills occurred, releasing 164,000 tonnes of oil, which was the least amount spilt in the last decades (ITOPF, 2021). Furthermore, the oil trade plays a vital role in economic development. There is a rising in the maritime transportation of the volume of oil. The oil tankers are responsible for around 90% of the oil transported worldwide (Chen et al., 2019).

In risk assessment of oil tanker spills, understanding oil spill behavior is fundamental to assess the spatial and temporal extent of the impact and which vulnerable ecosystems will suffer the consequences. Oil spill simulation models have been developed for simulating oil slick trajectories and fates (fate and transport models) under actual environmental conditions from spills created from marine traffic, oil production or other sources (Spaulding, 2017; Keramea et al., 2021). Therefore, oil spill risks are determined by the potential hazard of oil pollution and the environmental characteristics, e.g. ocean currents, winds, waves, and sea surface temperature (SST).

The simulation of the transport, diffusion and transformation of spilled oil in the ocean can be done using a Lagrangian formalism coupled with Eulerian circulation models. The Lagrangian formalism can track mass elements, such as the droplets of oil in the water; while in the Eulerian approach, the focus is on the flow properties in a specified point in space as a function of time so that one can model the currents and wind fields (Fox et al., 2014). Thus, with the Lagrangian approach, the model can compute the oil concentration on the water (De Dominicis, Pinardi, Zodiatis and Lardner, 2013).

Some of the Lagrangian operational models are COZOIL (Reed et al., 1989), SINTEF OSCAR (Reed et al., 1995), OILMAP (Spaulding et al., 1994), GULFSPILL (Al-Rabeh et al., 2000), ADIOS (Lehr et al., 2002), MOTHY (Daniel et al., 2003), MOHID (Carracedo et al., 2006), POSEIDON OSM (Annika et al., 2001; Nittis et al., 2006), OD3D (Hackett et al., 2006), the Seatrack Web SMHI model (Ambjorn, 2006), MEDSLIK (Lardner et al., 2006; Zodiatis et al., 2008), GNOME (Zelenke et al., 2012), OILTRANS (Berry et al., 2012), and MEDSLIK-II (De Dominicis et al., 2013a, b).

The development of models for risk assessment in oil spills is an ongoing research area due to the impacts of marine oil pollution. Some works integrate with the assessment fate and transport models. For instance, Amir-Heidari & Raie (2018) and Amir-Heidari et al. (2019) have used GNOME (Zelenke

DOI: 10.1201/9781003320289-29

et al., 2012) and ADIOS (Lehr et al., 2002) to simulate the spills and compute the oil concentrations on defined points. We can also cite Guo (2017), who integrated into the risk assessment a wave-current coupled model, coupling SELFE (Zhang and Baptista, 2008) and SWAN (Booij et al., 1999).

In this research, we used a Lagrangian fate and transport model, the MEDSLIK-II (De Dominicis et al., 2013a, b), to simulate oil spills trajectories and transformations in the ocean environment near the Fernando de Noronha Archipelago (FNA). Input data from oceanographic (i.e. currents and sea surface temperature) and atmospheric Eulerian models used in the MEDSLIK-II model solve the advection-diffusion and weathering process, as illustrated in Figure 1. These results are fundamental to quantify the impacts of potential oil spills that may occur due to oil tankers accidents that navigate nearby the FNA when integrated into a Quantitative Ecological Risk Assessment (Duarte et al., 2013, 2019; Duarte & Droguett, 2015).

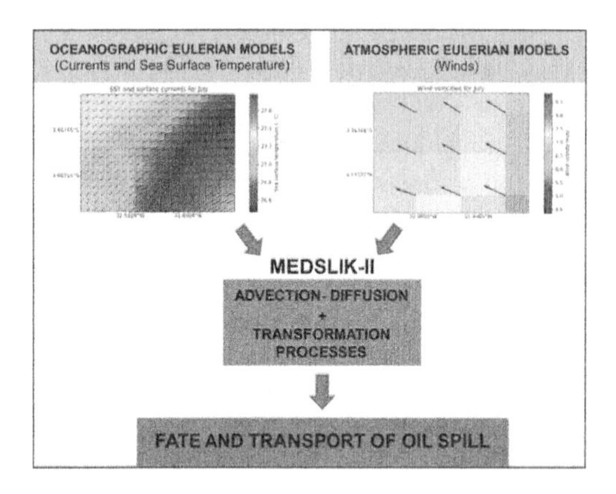


Figure 1. Simplified scheme of the MEDSLIK-II with the main inputs (oceanographic and atmospheric Eulerian models) and the output (fate and transport of oil spill).

The remainder of this work is structured as follows. First, we describe the FNA and its ecological importance, the meteo-oceanographic conditions around the archipelago (i.e. ocean currents, sea surface temperature and wind speed), and the main ship routes. Second, we detail the dataset used, define the spillage points on the most critical route and describe the fate and transport model used, i.e. the MEDSLIK-II. Third, we present and discuss the results regarding the oil spill simulations. Lastly, we present closing remarks about the simulations performed, comment the advantages and limitations and suggestion for future works.

## 2 PROBLEM DESCRIPTION

Oceanic islands are hotspots of biodiversity that host a large number of endemic species with unique biological value, and their isolation enables them to be a repository of threatened species being priorities regions for legal conservation acts (Gillespie, 2001; Whittaker and Fernández-Palacios, 2007; Gove et al., 2016; Kueffer and Kinney, 2017). In the Brazilian context, the Fernando de Noronha Archipelago (FNA) is the best-studied oceanic island. It possesses many marine and terrestrial species, attributed to its extension and habitat's heterogeneity. The FNA is located between latitudes 03°45′S and 03°57′S and longitudes 32°19′W and 32°41′W, in the Western Equatorial South Atlantic. It is situated approximately 345 km away from the northeast coast of Brazil and it is composed of 21 islands and islets with a total area of 26 km² (Figure 2) (IBAMA, 2005).

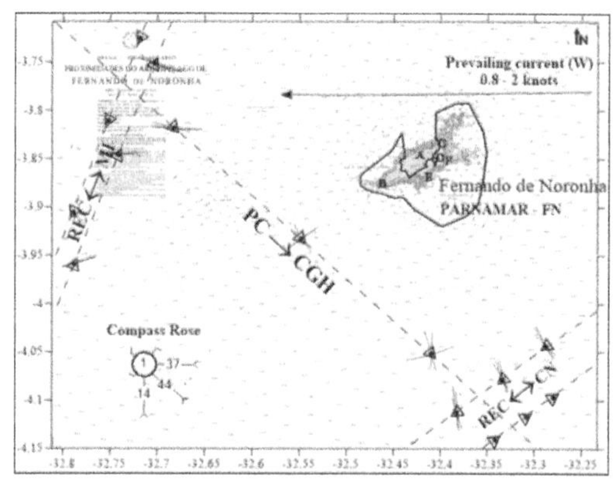


Figure 2. Main routes near FNA: Ponce and Colón – Cape of Good Hope (PC – CGH, going); Recife – Madeira Island (Rec – MI, going and return); Recife – Cape Noaudhibou (REC –CN, going and return). Scale 1:4222. (From: (Duarte and Droguett, 2015)).

The FNA is in a warm tropical region. The air temperature on average is 25°C and a well-defined dry season between August and February, and a rainy season between March and July, averaging 1400 mm rainfall (Serafini and França, 2010). The prevailing winds are the southeast trade winds. The greater intensity occurs between July and August (Tchamabi et al., 2017). The highest sea surface temperatures (SST) occur between March and June, typically exceeding 28°C due to the occurrence of the southwestern tropical Atlantic warm pool (Cintra et al., 2015) and the lowest between August and November (SST ~26.6°C) (Silva et al., 2009; Hounsou-gbo et al., 2015; Tchamabi et al., 2017). On the ocean surface, the central branch of the South Equatorial Current (cSEC) flows westward until it reaches the North Brazil Current (NBC) near the coast (Stramma and Schott, 1999; Lumpkin and Garzoli, 2005). The cSEC is stronger between March and July, and weaker between August and February (Lumpkin and Johnson, 2013; Tchamabi et al., 2017).

The FNA has a Conservation Unit status, protecting the endemic species and maintaining a healthy island ecosystem (Serafini & França, 2010). Two-thirds of the FNA consist of the Marine National Park of FN (PARNAMAR-FN), a marine protected area (MPA) that reaches the 50-meter isobathic line (ICMBio, 2013). The fundamental objective of creating PARNAMAR-FN is to preserve natural ecosystems with great ecological significance and scenic beauty, enabling scientific research, activities of environmental education, recreation and ecotourism (BRASIL, 2000).

Most projects devised and developed for preserving PARNAMAR-FN focus on the conservation of a single representative species (e.g. spinner dolphin, turtles, sharks, coral reefs) (Maida and Ferreira, 1995; Garla, 2004; Ferreira and Maida, 2006; TAMAR, 2006; Amaral et al., 2009; Lira et al., 2009; Silva, 2010). However, managers need to assess and manage ecological risks caused by routine (i.e. high frequency/low consequence) human activities within PARNAMAR-FN. These assessments should also contemplate improbable, significant events (i.e. low frequency/high impact). Although there is no evidence of large spills in recent years, it remains a latent threat (IUCN, 2020), and recently oil blobs have been reported to reach the FNA coast (ICMBio, 2021). On average, 75 ships, mainly oil tankers, navigate daily on routes near FN (Medeiros, 2009), using landmarks to determine the ship's position at sea more precisely and consistently.

The main ship routes nearby the FNA are presented in Figure 2. The routes were identified by Duarte & Droguett (2015) on Pilot Charts, which show the most recommended routes for navigation (i.e. those taking best advantage of currents, winds and possible nearby landmarks to help determine the ship's position) for each month of the year based on meteo-oceanographic data collected by the Brazilian Navy from 1951 to 1972 (BRASIL, 1993) The three main routes are:

1. Ponce and Cólon – Cape of Good Hope (PC-CGH – November – going), minimum distance to the FNA: 4.96 nautical miles (nm)
2. Recife – Madeira Island (REC – MI – Augusto – going and return), minimum distance to the FNA: 16.38 nm;
3. Recife – Cape Noaudhibou (REC – CN – March – going and return), minimum distance to the FNA: 12.58 nm.

## 3 METHODOLOGY

### 3.1 *Data source*

The global bathymetry data used is provided from GEBCO, which is a terrain model for ocean and land, providing elevation data (in meters) on a 15 arc-second interval grid (GEBCO, 2021). The coastline data is based on the Global Self-Consistent Hierarchical High-resolution Geography (GSHHG) from the National Oceanic Atmospheric Administration (NOAA) (NOAA, 2018).

The global oceanographic inputs are from the Global Ocean 1/12° Physics Analysis and Forecast provided by CMEMS that includes data regarding temperature (SST), salinity, currents, sea level, mixed layer depth and ice parameters from the top to the bottom over the global ocean (CMEMS, 2021). For this study, only the SST and currents data were retrieved. The atmospheric data (i.e. wind velocities) were retrieved from the ERA-Interim atmospheric fields, provided by the ECMWF (Berrisford et al., 2011; ECMWF, 2019).

### 3.2 *Spillage scenarios*

The primary forces that act on the oil spill are the currents and winds. We chose the monthly averaged currents and winds for March and July. The rationale is that the cSEC starts to intensify in March, but with a lower wind speed, reaching its highest intensity in July, with the intensification of the southeast trade winds (Molinari, 1982; Lumpkin & Garzoli, 2005).

The route that possesses a significant hazard is the REC – CN, which passes from south to east of the FNA (Duarte & Droguett, 2015). It is conceptually possible that an oil spill on this route would be transported to the archipelago coast due to prevailing winds and currents. We defined three initial spill releasing points on this route. These points were determined on the eastern side of the FNA, also known as the Windward Side (WS), that faces the open ocean and where the sea is more exposed to the action of winds and ocean currents (Ivar do Sul et al., 2009; Assunção et al., 2016). Moreover, these release points were selected based on traffic density on the route from the archipelago of the main vessel tracks registered in the Marine Traffic website (Queiroz et al., 2019; MARINE-TRAFFIC, 2021). The points are: P1 (3°31.3′S; 31°33′W), P2 (3°44.3′S; 31°51′W) and P3 (3°50.87′S; 31°58.8′W) (Figure 3).

We also assumed an instantaneous volume of oil spilled equal to 9100 tonnes, based on the capacity of a single typical Suezmax cargo tank (IMO, 2008).The oil used on the simulations is the 28° API (intermediate oil type, density approximately 886.6 kg/m$^3$) which is the most used in the Brazilian oil and gas exploratory activities (ANP, 2021). Once the spill starts, we simulate the fate and transport of the oil during 48h and store the main results every half hour. Therefore, we simulated the oil spill cases considering the weathering conditions for each month at each point of release, resulting in six scenarios. The names of the scenarios result from the combination of the releasing point and the month of the spill, summarized in Table 1.

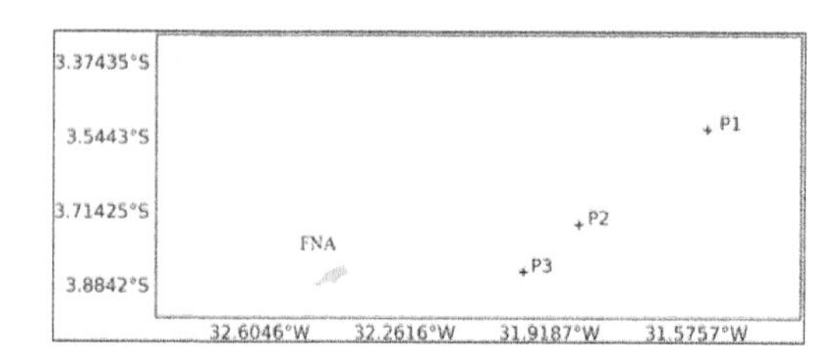


Figure 3. Location of FNA and coordinates of the release points P1, P2 and P3 for the oil spill simulations.

Table 1. Summary of the oil spill simulation scenarios.

| | Releasing points | | |
|---|---|---|---|
| | P1 | P2 | P3 |
| March | P1M | P2M | P3M |
| July | P1J | P2J | P3J |

### 3.3 *Lagrangian model: MEDSLIK-II*

The MEDSLIK-II is a Lagrangian oil model designed to simulate oil slick transport and transformation processes for spills on the maritime surface (Figure 1) (De Dominicis et al., 2013a, b). When an oil spill occurs, the wind and surface currents are the primary forces for oil transport in the aquatic environment (Spaulding, 2017). Therefore, the oil transport in the ocean is primarily attributed to advection by the large-scale flow field, and turbulent flow components cause the dispersion. Beyond transportation, oil spills impact depends mainly on the environmental conditions that control the weathering processes at the site of the spill (e.g. currents, climate, waves) and the time required to engage mitigation operations (NRC, 2003; Marta-Almeida et al., 2013; Lee et al., 2015). Weathering is a general definition for changes in oil properties due to physical, chemical and biological processes that occur when the spill is exposed to environmental conditions (e.g. in aquatic systems). The main weathering processes are illustrated in Figure 4 and defined as follows:

1. Spreading: The most dominant process in the first stage of a spill is spreading low pour point (i.e. the temperature below which the oil loses its liquid properties) oil on water. The spreading strongly influences late processes, such as evaporation and dispersion (Sebastião & Guedes Soares, 1995). Spreading consists of two processes: the first is the area lost due to oil converted from the thick to the thin slick, and the second is due to Fay's gravity-viscous phase of spreading (Al-Rabeh et al., 2000).
2. Evaporation: The primary initial process involved in the removal of oil from the sea. The evaporation rate is determined by the oil's physicochemical properties and is increased by spreading, high water temperatures, strong winds, and rough seas. By evaporation, low boiling components will rapidly be removed, thus reducing the remaining slick volume. For many oils, evaporation from the surface slick is the most critical mass loss process during the first hours of an oil spill (Sebastião & Guedes Soares, 1995).
3. Emulsification: Refers to the process by which water becomes mixed with the oil in the slick. The emulsification result is a significant increase in volume (3 or 4 times the volume of the original stabilized oil), a considerable increase in density, and a substantial increase in viscosity (Sebastião & Guedes Soares, 1995).
4. Dispersion: The wave action on the oil drives it into the water, forming a cloud of droplets beneath the spill. The droplets can be classified as large droplets that quickly rise and coalesce again with the surface spill or tiny droplets that grow more slowly and can be immersed long enough to diffuse into the lower water columns layers. In the latter case, the droplets are lost from the surface spill and considered to be permanently dispersed.

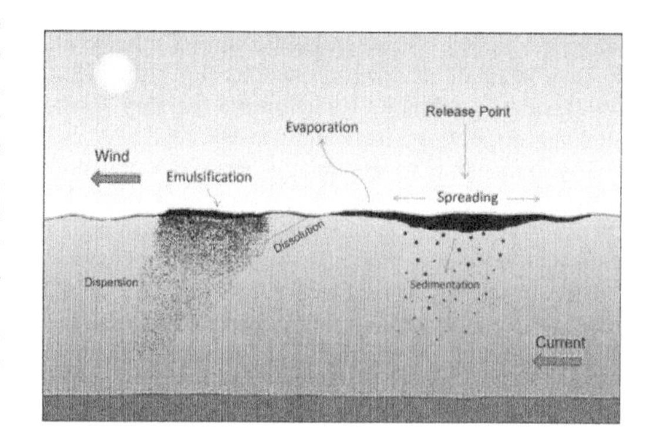


Figure 4. Main weathering processes acting on an oil spill. Adapted from (TRB; NRC, 2014)

The general equation for a tracer concentration (C(x, y, z, t)) with units of mass per volume, mixed in the marine environment is given by De Dominicis et al. (2013):

$$\frac{\partial C}{\partial t} + U \cdot \nabla C = \nabla \cdot (K \nabla C) + \sum_{j=1}^{M} r_j(x, C(x,t), t). \quad (1)$$

Where $\partial/\partial t$ is the local time-rate-of-change operator, **U** is the three-dimensional distribution of the horizontal ocean current components **u** and **v**, **K** is the turbulent diffusivity tensor, which parameterizes the sub-grid scale processes; the position vector (x, y, z) is denoted by **x**; and $r_i(\mathbf{x}, C(\mathbf{x}, t), t)$ are the

M transformation rates that modify the tracer concentration through the weathering processes (i.e. physical and chemical transformations).

Following the Lagrangian approach, the oil slick is constituted of oil particles that move like water parcels. However, the weathering processes acting on the entire slick instead of on the single-particle properties. Thus, the active tracer equation can be effectively split into two component equations:

$$\frac{\partial C_1}{\partial t} = \sum_{j=1}^{M} r_j(x, C_1(x,t), t). \quad (2)$$

$$\frac{\partial C}{\partial t} = -U \cdot \nabla C_1 + \nabla \cdot (K \nabla C_1). \quad (3)$$

Where $C_1$ is the oil concentration due to the weathering processes, while the final time rate of change of $C$ is given by the advection-diffusion acting on $C_1$. The model first solves (2) by considering the weathering processes acting on the total oil slick volume.

Then, the Lagrangian formalism is applied to solve (3), the advection-diffusion part of the equation, discretizing the surface oil slick in particles with position increments given by:

$$dx(t) = U(x,t)dt + Z\sqrt{2Kdt}. \quad (4)$$

Where dt is the model time step; **Z** are independent random vectors normally distributed, i.e. $Z \in N(0,1)$; and **K** is the turbulent diffusion diagonal tensor. The first part of the right side of the equation is the deterministic part of the flow field, while the second is the stochastic term, which characterizes random motion. Finally, the oil concentration is computed by assembling the particles with their associated properties.

## 4 RESULTS

### 4.1 *Oceanographic and atmospheric results*

The meteo-oceanographic results are fundamental to describe the fate and transport of the oil on the ocean. The average wind velocity is 2.86 m/s in the southwest direction in March and 7.52 m/s in the northwest direction in July (Figure 5).

The mean SST was higher in July, averaging 29.3° C, while for March the temperature averaged 27.01°C (Figure 6). Regarding the surface current, the most intense occurred in July, averaging 0.54 m/s southwestward. In March, the current presents a little deviation to the northeast, with a mean intensity equal to 0.17 m/s (Figure 6). The oil transport was mainly westward due to the surface current direction. The influence of the wind affects the direction of oil transport. Therefore, the northeast direction in July and the greater wind intensity contributed to carry the oil plume to FNA.

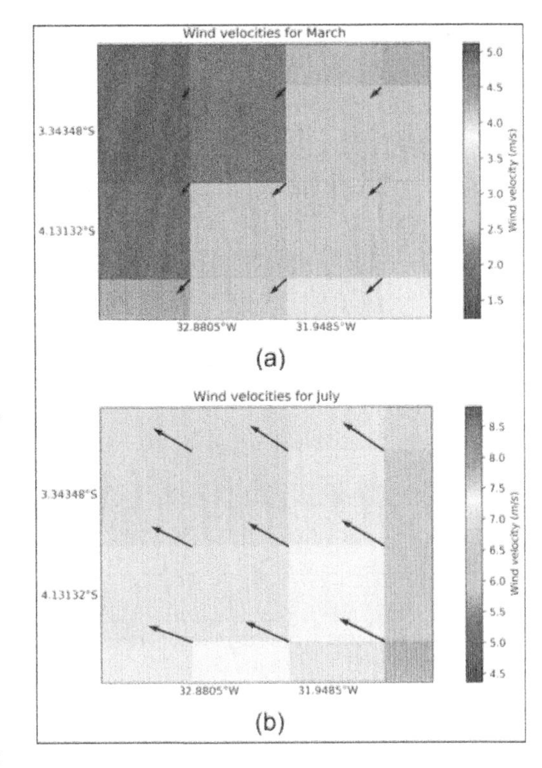


Figure 5. Average wind speed distribution (colors, m/s) and direction (arrows) for March (a) and July (b).

### 4.2 *Oil spill simulation*

The primary simulation for each scenario results is the percentages of the original oil volume that: evaporated, remained in the surface, dispersed in the water columns, and sedimented on the sea bottom and the coast. These results for the final time-step of the simulation are summarized in Table 2. In March, it was observed that the maximum percentage of evaporation was reached after 5.5 hours, while in July, the maximum was achieved within 3 hours. Also, the maximum rate of evaporated oil was similar in each month: approximately 35.84% in March and 35.83% in July (Table 2).

The density and viscosity increase is due to the emulsification of the water in the oil. For the spills simulated March, the density and the viscosity of the water-oil emulsion did not stabilize until the end of the simulation. The viscosities started at 220.02 Pa.s and increase to 2306.57 Pa.s for $P_1$, 2301.39 Pa.s for $P_2$ and 2296.82 Pa.s for $P_3$. Similarly, for the emulsion density, that increased from 885 kg/m³ to 996.65 kg/m³ for the releases that originated from points $P_1$ and $P_2$ and 996.68 kg/m³ for $P_3$. None of the spills simulated in March reached the FNA.

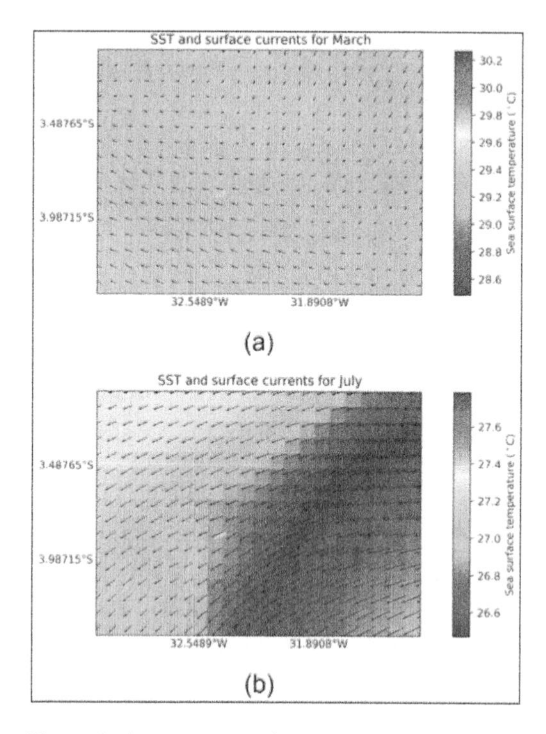


Figure 6. Average sea surface temperature distribution (colors, °C) and surface current direction (arrows) for March (a) and July (b).

For the simulations in July, the density stabilized around 22 hours after the start of the simulation, with the final value being 996.75 kg/m³ and remained equal until the end of the simulation for the releases from all three points. Similarly, to what happened in March, the variation in viscosity reached a stable value simultaneously with the density. The maximum emulsion viscosities are 2727.4 Pa.s for $P_1$, 2721.91 Pa.s for $P_2$, and 2714 Pa.s for $P_3$. After the evaporation process ends, around 64.17% of the oil remains on the surface. The final dispersed percentages can be seen in Table 2.

The oil reached the shore in two of the spillage scenarios in July: P1J and P2J. In scenario P1J, a small amount of oil that spread from the slick reached the FNA. The final amount of oil on the shore and the sea bottom after 48 hours was equivalent to 0.06% of the initial volume of the spill (~5.46 t). For scenario P2J, the oil reached the shore after 20 hours, initially with 0.5% (~45.5 t) of the total volume spilt. By the end of the 48h simulation time, the total oil fixed on the coast and at the bottom of the sea is approximately 63.94% (~5819 t). This oil portion is a substantial amount that can cause severe ecological damage and require cleanup efforts on a large scale.

The oil spill trajectory and concentrations on the sea surface for the release point $P_2$ for 10, 20 and 30 hours are presented in Figure 7. The illustration is helpful to show how the meteo-oceanographic conditions influence the oil slick trajectory.

## 5 DISCUSSION

The oil evaporation rate in the MEDSLIK-II depends on the physicochemical properties of the oil, spreading rate, SST and winds intensity (Sebastião and Guedes Soares, 1995; De Dominicis et al., 2013). Our simulation results showed that more than one-third of the volume spilt evaporates in the first few hours after the spill, i.e. 5.5h for the spills occurring in March, and 3h for the ones in July (Table 2). The emulsification formation varied with the month and happened after the portion evaporated reached its maximum. Since the winds are more potent in July (Figure 5), the emulsification occurs faster (within ~22h), as they provide energy for emulsification, but not enough to disperse it again (Lee 2015). In March, the wind intensity was lower, and the emulsion viscosity was still increasing within the simulation time. Still, the oil formed a stable and high viscosity emulsion that prevented the oil from spreading.

Table 2. Summary for each scenario of the percentage of oil that evaporated, remained on the sea surface, dispersed on the water column and fixed on the coast and the sea bottom.

| Scenario | Evaporated (%) | On the sea surface (%) | Dispersed on the water column (%) | Fixed on the coast and the sea bottom (%) |
|---|---|---|---|---|
| P1M | 35.84 | 63.89 | 0.27 | 0 |
| P2M | 35.84 | 63.90 | 0.26 | 0 |
| P3M | 35.84 | 63.90 | 0.26 | 0 |
| P1J | 35.83 | 63.18 | 0.93 | 0.06 |
| P2J | 35.83 | 0.02 | 0.21 | 63.94 |
| P3J | 35.83 | 63.29 | 0.88 | 0 |

For lighter oils (API > 30°), the evaporation is more significant, and thus the impacts of a spill reaching an ecosystem would be reduced. On the other hand, heavier oils (API > 22°), rich in asphaltenes and wax, would form more stable mousses with a low evaporative rate, therefore preventing the spreading. Regarding intermediate oil types as the one simulated, the evaporative loss was not significant since a considerable volume of oil reached the FNA in P2J (~5819 t). Moreover, the low percentage of oil dispersed and sedimented indicates that advective processes are more considerable than buoyancy effects.

The natural dispersion reduces the volume of the slick at the surface and reduces the evaporative loss; however, it does not alter the physicochemical properties of the oil (Sebastião & Guedes Soares, 1995). The wind speed and viscosity directly impact the percentage dispersed in the water columns, i.e. the higher the wind speed and the oil density, the more oil will be incorporated into the water column. The oil density did not vary significantly for both months, and the average wind speed in July (7.52 m/s) is higher than in March (2.86 m/s); thus, the oil dispersed in July is also higher, 0.93% (~8.43 t) in July against 0.27% (~2.46 t). The amount of oil dispersed into the water have serious consequences to the environment, since toxic components to marine life from the oil are mixed on the water (e.g. benzene, toluene, ethylbenzene and xylene) (Lewis & Pryor, 2013).

The combination of current and wind direction being westwards is problematic if an oil spill occurs on the east side of the FNA. The westward current and the predominant northeast wind in July intensifies the oil transport towards the FNA (Figure 7). It is important to note that the wind dictated the direction of the plume trajectory while the current dominated the flow. The southeast direction of the winds in March prevented the oil from reaching the archipelago. Still, the advective process predominates over the degradation effects (e.g. evaporation) in all scenarios. Thus, oil spills in July are of concern regarding catastrophic pollution in the archipelago.

The meteo-oceanographic conditions and physico-chemical processes brought a high parcel of the oil that first reaches the FNA within 20h. A total amount of around 63.95% of oil (~5819 t) of oil is on the archipelago shore or sedimented on the shallow water depths by the end of the simulation. This is illustrated in Figure 7F, when little oil is still being transported. The advective transport makes the time window for an effective response very narrow, potentially intensifying the impacts. Furthermore, mitigation actions such as the use of dispersants are very toxic to the environment (Shafir et al., 2007). Thus, preventive actions should be prioritized instead of mitigating measures.

A highly uniform oil slick reaches the FNA, with little spreading and 0.5 t/km² of oil concentration.

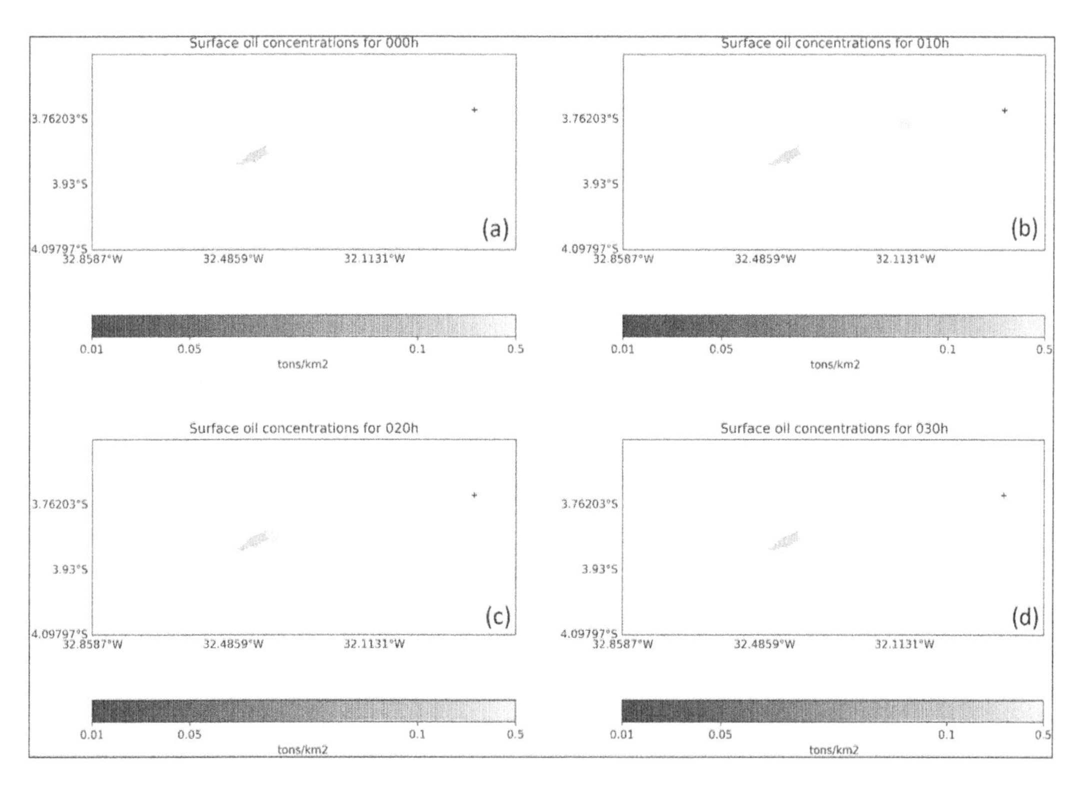


Figure 7. Oil concentration on the water (t/km²) for scenario P2J after (a) 0 hours, (b) 10 hours, (c) 20 hours and (d) 30 hours from the start of the spill.

The regions affected encompass the Rata islet to the Sueste bay. They are susceptible areas with rich ecological hotspots. For instance, the coral reefs of the FNA are of fundamental importance to the archipelago. They are sessile and sensitive to pollutants; the corals serve as food and shelter to many types of animals such as worms, crustaceans, sponges, sea urchins and many fishes (Yender et al., 2010). The loss of coral will affect humans and marine and terrestrial organisms since they protect the shoreline, support tourism, and contribute to fisheries.

Some species are essential to ecological tourism, such as sharks, sea turtles and the spinner dolphin (*Stenella longirostris*) (Queiroz et al., 2019). Moreover, the FNA presents several threatened fish species. The presence of such endangered species implies more conservation efforts. A recent assessment revealed that 17% of all fish species registered are threatened or near-threatened according to the International Union for the Conservation of Nature (IUNC) and the Chico Mendes Institute for Biodiversity Conservation (ICMBio) conservation status (SCHMID et al., 2020).

The damages suffered from the oil spills by the marine and terrestrial life may impact the ecosystem and the socioeconomic aspects of FNA. The ecological tourism and the domestic fishery will be fairly affected, thus placing the FNA inhabitants in a critical situation. The fishermen income decreases as the marine environment is affected by external factors such as oil spills. The presence of oil in the environment could result in the contamination of fishes due to oil ingestion, making them unfit for human consumption, besides raising the mortality rate of marine life.

## 6 CONCLUSIONS

Simulations of fate and transport are critical in a risk assessment for oil spills. This work simulated hypothetical oil spills with oil tankers that navigate near the Fernando de Noronha Archipelago. These simulations aimed to identify which of the hypothetical spills would affect the coast of the FNA in different seasons of the year. For this purpose, six different scenarios were proposed considering: three releasing points defined along the routes of ships passing near the archipelago; and the months of March and July, when the sea surface temperature, wind speed and currents intensity are stronger. The model simulated the oil trajectory and the weathering processes that act on the spill, giving as results the fractions of oil affected by each of the mentioned processes and the portions that remained on the coast and the sea surface.

The simulations showed that only in July the oil reached the coast of the FNA. This was due to the stronger wind and current conditions in this month, and thus, favorable to transport the oil plume to the archipelago. In the occurrence of such event, a high volume of oil would reach the FNA, causing considerable damage to the environment. The archipelago's ecosystem and the social activities (e.g. tourism and fisheries) would be negatively impacted. The oil slick reaches the FNA in a short time window due to the strong advective forces, and leaving little time to prepare mitigation actions.

Our simulation reinforces the need for preventive measures and mitigation strategies when facing the possibility of oil spills. These results can be integrated into a risk assessment methodology for quantifying ecological risks from oil spills. Hence, providing valuable information for decision makers regarding the risks to the environment which are vital for developing and choosing the most viable control strategies.

Although oil tankers accidents are infrequent, there is considerably high traffic of these ships nearby the FNA. The cases simulated in this work considered the most common Brazilian flag oil tanker that navigates in the FNA region and assumed the volume spilt from only one tank. As a proposal for future works, new simulations could be performed considering foreign oil tankers that navigate nearby the FNA and the oil volume that could be spilt.

## ACKNOWLEDGEMENTS

The authors thank the Human Resources Program (PRH 38.1) entitled "Risk Analysis and Environmental Modeling in the Exploration, Development and Production of Oil and Gas", managed by the Brazilian Agency for Petroleum, Natural Gas and Biofuels (ANP) and the Brazilian Funding Authority for Studies and Projects (FINEP), process number 044819, and the Brazilian National Agency for Research (CNPq), grant number 153238/2021-6, for the financial support in this research. This study was also financed in part by the Coordenação de Aperfeiçoamento de Pessoal de Nível Superior – Brasil (CAPES), Fincance Code 001.

## REFERENCES

Al-Rabeh, A. H., Lardner, R. W. & Gunay, N. (2000) 'Gulfspill Version 2.0: a software package for oil spills in the Arabian Gulf', *Environmental Modelling & Software*, 15(4), pp. 425–442.

Amaral, F. M. D. et al. (2009) 'Checklist & morphometry of benthic cnidarians from the Fernando de Noronha archipelago, Pernambuco, Brazil', *Cah. Biol. Mar.*, 50(3), pp. 227–290.

Ambjorn, C. (2006) 'Seatrack Web, forecasts of oil spills, a new version', in *IEEE US/EU Baltic International Symposium*. Institute of Electrical & Electronics Engineers, pp. 1–7.

Amir-Heidari, P. et al. (2019) 'A state-of-the-art model for spatial and stochastic oil spill risk assessment: A case study of oil spill from a shipwreck', *Environment International*, 126, pp. 309–320.

Amir-Heidari, P. & Raie, M. (2018) 'Probabilistic risk assessment of oil spill from offshore oil wells in Persian Gulf', *Marine Pollution Bulletin*, 136, pp. 291–299.

Annika, P. et al. (2001) 'The Poseidon Operational Tool for the Prediction of Floating Pollutant Transport', *Marine Pollution Bulletin*, 43(7–12), pp. 270–278.

ANP (2017) *Boletim Mensal da Produção de Petróleo e Gás Natural*. Available at: http://www.anp.gov.br/publicacoes/boletins-anp/2395-boletimmensalda-producao-de-petroleo-e-gas-natural.

Assunção, R. V. et al. (2016) 'Spatial-Temporal Variability of the Thermohaline Properties in the Coastal Region of Fernando de Noronha Archipelago, Brazil', *Journal of Coastal Research*, 75(sp1), pp. 512–516.

Berrisford, P. et al. (2011) *The ERA-Interim Archive Version 2.0*. Shinfield Park. Available at: https://www.ecmwf.int/en/elibrary/8174-era-interim-archive-version-20.

Berry, A., Dabrowski, T. & Lyons, K. (2012) 'The oil spill model OILTRANS and its application to the Celtic Sea', *Marine Pollution Bulletin*, 64(11), pp. 2489–2501.

Booij, N., Ris, R. C. & Holthuijsen, L. H. (1999) 'A third-generation wave model for coastal regions: 1. Model description and validation', *Journal of Geophysical Research: Oceans*, 104(C4), pp. 7649–7666.

BRASIL (1993) *Atlas de Cartas Piloto: Oceano Atlântico - De Trinidad ao Rio da Prata*. 2ªedição. Rio de Janeiro: DHN.

BRASIL (2000) 'Lei N°9.985, of 18 July 2000'.

Carracedo, P. et al. (2006) 'Improvement of pollutant drift forecast system applied to the Prestige oil spills in Galicia Coast (NW of Spain): Development of an operational system', *Marine Pollution Bulletin*, 53(5–7), pp. 350–360.

Chen, J. et al. (2019) 'Oil spills from global tankers: Status review and future governance', *Journal of Cleaner Production*, 227, pp. 20–32.

Cintra, M. M. et al. (2015) 'Physical processes that drive the seasonal evolution of the Southwestern Tropical Atlantic Warm Pool', *Dynamics of Atmospheres and Oceans*, 72, pp. 1–11.

CMEMS (2021) *Global Ocean 1/12° Physics Analysis and Forecast updated Daily, Copernicus Marine Service*. Available at: https://resources.marine.copernicus.eu/product-detail/GLOBAL_ANALYSIS_FORECAST_PHY_001_024/INFORMATION (Accessed: 25 August 2021).

Daniel, P. et al. (2003) 'Improvement of Drift Calculation in Mothy Operational Oil Spill Prediction System', *International Oil Spill Conference Proceedings*, 2003(1), pp. 1067–1072.

De Dominicis, M., Pinardi, N., Zodiatis, G. & Lardner, R. (2013) 'MEDSLIK-II, a Lagrangian marine surface oil spill model for short-term forecasting – Part 1: Theory', *Geoscientific Model Development*, 6(6), pp. 1851–1869.

De Dominicis, M., Pinardi, N., Zodiatis, G. & Archetti, R. (2013) 'MEDSLIK-II, a Lagrangian marine surface oil spill model for short-term forecasting – Part 2: Numerical simulations and validations', *Geoscientific Model Development*, 6(6), pp. 1871–1888.

Duarte, H. de O. et al. (2013) 'Quantitative Ecological Risk Assessment of Industrial Accidents: The Case of Oil Ship Transportation in the Coastal Tropical Area of Northeastern Brazil', *Human and Ecological Risk Assessment*, 19(6), pp. 1457–1476.

Duarte, H. O. et al. (2019) 'A novel quantitative ecological and microbial risk assessment methodology: theory and practice', *Human and Ecological Risk Assessment*, pp. 1–24.

Duarte, H. O. & Droguett, E. L. (2015) 'Quantitative Ecological Risk Assessment of accidental oil spills on ship routes nearby a marine national park in Brazil', *Human and Ecological Risk Assessment: An International Journal*, p. 0.

ECMWF (2019) *ERA Interim, Daily, European Centre for Medium-Range Weather Forecasts*. Available at: https://apps.ecmwf.int/datasets/data/interim-full-daily/levtype=sfc/ (Accessed: 25 August 2021).

Ferreira, B. P. & Maida, M. (2006) *Monitoramento dos Recifes de Coral do Brasil: Situação Atual e Perspectivas*. Brasília: MMA.

Fox, R. W. et al. (2014) *Introdução à Mecânica dos Fluidos*. Minas Gerais: LTC.

Garla, R. C. (2004) *Ecologia e conservação dos tubarões do Arquipélago de Fernando de Noronha, com ênfase no tubarão-cabeça-de-cesto Carcharhinus perezi (Poey, 1876) (Carcharhiniformes, Carcharhinidae)*. Rio Claro: Programa de Pós-Graduação de Ciências Biológicas da UNESP - Tese de doutorado em Zoologia.

GEBCO (2021) *Gridded Bathymetry Data, General Bathymetric Chart of the Oceans*. Available at: https://www.gebco.net/data_and_products/gridded_bathymetry_data/ (Accessed: 25 August 2021).

Gillespie, R. G. (2001) 'Oceanid Islands: Models of Diversity', *Enciclopedia of Biodiversity*, pp. 1–13.

Gove, J. M. et al. (2016) 'Near-island biological hotspots in barren ocean basins', *Nature Communications*, 7(1), p. 10581.

Guo, W. (2017) 'Development of a statistical oil spill model for risk assessment', *Environmental Pollution*, 230, pp. 945–953.

Hackett, B., Breivik, Ø. & Wettre, C. (2006) 'Forecasting the Drift of Objects and Substances in the Ocean', in *Ocean Weather Forecasting*. Berlin/Heidelberg: Springer-Verlag, pp. 507–523.

Hounsou-gbo, G. A. et al. (2015) 'Tropical Atlantic Contributions to Strong Rainfall Variability Along the Northeast Brazilian Coast', *Advances in Meteorology*, 2015, pp. 1–13.

IBAMA (2005) *Plano de Manejo da APA Fernando de Noronha - Rocas - São Pedro e São Paulo: Resumo Executivo*. Available at: http://www.icmbio.gov.br/portal/images/stories/imgs-unidadescoservacao/ResumoExecutivo_f.pdf.

ICMBio (2013) 'PARNAMAR - Parque Nacional Marinho de Fernando de Noronha'. Available at: http://www.parnanoronha.com.br/paginas/91-o-parque.aspx.

ICMBio (2021) *Operação Emergencial ao Aporte Atípico de Fragmentos de Óleo e Lixo Marinho nas Localidades do Mar de Fora de Fernando de Noronha*.

IMO (2008) *Formal Safety Assessment FSA-Crude Oil Tankers*. London: International Maritime Organization.

ITOPF (2021) *Oil Spill Tanker Statistics 2020*. London, UK.

IUCN (2020) *Brazilian Atlantic Islands: Fernando de Noronha and Atol das Rocas Reserves - 2020 Conservation Outlook Assessment, World Heritage Outlook*. Available at: https://worldheritageoutlook.iucn.org/.

Ivar do Sul, J. A., Spengler, Â. & Costa, M. F. (2009) 'Here, there and everywhere. Small plastic fragments and pellets on beaches of Fernando de Noronha (Equatorial Western Atlantic)', *Marine Pollution Bulletin*, 58(8), pp. 1236–1238.

Keramea, P. et al. (2021) 'Oil Spill Modeling: A Critical Review on Current Trends, Perspectives, and Challenges', *Journal of Marine Science and Engineering*, 9(2), p. 181.

Kueffer, C. & Kinney, K. (2017) 'What is the importance of islands to environmental conservation?', *Environmental Conservation*, 44(4), pp. 311–322.

Lardner, R. et al. (2006) 'An operational oil spill model for the Levantine Basin (Eastern Mediterranean Sea)', in *International Symposium on Marine Pollution*. European Communities.

Lee, K. et al. (2015) *The Behaviour and Environmental Impacts of Crude Oil Released into Aqueous Environments*. Ottawa. The Royal Society of Canada.

Lehr, W. et al. (2002) 'Revisions of the ADIOS oil spill model', *Environmental Modelling & Software*, 17(2), pp. 189–197.

Lewis, M. & Pryor, R. (2013) 'Toxicities of oils, dispersants and dispersed oils to algae and aquatic plants: Review and database value to resource sustainability', *Environmental Pollution*, 180, pp. 345–367.

Lira, S. M. A., Amaral, F. M. D. & Farrapeira, C. M. R. (2009) 'Population growth by the white sea-urchin Tripneustes ventricosus (Lamarck, 1816) (Echinodermata) at the Fernando de Noronha Archipelago Brazil', *Pan-American Journal of Aquatic Sciences*, 4, pp. 1–2.

Lumpkin, R. & Garzoli, S. L. (2005) 'Near-surface circulation in the Tropical Atlantic Ocean', *Deep Sea Research Part I: Oceanographic Research Papers*, 52(3), pp. 495–518.

Lumpkin, R. & Johnson, G. C. (2013) 'Global ocean surface velocities from drifters: Mean, variance, El Niño--Southern Oscillation response, and seasonal cycle', *Journal of Geophysical Research: Oceans*, 118(6), pp. 2992–3006.

Maida, M. & Ferreira, B. P. (1995) 'Preliminary evaluation of Suest Bay reef, Fernando de Noronha, with emphasis on the scleractinian corals', *Boletim Técnico Científico do CEPENE*, 3(1), pp. 37–47.

MARINETRAFFIC (2021) *MarineTraffic: Global Ship Tracking Intelligence | AIS Marine Traffic*. Available at: http://www.marinetraffic.com.

Marta-Almeida, M. et al. (2013) 'Efficient tools for marine operational forecast and oil spill tracking', *Marine Pollution Bulletin*, 71(1–2), pp. 139–151.

Medeiros, R. C. (2009) 'O Arquipélago de Fernando de Noronha e a presença militarnaval: uma condicionante Estratégia (I)'. Sagres. Available at: http://www.sagres.org.br/artigos/marinha_afn.pdf.

Molinari, R. L. (1982) 'Observations of eastward currents in the tropical South Atlantic Ocean: 1978–1980', *Journal of Geophysical Research*, 87(C12), p. 9707.

Nittis, K. et al. (2006) 'Operational monitoring and forecasting for marine environmental applications in the Aegean Sea', *Environmental Modelling & Software*, 21 (2), pp. 243–257.

NOAA\ (2018) *Global Self-Consistent Hierarchical High-resolution Geography, GSHHG, National Oceanic Atmospheric Administration*. Available at: https://www.ngdc.noaa.gov/mgg/shorelines/data/gshhg/latest/ (Accessed: 25 August 2021).

NRC (2003) *Oil in the Sea III: Inputs, Fates and Effects*. Edited by T. N. A. Press. Washing, DC.

Queiroz, S. et al. (2019) 'Simulation of Oil Spills Near a Tropical Island in the Equatorial Southwest Atlantic', *Tropical Oceanography*, 47(1), pp. 17–37.

Reed, M., Aamo, O. M. & Daling, P. S. (1995) 'Quantitative analysis of alternate oil spill response strategies using OSCAR', *Spill Science & Technology Bulletin*, 2(1), pp. 67–74.

Reed, M., Gundlach, E. & Kana, T. (1989) 'A coastal zone oil spill model: Development and sensitivity studies', *Oil and Chemical Pollution*, 5(6), pp. 411–449.

Sebastião, P. & Guedes Soares, C. (1995) 'Modeling the Fate of Oil Spills at Sea', *Spill Science & Technology Bulletin*, 2(2/3), pp. 121–131.

Serafini, T. Z. & França, G. (2010) 'Ilhas oceânicas brasileiras: biodiversidade conhecida e sua relação com o histórico de uso e ocupação humana', *Revista de Gestão Costeira Integrada/Journal of Integrated Coastal Zone Management*, 10(3), pp. 281–301.

Shafir, A., Van-Rijn, J. & Rinkevich, B. (2007) 'Short and long term toxicity of crude oil and oil dispersants to two representative coral species', *Environ. Sci. Technol.*, 41, pp. 5571–5574.

Silva, J. M. (2010) *The Dolphins of Noronha*. São Paulo: Bambu.

Silva, M. et al. (2009) 'Circulation and heat budget in a regional climatological simulation of the Southwestern Tropical Atlantic.', *Tropical Oceanography*, 37(1–2).

Spaulding, M. L. et al. (1994) 'Application of three-dimensional oil spill model (WOSM/OILMAP) to Hindcast the Braer spill', *Spill Science & Technology Bulletin*, 1(1), pp. 23–35.

Spaulding, M. L. (2017) 'State of the art review and future directions in oil spill modeling', *Marine Pollution Bulletin*, 115(1–2), pp. 7–19.

Stramma, L. & Schott, F. (1999) 'The mean flow field of the tropical Atlantic Ocean', *Deep Sea Research Part II: Topical Studies in Oceanography*, 46(1–2), pp. 279–303.

TAMAR (2006) *Life in the deep blue*. 1st edn. São Paulo: Bambu.

Tchamabi, C. C. et al. (2017) 'A study of the Brazilian Fernando de Noronha island and Rocas atoll wakes in the tropical Atlantic', *Ocean Modelling*, 111, pp. 9–18.

Ung, S.-T. (2019) 'Evaluation of human error contribution to oil tanker collision using fault tree analysis and modified fuzzy Bayesian Network based CREAM', *Ocean Engineering*, 179, pp. 159–172.

Whittaker, R. J. & Fernández-Palacios, J. M. (2007) *Island Biogeography: Ecology, Evolution and Conservation*. 2nd edn. Great Britain: Oxford University Press.

Yender, R. A. et al. (2010) *Oil Spills in Coral Reef: Planning and Response Considerations*. Florida: National Oceanic and Atmospheric Administration - NOAA.

Zelenke, B. et al. (2012) 'General NOAA Operational Modeling Environment (GNOME) technical documentation'. Edited by N. O. S. United States Office of Response and Restoration. U.S. Dept. of Commerce, National Oceanic and Atmospheric Administration, National Ocean Service, Office of Response & Restoration (NOAA technical memorandum NOS-OR&R 40). Available at: https://repository.library.noaa.gov/view/noaa/2620.

Zhang, Y. & Baptista, A. M. (2008) 'SELFE: A semi-implicit Eulerian–Lagrangian finite-element model for cross-scale ocean circulation', *Ocean Modelling*, 21(3–4), pp. 71–96.

Zodiatis, G. et al. (2008) 'The Mediterranean oil spill and trajectory prediction model in assisting the EU responde agency', in *Congreso Nacional de Salvamento en la Mar*. Cadiz, pp. 535–547.

*Trends in Maritime Technology and Engineering – Guedes Soares & Santos (eds)*


# Evaluating risk during evacuation of large passenger ships: A smart risk assessment platform for decision support

N.P. Ventikos, N. Themelis, K. Louzis, A. Koimtzoglou, A. Michelis & M. Koimtzoglou
*School of Naval Architecture and Marine Engineering, National Technical University of Athens, Greece*

A. Ragab
*Maritime and Logistic Department, Jade University, Elsfleth, Germany*

ABSTRACT: The PALAEMON Smart Risk Assessment Platform (SRAP) aims to provide dynamic, real-time risk assessment, with respect to the safety of passengers and crew members and to support the Bridge Command Team throughout all the phases of ship evacuation. The objective of this paper is to describe a Bayesian Network risk model, which was developed for SRAP for supporting the decision to sound the General alarm following an incident. The employed methodology includes: 1) identifying risk factors from a survey of the relevant literature and a qualitative analysis of evacuation-related maritime accidents, 2) developing the model structure, 3) quantification based on expert judgement, and 4) validation based on feedback from expert interviews and by testing the output of the models in selected case studies relevant to fire and flooding cases. The case studies are also used to demonstrate the functionality of the risk models and evaluate their sensitivity in different conditions.

## 1 INTRODUCTION

A broadly applicable definition of risk, as provided by ISO (2009), is the "effect of uncertainty on objectives", while risk assessment can be used for supporting decisions regarding actions and priorities during emergency situations. Ship evacuation is such a situation and a dynamic and high-risk process involving uncertainty, deviation from standard/best practices, and risk trade-offs. Risk assessment and analysis, supported by structured engineering judgment, may have a major impact on decisions in such a process (HSE 2002).

The success of ship evacuation strongly depends on the performance of the human element i.e. the crew, under extremely challenging and stressful conditions. Probably the most contributing factors for a positive outcome are the decisions to be made and exercised by the Master and the Bridge Command Team, which need to be taken timely and be the most appropriate for managing the incident effectively and with the minimum possible losses. In this respect, any decision-making tool that will improve the perception and the comprehension of the Master and his team regarding an accident onboard is of primary importance. The goal should be to support the human element effectively in terms of the necessary vigilance, early risk identification and adequate situation evaluation, in order to minimize human error and the response time during accidents.

The purpose of this paper is to present the concept of the Smart Risk Assessment Platform (SRAP), which is a dynamic risk assessment digital tool that intends to support the decisions of the Master and the Bridge Command Team throughout all phases of the evacuation process given the occurrence of an accident. The output of SRAP will be available in real-time to the Master and Bridge Command Team as an appropriate risk level indication regarding the following aspects: sounding the General Alarm (GA); monitoring the progress of the mustering process and taking any additional actions if required; and abandoning the ship or not.

Emphasis will be given on the form of the risk information so that it will be comprehensible for the decision-makers, and to the reasoning of the risk evaluation to help them identify both the facts and the assumptions that influence the risk level.

SRAP consists of the following three modules that correspond to the main evacuation phases:

- Situation assessment module;
- Mustering process assessment module; and
- Pre-abandonment assessment module.

This paper will focus only on analysing the situation assessment module which is activated after the occurrence of an accident. The risk information evaluated by SRAP will be available to the Master and the Bridge Command Team through the PALAEMON

DOI: 10.1201/9781003320289-30

(Koimtzoglou, et al. 2021) Decision Support System (DSS), which will generate alerts, prioritize unfolding events in terms of their criticality, recommend actions to be performed, etc. in real time. Its main aim is to evaluate whether raising the GA and initiate the evacuation process is needed. The assessment is based on the ship's safety status and the level of passenger's exposure to hazards. The situation assessment module is active during all phases of the evacuation process, until the successful containment of the accident or the abandonment of the ship.

The paper is structured as follows: Section 2 presents the relevant literature regarding the concept of risk assessment within the evacuation process examined in this work. Section 3 outlines the methodology applied to develop the SRAP model. Section 4 describes the prevailing conditions and the critical factors for each stage of the evacuation process that resulted from the study of five ship evacuation cases as derived from accident reports. Section 5 describes the BN risk model that was developed. After describing the risk factors that have considered, the structure of the developed BN model is presented, followed by a description of the rationale that was employed for quantification Section 6 presents the results of the applied validation process, which includes demonstrating the functionality of the risk model in specific fictional case studies. The paper concludes with the expected benefits of SRAP for ship evacuation and an overview of future research steps. In the Annex a table with a detailed description of the individual nodes of the BN model and the assumptions that have been employed in the development is presented.

## 2 BACKGROUND ON RISK ASSESSMENT FOR MARINE EVACUATION

### 2.1 *Risk assessment for escape and evacuation*

Many researchers have studied methods of assessing the risk of escape and evacuation on offshore installations during the occurrence of an incident (Norazahar et al. 2018). Several of them address the human element contribution, by incorporating in their modelling approach risks associated with personnel performance.

In recent years, Bayesian Network (BN) methodology, which represents both probabilistic and causal dependencies between risk factors in a network format, is increasingly utilized as an approach for risk analysis in the maritime domain. According to Zhang et al. (2018), the benefits of BNs include the incorporating expert's knowledge when statistical data is limited or unavailable, making dynamic updates when new observations are made, and including human and organizational factors. In addition, BNs can be considered as a way to formally represent domain knowledge and reason under uncertainty (Fenton and Neil 2012).

Eleye-Datubo (2006) used a BN technique to model a typical ship evacuation. Montewka et al. (2014) developed a framework for risk assessment for maritime transportation with the use of BNs. The framework focuses on open sea ship-to-ship collisions with a RoPax vessel being considered as the struck ship, and includes variables directly linked to the evacuation of the vessel. Norazahar et al. (2018) applied Bayesian analysis and binomial distribution for demonstrating the impact of harsh environmental conditions on personnel performing escape and evacuation on offshore installations. Ping et al. (2018) integrated BN and fuzzy Analytical Hierarchy Process (AHP) to estimate the probability of success of escape, evacuation, and rescue on offshore platforms.

Sarshar et al. (2013) proposed a BN model to analyse the passenger ship evacuation process during a fire accident to reveal the most influential factors that affect evacuation time. To build the model, they first identified the most important factors causing the evacuation time to exceed the standard time during a fire and then they associated them with the passengers' condition, the availability of evacuation information, the passengers' panic, the passengers' travel time, the awareness (reaction) time, the embarkation time, the launching time, and the evacuation time.

### 2.2 *Challenges and data availability*

A dynamic risk analysis seems to be the appropriate approach for the evacuation process, which is a highly dynamic process. Yang, Haugen, and Paltrinieri (2018) explain that the word "dynamic" is used to indicate that this analysis is frequently updated compared to the traditional static Quantitative Risk Assessments (QRAs). Furthermore, they designate that dynamic may refer to both updating of the risk model parameters and updating of the model itself, although most of the work that has been performed to date has been related to parameter updating. A dynamic risk assessment should be able to take any new information into account and adjust to the dominant dynamic environment (Khakzad, Khan, and Amyotte 2012).

Kristiansen (2005) provides an overview of the limitations of risk analysis in the maritime domain in general, especially in relation to quantitative analysis. The most significant and common limitation is the lack of statistical data relevant to each application/activity. This data forms the basis of the analysis and gives an indication of the frequency and the most likely consequences of a relevant accident scenario under consideration. The insufficiency of statistical data results in large uncertainties in the outcomes of the analysis and the need to make simplifying assumptions in the modelling process. It should be noted that BNs can exploit expert's judgement for quantification to address the limited availability of statistical data. However, expert judgement affects the reliability of the risk model due to the inherent subjectivity and related uncertainties (Guizhen Zhang et al. 2018; Hänninen 2014).

### 2.3 *Advantages of Bayesian Networks*

According to Khakzad, Khan and Amyotte (2011), although conventional risk assessment techniques such as Fault Tree (FT) have been used widely and effectively for dynamic safety analysis, their structure is static and have limitations with respect to uncertainty handling. Although Dynamic Fault Trees (DFTs) have been developed for modeling the dynamic behaviour of systems (see, for example, Hamza and Abdallah, 2015), they are mostly applied for analysing system reliability, while their quantification depends on the availability of failure data.

Some of the benefits of BNs, include explicit representation of the dependencies among events, updating probabilities, and coping with uncertainties (Khakzad, Khan and Amyotte, 2011). Bobbio, et al. (2001) state that BNs are more relevant compared to FTs in terms of representing complicated dependencies between components and incorporating uncertainty. BNs facilitate a flexible way to exploit expert knowledge by providing the ability to quantify Conditional Probability Tables (CPTs) even with partial information (Khakzad, Khan and Amyotte, 2011). Moreover, BNs have been used to dynamically update probabilities by continually take into account new information (Hamza and Abdallah, 2015).

## 3 METHODOLOGY

The employed methodology for the development of the SRAP risk models consists of the following steps (Figure 1):

1) Identifying factors and parameters that affect risk during the evacuation process on a passenger ship, based on information collected from the relevant literature (Section 2) and the accident analysis (Section 4).
2) Representing risk factors and their relationships in risk models, based on expert judgement, and quantifying the models,
3) Validating the developed risk models, by engaging experts in interviews,
4) Reviewing the risk models based on the feedback collected from the expert interviews.

The risk models in SRAP have been developed as discrete, static BNs, which are Directed Acyclic Graphs (DAGs) that include random variables (nodes) connected with edges that represent probabilistic and causal relationship (F. V. Jensen and Nielsen 2007). Nodes are distinguished into parents, which have no preceding connecting nodes, and children, which are conditionally dependent on parent nodes. Computationally, BN exploits Bayes' theorem for conditional probabilities, which is used for updating beliefs given the observation of evidence. BN modelling generally consists of the following steps (see Fenton and Neil 2012): 1) identifying variables that are important to the problem, 2) structuring the relationships between random variables, and 3) quantifying the strength of the relationship by specifying CPTs.

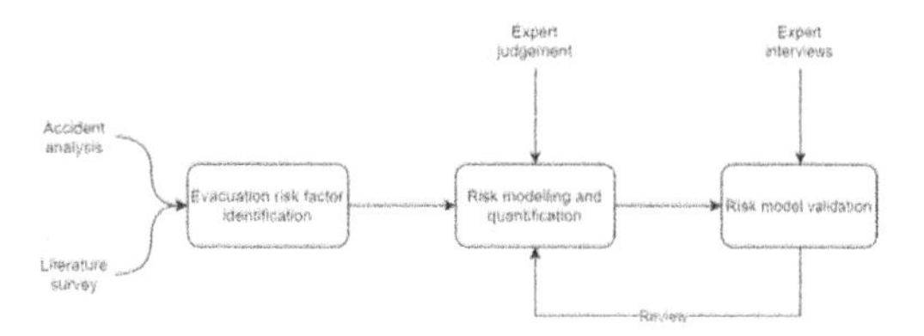


Figure 1. SRAP risk modelling methodology.

Considering that the developed BN risk models have been elicited from expert knowledge, we implement a validation approach similar to the framework proposed by Pitchforth and Mengersen (2013), which includes:

1) nomological validity to establish that the problem examined fits into a wider domain in the literature,
2) face validity to the relationships between the random variables reflect the problem,
3) content validity to establish that all important parameters that affect the problem have been included in the model,
4) concurrent validity to determine whether a network or one of its sections behaves similarly to a section of another network,
5) convergent validity to determine the similarity of the model (i.e. in terms of structure, discretization, and parameterization) to other models for similar purposes,
6) discriminant validity that reflects the degree to which the model is different compared to other models, and
7) predictive validity to establish that model behaviour and output are fit-for-purpose.

The developed BN models were validated in terms of face and content validity, by conducting a discussion with an expert (crew member of Ro-Pax ferry), and in terms of predictive validity through the case studies presented in Section 6. In addition, to verify the quantification rationale described in Section 4, the results of a strength of influence analysis, which reflect the distance between the conditional probability distributions of a child node conditional on the states of the parent nodes, are presented.

## 4 ANALYSIS OF EVACUATION-RELATED ACCIDENT CASES

In the context of this study, five accidents were studied involving passenger ships where an evacuation was conducted. These cases are the accidents of Sorrento (MIT 2015), Viking Sky (AIBN 2019), Carnival Liberty (NTSB 2015), Costa Concordia (MCIB 2012) and Queen of the North TSB 2008). The selected accidents occurred after the year 2000 and include cases where the events following the accident did not unfold rapidly (i.e. there was enough time for the crew to assess the situation and carry out the evacuation process). The objective of the analysis was to identify the key contributing risk factors and the prevailing conditions that affected the evacuation process in each case. Towards this end, the following information from the accident reports was considered: incident type and location, the time of the day they occurred, the way they affected the vessel status and the mitigative actions that were taken by the crew.

The analysis of the accident reports resulted in identifying parameters that affect the decision-making process with respect to initiating the evacuation process. The most influential factor, which affects the decision to initiate the evacuation process, seems to be the severity of the incident and the resulting damage to the vessel. Another important parameter is the time (day or night) the incident occurred. During night-time, the reaction of passenger and crew members may be delayed and darkness complicates the mustering process. In the Queen of the North accident, most passengers were asleep while the ship grounded and were alerted by the striking noise and vessel motion. In addition, an incorrect assessment of the situation and of the effectiveness of mitigation actions could be a factor of major importance. For example, if a fire gets out of control it may damage the Life-Saving Appliances (LSAs), which was the case in the Sorrento accident. Considering that the decision to evacuate is made after efforts to mitigate the incident are unsuccessful, incorrect assessment may also lead to actions taken too late (e.g. sounding the GA). In the Sorrento accident, the distress signal was launched after trying to mitigate the fire which broke out with the assistance of the fixed extinguishing system-drencher. Given that the fire became uncontrollable in a short amount of time, it is conceivable that the time devoted to mitigating a situation that would have escalated very quickly may have had an impact on the time efficiency of the evacuation.

# 5 SITUATION ASSESSMENT MODEL

This section describes the BN risk model that was developed for supporting the Master's decision to sound the GA while assessing the situation after the incident has occurred. After describing the risk factors that have considered, the structure of the developed BN model is presented, followed by a description of the rationale that was employed for quantification.

## 5.1 *Risk factors*

The aim of the situation assessment model is to provide inference on the post-accident severity of the situation, which depends on whether the incident can be contained, whether passengers are directly exposed to hazardous conditions, and whether the vessel may provide safe conditions to the passengers and crew. The risk factors that have been considered in the model are categorised as described below.

**Floatability, stability and watertight integrity**. To assess whether the damage (e.g. due to grounding or collision) has rendered the ship unsafe for the passengers, the following factors are considered: the new floating position, the restoring capability, the ship motions due to the prevailing weather conditions, the status of damage control systems, and whether the hull remains watertight. The critical conditions whose occurrence shall be examined at specific time instances are:

- whether the ship will capsize for the given damage, loading and weather conditions, considering progressive flooding; and
- whether the critical roll and pitch angle for the safe launching of the ship's Life Saving Appliances (LSAs) will be surpassed.

**Structural integrity**. Upon the occurrence of an accident, structural failure could occur due to the development of excessive global stresses in critical locations of the ship's structure. Such extreme loading of the structure could occur as an effect of the rapid development of stress due to a collision or an explosion. Additionally, progressive loading of the structure can occur in the case of flooding, from the thermal loading due to high temperatures developed in an uncontrolled fire or from the fire-fighting system. Thus, the critical conditions refer to surpassing the stress limits that could lead to collapsing of the structure.

**Fire safety integrity**. A fire can be ignited in any space of the ship and for a variety of reasons (e.g. electrical failure, human error, explosion, etc.). The fire effluents such as toxic gases and smoke, as well as the high temperatures and the heat radiation, can have a direct impact on passengers' health from the first phases of the development of a fire. The location of the fire incident is also important as it determines the type and the amount of the available fire load, its proximity to passengers and crew, as well as the potential impact on critical ship's systems (e.g. engine room). Fire detection and extinguishing systems (including manual suppression) are important for controlling a fire, especially when activated in the early phases of fire development. Moreover, the status of the fireproof doors status is also important. In the case of the failure of these systems and depending on the available fire load and the availability of oxygen, a failure of the passive containment will result in the spreading of fire outside or away from the space of fire origin and even of the

Main Vertical Zone (MVZ). Therefore, the following parameters determine the critically of the situation:

- The location and type of space of fire origin;
- Whether the fire safety systems (automatic and manual) are working properly;
- Whether the fire is spreading; and
- Whether passengers are in the proximity of the space of fire.

**Status of critical systems for ship's controllability and navigation**. An uncontrollable ship could increase the risk of grounding or collision. Furthermore, based on the Safe Return to Port requirements (IMO MSC.216 (82)), navigational and controllability systems are critical upon the occurrence of an accident. Thus, the status of the following systems will affect the situation assessment:

- Propulsion system and specifically the ability of the ship to move with a minimum speed at specific weather conditions;
- Steering system, which provides the required controllability; and
- Systems that support safe navigation.

**Status of critical systems for communication (internal and external)**. All radio-communications necessary to maintain the communications either inside the ship between the crew members (e.g. between the bridge and a response team) or between ship with other ships or land-based stations shall be examined for their operation.

**Passenger's exposure and readiness**. In the initial stages of the accident, it is possible for passengers to be directly and immediately affected (e.g. presence of toxic gases, flames, water, etc.). The exposure of the passengers to hazards is an important element that affects the decision to initiate the evacuation process and muster the passengers towards safety. Moreover, the readiness of the passengers affects their response time with respect to following cues and thus it is considered in the situation assessment model.

### 5.2 *Bayesian Network*

Figure *2* shows the relationships between the identified risk factors that were described in the previous section. A detailed description of the individual nodes of the BN model and the assumptions that have been employed is presented Table 5 in the Annex.

The containment of an incident that reflects vessel's integrity against fire or flooding depends on active and passive mitigation measures. The exposure of the passengers to hazardous conditions depends on how effectively the incident can be contained and from the characteristics of the incident (e.g. in which ship space has occurred, how severe is the damage etc.). Vessel integrity depends on the watertight and structural integrity of the hull, the operational status of the system's that are critical to the vessel's operation and for supporting the passengers and crew, and on whether the incident can be contained (e.g. a fire that is spreading uncontrollably could endanger the structural integrity of the vessel).

### 5.3 *Quantification*

A qualitative rationale was applied to develop the CPTs of the model's nodes by exploiting information collected from the the accident analysis (see Section 4), the relevant literature and expert judgment through a workshop conducted with the involved partners and stakeholders of the PALAEMON project. This approach was based on assessing the direction of the effect of the parent nodes on the probabilities of the

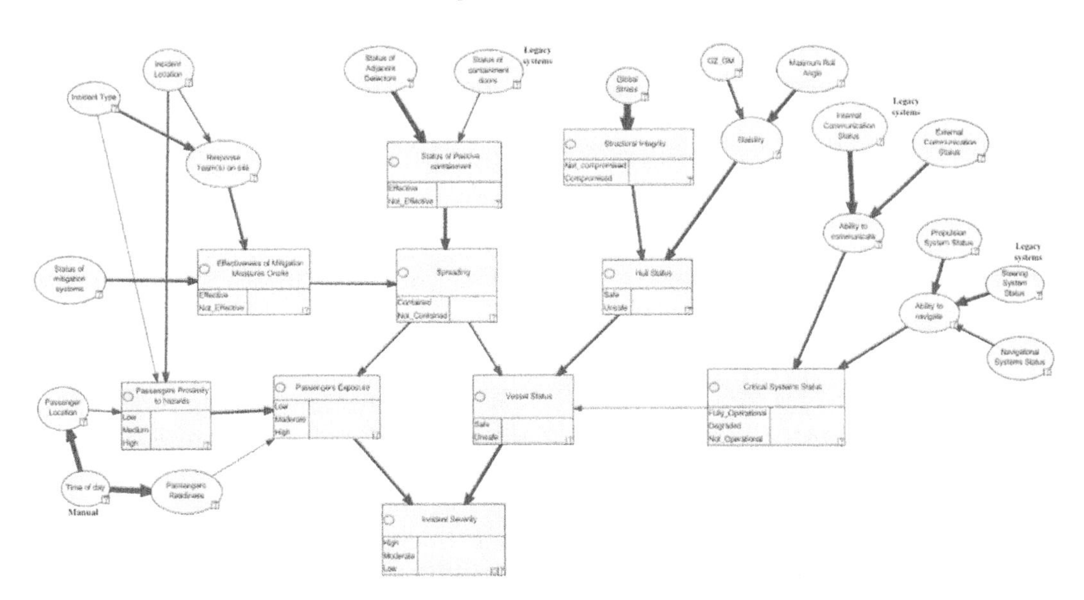

Figure 2. Situation assessment BN and strength of influence analysis.

states on the children nodes. For example, with increasing passenger proximity to hazards their probability of their exposure is increased. The magnitude of the effect was assessed by qualitatively weighing the contribution of each risk factor, as is described briefly below, which was subsequently reflected in the CPTs.

**Proximity to hazards (Incident type – Incident location)**. Flooding incidents are caused by hull breaches below the waterline. In cruise ships and Ro-Pax ferries, these spaces are mainly utilized as crew members accommodation, tanks, storage spaces, workshops, and machinery spaces. Therefore, it is assumed that in flooding incidents, passenger's proximity to hazards will be lower compared to fire incidents.

**Status of passive containment (Status of adjacent detectors – Status of containment doors)**. For capturing the fast spreading of an incident, the status of adjacent detectors is considered more vital compared to the status of the containment doors, because the status of the detectors present an indication of the spreading even if the containment doors are closed.

**Hull status (Structural integrity – Stability)**. Cruise ships and Ro-Pax ferries are vessels that usually do not present major structural integrity issues due to their design (increased number of decks) and their cargo (mainly passengers and vehicles). Hence, stability was considered more significant compared to structural integrity for assessing the hull status.

**Ability to navigate**. The status of the main propulsion and the steering system were weighted as more important compared to the condition of the navigational systems.

**Vessel status (Critical systems status – Hull status – Spreading)**. The order of significance that is assumed is: 1) the condition of the hull status, 2) the spreading, 3) the critical system status. During emergencies, the ship is considered the safest place to be. However, in case the hull is not safe for any reason (i.e. watertight integrity has been compromised), evacuation should be ordered immediately.

**Passengers exposure (Passengers proximity to hazards – Spreading)**. Passengers' proximity to hazards is assumed more critical compared to the level of spreading. If more passengers are closer to the incident, there is a higher risk level compared to an incident that may be spreading but passengers are not present in its vicinity.

**Incident severity (Passengers exposure -Vessel status)**. The main priority on cruise ships and Ro-Pax ferries during emergencies is the protection of life for the people on-board (i.e. passengers and crew members). Therefore, passengers' exposure is considered more important compared to the vessel status due to the need for their immediate protection.

Figure 2 shows the results of the strength of influence analysis for the situation assessment BN model. The results indicate that the developed CPTs adequately reflect the relative weights among the risk factors that have been described in this section. The thicker arrows show a greater influence of the parent to the connected child node with respect to their probabilities.

## 6 VALIDATION AND CASE STUDY RESULTS

This section describes the results of the face and content validation process based on the comments obtained from the expert interview regarding each node of the model. The section also presents the results from the application of the BN risk model to specific case studies. The objective of the analysis is to test the functionality of the situation assessment BN model and to illustrate the sensitivity of the model. The case studies include three fire and three flooding related cases. For each case, a brief description of the scenario and the result from the application of the situation assessment model are provided.

It is noted that the result of the risk model is the node "Incident severity" with Low, Moderate, and High states that are given as percent probabilities. The decision to sound the GA or not is supported by this result based on the following rationale that implies a risk-averse approach. Sounding the GA would be suggested if: 1) the probability of high incident severity is greater than the probabilities of low and medium severities, and 2) the probabilities of high and moderate incident severity are greater than the probability of low severity.

### 6.1 *Validation results*

Two comments of major importance are that night makes the handling of any incident more difficult and that fire is considered to need more immediate action for passenger's protection compared to flooding. With respect to the incident location, if it is not a public space and depending on the severity of the incident, the crew members will attempt to bring the situation under control without sounding the GA. In case the Response Team reports that the incident cannot be contained this will lead to the immediate sound of the GA. It was emphasized that the feedback provided by the Response Team(s) is very important for the assessment of the situation.

It was also highlighted that cruise ships and Ro-Pax Ferries have adequate strength and structural integrity is usually not an issue even in case of grounding and collision. Their main issue is the stability of the ship after an accident (i.e. low GM issues). Regarding stability, free surface moments should also be taken into account especially in fire accidents in Ro-Pax Ferries and within the car deck area due to water accumulation from firefighting systems.

Based on the expert comments on the structure of the BN model, it is considered that it adequately reflects the situation and that the most important risk factors and parameters have been considered.

### 6.2 *Case studies results - Fire cases*

For all fire cases, Table 1 presents the states of the parent nodes and Table 2 presents the states of the children nodes given the evidence for the states of the parent nodes of the situation assessment model.

6.2.1 *Fire case 1: Fire in cabin in daytime*

At 11.00 am, a fire breaks out in the balcony of a cabin and spreads inside the cabin. The cabin smoke detector generates an alarm and the fire sprinklers in the cabin are activated. The response team arrives on site. The smoke detectors in the adjacent corridor and cabins have not been activated, while the door of the incident cabin, as well as the doors of the neighbouring cabins are closed. All relevant ship systems are assumed to be working properly.

The stability and the structural integrity of the ship are not affected by the fire, therefore the node "Hull status" has an output of "Safe" with 93% probability. All critical systems are fully operational, and the "Spreading" node is "Contained" (68%). As expected, "Vessel status" is characterised "Safe" because the fire is not spreading (sprinklers are activated, and adjacent detectors remain deactivated) and there are no other issues for the ship's hull. On the contrary, as the fire breaks out in the passengers' cabin area, the node "Passengers proximity to hazards" has mainly an outcome of "Medium" (38%) and "High" (36%), as it assumed that the majority of the passengers are not in their cabin at the time of the incident. This reduces the passengers' exposure, but the moderate and high percentages are still considered enough to result in a high severity with 45% probability.

Based on this result, the SRAP would provide an indication to the Master to sound the GA. This is an expected outcome as any fire incident in the accommodation area, even with a small number of passengers, will initiate the evacuation and mustering process.

6.2.2 *Fire case 2: Fire in cabin in night-time*

The scenario is similar to Fire Case 1, but the incident initiates during the night. Initially, the cabin smoke detector is activated. The response team does not arrive on site. Soon, the corridor's and the adjacent cabins' smoke detectors are also activated.

In this scenario, although the "Stability" and the "Structural integrity" nodes of the ship are not affected by the incident, along with the critical systems, which are fully operational, the "Spreading" node is "Not contained" compared to Fire Case 1. This is attributed to the fact that the cabin's adjacent detectors are activated, and the response team is not on site. Therefore, the node "Vessel status" is characterised as "Unsafe" with 70% probability. Furthermore, because the fire breaks out in the passengers' cabin area at night, the node "Passengers proximity to hazards" is higher compared to Fire Case 1 as the majority of the passengers is assumed to be located in their cabin. As the "Spreading" node is "Not contained", it also affects the "Passengers exposure" which is assessed as "High" (64%).

The incident's severity is assessed as "High" with 79% probability and the SRAP would provide an indication to the Master to sound the GA.

Table 1. The parent nodes states for the fire cases.

| Node | Fire case 1 | Fire case 2 | Fire case 3 |
|---|---|---|---|
| Time of day | Day | Night | Day |
| Incident type | Fire | Fire | Fire |
| Incident location | Accommodation | Accommodation | Engine room |
| Response team onsite | Yes | No | Yes |
| Status of mitigation systems | Activated | Activated | Activated |
| Status of adjacent detectors | Not activated | Activated | Not activated |
| Status of containment doors | Opened | Opened | Closed |
| Global stresses | Not exceeded | Not exceeded | Not exceeded |
| Restoring capability (GZ/GM criteria) | Covered | Covered | Covered |
| Maximum roll angle | Not exceeded | Not exceeded | Not exceeded |
| Internal Communication | Operational | Operational | Operational |
| External Communication | Operational | Operational | Operational |
| Propulsion System | Operational | Operational | Operational |
| Steering System | Operational | Operational | Operational |
| Navigational Systems | Operational | Operational | Operational |

6.2.3 *Fire case 3: Fire in Engine Room*

The scenario is similar to Fire Case 1, but the fire breaks out in the Engine Room (ER).

The incident severity is "Low" (62%) and the SRAP would provide an indication to the Master not to sound the GA but to reassess the condition.

This is an expected outcome according to experts (see Section 6.1), where a case was described in which the passengers were not informed of a fire in the ER that was contained by the Response Team.

6.2.4 *Summary of fire cases*

The results from the fire cases show that the node "Spreading" is mainly affected by the status of the adjacent detectors, the presence of the Response Team on site and the status of the fireproof doors. Furthermore, the node "Time of day" affects all its connected children nodes. The "Incident severity" node is sensitive to the changes of its parent nodes, i.e. "Vessel status" and "Passengers exposure". In Fire Case 1, the "Vessel status" is "Safe" but the "Passengers exposure" is divided into Low 50%, Moderate 25% and High 25%. As these percentages are considered risky, the indication of the SRAP would lead to sound the alarm. In Fire Case 2, "Vessel status" is "Unsafe" and the "Incident severity" is significantly higher compared to Fire Case 1.

Finally, in Fire Case 3, the "Vessel status" is "Safe" and because the fire is located in the Engine Room and it is contained, the "Passengers exposure" node is significantly lower compared to the other cases. Hence, the indication of the SRAP would lead to reassess the situation. The condition of the nodes "Critical systems status", "Ability to communicate", "Ability to navigate" and "Stability" resulted to 100% (meaning "Sufficient" for the stability node and "Fully operational" for the three remaining nodes).

Table 2. The children nodes states for the fire cases.

| Node | Status | Fire case 1 | Fire case 2 | Fire case 3 |
|---|---|---|---|---|
| Effectiveness of mitigation measures onsite | Effective: | 90% | 50% | 90% |
| | Not effective: | 10% | 50% | 10% |
| Passengers proximity to hazards | Low: | 26% | 12% | 77% |
| | Medium: | 38% | 31% | 14% |
| | High: | 36% | 57% | 9% |
| Status of passive containment | Effective: | 70% | 5% | 95% |
| | Not effective: | 30% | 95% | 5% |
| Spreading | Contained: | 68% | 15% | 86% |
| | Not contained: | 32% | 85% | 14% |
| Structural integrity | Not compromised: | 95% | 95% | 95% |
| | Compromised: | 5% | 5% | 5% |
| Hull status | Safe: | 93% | 93% | 93% |
| | Unsafe: | 7% | 7% | 7% |
| Passengers exposure | Low: | 50% | 16% | 80% |
| | Moderate: | 25% | 20% | 13% |
| | High | 25% | 64% | 8% |
| Vessel status | Safe: | 70% | 34% | 82% |
| | Unsafe: | 30% | 66% | 18% |
| Incident severity | High: | 48% | 79% | 25% |
| | Moderate: | 18% | 11% | 13% |
| | Low: | 37% | 10% | 63% |

### 6.3 *Case studies results - Flooding cases*

For all flooding cases, Table 3 presents the states of the parent nodes and Table 4 presents the states of the children nodes given the evidence for the states of the parent nodes of the situation assessment model.

#### 6.3.1 *Flooding case 1: Grounding in night-time*

A grounding accident occurs during the night, resulting in a hull breach. The breach concerns the machinery space, where there is ingress of water in the ER while the steering system has been also affected by the impact. The response team reacts as planned, and automatic mitigation systems as well as other legacy systems are working properly. All watertight doors are closed. Sea state is calm and there is a 3-deg. roll inclination. The flooding of only one compartment does not create any structural stress issues.

As expected, the "Stability" and "Structural integrity" nodes are not affected by the accident and consequently the "Hull status" node is assessed as "Safe". On the contrary, the node "Critical systems status" is characterised as "Not operational", because one of the main systems that provide the ability to navigate (i.e. steering) is not operational. Combined with the fact that the watertight doors are closed and the "Spreading" node is "Contained", this results in the "Vessel status" to be assessed slightly "Safe" (52%). The "Passengers proximity to hazards" node is "Low", as the ER is a vessel compartment where access to passengers is restricted. Hence the "Passengers exposure" is also assessed as "Low". Finally, the "Incident severity" is assessed "High" (47%), hence the SRAP would provide an indication to the Master to sound the GA.

Sounding the GA to initiate the mustering process after a grounding accident even though there is not an immediate threat to the passengers, has been observed to be common, in order to ensure their safe mustering.

#### 6.3.2 *Flooding case 2: Grounding in night-time and blackout*

The scenario is similar to Flooding Case 1, but the ingress of water in the ER affects the Main Engine (ME) as well as the propulsion system, resulting in the ship being without power (i.e. blackout). The watertight doors in the ER are open and the roll angle is over 12 deg and steadily increasing. The flooding of only two compartments does not cause any immediate structural stress issues.

In this scenario the structural integrity of the vessel is slightly more affected compared to Flooding Case 1. The "Stability" node is highly affected by the increase of roll angle. This leads to "Hull status" being assessed as "Unsafe", combined with the states of the other parent nodes, such as "Critical system status", being assessed as "Not Operational". This attributed to the fact that two of the main systems that provide the ability to navigate (i.e. steering and propulsion systems) are not operational, which results in the "Vessel status" node to be "Unsafe" (91%). The nodes relating to the passengers remain the same as in the Flooding Case 1. Therefore, the "Incident severity" is assessed "High" with a higher percentage compared to Flooding Case 1 (i.e. 78% compared to 47%), due to the fact that the "Vessel status" is now "Unsafe".

The SRAP would provide an indication to the Master to sound the GA. Sounding the GA to initiate the mustering process after a grounding accident is common practice in such scenarios, given that the lack of power and steering may endanger people onboard.

6.3.3 *Flooding case 3: Grounding in night-time, steering intact*

This case is similar to the Flooding Case 1, but the ingress of water in the ER does not affect the steering system. The Response team, the automatic mitigation systems as well as all legacy systems are fully functional.

The "Incident severity" is characterized as "Low" compared to the other two Flooding Cases. This is attributed to the fact that the location of the accident is in the ER, where the Passengers Exposure is "Low", in combination with the "Spreading" node, which is assessed as "Contained". Regarding the "Vessel status", the condition of the critical systems, the structural integrity and the stability of the vessel have not been affected.

Based on these results, the SRAP would provide the indication to the Master not to sound the GA but to reassess the situation.

Table 3. The parent nodes states for the flooding cases.

| Node | Flooding case 1 | Flooding case 2 | Flooding case 3 |
|---|---|---|---|
| Time of day | Night | Night | Night |
| Incident type | Flooding | Flooding | Flooding |
| Incident location | ER | ER | ER |
| Response team onsite | Yes | Yes | Yes |
| Status of mitigation systems | Activated | Activated | Activated |
| Status of Adjacent Detectors | Not activated | Activated | Not activated |
| Status of containment doors | Closed | Opened | Closed |
| Global Stresses | Not exceeded | Not exceeded | Not exceeded |
| Restoring capability (GZ/GM criteria) | Covered | Covered | Covered |
| Maximum roll angle | Not exceeded | Exceeded | Not exceeded |
| Internal Communication | Operational | Operational | Operational |
| External Communication | Operational | Operational | Operational |
| Steering system | Not operational | Not operational | Operational |
| Propulsion | Operational | Not operational | Operational |
| Navigational | Operational | Operational | Operational |

6.3.4 *Summary of flooding cases*

The results from the Flooding Cases show that the "Vessel Status" node is sensitive to the maximum roll angle and therefore the "Stability" node. Specifically, in the Flooding Case 1, where the maximum roll angle is not exceeded, the "Hull status" is assessed as "Safe" (93%). In Flooding Case 2 the maximum roll angle gradually increases and therefore the "Hull Status" is assessed as "Unsafe" (69%). For both Flooding Cases, the "Passengers Exposure" node is mainly "Low", but the "Vessel Status" ranges from almost "Safe" (52%) for the Flooding Case 1 to "Unsafe" (84%) for the Flooding Case 2. For the Flooding Case 3 where the flooding is contained, the "Incident Severity" is "Low", and the indication of the model is to reassess the condition.

Table 4. The children nodes states for the flooding cases.

| Nodes | Status | Flooding case 1 | Flooding case 2 | Flooding case 3 |
|---|---|---|---|---|
| Effectiveness of mitigation measures onsite | Effective: | 90% | 90% | 90% |
| | Not effective: | 10% | 10% | 10% |
| Passengers Proximity to hazards | Low: | 79% | 79% | 79% |
| | Medium: | 16% | 16% | 16% |
| | High: | 5% | 5% | 5% |
| Status of passive containment | Effective: | 95% | 5% | 95% |
| | Not effective: | 5% | 95% | 5% |
| Spreading | Contained: | 86% | 22% | 86% |
| | Not contained: | 14% | 78% | 14% |
| Structural integrity | Not compromised: | 95% | 95% | 95% |
| | Compromised: | 5% | 5% | 5% |
| Critical systems status | Fully operational: | 0% | 0% | 100% |
| | Degraded: | 26% | 25% | 0% |
| | Not operational: | 74% | 75% | 0% |
| Stability | Sufficient: | 100% | 30% | 100% |
| | Not sufficient | 0% | 70% | 0% |
| Ability to communicate | Fully operational: | 100% | 100% | 100% |
| Ability to navigate | Fully operational: | 0% | 0% | 100% |
| | Degraded: | 20% | 0% | 0% |
| | Not operational: | 80% | 100% | 0% |
| Hull status | Safe: | 93% | 31% | 93% |
| | Unsafe: | 7% | 69% | 7% |
| Passengers exposure | Low: | 73% | 48% | 73% |
| | Moderate: | 16% | 32% | 16% |
| | High | 11% | 20% | 11% |
| Vessel status | Safe: | 52% | 9% | 82% |
| | Unsafe: | 48% | 91% | 18% |
| Incident severity | High: | 47% | 78% | 28% |
| | Moderate: | 15% | 15% | 14% |
| | Low: | 37% | 7% | 58% |

## 7 CONCLUSIONS

The Smart Risk Assessment Platform (SRAP) is a risk-based, digital tool that aims to assist the

Master and Bridge Command Team in the decisions that need to be taken after an incident with respect to evacuation.

The situation assessment risk model has been developed as a discrete, static BN that describes the probabilistic and causal relationships between the different risk factors. The development of this model in SRAP has been based on identifying critical factors that affect risk during marine evacuation from a survey of the relevant literature, an analysis of selected accident cases, and expert judgement. The BN has been quantified through a qualitative approach that has been verified by a strength of influence analysis on the BN nodes. In addition, the risk model has been initially validated in terms of face, content, and predictive validity based on comments obtained through an expert interview and the results of selected case studies respectively. The results of this process show that the model adequately reflects the situation, includes the most important risk factors, and produces reasonable results that would effectively achieve the objective of supporting the decisions for initiating the evacuation process.

The strength of influence analysis shows that the status of the vessel integrity (i.e. whether the vessel provides a safe place for passengers and crew) and the likelihood of the incident spreading are important factors that affect the decision to initiate the evacuation process. The results from the Fire and Flooding Cases show that the model is sensitive enough to capture different conditions that can typically be found in accidents where evacuating the passengers and crew is considered a possibility.

Future research steps include further validation activities, which can be based on real accident cases. The situation assessment risk model will be complemented with two other models that will be developed in the context of SRAP and be used in the phases of the evacuation process that follow sounding the GA. The models will support the decision-making process during mustering and giving the final order to abandon the vessel. The vision for SRAP is to build a comprehensive digital tool that will effectively support the Master and Crew and mitigate their stress throughout the evacuation process, ultimately leading to better decisions and increased safety.

## ACKNOWLEDGEMENTS

The work presented in this paper was supported by the Project "PALAEMON: A holistic passenger ship evacuation and rescue ecosystem", which has received funding from the European Union's Horizon 2020 research and innovation programme, EU.3.4. - SOCIETAL CHALLENGES - Smart, Green and Integrated Transport of the European Union under the Topic MG-2-2-2018 - Marine Accident Response (Grant Agreement number 814962).

## REFERENCES

AIBN. 2019. "Interim Report 12 November 2019 on the Investigation into the Loss of Propulsion and Near Grounding of Viking Sky, 23 March 2019." Norway: Accident Investigation Board Norway.

Bobbio, A, L Portinale, M Minichino, and E Ciancamerla. 2001. "Improving the analysis of dependable systems by mapping fault trees into Bayesian networks." Rliability Engineering & System Safety 249–260.

Eleye-Datubo, A. G., A. Wall, A. Saajedi, and J. Wang. 2006. "Enabling a Powerful Marine and Offshore Decision-Support Solution Through Bayesian Network Technique." *Risk Analysis* 26 (3): 695–721. https://doi.org/10/bvghhr.

Fenton, Norman, and Martin Neil. 2012. *Risk Assessment and Decision Analysis with Bayesian Networks*. CRC Press. https://doi.org/10.1201/9780367803018.

Hamza, Zerrouki, and Tamrabet Abdallah. 2015. "Mapping Fault Tree into Bayesian Network in safety analysis of process system." 2015 4th International Conference on Electrical Engineering (ICEE). Boumerdes, Algeria: IEEE.

Hänninen, Maria. 2014. "Bayesian Networks for Maritime Traffic Accident Prevention: Benefits and Challenges." *Accident Analysis and Prevention* 73: 305–12. https://doi.org/10.1016/j.aap.2014.09.017.

HSE. 2002. "Marine Risk Assessment." Offshore Technology Report 2001/063. Sudbury: Health & Safety Executive.

ISO. 2009. "ISO 31000:2009 Risk Management - Principles and Guidelines."

Jensen, Finn V., and Thomas Dyhre Nielsen. 2007. *Bayesian Networks and Decision Graphs*. 2nd ed. Information Science and Statistics. New York: Springer.

Khakzad, Nima, Faisal Khan, and Paul Amyotte. 2011. "Safety analysis in process facilities: Comparison of fault tree and Bayesian network approaches." Reliability Engineering and System Safety 925–932.

Khakzad, Nima, Faisal Khan, and Paul Amyotte. 2012. "Dynamic Risk Analysis Using Bow-Tie Approach." *Reliability Engineering and System Safety* 104: 36–44. https://doi.org/10.1016/j.ress.2012.04.003.

Koimtzoglou Alexandros, Konstantinos Louzis, Alexandros Michelis, and Marios-Anestis Koimtzoglou. 2021. "Mass Evacuation of Large Passenger Ships: A State-of-the-Art Analysis- Setting the foundations for the Intelligent Evacuation Ecosystem PALAEMON." Annual Conference of Marine Technology. Athens: Hellenic Institute of Marine Technology. 137–150.

Kristiansen, Svein. 2005. *Maritime Transportation: Safety Management and Risk Analysis*. Routledge.

MCIB. 2012. "Cruise Ship COSTA CONCORDIA Marine Casualty on January 13, 2012 Report on the Safety Technical Investigation." MINISTRY OF INFRASTRUCTURES AND TRANSPORTS - Marine Casualties Investigative Body.

MIT. 2015. "Fire on Board Ro-Ro Pax SORRENTO - Interim Report." MINISTRY OF INFRASTRUCTURE AND TRANSPORT, Directorate General for Rail and Marine Investigations, 3rd Division – Marine Investigations.

Montewka, Jakub, Soren Ehlers, Floris Goerlandt, Tomasz Hinz, Kristjan Tabri, and Pentti Kujala. 2014. "A Framework for Risk Assessment for Maritime Transportation Systems - A Case Study for Open Sea Collisions Involving RoPax Vessels." Reliability Engineering and System Safety 124: 142–57. https://doi.org/10.1016/j.ress.2013.11.014.

Norazahar, Norafneeza, Faisal Khan, Brian Veitch, and Scott MacKinnon. 2018. "Dynamic Risk Assessment of Escape and Evacuation on Offshore Installations in a Harsh Environment." Applied Ocean Research 79 (October): 1–6. https://doi.org/10/gf2sg5.

NTSB. 2015. "Marine Accident Brief. Engine Room Fire Aboard Cruise Ship Carnival Liberty." National Transportation Safety Board (NTSB).

Ping, Ping, Ke Wang, Depeng Kong, and Guoming Chen. 2018. "Estimating Probability of Success of Escape, Evacuation, and Rescue (EER) on the Offshore Platform by Integrating Bayesian Network and Fuzzy AHP." Journal of Loss Prevention in the Process Industries 54 (July): 57–68. https://doi.org/10/gdxm8d.

Pitchforth, Jegar, and Kerrie Mengersen. 2013. "A Proposed Validation Framework for Expert Elicited Bayesian Networks." *Expert Systems with Applications* 40 (1): 162–67. https://doi.org/10.1016/j.eswa.2012.07.026.

Sarshar, Parvaneh, Jaziar Radianti, Ole-christoffer Granmo, and Jose J Gonzalez. 2013. "A Dynamic Bayesian Network Model for Predicting Congestion During a Ship Fire Evacuation." *Proceedings of the World Congress on Engineering and Computer Science 2013 (WCECS 2013)* 1: 23–25.

TSB. 2008. "Striking and Subsequent Sinking. Passenger and Vehicle Ferry Queen of the North Gil Island, Wright Sound, British Columbia 22 March 2006." M06W0052. Marine Investigation Report. Transportation Safety Board of Canada.

Yang, Xue, Stein Haugen, and Nicola Paltrinieri. 2018. "Clarifying the Concept of Operational Risk Assessment in the Oil and Gas Industry." *Safety Science* 108: 259–68. https://doi.org/10/gf2sgj.

Zhang, Guizhen, Vinh V. Thai, Kum Fai Yuen, Hui Shan Loh, and Qingji Zhou. 2018. "Addressing the Epistemic Uncertainty in Maritime Accidents Modelling Using Bayesian Network with Interval Probabilities." *Safety Science* 102 (November 2017): 211–25. https://doi.org/10.1016/j.ssci.2017.10.016.

# ANNEX

Table 5 provides a detailed description of the individual nodes of the BN model.

Table 5. The description of nodes and the related assumptions of the situation assessment BN.

| Node | States | Description |
|---|---|---|
| Incident Type | Fire/Flooding | The type of incident determines the possibility for mitigating its effects and whether passengers can be directly exposed to hazards if it spreads. |
| Incident Location | Engine Room/ Accommodation/Public Space | The location where the incident has happened, which affects crew accessibility and whether passengers are likely to be at that location. |
| Response Team(s) on site | Yes/No | The presence of crew response teams at the incident site increases the likelihood of effectively mitigating the effects. |
| Status of mitigation systems | Activated/Not Activated | Indicates whether the systems at the incident location/source have been activated. |
| Effectiveness of mitigation measures on site | Effective/Not effective | Likelihood of mitigation measures being effective. If the incident is not suppressed at its source, it is more likely that it will spread. |
| Passenger location | In Cabins/Spread Out | Likelihood of passengers being mostly in their cabins or spread out in various parts of the ship. |
| Time of day | Day/Night | The time of the day the incident occurs determines the most probable locations for the passengers and how responsive they may be. |
| Passengers proximity to hazards | Low/Medium/High | Probability of passengers being close to danger, which determines the probability of being injured. |
| Passengers readiness | Immediate/Delayed | Probability of passengers being in a state of readiness to avoid immediate exposure to danger/hazards. |
| Passengers exposure | Low/Moderate/High | Probability of passengers being immediately exposed to danger/hazards. |
| Status of adjacent detectors | Activated/Not activated | Indicates whether detectors have been activated in areas adjacent to the location of the incident. |
| Status of containment doors | Opened/closed | Corresponds to the status of fireproof/waterproof doors that shall be activated to contain the incident. |
| Status of passive containment | Effective/Not Effective | Indicates whether passive containment of the incident is effective given the status of the detectors and the containment doors. |
| Spreading | Contained/Not Contained | Probability of incident spreading to other parts of the ship other (effectiveness mitigation measures). |
| Global Stress | Exceeded/Not exceeded | Indicates whether safe limits for global stress have been exceeded (compromised structural integrity issues). |
| Structural Integrity | Not Compromised/ Compromised | Probability of ship structural integrity being compromised (danger of collapsing). |
| Restoring criteria for GZ/GM | Criteria covered/ not covered | Indicates whether safe limits for righting lever moments have been exceeded (danger of capsizing). |
| Maximum roll angle | Exceeded/Not exceeded | Indicates whether safe limits for maximum roll angle have been exceeded (the safe launching of MEVs can be affected) |
| Stability | Sufficient/Not sufficient | Probability of ship stability being sufficient, which means that the ship remains in a stable position (either upright, or inclined). |
| Hull status | Safe/Unsafe | A safe state is one that is stable, i.e. not expected to become unsafe without warning. |
| Internal communication status | Operational/Not Operational | Indicates whether systems/equipment are operational for communication between crew members (influence on the crew's capability to mitigate the effects of the incident) |
| External communication status | Operational/Not Operational | Communication between the ship and the outside world (other ships, SAR services, Coast Guard, Shipping company) (e.g. request help etc.). |

(*Continued*)

Table 5. (Cont.)

| Node | States | Description |
|---|---|---|
| Ability to communicate | Fully operational/ Degraded/Not operational | Indicates the ship's ability to communicate, within the ship and with the outside world. Communication is essential for mitigating the effects of the incident and protect the safety of crew and passengers. |
| Propulsion system status | Operational/Not Operational | Indicates whether propulsion systems are damaged to the point of non-functionality (main engine, the shaft, and propeller). |
| Steering system status | Operational/Not Operational | Steering systems include the steering gear and the rudder and determine the ship's ability to change its course. |
| Navigational systems status | Operational/Not Operational | Indicates whether the navigational systems are damaged to the point of non-functionality. Navigational systems/equipment include AIS, radar, and ECDIS and determine the ship's ability to safely navigate (i.e. avoiding navigational hazards) and determine its position and course. |
| Critical systems status | Fully operational/ Degraded/Not operational | Critical systems are considered as defined in the IMO's Safe Return to Port regulation, and thus implying minimum functionality that will allow the ship to return to a haven. |
| Vessel status | Safe/Unsafe | A safe state is one that is stable and when the ship has at least partially some of its critical functionalities intact. |
| Incident severity | Low/Moderate/High | Probability of the incident resulting in adverse/unacceptable consequences for the crew and passengers. Depends on whether the ship can provide a safe environment for the crew and passengers and whether their direct exposure to hazards can be minimized. |

*Environmental conditions*

*Trends in Maritime Technology and Engineering – Guedes Soares & Santos (eds)*


# Long-term probabilistic identification of extreme sea-states as causes of coastal risk due to wave severity

G. Clarindo & C. Guedes Soares
*Centre for Marine Technology and Ocean Engineering (CENTEC), Instituto Superior Técnico, Universidade de Lisboa, Lisbon, Portugal*

ABSTRACT: In the present work, these techniques are applied to identify the expected extreme combinations of sea-states for different return periods. The extreme sea-state conditions are evaluated using a robust approach based on the environmental contours of the main variables such as significant height and spectral peak period (Hs-Tp). The computed values of combinations predicted by the contours are used to justify the exclusion of the wave climate as the main factor of dune erosion in the location. In other words, among the different agents that cause dune erosion, it is possible to consider that the severity of the wave conditions affects naturally, allowing the recovery of the dune over the years, something that does not happen with anthropogenic agents that cause disturbances in the fore-dune fields such as pedestrians paths. Nevertheless, the current work deals only with the long-term probabilistic prediction of wave climate conditions.

## 1 INTRODUCTION

Many coastal man-made or natural structures are exposed to the effect of severity induced by the sea environment, so identifying severe sea conditions as well the distribution of variables as waves, wind and current are crucial for an effective engineering practice in marine environments. Nowadays the probabilistic models describe the variability and can be used to extrapolate the extreme values. However, the prediction of wave conditions is given by wave models such as for example Silva et al. (2015 and 2016). They allow probabilistic assessments of the occurrence of different phenomena, as well as the quantification of the uncertainty of such predictions. Guedes Soares et al. (1988) describe how efficiently modelling the sea waves by probabilistic approach for engineering purposes. Applications of this approach for Portuguese coastal waters can be seen in Guedes Soares (1993 and 2003). Guedes Soares (2000) performed the WAVEMOD project in which a probabilistic approach was used to derive the design conditions for coastal engineering problems considering waves and current parameters. It is important to differentiate between a probabilistic and a deterministic approach. So, deterministic methods are in general more precise in that they describe the specific wave field but they are limited for some time scale and are not applicable for scales larger than four days, which is the present limit of meteorological models. Thus, for longer time scales and long-term prediction of the variables, assessments in the probabilistic sense are the most common and viable resource.

The environmental contours (EC) derived from the long-term probability density functions are often used in ocean engineering to identify extreme wave conditions. The time histories of environmental states are principally approximated by statistical models. The environmental contour concept, commonly applied in marine reliability design, permits the consideration of extreme environmental conditions independently of a particular natural or artificial structure. The contour lines correspond to a set of environmental states, which may be used to explore design options (ISSC, 2018).

The phenomenon of interaction between a natural (or artificial) structure and the environmental loads is considered random. Thus, the probabilistic approach is required to reliably assess this interaction. Moreover, information of this interaction is crucial for planning, design and development of coastal structures in the reliable sense. In order to correct works, the extreme conditions and its future projections are vital for clear decisions as well as avoid damage in the specific structure. Guedes Soares & Trovão (1991) focused the study on the influence of climate modelling on the long-term prediction of wave induced responses of ship structures. An overview of Probabilistic models of the wave environment for reliability assessment of offshore structures are provided by Bitner-Gregersen & Guedes Soares (1997). A natural structure as a sand beach with dunes and man-made structures have some probability to suffer damage due to extreme conditions and unwelcomed failure. To study examples of the interaction of wave severity on natural beaches, as

DOI: 10.1201/9781003320289-31

well as the probability of occurrence of extreme events that cause irreversible impacts on surrounding coastal structures, read Clarindo et al. (2018). Influence of the wave propagation in the entrance and inside of ports can be read in Rusu & Guedes Soares (2011) as well as Rusu & Guedes Soares (2014) respectively. As an example of the probability of the occurrence of severity of the waves in relation to structures designed by man, such as port structures, which are constantly at risk of overtopping and constant flooding, see Rusu & Guedes Soares (2013) and Santana-Ceballos et al. (2017). Some of this failure can be irreversible when taking into account the persistence and severity of the occurrence of these natural extreme events. Due to the random nature of this process, the probabilistic analysis plays a vital role in assessing the occurrence of severe conditions and is a powerful tool to develop effective coastal risks planning.

In this way, the environmental contours line method based on observed wave data for identifying the long-term extreme wave conditions are used. The work proposes to apply the extreme values computed by the contour lines as the most expected environmental load capable of producing structural damage. That in the scope of the work, is the irreversible erosion of the foredune of Caparica beach. The severe sea-states expected in this location are considered the most probable conditions to produce failure in the natural structure. The combinations of values are predicted based on various return periods and each of them with their own probability of failure (considered irreversible erosion). For implementing the proposed environmental contour line method, a conditional bivariate statistical scheme is used. Nowadays, there is a wide variety of environmental contour methods being the most popular: IFORM (Winterstein et al., 1993) and direct methods (Huseby et al., 2013). An application of direct method based on Monte Carlo sampling can be read in Clarindo & Guedes Soares (2020) as well the IFORM applied for wave energy converters in Clarindo & Guedes Soares (2021). A deep discussion including the uncertainty from the various EC methods is read in Haselsteiner et al. (2021). A comparative study for both approaches considering variance reduction found in Clarindo et al. (2021).

The results in this paper demonstrate that the performance of the method can be utilized as an effective tool for predicting the long-term extreme wave conditions to avoid coastal risk and damage in a natural structure. The coastal dunes fields generally include natural and human factors of disturbance. A simplified approach to assess normally not consider changes on the sea water levels as well the severity and occurrence of the sea-states. These kinds of factors could be included as important factors in the disturbance evaluation. Thus, the investigation of the local impact by the sea-states and its degree of severity can indicate some crucial reasons for coastal dune degradation due to occurrence of extreme sea conditions. The damage produced by anthropogenic factors such as pedestrians, cyclists, motorbikes among others in coastal dune vegetation has been recognized over the years and reported by: Bates (1935) as well Boorman & Fuller (1977) for vegetation of footpaths, sidewalks, cart tracks, and gateways. Baccus (1977) as well McDonnell (1981) studied the influence of recreational and user impact in the natural Parks. Andersen (1995), Hesp et al., (2010) and Fenu et al. (2013) dedicate the studies in human tramping in coastal zones, and finally García-Romero et al. (2019) investigate the processes in an arid transgressive dune field as indicators of human impact by urbanization. This kind of damage can be higher by its location within the dune field, however, the sections of coastal dunes next to the surf zone, especially embryo dunes and foredunes are subject to marine actions (waves and tides variability). This exposed situation of coastal dunes makes them more vulnerable to suffer erosion induced by wave loads. In fact, an assessment of erosion in which it considers the loads due to severe conditions is more realistic. Thus, it is important to identify the occurrence and intensity of these events, as well as predict the values of joint distribution for a given return periods. Consequently, this quantification may help to exclude total or partially severe conditions as the main agents of coastal erosion. In this section, the data considered to characterize the long-term wave climate for Caparica beach are described, as well as the most common variables employed to define the sea-states. The predicted joint distribution of the variables is based on N-years of return period. Moreover, the critical and severe conditions are identified using the EC method.

The paper is organized as follows. The sections 2 present the data and methods. Consists in a brief explanation of the main steps of EC derivation including the probabilistic statistical scheme as well as how IFORM contours can be constructed. The section 3 presents the results as well the discussions and the paper close with relevant conclusions in section 4.

## 2 DATA AND METHODS

### 2.1 *The sea waves measures*

The characterization of sea wave climate for the Caparica beach is executed based on data measured and managed by the Port of Lisbon Authority (APL) *in-situ* buoy, located at the entrance of the Tagus estuary (Figure 1). The TRIAXYS directional wave buoy is placed at 38º37'33,6" N; 9º23'16.8" W, at a water depth of 30m. The time series considers approximately 6 years (2007-2012) of sea wave records. In this study the relevant parameters such as significant wave height (Hs), peak period (Tp) and wave direction (Dir) are considered. Table 1 shows descriptive statistics of Hs, Tp and Dir data series recorded at the buoy location. Figure 2 shows the correlation of the variables recorded by the buoy device. Figure 3 describes the bivariate histogram of the variables defined by bins of occurrence of events. Figure 4 illustrates the directionality of incident waves recorded in the moored buoy.

Nowadays, *in situ* data from a buoy record is considered the utmost accurate way to measure the sea-states. In addition, a robust and realistic approach to modelling the distribution consists in describing the distribution 0conditioned in a multivariate way. So, the present work develops a conditional bivariate scheme of Hs and Tp variables based on suitable statistical distribution models.

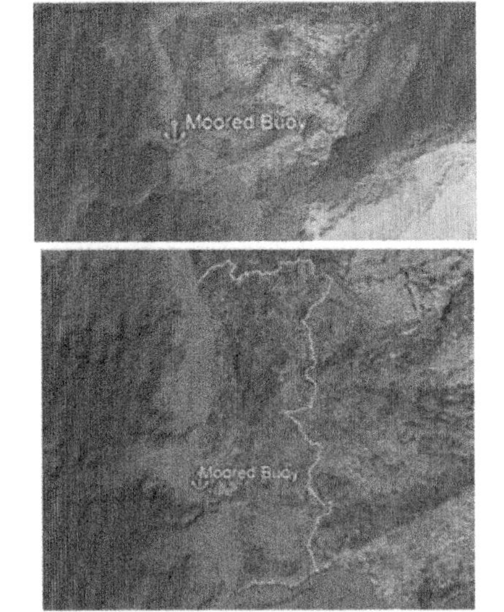


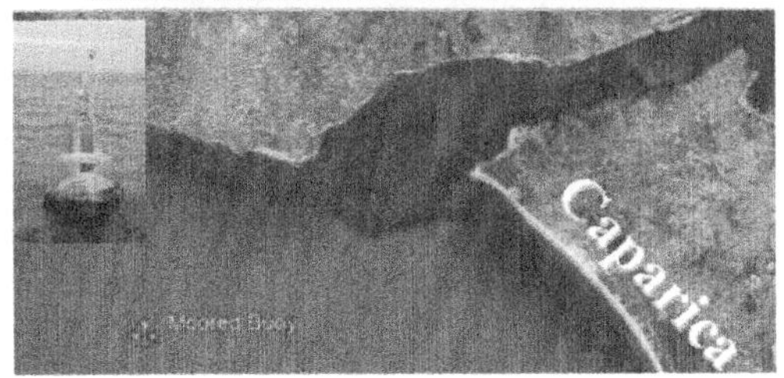


Figure 1. Location of the moored buoy in front of Caparica beach as well the device.

Table 1. Computed statistics of the wave observations.

| | Hs (m) | Tp (s) | Dir (°) |
|---|---|---|---|
| Mean | 1,26 | 10,70 | 280,0 |
| Median | 1,04 | 10,50 | 284,5 |
| Mode | 0,88 | 11,80 | 284,1 |
| Std Dev | 0,72 | 3,00 | 28,4 |
| Variance | 0,52 | 8,80 | 806,8 |
| Kurtosis | 4,60 | 0,02 | 29,2 |
| Asymmetry | 1,94 | 0,04 | -4,2 |
| Range | 5,91 | 18,10 | 359,6 |
| Minimum | 0,28 | 1,90 | 0,1 |
| Maximum | 6,19 | 20 | 359,7 |
| Number of Records | 29418 | 29418 | 29418 |

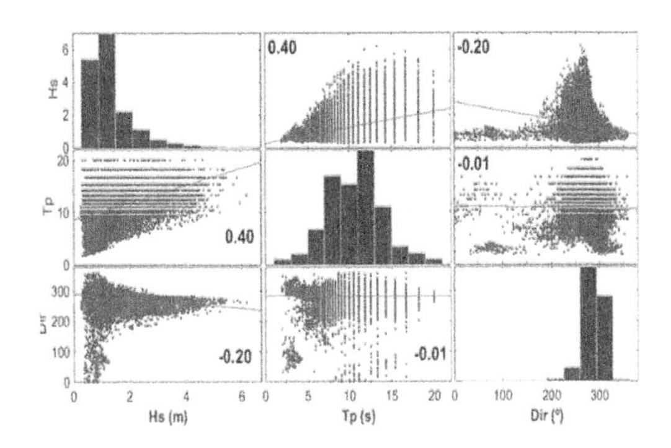


Figure 2. Matrix correlation of the variables considered, as well as the pairwise p-values for correlation.

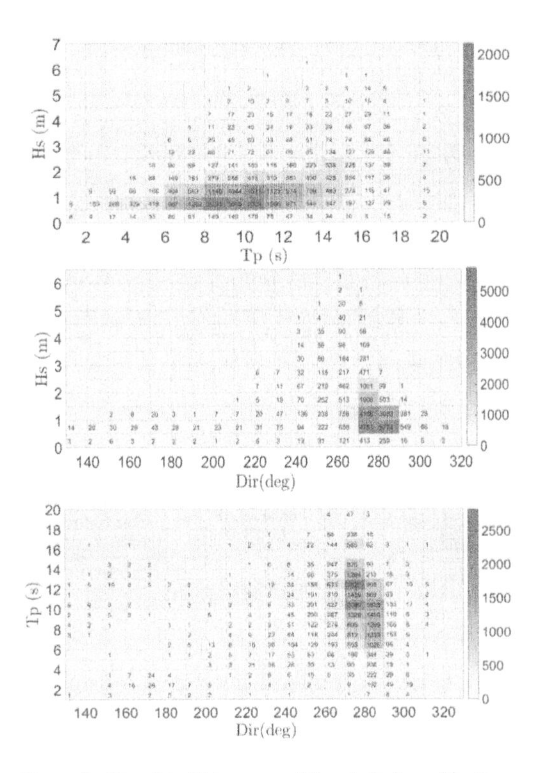


Figure 3. Bivariate histograms of the studied combinations of variables. Represents bins as colormap and indicates the number of events in each bin. The resolution for the bins is defined as: Hs = 0.5 (m); Tp = 1.0 (s); Dir = 20 (°).

## 2.2 *Methods*

The probabilistic study of the randomness of the sea waves characteristics is a concerning issue for reliable managing of coastal zones. In many marine applications, the simultaneous distribution of the most important met-ocean variables is required for calculations. Taking into account the risk and damage induced, some occurrences of the extremes values of these variables will be undesirable, so the prediction of their joint distribution for a given

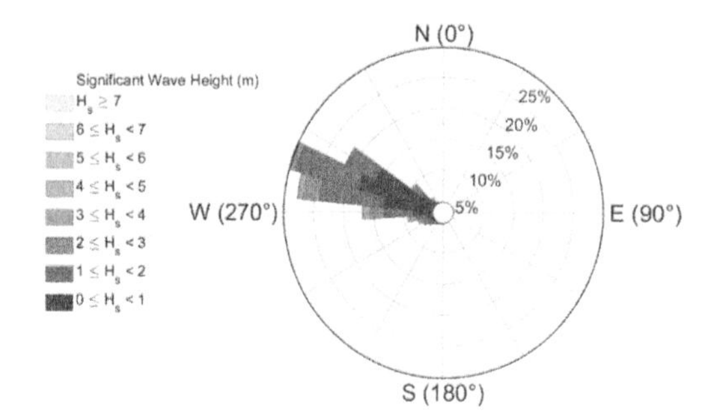


Figure 4. Directionality of incident waves.

return periods are crucial for effective management of the coastal environment. The long-term joint distribution of variables and its properties are computed using the EC methods that are briefly described next.

The EC concept permits the consideration of extreme environmental sea-states conditions. The EC methods generally involve three main stages: (i) statistical modelling: establishing a statistical model able to jointly characterize the environment based on sea-states observations, (ii) contour construction: computing the EC based on the previous established statistical model, that are constructed by IFORM approach, (iii) design conditions: selecting discrete points along the contour for design applications. (Figure 5).

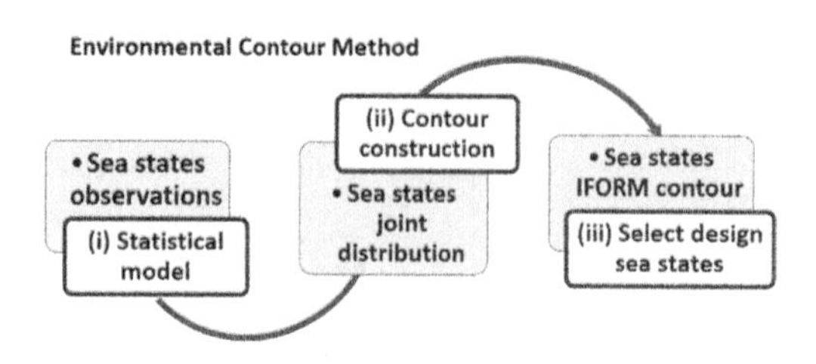


Figure 5. The general framework for environmental contour method.

*(i) Statistical modelling*

For the current work, the joint modeling of extreme sea waves conditions is based on the significant wave heights (Hs) and the corresponding conditional wave periods (Tp|Hs). The conditional modeling approach (CMA) have been shown efficiency in different applications as well to model different variables (Haver (1987), Guedes-Soares et al. (1988), Bitner-Gregersen & Haver (1991), Guedes Soares & Henriques (1996), Ferreira & Guedes Soares (2002), Jonathan et al. (2010), Bitner-Gregersen et al. (2015) and Li et al. (2015).

*Weibull*

Currently, perhaps the most used distribution to model significant wave height is the Weibull distribution. While some others authors use the 2-parameter Weibull distribution, most authors use the 3-parameter Weibull distribution for prediction of the waves. This distribution proved its versatility in different conditions of sea waves and usually fits well the majority of met-ocean data. The analytic expressions of Weibull cumulative density function (cdf) and probability density function (pdf) for a non-negative $x$ are, respectively:

$$F(x|\alpha,\beta,\mu) = 1 - e^{-\left(\frac{x-\mu}{\beta}\right)^{\alpha}}$$

$$f(x|\alpha,\beta,\mu) = \frac{\alpha}{\beta}\left(\frac{x-\mu}{\beta}\right)^{\alpha-1} e^{-\left(\frac{x-\mu}{\beta}\right)^{\alpha}}$$

The parameters, α, β and μ are known as the shape, scale and location parameters respectively. The model described above is applied to characterize the marginal distribution of the primary variable, that consists in distribution of Hs observations. The applied conditional scheme consists in modeling the Tp variables following the Lognormal distribution, which are briefly described in the next subsection.

*Conditional Lognormal*

Many coastal engineering applications require calculations of simultaneous distribution of the most important met-ocean variables. For this task a Conditional Modelling Approach (CMA) is considered. After fitting the marginal distribution of Hs by 3 parameters of Weibull, the conditional scheme is applied (see Eq. 1). The fitting of the conditional Lognormal distribution for the Tp dependent variable consists of fitting the function given by Equation 1a-b.

$$f_{T_p|H_s}(T_p|H_s) = \frac{1}{\sigma(H_s)T_p\sqrt{2\pi}}\exp\left(-\frac{(\ln T_p - \mu(H_s))^2}{2\sigma(H_s)^2}\right) \quad (1)$$

$$\mu(H_s) = E[\ln T_p|H_s] = a_1 + a_2 H_s^{a_3} \quad (1a)$$

$$\sigma(H_s) = Std[\ln T_p|H_s] = b_1 + b_2\, exp^{b_3 H_s} \quad (1b)$$

*(ii) Contour construction*

Certainly, the most common method to construct the environmental contour is the Inverse-FORM (Winterstein et al. 1993) that is also able to estimate the probability of occurrence of severe conditions based on periods of return. For more detail in how and why apply the environmental contour methods (including IFORM) see Ross et al. (2019) as well Haselsteiner et. al (2021). In the present work, the sea states to predict the severe conditions as factors of dune erosion are studied using the EC method constructed based on the IFORM approach.

*IFORM Environmental contour construction*

This step in the derivation procedure consists of constructing the EC lines based on the statistical model fitted on the previous step. The First Order Reliability Methods (FORM) are generally used in structural reliability problems to compute a multivariate probability integral. For a general problem definition let's denote $x = [x_1 \ldots x_n]$ the $n$ random variables represent the environmental quantities, the failure probability $P_f$ is formally described by the following integral:

$$P_f = \text{Prob}(g(x) \leq 0) = \int_{(g(x) \leq 0)} f(x)dx \quad (2)$$

where f(x) is the joint probability density function of x, and g(x) is the function defined such that g(x)≤0, being considered as the domain of integration and denotes the failure set, see Melchers (1987) for more detail. The FORM approach is considered an analytical approximation capable to compute the probability of exceedance of given threshold $P(Y>Y_0)$. In the standard FORM approach, a threshold value is considered and its likelihood is estimated in the standard normal space.

Nevertheless, the IFORM approach (Winterstein et al. 1993) is the opposite approach for mathematical considerations. It starts from the normal space and the probability of occurrence for a given period of return is defined by the contour's lines. These contour lines are then transformed into the environmental physical parameters space. First, the environmental sea waves variables $x = [x_1 \ldots x_n]$ are transformed into independent variables $U = [u_1 \ldots u_n]$ following a standard normal distribution. Giving the exceedance probability $P_f$, it is possible to calculate the safety index $\beta$. In order to obtain $\beta$, the probability of failure ($P_f$) is firstly estimated by:

$$P_f = \frac{T_{ss}}{(365 \times 24)T_{rt}} \quad (3)$$

where $T_{ss}$ is the sea-state duration in hours, remembering that the buoy data used consider a 1-hour for sea state durations. $T_{rt}$ is the return period in years, and 365 x 24 is a factor to convert the number of years to hours. Commonly, the relation between $\beta$ and $P_f$ is given by:

$$1 - P_f = \Phi(\beta) \quad (4)$$

The values of variables ($u_1$, $u_2$) in the normal space are given by:

$$u_1 = \beta \cos \theta \quad (5a)$$

$$u_2 = \beta \sin \theta \quad (5b)$$

where the safety index $\beta$ is the radius of the circle, $\Theta$ is the angle between the radius of the circle and $u_1$-axis varying between $\pi$ and $-\pi$. The relation between the physical space X and the normal space U is given by:

$$\Phi(u_1) = F_{H_{m0}}(h_s) \quad (6a)$$

$$\Phi(u_2|u_1) = F_{T_p|H_{m0}}(t_p|h_s) \quad (6b)$$

Therefore, for a given marginal distribution and the respective conditional distribution, the Rosenblatt procedure (Der Kiureghian & Liu, 1986) are performed to transform the data into the physical environmental parameter space, such as:

$$h_s = F_{H_{m0}}{}^{-1}(\Phi(u_1)) \quad (7a)$$

$$t_p|h_s = F_{T_p|H_{m0}}{}^{-1}(\Phi(u_2|u_1)) \quad (7b)$$

where Φ( ) denotes the cumulative distribution function of the standard Gaussian distribution. The safety index $\beta$ can be considered as the minimum distance from the origin to the limit of the safety domain (also named failure domain) in standardized normal space. Cunha & Guedes Soares (1999) discuss the choice of techniques for data transformation applied for Hs modelling. Nevertheless, the transformation process of the data to physical environmental parameter space is performed by Rosenblatt and illustrate in Figure 6.

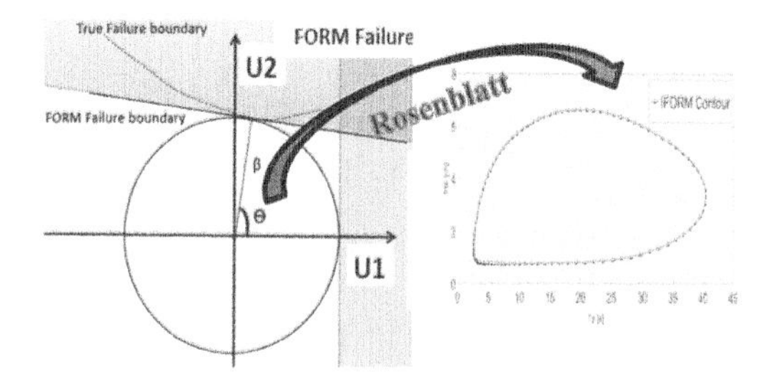


Figure 6. Illustration of the IFORM to compute environmental contours. The points along the U-circle are transformed by the Rosenblatt to the original Hs-Tp environmental variables.

## 3 RESULTS AND DISCUSSION

Table 2 describes the baseline used to assess the sea-states conditions near to Caparica beach as well as several return periods. The conditional lognormal fitted parameters by Least Square method are presented on Figure 7. In order to check visually the fitted values and to evaluate which zone can present exceedance around the contour. In addition, a dataset is generated artificially using the Monte Carlo technique based on the fitted statistical model. This procedure is important to verify the level of precision in the adjustment of the estimated parameters.

The bivariate distribution of the observed variables tends to follow the shape of the contour's lines, and in this way, the artificially generated data help to check this trend, in addition to providing information on which zones present excess values. This artificial dataset as well as the historical buoy data observations are then plotted overlaid with 100-years contour lines and shown in Figure 8. The Monte Carlo sample size is 1 million.

Table 3 describes the results of the fit procedure applied to the buoy data series considering the baseline defined previously. Summarizes the values of the statistical shape, scale and location parameters resulting from the best fit process following Weibull distribution, as well as the parameters from conditioned Tp distribution. As the Tp variable follows the Lognormal, then the parameters are conditioned by its mean and the standard deviations defined by Equation 1a-b.

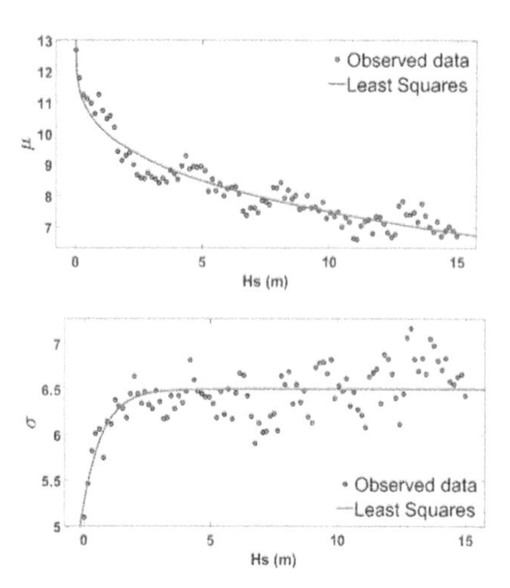


Figure 7. Nonlinear least square fit of μ [ln (Tp)] and σ [ln (Tp)] based on equations 1a-b.

Table 4 shows the extreme values of the return period for each variable considered, as well as the associated value for the secondary variable. These values are computed along the environmental contours constructed for several years, which allow us to observe probable long-term trends in the joint distribution of the variables that define the sea states. The environmental contours of Hs-Tp are shown in Figure 9. These combinations of variables are considered the most expected for each period

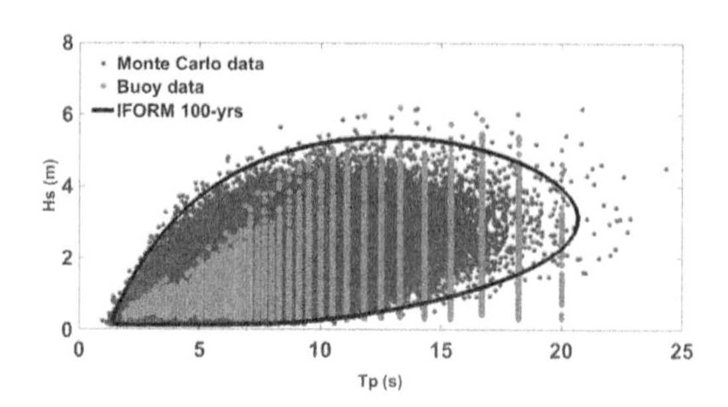


Figure 8. The 100-year sea wave contour line estimated by IFORM as well as the numeric simulated one million of sea-states generated by Monte Carlo overlaid with the historical buoy data.

Table 2. Summary of the baseline applied to identify long-term severe sea conditions.

| Variable | Hs | Tp |
|---|---|---|
| Distribution | 3P Weibull | Conditional Lognormal |
| Hs-Tp Return Period | 10 years<br>20 years<br>50 years<br>100 years | |

Table 3. Statistical parameters of fit procedure of sea states. (Eq. 1a-b).

| Marginal Hs (m) Weibull | α | β | γ |
|---|---|---|---|
| | 1.09 | 1.50 | 0.147 |

| Conditional Tp (s) Lognormal | | | |
|---|---|---|---|
| | $i = 1$ | $i = 2$ | $i = 3$ |
| $a_i$ | 0.66 | 1.03 | 0.36 |
| $b_i$ | 0.31 | -0.04 | 0.15 |

Table 4. Expected extremes values of sea states computed along the Hs-Tp environmental contour for various period of return.

| Return Period (Years) | Maximum Hs (m) and associated Tp (s) | | Maximum Tp (s) and associated Hs (m) | |
|---|---|---|---|---|
| 10 | 5.66 | 13.23 | 21.94 | 3.10 |
| 20 | 5.88 | 13.59 | 22.96 | 3.19 |
| 50 | 6.17 | 14.06 | 24.32 | 3.32 |
| 100 | 6.38 | 14.40 | 25.34 | 3.41 |

considered. Structurally, these combinations are more probable to generate damage due to the interaction between the dune-beach system versus the

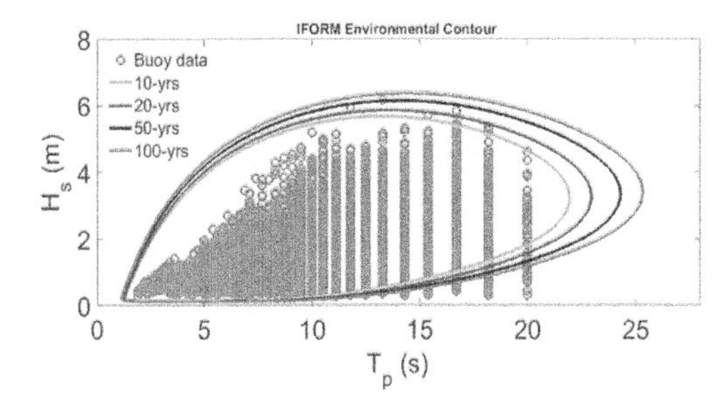


Figure 9. The Hs-Tp environmental contour derived by IFORM for various periods of return.

occurrence of severe marine conditions. The failure probability computed for each return period are: $Pf_{10} = 1.14e^{-5}$, $Pf_{20} = 5.70e^{-6}$, $Pf_{50} = 2.28e^{-6}$, $Pf_{100} = 1.14e^{-6}$.

## 4 CONCLUSIONS

Marine coastal research issues are extremely crucial and highly sensitive on a global and local scale. Adding complexity to the system when considering that life today is directly related to coastal marine resources. For many engineering purposes, it is vital to obtain reliable data defining correctly wave, wind and current and their joint probabilities for specific areas. The probabilistic models are efficiently applied to assess the distribution of sea-states variables as well the joint occurrence. For the problem of multivariate probabilistic analysis, a statistical conditional scheme is defined to assess the long-term expectation of sea-states variables. The statistical modeling of the observed data is considered the initial step, since it is soon coupled to the final step, which is the construction of the environmental contour. The combinations of predicted severe sea states are efficiently computed by environmental contours for different return periods.

In short, the probabilistic procedure followed to identify extreme sea-states considering observational data from shallow water moored buoy is useful for coastal engineering as it provides a clear overview of the long-term trend of wave dynamics, and can be applied for various purposes, such as coastal dune protection.

## ACKNOWLEDGEMENTS

The first author has the support of the Portuguese Foundation for Science and Technology (FCT), with the contract SFRH/BD/143567/2019. This work contributes to the Strategic Plan of the Lisbon Center for Naval and Ocean Engineering and Technology, which is also financed by the FCT under the contract UIDB/UIDP/00134/2020. The authors acknowledge the Administration of the Port of Lisbon (APL) for providing the buoys time series used in the work.

## REFERENCES

Andersen, U.V. 1995. Resistance of Danish coastal vegetation types to human trampling. *Biol. Conserv.* 71 3., 223–230.

Baccus, J. T. 1977. A study of beach and dunes floral and faunal interrelations as influenced by recreational and user impact on Padre Island National Seashore. *Southwest Region, National Park Service.* Santa Fe, New Mexico. 121 pp.

Bates, G. H. 1935. The vegetation of footpaths, sidewalks, cart tracks, and gateways. *J. Ecol* 23: 470–487.

Bitner-Gregersen E.M. & Haver S. 1991. Joint environmental model for reliability calculations. In: *Proceedings of the First International Offshore and Polar Engineering conference*. Edited by: S. Jin Chung. The International Society of Offshore and Polar Engineering ISOPE., August, Edinburgh, U.K., 11–16.

Bitner-Gregersen, E.M. & Guedes Soares, C. 1997. Overview of Probabilistic models of the wave environment for reliability assessment of offshore structures. In: Guedes Soares, C. Editor., *Advances in Safety and Reliability.* New York: Pergamon, pp. 1445–1456.

Bitner-Gregersen, E. M. 2015. Joint met-ocean description for design and operations of marine structures. *Appl. Ocean Res.* 51, pp. 279–292.

Boorman, L. A. & Fuller, R. M. 1977. Studies on the impact of paths on the dune vegetation at Winterton, Norfolk, England. *Biol. Cons.* 123.: 203–216.

Cunha, C. & Guedes Soares, C. 1999. On the choice of data transformation for modelling time series of significant wave height. *Ocean Engineering*, 26: 489–506.

Clarindo, G., Reis, M.T., Fortes, C.J.E.M. & Rodríguez, G. 2018. Risk assessment of coastal flood in a site of special scientific interest. *J. Coast. Conservation.* Vol. 22 – May 2018.

Clarindo, G. & Guedes Soares, C. 2020. Derivation of environmental contour by Direct Monte Carlo techniques. *Proceedings of the 5th International Conference on Maritime Technology and Engineering* Martech 2020., November 16-19, Lisbon, Portugal.

Clarindo, G. & Guedes Soares, C. 2021. Environmental wave contours for the west coast of Fuerteventura. *Developments in Renewable Energies Offshore* – Renew 2020. – 12-15 October 2020, Lisbon, Portugal.

Clarindo, G., Teixeira, A.P. & Guedes Soares, C. 2021. Environmental wave contours by Inverse FORM and Monte Carlo Simulation with variance reduction techniques. *Ocean Engineering*, vol. 228.

Fenu, G., Cogoni, D., Ulian, T. & Bacchetta, G. 2013. The impact of human trampling on a threatened coastal Mediterranean plant: the case of Anchusa littorea Moris Boraginaceae. *Flora-Morphol. Distrib. Funct. Ecol. Plants* 208 2., 104–110.

Ferreira, J.A. & Guedes Soares, C. 2002. Modelling bivariate distributions of significant wave height and mean wave period. *Applied Ocean Research* 24 1., 31–45.

García-Romero, L., Delgado-Fernández, I., Hesp, P.A., Hernández-Calvento, L., Hernández-Cordero, A.I. & Viera-Pérez, M. 2019. Biogeomorphological processes in an arid transgressive dunefield as indicators of human impact by urbanization. *Sci. Total Environ.* 650, 73–86.

Guedes Soares, C. & Trovao, M.E. 1991. Influence of wave climate modelling on the long-term prediction of wave induced responses of ship structures. *In: Price, W.C., and Memarel, P. Editors., Dynamics of Marine Vehicles*

*and Structures in Waves*, pp. 1–10. Amsterdam, The Netherlands: Elsevier Science Publishers.
Guedes Soares, C. 1993. Probabilistic Methodology for Coastal Site Investigation Based on Stochastic Modelling of Waves and Currents. *MAST Days*, Brussels, March 1993.
Guedes Soares C., Lopes, L.C. & Costa, M.D.S. 1988. Wave climate modelling for engineering purposes. *Computer modelling in ocean engineering*. Proceedings of an international conference on computer modelling in ocean engineering, Venice 19-23 September, pp. 169–175.
Guedes Soares, C. & Henriques, A.C. 1994. On the Statistical Uncertainty in Long Term Predictions of Significant Wave Height, *Proc. OMAE*, Vol. II, 1994, pp. 67–75.
Guedes Soares, C. 2003. Probabilistic Models of Waves in the Coastal Zone. *Advances in Coastal Modeling*. edited by V.C. Lakhan. Pp 159–187.
Haselsteiner, A.F., Coe, R.G., Manuel, L., Chai, W., Leira, B.J., Clarindo, G., Guedes Soares, C., Hannesdóttir, I., Dimitrov, N, Sander, A., Ohlendorf, J-H., Thoben, K.D., Hauteclocque, G. De, Mackay, E.; Jonathan, P., Qiao, C.; Myers, A., Rode, A., Hildebrandt, A., Schmidt, B., Vanem, E. & Huseby, A. B. 2021. A benchmarking exercise for environmental contours. *Ocean Engineering*, vol. 236.
Hesp, P.A., Schmutz, P., Martinez, M.M., Driskell, L., Orgera, R., Renken, K., Revelo, N.A.R. & Orocio, O.A. J. 2010. The effect on coastal vegetation of trampling on a parabolic dune. *Aeolian Res*. 2 2., 105–111.
Haver, S. 1987. On the joint distribution of heights and periods of sea waves. *Ocean Engineering*, 145., 359–376.
Huseby A.B., Vanem E. & Natvig B. 2013. A new approach to environmental contours for ocean engineering applications based on direct Monte Carlo simulations. *Ocean Engineering*, 60, March, 124–135.
Jonathan P., Flynn J. & Ewans K. C. 2010. Joint modelling of wave spectral parameters for extreme sea states. *Ocean Engineering*, 3711-12., 1070–1080.
Kiureghian, A.D. & Liu, P.L 1986. Structural Reliability Under Incomplete Probability Information. Journal Engineering *Mechanics*, 112, Pp. 85–104.
Li, L., Gao, Z. & Moan, T. 2015. Joint distribution of environmental condition at five European offshore sites for design of combined wind and wave energy devices. *Journal of Offshore Mechanics and Arctic Engineer-ing*: 1373.:031901: 16 pages.
Melchers, R. E. 1987. Structural Reliability: Analysis and Prediction. Ellis Horwood, Chichester, UK.
McDonnell, M.J. 1981. Trampling effects on coastal dune vegetation in the Parker River National Wildlife Refuge, Massachusetts, USA. *Biol. Conserv*. 21 4., 289–301.
ISSC 2018. Volume 1 – Technical Committee Reports & Volume 2 – Specialist Committee Reports. *International Ship and Offshore Structures Congress*.
Ross E., Astrup O., Bitner-Gregersen E., Bunn N., Feld G., Gouldby B., Huseby A., Liu Y., Randell D., Vanem E. & Jonathan P. 2019. On environmental contours for marine and coastal design. *Ocean Engineering*. 195, 106–194.
Rusu, E. & Guedes Soares, C. 2011. Wave modelling at the entrance of ports. *Ocean Eng*. 38:2089–2109.
Rusu, L. & Guedes Soares, C. 2013. Evaluation of a high-resolution wave forecasting system for the approaches to ports. *Ocean Eng*. 58: 224–238.
Rusu, E. & Guedes Soares, C. 2014. Influence of a new quay on the wave propagation inside the Sines harbor. *Maritime Technology and Engineering* – Guedes Soares & Santos Eds. Taylor & Francis Group, Lisbon, Volume: Vol 2, pp 1355–1364.
Santana-Ceballos, J., Fortres, C., Reis, M. & Rodriguez, G. 2017. Wave overtopping and flood risk assessment in harbors: a case study of the Port of Las Nieves, Gran Canaria. WIT Trans. Built Environ. *Coast. Cities their Sustain. Future*. II 170, 1–10
Winterstein, S. R., Ude, T. C., Cornell, C. A., Bjerager, P. & Haver, S. 1993. Environmental parameters for extreme response: Inverse FORM with omission factors. *In Proceedings of the 6th International Conference on Structural Safety & Reliability ICOSSAR*. August 9-13, Innsbruck, Austria: 551–557.

*Trends in Maritime Technology and Engineering – Guedes Soares & Santos (eds)*
 *ISBN 978-1-032-33583-4*

# Representing spectral changes in seasonal ocean wave patterns using interquartile ranges

G. Clarindo & C. Guedes Soares
*Centre for Marine Technology and Ocean Engineering (CENTEC), Instituto Superior Técnico, Universidade de Lisboa, Lisbon, Portugal*

G. Rodríguez
*Applied Marine Physics and Remote Sensing Group, Institute of Environmental Studies and Natural Resources (iUNAT), Department of Physics, Universidad de Las Palmas, Las Palmas, Spain*

ABSTRACT: This article presents the changes observed in the structure of wind-wave spectra due to wave period variability. A large set of wave data is analysed as well as a robust spectral representation of the sea states is applied for this purpose. After estimating and verifying the joint distribution of the sea states, a significant wave height interval is selected to then observe the climatic changes as the wave periods vary. For this fixed limit of wave height, three wave periods intervals that contain the maximum values of sea states occurrence are then selected. These different groups of data are also evaluated separately considering its seasonality by a robust statistical metrics. Furthermore, for each dataset these changes in seasonal patterns recorded by the spectra are presented using the inter quantiles ranges metrics. The resulting data from different ensemble of spectra are compared with the conventional approach that consider the maximum and minimum envelopes to represent the climatic spectrum with the robust representation (using inter quantile ranges) followed in the present work.

## 1 INTRODUCTION

Many marine structures are exposed to the effect of severity induced by the environment, so modeling the evolution and changes in the distribution of variables such as waves, wind and current are crucial for an effective engineering practice in marine environments. Guedes Soares et al. (1988) describe how to model the wave climate for engineering purposes. Initially, the wave spectral modeling studies addressed only single-peak spectra. However, to represent the physical processes realistically, the energy balance must include the interactions between local wind system as well swell systems from remote areas generated by different wind fields. This interaction produces mixed sea situations in which spectrally they can present two or more peaks.

Commonly the wave spectra in ocean are double peaked as consequence of remarkable dependence of meteorological and oceanographic factors (Lucas et. al (2021). Titov (1969), Thompson (1980), Cumming et al. (1981), Guedes Soares (1991), Guedes Soares & Nolasco (1992), Rodriguez & Guedes Soares (1999), Ewans et al. (2006) describes the high probability of occurrence of double peak spectra declaring that the two peaked spectra still occur with a relatively high frequency, which makes it important to use them in the design approach for oceanic and coastal areas. In this way, Ochi & Hubble (1976) and Guedes Soares (1984) proposed models to describe the combined situation of swell and wind systems.

Torsethaugen (1993) generalized the model proposed by Guedes Soares (1984) also adopting JONSWAP for each system, but now varying the peak value, as well as the enhancement factor, parameterizing the model based on conditions of the Norwegian coast. These studies were carried out even before the naval and offshore industry realized the importance of accounting for mixed sea situations in the mechanical-structural design approach. However, in recent years there has been an increasing interest in the evaluation of these types of cases in ocean engineering practices (see: Guedes Soares (1993), Yilmaz et al. (1998), Lawford et al. (2008), Rognebakke et al. (2008) and Teixeira et al. (2009)).

The wave climate data is mostly treated as long-term statistics of significant wave heights (Hs), peak (or mean) periods (Tp, Tm) and main directions (θ) for certain regions of the ocean. Boukhanovsky & Guedes Soares (2009) extended the application to model multi-peaked wave systems to the direction domain. Afterwards, Lucas et al. (2011) applied in modeling climate variability of directional wave spectra for the Portugal coast.

DOI: 10.1201/9781003320289-32

The basic procedure of wave climate description is defining the long-term distributions of the cited variables at the given location. Typically consists of compiling wave data during a long period of time to fit the observations to a suitable statistical model. This basic approach is initially formulated by Haver (1985) and detailed by Ferreira & Guedes Soares (2000).

Another specific approach is to adopt the techniques for evaluating the extreme distribution of significant heights, which is discussed in Ferreira & Guedes Soares (1998) using peaks over threshold, as well as Guedes Soares & Scotto (2004) using largest-order statistics. To fully describe a sea state, information regarding the period associated with these significant heights is needed. Both the spectral peak period (Tp) and the mean period (Tm) can be used to evaluate the distribution of the main variables that define a sea state, for requirements of more information see Bitner-Gregersen & Haver (1989) as well Ferreira & Guedes Soares (2002).

Generally, the problem of statistical analysis of wave spectra is associated with the multivariate statistical analysis. The simplest way of statistical spectral treating (see Scott, 1968 and Buckley, 1988) is based on averaging over the frequency and directional domain for different wave conditions (e.g., grouping by significant wave height and mean or peak wave period). This approach allows computation of the so called climatic wave spectra by averaging the climatic distribution of the wave energy on the frequencies and directions. However, the current paper only treat the densities distributed in frequencies. In other words, the description of the long-term wave climatic are based on significant wave height and spectral peak period.

This type of data grouping that joining events with the same wave parametric characteristics is performed in order to reduce the statistical variability and improve the climatic description of long-term patterns. Normally, the energy content, frequency composition and structure of the spectra derived from theses ensembles of sea states reveals the existence of patterns of various time scales, as well as the appearance of some episodic events related to the variability of the atmospheric conditions that generate diverse types of wave fields. However, it is necessary to present such information in a robust and practical way to better comprehend the climatic changes.

Various approaches can be used to reduce the large variability observed in the ensembles of wave spectra. For characterizing the wave climate, the present work group the individual wave spectra into homogeneous sets, which only include those elements with a certain degree of similarity generating several bins to evaluate the wave spectral structures.

The common approach to estimate the climatic wave spectrum from a given set of measured spectra is compute the average of the all-spectral density estimations associated to each frequency band. This procedure results in the well-known long-term average wave spectrum, defined by the International Maritime Organization as a spectrum measured over the ensemble of spectra (Buckley, 1993).

This study is organized as follows: Section 2 describe the methodology applied as well the dataset used. Section 3 presents the seasonal results obtained for each bin. Section 4 report the main conclusions.

## 2 METHODOLOGY

This section briefly describes the methods for obtaining homogeneous representative sets as well as the ensembles of wave spectra considered for the present work.

A classical approach to reach this goal and generate representative set of the data consists in classifying the whole collection of measured wave spectra in ranges of significant wave height and estimating a representative wave spectrum for each one of these subsets (Buckley (1988), Teng et al. (1994) and Teng (2001)). Grouping the sea states only by Hs is satisfactory but to compute the realistic description, it is necessary to categorize the sea states in terms of significant wave height and some characteristic wave period (Haver & Nyhus, 1986). Other natural way to decompose the ensembles of sea states collection is separate groups according to their climatic seasons.

Considering the previous mentioned and according to Nolasco & Guedes Soares (1992), firstly the distribution of sea state events is presented in bin arrangements scattered in the bivariate histogram, with intervals of 0.5m and 0.5s for Hs and Tp respectively (Figure 1). In second, the range of Hs = 1.0-1.5 meters is selected as the representative wave height to then vary the distribution of the periods. This range of Hs is selected as representative because it is the range that presents the highest occurrences of events, such as bin 1 (Hs=1.0-1.5m and Tp=4.5-5.0s) and bin 2 (Hs=1.0-1.5m and Tp= 5.0-5.5s) computing 4720 and 4285 events respectively. These mentioned bins are then selected to be decomposed into another groups of datasets that consider the seasonality. In this way, the bin 1 is subdived into four sets of data corresponding to spring, summer, autumn, winter. It can be said that each event is a computed unitary spectrum and this understanding extended for all bins. In this way it is possible to perform a type of mining/clustering from high to low resolution, which consequently reduce the statistical variability.

### 2.1 *Climatic wave spectrum estimation*

The traditional approach to estimate the climatic wave spectrum from a given set of measured spectra is to compute the average of the spectral density estimations for each frequency band and express its variability in terms of the envelops values and the standard deviation. The uncertainty of sea state parameters resulting from the methods of spectral estimation are studied by Rodriguez et al. (1999) as well as the slope of high frequency in Rodriguez and Guedes Soares (199b).

So, let $\eta_i$ be the $i$-th sea state in the sequence of $n$ successive sea states and $S_{\eta i}$ the associated spectral density function. Each one of the $n$ wave spectra represents a discrete function evaluated at $m$ specific frequencies. Conceptually, any particular wave spectrum, $S_{\eta i}$ can be interpreted as a realization of a random process and the complete collection of wave spectra, or realizations, is named the ensemble.

$$\{S_{\eta i}(f)\}\ i = 1, 2, 3, \ldots n$$

$f$ represents the frequency estimations.

After defining the ensemble of spectral data sets, it is necessary to extract relevant information to quantify their variability. Although details concerning the data are very important, it is possible and common to summarize this information using a few well-chosen parameters, easily calculated from the data that characterizes them for practical applications. This summary should include information about the range of the dataset (minimum and maximum), its location, spread and symmetry. The most used measures of location, which describe the central tendency of the data and dispersion, which assess the variability or degree of spread in the original data, are the sample mean and the standard deviation (or variance) respectively. Nevertheless, in the present work the metrics used to represent the seasonal climatic spectrum in each bin are those described by Rodriguez et al. (2016), in which basically consider the inter-quantile intervals instead of the maxima and minima as enveloping values.

### 2.2 *Representing significant and homogeneous wave datasets*

The climatic wave spectrum estimated through the conventional approach, as previous mentioned is given by the average value of spectral densities associated to each frequency and its statistical variability characterized by the variance. Though, this classical approach is based on the normality assumption for each one of the random variables, $S_\eta(f)$ that constitute the ensemble of wave spectra. Nonetheless, it is much more expected that a given ensemble is non-normal. Therefore, simple computation of parameters that describe a set of data completely only if the data are normal will not generally be satisfactory, because the distribution of non-normal variables may lack symmetry and may have extreme values or outliers.

The non-normal variables, (particularly those with extreme right or left tails), may be better summarized and exposed by means of robust and resistant statistical measures obtained from a small set of sample quantiles and related graphical techniques. In this way, an estimator is considered robust, generally, if it presents a lack of susceptibility for effects of incorrect assumptions regarding the assumed probability model, usually the effects of non-normality. On the other hand, an estimator is said to be resistant when a change in a few data will not substantially change the value of the estimate. Particularly, a resistant estimator is not unduly influenced by a small number of outliers. Amongst the large variety of robust and resistant estimators proposed in the statistical literature those based on sample quantiles are of particular interest due to its simple computation and direct interpretation, as well as no requirement of distributional assumptions. The q-th sample quantile ($x_q$) of a continuous random variable (x) for any q ($0 < q < 1$) is defined by $P(X \leq x_q) = q$. In other words, it represents a value that q% of the data lie below $x_q$. The quantile associated with $q = 0.5$ is called the median, or second quartile, denoted as $Q_2$. The quantiles associated with $q = 0.25$ and $q = 0.75$ are the lower and upper quartiles, denoted by $Q_1$ and $Q_3$, and those associated with $q = 0.1$ and $q = 0.9$ are the lower and upper deciles, denoted as $D_1$ and $D_9$, respectively. Therefore, it is defined in terms of the data values holding 25% and 75% of the values below them. That is, the interquartile range is defined as the difference between the upper and lower quartiles, $IQR = Q_3 - Q_1$.

Furthermore, skewed distributions affect the average values, making it less representative of the sample. Thus, the robust alternative approach proposed by Rodriguez et al. (2016), in which the median from IQR ($Q_2$) is considered as the main indicator of central tendency, or location, because it is fairly robust, remaining stable if is used to represent the wave spectra. Information provided by quartiles is used to generate a clear and compact visual representation of how the samples are distributed in the representative ensemble. The resulting wave spectra are strikingly summarized and displayed by the mentioned ranges that represents an important graphical tool for description of the data ensemble. This tactic allows us to identify long-term changes in the wave climate spectrum.

### 2.3 *Data*

This section describes the data used to obtain the homogeneous ensembles of spectra. The dataset is provided by the National Data Buoy Center (NDBC - NOAA) and measured by an *in-situ* discus buoy located in offshore of Gulf of Maine (43°12'2"N-69° 7'37"W) in the depth of 176.8 m. The total dataset consists of fifteen years (1996–2011) of hourly wave measurements from the station number 44005.

Having the measurements of the climatic spectra calculated, the first step in generating the homogeneous dataset is then performed. A bivariate histogram is created which includes the numbers of sea states for each sub-ensemble of samples. The division is completed considering intervals of 0.5m and 0.5s for Hs and Tp respectively.

Next, the range of Hs=1.0-1.5m is fixed and selected as representative to then extract information when the wave period (Tp) varies according the distribution of the maximum events int the bivariate histogram. Then, three bins in the period range (Bin 1, Bin 2 and Bin 3) are selected to be further subdivided into four new ensembles grouped now according to its seasonality. As indicated in the bivariate histogram, those bins are selected as the most representative, as they contain the highest number of sea state events of the entire histogram.

The idea is generated different and homogeneous wave spectrum ensembles. Then, the whole set of wave spectra for each bin has been split into four subsets, regarding climatic seasons. The individual spectra are distributed as follows: Spring (March, April, May), Summer (June, July, August), Autumn (September, October, November) and Winter (December, January, February). The Hs-Tp bivariate histogram for station 44005 is describe in Figure 1. The highlighted bins intervals are selected to exemplify the climatic spectra by its IQR. The Tp bins intervals for the fixed Hs range as well the total number of events and those considered for each season are described in Table 1.

In this way the statistical variability is presented compacted, comprehensive and informative, providing a clear distributional overview in terms of robust and resistant measures of the central tendency, dispersion, asymmetry, extremes, and tail lengths of the spectral estimates for each frequency range.

## 3 RESULTS

The individual wave spectra grouped according to the characteristics described in Table 1 are presented in the following figures. Firstly, the total number of events for the three different bins intervals are shown in Figure 2.

Naturally, maximum and minimum spectra represented in the left panel are heavily affected by the presence of outliers and, consequently, their representativeness is low when compared to the curves computed by IQR in the right panel of Figure 2. This fact is due to the large variability among the multiple measured spectra included in the examined set. Moreover, the envelope curve is defined by just one spectral estimation for every frequency band being the maximum and minimum, nevertheless, only the maximum spectrum is used in practice.

In second, the subsequent figures illustrate the bins subsets of spectra for each climatic season. These figures represent the all-individual wave spectra composed by the mean and envelopes values. Moreover, the IQRs ($D_1$, $Q_1$, $Q_2$, $Q_3$ and $D_9$) curves are also presented overlaid with the individual spectra.

It is observed that for longer energy period, the larger energy dispersion over the frequencies. Such changes are also noticeable seasonally, where the spectra are concentrating the energy in smaller frequencies as the period decreases. The IQR is able to identify these changes effectively.

The results provide a more intuitive idea of the variability associated to each frequency band and, hence, on the representativeness of the climatic (median) spectra. In principle, any of the four resulting data sets should be more homogeneous than the whole set showing more robustness and less variability.

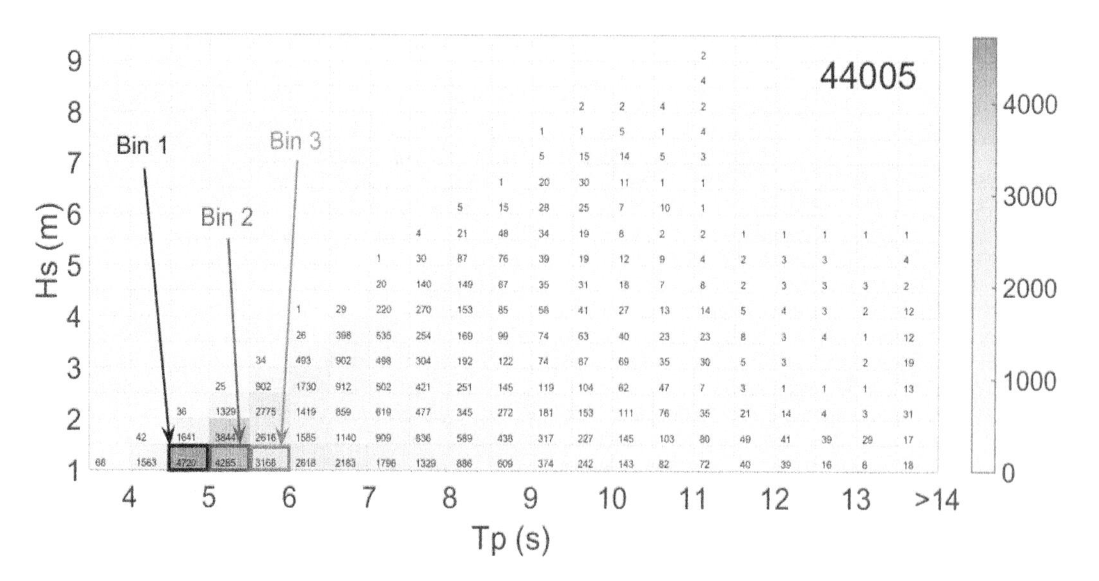


Figure 1. The bivariate histogram computed for station 44005 that represent several ensembles of wave spectra. The emphasized bins intervals exemplify the low Tp and high Tp values.

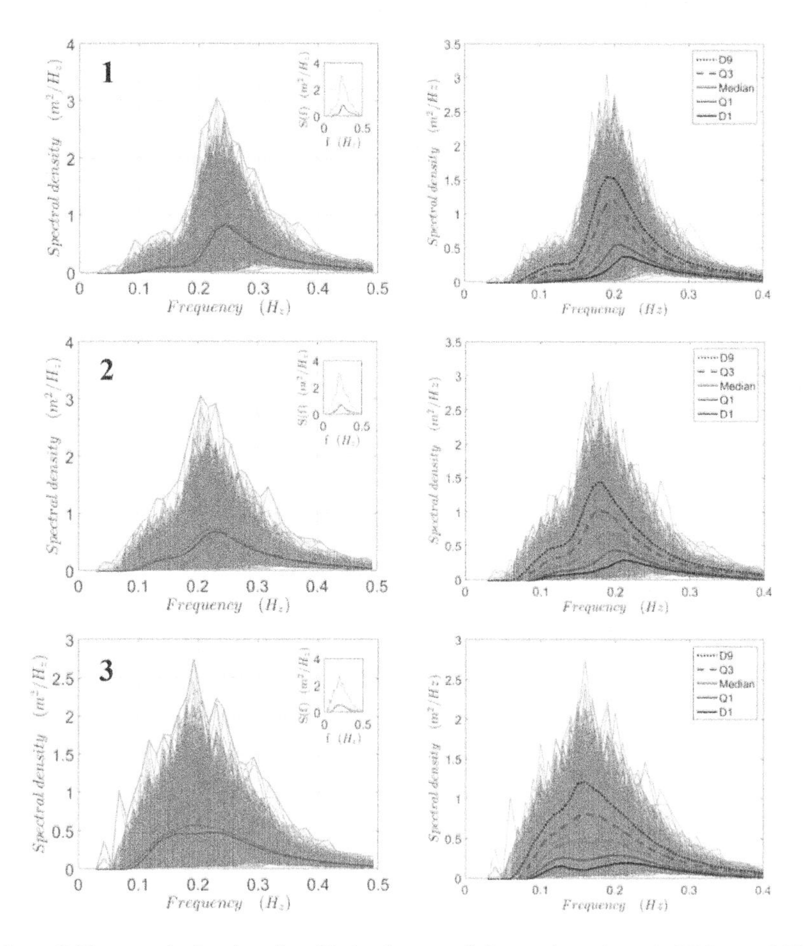


Figure 2. (Left panel) The mean (red) and median (blue) value as well the envelopes (green). (Right panel) The inter quantiles ranges for Bin 1, 2 and 3 considering the total events in each ensemble.

Table 1. Definition of bins intervals as well the number of events considered for each wave spectra ensembles including the seasonal events for the fixed representative Hs range.

| Hs=1.0-1.5 (m) | Tp (s) | Spring | Summer | Autumn | Winter | Total |
|---|---|---|---|---|---|---|
| Bin 1 | 4.5-5.0 | 972 | 1170 | 1368 | 1210 | 4720 |
| Bin 2 | 5.0-5.5 | 974 | 1321 | 1081 | 909 | 4285 |
| Bin 3 | 5.5-6.0 | 979 | 976 | 727 | 486 | 3168 |

## 4 CONCLUSIONS

A common and natural approach to reducing the variability for a given dataset is applied successfully to generate homogeneous and representatives wave ensembles. It consists of defining smaller ranges of data sets that have the same environmental characteristics. So, all observations are divided into smaller Hs-Tp groups named bins from the corresponding bivariate histogram.

Grouping multiple sea states, or wave spectra, in a relatively small bins of representative sea conditions that containing similar energy content, frequency composition, as well as the seasonality represents a robust approach to describe the changes for wave climatic patters. In this way, long-term changes are easily appreciated, even by visual inspection. To represent such climatic variability, it is suggested to implement the interquartile intervals for a given dataset instead of the maximum and minimum envelopes. Moreover, the variability expressed in terms of different range as the IQR has been shown to be robust and versatile in defining the climate variability. Compactly illustrates the variations

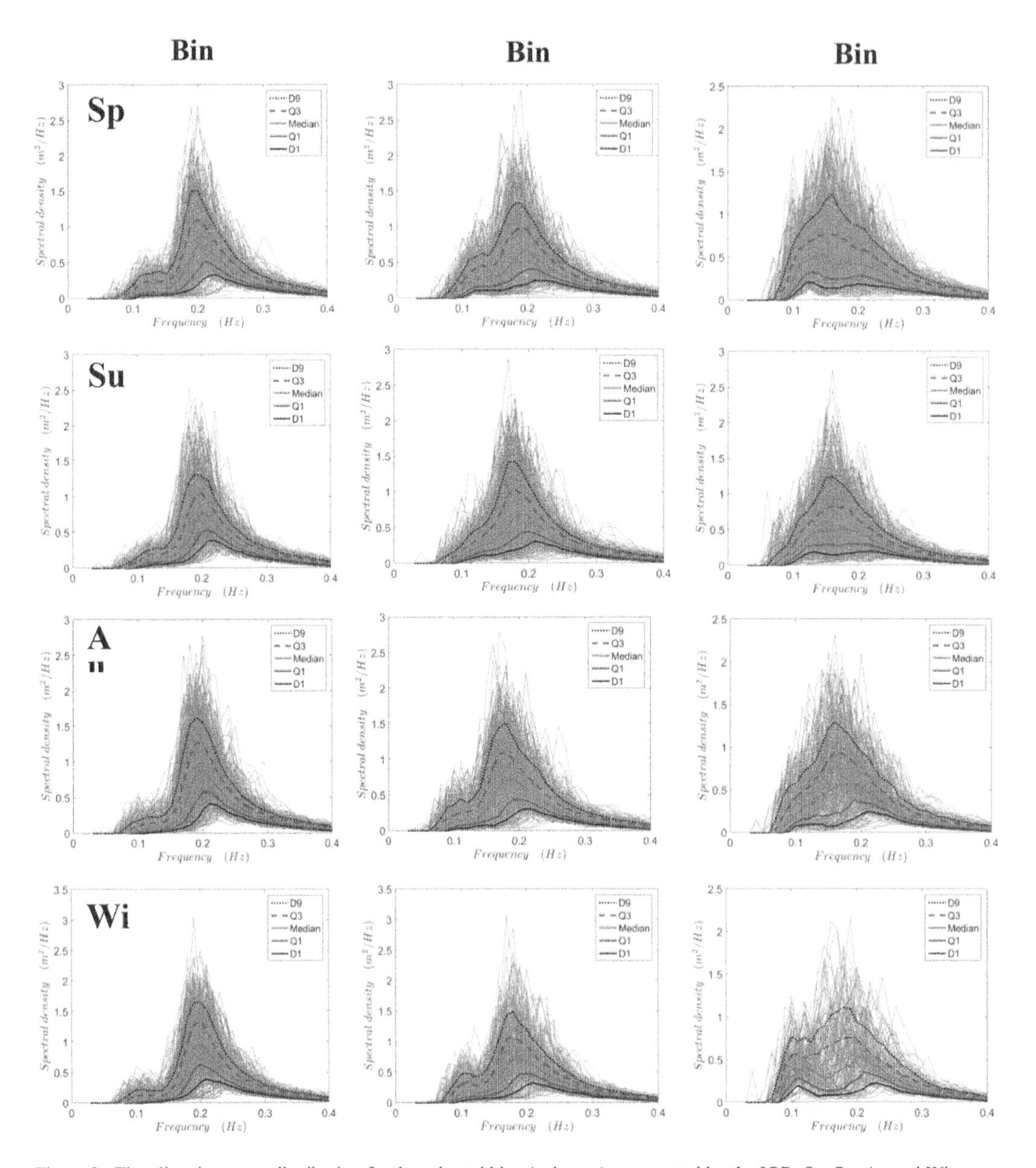


Figure 3. The climatic season distribution for the selected bins (columns) represented by the IQR. Sp, Su, Au and Wi represents spring, summer, autumn and winter respectively.

in energy distribution for several frequencies' bands. The dispersion in the data ensembles is better represented by the ranges of the quartiles, including those not normally distributed.

In summary, the followed procedure to represent the changes in the climatic wave spectrum from a set of measured wave spectra is usefulness in providing a clear, compact and informative overview of the spectral estimates for each frequency band.

## ACKNOWLEDGEMENTS

The first author had the support of the Portuguese Foundation for Science and Technology (FCT), with the contract SFRH/BD/143567/2019. This work contributes to the Strategic Plan of the Lisbon Center for Naval and Ocean Engineering and Technology, which is also financed by the FCT under the contract UIDB/UIDP/00134/2020.

## REFERENCES


Bitner-Gregersen, E.M. and Haver, S. (1989) Joint long-term description of environmental parameters for structural response calculation. *In: Proceedings of the 2nd International Workshop on Wave Hindcasting and Forecasting*, Canada.

Boukhanovsky, A. and Guedes Soares, C. (2009) Modelling of multi-peaked directional wave spectra. *Applied Ocean Research* 31 (2), 132–141.

Buckley, W.N. (1988) Extreme and climatic wave spectra for use in structural design of ships, *Naval Engineering Journal*, 100 (5), pp. 36–57.

Buckley, W.H. (1993) Design Wave Climates for the World-Wide Operations of Ships. Part 1: Establishments of Design Wave Climate. *Int. Maritime Organization (IMO)*, Selected Publications, October 1993.

Cumming, W.E., Bales, S.L. and Gentile, D.M. (1981) Hindcasting waves for engineering applications. *Proceedings of the international symposium on hydrodynamics in ocean engineering*, Trondheim, Norway, 70–89.

Ewans, K.C., Bitner-Gregersen, E.M., and Guedes Soares, C. (2006) Estimation of wind-sea and swell components in a bimodal sea state. *Journal of Offshore Mechanics and Arctic Engineering*, 128 (4), pp. 265–270.

Ferreira, J.A. and Guedes Soares, C. (1998) An application of the peaks over threshold method to predict extremes of significant wave height. *Journal of Offshore Mechanics and Arctic Engineering* 120 (3), 165–176.

Ferreira, J.A. and Guedes Soares, C. (2000) Modelling Distributions of Significant Wave Height. *Coastal Engineering*. 40(4):361–374.

Ferreira, J.A. and Guedes Soares, C. (2002) Modelling bivariate distributions of significant wave height and mean wave period. *Applied Ocean Research* 24 (1), 31–45.

Guedes Soares, C. (1984) Representation of double-peaked sea wave spectra. *Ocean Eng.* 11, pp. 185–207.

Guedes Soares C., Lopes, L.C. and Costa, M.D.S. (1988) Wave climate modelling for engineering purposes. *Computer modelling in ocean engineering*. Proceedings of an international conference on computer modelling in ocean engineering, Venice 19-23 September, pp. 169–175.

Guedes Soares, C. (1991) On the occurrence of double peaked wave spectra, *Ocean Engineering*, 18, pp. 167–171.

Guedes Soares, C. and Nolasco, M.C. (1992) Spectral modelling of sea states with multiple wave systems. *Journal of Offshore Mechanics and Arctic Engineering*, Vol. 114, pp. 278–284.

Guedes Scares, C. (1993) Long-Term Distribution of Non-Linear Wave Induced Vertical Bending Moments. *Marine Structures*, Vol. 6, 1993, pp. 475–483.

Guedes Soares, C. and Scotto, M.G. (2004) Application of the r largest-order statistics for long-term predictions of significant wave height. *Coastal Engineering* 51, 387–394.

Haver, S. (1985) Wave climate off northern Norway, *Applied Ocean Research*, 7(2): 85–92.

Haver, S. and Nyhus, K. A. (1986) A wave climate description for long term response calculations. *Proc. 5th Int Offshore Mechanics and Arctic Engineering Symposium*. ASME, Tokio.

Lawford, R., Bradon, J., Barberon, T., Camps, C. and Jameson, R. (2008) Directional wave partitioning and its applications to the structural analysis of an FPSO. *In: Proceedings of the 27th international conference on offshore mechanics and arctic engineering (OMAE).*

Lucas, C., Boukhanovsky, A. and Guedes Soares, C. (2011) Modelling the Climatic Variability of Directional Wave Spectra. *Ocean Engineering*. 38, 1283–1290.

Ochi, M.K. and Hubble, E.N. (1976) Six-parameter wave spectra. *In: Proc. 15th Intl Conf. on Coastal Engineering*, pp. 301–328.

Rodríguez, G., and Guedes Soares, C. (1999) Bivariate distribution of wave heights and periods in mixed sea states. *Journal of Offshore Mechanics and Arctic Engineering*, Vol. 121, pp. 102–108.

Rodriguez, G., Guedes Soares, C., and Machado, U. (1999) Uncertainty of the Sea State Parameters Resulting from the Methods of Spectral Estimation. *Ocean Engineering*. 26(10): 991–1002.

Rodriguez, G., and Guedes Soares, C. (1999) Uncertainty in the Estimation of the Slope of the High Frequency Tail of Wave Spectra. *Applied Ocean Research*. 21(4): 207–213.

Rodriguez, G., Clarindo, G. and Guedes Soares C. (2018) Robust estimation and representation of climatic wave spectrum. *Progress in Maritime Technology and Engineering*. 105–113.

Rognebakke, O., Andersen, O.J., Haver, S., Oma, N. and Sado, O. (2008) Fatigue assessment of side shell details of an FPSO based on non-collinear sea and swell. *In: Proceedings of the 27th international conference on offshore mechanics and arctic engineering (OMAE).*

Scott, J.R. (1968) Some average sea spectra. *Transactions Royal Institution of Naval Architects* 110 (2), 233–245.

Teixeira A.P. and Guedes Soares C. (2009) Reliability analysis of a tanker subjected to combined sea states. *Prob Eng Mech*. 2009; 24: 493–503.

Teng, C.C., Timpe, G. and Palao I. (1994) The Development of Design Waves and Wave Spectra for Use in Ocean Structure Design. *Transactions Society of Naval Architects and Marine Engineers*, 102: pp. 475–499.

Teng, C.C. (2001) Climatic and maximum wave spectra from long-term measurements. *Proc. Int. Conf. Ocean Wave Measurements and Analysis*, ASCE, San Francisco, USA.

Titov, L.F. (1969) Wind-driven waves. Jerusalem: *Israel Program for scientific Translations*, Vol 53.

Thompson, E.F. (1980) Energy spectra in shallow US coastal waters. *Technical paper*, 80-2, US Army corps of engineers, Coastal engineering research center.

Torsethaugen, K. (1993) A two peak wave spectrum model. *In: Proceedings of the 12th International Conference on Offshore Mechanics and Arctic Engineering (OMAE)*, Vol. 2, ASME, NY, USA, pp. 175–180.

Yilmaz, O. and Incecik, A. (1998) Effect of double-peaked wave spectra on the behavior of moored semi-submersibles. *Oceanic Eng Int*. Vol. 2, 54–64.

*Trends in Maritime Technology and Engineering – Guedes Soares & Santos (eds)*


# Extreme response analysis for TLP-type floating wind turbine using Environmental Contour Method

M.N. Sreebhadra, J.S. Rony & D. Karmakar
*Department of Water Resources and Ocean Engineering, National Institute of Technology Karnataka, Surathkal, Mangalore, India*

C. Guedes Soares
*Centre for Marine Technology and Ocean Engineering (CENTEC), Instituto Superior Técnico, Universidade de Lisboa, Lisbon, Portugal*

ABSTRACT: The reliability of structures against extreme loading conditions is a significant factor to be accounted for the design of marine structures. In Environmental Contour Method, using inverse reliability technique the most significant contributing environmental factors associated with the structure at the particular site can approximately give the long-term extreme response. In the present study, the extreme responses on five different configurations of TLP-type floating offshore wind turbine are analysed using the environmental contour method. The various responses including the maximum and minimum tower base bending moment loads at the blade root, tower base shear force is studied. The simulations are performed for the different wind speed and wave and a comparative study is made for the different TLP-type platforms. The estimation of extreme responses is obtained using the Environmental Contour Method and the present study will be helpful for the long-term load estimation of offshore structures.

## 1 INTRODUCTION

The wind energy has shown its significant contribution and demand over the recent years in the overall energy production. There has been a shift towards offshore wind turbines which has its advantages over the onshore site conditions. The wind energy extraction is even more significant as we move towards deeper ocean water depths. The shift from onshore to offshore wind farm sites are evident nowadays in many countries. The lack of suitable wind generation sites with the required magnitude of wind for feasible wind energy generation is one of the major factors that pushes towards the offshore floating wind generation. The failure probability of structures requires sufficient consideration and proper performance of the structures requires the failure probability to be small so that they can be designed without much conservation. Extreme responses of offshore wind turbine are of major importance and efforts are made to design the structure so that it can withstand such extreme conditions of loading during its service life. Using probabilistic design approaches for obtaining extreme limit states is a method that has been used by researchers over time but involves considerable number of simulations of wind turbine. Long-term extreme response for any type of offshore structure needs to be estimated for a given return period. Offshore structures use this design methodology of which the full long-term analysis yields correct results but is very time consuming and tedious process. In the structural analysis and design, the probabilistic approach is beneficial by using only limited environmental conditions to ensure sufficient capacity.

The Inverse First Order Reliability Method (IFORM), is suggested by Winterstein et al. (1993) which involves finding contours of environmental parameters that are independent on the structure with the variables decoupled with the response. Extreme sea states can be determined by this method and inverse FORM is introduced over First Order Reliability Method (FORM) method which proves to be beneficial. Winterstein (1998) presented the reliability-based prediction of design loads and responses for floating structures including Tension-Leg-platform and spar buoy which can be an alternative to the full long-term analysis. Environmental contour method is used by Christensen and Arnbjerg-Nielsen (2000) to obtain the design shear force and overturning moment at the seabed for an active stall-regulated wind turbine at two sites. Saranyasoontorn and Manuel (2004, 2005) demonstrated the environmental contour method to be applied to establish

DOI: 10.1201/9781003320289-33

ultimate wind turbine blade bending design loads for various wind turbines. The values obtained using the IFORM approach is compared with the solution by full integration over the failure domain and its accuracy is tested.

The long-term estimates of the most probable maximum values of motion to examine the safety of wind turbines against overturning when subjected to extreme waves and wind (Bagbanci et al., 2015). The reliability analysis of DeepCWind semi-submersible offshore floating wind turbine using 1-D model and 2-D model for 10-min mean wind speed is discussed in Karmakar et al. (2014). A brief comparison on the design loads of the floating wind turbine for I-D, 2-D and 3-D environmental contour method is analyzed in Karmakar et al. (2015) by performing reliability analysis on spar-type, DeepC-Wind and WindFloat semi-submersible floater. The estimation of the long-term joint probability distribution of extreme loads for different types of offshore floating wind turbines is studied using the environmental contour (EC) method as in Karmakar et al. (2016). The out-of-plane bending moment loads at the blade root and tower base moment loads for the offshore floating wind turbine of different floater configuration like spar-type, DeepCWind and Wind-Float semi-submersible floater is presented. Sultania et al. (2018) studied the dynamic behavior of a coupled wind turbine spar-buoy platform and the statistics of tower and rotor loads as well as platform motions using time domain analysis. The 2-D and 3-D IFORM method is used considering wind speed and wave height for the environmental parameters for extreme loading to understand the comparison for the different methods used and its accuracy. The environmental contour method is used to obtain the extreme responses under certain misalignment conditions adopted and the uncertainties in the response are studied by giving a correction to two-dimensional EC method. Further, different mooring patterns are studied to know the characteristics under different aligned wind-wave conditions as well as the structural properties.

Liu et al. (2019) obtained the short-term response of an integrated system using environmental contour method on semi-submersible platform and the long-term response is extrapolated based on the short-term loads. The 3D approach is used taking into account of the uncertainties and is noted to give more accurate long-term response analysis. Raed et al. (2019) obtained the 1D and 2D environmental contours employing the inverse first-order reliability method to derive the extreme responses on a semi-submersible floating wind turbine. The results thus obtained by 1D and 2D environmental contour methods is compared by performing a full long-term analysis. Recently, Konispoliatis et al. (2021) performed the study of a multi-purpose floating tension leg platform (TLP) concept suitable for the combined offshore wind and wave energy resources exploitation, taking into account the prevailing environmental conditions. The environmental contours are developed by computing the joint distribution of environmental parameters using the statistical characteristics of wind and wave time series at the examined location.

The Inverse First Order Reliability Method (IFORM) and the method for developing environmental contours has been devised in various studies conducted by researchers. The comparative studies using the 1-D, 2-D and 3-D models of environmental contours with that of the full long-term analysis for different return periods is also performed. The Environmental Contour Method is performed for the spar-type and semi-submersible type floaters but the reliability studies of wind turbines mounted on TLP-type platforms using the environmental contour method is observed to be limited. In the present study, analysis of five different types of TLP wind turbine is carried out using one-dimensional and two-dimensional methods by developing environmental contours using Inverse First Order Reliability Method. The five TLP-types offshore wind turbine configurations as in Bachynski et al. (2012) are chosen for the analysis and contour points corresponding to 20-year return period are obtained. The 5MW reference wind turbine supported on TLP type floating wind turbine is considered for the study. The environmental contours are developed from the joint distribution of environmental parameters at the site. The wind speed data generated using the TurbSim code is used as input for the bending loads simulations using FAST code developed by NREL. The forces and moments developed at the base of the turbine tower is studied for different TLP-type floating wind turbine for the life time of the structure.

## 2 EXTREME LONG-TERM ANALYSIS

The reliability analysis problem can be expressed by the limit state function $g(x)$ which can be written as

$$g(x) = y_{capacity} - y(x), \tag{1}$$

where $y_{capacity}$ is the strength and $y(x)$ is the load acting on the structure.

### 2.1 *Full long-term analysis*

In order to obtain the long-term extreme loads on the structure, the short-term distributions are integrated with the joint probability density function. The cumulative distribution function of short-term extremes and the probability of occurrence of the environmental extreme conditions is combined to obtain the long-term cumulative distribution function. Considering the service life of turbines as $T$-years, the design load $L_T$ and the design load for a 10-min duration segment $L_T - 10\,\text{min}$ and the extreme loads for design is studied. The environmental variables considered for the probabilistic distribution is taken as 10-min mean wind speed and wave height $H_S$. The exceedance probability $P_f$ is given by

$$P_f = P[L_{10-\min} > L_T], \quad (2)$$

$$\int_0^\infty \int_0^\infty P[L_{10-\min} > L_T | v, h_s] f_{v,Hs}(v, h_s) dv dh_s. \quad (3)$$

The term $P[L_{10-\min} > L_T | v, h_s]$ is the short-term probability and the joint probability density function is given by $f_{v,Hs}(v, h_s)$. Figure 1 depicts the schematic representation of 5MW wind turbine supported on TLP type floating wind turbine.

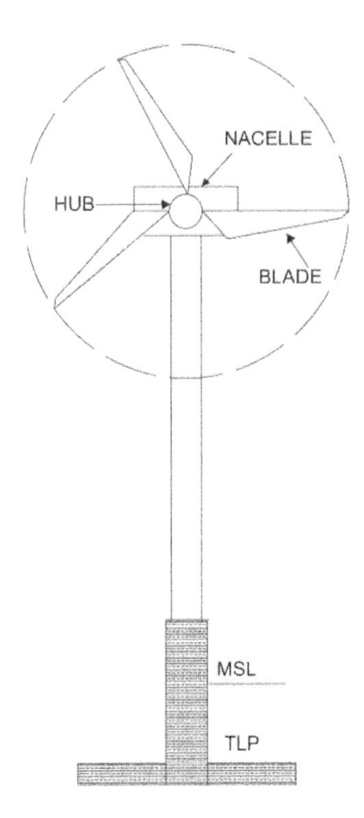


Figure 1. Schematic represenation of 5MW wind turbine supported on TLP type floating wind turbine.

It is evident from Eq. (3) that there is no coupling between the environmental random variable and the site-specific conditions. The environmental contour method is used to calculate the turbine T-year return period which proves to be more computationally simplified than other methods. Due to the fact that the full-long term analysis is time consuming and complex in nature, the First Order Reliability Method (FORM) and the Second Order Reliability Method (SORM) is studied. The reliability analysis on structures makes it helpful to obtain the probability of failure of structures based on finding the desired point by iteratively searching through the contour points for a particular target return period. The direct method of analysis are used, as well as methods involving the inverse reliability approach known as the Inverse First-Order Reliability Method (IFORM), is found to be efficient.

## 2.2 *Environmental contour method*

The Environmental Contour Method is a method by which the drawbacks of full long-term analysis can be avoided to obtain an approximate and efficient methodology. The method uses reliability approach where Rosenblatt transformation is used to transfer the environmental variables from the standard normal Uspace to the point related to the physical random variable which is based on inverse FORM method. The contour points in the environmental contour are obtained using the inverse reliability technique for a given return period to obtain the extreme responses of five different TLP configurations.

## 2.3 *Inverse first-order reliability method*

The methods for obtaining the probability of failure includes obtaining the combinations of all points that makes the limit state function $g(x) \leq 0$ by integrating the joint probability density function. The FORM method works on the principle of finding the design point on the failure surface in $U$ space, by making use of Rosenblatt transformation which transforms environmental variables to variables in standard normal $U$-space and then the probability of failure is determined by using the design point obtained in the failure surface. The design point is a point which is closest to the origin in $U$-space lies in the failure surface which has the highest probability of failure where the limit state function for the reliability analysis becomes negative.

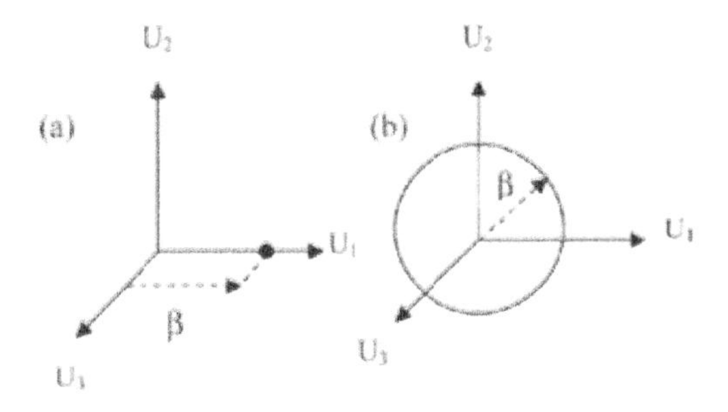


Figure 2. Illustration of 1D model and (b) 2D model (Karmakar et al., 2014).

The Inverse First-Order Reliability Method (IFORM) is based on the First-Order Reliability Method (FORM), which makes use of the reliability index $\beta$, which is used to obtain the points in the contour within the radius $\beta$ and the inverse Rosenblatt

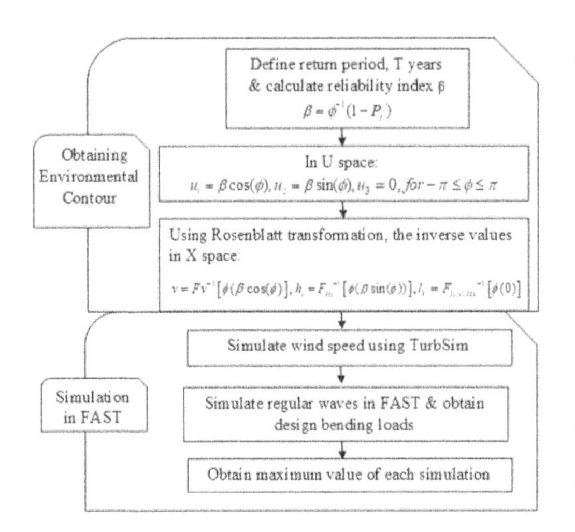


Figure 3. Methodology for the environmental contour method and simulation in FAST.

transformation is used to transform the variables to physical space. Figure 3 represents the procedure for Environmenal Contour Method and simuations in FAST.

### 2.3.1 *1-D environmental contour method*

In the present problem the joint probability density function of the two environmental variables, namely, the 10-min mean wind speed $V$ and the significant wave height $H_s$ are considered. Here, only the 10-min mean wind speed is modeled as random, while the wave height is considered at its median level. Knowing the reliability index $\beta$ associated with a particular return period and failure probability, the point corresponding to that failure probability in the standard normal space $U$ can be obtained by using the relation (Karmakar et al., 2014) given by

$$u_1 = \beta, u_2 = 0, u_3 = 0. \tag{3}$$

The point related to the design load in terms of the physical random variable may be obtained using the Rosenblatt transformation (Winterstein et al., 1993)

$$\phi(u_1) = F_v(v), \phi(u_2) = F_{Hs}(h_s), \ \phi(u_3) = F_{L_T}(l_T), \tag{4}$$

where $\phi$ is the standard normal cumulative distribution function and $F_v(v)$ is the standard normal cumulative distribution of mean wind speed $V$ which can be re-written in the form as

$$v = Fv^{-1}[\phi(\beta)], \ h_s = F_{Hs}{}^{-1}[\phi(0)], \ l_T = F_{l_T|v,Hs}{}^{-1}[\phi(0)]. \tag{5}$$

The reliability index $\beta$ used in the environmental contour method based on the target return period are considered for a particular exceedance probability, $P_f$ as 4.12 for the return periods of 20 years which is obtained from the relation

$$\beta = \phi^{-1}(1 - P_f), \tag{6}$$

$$\beta = \phi^{-1}\left(1 - \frac{Ts}{365 * 24 * Tr}\right), \tag{7}$$

where $Ts$ is the sea-state duration (10-min) and $Tr$ is the return period. The probability of exceedance in 10-min versus the mean wind speed is given for return period of 20 years is found to be $9.53 \times 10^{-7}$. Using 1D method, the mean wave heights corresponding to different return periods for various wind speeds is obtained.

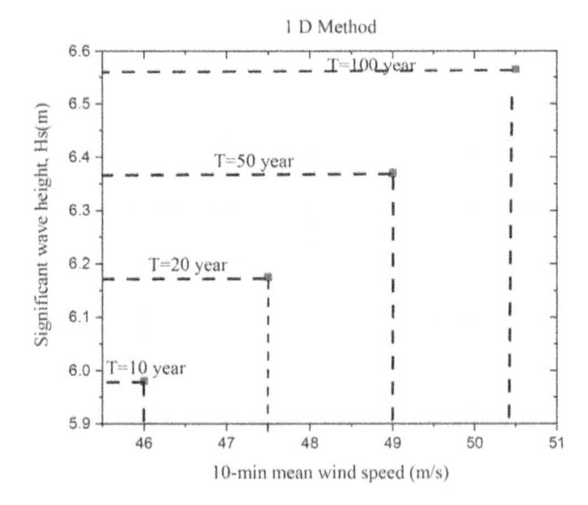


Figure 4. Mean wave height versus wind speed for different return periods.

### 2.3.2 *2-D method of environmental contour*

In 2-D Environmental Contour method, the environmental variables 10-min mean wind speed $V$ and wave height $H_s$ are considered random while the load variable $L_T$ is considered to be deterministic at its mean level. The probability of failure in $U$ space after determining reliability index $\beta$ is given by

$$u_1 = \beta\cos(\phi), \ u_2 = \beta\sin(\phi), \ u_3 = 0 \text{ for } -\pi \le \phi \le \pi. \tag{8}$$

The environmental contour in $U$ space can be represented by the points along the circle as illustrated in the Figure 1(b) with the radius value corresponding to the reliability index $\beta$ of a particular return period

$$u_1{}^2 + u_2{}^2 = \beta^2. \tag{9}$$

The transformations are carried out for the environmental variables from u-space to the physical space with the help of Rosenblatt transformation method

$$\phi(u_1) = F_v(v),\ \phi(u_2) = F_{H_s|V}(h_s),\ \phi(u_3) = F_{L_T|V,H_s}(l_T). \tag{10}$$

The standard normal cumulative distribution function of velocity $F_v(v)$ and the standard normal cumulative distribution function of wave height which is conditional on velocity $F_{H_s|V}(h_s)$, is rewritten in the form as

$$v = Fv^{-1}[\phi(\beta\cos\phi)],\ h_s = F_{Hs}{}^{-1}[\phi(\beta\sin\phi)] \text{ and } l_T = F_{l_T|v,Hs}{}^{-1}[\phi(0)]. \tag{11}$$

Using the above methodology and transformations, the environmental contours are developed based on 2D method for return period of 20-year. Each of the contour point shown in Figure 5 is characterized by a particular combination of mean wind speed and significant wave height.

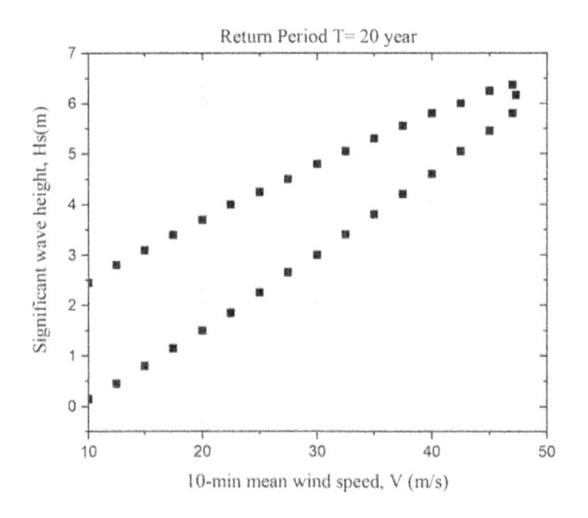


Figure 5. 2D Environmental Contour for 20-year return period.

Figure 6 shows the configurations of the five-TLP configurations and Table 1 indicates the dimensions for the TLP platforms taken from Bachynski et al. (2012). The configuration of TLP-1 with 4 legs has the highest displacement with dependence on center column for buoyancy. TLP-2 is a 3-legged platform characterized by 60% of displacement and 70% of initial tension of tendon as compared to first TLP and its 30% displacement is observed to have come from the pontoons. TLP-3, TLP-4 and TLP-5 are composed of 3 legs. TLP-3 has higher pretension per line compared to TLP-1. In case of TLP-4 and TLP-5, about 70% displacement is obtained from pontoons. TLP-4 is composed of 3 rectangular pontoons while TLP-5 has 3 round pontoons and also has twice the displacement and total pretension of TLP-5 (Bachynski et al.,2012).

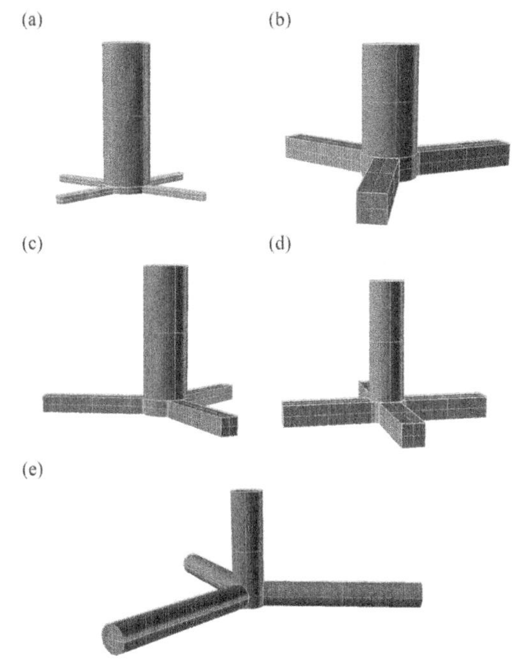


Figure 6. (a) TLP-1 (b) TLP-2 (c) TLP-3 (d) TLP-4 (e) TLP-5 configurations.

Table 1. Dimensions of various TLP configurations.

| | TLP-1 | TLP-2 | TLP-3 | TLP-4 | TLP-5 |
|---|---|---|---|---|---|
| Pontoons | Rect. | Rect. | Rect. | Rect. | Round |
| $D_1$ (m) | 18.0 | 14.0 | 14.0 | 6.5 | 6.5 |
| $D_2$ (m) | 18.0 | 14.0 | 14.0 | 10.0 | 6.5 |
| $h_1$ (m) | 52.6 | 40.0 | 26.0 | 33.0 | 23.0 |
| $h_2$ (m) | 2.4 | 5.0 | 6.0 | 6.0 | 5.0 |
| $r_p$ (m) | 27.0 | 32.0 | 28.0 | 25.0 | 32.5 |
| $d_p$ (m) | 2.4/2.4 | 5.0/5.0 | 6.0/6.0 | 6.0/6.0 | 5.0 |
| BF | 0.55 | 0.60 | 0.40 | 0.40 | 0.40 |
| $n_p$ | 4 | 3 | 3 | 4 | 3 |
| $z_s$ | -43.8 | -32.5 | -19.0 | -19.0 | -15.5 |
| $d_t$ (m) | 1.4 | 1.1 | 1.3 | 1.2 | 0.9 |
| $t_t$ (mm) | 46.2 | 36.3 | 42.9 | 39.6 | 29.7 |
| Steel mass (t) | 2322 | 1518 | 1293 | 859 | 505 |
| V(m3) | 11,866 | 7263 | 5655 | 4114 | 2320 |
| $T_t$(KN) | 6868 | 4963 | 8262 | 5556 | 3384 |

where $D_1$ and $D_2$ are the diameter of upper and lower TLPWT hull center column, $h_1$ and $h_2$ is the length of upper and lower TLPWT hull center column, $r_p$ is the pontoon radius measured from TLPWT hull centerline, $d_p$ is the height/width or diameter of pontoons, BF is ballast fraction, $n_p$ is the number of pontoons, $z_s$ is the vertical location of pontoons and $d_t$ is the tendon outer diameter.

## 3 RESULTS AND DISCUSSION

The reliability analysis of five-TLP type platforms is performed using environmental contour method. The 1-D and 2-D methods are used and contour plots are obtained for 20-year return period for analysis. The maximum and minimum values of tower base bending moments and tower base shear force values are obtained and the standard deviation of the time series data corresponding to the contour point for maximum value of tower-base bending moment is found. The design point corresponding to 20-year return period is observed at 10-min mean wind speed of 47.5 m/s with mean value of wave height of 6.17m.

Table 2. Maximum tower base forces and moments using 1D method.

| | TLP-1 | TLP-2 | TLP-3 | TLP-4 | TLP-5 |
|---|---|---|---|---|---|
| $M_x$ (KN-m) | 237000 | 124500 | 94010 | 84760 | 194300 |
| $M_y$ (KN-m) | 1009000 | 338500 | 301300 | 299300 | 878100 |
| $M_z$ (KN-m) | 74810 | 11800 | 10830 | 12580 | 32860 |
| $F_x$ (KN) | 36700 | 4286 | 3294 | 3258 | 13700 |
| $F_y$ (KN) | 5022 | 1217 | 1012 | 980.2 | 3698 |
| $F_z$ (KN) | 32750 | 7520 | 7521 | 7467 | 9239 |

Table 2 shows the results obtained for maximum amplitude of forces and moments based on 1D Environmental Contour Method for the 20-year return period for the five TLP configurations. It can be observed that TLP-1 with lower pontoon diameter has the highest fore-aft bending moment of 237 MN-m, side-to-side bending moment of 1009 MN-m, yaw moment of 74.81 MN-m. It is further observed that the side-to-side bending moment values are larger when compared to fore-aft and yaw moments. This is due to the fact that, as the reference wave direction has been taken in x-direction, the wave load in x-direction direction is dominating, further increasing the fore-aft shear force. The increase in forces in the fore-aft direction results in the dominance of side-to-side bending moment.

Table 3 depicts the results obtained for maximum amplitude of responses based on 2D Environmental Contour Method for 20-year return period. The TLP-1 is found to have maximum values of side-to-side bending moment of 237 MN-m at mean wind speed of 47.3 m/s with $H_s = 6.17$m followed by TLP-5, TLP-4, TLP-2 and TLP-3 respectively. It is observed that TLP-3 is least sensitive to extreme sea states which is due to its high pretension value even though its displacement value is lesser. It can be seen that the 3-legged TLP configuration which allows for easier installation, with pontoon radius ranging between 25m to 28 m, contributing to lesser tower base bending moment and shear force values which may be considered efficient to withstand the extreme design conditions. The minimum values as depicted in the results is helpful to determine the minimum values of the forces and moments that is essential for the wind turbine to restore to its original position when subjected to environmental loading. Further, the standard deviation of the time series data is obtained for mean wind speed and wave height that corresponds to the maximum values of tower base bending moment and shear force values for each of the contour point. The tower base bending loads and shear force values is observed higher for 2D model as compared to corresponding values in 1D model.

Table 3. Forces and moments at the tower base for 2D method.

| | | $M_x$ (KN-m) | $M_y$ (KN-m) | $M_z$ (KN-m) | $F_x$ (KN) | $F_y$ (KN) | $F_z$ (KN) |
|---|---|---|---|---|---|---|---|
| **TLP-1** | MAX | 237000 | 1009000 | 79440 | 36700 | 5669 | 32750 |
| | MIN | 3.155 | 4.02 | 0.0375 | 0.0423 | 0.00532 | 622.8 |
| | STDV | 12329.9 | 42166.3 | 3593.1 | 803.6 | 172.3 | 513.7 |
| **TLP-2** | MAX | 160500 | 632900 | 15550 | 9756 | 4406 | 8167 |
| | MIN | 0.5641 | 0.9007 | 0.0175 | 0.0211 | 0.0005 | 475.7 |
| | STDV | 4169.81 | 19459.1 | 642.16 | 296 | 91.31 | 126.8 |
| **TLP-3** | MAX | 136800 | 467900 | 16300 | 6995 | 3936 | 9074 |
| | MIN | 44.16 | 0.7066 | 0.0285 | 0.0174 | 0.4183 | 3755 |
| | STDV | 4952.54 | 35459.1 | 645.79 | 509.80 | 84.3 | 235.3 |
| **TLP-4** | MAX | 160700 | 499700 | 12100 | 7901 | 3994 | 9218 |
| | MIN | 103.7 | 46.24 | 0.0187 | 0.0004 | 0.0042 | 3462 |
| | STDV | 5243.13 | 25972.4 | 6839.6 | 367.4 | 94 | 82.5 |
| **TLP-5** | MAX | 542300 | 881600 | 37050 | 13720 | 9921 | 11120 |
| | MIN | 6.254 | 0.5346 | 0.0152 | 0.0458 | 0.0102 | 181.7 |
| | STDV | 2165.18 | 34366.9 | 1475.9 | 508.9 | 177.4 | 182.7 |

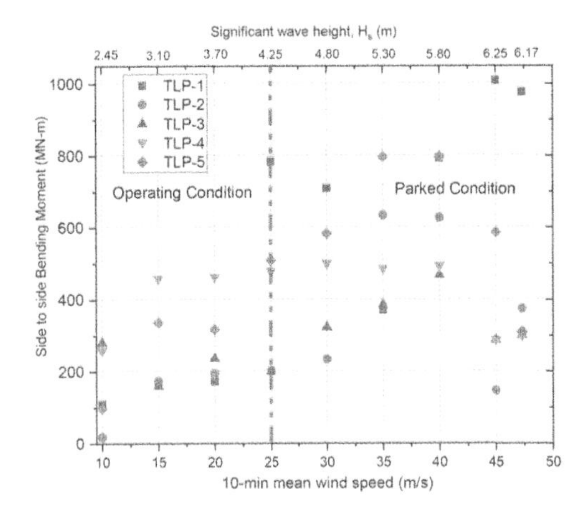


Figure 7. Maximum values of side-to-side bending moments at different mean wind speeds for various TLP configurations.

Figure 7 depicts the maximum values of side-to-side bending moments at different mean wind speeds for various TLP configurations considered in the present study. The corresponding values of significant wave height for the mean wind speed contour value is also depicted. In Figure 7, up to mean wind speed of 20m/s with $H_s = 3.7m$, TLP-4 is found to have higher fore-aft bending moment, while after cut-out wind speed (25 m/s), TLP-1 and TLP-4 shows higher values. It is seen that the peak values for TLP-2 and TLP-4 is at mean wind speed value of 40 m/s with $H_s = 5.8m$, 35 m/s with $H_s = 5.3m$ for TLP-3 and 47.3 m/s with $H_s = 6.17m$ for TLP-1. It is observed that the tower base bending moments in x-direction has lower values in the operating condition as the wind turbine is in working condition and consequent reduction in forces and the moments associated with it for higher values in the parked condition (after 25m/s) range. It may be due to the fact that there are no opposing forces or moments produced to counteract the extreme conditions of wind speed and wave height, in the parked state. In the parked condition, the TLP-1 is subjected to comparatively higher side-to-side bending moment value at each contour point which is due to its lower pontoon diameter. The side-to-side bending moment is observed to be reduced at the rated wind speed of 11.4 m/s as compared to the values after the cut-out wind speed of 25 m/s for the 5-MW wind turbine due to higher wind power absorption of the turbine at the rated wind speed. The comparative study with semi-submersible and Spar type floater geometry as in Karmakar et al. (2015), it is observed that the values of tower base moments for TLP-1, TLP-2 and TLP-5 are higher. An approximately 67%, 63% and 22 % increase in tower base bending moment values are observed for TLP-1, TLP-2 and TLP-5 respectively when compared with that of DeepCWind floater geometry, which may be related to the less stiff tendons leading to increased platform motions and the lesser wind power absorption of the wind turbine mounted on tension leg mooring when it is being subjected to higher wind and wave load combinations.

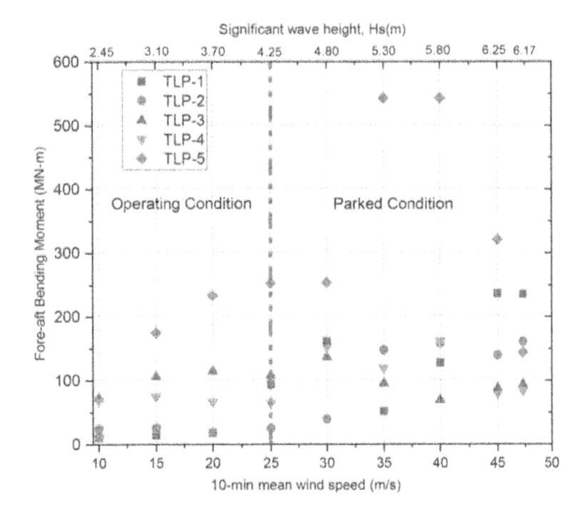


Figure 8. Maximum values of fore-aft bending moments at different mean wind speeds for various TLP configurations.

In Figure 8, the peak values of TLP-5 and TLP-4 is observed to occur at mean wind speed of 40 m/s with $H_s = 5.8m$ while for TLP-1, TLP-2 and TLP-3 the peak value occurs at 47.3 m/s with $H_s = 6.17m$.

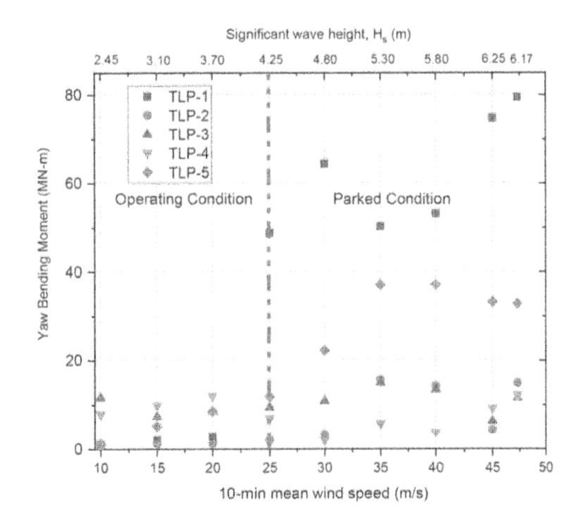


Figure 9. Maximum values of yaw moments at different mean wind speeds for various TLP configurations.

It is observed that TLP-5 is subjected to very large fore-aft bending moment at each contour point which may be due to its 3-legged round-shaped pontoon configuration having the least diameter and displacement as compared to other TLP configurations. Even though the least diameter made the structure least affected by wave action, the lower displacement characteristic made it more prone to extreme wave and wind condition.

In Figure 9, peak yaw moments are of the same range of 17 MN-m for TLP-2, TLP-3 and TLP-4 while around 80MN-m for TLP-1 and 37 MN-m for TLP-5. A drastic increase of values for higher mean wind speed in yaw moment values in case of TLP-1 and TLP-5 is noted, while for TLP-2, TLP-3 and TLP-4 has almost same range of values with increase in the mean wind speed. The yaw moment values for the five-TLP configurations are higher as compared with the values of semi-submersible floaters as given in Karmakar et al. (2014). The higher values of yaw moments for the five TLP configurations results in lower power absorption of the 5-MW wind turbine. The values obtained for the 20-year return period suggests that TLP-1 and TLP-5 configurations which are subjected to the higher bending loads may affect the design life of the platform. As the wind turbines are designed for pitch and yaw control, the turbine orients itself under extreme wind-wave conditions and thus the yaw moment values are much lesser than fore-aft and side-to-side moments.

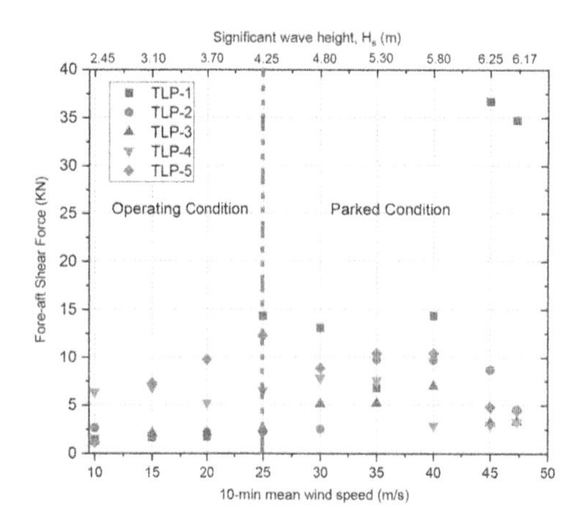


Figure 10. Maximum values of fore-aft shear force at different mean wind speeds for various TLP configurations.

The side-to-side bending moments in the wind turbine is associated with the fore-aft shear force. It can be observed from Figure 10, that the fore-aft shear force value shows maximum value at mean wind speed of 45 m/s with $H_s = 6.25m$ for TLP-1, 25 m/s with $H_s = 4.25m$ for TLP-5, 30 m/s for TLP-4, 40 m/s with $H_s = 5.8m$ for TLP-3, 35 m/s with $H_s = 5.3m$ for TLP-2.

In Figure 11, the side-to-side shear force has maximum values for TLP-1 and and TLP-2 at 47.3 m/s with $H_s = 6.17m$, TLP-5 has its maximum at 40 m/s with $H_s = 5.8m$ while TLP-3 and TLP-4 has its peak in the operating range, within 25 m/s. The wave force acting in x-direction is higher as it has been taken as the reference direction and thus the associated forces which is the fore-aft shear force values, are found to be higher compared with the side-to-side and axial forces acting on the wind turbine. As the wind loading is increased, the responses including the shear force values and associated bending moments are also increased as the wind power absorption by the wind turbine is reduced. In general, it is observed that the design point where the extreme force and moment responses occur are observed at contour value points which are closer to the highest wind-wave contour point.

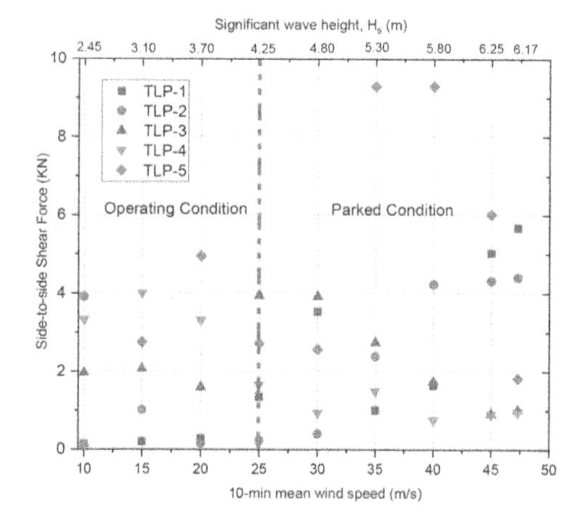


Figure 11. Maximum values of side-to-side shear force at different mean wind speeds for various TLP configurations.

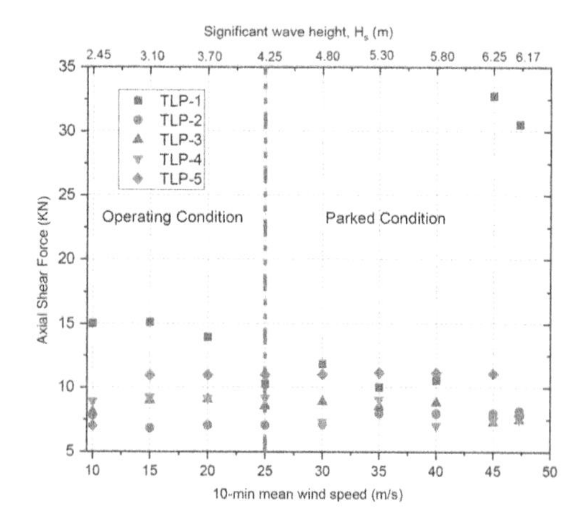


Figure 12. Maximum values of axial force at different mean wind speeds for various TLP configurations.

Figure 12 depicts the values of axial forces for the five different TLP configurations. It is observed that all the values lie around a range of 7KN to 10 KN with increase in mean wind speed, except for the case of TLP-1 which shows a drastic increase in peak value at 45 m/s with $H_s = 6.25m$. The wind turbines are designed for various controls including the yaw and pitch control, thus limiting the values of the axial force developed at the base of the turbine tower.

## 4 CONCLUSIONS

In the present study, the reliability analysis using Environmnental Contour Method is analysed for five different TLP type floating wind turbine configurations. The environmental contour points corresponding to 20-year return period is obtained and analysed. The contour points are further used to obtain the responses of the 5-MW wind turbine using aero-sevo-hydro-elastic simulations. The maximum, minimum and standard deviation of the maximum value for the time series data is obtained for the five different TLP configurations. Based on the study ferformed using Environmental Contour method, the conclusion drawn are as follows:

- The TLP-3 and TLP-4 having higher pontoon diameter/height and higher pretension values shows lower tower base bending moment as compared with other TLPWTs.
- The tower base bending moments and shear forces in parked condition is observed to have higher values than in the operating range which is due to the state of the wind turbine which is not opposing the huge bending loads caused due to the extreme wind and wave conditions.
- The tower base shear force and associated bending moments at the rated wind speed of 11.4 m/s is lesser as compared to the values after the cut-out wind speed of 25 m/s due to the higher wind power absorption at the rated wind speed.
- The design point where the extreme force and moment responses occur are observed at contour value points which are closer to the highest wind-wave contour point.
- The TLP-1 type floating wind turbine with the largest displacement has highest values of fore-aft and side-to-side shear forces and bending moments when compared with all other TLP configurations.
- The TLP-5 wind turbine is subjected to very large fore-aft bending moment at each contour point as the lower displacement characteristic made it more prone to extreme wave and wind condition even though the least diameter made the structure less affected by wave action.
- The 3-legged TLP configurations which allows for easier installation, with the pontoon radius ranging between 25m to 28m, contributing to lesser tower base bending moment and shear force values which may be considered efficient to withstand the extreme design conditions.
- The tower base bending loads and shear force values is observed higher for 2D environmental contour model as compared to 1D model.

The present study iis helpful in determining the wind-wave combination that leads to extreme responses, the maximum values of responses that the wind turbine may be subjected to in exrtreme conditions and thus the long-term design loads for the wind turbine is predicted without excessive computational efforts as in full long-term analysis.

## ACKNOWLEDGEMENTS

The authors acknowledge Science and Engineering Research Board (SERB), Department of Science & Technology (DST), Government of India for supporting financially under the research grant no. CRG/2018/004184, DST for India-Portugal Bilateral Scientific Technological Cooperation, Project grant no. DST/INT/Portugal/P-13/2017 and Vision Group for Science and Technology (VGST), Government of Karnataka, Project grant no VGST/RGS-F/GRD-901/2019-20/2020-21/198.

## REFERENCES

Bachynski, E.E. & Moan, T. (2012). Design considerations for tension leg platform wind turbines. Marine Structure. 29, 89–114.

Bagbanci, H., Karmakar, D. & Guedes Soares, C. (2015). Comparison of spar-type and semi-submersible type floaters concepts of offshore wind turbines using long term analysis, Journal of Offshore Mechanics and Arctic Engineering, Vol. 137 (6),061601-1-10.

Christensen, C.F. & Nielsen, T.A. (2000). Return period for environmental loads-combination of wind and wave loads for offshore wind turbines, *EFP 99*.

Karmakar, D. & Guedes Soares, C. (2014). Reliability based design loads of an offshore semi-submersible floating wind turbine, Developments in Maritime Transportation and Exploitation of Sea Resource - C. Guedes Soares & López Peña (Eds), Taylor & Francis Group, London, Vol-2, pp. 919–926.

Karmakar, D. & Guedes Soares, C. (2015). Extreme response prediction of offshore wind turbine using inverse reliability technique, Proceedings of 34th International Conference on Ocean, Offshore and Arctic Engineering, OMAE2015-42072.

Karmakar, D., Bagbanci, H. & Guedes Soares, C. (2016). Long-term extreme load prediction of spar and semisubmersible floating wind turbines using the Environmental Contour Method, Journal of Offshore Mechanics and Arctic Engineering, Vol. 138, 021601-1-10.

Konispoliatis, D.N., Katsaounis, G.M., Manolas, D.I., Soukissian, T.H., Polyzos, S., Mazarakos, T.P., Voutsinas, S.G. & Mavrakos, S.A., (2021). REFOS: A renewable energy multi-purpose floating offshore system. Energies, 14, 3126.

Liu, J., Thomas, E., Goyal, A. & Manuel, L. (2019). Design loads for a large wind turbine supported by a semi-submersible floating platform, Renewable Energy, 960–1481

Raed, K., Teixeira, A.P. & Guedes Soares, C. (2019). Assessment of long-term extreme response of a floating support structure using the environmental contour method. Advances in Renewable Energies Offshore, Taylor & Francis Group, London.

Saranyasoontorn, K. & Manuel, L. (2004). Efficient models for wind turbine extreme loads using inverse reliability, Journal of Wind. Engineering and Industrial Aerodynamics., 92, 789–804.

Saranyasoontorn, K. & Manuel, L. (2005). On assessing the accuracy of offshore wind turbine reliability-based design loads from environmental contour method. Journal of Offshore and polar Engineering, 15(2), 1–9.

Sultania, A. & Manuel, L. (2018). Reliability analysis for a spar-supported floating offshore wind turbine. Wind Engineering, Vol. 42(1) 51–65.

Winterstein, S.R., Ude, T.C., Cornell, C.A., Bjerager, P. & Haver, S. (1993). Environmental contours for extreme response: inverse form with omission factors, Proceedings of the International Conference on Structural Safety and Reliability, August 9–13, Innsbruck

Winterstein, S.R. & Engebretsen, K. (1998). Reliability based prediction of design loads and responses for floating ocean structures, Proceedings of 17$^{th}$ International Conference on Offshore Mechanics and Arctic Engineering, Lisbon, Portugal.

*Trends in Maritime Technology and Engineering – Guedes Soares & Santos (eds)*
*© 2022 copyright the Author(s), ISBN 978-1-032-33583-4*

# The effect of high-altitude wind forecasting models on power generation, structural loads, and wind farm optimisation

E. Uzunoglu, C. Bernardo & C. Guedes Soares
*Centre for Marine Technology and Ocean Engineering (CENTEC), Instituto Superior Técnico, Universidade de Lisboa, Lisbon, Portugal*

ABSTRACT: This paper provides a review of wind forecasting models and their applicability to high-altitude wind power generators such as large wind turbines and airborne devices. These devices are affected by factors such as turbulence and wind shear, which are majors factors characterizing wind flow in altitude. Therefore, as the altitude increases, the power law normally used for the vertical wind profile becomes invalid. Accordingly, accurate forecasting techniques that consider nonlinearity in wind properties are significant. As numeric models for such purposes are computationally demanding, data-driven methods are starting to gain prominence to overcome the shortcomings. Implementing these new methods will affect the estimation of power output, structural response, and farm planning for both wind turbines and airborne wind energy systems.

## 1 INTRODUCTION

As technologies in wind energy develop, research shifts towards larger turbines and airborne wind energy systems. The commonality between these two wind power systems is that they both operate in higher altitudes where the logarithmic law (i.e., power profile) does not apply. Additionally, recent studies show that the inherent fluctuation of wind power and the instability of output power can influence the performance of wind power systems. The variability of wind speed and the generated energy limits its utilisation to some extent. Therefore, it is important to increase the accuracy of wind energy forecasting approaches and reduce the reliance on simplifications.

Current wind forecasting approaches may be divided into two primary groups: probabilistic and deterministic forecasting (Liu *et al.* 2019). Probabilistic models eventually lead to a prediction of confidence levels of wind energy forecasts. Deterministic forecasting can be further divided into four categories: physical, statistical, intelligent, and hybrid. Physical models excel at forecasting using variables such as altitude and wind speed and can provide solutions detailing many variables. However, their use is difficult due to high computational demands.

Conversely, statistical and intelligent methods suggest forecasting wind as a stochastic process. They are data-driven models that leverage historical wind speed/power measurement and they can represent them in spectral form. They generally do well in simple time series prediction. However, they cannot process the nonlinearity present in realistic wind behaviour.

For the reasons given above and recent developments in machine learning and artificial intelligence, intelligent forecasting methods have garnered interest. Numerous comparison studies have shown that intelligent models have excellent forecasting performance. Compared with physical and statistical approaches, the intelligent forecasting model can achieve higher forecasting steps and better accuracy when coping with problems that cannot be analytically defined. These data-driven methods will also need to consider a method for removing noise from the measurements that they rely on (Wang *et al.* 2017).

Besides the single models listed above, hybrid models are an option that aims to bring advantages from the statistical methods and the intelligent methods. The definition of hybrid models is relatively vague. Generally, hybrid refers to the combination of two different algorithms or methods. The hybrid models combine the merits and characteristics of different methods, whose overall performances are better than single models in general (Ogliari *et al.* 2021). Thus, they are prime candidates to represent cases where nonlinearity and turbulence gain importance. Examples of such problems would include aerodynamics of large turbines, airborne wind energy, and effects of turbulence on structural components.

Numerous forecasting approaches have been used to predict wind speeds while research continues in this direction (Jiang *et al.* 2021) to develop a combined system and integrate several sub-models to provide

DOI: 10.1201/9781003320289-34

accurate point and interval forecasting performance. Despite the advances, there seems to be no accurate universal representation of wind behaviour valid for all scenarios and terrains. Developing a correct representation and realistic reflection of wind effects will improve turbine design, energy estimations, and wind farm location selection. This work aims to give a review of the topics and current practice.

## 2 WIND FORECASTING MODELS

### 2.1 *Power production*

The scarcity of available measurements and the assumptions of atmospheric stability and surface roughness result in limited wind forecasting models for cases such as airborne or offshore wind energy structures. The standard formulations only account for dependence on wind speed and include a neutral atmospheric stability assumption. However, the turbulence intensity (i.e., the ratio of standard deviation and the mean value of wind speed) also depends on the vertical atmospheric stability and the surface roughness. In other words, wind shear calculations need to account for additional factors to deliver realistic data as the accuracy of the applied turbulence intensities is affected twice by these assumptions.

Atmospheric stability can be characterized as stable, neutral, or unstable based on the "tendency of air particles to move vertically" relative to the temperature of their surroundings. In stable conditions, the air particles are cooler than the surrounding air, causing them to sink or remain where they are. This stratification leads to less mixing and typically results in lower turbulence intensities. In unstable conditions, the air particles are warmer than the surrounding air, which causes them to rise and leads to more vertical mixing. Hence, this is called "buoyancy generated" turbulence.

Studies show that a turbine can output different power between in-wake and out-of-wake scenarios (Zhang *et al.* 2014) despite operating at the same wind speed. The power production generally is higher in in-wake scenarios rather than out-of-wake scenarios, and it is sensitive to turbulence intensity even when the wind speed is higher than the turbine rated speed. A recent study that relies on the two years of observation at a location on the Dutch North Sea concluded that both surface roughness and atmospheric stability affect turbulence intensities (Caires *et al.* 2019). These findings are in line with the reports of (Kim *et al.* 2021) showing that the energy production of the Alpha Ventus wind farm in Germany, in Figure 1 (Wikipedia 2022), changed up to 20% depending on the atmospheric condition and the turbulence level despite being in the same wind speed bin of 10 m/s. Similarly, the wind turbine power in the offshore wind farm Egmond aan Zee (OWEZ) and North Hoyle was higher by approximately 10 – 20% in the strongly unstable atmospheric condition than in other atmospheric conditions.

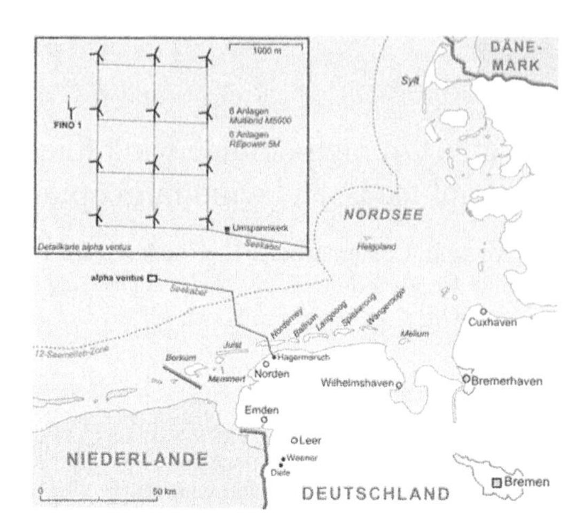


Figure 1. Location of the Alpha Ventus wind farm (Wikipedia creative commons license).

Hence, wind forecasting models included in the design codes need to be updated along with these findings to obtain realistic wind profiles. The IEC has also acknowledged that atmospheric stability characteristics of the installation site will need to be considered (Alblas *et al.* 2014). This case is more prominent given that turbines are getting larger and start operating at increasing hub heights. It would be expected that larger turbulent fluctuations will cause higher fatigue loads under unstable conditions, typically more dominant at offshore sites. However, stability conditions are largely dependent on the site location. Accordingly, it gains importance to expand the models to take such factors into account in wind forecasting models in the design of components and the selection of wind farm locations.

### 2.2 *Component level design*

While the International Electrotechnical Commission (IEC) standards recommend two turbulence models: the IEC Kaimal Spectra & Exponential Coherence Model and the Mann Spectral Tensor Model, neither of which consider the effect of atmospheric stability. However, recent studies examine alternatives such as using unstable spectra (e.g., Højstrup's 1981 Model) to show that high altitude and turbulence have a significant effect on both the tower and the blades (Knight & Obhrai 2019). The Højstrup model considers the effect of boundary layer height and buoyancy generated turbulence on the velocity spectra under unstable conditions, making it a candidate for evaluating effects on floating wind turbines.

The selection of the wind turbulence model can result in significant differences. Below rated speeds, the results show a high dependence on the turbulence class. However, as the wind speed increases and the wind turbine achieves the rated speed, the impact of the turbulence intensity is drastically reduced. The results also were apparent on mooring loads (Somoano *et al.* 2021). Aero-hydro-servo-elastic coupled simulation on a semisubmersible floating wind turbine showed wind shear to have limited global responses. However, its influence on individual blades was considerable. Comparatively, the turbine was found to be quite sensitive to turbulence intensity. In a wind field with high turbulence intensity, the platform motions become more violent, and the structural loads are increased substantially (Li *et al.* 2019).

On bottom fixed turbines, the effect of atmospheric stability on the loads and their relationship with fatigue loads has shown that atmospheric stability affects both the blade root bending and the tower base moments, which were the highest under unstable conditions. Similarly, for very unstable conditions, fatigue loads for the tower top torsion were 47% larger than neutral conditions for a SPAR platform and 30.4% larger for a semisubmersible (Knight and Obhrai 2019). In addition, (Kisvari *et al.* 2021) showed in a model that involved deep learning neural networks that even factors such as generator and gearbox temperature may affect the wind behaviour.

The case is similar to an onshore case described by (Liu & Stevens 2021), where the authors study turbulence caused by placing a wind farm behind a hill. Overall, the increased turbulence intensity by the hill results in increased velocity fluctuations along the blades. These stronger velocity fluctuations at the blades lead to higher fluctuations in the lift and drag forces experienced by the blades. For the neutral boundary layer case, the power fluctuations increased by 60%, while the increase in the thrust fluctuations is 75% due to the hill induced turbulence. The increased normalised lift and drag force fluctuations to which the blades are exposed suggested that the turbines experience higher fatigue loading. It was shown that the increase in the power fluctuations might not entirely reflect the increase in the local forces experienced by the blades. Any wear on the blades causes structural issues. Also, (Cappugi *et al.* 2021) shows that the annual energy loss from blade edge erosion will vary between 0.3 – 4 per cent as the edge erosion changes the flow on the blades estimated by a machine learning algorithm. Figure 2 (Cortés *et al.* 2017) shows an example of blade edge erosion over the lifespan of a wind turbine Accordingly, accurate modelling of atmospheric instabilities plays a vital role in better designs and maintenance strategies.

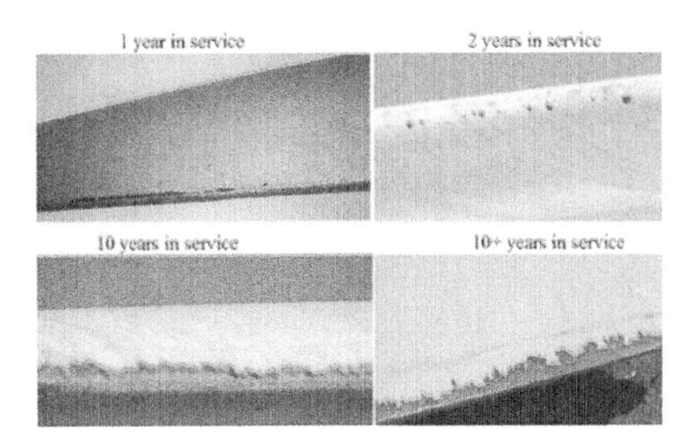


Figure 2. Blade edge erosion over the lifespan of a wind turbine (Cortés et al., 2017).

## 3 WIND RESOURCE ASSESSMENT

Numerical Weather Prediction (NWP) models have a wide range of applications in wind energy, from the assessment of wind resources necessary to the planning and implementation of a wind farm to the wind forecast fundamental to the operation and optimisation of the energy production. It is a method of weather forecasting that employs a set of equations that describe the flow of fluids. These equations are translated into computer code and use governing equations, numerical methods, parameterizations of other physical processes and combined with initial and boundary conditions before being run over a domain (geographic area). A set of parameters that can be considered in NWP is presented in Figure 3 (National Oceanic Atmospheric Administration 2022).

The weather research and forecast (WRF) model can be applied in the wind energy sector on different scales and purposes. It is a next-generation mesoscale numerical weather prediction system designed for atmospheric research and operational forecasting applications. It features two dynamical cores, a data assimilation system, and a software architecture supporting parallel computation and system extensibility. The model serves a wide range of meteorological applications across scales from tens of meters to thousands of kilometres. It results from the collaboration of several organisations to build a next-generation mesoscale forecast model and data assimilation (see Skamarock *et al.* (2021) for more details).

To assess the wind energy resources of one site, wind data for a minimum of one year is necessary, although a longer period (10 years) would increase the accuracy and allow the description of the long-term variability (Salvação et al. 2014). This data is traditionally measured locally; however, data from weather stations have several limitations: the high cost of installation and maintenance, the coarse resolution, and the frequent missing data.

Figure 3. Physical processes and variables that are typically parameterized for Numerical Weather Prediction (NWP), as they cannot be explicitly predicted in full detail in model forecast equations (National Oceanic Atmospheric Administration, 2022).

Besides this, most of the weather stations available are designed for meteorological purposes, and the wind is recorded at standard levels (10 meters). In contrast, wind energy applications require data at the turbine hub height, which ordinarily surpasses 80m (deCastro *et al.* 2019). NWP models have been used to overcome these constraints as they are less expensive, as the main cost is the processing computers, they can be run at very high resolutions, and they produce gridded data available at several vertical layers of the atmosphere. Besides this, depending on the computational power, they can provide a long period of data in a short time (Al-Yahyai *et al.* 2010).

The WRF is an NWP model that can be applied to wind resources assessment (Salvação & Guedes Soares 2018), showing a good performance in both offshore and onshore locations (Giannakopoulou & Nhili 2014, Nunalee & Basu 2014) . NWP models are also a valuable tool in stimulating the interaction between the wind farms and the Atmospheric boundary layer (Fischereit *et al.* 2022), as they can capture their impacts on the wind field, turbulence kinetic energy (TKE), temperature, humidity, clouds, and other meteorological or atmospheric parameters (Siedersleben *et al.* 2018).

The wind that flows through the wind farm decelerates as the kinetic energy is extracted from the mean flow to produce electrical energy, and the turbulence increases, due to the turbulent mixing of the turbine blades and the wind shear produced by the wake, which is generated within the layer of the turbine blades (Fitch *et al.* 2013). The WRF model can be applied to study this interaction, as it incorporates a wind farm parametrisation (WFP) which was developed by Fitch *et al.* (2012) and has been widely

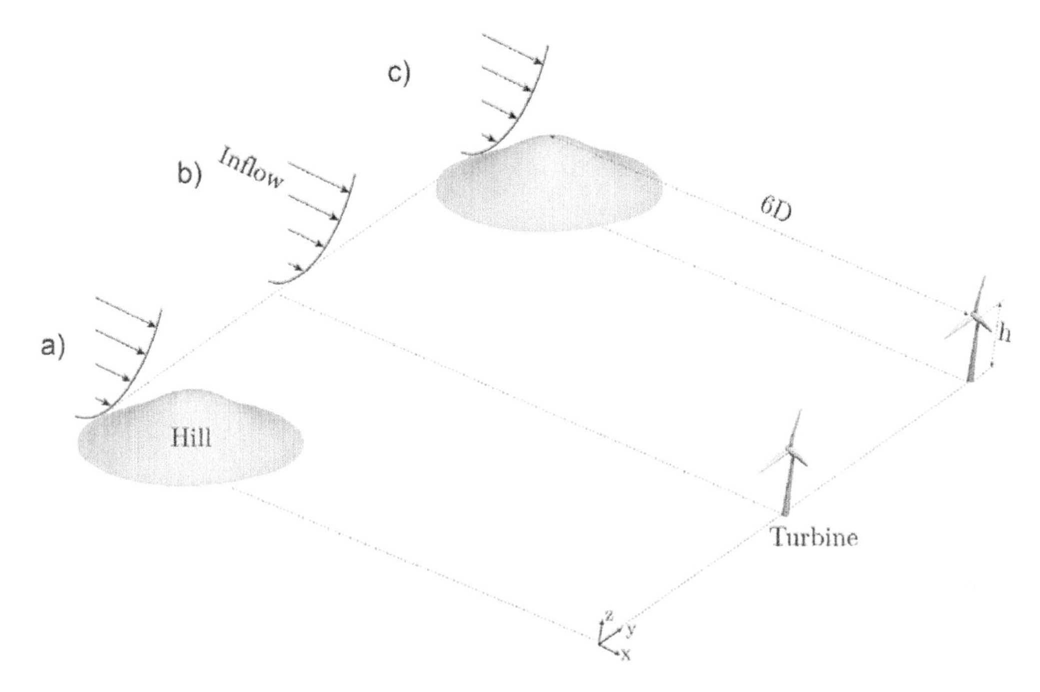


Figure 4. The farm optimization study in Liu and Stevens (2021), where the effect of atmospheric stability on power generation is studied. (a) a stand-alone hill and (b) a stand-alone turbine, and (c) the case in which the turbine is placed behind the hill. For each configuration, the authors considered a stable, neutral, and convective boundary layer case.

used to assess the impacts of wind farms on different spatial scales and different atmospheric stability regimes (Jiménez *et al.* 2015, Lee & Lundquist 2017, Xia *et al.* 2019).

The WFP represent the wind turbines as a turbulence source and a momentum sink in the vertical levels of the turbine rotor disk. The kinetic energy is extracted from the mean flow and added as (TKE) at the vertical levels of the turbine, depending on the thrust and power coefficients. The thrust coefficient describes the fraction of energy extracted from the mean flow, and the power coefficient is the fraction of energy converted into electrical energy, depending on the wind speed and wind turbine type. The difference between the thrust and power coefficients describes the fraction of energy converted into TKE. It neglects electrical and mechanical losses and assumes that all non-productive drag is converted into TKE (Siedersleben *et al.* 2020).

## 4 FARM OPTIMISATION AND SITE SELECTION PROCESSES

The previous sections discussed that recent studies show that IEC's commonly applied wind forecasting models do not capture wind instability and nonlinearity. Additionally, factors such as surface roughness on wind velocities at different altitudes need to be considered to accurately predict loads and energy production. It then becomes a natural question for site selection for wind turbines that consider all these factors for the efficiency of the farm. Accurate wind power estimation and its effect on components will eventually decrease economic uncertainties related to farm design and power production.

An example of placing a wind farm behind a hill is presented by (Liu & Stevens 2021). In the study, the distance between the hill and the turbine is six times the turbine diameter, and the hill height is equal to the hub height. The scenario shows that the hill wake reduces the power production of the downstream turbine by 35% for the convective boundary layer case under consideration. However, the wind turbine power production increases about 24% for the stable boundary layer case. It is shown that the hill wake also significantly increases the wind turbine power fluctuations, even when the turbine is not directly located in the hill wake, which is the case for the neutral boundary layer case. This study exemplifies the significance of the variables beyond the wind speed regarding power production.

Similarly, a 2021 study considers 88 existing wind sites and tries to formulate the effect of turbulence on wind turbines' power production and lifespan (Asadi & Pourhossein 2021). The researchers used an equivalent wind power density function, and the results revealed the significant effect of

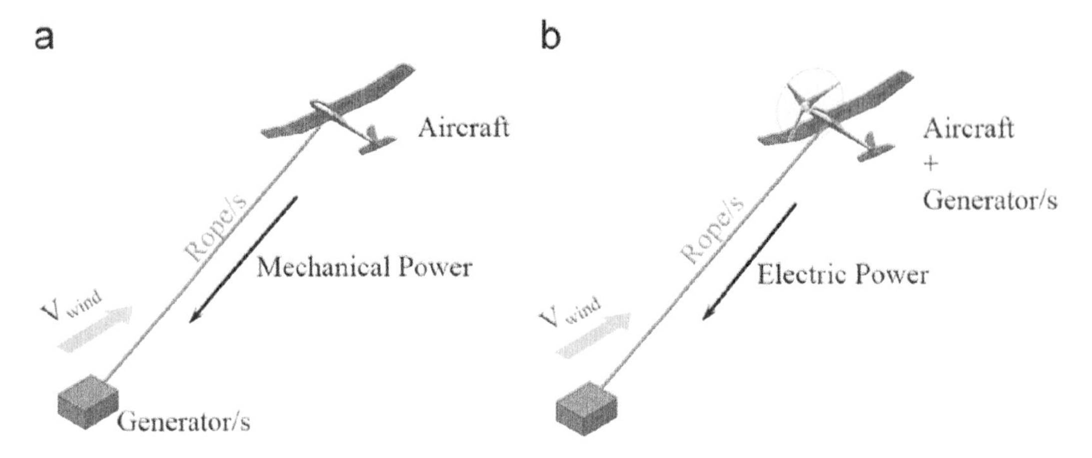


Figure 5. AWESs. Example of Ground-Gen (a) and Fly-Gen (b) AWESs (Cherubini *et al.* 2015).

turbulence on the site selection process. Some sites changed their rank by eight grades when turbulence intensity was considered in the selection process. Sensitivity analysis showed that, on average, the relative importance of turbulence intensity over wind power density could be approximated as 1/5, clarifying that it must be a factor to include in the calculations. When this factor is excluded (i.e., zero turbulence wind power), the annual energy output estimation changed by 50% (Bardal & Sætran 2017).

Additionally, the turbulence needs to be considered non-uniform for a correct representation of power output (Liu *et al.* 2021) and include the changes due to diurnal cycles. In (Radünz *et al.* 2021), the authors show that the time of the day can invert the power output between the front and back rows of a farm. The change can be up to 30%. Similar findings on multi-criteria decision making are reported by (Liu *et al.* 2020), where the authors stress the importance of correct forecasting techniques in the multi-criteria selection of wind farms. These forecasting techniques based on data-driven solutions also apply to offshore cases where the scarcity of measurements become an additional limiting factor. However, it is possible to obtain data onshore and then extend its use to the offshore case (Elshafei *et al.* 2021).

## 5 COMPARISON OF AIRBORNE WIND ENERGY AND WIND TURBINES

Airborne Wind Energy (AWE) is a new type of wind-harvesting energy technology that produces electricity using tethered wings that fly in a crosswind fashion. AWEs are generally made of two main components: a ground system and at least one aircraft that is mechanically connected (in some cases also electrically connected) by ropes (often referred to as tethers). Among the different AWES concepts, it is possible to classify Ground-Gen systems in which the conversion of mechanical energy into electrical energy takes place on the ground and Fly-Gen systems in which such conversion is done on the aircraft (Cherubini *et al.* 2015).

Two primary technologies are also referred to as pumping-mode and drag-mode systems. In the pumping-mode technology, the tethered wing generates power by reeling a tether from a ground-based winch connected to an electric generator. When reaching the maximum tether length, the wing is reeled back using a fraction of the energy generated during the reel-out phase. Instead, drag-mode systems generate power via turbines mounted on the tethered wings, transmitting the produced electricity via the tether to the ground station.

Comparing wind turbines and AWE regarding the exposure wind forces show that the wind field needs the relevant wind field is limited to around the height of the nacelle, i.e., about 100 m 180 m, depending on the type of turbine. AWE systems aim for winds at higher altitudes, typically up to 500 m, and tap into a larger vertical wind field that ranges from 50m to 500m during operation. This height invalidates the logarithmic law (i.e., power profile) assumption used for wind turbines (Malz, Verendel, *et al.* 2020). Hence, for detailed AWE analyses, such as an estimate of the annual power generation, it becomes necessary to use real wind-speed measurements at a specific location that may not be available.

Additionally, (Malz, Hedenus, *et al.* 2020) shows that the output can be similar to that of a wind turbine when it comes to airborne wind energy. However, it is also stated that, along with the wing design, the wind shear conditions at the installation site have the most significant impact on annual power production and the temporal distribution of the generation. The study states that wind shear is prominent between 200 – 1000 meters above sea level, especially at night-time (Bechtle *et al.* 2019). Hence, a forecasting methodology that can extrapolate existing data becomes a requirement for estimating the power generation of airborne turbines.

## 6 CONCLUSIONS

The assessment of wind energy output requires a correct assessment of the wind resource. In this regard, the nonlinearity of the wind seems to present the biggest challenge. The statistical approaches used by older models do not capture these. Therefore, data-driven methods will eventually replace these methods. The newer approaches will need to consider that wind energy systems need accurate representations of wind characteristics at higher altitudes.

Turbulence intensity, surface roughness, and wakes are evident factors affecting the amount of wind energy extracted. However, other models go further and include factors such as the gearbox temperature of a wind turbine. It may thus be necessary to identify the components that lead to a viable solution. These factors will have to be considered as a part of multi-criteria decision-making methods for farm selection.

Another issue stems from invalidating the power law in the calculation of wind shear. The behaviour of the wind changes with the altitude, and a universal solution that can apply to wind energy extraction devices of different altitudes (i.e., for both the wind turbines and airborne devices) does not seem to exist.

Structural loads on the blades and the tower base is reported to differ significantly with the calculated wind characteristics. Identification of fatigue loads at the component level is one of the topics to address. These studies may also be necessary for airborne wind energy systems, as the wind loads will be altered at high altitudes, causing different fatigue characteristics.

A final concern is the integration of data-driven methods into existing engineering software. Especially for wind turbines, the statistical methods are implemented into software such as NREL's TurbSim and FAST. As the methods to calculate the wind profile change, they must adapt to the new data. This integration will also be the case in calculations for farms and airborne wind energy.

## ACKNOWLEDGEMENTS

This work contributes to the Strategic Research Plan of the Centre for Marine Technology and Ocean Engineering (CENTEC), which is financed by the Portuguese Foundation for Science and Technology (Fundação para a Ciência e Tecnologia - FCT) under contract UIDB/UIDP/00134/2020.

## REFERENCES

Al-Yahyai, S., Charabi, Y., and Gastli, A., 2010. Review of the use of Numerical Weather Prediction (NWP) Models for wind energy assessment. *Renewable and Sustainable Energy Reviews*, 14 (9), 3192–3198.

Alblas, L., Bierbooms, W., and Veldkamp, D., 2014. Power output of offshore wind farms in relation to atmospheric stability. *Journal of Physics: Conference Series*, 555 (1).

Asadi, M. and Pourhossein, K., 2021. Wind farm site selection considering turbulence intensity. *Energy*, 236, 121480.

Bardal, L.M. and Sætran, L.R., 2017. Influence of turbulence intensity on wind turbine power curves. *Energy Procedia*, 137, 553–558.

Bechtle, P., Schelbergen, M., Schmehl, R., Zillmann, U., and Watson, S., 2019. Airborne wind energy resource analysis. *Renewable Energy*, 141, 1103–1116.

Caires, S., Schouten, J.J., Lønseth, L., Neshaug, V., Pathirana, I., and Storas, O., 2019. Uncertainties in offshore wind turbulence intensity. *Journal of Physics: Conference Series*, 1356 (1).

Cappugi, L., Castorrini, A., Bonfiglioli, A., Minisci, E., and Campobasso, M.S., 2021. Machine learning-enabled prediction of wind turbine energy yield losses due to general blade leading edge erosion. *Energy Conversion and Management*, 245 (July), 114567.

Cherubini, A., Papini, A., Vertechy, R., and Fontana, M., 2015. Airborne Wind Energy Systems: A review of the technologies. *Renewable and Sustainable Energy Reviews*, 51, 1461–1476.

Cortés, E., Sánchez, F., O'Carroll, A., Madramany, B., Hardiman, M., and Young, T.M., 2017. On the material characterisation of wind turbine blade coatings: The effect of interphase coating-laminate adhesion on rain erosion performance. *Materials*, 10 (10).

deCastro, M., Costoya, X., Salvador, S., Carvalho, D., Gómez-Gesteira, M., Sanz-Larruga, F.J., and Gimeno, L., 2019. An overview of offshore wind energy resources in Europe under present and future climate. *Annals of the New York Academy of Sciences*, 1436 (1), 70–97.

Elshafei, B., Peña, A., Xu, D., Ren, J., Badger, J., Pimenta, F.M., Giddings, D., and Mao, X., 2021. A hybrid solution for offshore wind resource assessment from limited onshore measurements. *Applied Energy*, 298 (February), 117245.

Fischereit, J., Brown, R., Larsén, X.G., Badger, J., and Hawkes, G., 2022. Review of Mesoscale Wind-Farm Parametrizations and Their Applications. *Boundary-Layer Meteorology*, 182 (2), 175–224.

Fitch, A.C., Olson, J.B., and Lundquist, J.K., 2013. Parameterization of Wind Farms in Climate Models.

Fitch, A.C., Olson, J.B., Lundquist, J.K., Dudhia, J., Gupta, A.K., Michalakes, J., and Barstad, I., 2012. Local and Mesoscale Impacts of Wind Farms as Parameterized in a Mesoscale NWP Model. *American Meteorological Society*, 140 (3017–3038).

Giannakopoulou, E.M. and Nhili, R., 2014. WRF model methodology for offshore wind energy applications. *Advances in Meteorology*, 2014.

Jiang, P., Liu, Z., Niu, X., and Zhang, L., 2021. A combined forecasting system based on statistical method, artificial neural networks, and deep learning methods for short-term wind speed forecasting. *Energy*, 217, 119361.

Jiménez, P.A., Navarro, J., Palomares, A.M., and Dudhia, J., 2015. Mesoscale modeling of offshore wind turbine wakes at the wind farm resolving scale: A composite-based analysis with the Weather Research and Forecasting model over Horns Rev. *Wind Energy*, 18 (3), 559–566.

Kim, D.Y., Kim, Y.H., and Kim, B.S., 2021. Changes in wind turbine power characteristics and annual energy production due to atmospheric stability, turbulence intensity, and wind shear. *Energy*, 214, 119051.

Kisvari, A., Lin, Z., and Liu, X., 2021. Wind power forecasting – A data-driven method along with gated recurrent neural network. *Renewable Energy*, 163, 1895–1909.

Knight, J.M. and Obhrai, C., 2019. The influence of an unstable turbulent wind spectrum on the loads and motions on floating Offshore Wind Turbines. *IOP Conference Series: Materials Science and Engineering*, 700 (1).

Lee, J.C.Y. and Lundquist, J.K., 2017. Observing and Simulating Wind-Turbine Wakes During the Evening Transition. *Boundary-Layer Meteorology*, 164 (3), 449–474.

Li, L., Liu, Y., Yuan, Z., and Gao, Y., 2019. Dynamic and structural performances of offshore floating wind turbines in turbulent wind flow. *Ocean Engineering*, 179 (February), 92–103.

Liu, H., Chen, C., Lv, X., Wu, X., and Liu, M., 2019. Deterministic wind energy forecasting: A review of intelligent predictors and auxiliary methods. *Energy Conversion and Management*, 195 (May), 328–345.

Liu, H., Li, Y., Duan, Z., and Chen, C., 2020. A review on multi-objective optimization framework in wind energy forecasting techniques and applications. *Energy Conversion and Management*, 224 (August).

Liu, L. and Stevens, R.J.A.M., 2021. Effects of atmospheric stability on the performance of a wind turbine located behind a three-dimensional hill. *Renewable Energy*, 175, 926–935.

Liu, Z., Peng, J., Hua, X., and Zhu, Z., 2021. Wind farm optimization considering non-uniformly distributed turbulence intensity. *Sustainable Energy Technologies and Assessments*, 43, 100970.

Malz, E.C., Hedenus, F., Göransson, L., Verendel, V., and Gros, S., 2020. Drag-mode airborne wind energy vs. wind turbines: An analysis of power production, variability and geography. *Energy*, 193.

Malz, E.C., Verendel, V., and Gros, S., 2020. Computing the power profiles for an Airborne Wind Energy system based on large-scale wind data. *Renewable Energy*, 162, 766–778.

National Oceanic Atmospheric Administration, U.S.D. of C., 2022. Numerical Weather Prediction (Weather Models) [online]. Available from: https://www.weather.gov/media/ajk/brochures/NumericalWeatherPrediction.pdf [Accessed 23 Feb 2022].

Nunalee, C.G. and Basu, S., 2014. Mesoscale modeling of coastal low-level jets: Implications for offshore wind resource estimation. *Wind Energy*, 17 (8), 1199–1216.

Ogliari, E., Guilizzoni, M., Giglio, A., and Pretto, S., 2021. Wind power 24-h ahead forecast by an artificial neural network and an hybrid model: Comparison of the predictive performance. *Renewable Energy*, 178, 1466–1474.

Radünz, W.C., Sakagami, Y., Haas, R., Petry, A.P., Passos, J.C., Miqueletti, M., and Dias, E., 2021. Influence of atmospheric stability on wind farm performance in complex terrain. *Applied Energy*, 282 (July 2020).

Salvação, N., Bernardino, M., and Guedes Soares, C., 2014. Assessing mesoscale wind simulations in different environments. *Computers and Geosciences*, 71, 28–36.

Salvação, N. and Guedes Soares, C., 2018. Wind resource assessment offshore the Atlantic Iberian coast with the WRF model. *Energy*, 145, 276–287.

Siedersleben, S.K., Platis, A., Lundquist, J.K., Djath, B., Lampert, A., Bärfuss, K., Cañadillas, B., Schulz-Stellenfleth, J., Bange, J., Neumann, T., and Emeis, S., 2020. Turbulent kinetic energy over large offshore wind farms observed and simulated by the mesoscale model WRF (3.8.1). *Geosci. Model Dev*, 13, 249–268.

Siedersleben, S.K., Platis, A., Lundquist, J.K., Lampert, A., Bärfuss, K., Cañadillas, B., Djath, B., Schulz-Stellenfleth, J., Bange, J., Neumann, T., and Emeis, S., 2018. Evaluation of a wind farm parametrization for mesoscale atmospheric flow models with aircraft measurements. *Meteorologische Zeitschrift*, 27 (5), 401–415.

Skamarock, W.C., Klemp, J.B., Dudhia, J.B., Gill, D.O., Barker, D.M., Duda, M.G., Huang, X.-Y., Wang, W., and Powers, J.G., 2021. A Description of the Advanced Research WRF Model Version 4.3. *NCAR Technical Note*, (July), 1–165.

Somoano, M., Battistella, T., Rodríguez-Luis, A., Fernández-Ruano, S., and Guanche, R., 2021. Influence of turbulence models on the dynamic response of a semi-submersible floating offshore wind platform. *Ocean Engineering*, 237.

Wang, H. zhi, Li, G. qiang, Wang, G. bing, Peng, J. chun, Jiang, H., and Liu, Y. tao, 2017. Deep learning based ensemble approach for probabilistic wind power forecasting. *Applied Energy*, 188, 56–70.

Wikipedia, 2022. Location of the Alpha Ventus wind farm [online]. *Creative commons licence*. Available from: https://de.wikipedia.org/wiki/Offshore-Windpark_alpha_ventus#/media/File:Windpark_alpha_ventus_Lagekarte.png.

Xia, G., Zhou, L., Minder, J.R., Fovell, R.G., and Jimenez, P.A., 2019. Simulating impacts of real-world wind farms on land surface temperature using the WRF model: physical mechanisms. *Climate Dynamics 2019 53:3*, 53 (3), 1723–1739.

Zhang, J., Chowdhury, S., and Hodge, B.M., 2014. Analyzing effects of turbulence on power generation using wind plant monitoring data. *32nd ASME Wind Energy Symposium*, (January).

*Renewable energy*

*Trends in Maritime Technology and Engineering – Guedes Soares & Santos (eds)*


# Numerical model of a WEC-type attachment of a moored submerged horizontal set of articulated plates

I.B.S. Bispo, S.C. Mohapatra & C. Guedes Soares
*Centre for Marine Technology and Ocean Engineering (CENTEC), Instituto Superior Técnico, Universidade de Lisboa, Lisbon, Portugal*

ABSTRACT: A 3D numerical model of a moored horizontal submerged structure, composed by a set of hinged plates attached to a vertical flap at its fore, intended to act as a WEC is presented. The horizontal submerged articulated float is composed of six hinged plates as to emulate a flexible horizontal plate. The numerical solution is calculated using BEM code ANSYS® AQWA to study the hydrodynamic behaviour of the structure by considering the rotational stiffness of the hinged plates and the linear stiffness of the mooring lines. The vertical displacements, wave excitation forces, and the horizontal responses of the flap are computed for a range of combinations of these stiffness parameters to estimate the power extraction potential. Present results show that the stability of the system is directly related to the rotational stiffness of the articulations, indicating that more power can be extracted from a stiffer structure for larger wavelengths. It is also shown that stability is dependent on the configuration of the design parameters and the number of mooring lines, as expected.

## 1 INTRODUCTION

Due to the growing interest in very large floating structures (VLFSs) and wave energy converters (WECs) in the field of marine structures and wave energy industry, the integration between the WECs (wave energy converters) and VLFSs (see Nguyen et al., 2020; Cheng et al., 2022). VLFSs can be used as an effective option of utilizing ocean space due to cost-effective, don't obstruct waves, environmentally friend, and flexibility nature of the structure whilst, the attachments are of lower costs for manufacturing and deployment, increases the reliability of the reduction of hydroelastic response, and act as a WEC for wave energy extraction.

However, there is no investigation of attaching a vertical plate at one end of the submerged horizontal flexible plate to act as a WEC (wave energy converters) and as well as an effective breakwater. The effectiveness of these structures is due to the fact that it does not block the incoming waves and can be properly tuned for attenuating the wave height and able to reflect, be cost-effective, and be environmentally friend.

Therefore, it is necessary to model a moored submerged structure attached with a submerged vertical plate to reduce the hydroelastic response and the vertical plate act as a WEC. The attachment of the vertical plate with submerged hinged plate brings several advantages, including (i) flexibility for expansion of the attachment; (ii) increasing reliability of the attachment as the system has higher redundancy; (iii) lower costs for manufacturing and deployment owing to the modularity in design, and (iv) the whole system be stable due to the greater number of mooring lines connected with the submerged plate. Motivated by the above advantages, this study presents a novel WEC-type attachment comprising floating horizontal plates and submerged vertical plates connected to the fore edge of moored submerged structure with hinges.

In this study, the performance of the WEC-type attachment is examined for various configurations which are specified by the numbers and locations of the submerged hinged horizontal plate with the auxiliary plate.

Recently, there has been considerable progress in the moored submerged horizontal flexible structures applications for the breakwater and wave energy absorption device using analytical methods. The effect of mooring lines and porous-effect parameters on the reflection, transmission, and dissipation coefficients as well as wave forces in a two-dimensional Cartesian coordinate system were analyzed in Mohapatra et al. (2018). Further, Mohapatra & Guedes Soares (2020) analyzed the wave energy absorption by the moored submerged flexible porous structure using the reduced wave equation along with Green's function technique and velocity decomposition method. Mohapatra & Guedes Soares (2021a) developed a three-dimensional Fourier expansion formula by formulating a 3D hydroelastic model associated with surface gravity wave

DOI: 10.1201/9781003320289-35

interaction with a submerged flexible porous structure and utilized to a real physical problem in finite water depth. Mohapatra & Guedes Soares (2019) studied the effect of submerged flexible plate combined with a moored floating elastic plate in three dimensions based on Fourier expansion and velocity decomposition method and in 2D (Mohapatra & Guedes Soares, 2016). Later, the effect of a moored submerged flexible porous plate on the floating flexible plate under a two-dimensional Cartesian coordinate system was analyzed in Mohapatra & Guedes Soares (2021b). Recently, Mohapatra & Guedes Soares (2021c) studied the effect of mooring lines on the hydroelastic response of a floating flexible structure through the structural deflection and reflection, transmission coefficients based on the boundary integral equation method.

Chandrasekaran & Sricharan (2020) proposed a floating WEC with the central buoy, circumferentially connected by a set of four, where the central buoy and floats are connected rigidly at one end while the other end is hinged and analyzed numerically in the frequency domain and time domain. Further, Sricharan & Chandrasekaran (2021) studied numerically a floating WEC consisting of a central buoy connected to a set of floats around it using a lever arm and is position restrained under a taut-mooring system in time-domain using an open-source code: WEC-Sim (Wave Energy Converter Simulator). Gao et al. (2011) investigated the hydroelastic response of the pontoon-type floating structures with a flexible line connection by adopting the modal expansion method and boundary element method based on the Mindlin plate theory in the frequency domain. Wang et al. (2010) presented a detailed review on the different methods of the structural arrangements for mitigating hydroelastic response of VLFS under wave action. Khabakhpasheva & Korobkin (2002) analyzed the hydroelastic behavior of compound floating plate being connected with elastically to the bottom based on normal mode method.

Ohmatsu (2000) presented the wave-induced hydroelastic response of a pontoon-type floating structure near a breakwater based on the three-dimensional eigenfunction expansion method and eigenmode expansion method and further results were compared with numerical and experimental results. Cheng et al. (2016) investigated the hydroelastic analysis of a pontoon-type floating structure edged with dual horizontal/inclined perforated plated based on boundary element method in the time domain and finite element method was adopted to present the deflection of the floating structure. Tay (2019) studied the energy extraction from an articulated plate anti-motion device that was connected to the VLFS by hinges by implementing a hybrid boundary element-finite element method. Nguyen et al. (2019) presented the use of modular draft (multiple independent auxiliary pontoons) WEC-type attachment at the fore edge of a rectangular VLFS for extracting wave energy and reduction of hydroelastic response of the VLFS in the frequency domain applying the hybrid Finite Element-Boundary Element (FE-BE) method. Zhang, et al. (2019) studied numerically the dynamic response and power capture performance of a VLFS composed of two hinged modules using the discrete-module-beam-bending based hydroelasticity method in the frequency domain.

Lu et al. (2019) studied the hydroelastic behaviour of VLFS consisting of multi-bodies using a numerical method based on multi-body hydrodynamics and Euler–Bernoulli beam assumption. Desmars et al. (2018) investigated the interaction of linear surface waves with a submerged WEC consisting of a submerged flexible plate actuated by one or more units of power take-off based on the Boundary Element Method. Zhang et al. (2018) proposed an embedded WEC installed in a super-scale modularized floating platform consisting of multiple blocks where on-top huge modular decks are flexibly supported by floating semi-submersible modules via elastic cushions and the adjacent blocks and neighboring decks are joined by rigid hinges based on a network modeling method. Nguyen et al. (2020) proposed a two-mode WEC-type attachment that comprises floating horizontal auxiliary plates and submerged vertical auxiliary plates connected to the fore-edge of the VLFS with hinges and linear PTO systems based on numerical methodology.

Bispo et al. (2021) reviewed the numerical approaches in the hydroelastic response of VLFSs. Recently, Bispo et al. (2022) employed the numerical approach using ANSYS® AQWA to simulate a moored articulated very large floating structure composed by a set of hinged plates, showing that this numerical model can be used to accurately obtain the reflection and transmission coefficients of incoming waves, besides providing insights on the influence of mooring and rotational stiffness to the reduction of vertical displacement.

In the present paper, a preliminary study of a three-dimensional numerical model of the WEC-type attachment with a moored articulated submerged horizontal set of plates is presented. The model is developed by employing ANSYS® AQWA and the results indicate that amount of power that can be generated is largely dependent on the mooring and rotational stiffness of the submerged articulated plates.

The paper is organized as follows: Section 2 presents the numerical model details and, briefly, the mathematical description of the problems being solved in frequency and time domains. Details of the implemented cases and the respective results are presented and discussed in Section 3. Finally, in Section 4, conclusions are drawn based on the previously presented results and the findings are highlighted, followed by the next steps to be developed in the current research.

## 2 NUMERICAL MODEL FORMULATION AND DESCRIPTION

### 2.1 *Numerical model*

A first implementation of the numerical model is made by adopting the approach of a Boundary Element Method solver, based on potential flow theory, using ANSYS® AQWA, a largely employed and well-known software throughout the industry when considering preliminary stages of development of floating or submerged structures. In this study, the mentioned software is employed to design and simulate the behaviour under regular waves, of a 3D model of an articulated structure, which is composed by a set of six horizontal plates and one vertical flap, joined by hinges with only one degree of freedom, around its (local) y-axis.

A common approach to determining the hydrodynamic forces and coefficients is to use the linear wave theory, assuming the waves are the sum of the incident, radiated, and diffracted wave components as it is later described in this section. AQWA employs one of the most common numerical tools to analyze the hydrodynamic behaviour of a large-volume structure in waves, a three-dimensional panel method, representing the structure surface by a series of diffraction panels. This method is based on potential flow theory, and it is assumed that: (i) the bodies have zero forward speed; (ii) the fluid is inviscid, incompressible and the fluid flow is irrotational; (iii) the incident regular wave train is of small amplitude compared to its length; (iv) motions are to the first order, hence, must be of small amplitude and all body motions are harmonic (ANSYS, 2015).

The adopted software solves the diffraction-radiation problem in the frequency domain to obtain the hydrodynamic coefficients, taking into account the body-to-body interaction. Nevertheless, this procedure does not account for the presence of mooring lines and articulations and, therefore, a second step is included to evaluate the effects that these structures might cause to the system dynamics. These constraints are added to the problem by solving the Cummins equation (Nolte & Ertekin, 2014), as is described later in this section. A time response analysis is performed in the same software as a way to provide time-series results of wave elevation, excitation forces and structures reaction forces, positions and velocities components for each degree-of-freedom.

The graphical representation of the system composed of the six plates, flap, hinges, and mooring lines is presented in Figure 1, where the origin of the coordinate system is located on the waterline, at mid-ship and keel line, and the motions follow the usual nomenclature and symbology found in the literature (X, Y, Z, RX, RY, RZ). Together with the structure geometry, the hinges for the articulation are represented by two-tones brown vertical cylinders between plates. Green spheres represent the centre of mass of each plate and flap. Black spheres are fixed positions used for all connections between plates or the mooring lines to the seabed.

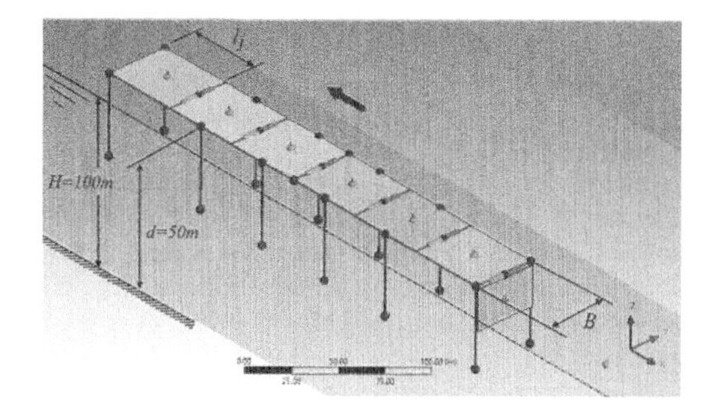

Figure 1. Schematic model definition – 3D numerical model.

Table 1. Design parameters of the articulated submerged horizontal and vertical multi-plate structure.

| Parameter (unit) | Symbol | Value |
|---|---|---|
| Overall structure | | |
| Length (m) | $L$ | 295 |
| Breadth (m) | $B$ | 40 |
| Thickness (m) | $D$ | 1 |
| Gap between plates (m) | $\Delta l$ | 1 |
| Number of diffracting panels | - | 17140 |
| Plates | | |
| Individual plate length (m) | $l_j$ | 48 |
| Mean density of Plate (kg/m³) | $\rho_P$ | 308 |
| Moments of Inertia (tons.m²) | $I_{xx}$ | 78989 |
| | $I_{yy}$ | 113722 |
| | $I_{zz}$ | 192612 |
| Flap | | |
| Height of Flap (m) | $F$ | 25 |
| Thickness of Flap (m) | $D_F$ | 1 |
| Mean density of Flap (kg/m³) | $\rho_F$ | 1366 |
| Moments of Inertia (tons.m²) | $I_{xx}$ | 316599 |
| | $I_{yy}$ | 89075 |
| | $I_{zz}$ | 227809 |

The structure is designed based on the parameters presented in Table 1. If the system is considered in equilibrium when no waves are acting on it, the vertical position of the centres of masses of each horizontal plate is positioned at half the height ($d$=50 m) of the water depth ($H$=100 m). Only waves coming from the positive $x$-axis ($\theta = -180°$) are considered at this stage of the research.

It may be noted that bending and torsional moments cannot be assessed by using AQWA with multiple bodies, unfortunately. Moments of Inertia are added in Table 1. The authors consider that assessing the effects of added moments of inertia in comparison to the mass moments of inertia without going into detail and presenting results for comparison would not bring enough depth that this subject deserves. On the other hand, a detailed discussion can derive a paper itself, as made by Tom et al (2015).

Mathematically, supposing the fluid as ideal and assuming the linear theory of waves, the time-dependent total velocity potential can be written as $\Phi = \phi e^{-i\omega t}$ where $\phi$ is the complex velocity potential, $\omega$ is the wave frequency in rad/sec and $t$ represents time. A Boundary Value Problem can be satisfied by the radiated potentials $\phi_R$ and diffracted potentials $\phi_D$:

$$\begin{cases} \nabla^2\phi = 0, & \text{in } \Omega \\ \omega^2\phi + g\frac{\partial\phi}{\partial z} = 0, & \text{on } S_f \\ \frac{\partial\phi}{\partial n_k} = \vec{V}_{S_k}.\vec{n}_k, & \text{on } S_k \\ \frac{\partial\phi}{\partial z} = 0, & \text{on } S_B \\ \lim_{r\to\infty}\sqrt{r}\left[\frac{\partial\phi}{\partial r} - \frac{i\omega^2}{g}\phi\right] = 0, & \text{on } S_\infty \end{cases} \tag{1}$$

where $g$ is the gravitational acceleration, $r$ is the radial distance from the origin and $\vec{n}_k$ is the inward unit vector at the $k$-th body surface. In this BVP, the fluid domain is represented by $\Omega$, $S_F$ is the free surface, $S_B$ is the bottom surface, $S_\infty$ represents the boundary at infinity, and the wetted surface of the body is given by $S_k$, with $k$ ranging from 1 to 7, representing each plate module (1 to 6) and the flap ($k$=7). The plates are connected by hinges with rotational stiffness $k_{rot}$ and the edges of the structure are connected by mooring lines with stiffness $k_t$, although the number of lines can also vary. The hinge that connects the flap to the first horizontal plate is considered as having no stiffness ($k_f = 0$).

For the $k$-th plate module, $\vec{n}_k$ is the unit vector normal to the wetted surface, pointing outwards, and $\vec{V}_{S_k}$ is the fluid velocity on the wetted surface of each plate. Once the velocity potential, $\phi$, is obtained, the added mass and the radiation damping can be computed. The velocity potential can be decomposed as a summation of the incident potential $\phi_I$, the diffraction potential $\phi_D$, and the radiation potential $\phi_R$:

$$\phi = \phi_I + \phi_D + \phi_R, \tag{2}$$

where

$$\phi_I = \sum_{m=1}^{\infty} \frac{igI_{m0}}{\omega} \frac{\cosh k_0(h-y)}{\cosh k_0 h} \cos\gamma_m z e^{-ip_{m0}x},$$

where $i = \sqrt{-1}$, $k_0$ is the wavenumber associated with the gravity wave dispersion relation and $p_{m0}$ is the wavenumber associated with plate covered region. Therefore, the excitation forces $F_{W_j}^{(k)}$, can be related to the incident and diffraction potentials for the $j$-th degree of freedom (DOF) and the $k^{th}$ body by:

$$F_{W_j}^{(k)} = \rho i\omega \iint_{\bar{S}_k} (\phi_I + \phi_D).\vec{n}_k dS, \tag{3}$$

in which $\rho$ is the fluid density, $\omega$ is the wave frequency, $\bar{S}_k$ is the mean wetted surface of each plate module $k$, and $i$ is the complex unit.

The generalized added mass and radiation damping are given by:

$$[a] + \frac{i}{\omega}[b] = \rho \iint_{\bar{S}_k} \phi_R.\vec{n}_k dS, \tag{4}$$

where $[a] = A_{jk}^{(mn)}$ and $[b] = B_{jk}^{(mn)}$ are the added mass and the radiation damping coefficients of the $j$-th mode of the $m$-th plate induced by $k$-th mode of the $n$-th plate. For six degrees of freedom, proceeding as in Jin et al. (2020), the equation of motion of the free multi-body floating structure is presented in the frequency domain and for unitary wave amplitude as:

$$[M + A(\omega)]\ddot{X} + B(\omega)\dot{X} + KX = F_w(\omega), \tag{5}$$

where $M$ is a matrix of masses for all bodies, in which its diagonal is composed by $6 \times 6$ matrices of masses for each plate $k$. The matrix $K$ is the hydrostatic stiffness matrix and it is defined similarly to the mass matrix, for each plate and 6 DOF. The added mass and radiation damping matrices, respectively represented by $A(\omega)$ and $B(\omega)$, are composed by $k \times k$ matrices of $6 \times 6$ size of added masses and radiation damping. The vector $X$ represents the displacement, from which follows that $\dot{X}$ and $X$ are its first- and second-time derivatives, respectively. Lastly, $F_W(\omega)$ is the vector of wave excitation forces for all $k$ bodies.

Once the hydrodynamic coefficients (wave excitation force, added mass and radiation damping) are obtained in the frequency domain by AQWA by solving the previously presented equations, the system can also be simulated in the time domain, assuming the linearity of wave theory. The system of differential equations employed in the time domain can be summarized by the Cummins equation:

$$(M + A_\infty)\ddot{x}(t) + (K + K_E)x(t) = f_w(t) + f_R(t) + f_C(t). \quad (6)$$

The structural mass matrix is denoted by $M$ and the fluid added mass matrix at infinite frequency is denoted by $A_\infty$. This equation includes the additional stiffness matrix, $K_E$, due to the connections of the plates. Displacement as a function of time is represented by $x(t)$, and likewise in the frequency domain, its first- and second-time derivatives are denoted by $\dot{x}(t)$ and $\ddot{x}(t)$, respectively. Forces actuating are the wave excitation force, $f_W(t)$, external forces such as the ones provided by the mooring, $f_E(t)$ and the so-called convolution forces, $f_C(t)$, such as the forces exerted by radiation of waves. This convolution force can be computed by:

$$f_C(t) = -\int_0^\infty R(\tau)\dot{x}(\tau - t)d\tau,$$

$$R(\tau) = \frac{2}{\pi}\int_0^\infty B(\omega)\cos(\omega t)d\omega,$$

where $R(t)$ is a retardation function, dependant on the radiation damping $B(\omega)$ computed in frequency domain. This function can also be included in the computation of the added mass at infinite frequency:

$$A_\infty = A(\omega) + \int_0^\infty R(t)\frac{\sin(\omega t)}{\omega}dt.$$

Details on the structures of these matrices and vectors are presented by Jin et al (2020). The solution of the system represented by Eq. (6) is performed by AQWA using a two-stage predictor-corrector numerical scheme to integrate it in time (ANSYS, 2015).

The effect of the mooring system on power absorption is studied by the introduction of a simple mooring model in AQWA, which works like non-linear springs, but neglects drag and inertia forces due to line motion. This simpler model seems like a reasonable approximation for a preliminary analysis of the WEC presented herein, although it is intended to further develop the numerical modelling of the mooring lines in the future.

Wave excitation forces, added masses and wave radiation damping are normalized accordingly to the following expressions before comparisons:

$$\overline{F_W(\omega)} = |F_W(\omega)|/\rho g,$$

$$\overline{A(\omega)} = A(\omega)/\rho,$$

$$\overline{B(\omega)} = B(\omega)/\rho\omega.$$

A hypothetical linear Power Take-Off (PTO) is assumed to be present in the system, capturing energy as the Flap moves. Since the aim is to verify the concept of the proposed WEC, no detailing of the PTO and its connections to the system are made at this moment of the research, it is only assumed that the horizontal force acting on the flap will move a linear piston and energy is transferred in this way. In order to quantify the power potential that can theoretically be produced by this WEC, the instantaneous power $P$ is computed by:

$$P = -F_{PTO} \cdot \dot{X}_{rel},$$

where $F_{PTO}$ is the horizontal force acting on the Flap in the direction of the $x$-axis and $\dot{X}_{rel}$ is the $x$ component of the relative velocity between the first horizontal plate and the Flap.

Time series extracted from the time response simulations are evaluated by computing the amplitudes of the corresponding spectrum using the Fast Fourier Transform (FFT) algorithm provided by MATLAB, version R2021b.

### 2.2 *Design parameters and comparison cases*

Eleven distinct configurations are tested in order to assess the effects of different values of mooring stiffness, the number of mooring lines and rotational stiffness of the hinges that join the horizontal plates. No rotational stiffness is considered, at this stage of the research, for the hinge that connects the vertical flap to the horizontal plate at the fore, since the aim is to verify which combinations can lead to a reasonable power output maintaining a stable condition of the structure when subjected to wave loads.

The design parameters comparison cases are presented in Table 2 below, in which nine out of eleven cases correspond to the employment of 12 mooring lines. This is due to the fact that if a lower rotational stiffness of the hinges is selected, the motion of the set of joined plates can rapidly become unstable, so the case would not converge, as it was noted during the stage of designing the numerical experiments. To avoid that, whenever the rotational stiffness is relatively small, more mooring lines are adopted.

Also, for higher values of rotational or mooring stiffness, it is considered a comparison case with less mooring lines.

Table 2. Cases considered for the design parameters.

| Case | $k_{rot}$ (N.m/°) | $k_t$ (N/m) | Number of Mooring Lines |
|---|---|---|---|
| A | $10^{10}$ | $10^{10}$ | 6 |
| B | $10^8$ | $10^8$ | 10 |
| C | $10^9$ | $10^8$ | 10 |
| D | $10^6$ | $10^{10}$ | 12 |
| E | $10^7$ | $10^6$ | 12 |
| F | $10^7$ | $10^8$ | 12 |
| G | $10^8$ | $10^8$ | 12 |
| H | $10^8$ | $10^{10}$ | 12 |
| I | $10^9$ | $10^6$ | 12 |
| J | $10^9$ | $10^8$ | 12 |
| K | $10^{10}$ | $10^{10}$ | 12 |

## 3 NUMERICAL RESULTS AND DISCUSSIONS

### 3.1 *Comparison between frequency and time domains*

Results from the time domain simulation are verified by comparing the normalized wave excitation force $\overline{F_W(\omega)}$ with the correspondent hydrodynamic coefficient obtained from the frequency domain simulation. These comparisons are performed considering the vertical component (Z) for the horizontal plates and X and RY components for the Flap. Examples of the comparison are presented in Figures 2-3.

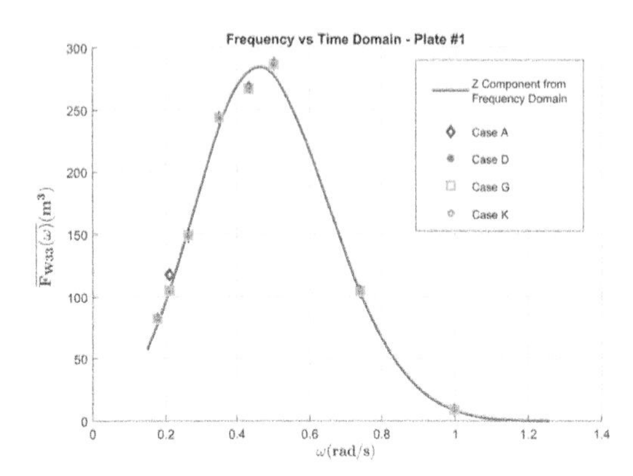


Figure 2. Comparison of frequency and time domain wave excitation forces – Z component.

Evidently, it is expected for the hydrodynamic parameters obtained from both methods to be very close to each other, since the time domain approach has inputs resulting from the frequency domain (and considering that the algorithms have been extensively tested for this commercial software). Nevertheless, it is worth mentioning that the effects of constraints are only considered in the time domain and that the wave force excitation is recomputed in this fashion, together with the other forces acting on the structure. Effects of body-to-body interactions are considered by default in the frequency domain in AQWA.

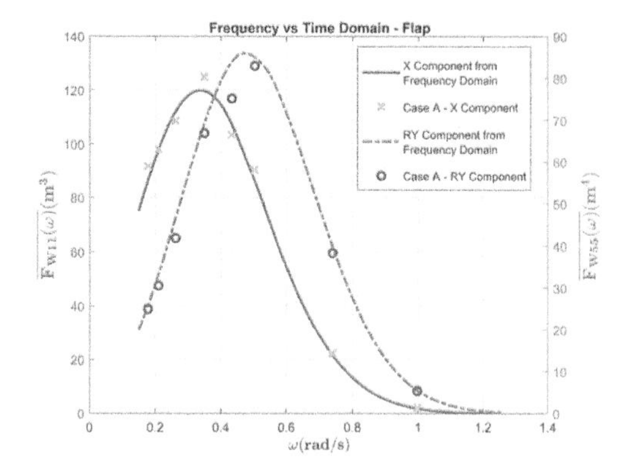


Figure 3. Comparison of frequency and time domain wave excitation forces – X and RY components.

Although there are slight deviations closer to the peak of the forces, such as the ones presented in Figures 2-3, respectively for Plate #1 and for the Flap, overall, the wave excitation forces components are showing a good agreement between the methods. From this, one can assume that the results of interest can be extracted with a reasonable degree of confidence.

### 3.2 *Numerical model simulations for different design parameters*

A preliminary analysis of the structure vertical response when subject to wave loads for a range of wave frequencies is presented along with Plate and Flap RAOs, and a first assessment of the power output potential, which are intended as indicators of viability for the proposed setup and design parameters configurations of the WEC. Results for particular wave frequencies where the system becomes unstable are not included in this analysis.

The proposed WEC acts like an Oscillating Surge Wave Energy Converter (OSWEC), so it is relevant to analyse its behaviour in this degree of freedom. To do so, the comparison of the RAO of Surge for the Flap obtained from frequency and from time domains is presented in Figure4. In this figure, it can be seen that for lower wave frequencies, the response is expected to be larger in surge and, in fact, data from the time domain simulation cases indicate that the coupling of the Flap to the horizontal plates, the stiffness of the joints of these plates and the stiffness of the mooring lines can have a significant impact on the response of the Flap where the larger the wave period, the larger the surge amplitude of motion.

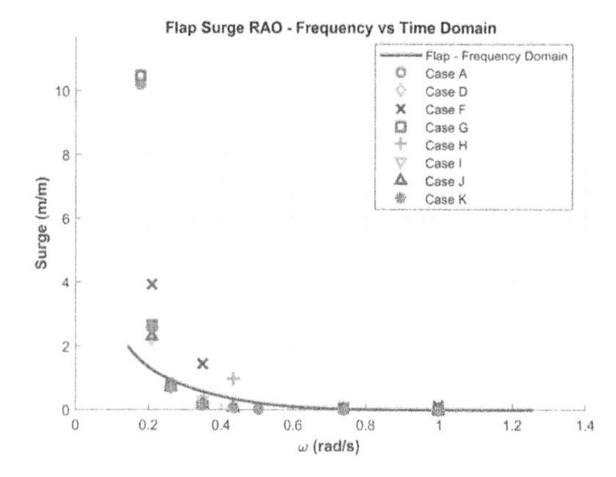


Figure 4. Comparison of frequency and time domain surge RAOs for the Flap.

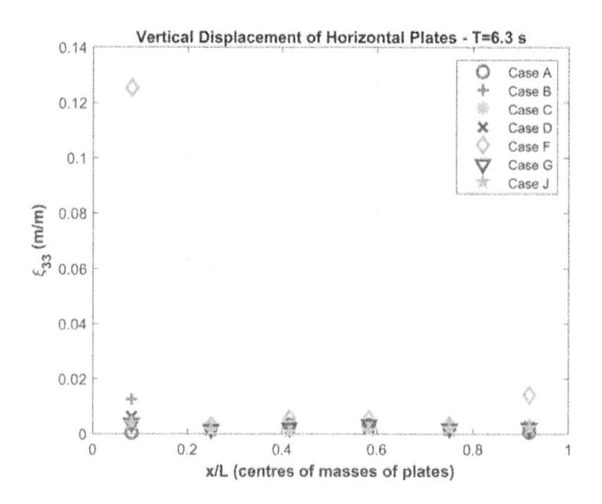


Figure 5. Comparison of vertical displacement of the plates for wave period of $T = 6.3$s.

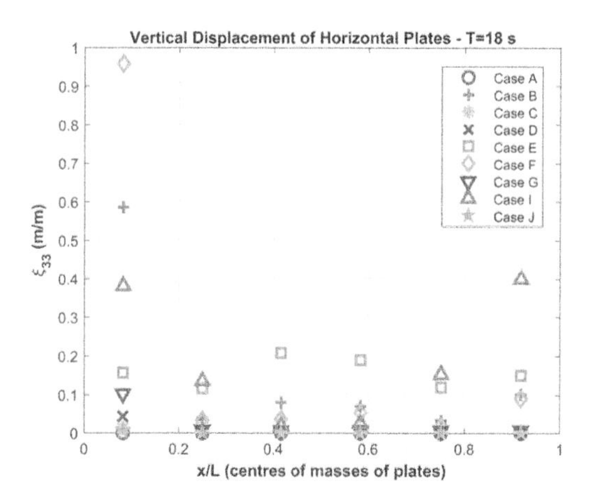


Figure 6. Comparison of vertical displacement of the plates for wave period of $T = 18$s.

The vertical displacements of the plates per wave amplitude are presented in Figures 5-7 for three-wave periods, showing the comparison among the cases previously described. It is worth noting that the horizontal position of the markers is relative to the centre of mass of each plate, according to the normalized length (*x/L*) of the structure, being the plate attached to the Flap the first one, on the left of the *x*-axis.

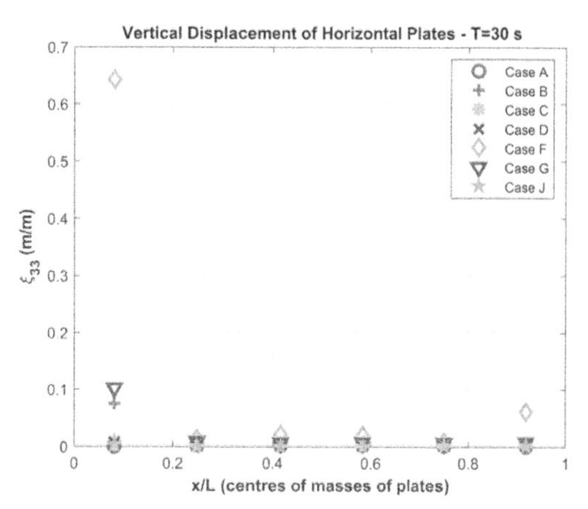


Figure 7. Comparison of vertical displacement of the plates for wave period of $T = 30$s.

According to these results, cases A through D show a relatively small vertical response for low and high frequencies (T24sand T $\leq$ 15s, respectively). Nevertheless, for intermediate wave periods, such as the example on Figure6, the vertical motion increases, especially for Plate #1, which is attached to the Flap. The cases where this happens are relative to values of $k_t$ that provide more freedom of vertical motion, allied to $k_{rot}$ values that can allow a more "flexible" structure, such as cases B, E, F, G and I, for example.

Unstable behaviour of the system is observed for cases E and I when considering low and high frequencies ($T \geq 24$s and $T \leq 15$s, respectively). This is due to the fact that mooring lines are too loose, with a lower value of $k_t$, which leads to very large vertical displacement, affecting its stability. If the structure is more "rigid" (larger value of $k_{rot}$), such as case I when compared to case E, the vertical response tends to be more uniform along with the structure and, therefore, the system will present a more stable behaviour along with a larger range of wave periods.

On the other hand, for cases with higher rotational and mooring stiffness, such as cases A, H and K, no significant vertical displacement is observed, which is expected, since the structure has little freedom to move vertically, and the plates also have a more solid connection between them.

Case D seems to be the one that presents a better trade-off between vertical displacement and stability of the system along the frequency range considered in this work. It presents more freedom for the plates to move in surge and pitch, but there is more restriction to heave, since the mooring stiffness is high. This means that the structure is more "flexible" while it is more rigidly fixed to the seabed.

The tested configurations of the proposed WEC are analysed in terms of the theoretical mean power output that can be generated by each setup represented by the cases from Table 2. Still, results for frequencies that exhibit unstable behaviour are not included in this analysis and, therefore, cases E and I are automatically excluded, since the excessive relative motion of Flap and Plate #1 can be misleading as higher values of power output are achieved.

Figure 8 presents the comparison of stable cases in terms of theoretical mean power output. A trend of higher values of mean power output can be observed for lower wave frequencies, indicating that more energy can be absorbed from waves with longer wave periods (wave amplitude is the same for all simulations, 1 m, thus the longer the frequency, the more energetic the waves are).

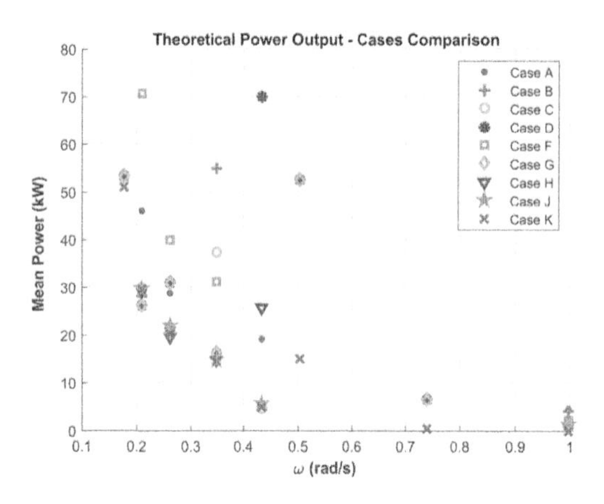


Figure 8. Theoretical mean power output estimated for the proposed WEC.

High-frequency waves show low mean power output since the relative motion for these waves is significantly reduced, both for the Flap and the horizontal part of the structure. Cases that have greater mooring stiffness and relatively higher values of rotational stiffness ($k_t \geq 10^8$ and $k_{rot} \geq 10^7$) seem to produce a steadier and smoother power output, although some outliers can be present, indicating intermediate frequencies with greater response than expected from the RAOs. These specific wave periods need further investigation.

## 4 CONCLUSIONS

A numerical model of a very large submerged, horizontal and moored anti-motion device, composed by a set of six plates joined by hinges and equipped with a Flap that acts as an Oscillating Surge Wave Energy Converter, was presented and analysed in terms of its theoretical power output, while its vertical response is intended to be kept in small ranges of amplitude.

The model was implemented in AQWA, a BEM code that can provide hydrodynamic coefficients in frequency and time domains, the later considering the presence of mooring lines and stiffened articulations between bodies. Such numerical modelling relies on mathematical models, and it is based on several assumptions and approximations. Therefore, it is acknowledged that the behaviour of an actual system could deviate from what is predicted by the numerical model. Even so, recalling that the primary objective of this study is to assess a first numerical implementation to simulate the behaviour of the proposed submerged WEC, the results are presented as a reasonable approximation and can be seen as preliminary. As mentioned by Babarit et al (2012), it is expected for the numerical model to overestimate the dynamic response of the system, so the results must be regarded with caution.

Comparisons of expected responses obtained from the frequency domain, within the framework of linear waves theory, were made against the respective ones computed in the time domain. Vertical displacement of the plates and horizontal responses of the Flap was considered to evaluate the effectiveness of the structure in reducing motion and also as a WEC for a variety of design parameters configurations, including mooring stiffness, rotational stiffness of plates articulations and the number of mooring lines. The concluding remarks of this study are summarized as:

1. The attenuation of the structure motion is indicated to be greater for larger values of mooring stiffness, as is expected. Nevertheless, the design parameters cases also indicate that the rotational stiffness of the hinges plays an important role in maintaining the stability of the system. A less significant but not negligible parameter is the number of mooring lines, since the less cables are present, the higher the demand for stiffness in each line.
2. When considering the horizontal motion necessary to produce a viable WEC coupled to the submerged horizontal structure, it was shown that intermediate values of rotational stiffness of the hinges present reasonable advantage since these configurations indicate a greater horizontal response and, therefore, can lead to a higher level of mean power output, specifically for longer wave periods.
3. Furthermore, the resulting theoretical mean power estimated for the presented cases has shown that by compromising in a small degree

the established trade-off between rotational and mooring stiffness, a higher level of mean power output can be attained from an intermediate range of wave periods.

4. Whilst it is expected for the performance of the WEC to be affected by the depth at which is deployed, as well as by the depth of the seabed since the effects for shallow waters can be significant on the responses of the structure, no analysis of such parameters was carried out in this work. Since it is an ongoing research topic, it is intended to be addressed in a future opportunity.
5. The future scope of the present research also includes: Addition of a very large floating structure at the free surface, above the submerged set of plates, intended to act as a breakwater; Analysis of the hydroelastic response of the floating flexible plate in the presence of the submerged compound; Specification of a PTO and computation of mean power output accordingly.

## ACKNOWLEDGEMENTS

The paper is performed within the project Hydroelastic behaviour of horizontal flexible floating structures for applications to Floating Breakwaters and Wave Energy Converters (HYDROELASTWEB), which is co-funded by the European Regional Development Fund (Fundo Europeu de Desenvolvimento Regional - FEDER) and by the Portuguese Foundation for Science and Technology (Fundação para a Ciência e a Tecnologia - FCT) under contract 031488_770 (PTDC/ECI-EGC/31488/2017). The second author has been contracted as a Researcher by the Portuguese Foundation for Science and Technology (Fundação para a Ciência e Tecnologia-FCT), through Scientific Employment Stimulus, Individual support under the Contract No. CEECIND/04879/2017. This work contributes to the Strategic Research Plan of the Centre for Marine Technology and Ocean Engineering (CENTEC), which is financed by the Portuguese Foundation for Science and Technology (Fundação para a Ciência e Tecnologia - FCT) under contract UIDB/UIDP/00134/2020.

## REFERENCES

ANSYS, 2015. Aqwa Theory Manual, ANSYS, Inc., Canonsburg, PA.

Babarit, A., Hals, J., Muliawan, M.J., Kurniawan, A., Moan, T., & Krokstad, J., 2012. Numerical benchmarking study of a selection of wave energy converters. *Renewable Energy* 41:44–63.

Bispo, I.B.S., Mohapatra, S.C., & Guedes Soares, C. 2022. Numerical analysis of a moored very large floating structure composed by a set of hinged plates. *Ocean Engineering*, 110785, doi: https://doi.org/10.1016/j.oceaneng.2022.110785

Bispo, I.B.S., Mohapatra, S.C., & Guedes Soares, C., 2021. A review on numerical approaches in the hydroelastic responses of very large floating elastic structures. In: Guedes Soares, C. & Santos, T. A. (Eds.), *Developments in Maritime Technology and Engineering,* Taylor & Francis Group, London, Vol 1, pp. 425–436.

Chandrasekaran, S. & Sricharan, V.V.S. 2020. Numerical analysis of a new multi-body mechanical wave energy converter with a linear power-take off system. *Renewable Energy* 159: 250–271.

Cheng, Y., Ji, C.Y., Zhai, G.J. & Gaidai, O. 2016. Hydroelastic analysis of oblique irregular waves with a pontoon-type VLFS edged with dual inclined perforated plates. Marine Structure 49: 31–57.

Desmars N, Tchoufag J, Younesian D, Alam MR. 2018. Interaction of surface waves with an actuated submerged flexible plate: Optimization for wave energy extraction. *Journal of Fluids Structures* 81:673–692.

Gao, R.P., Tay, Z.Y., Wang, C.M. & Koh, C.G. 2011. Hydroelstic response of very large floating structure with a flexible line connection. Ocean Engineering 38: 1957–1966.

Jin, C., Bakti, F.P., Kim, M.H., 2020. Multi-floater-mooring coupled time-domain hydro-elastic analysis in regular and irregular waves. Appl. Ocean Res. 101, 102276.

Khabakhpasheva, T.I & Korobkin, A.A. 2002. Hydroelastic behavior of compound floating plate in waves. *Journal of Engineering Mathematics* 44(1):21–40.

Lu, D., Fu, S., Zhang, X., Guo, F. & Gao, Y. 2019. A method to estimate the hydroelastic behaviour of VLFS based on multi-rigid-body dynamics and beam bending. *Ships Offshore Structures* 14(4): 354–62.

Mohapatra, S.C. & Guedes Soares, C. 2021b. Surface gravity wave interaction with a horizontal flexible floating plate and submerged flexible porous plate. *Ocean Engineering* 237, 109621.

Mohapatra, S.C. & Guedes Soares, C. 2021a. Hydroelastic behaviour of a submerged horizontal flexible porous structure in three-dimensions. *Journal of Fluids and Structures* 104, 103319.

Mohapatra, S.C. & Guedes Soares, C. 2020. Hydroelastic response of a flexible submerged porous plate for wave energy absorption. *Journal of Marine Science and Engineering* 8(9), 698,

Mohapatra, S.C. & Guedes Soares, C. 2019. Interaction of ocean waves with floating and submerged horizontal flexible structures in three-dimensions. *Applied Ocean Research* 83: 136–154.

Mohapatra, S.C. & Guedes Soares, C. 2021c. Effect of mooring lines on the hydroelastic response of a floating flexible plate using the BIEM approach. *Journal of Marine Science and Engineering* 9(9): 941.

Mohapatra, S.C., Sahoo, T. & Guedes Soares, T. 2018. Surface gravity wave interaction with a submerged horizontal flexible porous plate. *Applied Ocean Research* 78: 61–74.

Mohapatra, S.C. & Guedes Soares, C., 2016. Effect of submerged horizontal flexible membrane on moored floating elastic plate. *In*: Guedes Soares and Santos (Eds), *Maritime Technology and Engineering* 3. London: Taylor & Francis Group, pp.1181–1188, ISBN: 978-1-138-03000-8.

Nguyen, H.P., Wang C.M., Flocard, F. & Pedroso, D.M. 2019. Extracting energy while reducing hydroelastic responses of VLFS using a modular raft wec-type attachment. *Applied Ocean Research* 84: 302–316.

Nguyen, H.P., Wang, C.M. & Luong, V.H. 2020. Two-mode WEC-type attachment for wave energy extraction and reduction of hydroelastic response of pontoon-type VLFS. *Ocean Engineering* 197: 106875.

Nolte, J.D., & Ertekin, R.C., 2014. Wave power calculations for a wave energy conversion device connected to a drogue. *Journal of Renewable and Sustainable Energy*, 6: 013117.

Ohmatsu, S. 2000. Numerical calculation method for the hydroelastic response of a pontoon-type very large floating structure close to a breakwater. *Journal of Marine Science and Technology*-Japan 5: 147–160.

Sricharan, V.V.S. & Chandrasekaran, S. 2021. Time-domain analysis of a bean-shaped multi-body floating wave energy converter with a hydraulic power take-off using WEC-Sim. *Energy* 2021: 119985.

Tay, Z.Y. 2019.Energy extraction from an articulated plate anti-motion device of a very large floating structure under irregular waves. *Renewable Energy*. 130: 206–222.

Tom, N., Lawson M., Yu, Y-H., & Wright, A., 2015. Preliminary Analysis of an Oscillating Surge Wave Energy Converter with Controlled Geometry. 11$^{th}$ *European Wave and Tidal Energy Conference*. Nantes, France, *Sept. 6-11*.

Wang, C.M., Tay, Z.Y. & Utsunomiya, T. 2010. Literature review of methods for mitigating hydroelastic response of VLFS under wave action. Applied Mechanics Review 63(3): 1–18.

Zhang, X.T., Zheng, S.M., Lu, D. & Tian, X.L. 2019. Numerical investigation of the dynamic response and power capture performance of a VLFS with a wave energy conversion unit. *Engineering Structures* 195: 62–83.

Zhang, H.C., Xu, D.L., Zhao, H., Xia, S.Y. & Wu, Y.S. 2018. Energy extraction of wave energy converters embedded in a very large modularized floating platform. *Energy* 158: 317–329.

*Trends in Maritime Technology and Engineering – Guedes Soares & Santos (eds)*


# Standardisation of wind turbine SCADA data suited for machine learning condition monitoring

I.M. Black & A. Kolios
*Naval Architecture, Ocean and Marine Engineering Department, University of Strathclyde, Glasgow, UK*

ABSTRACT: Digital enabled asset management requires usable data, the measured information typically recorded on a wind farm requires processing for effective model development. As expected, a variety of different challenges are necessary in the pursuit of effective machine learning. This article provides a detailed overview of algorithms suited to time series analysis and techniques used in the data preparation. The pre-processing procedures involve feature selection, missing data treatment, and data projection. This is scrutinised against machine learning techniques: artificial neural network (ANN), Gaussian process regression (GRP), support vector machine (SVM), k-nearest neighbours (K-NN), and light gradient boosting (LGBM). The outcomes from this document highlight that pre-processing will increase the modal performance across 5 different machine learning algorithms. Secondly, if the process of organising SCADA data becomes standardised, model development can become modular and cooperative in the development of new tools. Lastly, the assessment of models will be more transparent.

## 1 INTRODUCTION

Europe installed 14.7 GW of new wind capacity in 2020, this is 6% less than in 2019 and 19% less than expected, but covid-19 struck. That is not to say that Wind energy in Europe is slowing down, quite the opposite. A realistic expectation for installed capacity across Europe over the next five years is 105 GW (Energy 2020). That is a doubling of the annual installation rate for offshore wind from 3 GW to 5.8 GW. The United Kingdom is expected to install a capacity of 18 GW where 15 GW will be offshore. Wind energy reduces the cost of energy (COE) and becomes a more desirable investment. A significant portion of the investment is operations and maintenance accruing to 30% of the overall cost (Crabtree et al.). This cost is decreasing, one reason being, trends towards direct-drive wind turbines. Wind turbines with no gears from the blades to the generator, consequently have zero gear losses and have the advantage of less maintenance and repairs but, they are larger (da Rosa 2013). Another incipient machinery fault detection before they become a catastrophic failure. Unexpected failure is directly related to wind turbine downtime and loss of revenue. Hence, effective condition monitoring (CM) for optimal maintenance actions is critical.

There are two main schools of thought for wind turbine maintenance: corrective, passive, or preventative, active. Corrective maintenance involves fixing issues on the occurrence of failure. Preventative maintenance is performed before the actualisation of failure. This can be further described by; scheduled maintenance, where fixed intervals based on the product's expected life are carried out. Another is Condition-based maintenance involves continual structural health monitoring of components to optimise maintenance schedules. Condition-based maintenance intends to prevent failures, reduce scheduled maintenance, and reduce maintenance costs (Verbert et al. 2017).

Condition-based maintenance can be carried out in the following ways: trending, damage modelling, alarm assessment, performance modelling, and clustering see (Black et al. 2021) for a breakdown of theses. Digital-enabled asset management performs the former tasks using computational methods. One requirement for computational intelligence is well-organised data. The handling of data in current IEC 61400 1 is limited only to the acquisition and implementation of condition monitoring but no clear standardisation of data for the purpose the implementation to computational methods is addressed to date. That is the purpose of this paper. To begin the discussion of standardising pre-processing for machine learning methods of wind turbine condition monitoring.

This paper aims to present a comprehensive process for which data can be effectively organised so that it can improve the accuracy of computational models for wind turbine condition monitoring. To highlight the increased accuracy of the pre-processing power, a trending condition monitoring model will be prepared using MET MAST information and the output power of an operational wind turbine.

DOI: 10.1201/9781003320289-36

Comparing the organised data with the raw. Section 1 introduces the scope, purpose, and motivation of this paper. Section 2 will provide some research background starting with current data acquisition and supervisory control (SCADA) standards for horizontal wind speed and power. Also, a detailed overview of trending condition fault detection. Section 3 preprocessing procedure. Section 4 details the machine learning models used for the trend condition monitoring. Section 5 demonstrates the improvement of the pre-processing to the raw data in a case study using SCADA data and METMAST data from a wind farm, applying the data sets to the fault detection models. Rounding things off in section 6 with a conclusion.

## 2 FORMULATION OF THE PROBLEM

Standardisation is predominantly from the IEC 61400 design requirements. Another contributor is the DNV GL, equally as important, but have a different philosophy. But both do not to this date, sufficiently address the concerns of modern machine learning model development for digital enabled asset management. The challenge of standardising preprocessing is broken down into two sections, firstly how SCADA is gained and secondly, how it is implemented in digital-enabled asset management.

### 2.1 *Supervisory control and data acquisition*

To perform effective digital-enabled asset management of a wind turbine requires information during the operational lifetime. Fortunately, a significant effort has been placed into gathering data throughout an operational wind turbine from oil temperature readings of the gearbox to measuring the vibrations throughout the span of the wind turbine blade. This has helped, those who have access to data, to apply machine learning methods for a range of purposes. Mostly with the aim to reduce the operational expenditure and increase the return on investment.

The magnitude of the power extraction is based on two aspects, the operational input, and the meteorological factors. The focus of this case study is on the pre-processing procedure and not how all data is gathered, only the wind speed and power are considered to highlight the process. However, for performance-based condition monitoring more meteorological and operational features could be included to improve the model. The acquisition of information is the same, just the instrument, and the validation setup varies. Current standards on gathering meteorological data are described in (IEC 2012b).

The set-up for wind speed measurements must be at the hub height, and it is recommended in (IEC 2012b)) to include wind shear and veer, since these are sources of uncertainty for horizontal flow. The rotor equivalent wind speed can be measured to reduce this uncertainty, but it may require another sensor configuration (IEC 2012b). The wind speed is calculated using a cup anemometer for all power performance measurements and can be mounted in a variety of ways. Depending on the position of the mounting there are specific requirements. A classical representation of the various sensors and where they are placed within the nacelle are highlighted in Figure 1. In this example the anemometer is places on the top of the nacelle.

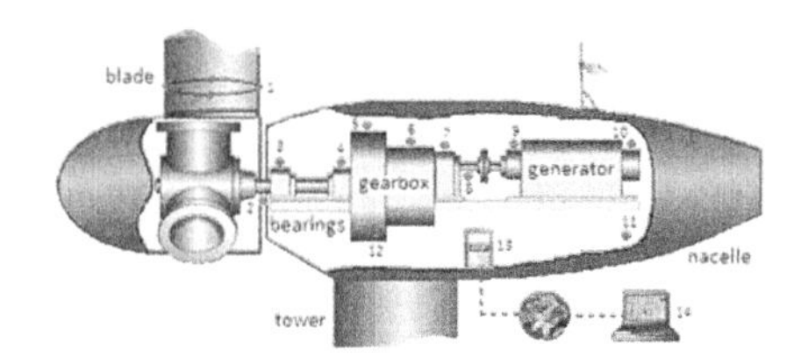


1 --- fibre optic transducers; 2, 8 --- speed transducers; 3, 4, 5, 6, 7, 9, 10, 11 --- accelerometers; 12 --- oil debris counter; 13 --- online CMS; 14 --- PC at control center.

Figure 1. A simplified graphic of classic sensor placement on the nacelle for a wind turbine (Yang, Tavner, Crabtree, Feng, & Qiu 2014).

An important aspect of all measurements from the supervisory control and data acquisition systems is the calibration. The sensor must be calibrated before installation in a wind tunnel to check that it can maintain the validity of its measurement period. If required, it is recommended in (IEC 2012b) to perform post calibration. Comparing initial results with the new information from the site and altering accordingly.

Certain features are more sensitive than others, for example, the air temperature does not fluctuate as much as say the wind, hence this is represented by higher fidelity measurements with wind speed being at least 1 Hz and the air temperature being of less importance. This information is aggregated into 10-minute periods derived from the continuous string of data. When the data is aggregated it must include statistical information such as mean, standard deviation, min, and max value. Some data rejection measures are included to ensure data obtained from the operation is only included, such as external conditions in particular wind speeds out of the operational range, manual shutdown, or specific atmospheric conditions.

After aggregating into 10-minute intervals, the data is normalised in five processes. These stages aim to improve the accuracy of the wind speed. The process is detailed in Figure 2. It includes references to each process. This entire process is to reduce the uncertainty from the wind turbulence in a wind farm.

An accurate estimation of the power is important for the power trending of the operational wind turbine. The net electric power measurement device is usually a power transducer. This establishes the voltage and current on each phase which can then be calculated into the active power. A class of 0.5 or higher must be implemented for MW wind turbines. Measuring the accuracy of the transducer can be

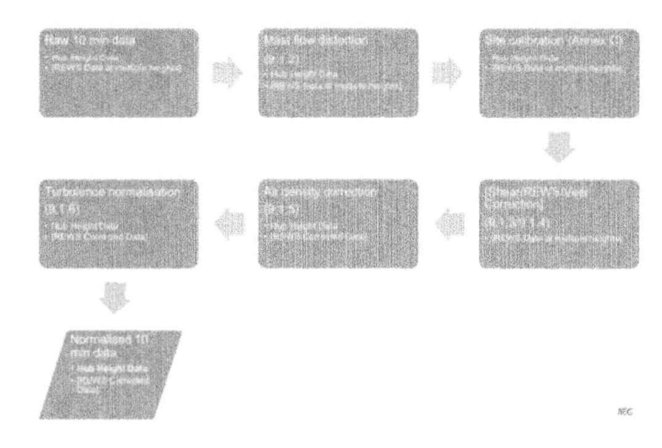


Figure 2. Process flow for wind speed pre-processing carried out in the (IEC 2012b).

calculated by flowing the procedure detailed the IEC standards (IEC 2017).

The data acquisition for all features is the same where higher frequency data is reduced to 10-minute intervals and the statistical attributes are calculated. The process is as follows, the sensor must be of a specific quality to ensure that the readings are of a certain standard. Once the sensor is placed on the operational wind turbine it will then be post-calibrated to ensure the readings are as expected. Recordings are taken, usually 1Hz, then the method of bins is applied. For complex parameters, a normalisation procedure will be included. Wind speed for example. For more information on SCADA data on mechanical loads and Acoustic noise signals see (IEC 2016) and (IEC 2012a) respectively.

### 2.2 *Trend condition monitoring models*

Trend monitoring systems evaluate the short-term and long-term trends in, the performance, oil temperatures, and the mechanical behaviour of wind turbines. Computational models are then used to determine these trends and they are compared against the actual value from the SCADA systems.

(Wilkinson et al. 2014), have successfully applied SCADA-based condition monitoring using an artificial neural network (ANN) applying the temperature of the oil in the drive train to reveal the patterns in the data. In this example, the data is just normalised. Another project carried out in (Catmull 2011) uses a self-organising map, this is an unsupervised learning ANN technique. There is no mention of the pre-processing procedure other than using the industry-standard 10-minute average data for 21 features. The self-organising map is used to determine the abnormality in the trends of, generator temperatures, reactive power, and gear winding temperatures.

The study from (Teng et al. 2018), performs a trend analysis on a direct drive wind turbine. In this instance the handling of the data is more complex, there is a variety of techniques used. Firstly, the bootstrap method is used for outlier detection, then the feature importance is determined using a random forest algorithm. After these methods are applied the most relevant features are used to train a deep neural network for the trend analysis.

SCADA monitoring can be carried out in a variety of ways. (Wilkinson et al. 2014), uses the torque/ rotational speed of the high-speed shaft, the RPM ratio from the low speed to the high-speed shaft, and the wind speed to determine the output power of the operational wind turbine using an adaptive ANN. In this project the data is; re-sampled to 1 Hz from 100Hz, clustered into regions using self-organised map algorithm, and then its split for training and testing purposes.

The Standards on trend analysis for wind turbines are limited. DNVGL-SE-0439 details the assessment on the validity of condition monitoring but no procedure. The IEC 61400-25 series on communications for monitoring and control of wind power plants offers insight into the transfer of information from a server to a client. In IEC 61400 25-3, it discusses a binning process again for vibrations on the power to aid in the process of alarms to deal with the no-linear data. Again, there is no beginning to end condition monitoring process describe.

Trend condition monitoring can be carried out in a variety of ways. Understandably, there are no standard models since the process of predicting values is rapidly changing with machine learning methods. The one area that is of concern is the handling of the data from server to model implementation. The articles discussed have included a pre-processing procedure and have not. How can one determine if that model is only suitable for that one data-set, a phenomenon called over-fitting, or in fact, is that model universally applicable? Standardising the pre-processing of dirty data and normalising it will aid in overcoming this phenomenon and streamline the condition monitoring model pipeline.

## 3 PRE-PROCESSING PROCEDURE

Computational algorithms require a presence of data in a mathematically feasible format, pre-processing is how that is achieved. Data processing techniques consist of data reduction, data projection, and missing data treatment. Data reduction aims to reduce the size of the data set through feature selection or case selection. Data projection intends to change the appearance of the data by scaling the features. Missing data treatment includes deleting missing values, and/or imputing them with estimates. These techniques will follow the process in Figure 3.

Effective SCADA data indexes are timestamped, and if it abides by the IEC standards it will be in 10-minute intervals at least. Depending on the provider the nacelle information, acoustics, mechanical loads, or METMAST data may be in different files. Concatenating all these together so that each row has the same timestamp ensures the relevant parameters are cohesive. In the situation where varying data sets are of different frequencies aggregating the entries such that the index is consistent.

Depending on the task knowing the property of the column may be important, for supervised learning, removing unnamed columns may be helpful, for unsupervised techniques discovering patterns and information is the main task, and removing columns will hinder this.

To deal with impurities (NANs, missing values, and anomalies) there are plenty of techniques (Myrtveit et al. 2001) discuss these in detail. Missing data treatment deletion methods,includes listwise or pairwise deletion. Pairwise the most popular according to (Huang et al. 2015), but this can lead to mass loss of information and may impact the results (Twala & Cartwright 2010). Alternatively, imputation methods, and according to (Strike et al. 2001) it is more effective since data is not lost. This can be carried out by; K- nearest neighbours, mean imputation, hot-deck imputation, cold deck imputation, regression imputation.

This paper includes the use of multivariate imputation using K-nearest neighbours, (Goldberger et al. 2004). The method applied implements the Minkowski distance to impute a value for the missing number or NaN for that feature. The Minkowski distance has the form:

$$\left(\sum_{i=1}^{n} |x_i - y_i|^p\right)^{1/p} \tag{1}$$

The Minkowski distance, also known as L2 when $p = 2$, is the distance between the two arbitrary points are xi – yi. Sensors do not have 100% up-time and are subject to failure and replacement. This technique deals with missing data and or NaN values.

Recently there has been significant work put into outlier detection, it can be carried out using; statistical methods, regression methods, or kernel-based methods. (Smiti 2020) provides an in-depth analysis of these techniques.

In this process, some insight has been included. For example, by setting hard limits on some of the variables such as wind speed or power. Another is knowing when a feature is a scalar and it can only have a positive value. Sensor errors are expected hence, removing these entries then applying the missing data treatment can improve the data quality.

Data projection is the process of normalising the data to exact, specified, and repeatable conditions. Scaling the values according to a defined rule such that all the features have the same degree of influence and thus the method is immune to units. Normally intervals of [0,1] or [-1,1] are used for the target.(Huang et al. 2015), has observed that [0,1] is mostly used and that is what has been implemented in this process. Calculated by:

$$[0,1]interval = \frac{x_i - min_x}{max_x - min_x} \tag{2}$$

Feature selection is the process of selecting a subset of properties listed in the data. The data-set may be of significance or irrelevance to the evaluation of the targeted outputs assuming that the data included has an impact on the accuracy of the model. In general, there are two methods filters and wrappers. Both evaluate the preset criteria independently before the machine learning method. Some of the most common methods comprise statistical methods such as Pearson's correlation or decision trees such as xgboost or LightGMB. To deal with the noisy data from an operational wind turbine A RandomeForest decision tree is implemented to perform the feature selection. The information gained from the RandomeForest is calculated by:

$$G(T, X) = Entropy(T) - Entropy(T, X) \tag{3}$$

Where $T$ is the target and $X$ is the feature split in its simplest form. The equation implemented is:

$$ni_j = w_j C_j - w_{left(j)} C_{left(j)} - w_{right(j)} C_{right(j)} \tag{4}$$

Where $ni_j$ is the impotence node, $wj$ is the weighted number of samples reaching $j$, $Cj$ is the impurity of the node, $C_{left}$ is the child node, and $C_{right(j)}$ is the child node from the right split. This will provide a number from 0-1 with the higher the gain the more significance that feature has on the desired output parameter.

The final technique for pre-processing is data splitting. This is one of the main procedures on how to evaluate the final model. Splitting the data into training, validation, and testing. The potion of these depends mostly on the size of the data-set.

The process flow for the implementation of all the techniques is addressed in Figure 3. This considers both supervised and unsupervised machine learning methods applied to operational wind turbines. There is scope for alternative techniques within the framework, as all standard models should. But the main takeaway from this standard process is how the data is concatenated, the known-known impurities removed, the known unknowns are controlled using feature selection and imputation. Finally, the data is scaled to provide a level playing field for all machine learning methods and continuity for other developers to work from.

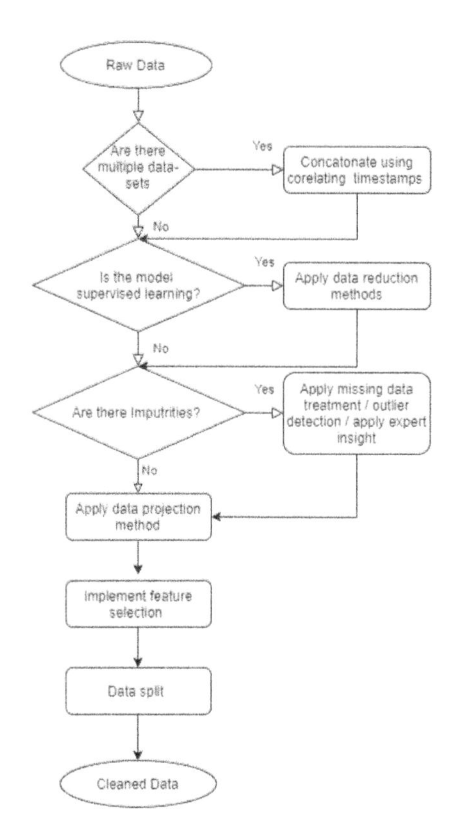


Figure 3. Flowchart detailing the procedure of taking raw data and applying techniques to clean data.

## 4 MACHINE LEARNING MODELS

There are vast and diverse amounts of regression machine learning methods. Starting with linear regression to more complex methods using transfer learning techniques. In this article 5 different methods have been included for their diversity, and different approaches to determine similar results to highlight that this process will aid in achieving improved results. Starting with; Artificial neural network, RandomeForrest, Gaussian process regression, xgboost, K-nearest neighbours, and finally Support vector Machine. This section will briefly describe these processes and the significant technique used for that process.

### 4.1 *Artificial neural network*

Artificial neural network (ANN) is a technique that takes inspiration from the functioning of the human brain. Where the neutrons work in parallel passing and storing information depending on the synaptic weights (activation function). The concept was first introduced by, (McCulloch & Pitts 1943)). Back-propagation (Rumelhart et al. 1986) revolutionised ANN's since it reduced computational time and efficiency.

More recently, the optimisation of activation functions improved, with different functions representing them, and the optimiser determining them. This article uses the adam method for stochastic optimisation for the activation functions described, (Kingma & Ba 2015). Adam, the adaptive learning method computes the learning rates for different parameters. Derived from the adaptive moment $m_n$ to the $n'h$ power,

$$m_n = E(X_n) \tag{5}$$

$X$ is a random variable. To estimate this, firstly, moving averages of the gradient-based on mini-batches are calculated using:

$$m_t = \beta_1 m_{t-1} + (1 - \beta_1)g_t \tag{6}$$

$$v_t = \beta_2 v_{t-1} + (1 -_2)g_t^2 \tag{7}$$

Where $v_t$ and $m_t$ are moving averages, $g_t$ is the gradient on the current mini-batch. $B_1, 2$ are the new additions from previous methods. These hyper-parameters have default values of 0.9 and 0.99 respectively. Since, $m_t$ and $vt$ are estimates of the first and second moments, they have the following property:

$$E(m_t) = E(g_t) \tag{8}$$

$$E(v_t) = E(g_t^2) \tag{9}$$

As you expand the value of $t$ a fewer number of values of the gradients contribute to the overall value. As they get multiplied by a smaller beta. This pattern is captured by:

$$m_t = (1 - \beta_1)\sum_{i=0}^{t} \beta_1^{t-i} g_i \tag{10}$$

Introducing a bias correction term for the momentum equations:

$$E(m_t) = E(g_i)(1 - \beta_1^{t-i}) + \varepsilon \tag{11}$$

This bias estimator is not just true for just Adam optimisation, this estimator holds for SGD with momentum and RMSprop among some methods.

### 4.2 *Support vector machine*

One of the most influential approaches to supervised learning is the support vector machine (Boser et al. 1992), (Cortes et al. 1995). One key innovation is the 'Kernel trick'. This consists of observing that many machine learning methods can be written as the dot product between examples. The kernel applied in this report is the radial basis function kernel (?):

$$K(x,x') = exp\frac{|x - x'|^2}{2\sigma^2} \quad (12)$$

Where two examples $x$ and $x'$, represented as feature vectors for an input space.

### 4.3 *K-nearest neighbours*

K-nearest neighbours the first of the non-parametric model implemented in this paper, where the complexity is a function of the training set size, unlike linear regression, for example, this has a fixed-length vector of weights, the nearest neighbour regression model simply stores the $X$ and $y$ from the training set. When asked to classify a test point $x$, the model looks up the nearest entry in the training set and returns the associated regression target. In other words:

$$\hat{y} = \sqrt{(x_i - y_i)^2 + (x_{i+1} - yi + 1)^2} \quad (13)$$

The algorithm can also be generalised to distance metrics other than the L 2 norm, such as learned distance metrics (Rasmussen & Williams 2005).

### 4.4 *Gaussian process regression*

The Gaussian process regression (GPR) is based on (Rasmussen & Williams 2005). This incorporates Bayesian interference, cross-validation, and gaussian noise. The kernel used is the squared exponential with Gaussian noise, as stated by, (Pandit et al. 2020) this is an effective method for extracting results for operational wind power regressions. It is described by:

$$K_{se}(x,x') = \sigma_f^2 exp\frac{|x - x'|^2}{2\sigma^2} + \sigma_f^2(x,x') \quad (14)$$

where $\sigma_f$ and $\sigma$ are defined as hyper-parameters represent the signal variance, and is a characteristic length scale that signifies how quickly the co-variance decreases with the distance between successive data points.

### 4.5 *Light gradient boosting decision tree*

The last method implemented is the light gradient boosting decision tree (LGBM), this method is developed based on the work from (Ke et al. 2017), on xgboost. But it differs in the growth of the trees. It is leaf-wise growth as appose to level-wise tree growth. The main principles of LGMB are to have the minimum objective loss, obj(0):

$$obj(0) = \sum_i^n l(y_i, \hat{y}_i) + \sum_{k=1}^K \Omega(f_k) \quad (15)$$

Where $k$ is the number of trees, $f$ is a function of the functional space $f$, and $f$ is all possible CARTs - a tree ensemble model consists of a set of classification and regression trees (CART). The loss function used in this model is theL2, with the regularisation term $\Omega$:

$$\omega(f) = \tau T + \frac{1}{2}\lambda\sum_{j-1}^T \omega_J^2 \quad (16)$$

Where $T$ is the number of leaves, $w$ is the vector score of the leaves, gamma and lambda are tuning hyper-parameters that require optimising. The growth of the tree structure is dependent on the gain:

$$gain = \frac{1}{2}\frac{G_L^2}{H_L - \lambda} + \frac{G_R^2}{H_R - \lambda} + \frac{(G_R + G_L)^2}{H_R + H_L + \lambda} - \tau \quad (17)$$

This formula can be decomposed as 1) the score on the new left leaf 2) the score on the new right leaf 3) The score on the original leaf 4) regularisation on the additional leaf. We can see an important fact here: if the gain is smaller than gamma, one would do better not to add that branch.

## 5 CASE STUDY

To validate the proposed framework for standardisation of SCADA data, a case study on an operational wind turbine is presented. The wind turbine considered is the Gamesa G97-2.0 MW IIA/III, designed for low to medium winds. The data-set of the wind turbine consists of 136 features, 39 of which are related to the meteorological mast data METMAST. The other 97 features consist condition monitoring sensor data, CMS data, parameters such as low-speed and high-speed generator RPMs. The data used in this report is of 10-minute internals with 102,082 time-steps (709 operational days). One can only assume that this data was taken from a higher frequency and processed into 10-minute averages as discussed in section 2.1. Only the power produced by the generator and the METMAST is included in this report for the power estimation model.

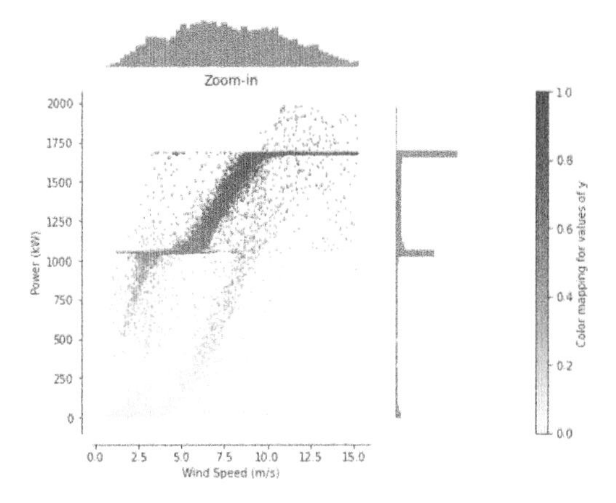


Figure 4. Scatter plot, with histograms on the x and y axis of unprocessed data, highlighting the issue of this data-set.

The data for this wind turbine has two significant modes operating around 1 MW and 2 MW, these two states are highlighted in the histogram distributed on the Y-axis. As expected, the wind speed distribution represents a standard normal distribution. The main concern is the distribution of the data points within the power. The power should have a similar distribution to the wind as they are inextricably linked however, various factors such as wind shear or operational philosophy may deter the distribution from that. In this case, it is unclear, with incredibly noisy data.

The variability of the trend in Figure 4 and the potential cause of outliers are discussed in detail in Section 2. The problem that needs addressing is how to implement this data such that the model created will can be effectively implemented for other wind turbines. The second aspect is, what strategy should be applied when there are multiple data streams of information being concatenated together. And lastly, how should one deal with the errors that are present in the transfer from sensor to the server and then to the client, some of the errors are highlighted in Table 1 for this data-set. The results section will answer all these issues and highlight the need for a consensus on how to deal with SCADA data for machine learning algorithms.

Table 1. Table of a few features from the SCADA data.

| Feature | Total Missing | Percent Missing(%) |
|---|---|---|
| Instantaneous air pressure 10M calc.(mmhg/mb) | 18658 | 18.28 |
| Instantaneous relative humidity 10M calc. (%) | 19338 | 18.94 |
| Instantaneous temperature 10M calc.(°C) | 19800 | 19.4 |
| Maximum direction 10M calc. (Height1) (°) | 18658 | 18.28 |
| Maximum horizontal speed 10M calc. (Height4) (m/s) | 18658 | 18.28 |
| Minimum horizontal speed 10M calc. (Height1)(m/s) | 18658 | 18.28 |
| Unnamed: 0 | 102082 | 100 |
| ...... | ...... | ...... |
| ...... | ...... | ...... |

### 5.1 *Error assessment*

In this case study the regression results are asses by using the R2 score, this represents the variance in the dependant variables that is predictable from the independent variables. This has the form:

$$R2 = 1 - \frac{\sum_i (y_i - f_i)^2}{\sum_i (y_i - \hat{y})^2} \tag{18}$$

Where, $y_i$ is the real real values, $f_i$ is the values associated to $y_i$, and $\hat{y}$ is the predicted value from the model.

## 6 RESULTS

The flow of this section will follow the process highlighted in Section 3 Figure 3. The trending monitoring model will both apply all the processes and exclude all of them to compare the results. The input parameters for the model are the METMAST data and the output data is the power.

The first set is addressing the issue of the separate two data-sets. Fortunately, this data set is time-stamped. And in this situation, the two data streams are taken every 10 minutes on every 10 minutes of the hour. Making this process a simple search, removing any entries that do not correlate, and concatenating them.

The trending condition monitoring technique used involves a regression, supervised learning machine learning model. Hence, unnamed features are removed.

Removing impurities, is the area of most creativity. In this procedure I have implemented specific techniques however, some alternatives might be more suitable to alternative data-sets. Also, new techniques may have been developed. But the process of dealing with impurities and the discussion of how they are implemented needs to be portrayed when discussing your pre-processing procedure.

The first step is dealing with missing entries, this report implements K-NN a multivariate imputation method. As appose to this, missing entries must be removed from the raw data. Resulting in a smaller data-set. For machine learning methods, the larger the data-set tends to improve the performance of the model.

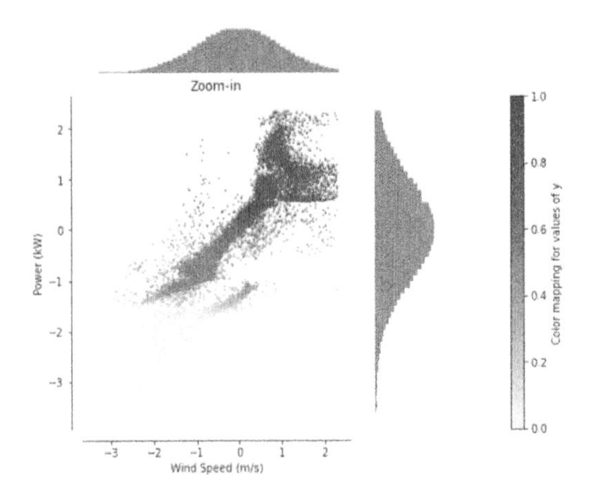


Figure 5. Scatter plot, with histograms on the x and y axis after processing.

In higher frequency models such as some financial trading models or vibration signals sub 1 Hz, the computational effort may not be practicable.

The last step is crucial for the evaluation. This is a mandatory step but the ratio of splitting the data should be dependent on the size of the data-set. The ratios implemented for training validation and testing are 0.7, 0.1, 0.2 respectively.

The purpose of standardising wind turbine SCADA data for machine learning is to transform the data into a format that can be used more effectively. Condition monitoring can be as simple as finding patterns in data that do not conform to normal behaviour. The Monte Carlo simulation is carried out over 100 iterations for each of the machine learning models to determine the r2 score variance. The preprocessing improves accuracy in which the data is observed across all models, highlighted in Figure 6.

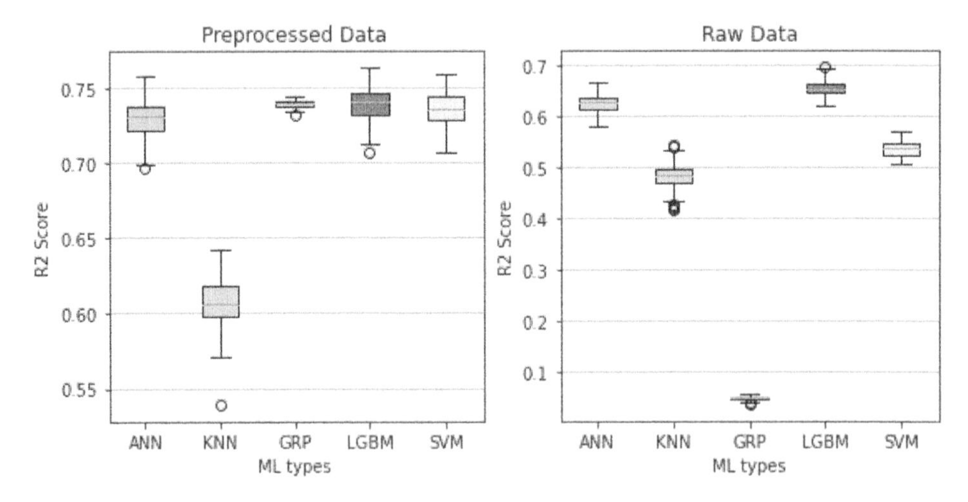


Figure 6. Direct comparison of raw vs clean data detailing the distribution of R2 scores after the Monte Carlo simulation.

Dealing with outliers is important. And it must be done with care, as one does not want to remove too much variance in the data such that the model overfits and cannot provide effective results from new, unseen data. The first procedure taking is removing all vector data from all the scalier features. Feature projection is especially important in the case of this data, with the power having to modes with significantly more data points. When training models this can lead to reduced performance. Figure 5 highlights how projection methods can redistribute data to improve performance. This technique is only applicable to this type of model since the data is in 10-minute intervals. When implementing this into a continuous deployment environment new data points can be incorporated into the previous data set and then converted to the quantile Gaussian distribution technique with ample time. The most significant improvement is the GRP improving from less than 10% accuracy to 73%.

## 7 CONCLUSION

Determining the power using a trending method for a wind turbine is made challenging with the stochastic nature of the wind and the operational states defined by the wind turbine. By creating a global model, this paper has presented a method of trend condition monitoring that uses METMAST data as input and power as output data. 5 machine learning techniques; ANN, K-NN, SVM, LGBM, and GPR have experienced improvements using the standardisation process with the accuracy and variance are improved dramatically for this data-set.

The work in this paper has focused on trending condition monitoring, the techniques implemented on the standardisation of SCADA data. This work will apply to machine learning methods within the wind turbine SCADA data realm.

This paper has highlighted that there is an inconsistency in the application of pre-processing SCADA data for operational wind turbines. This paper addresses this issue, the process should follow the predetermined process outlined in section 2 and this should be documented. This paper has highlighted a method from beginning to end that should be implemented as it will improve the accuracy of your results. But I do believe that this should not be the final form, the steps taken are necessary but the techniques in each process are interchangeable. The purpose of this procedure and study is to highlight the need for this process in standards so that the wind turbine industry has a consistent data pre-processing, and the focus can move onto other aspects, such as model development, removing doubt by highlighting the effective procedure. Collaborative machine learning model development, since there will be a standardised base for all model development. And lastly, transparency in model evaluation since the procedure of how the data is organised will be less ambiguous.

## REFERENCES


Black, I. M., M. Richmond, & A. Kolios (2021, 3). Condition monitoring systems: a systematic literature review on machine-learning methods improving offshore-wind turbine operational management. *International Journal of Sustainable Energy*.

Boser, B. E., I. M. Guyon, & V. N. Vapnik (1992). A training algorithm for optimal margin classifiers. ACM Press.

Catmull, S. (2011, 6). Self-organising map based condition monitoring of wind turbines.

Cortes, C., V. Vapnik, & L. Saitta (1995). Support-vector networks editor.

Crabtree, C. J., D. Zappalá, & S. I. Hogg. Wind energy: Uk experiences and offshore operational challenges.

da Rosa, A. (2013, 1). Wind energy.

Energy, W. (2020). Wind energy in europe 2020 statistics and the outlook for 2021-2025.

Goldberger, J., G. E. Hinton, S. Roweis, & R. R. Salakhutdinov (2004). Neighbourhood components analysis. *Advances in neural information processing systems 17*, 513–520.

Huang, J., Y. F. Li, & M. Xie (2015, 11). An empirical analysis of data preprocessing for machine learning-based software cost estimation. Volume 67, pp. 108–127. Elsevier.

IEC (2012a). 61400-11 acoustic noise measurement techniques.

IEC (2012b). 61869-2 instrument transformers - part 2: Additional requirements for current transformers.

IEC (2016). 61400-13 wind turbines. measurement of mechanical loads.

IEC (2017). 61400-12-1 power performance measurements of electricity producing wind turbines.

Ke, G., Q. Meng, T. Finley, T. Wang, W. Chen, W. Ma, Q. Ye, & T.-Y. Liu (2017). Lightgbm: A highly efficient gradient boosting decision tree. Volume 30. Curran Associates, Inc.

Kingma, D. P. & J. L. Ba (2015). Adam: A method for stochastic optimization.

McCulloch, W. & W. Pitts (1943). A logical calculus of the ideas immanent in nervous activity. *The Bulletin of Mathematical Biophysics 3*, 115–133.

Myrtveit, I., E. Stensrud, & U. Olsson (2001). Analyzing data sets with missing data: An empirical evaluation of imputation methods and likelihood-based methods. *IEEE Transactions on Software Engineering 27*, 999–1013.

Pandit, R. K., D. Infield, & A. Kolios (2020, 11). Gaussian process power curve models incorporating wind turbine operational variables. *Energy Reports 6*.

Rasmussen, C. E. & C. K. I. Williams (2005). *Gaussian Processes for Machine Learning (Adaptive Computation and Machine Learning)*. The MIT Press.

Rumelhart, D. E., G. E. Hinton, & R. J. Williams (1986, 10). Learning representations by back-propagating errors. *Nature 323*.

Smiti, A. (2020, 11). A critical overview of outlier detection methods. *Computer Science Review 38*.

Strike, K., K. E. Emam, & N. Madhavji (2001). Software cost estimation with incomplete data. *IEEE Transactions on Software Engineering 27*.

Teng, W., H. Cheng, X. Ding, Y. Liu, Z. Ma, & H. Mu (2018, 7). Dnn-based approach for fault detection in a direct drive wind turbine. *IET Renewable Power Generation 12*.

Twala, B. & M. Cartwright (2010, 5). Ensemble missing data techniques for software effort prediction. *Intelligent Data Analysis 14*.

Verbert, K., B. D. Schutter, & R. Babuška (2017, 3). Timely condition-based maintenance planning for multi-component systems. *Reliability Engineering and System Safety 159*, 310–321.

Wilkinson, M., B. Darnell, T. Delft, & K. Harman (2014, 5). Comparison of methods for wind turbine condition monitoring with scada data. *IET Renewable Power Generation 8*.

Yang, W., P. Tavner, C. Crabtree, Y. Feng, & Y. Qiu (2014, 05). Wind turbine condition monitoring: Technical and commercial challenges. *Wind Energy 17*.

*Trends in Maritime Technology and Engineering – Guedes Soares & Santos (eds)*


# Dynamic performance prediction of hywind floating wind turbine based on SADA method and full-scale measurement data

P. Chen & Z.Q. Hu
*Marine, Offshore and Subsea Technology Group, School of Engineering, Newcastle University, Newcastle Upon Tyne, UK*

C.H. Hu
*Research Institute for Applied Mechanics, Kyushu University, Kasuga, Fukuoka, Japan*

ABSTRACT: The highly coupled nonlinear performances of Floating Wind Turbines (FWTs) bring many challenges to the design and optimization of FWTs. This paper aims introduce a case study by using full-scale data through the SADA method and the full-scale data was collected by one of *Hywind* FWTs in Scotland. The methodology of the SADA method was first proposed by Chen and Hu, which consists of KDPs concepts, the *DARwind* program, and the application of AI algorithms. In this paper, the dynamic performance of *Hywind* FWT will be discussed in terms of platform motions, tower top, and blade tip deformation. The results show that SADA can predict the supporting floater motions with higher accuracy, though some design parameters are not accessible, and the numerical models are simplified. In summary, the SADA is a reliable and cost-effective method for dynamic performance analysis of FWTs, which can bring an innovative vision in engineering applications.

## 1 INTRODUCTION

Floating Offshore Wind Turbines (FWTs) have been acknowledged to be the workhorse of the next generation of wind energy harvest devices. However, the environmental loads applied on FWTs, and their dynamic performances are much more complicated than those of fixed-bottom wind turbines. There are still many gaps to be filled in FWTs design and optimization (Chen *et al.*, 2021b). Therefore, scholars and wind industry are constantly trying to overcome these challenges in FWTs for reliable analysis and predictions, and full-scale FWTs measurement is one of the useful tools

In recent years, full-scale FWTs projects are also rapidly progressing around the world, for example, the world's first floating wind farm Hywind Scotland pilot park (Equinor, 2019) has been successfully running for more than 4 years. Early researches on the comparison of measured data and simulated responses of the Hywind demonstration project can be found in reference by Hanson *et al.* (2011). More verifications of the Hywind prototype can be found in the works of literature (Skaare *et al.*, 2015; Driscoll *et al.*, 2016). Some other existing prototype projects of FWTs are also summarized in Table 1 (Chen *et al.*, 2020).

In addition, with the development of Artificial Intelligence (AI) technology, many cross-discipline technologies have been proposed in offshore engineering applications. The most extensive applications on a wind turbine are mainly in the two fields of power forecasting (Khan *et al.*, 2020) and condition monitoring (Stetco *et al.*, 2019). For FWTs design and optimization, there are fewer applications of AI technology so far.

This paper aims to demonstrate a case of study on the application of an AI-based method, named SADA, for the dynamic performance prediction of one Scotland Hywind FWT by using full-scale measurement data. The valuable full-scale measurement data of *Hywind* Scotland is provided by Equinor and ORE Catapult. The detail of *Hywind* FWT will be introduced in section 2. The SADA method was proposed by Chen *et al.* (2021a), which will be introduced in section 3 including the KDPs (Key Disciplinary Parameters) concept, *DARwind* program, and application of AI algorithm. Finally, the results of SADA prediction of platform motions, blade tip moment, and tower top deformation are discussed.

## 2 HYWIND SCOTLAND

This section introduces the floating wind farm of *Hywind* Scotland. More detail can be found in (Equinor, 2019)

*Hywind* Scotland is located off the east coast of Scotland, and there are 5 floating turbines and

DOI: 10.1201/9781003320289-37

Table 1. Some prototype projects of FWTs.

| Project | Company | Capacity and Site |
|---|---|---|
| WindFloat (Power, 2021) | Principle Power | 2MW in Aguçadoura, Portugal (2011) and Kincardine, Scotland (2018) |
| Damping Pool (Ideol, 2021) | Ideol | 2MW in Le Croisic, France (2018) and 3MW in Kitakyushu, Japan (2018) |
| Hywind (Equinor, 2019) | Equinor | 2.3MW in Karmøy, Norway (2009) |
| Blue H (Engineering, 2021) | Blue H Technologies | 2.4 MW in Brindisi, Italy (2009) |

a wave buoy at the site as shown in Figure 1. The full-scale measurement data is from FWT HS4, which has been circled in red in Figure 1.

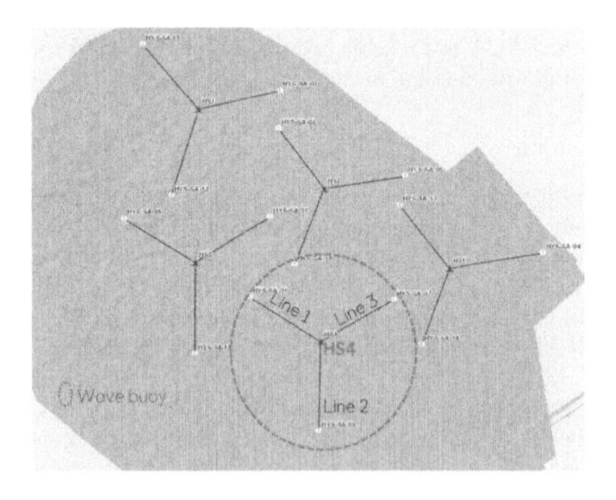


Figure 1. Overview of the Hywind Scotland and location of HS4 in the farm.

The main scantlings of the supporting Spar-type floater can be found in Table 2. The measurements from HS4 *Hywind* FWT for 3 operational cases are selected to conduct the dynamic performance prediction by SADA. The typical operational cases used in this paper with a range in wind speed and wave height are given in Table 3. The $H_s$ is significant wave height; $T_p$ is wave period; $V_w$ is the wind speed; $V_c$ is the current speed.

Table 2. Main scantlings of the floater.

| Z, bottom [m] | Z, top [m] | Length [m] | Diameter bottom [m] | Diameter top [m] |
|---|---|---|---|---|
| -77.6 | -19.6 | 58 | 14.4 | |
| -19.6 | -4.6 | 15 | 14.4 | 9.45 |
| -4.6 | 7.4 | 12 | 9.45 | |

The measured data used in this paper only includes 5DOF motions of the floater, and heave motion data is not provided. Besides, detailed structural parameters of the SWT-6.0-154 wind turbine is not available in the published materials. Therefore, there are several assumptions used in this investigation in this paper:

1) The wind, wave, and current are assumed to be in the same direction in the simulation.
2) Only steady wind and ocean current acting on the floating body is considered.
3) The parameters of the NREL 5MW Hywind wind turbine in OC3 (Jonkman, 2010) are used instead of the wind turbine of HS4.
4) The delta-connection of mooring lines is simplified in the numerical modelling which is directly connected to the hull, and an additional yaw stiffness is added.

Table 3. Cases matrix in measurement.

| Case No. | $H_s$ | $T_p$ | $V_w$ | $V_c$ |
|---|---|---|---|---|
| 1 | 2.1 | 6.5 | 15.5 | 0.27 |
| 2 | 3.2 | 9.3 | 13.9 | 0.09 |
| 3 | 3.0 | 7.9 | 16.6 | 0.33 |

## 3 METHODOLOGY

This section briefly introduces the SADA (**S**oftware-in-the-Loop combine **A**rtistical intelligence technology for **D**ynamic **A**nalysis of FWTs) method including the KDPs (Key Disciplinary Parameters) concept, *DARwind* program, and application of AI algorithm. In addition, the selected KDPs will also be given.

### 3.1 *SADA method*

SADA is a novel concept integrating the AI technology (Artificial Neural Networks and Deep Reinforcement Learning) and *DARwind* (a coupled aero-hydro-servo-elastic in-house program) for optimized design and dynamic performance prediction of FWTs. For more specific information about *DARwind*, please refer to the published literature (Chen *et al.*, 2019). Figure 2 shows the general introduction of the entire SADA algorithm. More detail of SADA can be found in the works of literature (Chen *et al.*, 2021a; Chen *et al.*, 2021b; Chen *et al.*, 2021c).

Firstly, the concept of KDPs is of most importance in SADA, working as a data transmission interface between AI technology and disciplines in numerical calculations. Specifically, the KDPs can be divided into three categories (Environmental KDPs, Disciplinary KDPs, Specific KDPs), which have their unique boundary conditions in SADA.

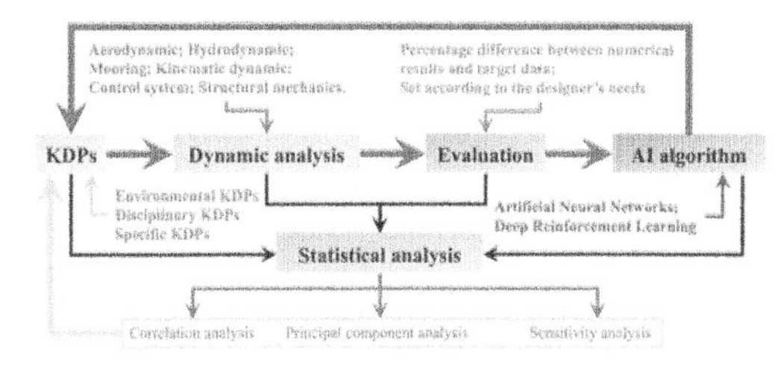


Figure 2. General introduction of the SADA method.

Secondly, based on the DRL (Deep Reinforcement Learning) framework, *DARwind* can be trained to be smarter and can give more accurate predictions for FWTs by self-learning on experimental or full-scale measured data. The specific notations and nouns combined FWTs and DRL in SADA are:

- **Agent:** *DARwind*
- **State(S):** The numerical analysis results from *DARwind*.
- **Action(A):** Adjusting the KDPs.
- **Reward(R):** Evaluation between numerical results and target data.

As an agent, *DARwind* will take continuous action through the SIL (Software-in-the-Loop) algorithm, i.e., adjusting KDPs appropriately to obtain more accurate prediction results and resilience to different environmental situations. At this point, the roles of the deep neural networks are amplified to record [state, action, reward, next state] in different situations through SIL. In addition, there will be percentage differences evaluation process to judge the current dynamic performance of FWTs. A good percentage differences definition can more accurately analyze the characteristics of KDPs, and it makes the optimization of SADA more efficient. In this paper, only the mean value is considered as the target value for each platform DOF motion. The variation of percentage differences ($P_d$) is defined as:

$$P_i = \left| \frac{O_t - O_i}{O_t} \right| \times 100\% \tag{1}$$

$$P_p = \left| \frac{O_t - O_w}{O_t} \right| \times 100\% \tag{2}$$

$$P_d = P_i - P_p \tag{3}$$

The $O_t$ is the output experimental physical quantity. The $O_i$ is the numerical results by initial KDPs by *DARwind*. $O_w$ is the AI training results by weighted KDPs by *DARwind*. The $P_d$ is used to measure whether the results of SADA are better than the initial KDPs. If the $P_d$ is positive, it means that the percentage differences between full-scale data and numerical simulation have decreased by SADA, otherwise the percentage differences have increased.

Thirdly, SADA can also build a data set to record the fluctuating of KDPs and FWTs' performance during AI training. Through main component analysis and correlation analysis, more deep-seated physical laws can be found in the dynamic performance of FWTs based on SADA's data set.

Table 4. Selected KDPs.

| No. | | KDPs |
|---|---|---|
| 1 | | Wind speed |
| 2 | Aero | *Glauert* correction |
| 3~13 | | Tower drag |
| 14 | | Cone Angle of blade |
| 15 | | Current speed |
| 16 | | Added static force |
| 17-22 | | Added linear viscous damping matrix |
| 23-28 | Hydro | Added quadratic viscous damping matrix |
| 29-34 | | Added linear restoring matrix |
| 35-40 | | Added linear mass matrix |
| 41 | | Significant wave height |
| 42 | | Peak period |
| 43 | | Shape factor |
| 44 | | Water depth |
| 45-47 | | Wet density |
| 48-50 | Mooring | Axial stiffness |
| 51-53 | | Length |
| 54-56 | | Diameter |
| 57 | | Hub mass |
| 58 | | Shaft angle of the rotor |
| 59 | | Nacelle mass |
| 60 | | Floater mass |
| 61 | | Floater volume |
| 62-64 | | Hub reference point. |
| 65-67 | | Hub dynamic reference point |
| 68-70 | | Hub inertia about the rotor axis. |
| 71-73 | Kinematics & Structural | Generator inertia about HSS |
| 74-76 | | Nacelle reference point |
| 77-79 | | Nacelle dynamic reference point |
| 80-82 | | Nacelle inertia |
| 83-85 | | Floater reference point |
| 86-88 | | Center of Floater mass |
| 89-91 | | Floater inertia |
| 92-96 | | Polynomial Flap 1$^{st}$ vibration modes |
| 97-101 | | Polynomial Flap 2$^{nd}$ vibration modes |
| 102-106 | | Polynomial Edge 1$^{st}$ vibration modes |
| 107 | | Generator torque constant |
| 108 | Servo | Gearbox ratio |

### 3.2 *KDPs selection*

In this paper, 108 KDPs are selected divided into categories. For example, the tower is divided into 11 station in aerodynamic simulation, so there are 11 KDPs of the drag coefficient of the tower. For the added viscous damping coefficient matrix and the restoring force coefficient matrix, only the diagonal parameters are considered in hydrodynamics. In summary, Table 4 shows the selected KDPs in this paper.

## 4 RESULT AND DISCUSSION

This section demonstrates and discusses the analysis results of SADA calculation in terms of platform motions, tower top, and blade tip deformation.

### 4.1 *Platform motions prediction*

In this part, SADA will use the average value of the 5 motions provided by the measured data of the Floater motions for AI training. Table 5 shows the percentage differences ($P_d$) of platform motions prediction. Among them, the percentage differences reduction of surge and pitch motion is the largest. Figure 3 shows the stacking diagram of the percentage differences variation under the 3 working conditions.

Table 5. Percentage differences of platform motions prediction.

| Case No | Case 1 | Case 2 | Case 3 |
|---|---|---|---|
| Surge | 29.77 | 33.17 | 41.33 |
| Sway | -4.18 | -3.96 | -23.89 |
| Roll | 11.27 | 9.22 | -30.16 |
| Pitch | 172.84 | -3.67 | -3.32 |
| Yaw | -68.64 | 8.27 | -35.6 |

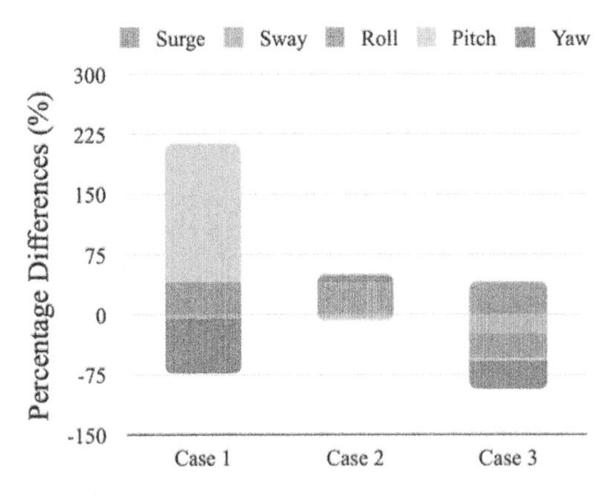


Figure 3. Stacked graph of percentage differences of platform motions.

Table 6 shows the amplitude changes of the five degrees of freedom under three working conditions. Combining Figure 4, it can be clearly found that although the percentage differences of some motions have increased but the amplitude change is particularly small. This is also due to the intelligent performance of the SADA method. By optimizing the typical motion response of *Hywind* FWT, physical quantities with little impact are discarded. Figure 5 shows the time history curve of the surge motion in case 3. It is clearly shown that after SADA simulation, the surge motion curve is much closer to that of full-scale measured data, compared with the non-trained *DARwind* program.

Table 6. Amplitude changes of platform motions.

| Case | Surge | Sway | Roll | Pitch | Yaw |
|---|---|---|---|---|---|
| **Case 1** | 0.94 | -0.08 | 0.01 | 0.37 | -0.11 |
| **Case 2** | 2.94 | -0.05 | 0.04 | -0.12 | 0.01 |
| **Case 3** | 4.58 | -0.04 | -0.05 | -0.16 | -0.06 |

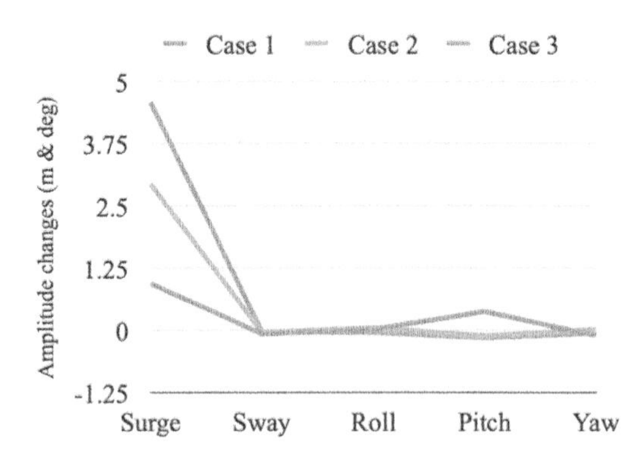


Figure 4. Amplitude changes of platform motions.

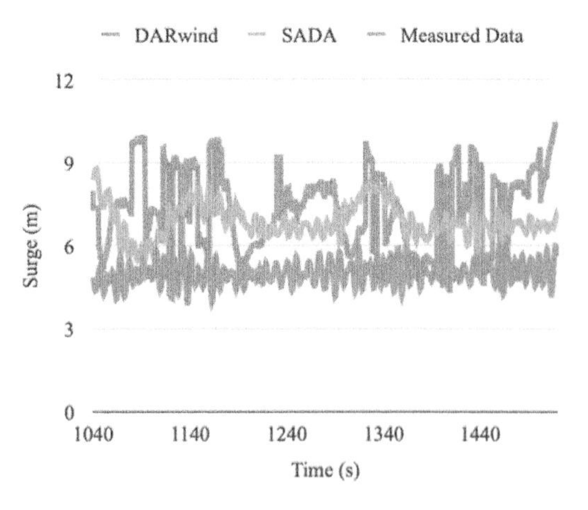


Figure 5. Time history of surge motions in Case 3.

## 4.2 *Tower top and blade tip deformation*

In this part, based on the 5DOF optimization of SADA, the prediction of other physical quantities is demonstrated, including the blade tip and tower top deformations. Figure 6 shows the blade deformations between models.

Table 7 shows the statistical results of the deformation of the tower top and blade tip in percentage differences. This statistical value is the result of subtracting the result of SADA without AI optimization. In the physical quantities in the table, the capital letter "T" stands for the tower, and "B1" stands for the first blade. The second lowercase letter represents the direction of deformation. The last small number represents the actual working condition number.

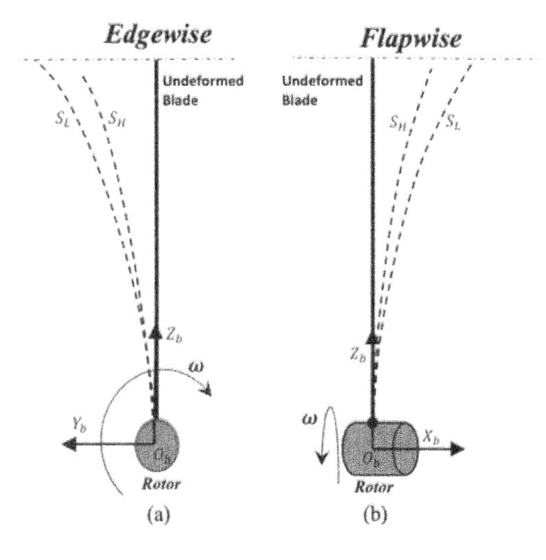


Figure 6. The blade deformations between models.

Table 7. Statistics of tower top and blade tip deformation between SADA and *DARwind*.

| Quantity | Avg | Std | Min | Max |
|---|---|---|---|---|
| Tx1 | -6.4E-03 | -2.2E-03 | 1.3E-03 | -1.2E-02 |
| Ty1 | -1.2E-03 | 5.4E-04 | -3.1E-03 | 1.2E-03 |
| B1x1 | -1.3E-01 | -9.8E-03 | -1.7E-01 | 1.0E-03 |
| B1y1 | -6.6E-03 | -4.5E-03 | 1.7E-03 | -8.8E-03 |
| B1z1 | 1.2E-02 | -1.6E-03 | 4.8E-04 | 1.2E-02 |
| Tx2 | -8.9E-04 | -1.9E-04 | 3.8E-04 | -2.5E-03 |
| Ty2 | 2.6E-04 | -2.6E-04 | 2.0E-03 | 6.0E-05 |
| B1x2 | -2.9E-01 | -2.2E-02 | -3.1E-01 | -2.1E-01 |
| B1y2 | -6.1E-03 | -6.3E-03 | 2.4E-03 | -1.9E-02 |
| B1z2 | 2.6E-02 | -4.3E-03 | 2.6E-02 | 1.8E-02 |
| Tx3 | -1.1E-03 | 1.3E-03 | -8.4E-04 | -2.0E-03 |
| Ty3 | -2.0E-03 | -4.8E-04 | -1.4E-03 | -1.7E-03 |
| B1x3 | -7.8E-03 | -1.4E-02 | -9.8E-03 | -1.1E-01 |
| B1y3 | 2.7E-03 | 1.2E-02 | -1.6E-02 | 2.7E-02 |
| B1z3 | 1.4E-05 | -5.6E-05 | 1.9E-04 | -2.0E-07 |

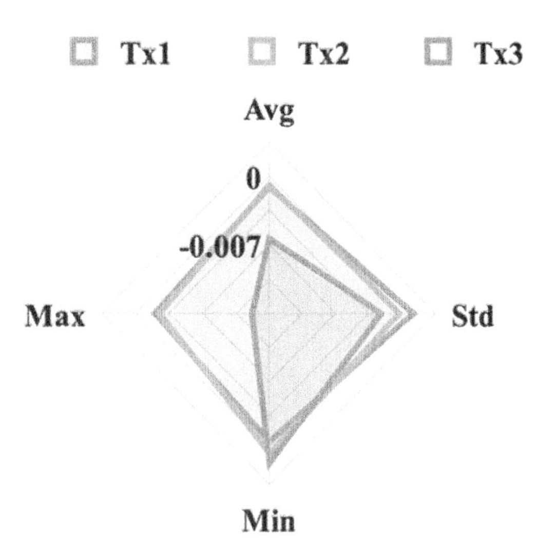


Figure 7. Radar chart of statistical changes of tower top.

From the perspective of the deformation of the tower top in the x-direction, Figure 7 shows the performance of the radar chart of the four statistical changes. The deformation at the top of the tower after SADA optimization is smaller than the result without AI optimization. The same conclusion can also be observed in Figure 8. Especially in case 3, the mean value of blade tip deformation decreases significantly in the x-direction.

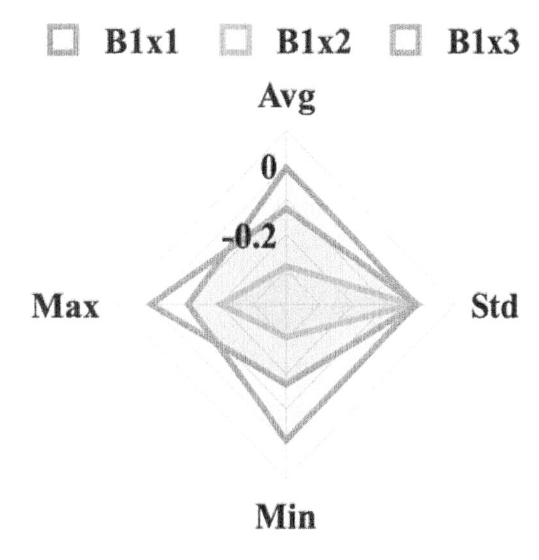


Figure 8. Radar chart of statistical changes of the blade tip.

Figures 9 and 10 show a time history of blade tip deformation variation. The red and black lines in the figure represent the average of the two curves respectively. It is not difficult to be pointed out from the figure that the SADA forecast of blade tip deformation is smaller than the forecast

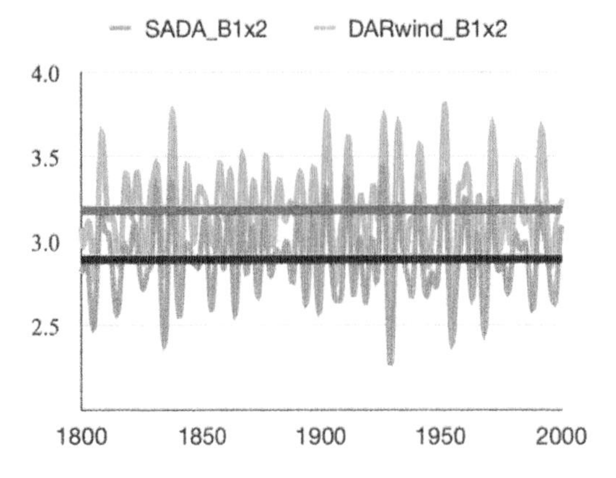


Figure 9. Time history of blade tip deformation in the x-direction.

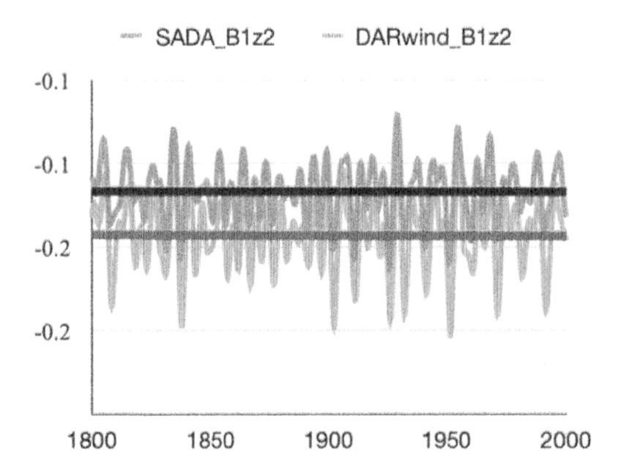


Figure 10. Time history of blade tip deformation in the z-direction.

results without AI optimization. Since SADA's platform motion is more accurate, it is believed that the results from AI-trained *DARwind* are more credible.

## 5 CONCLUSIONS

This paper presents a case of study to the dynamic performances prediction of FWTs by using full-scale data and a novel artificial intelligence (AI) technology-based method, named SADA. The results show that through the optimization of SADA, more accurate prediction results of platform motions can be obtained. The deformation of the tower top and blade tip of *Hywind* FWT is more reasonable than those without AI optimization. It is proved that SADA is a promising and reliable method for dynamic performance analysis tool and will benefit wind industry in the future.

## ACKNOWLEDGEMENTS

Thanks to Equinor for sharing operational data of Hywind Scotland through ORE Catapult, and the relevant link can be found at https://pod.ore.catapult.org.uk/. In addition, a great appreciation is given to Joint Usage/Research Center for Applied Mechanics of the Research Institute for Applied Mechanics, Kyushu University, for the kind support on this research (Project No. No.14_21RE-4).

## REFERENCES

Chen, J., Hu, Z., Liu, G. and Wan, D. (2019) 'Coupled aero-hydro-servo-elastic methods for floating wind turbines', *Renewable Energy*, 130, pp. 139–153.

Chen, P., Chen, J. and Hu, Z. (2020) 'Review of Experimental-Numerical Methodologies and Challenges for Floating Offshore Wind Turbines', *Journal of Marine Science and Application*, 19(3), pp. 339–361 10.1007/s11804-020-00165-z.

Chen, P., Chen, J. and Hu, Z. (2021a) 'Software-in-the-Loop Combined Reinforcement Learning Method for Dynamic Response Analysis of FOWTs', *Frontiers in Marine Sciense*, 7(1242) https://doi.org/10.3389/fmars.2020.628225.

Chen, P., Jia, C., Ng, C. and Hu, Z. (2021b) 'Application of SADA method on full-scale measurement data for dynamic responses prediction of Hywind floating wind turbines', *Ocean Engineering*, 239, p. 109814 https://doi.org/10.1016/j.oceaneng.2021.109814.

Chen, P., Song, L., Chen, J.-h. and Hu, Z. (2021c) 'Simulation annealing diagnosis algorithm method for optimized forecast of the dynamic response of floating offshore wind turbines', *Journal of Hydrodynamics*, 33(2), pp. 216–225 https://doi.org/10.1007/s42241-021-0033-9.

Driscoll, F., Jonkman, J., Robertson, A., Sirnivas, S., Skaare, B. and Nielsen, F.G. (2016) 'Validation of a FAST Model of the Statoil-hywind Demo Floating Wind Turbine', *Energy Procedia*, 94, pp. 3–19 https://doi.org/10.1016/j.egypro.2016.09.181.

Engineering, B.H. (2021) *Blue H Engineering*. Available at: http://www.bluehengineering.com/index.html.

Equinor (2019) *Equinor and ORE Catapult collaborating to share Hywind Scotland operational data*. Available at: https://www.equinor.com/en/news/2019-11-28-hywind-scotland-data.html (Accessed: 18/12).

Hanson, T.D., Skaare, B., Yttervik, R., Nielsen, F.G. and Havmøller, O. (2011) *Proceedings of the European Wind Energy Association Annual Event*.

Ideol (2021) *A FLOATING FOUNDATION THAT PUSHES THE BOUNDARIES OF OFFSHORE WIND*. Available at: https://www.ideol-offshore.com/en.

Jonkman, J. (2010) *Definition of the Floating System for Phase IV of OC3*. National Renewable Energy Lab. (NREL), Golden, CO (United States). [Online].

Khan, N.M., Khan, G.M. and Matthews, P. (2020) *IFIP International Conference on Artificial Intelligence Applications and Innovations*. Springer.

Power, P. (2021) *WindFloat*. Available at: https://www.prin-ciplepowerinc.com/en/windfloat (Accessed: 28/12).

Skaare, B., Nielsen, F.G., Hanson, T.D., Yttervik, R., Havmøller, O. and Rekdal, A. (2015) 'Analysis of measurements and simulations from the Hywind Demo floating wind turbine', *Wind Energy*, 18(6), pp. 1105–1122.

Stetco, A., Dinmohammadi, F., Zhao, X., Robu, V., Flynn, D., Barnes, M., Keane, J. and Nenadic, G. (2019) 'Machine learning methods for wind turbine condition monitoring: A review', *Renewable Energy*, 133, pp. 620–635 https://doi.org/10.1016/j.renene.2018.10.047.

*Trends in Maritime Technology and Engineering – Guedes Soares & Santos (eds)*


# Preliminary numerical optimization of the E-Motions wave energy converter

D. Clemente, P. Rosa-Santos & F. Taveira-Pinto
*Department of Civil Engineering – Faculty of Engineering of the University of Porto (FEUP), Porto, Portugal*
*Interdisciplinary Centre of Marine and Environmental Research (CIIMAR), Porto, Portugal*

P.T. Martins
*Portuguese Navy, Ministry of National Defence, Almada, Portugal*

ABSTRACT: This paper summarizes the preliminary outcomes of a recent numerical study towards the improvement of E-Motions. This is a wave energy converter capable of protecting sensitive equipment whilst converting energy from wave/wind induced roll oscillations of multipurpose offshore floating platforms. Firstly, E-Motions is presented and the preceding studies are briefly reviewed. Secondly, the numerical model calibration procedures are provided. Thirdly, the preliminary results on the numerical simulations, with regards to wave-platform-Power Take-Off interactions and power output, are discussed. Upgraded versions of previous E-Motions variants, namely with semi-cylindrical and semi-spherical hull designs, yield higher power ratios and exhibit roll response amplitude operator peaks above 40°/m. The estimated power outputs (average values of up to 30 kW) are further bolstered by improved combinations of Power Take-Off mass-damping values, adapted to each of the eight considered sea-states from two sites along the Portuguese coastline – Aguçadoura and São Pedro de Moel.

## 1 INTRODUCTION

As an unexplored and abundant resource, wave energy can significantly contribute towards improvement of the global energy mix. Given its sizeable theoretical power resource of about 2.1 TW, globally (Gunn & Stock-Williams, 2012), and greater density and predictability when compared with renewables such as solar or wind (Falcão, 2010; Falnes, 2007), wave energy exhibits all of the requirements needed towards achieving a significant share of the global energy market. The main hindrance to such goal derives from the low degree of maturity of existing technologies capable of harnessing the wave energy resource (Aderinto & Li, 2018; Pecher & Kofoed, 2017).

Thus far, out of the hundreds of existing and unique concepts, very few have reached a commercialization stage and virtually none demonstrated continued viability and deployment. Many factors contribute to this, including the challenges inherent to the hazardous ocean environment (*e.g.*, wave loads and bio-fouling), excessive attention given to technological fundamentals upgrade in detriment of key sub-systems (*e.g.*, moorings, cabling and control units), sizeable funding requirements and competition from more mature alternative energy sources, among others (Magagna et al., 2018; Ocean Energy Europe, 2021; Strang-Moran & Mountassir, 2018). Therefore, more pragmatic and versatile solutions are required, both in terms of wave energy converter (WEC) technology and in terms of target market selection (Clemente et al., 2021a).

The solution that is here proposed comes in the form of E-Motions, a WEC concept based on wave/wind induced roll oscillations of multipurpose offshore floating platforms. This device can convert wave energy into electricity via a sliding Power Take-Off (PTO), which operates in response to the height differential and (de)accelerations prompted by the rolling of floating platforms onto which it is installed (Clemente, 2015). E-Motions has, thus far, been subjected to a proof-of-concept study (Clemente et al., 2020), followed by an initial numerical modelling (NM) phase which assessed various hull geometry solutions and PTO configurations (Clemente et al., 2021b). The results from this latter phase supported a follow-up experimental study with three physical models, which involved preliminary tests (PTO free-fall, platform free decay and platform inclining tests), a parametric study with regular waves and an operational assessment for a set of irregular sea-states, based on two reference case studies from the Portuguese coastline (Silva et al., 2018).

A new NM study, aimed at further developing E-Motions, was recently carried out and focused on

DOI: 10.1201/9781003320289-38

assessing the performance of alternative hull configurations and PTO mass-damping combinations. In detail, the data on three E-Motions variants (*e.g.*, mass distribution, hull shape and mooring configuration), from the physical modelling (PM) phase, was used as input towards the calibration of the numerical model, which was reproduced in the Boundary Integral Element Method-based (BIEM) software ANSYS® Aqwa™ (ANSYS, 2018). The underlying code is based on the principles of Potential Flow Theory (PFT), which has been successfully applied by WEC researchers in numerous studies and for many decades (Al Shami et al., 2019; Bracco et al., 2011; Journée & Massie, 2001; Ma et al., 2020). This follows on standard practices of WEC development, namely through a composite modelling approach (Rosenberg & Mundon, 2016; Sutherland & Barfuss, 2011; Wu et al., 2019). It is worth noting that Aqwa had already been applied in the first numerical study of E-Motions (Clemente et al., 2021b).

The following paper is structured as follows:

- Section 2 presents and describes, in greater detail, the E-Motions concept, its mode of operation and summarizes the key capabilities of this device;
- Section 3 develops on the numerical model setup of the three E-Motions variants and the necessary steps towards a reliable implementation in Aqwa, to which adds a summary of the key calibration procedures and error assessments;
- Section 4 highlights and discusses the preliminary outcomes of the numerical study, focusing on the hydrodynamic response and power output estimates for each of the considered sub-variants and PTO mass-damping combinations;
- Lastly, the main conclusions are drawn in Section 5, including an overview of the work conducted with Aqwa and expected future developments towards further improvements of E-Motions.

## 2 E-MOTIONS WAVE ENERGY CONVERTER

As synthesized beforehand, E-Motions operates via a sliding PTO that generates electricity from wave and wind induced roll oscillations onto the floating platform it is integrated into. The superstructure, either at deck level or within the hull, isolates and protects the PTO from the surrounding environment. A schematic portray of the key sub-components and acting forces on the PTO is provided in Figure 1. The meaning of each symbol is provided in the following paragraphs.

The hydrodynamic behavior of E-Motions follows on the theoretical description of a standard floating structure, to which moorings and a sliding PTO are added. Starting with the platform component, the underlying mathematical formulation is based upon Newton's Second Law. Assuming that both the incoming wave $Z$ and body motion amplitudes $\zeta_j$ are

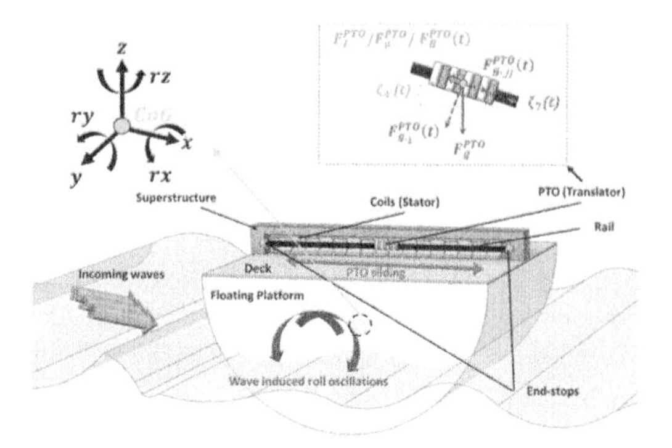


Figure 1. Schematic of the E-Motions and the sliding PTO, including the respective centers of gravity (CoG, orange markers).

relatively small, as well as that the non-linear effects are negligible (*e.g.*, viscous damping) the forces acting on a free-floating structure can be deducted, in the frequency-domain, through:

$$F_{e,i} + F_{h,i} + F_{r,i} = \sum_{j=1}^{6} m_{ij}\ddot{\zeta}_j, for\ i = 1,2 \ldots 6 \quad (1)$$

where $F_e$, $F_h$ and $F_r$ correspond to the excitation (Froude-Krylov and diffraction), hydrostatic restoration and radiation damping forces/moments, respectively. $m_{ij}$ regards the 6x6 mass/inertia matrix for the six degrees of freedom (DoFs): surge ($x$), sway ($y$), heave ($z$), pitch ($rx$), roll ($ry$) and yaw ($rz$).

These components can be further deducted which yields:

$$F_{e,i} = \sum_{j=1}^{6} \left[-\omega^2 \left(m_{ij} + A_{ij}\right) + i\omega B_{ij} + C_{ij}\right] \zeta_j, for\, i = 1,2 \ldots 6 \quad (2)$$

where $A_{ij}$, $B_{ij}$ and $C_{ij}$ are the 6x6 added (mass moment of) inertia, radiation damping and (linear) hydrostatic stiffness matrices, respectively, being the former two dependent upon the incoming wave frequency $\omega$.

Since the main mode of operation in E-Motions is associated with the platform's roll DoF $\zeta_4$, subscripts $_i$ and $_j$ can be replaced by 4 (coupling between DoFs is neglected, for simplification). By accounting for the PTO and mooring system, Equation 2 becomes:

$$F_{e,4} + F_{hn-l,4} + F_{PTO,4} = \left(-\omega^2 (m_{44} + A_{44}) + i\omega B_{44} + C_{44}\right)\zeta_4 = R_{44}\zeta_4 \quad (3)$$

where $F_{hn-1,4}$ and $F_{PTO,4}$ are the non-linear hydrostatic stiffness moment, which includes the influence of the mooring system, and the PTO moment, respectively. The condensed form of the main terms is represented by $R_{44}$.

Equation 3 can be re-written with regards to $\zeta_4$. If one divides it by the maximum wave amplitude $Z_0$, the corresponding response amplitude operator (RAO) can be obtained for each wave frequency:

$$RAO_4 = \frac{\zeta_4}{Z_0} \tag{4}$$

For a more suitable description, particularly of the non-linear $F_{PTO}, 4$ term, a time-domain approach can be employed instead of (or complementary to) the frequency-domain. To that end, the time term $t$ should be incorporated into the preceding equations. One important adjustment is required for the roll damping term, which should account for the time-history of the radiation patterns generated by the floating platform. To that end, this term is replaced by a "memory" impulse function $g_{44}$ of the time variables $t$ and $\tau$. The added mass moment of inertia term is included as the corresponding value at infinite frequency, resulting in the time-domain Cummins equation:

$$\begin{aligned} F_{e,4}(t) + F_{hn-l,4}(t) + F_{PTO,4}(t) \\ = \ddot{\zeta}_4(t)[m_{44} + A_{44}(\infty)] \\ + \int_{-\infty}^{t} g_{44}(t-\tau)\dot{\zeta}_4(\tau)d\tau + \zeta_4(t)C_{44} \end{aligned} \tag{5}$$

Determining the $F_{PTO}, 4$ component is a key milestone towards modelling the behavior of E-Motions. As seen in Figure 1, this component is composed of two parts: one dependent on the projected PTO force perpendicular to the sliding DoF $\zeta_7$ and another on the projected PTO force parallel to it. This latter sub-component, henceforth symbolized as $F_{PTO,7}$, is of direct interest for the estimation of the power output through the force scheme in Figure 1. Mathematically, it is described through:

$$\begin{aligned} F_{PTO,7}(t) = m_{PTO}\ddot{\zeta}_7(t) = F_g^{PTO}\sin(\zeta_4, t) \\ - \left[F_\mu^{PTO}(t) + F_B^{PTO}(t)\right] \end{aligned} \tag{6}$$

where $F_g^{PTO}$, $F_\mu^{PTO}$ and $F_B^{PTO}$ are the gravity, friction and damping forces acting on the PTO, respectively, while $m_{PTO}$ represents the PTO's mass.

By rewriting Equation 6 with regards to $F_B^{PTO}$, one obtains the PTO damping force, which can be applied towards estimating the instantaneous power output P by the PTO:

$$P(t) = \int_0^T F_B^{PTO}\dot{\zeta}_7 dt \tag{7}$$

E-Motions can, theoretically, be integrated into any suitable floating platform, so long as the roll oscillations are sufficient to generate the required energy but respect thresholds towards crew and/or cargo safety. This gives it a high degree of adaptability and enables cost reduction through equipment sharing and, if applicable, sub-system mitigation/elimination (*e.g.*, cabling and mooring). It also permits hybridization with other energy harvesting technologies, such as triboelectric nanogenerators (Clemente et al., 2021c; Rodrigues et al., 2020). Lastly, stacking of several rows of PTO can introduce a multiplication factor onto the energy output, although this ought to also impact the hydrodynamic response.

## 3 NUMERICAL MODEL SETUP

### 3.1 *Numerical approach*

As mentioned in the introductory section, the BIEM-based software ANSYS® Aqwa™ was employed during the latest phase of E-Motions development. Its employment requires the fulfillment of certain premises, namely that the fluid be considered as inviscid, irrotational and incompressible. Under such assumptions and by considering relatively small wave/body motion amplitudes, the initial Navier-Stokes equations which govern the fluid dynamics are simplified into the Laplacian and linearized Bernoulli equations (Guo & Ringwood, 2021; Papillon et al., 2020):

$$\nabla^2\phi = 0 \tag{8}$$

$$\frac{\partial\phi}{\partial t} + \frac{(\nabla\phi)^2}{2} + \frac{p}{\rho} + gz = constant \tag{9}$$

where $p$ and $\rho$ are the fluid pressure and water density, respectively, while $g$ is the acceleration of gravity. The symbol $\phi$ represents the velocity potential that satisfies the Laplacian equation as well as several other boundary conditions (ANSYS, 2019a).

Solving this velocity potential enables a suitable representation of the flow and the wave-structure interactions (in this case, the waves and floating platform), since the first order hydrodynamic pressure distribution can be computed from it and the linearized Bernoulli equation. This, in turn, permits the calculation of the acting fluid forces through integration over the wetted surface of the floating body. Solving $\phi$ implies computing its various terms, namely the potential associated with radiated waves from the six modes of body motion $\phi_r$, the incident wave $\phi_{inc}$ and the diffracted wave $\phi_d$. Through the linearization assumption, the superposition theorem enables a formulation of the total velocity potential as (ANSYS, 2019a):

$$\phi = \phi_{inc} + \phi_d + \sum_{i=1}^{6} \phi_{r,i}\dot{\zeta}_i \quad (10)$$

In Aqwa, this is achieved through a boundary integration approach that resorts to a pulsating Green's function $G$, which obeys the same boundary conditions as the velocity potential function. By implementing a source distribution over the mean wetted surface $S_0$, the velocity potential can be written as:

$$\phi = \frac{1}{4\pi} \int_{S_0} \sigma G dS \quad (11)$$

where $\sigma$ is the strength of each source determined by the hull surface boundary condition.

Numerically, this integral form of the velocity potential is solved, in Aqwa, via a discretization based on the Hess-Smith constant panel method, where the mean wetted surface is divided into a mesh composed of quadrilateral or triangular panels with a constant average source strength. Additional elements can be added, including frequency-independent added mass moment of inertia and hydrodynamic damping terms. Fenders and moorings can be equally used, with the latter ranging from linear cables to dynamic composite catenary lines composed of Morison elements. Lastly, articulations such as sockets and permit the establishment of connections between rigid bodies, which was required for the setup of the PTO and its sliding motions.

### 3.2 *Model configuration*

The three physical models used in the preceding experimental study – a half-cylinder (HC), a half-sphere (HS) and a trapezoidal prism (TP) – were reproduced in the numerical model through a computer assisted design (CAD) software, from which the geometries were imported into Aqwa. It should be stated that each geometry was transformed into a surface object and split at waterline level (*xy*-plane), as these are requirements for Aqwa to operate. An internal lid was generated to remove irregular frequencies, as recommended (ANSYS, 2019b). Data from the experimental study was used as input for configuring the physical properties of each E-Motions variant and the position of both the PTO and the floating platform (*e.g.*, vertical position and draught, respectively). This is summarized in Table 1. A fictitious articulation was generated between these two components, so that the PTO would slide in a realistic manner. This also enabled the introduction of a damping coefficient that could be calibrated. The amplitude of the sliding motions was limited between two high stiffness fenders on each flank of the floating platform, mimicking the rigid end-stops of the original physical models.

Table 1. Physical properties of E-Motions variants (prototype values).

| Geometry | HC | HS | TP |
|---|---|---|---|
| Mass (kg) | 452 051 | 103 202 | 332 943 |
| Radius of gyration (m) | | | |
| - roll | 7.284 | 5.676 | 6.040 |
| - pitch | 6.991 | 3.652 | 7.234 |
| - yaw | 8.797 | 5.836 | 7.518 |
| Centre of gravity, vertical coordinate from waterline (m) | 4.579 | 2.519 | 3.537 |
| Draught (m) | 2.495 | 2.315 | 1.622 |
| Transversal metacentric height (m) | 2.920 | 2.120 | 2.220 |

The mesh of each variant was selected through a heuristic process based on the layouts employed in the preceding numerical study with E-Motions (Clemente et al., 2021b) and the feedback from the Aqwa software. The number of elements and defeaturing tolerance was carefully adjusted to assure the mesh quality check requirements from Aqwa (ANSYS, 2019a), which resulted in a final number of panels equal to 10 996, 6 244 and 9 930 for the HC, HS and TP, correspondingly. This is exemplified for the HC's platform component in Figure 2.

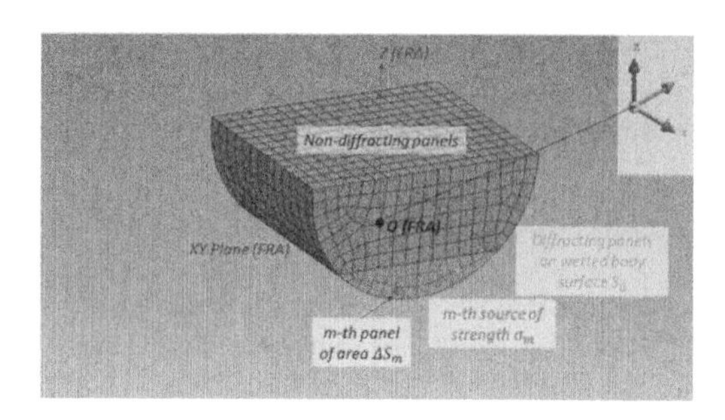


Figure 2. Meshing of the HC variant with regards to the fixed reference axis (FRA) coordinate system.

Figure 3 depicts the three numerical models prior to the execution of the Aqwa simulations.

For analysis purposes and based on the experience from the previous numerical study and references from literature (Clemente et al., 2021b; López et al., 2017; Ma et al., 2020), the wave direction range was set to -180° to 180°, at intervals of 45°. As for the wave frequency range, it was set from 0.02 Hz to 0.50 Hz, subdivided into about 100 intermediate frequency intervals. For the time-domain analysis, a time-step of 0.05 s and an output step of 0.10 s were set. Both the dynamic cable computation and the convolution integration approach were enabled,

a)

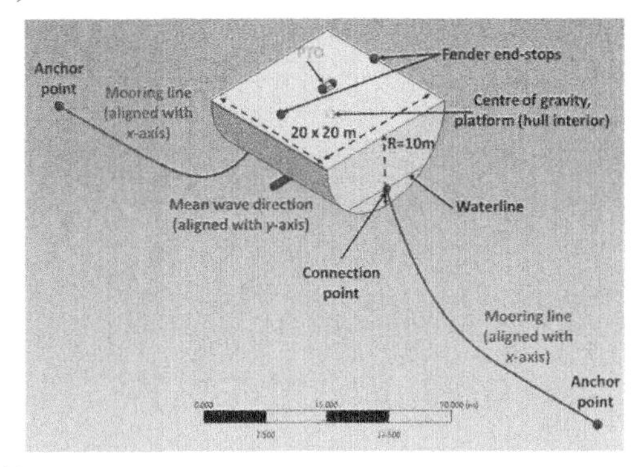


b)

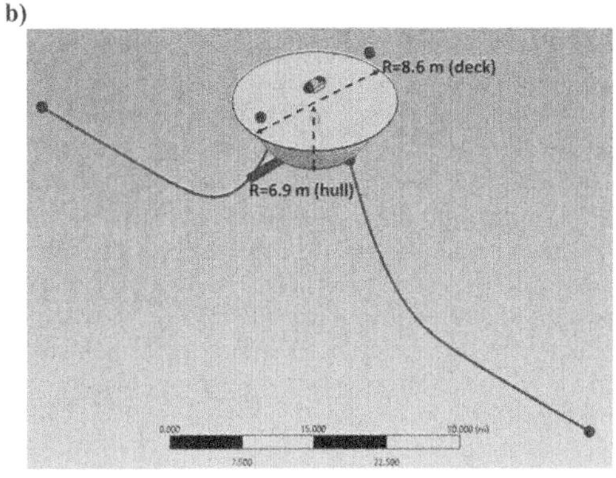


c)

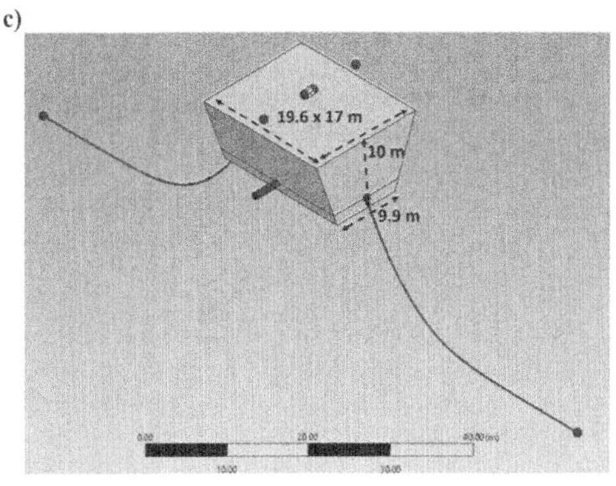


Figure 3. Numerical model of the a) HC, b) HS and c) TP variants of E-Motions in Aqwa.

as well as Wheeler Stretching with second order correction for a more accurate free surface elevation (FSW) reproduction.

Lastly, the mooring was configured in accordance with the original experimental setup. This required the setup of two catenary mooring lines extending 43.2 m in length and weighing 289 kg/m (prototype values). The remaining properties were obtained either through direct measurement, manufacturer catalogue or via estimates from literature references (Qiao et al., 2020; Xu et al., 2019). Each line had two connection points: one submerged anchor point at a depth of 16 m (original water depth) and a connection point on the bow or stern of each E-Motions variant, at waterline level.

### 3.3 *Calibration procedures and simulation plan*

Once the three numerical models were configured, a two-stage calibration procedure was performed. It implied an initial comparison of experimentally measured and numerically computed hydrostatic properties, for each variant, followed by a heuristic damping tuning associated with the hydrodynamic roll response and the PTO.

Starting with the physical and hydrostatic properties, the data from the original physical models was matched against the calculations carried out by Aqwa, as summarized in Table 2. Note that the metacentric height can be used as direct input to Aqwa, but it can also be calculated by the software. This latter approach was followed, so as to have an additional variable for comparison. There is a general good agreement found between the experimental and numerical values, with differences of 5% or less being reported.

Table 2. Experimental versus numerical values of the physical and hydrostatic properties for the HC/HS/TP variants (prototype values).

| Property | Experimental | Aqwa |
|---|---|---|
| Displaced volume ($m^3$) | 452.0/103.2/ 333.0 | 451.7/103.0/ 332.9 |
| Area moment of inertia ($m^4$) | 3 849/555/2 205 | 3 848/555/2 205 |
| Roll Stiffness (MN.m/rad) | 12.86/2.14/ 7.25 | 12.99/2.09/ 7.47 |
| Metacentric height (m) | 2.920/2.120/ 2.220 | 2.931/2.070/ 2.289 |
| Natural roll period, moored platform (s) | 9.12/7.69/ 9.75 | 9.10/7.72/ 9.41 |

The second step of the calibration procedures revolved around the calibration of the non-linear hydrodynamic roll damping, which is neglected, by default, in Aqwa. However, the software enables the introduction of a frequency-independent damping coefficient, which can be adjusted to improve the accuracy of the numerical model. Afterwards, an analogous procedure was followed for the definition for PTO damping coefficient. Even so, for this variable, a single value had to be selected, since the reproduced PTO in the experiments was the same in each physical model. It is worth noting that this step was carried out complementary to the previous one.

The heuristic definition of the two types of damping coefficients focused on the tests with regular waves for which experimental data was available, encompassing a total of 6 combinations of wave height, $H$, and wave period, $T$:

- Tests 1-7: fixed $H$=2 m, $T$=6.0/7.5/8.5/9.0/9.5/10.5 /12 s;
- Tests 8-10: $T$=9 s, $H$=1.0/3.0/4.0 m.

For each variant, different hydrodynamic roll damping ranges were initially considered, while a single range was assumed for the PTO damping coefficient. These were defined based on preliminary simulations and measured data from the preceding experimental campaign. A supporting peak-to-peak amplitude error minimization assessment was performed, which helped to further refine the values that ought to be attributed to each case. The error associated with each individual test was computed through:

$$NRMSE = \frac{\sqrt{\frac{\sum_{n=1}^{N}\left(\zeta_{04,n}^{exp}-\zeta_{04,n}^{num}\right)^2}{N}}}{\overline{\zeta_{04}^{exp}}} \times 100 \qquad (12)$$

where $\zeta_{04}^{exp}$ and $\zeta_{04}^{num}$ are, respectively, the peak-to-peak experimental and numerical roll amplitudes, $N$ the number of considered values, NRMSE the normalized root mean-square error and $\overline{\zeta_{04}^{exp}}$ the average of the experimental peak-to-peak amplitudes of each test (from 1 to 10). This approach was extended to the PTO sliding amplitudes, with the $\zeta_{04}$ values being replaced by the corresponding $\zeta_{07}$ and $\overline{\zeta_{04}^{exp}}$ by the average peak-to-peak PTO sliding amplitude $\overline{\zeta_{07}^{exp}}$, also from the experiments. The outcomes of this second stage are summarized in Figure 4 and Figure 5.

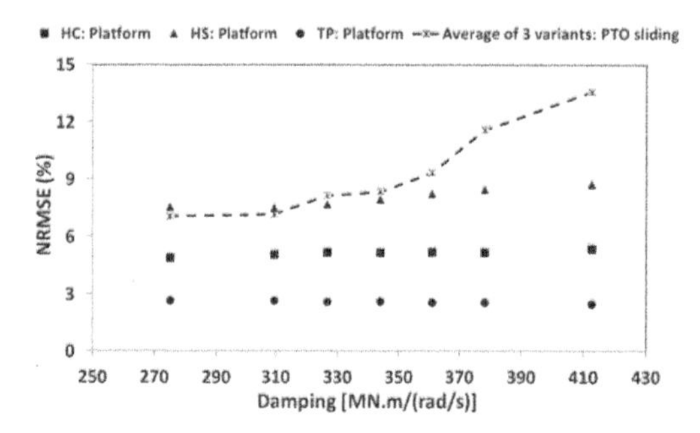


Figure 4. NRMSE values as a function of PTO damping coefficient.

Figure 4 Shows the NRMSE for each platform as a function of the PTO damping coefficient. This already reflects the corresponding hydrodynamic roll damping coefficients that yielded lowest NRMSE

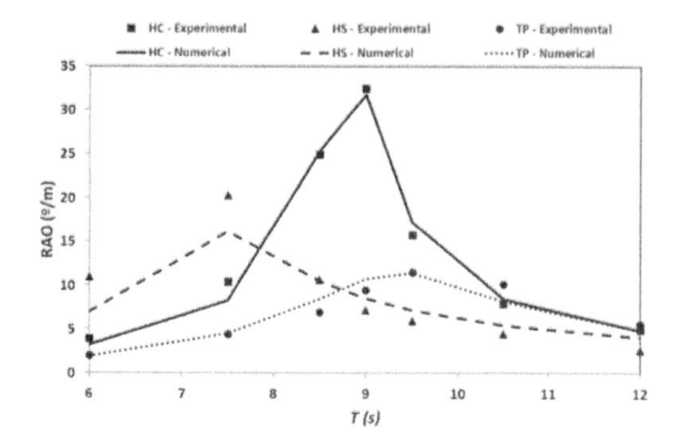


Figure 5. Experimental and numerical RAO curves for the moored HC, HS and TP variants of E-Motions, including the PTO.

values: 8.02×10$^5$, 2.72×10$^5$ and 19.8×10$^5$ N.m/(rad/s) for the HC, HS and TP (average NRMSE of 3.80%, 6.67% and 2.71%, respectively). On the PTO NRMSE values, the order of magnitude is generally low for the platform oscillations (below 10%) and the PTO sliding motions (below 15%). Overall, the lowest NRMSE values were obtained for a PTO damping coefficient of 2.75 × 10$^8$ N.m/(rad/s), which is the lower threshold of the considered range. Smaller values were discarded because, for one or more test cases, an excessive PTO peak-to-peak sliding amplitude deviation was detected. Given the regular nature of the roll oscillations' time series, the NRMSE served as a suitable approach when comparing the peak-to-peak amplitudes. Complementary, a general good agreement is found between the experimental values and the numerical RAO curves, albeit with some deviations attributed to the compromise with the PTO sliding calibrations and inaccuracies/limitations inherent to the numerical reproduction of the three E-Motions variants.

Following on the calibration procedures, the main simulation plan was defined. In order to further develop E-Motions, different hull sub-variants and PTO mass-damping combinations were studied in a two-stage analysis. The regular wave tests were replaced by 8 reference irregular sea-states, from the resource matrices associated with the case studies of Aguçadoura and São Pedro de Moel (Silva et al., 2018). Based on the local wave climate and the bin methodology recommended in (Pecher and Kofoed, 2017), the following representative significant wave height, $H_s$, and wave energy period, $T_e$, bins were selected, Table 3 (JONSWAP spectrum with peak enhancement factor of 3.3). The estimated probability of occurrence $Prob_{SS}$ and average wave power per metre of wave crest $P_w$ are also listed (deep water was assumed).

Several sub-variants were configured, namely in terms of hull dimensions and mass distribution. The alterations also resulted in similar unmoored natural roll periods of about 10s (prototype value). In detail, for the HC, HS and TP, this implied:

Table 3. Representative bins from the Aguçadoura and São Pedro de Moel case studies.

| BIN ID | $H_s$ (m) | $T_e$ (s) | $Prob_{SS}$ (%) | $P_w$ (kW/m) |
|---|---|---|---|---|
| SS1 | 1.33 | 8.00 | 7.21 | 6.94 |
| SS2 | 2.31 | 9.30 | 15.22 | 24.31 |
| SS3 | 2.36 | 10.24 | 15.05 | 27.92 |
| SS4 | 3.20 | 10.32 | 11.21 | 51.90 |
| SS5 | 3.32 | 11.21 | 10.30 | 60.61 |
| SS6 | 3.70 | 11.30 | 8.16 | 75.88 |
| SS7 | 3.83 | 12.17 | 4.56 | 87.37 |
| SS8 | 4.25 | 12.26 | 6.22 | 108.50 |

- HC: five sub-variants, MK-0 (original) to MK-4. Two sub-variants (MK-1 and MK-2) with a reduced length (75% and 50% of the initial value, respectively). The remaining two sub-variants, MK-3 and MK-4, were conceived as to assess the impact of a variation in the mass and hydrostatic stiffness. In detail, this implied a reduction to 75% and 50% of the MK-0 value for MK-3, respectively, and an increment of 50% and 7% for MK-4, correspondingly;
- HS: five sub-variants, MK-0 (original) to MK-4. MK-1, MK-2 and MK-3 were conceived with greater hull radii (10.0 m, 12.5 m and 8.5 m, by that order). MK-4 was also configured with a radius of 10.0 m, but as a lighter sub-variant with a reduced hydrostatic stiffness (75% of MK-1's value);
- TP: three sub-variants, MK-0 (original) to MK-2. Out of the three variants, TP exhibited, in general, comparably weaker power output ratios and hydrodynamic roll response. Hence, there is limited interest in deepening its improvement. Alike the HC's sub-variants, a reduced length was defined for TP MK-1 and MK-2 (15.0 m and 10.0 m, respectively). As a consequence, the mass and hydrostatic stiffness are reduced (*e.g.*, about 70% and 45% of MK-0's value on the latter parameter, correspondingly). This follows on the mind-set for low stiffness variants applied in HC's MK-3 and HS' MK-4.

Lastly, a new variant was included: an inverted enclosed frustum EF, Figure 6. Three sub-variants were considered for it:

- MK-0: reference case. Upper and lower base with radius of 10.0 m and 4.8 m, respectively, and a height of 10.0 m (cross-section angle between bases of 63.4°);
- MK-1: heavier version of MK-0 (additional 75%, stiffness 2.5 times greater). Bottom radius of 7.0 m, with conserved upper base and height (cross-section angle of 75.4°);
- MK-2: lighter version of MK-0 (nearly 75% and 60% of mass and hydrostatic stiffness, respectively), following on the example of HC MK-3, HS-MK4 and the TP sub-variants.

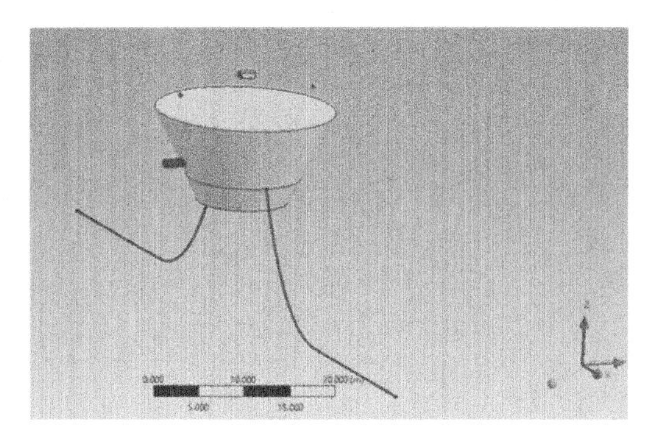

Figure 6. Enclosed Frustum variant (MK-0 sub-variant).

On the PTO, a total of 10 different mass-damping combinations were assessed. They range from the one found during the calibration procedures – V0 – to others with higher mass and damping values – V1 to V9, Figure 7.

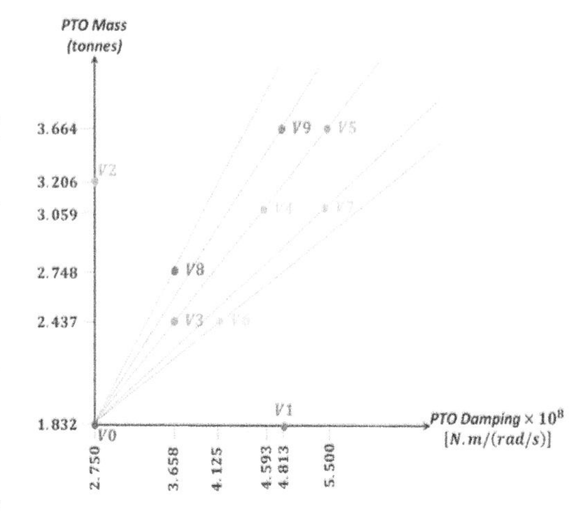


Figure 7. PTO mass-damping combinations.

## 4 RESULTS AND DISCUSSION

### 4.1 *Hull design and mass distribution assessment*

The comparison between the numerous sub-variants, in terms of hull dimensions and mass distribution, is discussed here. To that end, the estimates concerning the respective capture width ratios (CWR) are provided in Figure 8. This ratio was calculated for each of the 8 sea-states through:

$$CWR_{SS} = \frac{\overline{P_{SS}}}{P_w L} \times 100 \tag{13}$$

where $\overline{P_{SS}}$ and $L$ are the average power output, for a given sea-state, and the sub-variant's length or radius.

a)

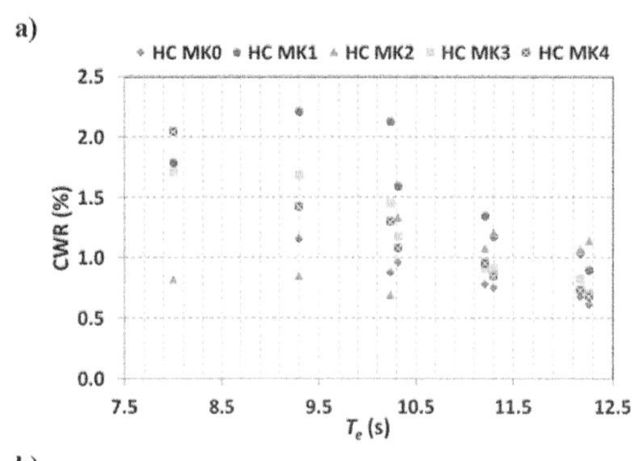


b)

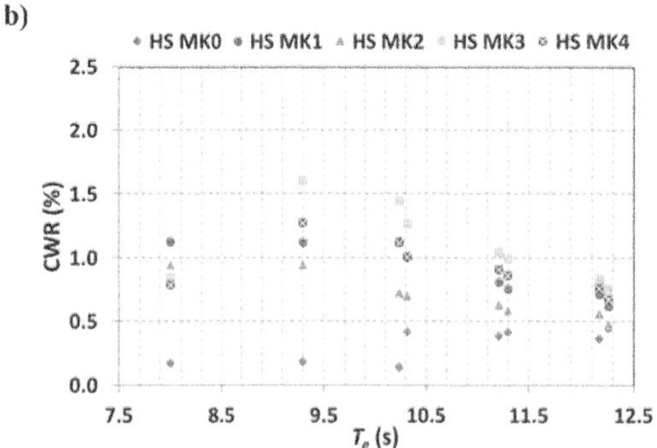


c)

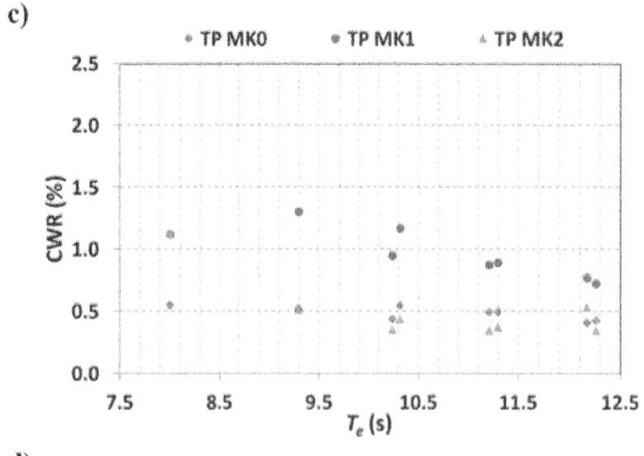


d)

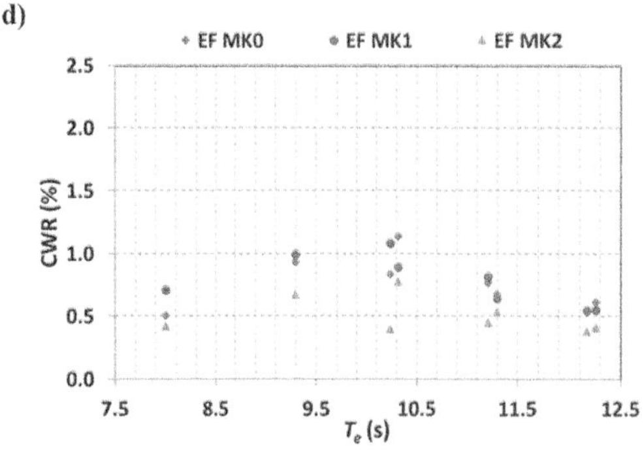


Figure 8. CWR for the a) HC, b) HS, c) TP and d) EF sub-variants for each sea-state, represented by the respective $T_e$.

At first glance, the order of magnitude of the CWRs seem relatively low when compared to literature references (Pecher and Kofoed, 2017). However, one must not forget that E-Motions is currently being developed towards marine niche markets and integration onto floating platforms, such as small boats. Therefore, the energy demand is expected not to be as high as that of the mainland power grid. Furthermore, these values represent the output of a single PTO unit, but, *a priori*, nothing prohibits the introduction of additional units onto a floating platform, apart from the availability of space. Therefore, these results should be interpreted relative to one another. On that regard, the HC and HS yield the highest CWR outputs, while TP and, even more so, EF exhibit the lowest values.

For HC, the MK-1 sub-variant shows, in general, higher CWR values over the considered irregular sea-states, which is also a consequence of the relatively smaller length in comparison with the original MK-0 sub-variant. Although the MK-2 sub-variant has an even lower length, the power output is reduced at a proportionally greater rate. The lighter MK-3 and stiffer MK-4 denoted intermediate values.

On the HS, MK-3 yields the highest CWR values, overall. The MK-1 and MK-2 denoted higher average power outputs, but lower CWRs due to the proportionally greater radius. MK-4, as a lighter version of MK-1, showed slightly better outputs than MK-1, in general, but still below those of MK-3.

The TP sub-variants exhibited, in accordance to the preceding experimental outcomes, lower CWRs than those of the curvilinear sub-variants, likely due to the higher hydrodynamic damping losses attributed to the sharp corners (Pecher and Kofoed, 2017). In relative terms, the MK-1 sub-variant provided the highest CWR values, while MK-2 demonstrated little to no improvements with regards to the original MK-0 sub-variant.

Lastly, the EF yielded CWRs of a similar to lower magnitude with regards to those obtained for TP. At a first glance, there seems to be a tie between the MK-0 and MK-1 sub-variants, but a closer look at the corresponding hydrodynamic roll responses, in the form of frequency-domain transfer functions, enables closure of this. As seen in Figure 9, the EF MK-0 exhibits a very high roll peak of up to 50°/m, but for the other two sub-variants this peak would be even greater, which led to capsizing issues, making them unviable as potential design solutions. This also resulted to the exclusion of the HC MK-4 sub-variant, since it also exhibited excessive roll oscillations (nearly 55°/m). Aside from EF MK-0, the roll transfer functions of other selected sub-variants – HC MK-1, HS MK-3 and TP MK-1 – are also displayed in Figure 9. It is worth noting that both the EF MK-0 and HC MK-1 are very close to the stability limits up from which capsizing is an issue (Pecher and Kofoed, 2017; Journée and Massie, 2001), which hints at the necessity of future studies with the sub-variants subjected to extreme wave conditions. This concern is extended to greater $H_s$ for a $T_e$ corresponding to the sub-variants' natural roll period, since for the 8 irregular waves conditions considered in this study the increment of $H_s$ would be compensated by an increase of $T_e$, thus further away from the resonance range (from BINs 2 and 3 onwards, particularly). Furthermore, all transfer function peaks are found in the vicinity of the sub-variants' natural roll period, which was constant for all sub-variants (10s, or the equivalent 0.10 Hz).

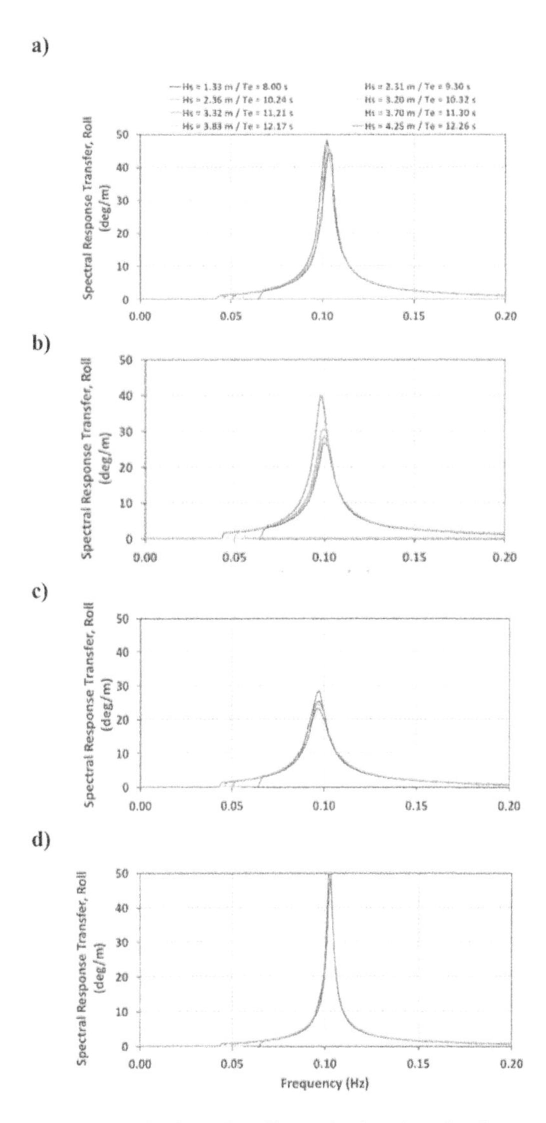


Figure 9. Hydrodynamic roll transfer functions for the a) HC MK-1, b) HS MK-3, c) TP MK-1 and d) EF MK-0 sub-variants.

### 4.2 *Power Take-Off mass-damping combinations comparison*

Following on the outcomes of the hull design analysis, the power output estimates are here presented and discussed. They also reflect the various PTO mass-damping combinations considered in this numerical study applied to the four selected sub-variants from the previous stage: HC MK-1, HS MK-3, TP MK-1 and EF MK-0. Because it was found that the different PTO mass-damping combinations did not significantly alter the hydrodynamic response of the respective sub-variants, the discussion will be centered on the power output estimates and the PTO's sliding motions.

Starting with the former, the CWRs obtained for each sub-variant, as a function of each sea-state and the PTO mass-damping combination, are presented in Figure 10, from which it is perceivable that the HC and HS sub-variants yield the highest values, once more.

In general, the CWR values tend to increase from SS8 to SS1, following mainly on the decrement of $H_s$ and the convergence towards the resonance range, with peaks for certain mass-damping combinations, depending on the considered sea-state. The HC sub-variant yields values as high as 3.57%, while the HS provides maximum CWR of 2.49%, given its slightly greater length and lower values. The TP and EF sub-variants, in contrast, barely surpass the 1.00% threshold due to a relatively low output and, in the case of EF, a greater length (20 m).

Certain PTO combinations share the same mass, which would permit a damping coefficient, for a specific set of wave conditions, if, in practice, a control unit were to be integrated into E-Motions. Given the potential of PTO damping tuning in the efficiency improvement of WEC devices (Giorgi et al., 2021; Ringwood, 2020), a simplified approach can be employed, here. Considering that, for each sea-state, the highest average power output of each viable combination – V3-V6, V4-V7 and V5-V9 – is taken (*e.g.*, the highest average power output of V5 or V9 for each BIN), then for HC, HS and TP, the V5-V9 mass-damping combination yields a total average power output greater than the standalone V5 or V9. The improvement reaches nearly 3.6%, 5.6% and 2.1% for the HC, HS and TP, respectively. These percentages are not to be confused with the previous CWR values. On the EF, the V4-V7 yields the highest change: an additional 9.0% to the maximum average standalone power value.

As seen in Figure 11, these outcomes are mainly justified by the PTO's velocity fluctuations (greater for HC and HS, mainly for SS1), since the available sliding stroke is always fully covered

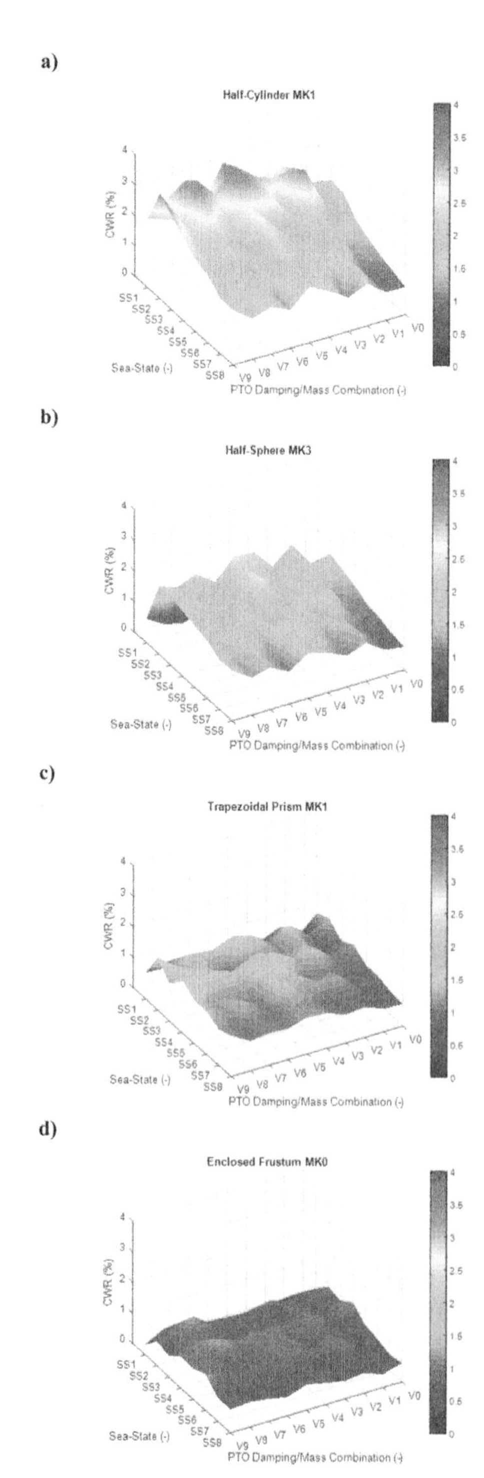


Figure 10. CWRs for the a) HC, b) HS, c) TP and d) EF sub-variants, for the 8 reference sea-states and V0 to V9 mass-damping combinations.

(14.4 m). The damping seems to have a limited effect on the PTO velocity, as well, in terms of order of magnitude.

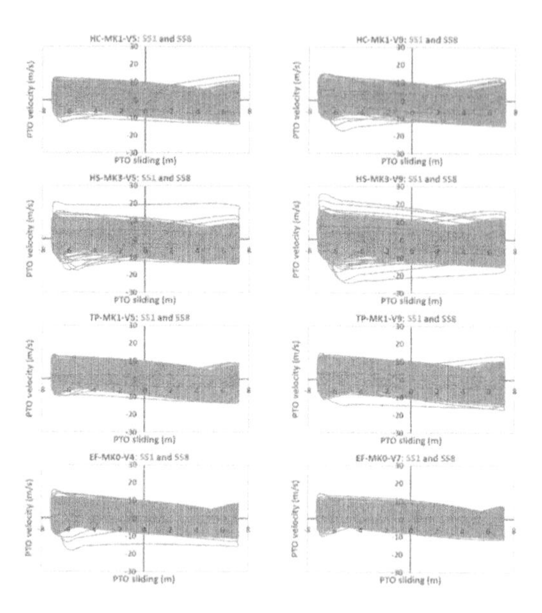


Figure 11. State spaces of PTO motions for the 4 E-Motions sub-variants, regarding SS1 and SS8.

## 5 CONCLUSIONS

The preliminary results from the numerical study of E-Motions have been presented and discussed along this paper, with the main outcomes being, now, summarized.

The calibration procedures were successfully carried out, both in terms of hydrostatic and physical properties, hydrodynamic roll response and PTO sliding motions. A generally good agreement was found between the experimental and numerical data, from which new sub-variants were considered towards the optimization of E-Motions.

Out of the proposed 16 sub-variants, 13 of which were updates of the original 3, the HC MK-1, HS MK-3, TP-MK1 and EF MK-0 yielded the highest CWR outputs, with values of nearly 2.5% being achieved for a single PTO unit under non-optimized mass-damping conditions. Significant hydrodynamic roll oscillations were reported (up to 50º/m), which led to the exclusion of 3 sub-variants due to high risk of capsizing: HC MK-4, EF MK-1 and EF MK-2.

Lastly, 10 different PTO mass-damping combinations were assessed for the 4 selected sub-variants. From said combinations, the HC and HS maintained their superiority in terms of CWR values, with maximum percentages of 3.57% and 2.49% being achieved, respectively. It was also demonstrated that specific mass-damping combinations could be merged by assuming a damping-tuning approach to each sea-state, which, in practice, would require the integration of a control unit. For all sub-variants the V5-V9 yielded improved power outputs, apart from the EF, for which the V4-V7 duo was more suitable.

Lastly, it was perceivable that the PTO velocity, especially for SS1 and "nearby" sea-states, would greatly influence the performance of each sub-variant.

Future studies will continue to develop on these findings whilst accounting for additional variables, such as the mooring system and alternative hull designs and PTO mass-damping combinations. It will also be pertinent to carry out a survivability assessment, given the very high roll oscillation peaks that were observed, which hint at the risk of capsizing.

## ACKNOWLEDGEMENTS

The authors acknowledge funding in the form of a Ph.D. scholarship grant by the FCT, co-financed by the EU's ESF through the NORTE 2020 program, with reference 2020.05280.BD. The authors further acknowledge the Interdisciplinary Centre of Marine and Environmental Research (CIIMAR) for its institutional support in attaining the aforementioned Ph.D. grant.

## REFERENCES

Aderinto, T., Li, H., 2018. Ocean Wave Energy Converters: Status and Challenges. Energies 11, 1250. http://dx.doi.org/10.3390/en11051250

Al Shami, E., Wang, X., Ji, X., 2019. A study of the effects of increasing the degrees of freedom of a point-absorber wave energy converter on its harvesting performance. Mechanical Systems and Signal Processing 133, 106281. http://dx.doi.org/10.1016/j.ymssp.2019.106281

ANSYS, 2019a. AQWA Theory Manual (Manual). ANSYS, Canonsburg, PA, USA.

ANSYS, 2019b. AQWA User's Manual (Manual). ANSYS, Canonsburg, PA, USA.

ANSYS, 2018. ANSYS AQWA. ANSYS, Inc, Canonsburg, PA.

Bracco, G., Giorcelli, E., Mattiazzo, G., 2011. ISWEC: A gyroscopic mechanism for wave power exploitation. Mechanism and Machine Theory 46, 1411–1424. http://dx.doi.org/10.1016/j.mechmachtheory.2011.05.012

Clemente, D., Rosa-Santos, P., Taveira-Pinto, F., 2021a. On the potential synergies and applications of wave energy converters: A review. Renewable and Sustainable Energy Reviews 135, 17. http://dx.doi.org/10.1016/j.rser.2020.110162

Clemente, D., Rosa-Santos, P., Taveira-Pinto, F., Martins, P., 2021b. Influence of platform design and power take-off characteristics on the performance of the E-Motions wave energy converter. Energy Conversion and Management 244, 15. http://dx.doi.org/10.1016/j.enconman.2021.114481

Clemente, D., Rosa-Santos, P., Taveira-Pinto, F., Martins, P., Paulo-Moreira, A., 2020. Proof-of-concept study on a wave energy converter based on the roll oscillations of multipurpose offshore floating platforms. Energy Conversion and Management 224, 19. http://dx.doi.org/10.1016/j.enconman.2020.113363

Clemente, D., Rosa-Santos, P., Taveira-Pinto, F., Rodrigues, C., Correia, J., Esteves, R., Ventura, J., 2021c. Experimental study combining a triboelectric nanogenerator solution with the E-Motions wave energy converter under irregular waves, in: Proc. of the 14th European Wave and Tidal Energy Conference. Presented at the 14th European Wave and Tidal Energy Conference, Plymouth, UK, p. 8.

Clemente, D.R.da S.P. , 2015. Energy production from the motions of offshore floating platforms (Master's Thesis). University of Porto, FEUP.

Falcão, A.F.de O. , 2010. Wave energy utilization: A review of the technologies. Renewable and Sustainable Energy Reviews 14, 899–918. http://dx.doi.org/10.1016/j.rser.2009.11.003

Falnes, J., 2007. A review of wave-energy extraction. Marine Structures 20, 185–201. http://dx.doi.org/10.1016/j.marstruc.2007.09.001

Giorgi, G., Faedo, N., Mattiazzo, G., 2021. Time-Varying Damping Coefficient to Increase Power Extraction from a Notional Wave Energy Harvester, in: Proc. of the 2021 International Conference on Electrical, Computer, Communications and Mechatronics Engineering (ICECCME). Presented at the 2021 International Conference on Electrical, Computer, Communications and Mechatronics Engineering (ICECCME), Mauritius, p. 9. http://dx.doi.org/10.1109/ICECCME52200.2021.9591020

Gunn, K., Stock-Williams, C., 2012. Quantifying the global wave power resource. Renewable Energy 44, 296–304. http://dx.doi.org/10.1016/j.renene.2012.01.101

Guo, B., Ringwood, J.V., 2021. Geometric optimisation of wave energy conversion devices: A survey. Applied Energy 297, 27. http://dx.doi.org/10.1016/j.apenergy.2021.117100

J Rosenberg, B., Mundon, T., 2016. Numerical and physical modeling of a flexibly-connected two-body wave energy converter. Presented at the Marine Energy Technology Symposium (METS), Washington DC, USA, p. 4.

Journée, J.M.J., Massie, W.W., 2001. Offshore Hydromechanics, First Edition. ed. Delft University of Technology, Delft, The Netherlands.

López, M., Taveira-Pinto, F., Rosa-Santos, P., 2017. Numerical modelling of the CECO wave energy converter. Renewable Energy 113, 202–210. http://dx.doi.org/10.1016/j.renene.2017.05.066

Ma, Y., Ai, S., Yang, L., Zhang, A., Liu, S., Zhou, B., 2020. Hydrodynamic Performance of a Pitching Float Wave Energy Converter. Energies 13, 27. http://dx.doi.org/10.3390/en13071801

Magagna, D., Margheritini, L., Alessi, A., Bannon, E., Boelman, E., Bould, D., Coy, V., Marchi, E.D., Frigaard, P.B., Soares, C.G., Golightly, C., Todalshaug, J. H., Heward, M., Hofmann, M., Holmes, B., Johnstone, C., Kamizuru, Y., Lewis, T., Macadre, L.-M., Maisondieu, C., Martini, M., Moro, A., Nielsen, K., Reis, V., Robertson, S., Schild, P., Soede, M., Taylor, N., Viola, I., Wallet, N., Wadbled, X., Yeats, B., 2018. Workshop on identification of future emerging technologies in the ocean energy sector: JRC Conference and Workshop Reports, in: E U R. Presented at the JRC Conference and Workshop: Workshop on identification of future emerging technologies in the ocean energy sector, European Commission * Office for Official Publications of the European Union, Ispra, Italy, p. 81. http://dx.doi.org/10.2760/23207

Ocean Energy Europe, 2021. Ocean Energy: Key trends and statistics 2020. Ocean Energy Europe, Brussels, Belgium.

Papillon, L., Costello, R., Ringwood, J.V., 2020. Boundary element and integral methods in potential flow theory: a review with a focus on wave energy applications.

J. Ocean Eng. Mar. Energy 6, 303–337. http://dx.doi.org/10.1007/s40722-020-00175-7
Pecher, A., Kofoed, J.P. (Eds.), 2017. Handbook of Ocean Wave Energy, Ocean Engineering & Oceanography. Springer International Publishing, Cham. http://dx.doi.org/10.1007/978-3-319-39889-1
Qiao, D., Haider, R., Yan, J., Ning, D., Li, B., 2020. Review of Wave Energy Converter and Design of Mooring System. Sustainability 12, 31. http://dx.doi.org/10.3390/su12198251
Ringwood, J., 2020. Wave energy control: status and perspectives 2020, in: Proc. of the IFAC World Congress. Presented at the IFAC World Congress, Berlin, Germany, p. 13.
Rodrigues, C., Nunes, D., Clemente, D., Mathias, N., Correia, J.M., Rosa-Santos, P., Taveira-Pinto, F., Morais, T., Pereira, A., Ventura, J., 2020. Emerging triboelectric nanogenerators for ocean wave energy harvesting: state of the art and future perspectives. Energy Environ. Sci. 27. http://dx.doi.org/10.1039/D0EE01258K
Silva, D., Martinho, P., Guedes Soares, C., 2018. Wave energy distribution along the Portuguese continental coast based on a thirty three years hindcast. Renewable Energy 127, 1064–1075. http://dx.doi.org/10.1016/j.renene.2018.05.037
Strang-Moran, C., Mountassir, O.E., 2018. Offshore Wind Subsea Power Cables (No. AP-0018). ORE Catapult, University of Hull, UK.
Sutherland, J., Barfuss, S.L., 2011. Composite Modelling, combining physical and numerical models. Presented at the 34th IAHR World Congress, Brisbane, Australia, p. 12.
Wu, J., Yao, Y., Sun, D., Ni, Z., Göteman, M., 2019. Numerical and Experimental Study of the Solo Duck Wave Energy Converter. Energies 12, 19. http://dx.doi.org/10.3390/en12101941
Xu, S., Wang, S., Guedes Soares, C., 2019. Review of mooring design for floating wave energy converters. Renewable and Sustainable Energy Reviews 111, 595–621. http://dx.doi.org/10.1016/j.rser.2019.05.027

*Trends in Maritime Technology and Engineering – Guedes Soares & Santos (eds)*


# Ferry ships: A cost/environmental comparison of innovative solutions for the electric power generation in port

T. Coppola, M. Fantauzzi, L. Micoli, L. Mocerino & F. Quaranta
*Department of Industrial Engineering, University of Naples "Federico II", Italy*

ABSTRACT: In order to cope with continuously increasing environmental demands, the maritime sector is investigating alternative solutions for the power generation on board. These solutions include both alternative fuel, such as Liquefied Natural Gas (LNG) and hydrogen, and green technologies, such as fuel cells (FC) and batteries. Although in-port emissions are responsible of a small percentage of the overall emissions, ports constitute sources of concentrated ship harmful emissions, potentially very dangerous for the human health. Moreover, the environmental regulations suggest switching present marine fuels as soon as possible. In this frame, this work investigates innovative solutions for the electric power generation in port for passenger ships in terms of cost and environmental benefits. These can be shore power solutions (such as cold ironing) and on-board solutions (such as dual fuel, fuel cell and others). The results show that some solutions could be more competitive than others according to specific operating conditions. The aim of the study is to identify the most eco-friendly and promising solutions. The cost analysis methodology takes into account the efficiency, fuel costs, operating and maintenance costs and the initial investment costs. The environmental comparison considers the emissions of nitrogen and sulphur oxides, carbon monoxide and dioxide, particulate matter and hydrocarbon.

## 1 INTRODUCTION

Exhaust gas emissions from ships represent a key issue for the environmental impact assessment of maritime transportation (Toscano et al., 2021). Diesel engines are used as the main power supply of marine vessels causing the emissions of carbon dioxide ($CO_2$), carbon monoxide (CO), nitrogen oxides (NOx), sulphur oxides (SOx), unburned hydrocarbons (HC), particulate matter (PM), Volatile Organic Compounds (VOC) and ash (Eyring et al., 2005). Ships emit pollutants during the cruise and manoeuvring phase and when mooring at berth. Although local emissions are a small fraction of global transport emissions, they could have serious effects on human health, especially in coastal areas and water cities. For instance, the exposure to toxic emissions is reported to cause cardiovascular and respiratory diseases, lung cancer, and premature deaths. Recent estimates are reported by the Fourth IMO Green House Gases (GHG) Study 2020 (International Maritime Organization, 2020) showing that GHG emissions of total shipping have increased from 977 Mt in 2012 to 1,076 Mt in 2018 (+9.6%), mostly due to a continuous increase of global maritime trade. This corresponds to an increase from 2.76% in 2012 to 2.89% in 2018 of the weight of shipping emissions on global anthropogenic GHG emissions (IMO, 2020).

As known, the key IMO Convention on this topic is the MARPOL (73/78). Further requirements, formulated in terms of the Energy Efficiency Existing Ship Index (EEXI), will enter into force in 2023 (MEPC 76). For NOx the reference is TIER II while the TIER III is to be considered valid only in the $NO_X$ ECA areas (MEPC 76). As regard to $SO_X$ instead, since 1st January 2020, the limit for sulphur in fuel oil used on board ships operating outside designated emission control areas is reduced to 0.50% w/w (from 1/1/2015 in the ECA areas, this limit is of 0.10%) (EUR-Lex). In order to reduce harmful emissions, several techniques have been introduced on board, most commonly adopted deal with the improvements of engine and ship performances (Altosole et al., 2021, 2018; De Luca & Pensa, 2012; Balsamo et al., 2012) the use of alternative or performed fuels which give better combustion process (Goldsworthy, 2012; Sun et al., 2017), after-burning treatments and the use of alternative power technologies (Altosole et al., 2020, 2014; Bucci et al., 2016). Most promising solutions are described as following.

### 1.1 *Innovative solutions*

Among conventional internal combustion engines (ICE), Dual Fuel Diesel engines (DF) offer a flexible solution that can be used with low infrastructures

DOI: 10.1201/9781003320289-39

cost and may include one or more generators fuelled by natural gas (NG), liquefied (LNG) or compressed (CNG). Switching from diesel oil to NG for internal combustion engines may significantly reduce exhaust gas emissions (Altosole et al., 2017). Furthermore, this solution could provide a significant reduction in fuel cost thanks to a lower consumption (Vávra et al., 2017). LNG is more suited in terms of storage, safety, and moving compared to CNG, mainly in maritime sector ships (Elkafas et al., 2021).

It must be taken into account that nowadays Fuel Cells (FCs) are considered a very promising technology in the field of power production. FCs are energy conversion devices that convert the chemical energy of a fuel and an oxidant into electrical energy (Moseley, 2001). This is an electrochemical process that shows several advantages over conventional ICE, such as the high electrical efficiency (typically 0.5 - 0.65), very low pollutant emissions, fuel flexibility and low noise. A first classification distinguishes between Low Temperature FCs (LT-FCs), operating in the range 60 - 120°C, and High Temperature FCs (HT-FCs), operating up to about 1000°C. LT-FCs include the Proton Exchange Membrane FC (PEMFC) technology, which is the most commercialized among the other FCs, but it requires pure hydrogen ($H_2$) as fuel. Application in the maritime sector of PEMFC has been demonstrated by several projects and studies worldwide (Sørensen & Spazzafumo, n.d.). On the other hands, applications of HT-FCs are more recent and refer mainly to the Solid Oxide FC (SOFC), but they seem to be promising thanks to the possibility of a heat recovery from the exhausts, which further increases the efficiency, and a fuel flexibility. Indeed, a SOFC does not require pure $H_2$ and can be fuelled by GNL, methanol or other light hydrocarbons.

In the last decades, the electrification of ships has contributed to increase the ships' efficiency as well, favouring the introduction on board of energy storage systems, such as batteries (B). There are generally full electric ships, in which B are used for propulsion, and hybrid propulsion, where B work just in ports and at low speeds. The designing of a hybrid ship deals with the choice of the type of B and their capacity, which is certainly influenced by weights, dimensions, charging times, costs, and cooling system. In 2018, a DNV GL study, commissioned by the European Maritime Safety Agency (EMSA) on the use of electrical storage systems in maritime sector, provides a state of art of the technology, ongoing projects, regulations and safety for maritime B applications According to this report, the most used technologies for cruise ships are Nickel Manganese Cobalt Oxide (NMC), Lithium Iron Phosphate (LFP), while Lithium Titanate Oxide (LTO) seems to be more suitable for yachts and high speed crafts ("IT - Italiano - IT - EMSA - European Maritime Safety Agency," n.d.). The main configuration of a hybrid propulsion consists of B modules integrated with a power system for electrical propulsion providing additional power to the large propulsion engines. This is a flexible solution where the vessel may use only generators, B stand-alone or both. This type of solution can facilitate the use of zero emission operation in port, reduce the noise and vibration level on the ship, and additionally, B will also smooth the load variations of the generator sets. In 2019, the Maritime Battery Forum published a review that analyses all the ferry (new or R&D projects) launched between 2008 and 2019 ("Maritime Battery Forum"). The review shows that 88% of applications consider the use of Li-ion B, while only in three different cases Lead-Gel (in 2008), Li-polymer (in 2014), and NiMH (in 2014) B were used. Most of these solutions are developed in the countries of northern Europe (Norway, Denmark, Finland, and Sweden) that are particularly sensitive to environmental issues. These hybrid solutions have a power range from 500 to 5000 kW and a number of passengers from 16 to 2000. The data updated to 01/03/2019 show that a major part of these ships (over 140) are car/passenger ferries followed by offshore supply ships.

The onshore power supply, also known as cold ironing (CI), represents another way to reduce local emissions while ships are berthed in port. It consists in providing power from shore-side electricity rather than using on board auxiliary generators (Innes & Monios, 2018; Spengler & Tovar, 2021). The system design can be broken into several sections such as a connection to the national grid carrying high voltage levels of electricity to a local shore-side substation, and on board installation used to connect the cable to a socket on board the vessel (Innes & Monios, 2018; Spengler & Tovar, 2021). Even though it has been proved that this solution could reduce the in-port emissions, particularly in countries where the energy generation from renewable resources is spread, it is implemented in few ports, which are mainly large ports with high total energy demand (Innes & Monios, 2018). This could be due mainly to the fact that shipowners are reluctant to pay the onshore electricity which is more expensive compared to the one generated on board (Coppola et al., 2016). According to WPCI (2022) only 28 ports in the world have installed cold ironing points: the main European ports are Rotterdam, Antwerp, Hamburg, and Civitavecchia. According to Zis (2019), there are in the world 43 ports with either cold ironing already installed or planned. According to Winkel et al. (2016) € 2.94 billion of health costs could have been saved and a potential $CO_2$ reduction of 800,000 tons could be estimated if all European ports had used cold ironing. In this framework, it is recalled that this solution is not adaptable to all vessels; for instance, adequate ships are Ro-Ro or cruise ships; on the other hand, obliviously only a specific logistic and geographical conformation of ports can accept this installation. Another challenge is that the AC on board ships always has a frequency of 60 Hz, while the national power grids in Europe generally operate at 50 Hz. While considering the potential alternative solutions for the electric power generation to reduce the in-port emissions, this work aims to compare them roughly in terms of cost and environmental benefits, assuming the case study of passenger ships with a power load from about 1000 to 3000 kW. Therefore, technologies considered are the auxiliary energy generator (AE) fuelled by MGO, assumed as reference case, the marine gas oil/LNG dual-fuel generating set (DF), batteries (B), PEMFC, SOFC and cold-ironing (CI).

## 2 METHODOLOGIES

### 2.1 *Cost evaluation*

Total cost for on-board generation of electricity depends mainly on the design of the ship's power supply system and the fuel used. Therefore, the comparison in terms of the overall cost ($C_{tot}$) of case studies considers the capex and opex, the last includes the operating, maintenance and fuel costs:

$$C_{tot} = Capex + Opex \tag{1}$$

Capex supposes an annual interest (i) of 5 % within 20 years (N) of ships working, it can be estimated according to the following equation (2).

$$Capex = P \cdot C_C \tag{2}$$

Where, P is the power (kW);$C_C$ is the specific capital cost ($/kW).

Opex is evaluated accordingly to equation (3):

$$Opex = P \cdot t \cdot \left(\sum_o C_O + \sum_m C_m\right) + C_F \tag{3}$$

Where, t is the berthed time (h); $\sum_o$ Co is the summation of specific operating costs ($/kWh); $\sum_m$ Cm is the summation of specific maintenance costs ($/kWh); CF is the fuel cost ($).

Parameters of the equation (2) and (3) are presented in Table 1 for each case study. These values are from those reported by other authors in recent literature. Specifically, AE plant cost of 276 $/kW and 345 $/kW for DF are reported by Coppola et al. (Coppola et al., 2016), authors assumed that DF is approximately 25 % higher than that of AE.

FC technologies are at the early stage of commercialization, as a result they show very high specific plant cost, which are 1200 $/kW and 5000 $/kW respectively (Kamarudin et al., 2006). It must be noted that FC plant costs include both the FC stack and the auxiliaries (BoP) required for the fuel supply, the electric energy production and distribution to the grid. The BoP of a FC system may be the main item of expenditure and it depends on the application and the power size of the plat. For instance, the US Department of Energy forecasts that direct hydrogen PEM stacks cost will be reduced nearly 45 $/kW, when manufactured at a volume of 500,000 units per year by 2025 (Inal & Deniz, 2020). This cost development indicates that FCs will compete with Diesel engines in the immediate future (Pratt and Klebanoff, n.d.). Specific B cost of 150 $/kW (Table 1) is an average value for Li-ion battery technologies cost (Roland et al., 2021). The value 160 $/kW for CI is reported by Anwar et al., (2020).

AE's maintenance cost (2.25 c$/kWh) is assumed by Coppola et al. (2016). PEM technologies for high power size application require a more complex system for the water and temperature management and often a coolant is needed (Gadducci et al., 2021; Sharaf & Orhan, 2014), as a consequence maintenance and operating costs are a relevant proportion of the PEM opex; they are assumed conventionally equal to 17.50 c$/kWh, since literature data often put together Co and Cm (Chen et al., 2015; Wu et al., 2008).

Main SOFC companies sell their products including the maintenance service cost in the installation plant contract, therefore the SOFC maintenance and operating costs are not reported in Table 1 (Cigolotti et al., 2021).

The specific fuel consumptions reported in Table 1 are assumed in average values according to a theoretical efficiency of the plant taken from literature and commercial data (Sharaf & Orhan, 2014; Vávra et al., 2017). As expected, these values vary significantly by the technology and the fuel considered. FC technology is well known for the high electrical efficiency, according to some commercial product values, most recent PEM products show an efficiency of 60 % (Dimitrovar and Nader, 2022; Özgür & Yakarуılmaz, 2018) while 80 % is for SOFC.

The last is generally made of an electrical efficiency of 55 % and a thermal efficiency of 25 % thanks to the heat recovery from the SOFC exhausts, thus reducing the fuel consumption (Alaswad et al., 2022; Cigolotti et al., 2021).

Table 1. Parameters' value for the capex and opex estimations.

| Parameter | Unit | AE | DF | FC 1* | FC 2* | B | CI |
|---|---|---|---|---|---|---|---|
| Specific plant cost, $C_C$ | $/kW | 276 | 345 | 1K | 1.2K | 150 | 160 |
| Specific maintenance cost, $\sum_m$ Cm | c$/kWh | 2.25 | 7.33 | 17.5 | - | 30 | 10.9 |
| Specific operating cost, $\sum_o$ Co | c$/kWh | 5.08 | 7.33 | 17.5 | - | 30 | 10.9 |
| Specific fuel consumption for FC | g/kWh | 220 | 190 | 50 | 90 | - | - |

* (FC 1= PEM, FC 2=SOFC)

A charging-discharging cycle for B assumes an average efficiency of 85 %, according to Patel (2021) (Patel, 2021). The present analysis considers ships having a range of power installed on board from 1250 to 3000 kW, which are typically passenger ships such as ferries or small high-speed crafts. It is assumed that such ships are berthed for 1500 hours annually (t).

Table 2 reports fuels' price and grid electricity price assumed in this work for the opex evaluation. It must be noted that fuel prices vary largely over time, place and by fuel quality, despite this, we considered average costs' referring to the present days. For instance, the price of the MGO has undergone a sharp decline at the beginning of the year 2020 reaching the price of about $ 300/metric tons; in the first months of 2021 reached an average value about $ 450/metric tons ("Ship & Bunker - Shipping News and Bunker Price Indications," n.d.). According to the European Union Natural Gas Import Price, the price of LNG, on the other hand, after reaching a minimum of $ 2/MMBtu around mid-2020, continued to rise up to the current $10/MMBtu ("European Union Natural Gas Import Price," n. d.). The electricity prices for households in Italy from 2010 to 2020 ranging from 0.23 $/kWh and 0.25 $/kWh. Hydrogen price depends also by the production process and capacity, besides the resource, as a consequence it varies significantly from about 2 to 20 c$/kWh.

Table 2. Resources' price.

| Resource | Unit | Cost |
|---|---|---|
| MGO | $/ton | 453 |
| LNG | $/m$^3$ | 110 |
| Hydrogen | c$/kWh | 10 |
| National grid electricity | $/kWh | 0,25 |

Estimation of the overall emissions is carried on by referring to the emission factors (EF) reported in recent literature data specifically for the technologies considered in this study (Ammar, 2019; "EMEP/EEA air pollutant emission inventory guidebook 2019 — European Environment Agency," n.d.; Inal and Deniz, 2020). EF refer to the route Tank-to-Propeller (TTP) for $CO_2$, CO, HC, NOx, SOx and PM; these are reported in average values in Table 3.

It is well known that emissions of an ICE depend strongly on the engine specifics and the operating conditions, which largely vary in consequence of the different applications (Mocerino et al., 2021), despite this, we assumed average values just for a rough evaluation and an easier comparison.

The amount of the specific pollutant emission is calculated according to the following equation (4):

Table 3. Emission factor for different pollutants and technologies (g/kWh).

| | | Pollutant | | | | | |
|---|---|---|---|---|---|---|---|
| System | Fuel | $CO_2$ | CO | NOx | PM | SOx | HC |
| AE | MGO | 698 | 1.68 | 12.6 | 0.23 | 2.56 | 0.53 |
| DF | LNG/ MGO | 430 | 0.79 | 1.57 | 0.025 | 0.003 | 3.00 |
| PEM | $H_2$ | 0 | 0 | 0 | 0 | 0 | 0 |
| SOFC | LNG | 343 | 0.002 | 0.005 | $\approx 0$ | 0 | $\approx 0$ |
| B | MGO | 698 | 1.68 | 12.60 | 0.23 | 2.56 | 0,53 |
| CI | - | 0 | 0 | 0 | 0 | 0 | 0 |

$$E_i = P \cdot t_p \cdot EF_i \tag{4}$$

Where: i is the generic pollutant ($CO_2$, CO, HC, NOx, SOx, PM), $E_i$ is the emission of *i*; P is the maximum power of the plant on board (kW); $t_p$ is the berthed time (h); $EF_i$ is the emission factor of i (g/kWh);

Assuming AE as reference case, the emission variation is calculated as follow:

$$\%\ variation = 100 \cdot \left[(E_i)_{AE} - (E_i)_j / (E_i)_{AE}\right] \tag{5}$$

Where, j refers to each case study, except AE.

## 3 RESULTS

Alternative solutions for the electric power generation on board ships require capex which vary significantly depending on the technology chosen, as reported in Table 1. It is quite evident that FC technologies (PEM, SOFC) have capex that are about one order of magnitude higher than others. This is due to the fact that they are at the early stage of commercialization, moreover, the TRL for the application in the maritime sector is still in the range 7-9 (Coppola et al., 2020; Se, 2011; van Biert et al., 2016). Therefore, it can be expected that FCs' capex will decrease appreciably in the next years.

Assuming that $C_o$, $C_m$, the price of MGO, LNG, $H_2$ and the electricity do not vary over the years, annual opex per kW can be calculated according to equation (3) and are shown separately in Figure 1.

As expected, fuel's and electricity energy's cost from the grid represent the main percentage of the opex for each solution considered. From the comparison between ICE technologies, it results that the amount of fuel cost (LNG) of DF (70 $/kW) is lower than AE (149 $/kW), on the other hand Cm and Co are higher. The SOFC is fuelled by LNG the same as DF, but thanks to a higher overall efficiency the

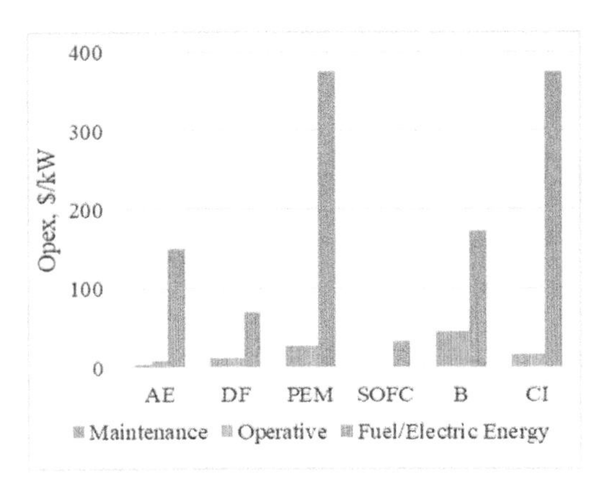


Figure 1. Annual opex ($/kW) for different plant solutions.

annual fuel cost is lower (33 $/kW). $H_2$ cost (375 $/kW) contributes to increase significantly the opex of the PEM solution, at the same time this technology has a relevant entry for the maintenance and operating cost (26 $/kW). In case of B solution, the MGO cost (172 $/kW) is slightly higher than those used for AE due to an efficiency factor (85 %) related to the charging-discharging cycle for recharging B.

Adding together opex and capex, Ctot per kW can be calculated for each solution and it is presented in Figure 2.

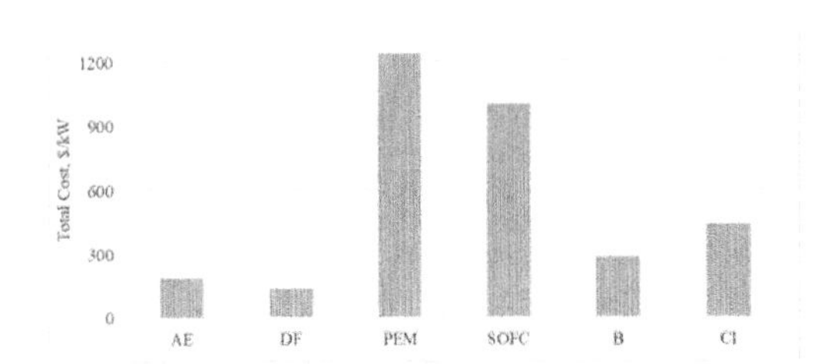


Figure 2. Annual total costs ($/kW) for different plant solutions.

PEM results the most expensive solution (1235 $/kW) due to the $H_2$ cost, while the SOFC cost (999 $/kW) remains high due to the capex, in spite of the lower fuel consumption. DF results the most affordable solution (134 $/kW) even than the reference case AE (183 $/kW). B's total cost (283 $/kW) results higher than that of AE and DF, but approximately in the same order of magnitude. In case of CI, the total cost (434 $/kW) is higher than AE, B and DF due to the electric energy cost from the port grid.

The costs analysis should be combined with the environmental impact of each solution taken into account. Assuming the traditional configuration (AE) as reference, Table 4 reports the emission variations in case of its replacing with an alternative electric power generation system (DF, PEM, SOFC, B, CI).

Table 4. Emission variation referred to AE (positive values correspond to a reduction).

| | Emission variation (%) | | | | | |
|---|---|---|---|---|---|---|
| System | $CO_2$ | CO | NOx | PM | SOx | HC |
| DF | 38.4 | 53.0 | 87.5 | 89.1 | 99.9 | -466 |
| PEM | 100 | 100 | 100 | 100 | 100 | 100 |
| SOFC | 50.9 | 99.9 | ≈100 | 100 | 100 | 100 |
| B | -15 | -15 | -15 | -15 | -15 | -15 |
| CI | 100 | 100 | 100 | 100 | 100 | 100 |

As well known, the use of LNG is spreading worldwide in the maritime sector thanks to the emission reduction allowing to comply the related rules, particularly referred to SOx emission (by 99.9% in this work). As reported in Table 4, DF contributes to reduce the emission of $CO_2$ by 38.4%, CO by 53.0%, NOx by 87.5% and PM by 89.1%. In spite of this, HC emission is significantly higher (466%) than AE mainly due to the methane slip phenomenon.

PEM using $H_2$ does not emit any pollutants in the route TTP, but it must be noted that if the $H_2$ is green, the PEM's emission still remains zero alongside the $H_2$ production chain, otherwise emissions must be taken into account in the route well–to–tank (WTT).

SOFC is a green technology fuelled by LNG, which is a fossil fuel; therefore, it allows to bring to almost to zero values the CO, NOx, PM, SOx and HC emission and to reduce significantly (about 50.9 %) the $CO_2$ emission thanks to its higher overall efficiency, which reduces the fuel consumption.

We assumed that B are charged by AE during the navigation; according to this hypothesis, emissions related to the case B are the same of AE raised considering the charging-discharging cycle efficiency (85 %). Therefore, B does not emit pollutants or GHG in ports, despite this, during the navigation all the emissions are 15 % higher than AE.

CI allows to cancel all the emissions from a ship in port, however this is true only in the case that the electric energy is produced from renewable resources, or else they must be taken into account.

## REFERENCES

Alaswad, A., Baroutaji, A., Rezk, A., Ramadan, M., Olabi, A.-G., 2022. Advances in Solid Oxide Fuel Cell Materials. *Encycl. Smart Mater.* 334–340. https://doi.org/10.1016/B978-0-12-803581-8.11743-6.

Altosole, M., Balsamo, F., Campora, U., Mocerino, L., 2021. Marine Dual-Fuel Engines Power Smart

Management by Hybrid Turbocharging Systems. *J. Mar. Sci. Eng.* 2021, Vol. 9, Page 663 9, 663. https://doi.org/10.3390/JMSE9060663.

Altosole, M., Benvenuto, G., Campora, U., Laviola, M., & Zaccone, R. (2017). Simulation and performance comparison between diesel and natural gas engines for marine applications. *Proceedings of the Institution of Mechanical Engineers, Part M: Journal of Engineering for the Maritime Environment, 231*(2), 690–704.

Altosole, M., Benvenuto, G., Campora, U., Silvestro, F., Terlizzi, G., 2018. Efficiency Improvement of a Natural Gas Marine Engine Using a Hybrid Turbocharger. *Energies* 2018, Vol. 11, Page 1924 11, 1924. https://doi.org/10.3390/EN11081924.

Altosole, M., Buglioni, G., Figari, M., 2014. Alternative Propulsion Technologies for Fishing Vessels: a Case Study. undefined 8, 296–301. https://doi.org/10.15866/IREME.V8I2.459.

Altosole, M., Campora, U., Vigna, V., 2020. Energy efficiency analysis of a flexible marine hybrid propulsion system. 2020 *Int. Symp. Power Electron. Electr. Drives, Autom. Motion, SPEEDAM* 2020 436–441. https://doi.org/10.1109/SPEEDAM48782.2020.9161873.

Ammar, N.R., 2019. ENVIRONMENTAL AND COST-EFFECTIVENESS COMPARISON OF DUAL FUEL PROPULSION OPTIONS FOR EMISSIONS REDUCTION ONBOARD LNG CARRIERS. Brodogr. Teor. i praksa Brodogr. i Pomor. Teh. 70, 61–77.

Anwar, S., Zia, M.Y.I., Rashid, M., De Rubens, G.Z., Enevoldsen, P., 2020. Towards Ferry Electrification in the Maritime Sector. *Energies* 2020, Vol. 13, Page 6506 13, 6506. https://doi.org/10.3390/EN13246506.

Balsamo, F., De Luca, F., Pensa, C., 2012. A new logic for controllable pitch propeller management, in: *Sustainable Maritime Transportation and Exploitation of Sea Resources*. pp. 639–648. https://doi.org/10.1201/b11810-97.

Bucci, V., Mauro, F., Marino, A., Bosich, D., Sulligoi, G., 2016. An innovative hybrid-electric small passenger craft for the sustainable mobility in the Venice Lagoon. 2016 *Int. Symp. Power Electron. Electr. Drives, Autom. Motion, SPEEDAM 2016*.

Chen, H., Pei, P., Song, M., 2015. Lifetime prediction and the economic lifetime of Proton Exchange Membrane fuel cells. *Appl. Energy* 142, 154–163. https://doi.org/10.1016/J.APENERGY.2014.12.062.

Cigolotti, V., Genovese, M., Fragiacomo, P., 2021. Comprehensive review on fuel cell technology for stationary applications as sustainable and efficient poly-generation energy systems. *Energies* 14. https://doi.org/10.3390/EN14164963.

Coppola, T., Fantauzzi, M., Miranda, S., Quaranta, F., 2016. Cost/benefit analysis of alternative systems for feeding electric energy to ships in port from ashore. AEIT 2016 - *Int. Annu. Conf. Sustain. Dev. Mediterr. Area, Energy ICT Networks Futur.* https://doi.org/10.23919/AEIT.2016.7892782.

Coppola, T., Micoli, L., Turco, M., 2020. State of the art of high temperature fuel cells in maritime applications, in: 2020 International Symposium on Power Electronics, *Electrical Drives, Automation and Motion, SPEEDAM 2020. pp. 430–435.*

De Luca, F., Pensa, C., 2012. Unconventional interceptors in still water and in regular waves: Experimental study on resistance reduction and dynamic instability, in: *NAV International Conference on Ship and Shipping Research*. p. 216369.

Dimitrovar, Z., Nader, W.B., 2022. PEM fuel cell as an auxiliary power unit for range extended hybrid electric vehicles. *Energy* 239. https://doi.org/10.1016/J.ENERGY.2021.121933.

Elkafas, A.G., Khalil, M., Shouman, M.R., Elgohary, M. M., 2021. Environmental protection and energy efficiency improvement by using natural gas fuel in maritime transportation. *Environ. Sci. Pollut.* Res. 28, 60585–60596. https://doi.org/10.1007/S11356-021-14859-6/TABLES/7.

EMEP/EEA air pollutant emission inventory guidebook 2019 — European Environment Agency, n.d. URL https://www.eea.europa.eu/publications/emep-eea-guidebook-2019 (accessed 1. 26.22).

European Union Natural Gas Import Price, n.d. URL https://ycharts.com/indicators/europe_natural_gas_price (accessed 1.20.22).

Eyring, V., Köhler, H.W., Van Aardenne, J., Lauer, A., 2005. Emissions from international shipping: 1. The last 50 years. J. Geophys. *Res. Atmos.* 110, 171–182. https://doi.org/10.1029/2004JD005619.

Gadducci, E., Lamberti, T., Bellotti, D., Magistri, L., & Massardo, A. F. (2021). BoP incidence on a 240 kW PEMFC system in a ship-like environment, employing a dedicated fuel cell stack model. *International Journal of Hydrogen Energy*, 46(47),24305–24317.

Goldsworthy, L., 2012. Combustion behaviour of a heavy duty common rail marine Diesel engine fumigated with propane. *Exp. Therm. Fluid Sci.* 42, 93–106.

Inal, O.B., Deniz, C., 2020. Assessment of fuel cell types for ships: Based on multi-criteria decision analysis. *J. Clean. Prod.* 265. https://doi.org/10.1016/J.JCLEPRO.2020.121734.

Innes, A., Monios, J., 2018. Identifying the unique challenges of installing cold ironing at small and medium ports – The case of aberdeen. *Transp. Res. Part D Transp. Environ.* 62, 298–313. https://doi.org/10.1016/J.TRD.2018.02.004.

IMO, 2020. Fourth Greenhouse Gas Study 2020.

Kamarudin, S.K., Daud, W.R.W., Md, Som, A., Takriff, M. S., Mohammad, A.W., 2006. Technical design and economic evaluation of a PEM fuel cell system. *J. Power Sources* 157, 641–649.

*Leading platform for distributed generation of electricity and hydrogen* | Bloom Energy n.d. URL https://www.bloomenergy.com/ (accessed 1. 26.22).

Maritime Battery Forum, n.d. URL https://www.maritimebatteryforum.com/ (accessed 12. 16.21).

Mocerino, L., Soares, C.G., Rizzuto, E., Balsamo, F., Quaranta, F., 2021. Validation of an Emission Model for a Marine Diesel Engine with Data from Sea Operations. *J. Mar. Sci. Appl.* 20, 534–545. https://doi.org/10.1007/S11804-021-00227-W/FIGURES/13.

Moseley, P.T., 2001. Fuel Cell Systems Explained. J. Power Sources. https://doi.org/10.1016/s0378-7753(00)00571-1.

Özgür, T., Yakaryılmaz, A.C., 2018. A review: Exergy analysis of PEM and PEM fuel cell based CHP systems. Int. J. Hydrogen *Energy* 43, 17993–18000. https://doi.org/10.1016/J.IJHYDENE.2018.01.106.

Patel, M.R., 2021. *Shipboard Electrical Power Systems*. Shipboard Electr. Power Syst. https://doi.org/10.1201/9781003191513.

Pratt, J.W., Klebanoff, L.E., n.d. *Feasibility of the SF-BREEZE: a Zero-Emission, Hydrogen Fuel Cell, High-Speed Passenger Ferry.*

Roland, Z., Máté, Z., Győző, S., 2021. Comparison of alternative propulsion systems-a case study of a passenger ship used in public transport. *Brodogradnja* 72, 1–18. https://doi.org/10.21278/BROD72201.

Se, G.L., 2011. FUEL CELLS IN MARITIME APPLICATIONS CHALLENGES, CHANCES AND EXPERIENCES, in: *4th International Conference on Hydrogen Safety.*

Sharaf, O.Z., Orhan, M.F., 2014. An overview of fuel cell technology: Fundamentals and applications. *Renew. Sustain. Energy* Rev. https://doi.org/10.1016/j.rser.2014.01.012.

Ship & Bunker - Shipping News and Bunker Price Indications n.d. https://www.shipandbunker.com/ (accessed 12.16.21).

Sørensen, B., Spazzafumo, G., n.d. *Hydrogen and fuel cells : emerging technologies and applications.*

Spengler, T., Tovar, B., 2021. Potential of cold-ironing for the reduction of externalities from in-port shipping emissions: The state-owned Spanish port system case. *J. Environ. Manage.* 279, 111807.

Sun, X., Liang, X., Shu, G., Wang, Yajun, Wang, Yuesen, Yu, H., 2017. Effect of different combustion models and alternative fuels on two-stroke marine diesel engine performance. *Appl. Therm. Eng.* 115, 597–606.

Toscano, D., Murena, F., Quaranta, F., Mocerino, L., 2021. Assessment of the impact of ship emissions on air quality based on a complete annual emission inventory using AIS data for the port of Naples. *Ocean Eng.* 232, 109166. https://doi.org/10.1016/J.OCEANENG.2021.109166.

Van Biert, L., Godjevac, M., Visser, K., Aravind, P. V., 2016. A review of fuel cell systems for maritime applications. *J. Power Sources.*

Vávra, J., Bortel, I., Takáts, M., Diviš, M., 2017. Emissions and performance of diesel–natural gas dual-fuel engine operated with stoichiometric mixture. *Fuel.* https://doi.org/10.1016/j.fuel.2017.07.057.

Winkel, R., Weddige, U., Johnsen, D., Hoen, V., Papaefthimiou, S., 2016. Shore Side Electricity in Europe: Potential and environmental benefits. *Energy Policy* 88, 584–593. https://doi.org/10.1016/J.ENPOL.2015.07.013.

World Ports Climate Action Program – World Port Sustainability Program, n.d. URL https://sustainableworldports.org/wpcap/ (accessed 1.26.22).

Wu, J., Yuan, X.Z., Martin, J.J., Wang, H., Zhang, J., Shen, J., Wu, S., Merida, W., 2008. A review of PEM fuel cell durability: Degradation mechanisms and mitigation strategies. *J. Power Sources* 184, 104–119. https://doi.org/10.1016/J.JPOWSOUR.2008.06.006.

Zis, T.P.V., 2019. Prospects of cold ironing as an emissions reduction option. *Transp. Res. Part A Policy Pract.* 119, 82–95. https://doi.org/10.1016/J.TRA.2018.11.003.

*Trends in Maritime Technology and Engineering – Guedes Soares & Santos (eds)*


# CFD analysis of the PTO damping on the performance of an onshore dual chamber OWC

J. Gadelho & C. Guedes Soares
*Centre for Marine Technology and Ocean Engineering (CENTEC), Instituto Superior Técnico, Universidade de Lisboa, Lisbon, Portugal*

G. Barajas & J.L. Lara
*IHCantabria - Instituto de Hidráulica Ambiental de La Universidad de Cantabria, Santander, Spain*

ABSTRACT: Three numerical approaches using CFD are used to model the PTO damping of an onshore dual chamber OWC tested in a scaled wave flume: In the first case, the PTO is modelled with the same geometrical dimensions of the experimental setup; in the second case, the PTO is modelled as a porous medium, while in the third case, it is modelled by a drag force applied directly to the momentum equation. Measurements of air pressure and free surface elevations inside the chambers of the OWC are compared between CFD approaches and the physical model. They show that the 3 CFD models reproduce with relative accuracy the physical experiments. Results also show that the simulations are faster around 25% up to 40% using numerical damping regions comparing to the PTO real geometry.

## 1 INTRODUCTION

Oscillating Water Columns (OWC) are one of the main Wave Energy Converters (WEC) systems widely used to capture energy, both onshore and offshore. A wide variety of OWC devices have been developed over the years (Falcão, 2015) and several numerical models have been build up around the subject (Iturrioz et al., 2015; Rezanejad et al., 2019; Shalby et al., 2019).

Due to its working principle, the Power Take-Off (PTO) system is usually modelled, in scaled experiments, as an air opening in the pneumatic chamber. The same principle can be applied to Computational Fluid Dynamics (CFD) studies. An air opening can be used to model the PTO. In some cases, the opening is small compared to the global mesh, therefore the simulations take excessive time, and the numerical model stability is not guaranteed.

Kamath et al. (2015), used the CFD software package REEF3D to model a single chamber OWC using Darcy's law for flow through porous media to model the PTO damping. They used an equivalent area of porous media 10x bigger than the area of the hole used in the experiments. They conclude that the PTO damping has a large influence on the hydrodynamics of an OWC and this can be used to attain the maximum possible hydrodynamic efficiency for a given incident wavelength.

Dimakopoulos et al. (2015), validated an OpenFOAM model to characterize the key aspects of the flow of a OWC wave energy converter. They conclude that the numerical model compares very well with physical model data, when the air is characterized as incompressible, as a low damping coefficient is used at the PTO.

Xu & Huang (2019), used a 3D CFD to model a circular OWC with a nonlinear power-takeoff. They validate the model and discuss on the resonant sloshing inside the pneumatic chamber and conclude that the CFD model setup can simulate the key physical processes involved: free surface elevations inside and outside the circular OWC and air pressures inside the chamber.

The DCOWC concept model used in this work is presented by Rezanejad et al. (2020). They used a boundary element code to model the onshore DCOWC and found out that the hydrodynamic performance is significantly improved up to 75% in a broad range of wave frequencies compared to the equivalent single chamber device. As explained by Rezanejad et al. (2015), the use of two chambers permits one of the chambers to extract more efficiently the power, for some wave periods, when the other is more efficiency for others, and vice-versa. Also, the resonance mechanisms and the wave interaction with bottom steps seem to play an important role to improve the global efficiency of the device (Rezanejad and Guedes Soares, 2018).

The novelty of this work is to validate and analyze, with the numerical model OpenFOAM, the PTO

DOI: 10.1201/9781003320289-40

damping on the performance of an onshore dual chamber OWC. The biggest challenge to model numerically dual chamber OWCs is to reproduce accurately the phenomena of the water flow exchange between chambers as well as its repercussion in the pneumatic processes.

Three numerical approaches are used to model the PTO: In the first case, the PTO is modelled with the same geometrical dimensions of the experimental setup; in the second case, the PTO is modelled as a porous medium (Romano et al., 2020); while in the third case, it is modelled by a drag force applied directly on the momentum equation.

The Free Surface Elevations (FSE) and air pressures inside the chambers are evaluated and compared to the experimental results. These results are relevant to estimate the global efficiency of a DCOWC installed onshore.

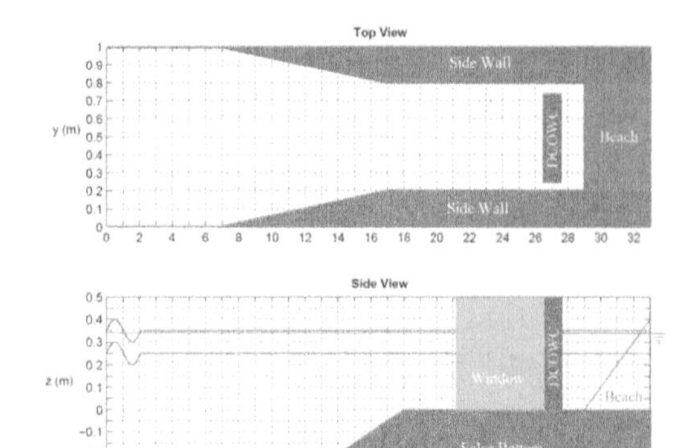


Figure 1. COI3 wave flume geometrical characteristics with minimum and maximum water depth representation. In the experiments the water depth of h = 0.384 m was used . Not to scale.

## 2 PHYSICAL MODEL

The experiments were performed at the COI3 irregular wave flume located at the National Laboratory of Civil Engineering (LNEC), Portugal. The wave flume is approximately 30.0 m long, and its cross-section varies from 1.0 m in the vicinity of the piston-type wave maker down to 0.6 m in the beach side. The bottom is irregular due to the inclusion of a false bottom to reduce the depth on the region of interest. The wave flume geometrical characteristics is presented in Figure 1. The geometrical scale of 1/25 is used in the experiments and in the numerical simulations.

The Dual Chamber Oscillating Water Column (DCOWC) is fixed in the center of the wave flume (top view) at approximately 25.0 m from the wave maker with a cross-section of 0.5 m and water depth in its vicinity of 0.384 m. Its geometry is represented in Figure 3a) and its dimensions is represented in Figure 2. There is one circular hole in each chamber with a diameter of 0.036 m to model the PTO pneumatic damping. Chamber 1 is the first facing the waves, while Chamber 2 is the one next to the beach side.

There is a gap of 0.05 m between the DCOWC and the wave flume walls in each side. In the back of the DCOWC there is an absorption beach. The piston type wave maker does not have an active absorption system and the generated waves were regular resulting from a combination of two wave heights H = 0.04 m and H = 0.08 m, and 13 periods ranging from T = 0.8 s up to T = 3.2 s, with an interval of 0.2 s.

Several sensors to measure Free Surface Elevations (FSE) and air pressures (P) were used in the experiments: Seven resistive Wave Gauges (WG) are located in the wave flume. These WGs are important to understand if the numerical model can generate and propagate the waves in the same way that the

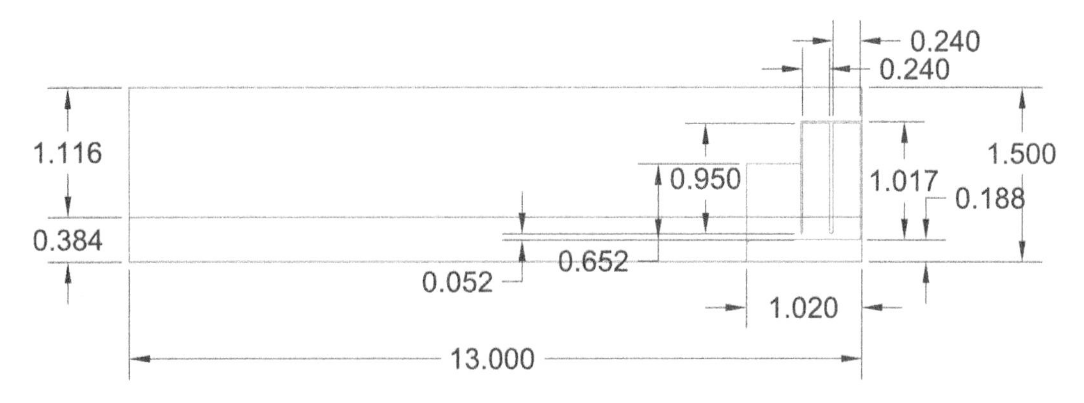


Figure 2. Computational Domain with the dimensions of the DCOWC and its location. Wave flume and DCOWC cross-section is 0.5 m. Dimensions in m and figure is not to scale. Thickness of DCOWC walls is 0.015 m, except central plate that is 0.030 mm thick.

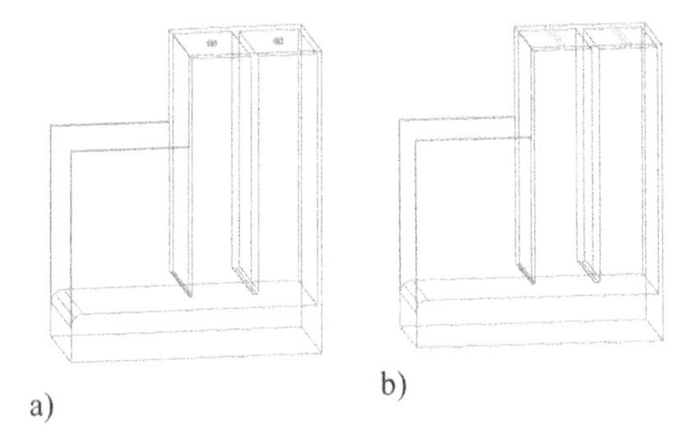


Figure 3. DCOWC with the PTO configurations: a) PTO modeled by holes with diameter $\phi = 0.036$ m; b) PTO modeled by a porous media/velocity damping.

observed in the physical model. Two capacitive WGs and two digital air pressure sensors were installed inside the chambers. Its synchronized data is important to evaluate the primary efficiency of the device.

## 3 NUMERICAL MODEL

The CFD software used in this work is release version 2012 of the open source software OpenFOAM (Open Source Field Operation and Manipulation) (*OpenFOAM*, 2022). The main solver is interFoam, which solves incompressible free surface Newtonian flows using the Navier–Stokes equations coupled with a volume of fluid method. The Navier-Stokes equations in vector form are:

$$\rho\left(\frac{\partial v}{\partial t} + v \cdot \nabla v\right) = -\nabla p + \mu \nabla^2 v + \rho g \quad (1)$$

where $\rho$ is the density of the fluid, $v$ is the velocity, $p$ is the pressure, $g$ is the acceleration of gravity, $\mu$ is the dynamic viscosity, $t$ represents the time and $\nabla^2$ is the vector Laplacian. The continuity equation is:

$$\nabla \cdot v = 0 \quad (2)$$

The wave generation and absorption, and the flow through the porous medium are guaranteed by the OpenFOAM suite named IHFoam (Higuera et al., 2014a, 2014b; Romano et al., 2020). In all simulations a k-ω SST turbulent model was used, enhanced by the correction of Larsen & Fuhrman (2018). The quasi-3D computational domain is represented in Figure 2. The base mesh was generated with blockMesh with a cell size of $\Delta x = 0.064$ m; $\Delta y = 0.02$ m; $\Delta z = 0.05$ m, where y represents the cross-section and is defined with 10 cells. The snappyHexMesh tool was used to define the DCOWC and to produce an extra refinement in the vicinity of the FSE and geometry. The final mesh has around 470 000 cells. A portion of the mesh is represented in Figure 4.

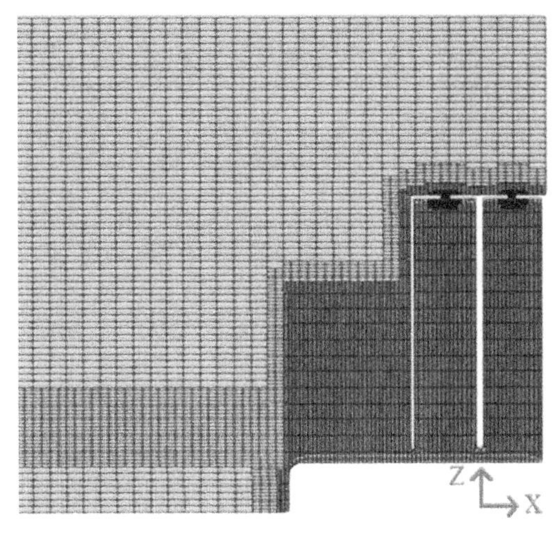


Figure 4. Lateral view of the mesh in the vicinity of the DCOWC. Extra refinement was used around the FSE, DCOWC and holes.

## 4 CALIBRATION

The calibration of the model consisted in a sensitive analysis for two parameters: computational domain and mesh refinement for the wave flume with the DCOWC geometry using holes to simulate the PTO.

To study the computational domain sensitivity, two numerical setups were used with almost the same mesh resolution. The first numerical domain consisted in modeling the entire experimental wave flume, including the false bottom ramp (Figure 1). This means that the gap between the DCOWC and the wave flume walls and a porous beach behind the structure were also modeled. A snapshot of a simulation can be seen in Figure 5. The length of the domain is 33.0 m and mesh has around 3 million cells.

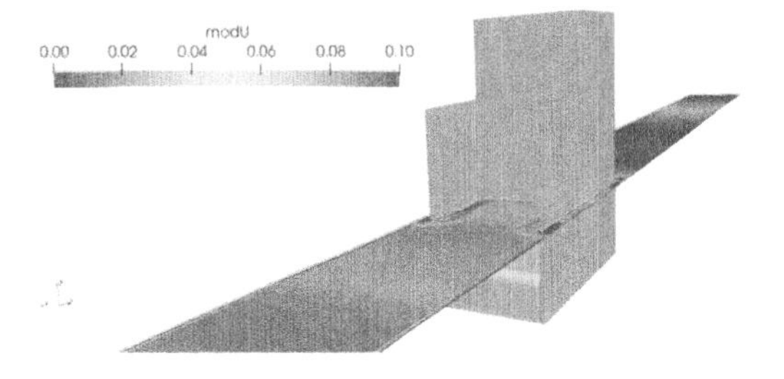


Figure 5. Snapshot of a CFD simulation for the full domain approach at t = 17.4s. H = 0.04 m and T = 2.0S.

The second domain consisted in modeling the wave flume cut in the inlet, and considering only its regular bottom part, and reducing its cross-section to 0.5 m, so the DCOWC occupies the entire outlet. This configuration is represented in Figure 2 and the mesh configuration can be seen in Figure 4. Its length is 13 m.

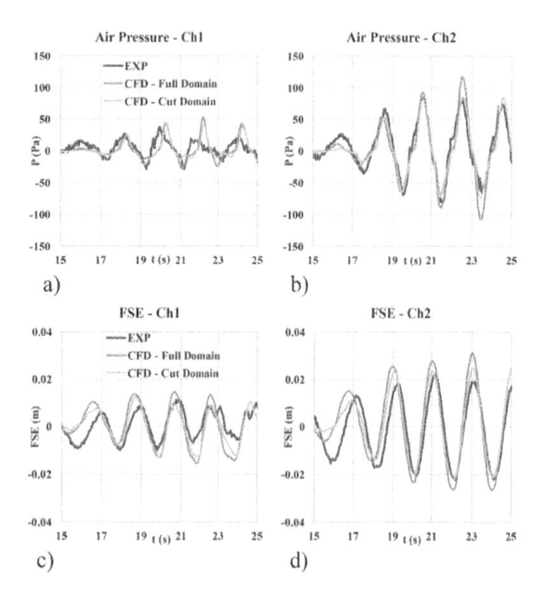


Figure 6. FSE and air pressures inside the chambers for H = 0.04 m and T = 2.0 s.

The results of FSEs and air pressures time series inside both chambers, for both domains, are compared with the experimental results with regular waves (H = 0.04 m and T = 2.0 s) and depicted in Figure 6. It is important to analyze the first instants of the results because is when the first waves arrive, and the numerical model has more difficult to reproduce the transient phenomena.

Observing the air pressure time series in chambers 1 and 2, in Figure 6 a) and b), respectively, it is visible that the numerical model can reproduce with higher accuracy the shape of the pressure signal in chamber 2 than in chamber 1. In chamber 1, the first facing the waves, the signal of the experiments is identical to the numerical model up to 21 s. Then the maximum experimental air pressure becomes smaller, which might indicate a loss of pressure due to the high nonlinear phenomenon of wave sloshing. The FSE inside chamber 1, Figure 6 c), also denote some irregularities after T = 21 s. This means that, while both domains can reproduce the hydro-aerodynamics in chamber 2, for chamber 1, the model cannot reproduce entirely all the high nonlinear interactions.

Globally, the cut domain present better results, for the same incident wave conditions, and it is faster to simulate. This means that the cut domain is adopted for all the rest of the simulations in this work.

The sensitive analysis of the mesh refinement consists in the analysis of the FSE envelope for regular waves with H = 0.02 m and T = 2.0 s, along the numerical wave flume. The simulations run for 60 s for three meshes described in Table 1. The refinement was made based on Mesh0 multiplied in x and z directions by a given factor.

Table 1. Mesh characteristics.

| Mesh | Factor | N. Cells |
|---|---|---|
| Mesh0 | 1 | 470k |
| Mesh1 | $\sqrt{2}$ | 800k |
| Mesh2 | 2 | 1400k |

The results of the FSE envelope for the three meshes is shown in Figure 7. It is visible the presence of nodes and anti-nodes, a clear indication of wave reflection from the DCOWC.

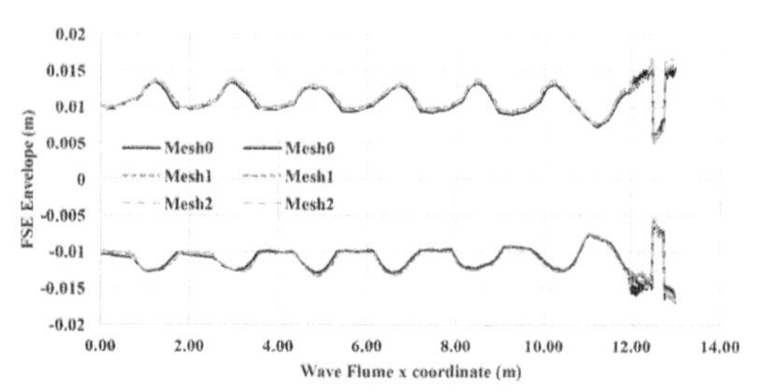


Figure 7. FSE envelope along the wave flume for the three meshes. Regular waves with H = 0.02 m and T = 2.0 s.

The FSE envelopes for the three meshes are identical, which means that even the coarser mesh is able to generate and propagate the waves without losing accuracy. The FSEs and air pressures inside the chambers were also evaluated and the biggest uncertainty of results is connected to chamber 1, where the air pressure had a maximum difference of 6% and 2% for the FSEs. In chamber 2, the maximum deviation was less than 1% for both measurements. This means that the air pressure inside chamber 1 is more sensitive to the mesh refinement than the other quantities inside the chambers. The observed differences inside the chambers might be connected to the hole geometry definition that the coarser mesh might fail to give a reasonable mesh density in this area.

Based on the previous results, the coarser mesh revealed to be accurate enough for the desired uncertainty in this study, so Mesh0, with extra care in the geometric holes definition, was adopted for the rest of the simulations.

## 5 PTO MODELING

One of the novelties of this work is to model the PTO damping with different techniques and simulate the key physical processes inside the DCOWC chambers. To do this, three techniques were adopted: one that is a simple replication of the physical holes used in the experiments; and two that adopts different flow rules in a defined numerical region. These numerical regions have a square area that is 10x bigger than the equivalent circular holes area (Figure 3).

The advantages of using damping numerical regions, instead of using the real geometry are mostly connected to computational savings. For example, if there is a need to correct the PTO damping of one chamber, the geometry must be changed and remeshing the entire case can be a very long process and generally needs a new validation.

In the case of using damping numerical regions, the user only needs to change a certain range of parameters, as an iterative process, until obtain the desired results. On the other hand, these types of approaches cannot replicate the flow in the vicinity of the PTO. Results show that, in terms of CPU time consumption, exists a reduction of around 25% up to 40% using Case 2 and Case 3, respectively, comparing to Case 1.

### 5.1 *Case 1 - PTO simulated with circular holes*

The first case is defined using holes in the top of the chambers to reproduce geometrically the same openings used in the experimental setup. Its geometry can be seen in Figure 3 a). The holes must be defined with a reasonable mesh resolution to permit a correct air flow through the orifice. In this case, the holes have approximately 4 cells per diameter. In Figure 8 it is visible the mesh used in the vicinity of the holes, and an intermediate mesh refinement between the refined part and the coarser part, to make the transition as smooth as possible. The diameter of the holes is $\phi = 0.036$ m.

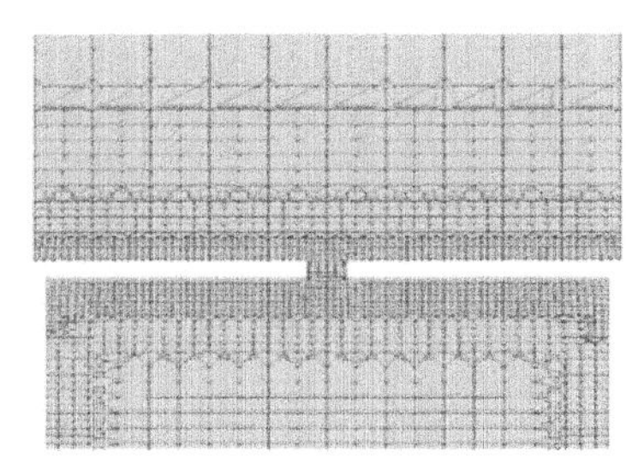

Figure 8. Mesh in the vicinity of the holes used in Case 1.

### 5.2 *Case 2 - PTO simulated with porous media*

A macroscopic approach has been used to describe the porous media flow in the numerical simulations (Romano et al., 2020), which calculates a mean behavior within the porous zone (avoiding the disadvantages of microscopic approaches). Due to the complexity and the lack of information of the internal structure of porous media, linear and nonlinear drag forces are modelled following Forchheimer's formula, (1901):

$$\frac{\partial \bar{u}_i}{\partial x_i n} = 0 \tag{3}$$

$$(1+c)\frac{\partial \rho \bar{u}_i}{\partial t n} + \frac{\bar{u}_j \partial \rho \bar{u}_i}{n \partial x_j n} = -g_j x_j \frac{\partial \rho}{\partial x_i} - \frac{\partial p^*}{\partial x_i} - f_{\sigma i} - \frac{\partial}{\partial x_j} \mu_{eff}\left(\frac{\partial \rho \bar{u}_i}{\partial x_j n} + \frac{\partial \rho \bar{u}_i}{\partial x_i n}\right) + -A\bar{u}_i - B|\bar{u}_i|\bar{u}_i \tag{4}$$

$$\frac{\partial \alpha}{\partial t} + \frac{\partial \bar{u}_i \alpha}{\partial x_i n} + \frac{\partial \bar{u}_{ic}\alpha(1-\alpha)}{\partial x_i n} = 0 \tag{5}$$

The Forchheimer parameters can be expressed following (Gent, 1995) expressions:

$$A = a\frac{(1-n)^2 \mu}{n^3 D_{50}^2} \tag{6}$$

$$B = b\left(1 + \frac{7.5}{KC}\right)\frac{(1-n)\rho}{n^3 D_{50}} \tag{7}$$

$$C = \gamma\frac{1-n}{n} \tag{8}$$

$u$ is the velocity, $\mu$ is the dynamic viscosity, $p$ is the pressure, $\rho$ density and $g$ gravity. $\alpha$ is the linear friction coefficient, $\beta$ is the non-linear friction coefficient, c is the added mass coefficient, $n$ is the porosity of the material, $D_{50}$ is the mean nominal diameter of the porous material and $KC$ is the Keulegan-Carpenter number.

After an iterative process of the calibration, the values used on the equations are listed in Table 2.

Table 2. value of the constants used in Case 2.

| constant | value |
|---|---|
| a | 2 |
| b | 2 |
| c | 0.34 |
| $D_{50}$ | 0.0159 |
| porosity | 0.25 |

### 5.3 *Case 3 - PTO simulated with velocity damping*

For the case where, the PTO is simulated with a velocity damping, the previous equations presented for porous media flow are adapted neglecting the role played by porosity on fluid acceleration (both local and convective).

$$\frac{\partial u_i}{\partial x_i} = 0 \tag{9}$$

$$\frac{\partial \rho u_i}{\partial t} + \frac{u_j \partial \rho u_i}{\partial x_j} - \frac{\partial}{\partial x_j}\left(\mu_{eff} \frac{\partial \rho u_i}{\partial x_j}\right) = \frac{\partial p^*}{\partial x_i} - g_j x_j \frac{\partial \rho}{\partial x_j} - F_{drag} \tag{10}$$

$$F_{drag} = a\|U\| + b\|U\| * \|U\| \tag{11}$$

where a and b are constants and ||U|| is the velocity magnitude. These constants must be calibrated and after an iterative process, the values used are a = 5 000; b = 10 000.

## 6 RESULTS

The present CFD results were separated into two categories. First, the results of the three PTO modeling techniques are presented and compared with the experimental results. Secondly, the key physical processes inside the chambers (air pressures and free surface elevations), for a wide range of wave periods, are explored and discussed.

### 6.1 *PTO damping results*

The results of air pressures and FSEs inside the chambers of the three PTO modeling techniques were compared with experimental results. Two regular wave cases were used:

- wave7: H = 0.035 m and T = 2.0 s
- wave10: H = 0.039 m and T = 2.6 s.

The results of the air pressures inside the chambers for wave7 are presented in Figure 9. Results show that the modeling of the PTO with porous media (Case2) present the best results in terms of pressure amplitude in the chamber 1. The other two cases overpredict the peak value. All the approaches fail to reproduce entirely the signal form obtained in the experiments. Probably this means that the model cannot reproduce the high nonlinear phenomenon of wave sloshing inside the first chamber. Also, the air compressibility, that is neglected in the numerical model, might have an important role in the overall pneumatic performance.

a)

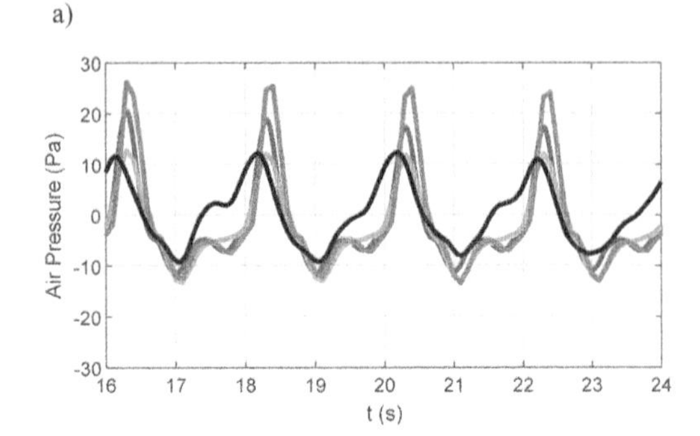


b)

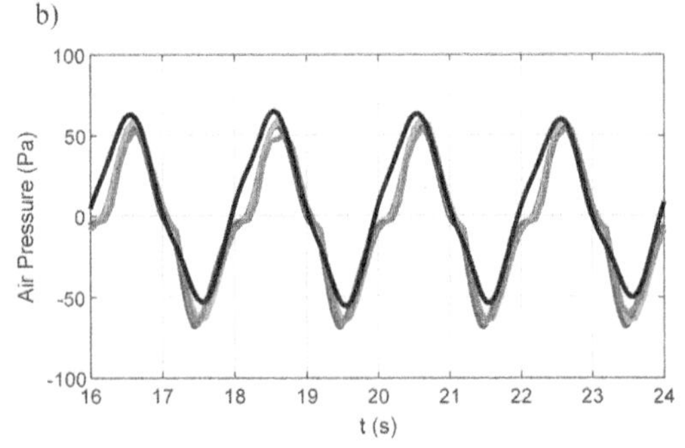


Figure 9. Air pressures inside the chambers for wave7: a) chamber 1; b) chamber 2. Black: Experimental; Blue: Case1; Green: Case2; Red: Case3.

For chamber 2, wave 7, the results for the air pressures time series are presented in Figure 9 b). They should a similar signal form for all the PTO modeling approaches. Comparing with the experimental results, numerical results show a more pronounced local attenuation when passing by zero pressure and underpredicts the minimum pressure.

For the case wave10, the results of the air pressures inside both chambers are presented in Figure 10. For chamber 1,

Figure 10 a), the three CFD results show almost the same signal form but cannot reproduce entirely the physical model, all of them overpredict the maximum pressure. The results of Case 2 seem to be the best in the negative pressure part. For the chamber 2,

Figure 10 b), the three models fail to give a correct signal form.

Snapshots of the CFD simulations for the three PTO modeling approaches can be seen in Figures 11 to 13. Apparently, the hydrodynamics is the same in all the approaches. As expected, the velocities around the PTO are approximately 10x bigger for Case1, where the area is 10x smaller.

### 6.2 *Physical processes inside the chambers*

The key physical processes inside the chambers (air pressures and free surface elevations) were evaluated. The analysis of these quantities is important to estimate the primary efficiency of the DCOWC.

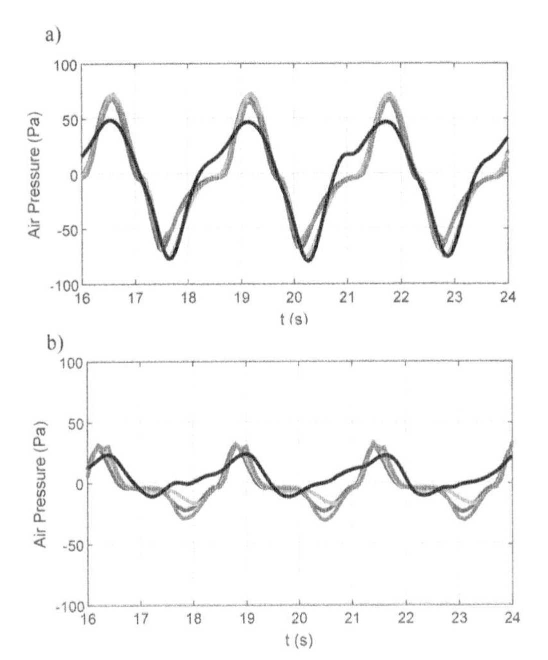


Figure 10. Air pressures inside the chambers for wave10: a) chamber 1; b) chamber 2. Black: Experimental; Blue: Case1; Green: Case2; Red: Case3.

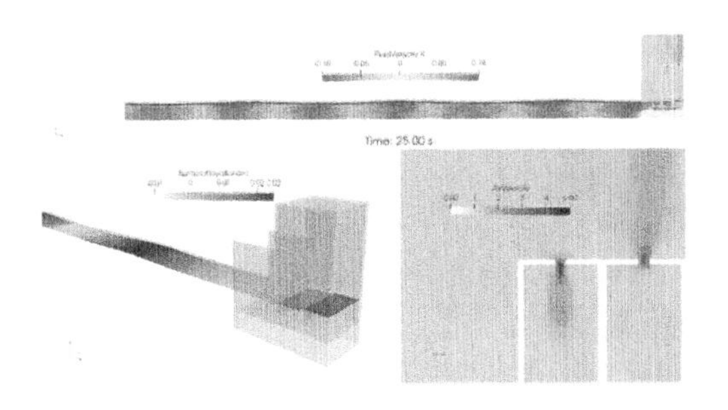

Figure 11. – Snapshots of the CFD results for the PTO damping Case 1 with T = 2.0 at 25.00 s.

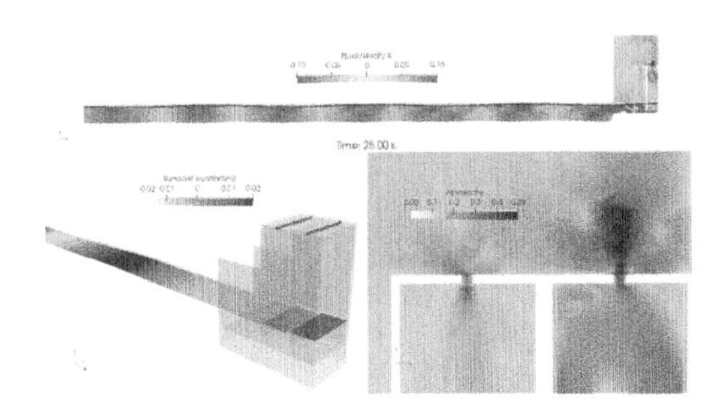

Figure 12. Snapshots of the CFD results for the PTO damping Case 2 with T = 2.0 at 25.00 s.

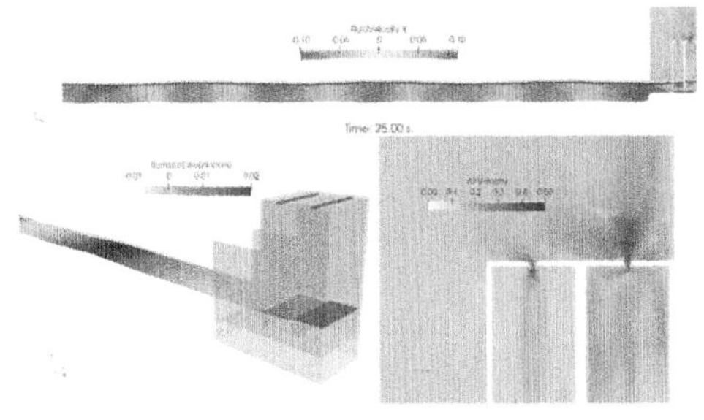

Figure 13. Snapshots of the CFD results for the PTO damping Case 3 with T = 2.0 at 25.00 s.

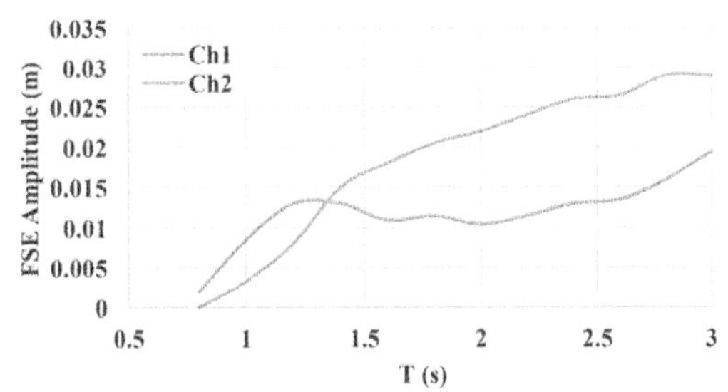


Figure 14. Evolution of the free surface elevations amplitude inside the chambers.

Twelve simulations with H = 0.04 m and with periods ranging from 0.8 s up to 3.0 s were made for Case1 (PTO is modeled with an opening).

The time series of each quantity inside the chambers is normalized, the first seconds before the system reach a steady state condition are excluded and in the end its amplitude is calculated.

The evolution of the FSE amplitude inside the chambers is presented in Figure 13. Results show that the amplitude of the FSE inside chamber 1 increases up to 0.025 m at T = 1.2 s and maintains its amplitude until T = 2.5 s, where it experiences a jump up to 0.04 m. For chamber 2, the amplitude of the FSE jumps from 0 m up to 0.03 m at 1.4 s, and then increases up to 0.0 6 m at T = 2.8 s. This means that in this chamber the wave amplitude is amplified up to 1.5x the incident wave amplitude due to a resonance phenomenon.

The evolution of the air pressure inside both chambers is shown in Figure 13. It follows almost the same tendence than the FSEs. In chamber 1, the air pressure jumps from 0 Pa up to 65 Pa at T = 1.2 s and then drops down to 30 Pa at T = 3.0 s. In chamber 2, the air pressure reaches 80 Pa at T = 1.8 s and remains high up to T = 3.0 s.

It can be concluded from the numerical analysis that chamber 1 might be more efficient in the lower periods, while for the higher periods chamber 2 might be the one that captures more pneumatic power from the waves.

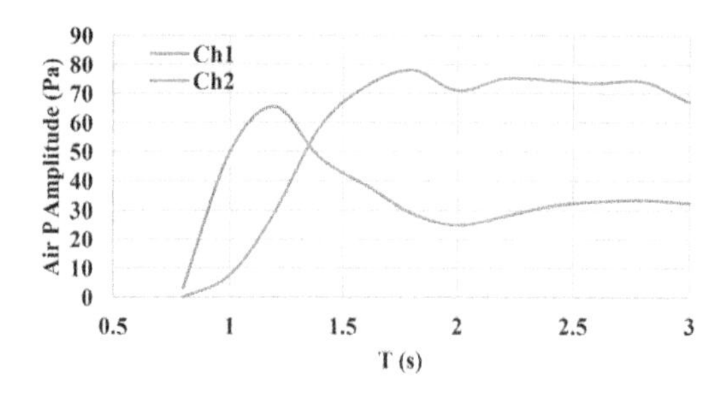


Figure 15. Evolution of the air pressures amplitude inside the chambers.

## 7 CONCLUSIONS

A novel dual chamber oscillating water column is studied in CFD, with 3 different approaches to model the PTO damping and simulate the key physical processes inside the DCOWC chambers. To do this, three techniques were adopted: one that is a simple replication of the physical holes used in the experiments; and two that adopts different flow rules in a defined numerical region.

Results show that the 3 PTO model can reproduce reasonable accuracy the air pressures and free surface elevations inside the chambers, but fail to reproduce the air pressure form, probably due to the impossibility of reproducing the high nonlinear phenomenon of wave sloshing inside the first chamber.

The use of damping numerical regions, proved to be faster 25% up to 40% using Case 2 and Case 3, respectively, comparing to Case 1, where real holes were used. On the other hand, damping numerical regions cannot replicate the flow in the vicinity of the PTO. Results show that the recorded velocities in the vicinity of the hole were 10x higher for the real geometry.

Results of the evolution of the air pressures and free surface elevations inside the chambers related to the wave periods, show that, in global, chamber 1 might be more efficient in the lower periods, while for the higher periods chamber 2 might can extract more pneumatic power from the waves. Results also show that in chamber 1, the wave amplitude is amplified up to 1.5x the incident wave amplitude due to a resonance phenomenon.

In future, a detailed sensitive analysis over the input parameters of the damping numerical regions cases (Case 2 and Case 3) might be important to calibrate the CFD numerical model.

## ACKNOWLEDGEMENTS

The first author has been funded by the University of Lisbon, and CENTEC within a PhD grant.

The authors would like to thank Dr. Kourosh Rezanejad and Eng. Gael Anastas who performed the experimental work and helped processing the data.

The paper is done within the project "Harbour protection with dual chamber oscillating water column devices", (HARBOUR OWC), which is co-funded by the European Regional Development Fund (Fundo Europeu de Desenvolvimento Regional - FEDER) and by the Portuguese Foundation for Science and Technology (Fundação para a Ciência e a Tecnologia - FCT) under contract PTDC/EME-REN/30866/2017.

This work contributes to the Strategic Research Plan of the Centre for Marine Technology and Ocean Engineering (CENTEC), which is financed by the Portuguese Foundation for Science and Technology (Fundação para a Ciência e Tecnologia - FCT) under contract UIDB/UIDP/00134/2020.

This work was produced with the support of the Portuguese INCD (Infraestrutura Nacional de Computação Distribuída) funded by FCT and FEDER under the project 01/SAICT/2016 nº 022153.

## REFERENCES

Dimakopoulos, A. S., Cooker, M. J., Lopez, E. M., & Longo, D. (2015). Flow characterisation and numerical modelling of OWC wave energy converters. *Proceedings of the 11th European Wave and Tidal Energy Conference*, 8.

Falcão, A. F. O., Developments in oscillating water column wave energy converters and air turbines, in: Guedes Soares, C. (Ed.), Renewable Energies Offshore. Taylor & Francis Group, London, pp. 3–11.

Forchheimer, P. Z. (1901). Wasserbewegungdurch Boden. *Zeitschrift Des Vereins Deutscher Ingenieure*, *45*, 1781–1788.

Gent, M. R. A. van. (1995). *Wave interaction with permeable coastal structures*. Doctoral Thesis. ISBN 978-90-407-1182-4.

Higuera, P., Lara, J. L., & Losada, I. J. (2014a). Three-dimensional interaction of waves and porous coastal structures using OpenFOAM®. Part I: Formulation and validation. *Coastal Engineering*, *83*, 243–258. http://dx.doi.org/10.1016/j.coastaleng.2013.08.010

Higuera, P., Lara, J. L., & Losada, I. J. (2014b). Three-dimensional interaction of waves and porous coastal structures using OpenFOAM®. Part II: Application. *Coastal Engineering*, *83*, 259–270. http://dx.doi.org/10.1016/j.coastaleng.2013.09.002

Iturrioz, A., Guanche, R., Lara, J. L., Vidal, C., & Losada, I. J. (2015). Validation of OpenFOAM® for Oscillating Water Column three-dimensional modeling. *Ocean Engineering*, *107*, 222–236. http://dx.doi.org/10.1016/j.oceaneng.2015.07.051

Kamath, A., Bihs, H., & Arntsen, Ø. A. (2015). Numerical modeling of power take-off damping in an Oscillating Water Column device. *International Journal of Marine Energy*, *10*, 1–16. http://dx.doi.org/10.1016/j.ijome.2015.01.001

Larsen, B. E., & Fuhrman, D. R. (2018). On the over-production of turbulence beneath surface waves in Reynolds-averaged Navier–Stokes models. *Journal of Fluid*

*Mechanics*, *853*, 419–460. http://dx.doi.org/10.1017/jfm.2018.577
*OpenFOAM* (Version 2012). (2022). [Computer software]. ESI Group. https://www.openfoam.com/
Rezanejad, K. and Guedes Soares, C. (2018). Enhancing the primary efficiency of an oscillating water column wave energy converter based on a dual-mass system analogy. Renewable Energy. 123:730–747.
Rezanejad, K., Abbasnia, A., & Guedes Soares, C. (2020). Hydrodynamic performance assessment of dual chamber shoreline oscillating water column devices. In C. Guedes Soares (Ed.), *Developments in Renewable Energies Offshore*, Taylor and Francis Group, pp. 188–196. http://dx.doi.org/10.1201/9781003134572-23
Rezanejad, K., Bhattacharjee, J., & Guedes Soares, C. (2015). Analytical and numerical study of dual-chamber oscillating water columns on stepped bottom. *Renewable Energy*, *75*, 272–282. http://dx.doi.org/10.1016/j.renene.2014.09.050
Rezanejad, K., Gadelho, J. F. M., & Guedes Soares, C. (2019). Hydrodynamic analysis of an oscillating water column wave energy converter in the stepped bottom condition using CFD. *Renewable Energy*, *135*, 1241–1259.
Romano, A., Lara, J. L., Barajas, G., Di Paolo, B., Bellotti, G., Di Risio, M., Losada, I. J., & De Girolamo, P. (2020). Tsunamis Generated by Submerged Landslides: Numerical Analysis of the Near-Field Wave Characteristics. *Journal of Geophysical Research: Oceans*, *125*(7). http://dx.doi.org/10.1029/2020JC016157
Shalby, M., Elhanafi, A., Walker, P., & Dorrell, D. G. (2019). CFD modelling of a small–scale fixed multi–chamber OWC device. *Applied Ocean Research*, *88*, 37–47. http://dx.doi.org/10.1016/j.apor.2019.04.003
Xu, C., & Huang, Z. (2019). Three-dimensional CFD simulation of a circular OWC with a nonlinear power-takeoff: Model validation and a discussion on resonant sloshing inside the pneumatic chamber. *Ocean Engineering*, *176*, 184–198. http://dx.doi.org/10.1016/j.oceaneng.2019.02.010.

*Trends in Maritime Technology and Engineering – Guedes Soares & Santos (eds)*


# Review on hardware-in-the-loop simulation of wave energy converters

J.F. Gaspar, R.F. Pinheiro, M. Kamarlouei & C. Guedes Soares
*Centre for Marine Technology and Ocean Engineering (CENTEC), Instituto Superior Técnico, Universidade de Lisboa, Lisbon, Portugal*

M.J.G.C. Mendes
*Instituto Superior de Engenharia de Lisboa (ISEL), Instituto Politécnico de Lisboa, Lisboa, Portugal*
*Centre for Marine Technology and Ocean Engineering (CENTEC), Instituto Superior Técnico, Universidade de Lisboa, Lisbon, Portugal*

ABSTRACT: This paper presents a review of the state-of-the-art of the Hardware-In-The-Loop (HIL) technique applied to the development of wave energy converters (WEC) and power take-off systems (PTO). The presentation discusses existing frameworks and the technologies involved. In addition, a case study is discussed in order to exemplify the HIL concept applied to a WEC-PTO.

## 1 INTRODUCTION

The hardware-in-the-loop (HIL) technique is a hybrid co-simulation approach that joins virtual and real physical components. The virtual components are, for example, numerical models of sea waves and Wave Energy Converter (WEC) hydrodynamics, while the real components are the Power Take-Off (PTO) and controller. One of the objectives of HIL is to verify the compliance of the real component to the functional requirements, such as verifying if the PTO controller is operating. This approach has a dramatic reduction on the testing cost in comparison to tests performed in the tank or at open sea. Thus, HIL has been considered as a cost-effective approach for the development of WEC and PTO controllers, analysis and validation of PTO dynamics, performance, efficiency, and calibration of mathematical models. HIL research on WECs has been published with some regularity, however a review of the state-of-the-art has not been found in the research literature.

Hardware-In-The-Loop (HIL) have been increasingly used in the field of technological development in several areas. In this simulation procedure, some of the virtual part, in the numerical model of the system under simulation, are replaced with real ones (Bracco et al., 2015; Zhang et al., 2016). The goal is to avoid the complication of some physical components or phenomena, which are hard to model and simulate precisely or in an acceptable time frame (e.g. flow of water around ship hulls and propellers, generator complex electro-visco-elastic behaviour), while other components may be numerically simulated in a precise and cost-effective way

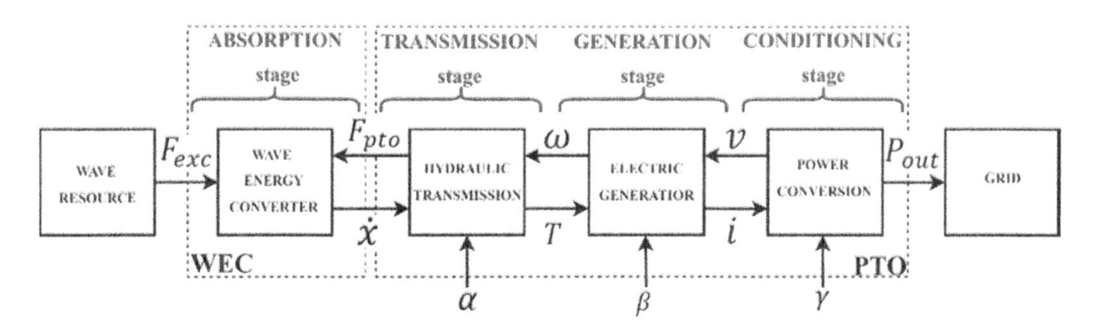


Figure 1. WEC-PTO with four conversion stages. The absorption stage is performed by the WEC prime mover while the transmission, generation and conditioning stages are performed by the PTO. Adapted from (Penalba & Ringwood, 2016).

DOI: 10.1201/9781003320289-41

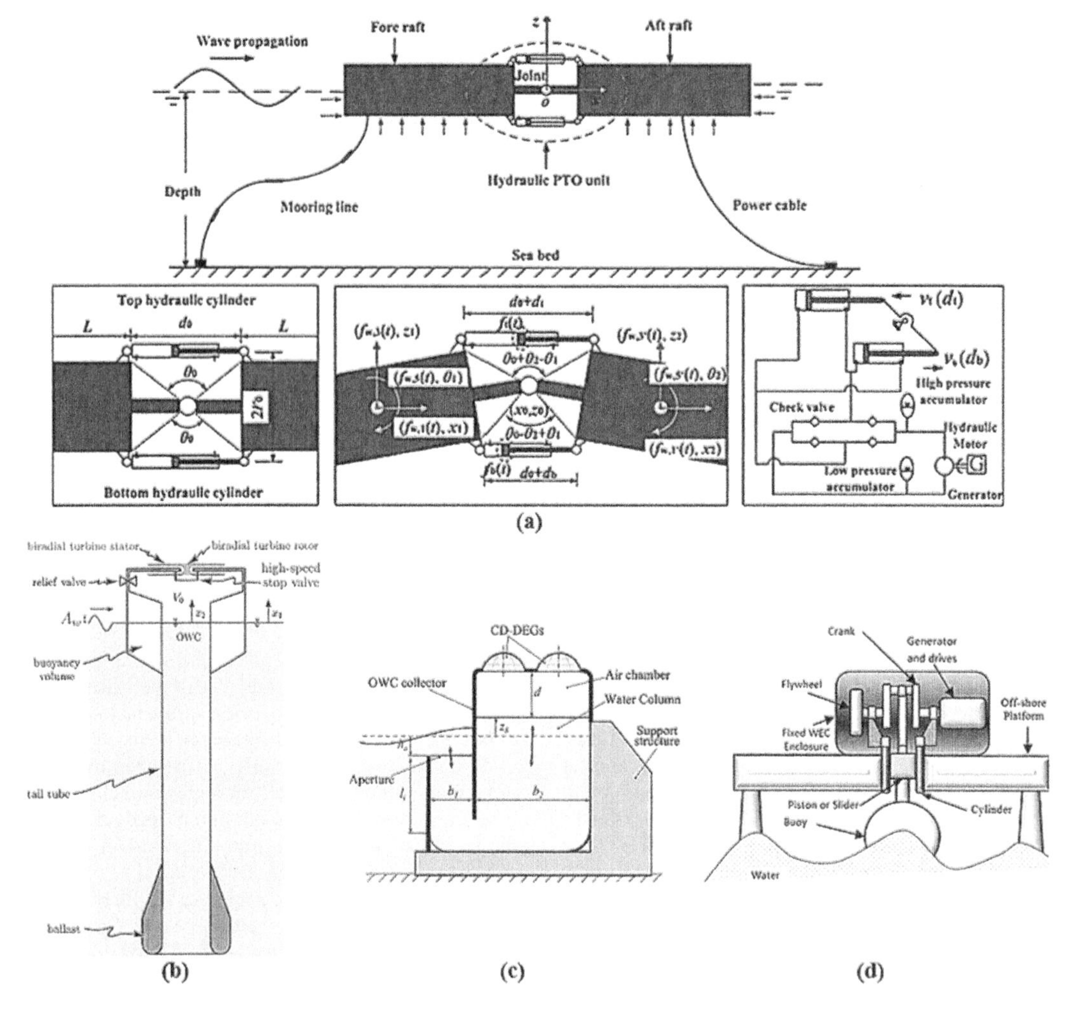

Figure 2. WECs. (a) Two-raft-type with oil-hydraulic PTO, (b) oscillating-water-column (OWC) with biradial turbine PTO (c) OWC with circular diaphragm dielectric elastomer generator (CD-DEGs), (d) point floater with slider-crank PTO. Adapted from (Henriques et al., 2016; Hollis et al., 2018; Liu et al., 2018; Moretti et al., 2021).

(Delmonte et al., 2020; Gaspar et al., 2016; Huijgens et al., 2021; Moretti et al., 2021; Vince et al., 2020). Moreover, HIL permits testing in a secure, controlled and replicable test environment and give a radical decrease of operational risks, price and execution time, compared to the testing of a real system operating in real conditions (e.g. sea or wave tank) (Bacelli et al., 2017; Delmonte et al., 2020; Neshati et al., 2020; Pedersen et al., 2016; Sancar et al., 2015).

The use of HIL systems began in the aeronautical field (flight simulators) and prospered in the military, aerospace, and automotive industries. An very widespread and state-of-the-art method in many fields of activity It is now considered to be a well-proven method (Bouscayrol et al., 2006; Bracco et al., 2015; Castellini et al., 2021; Signorelli et al., 2011; Skjetne & Egeland, 2006; Vince et al., 2020).

In the area of renewable energies, in specific in research and development of wave energy converters (WEC), HIL has been increasingly used (Aderinto & Li, 2018; Ahamed et al., 2020; Penalba & Ringwood, 2016; Xie & Zuo, 2013; Xu et al., 2019) and wind turbine systems (the later also verified and certified with HIL method) (Bracco et al., 2015; Curioni, 2019; Neshati et al., 2020; Signorelli et al., 2011; Vittori et al., 2021). WECs are designed to harness and absorb the low frequency motions of ocean waves and include power take-offs (PTOs) to transform the absorbed wave energy into usable electricity (Ahamed et al., 2020). Figure 1 presents a general form of the WEC known as the wave-to-wire (W2W) model.

The absorption stage catches the power of the wave source with the WEC prime mover (e.g., heave point floater or a surge flap) and transform it into mechanical energy (oscillatory motions) are presented in Figure 1. Via oil-hydraulic technology, the transmission stage rectifies the oscillating power and provide it to the generator stage. Subsequently, the generation stage transforms the oil-hydraulic in electrical power, whichever is suited to the mesh in the conditioning stage. In cases where the WEC is directly connected to the generator, the transmission stage can be not appear (e.g. linear generator) (Ahamed et al., 2020; Penalba & Ringwood, 2016; Zhang et al., 2021). The PTO inputs (α, β and γ inputs) have objective to control the WEC, in the sense of improving the efficiency.

Generally, WECs can be made of various types of absorber prime, i.e. two articulated barges coupled to the interface of the oil-hydraulic transmission PTO (Figure 2a) (Liu et al., 2018), a compartment where the oscillations of the water column pressurize and depressurize the caught air, which is constrained via the pneumatic PTO. (air turbine in Figure 2b (Henriques et al., 2016) and annular membrane in Figure 2c (Moretti et al., 2021)) or a point absorber WEC (Shami et al., 2018; Xie & Zuo, 2013) connected to a mechanical transmission PTO (Figure 2d) (Hollis et al., 2018).

Nevertheless, the HIL procedure can virtually emulate the wave generated motions of the WEC absorber (wave resource in Figure 1), in what is known as a dry-test of the WEC (Bracco et al., 2015). Therefore, the HIL technique has been reputed as an effective and efficient method for designing, optimization and testing of WEC controller and actuator systems, and, on sophisticated PTO systems (Bacelli et al., 2019; Delmonte et al., 2020; Lasa et al., 2012; Moretti et al., 2021; Signorelli et al., 2011). HIL has been implemented to the analysing and validation of PTO dynamics, control component, control algorithms, performance, effectiveness and mathematic model gauging standard operating conditions (Bacelli et al., 2019; Carrelhas et al., 2021; Hansen et al., 2019; Lasa et al., 2012; Liu et al., 2018; Signorelli et al., 2011).

Regarding HIL research on WEC, there are some publications, however a specific state-of-the-art review is not found in the literature. Thus, the goal of this paper is to give such review. Beginner researchers can carry this review as a helpful guidance to start their HIL works while expert researchers to updating their knowledge on HIL test. The review may be also utilized as a base-line for subsequent reviews on HIL testing of WECs.

This work is structured in 6 sections. Sections 2 and 3 one presents the HIL frameworks and taxonomies. The HIL technological systems are presented in Section 4. Section 5 discusses the review results, which are summarized in the conclusion.

## 2 FRAMEWORKS

The HIL procedure substitutes in part or entirely one or more W2W stages with real ones (Figure 1). The determination about the stages that should be virtual or real is fixed by the implementation rules and the exchange between costly tests on real systems and estimated numeric simulations (Delmonte et al., 2020; Signorelli et al., 2011). The distinction between HIL stages or element is not connected to the distinction between software or hardware, but preferably between the ones that are real in the WECs (e.g. controller software) from those numerically simulated (Signorelli et al., 2011). This procedure is illustrated by the HIL simulation of the Inertial Sea WEC (ISWEC) (Figure 3). The setup is made of a dry part, the WEC PTO (Figure 3a), and a moist one, the WEC hull (Figure 3b). The pitch motions of the hull are absorbed by the PTO inertial spin flywheel (left flank of the upper structure in Figure 3a) and then mechanically communicated to the generator for electrical generating (right flank of the upper structure in Figure 3a). The hull pitch motions caused by the waves are simulated with the rock platform positioned under the PTO. The time and cost of the wave tank are reduced by HIL simulation, and it is validated by real tests of the tank with an error as small as 10%. (Bracco et al., 2015).

The logical illustration of the HIL method is introduced in Figure 3c. The wave force (wave re-source stage) and hull dynamics (absorber stage) are numerically simulated while the test rig sustains and give the interface between the simulated hull dynamics (absorber stage) and the real part of the ISWEC (transmission and generation stages). The PTO pitching reference setting is fixed in real time on the hull dynamics block, in function of the wave and gyro loads, and is transmitted to the rock platform control hardware.

Other great level illustration of the HIL method, known as architectures (Delmonte et al., 2020; Grega, 1999; Liu et al., 2018; Signorelli et al., 2011), layouts (Curioni, 2019), configurations (Henriques et al., 2016), overviews (Henriques et al., 2016; Hollis et al., 2018), block diagrams (Neshati et al., 2016), system structure (Puleva et al., 2016; Sancar et al., 2015) and

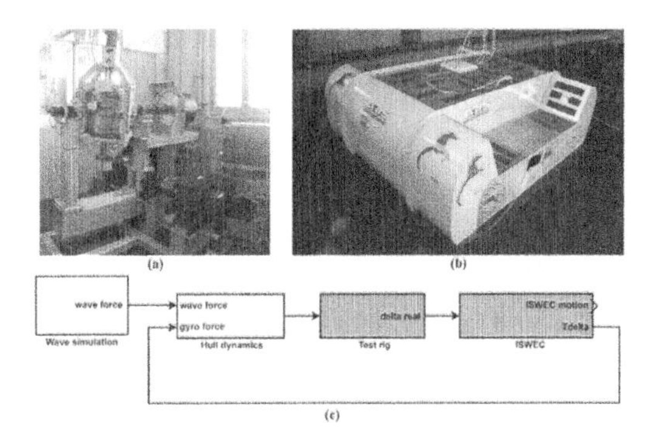


Figure 3. ISWEC HIL simulation. (a) test rig with PTO assembled on top of the rocking platform, (b) hull in the wave tank, (c) logic diagram with virtual and real parts of the WEC, identified with white and non-white coloured blocks, respectively. Adapted from (Bracco et al., 2015).

frameworks (Hayrettin Bora Karayaka et al., 2021) Are identified in the investigation literature.

A few of these representations are connected to the wind turbine research area (Curioni, 2019), given that some of their HIL test rigs can be utilized for WECs as well (Gaspar et al., 2016; Gaspar et al., 2017). These representations are specified a single definition in this review. They are all reputed as frameworks, however, with distinct progression of actionable and abstract information.

The actionable or actionability term is specified as the "ability of information to indicate specific actions to be taken in order to achieve the desired objective" (Shocker & Srinivasan, 1979; van Kleef et al., 2005). This term is utilized in the product development process where the client demand, generally more abstract and subjective, are translated into product specifications (i.e., actionable information). Therefore, the more actionable the information is, the easier the method characterization of the product, and the more abstract the information is, the less actionable is for technical realization. Frameworks can be in one or in between these hardliner, total actionable vs. total abstractness.

More WEC HIL frameworks are offered in Figure 4. The first in Figure 4a is more complete than the one in Figure 3c (i.e., grittier). It also covers and inside feedback loop that with the sum block (∑) is utilized to determine the mistake of the PTO relative position, which is then transmitting to the simulator hardware (this component is not shown in Figure 3c). The particulars go further to the analogical to digital converters (ADC) and integrator (1/s), thus viewing some actionability of this framework.

The HIL framework of the Two-raft-type WEC (Figure 4b) decomposes the "Simulated parts" domain into "Software model and controller" and "Hydraulic power pack system" subdomains. The second is not in the model of the WEC (presented in Figure 2a) as it just offers the interface between simulated and real parts. Thus, the classification of the "Hydraulic power pack system" as pertaining to the simulated component domain relates to the operability it has in the HIL framework, which is to occur the effects of the WEC simulated component on the real ones. The power pack and controller are real in the sense of the hardware and software, nevertheless, not real in the context of the WEC model.

The HIL framework, presented in Figure 5, simulates the OWC WEC (Figure 2b) and is organized in two different domains, the "HIL simulation and data logging (prototype scale)" and the "Tecnalia test rig (model scale)". Each one is a mixture of hardware and software. The test rig domain contains the simulator of the biradial turbine torque (Grid, Power converter and motor), the generation (Generator, PLC - Programmable Logic Computer and control law), conditioning (Power converter) and grid stages. The turbine simulator is controlled to deliver the real-time desired torque, computed from the WEC and wave models simulated in the xPC board computer. However, the simulator is not a simulated or real part of the WEC model while the generator, PLC, control law (software), power converter and the grid are real parts of the model. Motionless, they are all real in the test rig. This framework is placed at a lower, more functional or actionable level than the preceding frameworks (Figures 3 and 4).

Figure 4b Presents the framework that is located in between the frameworks presented in Figure 4a (logical) and Figure 5 (technological), as respects to the actionability level. Therefore, it takes logical and technological components. The same framework (Figure 4b) presents the numerically simulated wave resource ("wave excitation force") and absorber ("motion equation") stages, and the real transmission ("Hydraulic PTO unit") and generator ("Power generation system") stages.

Figure 6 Presents the HIL framework of the OWC with CD-DEGs (Figure 2c). It is less actionable than the ones in Figures 4 and 5, but more actionable than the framework presented in Figure 3c, as it gives some direction about technical employment (e.g., cylinder, piston, power electronics, DEG voltage). The parts are organized as belonging to one of three domains: "CD-DEG PTO", "Mechanical driver" and "Software". The "CD-DEG PTO" comprehends the physical parts, i.e., CD-DEGs, power electronics, pressure sensor and air chamber (Figure 2c).

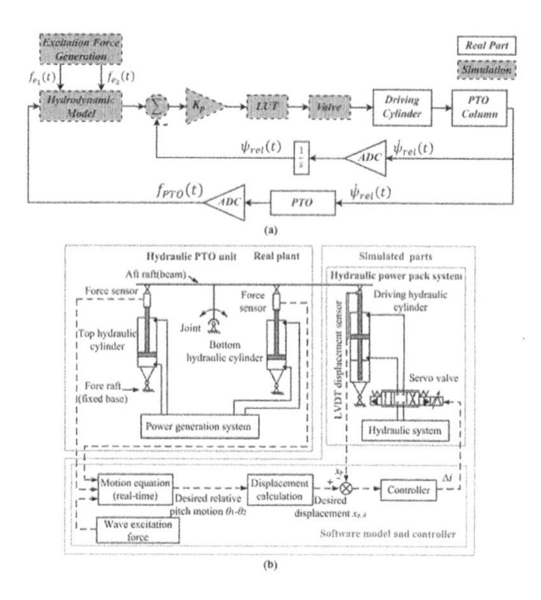


Figure 4. WEC HIL frameworks. (a) Heaving point absorber WEC with simulated (grey coloured) and real (white coloured) components (Signorelli et al., 2011), (b) Two-raft-type WEC with domain zones separating real from simulated parts (Liu et al., 2018).

Figure 7 Presents the HIL framework of the point floater WEC with slider-crank PTO (Figure 2d). The actionability level is similar to the one in Figure 6. Nevertheless, the parts are organized as belonging to one of two domains, "Simulation" and "Hardware" (Hollis et al., 2018; Karayaka et al., 2020).

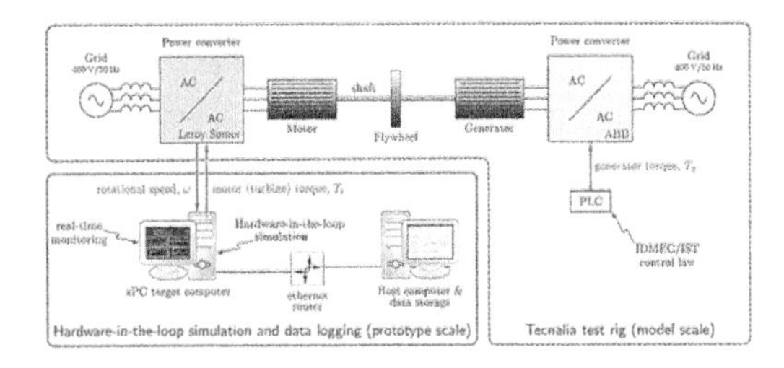


Figure 5. HIL framework of the OWC WEC (Figure 2b). The wave energy resource and WEC absorber stages are numerically simulated in the xPC target computer while the generator (including PLC and control law), conditioning and grid stages are the real parts of the W2W model. Adapted from (Henriques et al., 2016).

The parts in the "Simulation" domain are the usual simulated parts of the WEC model – wave resource and absorber stages – plus the control algorithms of the generator and motor.

The torques simulated by motor are calculated in the usual way and then sent to the motor control algorithm, which then sends control signals to the motor power electronics. On the other hand, the parts in the "Hardware" domain are the PTO flywheel, generator, motor and the power electronics ("Electric Machine Drives Board"). The later receives control signals from algorithms to control the generator and motor, respectively. Thus, Hollis et al. (2018) defines an interface between simulation and hardware domains, the first containing the simulated parts of the WEC model and control algorithms of the crank torque simulator while the second the real parts of the model (power electronics, generator and inertia wheel) and the crank simulator (power electronics and motor).

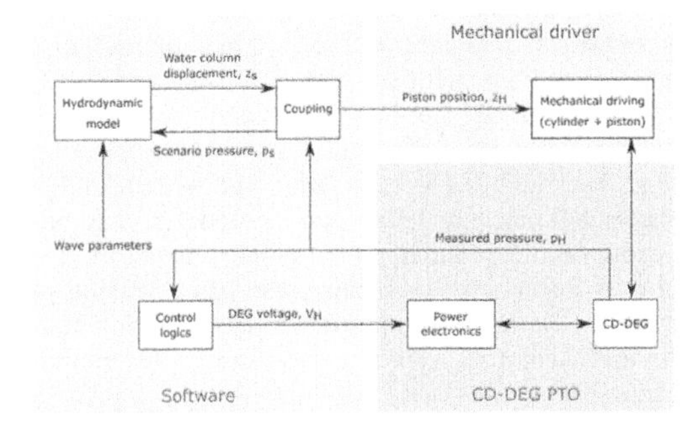


Figure 6. HIL framework of the OWC with CD-DEGs (Figure 2c). Adapted from (Moretti et al., 2021).

## 3 GENERIC FRAMEWORKS AND TAXONOMY

The HIL frameworks at different levels of actionability and abstractness, an unclear separation between WEC simulated and real parts, and blurred separation between these parts and the parts not belonging to the WEC are presented in this Section. Furthermore, diverse terms are specified to the same objects and the initial analysis indicates that a general framework might be removed and modified to the case studies. Therefore, a comparative review of the research literature is performed to understand which terms are used across frameworks, join the ones that share similar definitions and organize them hierarchically and horizontally. The generic HIL framework can be organized in three main domains: S – W2W (W2W Simulated parts), R – W2W (W2W Real parts) and S – R W2W interface (parts of the interface located between W2W simulated and real parts).

The S – W2W domain is related to the numerical simulation of the W2W stages (e.g., hardware and control algorithms). These stages are numerically modelled and simulated with integrated software running on a host computer. The software includes a compiler and drivers to convert the developed simulation model into a real time (RT) model. The RT model is made of code executable in a dedicated real time simulator (RTS) machine (RTSM). The RTSM (e.g., Input/Output board) is integrated in the host computer or in a second computer that interacts with the host computer (in this case denominated as the Master computer). However, the RTSM might be a standalone unit as well. The software also allows RT communication with the RTSM to collect data for visualization and tuning of the RT model parameters.

The R – W2W domain is related to the embedded subsystems or parts of the W2W stages (e.g., hardware, controller and control algorithms). These are the real physical parts of the W2W stages subjected to experimental testing, thus named as devices under test (DuT) or target hardware, controller or control algorithm. Development software, installed in a development computer, is used to develop the target control algorithm, compile and load it to the target controller (e.g., PLC or I/O board).

The S – R W2W domain is related to the system that physically simulates the action of the numerically simulated parts on the embedded ones. This system is defined as the emulator interface and is composed by a controller (e.g., I/O board), interface control algorithm, power electronics and electrical drive. The controller receives signals from the RT model and implements them on the target hardware through power electronics and electrical driver. The accurate implementation of these control signals is performed by a dedicated control algorithm of the emulator. However, a compensator may be added to adjust the differences between the target and test rig hardware specifications and to avoid the influence of the emulator dynamics on the simulation.

The RTS and control of the emulator interface are usually performed in the same controller hardware.

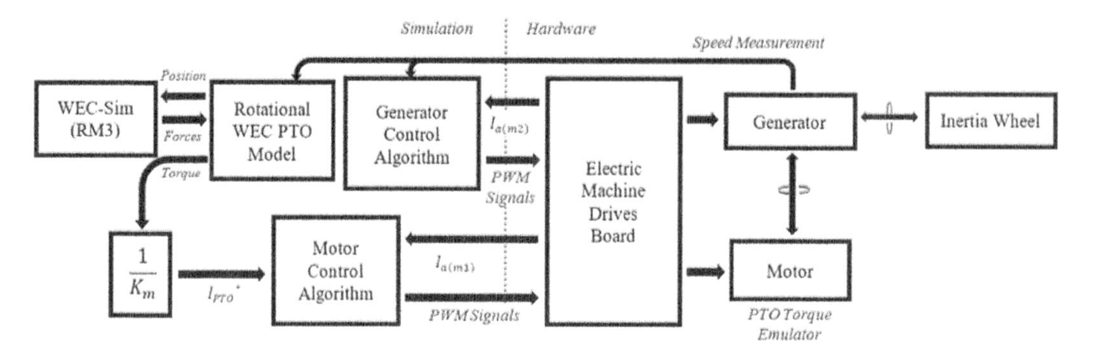


Figure 7. HIL framework of the slider-crank PTO at a more abstract level. Adapted from (Hollis et al., 2018).

However, this hardware may be also used as a target controller. Thus, different combinations may be achieved, depending on experimental complexity, specifications, and testing costs.

The clarification of this taxonomy is supported and articulated in Section 4, by presenting HIL technologies and simulation platforms collected from a second review pass on the research literature.

## 4 TECHNOLOGY

One case study for a better clarification of the generic framework and the technological challenges for HIL implementation is presented in this Section. Figure 8 presents the case study, the HIL setup of the point floater with slider-crank PTO (Figure 2d).

The electrical drive emulates the crank torque produced by the floater heaving motions and transmitted to the generator by the connecting shaft. The HIL target hardware (PTO generator attached to a flywheel) is a Direct Current (DC) machine (bottom-left side of Figure 8). It is attached to a second DC machine that works as the electrical drive of the emulator interface. The DC machines are powered by an electric machine drives board (power electronics) which is attached to the I/O board by a terminal board. The interface control algorithm runs in the I/O board.

The same hardware (I/O, terminal, and electric machine drive boards) is used to control the generator resistance torque, in other words, to test the target generator and control algorithm.

The RT wave and WEC hydrodynamic models run inside the same I/O board. Thus, the I/O board, simulation and development software are all located inside the same computer, i.e., the host computer is also used to develop the target control algorithm. The setup presented in Figure 8 is cost effective because uses the same technological platform to run the RT model of some W2W stages (S – W2W), emulate their physical action on the target generator (S – R W2W) and control the target generator (R – W2W).

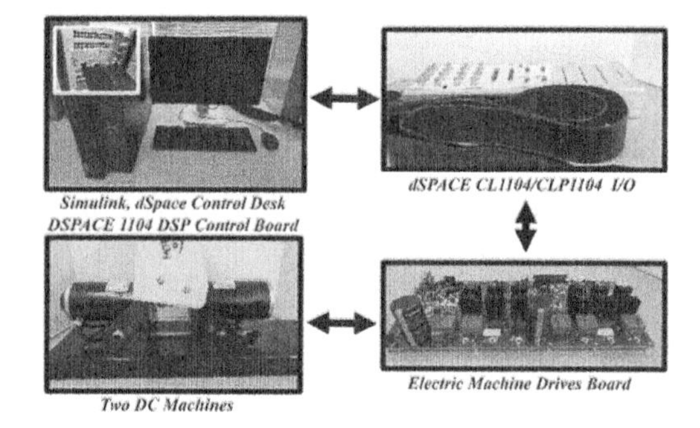


Figure 8. HIL setup of the slider-crank PTO. Top-left side: host computer with simulation software (Simulink, dSPACE Control Desk) and I/O board (dSPACE 1104 DSP control board). Top-right side: I/O terminal board (dSPACE CL1104/CLP1104 I/O). Bottom-right side: power electronics (Electric Machine Drives Board). Bottom-left side: emulator drive and target hardware (two DC machines). Adapted from (Hollis et al., 2018).

Important challenges are related to complex simulations and interactions between PTO and WEC (Signorelli et al., 2011), the difficulty in repeating the exact properties of a continuous system operation (e.g., hydraulic fluid degradation and system faults) and model inaccuracies caused by model oversimplification aiming computational efficiency and utilization of linear analysis techniques (e.g., omission of cylinder friction, oil compressibility, saturation constraints in force, velocity and power) (Signorelli et al., 2011).

The scaling of the models also adds undesirable effects that may affect the simulation accuracy. The testing of large-scale PTOs onshore is usually simplistic and do not care for realistic sea conditions, thus not allowing a full understanding of the system dynamics. As a result, the system design is more complex, overengineered, sub-optimal regarding power harvesting performance and is subjected to increased risks (Signorelli et al., 2011). On the other hand, HIL at model scale may add mechanical and hydrodynamic scale effects that distort the dynamic behaviours of the emulated system. Thus, the numerical model may behave differently than the downscaled real system. However, accurate simulations are achievable if these effects are corrected with dedicated solutions to provide a consistent dynamic equivalence between simulated and hardware systems (Bacelli et al., 2017; Curioni, 2019; Huijgens, 2021; Moretti et al., 2021), i.e. the interface which can also solve problems of interface between the numerical model and the physical components.

## 5 DISCUSSION

The results of this review show that knowledge in HIL simulation has been developed in the research field of to WECs and PTO testing. But, unlike languages and styles for the representation of the HIL frameworks are found, which generates complications in clarifying the method of the HIL simulation for a wider set of beginners and expert researchers. However, significant forms are found across HIL frameworks: diverse levels of framework actionability and abstractness, which depend on the objective at hand, such as having a wider view of the test setup to have an instantaneous understanding of the methodology (more abstract and less actionable) or a narrower one to understand how is implemented (less abstract and more actionable); in the literature is shown that one or more actionability levels of the framework are used in the same case study; existence of an interface between the simulated and real parts of the WEC that is not clearly defined, i.e. hard to define its boundaries in relation to the simulated and real parts of the WEC; existence of diverse terms for the same meanings; not easy to identify which software or hardware is the target for testing or interfacing between simulated and real parts of the WEC.

## 6 CONCLUSIONS

This paper, besides providing a state-of-the-art review on HIL testing, it gives a universal framework and terminology that may help a better comprehension of the HIL approach by researchers. It is expected that these review, framework and terminology should work as a baseline for further reviews on the state-of-the-art of HIL testing of wave energy converters. However, its articulation and review in other fields of research may be considered as well.

## ACKNOWLEDGEMENTS

This work was performed within the Strategic Research Plan of the Centre for Marine Technology and Ocean Engineering, which is financed by Portuguese Foundation for Science and Technology (FCT) under contracts: UIDB/UIDP/00134/2020, of the project "Experimental simulation of oil-hydraulic Power Take-Off systems for Wave Energy Converters"; PTDC/EME-REN/29044/2017, of the project "Variable geometry Wave Energy Conversion system for floating platforms"; and under contract PTDC/EME-REN/0242/2020.

## REFERENCES

Aderinto, T., & Li, H. 2018. Ocean Wave Energy Converters: Status and Challenges. *Energies*, *11*(5), 1250. https://doi.org/10.3390/EN11051250

Ahamed, R., McKee, K. & Howard, I. 2020. Advancements of wave energy converters based on power take off (PTO) systems: A review. *Ocean Engineering*, *204*, 107248. https://doi.org/10.1016/j.oceaneng.2020.107248

Bacelli, G., Spencer, S. J., Coe, R. G., Mazumdar, A., Patterson, D. & Dullea, K. 2017. Design and Bench Testing of a Model-Scale WEC for Advanced PTO Control Research. *European Wave and Tidal Energy Conference (EWTEC)*.

Bacelli, G., Spencer, S. J., Patterson, D. C. & Coe, R. G. 2019. Wave tank and bench-top control testing of a wave energy converter. *Applied Ocean Research*, *86*, 351–366. https://doi.org/10.1016/J.APOR.2018.09.009

Bouscayrol, A., Guillaud, X., Teodorescu, R., Delarue, P. & Lhomme, W. 2006. Hardware-in-the-loop simulation of different wind turbines using Energetic Macroscopic Representation. *IECON 2006 - 32nd Annual Conference on IEEE Industrial Electronics*, 5338–5343. https://doi.org/10.1109/IECON.2006.347616

Bracco, G., Giorcelli, E., Mattiazzo, G., Orlando, V. & Raffero, M. 2015. Hardware-In-the-Loop test rig for the ISWEC wave energy system. *Mechatronics*, *25*, 11–17. https://doi.org/10.1016/J.MECHATRONICS.2014.10.007

Carrelhas, A. A. D., Gato, L. M. C., Falcão, A. F. O. & Henriques, J. C. C. 2021. Control law design for the air-turbine-generator set of a fully submerged 1.5 MW mWave prototype. Part 2: Experimental validation. *Renewable Energy*, *171*, 1002–1013. https://doi.org/10.1016/J.RENENE.2021.02.128

Castellini, L., Gallorini, F., Alessandri, G., Alves, E., Montoya, D. & Tedeschi, E. 2021. Performance Comparison of Offline and Real-Time Models of a Power Take-Off for Qualification Activities of Wave Energy Converters. *2021 Sixteenth International Conference on Ecological Vehicles and Renewable Energies (EVER)*, 1–8. https://doi.org/10.1109/EVER52347.2021.9456630

Curioni, G. 2019. Robust Drive-Train Test Bench Control Framework via Hardware-in-the-Loop with Mechanical Inertia Emulation Capability. *2019 IEEE 58th Conference on Decision and Control (CDC)*, 1931–1936. https://doi.org/10.1109/CDC40024.2019.9030167

Delmonte, N., Robles, E., Cova, P., Giuliani, F., Faÿ, F. X., Lopez, J., Ruol, P., Martinelli, L., Fay, F. X., Lopez, J., Ruol, P. & Martinelli, L. 2020. An Iterative Refining Approach to Design the Control of Wave Energy Converters with Numerical Modeling and Scaled HIL Testing. *Energies*, *13*(10), 2508. https://doi.org/10.3390/en13102508

Gaspar, J. F., Kamarlouei, M., Sinha, A., Xu, H., Calvário, M., Faÿ, F. X., Robles, E. & Guedes Soares, C. 2016. Speed control of oil-hydraulic power take-off system for oscillating body type wave energy converters. *Renewable Energy*, *97*, 769–783. https://doi.org/10.1016/j.renene.2016.06.015

Gaspar, J. F., Kamarlouei, M., Sinha, A., Xu, H., Calvário, M., Faÿ, F. X., Robles, E. & Guedes Soares, C. 2017. Analysis of electrical drive speed control limitations of a power take-off system for wave energy converters. *Renewable Energy*, *113*, 335–346. https://doi.org/10.1016/J.RENENE.2017.05.085

Grega, W. 1999. Hardware-in-the-loop simulation and its application in control education. *FIE'99 Frontiers in Education. 29th Annual Frontiers in Education* Conference. Designing the Future of Science and Engineering Education., *2*, 12B6/7-12B612. https://doi.org/10.1109/fie.1999.841594

Hansen, A. H., Asmussen, M. F. & Bech, M. M. 2019. Hardware-in-the-Loop Validation of Model Predictive Control of a Discrete Fluid Power Power Take-Off System for Wave Energy Converters. *Energies*, *12*(19), 3668. https://doi.org/10.3390/en12193668

Henriques, J. C. C., Gomes, R. P. F., Gato, L. M. C., Falcão, A. F. O., Robles, E. & Ceballos, S. 2016. Testing and control of a power take-off system for an oscillating-water-column wave energy converter. In *Renewable Energy* (Vol.85, pp. 714–724). https://doi.org/10.1016/j.renene.2015.07.015

Hollis, T., Karayaka, H. B., Yu, Y. H. & Yan, Y. 2018. Hardware-in-the-Loop Simulation for the Proposed Slider-Crank Wave Energy Conversion Device. *OCEANS 2018 MTS/IEEE Charleston*, 1–7. https://doi.org/10.1109/OCEANS.2018.8604770

Huijgens, L. J. G. 2021. *Replication Data for: hardware in the loop emulation of ship propulsion systems at model scale* [TRAIL Research School]. https://doi.org/10.4233/uuid:24c5867a-6fbb-4e6a-9785-2c9362fada91

Huijgens, L., Vrijdag, A. & Hopman, H. 2021. Hardware in the loop experiments with ship propulsion systems in the towing tank: Scale effects, corrections and demonstration. *Ocean Engineering*, *226*, 108789. https://doi.org/10.1016/j.oceaneng.2021.108789

Karayaka, H. Bora, Yu, Y.-H., Tom, N. & Muljadi, E. 2020. Investigating the Impact of Power-Take-Off System Parameters and Control Law on a Rotational Wave Energy Converter's Peak-to-Average Power Ratio Reduction. *ASME 2020* 39 Th International Conference on Ocean,Offshore and Arctic Engineering, *9*. https://doi.org/10.1115/OMAE2020-18961

Karayaka, Hayrettin Bora, Yu, Y.-H. & Muljadi, E. 2021. Investigations into Balancing Peak-to-Average Power Ratio and Mean Power Extraction for a Two-Body Point-Absorber Wave Energy Converter. *Energies*, *14* (12), 3489. https://doi.org/10.3390/EN14123489

Lasa, J., Antolín, J. C., Angulo, C., Estensoro, P., Santos, M. & Ricci, P. 2012. Design, construction and testing of a hydraulic power take-off for wave energy converters. *Energies*, *5*(6), 2030–2052. https://doi.org/10.3390/en5062030

Liu, C., Yang, Q. & Bao, G. 2018. Influence of hydraulic power take-off unit parameters on power capture ability of a two-raft-type wave energy converter. *Ocean Engineering*, *150*, 69–80. https://doi.org/10.1016/J.OCEANENG.2017.12.063

Moretti, G., Scialò, A., Malara, G., Muscolo, G. G., Arena, F., Vertechy, R. & Fontana, M. 2021. Hardware-in-the-loop simulation of wave energy converters based on dielectric elastomer generators. *Meccanica*, *56*(5), 1223–1237. https://doi.org/10.1007/S11012-021-01320-8

Neshati, M., Feja, P., Zuga, A., Roettgers, H., Mendonca, A. & Wenske, J. 2020. Hardware-in-the-loop Testing of Wind Turbine Nacelles for Electrical Certification on a Dynamometer Test Rig. *Journal of Physics: Conference Series*, *1618*, 032042. https://doi.org/10.1088/1742-6596/1618/3/032042

Neshati, M., Zuga, A., Jersch, T. & Wenske, J. 2016. Hardware-in-the-loop drive train control for realistic emulation of rotor torque in a full-scale wind turbine nacelle test rig. *2016 European Control Conference (ECC)*, 1481–1486. https://doi.org/10.1109/ECC.2016.7810499

Pedersen, H. C., Hansen, R. H., Hansen, A. H., Andersen, T. O. & Bech, M. M. 2016. Design of full scale wave simulator for testing Power Take off systems for wave energy converters. *International Journal of Marine Energy*, *13*, 130–156. https://doi.org/10.1016/j.ijome.2016.01.005

Penalba, M. & Ringwood, J. V. 2016. A Review of Wave-to-Wire Models for Wave Energy Converters. *Energies*, *9*(7), 506. https://doi.org/10.3390/EN9070506

Puleva, T., Rouzhekov, G., Slavov, T. & Rakov, B. 2016. Hardware in the loop (HIL) simulation of wind turbine power control. *Mediterranean Conference on Power Generation, Transmission, Distribution and Energy Conversion (MedPower 2016)*, 1–8. https://doi.org/10.1049/cp.2016.1053

Sancar, U., Onol, A. O., Onat, A. & Yesilyurt, S. 2015. Hardware-in-the-loop simulations and control design for a small vertical axis wind turbine. *2015 XXV International Conference on Information, Communication and Automation Technologies, (ICAT)*, 1–7. https://doi.org/10.1109/ICAT.2015.7340497

Shami, E. Al, Zhang, R. & Wang, X. 2018. Point Absorber Wave Energy Harvesters: A Review of Recent Developments. *Energies*, *12*(1), 47. https://doi.org/10.3390/EN12010047

Shocker, A. D. & Srinivasan, V. 1979. Multiattribute Approaches for Product Concept Evaluation and Generation: A Critical Review. *Journal of Marketing Research*, *16*(2), 159. https://doi.org/10.2307/3150681

Signorelli, C. D., Villegas, C. & Ringwood, J. V. 2011. Hardware-In-The-Loop Simulation of a Heaving Wave Energy Converter. *Proceedings of the 9th European Wave and Tidal Energy Conference*.

Skjetne, R. & Egeland, O. 2006. Hardware-in-the-loop testing of marine control systems. *Modeling, Identification and Control*, *27*(4), 239–258. https://doi.org/10.4173/mic.2006.4.3

van Kleef, E., van Trijp, H. C. M. & Luning, P. 2005. Consumer research in the early stages of new product development: A critical review of methods and techniques. *Food Quality and Preference*, *16*(3), 181–201. https://doi.org/10.1016/j.foodqual.2004.05.012

Vince, T., Beres, M., Makis, S. & Mamchur, D. 2020. PLC Universal Hardware in the Loop System Based on ATmega2560. *2020 IEEE International Conference on Problems of Automated Electric Drive. Theory and Practice (PAEP)*, 1–4. https://doi.org/10.1109/PAEP49887.2020.9240867

Vittori, F., Pires, O., Azcona, J., Uzunoglu, E., Guedes Soares, C., Zamora Rodriguez, R. & Souto-Iglesia. A. 2021. Hybrid scaled testing of a 10MW TLP floating wind turbine using the SiL method to integrate the rotor thrust and moments. Guedes Soares, C., (Ed.).

*Developments in Renewable Energies Offshore*. London, UK: Taylor and Francis; 417–423.
Xie, J. & Zuo, L. 2013. Dynamics and control of ocean wave energy converters. *International Journal of Dynamics and Control*, *1*(3), 262–276. https://doi.org/10.1007/S40435-013-0025-X
Xu, S., Wang, S. & Guedes Soares, C. 2019. Review of mooring design for floating wave energy converters. *Renewable and Sustainable Energy Reviews*, *111*, 595–621. https://doi.org/10.1016/J.RSER.2019.05.027
Zhang, H., Zhang, Y. & Yin, C. 2016. Hardware-in-the-Loop Simulation of Robust Mode Transition Control for a Series-Parallel Hybrid Electric Vehicle. *IEEE Transactions on Vehicular Technology*, *65*(3), 1059–1069. https://doi.org/10.1109/TVT.2015.2486558
Zhang, Y., Zhao, Y., Sun, W. & Li, J. 2021. Ocean wave energy converters: Technical principle, device realization, and performance evaluation. *Renewable and Sustainable Energy Reviews*, *141*, 110764. https://doi.org/10.1016/j.rser.2021.110764

*Trends in Maritime Technology and Engineering – Guedes Soares & Santos (eds)*


# Time domain analysis of a conical point-absorber moving around a hinge

T.S. Hallak, M. Kamarlouei, J.F. Gaspar & C. Guedes Soares
*Centre for Marine Technological and Ocean Engineering (CENTEC), Instituto Superior Técnico, Universidade de Lisboa, Lisbon, Portugal*

ABSTRACT: This paper presents numerical time domain analyses for a conical point-absorber moving around a hinge, interconnected through a rigid arm. The power take-off system is simulated by implementing a rotational damper aligned with the hinge. The numerical model is developed in time domain and predicts the floater's motion and power performance parameters, e.g. the capture width, and absorbed power. The 1 dof hydrodynamic formulation is applied for the generalized mode. The numerical results are compared with experimental results, and frequency domain results. The paper concludes that non-linearities of motion may frequently be present in point-absorber dynamics. The irregular waves results show almost identical amplitudes of motion in comparison to the results based on response amplitude operators, also comparing well with the experiments, validating the code. In regular waves, the time domain solver presents a more sensitive response around the resonance, depending upon the damping force of the power take-off system.

## 1 INTRODUCTION

In the last couple of years, the COVID-19 pandemic has affected the most diverse global markets, the energy market included. The first observation is that global $CO_2$ emissions have dropped by around 8.8% in the first year of pandemic, triggered by reduced economic production (Liu et al. 2020). As the demand for energy returns to increase, renewable energies, including Offshore Renewable Energies (ORE), are considered as prospective solutions for such demand issues, also working in favor for the United Nations Sustainable Development Goals (UNs' SDGs).

The amount of mechanical energy contained in the ocean is undeniably impressive, and shall be consistently explored in the future. In the forms of current, tide, and wave energy, they shall be exploited depending upon the location and season. So far, however, the associated Levelized Cost of Energy (LCoE) is relatively high for investors. Moreover, the normal operation of devices may be affected by the lack of personnel, or rather expensive maintenance (Xiaojing et al. 2015).

In fact, technology for Wave Energy Converters (WECs), is still a few steps below the commercial scale. The Technology Readiness Level (TRL) of WECs is usually at TRL 7, or lower, for there is no device that delivers energy out of the prototype sphere, even though some do operate at this level, or did, as real scale prototypes, e.g. the WaveDragon, the WaveRoller, and the Wavestar, just to cite a few in various configurations. Other concepts are located at inferior TRL, e.g. the floating Oscillating Water Columns (OWCs), which are at most TRL 6. In fact, it is technically and economically challenging to extract energy from sea waves due to the slow, multi-directional and irregular motions of the particles and bodies (Hansen et al. 2013). In the case of heaving point-absorber WECs, as the one considered in this paper, TRL is 7, for projects such as the Wavestar (Denmark), and Usina de Ondas (Brazil) operated at this level.

That said, in the present study, a numerical time-domain code is developed to simulate the motions of the conical point-absorber WEC which has 1 degree of freedom (dof) and moves around a hinge. The results are obtained for regular and irregular waves, with and without a rotational damper. The results are then compared with peer-reviewed results, namely frequency domain results (Hallak et al. 2021), and experimental results obtained during a test campaign in wave basin (Kamarlouei et al. 2021). The aim of this study is to analyze the motions of the WEC, and to draw improved estimates of power performance.

## 2 METHODOLOGY

### 2.1 *Experimental setup*

The 1:27 scale model WEC is the conical point-absorber of diameter 18.5 cm (5.0 m in real scale), with 90° of opening angle. The WEC is connected to a hinge through a rigid arm of length 45.5 cm, and the initial angle between the arm and the horizontal

DOI: 10.1201/9781003320289-42

plane is 20°. Figure 1 shows the WEC in the test basin, at the Lir NOTF Facility (Kamarlouei et al. 2020), in Cork (Ireland), which is a 25.0 m long and 17.0 m wide wave tank. The basin has a moveable floor, so water depth was set at 1.10 m. Both arm and hinge were, at all times, dry. The initial draft of the body was 7.0 cm, and the initial distance between waterline and deck, 5.0 cm. In most of the wave scenarios, there was considerable interaction between bottom and water waves (shallow water condition).

Figure 1. Dry hinged-arm connection and WEC in view.

The WEC works as a heaving point-absorber device, even though the actual mode from which energy is obtained is the rotation around the uniaxial hinge. Indeed, because the pivot connecting the WEC with the arm is rigid, the system consists on a single dof mechanical system.

## 2.2 *Numerical model*

The wave-structure interaction model is based on linear potential flow formulation. The diffraction/radiation code Wadam, by DNV's SESAM HydroD® was used when obtaining the hydrodynamic coefficients of the hull. The post-processor was coded in Matlab® to perform time-integration using the Cummins equation, also using the Runge-Kutta 4$^{th}$ Order Method. Because the developed code is a linear solver, non-linear damping is accounted in a linearized way, done by assigning damping ratios for the vertical modes of the device, as shown in Table 1.

The Matlab code generates the wave elevation and wave excitation force series. For the irregular waves, series are generated by constructing the wave elevation array as sums of several random sinusoidal components, and by matching the wave energy in each frequency bands. Regular waves were generated considering the Airy wave model. Simulations were performed with 20 sec ramp duration.

Table 1. Assigned damping ratio.

| Device | Heave | Roll | Pitch |
|---|---|---|---|
| WECs | 25% | 5% | 5% |

The numerical model and basic formulae applied to obtain the motion response of the WEC is the generalized dof is presented in detail by (Hallak et al. 2020).

## 2.3 *Simulation conditions*

The wave scenarios considered in the test campaign are summarized in Table 2. The model tests were performed with and without a rotational damper at the hinge, with nominal value $B_{PTO}$ = 0.50 Nms (1381 kNms in real scale), which has been calibrated before the tests. It is worth mentioning that the wave periods of conditions RW 12 and IW 6 are close to the physical limits of the basin. It was not possible, even though it was desired, to simulate waves with higher periods.

Table 2. Regular and irregular wave data – Mode scale.

| Regular waves | Wave period [s] | Wave height [cm] | |
|---|---|---|---|
| RW 1 | 0.5 | 2.0 | |
| RW 2 | 0.8 | 2.0 | |
| RW 3 | 1.1 | 2.0 | |
| RW 4 | 1.4 | 2.0 | |
| RW 5 | 1.7 | 2.0 | |
| RW 6 | 2.0 | 2.0 | |
| RW 7 | 2.3 | 2.0 | |
| RW 8 | 2.6 | 2.0 | |
| RW 9 | 2.9 | 2.0 | |
| RW 10 | 3.2 | 2.0 | |
| RW 11 | 3.5 | 2.0 | |
| RW 12 | 3.8 | 2.0 | |
| **Irregular waves** | **Peak period [s]** | **Sig. wave height [cm]** | **γ-parameter [-]** |
| IW 1 | 1.27 | 6.25 | 1 |
| IW 2 | 1.39 | 7.39 | 1 |
| IW 3 | 1.49 | 8.44 | 1 |
| IW 4 | 1.60 | 9.65 | 1 |
| IW 5 | 1.70 | 10.8 | 1 |
| IW 6 (storm) | 1.70 | 14.0 | 3.3 |

In the numerical sphere, simulations were repeated with a hypothetical damping parameter, $B_{PTO}$ = 6.0 Nms, because it has been proved that the system could extract more wave energy with increased damping values (Hallak et al. 2020). As explained in the same reference, the "actual damping value was higher than the nominal value of the item, most likely due to manufacturing/testing issues, for it is very hard to control the damping moment of such a small device.

Simulation time was set at 20 min (real scale) for all wave scenarios. In addition to Table 2, a few extra wave periods were considered numerically in order observe the dynamics of the WEC next to its resonant period, namely at 0.2, 0.3, and 0.4 seconds.

### 2.4 *Power performance formulation*

Performance parameters are evaluated, namely the capture width, and the total wave absorbed power. Because planar waves are supposed, the capture width may achieve values higher than 1, meaning that the WEC absorbs more energy than what is contained in a wavelength of sea (in the direction orthogonal to the wave propagation).

To evaluate the capture width $L_w$. the formulation presented by (Gomes et al. 2015) is considered. The capture width is defined as:

$$L_w(\omega) = P(\omega)/P_w(\omega) \tag{1}$$

where the time-averaged absorbed power through the PTO is:

$$P(\omega) = 0.5 \cdot B_{PTO}(\omega|\psi|)^2, \tag{2}$$

and the time-averaged wave energy flux is, for regular waves,

$$P_w(\omega) = \frac{\rho g \omega A_w^2}{4k}\left(1 + \frac{2kh}{\sinh(2kh)}\right), \tag{3}$$

where $\rho$ is water density; $g$ is the acceleration due to gravity; $k$ is the wave-number and $h$ is the local water depth.

Alternatively, in the case of irregular waves, the following equation may be applied (Reguero et al. 2019):

$$P_w(\omega) = \frac{\rho g^2}{64\pi} T_C H_S^2, \tag{4}$$

where $T_c$ is the energy period, which is evaluated based on the spectral moments.

## 3 RESULTS

### 3.1 *Motion in regular waves*

Figure 2 shows the heave RAO of the WEC according to the experiments and the simulations. The results are also compared with the frequency domain RAOs. All experimental values of amplitude were evaluated as the averaged half-difference between maxima and minima after stationary regime was achieved. The results compare well in order of magnitude; however, they find different trends around the resonant period.

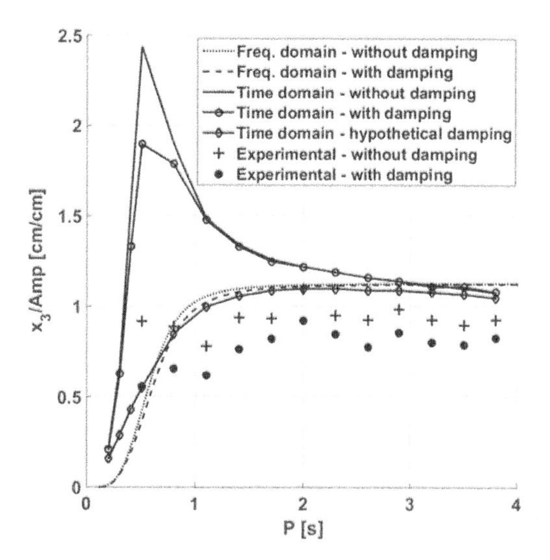


Figure 2. WEC's heave RAO in different damping scenarios.

Clearly, the time domain RAOs are very sensitive to the PTO damping value next to the heave resonant period (0.42 s). For instance, only the time domain simulations observed stronger resonant behavior – even though the damping ratio is not small, and has been double checked for the time-domain simulations. In the other hand, the frequency domain solver is much less sensitive to the PTO damping value, as can be observed by comparing dashed lines with solid lines.

Another clear result is that the experimental values are lower than the numerical values, even for the high periods when the WEC is practically riding the wave. This is mostly due to non-linearities and shall be explored in the future, for the underwater volume of the WEC may vary considerably. The time domain solution at the riding wave regime keeps approaching the experimental values as the wave period increases – instead of stabilizing at a particular amplitude of motion, as one could expect from a frequency domain analysis.

### 3.2 *Motion in irregular waves*

Figures 3 and 4 show the generated IW's series. Only the first 3 (out of 20) minutes of simulation are shown, to make it more visible. The wave significant height from JONSWAP is also shown in comparison with the average height of the one third highest waves, or $H_{1/3}$. The generated wave series contains individual waves of most varying heights and periods. Due to the presence of the smaller waves, $H_{1/3}$ is around 25-30% smaller than the inputted wave significant height from JONSWAP spectrum.

Figures 5 and 6 then present the heave response series of the WEC for the same waves of Figures 4 and 5, and including $B_{PTO}$ = 0.5 Nms. The first observation is that there are less individual heave oscillations then individual waves. Also, the variance of heave oscillations is higher than the variance of the water waves: around 33.6% in the cases without damping, and between 24.3% and 31.2% in the case with $B_{PTO}$ = 0.5 Nms. In the hypothetical scenario with $B_{PTO}$ = 6.0 Nms, the variance of heave oscillations is between 10.1% and 42.0% lower in comparison with the waves. The mitigation of heave is done by the PTO and outputs energy.

The significant amplitude of motion of the WEC is relatively close to the RAO-based significant amplitude of motion. Only in Sea State 6, or storm sea state, the difference reached the value of 20%. The experimental results are also shown for comparison in Figures 5 and 6. The results compare well, thus the code may be considered validated. The '+' markers represent experimental values measured without the consideration of PTO damping, and are around 25% to 30% lower in comparison to the numerical results obtained with the time domain solver. It is clear from the figures that, in general, results are in accordance, even though the overestimation of

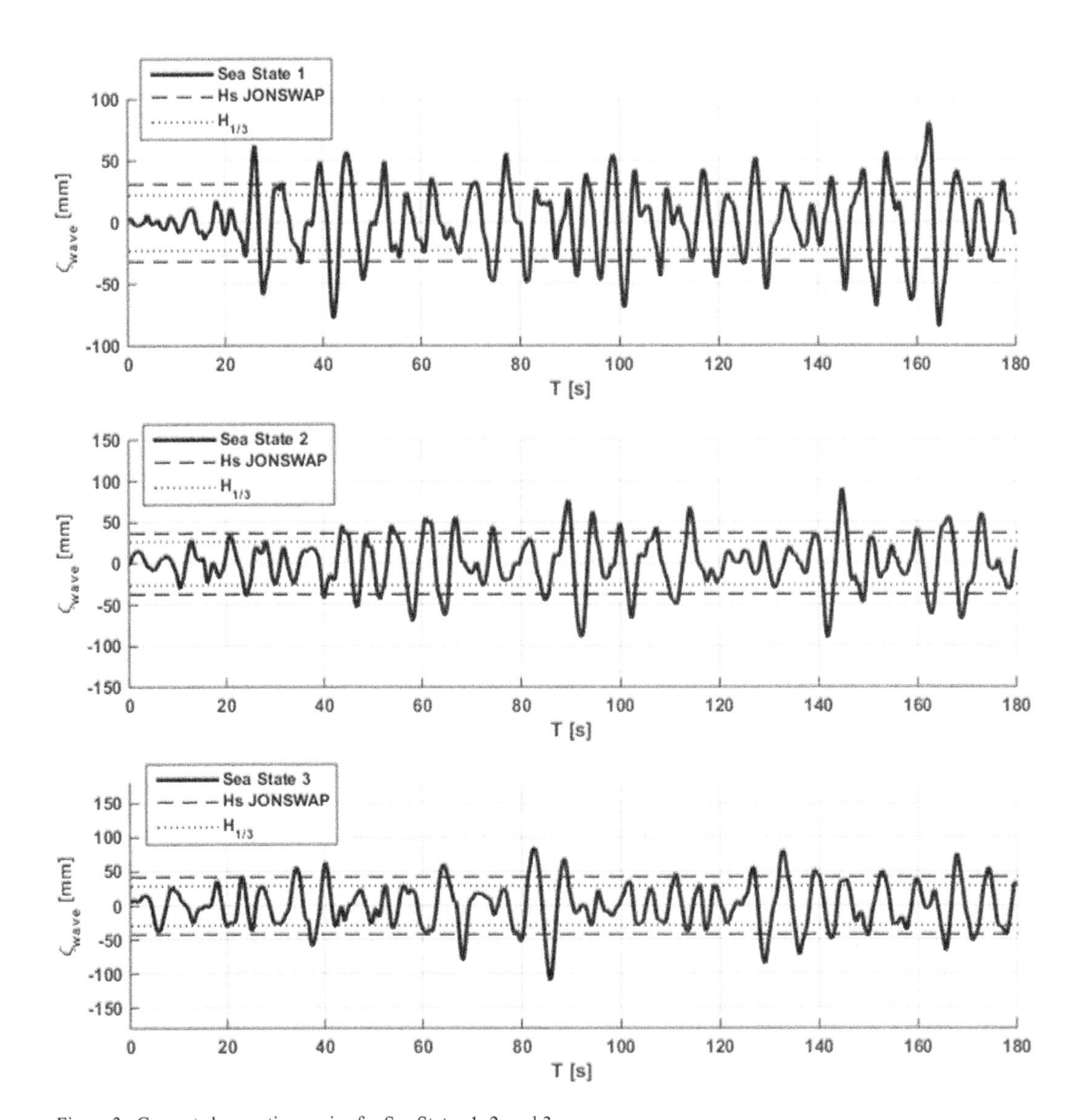


Figure 3. Generated wave time series for Sea States 1, 2, and 3.

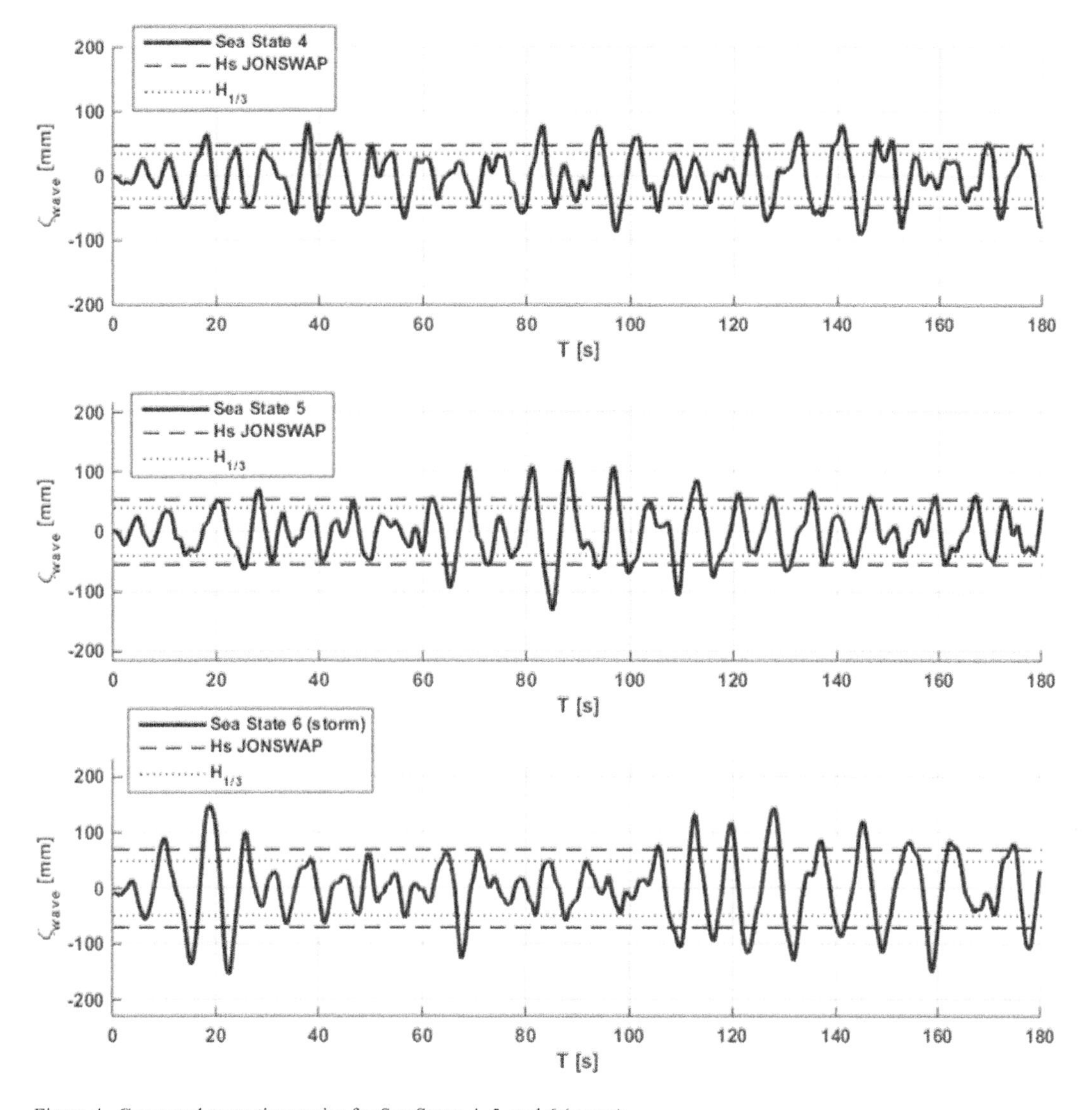


Figure 4. Generated wave time series for Sea States 4, 5, and 6 (storm).

motions found in this section shall lead to an over-estimation of absorbed power as well.

### 3.3 *Power performance*

The power performance of the real-scale WEC is measured based on the estimates of total wave absorbed power and capture width. Tables 3 and 4 present the results with $B_{PTO}$ = 0.5 Nms, and $B_{PTO}$ = 6.0 Nms, respectively, including both regular waves and irregular waves.

It is clear from Table 3 and 4 that the tested regular waves are not sufficient in order to produce reasonable amounts of power – around 6.0 kW maximum for both PTOs, which is mainly due to the relatively small waves, which are only 0.54 meters high. However, the capture width achieves great values in the low period range, which represent good wave energy absorption. The values of ~10 in the lowest periods basically mean that the WEC is absorbing energy from a planar wave that is the size of the WEC itself. Moreover, it is clear from Table 4 that the extra damping in the PTO is working in order to increase the frequency range of good wave energy absorption.

Table 3 suggests that the WEC may not absorb reasonable amounts of power in typical sea states with such a small damping. In Table 4, however, with the further enhancement of the PTO damping, the total wave absorbed power reaches the order of magnitude of the prototype WaveStar (Hansen 2013), actually surpassing it in non-dimensional values, for the tested WEC has a smaller diameter.

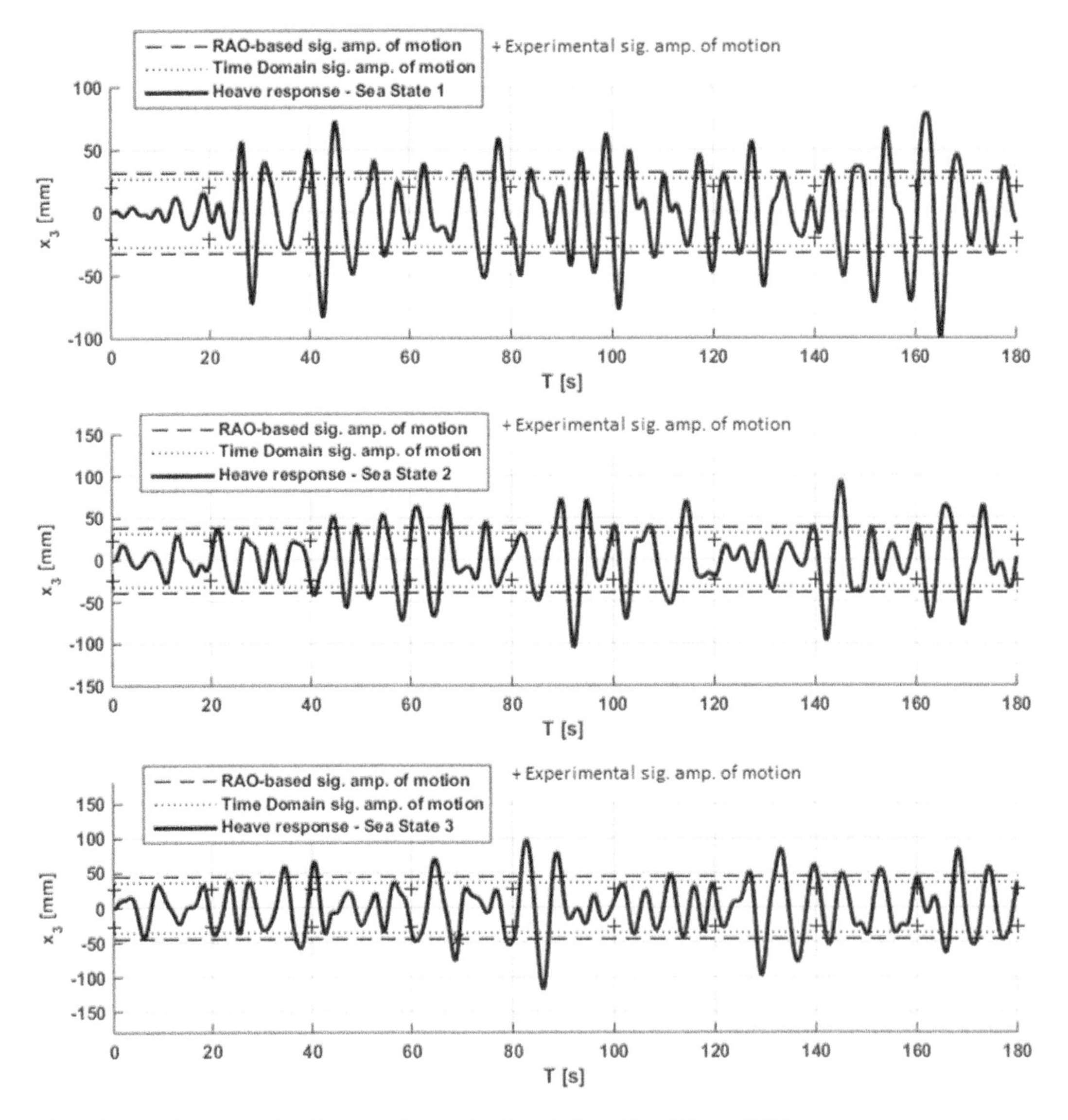


Figure 5. Numerical time series of heave motion for Sea States 1, 2, and 3, and $B_{PTO}$ = 0.5 Nms.

The values presented in Tables 3 and 4 are higher in comparison to the frequency domain results (Hallak et al. 2020). The discrepancy of the results may be explained from the fact that the frequency domain solver evaluates power performance from spectral values, without any kind of model for the motion dynamics inherently present in the phenomenon of wave energy absorption. Thus, the time domain solver is more reliable than the frequency domain solver, in terms of power performance.

### 3.4 *WEC's submergence and non-linear motion*

In Figure 6, the 'x' markers plotted above the heave motion time series for Sea States 5 and 6 indicate that the WEC is completely submerged. Basically, the code evaluates the relative heave in regards to the free surface elevation, and if its more negative than the free board, it adds a marker to the plot, meaning that the WEC is being fully submerged by the wave.

In total, there were 30 observations of numerical submergence. The amount of occasions increases with sea state, starting with 2 submergences in Sea State 1 up to 10 submergences in the stormy Sea State 6. Also the duration of submergence is evaluated by the code: in Sea States 1 to 3, the submergences usually last a single time step (~1.0 second in real scale), whereas in Sea State 6 some submergences lasted for more than 4.0 seconds. The most interesting is that these submergences were indeed observed during the test campaign. In the basin, however, they started to occur normally in Sea States 3 and 4, and were happening often only in the stormy Sea State 6.

Even though the time domain solver predicts more occasions of submergence, the orders of magnitude are consistent consistent. These results also

Table 3. WEC's power performance with $B_{PTO} = 0.5$ Nms.

| Scenario | Absorbed Power [kW] | Capture Width |
|---|---|---|
| RW 1 | 6.04 | 11.88 |
| RW 2 | 1.88 | 3.466 |
| RW 3 | 0.729 | 1.317 |
| RW 4 | 0.357 | 0.640 |
| RW 5 | 0.211 | 0.376 |
| RW 6 | 0.158 | 0.280 |
| RW 7 | 0.100 | 0.178 |
| RW 8 | 0.089 | 0.158 |
| RW 9 | 0.065 | 0.114 |
| RW 10 | 0.052 | 0.091 |
| RW 11 | 0.043 | 0.075 |
| RW 12 | 0.030 | 0.052 |
| IW 1 | 7.09 | 0.909 |
| IW 2 | 8.45 | 0.706 |
| IW 3 | 9.02 | 0.541 |
| IW 4 | 11.3 | 0.485 |
| IW 5 | 12.5 | 0.404 |

Table 4. WEC's power performance with $B_{PTO} = 6.0$ Nms.

| Scenario | Absorbed Power [kW] | Capture Width |
|---|---|---|
| RW 1 | 6.09 | 11.96 |
| RW 2 | 5.58 | 10.28 |
| RW 3 | 4.10 | 7.403 |
| RW 4 | 2.83 | 5.066 |
| RW 5 | 2.16 | 3.861 |
| RW 6 | 1.52 | 2.711 |
| RW 7 | 1.12 | 1.998 |
| RW 8 | 0.799 | 1.415 |
| RW 9 | 0.638 | 1.125 |
| RW 10 | 0.595 | 1.045 |
| RW 11 | 0.401 | 0.699 |
| RW 12 | 0.336 | 0.580 |
| IW 1 | 30.1 | 3.872 |
| IW 2 | 33.5 | 2.816 |
| IW 3 | 56.4 | 3.386 |
| IW 4 | 61.0 | 2.501 |
| IW 5 | 77.3 | 2.301 |

indicate that non-linearities of motion are already playing an important role in the WEC dynamics, because the underwater volume is changing considerably. As a matter of fact, the displacement volume of the fully submerged WEC is more than 150% higher in comparison to the operational displacement volume. The code shall be further developed in order include such non-linearities straightforwardly and iteratively, for the geometry of the hull allows to represent all hydrodynamic and restoration forces in non-linear fashion.

## 4 CONCLUSIONS

The present study investigates the motion dynamics and power performance of a conical point-absorber WEC with 1 dof moving around a hinge. The numerical results are obtained in time domain using an in-house code and are compared with peer-reviewed results in both frequency domain and from experimental campaign.

In general, the results compare well, and the code is validated after accomplishing the numerical-experimental comparison. The main discrepancy between frequency domain and time domain results is that the time domain solver presents more sensitiveness in regards to the damping next to the natural period. Basically, the time domain predicts resonant motion in the lower damping cases, whereas the frequency domain solver indicates that the resonant motion is already mitigated by the potential damping alone, after all, the damping ratio of the hull is definitely not small. The results also differ in terms of total wave absorbed power, for the frequency domain results use spectral analysis, which is not reliable for power performance, and thus wave energy absorption in irregular waves can only be correctly modelled in time domain.

Another result obtained is that the WEC may indeed capture reasonable amounts of power, especially with increased PTO damping values. If the dimensions are also modified in order to match wave energy period with WEC's natural period, then the WEC shall absorb amount of wave energy comparable to the most efficient point-absorber concepts known so far.

Last but not least, it is concluded from the time series of WEC's motion in irregular waves that non-linear effects may play an important role in the dynamics even in the common sea states, for the underwater volume of the WEC varied considerably in most simulation cases. Regarding the future development of the 1 dof solver, the next step is to include the non-linearities of WEC's motion within the iterative time-integration process.

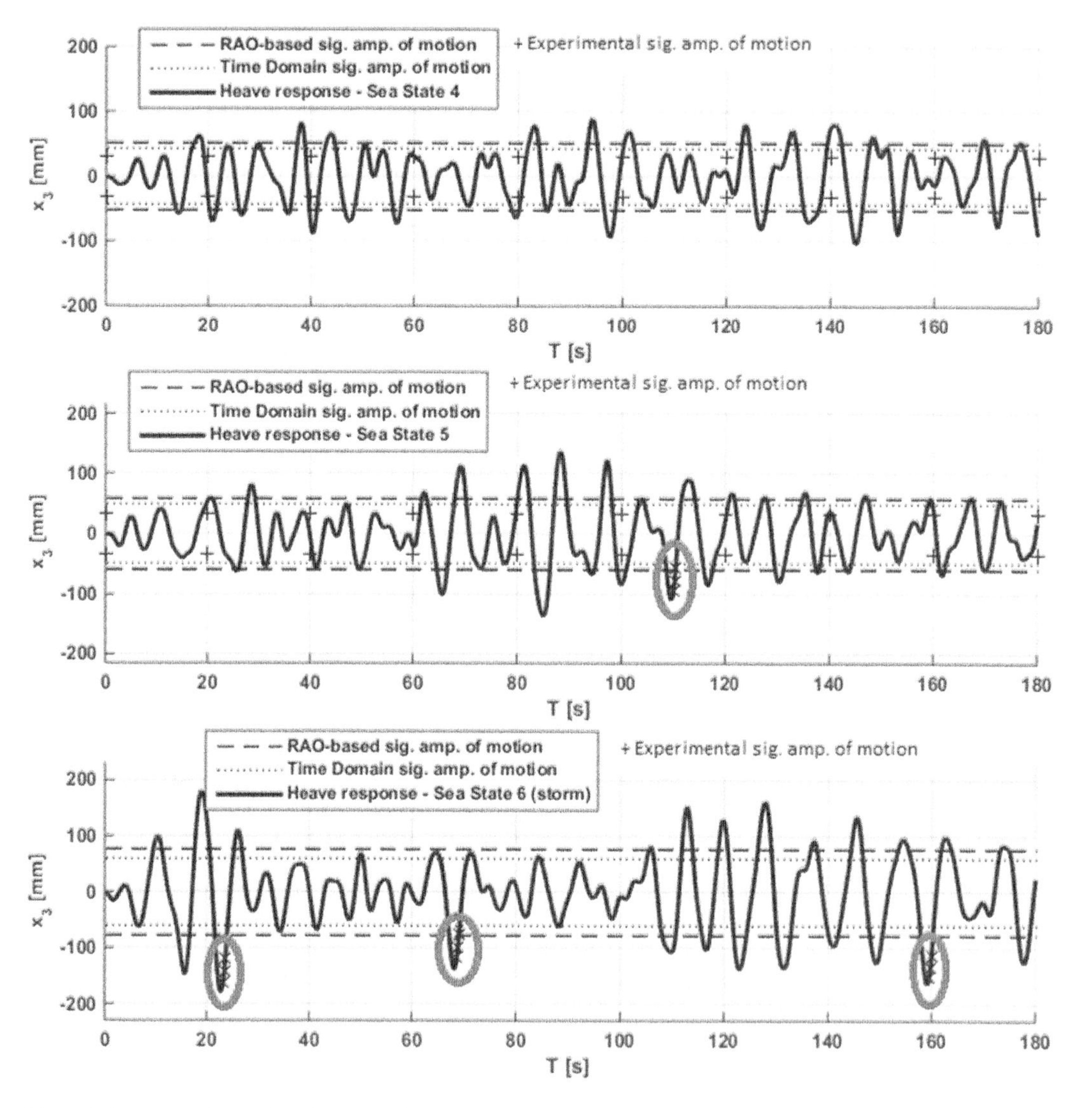


Figure 6. Numerical time series of heave motion for Sea States 4, 5, and 6 (storm), and $B_{PTO} = 0.5$ Nms.

## ACKNOWLEDGMENTS

This work is performed within the project WEC4MUP - Variable geometry Wave Energy Conversion system for floating platforms, which is funded by the Portuguese Foundation for Science and Technology (Fundação para a Ciência e a Tecnologia - FCT) under contract 031488-770 (PTDC/EME-REN/0242/2020). The first author of this work has a PhD Scholarship by the Portuguese Foundation for Science and Technology (Fundação para a Ciência e Tecnologia – FCT), under contract No. SFRH/BD/145602/2019. This work also contributes to the Strategic Research Plan of the Centre for Marine Technology and Ocean Engineering (CENTEC), which is financed by the FCT under contract No. UIDB/UIDP/00134/2020.

## REFERENCES

Gomes R. P. F., Lopes M. F. P., Henriques J. C. C., Gato L. M. C., Falcão A. F. O. (2015). The dynamics and power extraction of bottom-hinged plate wave energy converters in regular and irregular waves. *Ocean Engineering* **96**, 2015, pp. 86–99.

Hallak T. S., Gaspar J. F., Mojtaba K., Guedes Soares C. (2021). Numerical and experimental analyses of a conical point-absorber WEC moving around a hinge. Guedes Soares, C. & Santos T. A., (Eds.). *Developments in Maritime Technology and Engineering*, 2021. Vol 2, pp. 587–596.

Hansen R. H. (2013). *Design and control of the power take-off system for a wave energy converter with multiple absorbers*. PhD diss., Faculty of Engineering & Science, Aalborg University, Denmark. October, 2013.

Hansen R. H., Kramer M & Vidal E. (2013). Discrete displacement hydraulic power take-off system for the Wavestar Wave Energy Converter. *Energies* **6**, 2013, pp. 4001–4044.

Kamarlouei M., Gaspar J. F., Calvário M., Hallak T. S., Mendes M. J. G. C., Thiebaut F., Guedes Soares C. (2020). Experimental analysis of Wave Energy Converters concentrically attached on a floating offshore platform. *Renewable Energy* **152**, 2020, 1171–1185.

Kamarlouei M., Hallak T. S., Gaspar J. F. & Guedes Soares C. (2021). Evaluation of the negative stiffness mechanism on the performance of a hinged Wave Energy Converter, *Proceedings of the 40$^{th}$ International Conference on Ocean, Offshore and Arctic Engineering*. Vol. 9. V009T09A015, http://dx.doi.org/10.1115/OMAE2021-63748

Liu Z., Ciais P., Deng Z., et al. Near-real-time monitoring of global $CO_2$ emissions reveals the effects of the COVID-19 pandemic. *Nat Commun* **11**, 5172 (2020).

Reguero B. G., Losada I. J., Mendes F. J. (2019). A recent increase in global wave power as a consequence of oceanic warming. *Nat Commun* **10**, 205, 2019.

Xiaojing S., Huang D., Wu G. (2012). The current state of offshore wind energy technology development. *Energy* **41**, pp. 298–312.

*Trends in Maritime Technology and Engineering – Guedes Soares & Santos (eds)*


# Influence of platform configuration on the hydrodynamic performance of semi-submersibles for offshore wind energy

M. Hmedi, E. Uzunoglu & C. Guedes Soares
*Centre for Marine Technology and Ocean Engineering (CENTEC), Instituto Superior Técnico, Universidade de Lisboa, Lisbon, Portugal*

ABSTRACT: This work presents a comparison between the hydrodynamic behavior of three and four-column semi-submersibles. The platforms are the NAUTILUS and the OO-star semi-submersibles, with the DTU 10MW wind turbine. Both models are described, and simulations in wind and wave conditions are made using OpenFast. The natural frequencies and the response amplitude operators are calculated. The motions of the platform, the tower bending moment, and the nacelle acceleration are evaluated. These results are assessed to understand whether a particular geometry presents advantages over another. It is concluded that the platform geometry has a slight effect on the platform pitch motion, and economic factors should be considered for model selection.

## 1 INTRODUCTION

Fossil fuels, oils, and natural gases are finite energy resources, and the global energy demand is ever-increasing. In addition, fossil fuel-based power plants have negative environmental effects, emphasizing global warming. With the increase in energy consumption, atmospheric pollution from these energy sectors will intensify. On the other hand, wind energy is a sustainable, clean, and renewable energy resource. For several years, wind energy was the fastest-growing energy source in the world (Lowe & Drummond 2022).

Higher wind speeds and lower turbulence are driving factors to move offshore. Moreover, the visual and audible impact of wind turbines observed inland vanishes as the distance from the coastline increases. Thus, offshore wind turbines become advantageous, even more so in deeper waters and floating turbines. Another advantage of moving offshore is the possibility of implementing larger turbines (Bagbanci et al. 2012).

For the reasons given above, the move offshore has been prominent in recent years (Díaz & Guedes Soares 2020). Most offshore wind turbine floaters make use of the platform ideas from the oil and gas (O&G) sector. Among these platforms, the semi-submersible is one of the most stable platform types used by the industry (Uzunoglu et al. 2016). It consists of a deck, multiple columns, and pontoons.

Regarding their use in the O&G sector, there have been several hull configurations consisting of three and four columns. For instance, the SEDCO 135 was a three-column structure (Watts & Faulkner 1968) and Neptune Pentagon 81 was a five-column design (Halkyard 2005). Eventually, offshore O&G settled on four-column designs since 1961 (e.g. Blue Water Rig No.1) (Leffler et al. 2011). This configuration brought several advantages such as better deck support, sufficient stability, and ease of production. Similarly, the use of cylindrical columns gave way to rectangular columns due to ease of production and better support for the deck.

However, the requirements from the deck space are reduced when it comes to wind. For this reason, the smaller hull forms using three columns are often seen as in (Roddier et al. 2010, Robertson et al. 2014, Karimirad & Michailides 2015). In the case of a floating wind turbine platform, the floater's primary purpose is to support the wind turbine while ensuring a minimal response to wind and wave loads. This target needs to be met while keeping a low mass as it is one of the primary economic drivers (James & Costa Ros 2015, Uzunoglu & Guedes Soares 2019). Therefore, the geometry needs to suit both purposes regarding reduced mass and beneficial motion behaviour.

Several authors studied geometrical alternatives that meet these goals. Examples of geometrical variations can be given as V-shaped (Karimirad & Michailides 2015), the OC4 semi-submersible (Robertson et al. 2014), and the WindFloat (Roddier et al. 2010). These models were designed to support the NREL 5MW wind turbine (Jonkman et al. 2009). Two of the semi-submersibles designed and evaluated for 10 MW turbines during the LIFES50+ project (Wei et al. 2018) are the NAUTILUS and the OO-star semi-submersibles. The NAUTILUS utilizes four stability columns, and the tower is attached to the deck at a transition piece. The OO-star's

DOI: 10.1201/9781003320289-43

stability relies on three outer columns and a central column supports the tower. They were both designed to support the DTU 10 MW wind turbine (Bak et al. 2013), thus, comparison due to geometrical differences is possible.

However, the NAUTILUS and OO-star semi-submersibles were studied and tested separately during the LIFES50+ project. No comparison was made to highlight the benefits of each geometry. Given that the turbine is identical in both designs, the motion characteristics stem from the hull geometry. Thus, this paper compares the performance of these two models to identify the differences in their dynamic behaviour that affect the turbine's performance. The work is divided into three main parts. The first section presents the methodology and the tests made to analyse the models' responses. The second section describes the floaters and the turbine. The third section displays the simulation's results against different environmental conditions and compares the models' performance.

## 2 METHODOLOGY

This section consists of a background review of the tests and the formulas used in the analysis sections. Different tests were made to investigate the floating turbines' behaviour. Simulations are carried using OpenFast 3.0.0 (Jonkman 2013). They are divided into three main categories detailed below. The degrees of freedom (DOFs) of interest are the surge, heave, and pitch, as only headwind and waves are applied.

### 2.1 *Decay tests*

The main objective of these tests is to identify the natural periods of the models in the main DOFs. The platform natural frequencies must be far away from the excitation loads frequencies to avoid having resonance. This is an important phenomenon to consider when designing a model, as it damages the floating turbines and reduces their operating life. Decay tests are made by displacing the platform along a DOF direction from its static position, in still water and no wind conditions, and then releasing it to oscillate freely. Considering $N$ oscillations in a $\Delta T$ duration, the natural period $T_0$ is classically estimated as $T_0 = \Delta T/N$ and the natural frequency $f_0 = 1/T_0$. $f_0$ can be also estimated using the frequency analysis discussed in the next section. For a decay test, the motion spectrum's peak frequency corresponds to the system's natural frequency.

### 2.2 *Tests in wave conditions*

The examination of the platform's response in waves is an essential practice to investigate the system behaviour against wave excitation. The response amplitude operator (*RAO*) is a function facilitating the understanding of the system response. The *RAO* provides the frequency response of the platform to the wave excitation where the motion's amplitude is normalized by the wave height. In other words, *RAO* is a transfer function relating the motions of the body to the wave amplitude. It mainly depends on the frequency of the excitation wave and its direction. The study here is limited to head waves only, so the effect of the direction is not discussed. The model with the lowest peaks number and amplitude in its *RAO* curves is less excited by the waves and is considered to have better dynamics. Assuming a linear motion of the floating platform, the equation of motion is therefore written as follows (Newman 1977):

$$[M + A(\omega)]\ddot{x} + B(\omega)\dot{x} + [C_h + C_m]x = F(\omega) \quad (1)$$

where $x$ is the body motion, $\omega$ is the oscillation frequency, $M$ is the mass and inertia matrix, $A(\omega)$ is the added mass matrix, $B(\omega)$ is the damping matrix, $C_h$ is the hydrostatic stiffness matrix, $C_m$ is the mooring stiffness and $F(\omega)$ is the excitation force proportional to the incoming wave height $\eta$ by an $F_0$ factor, the latest being usually harmonic $\eta = \eta_0 e^{i\omega t}$. Assuming $x = ae^{i\omega t}$, hence *RAO* is defined by:

$$RAO = \frac{x}{\eta} = \frac{a}{\eta_0} = \frac{F_0}{(C_h + C_m) - [M + A(\omega)]\omega^2 + i\omega B(\omega)} \quad (2)$$

Note that the *RAO* is a complex function, and its magnitude (absolute value) is considered in the analysis. Its phase corresponds to the phase shift between the platform response and the upcoming wave.

In this work, an irregular wave simulation with no wind is made to find the *RAO* of both systems over the wave's excitation frequency bandwidth. The wave has a significant height $Hs = 4$ m, a peak period $Tp = 8$s, and a JONSWAP peak parameter $\gamma = 2.87$ with a frequency bandwidth between 0.05 and 0.25Hz. The JONSWAP spectrum was used as it provides an acceptable representation of the wave conditions mainly utilized in the calculation of the wave loads associated with the first-order wave frequency (Veldkamp & van der Tempel 2005). The *RAOs* are estimated by the spectral density analysis of the motions and wave output signals calculated by OpenFast. Briefly, *RAO* is calculated as following (Hee 2004):

$$RAO = (S_m(\omega)/S_w(\omega))^{1/2} e^{iArg(S_c(\omega))} \quad (3)$$

Where $S_m(\omega)$ is the motion spectrum density, $S_w(\omega)$ is the wave spectrum density and $S_c(\omega)$ is the cross-spectrum power density. A coherence threshold of 0.8 is set as a limitation of this formula, as the coherence measures the linear interdependence of the wave and motions signals. The coherence between both signals is:

$$C = \left( \frac{S_c^2(\omega)}{S_m(\omega)S_w(\omega)} \right)^{1/2} \tag{4}$$

### 2.3 *Tests in waves and wind conditions*

The simulations in wind and wave environments are essential to evaluate the functionality of the floating turbine at different operation points ("A" - below, "B" - at, and "C" - above the rated power speed). Table 1 summarizes the environmental conditions of each simulation. The wave spectrum used is the JONSWAP spectrum with a peak parameter $\gamma = 2.87$, and the values of the significant wave height ($H_s$) and the peak period ($T_p$) are chosen according to the Gulf of Maine (GoM) site conditions applied in (Pegalajar-Jurado et al. 2018).

Table 1. Wind and wave conditions of the operation points.

| Condition | $H_s$ (m) | $T_p$ (s) | $U_{ref}$ (m/s) | TI (%) |
|---|---|---|---|---|
| A | 1.67 | 8.0 | 8.0 | 18 |
| B | 3.04 | 9.5 | 11.4 | 18 |
| C | 4.29 | 10 | 17.9 | 18 |

The wind files are generated using TurbSim 2.00 (Jonkman & Buhl 2005) and by applying the Kaimal turbulence model following the IEC 61400-1 Ed. 2 standards (IEC 1999) with an "A" class turbulence characteristic corresponding to a turbulence intensity (TI) of 18%. The mean wind speed at the hub height is denoted $U_{ref}$. These tests under combined wind and wave excitation loads demonstrate the stability and the durability of the system settled in an intricate environment. Due to the wind excitation, a thrust force will be applied to the turbine influencing mainly the pitch motion and generating a tower base moment (Uzunoglu & Guedes Soares 2015). The motions signals, the blade's deflection, the tower base moments, and the nacelle acceleration are studied. The same blade pitch controller is used for both systems, so the controller's performance is not discussed in this work. The analysis is done by comparing the responses' maximum and standard deviation of both models.

## 3 MODELS' DESCRIPTION

This section presents a summary of the models used in the simulations. More details regarding the floaters and the full description of the turbine can be found in (Wei et al. 2018) and (Bak et al. 2013) respectively. Both models are available for public use in (Borg et al. 2018); these files are adaptable with OpenFast 1.0.0 and were adjusted in this work to be compatible with the newest version 3.0.0. Both models were designed for the GoM location having 130 m depth, and details about the wind and wave characteristics of the site can be found in (Gómez et al. 2015).

As previously mentioned, both floaters are equipped with the DTU 10MW Rotor-Nacelle Assembly (RNA). Each model has its own designed tower where the RNA is placed, ensuring a hub height of 119m above the mean sea level (MSL). Table 2 presents several key parameters of the turbine. The controller utilized in the simulations is the same one used in (Uzunoglu & Guedes Soares 2020).

Table 2. Key parameters of the DTU 10MW wind turbine.

| | Unit | Value |
|---|---|---|
| Rated power | MW | 10 |
| Rated wind speed | m/s | 11.4 |
| Cut out speed | m/s | 25 |
| Rotor diameter | m | 178.3 |
| Hub height | m | 119 |
| Maximum rotor speed | rpm | 9.6 |

### 3.1 *NAUTILUS semi-submersible*

The NAUTILUS support structure (Figure 1) consists of the tower, the floating platform, and the mooring lines. The mooring system is composed of four catenary lines anchored to the seabed. The semi-submersible is symmetrical, involving four columns connected at the bottom plane by a square-shaped ring pontoon and an X-shaped main deck at the top of the columns, and it is mainly made of steel. The tower is a single-piece conical design having a total length of 107m with a linearly varying thickness, and it is connected to the deck centre by a transition piece . The tower base outer diameter is 10.5m, and the top one is 5.5m to fulfil the RNA requirements. The tower is made of steel, and the material density (in the numerical model) was increased to 8,500 kg/m$^3$ to account for the mass of additional structural components such as flanges, bolts, secondary structures, and painting. The stability is attained by ballasting the ring pontoon with 3885 tons of concrete.

The NAUTILUS floating support structure includes a platform trim system (PTS) measuring the pitch and roll motions and the draft, then activating a pumping system that extracts seawater into/out of each column. The configuration used in this work is with full water ballast. The PTS is not included in the real-time simulation in OpenFast as

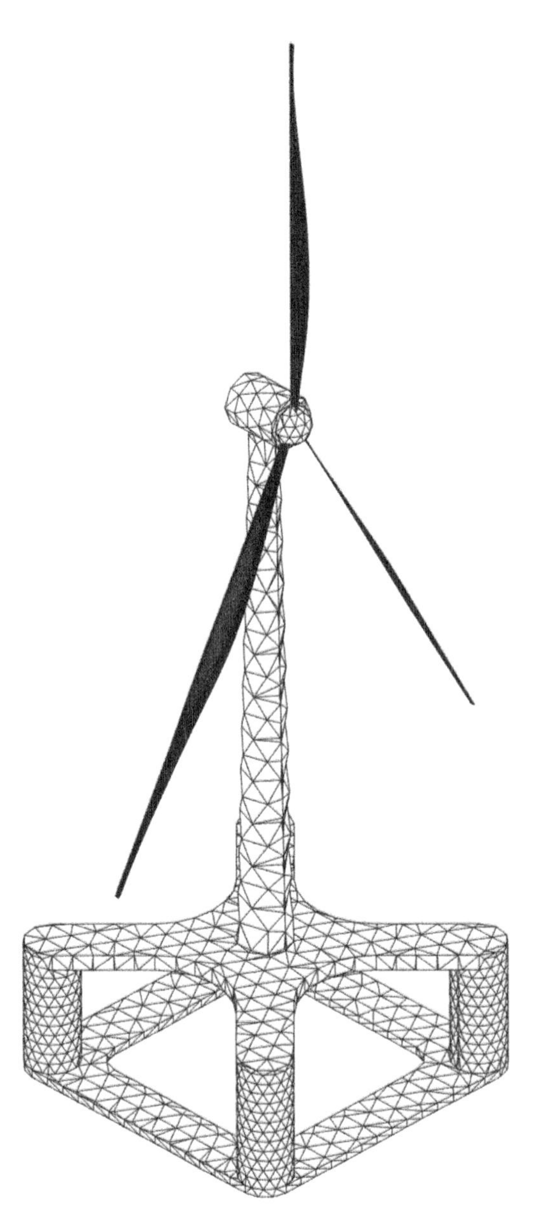

Figure 1. NAUTILUS support structure with DTU 10MW turbine.

it takes 30 minutes to take off the water from the tanks. The model parameters and dimensions are summarized in Table 3, and further details can be found in (Wei et al. 2018) and (Galván et al. 2018) concerning the real and the OpenFast numerical model. The potential flow model, including the additional hydrodynamic damping (ADH) coefficients, is utilized; it is concluded to be faster and presents a good agreement compared to the Morison elements (ME) method for implementing the viscous effects (Galván et al. 2018). The ADH coefficients are summarized in Table 4; they were obtained by upscaling the NAUTILUS-NREL 5 MW model experimentally tested in (Nava et al. 2014).

Table 3. Some parameters for the NAUTILUS semi-submersible.

| | |
|---|---|
| Transition piece diameter | 10.50m |
| Column diameter | 10.50m |
| Column height | 26.00m |
| Pontoon width | 10.50m |
| Pontoon height | 1.50m |
| Equilibrium draft | 18.33m |
| Displaced water volume | 9280.96m$^3$ |
| Tower top elevation | 114.667m |

Table 4. Linear and quadratic ADH for the NAUTILUS-DTU10 MW system (Galván et al. 2018).

| | Linear ADH terms | | Quadratic AHD terms | |
|---|---|---|---|---|
| DOF | Values | Units | Values | Units |
| Surge | 0 | [N.s/m] | 1100985 | [N.s$^2$/m$^2$] |
| Sway | 0 | [N.s/m] | 827308 | [N.s$^2$/m$^2$] |
| Heave | 335479 | [N.s/m] | 5637998 | [N.s$^2$/m$^2$] |
| Roll | 211.97E6 | [N.m.s/rad] | 38515.2E6 | [N.m.s$^2$/rad$^2$] |
| Pitch | 222.17E6 | [N.m.s/rad] | 41617.9E6 | [N.m.s$^2$/rad$^2$] |
| Yaw | 22.56E6 | [N.m.s/rad] | 7066.5E6 | [N.m.s$^2$/rad$^2$] |

### 3.2 *OO-star semi-submersible*

The OO-star semi-submersible consists of three outer columns and a central column mounted on a three-legged star-shaped pontoon with a bottom slab (Figure 2). The floating structure was designed by (Olav Olsen AS 2011). The station-keeping system is composed of three catenary mooring lines, each suspended with a clump weight. The floater's main material is post-tensioned concrete, it has a total mass of 21709 tons including the ballast. The tower is defined in segments made of steel with a constant wall thickness. The steel density (in the numerical model) is increased to 8243 kg/m$^3$ to consider the additional structural components. Table 5 recaps some of the main dimensions of the model. The potential flow model with the Morison elements method is applied to include the viscous effects. Table 6 summarizes the drag coefficients (CD) of some of the floater components. More details can be found in (Wei et al. 2018) and (Pegalajar-Jurado et al. 2018) concerning the real and the numerical models.

Table 5. Some parameters for the OO-star semi-submersible.

| | |
|---|---|
| Distance from the central column to the outer columns | 37.00m |
| Central column top diameter | 12.05m |
| Outer columns top diameter | 13.40m |
| Pontoon width | 16m |
| Pontoon height | 7m |
| Equilibrium draft | 22m |
| Displaced water volume | 23509m$^3$ |
| Tower top elevation | 115.63m |

Table 6. The drag coefficients for the OO-star semi-submersible (Pegalajar-Jurado et al. 2018).

| Components | Diameter | CD |
|---|---|---|
| Central vertical column upper section | 12.05m | 0.729 |
| Central vertical column bottom section | 12.05m | 0.729 |
| Central vertical column bottom section | 16.20m | 0.704 |
| Outer vertical column upper sections | 13.40m | 0.720 |
| Outer vertical column bottom sections | 13.40m | 0.720 |
| Outer vertical column bottom sections | 15.80m | 0.706 |
| Circular ends of the pontoon's legs | 15.80m | 0.706 |
| Outer heave plates | - | 10.00 |
| Pontoon's legs | - | 2.050 |

From the above description, both models have different floater, tower, and moorings designs. The floaters are also made of different materials, and the OO-star is heavier than the NAUTILUS. The NAUTILUS semi-submersible has four columns with a closed squared pontoon configuration while the OO-star has three stability columns with a star-shaped pontoon. The following sections present the models' performance to see the advantages of each geometry as both are mounted with the same turbine, and to check which design has better dynamics.

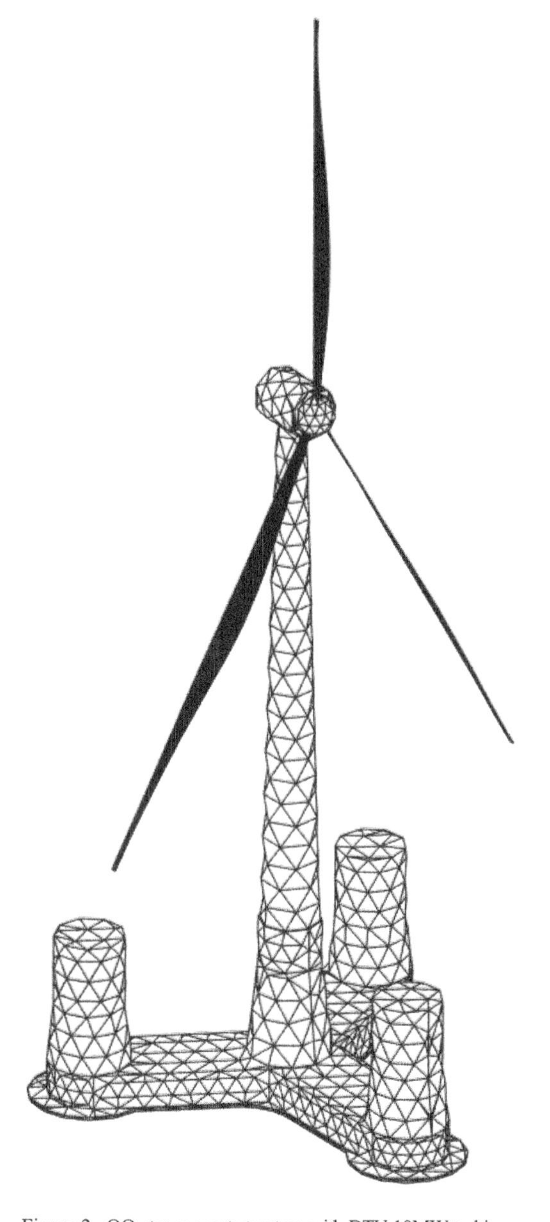

Figure 2. OO-star support structure with DTU 10MW turbine.

## 4 COMPARISON OF THE PLATFORMS' DYNAMICS

This section presents the simulations results of both models using OpenFast 3.0.0 and illustrates the difference in their behaviour for the same environmental conditions. Certain outputs are compared with the reference results obtained using FAST v8.16.00 documented in (Pegalajar-Jurado et al. 2018). The first simulation made is a static equilibrium test to verify the static offsets of the models. Table 7 outlines the offsets, which are identical to the reference ones. The offsets are calculated by averaging the output signals of a 2000 second simulation and skipping the first 600s. These offsets are used as initial conditions for the later simulations.

Table 7. Static offsets comparison.

| | | Surge | Heave | Pitch |
|---|---|---|---|---|
| NAUTILUS | Simulation | 0.06 m | -0.045 m | -0.26 deg |
| | Reference | 0.06 m | -0.045 m | -0.24 deg |
| OO-star | Simulation | 0.11 m | 0.054 m | -0.23 deg |
| | Reference | 0.11 m | 0.054 m | -0.23 deg |

### 4.1 *Decay results*

Decay tests serve to identify the natural frequency of the system. Each test is run for a sufficient duration to have enough oscillations to estimate the natural period of the models. The time-domain signals, as well as their spectral density, are depicted in Figure 3. The values of the natural frequency are introduced in Table 8. It is deduced that both models' natural frequencies help avoid severe resonance phenomenon in the operational wave conditions of the site. The NAUTILUS semi-submersible has a more damped response than the OO-star platform in the surge, heave, and yaw motions, while it is the opposite case for the pitch motion.

Table 8. Natural frequency results and comparison.

| | | Surge | Heave | Pitch | Yaw |
|---|---|---|---|---|---|
| NAUTILUS | Simulation | 0.0085 Hz | 0.052 Hz | 0.033 Hz | 0.011 Hz |
| | Reference | 0.0085 Hz | 0.052 Hz | 0.034 Hz | 0.011 Hz |
| OO-Star | Simulation | 0.0054 Hz | 0.048 Hz | 0.032 Hz | 0.009 Hz |
| | Reference | 0.0054 Hz | 0.048 Hz | 0.032 Hz | 0.009 Hz |

### 4.2 *Response in wave operational condition*

The two platforms are subjected to irregular waves of a significant height $H_s = 4$m, a peak period $T_p = 8$s, and the JONSWAP peak parameter $\gamma = 2.87$ with a frequency bandwidth between 0.05 and 0.25Hz to estimate their RAOs in the main DOF. Results are presented in Figure 4, and the applied frequency limit for the plots corresponds to the first-order wave frequency excitation range (DNV 2017). The NAUTILUS platform is more excited by the waves and presents multiple peaks in its RAOs plots. Even though these peaks are small, they can affect the life of the floating structure as they are located inside the most occurrent waves frequency range (0.1 to 0.2 Hz). In other words, the peaks correspond to a relatively significant displacement in the main DOFs, which might affect other components such as the moorings and the tower bending moments, intensifying the structural fatigue

### 4.3 *Simulations in wind and waves conditions*

Three operational conditions (Table 1) are selected to investigated and compare the behaviour of both platforms. A time window of 5000 seconds is selected for the analysis while skipping the first 600 seconds of the 5600 seconds simulation as transient mode. The difference in the viscous effects modelling between both models is assumed insignificant as the ADH model used with the NAUTILUS platform presents similar results when the ME method is applied (Galván et al. 2018). Figures 5 and 6 illustrate the maximum and standard deviation of the main DOFs, the nacelle accelerations, the tower base bending moment, and the blade out-of-plane deflection. The absolute value of the data is used to calculate the maximums.

Both models have roughly the same performance, and the main difference is seen in the platform pitch

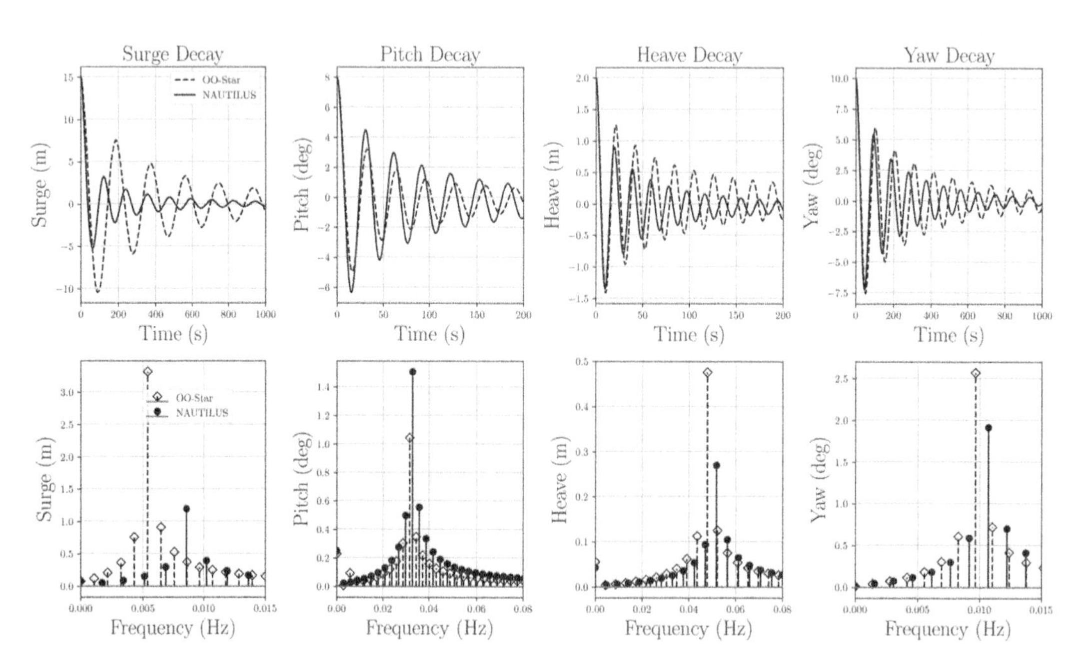


Figure 3. Time domain signals (top row) and spectrum (bottom row) of the decay test simulations.

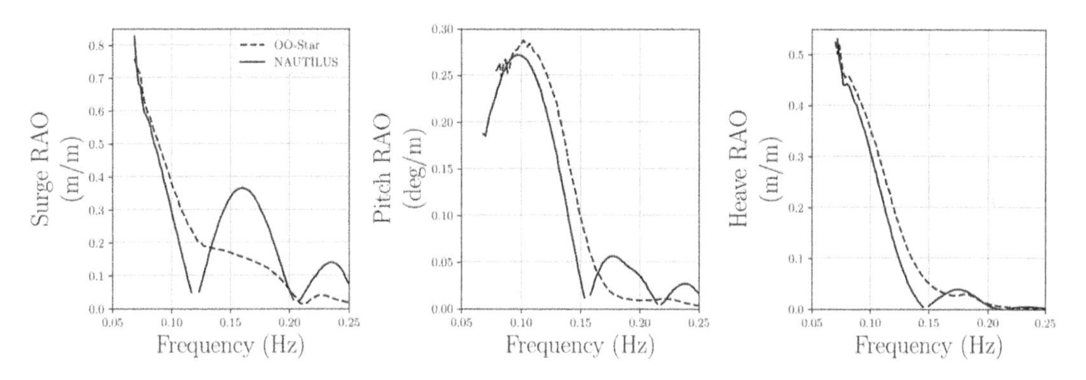


Figure 4. RAO plots of the main DOFs for both models.

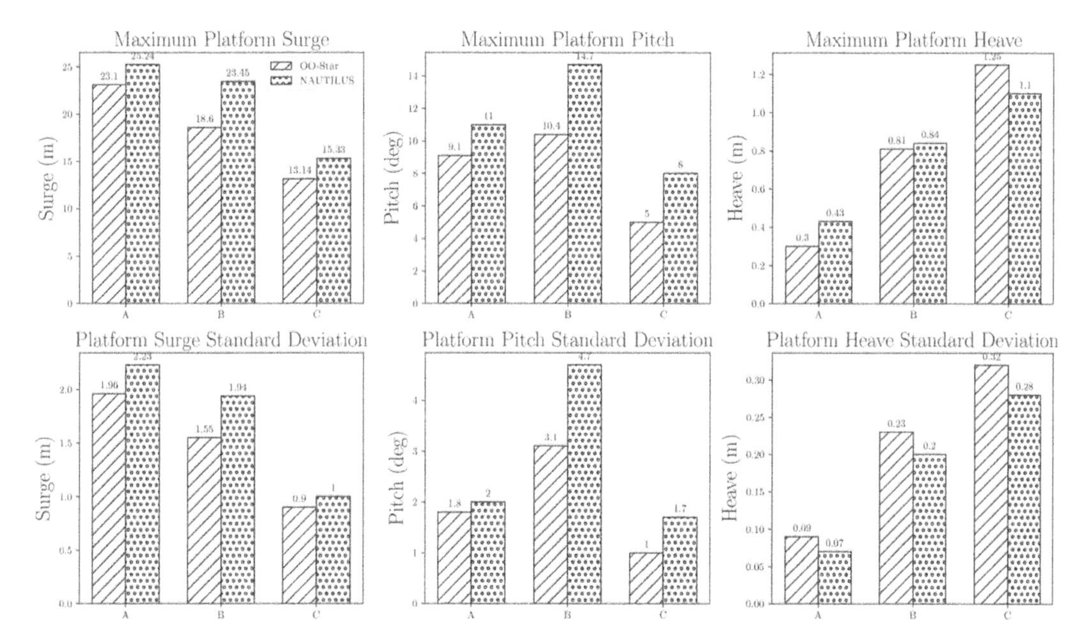


Figure 5. The maximum (top row) and the standard deviation (bottom row) of the main motions for the load cases A, B and C for both models.

motion. The NAUTILUS has a higher platform pitch response with a maximum of 14 degrees and a standard deviation of 4.7 degrees at the rated wind speed. The high standard deviation reflects high oscillations around the mean value, which intensifies the fatigue and reduces the platform's life. It is concluded in (Tran & Kim 2015) that the platform pitch motion of a FWT has a significant effect on the aerodynamic performance of the turbine, especially on the accuracy of the predicted aerodynamic power and thrust at high pitching motion. The fluctuations of the generated power, the main shaft bearing forces, and the tilting moment increase with the increase of the platform pitch amplitude (Li et al. 2021). The highest surge motion is observed in condition "A" for both NAUTILUS and OO-Star platforms, respectively 25.24 and 23.1m, and the smallest (respectively 15.33 and 13.14m) in condition "C". The heave response of both designs is small with a maximum of 1.25m in condition "C" reached by to OO-Star platform. Having the highest pitch response, the NAUTILUS platform is also subjected to the greatest nacelle acceleration and a slightly higher tower base bending moment compared to the OO-Star platform. The blade deflections are almost identical.

The OO-Star model presents slightly better performance than the NAUTILUS semi-submersible for the same environmental conditions. However, the PTS of the NAUTILUS was not used in the simulations. The PTS might reduce the platform pitching response. Although, the mean pitching response was found to be

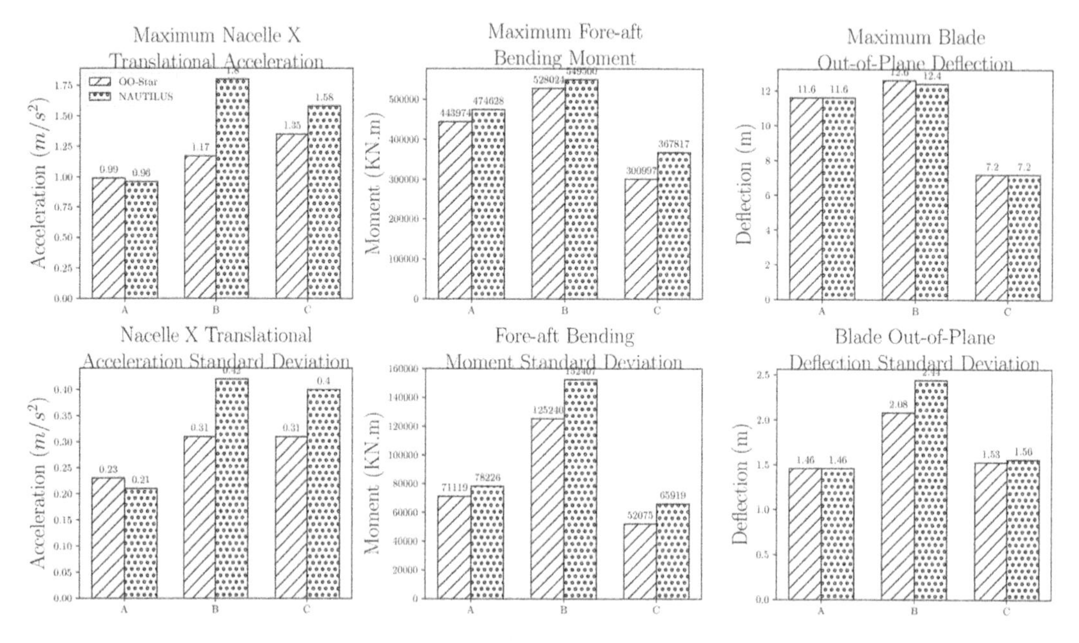


Figure 6. The maximum (top row) and the standard deviation (bottom row) of some parameters for the load cases A, B and C for both models.

12.7 seconds while the active ballast requires 30 minutes to respond to the motion. From these data, it is understandable that the mean pitching angle can be set using the active ballasting system. However, regarding the mean values, a further study should be carried to evaluate the PTS's effect. The PTS can still bring flexibility to vary the platform draft and preventing the waves from slamming over the deck in extreme conditions. It may also simplify the inspection and maintenance tasks. Another advantage of the PTS system would be the presence of pumps to remove seawater in the case of a damaged structure (Galván et al. 2018).

Choosing the most beneficial floater concept is not only related to the performance. Other factors related to the cost, construction, and transportation have an impact on the selection. The PTS of the NAUTILUS allows a variation of the draft offering access to more shipyards and ports (Wei et al. 2018). The NAUTILUS platform is simple to manufacture for being symmetrical with a squared ring pontoon that offers superior strength for a permanently sited system, helping to handle racking loads, and eliminating the need for bracing (Halkyard 2005). The study carried in (Liu et al. 2016) shows that the pontoon configuration of four-column semi-submersibles has a significant effect on the vortex-induced motions (VIM) where the four-pontoon system generates smaller lift forces in comparison with the two-pontoon system. Both designs don't use braces to connect the columns; this reduces their cost, simplifies their construction, and eliminates the fatigue potential the braces represent. Further study should be made to better highlight the performance of the star pontoon.

## 5 CONCLUSION

This paper presented a comparison of the dynamic behaviour between two semi-submersibles of three and four columns supporting the DTU 10MW wind turbine. The RAOs of the main DOFs were estimated using the spectral analysis of the motions and wave signals calculated by OpenFast in irregular wave conditions. The NAUTILUS platform is shown to be more excited by the waves and presented multiple peaks in its RAO curves.

The simulations in wind and wave conditions showed a slightly better performance of the OO-star platform compared to the NAUTILUS. The number of columns and their configuration mainly affected the platform pitch response. The other responses such as the heave motion, nacelle acceleration, and blade deflections were found to be roughly similar. However, considering that the platform pitch motion affects the turbine aerodynamic performance, it may be necessary to evaluate the aerodynamics further. This may require an extended study of controller behaviour to adapt it to each platform design.

The PTS of the NAUTILUS was not used in the simulations; the PTS would give the NAUTILUS platform several benefits. However, the mean pitching period was found 12.7 seconds which makes the PTS less likely to respond in time to smaller motions. The PTS may bring advantages to NAUTILUS over the OO-Star in terms of transportation from the port and response to extreme conditions. In addition, the cost and the ease of manufacturing are driving factors for a model selection.

## ACKNOWLEDGMENT

The first author has been funded by the project Hydroelastic behaviour of horizontal flexible floating structures for applications to Floating Breakwaters and Wave Energy Converters (HYDROELAST-WEB), which is funded by the Portuguese Foundation for Science and Technology (Fundação para a Ciência e a Tecnologia - FCT) under contract 031488_770 (PTDC/ECI-EGC/31488/2017). The second author has been funded by the project ARCWIND - Adaptation and implementation of floating wind energy conversion technology for the Atlantic region, which is co-financed by the European Regional Development Fund through the Interreg Atlantic Area Programme under contract EAPA 344/2016. This work contributes to the Strategic Research Plan of the Centre for Marine Technology and Ocean Engineering (CENTEC), which is financed by the Portuguese Foundation for Science and Technology (Fundação para a Ciência e Tecnologia - FCT) under contract UIDB/UIDP/00134/2020.


## REFERENCES


Bagbanci, H., D. Karmakar, & C. Guedes Soares (2012). Review of offshore floating wind turbines concepts. In C. Guedes Soares, Garbatov, Y., Sutulo, S., & Santos, T. A., (Eds.), *Maritime Engineering and Technology* (1 ed.)., pp. 553–562. London, UK: Taylor & Francis Group.

Bak, C., F. Zahle, R. Bitsche, T. Kim, A. Yde, L. C. Henriksen, M. H. Hansen, J. P. A. A. Blasques., M. Gaunaa, & A. Natarajan (2013). The DTU 10-MW Reference Wind Turbine. Technical report, Technical University of Denmark.

Borg, M., A. Pegalajar-Jurado, & F. Madsen (2018, June). FAST models of floating wind turbines.

DNV (2017). DNVGL-RP-F205 Global performance analysis of deepwater floating structures. Technical Report June, DNV GL AS.

Díaz, H. & C. Guedes Soares (2020, August). Review of the current status, technology and future trends of offshore wind farms. *Ocean Engineering 209*, 107381.

Galván, J., M. J. Sánchez-Lara, I. Mendikoa, G. Pérez-Morán, V. Nava, & R. Rodríguez-Arias (2018, October). *NAUTILUS-DTU10* MW Floating Offshore Wind Turbine at Gulf of Maine: Public numerical models of an actively ballasted semisubmersible. *Journal of Physics: Conference Series 1102*, 012015.

Gómez, P., G. Sánchez, A. Llana, & G. Gonzalez (2015, October). D1.1 Oceanographic and meteorological conditions for the design. Technical report, Iberdrola.

Halkyard, J. (2005, January). Chapter 7 - Floating Offshore Platform Design. In S. K. Chakrabarti (Ed.), *Handbook of Offshore Engineering*, pp. 419–661. London: Elsevier.

Hee, J. (2004, August). Signal Analysis Formulas.

IEC (1999, February). IEC 61400-1:1999: Wind turbine generator systems - Part 1: Safety requirements.

James, R. & M. Costa Ros (2015, June). Floating Offshore Wind: Market and Technology Review.

Jonkman, B. J. & M. L. Buhl, Jr. (2005, September). Turb-Sim User's Guide. Technical Report NREL/TP-500-36970, 15020326, NREL.

Jonkman, J. (2013, January). The New Modularization Framework for the FAST Wind Turbine CAE Tool. In *51st AIAA Aerospace Sciences Meeting including the New Horizons Forum and Aerospace Exposition*, Grapevine (Dallas/Ft. Worth Region), Texas. American Institute of Aeronautics and Astronautics.

Jonkman, J., S. Butterfield, W. Musial, & G. Scott (2009, February). Definition of a 5-MW Reference Wind Turbine for Offshore System Development. Technical Report NREL/TP-500-38060, 947422, NREL.

Karimirad, M. & C. Michailides (2015, November). V-shaped semisubmersible offshore wind turbine: An alternative concept for offshore wind technology. *Renewable Energy 83*, 126–143.

Leffler, W., R. Pattarozzi, & G. Sterling (2011). *Deepwater Petroleum Exploration & Production: A Nontechnical Guide* (2 ed.). PennWell Corporation.

Li, Z., B. Wen, X. Dong, X. Long, & Z. Peng (2021, July). Effect of blade pitch control on dynamic characteristics of a floating offshore wind turbine under platform pitching motion. *Ocean Engineering 232*, 109109.

Liu, M., L. Xiao, H. Lu, & J. Shi (2016, September). Experimental investigation into the influences of pontoon and column configuration on vortex-induced motions of deep-draft semi-submersibles. *Ocean Engineering 123*, 262–277.

Lowe, R. & P. Drummond (2022, January). Solar, wind and logistic substitution in global energy supply to 2050 – Barriers and implications. *Renewable and Sustainable Energy Reviews 153*, 111720.

Nava, V., G. Aguirre, J. Galvan, M. Sánchez-Lara, I. Mendikoa, & G. Perez-Moran (2014, November). Experimental studies on the hydrodynamic behavior of a semi-submersible offshore wind platform. In C. Guedes Soares (Ed.), *Renewable Energies Offshore* (1 ed.). London, UK: Taylor & Francis Group.

Newman, J. N. (1977). *Marine Hydrodynamics*. MIT Press.

Olav Olsen AS,. (2011). Dr. techn. Olav Olsen AS.

Pegalajar-Jurado, A., H. Bredmose, M. Borg, J. G. Straume, T. Landbø, H. S. Andersen, W. Yu, K. Müller, & F. Lemmer (2018, October). State-of-the-art model for the LIFES50+ OO-Star Wind Floater Semi 10MW floating wind turbine. *Journal of Physics: Conference Series 1104*, 012024.

Pegalajar-Jurado, A., F. Madsen, M. Borg, & H. Bredmose (2018, April). D4.5 State-of-the-art models for the two LIFES50+ 10MW floater concepts. Technical report, Technical University of Denmark.

Robertson, A., J. Jonkman, M. Masciola, H. Song, A. Goupee, A. Coulling, & C. Luan (2014, September). Definition of the Semisubmersible Floating System for Phase II of OC4. Technical Report NREL/TP-5000-60601, 1155123, NREL.

Roddier, D., C. Cermelli, A. Aubault, & A. Weinstein (2010, May). WindFloat: A floating foundation for offshore wind turbines. *Journal of Renewable and Sustainable Energy 2*(3), 033104.

Tran, T.-T. & D.-H. Kim (2015, July). The platform pitching motion of floating offshore wind turbine: A preliminary unsteady aerodynamic analysis. *Journal of Wind Engineering and Industrial Aerodynamics 142*, 65–81.

Uzunoglu, E. & C. Guedes Soares (2015, September). Comparison of numerical and experimental data for a DeepCwind type semi-submersible floating offshore wind turbine. In C. Guedes Soares (Ed.), *Renewable Energies Offshore* (1 ed.)., pp. 748–754. CRC Press.

Uzunoglu, E. & C. Guedes Soares (2019, January). A system for the hydrodynamic design of tension leg platforms of floating wind turbines. *Ocean Engineering 171*, 78–92.

Uzunoglu, E. & C. Guedes Soares (2020, February). Hydrodynamic design of a free-float capable tension leg platform for a 10 MW wind turbine. *Ocean Engineering 197*, 106888.

Uzunoglu, E., D. Karmakar, & C. Guedes Soares (2016). Floating Offshore Wind Platforms. In L. Castro-Santos & V. Diaz-Casas (Eds.), *Floating Offshore Wind Farms*, Green Energy and Technology, pp. 53–76. Cham: Springer International Publishing.

Veldkamp, H. F. & J. van der Tempel (2005, January). Influence of wave modelling on the prediction of fatigue for offshore wind turbines. *Wind Energy 8*(1), 49–65.

Watts, J. & R. Faulkner (1968, April). A Performance Review of the SEDCO 135-F Semi-Submersible Drilling Vessel. *Journal of Canadian Petroleum Technology 7*(02), 67–77.

Wei, Y., K. Müller, & F. Lemmer (2018, April). D4.2 Public Definition of the Two LIFES50+ 10MW Floater Concepts. Technical report, University of Stuttgart.

*Trends in Maritime Technology and Engineering – Guedes Soares & Santos (eds)*


# Review of hybrid model testing approaches for floating wind turbines

M. Hmedi, E. Uzunoglu & C. Guedes Soares
*Centre for Marine Technology and Ocean Engineering (CENTEC), Instituto Superior Técnico Universidade de Lisboa, Lisbon, Portugal*

ABSTRACT: This paper presents a review of the hybrid model testing strategies for floating wind turbines. It examines the method's accuracy, advantages, and weaknesses. The hybrid approach refers to tests that combine physical testing and numerical simulation in wave basin or wind tunnels. The coupling between the physical and numerical part is done via actuators. The different forms of hybrid testing adopted by researchers – such as robots, cable, propeller's actuators – are discussed and evaluated. A guidance is given for selecting the hybrid style suitable for model testing. It is concluded that hybrid testing is cost-efficient and versatile but its fidelity relays on the accuracy of the numerical tool. Also, the test's objectives are a key factor for choosing the suitable hybrid testing methodology.

## 1 INTRODUCTION

Energy demand is increasing, and the need of new resources is a reality. Fossil fuel-based power plants are environmentally harmful and lead to global warming. Meanwhile, wind energy presents a sustainable and renewable energy resource which is growing rapidly (Lowe & Drummond, 2022). In this regard, offshore wind turbines (WT) can be more efficient offshore due to higher wind speed and lower turbulence. Moreover, the audio-visual impact of the WT is reduced when placed far from the coastline. Thus, implementing larger turbines at offshore sites is more reasonable (Bagbanci et al., 2012).

Accordingly, the move offshore has been conspicuous in the recent years (Díaz & Guedes Soares, 2020) and for water depths greater than 50 meters, floating structures are more adequate. Regarding floaters, there are several options such as Tension leg platforms (TLP), semi-submersibles, and spar type structures (Uzunoglu et al., 2016) that all show differences in motion dynamics that needed to be accounted for in the design of wind energy structures.

Several scaled prototypes are already installed to evaluate the performance of floating wind platforms. For instance, Blue H (2007) is the first installed prototype of FWT (Collu & Borg, 2016). It is a TLP mounting 80 KW WT at 108 meters depth in Italy. Statoil's Hywind Spar (2009) in Norway (Skaare et al., 2015) and WindFloat semisubmersible (2011) in Portugal (Roddier et al., 2010) are other examples of installed prototypes.

However, building prototypes to validate theoretical designs is expensive and time consuming. It requires years to collect sufficient data for improvements. In contrast, experimental testing and numerical simulations are less expensive. Experimental testing requires more time than numerical simulations. However, numerical tools could be computationally costly and might disregard some non-linear effects. Inaccuracies might be present in the simulation results due to assumptions made in theoretical calculations; for example, excluding the second-order wave forces or neglecting viscous effects for potential flow calculations (Chen et al., 2020).

Thus, experimental testing allows investigating the model's performance under a more realistic environment as the loads and phenomena excluded by some numerical tools are now included. It also provides the desirable data to validate and improve the numerical models. This requires building a scaled model to replicate the full-scale system's performance. Customary, Froude scaling is applied to conserve the gravitational wave loads (Faltinsen, 1999). Contrarily, conserving Froude number yields to a lower Reynolds number and the aerodynamics loads are therefore out of scale (Müller et al., 2014).

Another significant requirement for a WT is adapting the blade pitch angle and generator torque (Pao & Johnson, 2011). This adaptation by the blade pitch controller affects the system motions by alternating the aerodynamics loads and rotor torque. Accordingly, the coupling effects need to be considered during a model testing. For instance, onshore WT controller may induce negative damping when used by FWT (Jonkman, 2008). Thus, efforts were made to test new control strategies such as in (Yu et al., 2017).

To surpass these challenges, mainly the scaling conflict, researchers applied different methodologies for testing FWTs. Static cables were applied

DOI: 10.1201/9781003320289-44

to emulate steady wind forces (Utsunomiya et al., 2009). Another method is using a drag disk at the tower's top to reproduce an approximate thrust force (Roddier et al., 2010). Froude scaled rotor with an increased wind speed to match the correct thrust force were utilized in (Robertson et al., 2013). Redesigned rotors, having the same performance as the full-scale rotor at a lower Reynolds number, were used in (Fowler et al., 2013; de Ridder et al., 2014; Duan et al., 2016) to correctly generate the rotor's aerodynamic loads. Another technique is the hybrid testing which consists of combining physical testing and numerical simulation in real time. An actuation system provides the coupling between the simulated WT and the physical floater in the wave basin (Azcona et al., 2014; Sauder et al., 2016) or inversely in wind tunnel (Bayati et al., 2013, 2014).

In hybrid tests, a physical part of the model is substituted by an actuator connected to a numerical tool. The approach can be applied in wind tunnels or wave basins. The hybrid approach allows testing FWT, under combined wind and wave conditions, in wave basins without a wind generation system. The actuators substitute the rotors and generate the required aerodynamic loads. This reduces the cost of the tests, as most wind generation installations require a high amount of energy to run (Otter et al., 2021). In addition to basin tests, the aerodynamics loads and wakes of FWTs can be studied in wind tunnels, where actuators mimic the system's motion. Besides these benefits, controlling the actuators can be complex.

Given the progress above, this paper reviews the hybrid systems used in previous testing campaigns. The advantages and weak points of each system are presented. The first section describes the hybrid approach. The second section reviews the common actuators employed in hybrid testing. The advantages of each actuation system are presented, and a comparison is given.

## 2 DEVELOPMENT OF HYBRID TESTING METHODOLOGIES

Several wave tank campaigns have been made to test FWTs (Stewart & Muskulus, 2016a) under different setups. In some projects, the models were tested in physical wind and wave environment (Robertson et al., 2013; Fowler et al., 2013; de Ridder et al., 2014; Duan et al., 2016). These cases can be categorized as full physical testing which requires the presence of wind and wave generating systems. The concept of the hybrid testing is quite different. It is a combination of physical testing and numerical simulation. This approach was initially used back to the 70's in earthquake engineering (Carrion & Spencer, 2007). There, to test the dynamic response of a structure in an earthquake condition, the structure's displacements were calculated on a computer and then applied to the tested model using several actuators.

The FTW tests borrow the same idea. The system's motions can be replicated using a robot (Bayati et al., 2014) in physical wind environment or the aerodynamic loads are emulated on the tower's top in physical waves conditions (Chabaud et al., 2013; Vittori et al., 2020) using actuators (cables, propellers, ducted fan). The easiest version is to emulate a recorded time-series of the motions or the aerodynamic loads via the actuation system. However, synchronizing the waves elevation and the emulated aerodynamic loads (or winds and robot motions) is challenging. Also, the controller effects are not replicated in real-time. In addition, the fidelity of these tests is low as the simulation does not consider the actual platform's motion in the calculations.

For these reasons, a higher fidelity approach with higher complexity was developed. It consists of combining the physical testing and the numerical simulation in real-time. For tests where the rotor is substituted by an actuator, the motions of the platform are tracked and sent to a numerical tool. The simulation gives the actual aerodynamic loads at the corresponding time step, which are then scaled and replicated by the actuator (Azcona et al., 2014). A study made by (Stewart & Muskulus, 2016b) compares using precalculated aerodynamic loads time series and real-time testing. The results demonstrate that using predefined loads leads to a higher difference in loads and motions when comparing simulation to experiment. The reason is the absence of a feedback between the system's motions and the predefined aerodynamic loads. However, real-time testing requires dealing with delays, whose compensation is problematic (Chabaud et al., 2013).

This methodology is referred to with different names such as "Software in the Loop" (SIL), "Hardware in the Loop" (HIL), and "Real-Time Hybrid Model" (ReaTHM) testing. They all denote tests that combine physical testing and numerical simulation. In general, this combination facilitates testing complex structures as FWT; however, it requires compensating delays for a rapid execution of the actuators in small time steps. The following recaps the hybrid testing campaigns of FWT and highlights the advantages of each actuation systems employed in these tests.

## 3 HYBRID TESTING METHODS

### 3.1 *Propeller actuators*

When it comes to actuators, there are several options such as propellers, ducted fans, and cables used in hybrid testing. Propellers and ducted fans are more commonly used by researchers. In 2014, researchers from CENER (Azcona et al., 2014) used a ducted fan at the tower top, more specifically at the hub height, to emulate the rotor's thrust. The thrust was calculated in real-time by the numerical tool FAST (Jonkman &

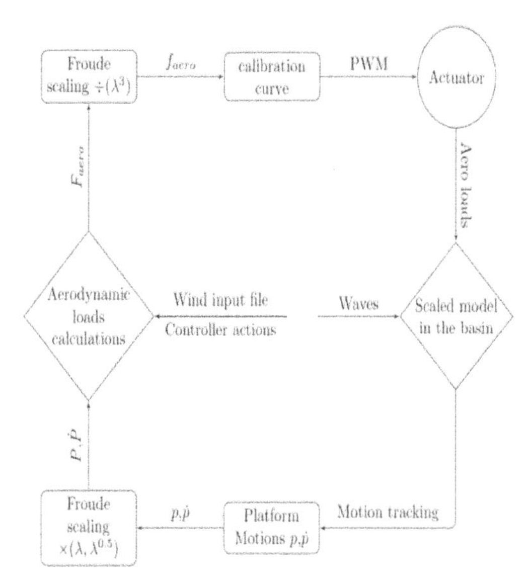


Figure 1. SIL diagram.

Buhl Jr., 2005) which receives the platform motions from the basin measurement (Figure 1). The technique was applied to test 1:40 scale-model of the OO-Star semi-submersible (Pegalajar-Jurado et al., 2018) with a 6MW wind turbine at the ECN wave basin. This campaign proved the capability of the SIL to include the aerodynamic damping.

The CENER team reapplied the same principle again at the ECN basin to test a 1:45 scaled model of the DeepCwind semi-submersible with the NREL 5MW WT (Vittori et al., 2018; Azcona et al., 2019) and with the DTU 10MW WT (Azcona et al., 2016). The improvement made was including the second-order hydrodynamics in the numerical model, which increased the accuracy of tests. As the previous system emulates only the thrust force, the CENER team developed a new multi-propellers system capable of reproducing the rotor moments (Pires et al., 2020). The new concept improved the SIL system by including the rotor moments generated by the aerodynamic and gyroscopic effects. During the ARCWIND project, CENER's multi-propellers system was used to test the CENTEC-TLP (Uzunoglu & Guedes Soares, 2020; Vittori et al., 2020).

In addition to the SIL systems developed by CENER, the hybrid approach using ducted fans or propellers have been widely applied by many researchers. A simple version of the hybrid testing was used by Desmond et al. (2019) to test the SCDnezzy[2] semi-submersible in Ireland. The tested model supports two 7.5MW turbines, thus, two ducted fans substituted the two rotors during the tests. The fans emulated steady wind loads only, providing up to 10N thrust on each tower. Similarly, one ducted fan generating steady thrust was used by (Wright et al., 2015) to test a hexagonal TLP design at two different scales. A ducted fan emulating recorded time-series of the thrust was used in (Andersen, 2016) to test a three-column semi-submersible. Nevertheless, as mentioned in section 2, it is a low fidelity test compared to SIL, as the controller effects are not replicated in real-time, and the instantaneous motions are not considered in the aerodynamic loads' simulation.

A comparison between hybrid testing with and without SIL was made in (Matoug et al., 2020). The tested platforms were the NAUTILUS semi-submersible (Galván et al., 2018) with the DTU 10MW (Bak et al., 2013) and the WindQuest 10MW (Achard et al., 2017). It was concluded that the SIL improved the cross-correlation between the numerical simulation and the experimental testing by 14%.

Following the previous conclusions, single propeller actuator with SIL was widely applied in many campaigns. A 1:36.67 scale-model of the Iberdrola TLP (Zamora-Rodriguez et al., 2014) was tested at the Kelvin Hydrodynamics Laboratory (KHL) at the University of Strathclyde (Oguz et al., 2018). The model was mounted with a single ducted fan and used the same SIL concept described by (Azcona et al., 2014). The same approach was used to test the SOFTWIND Spar (Arnal, 2020; Arnal et al., 2019) at ECN. Although, the single fan actuator aims to reproduce the thrust force only. The other aerodynamic loads and gyroscopic moments are not emulated by using a single actuator. Hence, efforts have been made to develop multi-actuator system capable to reproduce the full aerodynamic tensor.

An actuator of six drone propellers was developed at Instituto Hidraulica Cantabria (IHC) (Urbán & Guanche, 2019) and used to test the TELWIND FWT (Armesto et al., 2018). The propellers were calibrated and tuned to consider the aerodynamic influence resulting from the proximity of the propellers. The results were encouraging and demonstrated small deviations compared to the theoretical values of the loads. Another six propellers actuator was developed by (Otter et al., 2020) at MaREI - Ireland and showed a good performance in emulating the thrust and torque. An actuator with three propellers was developed and calibrated by (Hmedi, 2020) to emulate the thrust and the aerodynamic moments. The system performed well except for emulating the high frequency content of the loads. In addition to testing horizontal axis WT, a multi-propeller system was used to test a vertical axis wind turbine (Kanner et al., 2016).

### 3.2 *Cable winch actuators*

Cable winches were also used as actuators for hybrid testing. In early testing campaign, static lines (cables) were used to reproduce a static thrust obtained from the thrust curve for the given turbine (Utsunomiya et al., 2009). During that campaign, several tests were made including decay, regular, and irregular waves tests, as well as regular wave tests with a steady thrust force applied at the tower's top via the cables. Although, the other aerodynamic loads and interactions (turbulent wind, aerodynamic damping, controller effects, gyroscopic moments) were not considered.

This method is effortless and quick as it requires a simple mechanical system consisting of cables that apply static tensions at the tower's top. Thus, static loads are easily emulated by this technique. However, it lacks the control and variability as only static loads are reproduced. In addition, using the WT's thrust curve does not consider the platform response for the aerodynamic load calculation and no gyroscopic effects can be emulated. Moreover, this technique emulates only the thrust force and neglects the other aerodynamic loads components. In other words, testing with static cables is limited to steady wind cases where the thrust force is roughly constant.

To surpass these limitations, a radical improvement for the static cable method consists of applying the tensions instantaneously to the model according to the load simulation in real-time. The concept is the same as the SIL system with propellers actuators, except of using cable winches to apply the loads calculated by the numerical tool.

The first real-time hybrid system with cable winches was built at SINTEF - Norway. The system consists of six cables attached to a square frame at the hub height. SINTEF's researchers named their approach as ReaTHM testing and used it to test the NOWITECH model mounting the NREL 5MW (Bachynski et al., 2016; Berthelsen et al., 2016; Sauder et al., 2016). They also made a sensitivity analysis (Bachynski et al., 2015) and a reduced number of aerodynamic loads was emulated following the sensitivity analysis's results. The force allocation and a theorical overview of the system is presented in (Chabaud et al., 2018).

As limited number of loads was emulated, the system was improved by (Thys et al., 2018) to include the wind direction's variability and the tower loads. The frequency bandwidth was also extended up to the 3P frequency and the first tower bending frequency. The enhanced system was used to test the NAUTILUS semi-submersible with the DTU 10MW WT. A comparison between the enhanced and the old system was made by (Chabaud et al., 2018). It was concluded that the new system is more flexible as it can mimic more loads for the same number of cables. Meanwhile, a higher cable tensions was required.

A similar hybrid system with cables was used in (Antonutti et al., 2020). The tested model is a semi-submersible mounting the Haliade 6MW WT. The system was composed by five cables enabling the actuation of the surge, pitch, and yaw DoFs. The results showed a good matching between the experimental and numerical model.

The hybrid methodology with cable actuators was applied to test a 1:50 model scale of the DeepCwind semi-submersible at the University of Maine – USA (Hall et al., 2014; Hall & Goupee, 2018). Rather than emulation multiple loads, the system was developed to emulate the thrust force only. Therefore, the actuator was formed by two opposing cables attached to the turbine at the hub height. The results of the hybrid testing were compared to the ones obtained by full physical testing with a redesigned rotor (Fowler et al., 2013). Both results showed a good agreement especially for the motions and moorings tensions. It was also demonstrated that the hybrid approach was clearly able to reproduce the effect of the aerodynamic damping.

### 3.3 *Wind tunnel tests*

The hybrid tests with cables and propellers were carried in wave basins. Their objective is studying the hydrodynamics of the tested platform, where the actuators emulate the aerodynamic loads instead of subjecting a scaled rotor to a wind flow. Another class of hybrid testing is performed in wind tunnels and aims to study the aerodynamic loads, wakes, and turbines' interactions. Rather than calculating the aerodynamic loads, the motions of the FWT model are simulated by the numerical tool. The model with a physical rotor is placed in the wind tunnel and the FWT's motions are then mimicked via actuators.

Researchers from Politecnico de Milano – Italy developed a two DoF robot to mimic the surge and pitch motions in real time (Bayati et al., 2013). The hybrid methodology was called hardware in the loop (HIL) and technical details about the system can be found in (Ambrosini et al., 2020). The robot was used to impose only the surge motion to a 1:75 scale-model of the DTU 10 MW during wind tunnel tests. The study was made to investigate the wakes in steady and unsteady wind conditions (Bayati et al., 2017). The robot was improved during later campaigns to mimic more DoFs arising a six DoF robot named "HexaFloat" (Bayati et al., 2014). The new robot was again used during wind tunnel testing of the DTU 10MW mounted on the OO-star semi-submersible (Bayati et al., 2018; Belloli et al., 2020).

A similar approach was adopted in (Schliffke et al., 2020) to study the influence of the surge motion on the wake characteristics. The tested model is the FLOATGEN barge with a 2MW WT. A porous disc concept was adopted instead of using a scaled rotor. A linear motor controlled by the manufacturer's software imposed the surge displacement of the model. A nearly constant delay between the assigned and emulated motion was noticed. Nevertheless, as the study of the motion dynamics is not taking place, the delay was considered unimportant, especially it was independent of the imposed frequency and amplitude. In addition to the impact of the surge motion on wakes, some hybrid tests were conducted to study the effect of the pitch motion on FWT's wakes and the effect of wakes interactions of FWT (Rockel et al., 2014, 2016).

## 4 ADVANTAGES AND DISADVANTAGES

### 4.1 *Compared to full physical testing*

Hybrid testing in wave basins does not require physical wind generating system. In addition to the measured motions, a wind input file is sent to the numerical simulation, which allows implementing the wind turbulence in the tests. Moreover, the turbine controller's and aerodynamic damping effects can be emulated. Contrarily to scaling the rotors, the hybrid approach is not limited to a unique turbine and scale. The actuators can be used to reproduce the aerodynamic loads of any turbine. This makes the hybrid technique cheaper and quicker compared to full physical testing, as full physical testing requires building a scaled rotor for each turbine design and scaling coefficient. Hybrid testing campaigns have shown good results in emulating the thrust force, in addition to torques and gyroscopic moments that can also be reproduced and repeated in multiple tests.

However, hybrid testing requires dealing with delays that might affect the fidelity of the tests. The actuators should have enough bandwidth to emulated the assigned load (or motions) during an appropriate time-step (Otter et al., 2021). Moreover, the accuracy of the tests is related to several factors, such as the motion tracking measurements and the simulation tool. It is certain that aerodynamic phenomena (for hybrid testing in wave basins) and hydrodynamic response (for hybrid testing in wind tunnel) neglected in the numerical model are not emulated during the experiments. Also, calibration of the actuators might be challenging, especially for multi-actuator system. Having several actuators allows reproducing more loads (for basin tests) or mimicking multiple DoFs displacements (for tunnel tests). Although, this increases the complexity and sources of uncertainties of the system.

### 4.2 *Alternative hybrid testing methods*

Both propellers and cable actuators can successfully reproduce the thrust force. However, the propeller systems face difficulties in reproducing the aerodynamic moments (Otter et al., 2020), and considering the propellers' torques in the force allocation is challenging (Hmedi, 2020). Meanwhile, the cable actuators are capable to reproduce the moments. Although, the control of the cables is complex compared to propellers. Also, the cable's frame requires a wide space on the tower top, unlike the propellers which occupy smaller space. The cables' pretension slightly affect the results of free decay tests (Sauder et al., 2016), while it is simpler to make free decay tests with turned off propellers actuators.

Propellers rotate at a high rotational speed inducing undesirable vibrations to the model, while the cables induce less vibrations. A careful attention should be carried concerning the temperature of the active emulator to avoid burning the propeller's electrical motor or demagnetize it. For multi-actuator systems, the interaction between the propellers needs to be considered when tuning them. In fact, adjacent propeller's wakes modify the air inflow of the other ones which can change their performance up to 10% (Urbán & Guanche, 2019).

The selection between hybrid testing in wind tunnel or wave basin depends on the investigated phenomena. If the study of wakes and turbines interaction is investigated, hybrid testing in wind tunnels is of interests. Meanwhile, hybrid testing with propellers or cables in wave basins is used when the hydrodynamics are studied. If thrust loading is considered the dominating load for the tests, a single propeller (or a ducted fan) is recommended. Multi-cable winches system is more suitable when the aerodynamic and gyroscopic moments should be taken in the analysis.

## 5 CONCLUSIONS

This paper presented an overview of the hybrid technique applied for testing FWT. The hybrid approach was first defined as a combination of physical testing and numerical simulation, in wave basins or wind tunnels, where the coupling between both is made through actuators. This technique simplified the experimental tests in wave basins where wind generation systems are not present. It also allowed studying the aerodynamics of FWT in wind tunnels. A review of previous hybrid testing campaigns was made, and commonly used actuators such as cables and propellers were analysed.

The advantages and deficiencies of the hybrid technique were discussed. It was concluded that hybrid testing is cheap and flexible, allowing the assessment of several turbine at different scale. However, the fidelity of tests depends on the accuracy of the numerical simulation, the tracked motions, and the calibration of the actuators. The actuation systems were then assessed discussing the simplicity of calibrating propellers compared to cables system, and the capacity of cable actuators to correctly emulate the moments. As a final point, the choose of actuators depends on the investigated phenomena and the objectives of the tests.

## ACKNOWLEDGEMENTS

The first author has been funded by the project Hydroelastic behaviour of horizontal flexible floating structures for applications to Floating Breakwaters and Wave Energy Converters (HYDROELAST-WEB), which is funded by the Portuguese Foundation for Science and Technology (Fundação para a Ciência e a Tecnologia - FCT) under contract 031488_770 (PTDC/ECI-EGC/31488/2017). The second author has been funded by the project ARCWIND -

Adaptation and implementation of floating wind energy conversion technology for the Atlantic region, which is co-financed by the European Regional Development Fund through the Interreg Atlantic Area Programme under contract EAPA 344/2016. This work contributes to the Strategic Research Plan of the Centre for Marine Technology and Ocean Engineering (CENTEC), which is financed by the Portuguese Foundation for Science and Technology (Fundação para a Ciência e Tecnologia - FCT) under contract UIDB/UIDP/00134/2020.

## REFERENCES

Achard, J.-L., Maurice, G., Balarac, G., & Barre, S. (2017). Floating vertical axis wind turbine — OWLWIND project. *2017 International Conference on ENERGY and ENVIRONMENT (CIEM)*, 216–220.

Ambrosini, S., Bayati, I., Facchinetti, A., & Belloli, M. (2020). Methodological and Technical Aspects of a Two-Degrees-of-Freedom Hardware-In-the-Loop Setup for Wind Tunnel Tests of Floating Systems. *Journal of Dynamic Systems, Measurement, and Control*, *142*(6), 061002.

Andersen, M. T. (2016). *Floating Foundations for Offshore Wind Turbines*. Aalborg Universitetsforlag. Ph.d.-serien for Det Teknisk-Naturvidenskabelige Fakultet, Aalborg Universitet

Antonutti, R., Poirier, J.-C., & Gueydon, S. (2020). Coupled Testing of Floating Wind Turbines in Waves and Wind Using Winches and Software-in-the-Loop. *Day 2 Tue, May 05, 2020*, D021S019R001.

Armesto, J. A., Jurado, A., Guanche, R., Couñago, B., Urbano, J., & Serna, J. (2018). TELWIND: Numerical Analysis of a Floating Wind Turbine Supported by a Two Bodies Platform. *Volume 10: Ocean Renewable Energy*, V010T09A073.

Arnal, V. (2020). *Experimental Modelling of a floating wind turbine using a 'software-in-the-loop' approach*. Ecole Central Nantes.

Arnal, V., Bonnefoy, F., Gilloteaux, J.-C., & Aubrun, S. (2019). Hybrid Model Testing of Floating Wind Turbines: Test Bench for System Identification and Performance Assessment. *Volume 10: Ocean Renewable Energy*, V010T09A078.

Azcona, J., Bouchotrouch, F., González, M., Garciandía, J., Munduate, X., Kelberlau, F., & Nygaard, T. A. (2014). Aerodynamic Thrust Modelling in Wave Tank Tests of Offshore Floating Wind Turbines Using a Ducted Fan. *Journal of Physics: Conference Series*, *524*, 012089.

Azcona, J., Bouchotrouch, F., & Vittori, F. (2019). Low-frequency dynamics of a floating wind turbine in wave tank–scaled experiments with SiL hybrid method. *Wind Energy*, *22*(10), 1402–1413.

Azcona, J., Lemmer, F., Matha, D., Amann, F., Bottasso, C. L., Montinari, P., Chassapoyannis, P., Diakakis, K., Voutsinas, S., Pereira, R., Bredmose, H., Mikkelsen, R., Laugesen, R., & Hansen, A. M. (2016). *D4.2.4: Results of wave tank tests* (Deliverable D4.2.4; Deliverable Reports on INNWIND.EU).

Bachynski, E. E., Chabaud, V., & Sauder, T. (2015). Real-time Hybrid Model Testing of Floating Wind Turbines: Sensitivity to Limited Actuation. *Energy Procedia*, *80*, 2–12.

Bachynski, E. E., Thys, M., Sauder, T., Chabaud, V., & Sæther, L. O. (2016). Real-Time Hybrid Model Testing of a Braceless Semi-Submersible Wind Turbine: Part II — Experimental Results. *Volume 6: Ocean Space Utilization; Ocean Renewable Energy*, V006T09A040.

Bagbanci, H., Karmakar, D., & Guedes Soares, C. (2012). Review of offshore floating wind turbines concepts. In C. Guedes Soares, Garbatov, Y., Sutulo, S., & Santos, T. A., (Eds.), *Maritime Engineering and Technology* (1st ed., pp. 553–562). Taylor & Francis Group.

Bak, C., Zahle, F., Bitsche, R., Kim, T., Yde, A., Henriksen, L. C., Hansen, M. H., Blasques., J. P. A. A., Gaunaa, M., & Natarajan, A. (2013). *The DTU 10-MW Reference Wind Turbine*. Sound/Visual production (digital).

Bayati, I., Belloli, M., Bernini, L., & Zasso, A. (2017). Wind Tunnel Wake Measurements of Floating Offshore Wind Turbines. *Energy Procedia*, *137*, 214–222.

Bayati, I., Belloli, M., Facchinetti, A., & Giappino, S. (2013). Wind Tunnel Tests on Floating Offshore Wind Turbines: A Proposal for Hardware-in-the-Loop Approach to Validate Numerical Codes. *Wind Engineering*, *37*(6), 557–568.

Bayati, I., Belloli, M., Ferrari, D., Fossati, F., & Giberti, H. (2014). Design of a 6-DoF Robotic Platform for Wind Tunnel Tests of Floating Wind Turbines. *Energy Procedia*, *53*, 313–323.

Bayati, I., Facchinetti, A., Fontanella, A., Giberti, H., & Belloli, M. (2018). A wind tunnel/HIL setup for integrated tests of Floating Offshore Wind Turbines. *Journal of Physics: Conference Series*, *1037*, 052025.

Belloli, M., Bayati, I., Facchinetti, A., Fontanella, A., Giberti, H., La Mura, F., Taruffi, F., & Zasso, A. (2020). A hybrid methodology for wind tunnel testing of floating offshore wind turbines. *Ocean Engineering*, *210*, 107592.

Berthelsen, P. A., Bachynski, E. E., Karimirad, M., & Thys, M. (2016). Real-Time Hybrid Model Tests of a Braceless Semi-Submersible Wind Turbine: Part III — Calibration of a Numerical Model. *Volume 6: Ocean Space Utilization; Ocean Renewable Energy*, V006T09A047.

Carrion, J. E., & Spencer, B. F., Jr. (2007). *Model-based Strategies for Real-time Hybrid Testing*.

Chabaud, V., Eliassen, L., Thys, M., & Sauder, T. (2018). Multiple-degree-of-freedom actuation of rotor loads in model testing of floating wind turbines using cable-driven parallel robots. *Journal of Physics: Conference Series*, *1104*, 012021.

Chabaud, V., Steen, S., & Skjetne, R. (2013). Real-Time Hybrid Testing for Marine Structures: Challenges and Strategies. *Volume 5: Ocean Engineering*, V005T06A021.

Chen, P., Chen, J., & Hu, Z. (2020). Review of Experimental-Numerical Methodologies and Challenges for Floating Offshore Wind Turbines. *Journal of Marine Science and Application*, *19*(3), 339–361.

Collu, M., & Borg, M. (2016). Design of floating offshore wind turbines. In *Offshore Wind Farms* (pp. 359–385). Elsevier.

de Ridder, E.-J., Otto, W., Zondervan, G.-J., Huijs, F., & Vaz, G. (2014). Development of a Scaled-Down Floating Wind Turbine for Offshore Basin Testing. *Volume 9A: Ocean Renewable Energy*, V09AT09A027.

Desmond, C., Hinrichs, J.-C., & Murphy, J. (2019). Uncertainty in the Physical Testing of Floating Wind Energy Platforms' Accuracy versus Precision. *Energies*, *12*(3), 435.

Díaz, H., & Guedes Soares, C. (2020). Review of the current status, technology and future trends of offshore wind farms. *Ocean Engineering*, *209*, 107381.

Duan, F., Hu, Z., & Niedzwecki, J. M. (2016). Model test investigation of a spar floating wind turbine. *Marine Structures*, *49*, 76–96.

Faltinsen, O. M. (1999). *Sea loads on ships and offshore structures*. Cambridge University Press.

Fowler, M. J., Kimball, R. W., Thomas, D. A., & Goupee, A. J. (2013). Design and Testing of Scale Model Wind Turbines for Use in Wind/Wave Basin Model Tests of Floating Offshore Wind Turbines. *Volume 8: Ocean Renewable Energy*, V008T09A004.

Galván, J., Sánchez-Lara, M. J., Mendikoa, I., Pérez-Morán, G., Nava, V., & Rodríguez-Arias, R. (2018). *NAUTILUS-DTU10* MW Floating Offshore Wind Turbine at Gulf of Maine: Public numerical models of an actively ballasted semisubmersible. *Journal of Physics: Conference Series*, *1102*, 012015.

Hall, M., & Goupee, A. J. (2018). Validation of a hybrid modeling approach to floating wind turbine basin testing: Validation of a hybrid modeling approach to floating wind turbine basin testing. *Wind Energy*, *21*(6), 391–408.

Hall, M., Moreno, J., & Thiagarajan, K. (2014). Performance Specifications for Real-Time Hybrid Testing of 1:50-Scale Floating Wind Turbine Models. *Volume 9B: Ocean Renewable Energy*, V09BT09A047.

Hmedi, M. (2020). *Development of a Wind Turbine Emulator for the Experimental Testing of Floating Offshore Wind Turbine in a Wave Tank*. Ecole Central Nantes.

Jonkman, J. (2008, January 7). Influence of Control on the Pitch Damping of a Floating Wind Turbine. *46th AIAA Aerospace Sciences Meeting and Exhibit*. 46th AIAA Aerospace Sciences Meeting and Exhibit, Reno, Nevada.

Jonkman, J., & Buhl Jr., M. L. (2005). *FAST User's Guide* (NREL/TP-500-38230). National Renewable Energy Laboratory (NREL).

Kanner, S., Yeung, R. W., & Koukina, E. (2016). Hybrid testing of model-scale floating wind turbines using autonomous actuation and control. *OCEANS 2016 MTS/IEEE Monterey*, 1–6.

Lowe, R. J., & Drummond, P. (2022). Solar, wind and logistic substitution in global energy supply to 2050 – Barriers and implications. *Renewable and Sustainable Energy Reviews*, *153*, 111720.

Matoug, C., Augier, B., Paillard, B., Maurice, G., Sicot, C., & Barre, S. (2020). An hybrid approach for the comparison of VAWT and HAWT performances for floating offshore wind turbines. *Journal of Physics: Conference Series*, *1618*(3), 032026.

Müller, K., Sandner, F., Bredmose, H., Azcona, J., Manjock, A., & Pereira, R. (2014). *Improved tank test procedures for scaled floating offshore wind turbines*.

Oguz, E., Clelland, D., Day, A. H., Incecik, A., López, J. A., Sánchez, G., & Almeria, G. G. (2018). Experimental and numerical analysis of a TLP floating offshore wind turbine. *Ocean Engineering*, *147*, 591–605.

Otter, A., Murphy, J., & Desmond, C. J. (2020). Emulating aerodynamic forces and moments for hybrid testing of floating wind turbine models. *Journal of Physics: Conference Series*, *1618*, 032022.

Otter, A., Murphy, J., Pakrashi, V., Robertson, A., & Desmond, C. (2021). A review of modelling techniques for floating offshore wind turbines. *Wind Energy*, we. 2701.

Pao, L. Y., & Johnson, K. E. (2011). Control of Wind Turbines. *IEEE Control Systems*, *31*(2), 44–62.

Pegalajar-Jurado, A., Bredmose, H., Borg, M., Straume, J. G., Landbø, T., Andersen, H. S., Yu, W., Müller, K., & Lemmer, F. (2018). State-of-the-art model for the LIFES50+ OO-Star Wind Floater Semi 10MW floating wind turbine. *Journal of Physics: Conference Series*, *1104*, 012024.

Pires, O., Azcona, J., Vittori, F., Bayati, I., Gueydon, S., Fontanella, A., Liu, Y., de Ridder, E. J., Belloli, M., & van Wingerden, J. W. (2020). Inclusion of rotor moments in scaled wave tank test of a floating wind turbine using SiL hybrid method. *Journal of Physics: Conference Series*, *1618*(3), 032048.

Robertson, A., Jonkman, J. M., Goupee, A. J., Coulling, A. J., Prowell, I., Browning, J., Masciola, M. D., & Molta, P. (2013). Summary of Conclusions and Recommendations Drawn From the DeepCwind Scaled Floating Offshore Wind System Test Campaign. *Volume 8: Ocean Renewable Energy*, V008T09A053.

Rockel, S., Camp, E., Schmidt, J., Peinke, J., Cal, R., & Hölling, M. (2014). Experimental Study on Influence of Pitch Motion on the Wake of a Floating Wind Turbine Model. *Energies*, *7*(4), 1954–1985.

Rockel, S., Peinke, J., Hölling, M., & Cal, R. B. (2016). Wake to wake interaction of floating wind turbine models in free pitch motion: An eddy viscosity and mixing length approach. *Renewable Energy*, *85*, 666–676.

Roddier, D., Cermelli, C., Aubault, A., & Weinstein, A. (2010). WindFloat: A floating foundation for offshore wind turbines. *Journal of Renewable and Sustainable Energy*, *2*(3), 033104.

Sauder, T., Chabaud, V., Thys, M., Bachynski, E. E., & Sæther, L. O. (2016). Real-Time Hybrid Model Testing of a Braceless Semi-Submersible Wind Turbine: Part I — The Hybrid Approach. *Volume 6: Ocean Space Utilization; Ocean Renewable Energy*, V006T09A039.

Schliffke, B., Aubrun, S., & Conan, B. (2020). Wind Tunnel Study of a "Floating" Wind Turbine's Wake in an Atmospheric Boundary Layer with Imposed Characteristic Surge Motion. *Journal of Physics: Conference Series*, *1618*(6), 062015.

Skaare, B., Nielsen, F. G., Hanson, T. D., Yttervik, R., Havmøller, O., & Rekdal, A. (2015). Analysis of measurements and simulations from the Hywind Demo floating wind turbine: Dynamic analysis of the Hywind Demo floating wind turbine. *Wind Energy*, *18*(6), 1105–1122.

Stewart, G., & Muskulus, M. (2016a). A Review and Comparison of Floating Offshore Wind Turbine Model Experiments. *Energy Procedia*, *94*, 227–231.

Stewart, G., & Muskulus, M. (2016b). Aerodynamic Simulation of the MARINTEK Braceless Semisubmersible Wave Tank Tests. *Journal of Physics: Conference Series*, *749*, 012012.

Thys, M., Chabaud, V., Sauder, T., Eliassen, L., Sæther, L., & Magnussen, Ø. (2018). *Real-Time Hybrid Model Testing of a Semi-Submersible 10MW Floating Wind Turbine and Advances in the Test Method*.

Urbán, A. M., & Guanche, R. (2019). Wind turbine aerodynamics scale-modeling for floating offshore wind platform testing. *Journal of Wind Engineering and Industrial Aerodynamics*, *186*, 49–57.

Utsunomiya, T., Sato, T., Matsukuma, H., & Yago, K. (2009). Experimental Validation for Motion of a SPAR-Type

Floating Offshore Wind Turbine Using 1/22.5 Scale Model. *Volume 4: Ocean Engineering; Ocean Renewable Energy; Ocean Space Utilization, Parts A and B*, 951–959.

Uzunoglu, E., & Guedes Soares, C. (2020). Hydrodynamic design of a free-float capable tension leg platform for a 10 MW wind turbine. *Ocean Engineering*, *197*, 106888.

Uzunoglu, E., Karmakar, D., & Guedes Soares, C. (2016). Floating Offshore Wind Platforms. In L. Castro-Santos & V. Diaz-Casas (Eds.), *Floating Offshore Wind Farms* (pp. 53–76). Springer International Publishing.

Vittori, F., Bouchotrouch, F., Lemmer, F., & Azcona, J. (2018). Hybrid Scaled Testing of a 5MW Floating Wind Turbine Using the SiL Method Compared With Numerical Models. *Volume 10: Ocean Renewable Energy*, V010T09A082.

Vittori, F., Pires, O., Azcona, J., Uzunoglu, E., Guedes Soares, C., Zamora Rodríguez, R., & Souto-Iglesias, A. (2020). Hybrid scaled testing of a 10MW TLP floating wind turbine using the SiL method to integrate the rotor thrust and moments. In C. Guedes Soares (Ed.), *Developments in Renewable Energies Offshore* (1st ed., pp. 417–423). CRC Press.

Wright, C., O'Sullivan, K., Murphy, J., & Pakrashi, V. (2015). Experimental Comparison of Dynamic Responses of a Tension Moored Floating Wind Turbine Platform with and without Spring Dampers. *Journal of Physics: Conference Series*, *628*, 012056.

Yu, W., Lemmer, F., Bredmose, H., Borg, M., Pegalajar-Jurado, A., Mikkelsen, R. F., Larsen, T. S., Fjelstrup, T., Lomholt, A. K., Boehm, L., Schlipf, D., Armendariz, J. A., & Cheng, P. W. (2017). The Triple Spar Campaign: Implementation and Test of a Blade Pitch Controller on a Scaled Floating Wind Turbine Model. *Energy Procedia*, *137*, 323–338.

Zamora-Rodriguez, R., Gomez-Alonso, P., Amate-Lopez, J., De-Diego-Martin, V., Dinoi, P., Simos, A. N., & Souto-Iglesias, A. (2014). Model Scale Analysis of a TLP Floating Offshore Wind Turbine. *Volume 9B: Ocean Renewable Energy*, V09BT09A016.

*Trends in Maritime Technology and Engineering – Guedes Soares & Santos (eds)*


# A machine visual-based ship monitoring system for offshore wind farms

A. Huiting Ji
*Jimei University, Xiamen, Fujian, China*

B. Qing Yu
*Jimei University, Xiamen, Fujian, China, Hubei Key Laboratory of Inland Shipping Technology, Wuhan, China*

C. Chen Wei
*City University of Macau, Macau, China*

D. Yutian Hu & Tingting Lin
*Jimei University, Xiamen, Fujian, China*

ABSTRACT: Ship allision is a crucial topic for offshore wind farms (OWFs) due to catastrophic consequences. To safeguard the OWF facilities, operators adopt CCTV to detect and monitor passing ships in the waters near OWFs manually. This paper proposed an automatic ship detection system for offshore wind farms on the basis of machine vision technologies. In the system, a YOLOv3 algorithm is used to track and identify ships. A stereo vision algorithm is applied to locate ship positions. The system is tested in a real scenario and validated by comparing the visual location results and automatic identification system data. Extensive experiments demonstrate that the detection system of ships can automatically and accurately detect and locate ships. The potential uses of the system include ship invasion and allision avoidance.

## 1 INTRODUCTION

Numerous offshore wind farms (OWFs) are installed worldwide, not only proving their advantages of higher wind resources but also creating a potential danger for passing ships in the nearby waters. Recently, some collision accidents have been reported in Germany, UK and China, leading to shipping hull and turbine damage, electric power loss, especially for construction and fishing ships.

Currently, automatic identification systems (AIS), radar and video surveillance are the most popular approaches to detect ships in the waters near OWFs. In current studies, AIS can be considered as an efficient method to monitor surrounding ships. However, it has some limitations. For instance, some small ships (e.g., fishing activity, entertainment) are not obligated to install AIS facilities; However, under some unpredict circumstance, the vessel cannot be detected via AIS, such as turn off the AIS equipment, transmission interference etc. The radar detects objects in a physical way by using returning echoes but fails to prove detailed information such as ship type, ship size and ship states. Benefitting from machine visual technology, remote video surveillance becomes a novel way in maritime fields. The video system can compensate for further information that unrecognized by radar and AIS. For instance, video surveillance is able to prove real-time video for users to better understand the surrounding environments. It is an aid for invasion warning in OWFs, highly relay on operator experience and hard to remain 24 hours monitoring.

To this end, this study aims to propose a ship detecting and locating system that is purely based on real-time videos from OWFs. This system uses stereo-pair cameras to capture objects, identifies and tracks objects with a "YOLOv3" machine visual technology (Redmon & Farhadi, 2018) and locates objects with stereo vision approaches.

The remainder of this paper is structured as follows: previous studies are reviewed in Section 2. The ship detecting approaches, locating approaches in the system are introduced in Section 3 and Section 4, respectively. Section 5 drew the conclusion of the study.

## 2 RELATED WORKS

In the maritime field, ships can be determined and located with machine visuals when the real-time video is available. There are many traditional ship detection algorithms that can be considered for ship

DOI: 10.1201/9781003320289-45

detection. Such as constant false-alarm rate (CFAR), which bases on the intensity echoes from differences between ships and sea clutter (Frery et al., 1997; Liu et al., 2019; T. Zhang et al., 2020), but it is difficult for CFAR to efficiently process large collections of remote sensing imagery (Cui et al., 2012; Wang et al., 2013), fails to incorporate spectra and textures and leads to lower detection effects. In recent years, object detection based on deep learning methods becomes a hot research topic. It can extract deeper features and richer semantic information from video images. There are two major branches of object detection based on deep learning. The first branch is a two-stage detection algorithm: generating potential bounding boxes through the window sliding or regional proposal networks (RPN) and then completing the prediction of the target location and category through the detection network. For example, region-based CNN (R-CNN) (Girshick et al., 2014), fast R-CNN (Girshick, 2015), faster R-CNN (Ren et al., 2015). However, the CNN based ship detection models are less time efficient due to complicated data processing. To improve it, one-stage detection models are becoming more popular in the domain. They reframe object detection as a single regression problem. You only look once (YOLO) (Redmon et al., 2016; Redmon & Farhadi, 2017, 2018) is the most representative work with real-time speed which divides the image into equal grids and makes multi-class and multi-scale predictions for each grid cell. YOLOv2 (Redmon & Farhadi, 2017) and YOLOv3 (Redmon & Farhadi, 2018) are presented in succession by researchers to improve the performance of YOLO. They generate detection results (include ship positions, category) with a deeper convolutional model. Compared with other region-based methods, it has faster speed and better ability for feature extraction.

Unlike the techniques based on the sliding window and region proposal method, YOLO sees the entire image during detection, making it has fewer background errors. Furthermore, each individual component must be trained separately in a two-stage detection algorithm which leads to slow speed and difficulty optimizing. However, it is important to improve the detection speed in addition to accuracy for ship detection. In terms of the relatively balanced performance in detection accuracy and speed, YOLOv3 is fast and accurate object detection. Therefore, this paper considers applying the YOLOv3 model to the detection of ships.

The uses of stereo vision can be found in varies fields (Faugeras & Faugeras, 1993; Hartley & Zisserman, 2013; Jain et al., 1995; Trucco & Verri, 1998). Specific to the maritime domain, some researchers apply the real-time, stereo camera to inspect ship hull (Schattschneider et al., 2011), automatic berthing system (Nomura et al., 2021) or to assist marine automatic loaders. For instance, Mi et al. (2015) use projection positioning methods with the aid of stereo vision technologies develop an automated ship loader (Mi et al., 2015). In terms of ship navigation, Zheng et al.(2020) introduce a vision-based system to optimize the unmanned ships environment awareness. A difficult of obtaining the ship position through the image is highlighted when using the a single-vision system (Martins et al., 2007; Shimpo et al., 2005). To overcome such difficulties, the stereo vision system is proposed to detect and localize the ship by Yamamoto & Win (2006), and also be used in ship tracking (Anon, 2013).

As the advantages of using stereo vision are highlighted in the previous studies, a ships detection and locating system can be developed with aids of stereo vision and YOLOv3 algorithms.

## 3 SHIP DETECTING

This section dedicates the dataset and the methods used to develop the ship detection system. Section 3.1 explains the dataset used in the system. And Section 3.2 introduces the algorithm and the methods used in the system.

### 3.1 *Dataset*

The convolutional layers need to be pretrained on the dataset. YOLOv3 requires a large dataset of images to provide generalization and robust performance. The dataset contains a total of 7000 images, covers 6 distinct ship categories of ore carrier, bulk cargo carrier, general cargo ship, container ship, fishing boat and passenger ship. Each category holds 1167 images and is labelled (e.g., categories, coordinate position in the images) that constitute the YOLO model training process.

A tuple contains five attributes, which are categories, $X_{min}$, $X_{max}$, $Y_{min}$ and $Y_{min}$. $X_{min}$, $X_{max}$,$Y_{min}$ and $Y_{min}$ are the labeled box coordinates of a ship with the x-axis and y-axis values in the images. The categories are assigned by using the above-mentioned 6 ship categories. Table 1 gives some examples of labelling ships under the YOLO-configuration format.

Table 1. Examples of YOLO-labeled data file.

| | $X_{min}$ | $Y_{min}$ | $X_{max}$ | $Y_{max}$ |
|---|---|---|---|---|
| Setting | pixel | pixel | pixel | pixel |
| ore carrier | 633 | 467 | 944 | 510 |
| bulk cargo carrier | 411 | 498 | 1023 | 546 |
| general cargo ship | 622 | 508 | 1720 | 605 |
| container ship | 106 | 512 | 1454 | 546 |
| fishing boat | 1268 | 486 | 1920 | 564 |
| passenger ship | 518 | 512 | 1017 | 574 |

### 3.2 *Ship detection methods*

YOLOv3's structure can be divided into two main parts: a backbone network and a detection network as shown in Figure 1.

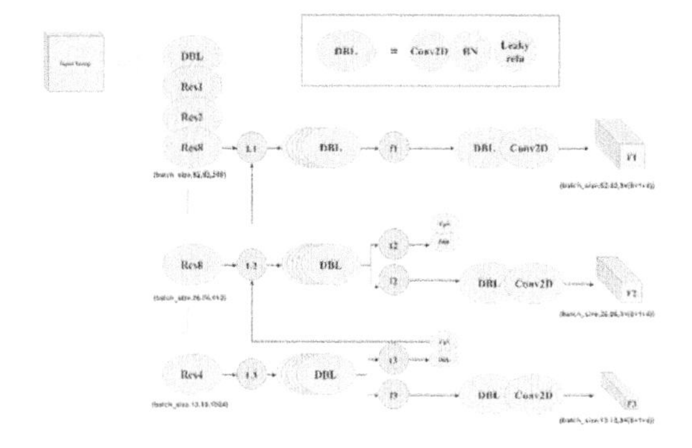


Figure 1. The network structure of YOLOv3.

The backbone network of YOLOv3 is called DarkNet-53 which is composed of convolution layers. Each convolution layer is followed by a batch normalization (BN) layer and an activation layer. Darknet-53 outputs three different scales of feature layers to describe the feature of the objects, $L_1$ is the middle layer, $L_2$ is the middle and lower layer and $L_3$ is the bottom layer. The layers can then be used in the detection network.

With the DarkNet-53, the detection network will obtain three different scales of feature layers. The feature layers of the three scales are $13\times 13$ (large target), $26\times 26$ (middle target) and $52\times 52$ (small target), respectively. Then the feature parameters $(L_1, L_2, L_3)$ are converted into prediction results during the reinforce feature extraction process. $L_3$ is divided into two parts of $f_3$ and $t_3$ after extraction, the $f_3$ is used to adjust the predict result $F_3$ after two convolutions while $t_3$ splices the $L_2$ through convolution and up-sampling process as shown in Figure 1 Similarly, $L_1$ and $L_2$ can be used to acquire their prediction results $F_1$ and $F_2$ through a series of convolutions and up-samplings.

As above, by using a similar concept of feature pyramid network (FPN), the predict results $F_1$, $F_2$, $F_3$ are used to encode the bounding box of the detected objects with aids of calculating their 3D tensor. In the meantime, the object categories can be predicted. In this study, we defined the size of the 3D tensor is $N\times N\times[3\times(6+1+4)]$, which means the tensor consists of 4 bounding box offsets ($c_x$, $c_y$, $p_w$, $p_h$), 1 prediction and 6 category predictions. $c_x$, $c_y$ is the offset distance between midpoint of the ship and the benchmark point (i.e., the Upper-left point in the image), $p_w$ and $p_h$ define the size of the anchor prior boxes. The predicted results can be then decoded as follows:

$$b_x = \sigma(t_x) + c_x \tag{1}$$

$$b_y = \sigma(t_y) + c_y \tag{2}$$

$$b_w = p_w e^{t_w} \tag{3}$$

$$b_h = p_h e^{t_h} \tag{4}$$

Where ($c_x$, $c_y$) represents the benchmark point and $p_w$, $p_h$ are the width and height of the prior anchor box. The ship can be therefore detected with their predicted result ($b_x$, $b_y$) and $b_w$, $b_h$.

After obtaining the final prediction results, non-maximal suppression screening should be carried out to fix the multiple detections, it is a common part of all target detections. In general, we detected each frame of the video, obtained the position of the ship in the picture and framed it in the part of detection.

## 4 SHIP LOCATING

This study uses stereoscopic vision technology to achieve ship locating. Stereoscopic vision is a machine vision technique that is used to deduce an object from two or more images. The internal and extrinsic parameters of two cameras can be obtained by calibrating the left and right cameras, simultaneously (T. Zhang et al., 2020). Thereby the positions of the ship are identified and captured. It is noted that we applied a rectangle prediction box in the system, the ship position is located in the midpoint of the bottom edge in the box. The location process is introduced as follows.

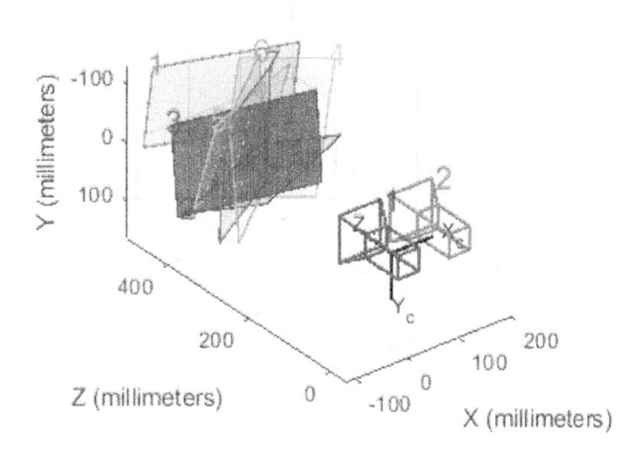


Figure 2. Checkerboard calibration.

### 4.1 *Camera parameters*

In order to obtain the relative position relations p between any two coordinate systems, the rotation matrix ***R*** and the translation vector ***T*** are need to be acquired by calibrating the left and right cameras, simultaneously. The checkerboard is rotated to different angles as shown in Figure 2, then the parameters are obtained by the "stereoCameraCalibrator"

Table 2. Calibration parameters.

| | Intrinsic Matrix | Rotation | Translation | Distortion |
|---|---|---|---|---|
| Setting | pixel | m | m | m |
| Left camera | $\begin{bmatrix} 534.31 & 0.00 & 342.64 \\ 0.00 & 534.31 & 234.42 \\ 0.00 & 0.00 & 1.00 \end{bmatrix}$ | $\begin{bmatrix} 1.00 & 1.00 & 1.00 \\ 1.00 & 1.00 & 1.00 \\ 1.00 & 1.00 & 1.00 \end{bmatrix}$ | $\begin{bmatrix} 1.00 \\ 1.00 \\ 1.00 \end{bmatrix}$ | $\begin{bmatrix} -0.29 \\ 0.11 \\ 0.00 \end{bmatrix}$ and $\begin{bmatrix} 0.00 \\ 0.00 \end{bmatrix}$ |
| Right camera | $\begin{bmatrix} 537.39 & 0.00 & 326.62 \\ 0.00 & 537.01 & 250.58 \\ 0.00 & 0.00 & 1.00 \end{bmatrix}$ | $\begin{bmatrix} 1.00 & 0.00 & 0.01 \\ -0.00 & 1.00 & 250.58 \\ -0.01 & 0.01 & 1.00 \end{bmatrix}$ | $\begin{bmatrix} -99.72 \\ 1.27 \\ 0.05 \end{bmatrix}$ | $\begin{bmatrix} -0.29 \\ 0.10 \\ 0.00 \end{bmatrix}$ and $\begin{bmatrix} 0.00 \\ 0.00 \end{bmatrix}$ |

tools of MATLAB. In this study, we assume the camera parameters as shown in Table 2.

### 4.2 *Distortion correction*

During the distortion correction process, Brown-Conrady model is applied (Clarke & Fryer, 1998; Duane, 1971). In the Brown-Conrady model, the lens distortion is classified as a radial distortion and a tangential distortion, in which the radial component is predominant for lens distortion and be used in this paper.

Then the coordinates used in the detection process are transformed to an image coordinate system by using the following equations:

$$y = \frac{u - u_0}{f_y} \tag{5}$$

$$x = \frac{v - v_0}{f_x} \tag{6}$$

where $(u, v)$ are the ship coordinates obtained by the detection part, $u_0$ and $v_0$ are the actual positions of the camera main points, $f_x$ and $f_y$ are the unit length of focal length on the $x$and $y$axes.

Then the transformed coordinates can be formulated as follows:

$$x_{corrected} = x\left(1 + k_1 r^2 + k_2 r^4 + k_3 r^6\right) \tag{7}$$

$$y_{corrected} = y\left(1 + k_1 r^2 + k_2 r^4 + k_3 r^6\right) \tag{8}$$

$$r^2 = x^2 + y^2 \tag{9}$$

where $r$ is the radius coordinates in polar coordinates.

After transformed the lenses distortions, a pixel coordinate system is developed:

$$u_{corrected} = x_{corrected} \cdot f_x + u_0 \tag{10}$$

$$v_{corrected} = y_{corrected} \cdot f_y + v_0 \tag{11}$$

where $u_{corrected}$ and $v_{corrected}$ are the corrected pixel coordinates.

### 4.3 *Ship locating by using stereo vision*

This study assumes that the camera are well verifies and their perspective transformation are satisfied with the center $V_C$, so an orthonormal camera coordinate system ($V_C$, $I_C$, $J_C$, $K_C$) is developed, in which $V$ is the optical center of the camera, $I_C$ and $J_C$ are the unit vectors of the image plane in the line and column directions, respectively, $K_C$ is the optical axis of the camera and $(O, I, J, K)$ is the world coordinate system.

The rotation matrix $\boldsymbol{R}$ and translation vector $\boldsymbol{T}$ are famulated:

$$\begin{bmatrix} X_w \\ Y_w \\ Z_w \end{bmatrix} = R \begin{bmatrix} X_c \\ Y_c \\ Z_c \end{bmatrix} + T \tag{12}$$

where ($X_W$, $Y_W$, $Z_W$) is a certain point in the world coordinate system, ($X_C$, $Y_C$, $Z_C$) is the same point in the camera coordinate system.

The relationship between any world point $M$ of coordinates $(x, y, z)$ and its corresponding image $m$ of coordinates $(x', y', f_c)$ can be formulated as follows:

$$x' = f_c \frac{x}{z} \tag{13}$$

$$y' = f_c \frac{y}{z} \tag{14}$$

where $f_c$ denotes the focal length of the camera. Then,

$$u = \frac{x}{dx} + u_0 \quad (15)$$

$$v = \frac{y}{dy} + v_0 \quad (16)$$

where $(u, v)$ are $m$ coordinates expressed in the pixel system, $d_x$, $d_y$ is the focal length and $(u_0, v_0)$ is a point in the image coordinate system. Note that the origin of the image coordinate system is the intersection of the camera optical axis and the imaging plane (Zhang, 2000).

As above, the relationship between the coordinate $(u, v)$ in the pixel coordinates system and the coordinate $(X, Y, Z)$ in the world coordinate system can be formulated as follows:

$$s\begin{bmatrix} u \\ v \\ 1 \end{bmatrix} = \begin{bmatrix} \frac{1}{dx} & 0 & u_0 \\ 0 & \frac{1}{dy} & v_0 \\ 0 & 0 & 1 \end{bmatrix} \cdot \begin{bmatrix} f & 0 & 0 & 0 \\ 0 & f & 0 & 0 \\ 0 & 0 & 0 & 1 \end{bmatrix} \cdot \begin{bmatrix} R & T \\ 0 & 1 \end{bmatrix} \cdot \begin{bmatrix} X \\ Y \\ Z \end{bmatrix} \quad (17)$$

where $s$ is an arbitrary scale factor.

As the camera is setting on offshore wind power plants, it is assumed that the origin of the world coordinate system at the light center of the left camera. As results, the camera coordinate system of the left camera is selected as the world coordinate system. The coordinates of the ship in the pixel coordinate system are obtained from the detection part, after distortion correction, the coordinate of the ship in the left camera and right camera are $(u_l, r_l)$ and $(u_r, r_r)$, the coordinate of the ship in the world coordinate system $(X, Y, Z)$ is obtained through the following calculation:

$$s\begin{bmatrix} u_l \\ v_l \\ 1 \end{bmatrix} = K_l \begin{bmatrix} R_{lw} & T_{lw} \\ \vec{0} & 1 \end{bmatrix} \begin{bmatrix} X \\ Y \\ Z \end{bmatrix} \quad (18)$$

$$s\begin{bmatrix} u_r \\ v_r \\ 1 \end{bmatrix} = K_r \begin{bmatrix} R_{rw} & T_{rw} \\ \vec{0} & 1 \end{bmatrix} \begin{bmatrix} X \\ Y \\ Z \end{bmatrix} \quad (19)$$

where, in general:

$$K = \begin{bmatrix} f_x & 0 & u_0 \\ 0 & f_y & v_0 \\ 0 & 0 & 1 \end{bmatrix} \quad (20)$$

The values of $u_l$, $v_l$, $u_r$ and $v_r$ are the points projected on the image plane, $\boldsymbol{R_{lw}}$ is a rotation matrix from left camera coordination system to world coordination system, so it is equivalent to the $\boldsymbol{0}$ matrix. $\boldsymbol{T_{lw}}$ is the translation vector from left camera coordination system to world coordination system, so it is equivalent to the $\boldsymbol{E}$ matrix. $\boldsymbol{R_{rw}}$ is a rotation matrix from right camera coordination system to world coordination system, $\boldsymbol{T_{rw}}$ is the translation vector from right camera coordination system to world coordination system. $\boldsymbol{K}$ is the internal parameter matrix obtained in Section 4.1.

By following the abovementioned process, the ship position can be well located through a stereoscopic version.

## 5 CONCLUSIONS

This paper introduces a ship detecting and positioning system based on the YOLOv3 algorithm and stereo vision technologies. The framework and the detailed methods used in the system are introduced. Although further tests are required to verify the accuracy of ship detection and locating, for instance, using AIS data, the novel system shares an idea of using machine version technology for ship allision prevention. Not only ease the OWF operators' workload during the CCTV monitoring but also delegate a possible way for ship traffic management in the water in the vicinity of OWFs.

## ACKNOWLEDGEMENTS

The work is supported by the Fund of Hubei Key Laboratory of Inland Shipping Technology (NHHY2021001) and the National Natural Science Foundation of China (NSFC) (Grant No. 52031009).

## REFERENCES

Clarke, T. A., & Fryer, J. G. (1998). The development of camera calibration methods and models. *The Photogrammetric Record*, *16*(91), 51–66.

Cui, Y., Yang, J., Yamaguchi, Y., Singh, G., Park, S.-E., & Kobayashi, H. (2012). On semiparametric clutter estimation for ship detection in synthetic aperture radar images. *IEEE Transactions on Geoscience and Remote Sensing*, *51*(5), 3170–3180.

Detection and tracking of ships using a stereo vision system. (2013). *Scientific Research and Essays*, *8*(7), 288–303. http://dx.doi.org/10.5897/SRE12.318

Duane, C. B. (1971). Close-range camera calibration. *Photogramm. Eng*, *37*(8), 855–866.

Faugeras, O., & Faugeras, O. A. (1993). *Three-dimensional computer vision: a geometric viewpoint*. MIT press.

Frery, A. C., Muller, H.-J., Yanasse, C. da C. F., & Sant'Anna, S. J. S. (1997). A model for extremely heterogeneous clutter. *IEEE Transactions on Geoscience and Remote Sensing*, *35*(3), 648–659.

Girshick, R. (2015). Fast r-cnn. *Proceedings of the IEEE International Conference on Computer Vision*, 1440–1448.

Girshick, R., Donahue, J., Darrell, T., & Malik, J. (2014). Rich feature hierarchies for accurate object detection and semantic segmentation. *Proceedings of the IEEE Conference on Computer Vision and Pattern Recognition*, 580–587.

Hartley, R., & Zisserman, A. (2013). Multiple view geometry in computer vision (cambridge university, 2003). *C1 C3, 2*.

Jain, R., Kasturi, R., & Schunck, B. G. (1995). *Machine vision* (Vol. 5). McGraw-hill New York.

Liu, T., Yang, Z., Yang, J., & Gao, G. (2019). CFAR ship detection methods using compact polarimetric SAR in a K-Wishart distribution. *IEEE Journal of Selected Topics in Applied Earth Observations and Remote Sensing, 12* (10), 3737–3745.

Martins, A., Almeida, J. M., Ferreira, H., Silva, H., Dias, N., Dias, A., Almeida, C., & Silva, E. P. (2007). Autonomous surface vehicle docking manoeuvre with visual information. *Proceedings 2007 IEEE International Conference on Robotics and Automation*, 4994–4999.

Mi, C., Shen, Y., Mi, W., & Huang, Y. (2015). Ship identification algorithm based on 3D point cloud for automated ship loaders. *Journal of Coastal Research, 73 (10073)*, 28–34.

Nomura, Y., Yamamoto, S., & Hashimoto, T. (2021). Study of 3D measurement of ships using dense stereo vision: towards application in automatic berthing systems. *Journal of Marine Science and Technology (Japan), 26*(2), 573–581. http://dx.doi.org/10.1007/s00773-020-00761-2

Redmon, J., Divvala, S., Girshick, R., & Farhadi, A. (2016). You only look once: Unified, real-time object detection. *Proceedings of the IEEE Conference on Computer Vision and Pattern Recognition*, 779–788.

Redmon, J., & Farhadi, A. (2017). YOLO9000: better, faster, stronger. *Proceedings of the IEEE Conference on Computer Vision and Pattern Recognition*, 7263–7271.

Redmon, J., & Farhadi, A. (2018). *YOLOv3: An Incremental Improvement*. http://arxiv.org/abs/1804.02767

Ren, S., He, K., Girshick, R., & Sun, J. (2015). Faster r-cnn: Towards real-time object detection with region proposal networks. *Advances in Neural Information Processing Systems*, 28.

Schattschneider, R., Maurino, G., & Wang, W. (2011). Towards stereo vision slam based pose estimation for ship hull inspection. *OCEANS'11 MTS/IEEE KONA*, 1–8.

Shimpo, M., Hirasawa, M., & Oshima, M. (2005). Detection and tracking method of moving ships through navigational image sequence. J. Japan Inst. *Navigation, 113*, 115–126.

Trucco, E., & Verri, A. (1998). *Introductory techniques for 3-D computer vision* (Vol. 201). Prentice hall Englewood Cliffs.

Wang, C., Jiang, S., Zhang, H., Wu, F., & Zhang, B. (2013). Ship detection for high-resolution SAR images based on feature analysis. *IEEE Geoscience and Remote Sensing Letters, 11*(1), 119–123.

Yamamoto, S., & Win, M. T. (2006). A feasibility study on ship detection using stereo vision. *Proc. of Techno-Ocean 2006/19th JASNAOE Ocean Engineering Symposium*, 38.

Zhang, T., Yang, Z., Gan, H., Xiang, D., Zhu, S., & Yang, J. (2020). PolSAR ship detection using the joint polarimetric information. *IEEE Transactions on Geoscience and Remote Sensing, 58*(11), 8225–8241.

Zhang, Z. (2000). A Flexible New Technique for Camera Calibration. In *IEEE TRANSACTIONS ON PATTERN ANALYSIS AND MACHINE INTELLIGENCE* (Vol. 22, Issue 11).

Zheng, X., Wang, X., Luo, Z., & Guo, B. (2020). Camera Imaging Calibration Optimization for Stereo Vision System of Marine Unmanned Ship. *Journal of Coastal Research, 111*(SI), 288–292.

*Trends in Maritime Technology and Engineering – Guedes Soares & Santos (eds)*


# Uncertainty analysis in the frequency domain simulation of a hinged wave energy converter

M. Kamarlouei & C. Guedes Soares
*Centre for Marine Technology and Ocean Engineering (CENTEC), Instituto Superior Técnico, Universidade de Lisboa, Lisbon, Portugal*

ABSTRACT: This paper presents the uncertainty analysis of a numerical approach applied for evaluating the power absorption of a hinged wave energy converter. The main goal of this paper is to uncover the effects of uncertainties in the absorbed power as the final numerical model output. This model is used for the validation of the experimental study. The numerical model is created according to the size and scale of a model used in a test campaign using boundary element method software to calculate the hydrodynamic forces and moments in surge, heave, and pitch degrees of freedom. The results are then transformed to the pitch motion around hinged point. This process includes uncertainties associated with various geometrical and testing parameters. Type A and Type B uncertainties are identified in the model inputs based on to their sources. Then, Monte Carlo simulation approach is utilized to analyse the propagation of uncertainty in the input parameters to the model output which is the average absorbed power of the device.

## 1 INTRODUCTION

Uncertainty on the performance of Wave Energy Converters (WECs) may significantly affect the financing opportunities for projects and increases cost of capital for commercialized ones. One of the main objectives of tank testing of WECs is to reasonably assess the power performance, survivability, and reliability of a particular concept. For instance, the performance assessment and survivability evaluation are the critical inputs to evaluate potential revenues and levelized cost of energy (LCoE) for wave energy devices.

Going to the higher scale of testing, up to open-sea testing stage, is expensive and concepts at this level should have promising numerical and low scale test results. Thus, the uncertainty of achieved results should be studied considering all the possible errors in measurements, prototyping, model imperfections, and mathematical assumptions. These types of uncertainty analysis are performed in several projects (Berque et al., 2016; Hiles et al., 2016; Orphin et al., 2018).

On the other hand, numerical modelling of WEC arrays, especially when combined with floating offshore wind turbines, is very complex and time consuming. The complexity is mainly related to the simulation of interactions between the floating bodies and incident waves. Thus, initial study of concepts may start with the small-scale physical model tests in the basin emulators. In parallel, smaller elements of the model may be validated using low – or – high fidelity numerical simulators (Hallak et al., 2018).

For instance, Figure 1(a) shows the small-scale floating platform model including the wind turbine and point absorber WECs (Gaspar, et al., 2021; Kamarlouei et al., 2020). This model is tested in the basin and intended to study the effect of WECs and wind turbine interactions with the dynamics of platform in different modes of operation. Numerical simulation of such system using available open-source tools, requires high number of assumptions and simplifications. Thus, it is more recommended to build a physical model and test it the ocean basin. However, numerical simulation of smaller components such as WEC module, platform structure, or wind turbine is still viable and provides better understanding of involved nonlinearities in the model test. In this case, various studies are available on the dynamics of point absorbers of different shapes (Berenjkoob et al., 2021; Sinha et al., 2015).

The uncertainty can be the model uncertainty (Guedes Soares, 1997) or it can be measurement uncertainty as specified in the ITTC documents. The model uncertainty is applied to the prediction of the calculation models and the measurement uncertainty to the laboratory measurements.

The uncertainty of any floating structure depends on the uncertainty of the excitation which is normally described by a wave spectrum (Guedes Soares, 1990), and the uncertainty of the response which is often represented by a transfer function (Guedes Soares, 1991). Combining both will lead to the short-term uncertainty, which reflects the situation in one sea state. For the lifetime of the structure a long-term formulation needs

DOI: 10.1201/9781003320289-46

to be developed so that all possible sea states are considered (Guedes Soares & Trovao, 1991, 1992).

(a)

(b)

Figure 1. Floating platform – six identical WECs are attached to the hybrid platform with a wind turbine in the centre (a) single WEC attached to the wave basin bridge and the PTO is tested with different combinations of rotational dampers and springs (b).

Other examples of model uncertainty in floaters are (Jafaryeganeh et al., 2016; Raed et al., 2015) while there are specific examples in the area of offshore renewable energy. For instance, Rezanejad studied the effect of spectral shape uncertainty in short-term performance of a Oscillating Water Column (OWC) device (Rezanejad & Guedes Soares, 2015). (Hiles et al., 2016) performed their uncertainty research on the mean annual energy production (MAEP) of a two-body WEC and analysed the WEC power data to quantify the key contributing factors to the MAEP uncertainty associated with the performance matrix approach. In their study, a Monte Carlo Simulation (MCS) approach is proposed for calculating the variability of MAEP estimates and is used to explore the sensitivity of the calculation. (Orphin et al., 2018) applied MCS method to propagate the uncertainty in a complex model of OWC WEC with time-varying variables and data reduction equations.

The wave energy research community has provided guidance for its members to perform uncertainty analysis of WECs following the ITTC guideline 7.5-02-07-03.7, entitled "Wave Energy Converter Model Test Experiments" (ITTC, 2017). Also, the Joint Committee for Guides in Metrology (JCGM) provides a complement to the "Guide to the expression of uncertainty in measurement" (GUM) (JCGM, 2008) concerned with the propagation of uncertainty using the MCS approach. The MCS is a practical alternative to the GUM uncertainty framework based on the law of propagation of uncertainty (Farrance & Frenkel, 2014; Guimaraes Couto et al., 2013). The MCS method is adequate for uncertainty propagation when: I- The uncertainties on the input parameters are not large, II- The values of the input parameters follow non-normal pdfs, III- The models are complex and non-linear.

The objective of this paper is to investigate the uncertainties in numerical modelling of a hinged WEC based on the parameters used in the model test in the ocean basin. The paper is organized into 4 sections. In section 2 the methodologies on uncertainty modelling as well as the mathematical model are presented. In Section 3 the uncertainty associated with the measurands are discussed. Then, the Monte Carlo simulation results are presented and discussed in Section 4 and then summarized in the conclusion.

## 2 METHODOLOGY

In this section the modelling of uncertainty in parameters and the MCS approach as well as the governing equations of the WEC hydrodynamic analysis problem are discussed in two separate sub-sections. Furthermore, this section explains the necessity of using MCS approach for uncertainty analysis of complex numerical simulation of hinged WEC.

### 2.1 *Uncertainty analysis*

The uncertainty evaluation of a process (e. g., experiment or numerical algorithm) can be carried out in the following main steps (Ellison & Williams, 2012):

1. Defining the outputs and expected outcomes of the process and determining the necessary measurements and calculations to produce the outcome.
2. Performing the required measurements.
3. Estimating the uncertainty of each input quantity that feeds into the result and stating the uncertainties in comparable terms.
4. Studying the dependency of input errors.
5. Calculating measurement result including any known corrections for processes like calibration.
6. Finding the combined standard uncertainty from all the individual aspects.

7. Expressing the uncertainty in terms of a coverage factor, together with a size of the uncertainty interval, and stating the level of confidence.
8. Concluding the evaluation by stating the measurement results and their uncertainties.

In this study, the main output of the experiment and numerical calculations is the mean power extraction of point absorber WEC. The measurements of input variables are carried out in an ocean basin lab (details are explained in next sub-section).

The uncertainty in results may arise from various possible sources such as inadequate definition of the measurand, sampling and data collection, matrix effects and parameter interventions, environmental conditions, uncertainties in measurements of masses, volumes, and reference values, approximations and assumptions related with the measuring procedures and methods, and random variation.

Evaluating uncertainty in measurement is categorized in three different classes: I- when expressed as a standard deviation, an uncertainty component of measurand $y$ is known as a Standard Uncertainty expressed by $u_s(y)$, II- the total uncertainty is known as Combined Standard Uncertainty and denoted by $u_c(y)$, is an estimated standard deviation equal to the positive square root of the total variance obtained by combining all the uncertainty components, and III- the Expanded Uncertainty denoted by $U$, provides an interval within which the value of the measurand is believed to be false with a higher level of confidence. $U$ is obtained by multiplying $u_c(y)$, the combined standard uncertainty, by a coverage factor $k$. The selection of the factor $k$ depends on the desired confidence level.

According to the GUM, the standard uncertainty of the input sources can be categorized in two main types: Type A, related to the sources of uncertainties from statistical analysis, such as the standard deviation obtained in a repeatability study; and Type B, related to any other source of information, such as a calibration certificate or obtained from limits deduced from personal experience. Type A and B uncertainties are also known as random and systemic uncertainty, respectively (ITTC, 2017; JCGM, 2008).

Type A uncertainties related with the repeatability studies are estimated by the GUM as the standard deviation of the mean value calculated by repetitive measurements. For instance, the uncertainty $u(x)$ due to repeatability of a set of $n$ measurements of the quantity $x$ can be expressed by $s(\bar{x})$ as follows:

$$u_{s-A}(x) = s(\bar{x}) = \frac{s(x)}{\sqrt{n}} \tag{1}$$

where $\bar{x}$ is the mean value of the repeated measurements, $s(x)$ is its standard deviation and $s(\bar{x})$ is the standard deviation of the mean. Also, it is important to note that the estimation of uncertainties of the Type B input sources must be based on precise assessment of observations or in an accurate scientific judgment, using all available information about the measurement and calibration procedure. Type B uncertainty in this experiment was evaluated primarily using instrument technical documents and expert evaluation, by performing a regression analysis:

$$u_{s-B} = \sqrt{\frac{(y_i - \hat{y}_i)^2}{N-2}} \tag{2}$$

where $N$ is the number of calibration points and $y_i - \hat{y}_i$ is the difference between the calibrated data point and the fitted value.

Therefore, the standard uncertainty ($u_s$) is calculated by the following equation:

$$u_s = \sqrt{(u_{s-A})^2 + (u_{s-B})^2} \tag{3}$$

According to the law of propagation of uncertainty (LPU), the combined uncertainty $u_c(y)$ is calculated by expanding the measurand model in a Taylor series and simplifying the expression by considering only the first order terms. This approximation is viable as uncertainties are very small numbers compared with the values of their corresponding quantities. Thus, the measurand $y$ is expressed as a function of $N$ variables $x_1, \ldots, x_N$:

$$y = f(x_1, \ldots, x_N) \tag{4}$$

leads to a general expression for propagation of uncertainties (Equation 5).

$$u_c^2 = \sum_{i=1}^{N} \left(\frac{\partial f}{\partial x_i}\right)^2 u_{x_i}^2 + 2\sum_{i=1}^{N-1} \sum_{j=i+1}^{N} \left(\frac{\partial f}{\partial x_i}\right)\left(\frac{\partial f}{\partial x_j}\right) u_s(x_i, x_j) \tag{5}$$

where $u_c$ is the combined standard uncertainty for the measurand y and $u_{xi}$ is the uncertainty for the $i^{th}$ input quantity. The second term of Eq. 5 is related to the correlation between the input quantities. If there is no supposed correlation between the inputs, this equation can be further simplified as:

$$u_c^2 = \sum_{i=1}^{N} \left(\frac{\partial f}{\partial x_i}\right)^2 u_{x_i}^2 \tag{6}$$

where $u_c$ corresponds to an interval that includes only one standard deviation. To have a better confidence level for the result, the GUM approach expands this interval by assuming a Student's t-Distribution for the measurand. The expanded uncertainty is then evaluated by multiplying the $u_c$ by a coverage factor $k$ that expands it to a coverage interval delimited by the t-Distribution table.

$$U = ku_c \tag{7}$$

As mentioned in the introduction section, when the GUM approach fails due to some reasons such as nonlinearities in the model, the Monte Carlo simulation can be used for the propagation of distributions (Guimaraes Couto et al., 2013). This process is illustrated in Figure 2 in comparison with GUM method.

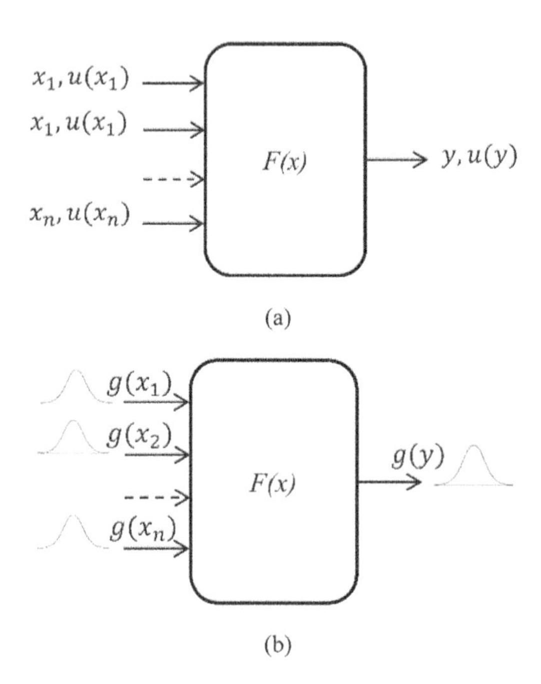


Figure 2. Propagation of uncertainties in the input quantities (a) and propagation of input distributions (b).

The initial steps of MCS method are comparable to the GUM method, however, the most appropriate probability density functions (or PDFs) should be selected for each of the input quantities. Therefore, the maximum entropy principle introduced in the Bayesian theory should be used, while the most generic distribution, for the level of information that is known about the input source, should be applied. Meanwhile, the PDFs that do not transmit more information than that which are known about the parameters, should not be selected. For instance, if the only data available on an input parameter is the maximum and minimum limit, the normal PDF should be applied. Then, after defining all the PDFs, the number of Monte Carlo trials (M) should be selected. Generally, the number of simulation trials has a direct relation with the convergence of the results. According to GUM recommendations (in Supplement 1) the number of trials should be greater than 200000 for the conversion interval of 95%. Figure 3 shows the flow chart of MCS method adapted in this paper.

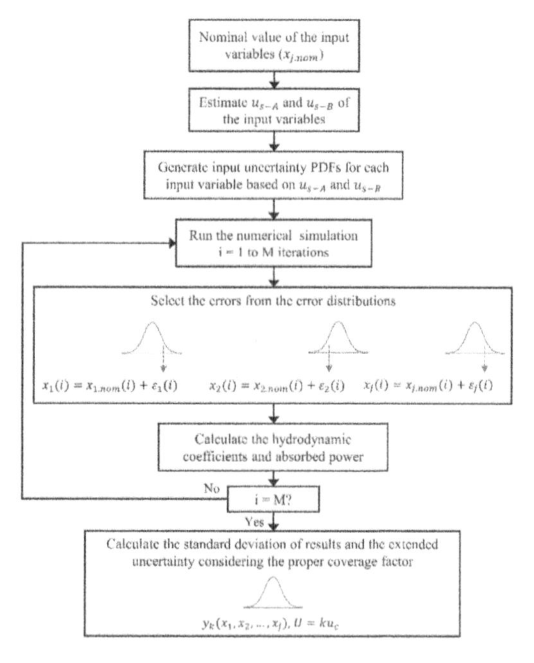


Figure 3. Flow chart of MCS method for the propagation of uncertainty (redrawn from (Orphin et al., 2018)).

In addition, for the time saving purposes, the following process is applied for verifying the number of trials in this paper:

- 10 Monte Carlo simulations are performed with 5 trials each while the standard deviation of the measurand (in this case, average extracted power) is calculated. Then, the standard deviation for each of the 10 simulations are calculated.
- This process of 10 simulations is repeated with 10, 100, and 1000 trials each. In each case the standard deviation of the measurand is recorded.

This procedure is terminated when the variation of the standard deviation is below the significant level defined for the study. Figure 4 illustrates this approach for deciding about the required number of MCS trials (M in Figure 3). The convergence interval of 95% is reached after 4000 trials.

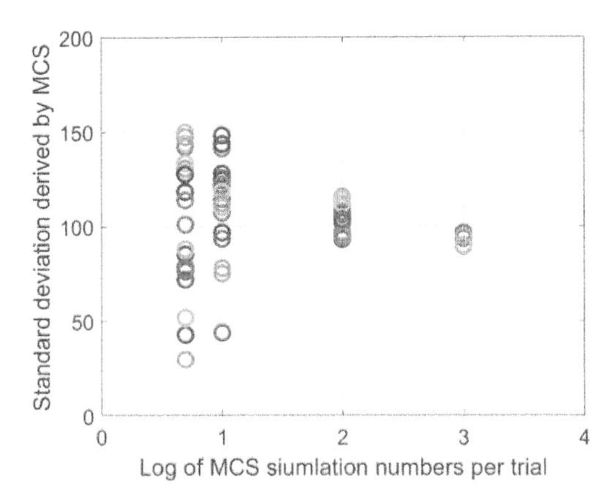


Figure 4. Variation of the standard deviation for extracted power calculated by 5, 10, 100, and 1000 MCS trials per simulation on 20 separate occasions.

## 2.2 *Mathematical model of hinged WEC*

The WEC concept has a cone shape prime mover with an apex angle of 90° that is attached to a rigid arm with a fixed angle shown in Figure 1(a). The arm is hinged to a fixed reference with an equilibrium angle of 20 degrees. Rotational dampers are applied on the hinge point simulating the PTO damping. Different types and combinations of mechanical springs are used as the negative stiffness module. Figure 5 shows schematically the main mechanical elements of the WEC concept, the rotational damper, spring, arm, and prime mover.

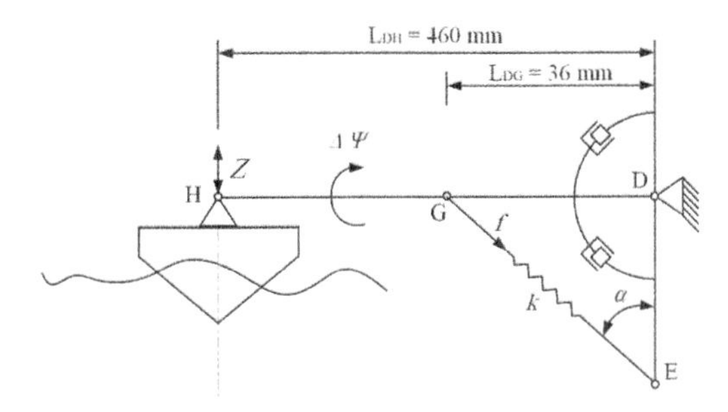


Figure 5. Schematic of the WEC concept. 1:27 scale.

In this schematic (Figure 4) the spring mechanism is considered by modelling the stiffness force ($f$) as a function of spring length and stiffness coefficient ($k$). Two unidirectional dampers are used to operate in different rotation directions, thus providing bidirectional damping.

The numerical study of the cone floater is performed using a NEMOH BEM code (Babarit & Delhommeau, 2015) considering that the energy absorption is performed by a single Degree of Freedom (single-DoF) which is the angle $\psi$ presented in Figure 3. Thus, the 6-DoF hydrodynamic coefficients of the floater are adapted to the hinged WEC using formulations discussed in this section.

The hydrodynamics problem of wave diffraction/radiation in the frequency domain on a free-floating rigid body is modelled as:

$$[M + M_A(\omega)]\{\ddot{x}(t)\} + [B_{PTO} + B(\omega)]\{\dot{x}(t)\} + K_{hyd}\{x(t)\} + F_{NS}(t) = \{f_{ex}(\omega;t)\} \tag{8}$$

where $M$ is the mass matrix of the rigid body; $M_A(\omega)$ is the frequency-dependent hydrodynamic added mass; $B(\omega)$ is the frequency-dependent hydrodynamic damping; $B_{PTO}$ is the PTO damping matrix; $K_{hyd}$ is the hydrostatic restoring matrix; $F_{NS}$ is the negative stiffness force defined in Eq. (8); $x$ represents body motion and $f_{ex}(\omega)$ the frequency-dependent wave excitation force.

$$F_{NS} = K_N(x(t) + L_S) \tag{9}$$

where $K_N$ is the spring stiffness coefficient and $L_S$ is the pretension length of the spring. After a series of conversions applied on the surge, heave, and pitch DoFs, a form for $\psi$ motion is achieved by (Kamarlouei et al., 2021):

$$\left[M_{\psi} + M_{A_{\psi}}(\omega)\right]\{\ddot{\psi}(t)\} + \left[B_{PTO_{\psi}} + B_{\psi}(\omega)\right]\{\dot{\psi}(t)\} + K_{hyd_{\psi}}\{\psi(t)\} - F_{NS_{\psi}}(t) = \left\{f_{ex_{\psi}}(\omega;t)\right\} \tag{10}$$

Solving this equation provides the $\psi$ motion which is applied in the average absorbed power formulation:

$$P_{abs} = \int_0^{\infty} B_{PTO}\omega^2 \left(\frac{\psi_A}{\zeta_A}\right)^2 S_{\zeta}(\omega)d\omega \tag{11}$$

where, $S_{\zeta}(\omega)$ is the irregular wave spectrum and $\psi_A/\zeta_A$ is the RAO of the $\psi$ DoF.

# 3 CASE STUDY

In this case study, the absorbed power will be considered as the measurand, using Equation 11. The values and sources of uncertainty of inputs have been estimated from data in the literature (Gaspar, et al., 2021). Figure 6 shows the cause-effect diagram for the hinged WEC power absorption uncertainty estimation. The input uncertainty sources are from Type A and B, and they are summarized in Table 1. The main sources of the uncertainty are the prototyping errors, estimations, calibration, and repeatability.

The prototyping process of 12 floaters has been done in the mechanical workshop using 3D printer and plaster moulds which are filled by injecting expandable fumes (Kamarlouei et al., 2020). The rotational arms are built using aluminium bars while the spring and damper fixtures are built manually. Thus, the repeatability of model is associated with measurement errors and Type A uncertainty in the floater apex angle and CoG, arm length and its mass. The outlier models were excluded from the test campaign.

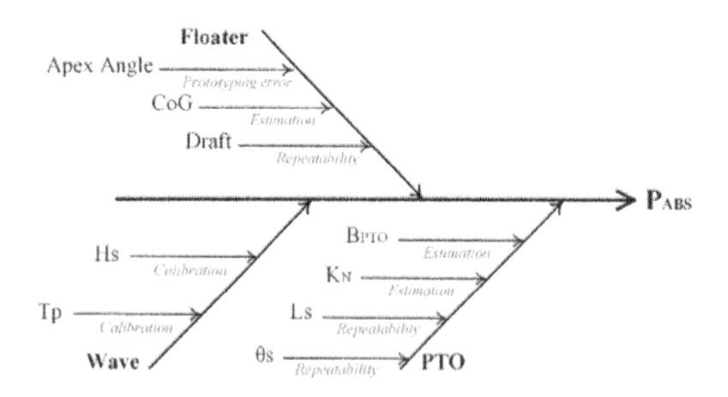


Figure 6. Cause-effect diagram for the uncertainty estimation of the case study hinged WEC power absorption.

Table 1. Input sources and their associated uncertainties.

| Parameter | Mean value | Standard Deviation |
|---|---|---|
| **Floater** | | |
| Apex angle [°] | 90 | 0.2 |
| Draft [mm] | 80 | 0.5 |
| CoG [mm] | 80 | 2 |
| **Arm** | | |
| Length [mm] | 460 | 2.3 |
| Mass [g] | 400 | 3.2 |
| Initial angle [°] | 70 | 1 |
| **PTO and Negative stiffness** | | |
| Damping ratio ($B_{PTO}$) [Nms] | 0.5 | 0.005 |
| Initial spring length ($L_s$) [mm] | 80 | 0.5 |
| Spring angle ($\theta_s$) [°] | 30 | 1 |
| Spring stiffness ($K_N$) [N/m] | 6.8 | 0.04 |
| **Wave** | | |
| Significant Height ($H_s$) [cm] | 10.8 | 0.005 |
| Peak Period ($T_p$) [s] | 1.70 | 0.04 |

The stiffness coefficients of springs ($K_N$) are measured in the laboratory, and so, with Type A uncertainties that are estimated based on the number of measurements and Type B uncertainties of applied equipment, i.e., calliper and standard masses. The uncertainty in damping ratio ($B_{PTO}$) is of Type B and is estimated based on the provided datasheets (ACE Controls Inc, 2021) and equipment aging estimations by experts.

The measurement of floater draft, initial arm angle, initial spring length, and spring angle are

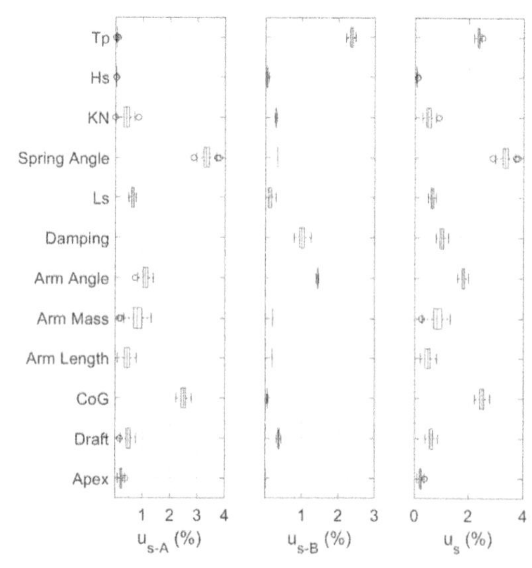


Figure 7. Uncertainty sources including Type A, B, and combined values in the input parameters.

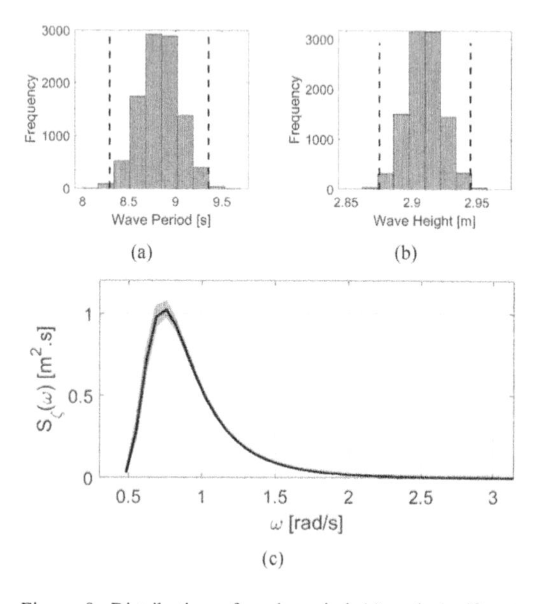


Figure 8. Distribution of peak period (a) and significant wave height (b) uncertainties and their propagation in the generated irregular wave spectrum (c).

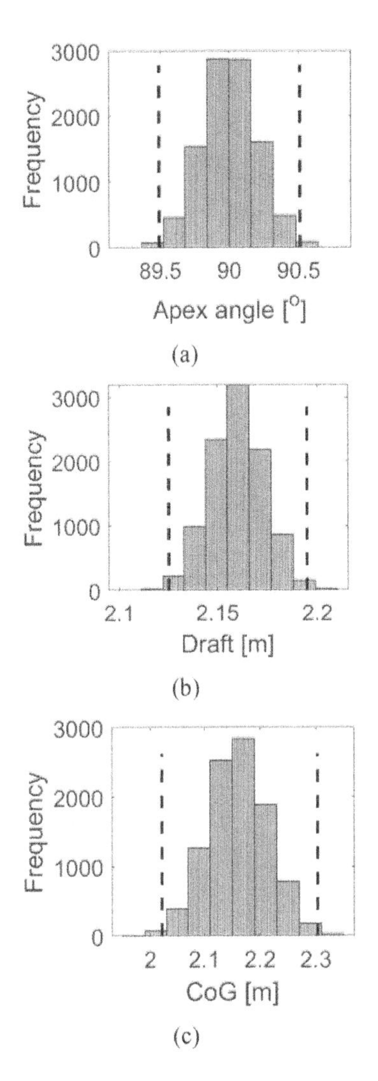


Figure 9. Distribution of apex angle (a), draft (b), and CoG (c) uncertainties.

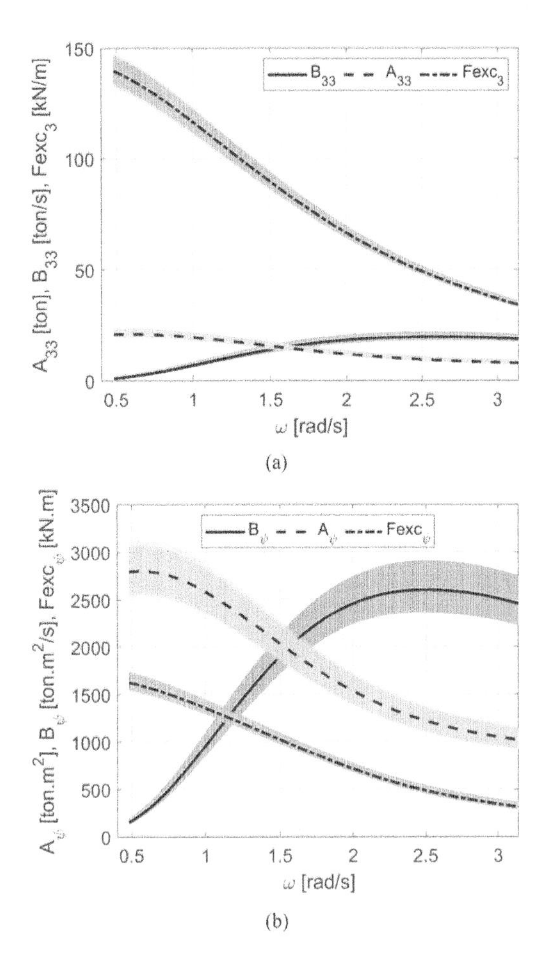


Figure 10. Uncertainty propagation in the hydrodynamic coefficients of cone shape floater in heave DoF (a) and $\psi$ DoF (b).

performed using standard devices in the ocean basin before each test. Thus, their uncertainties are of both Type A and Type B.

Finally, Type A uncertainties within the wave parameters ($H_s$ and $T_p$) are calculated based on the wave height and period data of individual waves, identified by zero-up crossing. Despite of the first few waves (i.e., 3 to 5 waves depending on the sea state) Type A uncertainty is found to be negligible in the $H_s$ and $T_p$. However, the Type B uncertainties are considered according to the wave gauges calibration logs. Figure 7 shows the Type A and B uncertainties in the input variables of Table 1.

## 4 RESULTS AND DISCUSSION

In this section, the MCS for the hinged WEC is presented. The first step in the uncertainty analysis of the hinged WEC is to perform the hydrodynamic analysis of the device in the frequency domain. This study may start with the propagation of wave uncertainties in the generated spectrum. In this study, sea state 5 (with $H_s$ and $T_p$ values presented in Table 1) is considered for the study of WEC power performance. Figure 8(a-c) show the distribution of uncertainty in wave height and period and their propagation in the irregular wave spectrum, respectively. It should be indicated that Pierson Moskowitz spectrum is used for the generation of the irregular waves. The maximum uncertainty is associated with the pick period with $U = 2.0\%$.

Figure 9 shows the uncertainty distribution of floater geometrical parameters including apex angle, draft, and CoG.

Figure 10(a and b) show the uncertainty propagation in the hydrodynamic coefficients of the cone shape floater in heave and $\psi$ DoFs.

The expanded uncertainty of heave damping, added mass, and excitation force have the following ranges, respectively:

$U_{B_{33}} = 5.5\%$ in $\omega = 0.48$ and 3.3% in $\omega = 3.14$,
$U_{A_{33}} = 4.3\%$ in $\omega = 0.48$ and 4.0% in $\omega = 3.14$,
$U_{Fexc_{33}} = 2.7\%$ in $\omega = 0.48$ and 1.7% in $\omega = 3.14$.

The hydrodynamic coefficients in the $\psi$ DoF are calculated using the surge, heave and pitch coefficients, transformation functions (Kamarlouei et al., 2021), and the arm geometrical parameters. Thus, the uncertainties in the $\psi$ DoF are presented as follows:

$U_{B_{\psi}} = 6\%$ in $\omega = 0.48$ and 5% in $\omega = 3.14$,
$U_{A_{\psi}} = 5.3\%$ in $\omega = 0.48$ and 6.1% in $\omega = 3.14$,
$U_{Fexc_{\psi}} = 3.2\%$ in $\omega = 0.48$ and 5% in $\omega = 3.14$.

As it is seen in Figure 10(b) the uncertainties in $\psi$-DoF hydrodynamic coefficients are considerably higher than those in heave DoF. These observation shows the effect of uncertainty in the arm length, initial angle, and mass (presented in Table 1 and Figure 7) which are mainly imposed by Type A uncertainties.

Figures 11(a-c) respectively present the uncertainty distribution in the significant amplitude of the WEC in $\psi$ DOF ($A_{\psi}$), the negative stiffness ($K_N$), and PTO damping ($B_{PTO}$). Finally, in Figure 11(d) the propagation of uncertainty in the average absorbed power is presented. The expanded uncertainty in $P_{ABS}$ is estimated as $U_{PABS} = 3.7\%$. Table 2 illustrates more details about the statistical parameters and uncertainties in $P_{ABS}$.

One may notice that the combined standard uncertainty of the $P_{ABS}$ is 0.093. This result is obtained from 10000 trials and 95% coverage probability (coverage factor of 1.96). The skewness and kurtosis coefficients are approximately 0.096, and 3.007, respectively. Ideally, with the defined number of trials and coverage interval for a normal distribution the skewness should be 0.0 and Kurtosis should be 3.0. However, the obtained values match with the observation in Figure 11(d) where the tail on the right side is slightly longer and the Kurtosis coefficient proves that the distribution is quite normal.

Figure 12 shows the scatter diagram of 10 input parameters versus the $P_{ABS}$. It is noticeable that the average absorbed power is more sensitive to

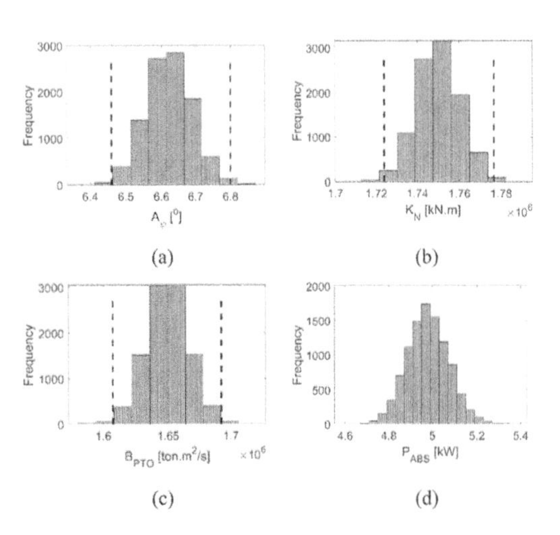


Figure 11. Distribution of $\psi$ significant amplitude (a), negative stiffness (b), and PTO damping (c) uncertainties, and the propagation uncertainty in the average absorbed power (d).

Table 2. Results obtained for the extracted power uncertainty estimation using the MCS approach, with a coverage probability of 95%.

| Parameter | Value |
|---|---|
| Mean (kW) | 4.977 |
| $u_c$ (standard deviation) | 0.093 |
| Low end point (95%) | 4.748 |
| High end point (95%) | 5.227 |
| Coverage factor ($k$) | 1.96 |
| Skewness coefficient | 0.096 |
| Kurtosis coefficient | 3.007 |
| Expanded uncertainty ($U_{PABS}$) | 0.183 |

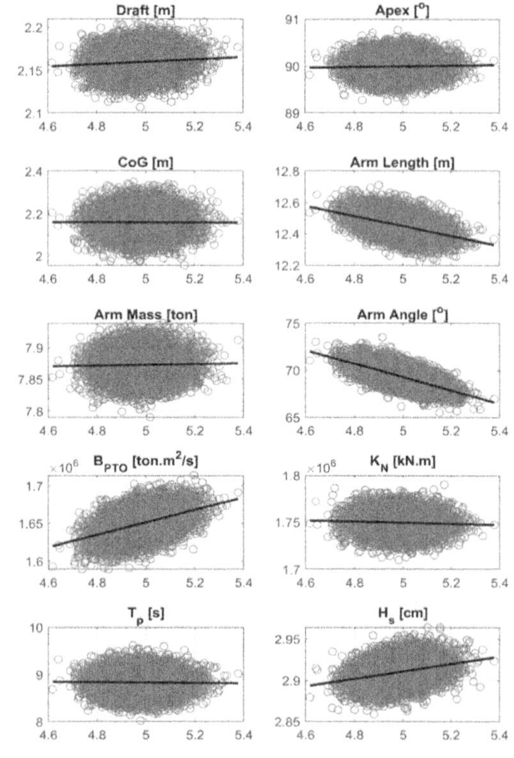


Figure 12. Scatter diagram and polynomial trend of different input uncertainties (Y-axis) vs the uncertainty in the average absorbed power $P_{ABS}$ in kW (X-axis).

uncertainties in the draft, arm length, initial arm angle (also affecting the negative spring), and $B_{PTO}$, and $H_s$. The first three are mainly affected by Type A uncertainty and repeatability in the measurements. However, uncertainties in the last two inputs are related to the damper and wave generator calibrations (Type B).

## 5 CONCLUSIONS

In this paper the Monte Carlo Simulation method is applied for the propagation of uncertainties in the geometrical and hardware specifications that were applied in the experimental analysis of a hinged WEC, using a numerical approach.

Standard uncertainties of Type A and Type B in the wave, floater, arm, and PTO were propagated through a series of functions including those applied by linear frequency domain BEM code, transformation functions for the generalized DoF $\psi$, and power absorption formula. The expanded uncertainty of the average absorbed power is estimated as 3.7% with the confidence interval of 95% and coverage factor of 1.96. It is also concluded that the draft, arm length, initial arm angle, PTO damping, and significant wave height have the higher influence on the obtained overall uncertainty in $P_{ABS}$. Thus, low precision tools for the measurement of first three variables (causing Type A uncertainty) may impose higher errors in the numerical simulation of the WEC. Also, Type B uncertainties in the generated waves and PTO dampers will significantly affect the uncertainty in the estimated output power.

## ACKNOWLEDGEMENTS

This work was performed within the project "Experimental simulation of oil-hydraulic Power Take-Off systems for Wave Energy Converters", funded by the Portuguese Foundation for Science and Technology (FCT), under contract PTDC/EME-REN/29044/2017. The experimental work has received support from MaRINET 2, a Marine Renewable Infrastructure Network for Enhancing Technologies 2 under H2020-EU.1.4.1.2 "Integrating and opening existing national and regional research infrastructures of European Interest", project ID 731084. The work contributes to the Strategic Research Plan of the Centre for Marine Technology and Ocean Engineering, which is financed by Portuguese Foundation for Science and Technology (FCT), under contract UIDB/UIDP/00134/2020.

## REFERENCES

ACE Controls Inc. (2021). *Rotary Dampers FFD-25FS-L502*. https://www.acecontrols.com/us/products/motion-control/rotary-dampers/ffd/ffd-fs-l/ffd-25fs-l502.html

Babarit, A., & Delhommeau, G. (2015, September 6). Theoretical and numerical aspects of the open source BEM solver NEMOH. *11th European Wave and Tidal Energy Conference*. https://hal.archives-ouvertes.fr/hal-01198800/

Berenjkoob, M. N., Ghiasi, M., & Guedes Soares, C. (2021). Influence of the shape of a buoy on the efficiency of its dual-motion wave energy conversion. *Energy*, *214*. https://doi.org/10.1016/j.energy.2020.118998

Berque, J., Armstrong, S., Sheng, W., Crooks, D., de Andres, A., Medina-Lopez, E., Weller, S., & Lazkano, U. (2016). *Wave Energy Measurement Methodologies for IEC Technical Specifications. Open Sea Operating Experience to Reduce Wave Energy Costs Deliverable D5.1*. University College Cork. https://doi.org/10.25607/OBP-532

Ellison, S. L. R., & Williams, A. (Eds.). (2012). *Quantifying Uncertainty in Analytical Measurement, 3rd Edition*. Eurachem/CITAC. https://doi.org/10.25607/OBP-952

Farrance, I., & Frenkel, R. (2014). Uncertainty in Measurement: A Review of Monte Carlo Simulation Using Microsoft Excel for the Calculation of Uncertainties Through Functional Relationships, Including Uncertainties in Empirically Derived Constants. *The Clinical Biochemist Reviews*, *35* (1),37. /pmc/articles/PMC3961998/

Gaspar, J. F., Kamarlouei, M., Calvário, M., & Guedes Soares, C. (2021). Design of power take – offs for combined wave and wind harvesting floating platforms. In C. Guedes Soares & T. A. Santos (Eds.), *Developments in Renewable Energies Offshore - Proceedings the 4th International Conference on Renewable Energies Offshore, RENEW 2020* (pp. 785–793). CRC Press.

Gaspar, J. F., Kamarlouei, M., Thiebaut, F., & Guedes Soares, C. (2021). Compensation of a hybrid platform dynamics using wave energy converters in different sea state conditions. *Renewable Energy*, *177*, 871–883. https://doi.org/10.1016/J.RENENE.2021.05.096

Guedes Soares, C. (1997). *Quantification of Model Uncertainty in Structural Reliability* (C. Guedes Soares (Ed.); pp. 17–37). Springer, Dordrecht. https://doi.org/10.1007/978-94-011-5614-1_2

Guimaraes Couto, P. R., Carreteiro, J., & de Oliveir, S. P. (2013). Monte Carlo Simulations Applied to Uncertainty in Measurement. *Theory and Applications of Monte Carlo Simulations*. https://doi.org/10.5772/53014

Hallak, T. S., Gaspar, J. F., Kamarlouei, M., Calvário, M., Mendes, M. J. G. C., Thiebaut, F., & Guedes Soares, C. (2018). Numerical and Experimental Analysis of a Hybrid Wind-Wave Offshore Floating Platform's Hull. *Proceedings of the International Conference on Offshore Mechanics and Arctic Engineering - OMAE, 11A*. https://doi.org/10.1115/OMAE2018-78744

Hiles, C. E., Beatty, S. J., & de Andres, A. (2016). Wave Energy Converter Annual Energy Production Uncertainty Using Simulations. *Journal of Marine Science and Engineering 2016, 4*(3), 53. https://doi.org/10.3390/JMSE4030053

ITTC. (2017). *ITTC-Uncertainty Analysis for a Wave Energy Converter.*

Jafaryeganeh, H., Teixeira, A. P., & Guedes Soares, C. (2016). Uncertainty on the bending moment transfer functions derived by a three-dimensional linear panel method. In C. Guedes Soares & T. A. Santos (Eds.), *Maritime Technology and Engineering* (pp. 295–302). Taylor & Francis Group.

JCGM. (2008). *Evaluation of measurement data - Guide to the expression of uncertainty in measurement (GUM) JCGM*. Joint Committee for Guides in Metrology. http://www.bipm.org/en/publications/guides/gum.html

Kamarlouei, M., Gaspar, J. F., Calvario, M., Hallak, T. S., Mendes, M. J. G. C., Thiebaut, F., & Guedes Soares, C. (2020). Experimental analysis of wave energy converters concentrically attached on a floating offshore platform. *Renewable Energy*, *152*, 1171–1185. https://doi.org/10.1016/j.renene.2020.01.078

Kamarlouei, M., Gaspar, J. F., Hallak, T. S., Calvário, M., Guedes Soares, C., & Thiebaut, F. (2021). Experimental analysis of wind thrust effects on the performance of a wave energy converter array adapted to a floating offshore platform. In C. Guedes Soares & T. A. Santos (Eds.), *Developments in Maritime Technology and Engineering* (pp. 597–605). Taylor & Francis Group, CRC Press. https://doi.org/10.1201/9781003216599-63

Kamarlouei, Mojtaba, Hallak, T. S., Gaspar, J. F., & Guedes Soares, C. (2021). Evaluation of the Negative Stiffness Mechanism on the Performance of a Hinged Wave Energy Converter. *Proceedings of the International Conference on Offshore Mechanics and Arctic Engineering - OMAE*, *9*. https://doi.org/10.1115/OMAE2021-63748

Orphin, J., Penesis, I., & Nader, J.-R. (2018). Uncertainty Analysis for a Wave Energy Converter: the Monte Carlo Method. *AWTEC 2018 Proceedings, September*, 444.

Raed, K., Karmakar, D., & Guedes Soares, C. (2015). Uncertainty modelling of wave loads acting on semi-submersible floating support structure for offshore wind turbine. In C. Guedes Soares (Ed.), *Renewable Energies Offshore* (pp. 945–951). Taylor & Francis Group.

Rezanejad, K., & Guedes Soares, C. (2015). Effect of spectral shape uncertainty in short term performance of a Oscillating Water Column device. In C. Guedes Soares (Ed.), *Renewable Energies Offshore* (pp. 479–487). Taylor & Francis Group.

Sinha, A., Karmakar, D., & Guedes Soares, C. (2015). Effect of floater shapes on the power take-off of wave energy converters. In C. Guedes Soares (Ed.), *Renewable energies offshore* (pp. 375–382). Taylor and Francis Group, London, UK.

*Trends in Maritime Technology and Engineering – Guedes Soares & Santos (eds)*


# Floating offshore wind turbine stability study under self-induced vibrations

S. Piernikowska & M. Tomas-Rodriguez
*The City, University of London, London, UK*

M. Santos Peñas
*Institute of Knowledge Technology, University Complutense of Madrid, Madrid, Spain*

ABSTRACT: Floating Offshore Wind Turbines (FOWT), due to their geographical location are subjected to strong dynamic loadings originated from external sources such as wind and waves. Consequently, these systems are highly sensitive to a specific range of operational frequencies. If a dynamical system is under resonance conditions, a significant amplification of the dynamic response amplitude is prone to occur. The aim of this work is to study the response of a particular FOWT model with an externally applied periodic force so that the system may reach resonance conditions. The authors make use of a dynamic model described by a set of second order differential equations. Time and frequency domain analysis are performed with a major focus on the response of the system.

## 1 INTRODUCTION

Renewable energies and in particular, Wind turbine technology became crucial in the last decade due to the increasing possibilities of fully exploiting natural energy resources. Wind as a source of energy has many advantages: it is cost-effective and truly clean and in addition to this, it does not contribute to $CO_2$ footprint increase in the generation process. Wind energy has enough potential to replace classically used energy sources (e.g., coal or fuel) and it is predicted to produce the most significant amount of the electric power from all industrially available renewable sources. Wind onshore technology is highly reliable and environmentally friendly. However, it naturally carries some well-known disadvantages (Leithead, 2006). On the other hand, offshore wind technology offers possible solutions which overcome the drawbacks of onshore technology. Offshore wind turbines are mounted off land, usually in seawater or in freshwater, at different depths. Offshore location allows these systems to take full advantage of stronger and more constant winds than those fixed inland, hence, to extract more energy from the wind. Geographical location of wind farms is also an important factor when determining the environmental impact. Ideally, visual and noise pollutions should be kept to a minimum. From very early stages of wind technology, EU countries have been leading the development of this source of energy. Denmark, started to harvest electrical energy from the wind in early 1970s. Nowadays, at least 40% of their energy production comes from the wind. Other EU countries which also have devoted resources and research to implement this type of clean energy are Germany, Belgium, Netherlands, France, or Italy. Outside of Europe, China is the current leading country in wind technology developments- as an example, the very large wind farm at the Gobi Desert (Jha, 2010). Currently, United Kingdom is becoming a leader in offshore power harvesting. UK owns over 34% of the total global installations with the largest offshore wind farm build in England (Broom, 2020).

Vibrations in mechanical systems are often a response to disturbances from their equilibrium point. One of the factors that could induce oscillations in FOWT systems is the presence of loadings originated by the wind itself or waves/currents. Also, the rotating nature of the rotor blades do transfer vibrations along the system's tower into the barge/platform.

Self-induced vibrations are a natural phenomenon often experienced by mechanical systems of rotating nature (Dinh & Basu, 2015) and in the case of wind turbines, this is a challenge faced by floating turbines due to the lack of fixed foundation.

In the case of FOWT, above certain windspeed, maximum energy extraction is achieved by means of blade-pitch control routines that can induce increasing amplitude barge pitch motions, this is what appears in the literature as "self-induced unstable vibrations" (Jonkman, 2007). Consequently, a significant reduction in power generation could happen.

Moreover, in FOWT, structural costs are primarily associated to the floating foundation stabilization (Roddier, 2010). Hence, the main control challenge

DOI: 10.1201/9781003320289-47

is focused on reduction of the unwanted oscillation to extend the operation life of the structure.

## 2 MODEL DESCRIPTION

FOWTs are highly complex mechanical structures composed of three main subsystems, platform (barge), tower and nacelle (Figure 1).

As mentioned in the introduction, these structures are exposed to challenging environmental conditions due to the geographical location of the wind farms.

By means of mathematical simplifications it is possible to represent a FOWT system as a simplified rigid-body model composed by three Degrees Of Freedom (DOF). The approach followed in this work uses the benchmark model used in (Jonkman, 2008) which is a widely used approach in several studies i.e. (Jonkman 2009, Steward et al. 2011, He et al. 2017, Tomás-Rodríguez & Santos 2018) just to cite a few.

The used model is the 5MW wind turbine developed by National Renewable Energy Laboratory (NREL) and studied by (Jonkman, 2009). The model is a three bladed upwind system with a 90 m hub height and a 126 m rotor diameter. In (Jonkman, 2008), this model was described by a set of limited degrees-of-freedom differential equations.

Figure 2 illustrates the simplified system as a diagram where each component of the structure is represented as an individual block with springs and dampers characterizing various flexibilities and torsion properties. For the simplicity, the subscripts 'p', 't' and 'T' indicate the platform, tower, and nacelle. The *k* elements, appear on Figure 2 as $k_p$, $k_t$, $k_T$, represent the rotational and linear stiffnesses. Similarly, *d* terms ($d_p$, $d_t$, $d_T$) describe the rotational and linear damping constants. The center of masses of each of the subsystem's elements are represented by $m_pg$, $m_tg$, $m_Tg$ (Stewart & Lackner, 2011).

Another important factor is the relative motion (degree of freedom) between components. Hence, the existing rotation between the tower and the barge ($\theta_t$) and barge and the surface of the water ($\theta_p$) are also considered in this model. The existing motion between the tower and the TMD mass in the nacelle can be assumed as translational ($x_T$) due to the restricted displacement of the TMD inside the nacelle. Based on the simplified model, a mechanical problem can now be fully described by a set of 3 single-degree-of-freedom second order differential equations presented in Equation 1, 2 and 3.

$$
\begin{aligned}
I_p\ddot{\theta}_p = -d_p\ddot{\theta}_p - k_p\theta_p - m_pgR_p\theta_p \\
+k_t(\theta_t - \theta_p) + d_t(\dot{\theta}_t - \dot{\theta}_p)
\end{aligned} \quad (1)
$$

$$
\begin{aligned}
I_t\ddot{\theta}_t = & - m_tR_t\theta_t - k_t(\theta_t - \theta_p) - d_t(\dot{\theta}_t - \dot{\theta}_p) \\
& - k_TR_T(R_T\theta_t - x_T) - d_TR_T(R_T\dot{\theta}_t - \dot{x}_t) \\
& - m_Tg(R_T\theta_t - x_T)
\end{aligned} \quad (2)
$$

$$
m_T\ddot{x}_T = k_T(R_T\theta_t - x_T) + d_T(R_T\dot{\theta}_t - \dot{x}_T) + m_Tg\theta_t \quad (3)
$$

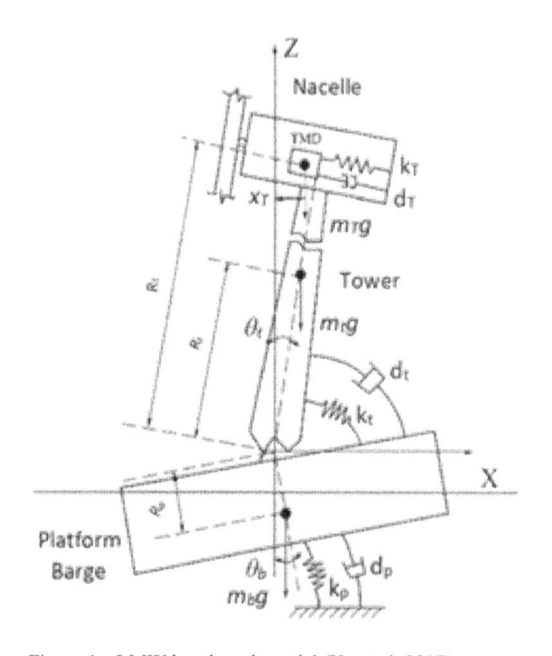

Figure 1. 5 MW benchmark model (He et al. 2017).

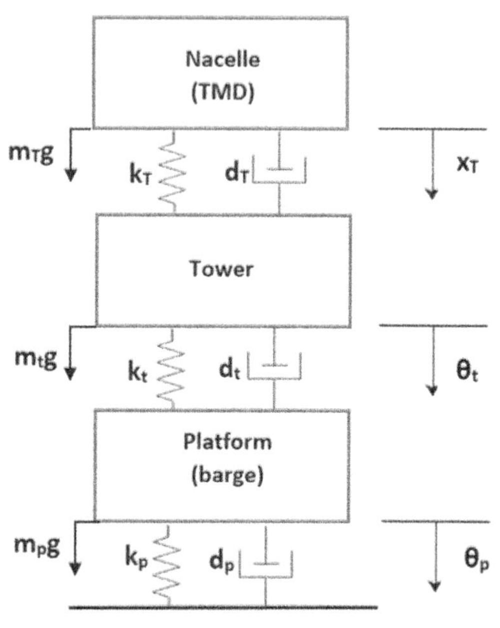

Figure 2. Rigid-body platform-pitch model of FOWT.

where $I$ = rotational inertia; $R$ = various distances of the structural elements from its centers of the mass. For further details on this model and how to obtain Equations 1-3, the reader is referred to (Hen et al. 2017).

The dynamic equations of motion (Eq. 1-3) are derived based on the simplified model (Figure 2) and full details on how to obtain them can be found in (He at al. 2017) In our study, environmental factors such as wind and wave loading can initially be omitted as structural dynamics of the system are independent of those type of forcing. The control of the blade-pitch for above-rated wind speeds is critical for the overall system's stability. It has been demonstrated how blade pitching for maximum power extraction can lead to an overall negative damping of the tower and floating platform (Jonkman 2008, Jose et al. 2018, Nielsen et al. 2008, Larsen et al. 2007).

Hence, the overall damping of the barge-pitch mode needs to be kept positive. This barge-pitch damping problem can be further analyzed by considering the rigid-body platform-pitch subsystem as a second order single DOF system as in (Jonkman, 2008). Following this approach, the platform equation of motion (Eq. 1) can be re-written as a second order differential equation as follows:

$$(I_{mass} + A_{Radiation})\ddot{\xi} + (B_{Radiation} + B_{Viscous})\dot{\xi} + \\ + (C_{Hydrostatic} + C_{Lines})\xi = L_{HH}T \quad (4)$$

where $\xi$ = platform-pitch angle (rotational displacement); $I_{mass}$ = pitch inertia associated with wind turbine and barge mass; $A_{Radiation}$ = added inertia (added mass) associated with hydrodynamic radiation in pitch; $B_{Radiation}$ = damping associated with hydrodynamic radiation in pitch; $B_{Viscous}$ = linearized damping associated with hydrodynamic viscous drag in pitch; $C_{Hydrostatic}$ = hydrostatic restoring in pitch; $C_{Lines}$ = linearized restoring in pitch from all mooring lines (can be compared to the stiffness induced by a spring); $T$ = aerodynamic rotor thrust; $L_{HH}$ = hub height (i.e., rotor-thrust moment arm).

According to (Jonkman, 2008), in FOWT, the rotor thrust has an impact into the platform-pitch damping. This depends on the relative wind speed at the hub due to the motion of the hub, rotor speed, and blade-pitch angle. The translational displacement and platform-pitch angle can be linearly related to each other as $x = L_{HH} \times \xi$ by the assumption that the pitch angle is small enough. If the hub translation varies slowly, the response of the wake of the rotor is identical for changes in hub and wind speed. If only variation of the rotor thrust with the speed of the hub is considered, a first-order Taylor series can be applied to express the aerodynamic rotor thrust as shown in Equation 5.

$$T = T_0 - \frac{\partial T}{\partial V}\dot{x} = T_0 - \frac{\partial T}{\partial V}L_{HH}\dot{\xi} \quad (5)$$

where $T_o$ = rotor thrust at a linearization point; $V$ = rotor- disk-averaged wind speed.

The equation of motion of the platform-pitch mode presented in Equation 4 can now be expressed in terms of the translational motion of the hub (Eq. 5) and this is shown in Equation 6:

$$\underbrace{\left(\frac{I_{mass} + A_{Radiation}}{L_{HH}^2}\right)}_{M_x}\ddot{x} + \underbrace{\left(\frac{B_{Radiation} + B_{Viscous}}{L_{HH}^2} + \frac{\partial T}{\partial V}\right)}_{C_x}\dot{x} + \\ + \underbrace{\left(\frac{C_{Hydrostatic} + C_{Lines}}{L_{HH}^2}\right)}_{K_x}x = T_0 \quad (6)$$

Therefore, the isolated rigid-body platform-pitch model will respond as a second-order system with certain natural frequency and damping ratio as presented in Equations 7 and 8. The overall damping coefficient can be extracted from Equation 6 and is shown in Equation 9 as $C_x$.

$$\omega_{xn} = \sqrt{\frac{K_x}{M_x}} \quad (7)$$

$$\varsigma_x = \frac{C_x}{2\sqrt{K_x M_x}} \quad (8)$$

$$\underbrace{\left(\frac{B_{Radiation} + B_{Viscous}}{L_{HH}^2} + \frac{\partial T}{\partial V}\right)}_{C_x} \quad (9)$$

where $\omega_{xn}$ = natural frequency; $\varsigma_x$ = damping ratio; $C_x$ = overall damping coefficient.

Just above rated wind speeds, the damping coefficient could become negative. In consequence, the platform may experience increasing amplitude of oscillations in its pitch mode. The high sensitivity of the platform-pitch to loads originated on the nacelle/rotor can be seen. For future control development, it is important to analyze the stability of the system in this disturbed state.

In this paper, the stability of the FOWT is examined by application of BIBO stability criterion and by analysis of the time and frequency domain performances.

Equations 1-3 were used to derive a set of transfer functions describing the model. The transfer function indicates the relationship between the input and the output of the system in the Laplace domain.

$$TF_1 = \frac{\theta_p}{x_T} = \frac{s^2(I_p m_T g + k_T R_T) + s(m_T g d_t + m_T g d_p + k_T R_T d_t) + (m_T g k_t + m_T g k_p + m_p g^2 R_p m_T)}{s^3(m_T d_t) + s^2(d_T d_t + m_T k_t) + s(d_t k_T + d_T k_t) + (k_t k_T)} \tag{10}$$

$$TF_3 = \frac{\theta_t}{\theta_p} = \frac{s d_t + k_t}{s^2 I_p + s(d_t + d_p) + (k_t + k_p + m_p g R_p)} \tag{11}$$

$$TF_3 = \frac{\theta_t}{x_T} = \frac{(m_t g + k_t R_t)}{s^2 m_T + s d_T + k_T} \tag{12}$$

where *TF1* = transfer function 1 (motion of the platform with respect to the mass M translation located in the nacelle); *TF2* = transfer function 2 (rotation of the tower with respect to the platform's rotation); *TF3* = transfer function 3 (rotation of the tower with respect to the mass M translation located in the nacelle).

## 3 SIMULATIONS

The fundamental analysis of the behavior of any system involves the study of the denominator of its transfer functions. The operation provides the information about the BIBO stability of the computed model. MATLAB was used to apply classic control theory. Figure 3 and Figure 4 show the pole-zero mapping and root locus of the platform-pitch, respectively.

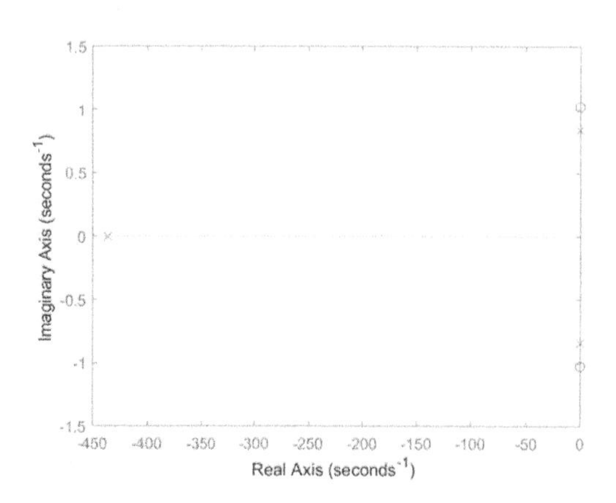


Figure 3. Pole-zero map platform-pitch mode (TF1).

Figure 3 is the representation of the pole-zero map computed from the TF1 relating the rotation of the platform with respect to the translation of the TMD mass M located in the nacelle. There is a pair of pure complex conjugate modes indicating the oscillatory nature of this subsystem and one stable

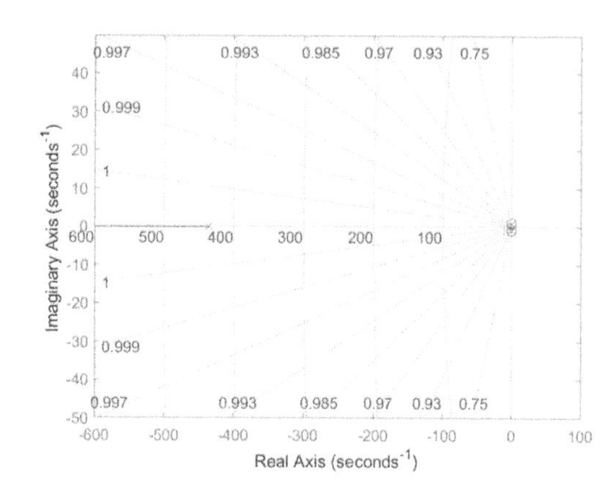


Figure 4. Root Locus of platform-pitch mode (TF1).

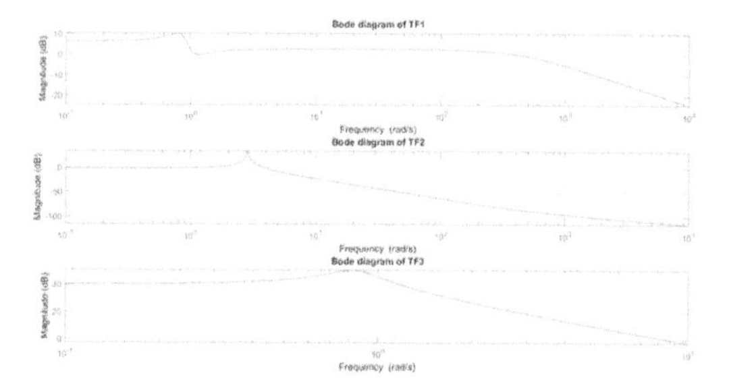


Figure 5. Magnitude-Bode plots of the system.

single pole. Figure 4 represents the roots location of the closed-loop poles obtained as a solution of the characteristic polynomial. The plot shows one real pole position on the real axis looking for its zero in infinity. Also, there is a complex conjugate pair with zero real parts. The presence of the purely imaginary complex conjugate pair indicates the oscillation with constant amplitude. However, due to the location of the third pole, the system behaves an underdamped system with a decaying amplitude of oscillation.

The frequency domain studies the behavior of the system within certain range of operational frequencies and describes how the system energy is dissipated. Any possible peak in its magnitude would indicate near resonance conditions.

Figure 5 shows the resultant Bode plots obtained for this model's transfer functions. The barge-pitch mode is the one with the lowest resonance frequency. It can be deducted that generally; floating wind turbines experience low structural frequencies. Therefore, there is a risk that the system could reach its resonance condition and as result to it, the oscillations amplitude would significantly increase possibly causing damages to the system.

## 4 RESULTS

The magnitude-Bode plot for the barge-pitch mode was used to extract the value of the resonance condition of the model. The resonance frequency was established as 0.829 Hz and further applied into the time domain analysis as a reference value for the range of the operational frequencies.

The mathematical model has been also validated by the theoretical calculations of the resonance frequency. The characteristic polynomial describing the dynamic behavior of the system can be expressed in term of natural angular frequency and the damping ratio as shown in Equation 13. The theoretical resonance frequency was calculated as 0.8486 Hz which falls within the numerical percentage error of ±2.31%.

$$m\ddot{x} + b\dot{x} + kx = 0 \rightarrow \ddot{x} + 2\varsigma\omega_n\dot{x} + {\omega_n}^2 x = 0 \quad (13)$$

Figure 6 shows the oscillation of the barge for different cases of TMD mass oscillation inside the nacelle. The model was tested with a sinusoidal input at different frequencies $w = \{0.6,\ 0.829,\ 1\}$Hz. This periodic input was introduced to simulate a periodic blade-pitch change as it is usually done to carry out blade control in above-rated wind conditions. It can be seen how at the resonance condition; the system displays the highest amplitude of oscillation. When the system is subjected to lower/higher frequencies than resonance, it oscillates with similar amplitude.

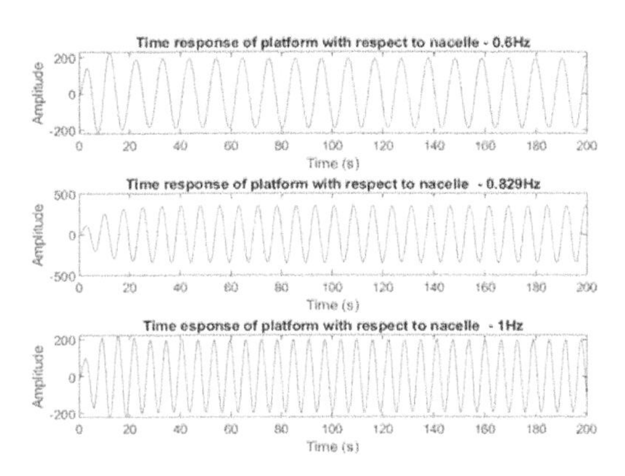


Figure 6. Time domain analysis.

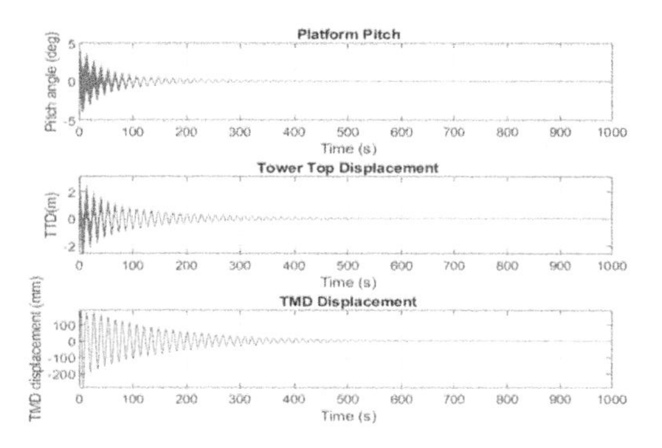


Figure 7. Time response of the model (no external force).

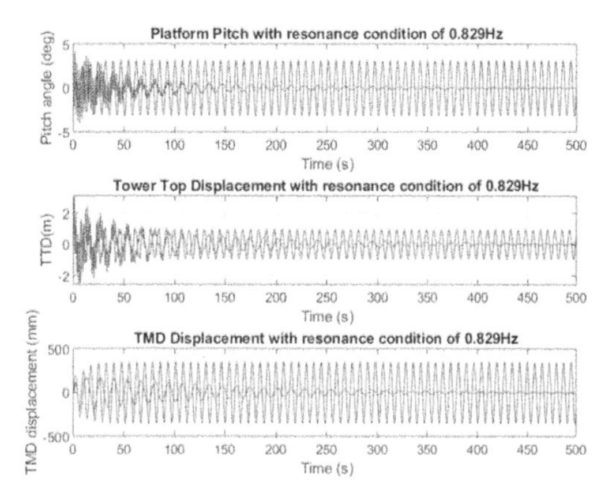


Figure 8. Comparison of system's response with (red line) and without (blue line) periodic external force.

Figure 7 shows the behavior of the barge, tower, and TMD mass when no external force is applied. The Figure 8 shows the comparison of dynamic responses of the model both in the presence of a sustained external excitation and in the case of an initial natural disturbance from the equilibrium position. The displacement plot represents the transient response of the system subjected to the disturbance from its equilibrium position.

The model will vibrate after excitation until it reaches its initial state and then it goes to rest. The sinusoidal time response describes reaction of the model exposed to the external force. Again, the system will oscillate with the frequencies introduced by input.

Figure 8 shows the system's response when conditions simulate periodic blade-pitch changes that could induce periodic rotor thrust changes, therefore, this could lead to potential instabilities of the system if the overall coefficient $C_x$ is negative. In here, unstable conditions have not been reached, but it is clear to see that in the absence of external forcing the system's oscillations tend to zero after certain time, whilst in the case of an external periodic force with frequency near the resonant frequency of the system, the amplitude grows and remains as such for the entire simulation time interval.

## 5 CONCLUSIONS

In this article, the authors have studied potential instabilities as a result of self-induced vibrations. The mathematical modelling and computer simulations allow the analysis of FOWs. The operating model is a simplification of the complex turbine structure into a rigid- body platform-pitch mode by isolation from the environmental factors. It has been shown that the system can be fully described by a set of second-order differential equations upon such simplification.

The main aim of this work was to study the oscillatory behavior of FOWTs when this might have been naturally induced by the rotor during the blade-pitch control routine at above-rated wind speeds. In order to do so, a simplified simulation model has been used. The platform oscillatory motion has been studied for various frequencies of the input (applied to the nacelle). It has been shown by analysis of the pole-zero map and root locus, that the model is behaving as an underdamped system when no external force is applied.

By means of a simplified model and with application of the Bode plot, the response of the system to the externally applied force with the frequency around its resonance condition has been studied. It has been proven that the amplitude of oscillation of the model significantly increased and such behavior remain through all simulation time.

It can be concluded that blade-pitch control introduces instabilities to FOWTs which results in appearance of unwanted oscillation on the structure.

## REFERENCES


Broom, D. 2020. *These 3 countries are global offshore wind powerhouses*, 7th April 2020. https://ww.weforum.org/agenda/2019/04/these-3-countries-are-global-offshore-wind-powerhous es/K. Elissa.

Dinh, V. & Basu, B. 2015. Passive control of floating offshore wind turbine nacelle and spar vibrations by multiple tuned mass dampers. *Structural Control and Health Monitoring*. vol 22. 10.1002/stc.1666.

He, E.G. & Hu, Y. & Zhang, Y. 2017. Optimization design of tuned mass damper for vibration suppression of a barge-type offshore floating wind turbine. Proceedings of the Institution of Mechanical Engineers, Part M. *Journal of Engineering for the Maritime Environment* vol.231(1): 302–315

Jha, A.R. 2010. *Wind Turbine Technology:* 1-30, 79–136. CRC Press.

Jonkman, J.M. 2007. Dynamics modeling and loads analysis of an offshore floating wind turbine. No. NREL/TP-500-41958, National Renewable Energy Lab. (NREL), Golden, University of Colorado.

Jonkman, J.M. 2008. Influence of Control on the Pitch Damping of a Floating Wind Turbine. *2008 ASME Wind Energy Symposium*. January 7-10, 2008. Reno, Nevada.

Jonkman, J.M. & Butterfield, S. & Musial, W & Scott, G. 2009. Definition of a 5-MW Reference Wind Turbine for Offshore System Development. Technical Report NREL/TP-500-38060. National Renewable Energy Lab. (NREL). Golden, CO (United States).

Jose, A. & Falzarano, J. & Wang, H. 2018. A Study of Negative Damping in Floating Wind Turbines Using Coupled Program FAST-SIMDYN. *Proceedings of the ASME 2018 1st International Offshore Wind Technical Conference. ASME 2018 1st International Offshore Wind Technical Conference.* November 4–7, 2018. San Francisco, California, USA.

Lackner, M.A. & Rotea, M. 2011. Structural control of floating wind turbines. *Mechatronics* 21(4): 704–719.

Larsen, T.J. & Hanson, T.D. 2007. A method to avoid negative damped low frequent tower vibrations for a floating, pitch controlled wind turbine. *Journal of Physics: Conference Series:* 75 012073, *Volume 75.* August 2007. Technical University of Denmark. Denmark, The Netherlands.

Leithead, W. 2006. *Wind Energy, Philosophical transactions of the Royal Society* 363(1853): 957–970.

Nielsen, FG. & Hanson, T.D. & Skaare, B. 2008. Integrated Dynamic Analysis of Floating Offshore Wind Turbines. *Proceedings of the 25th International Conference on Offshore Mechanics and Arctic Engineering. Volume 1: Offshore Technology; Offshore Wind Energy; Ocean Research Technology; LNG Specialty Symposium.* June 4–9, 2006. vol 1: 671–679. Hamburg, Germany.

NREL, https://www.nrel.gov/

Stewart, G.M. & Lackner, M. A. 2011. The effect of actuator dynamics on active structural control of offshore wind turbines. *Engineering Structures* vol.33: 1807–1816

Tomás-Rodríguez, M. & Santos, M. 2018. Floating Offshore Wind Turbines: Controlling the impact of vibrations. *7th International Conference on System and Control, ICSC'18- October 24-26, 2018.* Valencia, Spain.

Roddier, D. & Weinstein, J. 2010. Floating Wind Turbines. *ASME. Mechanical Engineering* 132(4): 28–32.

*Trends in Maritime Technology and Engineering – Guedes Soares & Santos (eds)*


# AEP assessment of a new resonant point absorber deployed along the Portuguese coastline

V. Piscopo & A. Scamardella
*Department of Science and Technology, The University of Naples "Parthenope", Naples, Italy*

ABSTRACT: The paper focuses on the assessment of the Annualised Energy Production of a new resonant point absorber, equipped with a fully submerged toroidal shape, whose main dimensions can be properly tuned depending on the met-ocean conditions at the deployment site. After briefly discussing the hydrodynamic model in the time-domain, a parametric study is performed to investigate the incidence of the main dimensions of the fully submerged mass on the yearly energy production of the new WEC device, considering two candidate deployment sites located along the Portuguese coastline, the former between Viana do Castelo and Porto, the latter between Nazaré and Peniche. Current results are fully discussed and a comparative analysis with the energy production of a conventional floating buoy, with the same main dimensions but without the toroidal shape, is also provided to prove the effectiveness of the new WEC device, in terms of power production.

## 1 INTRODUCTION

Marine renewables represent one of the most promising sources to reduce the global warming and reach the goals fixed by the Paris Agreement international treaty on climate change. In this respect, the European "Ocean Energy Implementation Plan" (European Commission, 2018) provides a strategic roadmap, with a set of key-actions devoted to increase the technology readiness level of wave energy conversion (WEC) devices and move the marine renewable energy sector towards the commercialization phase.

Among the variety of WEC devices, point absorbers represent one of the most mature and well-established technologies, due to the relatively simple working principle and the low technological effort required for both the construction phase and the maintenance operations. In this respect, starting from the pioneering works by Budal & Falnes (1975), consistent improvements have been achieved in the last decade, thanks to the wide research activities carried out throughout the world and mainly devoted to improving the hydrodynamic performances of this WEC device (Piscopo et al., 2016, 2017; Do et al., 2018; Kolios et al., 2018) and enhancing the survivability in harsh weather conditions (Zheng et al., 2020; Piscopo et al., 2020).

In this paper, the resonant point absorber, with spar-type configuration, recently developed at the University of Naples "Parthenope" (Piscopo & Scamardella, 2019, 2021), and currently covered by a patent demand, released by the Italian Ministry of Economic Development, is assumed as reference WEC device. The point absorber is equipped with a gravity-based Permanent Magnet Linear Generator (PMLG) lying on the seabed (Ulvgård et al., 2016; Sjökvist & Göteman, 2017) that allows the direct conversion of the kinetic energy, due to the heave motion of the floating buoy, into electrical power. The main advantage of this WEC device is mainly due to the presence of the fully submerged toroidal shape, whose main dimensions can be properly tuned depending on the met-ocean conditions at the deployment site. Particularly, the correct selection of the fully submerged mass main dimensions allows properly tuning the heave natural period of the floating buoy, so as it works near to the resonance condition with the prevailing sea states at the deployment site, with a positive impact in terms of power production and Levelised Cost of Energy (LCoE).

Based on previous remarks, the hydrodynamic performances of the WEC device are assessed at two candidate deployment sites, located along the Portuguese coastline, the former between Viana do Castelo and Porto, the latter between Nazaré and Peniche. These deployment sites are characterized by appreciable values of the mean annual energy flux, and they were widely investigated in the past, in order to carefully assess the relevant wave climate (Rusu et al., 2009; Silva et al., 2015; Silva et al., 2018). The main dimensions of the fully submerged mass are systematically increased to investigate the effect of the heave natural period of the WEC device on the relevant hydrodynamic performances, as well as on the Annualised Energy Production (AEP). The paper is structured as follows. Section 2 provides the main features of the new WEC device, together with the relevant nonlinear hydrodynamic model in the

DOI: 10.1201/9781003320289-48

time-domain and the basics for the AEP assessment. The met-ocean conditions at the two candidate deployment sites are provided in Section 3, together with the main data of the WEC device and the reference Power Take-Off (PTO) unit, lying on the seabed. Section 4 deals with the AEP assessment at the two deployment sites, while Section 5 discusses the main results of the analysis. Finally, the main outcomes of the study and some suggestions for future works are provided in Section 6. All calculations are performed by a code developed in Matlab (MathWorks, 2018), while the hydrodynamic properties of the floating buoy are assessed by the open-source code NEMOH (Babarit & Delhommeau, 2015).

## 2 THEORETICAL BACKGROUND

### 2.1 *Layout of the resonant point absorber*

The layout of the new resonant point absorber, equipped with a fully submerged toroidal shape, is depicted in Figure 1. The WEC device consists of a hemispherical floating buoy with diameter $D$, connected by means of a tapered vertical cylinder with spar-type configuration, equipped with 8 equally spaced bracings, to a fully submerged toroidal shape, with height $h$, submergence $s$ and inner (outer) diameter $D_{inn}$ ($D_{out}$).

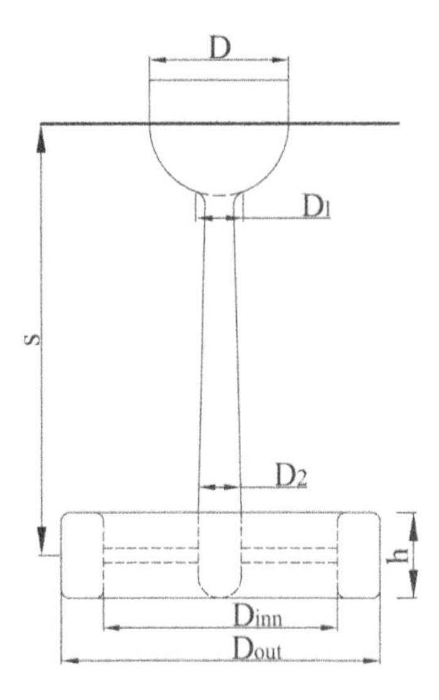


Figure 1. Layout of the WEC device.

The point absorber is made of ordinary steel, apart from the fully submerged mass that is made of marine concrete, to reduce the assembly costs of the WEC device, in view of mass production. The floating buoy is connected, by means of a tension line, to a PMLG lying on the seabed. In this respect, the PTO unit is also equipped with a marine concrete foundation that behaves as stationkeeping system of the WEC device.

The typical layout of the PTO unit is provided in Figure 2. Power production is due to the relative motion between the translator mass and the stator. The PMLG is also equipped with upper and lower end-stop springs. The overall layout of the PTO unit is like the PMLG developed at Uppsala University (Ulvgård et al., 2016; Sjökvist & Göteman, 2017), with the only exception of the upper free stroke length that is increased, to avoid sudden reaction forces in extreme sea state conditions, during the upward motion of the floating buoy, when the translator mass hits the upper end-stop spring. Besides, the translator mass is heavy enough to keep the line cable tensioned during the downward motion of the floating buoy, with no reductions of power production.

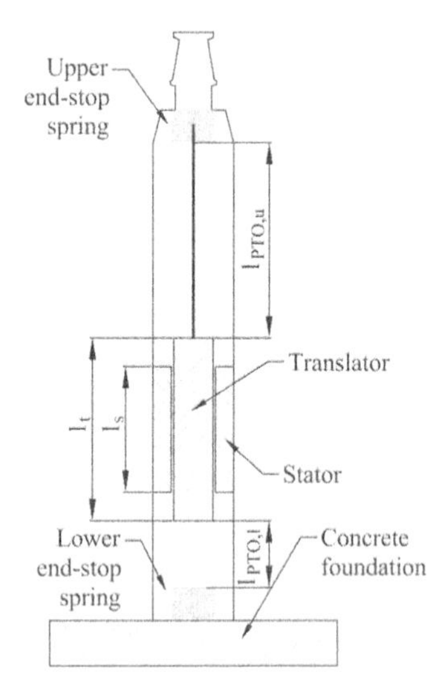


Figure 2. Layout of the PMLG.

### 2.2 *The hydrodynamic model*

The nonlinear time-domain model shall account for the hydrodynamic and drag forces exerted by the floating buoy, as well as for the additional forces exerted by the tensioned line $F_l$, by the upper/lower end-stop springs $F_{s,u}/F_{s,l}$, by the nonlinear viscous drag forces $F_d$ and by the full/partial overlapping between the translator mass and the stator, according to eq. (1), where $\zeta_1$, $\zeta_3$ and $\zeta_5$ denote the surge, heave and pitch motions, respectively. In eq. (1) $m_b$ and $I$ are the mass and the pitch moment of inertia of the floating buoy; $a_{ij,\infty}$ is the added mass of the i-j*th* mode at infinite frequency; $K_{ij}$ is the impulse response function, to be determined by the inverse Fourier transform of the radiation coefficient, according to the Cummins (1962) approach. Further details about the hydrodynamic model can be found in Piscopo & Scamardella (2021). Particularly, the hydrodynamic model accounts for surge, heave and pitch motions of the floating buoy, as they are the most significant ones for point absorbers with dimensions that are negligible as regards the incoming waves (Falcão, 2010; Piscopo & Scamardella, 2021 among others). The yaw motion is not considered as it is not significant for point absorbers. The hydrodynamic model accounts for the main non-linear forces. Anyway, it still needs to be further validated. This topic will be the subject of future works, provided that the WEC device was recently developed, as proved by the patent demand released by the Italian Ministry of Economic development on 4th February 2021.

$$\left\{\begin{array}{c} (m_b + a_{33,\infty})\ddot{\zeta}_3 + \int_{-\infty}^{t} K_{33}(t-\tau)\dot{\varsigma}_3(\tau)d\tau \\ +\varrho g A_{wp}\varsigma_3 = F_e^3 - F_d^3 - F_l\cos\vartheta - m_b g + \rho g \nabla \\ (m_b + a_{11,\infty})\ddot{\zeta}_1 + \int_{-\infty}^{t} K_{11}(t-\tau)\dot{\varsigma}_1(\tau)d\tau \\ +a_{15,\infty}\ddot{\zeta}_5 + \int_{-\infty}^{t} K_{15}(t-\tau)\dot{\varsigma}_5(\tau)d\tau = F_e^1 - F_d^1 - F_l\sin\vartheta \\ (I + a_{55,\infty})\ddot{\zeta}_1 + \int_{-\infty}^{t} K_{55}(t-\tau)\dot{\varsigma}_5(\tau)d\tau + m_b g GM\varsigma_5 \\ +a_{51,\infty}\ddot{\zeta}_1 + \int_{-\infty}^{t} K_{51}(t-\tau)\dot{\varsigma}_1(\tau)d\tau = F_e^5 - F_d^5 - F_l\sin\vartheta(s - z_g) \\ m_t\ddot{\zeta}_t + A_{act}\gamma\dot{\zeta}_t + F_{s,u} - F_{s,l} + m_t g = F_l\cos\vartheta \end{array}\right. \tag{1}$$

The convolution integrals are resembled by a 5th order approximate state-space realization (Safonov & Chiang, 1989; Yu & Falnes, 1995) by the Hankel Singular Value Decomposition algorithm (Kung, 1978). The random wave force is provided by eq. (2):

$$F_e^i = \int_{-\infty}^{\infty} k_e^i(t-\tau)\eta(\tau)d\tau \tag{2}$$

that is based on the convolution product between the random sea surface elevation $\eta$ and the excitation kernel function $k_e$ for the i-*th* motion mode.

### 2.3 *AEP assessment*

The power produced by the PTO unit mainly depends on the electromagnetic damping $\gamma$ and the electric efficiency $\eta_e$ of the PMLG. Hence, it is determined as the mean value of the instantaneous power production over the time interval $T$ (Falcão, 2010) by eq. (3):

$$P_e = \frac{\gamma\eta_e}{T}\int_0^T A_{act}\dot{\zeta}_t^2 dt \tag{3}$$

In eq. (3) $\zeta_t$ denotes the vertical motion of the PMLG translator mass, while $A_{act}$ is the active area ratio that is determined by eq. (4):

$$A_{act} = \left\{\begin{array}{c} 0 \; if \; |\varsigma_t| \geq \frac{1}{2}(l_t + l_s) \\ 1 \; if \; |\varsigma_t| \leq \frac{1}{2}|l_t - l_s| \\ \frac{1}{\min\{l_t; l_s\}}\left[\frac{1}{2}(l_t + l_s) - |\varsigma_t|\right] \text{else} \end{array}\right. \tag{4}$$

where $l_t$ ($l_s$) is the translator (stator) length, as depicted in Figure 2. Hence, the AEP can be easily determined by eq. (5):

$$AEP = 8760\sum_{i=1}^{n_s} p_i P_{e,i} \tag{5}$$

having denoted by $p_i$ the probability of occurrence of the i-*th* sea state condition and by $n_s$ the number of sea states. Obviously, the AEP mainly depends on the wave statistics at the deployment site, which implies that the careful assessment of the wave climate is a key factor to assess the yearly energy amount produced by the point absorber.

## 3 MAIN DATA

### 3.1 *Met-ocean conditions at the deployment sites*

As previously said, two candidate deployment sites, located along the Portuguese coastline, are considered in the analysis. In fact, the entire Atlantic coastline of Europe presents a large amount of wave energy due to the exposure to Westerly winds blowing over the North Atlantic (Clément et al., 2002). Particularly the northern countries of Europe have the large wave energy amount, with a mean energy flux that reaches 70 kW/m off the Ireland coastline (Pontes, 1998). The western coasts of southern countries, instead, receive a wave energy flux equal to about 25 kW/m off the Canary Islands. Among the various deployment sites, the Portuguese coastline is one of the most promising areas, with an available wave energy flux equal to about 34 kW/m between Viana do Castelo and Porto, and between Nazaré and Peniche, considering all sea state conditions up to 5.0 m significant wave height that represents the cut-out sea state for the point absorber. Beyond this limit value, in fact, the WEC device switches to the survival mode and no power production occurs. By the analysis of the wave scatter diagrams at the two deployment sites, provided by Mota & Pinto (2014), Figure 3 provides the marginal distribution of the available energy at the first deployment site, between Viana do Castelo and Porto, based on 13 wave energy period classes, ranging from 4.5 up to 16.5 s, with 1.0 s step.

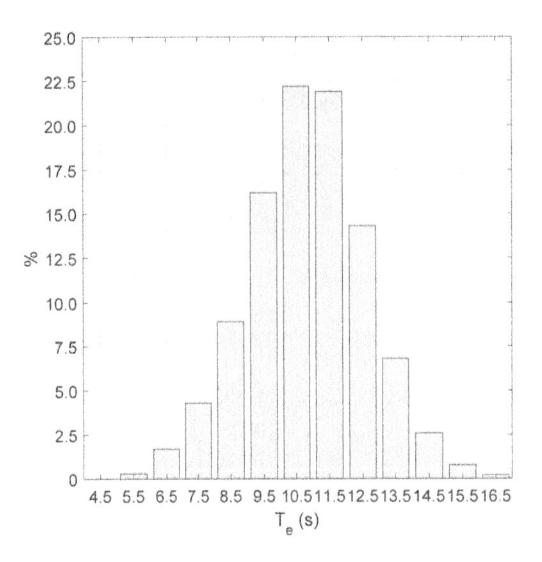


Figure 3. Wave energy marginal distribution based on wave period classes between Viana do Castelo and Porto.

Similarly, Figure 4 provides the same distribution at the second deployment site, between Nazaré and Peniche. By the marginal distributions of the wave energy at the two deployment sites, it is gathered that about one-half of the total wave energy up to 5.0 m significant wave height corresponds to wave periods ranging from 9.5 to 11.5 s, which implies that the heave natural period of the WEC device needs to be properly tuned, as far as possible, in order to match the prevailing sea states conditions and increase the power production.

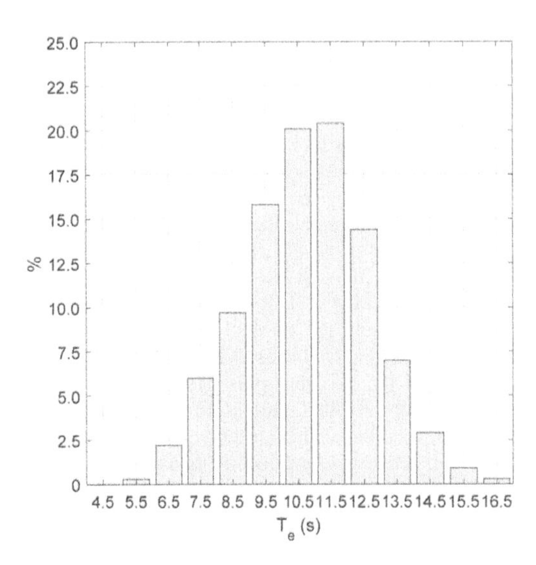


Figure 4. Wave energy marginal distribution based on wave period classes between Nazaré and Peniche.

### 3.2 *The WEC device*

Table 1 provides the main dimensions of the point absorber depicted in Figure 1. The diameter of the floating hemispherical buoy is 5.0 m, while the inner and outer diameters of the toroidal shape are equal to 7.0 and 9.0 m, respectively. The fully submerged mass is located 15.0 m below the sea water level to reduce, as far as possible, the hydrodynamic interference with the floating hemispherical body. The height of the fully submerged toroidal shape ranges from 1.0 to 2.0 m with 0.5 m step, so as three candidate layouts are analysed to investigate the effect of the fully submerged volume on the AEP of the WEC device. In addition, a reference point absorber, consisting of a similarly sized floating buoy without the fully submerged mass, is also investigated. It corresponds to the reference case $h$=0.0 m in current analysis.

Table 1. Main data of the point absorber.

| | | | |
|---|---|---|---|
| Floating buoy diameter | $D$ | 5.0 | m |
| Upper diameter of the vertical cylinder | $D_1$ | 1.0 | m |
| Lower diameter of vertical cylinder | $D_2$ | 1.5 | m |
| Submergence of toroidal shape | $s$ | 15.0 | m |
| Inner diameter of toroidal shape | $D_{inn}$ | 7.0 | m |
| Outer diameter of toroidal shape | $D_{out}$ | 9.0 | m |
| Height of toroidal shape | $h$ | 0.0-2.0 | m |

The reference PTO unit is the 20 kW PMLG, recently designed by Ulvgård et al. (2016) and analysed by Sjökvist & Göteman (2017). The only difference is the upper free stroke length that is increased to avoid sudden and very high reaction forces exerted by the upper end-stop spring in extreme sea state conditions that in current analysis correspond to a cut-out sea state with 5.0 m significant wave height. The main data of the PMLG are listed in Table 2.

Table 2. Main data of the permanent magnet linear generator.

| | | | |
|---|---|---|---|
| Nominal power | $P$ | 20 | kW |
| Electromagnetic damping | $\gamma$ | 40816 | Ns/m |
| Translator mass | $m_t$ | 6500 | kg |
| Upper free stroke length | $l_{PTO,u}$ | 10.0 | m |
| Lower free stroke length | $l_{PTO,l}$ | 2.0 | m |
| Translator length | $l_t$ | 3.0 | m |
| Stator length | $l_s$ | 2.0 | m |
| Stiffness of end-stop springs | $k_s$ | 250 | kN/m |
| Mean efficiency | $\eta_e$ | 0.87 | — |

### 3.3 *Hydrodynamics of candidate WEC devices*

As previously said, four WEC devices are considered in the analysis. The first one is a conventional floating buoy, while the remaining

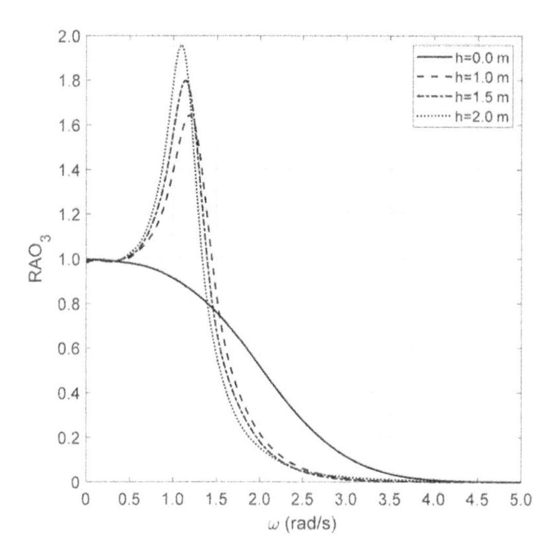


Figure 5. Heave RAOs of candidate WEC devices.

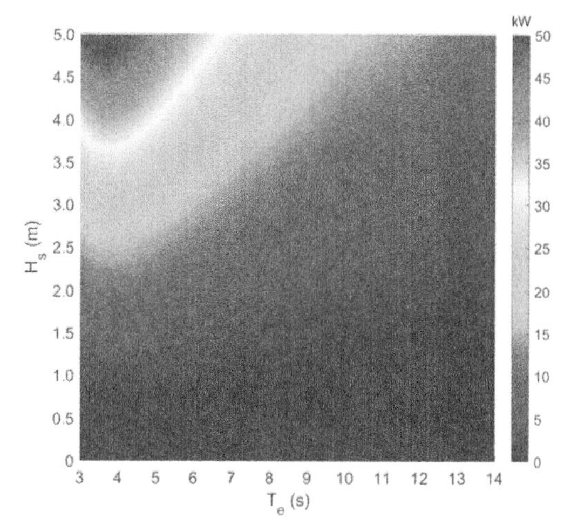


Figure 6. Power production bivariate distribution – $h$=0.0 m.

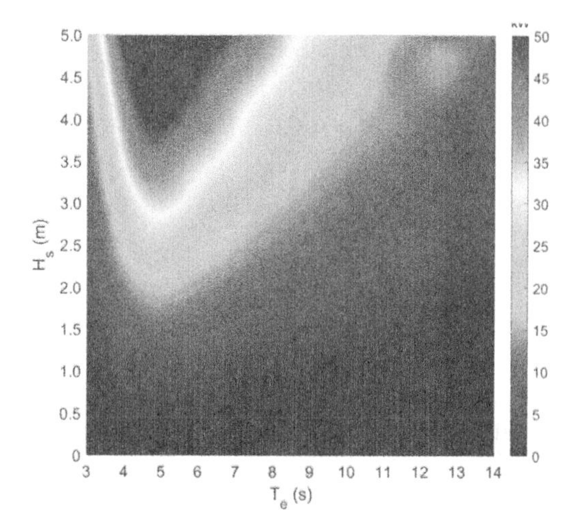


Figure 7. Power production bivariate distribution – $h$=2.0 m.

three ones are equipped with a toroidal shape with three different heights, namely 1.0, 1.5 and 2.0 m, respectively. The volume of the fully submerged mass significantly affects the hydrodynamic behaviour of the WEC device and, more specifically, the heave Response Amplitude Operator ($RAO_3$) depicted in Figure 5. Particularly, the continuous line refers to the conventional WEC device without the fully submerged mass, while the dashed, pointed-dashed and pointed lines refer to the new resonant point absorber, equipped with the toroidal shape with different heights.

As it can be gathered by Figure 5, the toroidal shape significantly affects the hydrodynamic behaviour of the floating buoy. In fact, the reference WEC device is characterized by a heave RAO that is typical of an overdamped floating buoy. On the contrary, the hydrodynamic behaviour of the three WEC devices, equipped with the fully submerged toroidal shape, is clearly underdamped, with a resonance circular wave frequency that depends on the volume of the fully submerged mass. In this respect, the fully submerged toroidal shape allows shifting the heave natural period of the point absorber up to about 6.0 s. Obviously, the volume of the fully submerged mass could be further increased until maximum power production occurs.

Based on previous remarks, the power production of the new WEC device will be considerable higher, as regards the conventional point absorber, with a positive impact on the AEP. These outcomes can be further stressed by the power production bivariate distributions provided in Figures 6 and 7 for the reference WEC device and for the new point absorber equipped with a 2.0 m high toroidal shape, respectively.

The fully submerged mass allows shifting the wave period and increasing the energy production, as regards the reference WEC device, in the frequency range where most of the available wave energy at the deployment site is recognized. The power production bivariate distributions are considered in the case study provided in Section 4, where the AEP assessment at the two candidate deployment sites is carried out.

## 4 CASE STUDY

### 4.1 *AEP assessment at the first deployment site*

The AEP assessment of the four candidate WEC devices at the first deployment site, located between Viana do Castelo and Porto, is provided in Figures 8 and 9, where the marginal distributions, based on significant wave height and wave period classes, are reported, respectively.

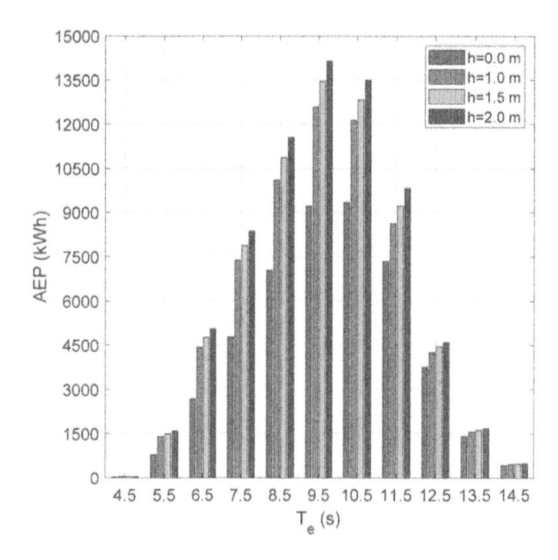


Figure 8. AEP marginal distribution between Viana do Castelo and Porto based on wave period classes.

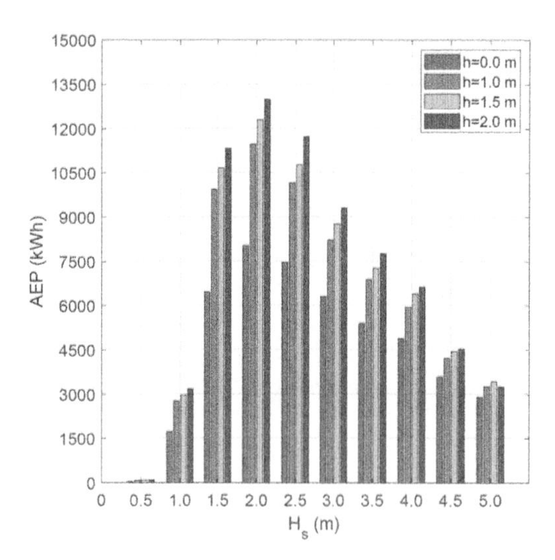


Figure 9. AEP marginal distribution between Viana do Castelo and Porto based on significant wave height classes.

By Figure 8 it is gathered that the new WEC devices allow significantly increasing the power production in the range between 7.5 and 11.5 s where most of the wave energy is detectable. As concerns the marginal distribution provided in Figure 9, an appreciable increase is recognized between 1.5 and 4.0 m significant wave height classes. Similar outcomes can be stressed by the AEP assessment reported in Table 3, from which it is gathered that the new WEC devices allow increasing the power production up to 51%, as regards the reference point absorber ($h$=0.0 m), depending on the height of the fully submerged toroidal shape.

Table 3. AEP assessment between Viana do Castelo and Porto.

| | h=0.0 m | h=1.0 m | h=1.5 m | h=2.0 m |
|---|---|---|---|---|
| AEP (kWh) | 46850 | 63040 | 67163 | 70884 |
| Δ (%) | | 34.56 | 43.36 | 51.30 |

## 4.2 *AEP assessment at the second deployment site*

The AEP assessment at the second deployment site, located between Nazaré and Peniche, is provided by the marginal distributions reported in Figures 10 and 11, as well as in Table 4, where the AEP assessment is provided. Almost the same outcomes, already stressed in Subsection 4.1, can be repeated with reference to the second candidate deployment site. Also in this case, the effect of the fully submerged mass is remarkable, and it leads to an appreciable increase of the power production, provided that the new resonant point absorber operates closer to the resonance condition with the prevailing sea states, if compared with the reference WEC device. Besides, the AEP between Nazaré and Peniche is always 4-5% higher than the relevant values at the first deployment site, even if the available wave energy is almost the same.

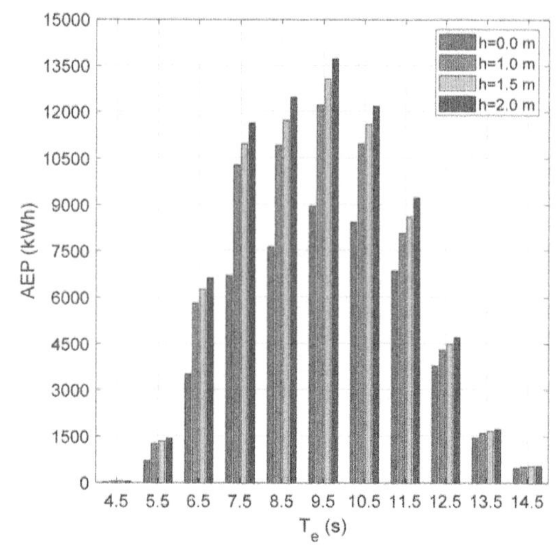


Figure 10. AEP marginal distribution between Nazaré and Peniche based on wave period classes.

This outcome confirms that the available wave energy itself is not the unique parameter that affects the power production of the WEC device and that the marginal distribution, based on wave period classes, plays a fundamental role, too.

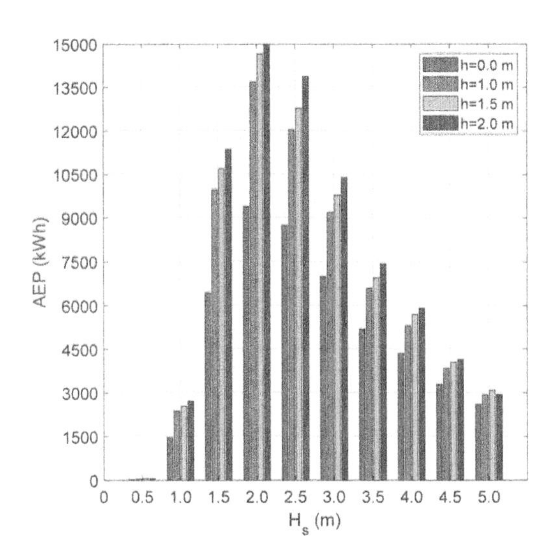


Figure 11. AEP marginal distribution between Nazaré and Peniche based on significant wave height classes.

Table 4. AEP assessment between Nazaré and Peniche.

| | h=0.0 m | h=1.0 m | h=1.5 m | h=2.0 m |
|---|---|---|---|---|
| AEP (kWh) | 48570 | 66045 | 70377 | 74376 |
| Δ (%) | | 35.98 | 44.90 | 53.13 |

## 5 DISCUSSION

The benchmark study reported in Section 4 clearly highlights that the new WEC device allows increasing the AEP, as regards a conventional one without the fully submerged toroidal shape. On the other side, it is conceivable that the installation costs of the resonant point absorber are higher, due to the presence of the vertical tapered cylinder, the 8 bracings and the marine concrete submerged mass. In this respect, a comparative analysis in terms of LCoE needs to be performed. This parameter allows estimating the cost of energy and comparing different renewable sources (IEA & NEA, 2015), according to eq. (6):

$$LCoE = \frac{1}{AVA \cdot AEP}\left[CAPEX\frac{r(1+r)^n}{(1+r)^n - 1} + OPEX\right] \quad (6)$$

having denoted by: AVA the availability of the WEC device, CAPEX the capital cost, OPEX the annual operating cost, r the discount rate and n the expected design lifetime of the wave farm. The availability is assumed equal to 98%, as power production is expected to be stopped one week a year for the scheduled maintenance operations. In this respect, it must be pointed out that the availability of the WEC device is set equal to 98% provided that the AEP already accounts for the only sea state conditions up to the cut-out sea state, as discussed in Subsection 3.1. The design lifetime is equal to 20 years, combined with a 12% discount rate (Magagna, 2019), so reflecting the high investment risk in the marine renewable energy sector, perceived by international investors. As concerns the annual operating costs, the assessment of the expenses, arising during the maintenance operations of the wave farm, is a very challenging issue, as there is not much experience in real environment and for very long periods. In this respect, the annual operating cost is roughly assumed equal to 5% of capital cost, that is the reference value for the ocean energy sector (Magagna, 2019). Finally, the CAPEX of the four WEC devices is taken from Piscopo & Scamardella (2021) who systematically investigated the relevant main cost items, due to: (i) the PTO unit, including the marine concrete foundation; (ii) the floating buoy, including the marine concrete toroidal shape (if present) and the tensioned wire rope; (iii) the connection of the marine substation to the ashore grid and (iv) the deployment of the WEC device.

Based on these remarks, Table 5 provides the main input parameters required to estimate the LCoE of the four WEC devices that is also reported in Figures 12 and 13 for the first and second deployment site, respectively. The same Figures also investigate the incidence of the discount rate that reduces from the initial value of 12% up to 3%, so matching the common values typical of other renewable sources on the international market. By the comparative analysis reported in Figures 12 and 13, it is gathered that the cost of energy is still high and not yet competitive on the international market, as regards other renewable sources. Anyway, significant savings can be gained if the discount rate diminishes as it is expected to occur in the next future, due to the increasing readiness level of all ocean technologies. Anyway, the EU target of 0.20 Euro/kWh is still far, but not impossible to reach. Really, additional concurrent actions need to be undertaken to reach the EU target that mainly involve: (i) the reduction of capital costs, that can be gained only thanks to the mass production of WEC devices and (ii) the increase of the AEP that can be reached by shifting the cut-out sea state, to harness additional wave energy. In this respect, the WEC devices with the fully submerged mass allow obtaining additional savings that are not so high as the power production increase, provided that the AEP is counterbalanced by the higher capital costs required to produce the resonant floating buoy, as predictable. Anyway, apart from these savings, the new WEC device allows more efficiently harnessing the available wave energy, with a positive impact in terms of sea areas to be occupied by wave farms.

Table 5. Main parameters for LCoE assessment.

| | h=0.0 m | h=1.0 m | h=1.5 m | h=2.0 m |
|---|---|---|---|---|
| CAPEX (k€) | 115.4 | 148.3 | 157.4 | 166.8 |
| OPEX (k€) | 5.8 | 7.4 | 7.9 | 8.3 |
| r | 12% | | | |
| n (years) | 20 | | | |
| AVA | 98% | | | |

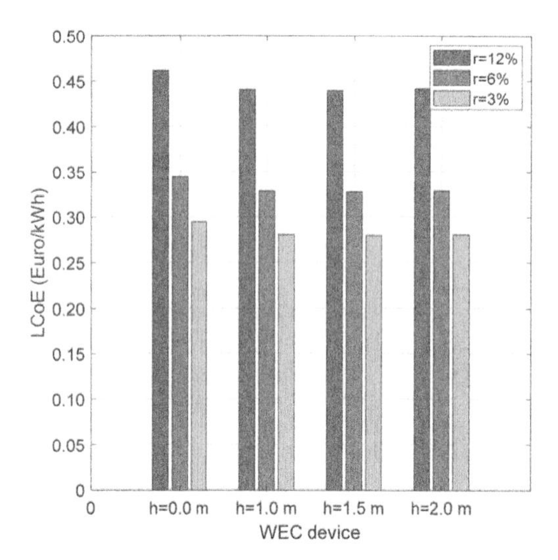


Figure 12. LCoE assessment between Viana do Castelo and Porto.

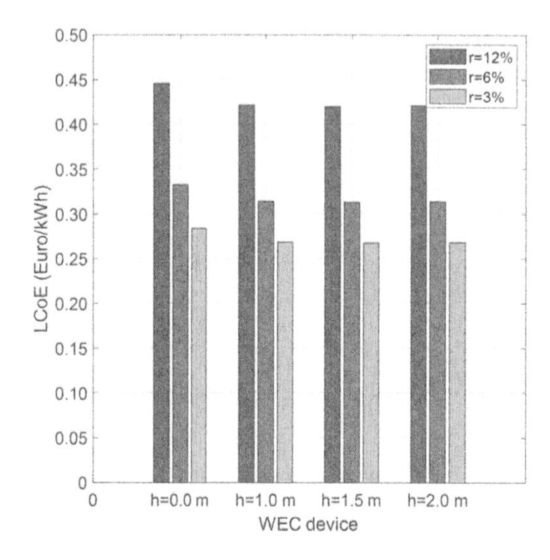


Figure 13. LCoE assessment between Nazaré and Peniche.

## 6 CONCLUSIONS

The paper focused on the AEP assessment of the resonant point absorber, equipped with a fully submerged toroidal shape, recently developed at the University of Naples "Parthenope". After a brief review about the hydrodynamic modelling in the time-domain, the main input data required to perform the AEP analysis are outlined. Two candidate deployment sites, located along the Portuguese coastline, are selected and the relevant $H_s$-$T_e$ bivariate distributions are provided. Four reference layouts are investigated, the former consisting of a hemispherical floating buoy, the remaining three ones equipped with a toroidal shape whose height ranges from 1.0 to 2.0 m, with 0.5 m step.

By the case study reported in Section 4, the new WEC device allows increasing the AEP up to 50%, as regards a conventional point absorber. This outcome is mainly due to the different hydrodynamic behaviour of the floating buoy that mainly depends on the heave RAO. Besides, by the preliminary assessment of the LCoE, it is gathered that the new layout allows reducing the cost of energy, even if the savings are not so high as the power production increase, provided that a balancing occurs between the increase of the AEP and the capital cost. Finally, by Figures 12 and 13 it is gathered that appreciable savings can be gained reducing the discount rate that is still high for the ocean energy sector, if compared with other well-established renewable sources. Anyway, the EU target to be reached by 2025 is not so far and it can be gained and even overpassed if the suggestions provided in Section 5 will be envisaged. This topic will be the subject of future works, in order to make the marine renewable energy sector more competitive on the international market and move it towards the commercialization phase.

## REFERENCES


Babarit, A., Delhommeau, G., 2015. Theoretical and numerical aspects of the open-source BEM solver NEMOH. Proceedings of the 11th European Wave and Tidal Energy Conference, Nantes, France.

Budal, K., Falnes, J., 1975. A resonant point absorber of ocean-wave power. *Nature* 256, 478–479.

Clément, A., McCullen, P., Falcão, A., Fiorentino, A., Gardner, F., Hammarlund K., Lemonis, G., Lewis, T., Nielsen, K., Petroncini, S., Pontes, M.-T., Schild, P., Sjöström, B.-O., Sørensen, H.C., Thorpe, T., 2002. Wave energy in Europe: current status and perspectives. *Renewable and Sustainable Energy Reviews* 6(5), 405–431.

Cummins, W.E., 1962. The impulse response function and ship motions. *Schiffstechnik* 9, 101–109.

Do, H-T., Dang, T.-D., Ahn, K.K., 2018. A multi-point absorber wave-energy converter for the stabilization of output power. *Ocean Engineering* 161, 337–349.

European Commission, 2018. Ocean Energy Implementation Plan – final version – 21st March 2018.

Falcão, A.F.O., 2010. Wave energy utilization: A review of the technologies. *Renewable and Sustainable Energy Reviews* 14, 899–918.

IEA and NEA, 2015. Projected Costs of Generating Electricity, Technical Report of the International Energy Agency and the Nuclear Energy Agency.

Kolios, A., Di Maio, L.F., Wang, L., Cui, L., Shen, Q., 2018. Reliability assessment of point-absorber wave energy converters. *Ocean Engineering* 163, 40–50.

Kung, S.Y., 1978. A New Identification and Model Reduction Algorithm via Singular Value Decompositions. Proceedings of the 12th Asilomar Conference on Circuits, Systems and Computers (Institute of Electrical and Electronic Engineers, 6-8 November 1978, New York, 705–714.

Magagna, D., 2019. Ocean Energy Technology Development Report 2018, EUR 29907 EN, European Commission, Luxembourg, 2019, JRC118296.

MathWorks, 2018. Matlab User Guide R2018b. Online version at www.mathworks.com

Mota, P., Pinto, J.P., 2014. Wave energy potential along the western Portuguese coast. *Renewable Energy* 71, 8–17.

Piscopo, V., Benassai, G., Cozzolino, L., Della Morte, R., Scamardella, A., 2016. A new optimization procedure of heaving point absorber hydrodynamic performances. *Ocean Engineering* 116, 242–259.

Piscopo, V., Benassai, G., Della Morte, R., Scamardella, A., 2017. Towards a cost-based design of heaving point absorbers. *International Journal of Marine Energy* 18, 15–29.

Piscopo, V., Benassai, G., Della Morte, R., Scamardella, A., 2020. Towards a unified formulation of time and frequency-domain models for point absorbers with single and double-body configuration. *Renewable Energy* 147, 1525–1539.

Piscopo, V., Scamardella, A., 2019. Design of a new point absorber with a fully submerged toroidal shape. *Ocean Engineering* 191, 106492:1–13.

Piscopo, V., Scamardella, A., 2021. Parametric design of a resonant point absorber with a fully submerged toroidal shape. *Ocean Engineering* 222, 108578: 1–17.

Pontes, M.T., 1998. Assessing the European wave energy resource. *Journal of Offshore Mechanics and Arctic Engineering* 120, 226–231.

Rusu, E., Guedes Soares, C., 2009. Numerical Modelling to estimate the spatial distribution of the wave energy in the Portuguese nearshore. *Renewable Energy* 34(6), 1501–1516.

Safonov, M.G., Chiang, R.Y., 1989. A Schur Method for Balanced Model Reduction. *IEEE Transactions on Automatic Control* 34(7), 729–733.

Silva, D.; Bento, A. R.; Martinho, P., Guedes Soares, C., 2015. High resolution local wave energy modelling for the Iberian Peninsula. *Energy* 91, 1099–1112.

Silva, D.; Martinho, P. and Guedes Soares, C., 2018. Wave energy distribution along the Portuguese continental coast based on a thirty-three years hindcast. *Renewable Energy* 127(4), 1067–1075.

Sjökvist. L., Göteman, M., 2017. Peak Forces on Wave Energy Linear Generators in Tsunami and Extreme Waves. *Energies* 10(9), 1323.

Ulvgård, L., Sjökvist, l., Göteman, M., Leijon, M., 2016. Line Force and Damping at Full and Partial Stator Overlap in a Linear Generator for Wave Power. *Journal of Marine Science and Engineering* 4(4), 1–14.

Yu, Z., Falnes, J., 1995. State-space modelling of a vertical cylinder in heave. *Applied Ocean Research* 17, 265–275.

Zheng, S, Zhang, Y., Iglesias, G., 2020. Concept and performance of a novel wave energy converter: Variable Aperture Point-Absorber (VAPA). *Renewable Energy* 153, 681–700.

*Trends in Maritime Technology and Engineering – Guedes Soares & Santos (eds)*


# Potential opportunities of multi-use blue economy concepts in Europe

S. Ramos, H. Díaz & C. Guedes Soares
*Centre for Marine Technology and Ocean Engineering (CENTEC), Instituto Superior Técnico, Universidade de Lisboa, Lisbon, Portugal*

ABSTRACT: A non-exhaustive review of existing or proposed combinations of marine exploitation activities in multi-use configurations, either in co-location, integration, or island structure setups is presented. The reviewed designs involve the synergic combination of the marine uses for marine renewable energies (wind, wave, solar, tidal, OTEC, hydrogen and gas), aquaculture, tourism and leisure, transport and ports, or desalinization. Potential multi-use opportunities in European regions have been presented and their financial, technical, regulatory, and environmental benefits or challenges analyzed. Further efforts requiring special attention to move towards the commercialization of multi-use have been highlighted.

## 1 INTRODUCTION

A context of global energetic transition has boosted the exploitation of various sources of renewable energy for the partial supply of the world energy demand during the past decades. Solar, wind, hydropower, geothermal and biomass, are some of the most popular renewable energies. Together, they account for about 26% of the annual average generated electricity worldwide (~25,619 TWh) (EIA 2020). Nonetheless, these forms of energy are associated with major issues, such as, resource intermittency, market liberalization, energy storage limitations, grid reconfiguration and conflicting use of inland space (Clemente et al. 2021). Marine renewable energy (MRE) sources (conformed by offshore wind, tidal and ocean currents, salinity gradients, ocean thermal energy conversion and wave energy) may contribute to defeating these challenges.

Offshore wind energy is currently commercialized to some extent and their installation costs have been reduced as its efficiency and learning curve improved. The European Union (EU) leads the global market, with about 90% of global recently concluded offshore wind farm projects (EU Commission 2017). Tidal and current energies are emerging, at the border between full-scale prototype demonstration and commercial exploitation; wave energy, salinity and thermal gradient are still in a prototype testing stage.

A review on the global state-of-the-art of the different marine energies is given in Esteban et al. (2019) and Nassar et al. (2020). Besides the emerging MRE, other marine industry sectors, such as maritime transport, fisheries, aquaculture, desalinization, mineral extraction, oil and gas production, and tourism, are also in an expanding process, because of increasing consumerism, market and trade globalization, a growing seafood demand and rising interest for the coastal tourism.

Despite a slight drop in 2020 due to the COVID19 crisis, international maritime transport has been constantly growing for the last few decades. The world maritime trade is expected to continue rising over the 2022–2026 period (UNCTAD 2021). The aquaculture industry has also raised in the last three decades, where the annual average production outstepped captures from the fishery (Billing et al. 2022) .

In such a context, the increasing competition for space in coastal seas has become an issue requiring prompt action. Consequently, there is currently a global motivation to find synergies between Blue Economy sectors and marine exploitation industries in multi-use configurations (EU Commission 2021).

Multi-use activities can be referred to as the simple sharing of the same marine space, but also to different activities built upon a single installation, often regarded as Multi-Purpose Offshore-Platforms (Aryai et al. 2021). The design of new floating multi-use platforms and subsea engineering is inspired by the offshore platforms of the oil and gas industries that date back to the 19th century.

The combination of different activities into the same platform or space contributes to lowering the infrastructure costs and makes the most of the marine space. As an example, a 5% decrease in the operations and management costs of a wind farm was found when integrating it into a mussel cultivation facility (Buck et al. 2010). Furthermore, cost savings were found to occur when exploitation devices share moorings and logistics in cases of integration between marine activities (Connolly & Hall 2019, Dalton et al. 2019). A spatially efficient

DOI: 10.1201/9781003320289-49

production system can remarkably enlarge the production rate. Moreover, an 8% increase in spatial efficiency was found by Griffin et al. (2015) with the combination of wind turbines and mussel farming. Moreover, multi-use may also contribute to offsetting the high intermittency of some MREs and add stability to the electric systems.

Seeking the advantages of multi-use, the European Commission has assisted the progress of the Blue Economy and Multi-use sector since the late 2000s, by funding different R&D programs and granting projects, such as the Cooperation Energy program (FP7 - Energy 2007), the Ocean for Tomorrow (FP7- Ocean 2013) the Horizon2020 (H2020 2014), or the EMFAF Flagship (EMFAF-Flagship 2021), among others.

This paper presents a review on existing or proposed configurations of multi-use platforms up to date, proposes potential multi-use opportunities in the European regions, analyzes their associated benefits and challenges and highlights further lines of research. The paper is structured as follows: Section 2 presents a review of existing multi-use configurations and projects. Section 3 proposes some potential opportunities for multi-use in European Atlantic regions and discusses the main advantages and challenges associated with the operation of multi-use activities, Section 4 present further research to move forward, and the last section states the main conclusions of the study.

## 2 STATE-OF-THE-ART

Several multi-use concepts have been proposed up to date, either in singular research publications or as a part of bigger scale projects with different funding sources. Main projects around multi-use in different European regions were: the TROPOS project, where different modular multi-use concepts were designed targeting the Mediterranean, tropical and subtropical regions (Hernández Brito 2015, TROPOS 2019); the MERMAID project, where multiple concepts were analyzed under different conditions at representative sites (MERMAID 2012, MERMAID Final Report 2015); H2Ocean which had a special target on the combination of wind and wave energy exploitation for hydrogen production (H2OCEAN 2011); the MARIBE project examine business opportunities for the most promising multi-use combinations (MARIBE 2015); and the MUSES project explores the opportunities for multi-use across five EU sea basins (Baltic Sea, North Sea, Mediterranean Sea, Black Sea and Eastern Atlantic) (MUSES 2018). The EU SCORES project, which assess the possibilities of combination of offshore wind, wave, and offshore solar PV energy, is currently ongoing (EU SCORES 2021).

Most of the proposed multi-use designs are in the concept and design stages, although some prototypes have already been tested in laboratories, demonstrated in real sea state conditions, and even commercially operated. This review classifies multi-use setups based on the connectivity between activities into two different approaches: co-location, integration, and island structure (Nassar et al., 2020).

In a co-location configuration, different marine exploitation activities are located close together and share the same marine space, maintenance, operation equipment and grid connections, although they do not share the same platform or foundations. It has been proved that a strategic configuration of sharing space can result in better structural and operational reliability (Aryai et al. 2021).

For instance, research performed by (Rusu & Guedes Soares 2013, Bento et al. 2014, Silva et al. 2018) demonstrated that the installation of wave energy converters in the close vicinity of a nearshore aquaculture farm has a positive sheltering effect in aquaculture cages, as it contributes to the dissipation of the energy associated with the incident wave front. Examples of co-location setups are given in and Table 2, and the graphical illustrations are shown in Figures 1, 2.

Conversely, an integration set-up is achieved when various marine resource exploration devices (such as wind turbines, solar panels, or aquaculture cages) are built upon the same shared platform and foundation that supports multiple uses. This configuration carries a few advantages, such as shared logistics, reduced use of the marine space, and potential internal energy supply (aquaculture installations can benefit from the supply of wind or wave energy from the same structural unit). The foundation of these structures can be floating, or bottom fixed and hybrid structures can be achieved by either building new platforms or adapting the operation of already existing structures (i.e. offshore oil and gas exploitation platforms). Transforming existing structures has the added value of lowering the decommissioning costs of old platforms (Aryai et al. 2021). Examples of integrated platforms are presented in Table 3, and a graphical idea of their characteristics is provided in Figure 3.

Lastly, the definition of island structures consists of large designs of multi-use platforms that act as artificial offshore islands and host a variety of marine-related synergic activities (either integrated into the platform itself or in its vicinity). Synergic activities can be the exploration of marine energy, aquaculture farming, seawater desalinization, or a variety of leisure activities, among others. The island concept became popular with the development of the TROPOS project, where three island settings were proposed: the sustainable service hub island, the green and blue island, and the leisure island (Table 4 and Figure 4).

A more in-depth review of existing multi-use concepts and projects can be found in (Nassar et al. 2020).

## 3 POTENTIAL MULTI-USE OPPORTUNITIES.

Not all the regions are equally suitable for the installation of the same multi-use concepts. The location

Table 1. Reviewed multi-use integrated systems.

| Concept | Ref. | Funding | Company | Tested/ Proposed Location | Status | Graphical Description | Synergic Marin Uses |
|---|---|---|---|---|---|---|---|
| SPACE@SEA platform | (SPACE@ SEA 2017) | H2020 | Marin, TUDelft, Nemos, etc. | Europe | Prototype test | Fig. 1A | |
| MPU | (MUSICA 2020) | H2020 | Polcan, Coral Ltd., Innosea etc. | Oinousses Island (Greece) | Concept & Design | Fig. 1B | |
| ECO Wave Energy | (ECO Wave Power 2011) | H2020 + Private | Eco Wave Power | Gibraltar (Spain); Jaffa Port (Israel) | Demonstration (grid-connected and off-grid) | Fig. 1C | |
| Penghu | (GIEC 2019) | Chinese Ministry of Natural Resources | Guangzhou Institute of Energy Conversion | China | Demonstration | Fig. 1D | |
| Satellite Unit | (Hernández Brito 2015, TROPOS 2019) | FP7-OCEAN.2011-1 | Abengoa, UPM, etc. | Mediterranean, Tropical and Sub-Tropical regions | Concept & Design | Fig. 1E | |
| WSA system | (Zheng et al. 2020) | China National Key Research Scheme 2016YFC0303706 | - | South China Sea | Concept & Design | Fig. 1F | |
| Monotowers | (Shell 2006) | Private | Royal Dutch Shell | North Sea | Commercial (2006) | Fig. 1I | |
| FOWT-SFFC | (Zhang et al. 2018, Lei et al. 2020) | - | - | - | Concept & Design | Fig. 1J | |

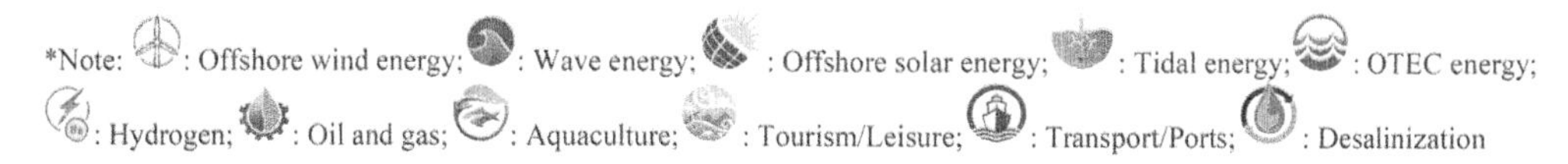
*Note: : Offshore wind energy; : Wave energy; : Offshore solar energy; : Tidal energy; : OTEC energy; : Hydrogen; : Oil and gas; : Aquaculture; : Tourism/Leisure; : Transport/Ports; : Desalinization

characteristics, such as the sea state conditions, population needs, existing infrastructure or the characteristics of the inland energy market determine which combinations of technologies are more appropriate. Moreover, despite the numerous benefits, the joint exploitation of different blue economy activities in the same space or platform is also associated with several challenges from technical, economic, regulatory, administrative, social, safety, spatial or environmental nature. Here, some potential opportunities of activity combination have been proposed, and their advantages and challenges discussed as per (Karmakar & Guedes Soares 2015, Schultz-Zehden et al. 2018, Aryai et al. 2021, Steins et al. 2021).

### 3.1 *MRE and aquaculture*

The combination of MRE with aquaculture entails the direct integration of aquaculture cages to offshore wind, wave, tide or solar energy foundations, or the co-location of aquaculture installations within the security zone of an MRE farm.

Although not abundant, one prototype of multi-use combining wind and solar energy with aqua culture currently exists in Cyprus, and few other prototypes have been considered for operations in Belgium, the Netherlands, Germany, Sweden, Denmark, and the UK. Most projects are focused the combination of offshore wind with mussel or seaweed cultivation (Schultz-Zehden et al. 2018, Bocci et al. 2019).

Table 2. Reviewed multi-use integrated systems (continuation).

| Concept | Ref. | Funding | Company | Tested/ Proposed Location | Status | Graphical Description | Synergic Marine Uses |
|---|---|---|---|---|---|---|---|
| Hex Box | (Ocean Aquafarms 2015) | - | Ocean Aquafarms | Norway/ China | Prototype test | Fig. 2K | |
| OIPS platform | (INNOVAKEME 2009) | IDERAM | Hexicon, EEM, AREAM, etc. | Porto Santo (Madeira) | Concept & Design | Fig. 1G | |
| Blue Growth Farm platform | (UniRC 2021) | H2020 | RINA, FINCOSIT, SAFIER, SAMS, etc. | Reggio Calabria (Italy) | Demonstration | Fig. 2L | |
| The FFP platform | (Floating Power Plant 2004) | EUDP + Private | Cefront, Siemens, DTU, etc. | - | Prototype test | Fig. 1H | |
| Poseidon, Two wave one wind, Cantabria Platform, Wave Treader | (Sarmiento et al. 2015) | FP7-OCEAN.2011-1 | Poseidon Floating Power, Green Ocean Energy Company, Enerocean | Kriegers Flak (Denmark); Gemini site (Netherlands); Santoña (Spain); Venice (Italy) | Concept & Design; Demonstration (grid-connected) | Fig. 2M, 2P, 2Q, 2S | |
| W2Power, OWC Array, Spar Tours Combination, and Semi-submersible Flap Combination | (MARINA 2010) | FP7 – Energy | Acciona, Pelagic Power, NTNU, etc. | Norway; Spain; Sardinia | Concept & Design; Prototype test | Fig. 2O, 2T, 2V | |
| WEGA Hybrid Coupling | (Sojo & Auer 2013) | Private | Sea for Life Lda. | - | Prototype test | Fig. 2R | |
| Ocean Hybrid Platform | (OHP 2021) | Private | SINN | Greece | Prototype test | Fig. 2N | |
| Seagen W-Shape | (Sarmiento et al. 2015) | UK gov; EU commission | Seagen, Simec Atlantis | Ireland | Prototype test | Fig. 2U | |

*Note: : Offshore wind energy; : Wave energy; : Offshore solar energy; : Tidal energy; : OTEC energy; : Hydrogen; : Oil and gas; : Aquaculture; : Tourism/Leisure; : Transport/Ports; : Desalinization

This combination represents a remarkable opportunity for territories where aquaculture or MRE exploitation areas are already functional, or at least projected for the medium term and included on the correspondent marine spatial plans. Such is the case of the Mediterranean, Norway, or France. Also, the north and west regions of the continental Spanish and Portuguese territories present interesting opportunities for the deployment of those concepts, as they are characterized by several locations devoted to offshore aquaculture farming. The social acceptance of this short of combination has been analyzed by Billing et al. (2022) for two potential offshore sites in regions of Italy and Scotland.

Advantages:

- Represents an opportunity to move aquaculture offshore to further exposed sites and create cost saving through shared commonalities.
- Reduces the competition of those two emerging uses of the ocean.
- Enables the use of wind, wave or solar energy generated onsite for the aquaculture operation demand. In

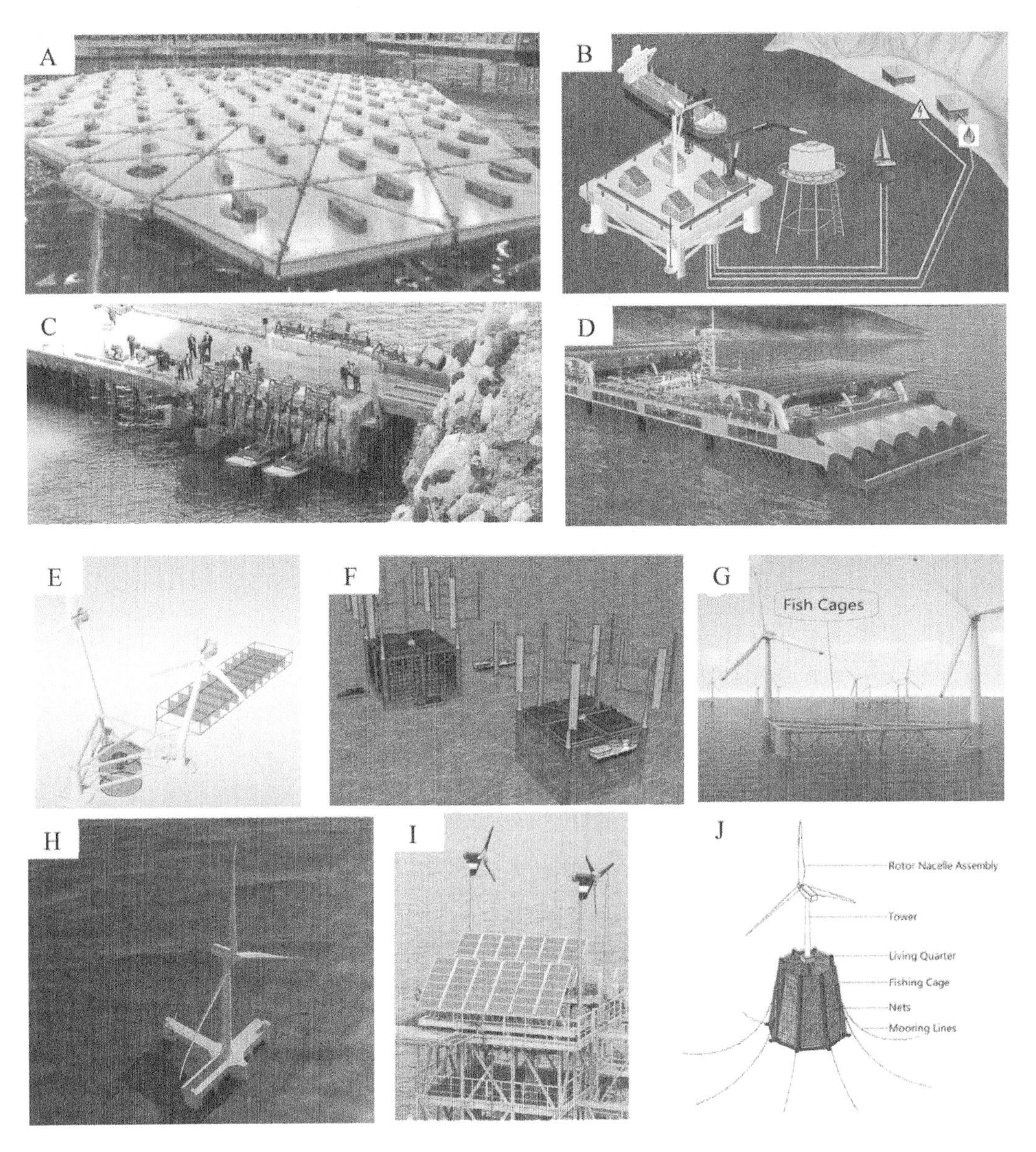


Figure 1. Graphical description of reviewed multi-use integrated systems: A) Space@Sea platform prototype (SPACE @SEA 2017). B) MPU platform (MUSICA 2020). C) ECO Wave Power (ECO Wave Power 2011). D) Penghu (GIEC 2019). E) Satellite Unit (Hernández Brito 2015). F) WSA system (Zheng et al. 2020). G) OIPS Platform (INNOVAKEME 2009). H) FFP platform (Floating Power Plant 2004). I) Monotowers (Shell 2006). J) FOWT-SFFC (Lei et al. 2020).

addition, electricity could also be conducted onshore, especially in isolated regions with high power costs.

- The systems can share the same export cable, anchors, moorings, vessels, and human resources.
- Reduced operation and maintenance costs throughout the operational lifetime.
- May reduce the wave loads in the aquaculture cages if several wave energy converters are installed immediately up wave.
- Structural foundations can have a positive impact on nature conservation, as they can act as artificial reefs for invertebrates and shelter for fish.

Challenges:

- Low technology readiness, limited knowledge, and uncertain economic return.
- Concerns about structural reliability of shared components or structures. The structural reliability of this king of combination and the most common failure models to assess it have been presented in (Aryai et al. 2021).

Figure 2. Graphical description of reviewed multi-use integrated systems (continuation): K) Hex Box (Ocean Aquafarms 2015). L) Blue Growth Farm platform (UniRC 2021). M) Poseidon Platform (MERMAID 2012). N) Ocean Hybrid Platform(OHP 2021). O) Wave2Power structure (Sojo & Auer 2013). P) Two Wave One Wind Platform (MERMAID 2012). Q) Cantabrian Platform (MERMAID 2012). R) WEGA Platform (Sarmiento et al. 2015). S) Wave Treade (Sarmiento et al. 2015)). T) OWC array (Sojo & Auer 2013). U) Seagen W-shape (Sarmiento et al. 2015). V) Semi-submersible Flap Combination and Spar Tours Combination (Sojo & Auer 2013).

- Small financial benefits of an aquaculture farm compared to potential increased risks.
- Aquaculture has high maintenance requirements, which increases traffic at the site, generating potential increased impacts on the environment and the MRE installation.
- May have collateral physical and chemical effects and be eventually harmful for the marine ecosystem. The impacts can be the modification of the wind patterns, acoustic pollution, or changes on the stratification the water column, and chemical pollution of the surrounding waters.

### 3.2 *MRE and fisheries*

This combination of activities implies marine renewable energy exploitation operations taking place in the same space than piscatorial activities, so that

Table 3. Reviewed multi-use co-located systems.

| Concept | Ref. | Funding | Company | Tested/ Proposed Location | Status | Graphical Description | Synergic Marine Uses |
|---|---|---|---|---|---|---|---|
| German Pilot | (UNITED 2020) | H2020 | R & D Centre Kiel University of Applied Sciences | Germany | Demonstration | Fig. 3A | |
| Dutch Pilot | (UNITED 2020) | H2020 | Seaweed Company and Oceans of Energy. | Netherlands | Demonstration | Fig. 3A | |
| Belgium Pilot | (UNITED 2020) | H2020 | Parkwind Farm, Jan de Nul logistics, Brevisco Aqua., UGhent, etc. | Belgium | Demonstration | Fig. 3A | |
| Danish Pilot | (UNITED 2020) | H2020 | Middelgrunden farm | Denmark | Demonstration | Fig. 3A | |
| Greek Pilot | (UNITED 2020) | H2020 | Skironis Aqua. Farm & Blue Planet Scuba Center | Greece | Demonstration | Fig. 3A | |
| H2OCEAN concept | (UNITED 2020) | FP7-OCEAN.2011-1 | - | North Atlantic, North Sea, Mediterranean Sea | Concept & Design | Fig. 3B | |

*Note: : Offshore wind energy; : Wave energy; : Offshore solar energy; : Tidal energy; : OTEC energy; : Hydrogen; : Oil and gas; : Aquaculture; : Tourism/Leisure; : Transport/Ports; : Desalinization

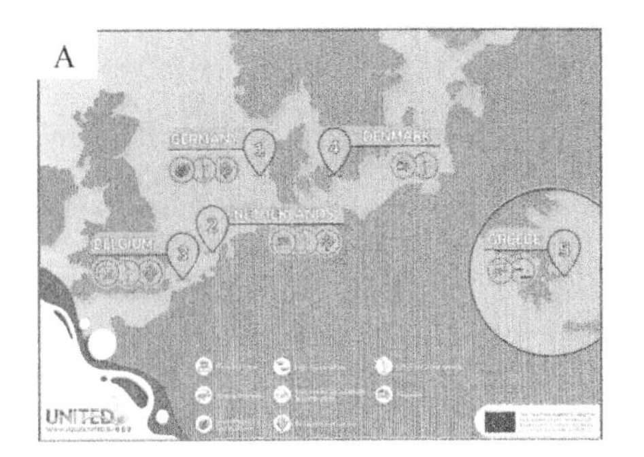


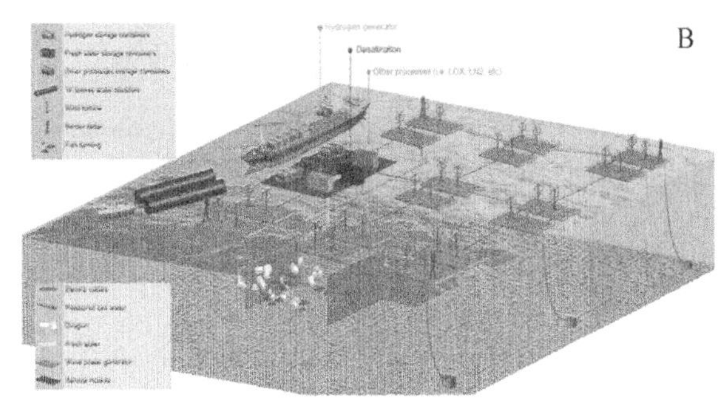


Figure 3. Graphical description of reviewed multi-use co-located systems: A) Multi-use pilot concepts within the UNITED project (UNITED 2020). B) H2Ocean concept (H2OCEAN 2011).

fisheries are not excluded either from the marine renewable energy exploitation area (in the case of offshore wind farms constitute a maximum 500 m safety zone during turbine operation) or along the offshore export power cable corridor (Díaz & Guedes Soares 2020).

This concept is especially relevant for those countries whose economy is driven by the fishing industries, and particularly if offshore MRE already exist. For instance, fishery is allowed within the offshore wind farm area during operation in Denmark and the UK. However, in Belgium and Germany, fishing is not allowed within the OWF safety zones (Schultz-Zehden et al. 2018).

Advantages:

- Potential resolution of the conflict between these two uses, facilitating social acceptance of MRE exploitation.

Table 4. Reviewed multi-use island structure systems.

| Concept | Ref. | Funding | Company | Tested/ Proposed Location | Status | Graphical Description | Synergic Marine Uses |
|---|---|---|---|---|---|---|---|
| Sustainable Service Hub Island | (TROPOS 2019) | FP7-TRANSPORT | Abengoa, UPM, etc. | North Sea | Concept & Design | Fig. 4D | |
| Green and Blue Crete | (TROPOS 2019) | FP7-TRANSPORT | Abengoa, UPM, etc. | Crete | Concept & Design | Fig. 4A | |
| Leisure Island | (TROPOS 2019) | FP7-TRANSPORT | Abengoa, UPM, etc. | Gran Canaria | Concept & Design | Fig. 4C | |
| Green and Blue Taiwan | (TROPOS 2019) | FP7-TRANSPORT | Abengoa, UPM, etc. | Taiwan | Concept & Design | Fig. 4B | |
| Offshore Container Terminal | (TROPOS 2019) | FP7-TRANSPORT | Abengoa, UPM, etc. | Panama | Concept & Design | Fig. 4F | |

*Note: : Offshore wind energy; : Wave energy; : Offshore solar energy; : Tidal energy; : OTEC energy; : Hydrogen; : Oil and gas; : Aquaculture; : Tourism/Leisure; : Transport/Ports; : Desalinization

- MRE foundations constitute particularly valuable fishing grounds that serve as artificial reefs.
- Sharing of human resources, vessels, port facilities or monitoring activities.

Challenges:

- Added environmental impacts and safety risks of fishing within the wind farms require exhaustive risk assessment and additional regulatory frameworks.
- Fishery mobile gear are generally not suitable for MRE areas, as bottom-contact gears might cause damage to submarine cables and to the fishing gear, carrying increased economic losses.

### 3.3 *MRE and desalinization*

The combination of MRE and desalinization (where the desalinization energy demand is directly supplied by solar, wind or wave renewable energy onsite) presents a promising concept in areas suffering from scarce freshwater sources and episodes of drought, which leads to permanent water shortage and the greatest supply of drinking water proceeding from desalinated seawater.

An existing, but currently not operational case of offshore wind and desalination multi-use lies in Irakleia Island and a potential case combining wind and solar energy is projected in Northern Mykonos Island, both located in Greece (Depellegrin et al. 2019). Locations such as the Canary Islands or some other coastal areas of the Mediterranean Sea could be a special target for those types of multi-use activities.

Advantages:

- The synergetic processes, where the desalinization energy demand is directly supplied by MRE onsite, constitutes a self-sustainable process with an added value to the local population.
- The combination constitutes reduced construction efforts and costs if MRE infrastructure already exists.
- Commonalities of certain infrastructures, such as cables or platforms.
- Operation and maintenance synergies have the potential of the overall reduction of costs.
- Being installed at sea, the floating desalinization platform has the potential to move and accommodate to the needs of desalinated water in different locations.

Challenges:

- Associated with high values of levelized cost of energy and large uncertainties about economic return.
- Concerns about the structural integrity of in harsh environments characterized by strong winds, intense currents, and wave interactions.
- May have collateral physical effects and eventually for the marine ecosystem, such as modification of the wind patterns, acoustic pollution or influence on the stratification and salinity of the water column.

### 3.4 *MRE and tourism*

Tourism and offshore MRE can easily share the sea space in a co-location manner. Synergetic activities include: sightseeing boat tours to offshore wind,

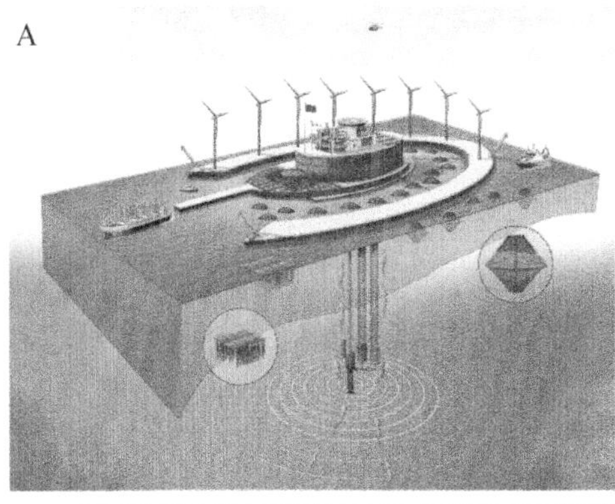


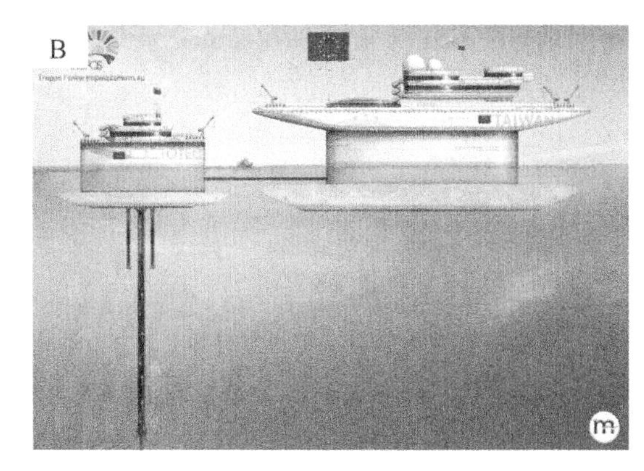

Figure 4. Graphical description of reviewed multi-use island structures: A) Blue and Green Island Crete. B) Blue & Green Island Taiwan C) Leisure Island Gran Canaria. D) Sustainable Service Hub North Sea. E) Offshore container Terminal Panamá (Hernández Brito 2014).

wave or solar energy farms; specially designed platforms can be set up around the MRE foundations (ie. wind turbines) and serve as facilities for the practice of water sports, or as offshore restaurants; boat tour operators can be engaged in MRE related monitoring or maintenance activities.

Examples of this multi-use combination already exist in most of the European countries where offshore wind farms have already been installed, mainly in the North and Baltic Seas. However, this combined activity is of prime interest for those all Atlantic and Mediterranean regions where planned projects for the development of MRE facilities exist (Schultz-Zehden et al. 2018).

Advantages:

- Minor structural modification in existing MRE infrastructure already exists.
- MRE infrastructure can constitute an added value to the landscape.
- Potential to increase of the social acceptance of MRE industries by coastal communities, as it offers socio-economic benefits.
- Potential positive impact on nature conservation, as MRE infrastructure can act as artificial reefs for invertebrates and shelter for fish.

Challenges:

- Concerning safety regarding hosting tourists on board of aquaculture or fishing vessels. Need of a complete risk assessment and insurance against accidents.
- Barriers related to distance from shore, weather, tide conditions and seasonality.
- Collateral impacts for the marine ecosystem, such as the modification of the wind patterns, acoustic

pollution, or changes on the stratification the water column.

### 3.5 *MRE and environmental monitoring and protection*

This combination implies the integration of monitoring equipment (such as passive acoustic, sonar, audio and visual on a platform or a vessel) in the offshore MRE structures. Examples of case studies about the potential implementation of this type of multi-use have been reported for in Scotland (Schultz-Zehden et al. 2018).

Advantages:

- Provides data required for the controlled and safe operation of a specific activities, to inform research or for environmental protection.

Challenges:

- Lack of subsides by public funds.

### 3.6 *MRE and shipping terminal*

This multi-use involves the generation of energy from marine renewable energy converters (offshore wind turbines, wave, or tidal devices), its transmission to a port substation and the use of that green energy to cover the consumption of the port. Often, a synergic benefit exist as those are the ports supporting the manufacturing, installation, operation, and managements of the offshore energy farms.

This combination is especially suitable where offshore marine energy converters already exist and are well-supported by a shipping terminal. That is the case of the North, Baltic, and Eastern Atlantic Sea Basins, where the offshore renewable energy industry is rapidly developing.

A case study about the potential combination of these to resources has been carried out for the West Coast of Scotland (Bocci et al. 2019). That case study showed that this combination could have potential for small docks to supply the demand of essential ferry routes between islands or remote mainland areas. Obsoleted port facilities could be used for the prototype testing.

Advantages:

- Green energy supply for shipping terminals and ports.
- Reduction of green-house gas emissions in the terminals and human health benefits.
- Inland grid connection terminal and infrastructure for offshore renewable energy farms.

Challenges:

- The connection of offshore energy to ports is associated with several technical barriers.
- Lack of a specific regulatory framework.
- Sufficient port infrastructure and vessel capacity for the support of assembly, installation, and operation of MRE devices.

### 3.7 *Synergies between MREs*

The combination between MRE sources (offshore wind, wave, tidal or solar energy) can be obtained by installing them on the same integrated platform, or just sharing the same marine space in a co-location scheme, where wave energy converters may for example fill the gaps between offshore wind turbines sharing the cable array and grid connections.

The North Sea represents a key spot for this kind of multi-use combination, as the offshore wind farm there is already in a well-established stage. Indeed, a pilot test hybrid of wind and wave energy converters has been projected in Scotland (Caithness), whose long-term aim is to deploy a commercial scale project. Within the scope of the MERMAID project, a feasibility study has been also performed in the Eastern Atlantic, in the Spanish region of Cantabria (Schultz-Zehden et al. 2018).

Advantages:

- Capacity to generate maximum energy per nautical space.
- Reduced costs due to the ability to share installation equipment, moorings, cable connections, monitoring systems, operation, and maintenance activities.
- Mitigates space conflict between MREs.
- The simultaneous production of different MREs can improve the quality of the power output and have a positive impact on levelling the power output delivered to the grid.
- The setup of wave energy converter arrays up wave offshore wind, tide or solar energy converters reduces the wave loads impacting them.
- MRE platforms may constitute a positive impact on nature conservation, acting as artificial reefs for invertebrates and shelter for fish.

Challenges:

- Need of separate permitting and regulatory processes and different tariff rates.
- High levels of uncertainty regarding the economic return.
- Potential effects on the ecosystem, such as modification of the wind patterns or acoustic pollution. Impacts can be suffered by communities of seabirds and bats, fish populations, marine mammals, etc.
- Concerns about the structural integrity of multi-use platforms in harsh environments characterized by strong winds, intense currents, and wave interactions.

### 3.8 *Oil and gas decommissioning*

Oil and gas offshore platforms, their functions and pipelines can take a second life before being totally remove, and serve as structures for other marine uses, such as: $H_2$ storage, generation of offshore wind power to gas or $H_2$, docking ports and fuel supply stations for recreational or industrial navigation, support for recreational activities (diving, recreational fishing, marinas, hotels, restaurants, etc.), platforms for ocean monitoring and research activities, structural support for aquaculture activities or renewable energy converters, and as artificial reefs.

This multi-use concept is a key opportunity for those regions where several oil and gas structures are expected to be decommissioned in the next years, such as the North Sea or the Northern Adriatic Sea (Schultz-Zehden et al. 2018).

Advantages:

- Significative reduced cost of decommissioning and of new installations for other uses.
- Can act as artificial reef causing a positive impact for the ecosystem once chemical contamination risk due to oil or gas discharge are suppressed.

Challenges:

- Lack of specific regulation and guidance specifying the ownership rules and liability during the reuse period.
- Poor public acceptance of activities linked to oil and gas companies.

### 3.9 *Tourism and aquaculture*

This multi-use concept propones the inclusion of touristic offers into aquaculture related activities such as visits to aquaculture sites, diving/snorkeling around the aquaculture installation or spearfishing next to the aquaculture installation.

Tourism and aquaculture combination is especially interesting in territories whose economy is mostly driven by the fishery and tourism sectors, and where marine-related leisure activities are also an important source of income (such as kayaking, snorkeling, spearfishing, surfing and Stand-Up Paddling). Ideally, multi-use configurations would involve the co-location of tourism and leisure services in locations where existing aquaculture infrastructure already exist.

Instances of this multi-use of the space currently exist in countries such as Italy, Slovenia, Greece, or Malta (Depellegrin et al. 2019). Some areas in the South of Portugal and northern Adriatic Sea, as well as the Madeira Islands, Canary Islands or Cyprus, gather these characteristics, which make them suitable for the combination of aquaculture activities, possibly fed by offshore solar energy, together with touristic, and leisure activities.

Advantages:

- The combination would constitute a minor modification process if aquaculture infrastructure already exists, resolving the competition for space among tourism and aquaculture.
- Constitutes increased economic benefits for aquaculture operators and promotes aquaculture acceptance and value of the locally produced fish.
- Only one environment impact assessment (EIA) and permitting process may be needed.

Challenges:

- Concerning safety regarding hosting tourists on board of aquaculture or fishing vessels. Need of a complete risk assessment and insurance against accidents.

### 3.10 *Tourism, fisheries & environmental protection*

Combination of small-scale fishing vessels with touristic tours where visitors can personally discover the fishing traditions. Pedagogical activities related to environmental protection and sustainability are to be applied during the experience.

Existing examples of this type of combination can be found in Mediterranean countries like France, Spain, Italy, Greece, or Malta. Moreover, other territories whose economy has been historically driven by the fishing industries, and characterized by a good-established touristic attraction, constitute great potential opportunities for this short of multi-use, such as Slovenia, Croatia, or Cyprus (Depellegrin et al. 2019). Case studies about the potential of this combinations were also assessed for the South Coast of Portugal, the Archipelago of Azores, the North Adriatic, and the Aegean Sea (Bocci et al. 2019).

Advantages:

- Added source of income for small fishermen or fishing companies.
- Contribute to raising environmental protection and sustainability awareness.
- Shares vessels, dock infrastructures and logistics, reducing the conflict for the use of the marine space.

Challenges:

- Concerning safety regarding hosting tourists on board of fishing vessels. Need of a complete risk assessment and insurance against accidents.

### 3.11 *Tourism, underwater cultural heritage & environmental protection*

This multi-use concept is about linking touristic or recreational activities with the protection of underwater archaeology and its surrounding marine ecosystems. This can be performed through land-based museums about the underwater cultural heritage, onsite visits to marine locations of interest using glass bottom boats, or in situ access to scuba divers. Those practices are to be carried out by always transmitting a sense of marine environment protection and respect.

Instances of this synergetic activities already exist in some locations of the Baltic, the Mediterranean and Eastern Atlantic. In the Black Sea, the HERAS project promoted thus multi-use idea as well (HERAS Project 2014). Locations rich in underwater cultural heritage, warm temperatures, and clear waters, are particularly interesting for these kind of use (such as the Mediterranean Sea) (Schultz-Zehden et al. 2018, Depellegrin et al. 2019).

Advantages:

- Win-win situations for both tourism, cultural protection, and environmental protection, as it raises public awareness and appreciation.
- Added source of income for local communities.

Challenges:

- Potential impacts and disturbances for the marine ecosystem.
- Strict protection and resistance from authorities regarding tourist access to cultural heritage sites due to the risk of damage.

A total of 10 activities of the marine sector have been proposed for combination, with tourism and MRE (either offshore wind, wave, tidal or solar energy) being the more frequently activities proposed for multi-use set-ups. The aquaculture, fishing and environmental protection sectors were also considered by more than one potential combination.

When comparing the different possible combinations, it is evident that common advantages and challenges exist for all of them. On one hand, the mitigation of conflict and competition between marine users is one of the main shared benefits by all the multi-use combinations. The reduction of costs due to shared infrastructure, logistics and human resources is also a major driver for the implementation of all the concepts, especially if part of the infrastructure exists because of the active operation of one of the marine uses.

On the other hand, due to the still emerging technological state, with limited knowledge and just few existing concepts operating in real conditions, the combinations must deal with challenges associated to high economic return uncertainty, concerns about the structural integrity of the installations when operating in harsh marine environments, added environmental impacts, safety risks, and lack of a specific regulatory framework and guidance.

Nonetheless, some combinations are qualitatively associated with a positive balance between benefits and challenges. As an example, the use of existing MRE or disused oil and gas infrastructures to set up environmental monitoring devices, has the potential of providing valuable real marine data for multiple public users and entities, and it is just hindered by a lack of subsides and public funds.

Also, the development of touristic and environmental protection activities in underwater cultural heritage spots, which can constitute a great source of income for local communities, is associated with relatively small challenges that can be easily minimized (such as disturbance for the marine ecosystem and the interdicted access to underwater heritage areas due to excessive protection from the authorities). Similarly, creating touristic tours around MRE facilities, such as existing offshore wind farms, is mainly associated to regulatory issues and safety concerns for the tourists. Those challenges can be mitigated with the contract of special insurances and the proper elaboration of specific risk assessments.

Conversely, other multi-use combinations entail difficult challenges which can eventually surpass the benefits, and those require detail feasibility assessments before deployment. Such is the case of performing fishing or aquaculture activities within MRE exploitation areas. In exchange of small financial benefits, this multi-use is associated to high concerns about the structural reliability of shared components, and with increased safety risks and equipment damage due to the simultaneous operation of activities.

Similar conclusions were retrieved in a previous study performed by (Bocci et al. 2019) within the MUSES project. They computed specific scores to quantitatively rank the expected potential and effects of different multi-use opportunities, based on multiple stakeholders' experiences and opinions. The combinations scoring the highest positive potential and effects (meaning that benefits from drivers and added values exceed the drawbacks of the associated barriers and impacts) was found to be that of the MRE and monitoring, followed by the combination of tourism, underwater cultural heritage, and environmental protection, and that of MRE with desalinization. Conversely, the combinations scoring the worst balance were found to be the links between MRE and Aquaculture, and that combining shipping terminals with the MRE.

Despite that qualitative and quantitative classification, the decision on the most suitable combination can be significantly influenced by local conditions and should be ultimately made site-specific, by analyzing the local drivers, barriers, impacts, added values and site characteristics.

## 4 FUTURE RESEARCH

Overcoming the challenges hindering existing multi-use concepts and promoting the implementation of multi-use requires common cooperation between stakeholders and decision-makers in marine spatial planning processes. Yet, further research is required to identify the best technologies and configurations that minimize the impacts and optimize return, so it results attractive for the energy market. The following topics require further attention:

1. Policy, regulations, and marine spatial planning:

Multi-use of the marine space can be highly encouraged through the demarcation of specific marine areas and prioritizing multi-use instead of single use in the regional and national marine spatial planning. For that purpose, greater consideration should be taken from the stakeholders' priorities in the decision-making, as well as from the macro-regional needs.

Moreover, a minimization of regulatory and administrative barriers for multi-use must be promoted, and a consistent policy framework and guidelines (currently weak or non-existing) must be developed to harmonize the safe operation of multi-use. As an example, roadmaps might be needed for existing ports to get provided from the required infrastructures, vessels and human resources needed for the future support of marine multi areas.

2. Promotion of research, data monitoring and social awareness:

Support the production and sharing of marine data, as well as the access to measured economic, social, and environmental impacts of operational multi-use concepts, such as the resilience of marine ecosystems in a contexts of multi-use deployment. This generates greater knowledge for the safety operation of multi-uses, ensures environmental protection and improves social acceptance. Business models and case studies including possible deployment paths of the most promising concepts in the real sea are also crucial to attract financial investment.

3. Funding for pilot projects:

At the current stage of technological development of multi-use systems, still characterized by a high levelized cost of energy and uncertain economic return, government funding is essential to keep up in the development path until multi-use becomes competitive. Due to a still poorly attractive business for investors, public funding might be the only chance for the deployment of pilot projects and testing sites, ultimately essential for the compilation of knowledge around multi-use.

4. Technical design and structural reliability:

The structural reliability of a multi-use infrastructure design will be affected overtime, due to external loadings, deterioration phenomena, accidents, poor workmanship, or severe storms. Therefore, innovation around the technical design and structural reliability of the different elements of the multi-use structure (moorings, cable installation, fishing-friendly cable protection, connection of offshore energy to ports, etc.) is fundamental for moving forward towards the commercialization of the industry. Monitoring and collecting operational data is essential for the structural reliability analysis and to detect possible opportunities of improvement.

## 5 CONCLUSIONS

This paper reviewed existing or projected combinations of marine exploitation activities, either in co-location, integration, or island structure setups. The reviewed designs involve the synergic use of the marine space for MRE exploitation (wind, wave, solar, tidal, hydrogen and gas), aquaculture, tourism/leisure, transport/ports, and/or desalinization. A great amount of those designs has been possible because of public funding granted to a variety of EU projects targeting the impulse of renewable energies, the encouragement of blue growth and the mitigation of climate change. Yet, several financial, technical, spatial, regulatory, and environmental challenges need to be addressed to make multi-use commercially viable.

Several potential opportunities of multi-use combination have been put forward in the scope of the European regions, considering location characteristics, population needs, existing infrastructure or the characteristics of the inland energy market. When assessing the advantages and challenges associated with each multi-use concept, it can be envisaged that some combinations are characterized by a positive balance between expected benefits and associated challenges, while others are associated with heavier barriers than advantages.

For successful multi-use future deployments, the importance to delimit specific areas for multi-use purposes in the national marine spatial planning is specially highlighted, by covering stakeholders' motivations and attitudes, and identifying feasible policy solutions. Also, when it comes to extend the research on multi-use concepts, it is essential to increase the knowledge and data base on the socio-economic, spatial, regulatory, and environmental aspects, apart from those related to the technical design.

As a concluding note, in the current stage of technological development and considering its associated challenges, governmental action is still needed to boost multi-use in the blue economy sectors, such as including of multi-use in the national targets and allocating special funding.


## ACKNOWLEDGEMENTS

This work contributes to the Strategic Research Plan of the Centre for Marine Technology and Ocean Engineering (CENTEC), which is financed by the Portuguese Foundation for Science and Technology (Fundação para a Ciência e Tecnologia - FCT) under contract UIDB/UIDP/00134/2020. The first author is financed by FCT under the grant 2020.06618.BD.


## REFERENCES


Aryai, V., Abbassi, R., Abdussamie, N., Salehi, F., Garaniya, V., Asadnia, M., Baksh, A.A., Penesis, I., Karampour, H., Draper, S., Magee, A., Keng, A.K., Shearer, C., Sivandran, S., Yew, L.K., Cook, D., Underwood, M., Martini, A., Heasman, K., Abrahams, J., & Wang, C.M. 2021. Reliability of multi-purpose offshore-facilities: Present status & future direction in Australia. *Process Safety & Environmental Protection*, 148, 437–461.

Bento, A.R., Rusu, E., Martinho, P., & Guedes Soares, C. 2014. Assessment of the changes induced by a wave energy farm in the nearshore wave conditions. *Computers and Geosciences*, 71 (1), 50–61.

Billing, S.-L., Charalambides, G., Tett, P., Giordano, M., Ruzzo, C., Arena, F., Santoro, A., Lagasco, F., Brizzi, G., & Collu, M. 2022. Combining wind power and farmed fish: Coastal community perceptions of multi-use offshore renewable energy installations in Europe. *Energy Research and Social Science*, 85, 102421.

Bocci, M., Sangiuliano, S.J., Sarretta, A., Ansong, J.O., Buchanan, B., Kafas, A., Caña-Varona, M., Onyango, V., Papaioannou, E., Ramieri, E., Schultz-Zehden, A., Schupp, M.F., Vassilopoulou, V., & Vergílio, M. 2019. Multi-use of the sea: A wide array of opportunities from site-specific cases across Europe. *PLoS ONE*, 14 (4).

Buck, B.H., Ebeling, M.W., & Michler-Cieluch, T. 2010. Mussel cultivation as a co-use in offshore wind farms: potential & economic feasibility. *Aquaculture Economics and Management*, 14 (4), 255–281.

Clemente, D., Rosa-Santos, P., & Taveira-Pinto, F. 2021. On the potential synergies and applications of wave energy converters: A review. *Renewable and Sustainable Energy Reviews*, 135 (July 2020), 110162.

Connolly, P. & Hall, M. 2019. Comparison of pilot-scale floating offshore wind farms with shared moorings. *Ocean Engineering*, 171, 172–180.

Dalton, G., Bardócz, T., Blanch, M., Campbell, D., Johnson, K., Lawrence, G., Lilas, T., Friis-Madsen, E., Neumann, F., Nikitas, N., Ortega, S.T., Pletsas, D., Simal, P.D., Sørensen, H.C., Stefanakou, A., & Masters, I. 2019. Feasibility of investment in Blue Growth multiple-use of space and multi-use platform projects; results of a novel assessment approach and case studies. *Renewable and Sustainable Energy Reviews*, 107, 338–359.

Depellegrin, D., Venier, C., Kyriazi, Z., Vassilopoulou, V., Castellani, C., Ramieri, E., Bocci, M., Fernandez, J., & Barbanti, A. 2019. Exploring Multi-Use potentials in the Euro-Mediterranean Sea space. *Science of the Total Environment*, 653, 612–629.

Díaz, H. & Guedes Soares, C. 2020. An integrated GIS approach for site selection of floating offshore wind farms in the Atlantic continental European coastline. *Renewable and Sustainable Energy Reviews*, 134.

ECO Wave Power, 2011. Available from: https://www.ecowavepower.com/ [Accessed 6 Feb 2022].

EIA, 2020. Key World Energy Statistics, 2020. *Int. Energy Agency*, 33 (August), 4649.

EMFAF-Flagship, 2021. Available from: https://www.flagshipproject.eu/ [Accessed 7 Feb 2022].

Esteban, M.D., Espada, J.M., Ortega, J.M., López-Gutiérrez, J.S., & Negro, V. 2019. What about marine renewable energies in Spain? *Journal of Marine Science and Engineering*, 7 (8).

EU Commission, 2017. *Report on the Blue Growth Strategy: Towards more sustainable growth and jobs in the blue economy.*

EU Commission, 2021. *Communication from the commission to the European parliament, the council, the Europe-an economic and social committee, and the committee of the regions.*

EU SCORES, 2021. Available from: https://euscores.eu/ [Accessed 23 Feb 2022].

Floating Power Plant [online], 2004. Available from: https://www.floatingpowerplant.com/ [Accessed 5 Feb 2022].

FP7 - Energy, 2007. Available from: https://cordis.europa.eu/programme/id/FP7-ENERGY [Accessed 7 Feb 2022].

FP7- Ocean, 2013. Available from: https://ec.europa.eu/research/bioeconomy/fish/research/ocean/infoday_en.htm [Accessed 7 Feb 2022].

GIEC, 2019. GIEC Successfully Built the First Semisubmersible Open Sea Aquaculture Platform "Penghu". Available from: http://english.giec.cas.cn/ns/rp/201908/t20190809_213996.html [Accessed 23 Feb 2022].

Griffin, R., Buck, B., & Krause, G. 2015. Private incentives for the emergence of co-production of offshore wind energy and mussel aquaculture. *Aquaculture*, 436, 80–89.

H2OCEAN, 2011. Available from: http://www.vliz.be/projects/mermaidproject/project/related-projects/h2ocean.html [Accessed 5 Feb 2022].

H2020, 2014. Available from: https://ec.europa.eu/programmes/horizon2020/en/what-horizon-2020 [Accessed 7 Feb 2022].

HERAS Project, 2014. Available from: https://maritime-spatial-planning.ec.europa.eu/projects/submarine-archaeological-heritage-western-black-sea-shelf [Accessed 18 Feb 2022].

Hernández Brito, J. 2014. *TROPOS project periodic report publishable summary.*

Hernández Brito, J., 2015. *TROPOS Project Report.*

INNOVAKEME, 2009. Available from: https://www.innovakeme.com/oips/ [Accessed 5 Feb 2022].

Karmakar, D. & Guedes Soares, C. 2015. Review of the present concepts of multi-use offshore platforms. *In*: C. Guedes Soares, ed. *Renewable Energies Offshore.* London: Taylor & Francis Group, 867–875.

Lei, Y., Zheng, X.Y., & Li, W. 2020. Effects of Fish Nets on the Nonlinear Dynamic Performance of a Floating Offshore Wind Turbine Integrated with a Steel Fish Farming Cage. *International Journal of Structural Stability and Dynamics*, 03 (2050042).

MARIBE, 2015. Available from: http://maribe.eu/ [Accessed 9 Feb 2022].

MARINA, 2010. Available from: https://maritime-spatial-planning.ec.europa.eu/projects/marina-platform [Accessed 5 Feb 2022].

MERMAID Final Report, 2015. *Go offshore - Combining food and energy production content.*

MERMAID, 2012. Available from: http://www.vliz.be/projects/mermaidproject/indexd39b.html [Accessed 5 Feb 2022].

MUSES, 2018. Available from: https://muses-project.com/ [Accessed 9 Feb 2022].

MUSICA, 2020. Available from: https://musica-project.eu/about-us/ [Accessed 5 Feb 2022].

Nassar, W.M., Anaya-Lara, O., Ahmed, K.H., Campos-Gaona, D., & Elgenedy, M. 2020. Assessment of multi-use offshore platforms: Structure classification and design challenges. *Sustainability (Switzerland).*

Ocean Aquafarms, 2015. Available from: https://www.oceanaquafarms.com/ [Accessed 5 Feb 2022].

OHP, 2021. Available from: https://www.sinnpower.com/platform [Accessed 5 Feb 2022].

Rusu, E. & Guedes Soares, C. 2013. Coastal impact induced by a Pelamis wave farm operating in the Portuguese nearshore. *Renewable Energy*, 58, 34–49.

Sarmiento, J., Guanche, R., Belloti, G., Cecioni, C., Cantu, M., Franceschi, G., & Sufferdini, R. 2015. *MER-*

*MAID Project Deliverable report: Integration of energy converters in multi-use offshore platforms.*

Schultz-Zehden, A., Lukic, I., Onwona Ansong, J., & Susanne Altvater, et al. 2018. *Ocean multi-use action plan.*

Shell, 2006. *Meeting the energy challenge. The Shell Sustainability Report 2006.*

Silva, D., Rusu, E., & Guedes Soares, C. 2018. The effect of a wave energy farm protecting an aquaculture installation. *Energies*, 11 (8).

Sojo, M. & Auer, G. 2013. *MARINA Project Report.*

SPACE@SEA, 2017. Available from: https://spaceatsea-project.eu/about-space-at-sea [Accessed 5 Feb 2022].

Steins, N.A., Veraart, J.A., Klostermann, J.E.M., & Poelman, M. 2021. Combining offshore wind farms, nature conservation and seafood: Lessons from a Dutch community of practice. *Marine Policy*, 126.

TROPOS, 2019. Available from: https://www.ed.ac.uk/sustainability/what-we-do/climate-change/case-studies/climate-research/tropos-project [Accessed 5 Feb 2022].

UNCTAD, 2021. *Review of maritime transport 2021.* United Nations.

UniRC, 2021. Progetto "The Blue Growth Farm". Experimental Campaign at the NOEL Laboratory of the Università degli Studi Mediterranea di Reggio Calabria Available from: https://www.unirc.it/en/news/24094/unirc-progetto-the-blue-growth-farm-experimental-campaign-at-the-noel-laboratory-of-the-universita-degli-studi-mediterranea-di-reggio-calabria [Accessed 7 Feb 2022].

UNITED, 2020. Available from: https://www.h2020united.eu [Accessed 5 Feb 2022].

Zhang, X., Lu, D., Guo, F., Gao, Y., & Sun, Y. 2018. The maximum wave energy conversion by two interconnected floaters: Effects of structural flexibility. *Applied Ocean Research*, 71, 34–47.

Zheng, X., Zheng, H., Lei, Y., Li, Y., & Li, W. 2020. An offshore floating wind–solar–aquaculture system: Concept design and extreme response in survival conditions. *Energies*, 13 (3).

*Trends in Maritime Technology and Engineering – Guedes Soares & Santos (eds)*


# Dynamic response analysis of a combined wave and wind energy platform under different mooring configuration

J.S. Rony & D. Karmakar
*Department of Water Resources & Ocean Engineering, National Institute of Technology Karnataka, Surathkal, Mangalore, India*

C. Guedes Soares
*Centre for Marine Technology and Ocean Engineering (CENTEC), Instituto Superior Técnico, Universidade de Lisboa, Lisbon, Portugal*

ABSTRACT: In the present study, a novel concept of combining a submerged tension leg platform (STLP) with six heaving type point absorbers WEC in circular pattern is presented for different mooring configurations. The tensioned tendons are used to fix the floating combined wave and wind energy system in position. The safety, stability and power production of the combined floating platform depends significantly on the integrity of the tendons. The combined wind and wave platform supported by four, five, eight and nine tendons are analysed for the operating conditions of the 5MW wind turbine under regular waves. Time domain numerical simulation tool FAST developed by NREL is used to perform the aero-servo-hydro-elastic simulation. The spectra of surge, sway, roll, pitch and yaw motion of the combined system under each mooring configuration is presented to analyze the behavior of the combined wave and wind energy system. Statistical results on the tension developed on each tendon for different mooring configurations is also presented to study the importance of mooring and the influence of mooring system on the dynamic responses of the combined floater. The study performed will be helpful in the design and analysis of possible configurations of mooring lines supporting the floating platform and improving the structural integrity of the combined floating concept.

## 1 INTRODUCTION

Limited fossil fuel resources and the effect of global warming has increased the demand for clean renewable energy resources. The offshore region contains a tremendous quantity of renewable energy in the form of winds and waves. Thus, the study of offshore wind turbines and wave energy converters has become more significant. In the last few decades, fixed structures such as jacket platforms and gravity platforms were used for offshore wind turbine support structure. In the deeper ocean, the winds blowing with a constant speed over a fetch area and non-breaking waves can increase the power production rate. As we move away from the shoreline to deeper oceans, the cost of fixed structures increases tremendously. Hence to reduce the cost of installation, the floating structures are considered in the deep ocean. In order to keep the floating structure in stable position under different environmental conditions, different mooring systems are used. In offshore regions, mooring system plays an important role in the dynamic stability along with design and analysis of the floating platform. Three types of mooring systems are commonly such as taut mooring system, semi-taut mooring system and catenary mooring system. An array of heaving point absorbers in circular and linear pattern is studied by Sinha et al. (2015) to understand the absorption of wave energy power. The efficiency of power absorption in irregular waves are examined for different floater shapes, wave heading angles and positioning of the floater. Shen et al. (2016) investigated the coupled responses of the mini-TLP supporting 5MW wind turbine (WT) using in-house code CRAFT (Coupled Response Analysis of Floating wind Turbine). The viscous drag force is observed to induce higher harmonic pitch resonance and the resonant pitch motion is reduced by aerodynamic damping. Crudu et al. (2016) analysed the moored offshore structure and evaluated the forces in the mooring lines to predict the motions of the floating offshore structure under different environmetal conditions and mooring forces acting on the structure. The analysis of the moored structure and the dynamic behaviour of mooring chain is considered for five different scales. Soeb et al. (2016) discussed the hydrodynamic analysis of the floating platform due to action of the sea

DOI: 10.1201/9781003320289-50

environmental condition. The spar platform is considered and the cost effectiveness along with good capability from mooring loads and dynamic loads on the structure is analyzed.

The concept of submerged tension leg platform (STLP) for moderate water depth of 70 - 150m which is self-stabilized during the transportation phase is proposed by Han et al. (2017). The study investigated the dynamic responses of the submerged system and the effect of second order wave loads on the dynamics of STLP. Karimi et al. (2019) proposed a frequency domain modelling approach for FOWT supporting wind turbine and mooring systems under irregular wave loads. The study analysed and compared the six DOF system responses of 5MW wind turbine supported on OC3-Hywind, MIT/NREL TLP and OC4-DeepCwind. Le et al. (2020) studied the coupled dynamic response of submerged FOWT for intermediate water depth of 50 -200m under different mooring conditions using aero-servo-hydro-elastic-mooring tool for irregular wave environments. The study observed that the tether length has significant effects on the motion responses of the surge, heave, pitch and yaw. Zhang et al. (2020) conducted the hydrodynamic analysis of V-shaped semi-submersible, brace less semi-submersible and OC4-DeepCwind semi-submersible supporting 5MW reference wind turbine. The study investigated the hydrodynamic responses of the platform for two different water depths to observe the influence of the second order wave loads on the platform motions and mooring tensions. The study revealed that the dynamic responses of the platform are over estimated when omitting the effect of Morison's drag on the platform columns.

A novel combined concept of TLP with heaving WEC is studied by Ren et al. (2020). The coupled dynamic responses of the combined system are analysed numerically using the simulation tool AQWA and was experimentally validated using 1:50 scale model. Two non-linear air dampers are used to simulate the PTOs and a rotating wind turbine is used to develop the mean thrust effect. Wang (2020) investigated the dynamic responses of the floating wind turbine (FWT) and specified an identification method for fitting space model based on mathematical procedures to approximate the convolution terms in the equation of motion. Sakaris et al. (2021) observed the damage diagnosis problem of damage detection, identification of damaged tendons and quantification of precise damage under various operating conditions for a 10MW multibody FOWT supported on platform having two rigid bodies supported by 12 tendons. The dynamic response of the WT under different damage states are obtained based on the stimulated FOWT. The damage diagnosis is obtained via functional model-based method to operate single response signal. Si et al. (2021) proposed a combined floating wind and wave energy converter by integrating a semi-submersible floating platform and three-point absorber WECs. The dynamic responses and the power output of the combined system under various environmental loads are investigated using the aero-servo-hydro-mooring simulation tool. The study observed that, the dynamics of different power take off (PTO) control strategies showed considerable influence on the motions of the combined system. Gaspar et al. (2021) experimentally investigated a floating platform supporting twelve WECs without a wind turbine. The study observed that the WECs on the upward side and the downward side of the platform have different roles in the compensation of the platform dynamics. Yang et al. (2021) investigated the dynamic response of a 10MW offshore wind turbine supported by a multi-body floating platform consisting of cylindrical platform supported by six tendons to observe the different tendon breakage scenario. Numerical simulation tool F2A based on AQWA and FAST is used to perform coupled analysis of FOWT. The study observed the influence of platform response on the health of tendons.

In the present study a submerged tension leg platform is combined with six heaving type cone-cylindrical shaped point absorber wave energy converters in circular pattern. The study considers four, five, eight and nine mooring lines for the taut mooring system in stabilizing the STLP-WEC system under regular sea wave conditions. The dynamic responses of the combined system for different operational conditions of the wind turbine are investigated using the numerical aero-servo-hydro-elastic simulation tool FAST. The responses are then studied in frequency domain to preserve the frequency dependent nature of the wave excitation responses of the system and to integrate the turbulent wind load on to the model. The study also analyses the tension developed on each mooring cable and the effect of addition of WECs on the tension developed on the mooring cables.

## 2 STLP-WEC CONCEPT WITH MOORINGS

The combined STLP-WEC concept is analysed considering different mooring configurations. The description of the model along with the coupled dynamic analysis of the combined wave and wind energy system under different mooring configuration is presented in detail.

### 2.1 *Description of the model*

An artistic illustration of the submerged TLP floater concept supporting 5MW wind turbine and six heaving type point absorber WECs stabilized using tensioned tendons is shown in Figure 1. The STLP floater is provided with four outer cylindrical pontoons and a central column. The outer pontoons are connected to each other using the horizontal rectangular pontoons and the outer pontoons are connected to the central column through inclined rectangular braces. Six numbers of cone-cylinder shaped heaving point absorber WEC is connected in circular pattern around the STLP to derive wind and wave energy simultaneously. The dimensions of the STLP and the WEC is same as is discussed in Rony et al. (2021). The model surface is subdivided into meshes representing the wetted surface below the water line. To provide a better continuation of the body surface, the surface is discretized by the meshes, a set of small elements called panels are defined with these meshes.

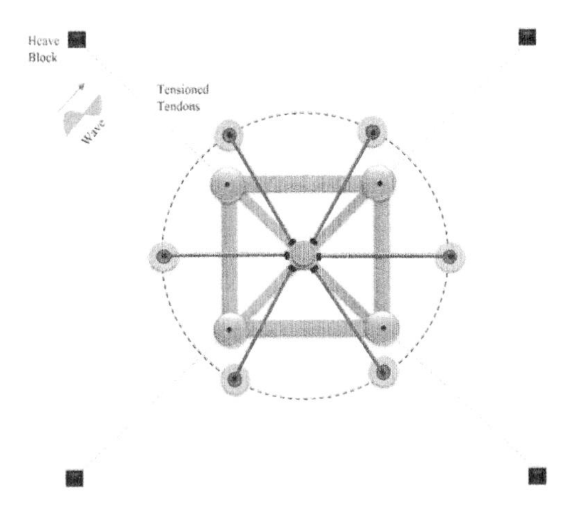


Figure 1. STLP with six WEC in circular pattern moored using the tensioned tendons.

The panel size can be modified depending on the accuracy requirements. The combined STLP-WEC is designed to support 5 MW NREL wind turbine (Jonkman, 2010) and is anchored at fairlead positions to the seabed by different numbers of taut mooring lines. The representation of the positioning of mooring lines for different arrangements of the taut mooring configurations (four, five, eight and nine) is shown in Figure 2. In the case of the four and eight mooring configurations, the tensioned cables are provided on the outer pontoons of the STLP. An additional tendon is provided to the central column along with the cables provided for the outer pontoons for the five and nine mooring configurations as shown in Figure 2.

The tensioned mooring cables are attached to the sea bed using heave blocks to further reduce the heave motion on the combined system. The properties of the mooring cables are shown in Table 1.

## 2.2 *Coupled dynamic analysis*

The aero-hydro-servo-elastic simulation tool FAST (Fatigue, Aerodynamics, Structures and Turbulence) is considered for analysis of floating platform under the wave, wind, mooring line and current forces. The generalized equation of the motion of the platform

(a)

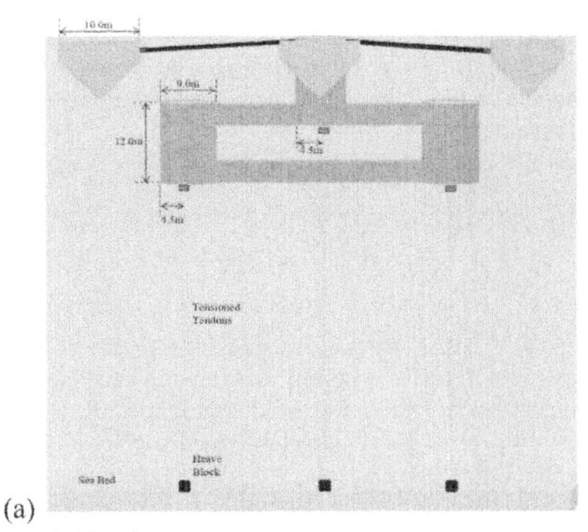


(b)

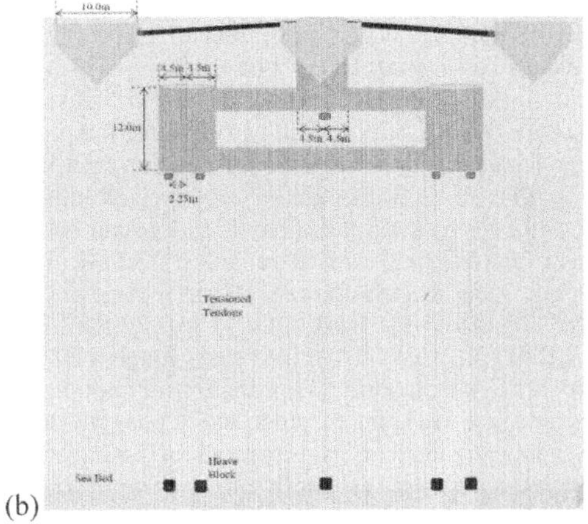


Figure 2. Illustration of position of mooring cables in case of (a) 4 and 5- mooring (b) 8 and 9-mooring.

Table 1. Properties of taut mooring system.

| Characteristics | Value |
|---|---|
| Unstretched mooring-line length | 130.978 m |
| Line diameter | 0.12 m |
| Line mass per unit length | 1159.248 kg/m |
| Line extensional stiffness | 31.359E9 N |
| Average steel density | 7850 kg/m$^3$ |
| Average concrete density | 2562.5 kg/m$^3$ |

for the model linearization as given in FAST code (Jonkman, 2007) is given by

$$M\ (q,u,t)\ddot{q} + f\ (q,\dot{q},\ddot{q},u,u_d,t) = 0, \qquad (1)$$

where $M$ is mass matrix, $f$ is non-linear forcing function, $q$ is the displacement vector of DOF, $\dot{q}$ is velocity vector of DOF, $\ddot{q}$ is the acceleration vector of DOF, $u$ is the control point vector and $u_d$ is the wind input disturbance vector. The equation of motion of the floating platform under regular wave is given by

$$[M+M_a(\omega)]\{\ddot{X}\}+[Q(\omega)]\{\dot{X}\}+[K]\{X\}=\{F(\omega)\}. \tag{2}$$

where $X$ is the structural displacement vector, $\dot{X}$ is structural velocity vector, $\ddot{X}$ is structural acceleration vector, $M$ is total mass of the platform, $K$ is the restoring force coefficient (stiffness matrix), $M_a(\omega)$ is the added mass, $Q(\omega)$ is the linear damping and $F(\omega)$ is the harmonic excitation force proportional to the incoming wave. The numerical simulation tool for FOWTs and WECs are quite different as they have distinct motion DOFs and environmental loadings even though they have similar physical domains. Hence, it is difficult to perform integrated time-domain analysis for the hybrid floating platform within the same framework. The numerical simulations for investigating the frequency domain behaviour of the WECs is carried out using WAMIT in the present study. On the other hand, various researchers used SIMO-TDHMILL3D, SIMO-RIFLEX-AERODYN, HAWC2-WASIM tools coupled FOWT and WEC simulations. The equation of motion for a single heaving point absorber and $N$ point absorber in frequency domain is presented in Sinha et. al (2016).

The mooring system made of cables (chains, steel, synthetic fiber) are attached to the floater at fairlead connections and the opposite end anchored to the sea bed. Restraining forces at the fairlead are established through tension developed on the mooring lines, which depends on buoyancy of the floater, weight of the cable in water, viscous-separation effects, elasticity of the cables and the geometric layout of the mooring system. The total load on the support platform for linear mooring system ignoring the mooring inertia and damping (Matha, 2009) is given by

$$F_i^{Lines}=F_i^{Lines,0}-C_{ij}^{Lines}q_j \tag{3}$$

where $F_i^{Lines,0}$ is the $i^{th}$ component of the total mooring system load acting on the support platform in its un-displaced position and $C_{ij}^{Lines}$ is the component of the linearized restoring matrix from all mooring lines. In the case of the catenary mooring lines $F_i^{Lines,0}$ takes the pre-tension at the fairlead from the weight of the cable not resting on the sea floor. If the catenary mooring lines are neutrally buoyant, then the pre-tension developed is zero. On the other hand, in the case of taut mooring system, $F_i^{Lines,0}$ is calculated based on the excess buoyancy in the tank in un-displaced position including the weight of the cable in water. $C_{ij}^{Lines}$ is the combination of elastic stiffness of the mooring lines and the effective geometric stiffness considering the weight of cables in water. However, the mooring system dynamics are not linear in nature as it includes non-linearities in force-displacement relationships, non-linear hysteresis effect due to the loss of energy as the lines oscillate with floater about mean position. FAST considers quasi-static mooring module to simulate the non-linear restoring loads from the mooring system of the floater. The module models the taut mooring lines taking into account the apparent weight in fluid, elastic stretching and the sea bed friction of each line, neglecting the individual line bending stiffness. Knowing the fairlead points for a given platform displacement at any instant in time, the mooring system solves for tensions within and configuration of each mooring line assuming that each cable is in static equilibrium at that instant. FAST then solves the dynamic equation of motion for acceleration of the rest of the system using additional tension and additional loading on the platform from aerodynamics and hydrodynamics. FAST then integrates in time domain to obtain new platform and fairlead positions at next time step and repeating the process. Independent analysis of each mooring line is necessary for any mooring system. Specifying the fairlead locations of each mooring line relative to the support platform, the anchor locations of each mooring line relative to the inertial frame, total upstretched length $L$, extensional stiffness EA, coefficient of sea-bed friction drag $C_B$, the apparent weight in fluid per unit length $\omega$ is related to the mass of the line per unit length $\mu_c$ is given by

$$\omega=\left(\mu_c-\rho\frac{\pi D_c^2}{4}\right)g. \tag{4}$$

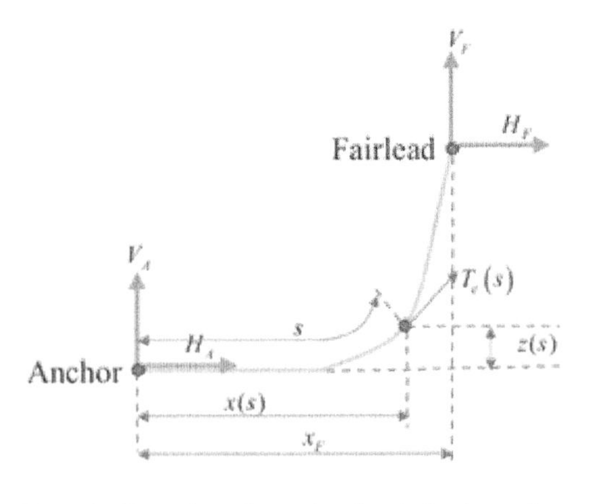


Figure 3. Illustration of mooring line in local coordinate system.

The equation for the horizontal vertical distance from the anchor point to the given point on the line when no portion of the mooring line is resting on the sea-bed are of the form

$$x(s) = \frac{H_F}{\omega}\left\{\ln\left[\frac{V_A+\omega s}{H_F}+\sqrt{1+\left(\frac{V_A+\omega s}{H_F}\right)^2}\right] - \ln\left[\frac{V_A}{H_F}+\sqrt{1+\left(\frac{V_A}{H_F}\right)^2}\right]\right\}+\frac{H_F s}{EA}, \quad (5)$$

$$z(s) = \frac{H_F}{\omega}\left[\sqrt{1+\left(\frac{V_A+\omega s}{H_F}\right)^2}-\sqrt{1+\left(\frac{V_A}{H_F}\right)^2}\right] + \frac{1}{EA}\left(V_A s+\frac{\omega s^2}{2}\right). \quad (6)$$

The effective tension is then calculated with the upstretched length $s$ using the relation

$$T_e(s) = \sqrt{H_F^2+(V_A+\omega s)^2}. \quad (7)$$

The present study considers different wind speeds representing the operational conditions of the wind turbine to analyse the motion amplitudes of the combined STLP-WEC in frequency domain.

Table 2. Met-ocean conditions (Rony et al. (2021)).

| Case No. | $V_{mean}(m/s)$ | $H_s(m)$ | $T_p(s)$ | Turbine status |
|---|---|---|---|---|
| LC-1 | 8 | 2.5 | 9.8 | Operating |
| LC-2 | 11.2 | 3 | 10 | Operating |
| LC-3 | 14 | 3.6 | 10.2 | Operating |
| LC-4 | 17 | 4.2 | 10.5 | Operating |

Table 2 shows the selected wind speeds and the correlated sea state conditions (expected values of $H_s$ and $T_p$) used to investigate the motion response of the system as well as the tension developed on the mooring cables for the different mooring configurations.

## 3 RESULTS AND DISCUSSION

The natural frequency of the STLP and STLP-WEC configuration are determined using free decay test. The submerged TLP configuration in the free decay test are released from certain offset for each degree of freedom (DOF) in calm water with the wind turbine in parked condition. The six-DOF natural frequencies for the two different configurations of the submerged TLP are tabulated and compared in Table 3. The addition of WECs to the STLP have minimum impact on the natural frequency of the submerged floater. The heave motion is observed to have slight variation with the addition of WECs when compared to the other five degrees of motion. Also, it is observed that the natural frequencies of horizontal plane motions (surge, sway and yaw) are lower than those on the vertical plane motions (heave, pitch and roll) which concludes that the stiffness for the submerged TLP is higher in the vertical plane.

Table 3. Natural frequency for STLP and STLP-WEC offshore floater using free decay test.

| Mode | STLP | STLP+6WEC |
|---|---|---|
| Surge | 0.06 | 0.06 |
| Sway | 0.06 | 0.06 |
| heave | 0.32 | 0.29 |
| Roll | 0.34 | 0.34 |
| Pitch | 0.34 | 0.34 |
| Yaw | 0.08 | 0.07 |

The frequency domain equation of motion is used to model the physical problem that consolidated regular and irregular waves along with turbulent wind load. On observing the frequency domain model outputs over the frequency range of $0 \leq \omega \leq 0.5$, time series responses for internal loads and motions can be analysed, which can further be used to observe the performance of the floater model. The responses are analysed using frequency domain analysis for the six WEC configuration in the case of four different operating wind speed conditions.

In Figure 4(a-d) the surge response is analysed for different mooring configurations of six WEC configuration under different wind speed and regular wave. The maximum value of surge response of about 0.74m/m is observed for LC4 for the frequency 0.09Hz for eight mooring configuration. In the case of four different load cases considered, it is observed that the responses tend to reduce with the addition of tensioned tendons at the central column. The maximum responses are observed to occur at a lower frequency range close to 0.5Hz for lower wind speed conditions in the case of four and five mooring configuration. Also, two peaks in the surge spectrum for the lower wind speed represents the surge natural frequency and the wave spectral peak frequency respectively. Further, for the eight and nine mooring configurations, the maximum value of surge response is observed to occur at 0.09Hz which is

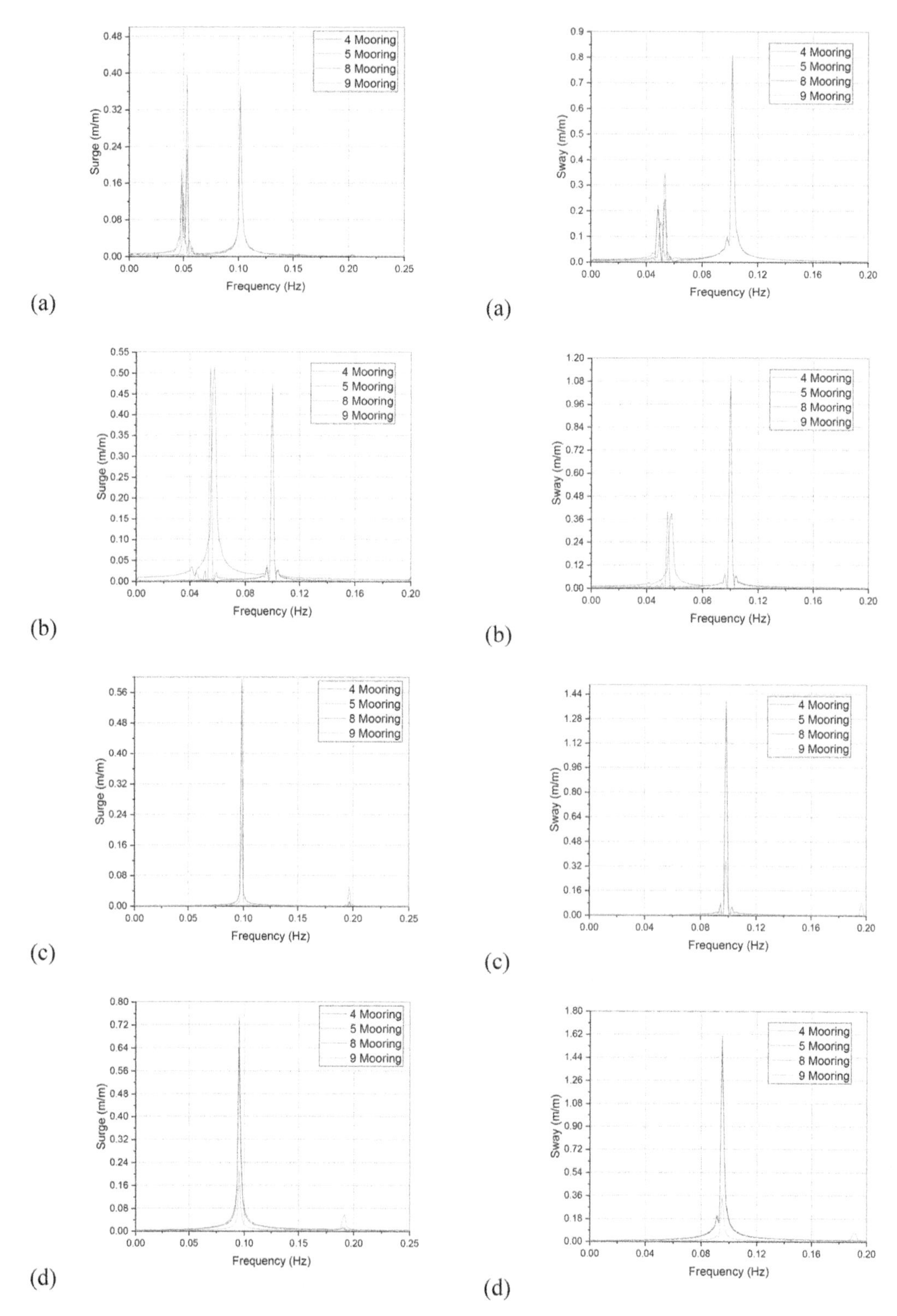


Figure 4. Surge response for (a) LC-1, (b) LC-2, (c) LC-3 and (d) LC-4 for different mooring configurations.

Figure 5. Sway response for (a) LC-1, (b) LC-2, (c) LC-3 and (d) LC-4 for different mooring configurations.

closer to the natural frequency of the STLP-WEC combined wave and wind energy system.

Figure 5(a-d) shows the sway response amplitude for different mooring configurations of six WEC configuration under different operating wind speed conditions. The sway motion amplitude is observed to increase with the increase in wind speed and wave height. The sway peak spectrum is much away from the natural frequency condition for higher wind speed conditions in the case of all mooring configurations. The peak value for the operating condition is observed to be 1.62m/m for 17 m/s wind speed at 0.11Hz. In addition, two peak spectrums are observed for lower wind speed conditions for four, five and eight mooring configurations similar to the surge response. Both peaks of the sway spectrum are much away from the natural frequency of the combined floater system.Also, the sway motion is observed to be higher than the surge motion as seen from Figure 4(a-d) and Figure 5(a-d). The heave motion for the combined floater is observed to be very small under regular wave conditions.

Figure 6(a-d) shows the roll motion spectrum for different mooring configurations of the six WEC configuration under different operating wind speed conditions. In Figure 6(a-d) it is observed that the peak frequency for all the four operating conditions in the case of roll motion is close to 0.11 Hz and the peak response value is less than 0.4deg/m, thus showing higher stability of the combined system against overturning. For higher wind speeds (Figure 6c-d) another dominant peak is observed near 0.21Hz frequency range. Both the peaks are observed away from the natural frequency, further suggesting that there is no possibility of occurrence of resonance. The platform roll for all the conditions is observed to be minimum for the zero-degree wave heading angle as the platform is symmetric about the wave heading direction.

Figure 7(a-d) shows the pitch motion amplitude for different mooring configurations of six WEC configuration under different operating wind speed conditions. In Figure 7(a,b) it is observed that four peaks in the case of pitch motion when the wind speed is less than the rated wind speed for the 5MW reference wind turbine. Further, the peak values for pitch motion tends to increase with the increase in the wind speed for all four operating conditions.

The pitch motion tends to increase with the increase in the number of mooring lines as the nine mooring lines configuration is observed to have highest response value for all operating conditions. It is also observed that the peak motion

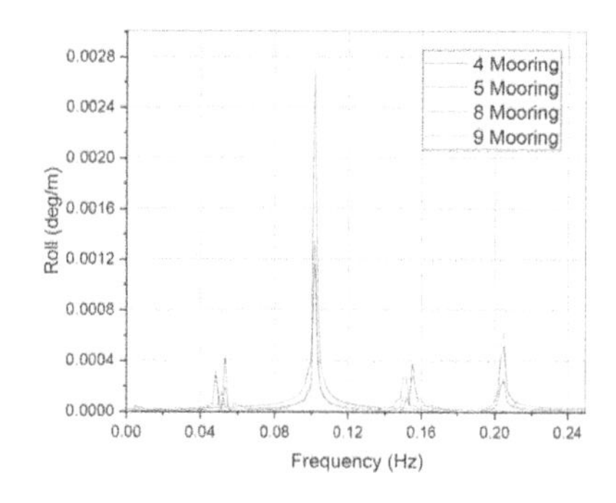


(a)

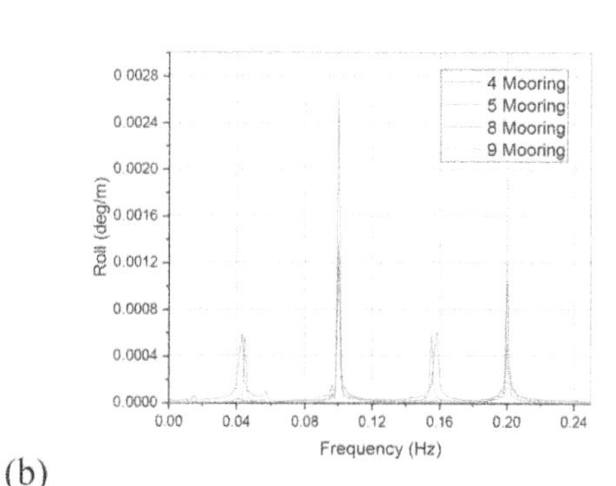


(b)

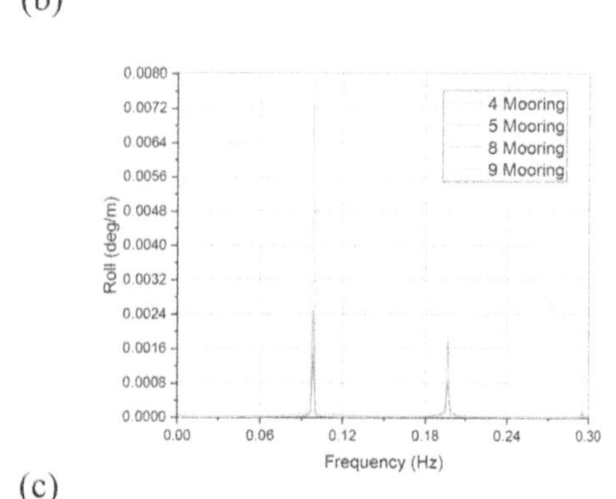


(c)

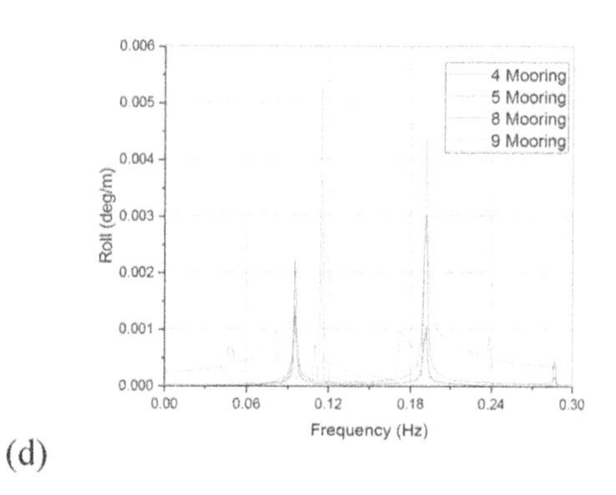


(d)

Figure 6. Roll response for (a) LC-1, (b) LC-2, (c) LC-3 and (d) LC-4 for different mooring configurations.

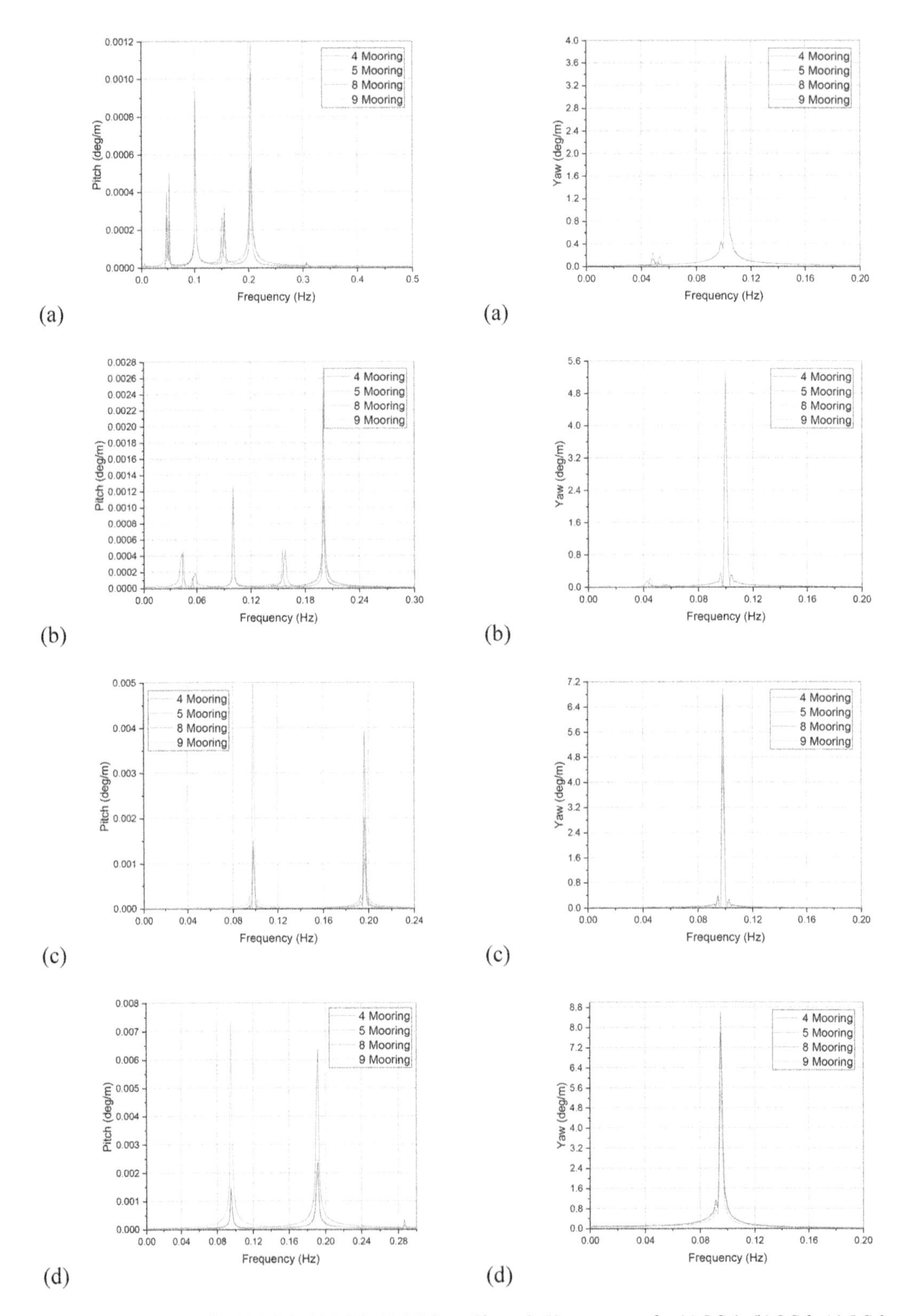


Figure 7. Pitch response for (a) LC-1, (b) LC-2, (c) LC-3 and (d) LC-4 for different mooring configurations.

Figure 8. Yaw response for (a) LC-1, (b) LC-2, (c) LC-3 and (d) LC-4 for different mooring configurations.

values are observed before the natural frequency of the floating system thus avoiding the occurrence of resonance for the floater. The surge and pitch motion are observed to be higher for the lower wave frequency, further suggesting proper orientation of the floater against wind for higher frequency of waves, further improving the wind power absorption.

Figure 8(a-d) shows the yaw motion amplitude for different mooring configurations of six WEC configuration under different operating wind speed conditions. The yaw motion amplitude is observed to be higher compared to all other translational and rotational motions. Also, the peak value for the yaw motion tends to increase with the increase in the wind speed. The peak value for the yaw motion is observed for the frequency range of $0.9 \leq \omega \leq 0.11$.

The peak value for the yaw motion is observed to be closer to the natural frequency of the floater and hence proper damping of the floater needs to be provided to prevent the floater against resonance.

Thus, from Figures 4-8, it is observed that the responses for the floater is increased for the 8-mooring and 9-mooring configurations. The peak response for the surge and pitch motion is closer to the natural frequency region in case of the 9-mooring configuration, suggesting that the position of the taut mooring for the 9-mooring configuration needs to be shifted towards the center of gravity (CG) of the outer pontoon of the floater. The 4-mooring configuration is observed to have minimum response values under the action of regular waves for all the four different load cases. The maximum value of fairlead tension developed on each mooring cable is studied for different load case conditions in Figure 9(a-d). The tension developed on each cable is considerably reduced with the increase in the number of mooring cables. In Figure 9(a) it is observed that under the operational conditions of the 5MW wind turbine, the tension developed on the mooring cables have minimum variation for the 4-mooring configuration. In the case of 5-mooring configuration, the tension developed on the cable provided on the central column is minimum compared all other cables provided for the outer pontoons. Further, in the case of 9-mooring configuration, the tendons on the leeward side are observed to have higher tension compared to tendons on the seaward side for wind speeds greater than the rated wind speed for the 5MW wind turbine. The study also observed that the addition of wave energy converters around the STLP floater has helped to reduce the tension developed on the mooring cables. The maximum value of tension developed on the mooring cables for the single STLP floater is presented in Figure 10(a-d). The reduced tension on the mooring cables ensures that the fatigue load developed on the tendons for the combined floating concept is minimum.

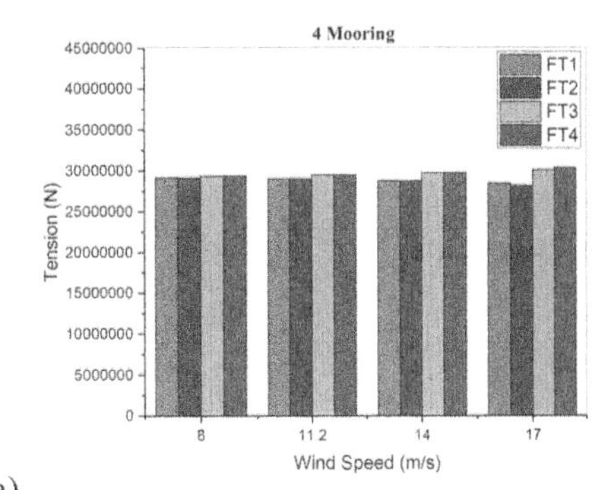


(a)

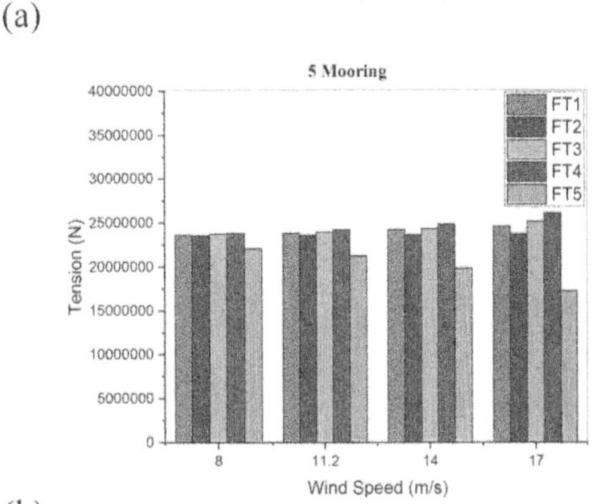


(b)

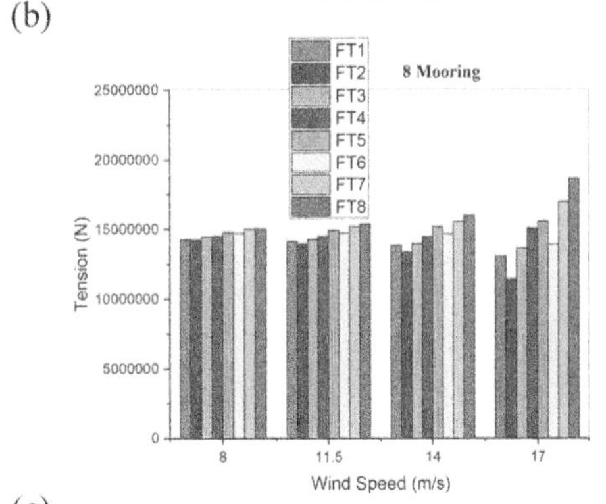


(c)

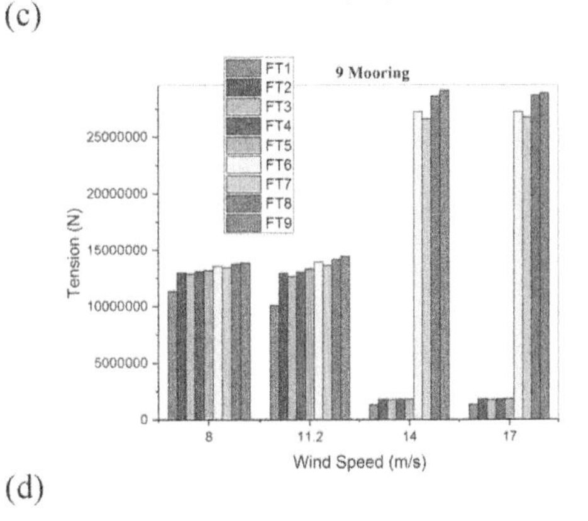


(d)

Figure 9. Maximum tension developed on the tendons for (a) 4-mooring, (b) 5-mooring, (c) 8-mooring and (d) 9-mooring for combined floater.

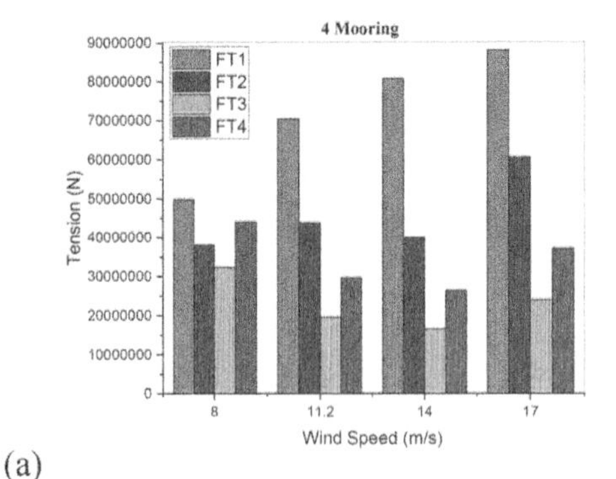


(a)

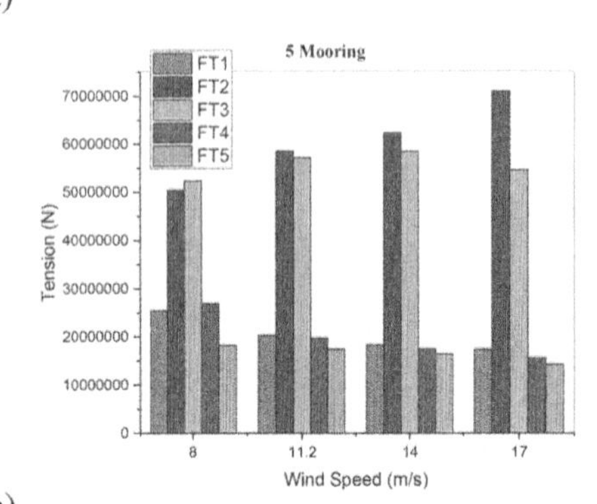


(b)

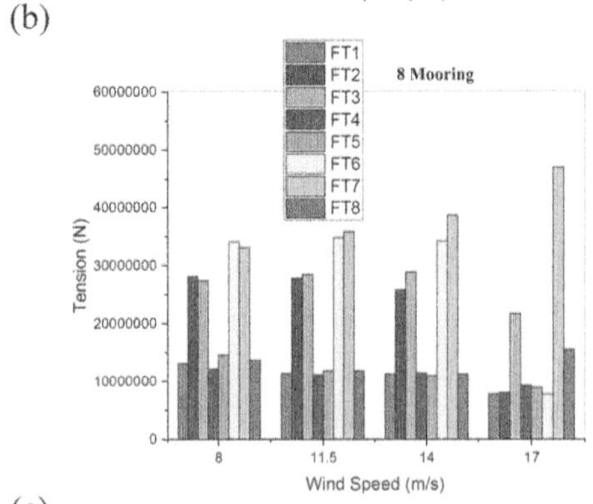


(c)

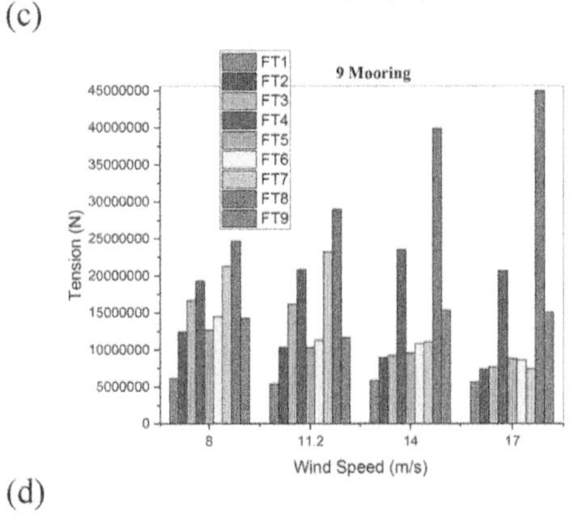


(d)

Figure 10. Maximum tension developed on the tendons for (a) 4-mooring, (b) 5-mooring, (c) 8-mooring and (d) 9-mooring for STLP floater.

## 4 CONCLUSIONS

The study presents a submerged tension-leg platform (STLP) combined with six WECs in circular pattern to support 5MW NREL wind turbine for moderate water depth of 150 m. The dynamic response characteristics of the combined floating system in frequency domain is studied for different wind speed conditions with four different taut mooring configurations under regular waves. Fully coupled dynamic simulations are carried out using the numerical simulation tool FAST to determine the structural responses of the structure. The amplitude of surge, sway and yaw motions are observed to be more sensitive to the wind-wave actions. The responses are observed to be higher for the 8-mooring and 9-mooring configurations as compared to the 4-mooring configuration. The higher response for surge and pitch for the 8-mooring and 9-mooring configurations suggests that the wind power absorbed for the combined system will be reduced under these mooring systems. The addition of the six WECs around the STLP is noted to reduce the tension developed on the mooring cables by almost 50%. The presence of higher number of the mooring cables helps the in reducing the tension developed on each cable. It is also noted that, the reduction of tension may not be significant on considering the cost of each taut mooring cable. The study thus concludes that the 4-mooring configuration is best suited for STLP with six WEC in circular pattern owing to its higher wind power absorption and with no significant reduction in the tension developed on the cables.

## ACKNOWLEDGEMENTS

The authors acknowledge Science and Engineering Research Board (SERB), Department of Science & Technology (DST), Government of India for supporting financially under the research grant no. CRG/2018/004184, DST for India-Portugal Bilateral Scientific Technological Cooperation, Project grant no. DST/INT/Portugal/P-13/2017 and Vision Group for Science and Technology (VGST), Government of Karnataka, Project grant no VGST/RGS-F/GRD-901/2019-20/2020-21/198.

## REFERENCES

Crudu, L., Obreja, D.C. & Marcu, O. (2016). Moored offshore structures – evaluation of forces in elastic mooring lines, International Conference on Advanced Concepts in Mechanical Engineering, pp.1–12.

Gaspar, J. F., Kamarlouei, M., Thiebaut, C. and Guedes Soares, C. (2021). Compensation of a hybrid platform dynamics using wave energy converters in different sea state conditions. Renewable Energy, 177, 871–883.

Han, Y., Le, C., Ding, H., Cheng, Z. & Zhang, P. (2017) Stability and dynamic response of a submerged tension-leg flatform for offshore wind turbines. Ocean Engineering 129, 68–82.

Jonkman, J.M. (2007). Dynamic modelling and loads analysis of an offshore floating wind turbine. National Renewable Energy Laboratory, Golden, CO, Technical Report No. NREL/TLP-500-41958.

Karimi, M., Buckham, B. and Crawford, C. (2019). A fully coupled frequency domain model for floating offshore wind turbines. Journal of Ocean Engineering and Marine Energy, 5, 135–158.

Le, C., Jian, Z., Hongyan, D., Puyang, Z. and Guilan, W. (2020). Preliminary design of a submerged support structure for floating wind turbines. Journal of University China, 19, 1265–1282.

Matha, D. (2009). Model Development and Loads Analysis of an Offshore Wind Turbine on a Tension Leg Platform, with a Comparison to Other Floating Turbine Concepts. National Renewable Energy Laboratory, Technical Report No. NREL/SR-500-45891.

Sakaris, S.C., Yang, Y., Bashir, M., Michailides, C., Wang, J., Sakellariou, S.J. & Li, C. (2021). Structural health monitoring of tendons in a multibody floating offshore wind turbine under varying environmental and operating conditions. Renewable Energy, 179, 1897–1914.

Shen, M., Hu, Z. & Liu, G. (2016) Dynamic response and viscous effect analysis of a TLP type floating wind turbine using coupled aero-hydro-mooring dynamic code. Renewable Energy 99, 800–812.

Sinha, A., Karmakar, D. & Guedes Soares, C. (2015). Numerical modelling of an array of heaving point absorbers. Renewable Energies Offshore, C. Guedes Soares (Eds), Taylor & Francis Group, London, 383–391.

Soeb, M.R, Islam, A.B.M.S. & Jumaat, M.Z., (2016). Hydrodynamic Response of Floating Coupled Spar in Deep Sea, International Conference on Marine Technology, MARTEC, pp.182–188.

Si, Y., Chen, Z., Zeng, W., Sun, J., Zhang, D., Ma, X. & Qian, P. (2021). The influence of power-take-off control on the dynamic response and power output of combined semi-submersible floating wind turbine and point-absorber wave energy converters. Ocean Engineering, **227**, pp. 1–23.

Ren, N., Ma, Z., Shan, B., Ning, D. & Ou, J. (2020). Experimental and numerical study of dynamic responses of a new combined TLP type floating wind turbine and a wave energy converter under operational conditions. Renewable Energy, 151, 966–974.

Rony, J.S., Karmakar, D. & Guedes Soares, C. (2021) Coupled dynamic analysis of spar-type floating wind turbine under different wind and wave loading. Marine Systems and Ocean Technology, 16, 169–198.

Rony, J.S., Karmakar, D. & C. Guedes Soares. (2021) Dynamic analysis of submerged TLP wind turbine combined with heaving wave energy converter. Developments in Maritime Technology and Engineering, C. Guedes Soares and T.A. Santos (Eds), Taylor & Francis Group, London, Vol-2, pp. 639–646.

Wang, Y. (2020). Efficient computational method for the dynamic responses of a floating wind turbine. Ships and Offshore Structures, 15, 269–279.

Yang, Y., Bashir, M., Michailides, C., Mei, X., Wang, J & Li, C. (2021). Coupled analysis of a 10MW multi-body floating offshore wind turbine subjected to tendon failure. Renewable Energy, 176, 89–105.

Zhang, L., Shi, W., Karimirad, M., Michailides, C. & Jiang, Z. (2020). Second-order hydrodynamic effects on the response of three semisubmersible floating offshore wind turbines. Ocean Engineering, 207, 1–22.

*Trends in Maritime Technology and Engineering – Guedes Soares & Santos (eds)*


# Modelling the hydrostatic stability characteristics of a self-aligning floating offshore wind turbine

D. Scicluna, T. Sant & C. De Marco Muscat-Fenech
*University of Malta, Msida, Malta*

G. Vernengo
*Universita di Genoa, Genoa, Italy*

Y.K. Demirel
*University of Strathclyde, Glasgow, Scotland, UK*

ABSTRACT: The offshore wind energy market is fast-growing, particularly in the EU where 102.6 GW are expected to be added by 2030 on top of the 220 GW currently installed. There exist vast areas of offshore wind resources with water depths greater than 100m necessitating floating structures. The paper explores the hydrostatic stability characteristics of a novel floating wind turbine concept, having a single point mooring system and self-aligning capabilities to eliminate a rotor yaw mechanism. The preliminary design supports a 8 MW horizontal-axis wind turbine with a custom floater that can be constructed within the shipyard facilities available in Maltese Islands. The hydro static stability analysis is carried out for different levels of ballast while considering the maximum axial thrust induced by the rotor during operation. It is shown that the entire floating structure exhibits favourable hydrostatic stability characteristics for both the heeling and pitching axes, meeting the requirements set by the DNV ST0019 standard.

## 1 INTRODUCTION

### 1.1 *Background*

An increase in energy demand and the evident consequences of climate change are now fuelling significant interest in alternative energy sources such as wind energy. Wind energy presents an opportunity to decrease the current energy dependence on fossil fuels by increasing the production of green energy. Wind energy has been used for centuries though the use of windmills which were later developed into onshore wind turbines. Technological developments have seen the rapid growth of wind energy industry, which now also includes offshore renewable energy.

The scope of this paper is twofold: (1) to present a brief overview of floating offshore wind turbines (FOWTs) and introduce self-aligning turbines using single-point mooring (SPM) technology and (2) to present a preliminary design of a self-aligning 8 MW floater and assess its hydrostatic stability characteristics when the turbine is operating at the rated wind speed.

### 1.2 *Offshore structures*

Offshore wind turbines provide a number of benefits over systems installed onshore. Although onshore turbines are cheaper and present less technological challenges to install and maintain, such installations are subject to spatial restrictions, an issue that is not present for offshore turbines due to the vast seas and oceans found all over the world. Offshore structures also create less visual and noise pollution than onshore, resulting in a positive public opinion associated with offshore wind energy structures. Moreover, given that wind speeds increase with distance from shore, offshore turbines encounter stronger and less turbulent winds, resulting in an opportunity for increased and steadier power generation.

Another important factor to consider is the fast growing interest in wind energy which caused a growing need and demand for larger turbines, with expected sizes of up to 15 – 20 MW, as highlighted by Sieros *et al.*, (2012). In order to accommodate the increasing turbine sizes, larger platforms and substructures are necessary. Larger structures are difficult to incorporate in onshore structures as stated by McKenna et al (2016), given the aforementioned spatial limitations associated with onshore wind energy (Mckenna & Fisher, 2016).

Offshore structures are significantly easier to modify to accommodate larger turbine sizes, making offshore wind energy a more viable option. Numerous studies, such as Ashuri et al (2016) and

DOI: 10.1201/9781003320289-51

Chaviaropoulos et al (2014) have investigated challenges related to technical, economic and design characteristics, including performance indicators and target values that are associated with the design of large-scale wind turbines. The main issue with upscaling, as presented in the aforementioned research papers, is the increase in rotor mass which may be mitigated through incorporating alternative lightweight components and novel design concepts.

A number of support systems currently exist in the offshore wind turbine (OWT) industry, which can be segregated into bottom-fixed structures and floating structures. As the name suggests, bottom-fixed structures are connected directly to the seabed. This imposes a limit to the sea depth the structures can be installed in.

In general, bottom-fixed structures are limited to a depth of 50 m as the costs of implementing bottom fixed-structures increase exponentially with increasing depth (IRENA, 2016). The increasing demands for wind energy have resulted in an increased use of FOWT structures to keep up. Floating structures connect to the seabed using mooring lines and anchors and can be installed in deeper waters without the high costs and environmental impacts that are associated with bottom-fixed OWT.

Floating platforms were first introduced in the O&G industry and throughout the years, the technology has been adapted and developed for the OWT industry. Both bottom-fixed and floating offshore wind energy platforms have similar design characteristics to structures presently used in the offshore O&G industry with modifications included to accommodate the wind turbine structure (Schneider & Senders, 2010).Three main types of floating offshore platforms: the spar-buoy, the tension leg platform (TLP) and the semi-submersible as shown in Figure 1 (IRENA, 2016).

Figure 1. (a) Spar-buoy (b) Semi-submersible (c) TLP (Leimeister, Kolios and Collu, 2018).

An important aspect for any floating structure is stability. FOWT structures make use of three main mechanisms to reduce the degree of motion and achieve stability: ballast-stabilised, buoyancy-stabilised and mooring-stabilised (Leimeister, et al, 2018).

Ballast stabilised structures use a large ballast at the bottom of the floating platform in order to place the centre of gravity (COG) of the system below the centre of buoyancy (COB). As the structure tilts, a stabilising righting moment occurs that opposes the displacement of the structure. Buoyancy-stabilised structures use the concept that a large water plane area generates a large second moment of area for the necessary righting moment that stabilises the structure during rotational motion. Mooring-stabilised structures make use of tensioned mooring lines to generate the necessary restoring force when the structure is heeled.

Table 2 presents the different types including the stabilising mechanism used in each case (Schneider & Senders, 2010; Leimeister, et al, 2018).

Table 1. FOWT stabilising mechanisms.

| FOWT | Description and stabilising mechanism |
|---|---|
| Spar-buoy | A long cylindrical structure with a low WPA which is ballast stabilised. |
| TLP | A central column connected to a submerged buoyant platform which is mooring stabilised. |
| Semi-submersible | Connected columns with braces, pontoons or platforms that are buoyancy stabilised. |

Although in general each structure type is presented as having only one type of mechanism for stability, in practice all floating platforms use more than one stabilising mechanism (Butterfield *et al.*, 2005).

In the offshore industry, self-alignment characteristics are defined as the ability of a floating platform to yaw freely with the varying wind and wave direction without the use of an active yaw system. In general, wind turbines make use of a yaw mechanism in order to continuously retain the position of the rotor plane to face the wind flow. Any misalignment present between the direction of wind flow and the rotor results in decreased efficiency. A self-aligning floating structure avoids the need for a complex yaw mechanism.

Through implementing self-alignment characteristics at the preliminary design stage, the overall costs and maintenance required can be reduced (Netzband, et al, 2020).

Mooring technology is being used to introduce self-aligning characteristics to the design of floating structures in the FOWT industry. Mooring costs contribute to a significant percentage of the overall costs of the floating offshore platforms, therefore the choice of mooring systems in FOWT structures is an important factor that needs to be considered carefully (Caro *et al.*, 2018).

Currently, the most common mooring system adopted in the FOWT industry is the multi-point mooring system (MPM) where multiple mooring lines are geometrically distributed around the floating structure and connected to anchors fix the structure remains in position without weathervaning (Mooring). Through technological developments, single point mooring (SPM) was then introduced in the FOWT industry. In SPM systems, the mooring system is connected to the floating platform at only a single point. SPM systems can use either a single mooring line or multiple mooring lines connected at the sea-bed that come from a single point on the floating platform (Wichers, 2013).

SPM systems were first referred to as passive weathervaning mooring systems as the floating structure was allowed to self-align with the prevailing wind and wave conditions. The weathervaning abilities presented by SPM systems reduce wear and damage on the mooring lines as well as minimising the need for tug assistance given that the structures do not experience large hydrodynamic and aerodynamic forces (Wichers, 2013).

Although the self-alignment properties offer a number of benefits to FOWT structures, in reality, a small degree of misalignment between the direction of wave propagation and direction of wind flow will always be present. Such a misalignment leads the turbine to experience a phenomena referred to as yaw error, wherein the turbine is operating at a misalignment with the wind, resulting in a loss of power.

Therefore, although the SPM systems offer a large degree of self-alignment abilities to FOWT, additional characteristics need to be implemented into the design to further supplement the structure's weathervaning characteristics and improve the possibility of self-alignment of the wind turbine with the wind direction as quickly as possible.

SPM systems are becoming increasingly popular, an example is the HyStOH concept (TUHH, 2020) produced as a collaboration project between CRUSE Offshore, the Hamburg University of Technology, DNV and more. The concept not only makes use of an SPM system for self-aligning characteristics but also includes an innovative tower design in the shape of an aerofoil. The tower design provides the necessary forces for self-alignment under prevailing wind conditions. In spite of the positive results obtained by the research project, the prevailing conclusion was that the self-alignment characteristics were limited by current velocities and yaw errors. Therefore including additional techniques to increase the self-alignment moments are necessary (Wichers, 2013; Netzband, Schulz and Abdel-Maksoud, 2020).

Alternative emerging designs that also make use of SPM systems are Eolink and X1 Wind. Eolink makes use of a pyramid-shaped superstructure with a square base to decrease the weight of the structure. Similarly, X1 Wind makes use of a triangularly-shaped structure, instead of the conventional tower structure to support the turbine, combined with a custom SPM system called PivotBuoy.

The design proposed in this paper consists of a semisubmersible structure making use of novel geometrical characteristics in order to promote self-aligning properties to the structure.

## 2 DESIGN STANDARDS FOR FLOATING TURBINES

The FOWT industry is rapidly developing, with a significant number of installations together with other additional FOWTs to be set up. In order to keep up with the fast FOWT development, there is a need for more detailed standards to be drafted to ensure safer, successful, and long-term operations.

The International Maritime Organisation (IMO) presents the IMO MSC 267(85), which is the parent of all standards produced by regulation organisations worldwide, and presents the intact stability criteria of various structure types (International Maritime Organization (IMO), 2008).

The stability of an offshore structure is found using righting moment and heeling moment curve, taken about the critical axis with free surface effects also taken into consideration (International Maritime Organization (IMO), 2008).

Parameters such as the minimum and maximum righting lever (GZ) area and GZ values, initial metacentric height (GM), and wind and rolling criteria are used to identify whether a structure is safe and ready for deployment (International Maritime Organization (IMO), 2008).

The DNV standards Intact OS-C301 and DNV ST-0119, based on the IMO standard, are commonly used for FOWT structures.

The main requirements for floating offshore structures can be seen in Table 2.

Table 2. Requirements set by standards (IMO, 2008; DNV, 2017, DNV, 2018).

| | IMO MSC267 (85) MODU | DNV OS-C301 Sect. 4.3 | DNV ST-0119 Sect. 10.2.3 |
|---|---|---|---|
| For Column Stabilised Units | ✓ | ✓ | ✓ |
| For Semi-submersibles | - | - | ✓ |
| Area under Righting moment curve to the second intercept or downflooding angle in excess of the area under the wind heeling moment curve | | | |
| | ≥ 30% | ≥ 40% | ≥ 130% |
| Righting moment curve should be positive over range of angles from upright to the second intercept. | | | |
| | ✓ | ✓ | ✓ |

Therefore, given that the model considered is a column-stabilised semi-submersible unit, DNV ST-0119 will be used to ensure that the required standards are met. The righting moment and heeling moment curves as defined by the DNV ST-0119 standard are shown in Figure 2.

The DNV ST-0019 for Floating wind turbine structures where it is stated that "*For evaluation of sufficient floating stability it is crucial to assess the wind loads. The wind loads consist of wind loads on the rotor in combination with wind loads on the tower and other wind-exposed parts of the support structure*" (DNV, 2018).

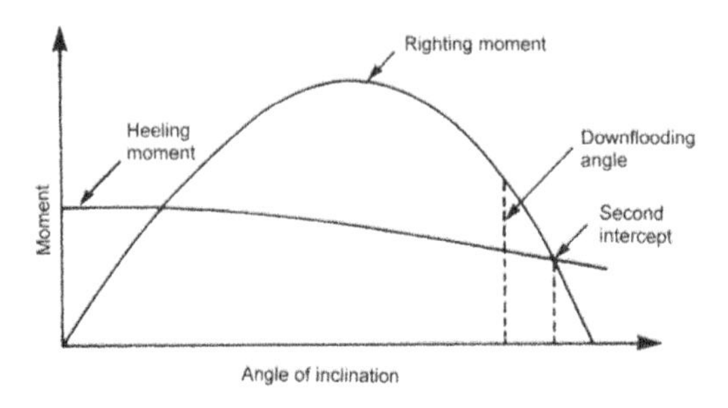


Figure 2. Righting moment and wind heeling moment curves (DNV, 2018).

The righting moment is dependent on the restoring capabilities of the structure. The wind heeling moment is caused by a force due to wind acting on the structure at a distance from the centre of rotation.

The wind loads on the rotor can be calculated using Equation 1.

$$F_T = 0.5\rho C_T A_{rotor} U^2 \tag{1}$$

Where $F_T$ is the thrust force
$\rho$ is the density of air
$C_T$ is the thrust coefficient of the rotor
$A_{rotor}$ is the swept area of the rotor
$U$ is the rated wind speed of the turbine

Based on DNVGL-0S-C301, the force of the wind on the structure can be calculated using Equation 2.

$$F_S = 0.5\rho C_S C_H A_P U^2 \tag{2}$$

Where $F_S$ is the wind loads on the structure
$\rho$ is the density of air
$C_S$ is the shape coefficient depending on the shape of the structural member exposed to the wind
$C_H$ is the height coefficient depending on the height above sea-level of the structural member exposed to the wind
$A_P$ is the projected area of all exposed surfaces in either the upright or heeled condition
$U$ is the wind speed

The wind heeling moment can then be calculated using Equation 3.

$$\mathrm{M_{wind}} = \mathrm{F_T} * (\mathrm{z_{hub}} - \mathrm{z_{COB}}) \cos^2(\theta) \tag{3}$$

Where $\mathrm{M_{wind}}$ is the wind heeling moment $\mathrm{z_{hub}}$ is the vertical distance from the base of the tower plus the freeboard of the structure $\theta$ is the pitch angle.

## 3 PROPOSED DESIGN

The proposed design consists of a semi-submersible structure making use of novel geometrical characteristics to promote self-aligning properties to the structure. The design considered can be divided into two different sections which will be referred to as the superstructure and the platform. The superstructure consists of the rotor nacelle assembly (RNA) assembly and the wind turbine tower, while platform consists of two pontoons, a deck, four columns and connecting beams.

The main dimensions of the structure were acquired through a controlled design spiral. Different design parameters which were varied until the ideal combination of dimensions were established for specific construction and operational requirements. The parametric analysis involved mass estimation, calculation of draught, centre of gravity for each section and overall structure, displaced volume and metacentric height.

The submerged section of the substructure is designed utilising two truncated Wigley Hull shape pontoons, similar to a catamaran structure, as can be seen in Figure 3. The novel shape of the pontoons promotes the self-aligning characteristics and the manoeuvrability of the structure. The pontoons are subdivided into eight compartments, for the required in-service ballasting options, and are connected to a rectangular deck using four aerofoil shaped columns, as seen in Figure 3. The aerofoil shaped columns were also chosen to further enhance the floater's self-alignment characteristics with the prevailing wind and wave conditions.

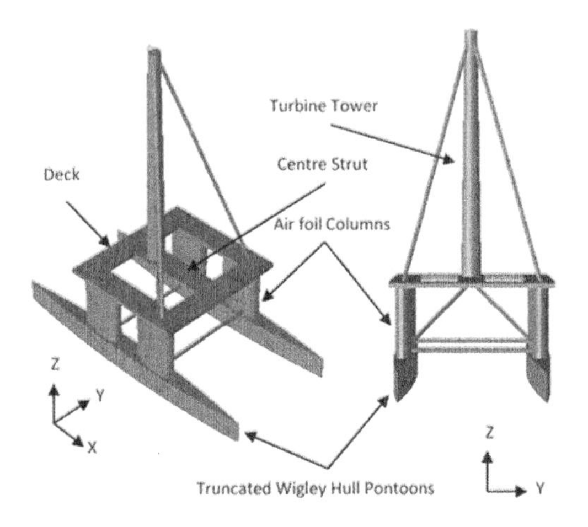


Figure 3. Proposed Design.

The tapered cylindrical turbine tower is fixed to the rectangular deck, designed for an 8 MW turbine. The tapered tower has diameters on the base of

7.7 m and 5 m at the top, below the nacelle. Connecting beams were placed strategically in the model to provide further necessary support to the components as well as to aid with load distribution. The thicknesses of the plates used in the structure were calculated using the DNV ST0119 standard.

In order to create the ideal FOWT design for Malta, the restrictions associated with the Maltese Archipelago needed to be carefully considered. The restrictions considered are mainly the spatial limitations due to the size of the ship building docks present in Malta and surrounding deep nearshore waters which impose limitations, not only on the size of the structure but also on the type of FOWT that can be constructed. Therefore, it is necessary that the finalised design adheres to the restrictions without compromising the weathervaning capabilities.

Before initialising the parametric study, the fundamental fixed parameters, such as the maximum width and the draught of the structure, were defined and set to remain within the limits of the maximum dimensions of the intended on-site construction ship-building dock found within the Maltese Islands. The maximum length, width and draught of the ship-building dock are 362 m, 62 m and 9.14 m respectively. For the parametric study, additional limitations as specified in DNV standards were also included in the study with a PASS/FAIL criteria to easily identify any design problems. The parameters of the structure can be seen in Table 3.

Table 3. Model parameters.

| Parameter | Value |
|---|---|
| Power Rating, rated wind speed | 8 MW, 12.5 m/s (Cian, 2016) |
| Rotor Orientation, configuration, diameter | Downwind, 3 Blades, 164 m |
| Tower height, base diameter, top diameter | 92.5 m, 7.7 m, 5 m, |
| Rotor diameter, Nacelle Diameter | 4 m, 7.5 m |
| Pontoon: Length, Width, Height | 121 m, 8 m, 9.9 m, |
| Column: height, number of columns | 25 m, 4 |
| Deck: height, length, breadth | 2 m, 55 m, 60 m |
| Total mass (no ballast), Displaced volume (no ballast) | 5254t, 5125.5 $m^3$ |
| Draught, | 4 m, |
| COG (w.r.t free water surface) | (1.6 m, 0 m, 23.3 m) |

Although when compared to other FOWT types, typical semisubmersibles are the largest structures, the proposed design boasts a lower overall mass than other structures as seen in Table 2. Moreover, the low mass of the structure allows for a smaller operating draught, ideal for the port limitations present in the ship-building docks of the Maltese Archipelago.

Figure 4 contains a schematic diagram of the top and front views of the pontoons which presents the

Table 4. Comparison of semi-submersible FOWT.

| Structure | MW | Mass (t) |
|---|---|---|
| Proposed Design | 8 | 5253 |
| Semi-submersible (Oh, 2018) | 2 | 5254 |
| Semi-submersible (Zhao, 2019) | 5 | 2403 |
| Semi-submersible (Karimirad, 2014) | 5 | 6954 |
| Semi-submersible (Iijima *et al.*, 2013) | 5 | 13968 |
| Semi-submersible (Zhao, 2019) | 10 | 4461 |
| Semi-submersible (Galvin, *et al.*, 2018) | 10 | 8137 |

reference point considered for the structure. After finalising the parameters of the model, the structure was created using a combination of plates and beams in DNV SESAM® Genie (DNV). The compartments required for the ballast tanks were also created. The model was then meshed accordingly for hydrostatic analysis in SESAM® HydroD (DNV).

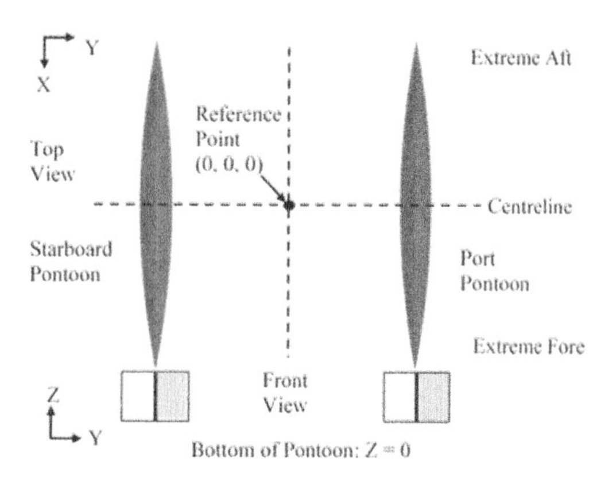


Figure 4. Schematic diagram of pontoons.

## 4 ANALYSIS LAYOUT

After finalising the model design in SESAM® Genie (DNV), the stability of the structure was analysed SESAM® HydroD (DNV) to ensure compliance with the necessary standards. Therefore, two different types of analysis, denoted as A1 and A2, were defined and carried out, as described in Table 4.

Table 5. Analysis Layout.

| | Ballast | Wind | Standard Check |
|---|---|---|---|
| A1 | 0 – 100% | No | DNVGL-ST-0119 Sect. 10.2.3 |
| A2 | 0 – 100% | Yes | DNVGL-ST-0119 Sect. 10.2.3 |

The scope of analysis A1 was to test the static stability of the structure when no wind is present. A1 was carried out in HydroD (DNV)where different loading conditions were specified with ballast ranging from 0% to 100%. The models created in SESAM® Genie (DNV) were imported and the necessary parameters for the analysis were set.

Analysis A2 was used to identify the effect of wind loads on the structure. In general, FOWT systems require a control system to help regulate the power generated. The continuous pitch and surge motions of a FOWT towards prevailing wind result in an increase in relative wind speed. As the wind speed increases, the power output gradually increases until a maximum value of generated power is reached at a specific wind speed, referred to as the rated wind speed. Analysis A2 investigated the effect of wind heeling moments on the proposed design at a rated wind speed obtained from (Cian, 2016) and defined in Table 3.

## 5 DISCUSSION OF RESULTS

The results of the analyses A1 and A2 are now presented and discussed.

### 5.1 *Analysis A1*

The hydrostatic values obtained from Analysis A1 can be seen in Table 6. As can be seen, the structure experiences a small initial trim angle of 0.5° towards the positive x-direction due to the weight of the support beams connecting the tower and the deck. At 80% ballast and continuing up to 100% ballast, the initial trim angle of the structure increases sharply from 0.5° to 15.2° and reaching up to 25.2° at 100% ballast.

Table 6. Hydrostatic values obtained from Analysis A1.

| Ballast | T (m) | Mass (t) | WPA ($m^2$) | Trim Angle (°) | $GM_L$ (m) | $GM_T$ (m) |
|---|---|---|---|---|---|---|
| 0% | 3.3 | 5273 | 1539 | 0.5 | 185 | 150 |
| 10% | 4.2 | 6562 | 1539 | 0.5 | 136 | 121 |
| 20% | 5.0 | 7851 | 1539 | 0.5 | 114 | 101 |
| 30% | 5.8 | 9140 | 1539 | 0.5 | 99 | 87 |
| 40% | 6.6 | 10428 | 1539 | 0.5 | 87 | 77 |
| 50% | 7.4 | 11717 | 1539 | 0.5 | 78 | 69 |
| 60% | 8.2 | 13006 | 1538 | 0.5 | 70 | 62 |
| 70% | 9.1 | 14295 | 1538 | 0.5 | 64 | 57 |
| 80% | 14.6 | 15584 | 884 | 15.2 | 17 | 28 |
| 90% | 17.6 | 16873 | 693 | 18.9 | 19 | 21 |
| 100% | 22.7 | 18162 | 497 | 25.2 | 14 | 13 |

At a ballast of 80% the draught exceeds the maximum height of the pontoon and the columns begin to submerge. Under these conditions there will be a reduction in the waterplane area (WPA). The reduced WPA causes the Centre of Floatation (COF) of the structure to shift further aft than the position of the tower, thereby causing the large trim angles. Furthermore, as can be seen in Table 6, both transverse and longitudinal GM values sharply decrease at 80% ballast, which is also attributed to the sudden decrease in WPA caused by the submerging of the pontoons.

The results of Analysis A1 conclude that the ballast compartments cannot be filled equally past 70%. To prevent the excess trim present between at 80% to 100% ballast, dynamic balancing must be used. The individual ballast tanks must be filled accordingly in order to offset the large trim obtained.

### 5.2 *Analysis A2*

Analysis A1 was repeated for the case in which wind is blowing at the rated conditions of a wind speed of 12.5 m/s, corresponding to the scenario where the wind turbine exerts the highest axial thrust on the floater across its operating envelope. The DNV standard ST-0119 flor floating wind turbines was used in conjunction with Equations 1-3.

The wind heeling moment was considered in two directions, for the wind direction parallel to the rotor, to consider pitching, and perpendicular to the rotor, for heeling.

The ST-0119 standard specifies that the wind heeling moment needs to be considered solely in the longitudinal (heeling) axis i.e. with the rotor plane perpendicular to the wind to ensure that the structure does not heel over. However, given that the structure is self-aligning and therefore it will align itself to the prevailing wind and wave conditions, realistically there is no scenario where the wind direction will be perpendicular to the structure while the turbine is in operation. In any case, during the towing operation of the FOWT, the structure will not be able to self-align and so it is possible that the wind direction will be perpendicular to the structure and there is a chance the structure will heel over.

Although it is not required by the ST-0119 that the structure meet all requirement for the transverse axis as well, i.e. when the wind direction is parallel to the turbine, it was still considered in this analysis. As mentioned previously, the dual-hull design included in the structure bears a significant resemblance to a catamaran structure. Research carried out by Deakin (2003) shows that the majority of catamaran capsize incidents were caused by either wind induced capsize or pitchpoling i.e. the capsizing by pitching forward.

Therefore, given that dual-pontoon structures have tend to capsize and considering that the structure needs to weathervane, verifying that the structure is also stable under wind heeling conditions in the transverse (pitching) axis is considered to be important.

The area underneath the wind heeling and righting moment graphs was evaluated using numerical techniques.

The GZ-curves for both the heeling and the pitching axes can be seen in Figures 5 and 6 respectively.

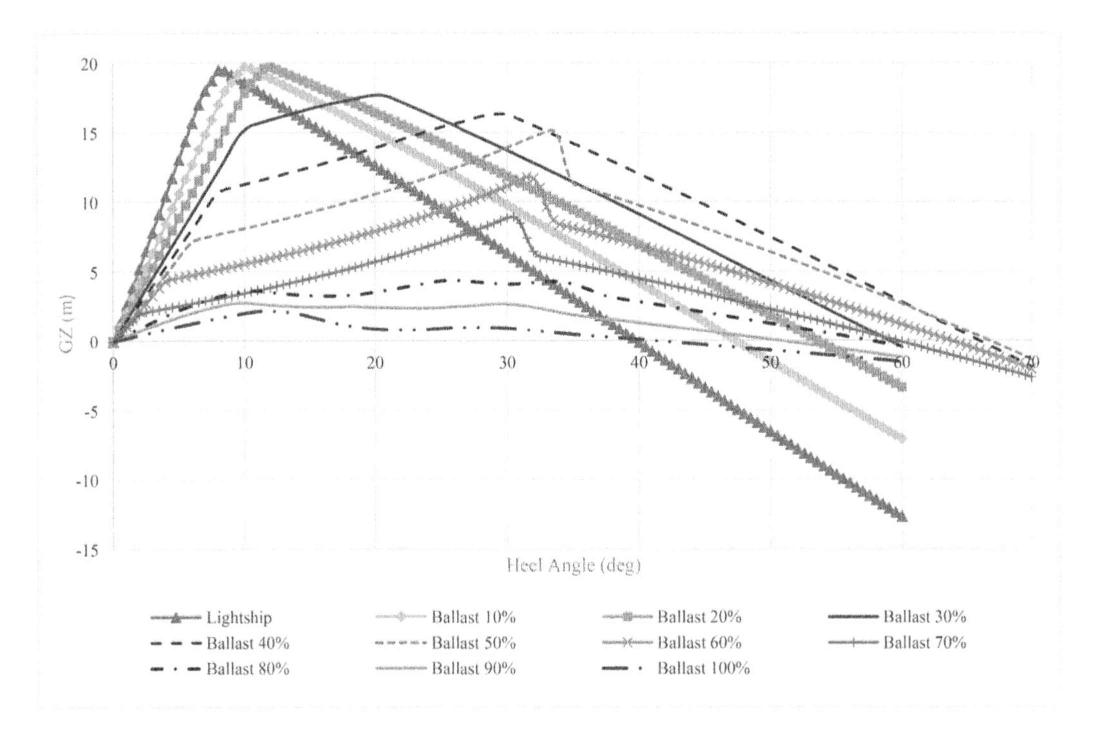


Figure 5. GZ curves for heeling.

From the GZ curves obtained for the different ballast conditions it can be seen that under the ballast conditions of 0% – 40%, the traditional shape of GZ curve is preserved.

On the other hand, when the ballast is increased between 50 -100%, the shape of the GZ curves deviate from the traditional curves, occurring most prominently between 80-100% ballast. The unusual shapes of the graphs can be attributed to the increased trim angle at those ballast conditions.

In the heeling axis, the graphs of 50% - 70% ballast can be seen to peak sharply before decreasing. The GZ curves obtained for the 80%-100% ballast show smaller peaks around the same heel angles. The peaks are due to the deck being immersed under water. The shape of the peaks shown in ballasts 80% to 100% are different due to the large trim angles present.

The GZ curves obtained for the pitching axis can be seen in Figure 6. The curves for ballast conditions 0%-50% encompass the traditional GZ curve shape shown in literature. From 60% ballast the GZ curve continuously flattens in shape until 100% ballast where the graph stops abruptly at a pitch angle of 44.5°, unlike the rest of the ballast conditions. Under full ballast conditions the structure submerges totally at 44.5°.

The results of analysis A2 with regards to the standard requirements can be seen in Table 7.

As shown in Table 7 the model passed all the necessary standards requirements both in the heeling and pitching axes. Furthermore, the values obtained of the area under the righting moment curve to the second intercept in excess of the area under the wind heeling moment curve for the pitching axis are significantly lower than those obtained for the heeling axis. Therefore, the results of Analysis A2 further reinforce the notion that dual-pontoon are more prone to instability in the pitching axis.

The positive static stability values obtained from Analysis A2 show that the structure is stable even under with 0% ballast. Therefore, there is no need to ballast the structure during the towing operations.

Moreover, the static stability results identify to which extent ballasting is possible on the structure

Table 7. Results of analysis A2.

| DNV ST-0119 Sect. 10.2.3 | | | |
|---|---|---|---|
| Ballast | Result | Heeling Axis | Pitching Axis |
| 0% | Pass | 48 > 1.3 | 5.6 > 1.3 |
| 10% | Pass | 87 > 1.3 | 6.0 > 1.3 |
| 20% | Pass | 135 > 1.3 | 6.3 > 1.3 |
| 30% | Pass | 221 > 1.3 | 6.3 > 1.3 |
| 40% | Pass | 279 > 1.3 | 6.4 > 1.3 |
| 50% | Pass | 300 > 1.3 | 6.4 > 1.3 |
| 60% | Pass | 271 > 1.3 | 5.7 > 1.3 |
| 70% | Pass | 899 > 1.3 | 4.5 > 1.3 |
| 80% | Pass | 274 > 1.3 | 8.4 > 1.3 |
| 90% | Pass | 451 > 1.3 | 3.7 > 1.3 |
| 100% | Pass | 136 > 1.3 | 2.9 > 1.3 |

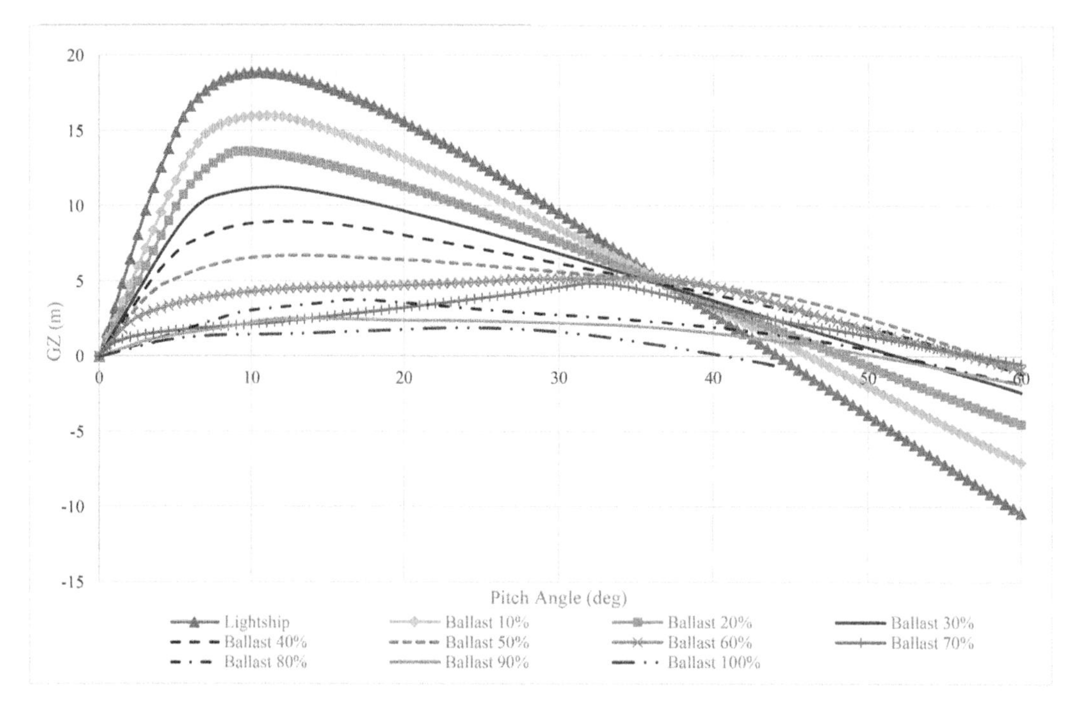


Figure 6. GZ Curves for pitching.

so that future work can incorporate dynamic stability in the structure.

## 6 CONCLUSION

The model being tested makes use of two innovative characteristics to promote weathervaning capabilities in a FOWT designed for Mediterranean conditions. By incorporating two truncated Wigley Hull shape pontoons and four airfoil shaped support columns in conjunction with an SPM system, the self-aligning characteristics of the structure are enhanced. Given the novelty of the design, significant importance was given to ensuring that the model adhered to all the necessary stability standards and that good static stability characteristics are present.

The structure shows positive results in both the heeling and pitching axes and passes all the stability criteria set by DNV ST-00019. Although in reality the structure would self-align itself to the varying wind and wave conditions, the model still complies with all criteria set by standards.

The model experiences drastic trimming between ballasts of 80% to 100% which necessitate dynamic balancing of the structure in order to offset the trim. The result obtained also show that the model has better stability characteristics in the heeling axis than the pitching axis which is common in dual pontoon structures.

The aim of this research is to assess the static stability of the structure in order to identify to which extent ballasting the structure is possible. For dynamic stability. Assessment of the dynamic stability and self-aligning characteristics of the proposed floater under wave action is beyond the scope of this paper and is a suggestion for future work.

More options for future work can include modifying the design to prevent the presence of excess trimming in higher ballast conditions; and modelling its self-aligning ability when moored with an SPM system.

## ACKNOWLEDGEMENTS

The presented work has been supported through the Maritime Seed Award 2020 and the collaborative programme of the VENTuRE (project no. 856887) EU funded H2020 project.

## REFERENCES

Butterfield, S. *et al.* (2005) 'Engineering Challenges for Floating Offshore Wind Turbines', in *e 2005 Copenhagen Offshore Wind Conference*, October 26–28,. Copenhagen, Denmark. Available at: https://www.nrel.gov/docs/fy07osti/38776.pdf.

Caro, R.. *et al.* (2018) 'A Review of Ship Mooring Systems', *Brodogradnja*, 69(1). doi: 10.21278/brod69108.

Cian, D. (2016) 'Description of an 8 MW reference wind turbine', *Journal of Physics: Conference Series*, 753(9). doi:10.1088/1742-6596/753/9/092013.

Deakin, B. (2003) 'Model Tests To Study Capsize and Stability of Sailing Multihulls', *The International Journal of Small Craft Technology*. doi: 10.3940/rina.ijsct.2003.b2.9031.

DNV (2017) *DNV- OS- C301: Stability and Watertight Integrity*. Available at: https://rules.dnv.com/docs/pdf/DNV/OS/2017-01/DNVGL-OS-C301.pdf.

DNV (2018) *DNV-ST-0119: Floating wind turbine structures*. Available at: https://rules.dnv.com/docs/pdf/DNV/ST/2018-07/DNVGL-ST-0119.pdf.

DNV (no date) *Software products overview*. Available at: https://www.dnv.com/software/products/sesam-products.html (Accessed: 30 January 2022).

Galvin, J., Sanchez-Lara, M. and Menfikoa, I. (2018) 'NAUTILUS-DTU10 MW Floating Offshore Wind Turbine at Gulf of Maine: Public numerical models of an actively ballasted semisubmersible', *Journal of Physics: Conference Series*, 1102. doi:10.1088/1742-6596/1102/1/012015.

Iijima, K. *et al.* (2013) 'Conceptual Design of a Single-Point Moored FOWT and Tank Test for its motion characteristics', in *32nd International Conference on Ocean, Offshore and Arctic Engineering*. doi: 10.1115/OMAE2013-11259.

International Maritime Organization (IMO) (2008) *Adoption of the International Code on Intact Stability (2008 IS CODE), Resolution*. Available at: https://wwwcdn.imo.org/localresources/en/KnowledgeCentre/IndexofIMOResolutions/MSCResolutions/MSC.267(85).pdf.

Karimirad, M. (2014) 'V-shaped semisubmersible offshore wind turbine: An alternative concept for offshore wind technology', *Renewable Energy*, 83, pp. 126–143. https://doi.org/10.1016/j.renene.2015.04.033.

Leimeister, M., Kolios, A. and Collu, M. (2018) 'Critical review of floating support structures for offshore wind farm deployment', in *Journal of Physics: Conference Series. EERA DeepWind'2018, 15th Deep Sea Offshore Wind R&D Conference*. EERA DeepWind'2018, 15th Deep Sea Offshore Wind R&D Conference.

Mckenna, R., P, O. and Fisher, L.W. (2016) 'Key challenges and prospects for large wind turbines', *Renewable and Sustainable Energy Reviews*, 53, pp. 1212–1221. https://doi.org/10.1016/j.rser.2015.09.080.

Mooring, A. (no date) *MOORING SYSTEMS*. Available at: http://abc-moorings.weebly.com/mooring-systems.html (Accessed: 26 December 2020).

Netzband, S., Schulz, C. W. and Abdel-Maksoud, M. (2020) 'Passive self-aligning of a floating offshore wind turbine', *Journal of Physics: Conference Series*. doi: 10.1088/1742-6596/1618/5/052027.

Oh, S. (2018) 'Numerical modelling and validation of a semisubmersible floating offshore wind turbine under wind and wave misalignment', *Journal of Physics: Conference Series*. doi: 10.1088/1742-6596/1104/1/012010.

Schneider, J. A. and Senders, M. (2010) 'Foundation Design: A Comparison of Oil and Gas Platforms with Offshore Wind Turbines', *Marine Technology Society Journal*, 44(1), pp. 35–51. https://doi.org/10.4031/MTSJ.44.1.5.

Sieros, G. *et al.* (2012) 'Upscaling wind turbines: Theoretical and practical aspects and their impact on the cost of energy', *Wind Energy*, 15(1), pp. 3–17. doi: 10.1002/we.527.

TUHH (2020) *HyStOH/Hydrodynamic and Structural Optimisation of a Semi-Submersible for Offshore Wind Turbines -*. Available at: https://www.tuhh.de/panmare/de/applications/research-projects/completed-projects/hystoh.html (Accessed: 29 January 2022).

Wichers, I. J. (2013) *Guide to Single Point Mooring*. WMooring.inc. Available at: http://www.wmooring.com/files/Guide_to_Single_Point_Moorings.pdf.

Zhao, Z. (2019) 'Analysis of Dynamic Characteristics of an Ultra-Large Semi-Submersible Floating Wind Turbine', *Journal of Marine Science and Engineering*, 7(6), p. 169. https://doi.org/10.3390/jmse7060169.

*Trends in Maritime Technology and Engineering – Guedes Soares & Santos (eds)*


# Hydrodynamic analysis of a dual-body wave energy converter device with two different power take-off configurations

J.C. Souza Filho, K. Rezanejad & C. Guedes Soares
*Centre for Marine Technology and Ocean Engineering (CENTEC), Instituto Superior Tecnico, Universidade de Lisboa, Lisbon, Portugal*

ABSTRACT: The study aims to compare two different power take-off systems installed in a dual body wave energy converter, first the power take-off in between the floater and the submerged body, and second in between the submerged body and the sea bottom. It is intended to prove that the second is more efficient. The analysis is carried out for three WEC's with different bodies, cylinder-sphere, cylinder-cylinder, and sphere-sphere, for the floater and submerged body. The analysis begins by modeling the bodies in NEMOH, a software that calculates hydrodynamical coefficients. After, a dynamical model is created in the frequency domain to calculate power absorbed and efficiency in regular and irregular waves for all geometries with both the first and the second PTO's configuration. The final analysis is comparing the efficiency in irregular waves for the first PTO configuration against the second, it shows that the second configuration has an improvement of 11%.

## 1 INTRODUCTION

A wave energy converter is a device that captures energy from the waves and converts it into electrical energy. Normally, the wave energy is harvested by the movements of the device. There are several different types of WECs and therefore they can be classified into different groups.

One of the possible classifications is by the distance to the shoreline, studied by Cruz (2008); Falcão (2010); Guedes Soares et al (2012), classifying them into onshore, nearshore, and offshore.

Another one is the one used by The European Marine Energy Center LTD (EMEC). In which the device is classified into 8 different types: attenuator, oscillating wave surge converter, oscillating water column, overtopping/terminator device, submerged pressure differential, bulge wave, rotating mass, point absorber, and others.

The exploration of wave energy started recently, R&D started around the 1970s, and demonstrations of wave energy converter devices around the 1980s, so there is still a lot of development to be done to optimize its efficiency and lower the costs of installation and operation.

Some research in the area has been performed by different authors, for instance, Al Shami et al. (2019) studied the effects of increasing the number of degrees of freedom through adding submerged bodies to a point absorber WEC in an attempt to capture more power while keeping the same total mass and volume. It was found that increasing the number of degrees of freedom through adding spherical submerged bodies increases the average captured power by 26% for the WECs going from 2 DOF to 4 DOF. Rezanejad et al. (2018) studied the beneficial effects of designing an oscillating water column WEC by associating the hydrodynamics of the oscillating water masses to a dual-mass system. It was proved that the implementation of the bottom step in the configuration of the OWC system will improve the performance of the device up to 150 and 60% in regular and random wave conditions, respectively.

Moreover, Mørk et al. (2010) has estimated that the global availability of gross power of wave energy is about 3.7 TW, while the installed capacity around the end of 2016 was only 12 MW, data on the Atlantic coasts of Europe are found in Guedes Soares et al (2014). To understand the dimension of the power availability, it would represent 32412 TWh in one year, ignoring losses system due to lack of perfect efficiency, and the world consumption of energy in 2019 was 173340 TWh, (Ritchie 2017).

Now it is possible to affirm that wave energy is hugely underused besides the potential to represent nearly 20% of the world's consumption. That means that any improvement in the field could significantly increase the world's power generation, something that will almost undeniably be one of the problems for future years.

Also, it is important to remember the environmental effect of using wave energy instead of nonrenewable ones. As it is renewable, it does not pollute the environment as strongly as some traditional energy

DOI: 10.1201/9781003320289-52

sources (as coal or oil for example), also, is not as dangerous to deal as nuclear energy.

It is possible to say now that there are several good reasons and needs to continue the studies in the field of wave energy exploration. Following that idea, the present study was developed inspired by the previous research study carried out by Rezanejad and Guedes Soares (2018) to improve the efficiency of a wave energy converter.

The objective of this study is to compare the performance of a two-body wave energy converter with different configurations of its PTO system. The first configuration consists of the PTO located between the two bodies, the second configuration consists of the PTO located between the submerged body and sea bottom.

The process begins with the optimization of both configurations varying their geometrically, PTO and mooring parameters in a range of values. Then, prove that the second configuration is better than the first one in terms of harvesting energy efficiency.

That is of paramount importance, as improving the efficiency of a Wave Energy Converter means in other worlds reducing the cost to generate energy. The reduction of the cost of the energy produced by Wave Energy Converters is one of the main challenges in the field today, as the technology to install it already exists.

Firstly, a common configuration for the two-body WEC (Power Take-Off unit placed between the two floaters) is analyzed and optimized. This analysis is carried out for three different shapes of the geometries of the bodies, the floater and the submerged body are respectively: cylinder-sphere, cylinder-cylinder, sphere-sphere.

The optimization was done for each one of the three cases (cylinder-sphere, cylinder-cylinder, sphere-sphere) by varying the geometrical parameters, radius, draft, height, of both bodies, and the parameters of the power take-off unit and mooring system. Then, the best combination of them in terms of power absorbing efficiency in irregular waves was chosen as the optimal one.

Secondly, using the same procedure described, a different configuration for the wave energy converter (Power Take-Off unit placed between sea bottom and the fully submerged body) is analyzed. Also, the best ones in terms of efficiency are pointed out.

Finally, the comparison between the two configurations is done for each one of the three cases, cylinder-sphere, cylinder-cylinder, sphere-sphere. In the end, it has been proven that the second configuration is the best in terms of efficiency in harvesting wave energy for all three cases analyzed, as was desired. Hence, the application of the second configuration would significantly reduce energy production costs, which is one of the main barriers nowadays to the industrial development of Wave Energy Converter devices. Therefore, this might be helpful to promote the application of this kind of PTO system in WECs.

## 2 PROCESS DESCRIPTION

The PTO in the present study, as in Al Shami et al (2019), is modeled as a combination of a damper and a spring, and the mooring/interconnection is modeled as a spring. Being them represented here by $c_{pto}, K_{pto}, K_m$.

The two different PTO configurations being studied can be seen in *Figure 1* and the description of the model is present in the next lines.

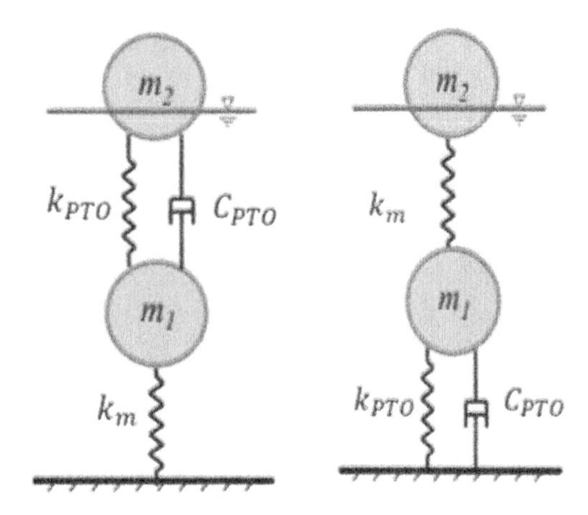


Figure 1. Dynamic model of first and second configurations of the PTO.

Firstly, three different combinations of geometrical shapes were defined to be analyzed: cylinder-sphere, cylinder-cylinder, and sphere-sphere for the floater and the submerged body respectively.

The main importance of considering these three geometries is to improve the reliability of the study, as by raising the number of cases considered the conclusion of the present study becomes more generic and valid not just in a specific case. Several different other authors used the same method, for instance Al Shami et al. (2019) considered the possible bodies as cylinder and sphere for both the floating and submerged body.

The geometries considered were chosen by selection as the most common ones used in the research of WEC.

All three combinations were analyzed for both the first configuration, PTO between the two bodies, and the second configuration, PTO between submerged body and sea bottom.

The process described in the next lines was carried out for all the three configurations combined with both the first and second configurations of the PTO, so it was carried out six times. For example, the process can be carried out for the case of

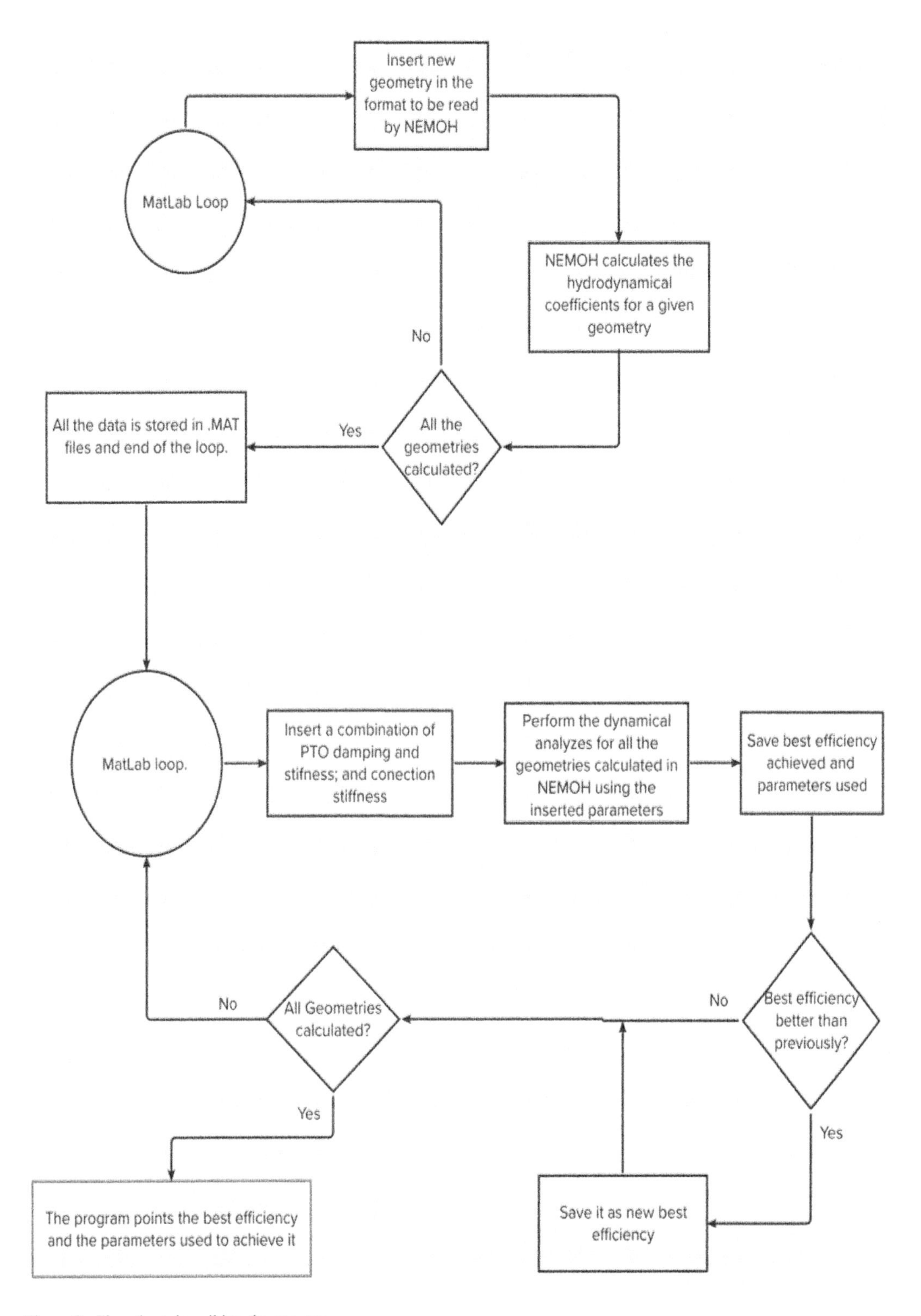


Figure 2. Flowchart describing the process.

cylinder-sphere using the first configuration of the PTO, cylinder-sphere using the second configuration of the PTO, cylinder-cylinder using the first configuration of the PTO, and so on.

The process consists in the creation of loops in MatLab involving both the NEMOH (Babarit, A. et al 2015), a software that allows the calculation of the hydrodynamical coefficients of the bodies, and codes created to perform the dynamical analysis and optimization of the system, resulting in the calculation of absorbed power and efficiency.

First, the shape of the bodies being analyzed are defined, for example, cylinder-sphere. Then, different mesh geometries are created, using a MatLab loop, in the correct file to be inputted input in NEMOH. NEMOH then outputs the hydrodynamical coefficients that are stored in a '. MAT' file by the MatLab loop.

The meaning of these different meshes created is to set different sizes of the bodies to be analyzed. So, for example, it is analyzed for the case cylinder-sphere (using the first PTO configuration) for different sizes of the cylinder combined with different sizes of the sphere, each one of these combinations generating a different output in NEMOH.

For example, one of these meshes could contain data about a cylinder-sphere WEC with a cylinder with 2 meters radius and 1-meter draft combined with a sphere with a 2 meter radius, another one of these meshes could contain a cylinder with a 1 meter radius and 0.5 meters of draft combined with a sphere with 1 meter radius, and so on.

Then, a range of values of PTO and mooring parameters $(c_{pto}, K_{pto}, K_m)$ are defined and will be used to optimize the efficiency.

After, a loop proceeds to the calculation of efficiency and power absorbed for all the different meshes (different combinations of bodies sizes) combined with all the different parameters $(c_{pto}, K_{pto}, K_m)$.

Finally, the code points which is the best geometrical configuration and mooring and PTO parameters $(c_{pto}, K_{pto}, K_m)$ that reached the maximum efficiency for the case analyzed.

For example, if the case been analyzed was cylinder-sphere using the first configuration of the PTO, the code would, in this stage, point which combination of mesh (sizes of the buoy and the submerged body) and mooring and PTO parameters $(c_{pto}, K_{pto}, K_m)$ generated the maximum efficiency, being this the optimum case.

A flowchart better explaining the way the process works can be seen at Figure 2.

For instance, the case decided to be analyzed could be sphere-sphere. So, the loop at the flowchart could be initialized with two different geometry sizes, sphere-sphere with both 1-meter radius and another one with 2 meters radius each. For that case, after the first loop finishes its calculations, two different groups of results with hydrodynamical coefficients would be stored in .MAT files.

Using each of that .MAT files, the second loop performs the dynamical calculation for both the first and second configurations of the PTO. Also, for each of these PTO's configurations, the system tries all the values of mooring and PTO parameters $(c_{pto}, K_{pto}, K_m)$ using a pre-defined range of values. The final output of this loop is the results of the best case for each scenario consider. So, the program would output the dynamical calculations (power and efficiency in regular and irregular waves) for both PTO configurations using the parameters that generated the maximum efficiency possible. Having these results, a final comparison between the two PTO configurations shows which one is better in terms of efficiency. For instance, following the example, the best case for the sphere-sphere with the first PTO configuration could've been both with 1 meter radius and generated 40% of efficiency; and for the second configuration with the PTO the one of both with 2 meters and generated 47% of efficiency. In that fictional example, the second configuration would've been considered better for the case sphere-sphere because it has a higher best efficiency.

## 3 MATHEMATICAL MODEL

The goal of this section is to describe a mathematical model using both a BEM (boundary element method) solver software, NEMOH, and a code created in MatLab that allows the calculation of the efficiency/Power Absorbed, both in regular and irregular waves, of the two models/configurations of the WEC's considered in the present study.

### 3.1 *NEMOH*

In order to create the dynamic model in frequency domain and carry out the necessary calculations, it is necessary to calculate the frequency-dependent hydrodynamic parameters of the two bodies (added mass, radiation damping, and wave forces).

There are some programs specialized in this calculation, for example, WAMIT and NEMOH, a comparison between the two of them was made by Penalba et al (2017) showing good agreement between them. So, NEMOH was the one chosen because is an open-source program. The program requires some inputs in order to work and proceed with the calculations, a mesh file with the discretization of the panels for both bodies, conditions of the sea (like water depth, wave height, distance between the bodies, and draft of the floating buoy) and values assumed for gravity and ρ. In order to properly perform the calculations, NEMOH makes some assumptions, as is described by the developer of the code (Babarit and Delhommeau 2015). The assumptions are inviscid fluid, incompressible and irrotational flow.

### 3.2 *First geometry*

Firstly, it is important to clarify that the movement of the dual body WEC is restricted to heave and the two bodies are coupled due to the PTO forces, so the system has two degrees of freedom: heave movement for the buoy and heave movement for the submerged body.

The model is based on the linear wave theory, meaning small wave amplitudes and small body motions were considered. As described before, the analysis was developed in frequency domain. The parameters $c_{pto}, K_{pto}, K_m$ are PTO damping and stiffness; and mooring/interconnection stiffness respectively. The hydrodynamic parameters (added mass, radiation damping and excitation force) for both the buoy and the submerged body were calculated using the BEM solver software NEMOH. In order to create the model, Newton's second law was applied for both bodies.

$$M\ddot{x} = \sum F_{external} \tag{1}$$

Then, as in (Al Shamiet al 2018); (Liang and Zuo 2016); (Al Shami et al 2019); (Cheng et al. 2015), it is possible to denote the matrices:

$$M = \begin{bmatrix} m_1 + A_{11} & A_{12} \\ A_{21} & m_2 + A_{22} \end{bmatrix} \tag{2}$$

$$C = \begin{bmatrix} b_{11} + b_{visc1} + c_{pto} & b_{12} - c_{pto} \\ b_{21} - c_{pto} & b_{22} + b_{visc2} + c_{pto} \end{bmatrix} \tag{3}$$

$$K = \begin{bmatrix} k_s + k_{pto} & -k_{pto} \\ -k_{pto} & k_{pto} - k_m \end{bmatrix} \tag{4}$$

$$X = \begin{bmatrix} X_1 \\ X_2 \end{bmatrix} \tag{5}$$

$$F = \begin{bmatrix} F_1 \\ F_2 \end{bmatrix} \tag{6}$$

It is possible now to model the system as:

$$Z(i\omega) = -\omega^2 M + i\omega C + K \tag{7}$$

Where Z is the impedance matrix and can be written

$$Z(i\omega) = \begin{bmatrix} Z_{11} & Z_{12} \\ Z_{21} & Z_{22} \end{bmatrix} \tag{8}$$

with:

$$Z_{11} = -\omega^2(m_1 + A_{11}) + i\omega(b_{11} + b_{visc1} + c_{pto}) + k_s + k_{pto} \tag{9}$$

$$Z_{12} = -\omega^2 A_{12} + i\omega(b_{12} - c_{pto}) - k_{pto} \tag{10}$$

$$Z_{21} = -\omega^2 A_{21} + i\omega(b_{21} - c_{pto}) - k_{pto} \tag{11}$$

$$Z_{22} = -\omega^2(m_2 + A_{22}) + i\omega(b_{22} + b_{visc2} + c_{pto}) + k_{pto} - k_m \tag{12}$$

The solution for the equations (15), (16), (17) and (18) can now be written as:

$$X = Z(i\omega)^{-1}F \tag{13}$$

### 3.3 *Second geometry*

Analogously, for the second geometry the impedance matrix became:

$$Z_{11} = -\omega^2(m_1 + A_{11}) + i\omega(b_{11} + b_{visc1}) + k_s + k_m \tag{14}$$

$$Z_{12} = -\omega^2 A_{12} + i\omega(b_{12}) - k_m \tag{15}$$

$$Z_{21} = -\omega^2 A_{21} + i\omega(b_{21}) - k_m \tag{16}$$

$$Z_{22} = -\omega^2(m_2 + A_{22}) + i\omega(b_{22} + b_{visc2} - c_{pto}) - k_{pto} + k_m \tag{17}$$

Finally, the solution is now calculated the same way of the first geometry:

$$X = Z(i\omega)^{-1}F \tag{18}$$

### 3.4 *Power and efficiency regular waves*

The efficiency for regular waves is defined as the ratio between the power absorbed by the WEC and the maximum power available to be absorbed in the income wave. The calculation is carried out for each different frequency of incoming waves.

$$\eta(\omega) = \frac{P_{avg}(\omega)}{P_{max\_wave}(\omega)} \tag{19}$$

The calculation of the power absorbed for each frequency of income wave can be calculated as in (Al Sham et al 2018); (Liang and Zuo 2016); (Al Shami et al 2019); (Cheng et al. 2015). It is important to notice that they are different for the first and second geometry, as the PTO (which generates the power) is placed in different places. For the first geometry it generates power in relation to the relative motion between the two bodies, for the second, in relation to the absolute motion of the second body.

For the first configuration:

$$P_{avg}(\omega) = \frac{1}{T}\int_0^T c_{pto}(\dot{x}_1 - \dot{x}_2)dt = 0.5\omega^2 c_{pto} abs(X_1 - X_2) \tag{20}$$

For the second configuration:

$$P_{avg}(\omega) = \frac{1}{T}\int_0^T c_{pto}(\dot{x}_2)dt = 0.5\omega^2 c_{pto} abs(X_2) \tag{21}$$

And the maximum power available in the wave to be absorbed by the WEC, according to Dean and Darimple (2010) can be calculated as the total energy per wave per unit width times the group velocity.

$$P_{max\_wave}(\omega) = Ec_g(\omega) \tag{22}$$

where:

$$E = 0.5\rho g H^2 L \tag{23}$$

and

$$c_g(\omega) = 0.5\frac{\omega}{\kappa}\left(1 + (2\kappa(\omega)D_{epth})/(\sinh 2\kappa(\omega)D_{epth})\right) \tag{24}$$

and $\kappa$ is the wave number, calculated by Dean and Darimple (2010) as:

$$g\kappa(\omega)\tanh\kappa(\omega) * D_{epeth} = \omega \tag{25}$$

The calculation of the wave number must be an iterative process, however as the MatLab software were used a simple command 'vpasolve' was enough to reach the desired results.

### 3.5 *Powe and efficiency irregular waves*

The calculation for irregular waves depends on the sea state of the place where the WEC is installed. The parameters taken into considerations are the significant wave high ($H_s$) and energy wave period ($T_e$). The efficiency can then be calculated as described in (Liang and Zuo 2016); (Cheng et al. 2015).

$$\eta = \frac{P_{avg\_irr}}{P_{max\_wave\_irr}} \tag{26}$$

It is possible to notice that unlike in regular waves, the final efficiency is one number independent on the income wave frequency. The calculation of absorbed and available power in irregular waves depends on the constructions of a sea spectra. The one chosen in this model was Pierson-Moskowitz spectra:

$$S(\omega) = 526H_s^2T_e^{-4}\omega^{-5}e^{-1054T_e^{-4}\omega^{-4}} \tag{27}$$

Now, both the absorbed and available power can be written as:

$$P_{irr} = \int_0^\infty P_{reg}(\omega)S(\omega)d\omega \cong \sum P_{reg}(\omega)S(\omega)\Delta\omega \tag{28}$$

## 4 PROCESS EXPLANATION

Firstly, in order to proceed to all calculations, the sea state spectrum of Pico-Azores is calculated. The considered Sea State is the one obtained by Matos et al (2015) and was used to obtain the spectrum.

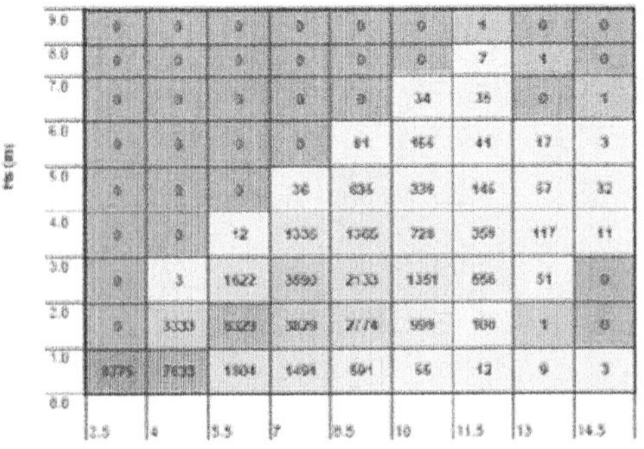

| Hs (m) \ TP (s) | 2.5 | 4 | 5.5 | 7 | 8.5 | 10 | 11.5 | 13 | 14.5 |
|---|---|---|---|---|---|---|---|---|---|
| 9.0 | 0 | 0 | 0 | 0 | 0 | 0 | 1 | 0 | 0 |
| 8.0 | 0 | 0 | 0 | 0 | 0 | 0 | 7 | 1 | 0 |
| 7.0 | 0 | 0 | 0 | 0 | 0 | 34 | 35 | 0 | 1 |
| 6.0 | 0 | 0 | 0 | 0 | 81 | 464 | 41 | 17 | 3 |
| 5.0 | 0 | 0 | 0 | 36 | 635 | 339 | 146 | 57 | 32 |
| 4.0 | 0 | 0 | 12 | 1336 | 1365 | 728 | 359 | 117 | 11 |
| 3.0 | 0 | 3 | 1622 | 3590 | 2133 | 1351 | 656 | 51 | 0 |
| 2.0 | [illegible] | 3333 | 6329 | 3829 | 2774 | 999 | 100 | 1 | 0 |
| 1.0 | 8775 | 7633 | 1804 | 1494 | 604 | 66 | 12 | 9 | 3 |
| 0.0 | | | | | | | | | |

Figure 3. Matrix of occurrence of each combination of Ts and Hs at Pico-Azores.

The range of values of mooring and PTO parameters ($c_{pto}, K_{pto}, K_m$) are the same for all the three cases and can be seen in Table 1.

Table 1. Range of values considered for PTO and mooring parameters.

| Mooring and PTO parameters | | | |
|---|---|---|---|
| Parameter | Starting Value | Ending Value | Step |
| PTO Damping [Ns/m] | 10000 | 300000 | 10000 |
| Equivalent PTO Stiffness [N/m] | 10000 | 140000 | 10000 |
| Mooring Stiffness/ Floaters Interconnection Stiffness [N/m] | 10000 | 140000 | 10000 |

4.1 *Cylinder-sphere*

To perform the calculations and the optimization process for the case cylinder-sphere, nine different combinations of geometrical parameters were chosen. These different combinations can be seen at Table 2.

Table 2. Configurations considered for the case cylinder-sphere.

| Configuration Number | Buoy Radius [m] | Buoy Draft [m] | Submerged Body Radius [m] |
|---|---|---|---|
| 1 | 1 | 0.5 | 1 |
| 2 | 2.5 | 0.5 | 1 |
| 3 | 2.5 | 1 | 1 |
| 4 | 2.5 | 1 | 3 |
| 5 | 4 | 1 | 3 |
| 6 | 6 | 1 | 3 |
| 7 | 4 | 1 | 1 |
| 8 | 6 | 1 | 1 |
| 9 | 6 | 2 | 3 |

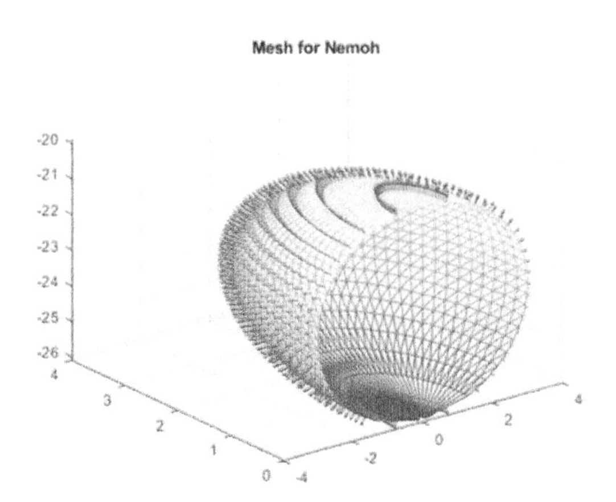


Figure 4. Mesh for NEMOH of the floating buoy.

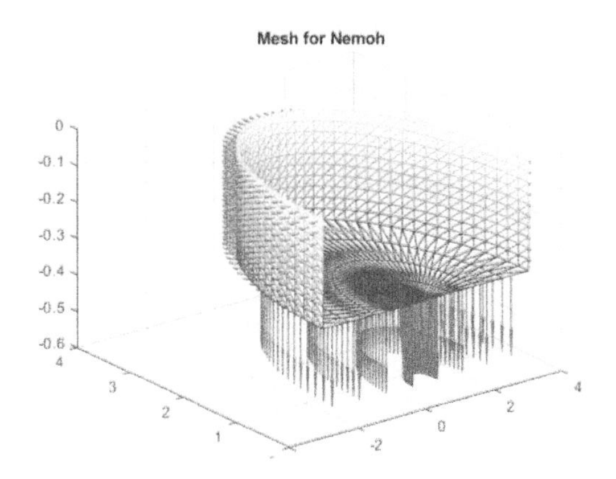


Figure 5. Mesh for NEMOH of the floating buoy.

The mesh files used as input and the efficiency curves are presented now. This will be presented just now as they are similar for all cases.

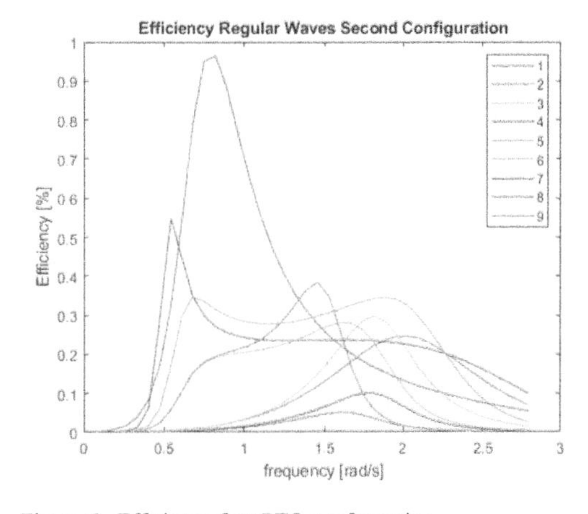


Figure 6. Efficiency first PTO configuration.

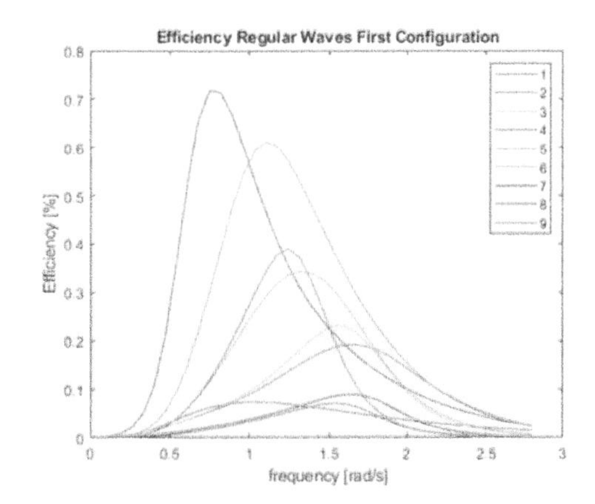


Figure 7. Efficiency second PTO configuration.

### 4.2 *Cylinder-cylinder*

To perform the calculations and the optimization process for the case cylinder-cylinder, eleven different combinations of geometrical parameters were chosen. These different combinations can be seen at Table 3.

Table 3. Configurations considered for the case cylinder-cylinder.

| Configuration Number | Buoy Radius [m] | Buoy Draft [m] | Submerged Body Radius [m] | Submerged Body Height [m] |
|---|---|---|---|---|
| 1 | 1 | 0.5 | 1 | 1 |
| 2 | 1 | 1 | 1 | 1 |
| 3 | 1 | 1 | 2.5 | 1 |
| 4 | 1 | 1 | 2.5 | 2 |
| 5 | 2.5 | 1 | 1 | 1 |
| 6 | 2.5 | 2 | 1 | 1 |
| 7 | 2.5 | 1 | 2.5 | 2 |
| 8 | 2.5 | 2 | 2.5 | 2 |
| 9 | 2.5 | 2 | 4 | 2 |
| 10 | 4 | 2 | 2.5 | 2 |
| 11 | 4 | 2 | 4 | 2 |

### 4.3 *Sphere-sphere*

To perform the calculations and the optimization process for the case sphere-sphere, seven different combinations of geometrical parameters were chosen. These different combinations can be seen at Table 4.

Table 4. Configurations considered for the case sphere-sphere.

| Configuration Number | Buoy Radius [m] | Submerged Body Radius [m] |
|---|---|---|
| 1 | 1 | 1 |
| 2 | 1 | 2 |
| 3 | 2 | 1 |
| 4 | 2 | 2 |
| 5 | 2 | 3 |
| 6 | 3 | 1 |
| 7 | 3 | 3 |

## 5 RESULTS

By running the model, the best efficiency for each case is achieved, the parameters chosen and results for each case can be seen at Table 5.

As it can be seen, all the efficiencies for the second configuration of the PTO are higher than the first configuration. This means that it costs less money to generate energy in Pico-Azores using the second configuration (PTO between submerged body and sea bottom) instead of the first one (PTO between buoy and submerged body).

The wave energy converter that reached the best efficiency was the cylinder-cylinder for both the first and the second configuration of the PTO. However, the second most efficient was the sphere-sphere, for the case of second PTO configuration, and cylinder-sphere for the case of first PTO configuration.

Table 5. Results for optimized configuration for both PTO configurations for each case.

| | First Configuration Optimized bodies | Second Configuration Optimized bodies |
|---|---|---|
| **Geometrical Configuration** | **Cylinder-Sphere** | |
| Buoy Radius [m] | 2.5 | 1 |
| Buoy Draft [m] | 1 | 0.5 |
| Submerged Body Radius [m] | 3 | 1 |
| Absorbed Power Irregular Waves [W] | 1.26E+04 | 5.91E+03 |
| Efficiency Irregular Waves | 47.07% | 55.43% |
| Variation in the Efficiency | 8.36 % | |
| **Geometrical Configuration** | **Cylinder-Cylinder** | |
| Buoy Radius [m] | 2.5 | 2.5 |
| Buoy Draft [m] | 2 | 1 |
| Submerged Body Radius [m] | 4 | 2.5 |
| Submerged Body Height [m] | 2.5 | 2 |
| Absorbed Power Irregular Waves [W] | 1.46E+04 | 1.86E+04 |
| Efficiency Irregular Waves | 54.86% | 65.67% |
| Variation in the Efficiency | 10.71 % | |
| **Geometrical Configuration** | **Sphere-Sphere** | |
| Buoy Radius [m] | 2 | 2 |
| Submerged Body Radius [m] | 3 | 1 |
| Absorbed Power Irregular Waves [W] | 9.96E+03 | 1.28E+04 |
| Efficiency Irregular Waves | 46.67% | 60.06% |
| Variation in the Efficiency | 13.39 % | |

Also, it is possible to see that the second configuration of the PTO is optimized always for smaller bodies (both floater and submerged body, exception made for the case of cylinder-cylinder in which the size of the buoy is the same) in comparison with the first configuration, this is even more noticed in the case of the submerged body.

For the case cylinder-sphere, the size of the submerged sphere goes from 3 meters radius (first configuration of the PTO) to 1 meter (second configuration of the PTO).

For cylinder-cylinder it goes from 4 meters of radius and 2.5 meters of height (for the first PTO configuration) to 2.5 meters radius and 2 meters height (second configuration of the PTO).

For the case sphere-sphere it goes from 3 meters radius (for the first PTO configuration) to 1 meter (for the second PTO configuration).

Now, analyzing all the optimized data together, it is possible to notice that the first configuration of the PTO produced an average efficiency of 49.53%, while the second configuration of the PTO produced an average efficiency of 60.39%. That means that using the second configurations instead of the first one generates an average improvement in the efficiency of the optimized data of 10.85%.

Comparing all the cases and efficiencies calculated (not just the optimized ones) in the present study it is possible to notice that the second configuration of the PTO still more efficient than the first one. The second configuration has an average efficiency of 24.62% while the first one has an efficiency of 18.17%. So, the second configuration is 6.45% more efficient than the first one considering all the cases analyzed.

Also, comparing just all the cases for each configuration (cylinder-sphere, cylinder-cylinder, and sphere-sphere) the second configuration of the PTO is still more efficient in all the cases. For cylinder-sphere 17.40% against 11.86%, for cylinder-cylinder 34.27% against 26.89% and for the case of sphere-sphere 22.19% against 15.77%. This means that the second configuration is indeed more efficient than the first one for all the cases analysis. This conclusion proofs the starting objective.

## 6 CONCLUSIONS

The present study was carried out in order to compare two different configurations of wave energy converters (PTO in between floater and submerged body; and PTO in between submerged body and sea bottom) and decide which one is better in terms of efficiency of harvesting energy (as having a higher efficiency means cheaper costs for generating energy).

The initial desire of the study was to prove that the second configuration is more efficient (and so produce energy in a cheaper way if all other costs are assumed constant) than the first configuration.

After the analysis was carried out and the results compared, it was concluded (as was desired) that the second configuration is more efficient than the first one for all three geometrical configurations. The average difference in efficiency between the two configurations is 10.85%. That means that, if considering all the other costs (for instance, installation) constant, the second configuration can generate energy with fewer costs, being this of paramount importance as the cost of energy is one of the main barriers for the usage of wave energy converters nowadays. This is the main barrier nowadays because wave energy is generally clean, so it doesn't pollute as much as conventional ways of generating energy (such as coal or fuel for instance). That means that if somehow this energy could be harvested in a cheaper way it would be way more used all over the globe and it would also prevent the excess of pollution caused by the conventional ways of generating energy (being this one of the biggest current problems of the world)

It was also noticed that the second configuration is optimized for smaller bodies than the first one. For example, in the case cylinder-sphere the first configuration is optimized for buoy radius of 2.5m, buoy draft of 1m, and submerged body of 3m; and the second configuration is optimized for buoy radius of 1m, buoy draft of 0.5m and submerged body of 1m.

The work presented in this study is expected to contribute to the understanding of the effect of the variation of the design variables of a WEC on its performance.

For future work, it is suggested to improve the model by taking into consideration some non-linear effects. Also, in future works, the analysis should be carried out for the same configurations but other areas in the ocean (as in the present study just the region of Pico-Azores was considered). By doing these the results are more reliable, once it is guaranteed that the second configuration is indeed more efficient than the first one no matter where in the ocean the analysis is being made.

## ACKNOWLEDGEMENTS

This work was performed within the Strategic Research Plan of the Centre for Marine Technology and Ocean Engineering, financed by the Portuguese Foundation for Science and Technology (Fundação para a Ciência e Tecnologia-FCT). The work has been funded by the Portuguese Foundation for Science and Technology (Fundação para a Ciência e Tecnologia - FCT) under contract number UIDB/UIDP/ 00134/2020.

## REFERENCES

Al Shami, E., Wang, X., Zhang, R. and Zuo, L. 2019. A pa-rameter study and optimization of two body wave energy converters. Renewable energy 131: 1–13.

Al Shami, E., Wang, X. and Ji, X., 2019. A study of the effects of increasing the degrees of freedom of a point-absorber wave energy converter on its harvesting performance. Mechanical Systems and Signal Processing, 133.

Babarit, A. and G. Delhommeau (2015). Theoretical and numerical aspects of the open source BEM solver NEMOH. 11th European Wave and Tidal Energy Conference (EWTEC2015).

Bozzi, S., Miquel, A.M., Antonini, A., Passoni, G. and Archetti, R., 2013. Modeling of a point absorber for energy conversion in Italian seas. Energies, 6(6)

Cheng, Z., Yang, J., Hu, Z. and Xiao, L., 2014. Frequency/time domain modeling of a direct drive point absorber wave energy converter. Science China Physics, Mechanics and Astronomy, 57(2): 311–320.

Guedes Soares, C., Bhattacharjee, J., Tello, M. and Pietra, L., 2012. Review and classification of wave energy converters. Maritime Engineering and Technology. London: Taylor & Francis Group, pp.585–594

Guedes Soares, C; Bento, A. R.; Goncalves, M.; Silva, D.; and Martinho, P., 2014. Numerical evaluation of the wave energy resource along the Atlantic European coast. Computers & Geosciences; 7137–49

Liang, C. and Zuo, L., 2016, September. On the dynamics and design of a two-body wave energy converter. In Journal of Physics: Conference Series (Vol. 744, No. 1, p. 012074). IOP Publishing.

Matos, A., Madeira, F., Fortes, C.J.E.M., Didier, E., Poseiro, P. and Jacob, J., 2015. Wave energy at Azores islands. Proc., SCACR.

Mørk, G., Barstow, S., Kabuth, A. K. 2010. Assessing the Global Wave Energy Potential. ASME 2010 29th International Conference on Ocean, Offshore and Arctic Engineering.

Penalba, M., Kelly, T. and Ringwood, J., 2017. Using NEMOH for modelling wave energy converters: A comparative study with WAMIT.

Rezanejad, K., Guedes Soares, C. 2014. Numerical study of a large floating oscillating water column device using a 2D boundary element method, Developments in Maritime Transportation and Exploitation of Sea Resources

Rezanejad, K., Guedes Soares, C. 2015. Hydrodynamic performance assessment of a floating oscillating water column, Maritime Technology and Engineering. London: Taylor & Francis Group, 1287–1296.

Rezanejad, K., Guedes Soares, C., 2018 Enhancing the primary efficiency of an oscillating water column wave energy converter based on a dual-mass system analogy, Renewable Energy, Volume 123, 2018, 730–747

Rezanejad, K. Gadelho, J. F. M. Xu, S. Guedes Soares, C. 2021. Experimental investigation on the hydrodynamic performance of a new type floating Oscillating Water Column device with dual-chambers, Ocean Engineering 234, 109307

Ruezga, A., 2019. Buoy Analysis in a Point-Absorber Wave Energy Converter. IEEE Journal of Oceanic Engineering, 45(2): 472–479.

Sang, Y., Karayaka, H.B., Yan, Y., Zhang, J.Z., Bogucki, D. and Yu, Y.H., 2017. A rule-based phase control methodology for a slider-crank wave energy converter power take-off system. International Journal of Marine Energy, 19, pp.124–144.

Shadman, M., Estefen, S.F., Rodriguez, C.A. and Nogueira, I.C., 2018. A geometrical optimization method applied to a heaving point absorber wave energy converter. Renewable energy, 115, pp.533–546.

Shadman, M., Avalos, G.O.G. and Estefen, S.F., 2021. On the power performance of a wave energy converter with a direct mechanical drive power take-off system controlled by latching. Renewable Energy, 169, pp.157–177.

Vicente, Pedro & Falcao, Antonio & Justino, Paulo. (2011). Optimization of Mooring Configuration Parameters of Floating Wave Energy Converters. Proceedings of the International Conference on Offshore Mechanics and Arctic Engineering - OMAE. 5. 10.1115/OMAE2011-49955.

Walton, R., 2019. Global electricity consumption to rise 79 percent higher by 2050, EIA says | Power Engineering. [online] Power Engineering. Available at: <https://www.power-eng.com/renewables/global-electricity-consumption-to-rise-79-percent-higher-by-2050-eia-says/> [Accessed 7 June 2020].

Wang, L., Lin, M., Tedeschi, E., Engström, J. and Isberg, J., 2020. Improving electric power generation of a standalone wave energy converter via optimal electric load control. Energy, 211, p.118945.

Zhao, X.L., Ning, D.Z., Zou, Q.P., Qiao, D.S. and Cai, S.Q., 2019. Hybrid floating breakwater-WEC system: A review. Ocean Engineering, 186, p.106126.

*Trends in Maritime Technology and Engineering – Guedes Soares & Santos (eds)*


# A preliminary evaluation of the performance parameters of point absorbers for the extraction of wave energy

J.B. Valencia & C. Guedes Soares
*Centre for Marine Technology and Ocean Engineering (CENTEC), Instituto Superior Técnico, Universidade de Lisboa, Lisbon Portugal*

ABSTRACT: This present paper preliminarily assesses the performance parameters of point absorber wave energy converter with selected bottom shapes in specific geographic locations and considering viscous effects. Three geometries of cylindrical, conical and hemispherical base are modelled as axi-symmetric bodies having common displacement and stiffness power take-off set to zero. In regular shallow water waves, each model is compared themselves at different ratios and angles as the case may be. The three bodies are scaled to the prototype size and under conditions of regular deep water waves, the hydrodynamic and energy performance parameters are compared and optimized. For an irregular waves marine environment, four geographic zones are chosen to evaluate the performance of the three WECs. For this purpose, the following metrics are calculated: mean annual power flux, mean annual energy production and the mean annual capture width, considering the scenarios: (i) variable power take-off damping (ii) optimal power take-off damping.

## 1 INTRODUCTION

Numerical modelling is applied from the initial stages of design of a wave energy converter to get an approximation of the hydrodynamic behavior that it has and so understanding how some parameters (the direction of the wave, water depth, wavelength) affect the performance of the mechanical power in the process definition of the concept designs. Numerous studies of single body point absorbers that comparing the absorbed power of axisymmetric geometries were carried out in the last decades. Pioneering research which analyzed cylindrical floating buoys with conical and hemispherical bottom in heave motion were done by (De Backer et al., 2007) simulated with WAMIT and (Pastor and Liu, 2014) using Ansys Aqwa. Both papers follow a similar methodology to calculate the energy absorption of the irregular waves. A hydrodynamic research of moored floating bodies which have their submerged part with shapes of cosine and spherical type was developed by (Berenjkoob et al., 2018) using Ansys Aqwa. The viscous effects were included by (Bhinder et al., 2011) using CFD code Flow 3D verifying for a fully submerged cylinder that there is a notable diminution of the power function and the mean annual energy production. (Tom and Yeung, 2013) investigated the differences in hydrodynamic performance between flat and hemispherical bottom floaters using a CFD inhouse code. A approach applied by (Zhou et al., 2020) to speed up the calculations in the frequency domain of the device converter performance in viscous conditions is to use the BEM codes to solve the motion equation in inviscid fluid and to add viscous corrections which can be obtained via the decay test. Considering this last focusing and with the objective of obtaining the performance parameters on the European Atlantic coast, in the present work a step-by-step study was developed in the frequency domain describing the procedure in detail with the necessary simplifications. The potential flow equations were solved using WAMIT. Albeit the response amplitude operator *RAO* can be obtained directly from the BEM solver, it was calculated from the heave body motion equation $\hat{z}$ divided by the incident wave amplitude $\zeta_a$ to be able to introduce the viscous term. Next, the WEC performance parameters of the geometries in study in regular and irregular waves are obtained and hydrodynamically compared. Although this domain gives an useful overview of the optimal dynamics of WEC, will be necessary to perform a study future in the time domain to introduce non linear variables and to compare that with the results of a CFD research.

## 2 THEORETICAL BACKGROUND

### 2.1 *Dynamic of the floater*

Using the linear wave theory, the frequency domain equation to the heaving free-floating body is derived:

DOI: 10.1201/9781003320289-53

$$\{-\omega^2(m+\mathrm{a})+i\omega(b)+(k)\}\hat{z}=\hat{f}_e \quad (1)$$

The coefficients added mass a, potential damping $b$ and hydrostatic restoring $k$ can be obtained analytically for simple geometries, but when complexity increases the BEM solvers are needed.

If it is considered the heaving device doesn't haves mooring, the PTO force $\hat{f}_{pto}$ and the viscous force $\hat{f}_v$ are linear and included with a velocity $\hat{v}=i\omega\hat{z}$:

$$\hat{f}_{pto}=-b_{pto}\hat{v}-k_{pto}\hat{z} \quad (2)$$

$$\hat{f}_v=-b_v\hat{v} \quad (3)$$

Then the motion equation is:

$$\{-\omega^2(m+\mathrm{a})+i\omega(b+b_v+b_{pto})+(k+k_{pto})\}\hat{z}=\hat{f}_e \quad (4)$$

Equation 4 represents the mass-spring-damper system model for a single DOF WEC body, see Figure 1.

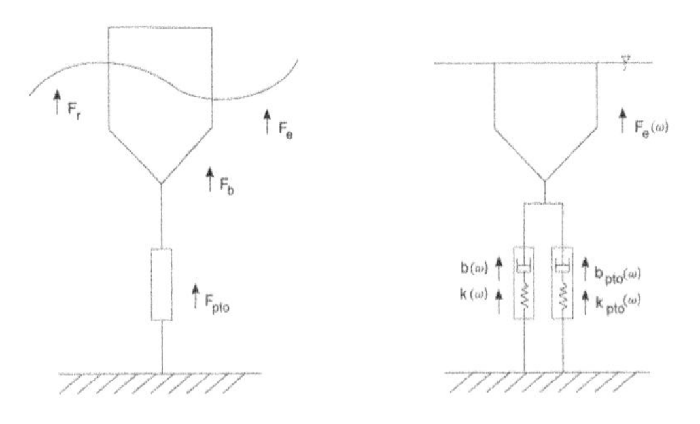


Figure 1. Heaving WEC model including radiation, exciting, buoyancy and PTO force.

A summary of the device performance equations used in regular waves is shown (Thomas, 2008):

Response amplitude operator [m/m]

$$RAO=|\hat{z}|/\zeta_a \quad (5)$$

Power absorbed [kW]:

$$\bar{P}_{a,reg}=\frac{1}{2}\omega^2 b_{pto}|\hat{z}|^2 \quad (6)$$

Power function [kW/m$^2$]

$$\bar{p}_{reg}=\frac{1}{2}\omega^2 b_{pto}RAO^2 \quad (7)$$

Capture width [m]

$$L_{w,reg}=\bar{P}_{a,reg}/\bar{P}_{w,reg} \quad (8)$$

Capture width ratio [-]

$$CWR_{reg}=L_{w,reg}/L_c \quad (9)$$

Also the wave energy flux per meter wave-front for regular waves (Shaw, 1982) is expressed in [kW/m].

Deep-water waves:

$$\bar{P}_{w,reg}=\frac{\rho g^2\zeta_a^2}{4\omega} \quad (10)$$

Shallow-water waves:

$$\bar{P}_{w,reg}=\frac{1}{2}\rho g^{1.5}\zeta_a^2\sqrt{h} \quad (11)$$

Similarly, the irregular version for deep waters [kW/m]:

$$\bar{P}_{w,irr}=\frac{\rho g^2}{64\pi}H_{m0}^2 T_e \quad (12)$$

where $H_{m0}$ and $T_e$ are the spectral moments: significant wave height and energy period respectively, derived from Jonswap spectrum.

In point absorbers, the maximum energy extraction usually will occur when system natural oscillations $\omega_o$ approach or equal the frequency of incoming waves $\omega_i$, be these of regular or irregular type. This is the resonance phenomenal and the frequency $\omega_r=\omega_o=\omega_i$ is only reached when the velocity of the oscillating body $\hat{v}$ is in phase with the excitation force $\hat{f}_e$, from which can be obtained the resonance frequency $\omega_r$ under two situations: the PTO force includes or omits, the spring coefficient $k_{pto}$.

$$\omega_r=\omega_o=\sqrt{\frac{k+k_{pto}}{m+\mathrm{a}}} \quad (13)$$

Damping passive control approach $k_{pto}=0$ and the maximization of mean power (equation 6) is affected considering that the average absorbed power only depends on the PTO damping $\bar{p}_{reg}=\bar{p}_{reg}(b_{pto})$. Thus, deriving $\frac{\partial\bar{p}_{reg}}{\partial b_{pto}}=0$, it results the optimal damping coefficient:

$$b_{opt}=b_{pto}=\sqrt{b^2+\frac{1}{\omega^2}[-\omega^2(m+\mathrm{a})+k]^2} \quad (14)$$

If one introduces equation 13 in equation 14, the optimal condition in resonance is:

$$b_{pto} = b \tag{15}$$

Direct consequents of the damping PTO control, are the expressions to calculate the maximum theoretical mean power of a single mode WEC:

$$P_m = \frac{\left|\widehat{\mathrm{f}}_e\right|^2}{8\mathrm{b}_{\mathrm{opt}}} \tag{16}$$

The final aim of the development of WEC concept and its numerical stages is the estimation of levelized cost of energy (LCoE), that involve a complex quantity of variables and uncertainties: (a) capital cost (device cost, installation, production) (b) operation cost (maintenance, insurance) (c) mean annual energy production MAEP, all them depend directly of the PTO system and in this study will be used a simplified linear version as described in equation 2.

## 2.2 *Performance of the WEC*

The scatter diagram represents the wave climate of a geographical area during a time period (usually tens of years) and considering the wave directionality, it is normally organized in a two-dimensional matrix $C(H_s, T_e)$. The vertical and horizontal references correspond to the wave significative heights $H_s$ and time periods $T_e$ respectively. Each cell (bin) of this matrix represents the relative frequency of occurrence $f_{H_s,T_e}$ of the respective combination ($H_s$,$T_e$). It both span in constant way, $H_s$ in meters, $\mathrm{T_e}$ in seconds and depend on the size of the zone in study. The matrix $C(H_s, T_e)$ meets the next condition: $\sum_{i=1}^{N} C_i = 1$, where $N$ is the number of sea-states.

The mean annual wave power flux (MAPF) per meter wave-front [kW/m], may be calculated by summing over all energy fluxes of an element-wise matrix multiplication between $\bar{\mathrm{P}}_{\mathrm{w,irr}}$ transported in each sea state defined in equation 12 and the probability of occurrence $C(H_s, T_e)$ (Beels et al., 2007).

$$MAPF = \sum_{H_s} \sum_{T_e} C(H_s, T_e) \odot \bar{\mathrm{P}}_{\mathrm{w,irr}}(H_s, T_e) \tag{17}$$

The performance matrix, is the representation of the mechanical power ideally extractable in [kW] by the WEC device. This involves all the components that participate actively on the process of primary conversion energy that varies with sea state and falls mainly on the PTO machine. Each element of the matrix is calculated by the following formula:

$$\bar{P}_{a,irr}(H_s, T_e) = 2\int_0^\infty \bar{p}_{reg}(\omega) S_\zeta(\omega) d\omega \tag{18}$$

where $\bar{p}_{reg}$ is the power function defined in equation 7, and $S_\zeta$ is the Jonswap energy spectrum. However, it can be considered an additional power matrix that represents the absorbed power of a specific area and to obtain it, $\bar{P}_{a,irr}$ is multiplied element-wise by the probability of each sea state.

$$PM_{H_s T_e} = C(H_s, T_e) \odot \bar{P}_{a,irr}(H_s, T_e) \tag{19}$$

The mean annual energy production (MAEP) is the total energy produced over an one-year period [kW-h], that can be estimated in its form simpler (Kofoed and Folley, 2016):

$$MAEP = \sigma \sum_{H_s} \sum_{T_e} PM(H_s, T_e) \tag{20}$$

where $PM(H_s, T_e)$ is the power matrix obtained in equation 19, $\sigma$ is the numerical factor 24x365 hours/year (Gregorian year) that assume the ideal conditions of work of generate energy uninterruptedly with 100% of availability.

The mean annual capture width (MACW) is the same concept of capture width [m] of the equation 8, but this time using the parameters MAEP and MAPF as variables:

$$MACW = \sigma^{-1}\frac{MAEP}{MAPF} \tag{21}$$

If this value is divided by the characteristic length $L_c$ in meters, a nondimensional parameter is obtained: mean annual capture width ratio (MACWR).

# 3 WAMIT MODELLING

To validate our subsequent hydrodynamics calculations, the results will be compared with numerical data from (Falnes, 2002) for a floating cylinder. WAMIT has two approaches for get the hydrodynamic coefficients: low order LO and high order HO analysis (WAMIT Inc, 2013).

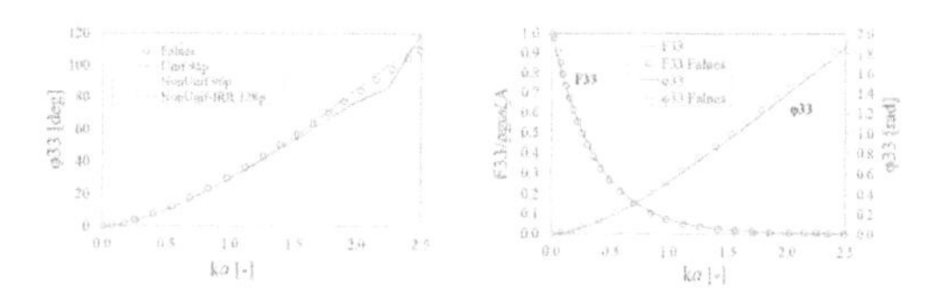

Figure 2. LO/HO discretizations (a) Phase lag (b) $F_{33}$ and phase.

For LO validation with Falnes curves, Figure 2(a) considers a discretization with uniform meshing Unif, non-uniform meshing (cosine spacing) NonUnif and non-uniform meshing under the irregular frequencies removal NonUnif-IRR. For a similar HO validation curves, Figure 2(b) considers a meshing with cosine space distribution with irregular frequency removal.

With a simple visual analysis in LO, it is clear that the non uniform version with irregular frequency removal is the best fit. It was also verified in the HO method. To quantify that, it was done a convergence study on heave mode with an incremental mesh refinement on a cylinder with a radius of 5 meters and heading angle of 45 degrees.

The low order convergence is soft and was done with discretization of 32 and 1568 panels, as minimal and maximum values by quadrant. See Figure 3(a).

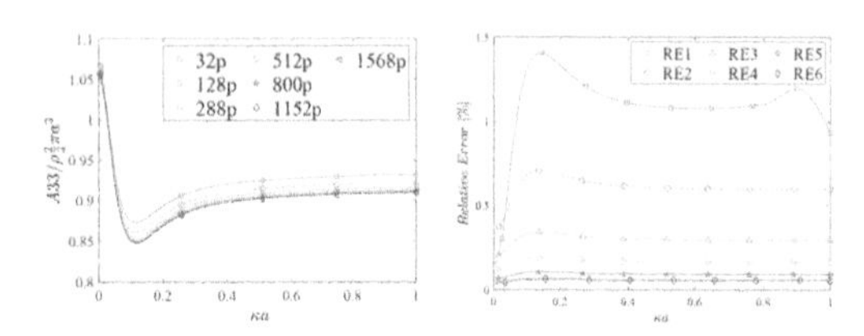


Figure 3. Heave convergence LO (a) Added mass (b) Relative error.

The relative error tends to the zero percent with the higher value of the panelization. See Figure 3(b).

A comparison among both approaches is made. In the Figure 4(a) is visible that HO convergence is fastest and in all cases the discretization ILOWHI-1 of 128 and 288 panels is under 0.5% relative error compared with the best value of the low order discretization. See Figure 4(b).

The value of 128 panels was chosen for subsequent calculations because in practical purposes the BEM solver runs faster.

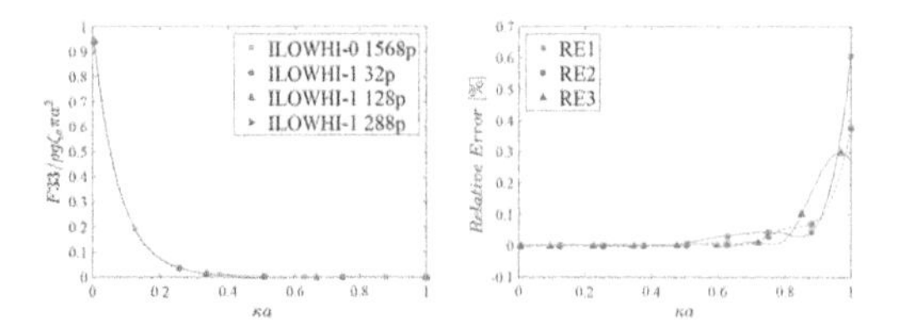


Figure 4. Heave convergence HO (a) Exciting force β=45° (b) Relative error.

## 4 PERFORMANCE IN REGULAR WAVES

The geometrical bodies in this study have as common base the ratio $r = a/d$. Figure 5(a) cylinder with a and $d$ as the *radius* and *draft* respectively from which the expressions are deducted for the *equivalents drafts* of the others WECs devices: Figure 5(b) cylinder with conical base $(d_1 + d_2)$ and the Figure 5(c) cylinder with hemispherical base $(d_1' + d_2')$

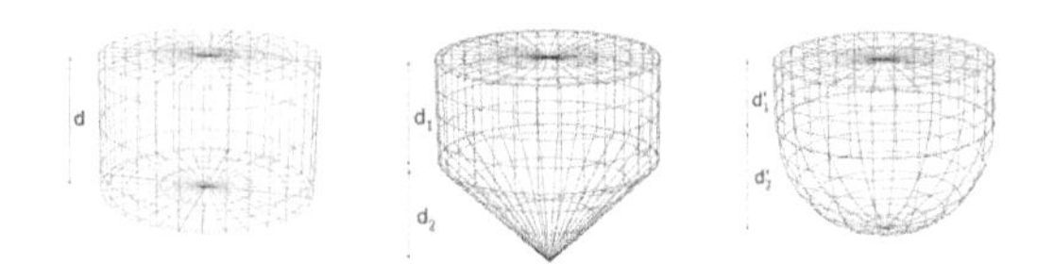

Figure 5. Axi-symmetrical bodies (a) Cylindrical (b) Conical (c) Hemispherical.

The draft formulas that correspond to each geometry with the condition d = 1 are shown in Table 1.

Table 1. Equivalent draft formulas.

| BCyl | BCone | BHemis |
|---|---|---|
| $d$ | $d_1 = d\left(1 - \frac{1}{3}\ \frac{r}{\tan(\propto)}\right)$ | $d'_1 = d(1 - \frac{2}{3}\ r)$ |
| | $d_2 = \frac{d \cdot r}{\tan(\propto)}$ | $d_2' = a$ |

### 4.1 *Dimensions and characteristics*

The input's parameters of design are: BCyl$(a, d)$, BCone$(a, d_1, d_2, \propto)$ and BHemis$(a, d_1', d_2')$ see Table 2, with which can be obtained derived parameters as water plane area, submerged volume, the hydrostatic restoring coefficient, among others.

Table 2. Model dimensions with scale=0.2.

| BCyl | r = 0.8 | r = 1.2 | r = 1.4 |
|---|---|---|---|
| *a* | 0.8000 | 1.2000 | 1.4000 |
| d | 1.0000 | 1.0000 | 1.0000 |
| BCone | | | |
| r = 0.8 | α = 45° | α = 60° | α = 75° |
| *a* | 0.8000 | 0.8000 | 0.8000 |
| d1 | 0.7333 | 0.8460 | 0.9285 |
| d2 | 0.8000 | 0.4619 | 0.2144 |
| BHemis | r = 0.8 | r = 1.2 | r = 1.4 |
| *a* | 0.8000 | 1.2000 | 1.4000 |
| d1 | 0.4667 | 0.2000 | 0.0667 |
| d2 | 0.8000 | 1.2000 | 1.4000 |

The values of the environmental constants used on the calculations are shown in the Table 3.

Table 3. Environment constants.

| | Gravity | $9.81\ m/s^2$ |
|---|---|---|
| | Shallow water | Deep water |
| Wave amplitude | 0.1 m | 1 m |
| Depth | 15 m | 500 m |
| Density | $1000\ kg/m^3$ | $1025\ kg/m^3$ |

The viscous correction data was obtained from (Zhou et al., 2020), see Table 4.

Table 4. Models' viscous damping [kg/s].

| BCyl | BvisT | Binv | Bvis | Fv |
|---|---|---|---|---|
| 0.8 | 1165 | 453 | 712 | 2.57 |
| 1.2 | 3378 | 1766 | 1612 | 1.91 |
| 1.4 | 4880 | 2912 | 1968 | 1.68 |
| **BCone 45°** | **BvisT** | **Binv** | **Bvis** | **Fv** |
| 0.8 | 558 | 468 | 90 | 1.19 |
| 1.2 | 2145 | 1967 | 178 | 1.09 |
| 1.4 | 3507 | 3435 | 72 | 1.02 |
| **BHemis** | **BvisT** | **Binv** | **Bvis** | **Fv** |
| 0.8 | 488 | 485 | 2.5 | 1.01 |
| 1.2 | 2048 | 1988 | 60 | 1.03 |
| 1.4 | 3675 | 3588 | 87 | 1.02 |

$B_{visT}$ is the total viscous damping, $B_{inv}$ is the inviscid damping, $B_{vis}$ is the correction damping and $F_v$ is the non-dimensional coefficient viscous damping correction.

The viscous damping was directly applied to the scale of the models and then scaled using the values $[B_{visT}, B_{inv}, B_{vis}]$ x scale$^{2.5}$ for the prototype body.

The expressions for the natural frequency $\omega_n$ of the three free floating bodies are given next and include $\mu_{33}$ the non-dimensional added mass, $g$ the gravity and the drafts before calculated.

Table 5. Natural frequency formulas.

| BCyl $\omega_n^2$ | BCone $\omega_n^2$ | BHemis $\omega_n^2$ |
|---|---|---|
| $\frac{g}{d[1+\mu_{33}(\omega_n)]}$ | $\frac{g}{\left[d_1+\frac{1}{3}d_2\right][1+\mu_{33}(\omega_n)]}$ | $\frac{g}{\left[d'_1+\frac{2}{3}d'_2\right][1+\mu_{33}(\omega_n)]}$ |

Natural frequency $\omega_n$ [rad/s], calculated iteratively for the scenarios simulated:

Table 6. Natural frequencies [rad/s] for models scale 1/5.

| BCyl | r | 0.8 | 1.2 | 1.4 |
|---|---|---|---|---|
| | $\omega_n$ | 2.607 | 2.449 | 2.381 |
| BCone | r=0.8/$\alpha$ | 45° | 60° | 75° |
| | $\omega_n$ | 2.773 | 2.712 | 2.658 |
| | r | 0.8 | 1.2 | 1.4 |
| BHemis | $\omega_n$ | 2.784 | 2.669 | 2.620 |

Table 7. Natural frequencies [rad/s] for prototypes.

| r=0.8 | BCyl | BCone 45° | BHemis |
|---|---|---|---|
| $\omega_n$ | 1.166 | 1.240 | 1.245 |

### 4.2 *Shallow water regular waves simulations*

#### 4.2.1 *WEC cylindrical base (BCyl)*

Simulations of *BCyl* models with ratios $r = \{0.8, 1.2, 1.4\}$ are shown. Using programming was possible to add the PTO force with a damping boot $b_{pto} = 400\mathrm{kg/s}$, and besides the viscous damping from Table 4.

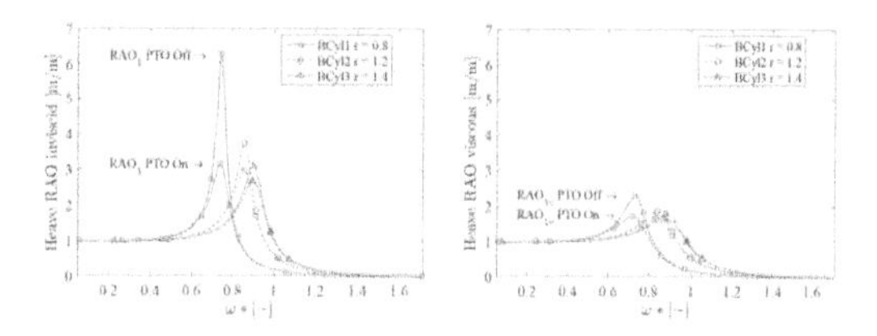


Figure 6. BCyl RAO with PTO on/off (a) Inviscid fluid (b) Viscous fluid.

It is visible a clear decrease of three times its peak value when the PTO is applied under the inviscid condition Figure 6(a). Then a new drop of almost two times the value of peak occurs, by the effect of the viscosity in the captor BCyl1 Figure 6(b). This situation of peaks falling is repeated in each geometry with smaller impact and it is evident that the viscous correction affects notably to the more slender body, i.e. the one with less ratio. Also, the calculated resonant frequencies of the captors in inviscid fluid $\omega = \{2.591,\ 2.440, 2.370\}$ [rad/s], decrease whenis considered the $\omega = \{2.530,\ 2.380, 2.320\}$ [rad/s].

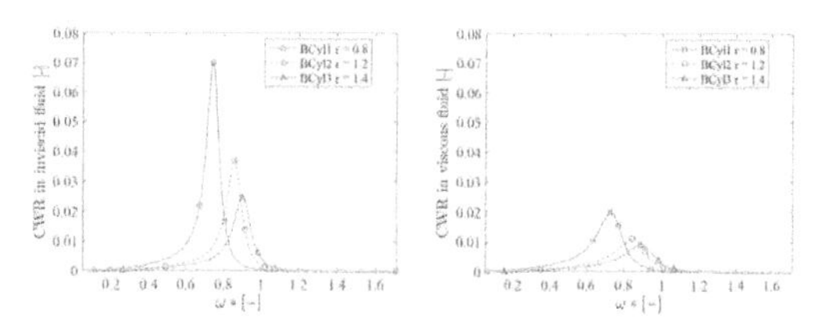

Figure 7. BCyl CWR (a) Inviscid fluid (b) Viscous fluid.

Figure 7(a) shows the capture width ratio CWR for the characteristic length $L_c = [0.8, 1.2, 1.4]$. The unevenness in the peaks is notable in ideal fluid due to the ratios of the bodies, and decreases abruptly under viscous regime Figure 7(b), being the most affected, the one that has less ratio.

As expected, for the low boot damping PTO used, the resonant frequencies coincide with those of the absorbed power, in inviscid or viscous fluid.

4.2.2 *WEC conical base (BCone)*

Simulations of the three BCone models, with ratios $r = \{0.8, 1.2, 1.4\}$ and half apex angles $\alpha = \{45°, 60°, 75°\}$ are carried out. (Plots with $r = \{1.2, 1.4\}$don't are shown.)

Figure 8(a) show the behavior of the RAO PTO-on curves for BCone with ratio: $r = \{0.8\}$ and different half apex angles $\alpha = \{45°, 60°, 75°\}$, in an inviscid fluid. Clearly is appreciated in the resonant frequency region, the shift of the peaks when the angle $\alpha$ rises, being the BCone 75°, the one with the highest response amplitude.

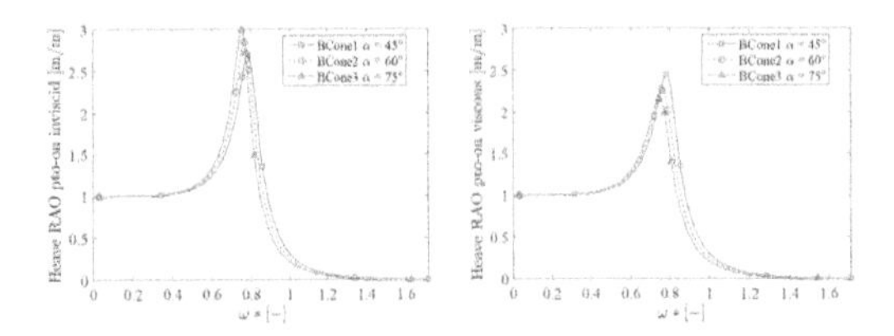

Figure 8. BCone RAO (a) Inviscid fluid (b) Viscous fluid.

Figure 8(b) shows the viscous version, where the order of the peaks is altered and the viscous correction affects notably the geometry with greater conical angle. When comparing the three bodies, it is verified that the fall of the peaks increases when the ratio also increases.

Figure 9(a) shows that Bcone1 75° has the greatest capture width ratio, still when there is a noticeable peak decrease related to ratio increase.

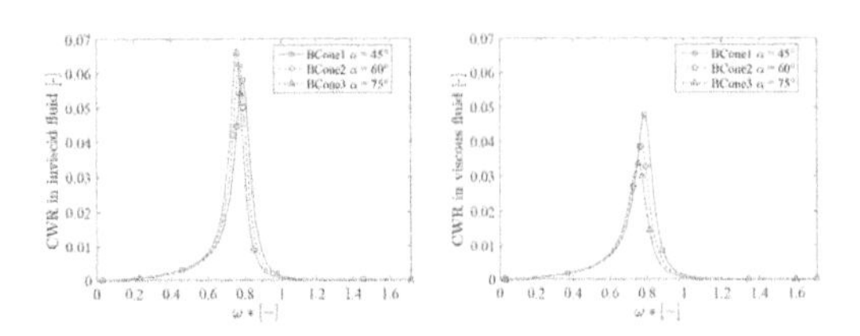

Figure 9. BCone CWR (a) Inviscid fluid (b) Viscous fluid.

In viscous conditions Figure 9(b), the smallest peak becomes the largest and the order of the peak's size is altered but maintaining frequencies coverage.

4.2.3 *WEC hemispherical base (BHemis)*

Simulations of the three *BHemis* models, with ratios $r = \{0.8, 1.2, 1.4\}$ are carried out.

Figure 10(a) shows that the more slender body has the highest response and that due to the viscous effects are very small, the curves almost overlap with the potential version.

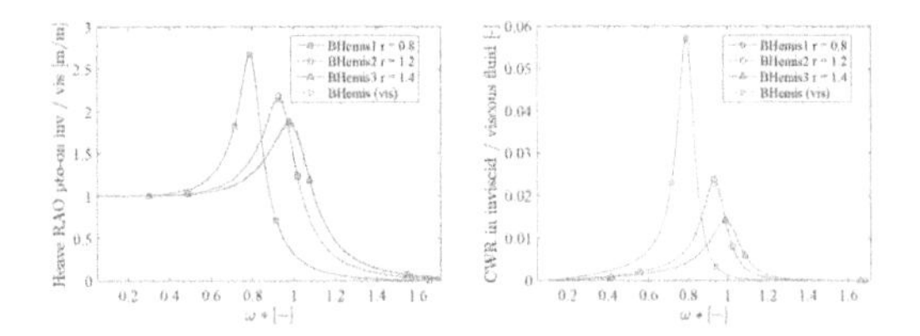

Figure 10. BHemis inviscid/viscous fluid (a) RAO (b) CWR.

Figure 10(b) similarly, due to that the viscous coefficient is close to 1 in each ratio, their effects are very smalls and the curves overlap. A small difference near to the peaks of BHemis2 and BHemis3 is shown. Like before, the slender body has the best performance in the energy's capture.

4.3 *Deep water regular waves simulations for prototypes bodies*

For study the influence of the geometry, three WEC prototypes are compared: *BCyl*, *BCone* ($\propto$= 45°), *BHemis*, with ratio $r = 4m$, Other scaled magnitudes are the PTO damping boot 2.236x10$^4$ kg/s, the hydrostatic restoring coefficient 5.053x10$^5$ N/m.

The inviscid response $Z$ of the vertical oscillation Figure 11(a) is more pronounced for the BCyl body (aprox. 3.15m). The other responses of BCone 45° and BHemis, overlap and are lower (aprox. 2.6m). A situation a bit more realistic is appreciated in the Figure 11(b) where the BCyl's peak decreases abruptly to almost half of its original value. BCone 45° decreases its peak by only some centimeters and BHemis keeps almost the same position due to the low viscosity for this geometry.

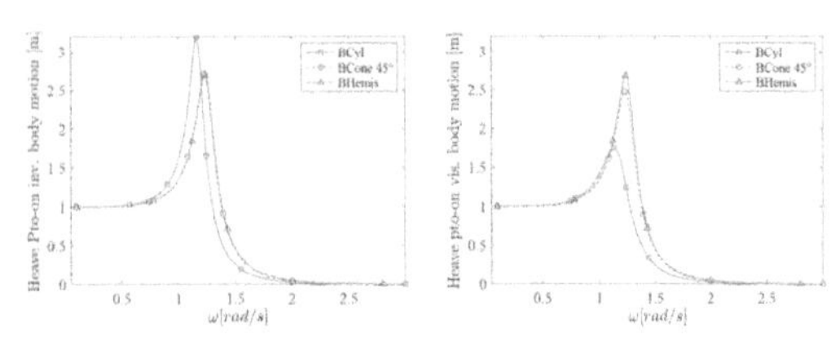

Figure 11. Three bodies, heave motion Z (a) Inviscid fluid (b) Viscous fluid.

Simulations of the theoretical maximal power of absorption $P_m$ and the absorbed power $\bar{P}_a$ together under inviscid and viscous fluid, are carried out.

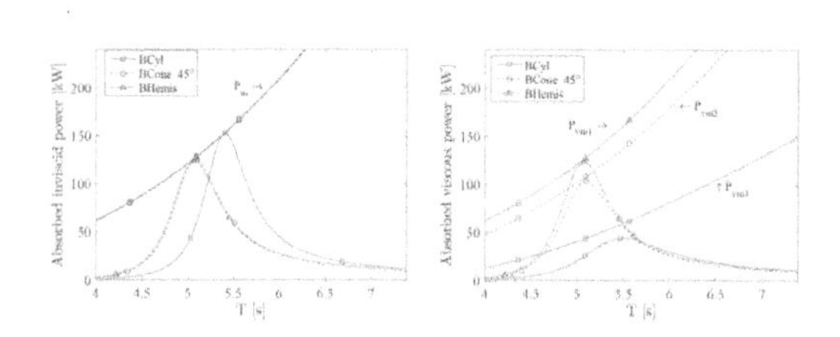


Figure 12. Three bodies, power absorbed $\bar{P}_a$(a) Inviscid fluid (b) Viscous fluid.

In Figure 12(a) is observed that the three bodies get to reach the maximal power, that is a common curve. Figure 12(b) shows that each of the three geometries has a different maximal power and that depending on the value of the viscous coefficient, the curves move away or closer.

The best performance is observed when the BCyl body is in an inviscid regime but in the viscous frame the best one is the BHemis body even than BCone 45°.

When the equation 15 is introduced in the power equations, it brings the optimized power version $\bar{P}_{opt}$.

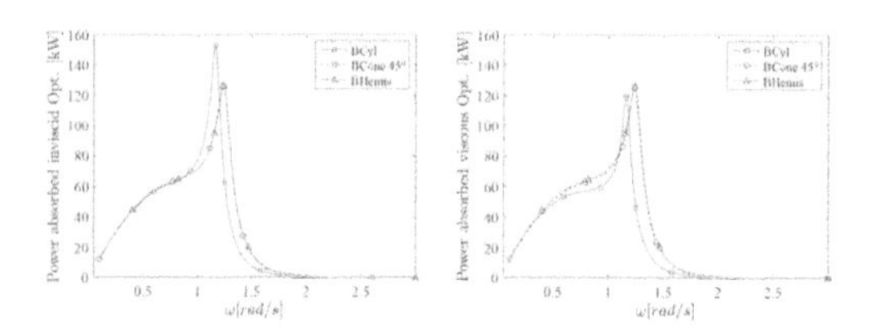


Figure 13. Three bodies, optimal power $\bar{P}_{opt}$ (a) Inviscid fluid (b) Viscous fluid.

At first sight, the Figure 13(a) and Figure 13(b), are similar to the previous plots near to the peaks but they are much different at the spectral low band, giving additional criteria to choose the best WEC. BHemis has slightly better performance in absorbed power over BCone under optimal conditions.

Figure 14(a) shows that the BCyl has the highest value of CWR but a smaller value of frequency range captured compared to the others bodies (under inviscid conditions).

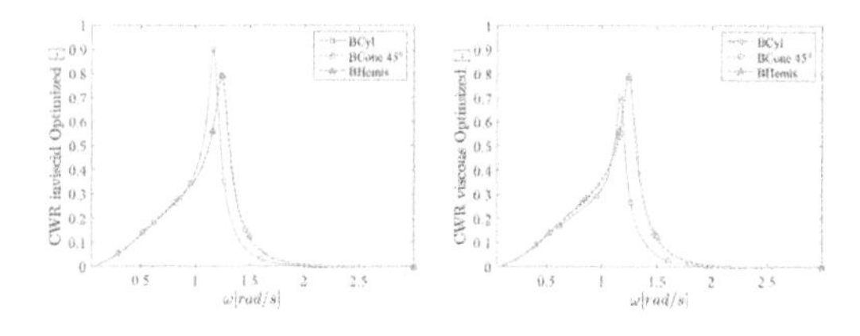


Figure 14. Three bodies, optimal capture width ratio $CWR_{opt}$ (a) Inviscid fluid (b) Viscous fluid.

Figure 14(b) shows that under viscous effects, the peak decreases strongly in the BCyl body and the BCone 45° and BHemis has a similar behavior, however the best performance pertain to the BHemis device because it is lightly better in captured frequencies range and the CWR parameter. Also it has the less viscous dissipation.

## 5 PERFORMANCE IN IRREGULAR WAVES

### 5.1 *Input environment's parameters*

The European Atlantic coast was chosen as the study area. Scatter tables of four geographical areas (Emec, Yeu, Lisbon, Belmullet) were processed and the data files obtained from (LHEEA, 2017). The observation period is assumed to be obtained in an annual period (Pontes, 1998).

The next step is to verify if they actually represent the correct values of the respective energy resource, (this validation is independent of the calculation of the power matrices) which implies the all have the energy spectrum calibrated. For to check so, simulations of some representative sea states of the north sea (De Backer et al., 2007) were done using the Jonswap spectrum.

| Sea State | Hs [m] | Tp [s] |
|---|---|---|
| 1 | 1.25 | 6.7 |
| 2 | 1.75 | 7.4 |
| 3 | 3.25 | 8.81 |

Figure 15. Typical Jonswap states for the north sea area.

Table 8. Sites location wave energy resource (kW/m).

| Country | Location | Water depth [m] |
|---|---|---|
| Scotland | 059° 00,000′N - 003° 66,000′W | 50 |
| France | 046° 40,000′N - 002° 25,000′W | 47 |
| Portugal | 039° 00,000′N - 012° 00,000′W | 100 |
| Ireland | 054° 00,000′N - 012° 00,000′W | 100 |

| Site | $\gamma = 1$ | $\gamma = 3.3$ | $\gamma = 7$ |
|---|---|---|---|
| Emec | 22.19 | 23.36 | 24.06 |
| Yeu | 25.78 | 27.13 | 27.96 |
| Lisbon | 36.17 | 38.09 | 39.25 |
| Belmullet | 77.75 | 81.88 | 84.38 |

Next, using the equation 17, it is calculated the mean annual wave power flux *MAPF* per meter wave front [kW/m] for each site location (see Table 8). They are compared to the results of (Babarit et al., 2012), that considers the three common values of frequency spreading factor $\gamma = [1, 3.3, 7]$, referred to the Jonswap spectrums [wind sea, typical, long Atlantic swell].

### 5.2 *WECs metrics with Bpto variable*

In this scenario (case i), the PTO damping coefficient *Bpto* that maximize the energy performance in each site is optimized. The chosen values are powers of ten multiplied by the damping boot. The PTO damping range is: [$10^{0}$ $10^{1}$ $10^{2}$ $10^{3}$ $10^{4}$ $10^{5}$] x -BptoZo kg/s, with BptoZo = 2.2 x$10^{4}$ kg/s.

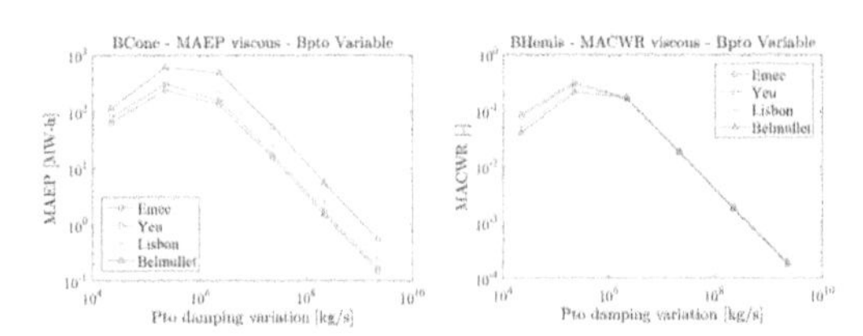


Figure 16. PTO damping variation (a) MAEP (b) MACWR.

It is observed that for the three energy devices, with the PTO damping 10 x BptoZo, the MAEP and MACWR maximum is obtained (Figure 16 (a) and (b) are in logarithmic scale).

The absorbed power matrix was calculated for the areas and geometries under study. These results were multiplied by the respective energy resource occurrence matrices, as indicated in equation 20 to obtain the MAEP of each energy converter device.

If the MAEP is divided between the energy resource and the characteristic length, as the equation 21 shows, the mean annual capture width ratio MACWR is obtained (expressed as a percentage for better visibility). The two performance parameters for the devices under study are shown in Table 9.

### 5.3 *WECs metrics with Bpto optimal*

For this scenario (case ii) the absorbed power for each sea state is optimized using the damping passive control of the equation 14. The results of the performance parameters obtained as in section 5.2 are shown in Table 10:

For the case (i), the energy extraction is higher for BHemis and BCone in all sites, if the viscous fluid is considered. In the case of inviscid fluid, the situation

Table 9. Performance metrics for WECs with Bpto variable.

| BCyl | Bpto variable | | | |
|---|---|---|---|---|
| | MAEP [MW-h/year] | | MACWR [%] | |
| | inv | vis | inv | vis |
| Emec | 234.3943 | 208.2511 | 28.6419 | 25.4473 |
| Yeu | 299.8000 | 266.0321 | 31.5314 | 27.9799 |
| Lisbon | 301.3851 | 273.8608 | 22.5819 | 20.5196 |
| Belmullet | 594.0930 | 547.2563 | 20.7073 | 19.0748 |
| BCone 45° | Bpto variable | | | |
| | MAEP [MW-h/year] | | MACWR [%] | |
| | inv | vis | inv | vis |
| Emec | 236.0516 | 235.7323 | 28.8444 | 28.8054 |
| Yeu | 301.7447 | 301.3333 | 31.7360 | 31.6927 |
| Lisbon | 301.8738 | 301.5451 | 22.6185 | 22.5939 |
| Belmullet | 593.8257 | 593.2731 | 20.6980 | 20.6787 |
| BHemis | Bpto variable | | | |
| | MAEP [MW-h/year] | | MACWR [%] | |
| | inv | vis | inv | vis |
| Emec | 236.5416 | 236.5399 | 28.9043 | 28.9041 |
| Yeu | 302.3687 | 302.3664 | 31.8016 | 31.8014 |
| Lisbon | 302.1745 | 302.1727 | 22.6410 | 22.6409 |
| Belmullet | 594.0799 | 594.0769 | 20.7069 | 20.7068 |

is similar except for Belmullet, by tenths, due to the statistical variability of the results. For the case (ii), BHemis and BCone continues to be the WEC with the best productivity in an inviscid and viscous fluid when compared in each site or in all sites, due to its less viscous effects.

In all cases, the performance of each device in the four sites projects an ascending linear relationship for the MAEP, as the energy resource increases Figure 17. Indeed, it seems reasonable to expect greater energy extraction in the sites with the highest energy resources, taking into account that only one technology is being modeled. Also can be inferred that there is a limit of absorbed power (Babarit and Hals, 2011).

Take into account that when two sites as Emec and Yeu, with similar level of depth and energy resource are compared, the second one is more productive. One explanation could be that around the island of Yeu, longer wavelengths are more common, whose magnitude is of the order of the square of the period ~ $1.56T^2$.

Table 10. Performance metrics for WECs with Bpto optimal.

| BCyl | Bpto optimal | | | |
|---|---|---|---|---|
| | MAEP [MW-h/year] | | MACWR [%] | |
| | inv | vis | inv | vis |
| Emec | 326.0474 | 276.7662 | 39.8415 | 33.8196 |
| Yeu | 405.8235 | 342.9868 | 42.6824 | 36.0736 |
| Lisbon | 451.0340 | 398.2983 | 33.7947 | 29.8433 |
| Belmullet | 924.7852 | 831.6154 | 32.2337 | 28.9862 |
| BCone 45° | Bpto optimal | | | |
| | MAEP [MW-h/year] | | MACWR [%] | |
| | inv | vis | inv | vis |
| Emec | 327.4168 | 326.7172 | 40.0088 | 39.9233 |
| Yeu | 407.8326 | 406.9366 | 42.8938 | 42.7995 |
| Lisbon | 454.5026 | 453.7786 | 34.0545 | 34.0003 |
| Belmullet | 933.6648 | 932.4134 | 32.5432 | 32.4996 |
| BHemis | Bpto optimal | | | |
| | MAEP [MW-h/year] | | MACWR [%] | |
| | inv | vis | inv | vis |
| Emec | 327.9996 | 327.9958 | 40.0800 | 40.0796 |
| Yeu | 408.6038 | 408.5990 | 42.9749 | 42.9744 |
| Lisbon | 454.9935 | 454.9895 | 34.0913 | 34.0910 |
| Belmullet | 934.3735 | 934.3667 | 32.5679 | 32.5677 |

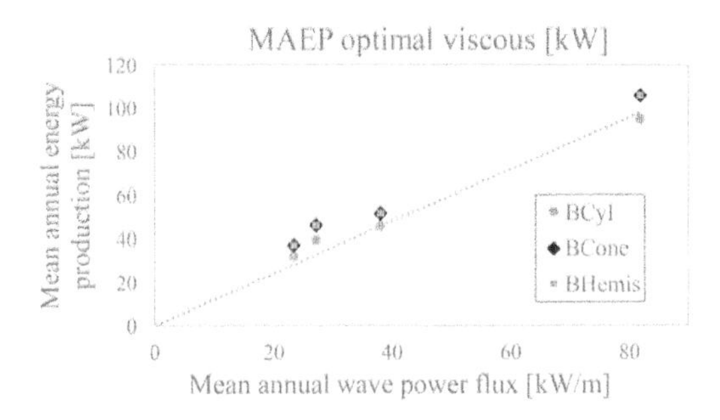


Figure 17. MAEP versus energy resource sites.

## 6 CONCLUSIONS

With numerical validation, it was concluded that WAMIT higher order method is the most efficient for evaluating geometries with symmetries, given its rapid implementation and precise convergence.

In regular waves, from the individual study of each axisymmetric geometry model considering constant damping PTO 400 kg /s, it was concluded that:

For BCyl and BHemis under inviscid conditions, the shift of the resonance frequencies towards the high frequency zone is directly related to the increase in the dimensions of the buoy. For BCone, the shift of the resonance frequencies towards the low frequency zone is directly related to the increase of the half apex angle when the displacement is kept constant. This is explained by the influence that the hydrodynamic parameters have on the resonance equation 13. Considering viscous conditions for BCyl and BHemis there is a natural decrease in the RAO maximums but a pronounced drop in the more slender bodies (less ratio) and a decrease in the resonance frequencies. For BCone there is a decrease of resonant responses that affects notably the geometry with greater conical angle and also the peaks order is altered, when the displacement held constant.

In regular waves, were compared the performance parameters of three axisymmetric geometries prototype optimized considering constant damping PTO ~ $10^4$ kg/s and their viscous corrections. After was concluded that: BCone and BHemis have the best capture width ratio CWR and power absorbed with the highest resonant frequencies although the second one body has slightly better performance.

The wave power available in regular deep water waves and shallow water waves has an evident impact on results obtained for CWR of the analyzed geometries having their origin in the celerity embedded in equation 10 and equation 11 where the first one has a wave length dependency and second one only depend of water depth.

It is evident a pronounced difference in the capture of energy from regular shallow waters and deep waters, mainly due to the first one doesn't depend of the wave frequency.

In irregular waves of four geographical areas were compared the performance parameters MAEP and MACWR of three axisymmetric geometries prototype under viscous effects, considering: (i) A variation of damping PTO seeking to maximize energy performance in each site. (ii) An optimized PTO damping coefficient seeking for the best performance for each sea state. After was concluded that:

Under optimal conditions, there is a linearity related to energy resources. That is, if the productivity of BCyl, BCone and BHemis is analyzed in all the sites, a linear correlation with the increase of the energy resource is visible and it can be inferred that there is a limit of absorbed power. The designs with the highest energy productivity are BCone 45° and BHemis, considering the latter has a slightly better performance, however there are factors that need to be verified for it to be a definitive result.

These factors should be developed in future research and include:

- Viscous corrections must be verified via CFD or laboratory experiments.
- The statistical variability of the scatter tables must be contrasted with time series data.
- Time domain model must be compared.
- Non-linear effects must be included in the numerical model.

## REFERENCES


Babarit, A., and Hals, J. (2011). On the maximum and actual capture width ratio of wave energy converters. in *10th European wave energy conference*, (Southampton).

Babarit, A., Hals, J., Muliawan, M. J., Kurniawan, A., Moan, T., and Krokstad, J. (2012). Numerical benchmarking study of a selection of wave energy converters. *Renew. Energy* 41, 44–63. doi:https://dx.doi.org/10.1016/j.renene.2011.10.002.

Beels, C., De Rouck, J., Verhaeghe, H., Geeraerts, J., and Dumon, G. (2007). Wave energy on the Belgian Continental Shelf. in *OCEANS 2007 - Europe*, 1–6.

Berenjkoob, M. N., Ghiasi, M., and Soares, C. G. (2018). Hydrodynamic analysis of different geometries of a wave energy absorber buoy. in *3rd International Conference on Renewable Energies Offshore (RENEW 2018)* Proceedings in Marine Technology and Ocean Engineering.

Bhinder, M., Babarit, A., Gentaz, L., and Ferrant, P. (2011). Assessment of viscous damping via 3D-CFD modelling of a Floating Wave Energy Device. in *9th European Wave and Tidal Energy Conference* (Southampton, United Kingdom).

De Backer, G., Vantorre, M., Banasiak, R., Beels, C., and De Rouck, J. (2007). Numerical Modelling of Wave Energy Absorption By a Floating Point Absorber System. in International Ocean and Polar Engineering Conference. (Lisbon, Portugal), 374–379.

Falnes, J. (2002). *Ocean Waves and Oscillating Systems: Linear Interactions Including Wave-Energy Extraction*. Cambridge University Press.

Kofoed, J. P., and Folley, M. (2016). "Chapter 13 - Determining Mean Annual Energy Production," in *Numerical Modelling of Wave Energy Converters*, ed. M. Folley (Academic Press), 253–266.

LHEEA (2017). LHEEA (last visited Feb. 2017). Available at: https://lheea.ec-nantes.fr/doku.php/emo/ewtec2015 shortcourse/start.

Pastor, J., and Liu, Y. (2014). Frequency and time domain modeling and power output for a heaving point absorber wave energy converter. *Int. J. Energy Environ. Eng.* 5, 101. doi:10.1007/s40095-014-0101-9.

Pontes, M. T. (1998). Assessing the European Wave Energy Resource. *J. Offshore Mech. Arct. Eng.* 120, 226–231. doi:10.1115/1.2829544.

Shaw, R. (1982). *Wave Energy - A Design Challenge*. Wiley.

Thomas, G. (2008). "The Theory Behind the Conversion of Ocean Wave Energy: a Review," in *Ocean Wave Energy: Current Status and Future Prespectives*, ed. J. Cruz (Berlin, Heidelberg: Springer Berlin Heidelberg), 41–91. doi:10.1007/978-3-540-74895-3_3.

Tom, N., and Yeung, R. W. (2013). Performance Enhancements and Validations of a Generic Ocean-Wave Energy Extractor. *J. Offshore Mech. Arct. Eng.* 135. doi:10.1115/1.4024150.

WAMIT Inc (2013). WAMIT v7.0 user manual.

Zhou, B., Hu, J., Sun, K., Liu, Y., and Collu, M. (2020). Motion Response and Energy Conversion Performance of a Heaving Point Absorber Wave Energy Converter. *Front. Energy Res.* 8. doi:10.3389/fenrg.2020.553295.

*Trends in Maritime Technology and Engineering – Guedes Soares & Santos (eds)*


# Stress distribution on the CENTEC-TLP in still water and rated wind speed

E. Zavvar, B.Q. Chen, E. Uzunoglu & C. Guedes Soares
*Centre for Marine Technology and Ocean Engineering (CENTEC), Instituto Superior Técnico, Universidade de Lisboa, Lisbon, Portugal*

ABSTRACT: This work presents a preliminary study of the stress distribution on the tension leg platform of a 10MW floating wind turbine. The 10 MW turbine combined with the platform hull are modelled and analysed by the finite element method. The design conditions considered for analysis are the platform in still water with the turbine producing its thrust at rated speed. Recommendations from Classification Societies are used to design the structure, in particular the stiffener dimensions, span, and spacing. The results of the analysis reveal that the platform's critical sections are the interaction of the lower columns with the upper columns, the connection of the tower of the DTU 10MW and central column, and the attachments of the tendons, where additional reinforcements should be considered. Furthermore, the leeward locations are more stressed than other sides.

## 1 INTRODUCTION

With the rising demand for electrical energy, several studies were conducted to determine methods of producing a significant quantity of power. As a result, there has been a shift of focus towards improving cleaner energy sources such as dams, wind turbines, and solar panels. Among these, wind turbines garnered interest as a viable alternative. Onshore wind has been in use for decades, and this experience is now serving as a basis for designing offshore wind turbines (Uzunoglu et al. 2016).

When it comes to the structural analysis of onshore designs, documents as early as IEC 61400-1 (2005) provide regulations and recommendations. Additionally, there have been several detailed studies on the topic. For example, Reyno et al. (2015) worked on the stress concentration on the door opening under design loading conditions. The investigation established that the typical value of stress concentration is 1.45, but by reinforcing, it was decreased 78%. In this study, stress distribution in the weld and the whole tower was not investigated.

Önder et al. (2011) investigated square cross-sections for wind turbine towers with different thicknesses under gravity, rotor forces, and wind loads/vortex loads. It was concluded that square cross-sections for wind turbine towers may be a viable alternative to the more typically employed circular cross-sections. However, the stress distribution was not considered in this study.

Comparing offshore and onshore structures, the platform or the support structure makes the difference, becoming additional factors that need to be checked for structural integrity. For instance, Tempel (2006) discussed the design of support structures for offshore wind turbines, while Teixeira et al. (2019) and Iliopoulos et al. (2014) studied the fatigue analysis of offshore wind turbines' towers.

Within the offshore category, fixed bottom and floating structures need to be evaluated separately as the loads that they encounter differ. For instance, Yeter et al. (2015) investigated a fixed offshore wind turbine support structure subjected to combined wave and wind-induced loading. It was concluded that the brace component experienced the most severe fatigue damage, and the estimated fatigue life was above 1000 years. However, in this study, the effects of different kinds of stiffeners on stress distribution were not investigated.

Regarding floating turbines, there are also works evaluating the platform types. Kim et al. (2018) investigated a support structure of a floating wind turbine to reflect the effect of wind speed in the stress transfer function by an artificial neural network. To minimize the number of simulations while increasing the accuracy of the findings, a correction factor for the stress transfer function produced by wind load was estimated using an artificial neural network. To increase the accuracy of the ANN model, a superposition model is developed. The overall stress spectrum was determined by adding the stress spectra caused by wind load from the ANN model and inertia load from motion analysis using linear wave theory. In another study on a TLP, Bachynski (2014) performed a study on the design and dynamic analysis of a single-column 5 MW tension leg platform wind turbine (TLPWT). The computational tool SIMO-RIFLEX-AeroDyn was used to test the design and analysis of TLPWTs. It was concluded that for

DOI: 10.1201/9781003320289-54

some TLPWTs, second-order sum-frequency wave forces are important for fatigue and extreme response calculations. Ringing forces of third-order were shown to be essential for TLPWTs with large diameters (14-18 m), particularly when the turbine was idle or parked. The works above show the interest in identifying possible improvements in the structural design of floaters. Similar to IEC 61400-3-1 (2019) and IEC 61400-3-2 (2020) that discuss fixed and offshore structures, respectively, rules and regulations are also developing in this regard. DNV-OS-J103 (2013), DNV-OS-J101 (2014), DNV-ST-0119 (2018), DNV-OS-C105 (2011), DNV-RP-C201 (2002), and DNV-RP-C202 (2013) are examples of this progress. However, there are significant geometric variations in wind turbine platforms. Hence, it becomes necessary to examine whether applying the regulations is sufficient for structural design purposes.

Accordingly, this work presents an initial effort to evaluate the CENTEC-TLP (Uzunoglu & Guedes Soares, 2020) from a structural point of view in still water. This platform is a design that resembles a barge and is self-stable for transportation purposes (Mas-Soler et al., 2021). Once installed, it functions as a TLP. Its hydrodynamic performance and the design methodology have been discussed throughout several works (Uzunoglu & Guedes Soares 2018, 2021). Here, the focus is only on the structural responses where the floater is assumed to be in still water with the wind thrust applied assuming the steady rated speed wind. The stress distribution in this condition is evaluated on the sections of the platform to identify the sections that require further reinforcements.

## 2 DESCRIPTION OF THE SYSTEM AND GUIDELINES

### 2.1 *Description of the system*

In this study, the CENTEC-TLP with the DTU 10MW is analyzed under hydrostatic pressure and reinforced with two kinds of stiffeners. Properties of the TLP are presented in Table 1 and Figure 1. The properties of the stiffeners are shown in Figure 2 and Table 2.

The DTU 10MW has a total mass (rotor, nacelle, and tower) of 1302 tonnes and a height of 129 meters. Further information on the DTU 10MW is available in Uzunoglu & Guedes Soares (2020), Bak et al. (2013), and Table 3. The Young's modulus and Poisson's ratio were taken to be 210 GPa and 0.3, respectively. Steel NV-550 with 550 MPa yield stress was used.

### 2.2 *Guidelines and standards for the TLP*

To get some basic information about the properties of stiffeners, DNV-RP-C203 (2016) and DNV-OS-C101 (2011) recommendations are used. According to the DNV-OS-C101 (2011), the minimum thickness should not be less than:

Table 1. Properties of the CENTEC-TLP (Uzunoglu & Guedes Soares 2020).

| Parameter | Value | Units |
|---|---|---|
| Overall length& beam | 49 | [m] |
| Pontoon diameters | 4 | [m] |
| Draft when installed | 20 | [m] |
| Lower column height | 7.5 | [m] |
| Lower column diameter | 10.5 | [m] |
| Height of the central column above the waterline | 10 | [m] |
| Distance between the keel and the pontoons | 1 | [m] |
| Steel thickness of components below installed draft | 0.04 | [m] |
| Steel thickness of components above installed draft | 0.04 | [m] |
| Steel density | 7850 | [kg/m] |

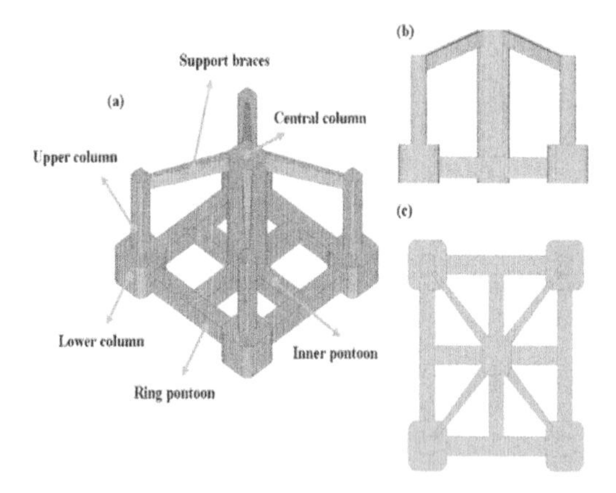


Figure 1. A model of the 10 MW CENTEC-TLP. a) A perspective view; b) A side view; c) An upper view.

$$t = \frac{14.3t_0}{\sqrt{f_{yd}}}(mm) \quad (1)$$

where $f_{yd}$ is design yield strength (N/mm2), and $t_0$ is the thickness of the structure based on the application categories of components (DNV-OS-C105 2011). For plates subjected to lateral pressure, the minimum thickness is determined as follows:

$$t = \frac{15.8k_a s\sqrt{p_d}}{\sqrt{\sigma_{pd1}k_{pp}}}(mm) \quad (2)$$

where $k_a, s, p_d, \sigma_{pd1}$, and $k_{pp}$ are defined as a correction factor for aspect ratio of plate field, stiffener spacing (m), design pressure (kN/m$^2$), design bending stress (N/mm$^2$), and fixation parameter for

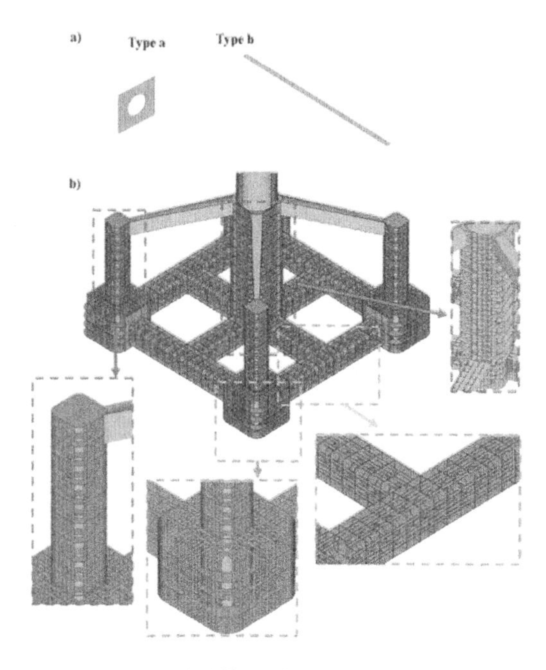


Figure 2. Designed stiffeners for the TLP; a) types of stiffeners, b) the arrangements of stiffeners inside the TLP.

Table 2. Properties of the stiffeners.

| Part | Type | No. Stiff | Thickness (m) | Height (m) | Mesh size (m) |
|---|---|---|---|---|---|
| TLP | - | - | 0.016 | - | 0.50 |
| Long | a | 13 | 0.016 | 1.00 | 0.50 |
| Pontoon | b | 3 | 0.016 | 0.50 | 0.50 |
| Short | a | 6 | 0.016 | 1.00 | 0.50 |
| Pontoon | b | 3 | 0.016 | 0.50 | 0.50 |
| Central | a | 14 | 0.016 | 2.00 | 0.50 |
| Column | b | 4 | 0.016 | 0.80 | 0.50 |
| Upper | a | 4 | 0.016 | 0.90 | 0.50 |
| Column | b | 3 | 0.016 | - | 0.50 |
| Lower Column | a | 3 | 0.016 | - | 0.50 |
| | b | 6 | 0.016 | 1.30 | 0.50 |

plate, respectively. According to the DNV-OS-C101 (2011), the minimum section module for stiffeners is defined as follows:

$$Z_s = \frac{l^2 s p_d}{k_m \sigma_{pd2} k_{ps}}, min\ 15.10^3 \left(mm^3\right) \qquad (3)$$

where $l, k_m, \sigma_{pd2}$, and $k_{ps}$ are defined as stiffener span (m), bending moment factor, design bending stress (N/mm$^2$), fixation parameter for stiffeners, respectively. The minimum shear area is defined as follows (DNV-OS-C101 2011):

Table 3. The DTU 10MW properties Bak et al. (2013).

| Parameter | Value | Units |
|---|---|---|
| Cut in wind speed | 4 | [m/s] |
| Cut out wind speed | 25 | [m/s] |
| Rated wind speed | 11.4 | [m/s] |
| Rated power | 10 | [MW] |
| Minimum/maximum rotor speeds | 6–9.6 | [rpm] |
| Rotor frequency (1P) | 6.5–10 | [s] |
| Blade passing frequency (3P) | 2.08–3.33 | [s] |
| Number of blades | 3 | [-] |
| Rotor diameter | 178.3 | [m] |
| Hub diameter | 5.6 | [m] |
| Hub height | 129 | [m] |
| Rotor mass | 227,962 | [kg] |
| Nacelle mass | 446,036 | [kg] |
| Tower mass | 628,442 | [kg] |

$$A_s = \frac{l s p_d}{2\tau_{pds}} 10^3 \left(mm^2\right) \qquad (4)$$

where $\tau_{pds}$ is the design shear stress (N/mm$^2$). The height of the web is calculated as follows (DNV-RP-C202 2013):

$$h \le 0.4 t_w \sqrt{E/f_y} \qquad (5)$$

where $t_w$ is the thickness of the web. Therefore, based on the above formulae, the minimum thickness for stiffeners with the spacing of 1 meter and the span of 2 meters is 16 millimeters, and the height of the web should be 420 millimeters.

## 3 DEVELOPMENT OF THE FEM MODEL

Shell 281, an element with eight nodes in the ANSYS, was used to model the TLP (Figure 3) and the stiffeners (Figure 4). To model the interaction of stiffeners and TLP, the ANSYS contact capability was utilized. In contact between two elements, the node-to-line method was established as the target and the contact. The hull of the TLP was defined as the target surface, and the edges of the stiffeners were introduced as the contact surface. The mesh size is 0.5 meters (Yeter et al. 2015) for the TLP and 1.0 meters for the tower. The mesh size is larger in the tower as the focus is on the stress distribution on the TLP. The number of elements in the tower and the TLP are 3165 and 33938, respectively. Figure 3 and Figure 4 present the meshed TLP and stiffeners.

The rotor and the nacelle weight are defined as a point mass, and the wind pressure is defined as a force on the top of the tower (Figure 5a). The rotor and nacelle masses are 230 tonnes and 450 tonnes, respectively. The wind force is 1500 kN. The moment of nacelle and the rotor are 40000 kN.m. The TLP is under hydrostatic pressure as the still water, as shown in Figure 5b. The TLP's draft after installation is 20 meters. Fixed boundary conditions are used in each corner of the model in the locations of tendons.

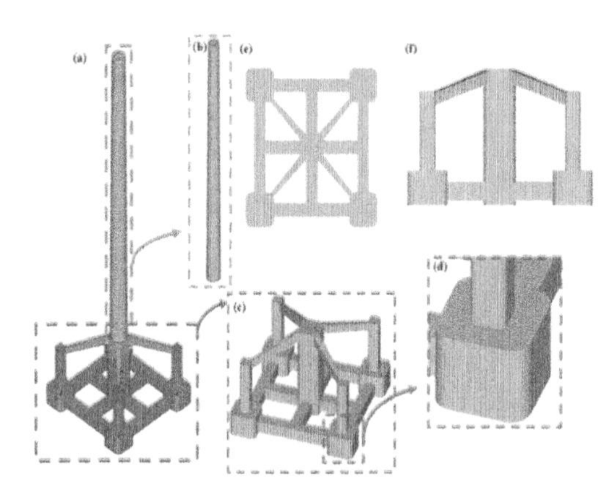

Figure 3. The meshed model of the CENTEC-TLP. a) A perspective of the TLP with the tower of the 10 MW wind turbine; b) The tower of the 10 MW wind turbine; c) the TLP; d) the connection of the upper column with the lower column; e) A upper view of the meshed TLP; f) A side view of the meshed TLP.

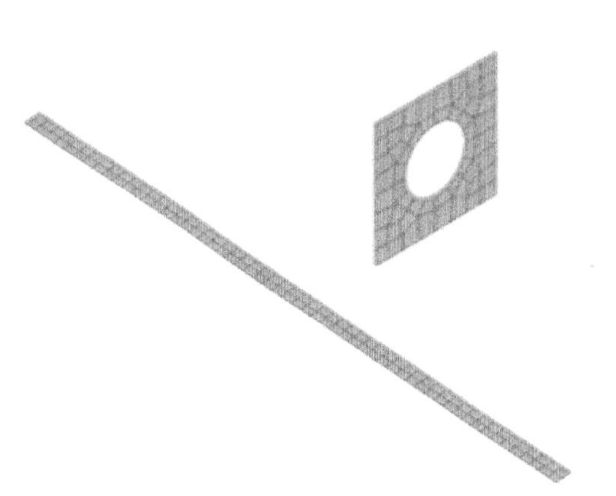

Figure 4. The meshed stiffeners used in the TLP.

## 4 STRESS DISTRIBUTIONS ON THE PLATFORM

The CENTEC-TLP is analyzed with the minimum thickness and height of stiffeners proposed by DNV-OS-C101 (2011) and DNV-RP-C202 (2013). Figure 6 shows the stress distribution, and Table 4 presents the von Mises stress of the reinforced-TLP.

According to Figure 6 and Table 4, it can be concluded that the thickness, the height, and the number of stiffeners, which are determined by DNV-OS-C101 (2011) and DNV-RP-C202 (2013), are not enough for some parts of the TLP, for the values of stress in some critical points, such as location 1, 2, 3, and 4, is higher than the yield stress; thus, a yielding would occur in the TLP before those stress levels would be reached. As a result, these locations need to be reinforced.

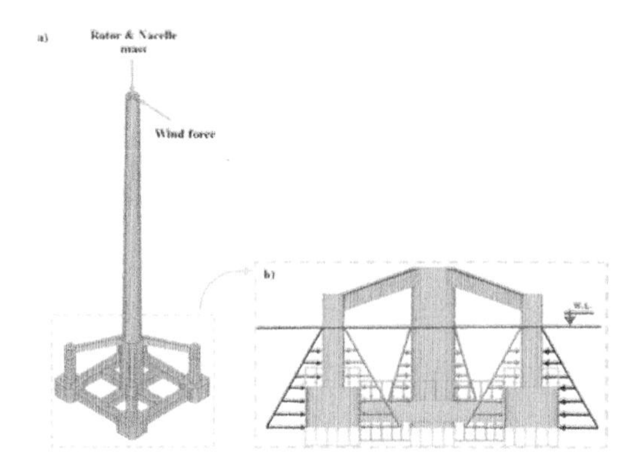


Figure 5. The loaded model; a) Location of the rotor and the nacelle mass and the wind pressure, b) the concentrated force applied on the top of the CENTEC-TLP.

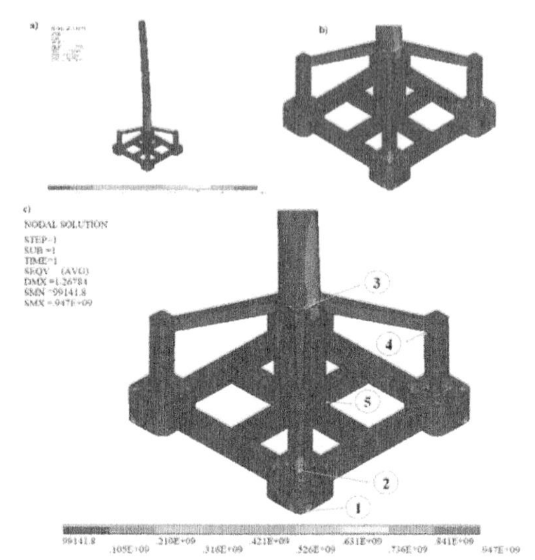


Figure 6. The stress distribution on the CENTEC-TLP with minimum stiffeners proposed by DNV-OS-C101 (2011) and DNV-RP-C202 (2013); a) the stress distribution on the TLP with tower, b) the stress distribution on the TLP, c) The TLP with marked maximum stresses in different points.

Table 4. The von Mises stress in the CENTEC-TLP.

| Location | Von Mises Stress (MPa) |
|---|---|
| 1 | 701 |
| 2 | 947 |
| 3 | 762 |
| 4 | 456 |
| 5 | 681 |

Locations presented in Figure 6.

These options could be stiffeners, FRP, bracket collar plates, doubler plates, or similar methods. Overall, the downwind of the TLP experienced higher stress than other parts. The increase of the thickness of the stiffeners would be one of the options to reinforce the TLP.

The results indicate that increasing the thickness of all stiffeners compared to the increase of the thickness of the lower, the upper, and the central column has negligible effects on maximum stresses of different locations. Thus, the thickness of the lower, the upper, and the central column just increased. Table 5 presents the effects of the stiffeners' thickness on von Mises stress in the TLP.

Table 5. The effects of the stiffeners' thickness of the lower, the upper, and the central column on von Mises stress in the CENTEC-TLP.

| | Von Mises Stress (MPa) | | | | |
|---|---|---|---|---|---|
| Location | 20 mm | 25 mm | 30 mm | 35 mm | 40 mm |
| 1 | 576 | 567 | 560 | 555 | 550 |
| 2 | 924 | 778 | 676 | 599 | 539 |
| 3 | 673 | 595 | 528 | 474 | 429 |
| 4 | 429 | 419 | 413 | 406 | 401 |
| 5 | 561 | 467 | 402 | 354 | 317 |

Locations presented in Figure 6.

According to Table 5, the increase of the stiffeners' thickness in the lower, the upper, and the central column leads to a decrease in the maximum stress in different locations. It can be concluded that instead of increasing the stiffeners' thickness in the mentioned three parts to 40 mm, it is better to increase the thickness of the lower column to 40 mm, the upper column to 30 mm, and the central column to 35. Table 6 presents the effects of the combination of different thicknesses on the von Mises stress in the CENTEC-TLP.

In the following subsections, the stress distribution in each part of the TLP will be discussed.

Table 6. The von Mises stress in the CENTEC-TLP by increased thicknesses.

| Location | Von Mises Stress (MPa) |
|---|---|
| 1 | 553 |
| 2 | 536 |
| 3 | 474 |
| 4 | 413 |
| 5 | 354 |

Locations presented in Figure 6.

## 4.1 *The pontoons*

The stress distribution on the pontoons in the TLP is presented in Figure 7. According to Figure 7d, and 7f, the pontoons connected to the central

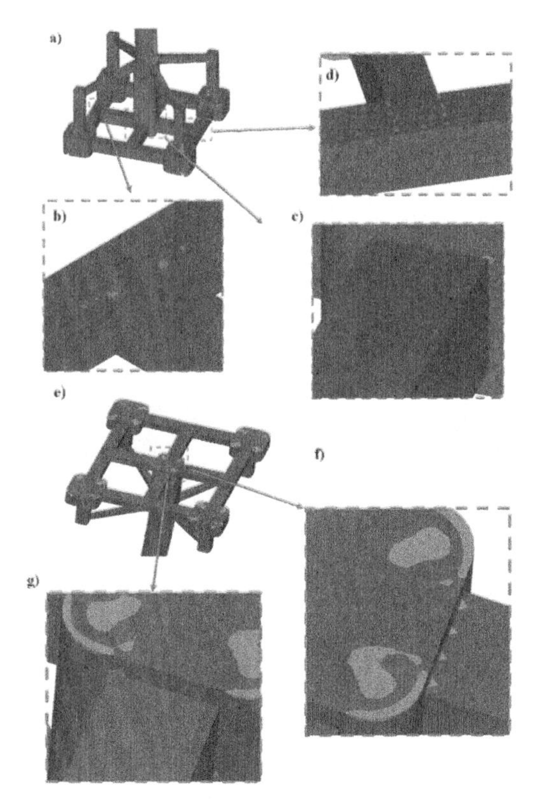


Figure 7. The stress distribution on the pontoons in the CENTEC-TLP; a) A close view of the TLP, b) The stress distribution on the pontoons in the opposite side of the TLP, c) The stress distribution on the connection of pontoons with the central column, d) The stress distribution on the downwind parts of the pontoon, e) The stress distribution on the bottom of the TLP, f) The stress distribution on the downwind parts of the pontoon in the bottom, g) The stress distribution on the connection of the pontoon with the central column at the bottom.

column in the downwind of the TLP has more stress than other parts of the pontoons. It can be seen that the bottom of the pontoons, especially pontoons that are in the leeward of the TLP, have more stress because of having both the hydrostatic pressure and the wind force together; therefore, that part should be reinforced by appropriate methods (Figure 7f).

### 4.2 *The lower column and the mooring connections*

The stress distribution on the lower column and the mooring connections in the TLP are presented in Figure 8. It can be seen that stress concentration appears in the mooring connections and the interaction of the lower column with the upper column, especially the corners that are close to the mooring lines. Figure 8, d indicates that the inside parts of the lower column get less stress than the outside parts. All corners of the lower columns and the mooring connections need extra support by stiffeners or other reinforcing methods.

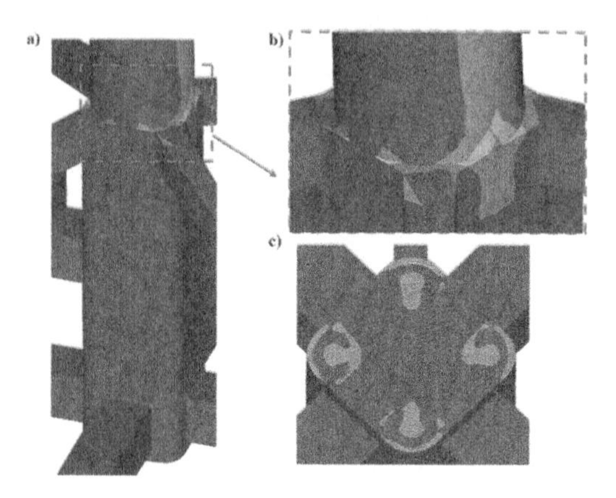


Figure 9. The stress distribution on the central column in the CENTEC-TLP; a) A close view of the central column, b) The stress distribution on the connection of the tower of the DTU 10MW and central column, c) The stress distribution on the bottom of the central column.

### 4.3 *The central column*

Figure 9 indicates the connection of the tower of the DTU 10MW and the central column. According to

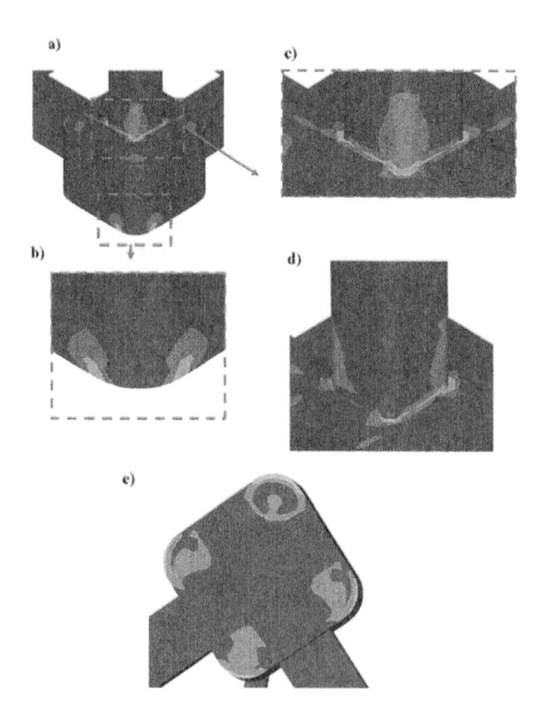


Figure 8. The stress distribution on the lower column and the mooring connections in the CENTEC-TLP; a) A close view of the lower column, b) The stress distribution on the mooring connections, c) The stress distribution on the interaction of the lower column with the upper column (outside), d) The stress distribution on the interaction of the lower column with the upper column (inner side), e) The stress distribution on the bottom of the lower column.

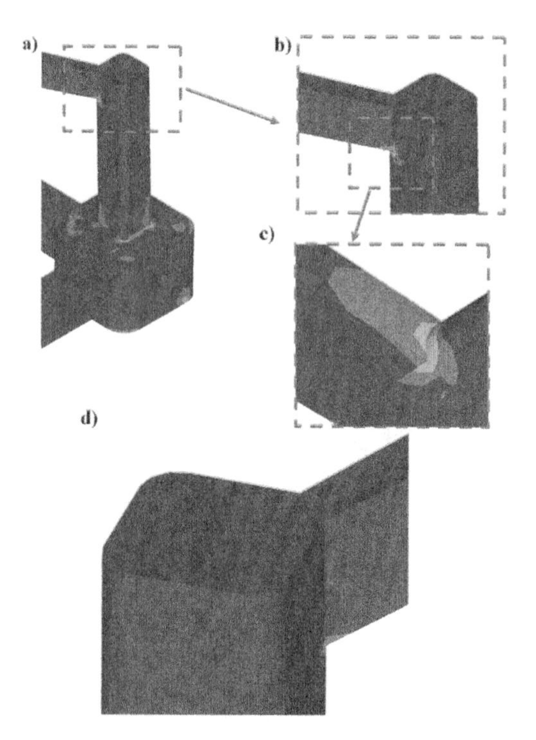


Figure 10. The stress distribution on the upper column and the support brace in the CENTEC-TLP; a) A close view of the upper column, b) The stress distribution on the connection of the support brace and upper column in the leeward, c) The stress distribution on the bottom of support brace in the leeward, d) The stress distribution on the bottom of support brace.

Figure 9c, the interaction of the tower with the central column and the bottom of the central column, all corners of the TLP should be reinforced.

### 4.4 *The upper column and the support brace*

The stress distribution on the upper column and the support brace in the TLP is presented in this section.

Figure 10 shows that stress concentration could be seen in the interaction of the lower column and support brace with the upper column in the leeward side of the TLP. According to Figure 10, c, the bottom of the support brace requires more stiffener tan than the top of the support brace. Figure 10, d indicates that other support braces which are not in the downwind of the TLP have less stress concentration.

## 5 CONCLUSIONS

The stress distribution on CENTEC-TLP at the design condition in still water and operational wind thrust is investigated. The wind is assumed to be linear. Two kinds of stiffeners are used to reinforce the TLP, and they are analyzed with ANSYS software. The main conclusions are summarized as follows:

- The connection of the pontoons with the central column and lower columns which are downwind of the TLP had more stress than other sides of the pontoons.
- The mooring connections showed higher stress. Therefore, the determined stiffeners were not enough for these locations, and they should be reinforced with increasing thickness of the stiffeners, number of stiffeners, or other reinforcing methods.
- The connection of the tower of the DTU 10MW and central column had a higher maximum stress than other parts, which is downwind of the TLP.
- The stress concentration was seen in the interaction of the lower column and support brace with the upper column in the leeward side of the TLP. The lower column, especially the corners close to the mooring lines, had higher stress values than other corners of the lower column.
- Overall, the leeward of the TLP had higher stress values, and therefore that side of the TLP needs more attention and precise design.
- Increasing the thickness of the stiffeners leads to a decrease in the maximum stress in the TLP.

## ACKNOWLEDGEMENTS

This work was conducted within the ARCWIND project–Adaptation and implementation of floating wind energy conversion technology for the Atlantic region (http://www.arcwind.eu/), which is co-financed by the European Regional Development Fund through the Interreg Atlantic Area Programme under contract EAPA 344/2016. This work contributes to the Strategic Research Plan of the Centre for Marine Technology and Ocean Engineering (CENTEC), which is financed by the Portuguese Foundation for Science and Technology (Fundação para a Ciência e Tecnologia - FCT) under contract UIDB/UIDP/00134/2020.

## REFERENCES

Bak, C., Zahle, F., Bitsche, R., Kim, T., Yde, A., Henriksen, L.C., Hansen, M.H., Blasques, J.P.A., Gaunaa, M., & Natarajan, A. 2013. The DTU 10-MW Reference Wind Turbine. DTU Wind Energy Report-I-0092.

Bachynski, E.E. Design and Dynamic Analysis of Tension Leg Platform Wind Turbines. PhD thesis. *Norwegian University of Science and Technology*. Norway.

Det Norske Veritas. 2010. Buckling strength of plated structures, DNV-RP-C201.

Det Norske Veritas. 2011. Structural design of TLPS (LRFD Method), DNV-OS-C105.

Det Norske Veritas. 2013a. Buckling strength of shells, DNV-RP-C202.

Det Norske Veritas. 2013b. Offshore standard design of floating wind turbine structures, DNV-OS-J103.

Det Norske Veritas. 2014. Design of offshore wind turbine structures, DNV-OS-J101.

Det Norske Veritas. 2016. Fatigue design of offshore structures, DNV-RP-C203.

Det Norske Veritas. 2018. Standard for floating wind turbines, DNV-ST–0119.

Det Norske Veritas. 2011. Design of offshore steel structures, general (LRFD method), DNV-OS-C101.

International Electrotechnical Commission 2005. Wind turbines – part 1: design requirements, IEC 61400–1.

International Electrotechnical Commission 2019. Design requirements for fixed offshore wind turbines, IEC 61400-3–1.

International Electrotechnical Commission 2020. Design requirements for floating offshore wind turbines, IEC 61400-3–2.

Iliopoulos, A.N., Devriendt, C., Iliopoulos, S.N., & Van Hemelrijck, D. 2014. Continuous fatigue assessment of offshore wind turbines using a stress prediction technique. *Health Monitoring of Structural and Biological Systems* 9064: 90640S.

Kim, H, Beom, J., Chang, K., & Yoon, H. 2018. Fatigue Analysis of Floating Wind Turbine Support Structure Applying Modified Stress Transfer Function by Artificial Neural Network. *Ocean Engineering* 149: 113–26.

Mas-Soler, J., Uzunoglu, E., Bulian, G., Guedes Soares, C., & Souto-Iglesias, A. 2021. An experimental study on transporting a free-float capable tension leg platform for a 10 MW wind turbine in waves. *Renewable Energy* 179: 2158–2173.

Reyno, H., Park, J., Kang, S., & Kang, Y. 2015. A numerical analysis for stress concentration of openings in offshore tubular steel tower under design loading condition. *Journal of the Korea Academia-Industrial Cooperation Society* 16 (2): 1516–23.

Teixeira, R., Nogal, M., O'Connor, A., Nichols, J., & Dumas, A. 2019. Stress-cycle fatigue design with

Kriging applied to offshore wind turbines. *International Journal of Fatigue* 125: 454–467.
Tempel, J.V.D. 2006. Design of support structures for offshore wind turbines. PhD thesis. *Delft University of Technology*. Netherlands.
Önder, U., Akbas, B., & Shen, J. 2011. Design issues of wind turbine towers. *Proceedings of the 8th International Conference on Structural Dynamics*: 1592–1598, *4–6 July 2011*, Leuven: Belgium.
Uzunoglu, E., & Guedes Soares, C. 2018. Parametric modelling of marine structures for hydrodynamic calculations. *Ocean Engineering* 160: 181–196.
Uzunoglu, E., & Guedes Soares, C. 2020. Hydrodynamic design of a free-float capable tension leg platform for a 10 MW wind turbine. *Ocean Engineering* 197: 106888.
Uzunoglu, E., & Guedes Soares, C. 2021. Response dynamics of a free-float capable tension leg platform for a 10 MW wind turbine at the northern Iberian peninsula. In *Developments in Renewable Energies Offshore*, edited by C Guedes Soares, 408–16. London: Taylor & Francis Group. https://doi.org/10.1201/9781003134572-47
Uzunoglu, E.; Karmakar, D., and Guedes Soares, C. 2016; Floating offshore wind platforms. L. Castro-Santos & V. Diaz-Casas (Eds.). Floating Offshore Wind Farms. Springer International Publishing Switzerland; pp. 53–76.
Yeter, B., Garbatov, Y., & Guedes Soares, C. 2015. Fatigue damage assessment of fixed offshore wind turbine tripod support structures. *Engineering Structures* 101: 518–528.

*Oil & gas*

*Trends in Maritime Technology and Engineering – Guedes Soares & Santos (eds)*


# Analysis of the basic causes of FPSO fluid releases

U. Bhardwaj, A.P. Teixeira & C. Guedes Soares
*Centre for Marine Technology and Ocean Engineering (CENTEC), Instituto Superior Técnico, Universidade de Lisboa, Lisbon, Portugal*

ABSTRACT: An analysis identifying the basic causes of fluid releases in Floating Production Storage and Offloading (FPSO) units is conducted based on 321 spills or releases events in FPSO units occurred from 1992 to 2005 in UK continental shelf. First, the components involved in spills or releases are identified. The frequencies of releases from this system level are presented with the objective of identifying the release prone areas and system elements. An analysis is performed highlighting the causal factors of releases from components in terms of design errors, components failures, operational and procedural errors. Statistics are presented in order to prioritize basic causes and to formulate relationships of releases with FPSO components. Such analysis may indicate the need for improvement of safety standards and system design pertaining to FPSOs.

## 1 INTRODUCTION

Among all floating units, Floating Production Storage and Offloading (FPSO) vessels are used extensively by the offshore industry and have become one of the primary methods for oil and gas production, processing and storage in more remote areas and deeper waters. FPSO vessels are equipped with hydrocarbon processing equipment for the separation and treatment of crude oil, water and gases arriving from sub-sea oil wells. Moreover, the storage capacity of FPSOs along with its offloading system allows transferring the crude oil to shuttle tankers for shipment to shore in alternative to export pipelines. With the development of oil and gas exploitation in the deep sea (Silva and Guedes Soares, 2019, 2021), the use of offshore floating installations operated in complex and high-risk environments increases the corresponding risk level significantly around the globe (Bhardwaj et al., 2019, 2022a).

Today over 270 FPSOs (including Floating Storage and offloading units - FSO) are in operation in intricate and high-risk environments (STF Analytics, 2018). The different operational modes of FPSOs and the complexity of their equipment and components (UKOOA, 2002) pose different risks compared to fixed platforms and commercial trading tankers, so their risks should be discussed separately.

According to the Health and Safety Executive's definition, a major accident in the offshore industry is understood to be an accident sequence that is out of control and that has the potential to cause five or more fatalities (HSE, 2015a). Though rare, still recently an accident with an FPSO involving 9 fatalities and many injuries and causing damage of 250 million dollars happened in Camarupim Field, Brazil (Marsh, 2016) that has started with a condensate leak during fluid transfer. Releases of hydrocarbons, chemicals and other fluids are the main contributors to major accidents in the oil and gas industries. In the upstream sector (oil and gas exploration/extraction industries), the 9 largest losses out of 25 are caused by releases of fluids (hydrocarbons or others) (Marsh, 2016).

Several incidents involving hydrocarbon releases from wells, subsea installations and pipelines have been recorded by various organizations around the world. These occurrences of hydrocarbons and other fluids pose risks to the environment and people working in the offshore installations, which should be as low as reasonably practicable, according to the Health and Safety Executive (HSE) (HSE, 2015a). The main hazards of releases in the offshore industry are:

- fire, after ignition of released hydrocarbons;
- explosion, after gas releases, formation and ignition of an explosive cloud;
- oil release on the sea surface or subsea;
- falling load, injury to the crew.

The occurrence of release events is appreciable, typically in the order of 2.5 releases per FPSO per year from 2000 to 2005 (Bhardwaj et al., 2017). It is therefore important to maintain awareness to prevent as far as possible the occurrence of such initiating events.

A vast amount of studies exists on the various aspects of releases from offshore installations, primarily focused on releases of hydrocarbons. Such studies have mainly paid attention to the statistics and circumstances of releases (Sklet, 2006, Aven et al., 2006 & Li, 2011). An assessment model for

DOI: 10.1201/9781003320289-55

predicting the frequency of process leaks occurring during all operation phases and topside leaks from the well system occurring during normal production has been proposed by Lloyd's Register Consulting (2018).

In terms of hydrocarbon release causes, many dimensions have been explored. For instance, Sklet et al. (2006) have analysed technical, human, operational and organizational factors. Kongsvik et al. (2011) have compared and verified that the safety climate dominates over technical indicators (such as installation age, weight and number of leakage sources). Okstad et al. (2012) have reviewed a limited set of 25 accident investigations and identified the causal factors involved, while Lootz et al. (2012) have focused on human factors associated with releases. Bergh et al. (2014) have addressed psychosocial risk factors. Olsen et al. (2015) have established relationships between work climate and hydrocarbon releases.

Several studies have discussed in detail other aspects of the circumstances of releases in Norway continental shelf (NCS) such as (Vinnem et al., 2007, 2010), Vinnem (2012, 2013) and Vinnem & Røed (2015). In particular, Vinnem & Røed (2015) conducted a root cause analysis on 78 hydrocarbon releases on the NCS. They have used six categories of root causes defined in the Barrier and Operational Risk Analysis of hydrocarbon releases (BORA-Release) project and highlighted types of systems and equipment involved in releases (Aven et al., 2006). Almost half of the releases were caused by technical faults mainly from small diameter pipes and the smallest equipment were associated with the highest number of releases. Other dimensions at the instance of releases such as activities, time and quantity have also been analysed. Vinnem & Røed (2015) have concluded that verification faults were dominating among installations and lack of compliance with the controlling documentation is the most common cause of releases.

To identify safety measures, Suardin et al. (2008) have assessed the consequences of releases, such as fires and explosions in FPSOs. Park et al. (2018) have also shown that LNG-FPSOs (Liquefied Natural Gas) or FLNGs (Floating Liquid Natural Gas) have a relatively high risk of release and explosion. Pipes, particularly of small diameter, are the main contributors to fluid releases. Keprate et al. (2017) have included some prevention mechanisms in their analysis of releases from the piping. This indicates that many research studies analyse offshore fluid releases from many different angles and use different approaches, but most of them are not specific for their applicability on particular installations. To the best of the authors' knowledge, basic cause analysis of releases specifically for FPSOs has not yet been conducted (Bhardwaj et al., 2021, 2022b).

A systematic analysis of dangerous events on FPSOs has been conducted by Bhardwaj et al., (2017, 2021) who compared risk scenarios in terms of events and their frequency for FPSO and other floating units. FPSO specific statistics have also been discussed in terms of types of events, their frequency, and operational modes. The consequences of these events have been seen in terms of injuries and terminal events. It has been concluded that spill/releases in FPSOs are the most recurring event of all events. This finding is further analysed in detail by Bhardwaj et al. (2018). This study has identified the module or systems, the equipment, and activities involved in the spill or releases, the type of fluid released and the number of shutdowns of the processing units.

The present work follows on from the previous study by Bhardwaj et al. (2018), where releases were perceived at two levels (module or system and equipment). This paper further extends release scenarios to components and establishes relations among equipment and components regarding releases in FPSOs. The basic causes of releases from a component level are also identified.

A basic cause is the main cause of an accident or incident that can be rationally identified and managed. The present study conducts a causal analysis of releases in FPSOs (including FSOs) with the main objective of presenting and discussing the circumstances that lead to the spill/release. In particular, the causes of such events, for instance, technical and operational, are characterised, using elements from several incident and accident reporting systems such as from HSE (2003, 2008, 2015b) for the UK offshore industry and Petroleum Safety Authority (PSA) for Norway (PSA, 2016). None of these reports emphasizes the type of installations, their complexity, and the equipment involved in the releases. It is believed that this analysis will provide information to the FPSO operators for reducing spill/releases and related risks by focusing on the critical components of the system.

## 2 HYDROCARBON RELEASE DATA

A large and descriptive set of UK Continental Shelf accident and incident data utilized in the present study has been obtained from HSE (2007). It is believed that such a compilation of information of FPSO accidents is the most comprehensive data publicly available. There exist other reports by HSE (e.g. HSE, 2015b) on hydrocarbon release data which are very useful for understanding the characteristics of hydrocarbon releases. Such reports however lack certain features and descriptions to distinguish FPSO among the offshore installations. The "real" number of accidents and incidents around the world might be much higher.

The dataset contains information on accidents and incidents that have occurred in floating offshore units engaged in the oil and gas activities in the UKCS over 25 years, from 1980–2005. A total of 16 FPSOs and 6 FSUs were active during these years with the first oil production in 1982 by Fulmar FSU. This corresponds to a total of 170 unit-years of FPSO operating from 1980 to 2005. The data consists of a description of 321 release events with the first occurrence recorded in 1992, defined as a situation that has the potential to result in an unwanted outcome. In terms of their severity the release events are characterised as follows: i) Accidents - events causing serious injury to crew members, structural damage and environmental pollution; ii) Incidents - events causing releases of dangerous fluid, minor injuries to crew and long shutdown of equipment; iii) Insignificant - events causing releases of oil, water and other less dangerous fluid, less pollution and no injury.

One accident or incident may comprise a chain of events, e.g. a blowout resulting in an explosion, fire and oil spill. This means that one single accident or incident may give rise to several events. The total number of events is thus much higher than the total number of accidents recorded. Fluid releases in the present context are defined as a gas release or oil release (including condensate), inert gas, water, lube oil, methanol etc. from the process flow, well flow, or flexible risers, unintentional or uncontrolled releases (heli-fuel, diesel).

## 3 COMPONENT AND EQUIPMENT INVOLVED IN RELEASES

The present analysis intends to highlight the vulnerabilities of components and to provide guidance on how to minimize their frequency. Every single release is analysed irrespective of its implication, as it can indicate local or system-related weaknesses and potential symptoms of larger system failures. Components defined here are the parts or elements of equipment (Table 1). It is impossible to incorporate all components individually. Hence the components discussed in this section are not necessary to be all, but the ones involved in releases.

Table 2 presents the components involved in releases. A pipe is the single largest source, contributing to 34.3% of the releases and also accounted for two accidents in the FPSO (see Figure 1). To avert such a fateful event and to enhance process safety is vital to maintaining the technical integrity of the topside and subsea piping. Valve and flange components follow the sequence. It is noteworthy that several accidents (like piper alpha, Cullen, 1990) have started with the failure of valves, mainly pressure safety valves.

Vinnem (2013) illustrated that failure or maloperation in coupling (flange) is a dominating cause of releases during maintenance work. However, due to the limitation of data, in 12% of the releases, the components involved were unknown. Statistics from HSE (2008) for offshore installations also indicate that the single largest contributor to releases is the topside piping (as pipework contributed to 56% of failures on offshore oil and gas platforms in the UK sector of the North Sea), followed by *valves*, *flanges*, and other pressure equipment.

One further step is conducted to analyse the relationship between components and equipment. For example, *compressors* work on very high pressure, contributing to a high number of releases from *valves* and *flanges* that are around them.

Table 1. Description of the components.

| Component | Description |
|---|---|
| Casing | It consists of the casing of pump volute casing, enclosure for engine and turbine |
| Seal | O ring of a compressor, extended seal in the high-pressure side of exchangers, seal on methanol pump, swivel bottom seal, HP barrier seal, valve seal vent, oil pump seal, water seal |
| Valve | Comprises valve body, stem and packer. Manual valve – Block (isolation, shut off and kill), bleed, choke and check valve Actuated valve – ESDV (Emergency shutdown valve), control (FCV or PCV), block, choke, blowdown, relief and pipe SSIV. Valve types including gate, ball, plug, globe, needle and butterfly(for check valves) |
| Flange | Comprises flange joint and gasket (where fitted, coupling, fittings, other connection between equipment, quick connect and disconnect (QCDC) coupling. Flange type includes ring type joint, compressed joint, spiral wound, clamp, hammer union. |
| Pipe | Gas, water and oil pipelines-steel or flexible, riser, flowline, hoses (flexible). |

Table 2. Components involved in Spill/releases.

| Type of component | Number of events | % |
|---|---|---|
| Pipe | 110 | 34.3 |
| Valve | 70 | 21.8 |
| Flange | 69 | 21.5 |
| Seal | 17 | 5.3 |
| Casing | 16 | 5 |
| Unknown | 39 | 12.1 |
| Total | 321 | 100 |

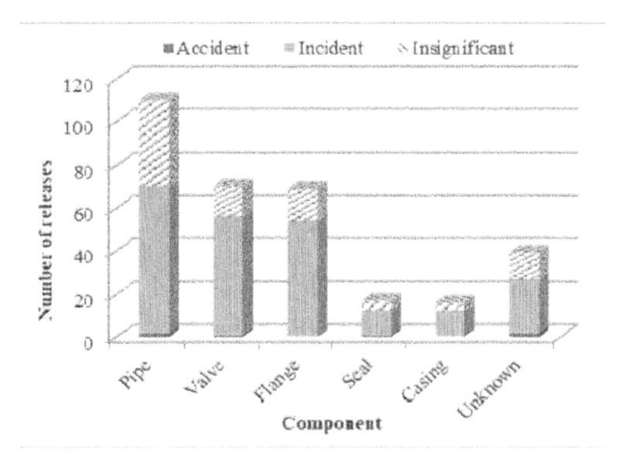


Figure 1. Spill/releases from components.

Table 3. Relationship of releases from equipment with components.

| | Component | | | | | | |
|---|---|---|---|---|---|---|---|
| Equipment | Casing | Flange | Pipe | Valve | seal | Unknown | Total |
| Compressor | 1 | 22 | 17 | 33 | 1 | 7 | 81 |
| Tank | 2 | 6 | 10 | 10 | 6 | 14 | 48 |
| Pump | 5 | 5 | 13 | 3 | 3 | 4 | 33 |
| Engine | 3 | 7 | 9 | 4 | | 2 | 25 |
| Flowline/riser | | 3 | 14 | 3 | | 1 | 21 |
| Heat Exchanger | 1 | 8 | 4 | 1 | 1 | 2 | 17 |
| Metering skid | | 3 | 3 | 5 | | 1 | 12 |
| Separator | | | 6 | 3 | 2 | | 11 |
| Turbine | 3 | 2 | 1 | 2 | | | 8 |
| Scrubber | 1 | 2 | 3 | | | | 6 |
| Boiler | | | 2 | | | 2 | 4 |
| Manifold | | 1 | 1 | 1 | | | 3 |
| Header | | 1 | 2 | | | | 3 |
| Umbilical | | | 2 | | | | 2 |
| Well | | | | 2 | | | 2 |
| Degasser | | | 2 | | | | 2 |
| Coalescer | | | 1 | | | | 1 |
| Filter | | 1 | | | | | 1 |
| Unknown | | | | | | | 41 |

A few important observations can be made from Table 3. When analysing the *compressor*, it is understood that overpressure, blowout, and vibration have caused the highest failure of components, especially *valves* and *flanges*. Releases around *tanks* are mainly from *valves* (10), *pipes* (10) and *flanges* (6). Major releases from *pipes* (13) are observed with the *pump*. Since population data were not available, it is not possible to calculate release rates or failure rates of the components and equipment.

*Valve* is a critical component in all equipment since it is subjected to high-pressure differences. *Valve* failures depend on an operating sector, processing condition, operating regime, type, and size. *Valve* seal failure accounts for 23% followed by cover failure. Failures in *valve* seals are mainly due to the degradation of material properties under high tension (Table 4).

One important parameter to *pipe* failure is the *pipe* diameter (Silva et al. 2019). Figure 2 shows that failure is predominant in small diameter (< 0.5 inches) pipelines and large diameter (> 2mm). Vinnem & Røed (2015) also suggested that any

Table 4. Spill/releases from subcomponent failures.

| Subcomponent | Number of failures | % |
|---|---|---|
| Valve seal | 16 | 22.9 |
| Cover | 8 | 11.4 |
| Distance piece | 2 | 2.9 |
| Stem | 1 | 1.4 |
| Other | 43 | 61.4 |
| Total | 70 | 100 |

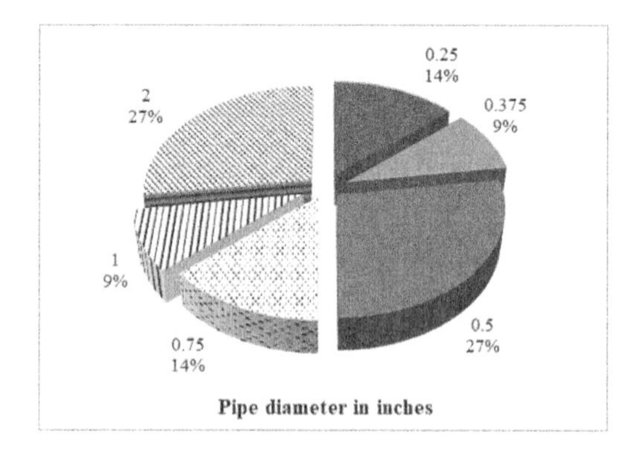


Figure 2. Pipe diameter involved in releases.

*pipes* with a diameter between 0.5 and 1 inch are the most frequent in the failures. To avoid piping failures nuclear and process industries widely use the leak before break (LBB) concept to design offshore piping (Matthews 2003). Visualizing the various stages of release events (from failure initiation to release event) through a metallurgic point of view is out of the scope of this paper.

## 4 CAUSES OF RELEASES

The present study uses a systematic approach in which the causes of releases from components are analysed. A root cause is the most basic cause of an accident or incident that can be reasonably identified and managed. Thus, first, a taxonomy of root causes must be established. A typical cognitive approach to root causes adopted in the BORA project classifies root causes as (Haugen et al., 2007 & Vinnem et al., 2007):

(i) Technical degradation of system (Category A);
(ii) Human intervention;
  (a) introducing latent error (delayed release); (Category B);
  (b) causing immediate release (Category C);
(iii) Process disturbance (Category D),
(iv) Inherent design errors (Category E);
(v) External events (Category F).

Offshore Safety Directive Regulator (OSDR) has proposed the following four main categories (causal factors) classified by HSE (2015c) that are adopted in the present study:

- Design failure;
- Component failures;
- Operational errors;
- Procedural errors.

Table 5 shows the distribution of the main causal factors in spill/releases. Due to limited descriptions available, around 36.8% of the causes remain unknown. 32.1 % of cases are due to operational errors, followed by component failures (27.1%). These causes are further explored in the following sections. Severity analysis (Figure 3) shows that the major accidental events are caused by operational errors.

Table 5. Causal Factors involved in spill/releases.

| Causal Factors | Number of events | % |
|---|---|---|
| Design Error | 4 | 1.2 |
| Component failure | 87 | 27.1 |
| Operational errors | 103 | 32.1 |
| Procedural errors | 9 | 2.8 |
| Unknown | 118 | 36.8 |
| Total | 321 | 100 |

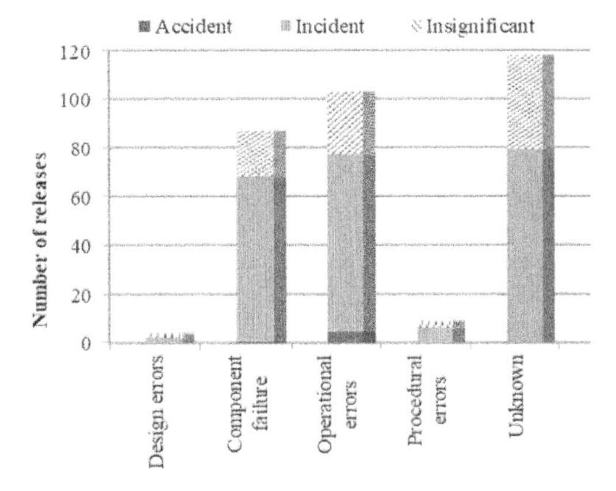


Figure 3. Causal factors involved in spill/releases.

## 4.1 *Design errors*

Design related failures are latent failures introduced during design, assembling and manufacturing that cause releases during normal operations. Four releases are due to substandard material in components.

## 4.2 *Component failure*

The failures in components can be attributed to design deficiencies, poor selection of materials, manufacturing defects, exceeding design limits and overloading, inadequate maintenance (Maleque & Salit 2013).

The design and manufacturing faults are considered in section 4.1, which incorporates the technical or physical failures of components leading to releases due to mechanical and material degradation caused by ageing, wear-out, corrosion, erosion. The highest number of component failures are caused by weld crack of component (16%) followed by erosion (11.5%) (see Table 6).

Table 6. Basic causes of component failures.

| Type of failure | Number of failures | % |
|---|---|---|
| Weld crack | 14 | 16.1 |
| Erosion | 10 | 11.5 |
| Vibration | 5 | 5.7 |
| Corrosion | 4 | 4.6 |
| Wear | 3 | 3.5 |
| Other Mechanical | 51 | 58.6 |
| Total | 87 | 100 |

## 4.3 *Operational errors*

Operational errors consist of a wide domain of errors and failures during normal operating conditions. Besides, human errors external events or loads like releases due to falling objects, collisions, bumping, etc. are also included in this category. Moreover, process upsets like overpressure, under pressure, overheating contributing to component failure is also

Table 7. Basic causes of operational errors.

| Type of equipment | Number of events | % |
|---|---|---|
| Fatigue fracture | 25 | 24.3 |
| Incorrect fitting | 25 | 24.3 |
| Overpressure | 21 | 20.4 |
| Left open | 19 | 18.4 |
| Overheating | 7 | 6.8 |
| Excess load | 2 | 1.9 |
| Collision | 2 | 1.9 |
| Deck movement | 1 | 1 |
| Dropped object | 1 | 1 |
| Total | 103 | 100 |

Table 8. Relationship between causal and basic factors of releases with components.

| Causal factors | Basic causes | Component | | | Total |
|---|---|---|---|---|---|
| | | Flange | Pipe | Valve | |
| Design error | Substandard material | 1 | 3 | | 4 |
| | Erosion | 2 | 6 | 1 | 9 |
| | Vibration | 3 | 1 | 1 | 5 |
| Component failure | Weld crack | 1 | 3 | 1 | 5 |
| | Corrosion | | 4 | | 4 |
| | Wear | | 2 | 1 | 3 |
| | Incorrect fitting | 17 | 1 | 5 | 23 |
| | Fatigue fracture | 4 | 11 | 9 | 24 |
| Operational | Overheating | 3 | 4 | 13 | 20 |
| | Overpressure | 2 | 8 | 3 | 13 |
| | Left open | 3 | 2 | 13 | 18 |
| | Excess loads | | 2 | | 2 |
| Procedural | Noncompliance | 2 | 5 | 2 | 9 |
| Other | | 10 | 16 | | 26 |

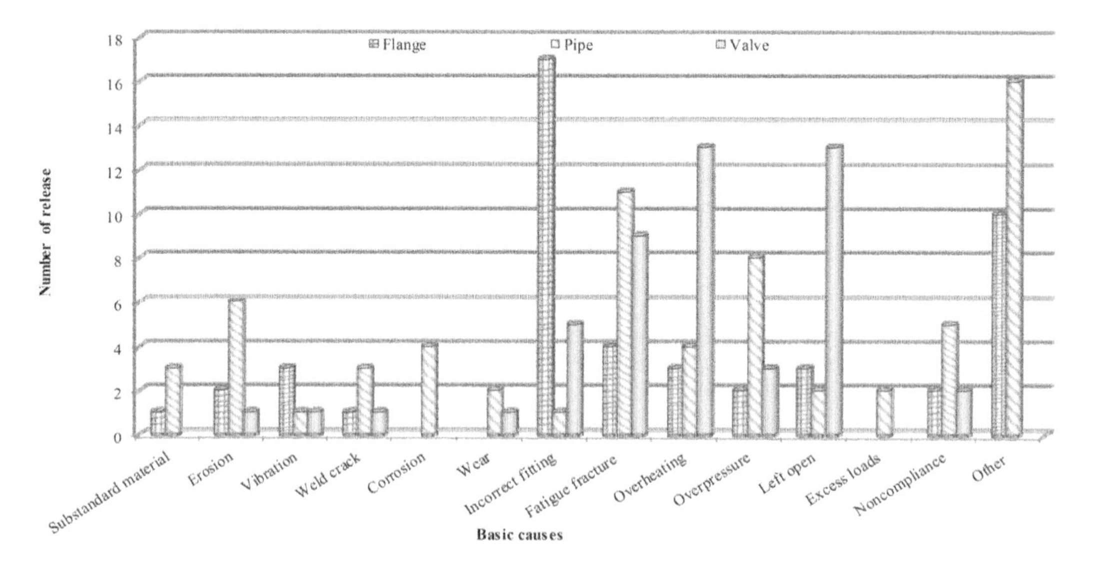


Figure 4. Distribution of basic causes with component.

considered. Table 7 breakdowns the types of operational errors leading to releases found in the present analysis. Operational errors are mainly caused by human interventions like incorrect fittings (24.3%), components left open after maintenance (18.4%). Fatigue failure also accounts for almost one-fourth of errors. Failure due to overpressure is the next appreciable operational error (20.4 %).

### 4.4 *Procedural error*

Nearly all FPSO operations are manually executed by platform personnel, therefore compliance with procedures and instructions is crucial. In the present dataset 9 releases are due to *non-compliance* with procedures.

It is interesting to note that most of the *operational* and *procedural errors* are a subset of human errors. However, the real share of human errors would have been much higher since 36.8% of releases causes are unknown (Table 5). Skogdalen & Vinnem (2011) suggested that human errors and organization factors must be modelled and considered as safety barriers for quantitative risk analysis in the offshore industry. When other investigation reports are scrutinized for human errors, a lack of compliance with procedures, instructions, and standards is observed.

## 5 COMPONENT-SPECIFIC BASIC CAUSES

In this section, an attempt is made to establish a relationship between the individual components and root/basic causes. As demonstrated in Table 8 and Figure 4, pipelines failures are mainly due to weld crack, fracture, overpressure and erosion. Releases due to erosion in pipes are typically caused by the production of sand from the reservoir. Although corrosion is an important phenomenon in offshore installations, only four cases have been observed in pipelines.

Releases from the valves are due mainly to overheating and being left open after maintenance. Flanges releases are associated typically with operational causes, mainly incorrect fittings (due to tightening with too low or too high, tension and/or misalignment). Releases from valves and flanges can be easily mitigated since most of their causes are attributed to human errors during maintenance.

## 6 CONCLUSIONS

*Spill/releases* of fluids (HC, diesel, chemical etc.) are the most hazardous events in FPSOs. The prevention of *spills or releases* is fundamental since they may lead to major accidents. Reports on hydrocarbon releases published by HSE and PSA show trends but not the basic causes of releases specific to FPSO installations.

A large and descriptive dataset of UK Continental Shelf accidents and incidents over 25 years, from 1980-2005 has been analysed. 321 release events in FPSOs have been investigated, with the first occurrence recorded in 1992. Understanding how the releases evolved at the component level of the FPSO system is relevant for risk management and maintenance planning.

The present paper provides information about the basic causes of spill/releases in FPSOs. An attempt has been made to quantify the contribution of components of equipment to the releases. Based on the analysis in the study the following observations are drawn:

- One third of the releases are from pipes, followed by valves (21.8%) and flanges (21.5%);
- Compressors have the highest releases from valves (33), flanges (22) and pipes (17);
- Releases around tanks are mainly from valves (10), pipes (10) and flanges (6);
- Major releases from pipes (13) are associated with pumps;
- Failure of valves are critical and 23% of failures are due to valve seal failures while 11% are from valve cover failures;
- Small (< 0.5 inches) and large diameter (> 2 inches) pipes are predominate in releases;
- Failures in seals are mainly due to the degradation of material properties under high tension.
- Operational errors (32.1%) and component failures (27.1%) are the main causal factors of releases. However, due to the limited description available in the dataset 36,8% of total releases causes are unknown.
- The basic causes of operational errors are mainly related to fatigue (24.3%), incorrect fitting (24.3%) and overpressure (20.4%).
- The basic causes of component failures are mainly associated with weld crack (16.1%) and erosion (11.5%);

Lastly, it is shown the relationship between causal and basic factors of releases with components:

- The main basic cause of flange failures is incorrect fitting;
- Pipes fail due to fracture then overpressure, erosion and corrosion;
- Valves fail primarily due to overheating or being left open after maintenance.

It is clear that the catastrophic outcomes of releases typically result from a long sequence of accidental events, so there are usually several opportunities to avoid accidents if the causes of releases are known. This work can be extended with the analysis of the sequence of events leading to critical accidental scenarios and of the safety barriers and design improvements necessary to prevent releases and mitigate their consequences.

Finally, it can be concluded that the main areas of concern of spill/releases are related to bolted joints, flexible hoses, small bore tubing, erosion and vibration management, as well as to human factors and awareness and communication issues.

In the future study, updated data will be collected to reflect the latest development of offshore engineering equipment.

## ACKNOWLEDGEMENTS

The first author was financed by the European Union Horizon 2020 research and innovation under the Marie Skodowska-Curie grant agreement No. 73088 RESET. This work contributes to the Strategic Research Plan of the Centre for Marine Technology and Ocean Engineering, which is financed by the Portuguese Foundation for Science and Technology (Fundação para a Ciência e Tecnologia-FCT) under contract UIDB/UIDP/00134/2020.

## REFERENCES

Aven, T., Sklet, S. & Vinnem J.E. 2006. Barrier and operational risk analysis of hydrocarbon releases (BORA-Release) Part I. Method description. *Journal of Hazardous Materials* 137: 681–691.

Bergh, L.I. V., Ringstad, A. J., Leka, S. & Zwetsloot, G. I. J. M. 2014. Psychosocial risks and hydrocarbon leaks: An exploration of their relationship in the Norwegian oil and gas industry. *Journal of Cleaner Production* 84: 824–830.

Bhardwaj, U., Teixeira, A. P. & Guedes Soares C. 2017. Analysis of FPSO accident and incident data. In C. Guedes Soares and Y. Garbatov (Eds.), *Progress in the Analysis and Design of Marine Structures*: 773–782, Taylor & Francis Group, London.

Bhardwaj, U., Teixeira, A. P. & Guedes Soares C. 2018. Analysis of FPSO accident and incident data. In C. Guedes Soares and Y. Garbatov (Eds.), *Maritime Transportation and Harvesting of Sea Resources*: 1121–1131, Taylor & Francis Group, London.

Bhardwaj, U., Teixeira, A. & Guedes Soares, C. 2019. Reliability prediction of a subsea separator. In *4th Workshop and Symposium on Safety and Integrity Management of Operations in Harsh Environments (CRISE4). St. John's, NL*: 1–9. Canada.

Bhardwaj, U., Teixeira, A.P., Guedes Soares, C., Ariffin, A. K. & Singh, S.S. 2021. Evidence based risk analysis of Fire and explosion accident scenarios in FPSOs. *Reliability Engineering and System Safety* 215: 107904.

Bhardwaj, U., Teixeira, A.P. & Guedes Soares, C. 2022a. Bayesian framework for reliability prediction of subsea processing systems accounting for influencing factors uncertainty. *Reliability Engineering and System Safety* 218: 108143.

Bhardwaj, U., Teixeira, A.P. & Guedes Soares, C. 2022b. Casualty analysis methodology and taxonomy for FPSO accident analysis. *Reliability Engineering and System Safety* 218: 108169.

Cullen, W.D. 1990. The Public Inquiry into the Piper Alpha Disaster, Department of Energy, London.

Haugen, S., Seljelid, J., Sklet, S. & Vinnem, J.E. 2007. BORA Generalization Report, *Preventer report 200254-07*, 29 Jan.

HSE 2003. Offshore hydrocarbon release statistics and analysis, *HID statistics report*, HSR 2002-002.

HSE 2007. Accident statistics for floating offshore units on the UKCS 1980-2005, *Det Norske Veritas/UK Health & Safety Executive*. HSE Research Report Series. Report No. RR567, www.hse.gov.uk/research/rrhtm/rr567.htm.

HSE 2008. Offshore hydrocarbon releases 2001–2008. *Research Report 672*.

HSE 2015a. The Offshore Installations (Safety Case) Regulations 2015. *Statutory Instrument 2015, no. 398*, HMSO, London.

HSE 2015b. Offshore Hydrocarbon Releases 1992–2015, www.hse.gov.uk/offshore/statistics.htm.

HSE 2015c. Guidance on reporting offshore hydrocarbon releases, Issued by the Offshore Safety Directive Regulator [OSDR]. HSE – Health & Safety Executive. Avail from http://www.hse.gov.uk/osdr/assets/docs/guidance-on-reporting-offshore-hydrocarbon-releases.docx

Keprate, A., Ratnayake, C. R.M. & Sankararaman, S. 2017. Minimizing hydrocarbon release from offshore piping by performing probabilistic fatigue life assessment. *Process Safety and Environmental Protection* 106: 34–51.

Kongsvik, T., Kjøs J., Svein A. & Sklet, S. 2011. Safety climate and hydrocarbon leaks: An empirical contribution to the leading-lagging indicator discussion. *Journal of Loss Prevention in the Process Industries* 24: 405–411.

Li, A. 2011. Cross industry hydrocarbon release analysis. In: SPE Paper 145449, *SPE Offshore Europe Oil and Gas Conference and Exhibition*, Aberdeen, UK, 6-8 September.

Lloyd's Register Consulting 2018. Process leak for offshore installations frequency assessment model—PLOFAM (2), Report no: 107566/R1.

Lootz, E., Hauge, S., Kongsvik, T. & Mostue, B.A. 2012. Human Factors and Hydrocarbon Leaks on the Norwegian Continental Shelf. *Contemporary Ergonomics and Human Factors*. Martin Anderson (Ed.), Taylor & Francis London.

Maleque, M. A. & Salit, M.S. 2013. Material selection & design, Springer, pp. 17–38.

Marsh 2016. The 100 Largest Losses 1974-2015 Large property damage losses in the hydrocarbon & Chemical industries, Marsh & Mclennan companies.

Matthews, C., 2003. *Handbook of mechanical in-service inspection: Pressure Systems and Mechanical Plant*, Professional Engineering Publishing Limited, London, UK.

Okstad, E., Jersin, E. & Tinmannsvik, R. K. 2012. Accident investigation in the Norwegian petroleum industry - Common features and future challenges, *Safety Science* 50: 1408–1414.

Olsen, E., S. Næss, and S. Høyland, 2015. Exploring relationships between organizational factors and hydrocarbon leaks on offshore platform. *Safety Science* 80: 301–309.

Park, S., Jeong, B., Lee, B.S., Oterkus S. & Zhou, P. 2018. Potential risk of vapour cloud explosion in FLNG liquefaction modules. *Ocean Engineering* 149: 423–437.

PSA 2016. Trends in Risk Level in the Petroleum Activity 2015, Annual report published by the Petroleum Safety Authorities, Stavanger, Norway, 28. April.

Silva, L. M. R. and Guedes Soares, C. An integrated optimization of the floating and subsea layouts. *Ocean Engineering*. 2019; 191:106557.

Silva, L. M. R. and Guedes Soares, C. Oilfield development system optimization under reservoir production uncertainty. *Ocean Engineering*. 2021; 225:108758.

Silva, L. M. R.; Teixeira, A. P., and Guedes Soares, C. A methodology to quantify the risk of subsea pipeline systems at the oilfield development selection phase. *Ocean Engineering*. 2019; 179:213–225.

Sklet, S. 2006. Hydrocarbon releases on oil and gas production platforms: Release scenarios and safety barriers. *Journal of Loss Prevention in the Process Industries* 19: 481–493.

Sklet, S., Vinnem, J.E. & Aven, T. 2006. Barrier and operational risk analysis of hydrocarbon releases (BORA-Release). Part II: Results from a case study. *Journal of Hazardous Materials* 137: 692–708.

Skogdalen, J. E. & Vinnem, J. E. 2011. Quantitative risk analysis offshore - Human and organizational factors, *Reliability Engineering & System Safety* 96: 468–479.

STF Analytics 2018. *Submarine Telecoms Forum Analytics. Market Sector Report: Offshore Oil & Gas Edition*, Submarine Telecoms Forum Analytics.

Suardin, J. A., Mcphate, A. J., Sipkema, A., Childs, M. & Mannan M. S. 2008. Fire and explosion assessment on oil and gas floating production storage offloading (FPSO): An effective screening and comparison tool. *Process Safety and Environmental Protection* 87: 147–160.

UKOOA 2002. UKOOA FPSO design guidance notes for UKCS service.

Vinnem, J. E., Hestad, J. A., Kvaløy, J. T. & Skogdalen, J. E. 2010. Analysis of root causes of major hazard precursors (hydrocarbon leaks) in the Norwegian offshore petroleum industry, *Reliability Engineering and System Safety* 95: 1142–1153.

Vinnem, J.E. & W. Røed, 2015. Root causes of hydrocarbon leaks on offshore petroleum installations. *Journal of Loss Prevention in the Process Industries* 36: 54–62.

Vinnem, J.E., 2012. On the analysis of hydrocarbon leaks in the Norwegian offshore industry. *J. Loss Prevention in the Process Industries* 25: 709–717.

Vinnem, J.E., 2013. On the development of failure models for hydrocarbon leaks during maintenance work in process plants on offshore petroleum installations. *Reliability Engineering and System Safety* 113: 112–121.

Vinnem, J.E., Seljelid, J., Haugen, S. & Husebø, T. 2007. Analysis of hydrocarbon leaks on offshore installations. *Risk, Reliability and Societal Safety*, T. Aven and J. E. Vinnem (Eds.): 25–27, Taylor & Francis Group, London.

*Trends in Maritime Technology and Engineering – Guedes Soares & Santos (eds)*


# Stochastic characterization of a petroleum reservoir

J.V. Saíde
*Centre for Marine Technology and Ocean Engineering (CENTEC), Instituto Superior Técnico, Universidade de Lisboa, Lisbon, Portugal*
*Centre of Studies in Oil and Gas Engineering and Technology (CS- OGET), Universidade Eduardo Mondlane, Mozambique*

A.P. Teixeira
*Centre for Marine Technology and Ocean Engineering (CENTEC), Instituto Superior Técnico, Universidade de Lisboa, Lisbon, Portugal*

ABSTRACT: Reservoir characterization plays a fundamental role in the oil and gas exploration and production industry. Stochastic methods are very useful in the early stages of a field's life for modelling reservoir heterogeneities when well data are too sparse. This paper provides an overview of reservoir characterization based on stochastic reservoir models applied to the assessment of an oilfield. The methods adopted consider the geological knowledge about the likely shape and size of the sand bodies and incorporate information about the relationships between properties at neighbouring locations. The methods allow building a stochastic model of an oil and gas reservoir to improve the estimation of reserves and support decision making, reflecting not only the knowledge of the internal characteristics of the reservoir but also the uncertainties associated. The modelling of Xiaermen Oilfield is used as case study.

## 1 INTRODUCTION

Petroleum reservoirs may contain oil, natural gas, or both. The reservoir characterization consists of an integrated process to generate an effective understanding of the reservoir main properties and hydrocarbon distribution using different data sources. Today, practically all aspects of oil and gas exploration and development use modelling for reservoir characterization (Islam, 2013), where available data are transferred into a computer-based mathematical representation (Bjørlykke, 2015).

Reservoir characterization involves the prediction of various parameters to visualize reservoir heterogeneities (Liu et al., 2021), as represented by the spatial variability of properties such as porosity, permeability, thickness, lithofacies types, fault orientations, among others (Mohaghegh et al., 1996).

Stochastic modelling is very useful in the early stages of a field's life, when well data are too sparse. This type of approach allows greater control in the quantification of the oil and gas volume, and it represents several equiprobable results, commonly used to assist in the representation of the heterogeneities of the reservoirs and the spatial variability of the variables, inferring the properties, reliable predictions follow trends of observation data (Ahn & Choe, 2022). To represent the reservoir heterogeneities geostatistical principles are applied to geological modelling (Normando et al., 2022). Most of the companies throughout the world face challenges during the production stage due to poor reservoir characterization. To minimize production challenges of any oilfield reservoir a clear 2D or 3D grid model must be constructed. These can be used to forecast and manage the future performance of the oilfield, which is important for better oilfield life cycle reservoir management.

This paper addresses the stochastic characterization of a petroleum reservoir using the Xiaermen Oilfield as case study. The paper is organised as follows. Section 2 discusses the geologic setting of the Xiaermen Oilfield. Section 3 presents the data and the approach adopted to build the stochastic model for reservoir characterization. The model results are presented in section 4, including the research findings and a discussion on the 3D geological model of different petrophysical and volume calculations. Finally, section 5 provides the conclusions of the modelled facies, petrophysical distribution and volume throughout the model analysed.

## 2 GEOLOGIC SETTING OF XIAERMEN OILFIELD

The Xiaermen Oilfield found in Biyang County, Henan Province, China is situated in the middle of a large northeast fault edge, to the east of Biyang

DOI: 10.1201/9781003320289-56

Sag, Nanxiang Basin. The oilfield reservoir is structured with brachy-anticline in east-west direction having four complicated major faults and many minor faults adjacent to the main oil-bearing area. There are small faults to the east which are emerged from the big fault. The extreme trap height is 275 m of the reservoir. The Xiaermen Reservoir is characterized as unconsolidated sandstone where its sand was deposited by deltaic fans. Considerable variation in sand thickness occurring over short lateral distances is characteristic of this deposit and thick layers with high permeability are patchy and isolated. The sediment underlying the project area is finely grained and relatively clean quartz sand. The average porosity is about 24% and the mean geometric permeability is 2 $\mu m^2$. The porosity and permeability are generally high, but heterogeneous. Xiaermen oilfield has gone into four stages of production up to 2009 reaching 31 years of production. The Xiaermen Oilfield has four main formation (1) H2II formation, (2) H2III formation (3) H2IV formation (4) H2V formation. In this regard, the paper is focused on the H2V formation data that has an oil-bearing area of 1.43 $Km^2$, primary geological reserves of $218x10^4$ t. The total number of well is 40 wells, the well density is 18.6 /$km^2$ and the average well spacing is 235m. The Xiaermen oilfield has sediments such as small trough cross bedding (well named T5-241 at depth of 1352.2m), lenticular bedding present at depth of 1343.7 m in well T5-241, progressive bedding (gas 1 well at the depth of 1162.4m).

## 3 DATA AND METHODS

### 3.1 *Data*

This paper builds a geological model using 40 well log data of the H2V formation of the Xiaermen oil field.

### 3.2 *Deterministic geostatistical modelling*

Geostatistical methods for reservoir characterization and modelling can be divided into deterministic and stochastic methods. In deterministic methods, all the conditions that can influence the predictions have to be completely known. Deterministic results can be unambiguously described by the completely known finite conditions, so deterministic analysis can only offer one solution. The true geological model in the subsurface is singular, but since the description of the subsurface is based on well data (point data), it is not possible to be certain that the solution obtained with geostatistical methods is the correct one. Therefore, all geostatistical models contain some uncertainty. They are a way of estimating the sub-surface conditions and they can provide only the most probable solution, in other words the closest to the real conditions (Zelenika et al., 2017).

### 3.3 *Stochastic geostatistical modelling*

Deterministic methods are still commonly used for reservoir characterization and modelling, even though the knowledge about the geological processes is far from complete. Stochastic modelling is a standard part of mathematical geology and are increasingly being used as an integral part of the tools for hydrocarbon reservoir characterizations worldwide. Stochastic realizations provide a different number of solutions for the same input data set. Those solutions can be very similar but never identical.

### 3.4 *Reservoir modelling workflow*

Several frameworks are required to build a numerical reservoir model such as: 1) the structural and reservoir framework; 2) the depositional and geostatistical framework and 3) the reservoir flow simulation framework. Under the structural and reservoir framework stage, the general structural design of the reservoir is shown. Depositional and geostatistical framework shows how facies and petrophysical information are distributed and the reservoir flow simulation framework deals with production and Pressure -Volume - Temperature (PVT) data. This paper has been carried out by integrating static data such as seismic, well logs, core description, borehole images, tectonic history, geological structures, among others (Azim, 2016). Then, the Sequential Indicator Simulation (SIS) technique for categorical variables is used to model facies or rock types, complexity of depositional history.

### 3.5 *Property modelling*

Properties modelling allow the distribution of properties with the available wells data. The process involves showing how the well data are matching to each other as well as the heterogeneity of the reservoir. Firstly, well data is scaled up with the intension to distribute the available data to nearby grid cells where the well penetrates. For far away cells, property modelling is done to show the distribution of a specific property to the whole reservoir grid cell. Secondly, data analysis is conducted using different variograms, according to the modelled property. Data analysis process is essentially used to check and interpret data based on their similarities and flow trend of the given well logs. For inferring these properties in areas where there are no data, semi-variograms are used to quantify their spatial dependence, as:

$$\gamma(L) = \frac{1}{2N(L)} \sum_{i=1}^{N(L)} [Z(X_1) - Z(X_i + L)]^2 \quad (1)$$

where $h$ is the distance between both locations, $x_i$ and $x_i + L$, and $N(L)$ is the number of pairs for $x_i$ and $x_i + L$.

Gamma Ray (GR) log is a measurement of natural radioactivity of the formation. The log normally reflects the shale contents in a sedimentary formation. This is the radioactive concentration of elements in clay and shales. A low level of radioactive elements indicates a sandstone formation unless radioactive contaminants such as volcanic ash or granite ash are present. Meanwhile, a high level of radioactive elements indicates a shale formation. The total gamma ray level is recorded and plotted in API units which vary from 0 to 150 API. The well log data are loaded to Schlumberger Petrel software for data processing. Reservoirs (sandstones) are identified and correlated across the loaded wells. Petrophysical properties such as porosity (($\Phi$)), water saturation ($S_w$), hydrocarbon saturation ($S_h$) and shale volume ($V_{sh}$) are determined to characterize each reservoir according to its ability to hold hydrocarbon.

### 3.6 *Determination of total porosity*

Porosity is one of the petrophysical parameters that determine the amount of fluids that the pore spaces within a rock could hold. It could be categorized as total or effective. Total porosity determines the viability of all the interconnected and isolated pore spaces for fluid accumulation, while effective porosity only accounts for the interconnected pore spaces. Density porosity ($\varphi_{den}$) can be determined using Eq. (2) as given by:

$$\varphi_{den} = \frac{\rho_{ma} - \rho_b}{\rho_{ma} - \rho_f} \tag{2}$$

where $\rho_{ma}$ is the matrix density, $\rho_b$ or RHOB is the bulk density and $\rho_f$ is the fluid density. To determine the $\varphi_{den}$, $\rho_{ma}$ and $\rho_f$ must be known. For estimation of total porosity ($\varphi_t$), $\rho_{ma}$ is replaced as particle density, which is assumed as 2.65 $gcm^{-3}$ for sandstone, while $\rho_f$ for oil, water and gas are 0.87, 1.00 and 0.65 $gcm^{-3}$, respectively. However, $\varphi_t$ can be determined by Eq. (3).

$$\varphi_t = \frac{(2.65 - \rho_b)}{(2.65 - 0.87)} \tag{3}$$

In Eq. (2), $\rho_f = 0.87$. This condition is valid if the two porosity (RHOB and neutron (PHIN)) curves for a hydrocarbon reservoir separate from each other and RHOB log value is less than PHIN log value, provided that the log matrix lithology is known. Determination of effective porosity ($\varphi_e$) is as shown in Eq. (4).

$$\varphi_e = \varphi_t \times (1 - V_{sh}) \tag{4}$$

### 3.7 *Using variogram in modelling*

Basically, data trends are identified in the data analysis process. Then property modelling processes are used to model the random variation away from that trend. A variogram under property modelling is used to describe variations of the property. Two points close together typically have similar property values than points far away from each other.

### 3.8 *Water saturation determination*

The evaluation of the amount of the hydrocarbons present in the reservoir is based on estimation of the volume of water present in the pore spaces. When a hydrocarbon is present in the core, the nonconductive hydrocarbon reduces the cross-sectional area and blocks flow-paths in the rock. This effect increases the resistivity of the rock. In an oil-wet system, resistivity will decrease at a great rate (with respect to brine saturation) than it does in a water-wet system. Water saturation ($S_w$) is the ratio of the volume of water in the pore space to the total volume of the pore spaces in a rock. By using Archie's relationship as presented in Eq. (5), $S_w$ could be determined.

$$S_w^n = \frac{R_w}{\varphi^m \times R_t} \tag{5}$$

where $n$ and $m$ are the saturation and cementation exponents, $n$ varies from 1.8 to 4.0, while $m$ varies from 1.7 to 3.0, but the default value for the two exponents is usually 2.0. $R_w$ is the formation water resistivity and $R_t$ is the true resistivity of the formation.

Hydrocarbon saturation ($S_h$) is the fraction of pore volume occupied by hydrocarbon. It can be estimated by Eq. (6)

$$S_h = 1 - S_w \tag{6}$$

To compute the $V_{sh}$, the GR index ($I_{GR}$) needs to be determined, as shown in Eq. (7). Shale contents are mostly present within the hydrocarbon-bearing sandstone reservoir. The presence of shale in the productive zone has a severe impact on the petrophysical properties and can cause a reduction in the effective and total porosity, as well as permeability. Moreover, it also poses problem in the interpretation of wireline well logs and can affect proper and effective estimation of hydrocarbon. Hence, its determination is crucial for solving the problem stated earlier. Various methods exist for shale volume estimation, such as gamma ray log, spontaneous potential log or porosity-neutron log. In this study, the gamma ray log technique is used for the shale volume estimation by first estimating the gamma ray index ($I_{GR}$) (Eq. 7).

$$I_{GR} = \frac{GR_{log} - GR_{min}}{GR_{max} - GR_{min}} \tag{7}$$

where $GR_{log}$ represents the gamma ray reading at the depth of interest, $GR_{min}$ and $GR_{max}$ represent the minimum and maximum gamma ray values of the clean and shale formation, respectively. $GR_{min}$ is the minimum GR (i.e., sand) and $GR_{max}$ is the maximum GR (i.e. shale).

$$V_{sh} = 0.83(2(3.7 \times I_{GR}) - 1.0) \tag{8}$$

The net-to-gross ratio is the ratio of the total thickness of the subsurface productive pay zone to that of the total thickness of the subsurface reservoir interval for a vertical well, given by:

$$Net - To - Gross\ ratio = 1 - V_{sh} \tag{9}$$

### 3.9 *Volumetric estimation technique*

The hydrocarbon volumes are estimated using the volumetric method. Firstly, the hydrocarbon pore volume is calculated using the reservoir simulation and modelling software prior to the computation of the original oil in place at surface condition by means of the formation volume factor. The petrophysical parameters such as porosity and water saturation, as well as net-to-gross ratio and reservoir thickness are used to estimate the reservoir fluid volume. These parameters are imputed into the following formula.

$$\text{STOIIP} = \frac{V_b \times \emptyset_{eff} \times NTG \times (1 - S_w)}{B_{oi}} \tag{10}$$

and

$$V_b(bbl) = 7758Ah \tag{11}$$

where $V_b$ = Bulk reservoir in volume, A = Reservoir cross-sectional area; acres (obtained from map data) and h = Reservoir thickness (pay zone; ft) (obtained from log data). $\phi$ = Formation porosity (fraction) (obtained from log data). NTG = Net-To-Gross ratio. $S_w$ = Formation water saturation (fraction) (obtained from log data) and $B_{oi}$ = Oil formation volume factor (1.2 bbls/stb).

## 4 RESULTS AND DISCUSSION

### 4.1 *Facies modelling*

Surfaces control facies and property modelling is a powerful way to control the modelling and ensure that the property distribution agrees with the geological conceptual model, (Figure 1).

Using surfaces to control facies and structure model (Figure 2) the properties are modelled in this representation to ensure that the property distribution agrees with your geological conceptual model.

Figure 2 a) shows the flow of Sand and conglomerate together with fine sand from North to South direction within a 3D Model. A depositional model that shows a Alluvial fan-Delta sedimentary system in the Xiaermen oilfield with trend from north to south direction is illustrated in Figure 2 b).

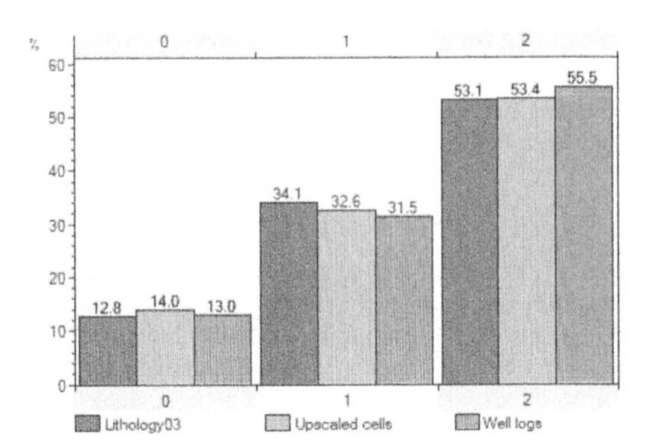


Figure 1. Facies Well logs and Upscale log graph.

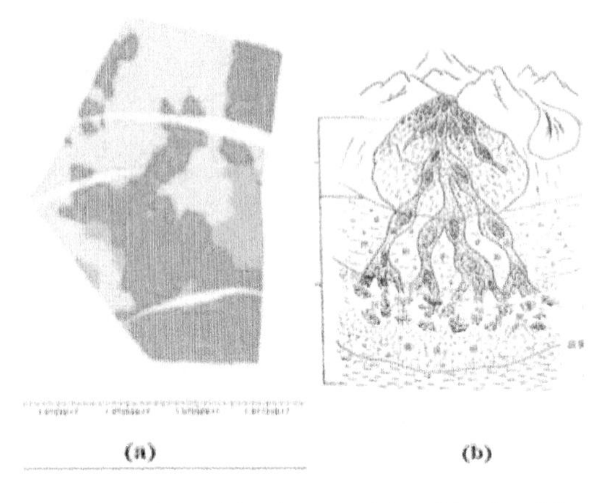


Figure 2. Facies model and proposed depositional model.

### 4.2 *Variogram analysis*

For each of the variables experimental variograms were constructed, but only lithofacies are presented here. Figure 3 shows the variation trend of facies modelling data in a) vertical b) major and c) minor directions. The vertical and major direction trends have a good fitting since there were a lot of well data. Minor trend fitting was not good compared to vertical direction since there were few well data.

### 4.3 *Porosity*

The porosity model (Figure 4) shows the distribution over the whole model with an average mean value of porosity of 17.3%. It shows clearly that there is a high porosity value. This implies that the pore spaces have enough space for reservoir fluid to be accommodated.

From lithology identification, a reservoir contains Sand and conglomerate together with fine sand, (Figure 5), hence it is possible for the reservoir to encounter high porosity distribution in some areas.

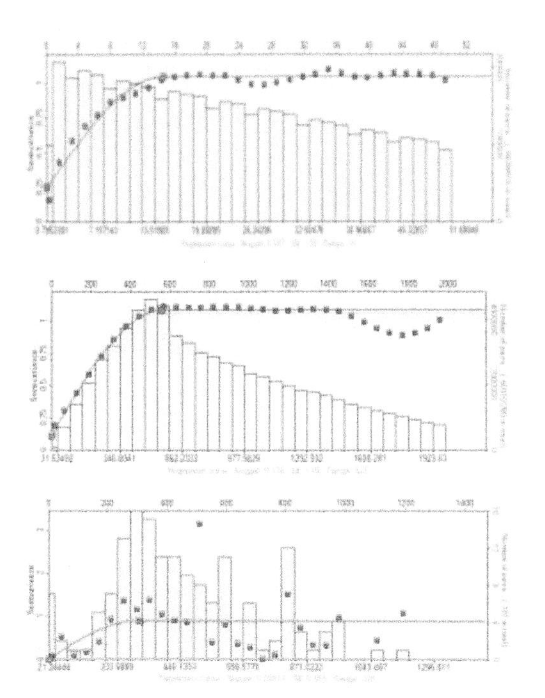

Figure 3. Experimental and model variograms of lithofacies in a) vertical b) major and c) minor directions.

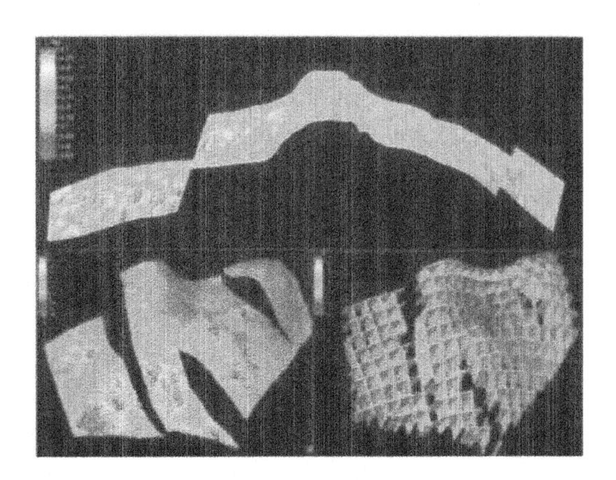

Figure 4. Porosity model.

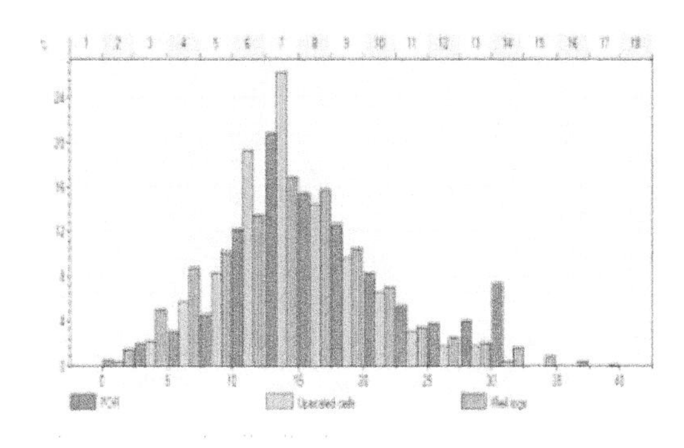


Figure 5. Porosity distribution.

### 4.4 *Permeability*

Permeability values range between of 1mD to 681mD, while the mean permeability value is 160 mD for the modelled H2V reservoir. The 3D geological model of permeability distribution results (Figures 6 and Figure 7) show that permeability concentration varies from North to south as it is the same way as in a lithology model flow, where the area with high permeability is the same where there is Sand and conglomerate together with fine sand facies rock type.

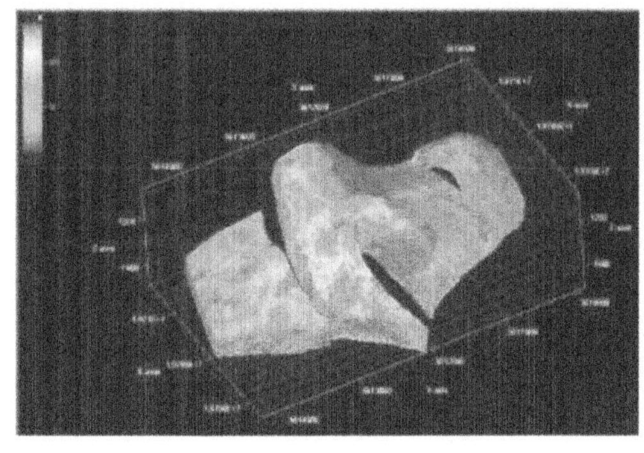

Figure 6. 3D perspective view of the permeability model.

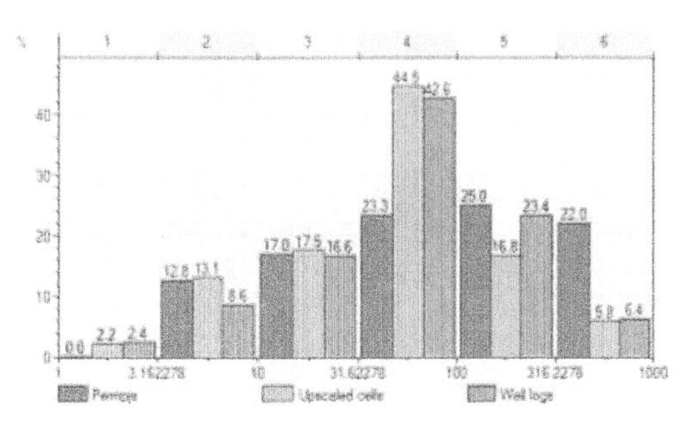


Figure 7. Permeability distribution.

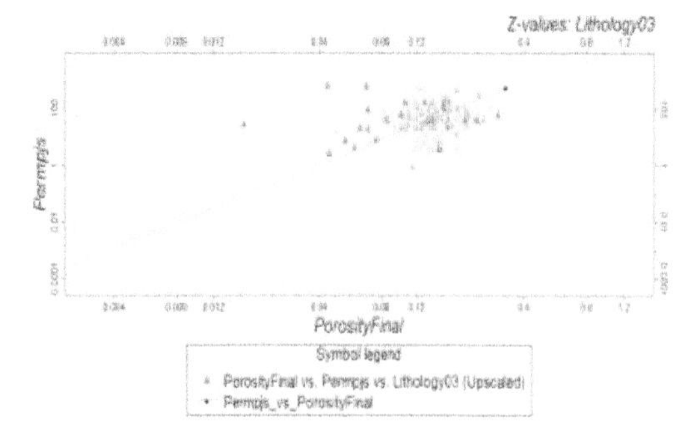


Figure 8. Relationship between porosity, permeability, and lithology.

### 4.5 *Relationship between porosity, permeability, and lithology*

Porosity and permeability are the most important physical properties of these reservoirs. Figure 8 shows the relationship between porosity, permeability, and lithology. It shows areas where there are high permeability and porosity. The areas where these properties have high concentration reflect the actual behaviour of a good reservoir.

For the best results these poroperm cross-plots should be constructed for clearly defined lithologies or reservoir zones. Figure 9 shows the porosity and permeability cross-plots in sandstone and carbonate reservoir.

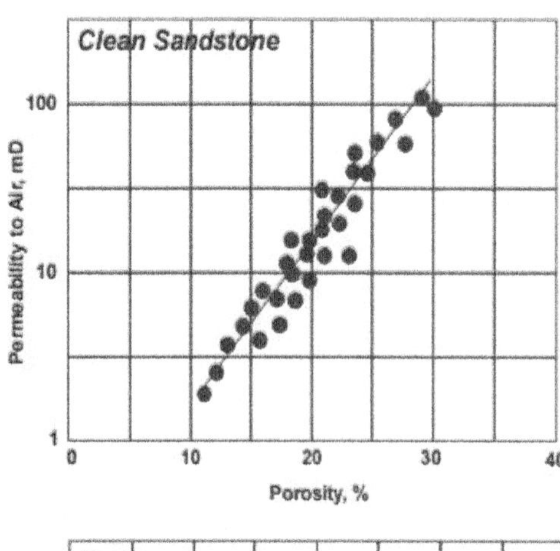


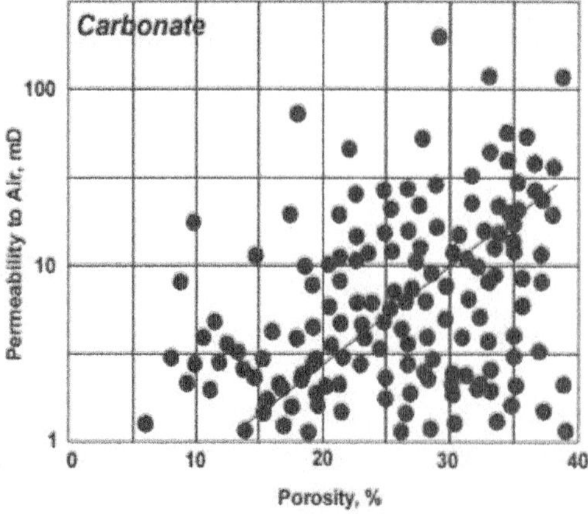


Figure 9. Porosity and permeability in sandstone and carbonate reservoir.

### 4.6 *Oil water contact model*

Water Oil contact was set using resistivity logs to differentiate water and hydrocarbon zones. The depth of the contact was different depending on the location of the well as demonstrated in Figure 10.

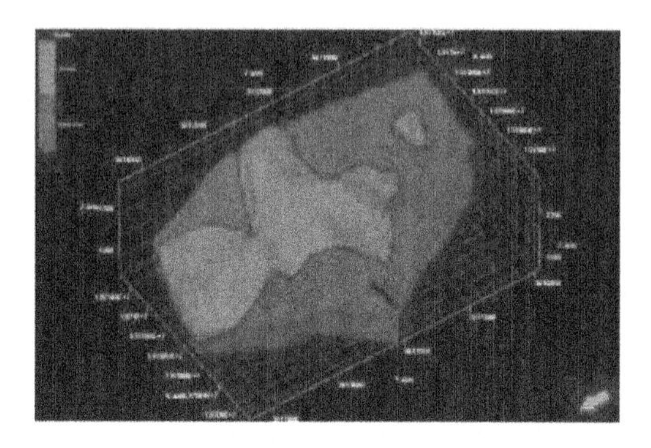

Figure 10. Oil water contact model.

### 4.7 *Net to Gross (NtG) property model*

The net-pay model quantifies a reservoir rock/productive rock and eliminate non-productive/non-reservoir rock, based on well data, as shown in the (Figure 11). The resulting 3D NTG model of a H2V reservoir provides much information on the quantity of the hydrocarbons-in-place. The model shows the most paying zone, which has great probability of accommodating hydrocarbons in a reservoir.

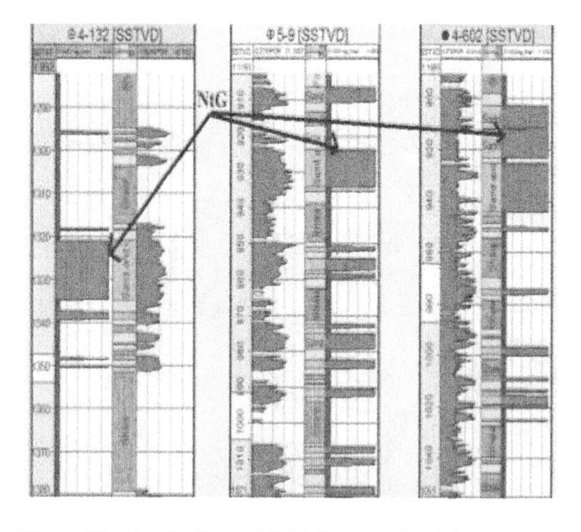


Figure 11. Net To Gross (NTG) Property Model.

### 4.8 *Water saturation model*

Figure 12 shows. The resulting constructed 3D model of Water saturation shows the distribution of water in the H2V formation of the Xiaermen oilfield. It has an average value of 0.5.

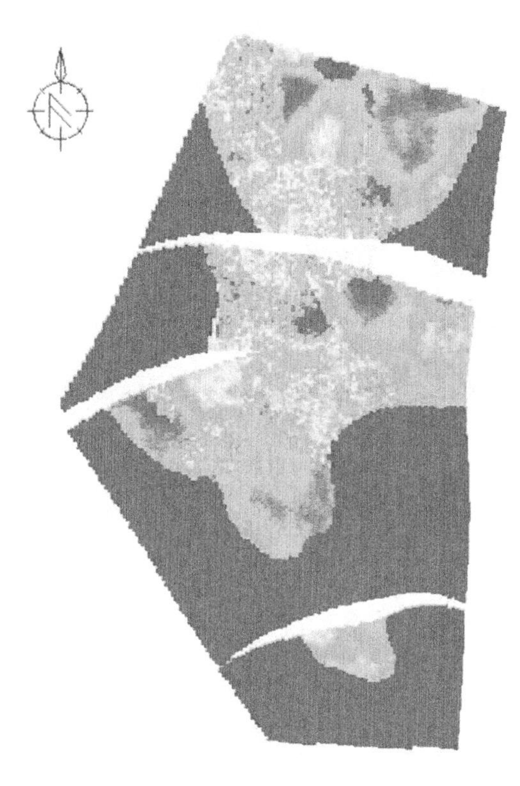

Figure 12. Water saturation distribution and oil distribution throughout the model.

The model is created using 16 well logs data as secondary interpretation data. Areas with high water saturation have a high probability of low content of oil and vice versa. From the model, the flow is high from north part to south of the model. The middle part and first part from north showed to have high concentration of oil.

### 4.9 *Reservoir volumetric results*

A geological model was created based on well data of the H2V formation in Xiaermen oilfield. Table 1 shows that the total bulk volume of reservoir was $901\times10^{6}$ $m^{3}$ in a model and the STOIIP for all 14 zones was $229.3\times10^{4}$ t. Finally, the reserve contained in the H2V formation is economical feasible and the model reserve results are relating to the primary/original geological reserve ($218\text{x}10^{4}$t). Thus, the reservoir model could be used for further production plan of a reservoir. It is ready to be used as input for simulation and optimization study of the reservoir, to solve different production challenges. Figure 13 shows the oil distribution map. There is highest amount of oil in the middle of area of a reservoir, the distribution is from northern part to southern part of the reservoir.

Table 1. Total reserve in H2V formation.

| BULK VOLUME [$10^{6}$ $m^{3}$] | NET VOLUME [$10^{6}$ $m^{3}$] | STOIIP [$10^{6}$ $sm^{3}$] | STOIIP [$10^{4}$tons] | PRIMARY STOIIP [$10^{4}$t] |
|---|---|---|---|---|
| 901 | 180 | 2.6291 | 229.3 | 218 |

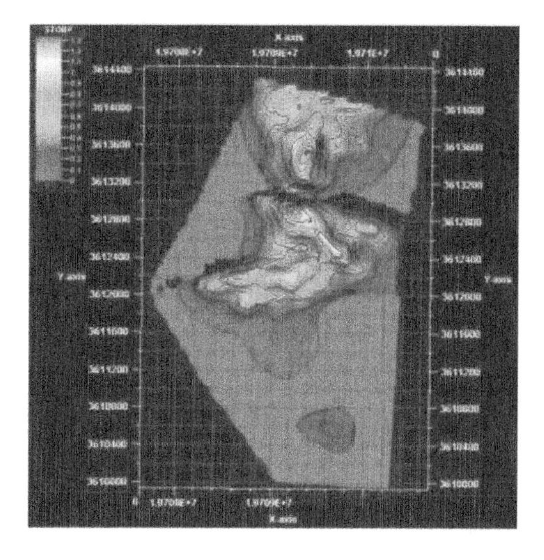

Figure 13. Oil distribution map.

## 5 CONCLUSIONS

A 3D reservoir model of the H2V formation of Xiaermen oilfield is developed. The results of the modelled facies and petrophysical distribution throughout the model are analysed from 40 well data. The created model showed its capability to calculate the trend of the properties in a defined depth. Consequently, the constructed 3D geological model of the H2V formation of Xiaermen oilfield reservoir is very promising with an average porosity value of 17.3%, while permeability model has an average value of 160mD. Finally, from the calculated volumetric parameter (STOIIP) of a reservoir fluid content of the H2V formation of the Xiaermen oilfield, the results are compared with primary geological reserves. The Xiaermen oilfield had an original geological reserve of $218\text{x}10^{4}$t while the calculated geological reserve obtained from the developed model is

$229x10^4$t. Hence, the reserve estimates showed a very close relation to the original/primary reserves. Therefore, the 3D geological model accuracy is adequate and recommended to be used for further field development plans.

## ACKNOWLEDGEMENTS

This work also contributes to the Strategic Research Plan of the Centre for Marine Technology and Ocean Engineering (CENTEC), which is financed by the Portuguese Foundation for Science and Technology under contract UIDB/UIDP/00134/2020, and Centre of Studies in Oil and Gas Engineering and Technology (CSOGET), Universidade Eduardo Mondlane – Mozambique.

## REFERENCES

Ahn, Y., & Choe, J. (2022). Reliable channel reservoir characterization and uncertainty quantification using variational autoencoder and ensemble smoother with multiple data assimilation. *Journal of Petroleum Science and Engineering*, *209*. https://doi.org/10.1016/j.petrol.2021.109816.

Azim, R. A. (2016). Integration of static and dynamic reservoir data to optimize the generation of subsurface fracture map. *Journal of Petroleum Exploration and Production Technology*, *6*(4), 691–703. https://doi.org/10.1007/s13202-015-0220-8.

Bjørlykke, K. (2015). *Petroleum Geoscience: From Sedimentary Environments to Rock Physics*. 2nd Edition. Springer-Verlag Berlin Heidelberg, https://pt1lib.org/book/2571001/1a9d5b.

Islam, A. R. M. T. (2013). Interpretation of wireline log data for reservoir characterization of the Rashidpur Gas Field, Bengal Basin, Bangladesh. *IOSR Journal of Applied Geology and Geophysics*, *1*(4), 47–54. https://doi.org/10.9790/0990-0144754.

Liu, X., Ge, Q., Chen, X., Li, J., & Chen, Y. (2021). Extreme learning machine for multivariate reservoir characterization. *Journal of Petroleum Science and Engineering*, *205*. https://doi.org/10.1016/j.petrol.2021.108869.

Mohaghegh, S., Arefi, R., Ameri, S., Aminiand, K., & Nutter, R. (1996). Petroleum reservoir characterization with the aid of artificial neural networks. *Journal of Petroleum Science and Engineering*, *16*(4), 263–274. https://doi.org/10.1016/S0920-4105(96)00028-9.

Normando, M. N., do Nascimento Junior, D. R., de Souza, A. C. B., Leopoldino Oliveira, K. M., Filho, F. N., da Silva Barbosa, T. H., Salgueiro, A. R. G. N. L., & Maia de Almeida, N. (2022). A proposal for reservoir geostatistical modeling and uncertainty analysis of the Curimã Field, Mundaú Sub-Basin, Ceará Basin, Brazil. *Journal of South American Earth Sciences*, *114*, 103716. https://doi.org/10.1016/j.jsames.2022.103716

Zelenika, K.N., Vidaček, R., Ilijaš, T., & Pavić, P. (2017). Application of deterministic and stochastic geostatistical methods in petrophysical modelling – a case study of upper pannonian reservoir in sava depression. *Geologia Croatica*, *70*(2), 105–114. https://doi.org/10.4154/gc.2017.10.

*Trends in Maritime Technology and Engineering – Guedes Soares & Santos (eds)*



# A stochastic programming model for designing offshore production systems

L.M.R. Silva & C. Guedes Soares
*Centre for Marine Technology and Ocean Engineering (CENTEC), Instituto Superior Técnico, Universidade de Lisboa, Lisbon, Portugal*


ABSTRACT: A two-stage stochastic model is presented aiming to enhance the decision process of designing a production system. The optimization methodology integrates the production planning with the decisions of surface and subsea systems, screening out the initial capital expenditures of the main facilities and the equipment, the project revenue, and the system concept. The probabilistic knowledge of three scenarios of oil production allows to change the nature of the previous discrete model to an uncertainty decision-dependent model. The efficiency of the proposed stochastic mixed integer linear programming model is demonstrated using a realistic case study and a validation process.


## 1 INTRODUCTION

The production system of an offshore oilfield is responsible for the hydrocarbon drainage from the reservoir to the seabed; the collection and transportation of hydrocarbons from the wellhead to the surface facility (platform or central collection point); and the pretreatment of fluids for storage and transfer to the refinery. The main strategic decisions to be taken when designing an offshore production system are related to the proper subsea arrangement of the equipment and components, the network of pipelines and the intended surface capacity and the location that drives the facility design.

Most of the works presented in the literature treat these decisions as independent research topics (Devine & Lesso, 1972; Wang et al., 2014, 2012; Zhang et al., 2017). Moreover, few approaches combine the variety of decision within a single model (Iyer & Grossmann, 1998; Rodrigues et al., 2016; Silva & Guedes Soares, 2019).

Meanwhile, the high degree of uncertainty is an important factor during the early phases of an oilfield development that need to be considered into the decision process. The information and data related to the reservoir characteristics and properties are scarce impacting on the accuracy of the assessment of the reservoir production during the oilfield life cycle.

The earlier approach of Haugen (1996) adopted a general production performance to represent the production profile attempting to investigate the investment planning decision of pipeline system under production uncertainty. The uncertainty is inserted into the model as probability density associated to each phase of the production life. Gupta & Grossmann (2014) developed a decision approach that considers the decisions of surface and pipeline systems dependent of oilfield size, oil deliverability, water-oil ratio, and gas-oil ratio uncertainties. However, these research works do not integrate all the main decisions of a production system.

Silva & Guedes Soares (2021a) proposed a deterministic model and uncertainty methodology to investigate the impact of the production uncertainty into the offshore production system. The model considers the oil, gas, and water productions during the production life as dynamic metrics into the optimization model that defines the surface and subsea production system.

The present model converts the deterministic model provided by Silva & Guedes Soares (2021a) into a stochastic model. The main advantage is that instead of formulating the problem with known expected values, the stochastic formulation considers a finite number of possible scenarios with respective probabilities. Therefore, the new model aims to enhance the decision process of designing a production system. The model attempts to define the most feasible production system that maximize the project revenue and simultaneously minimize the initial capital expenditures of the facilities and equipment.

## 2 PRODUCTION UNCERTAINTY METHODOLOGY

To perform the estimation of the oil production, the production decline curve (PDCA) is adopted as the analytical method used to establish the production curve of the well Moreover, to assess the production uncertainty, a sample of data of production history needs to be collected. Thus, the database used to define the statistical relationship necessary for this

DOI: 10.1201/9781003320289-57

work is the same as the one presented by Silva & Guedes Soares (2021a, 2021b).

First, a short period of each production system is used to optimize the PDCA parameters. The optimization approach searches for the values of the parameters that minimize the error between the predicted data and the observed data. Afterwards, the optimized parameters are used to forecast the production curve that best fits the entire production history. As a final step, the comparison between the estimated production curve and the production history allows evaluating the mean absolute percentage error (MAPE) point-to-point. As a result of the uncertainty assessment, beyond the predicted scenario, two possible production scenarios were established: optimistic and pessimistic. See Silva & Guedes Soares (2021a) for a more detailed description and discussion regarding the methodology.

Now, extending the described methodology, the evaluation of the probability of occurrence of each production scenario can be performed by comparing the final real production and the final estimated production. This comparison allows identifying when the real production is under, above or equal to the estimated production. Therefore, this procedure allows the experts to not only elicit opinions on the future but to define the probabilities based on data.

## 3 PRODUCTION SYSTEM OPTIMIZATION

The production system optimization integrates the production planning under uncertainty, the investment planning and the facility location-allocation problem taking into consideration the subsea and surface production design. The stochastic mixed-integer linear programming model (SMILP) accomplish the need to use continuous and integer variables, assuming values and decisions subject to linear equations, and the need to capture the uncertainty.

The optimization problem is modelled as two-stage stochastic programming. The first stage represents the decisions that are not directly affected by the occurrence of a random event, and the decisions that are dependent on the uncertain data represent the second stage. The main idea of the two-stage stochastic programming model is to take the first stage of decisions considering the decision costs of the second stage that need to be made later. The discrete MILP model provided by Silva & Guedes Soares (2021a) concluded that due to the structure of the problem only the surface system is impacted by the production uncertainty. Therefore, the first stage accomplished the subsea production decisions. Thus, the second stage regards the decisions of the surface production system.

The main decisions taken at the first stage are the number and location of the surface platforms that minimize the cost of the pipelines connecting the wells to the platforms; the number, size, and location of the manifolds to be installed when the model decides to connect the wells by a manifold to the platforms minimizing the pipeline cost; and the proper assignment of each well, directly to the platform or by a manifold. Moreover, due to the insertion of the reservoir dynamic metrics to represent the behaviour of the fluids (oil, gas, and water) into the model, the oil, gas, and water flowrates at the manifold for each production time are also defined.

Thus, the second stage regards the surface production system decisions and the estimation of the project revenue. The project defines the necessary platform capacity to store the total production, and the cumulative production of oil, gas and water received at the installed platform. The total oil production allows the estimation of the project revenue. The second stage data is modelled as a random vector with a finite number of production scenarios with respective probabilities.

### 3.1 *Model assumptions*

In a mathematical model, a set of approximations and assumptions about the way the system will work are necessary to reduce the size of the combinatorial problem allowing the model to be evaluated numerically. The main model assumptions are summarized below:

1. The number, flowrate, and location of the wells in x-y axis are pre-defined data.
2. The separated geographic location of the wells allows assuming that the model can handle one or more reservoirs.
3. The total oil production received at each installed platform cannot exceed the standardized maximum capacity.
4. The maximum number of the wells that can be connected to each manifold is pre-defined data.
5. The installation of the pipelines and manifolds are not impacted by topographic limitations.
6. The pipeline routes are considered as straight lines and, to avoid the crossing effect of the pipelines, pipe supports can be installed.

### 3.2 *Conceptual model*

The problem of optimizing the offshore production system is formulated in the next sub-section, but, schematically, it can be stated as it follows below:

- *Given* the number and location of the wells; the reservoir volume-in place of hydrocarbons; the minimum flowrate that each manifold can receive; the constant price of barrel price; the length of the risers; the manifolds slot sizes; the platform capacities; the cost of pipelines; the cost of each size of the platform and manifold slots; the oil production of each well during the reservoir production time; the gas production of each well during the reservoir production time and the water production of each well during the reservoir production time.

- *Obtain* the distance between the touchdown point of the platforms and the wells; the distance between the manifolds and the wells; the distance between the touchdown point of the platforms and the manifolds; and the probability of occurrence of each production scenario.
- *Subject to* well assignment constraints; manifold installation and assignment constraints; flowrate conservation constraints; cumulative production constraints and platform installation, assignment and received production constraints.

The solution provides the optimal technology selection, system elements, subsea layout concept, surface design, initial capital expenditures of facilities and equipment, and the project revenue that maximizes the project profit.

### 3.3 *Mathematical model*

The sets, parameters and variables following the notation presented in Tables 1– 3. The objective function aims at maximizing the tradeoff between the second and first stage of decisions representing the profit of the project, following:

$$\text{maximize } \{SecondStage - FirstStage\} \quad (1)$$

being:

$$\text{SecondStage}: \{Revenue - CostOfPlatform\} \quad (2)$$

$$\text{FirstStage}: \{CostOfPipelines + CostOfRisers + CostOfManifolds\} \quad (3)$$

where:

$$\text{Revenue}: \left\{\sum_c to_c \times BP \times PROB_C\right\} \quad (4)$$

$$\text{CostOfPlatform}: \left\{\sum_P \sum_S PLATCOST_{PS} \times y_{ps}\right\} \quad (5)$$

$$\text{CostOfPipelines}: \left\{\sum_p \sum_w DIST_{mw} \times x_{pw} + \times PIPECOST\left(\sum_m \sum_w DIST_{mw} \times x_{mw} + \sum_p \sum_m DIST_{pm} \times x_{pm}\right) \times PIPECOST\right\} \quad (6)$$

$$\text{CostOfRisers}: \left\{\sum_p \sum_w RISERLENGTH \times x_{PW} \times PIPECOST + \sum_p \sum_m RISERLENGTH \times x_{Pm} \times PIPECOST\right\} \quad (7)$$

$$\text{CostOfManifolds}: \left\{\sum_p \sum_m y_m \times MCOST_m\right\} \quad (8)$$

Table 1. Sets index.

| Index | Description |
|---|---|
| w ϵ W | Set of wells |
| m ϵ M | Set of possible manifolds |
| p ϵ P | Set of possible platforms |
| s ϵ S | Set of size of platforms |
| t ϵ T | Set of time |
| r ϵ R | Set of reservoirs |
| c ϵ C | Set of production scenarios |

Table 2. Parameters.

| Parameters | Description |
|---|---|
| RISERLENGTH | length of risers; |
| MAXWELLS | maximum number of wells; |
| $OHIP_r$ | volume of original hydrocarbon-in-place of each reservoir; |
| QMAX | maximum flowrate that manifolds can receive from the wells; |
| BP | a constant price of oil barrel; |
| PIPECOST | cost of pipelines; |
| $PLATCOST_{ps}$ | cost of platform *p* of size *s*; |
| $MCOST_m$ | cost of manifold *m* according to the slot capacity; |
| $DIST_{pw}$ | distance between the touchdown point of platform *p* and well *w*; |
| $DIST_{mw}$ | distance between manifold *m* and well *w*; |
| $DIST_{pm}$ | distance between the touchdown point of the platform *p* and manifold *m*; |
| $SLOT_m$ | manifold slot size; |
| $CAPACITY_{ps}$ | maximum platform capacity size; |
| $QO_{wtc}$ | oil flowrate of well *w* of time *t* in the scenario *c*; |
| $QG_{wtc}$ | gas flowrate of well *w* of time *t* in the scenario *c*; |
| $QW_{wtc}$ | water flowrate of well *w* of time *t* in the scenario *c*; |
| $PROB_c$ | Probability of occurrence of production scenario *c*; |

Eqs. (9)-(11) must guarantee that the wells cannot be simultaneously assigned to both platform and manifold:

$$\sum_{p} x_{pw} \leq 1 \forall w \in W \tag{9}$$

$$\sum_{m} x_{mw} \leq 1 \forall w \in W \tag{10}$$

Table 3. Variables.

| Variables | Description |
|---|---|
| $y_{ps} \in \{0,1\}$ | assumes the value 1 if the platform *p* of size *s* is installed; |
| $x_{pw} \in \{0,1\}$ | assumes the value 1 if the well *w* is assigned directly to the touchdown point of the platform *p*; |
| $x_{mw} \in \{0,1\}$ | assumes the value 1 if the well *w* is assigned to the manifold *m*; |
| $x_{pm} \in \{0,1\}$ | assumes the value 1 if the manifold *m* is assigned to the touchdown point of the platform *p*; |
| $y_{m} \in \{0,1\}$ | assumes the value 1 if the manifold *m* is installed; |
| $qo_{mtc} \in IR^{+}$ | oil flowrate received at the manifold *m* at time *t* in the scenario *c*; |
| $qo_{ptc} \in IR^{+}$ | oil flowrate from well *w* received at the platform *p* at time *t* in the scenario *c*; |
| $qot_{ptc} \in IR^{+}$ | total oil flowrate received at platform *p* at time *t* in the scenario *c*; |
| $zo_{pmc} \in IR^{+}$ | auxiliary variable used for the linearization of the oil flowrate from the manifold *m* to the platform *p* if the manifold *m* is assigned to the platform *p* at time *t* in the scenario *c*; |
| $no_{tc} \in IR^{+}$ | cumulative oil production at time *t* in the scenario *c*; |
| $to_{c} \in IR^{+}$ | auxiliary variable adopted to capture the total oil production of scenario *c*; |
| $qg_{mtc} \in IR^{+}$ | gas flowrate received at the manifold *m* at time *t* in the scenario *c*; |
| $qg_{ptc} \in IR^{+}$ | gas flowrate received at the platform *p* at time *t* in the scenario *c*; |
| $qgt_{ptc} \in IR^{+}$ | total gas flowrate received at platform *p* at time *t* in the scenario *c*; |
| $zg_{pmc} \in IR^{+}$ | auxiliary variable used for the linearization of the gas flowrate; |
| $ng_{tc} \in IR^{+}$ | cumulative gas production at time *t* in the scenario *c*; |
| $qw_{mtc} \in IR^{+}$ | water flowrate received at the manifold *m* at time *t* in the scenario *c*; |
| $qw_{ptc} \in IR^{+}$ | water flowrate received at the platform *p* at time *t* in the scenario *c*; |
| $qwt_{ptc} \in IR^{+}$ | total water flowrate received at platform *p* at time *t* in the scenario *c*; |
| $zw_{pmc} \in IR^{+}$ | auxiliary variable used for the linearization of the water flowrate; |
| $nw_{tc} \in IR^{+}$ | cumulative water production at time *t* in the scenario *c*; |

$$\sum_{p} x_{pw} + \sum_{m} x_{mw} = 1 \forall w \in W \tag{11}$$

Eq. (12) guarantees that the total number of wells assigned to the facilities cannot exceed the maximum number of wells:

$$\sum_{p}\sum_{w} x_{pw} + \sum_{m}\sum_{w} x_{mw} = MAXWELLS \tag{12}$$

A manifold must be installed once (Eq. (*13*)), and whether the manifold is installed it must be assigned to one platform (Eq. (*14*)). Thus, the Eq. (15) guarantees that the installed manifold can only be connected to one platform.

$$\sum_{m} y_{m} \leq 1 \tag{13}$$

$$y_{m} \leq \sum_{p} x_{pm} \forall m \in M \tag{14}$$

$$\sum_{p} x_{pm} \leq 1 \forall m \in M \tag{15}$$

Eq. (16) assures that the maximum number of slots of each manifold cannot be exceeded:

$$\sum_{w} x_{mw} \leq SLOT_{m} \times y_{m} \in M \tag{16}$$

Eq. (17) changes the even number of manifold slots to an odd number to give flexibility to the design layout sparing one slot for future tieback wells. It is important to note that Eq. (17) is a substitute for Eq. (16):

$$\sum_{w} x_{mw} \leq (SLOT_{m} - 1) \times y_{m} \in M \tag{17}$$

The summation of the production received cannot exceed the volume of original hydrocarbon-in-place of each reservoir:

$$\sum_{p}\sum_{w}\sum_{t}\sum_{c} x_{pw} \times (QO_{wtc} \times QG_{wtc}) + \sum_{m}\sum_{w}\sum_{t}\sum_{c} x_{mw} \times (QO_{wtc} \times QG_{wtc}) \leq \sum_{r} OHIP_{r} \tag{18}$$

Eq.(19) assures the directly flowrate from the wells to the manifold, for all scenarios:

$$qo_{mtc} = \sum_{w} x_{mw} \times QO_{wtc} \forall \mathrm{m} \in \mathrm{M}, \mathrm{t} \in \mathrm{T}, \mathrm{c} \in \mathrm{C} \quad (19)$$

Eqs. (20)-(22) represent the group of constraints necessary to perform the linearization of the flowrate from the manifold to the platform, for all scenarios:

$$-qo_{mtc} + zo_{pmtc} \leq 0 \forall \mathrm{m} \in \mathrm{M}, \mathrm{p} \in \mathrm{P}, \mathrm{t} \in \mathrm{T}, \mathrm{c} \in \mathrm{C} \quad (20)$$

$$-(x_{pm} \times QMAX) + zo_{pmtc} \leq 0 \forall \mathrm{m} \in \mathrm{M}, \mathrm{p} \in \mathrm{P}, \mathrm{t} \in \mathrm{T}, \mathrm{c} \in \mathrm{C} \mathrm{c} \in \mathrm{C} \quad (21)$$

$$qo_{mtc} + (x_{pm} \times QMAX) - zo_{pmtc} \leq QMAX \forall \mathrm{m} \in \mathrm{M}, \mathrm{p} \in \mathrm{P}, \mathrm{t} \in \mathrm{T}, \mathrm{c} \in \mathrm{C} \quad (22)$$

The received gas flowrate at the manifold, for all scenarios, and the linearization constraints, are represented by Eqs. (23)-(26):

$$qg_{mtc} = \sum_{w} x_{mw} \times QG_{wtc} \forall \mathrm{m} \in \mathrm{M}, \mathrm{t} \in \mathrm{T}, \mathrm{c} \in \mathrm{C} \quad (23)$$

$$-qg_{mtc} + zg_{pmtc} \leq 0 \forall \mathrm{m} \in \mathrm{M}, \mathrm{p} \in \mathrm{P}, \mathrm{t} \in \mathrm{T}, \mathrm{c} \in \mathrm{C} \quad (24)$$

$$-(x_{pm} \times QMAX) + zg_{pmtc} \leq 0 \forall \mathrm{m} \in \mathrm{M}, \mathrm{p} \in \mathrm{P}, \mathrm{t} \in \mathrm{T}, \mathrm{c} \in \mathrm{C} \quad (25)$$

$$qg_{mtc} + (x_{pm} \times QMAX) - zg_{pmtc} \leq QMAX \forall \mathrm{m} \in \mathrm{M}, \mathrm{p} \in \mathrm{P}, \mathrm{t} \in \mathrm{T}, \mathrm{c} \in \mathrm{C} \quad (26)$$

The received water flowrate at the manifold, for all scenarios, and the linearization constraints, are represented by Eqs. (27)-(30):

$$qw_{mtc} = \sum_{w} x_{mw} \times QW_{wtc} \forall \mathrm{m} \in \mathrm{M}, \mathrm{t} \in \mathrm{T}, \mathrm{c} \in \mathrm{C} \quad (27)$$

$$-qw_{mtc} + zw_{pmtc} \leq 0 \forall \mathrm{m} \in \mathrm{M}, \mathrm{p} \in \mathrm{P}, \mathrm{t} \in \mathrm{T}, \mathrm{c} \in \mathrm{C} \quad (28)$$

$$-(x_{pm} \times QMAX) + zw_{pmtc} \leq 0 \forall \mathrm{m} \in \mathrm{M}, \mathrm{p} \in \mathrm{P}, \mathrm{t} \in \mathrm{T}, \mathrm{c} \in \mathrm{C} \quad (29)$$

$$qw_{mtc} + (x_{pm} \times QMAX) - zw_{pmtc} \leq QMAX \forall \mathrm{m} \in \mathrm{M}, \mathrm{p} \in \mathrm{P}, \mathrm{t} \in \mathrm{T}, \mathrm{c} \in \mathrm{C} \quad (30)$$

Eq. (31) assures the direct oil flow rate from the wells to the platform, for all scenarios. Thus, Eq.(32) guarantees that the total oil flowrate received at the platform, for all scenarios, is the summation of the direct flowrate from the wells and the flowrate from the manifolds:

$$qo_{ptc} = \sum_{w \in W} x_{pw} \times (QO_{wtc}) \forall p \in P, t \in T, \mathrm{c} \in \mathrm{C} \quad (31)$$

$$qot_{ptc} = \sum_{m} zo_{pmtc} + qo_{ptc} \forall p \in P, t \in T, \mathrm{c} \in \mathrm{C} \quad (32)$$

Eqs. (33)-(34) represent the direct and the total gas flowrate received at the platform, for all scenarios:

$$qg_{ptc} = \sum_{w} x_{pw} \times (QG_{wtc}) \forall p \in P, t \in T, \mathrm{c} \in \mathrm{C} \quad (33)$$

$$qgt_{ptc} = \sum_{m} zg_{pmtc} + qg_{ptc} \forall p \in P, t \in T, \mathrm{c} \in \mathrm{C} \quad (34)$$

Eqs. (35)-(36) represent the direct and the total water flowrate received at the platform, for all scenarios:

$$qw_{ptc} = \sum_{w} x_{pw} \times (QW_{wtc}) \forall p \in P, t \in T, \mathrm{c} \in \mathrm{C} \quad (35)$$

$$qwt_{ptc} = \sum_{m} zw_{pmtc} + qw_{ptc} \forall p \in P, t \in T, \mathrm{c} \in \mathrm{C} \quad (36)$$

Eqs. (37)-(39) assure the cumulative production of oil, gas, and water, for all scenarios:

$$no_{tc} = \sum_{p} qot_{ptc} \forall p \in P, t \in T, \mathrm{c} \in \mathrm{C} \quad (37)$$

$$ng_{tc} = \sum_{p} qgt_{ptc} \forall p \in P, t \in T, \mathrm{c} \in \mathrm{C} \quad (38)$$

$$nw_{tc} = \sum_{p} qwt_{ptc} \forall p \in P, t \in T, \mathrm{c} \in \mathrm{C} \quad (39)$$

The oil production is input data. However, the total oil production is adopted in this model as an auxiliary variable to guarantee that the flowrate balance constraints are satisfied in terms of volume conservation. The total oil production received represents the summation of each well flowrate during time $t$, for all scenarios $c$ Eq.(40):

$$to_{c} = \sum_{t} no_{tc} \forall \mathrm{c} \in \mathrm{C} \quad (40)$$

A platform must be installed once (Eq. (41)). Whether a platform is installed the total number of assignments to each platform cannot exceed the maximum number of wells (Eq.(42)). Thus, the total flow rate received at each installed platform cannot exceed the platform capacity (Eq. (*43*)).

$$\sum_{s} y_{ps} \leq 1 \forall p \in P \quad (41)$$

$$\sum_{w} x_{pw} + \sum_{m} x_{pm} \leq MAXWELLS \times \sum_{s} y_{ps} \forall p \in P \quad (42)$$

$$\sum_{C} qot_{ptc} \times PROB_C \leq \sum_{s} CAPACITY_{ps} \times y_{ps} \forall p \in P, t \in T, c \in C \quad (43)$$

## 4 CASE STUDY

A hypothetical offshore oilfield based on the ones located in Campos's basin-Brazil is presented in this section, which is similar to the one presented by Silva & Guedes Soares (2021a). The estimated volume of the original hydrocarbon-in-place is 896,170,000 bbl and the pre-defined number of wells is 10. The well clustering division is evaluated for manifolds with 2 and 4 slots. Moreover, only one possible platform location is considered. There are three possible platform capacity, 40,000, 50,000 and 70,000 bbl/ d. The location of the wells and the facilities is presented in Figure 1.

The mean water depth of the oilfield location is 95m and the respective riser length is 114m (1.2 times the value of the mean water depth was used for safety of all risers). The gross cost estimations of each floating and subsea infrastructure are presented in Table 4.

Table 4. Cost estimates for the production system components.

| Production System Components | Costs | Unit |
|---|---|---|
| Platform Cost (40,000 bbl/d) | 50 $\times 10^{+06}$ | US$ per unit |
| Platform Cost (50,000 bbl/d) | 65 $\times 10^{+06}$ | US$ per unit |
| Platform Cost (70,000 bbl/d) | 80 $\times 10^{+06}$ | US$ per unit |
| Manifold Cost (2 Slots) | 1,500.00 | US$ per unit |
| Manifold Cost (4 Slots) | 4,000.00 | US$ per unit |
| Pipeline Cost | 270.00 | US$ per meter |

As a result of the uncertainty assessment methodology presented by Silva & Guedes Soares (2021a), three production scenarios are established: predicted, optimistic, and pessimistic production curves. Attempting to reduce the computational effort required to solve the problem, only 12 years of production of each decline curve are used as production data. Figure 2 presents the three estimated decline curves.

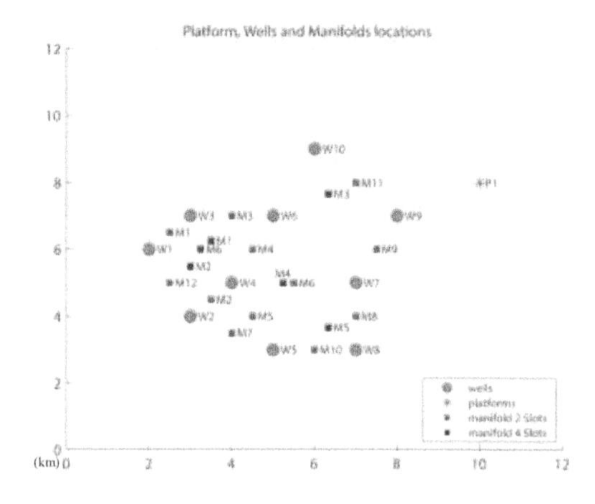


Figure 1. Spatial locations of the wells and the facilities.

The probability of occurrence of each production scenario is established based on Campos's database. The comparison of the final real production and the estimated production of 25 oilfields allows the evaluation of each probability. The production history of 11 oilfields is similar to the estimated production curve; 9 oilfields' production histories are above the estimated production curve, and 5 oilfields' production histories are under the estimated production. As a result, the probability of the predicted scenario is 0.44, the probability of the optimistic scenario is 0.36 and the probability of the pessimistic scenario is 0.2.

The problem was implemented at the IBM ILOG CPLEX Studio Optimization environment and solved through the branch-and-bound technique on a computer with Intel(R) Core (TM) i7 of 2.30 GHz and 16GB RAM.

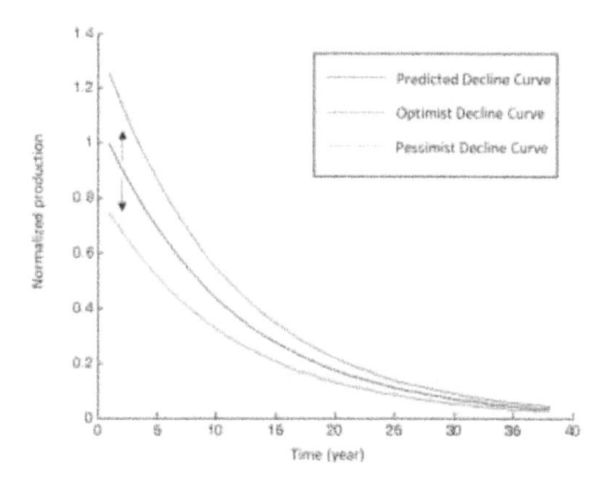


Figure 2. The estimated predicted, optimistic and pessimistic production decline curves.

Table 5. The stochastic solution and the deterministic solutions of the predicted, optimistic, and pessimistic scenarios.

| | Stochastic Two Stages P (0.44, 0.36, 0.2) | Deterministic solution - predicted scenario | Deterministic solution - optimistic scenario | Deterministic solution - pessimistic scenario |
|---|---|---|---|---|
| Obj. function | 1.69E+07 | 1.58E+07 | 2.25E+07 | 9.10E+06 |
| Revenue | 2.75E+07 | 2.64E+07 | 3.31E+07 | 1.96E+07 |
| Total Costs | 1.06E+07 | 1.05E+07 | 1.06E+07 | 1.05E+07 |
| Plat. Capacity | 70,000.00 (bbl/d) | 50,000.00 (bbl/d) | 70,000.00 (bbl/d) | 40,000.00 (bbl/d) |
| Subsea layout | 2 manifolds of 4 slots clustering 8 wells and 2 wells directly assigned to Platform P1 | 2 manifolds of 4 slots clustering 8 wells and 2 wells directly assigned to Platform P1 | 2 manifolds of 4 slots clustering 8 wells and 2 wells directly assigned to Platform P1 | 2 manifolds of 4 slots clustering 8 wells and 2 wells directly assigned to Platform P1 |

## 4.1 *Validation process*

In stochastic programming, it is a good common sense to examine the quality of the solution provided by the model, comparing the stochastic solution and the expectation of the deterministic solution, assessing the value of the stochastic solution (VSS).

Birge & Louveaux(2011), determines that the VSS is the difference of the expectation of the expected value problem (EEV) and the respective stochastic solution, Eq. (44). Thus, the EEV is the summation of the deterministic solution of each scenario multiplied by its probability of occurrence, Eq.(45).

$$VSS = EEV - RP \tag{44}$$

$$EEV = p_1(ExpOptValue_{sc1}) + p_2(ExpOptValue_{sc2}) + p_3(ExpOptValue_{sc3}) \tag{45}$$

where, "p" represents the probability of each scenario, "ExpOptValue" means the expected optimal value of each deterministic scenario, and "sc" the respective scenario.

## 4.2 *Results and discussions*

After the explanation of the indicator of the quality of the stochastic solution, the new SMILP model is allowed to solve the case study described in section 4, adopting the estimated probabilities of occurrence of the three production scenarios. Table 5 summarizes the main features of the solutions provided by the stochastic model and the deterministic model evaluated for each possible production scenario (predicted, optimistic and pessimistic). For a better presentation of the results allowing the comparison between solutions, the stochastic result of the platform cost is added to the total cost of the manifolds, the pipelines, and the risers.

At first glance, the comparison of the stochastic solution and the discrete solutions of the deterministic model provided by Silva & Guedes Soares (2021a), identifies that the subsea layout is the same for all solutions of the production system (Figure 3). Moreover, the total EEV and the EEV for each scenario are presented in Table 6.

Adopting the approximate deterministic strategy (EEV), it is possible to assess the value of the stochastic solution (VSS), where, for the present case, is zero. Meaning that the gain in considering the stochastic problem instead of the EEV problem is negligible and, therefore, the approximate expected value strategy can be used.

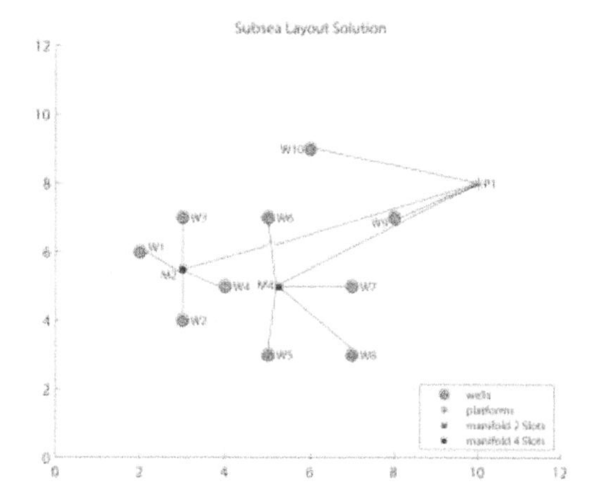


Figure 3. Subsea layout design, for all the solutions.

Table 6. Expectation of the expected value problem.

| Scenarios | EEV |
|---|---|
| Predicted | $6.95 \times 10^{+06}$ |
| Optimist | $8.10 \times 10^{+06}$ |
| Pessimist | $1.82 \times 10^{+06}$ |
| Total | $1.69 \times 10^{+07}$ |

## 5 CONCLUSIONS

The paper presents a stochastic mixed-integer linear programming that integrates the production planning under uncertainty with the decisions of surface and subsea systems screening out the initial capital expenditures of the main facilities and the equipment, the project revenue, and the system concept. The main contribution of the present model is the determination of two stages of decisions. The cost of the subsea production system is the first stage of the decision. Thus, the surface decisions and the estimation of the project revenue are considered as the second stage of decisions.

The value of the stochastic solution is adopted to examine the quality of the solution provided by the model, comparing the model solution and the expectation of the deterministic solution. A realistic case study is also provided to demonstrate the efficiency of the proposed stochastic mixed-integer linear programming model. As conclusion, the stochastic model simultaneously satisfies all the scenarios that can happen. But there is no gain in considering the stochastic problem instead of the approximate deterministic strategy. This result shows that the methodology developed by Silva & Guedes Soares (2021a) is efficient and can be adopted avoiding having to model the uncertainty in the model without compromising the quality of the solution.

## ACKNOWLEDGEMENTS

This work was performed within the scope of the Strategic Research Plan of the Centre for Marine Technology and Ocean Engineering (CENTEC), which is financed by the Portuguese Foundation for Science and Technology (Fundação para a Ciência e Tecnologia - FCT) under contract UIDB/UIDP/00134/2020.

## REFERENCES

Birge, J. R., & Louveaux, F. (2011). *Introduction to Stochastic Programming* (T. V. Mikosch, S. I. Resnick, & S. M. Robinson (eds.); Second Edition). http://www.springer.com/series/3182

Devine, M. D., & Lesso, W. G. (1972). Models for the Minimum Cost Development of Offshore Oil Fields. *Management Science*, *18*, 378–387.

Gupta, V., & Grossmann, I. E. (2014). Multistage stochastic programming approach for offshore oilfield infrastructure planning under production sharing agreements and endogenous uncertainties. *Journal of Petroleum Science and Engineering*, *124*, 180–197. http://dx.doi.org/10.1016/j.petrol.2014.10.006

Haugen, K. K. (1996). A stochastic dynamic programming model for scheduling of offshore petroleum fields with resource uncertainty. *European Journal of Operational Research*, *88*, 88–100. http://dx.doi.org/10.1016/0377-2217(94)00192-8

Iyer, R. R., & Grossmann, I. E. (1998). Optimal Planning and Scheduling of Offshore Oil Field Infrastructure Investment and Operations. *Industrial & Engineering Chemistry Research*, *37*(4), 1380–1397.

Rodrigues, H. W. L., Prata, B. A., & Bonates, T. O. (2016). Integrated optimization model for location and sizing of offshore platforms and location of oil wells. *Journal of Petroleum Science and Engineering*, *145*, 734–741. http://dx.doi.org/10.1016/j.petrol.2016.07.002

Silva, L. M. R., & Guedes Soares, C. (2019). An integrated optimization of the floating and subsea layouts. *Ocean Engineering*, 191. 106557. http://dx.doi.org/10.1016/j.oceaneng.2019.106557

Silva, L. M. R., & Guedes Soares, C. (2021a). Oilfield development system optimization under reservoir production uncertainty. *Ocean Engineering*, *225*, 108758. http://dx.doi.org/10.1016/j.oceaneng.2021.108758

Silva, L. M. R., & Guedes Soares, C. (2021b). Statistical analysis of the oil production profile of Campos' basin in Brazil. In C. Guedes Soares & Santos T. A. (Eds.), *Developments in Maritime* Technology *and* Engineering (Volume 2, pp. 775–784). Taylor and Francis Group. http://dx.doi.org/10.1201/9781003216599-83

Wang, Y., Duan, M., Xu, M., Wang, D., & Feng, W. (2012). A mathematical model for subsea wells partition in the layout of cluster manifolds. *Applied Ocean Research*, *36*, 26–35. http://dx.doi.org/10.1016/j.apor.2012.02.002

Wang, Y., Wang, D., Duan, M., Xu, M., Cao, J., & Estefen, S. F. (2014). A mathematical solution of optimal partition of production loops for subsea wells in the layout of daisy chains. *Journal of Engineering for the Maritime Environment*, *228*(3), 211–219. http://dx.doi.org/10.1177/1475090213506186

Zhang, H., Liang, Y., Ma, J., Qian, C., & Yan, X. (2017). An MILP method for optimal offshore oilfield gathering system. *Ocean Engineering*, 141, 25–34. http://dx.doi.org/10.1016/j.oceaneng.2017.06.011

*Trends in Maritime Technology and Engineering – Guedes Soares & Santos (eds)*


# Intuitionistic fuzzy-MULTIMOORA-FMEA for FPSO oil and gas processing system

Peijie Yang
*Ocean Engineering Technology Research Center, School of Naval Engineering, Harbin Engineering University, Harbin, China*

Cong Yi
*Engineering Research and Design Department, CNOOC Research Institute, Beijing, China*

Zhuang Kang & Jichuan Kang
*Ocean Engineering Technology Research Center, School of Naval Engineering, Harbin Engineering University, Harbin, China*

ABSTRACT: To tackle the inadequacies of failure mode and effects analysis (FMEA) in the risk assessment of Floating Production, Storage and Offloading (FPSO) oil and gas handling system - the evaluation of risk factors, weight assignment problem and risk ranking problem. This paper proposes an improved FMEA method based on the intuitionistic fuzzy MULTIMOORA method. This method is based on intuitionistic fuzzy set theory and intuitionistic fuzzy priority weighted average (IFPWA) operator, and introduces the improved MULTIMOORA method to optimize risk ranking. The effectiveness and feasibility of the proposed method is verified by applying it to the FPSO risk assessment scenario, and the superiority of the proposed method is demonstrated by comparing with the conventional FMEA.

## 1 INTRODUCTION

The risk of offshore oil and gas development mainly comes from the oil and gas leakage events caused by the defects of facilities and human errors. More than 70% of offshore platform accidents originate from the highly destructive oil and gas explosion and fire (Paik et al., 2011). The process equipment, pressure vessels and pipelines used to process, store or transport hazardous media in floating production system (FPSO) may cause oil and gas leakage due to perforation, crack or even fracture under the long-term action of harsh marine environment. Once the ignition source appears, it is extremely prone to fire, explosion and other major accidents. According to the statistics of RIDDOR (Muncer, 2003), there were 316 FPSO dangerous events from 1996 to 2002, including 157 oil and gas leakage events, accounting for 50%, 27 fire and explosion events, accounting for 8%. According to the statistics of 135 FPSO operation accidents in British waters (Paik et al., 2011), the oil and gas leakage events overall show an obvious growth trend, and the annual frequency of small leakage events, serious leakage events and major leakage events are 2.081, 2.330 and 0.100 respectively. The chain risk of fire and explosion caused by oil and gas leakage in FPSO production has become a safety problem that cannot be ignored in offshore oil and gas development.

Analyze the failures causes and modes of the equipment, which contributes to the reliability of the equipment. In the standard promulgated by the International Standard Organization (ISO) on risk management in offshore oil and gas production, the reliability or failure analysis techniques that can be utilized for offshore oil and gas production equipment, including fault tree analysis, failure mode and effect analysis (FMEA) (Liu et al., 2015), and Failure Modes, Effects, and Criticality Analysis (FMECA). Among them, FMEA, as an important safety and reliability analysis tool, has been widely used in various industries.

FMEA is a reliability assessment method widely used to define, identify and eliminate potential failures and consists of three main conceptual phases: the first phase is the qualitative analysis phase, by identifying all potential failure modes and their causes and effects; the second phase is the quantitative analysis phase, through risk ranking number (RPN) for assessment; and the third phase is the modified analysis phase, where risk level reduction is achieved by implementing improvement measures. The RPN is the multiplication of three risk factors:

DOI: 10.1201/9781003320289-58

occurrence (O), severity (S), and detectability (D). The three risk factors are scored according to their respective scales (1 to 10), and the RPN score ranges from 1 to 1000, with higher RPN scores representing larger the risk, and the failure mode should be given priority.

However, while the traditional FMEA method has advantages such as ease of use and generality, it still has many defects: ① the accurate quantitative value cannot accurately describe the real fault evaluation information provided by experts; ② the same weight is given to the three risk factors, ignoring the relative importance of the risk factors; ③ due to the lack of a comprehensive scientific basis for the RPN calculation formula, the RPN value is very sensitive to changes in risk factors, and it is susceptible to the same RPN value for failure modes, so that it is difficult to judge the risk order.

Information vehicles such as rough sets (Kutlu et al., 2012), fuzzy rough sets (An et al., 2016) and hesitant fuzzy sets (Geng et al., 2017) were employed to represent expert evaluation information in FMEA. Nie et al. (2018) expressed the assessment of system risk with multi-gram linguistic term sets; Wang et al. (2018) introduced rough numbers to deal with the subjectivity and ambiguity of expert evaluation in FMEA. In addition, since intuitionistic fuzzy sets (Atanassov, 2017) consider more non-subordinate information compared to fuzzy sets, the expressions are more flexible and practical. The results are better when applied to fuzzy evaluation. The above study features a combination of FMEA methods with other methods to overcome the ambiguity of expert evaluation information. The comprehensive weighting method to obtain the weights of risk factors can not only override the shortcomings of experts' empirical judgment, but also can fully consider the objective factors of the expert evaluation information itself. For example, Song et al. (2014) constructed a comprehensive weighting method combining expert evaluation method and entropy weighting method to determine the weights of risk factors.

For risk ranking problems, FMEA can also be seen essentially as a multi-criteria decision making (MCDM) process. Thus, the Vlsekriterijumska Optimizacija I Kompromisno Resenje (VIKOR) (Tian et al., 2018), Decision Experimentation and Evaluation Laboratory (DEMATEL) and the Technique for Order Preference by Similarity to Ideal Solution (TOPSIS) (Kang et al., 2017). However, the above multi-attribute decision methods have a single decision approach and their ranking robustness is yet to be improved. Brauers et al. (2015) introduced the full multiplicative model into the Multi-Objective Optimization by Ratio Analysis (MOORA) method, and then proposed the MOORA combined with the full multiplicative form (MULTIMOORA) method, which has been widely used in economic, weather and electronic decision-making research due to its simplicity and robustness.

In this paper, an improved FMEA method based on intuitionistic fuzzy MULTIMOORA is proposed. First, the method incorporates a fuzzy language, which is adapted to the expression habits of experts with different knowledge backgrounds and experience levels. After that, considering both the subjective experience knowledge of experts and objective information of fault assessment, a comprehensive assignment method combining IFPWA operator and entropy weight method is implemented to determine the risk factor weights; finally, based on intuitionistic fuzzy MULTIMOORA, the total evaluation value of risks is obtained and the risks are ranked by this method.

## 2 METHODS

In order to solve the insufficiency of the traditional FMEA model, this paper proposes an improved FMEA method based on intuitionistic fuzzy probability language. The method mainly includes three stages: ① failure mode risk assessment based on intuitionistic fuzzy probability language; ② Subjective and objective weights of risk factors determined by IFPWA (Intuitionistic Fuzzy Priority Weighted Average) operator and entropy method; ③ Prioritization of failure modes based on Intuitionistic fuzzy MULTIMOORA. The framework flowchart of the improved FMEA method is shown in Figure 1.

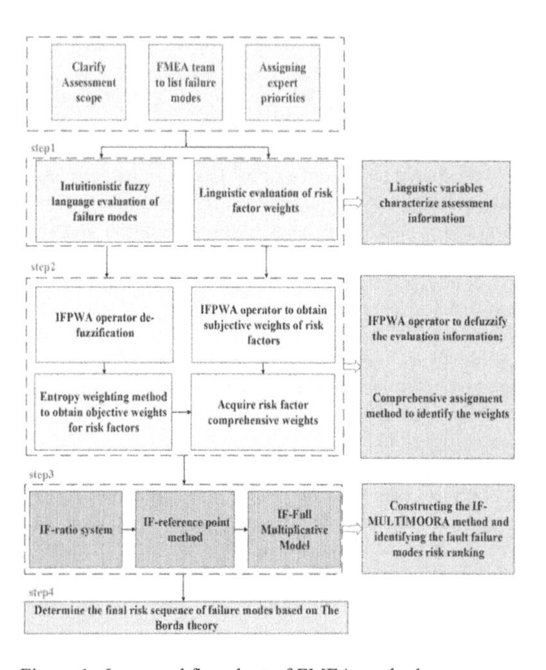


Figure 1. Improved flowchart of FMEA method.

### 2.1 *Intuitionistic fuzzy set theory*

Let $X = \{x_1, x_2, \cdots, x_n\}$ be a non-empty theoretical domain, the intuitionistic fuzzy set $A$ $X$ on Definition $A = \{\langle x, \mu_A(x), \upsilon_A(x)\rangle | x \in X\}$.
where $\mu_A(x) : X \to [0,1]$ and $\upsilon_A(x) : X \to [0,1]$ are the subordinate and unaffiliated functions, respectively, with $x$ respect to $A$ and satisfying $\forall x \in X: \mu_A(x) + \upsilon_A(x) \leq 1$.

In the intuitionistic fuzzy set, since the sum of the affiliation function and the non-affiliation function is not greater than 1, another function can be derived, that is the hesitation function. Called $\pi_A(x) : X \to [0,1]$ as the hesitation function of $x$ relative to $A$, the hesitation function $\pi_A(x)$ is denoted as: $\pi_A(x) = 1 - \mu_A(x) - \upsilon_A(x)$.

The hesitancy function $\pi_A(x)$ represents a measure of the hesitation of an element $x$ belonging to $A$ and $\forall x \in X, \pi_A(x) \in [0,1]$.

LetA $= (\mu_1, \upsilon_1)$ and $B = (\mu_2, \upsilon_2)$ be two intuitionistic fuzzy numbers, then the Hamming distance between $A$ and $B$ is:

$$d(A,B) = \frac{1}{2}(|\mu_1 - \mu_2| + |\upsilon_1 - \upsilon_2| + |\pi_1 - \pi_2|) \quad (1)$$

Let $A = \langle \mu_A(x), \upsilon_A(x)\rangle \in IFV(X)$, then.

(1) $S(A) = \mu_A(x) - \upsilon_A(x)$ is called the intuitionistic fuzzy score function of the intuitionistic fuzzy set $A$.

(2) $H(A) = \mu_A(x) + \upsilon_A(x)$is called the intuitionistic fuzzy exact function of the intuitionistic fuzzy set $A$.

Obviously, $H(A) + \pi_A(x) = 1$, the larger the value of hesitation function, the lower the intuitionistic fuzzy accuracy. On the contrary, the higher the intuitionistic fuzzy accuracy.

Since the IFPWA operator is to be applied in the text to aggregate the expert assessment information, which $s(A_j)$must not be a negative number, the modified scoring function is (Xu et al. 2011).

$$s(A_j) = \mu + \frac{1}{2}(1 - \mu - \upsilon) \quad (2)$$

Suppose $A_i = \langle \mu_{Ai}(x), \upsilon_{Ai}(x)\rangle \in IFV(X), i = [1,n]$, define the intuitionistic fuzzy operator $IFWA : X^n \to X$, if:

$$IFWA_\omega(A_1, A_2, \cdots, A_n) = \omega_1 A_1 + \omega_2 A_2 + \cdots + \omega_n A_n = \sum_{i=1}^{n} \omega_i A_i \quad (3)$$

$IFWA$ is called the intuitionistic fuzzy weighted average operator, where $\omega_1, \omega_2, \cdots, \omega_n$ are the exponential weighting coefficients of $A_1, A_2 \ldots A_n$ respectively, and $\omega_i \in [0,1], \sum_{i=1}^{n} \omega_i = 1$. In particular, $IFWA$ is called the intuitionistic fuzzy average operator when $\omega_i \equiv 1/n$.

According to Eq. (3), the $IFWA$ result of the integration operation is still an intuitionistic fuzzy number, and

$$IFWA_\omega(A_1, A_2, \cdots, A_n) = \left\langle 1 - \prod_{i=1}^{n}(1 - \mu_{Ai}(x))^{\omega_i}, \prod_{i=1}^{n}(\upsilon_{Ai}(x))^{\omega_i} \right\rangle \quad (4)$$

Suppose $A_i = \langle \mu_{Ai}(x), \upsilon_{Ai}(x)\rangle \in IFV(X), i = [1,n]$, define the intuitionistic fuzzy operator $IFWG : X^n \to X$, if:

$$IFWG_\omega(A_1, A_2, \cdots, A_n) = \prod_{i=1}^{n} \omega_i A_i = \left\langle \prod_{i=1}^{n}(\mu_{Ai}(x))^{\omega_i}, 1 - \prod_{i=1}^{n}(1 - \upsilon_{Ai}(x))^{\omega_i} \right\rangle \quad (5)$$

$IFWG$ is called the intuitionistic fuzzy weighted geometric operator, where $\omega_1, \omega_2, \cdots, \omega_n$ are the exponential weighting coefficients of $A_1, A_2 \ldots A_n$ respectively and $\omega_i \in [0,1], \sum_{i=1}^{n} \omega_i = 1$.

### 2.2 *Improved FMEA method based on intuitionistic fuzzy MULTIMOORA*

#### 2.2.1 *Intuitive fuzzy probabilistic language evaluation*

Due to the uncertainty and incompleteness of information related to failure modes in practice, it is difficult for experts to accurately evaluate from the real number form of 1 to 10, and they are more inclined to use linguistic variables to evaluate information, so the enterprise organizes $s$ evaluation experts Using linguistic variables, and converting linguistic variables into corresponding intuitionistic fuzzy numbers according to certain rules (Table 1), the failure mode intuitionistic fuzzy evaluation matrix is obtained. This article follows the traditional FMEA method of occurrence, severity and detectability of three risk factors.

Assuming that the FMEA expert team is composed of $s$ experts $DM_k$ $(k = 1, 2, \cdots, s)$, the experts are divided into $s$ priority levels according to their knowledge structure and field experience, that is, the knowledge structure of $DM_1$ is closer to the FMEA evaluation object and the field experience is more abundant, with the highest priority Grade, the risk assessment gives priority to its assessment information, and $DM_s$ has the lowest priority.

The expert team evaluates the actual performance of $m$ potential failure modes $FM_i$ $(i = 1, 2, \cdots, m)$ under $n$ risk factors $RF_j$ $(j = 1, 2, \cdots, n)$, and the importance of risk factors to the ranking results through language variables. The evaluation results can be converted into corresponding intuitionistic fuzzy numbers, where: $R_k = (A_j^k)_{m\times n}$ represents the failure mode intuitionistic fuzzy evaluation matrix given by $DM_k$, $A_j^k = \langle \mu_{Aj}^{(k)}, \upsilon_{Aj}^{(k)} \rangle$ represents the intuitionistic fuzzy number transformed by $DM_k$ for the language evaluation information of $RF_j$ on $FM_i$. $W_k = (A_j^k)_{1\times n}$ represents the intuitionistic fuzzy evaluation matrix of risk factor weight given by $DM_k$, and $A_j^k = \langle \mu_j^{(k)}, \upsilon_j^{(k)} \rangle$ represents the intuitionistic fuzzy number transformed by $DM_k$ for the language evaluation information of the significance of $FM_i$.

Table 1. Failure mode language variable levels.

| Language Variable | Intuitionistic Fuzzy Number |
|---|---|
| Extremely Low(EL) | (0.025,0.900) |
| Very Low (VL) | (0.075,0.850) |
| Low (L) | (0.150,0.750) |
| Medium Low (ML) | (0.350,0.550) |
| Medium (M) | (0.500,0.500) |
| Medium High (MH) | (0.550,0.350) |
| High (H) | (0.750,0.150) |
| Very High (VH) | (0.850,0.075) |
| Extremely High (EH) | (0.900,0.025) |

### 2.2.2 *Comprehensive weighting method to determine risk factor weights*

Determining expert weights is one of the key steps in the process of collecting information for group evaluation. The subjective weights in the proposed integrated weighting method are defined by the IFPWA (intuitionistic fuzzy priority weighted average) operator. Specific steps are as follows:

①The $s$ experts are organized to evaluate the three risk factors O, S, and D of the failure mode with Table 1, and the risk factor weight intuitive fuzzy evaluation matrix $W_k = (A_j^k)_{1\times n}$ is obtained.

②Aggregate the expert risk factor weight information by IFPWA, that is.

$$IFPWA\left(A_j^{(1)}, A_j^{(2)}, \cdots, A_j^{(s)}\right) = \frac{T_1}{\sum_{k=1}^{s} T_k} A_j^{(1)} \oplus \frac{T_2}{\sum_{k=1}^{s} T_k} A_j^{(2)} \oplus \cdots \oplus \frac{T_s}{\sum_{k=1}^{s} T_k} A_j^{(s)}$$
$$= \left(1 - \prod_{k=1}^{s} \left(1 - \mu_{Aj}^{(k)}\right)^{\frac{T_k}{\sum_{k=1}^{s} T_k}}, \prod_{k=1}^{s} \left(\upsilon_{Aj}^{(k)}\right)^{\frac{T_k}{\sum_{k=1}^{s} T_k}}\right) \tag{6}$$

In the formula, $A_j^k = \langle \mu_{Aj}^{(k)}, \upsilon_{Aj}^{(k)} \rangle (k = 2, 3, \cdots, s)$ is a set of failure mode intuitionistic fuzzy number evaluation information given by experts, where the experts are classified into $s$ priority levels, and $T_k = \prod_{t=1}^{k-i} s(A_j^{(t)}) (k = 2, 3, \cdots, s), T_1 = 1$. This step accomplishes the defuzzification of the matrix $W_k = (A_j^k)_{1\times n}$.

③The values obtained in the previous step are processed with Eq. (4) to obtain the risk factor weight information scoring function, and the subjective weights of the risk factors are determined according to Eq. (7).

$$\omega_j^s = \frac{s(A_j)}{\sum_{j=1}^{n} s(A_j)} \tag{7}$$

In this paper, the entropy weighting method is used to solve the objective weights of risk factors. The basic steps are as follows:

④The population failure assessment matrix $R_k = (A_j^k)_{m\times n}$ processed by the IFPWA operator is normalized.

$$Y = (yij)m \times n = (gij/\sum_{i=1}^{m} gij)m \times n. \tag{8}$$

⑤Calculate the entropy value of each risk factor.

$$Ej = -1/\ln n \sum_{i=1}^{m} yij \ln yij \tag{9}$$

⑥Calculate the objective weight of each risk factor.

$$\omega_j^o = (1 - Ej)/(n - \sum_{j=1}^{n} Ej). \tag{10}$$

Integrating the subjective and objective weights derived from the IFPWA operator and the entropy weight method, the comprehensive weight of each risk factor can be calculated as:

$$\omega_j = \gamma\omega_j^S + (1 - \gamma)\omega_j^O \tag{11}$$

where $\gamma$ is the subjective weight coefficient, reflecting the proportion of the subjective judgment of the expert in the weight of the risk factor, taking the value range of $[0, 1]$.

### 2.2.3 *Risk ranking of failure modes based on intuitionistic fuzzy MULTIMOORA*

In this paper, the IFWA operator and the IFWG operator are introduced into the ratio system and the fault mode information fusion of the full multiplicative model in the MULTIMOORA method, respectively, in order to avoid information loss. The improved Hamming distance between the failure mode information and the reference point in the computational reference point method is implemented to enhance the ranking robustness. At last, the final risk ranking of failure modes is determined based on Borda theory (Liu et al. 2021). The steps of the risk ranking method based on Intuitionistic fuzzy MULTIMOORA are as follows:

⑦IF-ratio system. According to Eq. (4), the integrated utility value $FM_i$is calculated with the intuitionistic fuzzy integrated evaluation matrix $R = (A_i)_{m\times n}$ and the risk factor weight vector $\omega_j = \{\omega_1, \omega_2, \cdots, \omega_n\}$ of failure modes.

$$y_i = IFWA(A_1, A_2, \cdots, A_n) = \overset{n}{\underset{j=1}{\oplus}} \omega_j A_i = \left\langle 1 - \prod_{j=1}^{n} (1 - \mu_{Ai}(x))^{\omega_j}, \prod_{j=1}^{n} (\upsilon_{Ai}(x))^{\omega_j} \right\rangle \tag{12}$$

where $y_i$ denotes the combined utility value $FM_i$ under all risk factors. The larger the $y_i$, the higher the risk ranking.

⑧IF-reference point method. Introducing the risk factor weight vector $\omega_j=\{\omega_1, \omega_2, \cdots, \omega_n\}$ as the significance coefficient, combined with Eq. (1), the Tchebycheff distance of the reference point under different risk factors is calculated.

$$d\left(\tilde{\beta}_j, FM_i\right) = \max_{1\le j\le n} d\left(\tilde{\beta}_j, \tilde{a}_{ij}\right) = \max_{1\le j\le n} \frac{\omega_j}{2}\left(\left|\mu_{ij} - 1\right| + \left|\upsilon_{ij}\right| + \left|\pi_{ij}\right|\right) \tag{13}$$

The smaller the Tchebycheff distance of the failure mode, the higher the risk ranking.

⑨IF-Full Multiplicative Form model. According to Eq. (5), the multiplicative utility value for the failure mode is calculated by using the failure mode intuitionistic fuzzy integrated evaluation matrix $R = (A_j)_{m\times n}$and the risk factor weight vector $\omega_j = \{\omega_1, \omega_2, \cdots, \omega_n\}$.

$$U_i = IFWG(A_1, A_2, \cdots, A_n) = \overset{n}{\underset{j=1}{\otimes}} A_i^{\omega_j} = \left\langle \prod_{j=1}^{n} (\mu_{Ai}(x))^{\omega_j}, 1 - \prod_{j=1}^{n} (1 - \upsilon_{Ai}(x))^{\omega_j} \right\rangle \tag{14}$$

where the larger the $U_i$, the higher the risk ranking.

⑩Borda theory. Combining Borda theory, the generalized dominance relationships in a ternary array of 3 risk orders for each failure mode derived from these three parts are compared, as well as determine the final risk order of the failure modes.

## 3 FPSO OIL AND GAS PROCESSING SYSTEM RISK ASSESSMENT

The risk assessment of the FPSO is performed according to the proposed Intuitionistic Fuzzy -MULTIMOORA-FMEA method. The main failure modes of oil and gas processing equipment are shown in Appendix Table A.

### 3.1 *Risk assessment based on intuitionistic fuzzy MULTIMOORA*

First, the FMEA expert team consists of three members with various priorities based on their knowledge structure and domain experience, classified as priority DM1, DM2 and DM3. By combining the linguistic variables in Table 1, the expert group assessed the actual performance of the 71 potential failure modes in Appendix Table A under the three risk factors O, S and D, as well as the importance of the risk factors themselves, as shown in Appendix Table B. Then the linguistic variables assessment information was converted into the corresponding intuitionistic fuzzy numbers to compose the failure mode intuitionistic fuzzy assessment matrix $R = \left(A_i^k\right)_{m\times n}$. At last, the failure mode intuitionistic fuzzy comprehensive evaluation matrix $R = (A_i)_{m\times n}$ is obtained by aggregating the expert evaluation information according to the IFPWA operator, as shown in Appendix Table C.

Second, the integrated weight method was applied to solve the risk factor weight information. According to Eq. (6), the intuitive fuzzy integrated evaluation matrix of risk factor weights and the corresponding scoring function values is obtained as in Table 2. From Eq. (7), the subjective weight vector of risk factors is obtained as $\omega_j^s=\{0.330, 0.352, 0.318\}$. From Eqs. (8) ~ (10), the objective risk factor weight vector is obtained as $\omega_j^o=\{0.527, 0.166, 0.307\}$. Taking $\gamma=0.5$, the risk factor weight information is determined as $\omega_j = \{0.429, 0.259, 0.312\}$.

Finally, the risk ranking of each failure mode under IF-ratio system, IF-reference point method and IF-full multiplication form model are obtained through utilizing Eqs. (12) ~ (14), and then the final risk ranking of each failure mode is determined by Borda theory, as shown in Appendix Table D. Figure 2 shows the three risk sequences under the intuitionistic fuzzy MULTIMOORA method. It can be seen that FM49 has the highest risk.

### 3.2 *Comparative analysis*

In order to verify the effectiveness of the FMEA method proposed in this paper, the risk ranking results are compared with the RPN ranking results calculated from the traditional FMEA, The O, S and D values of the traditional FMEA are shown in Figure 3

As can be seen from Figure 4, the top 10 failure modes are basically the same com-pared to the conventional FMEA, and the risk ranking of FM49 is the highest of the two results. That is due to the fact that these 10 failure modes are prone to oil and gas leaks,

Table 2. Relevant parameters of expert evaluation.

| Risk factors | $A_j^1$ | $A_j^2$ | $A_j^3$ | $A_j$ | $S(A_j)$ |
|---|---|---|---|---|---|
| O | (0.850,0.075) | (0.850,0.075) | (0.750,0.150) | (0.825,0.092) | 0.867 |
| S | (0.900,0.025) | (0.850,0.075) | (0.900,0.025) | (0.886,0.035) | 0.926 |
| D | (0.850,0.075) | (0.750,0.150) | (0.750,0.150) | (0.789,0.119) | 0.835 |

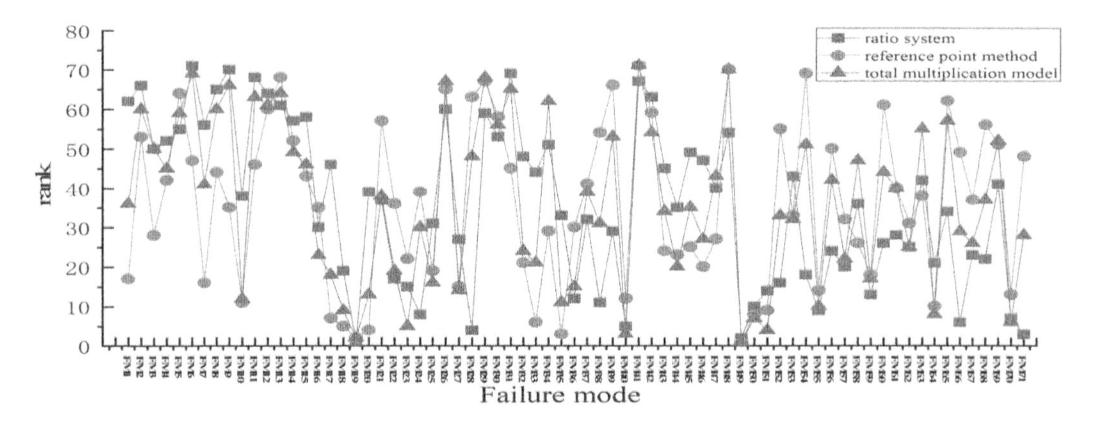


Figure 2. Three risk sequences under Intuitionistic Fuzzy MULTIMOORA method.

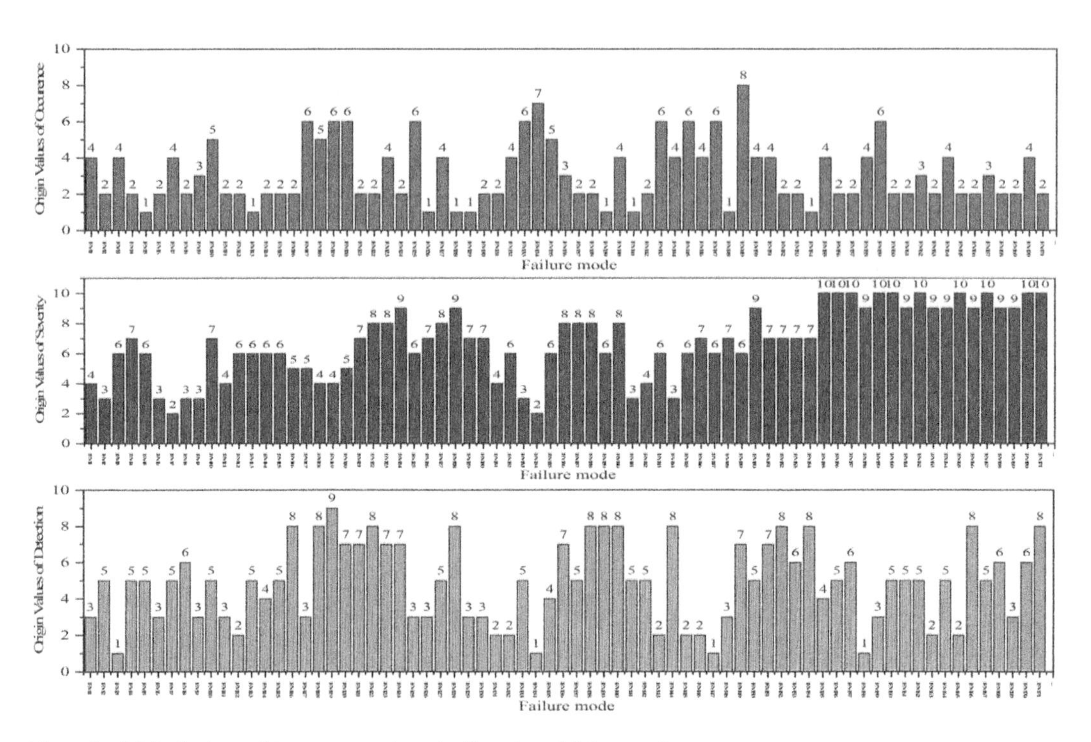


Figure 3. Original values of the occurrence/severity/detection of failure modes.

fire and explosion hazards and have a high hazard, and this ranking illustrates the effectiveness of the proposed method. The risk ranking of other failure modes has changed somewhat, taking FM35 and FM47 as examples. FM35 is wire insulation aging, which has a higher incidence of failure and is not susceptible to detection, and is vulnerable to the potential danger of leakage, as well as fire and explosion when encountering combustible materials. FM47 is valve corrosion damage, which is due to the large number of related valve parts in the oil and gas treatment system, and the corrosion dam-age of a valve part is not vulnerable to detection, and once it occurs, it is susceptible to oil and gas leakage accident. From these two failure accidents, we can observe that the potential hazards prone to be induced after the failure are higher, so the risk ranking of FM35 and FM47 should be higher. Comparing the results of the two methods, the proposed method in this paper is more in compliance with the results, which confirms that the proposed method is better compared with the traditional FMEA analysis, and the same risk priority values appear in the traditional FMEA analysis (e.g., FM50, FM59, FM64), which makes it complicated to evaluate the risk level.

The method adopted in this paper was optimized for each step of the FMEA process compared to the traditional method. The results obtained can provide decision makers with a more accurate reference basis for risk management.

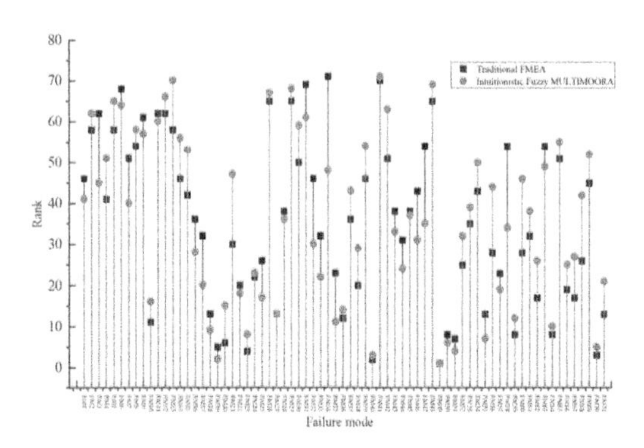


Figure 4. Compare and analyze the results with traditional FMEA.

## 4 CONCLUSIONS

This paper presents an intuition-based fuzzy MULTIMOOR approach to FMEA risk assessment. Compared with existing FMEA methods, this paper has the following advantages.

The evaluation information represented by language variables is transformed into corresponding intuitionistic fuzzy numbers, thus effectively dealing with the uncertainty of the expert assessment information; a comprehensive weighting method combining expert evaluation method and entropy weighting method is presented to distinguish the relative importance of risk factors and overcome the one-sided disadvantages of subjective weighting method or objective weighting method. By introducing IFWA operator, improved Hamming distance and IFWG operator, the intuitionistic fuzzy MULTIMOORA method is proposed. The risk ranking of failure modes improves the robustness of existing FMEA results.

Finally, FMEA risk assessment of FPSO oil and gas handling system is taken as an example, and it is concluded that valve damage is the most harmful to FPSO oil and gas handling system, which verifies the feasibility and effectiveness of the method.

## ACKNOWLEDGMENTS


This work is supported by the Specialized Research Project for LS17-2 Semi-submersible Production Platform (LSZX-2020-HN-05-0405), the National Natural Science Foundation of China (Grant No. 52101305), the Fundamental Research Funds for Central Universities (Grant No. 3072021CF0101).


## REFERENCES


An X, Yu J, Cai W, 2016. Fuzzy rough FMEA evaluation method based on hybrid multi-attribute decision and correlative analysis. Computer Integrated Manufacturing Systems. 22(11):2613–2621.

Atanassov K, 2017. Type-1 Fuzzy Sets and Intuitionistic Fuzzy Sets. Algorithms. 10(3): 12.

Brauers W, Zavadskas EK, Ginevicius R, 2015. Multi-objective decision making in macro- and micro-economics with application of the MULTIMOORA method. International Journal of Applied Nonlinear Science. 2(1/2):75.

Geng X, Zhang Y, 2017. Improved FMEA approach for risk evaluation based on hesitant fuzzy set. Computer Integrated Manufacturing Systems. 23 (2):340–348.

Kang J, Sun L, Sun H, Wu C, 2017. Risk assessment of floating offshore wind turbine based on correlation-FMEA. Ocean Engineering.129:382–388.

Kutlu A, Ekmekcioglu E, 2012. Fuzzy failure modes and effects analysis by using fuzzy TOPSIS-based fuzzy AHP. Expert Systems with Applications. 39 (1): 61–67.

Liu HC, You JX, Ding XF, Su Q, 2015. Improving risk evaluation in FMEA with a hybrid multiple criteria decision making method. International Journal of Quality & Reliability Management. 32(7): p. 763–782.

Liu PD, Gao H, Fujita H, 2021. The new extension of the MULTIMOORA method for sustainable supplier selection with intuitionistic linguistic rough numbers. Applied Soft Computing. 99:106893.

Muncer P, 2003. Analysis of accident statistics for floating monohull and fixed installations.

Nie RX, Tian ZP, Wang XK, et al., 2018. Risk evaluation by FMEA of supercritical water gasification system using multi-granular linguistic distribution assessment. Knowledge-Based Systems. 162: p. 185–201.

Paik J, Czujko J, Asme, 2011. ASSESSMENT OF HYDROCARBON EXPLOSION AND FIRE RISKS IN OFFSHORE INSTALLATIONS: RECENT ADVANCES AND FUTURE TRENDS. Omae2011: Proceedings of the Asme 30th International Conference on Ocean, Offshore and Arctic Engineering, Vol 2: Structures, Safety and Reliability. 313–323.

Paik J, Czujko J, Kim BJ, et al., 2011. Quantitative assessment of hydrocarbon explosion and fire risks in offshore installations. Marine Structures.24(2): p. 73–96.

Song WY, Ming XG, Wu ZY, Zhu BT, 2014. A rough TOPSIS Approach for Failure Mode and Effects Analysis in Uncertain Environments. Quality and Reliability Engineering International. 30(4): p. 473–486.

Tian ZP, Wang JQ, Zhang HY, 2018. An integrated approach for failure mode and effects analysis based on fuzzy best-worst, relative entropy, and VIKOR methods. Applied Soft Computing. 72: p. 636–646.

Wang Z, Gao JM, Wang RX, et al., 2018. Failure Mode and Effects Analysis by Using the House of Reliability-Based Rough VIKOR Approach. IEEE Transactions on Reliability. 67(1): p. 230–248.

Xu YJ, Sun T, Li DF, 2011. Intuitionistic fuzzy prioritized OWA operator and its application in multicriteria decision-making problem. Control & Decision. 26 (1):129–132.

## APPENDIX

Table A. Oil and gas processing system failure mode table

| Failure mode | Failure factor | Failure mode | Failure factor | Failure mode | Failure factor |
|---|---|---|---|---|---|
| FM1 | Scaling not cleaned | FM25 | Pressure gauge failure | FM49 | Valve damaged |
| FM2 | Excessive corrosives or impurities in hot oil | FM26 | Nitrogen pressure control valve failure | FM50 | Control failure. |
| FM3 | High temperature of heat exchange oil | FM27 | Liquid level alarm failure | FM51 | Lax flange connection |
| FM4 | Failure of safety valve of heat exchange pipe | FM28 | Personnel error | FM52 | Flange mis-match |
| FM5 | The heat exchange tube is corroded | FM29 | Poor sealing | FM53 | Does not match the pipeline pressure level |
| FM6 | Falling object collision | FM30 | Corrosion | FM54 | Personnel error |
| FM7 | High concentration of corrosive substances in crude oil | FM31 | Pumping vacuum. | FM55 | Oil and gas leaks. |
| FM8 | Heat exchanger rupture | FM32 | Cavitation erosion | FM56 | Failure to cut off the medium flow in time |
| FM9 | External anti-corrosion failure | FM33 | Serious bearing wear | FM57 | Fire and combustible gas detection failure. |
| FM10 | Failure of shell side safety valve | FM34 | Bolt loose. | FM58 | Crude oil accumulation on the ground. |
| FM11 | Seal failure | FM35 | Wire insulation aging. | FM59 | Presence of ignition source |
| FM12 | Sheet corrosion | FM36 | Cable breakage. | FM60 | Smoking in the work area |
| FM13 | Weld damage | FM37 | Illegal wiring | FM61 | Use metal tools to knocked. when inspecting. |
| FM14 | Damaged safety valve leakage | FM38 | Circuit failure due to mis-operation | FM62 | Electrical equipment leakage |
| FM15 | Leakage of orifice valves | FM39 | Material fatigue fracture. | FM63 | Electric spark or arc is generated when switching |
| FM16 | Internal corrosion damage | FM40 | Overpressure | FM64 | Generate static electricity. |
| FM17 | External corrosion damage | FM41 | Material defects. | FM65 | Struck by lightning. |
| FM18 | Stress corrosion damage | FM42 | Pipeline weld damage | FM66 | High temperature ignition |
| FM19 | Third party damage | FM43 | Vibration breakage. | FM67 | Fire suppression system failure. |
| FM20 | Falling object collision, tank damage | FM44 | Collision. | FM68 | Combustible gas accumulation |
| FM21 | Outlet blockage | FM45 | Pipeline corrosion damage. | FM69 | Low explosion-proof level of equipment. |
| FM22 | The safety valve is blocked or damaged | FM46 | Connection loose | FM70 | Long-term overpressure of equipment |
| FM23 | Pressure alarm failure | FM47 | Valve corrosion damage | FM71 | Fire inside the equipment. |
| FM24 | Emergency shut-off valve failure | FM48 | Valve seat rusted | | |

Table B. evaluation information of expert language variables

| Risk factors | O | | | S | | | D | | | Risk factors | O | | | S | | | D | | |
|---|---|---|---|---|---|---|---|---|---|---|---|---|---|---|---|---|---|---|---|
| expert | $DM_1$ | $DM_2$ | $DM_3$ | $DM_1$ | $DM_2$ | $DM_3$ | $DM_1$ | $DM_2$ | $DM_3$ | expert | $DM_1$ | $DM_2$ | $DM_3$ | $DM_1$ | $DM_2$ | $DM_3$ | $DM_1$ | $DM_2$ | $DM_3$ |
| Importance of risk factors | VH | VH | H | EH | VH | EH | VH | H | H | Importance of risk factors | VH | VH | H | EH | VH | EH | VH | H | H |
| $FM_1$ | ML | M | M | L | L | M | ML | L | L | $FM_{37}$ | VL | L | L | VH | VH | VH | M | MH | MH |
| $FM_2$ | VL | L | VL | L | L | L | M | M | ML | $FM_{38}$ | VL | VL | L | VH | H | VH | H | VH | VH |
| $FM_3$ | M | ML | ML | H | MH | MH | EL | VL | EL | $FM_{39}$ | EL | EL | VL | MH | MH | H | VH | H | VH |
| $FM_4$ | VL | L | L | MH | H | H | M | ML | M | $FM_{40}$ | M | ML | ML | VH | H | H | H | VH | H |
| $FM_5$ | EL | VL | EL | MH | MH | H | M | M | ML | $FM_{41}$ | EL | EL | EL | L | ML | ML | M | ML | M |
| $FM_6$ | VL | L | L | L | L | VL | L | ML | L | $FM_{42}$ | VL | VL | L | ML | L | ML | M | MH | M |
| $FM_7$ | ML | M | M | VL | VL | L | M | M | MH | $FM_{43}$ | MH | MH | M | M | MH | MH | VL | L | VL |
| $FM_8$ | VL | L | L | L | VL | VL | M | MH | M | $FM_{44}$ | ML | ML | L | L | ML | L | VH | H | H |
| $FM_9$ | L | VL | L | L | VL | L | L | L | VL | $FM_{45}$ | MH | M | MH | M | MH | MH | VL | VL | L |
| $FM_{10}$ | M | ML | M | MH | MH | H | MH | MH | M | $FM_{46}$ | ML | M | ML | H | MH | MH | L | VL | L |
| $FM_{11}$ | VL | L | L | M | ML | M | L | L | L | $FM_{47}$ | MH | MH | H | MH | H | H | EL | VL | VL |
| $FM_{12}$ | VL | VL | L | MH | MH | H | L | VL | VL | $FM_{48}$ | EL | EL | EL | H | VH | MH | L | L | VL |
| $FM_{13}$ | EL | EL | VL | M | MH | MH | M | M | ML | $FM_{49}$ | VH | H | H | MH | M | MH | H | MH | H |
| $FM_{14}$ | VL | L | VL | MH | MH | MH | M | M | ML | $FM_{50}$ | M | M | ML | EH | VH | EH | M | M | ML |
| $FM_{15}$ | VL | L | L | M | MH | MH | M | MH | M | $FM_{51}$ | M | M | ML | H | H | VH | H | MH | H |
| $FM_{16}$ | L | VL | L | M | M | M | VH | VH | H | $FM_{52}$ | VL | VL | L | H | VH | H | VH | VH | H |
| $FM_{17}$ | M | MH | MH | ML | M | M | ML | L | L | $FM_{53}$ | L | L | VL | H | H | H | MH | MH | M |
| $FM_{18}$ | M | M | MH | ML | ML | M | VH | VH | H | $FM_{54}$ | EL | EL | EL | H | H | VH | VH | H | VH |
| $FM_{19}$ | MH | H | H | M | ML | M | VH | VH | EH | $FM_{55}$ | M | ML | ML | EH | EH | EH | M | M | ML |
| $FM_{20}$ | M | MH | MH | ML | ML | M | MH | MH | H | $FM_{56}$ | VL | L | VL | EH | EH | EH | M | M | M |
| $FM_{21}$ | VL | VL | L | H | MH | H | H | MH | H | $FM_{57}$ | L | L | L | EH | EH | EH | M | MH | M |
| $FM_{22}$ | L | VL | VL | VH | H | H | VH | H | H | $FM_{58}$ | ML | M | M | VH | VH | VH | EL | VL | EL |
| $FM_{23}$ | ML | ML | M | VH | H | VH | H | MH | H | $FM_{59}$ | MH | M | MH | EH | EH | EH | L | L | ML |
| $FM_{24}$ | VL | L | L | EH | VH | EH | H | H | H | $FM_{60}$ | VL | EL | VL | EH | EH | EH | M | M | ML |
| $FM_{25}$ | MH | H | H | MH | MH | H | L | L | ML | $FM_{61}$ | VL | L | L | VH | EH | VH | M | MH | M |
| $FM_{26}$ | EL | VL | EL | H | MH | H | L | ML | L | $FM_{62}$ | L | L | ML | EH | EH | EH | ML | ML | M |
| $FM_{27}$ | ML | M | M | VH | H | VH | M | ML | M | $FM_{63}$ | L | VL | VL | VH | VH | VH | VL | VL | L |
| $FM_{28}$ | EL | VL | VL | EH | VH | EH | VH | H | VH | $FM_{64}$ | M | ML | M | VH | VH | VH | M | M | MH |
| $FM_{29}$ | EL | EL | VL | MH | H | VH | L | ML | ML | $FM_{65}$ | VL | EL | VL | EH | EH | EH | L | L | VL |
| $FM_{30}$ | VL | VL | L | H | H | VH | L | L | ML | $FM_{66}$ | VL | L | VL | VH | EH | EH | H | H | VH |
| $FM_{31}$ | VL | L | L | ML | M | M | L | L | VL | $FM_{67}$ | L | VL | VL | EH | EH | EH | M | ML | M |
| $FM_{32}$ | M | ML | M | MH | MH | H | L | VL | VL | $FM_{68}$ | VL | VL | L | VH | EH | EH | MH | MH | H |
| $FM_{33}$ | MH | MH | MH | L | L | ML | M | M | M | $FM_{69}$ | VL | L | VL | VH | EH | VH | L | ML | L |
| $FM_{34}$ | MH | H | H | VL | L | VL | EL | EL | VL | $FM_{70}$ | M | ML | ML | EH | EH | EH | MH | M | MH |
| $FM_{35}$ | MH | M | MH | MH | H | H | ML | M | M | $FM_{71}$ | VL | L | VL | EH | EH | EH | H | H | VH |
| $FM_{36}$ | L | ML | L | VH | H | VH | H | H | VH | | | | | | | | | | |

Table C. evaluation information of expert language variables

| Failure mode | $O$ | $S$ | $D$ | Failure mode | $O$ | $S$ | $D$ | Failure mode | $O$ | $S$ | $D$ |
|---|---|---|---|---|---|---|---|---|---|---|---|
| $FM_1$ | 0.411,0.531 | 0.164,0.740 | 0.291,0.608 | $FM_{25}$ | 0.668,0.225 | 0.596,0.300 | 0.157,0.743 | $FM_{49}$ | 0.795,0.115 | 0.535,0.392 | 0.693,0.202 |
| $FM_2$ | 0.083,0.840 | 0.150,0.750 | 0.481,0.507 | $FM_{26}$ | 0.028,0.897 | 0.693,0.202 | 0.185,0.715 | $FM_{50}$ | 0.481,0.507 | 0.885,0.036 | 0.481,0.507 |
| $FM_3$ | 0.443,0.520 | 0.652,0.241 | 0.028,0.897 | $FM_{27}$ | 0.411,0.531 | 0.821,0.095 | 0.460,0.514 | $FM_{51}$ | 0.481,0.507 | 0.781,0.125 | 0.693,0.202 |
| $FM_4$ | 0.084,0.837 | 0.668,0.225 | 0.460,0.514 | $FM_{28}$ | 0.028,0.897 | 0.885,0.036 | 0.821,0.095 | $FM_{52}$ | 0.076,0.849 | 0.788,0.120 | 0.826,0.092 |
| $FM_5$ | 0.028,0.897 | 0.596,0.300 | 0.481,0.507 | $FM_{29}$ | 0.025,0.900 | 0.705,0.192 | 0.198,0.701 | $FM_{53}$ | 0.148,0.753 | 0.750,0.150 | 0.367,0.545 |
| $FM_6$ | 0.084,0.837 | 0.148,0.753 | 0.185,0.715 | $FM_{30}$ | 0.076,0.849 | 0.781,0.125 | 0.157,0.743 | $FM_{54}$ | 0.025,0.900 | 0.781,0.125 | 0.821,0.095 |
| $FM_7$ | 0.411,0.531 | 0.076,0.849 | 0.507,0.475 | $FM_{31}$ | 0.084,0.837 | 0.411,0.531 | 0.148,0.753 | $FM_{55}$ | 0.443,0.520 | 0.900,0.025 | 0.481,0.507 |
| $FM_8$ | 0.084,0.837 | 0.137,0.767 | 0.514,0.453 | $FM_{32}$ | 0.460,0.514 | 0.596,0.300 | 0.137,0.767 | $FM_{56}$ | 0.083,0.840 | 0.900,0.025 | 0.500,0.500 |
| $FM_9$ | 0.138,0.766 | 0.138,0.766 | 0.148,0.753 | $FM_{33}$ | 0.550,0.350 | 0.157,0.743 | 0.500,0.500 | $FM_{57}$ | 0.150,0.750 | 0.900,0.025 | 0.514,0.453 |
| $FM_{10}$ | 0.460,0.514 | 0.596,0.300 | 0.541,0.374 | $FM_{34}$ | 0.668,0.225 | 0.083,0.840 | 0.025,0.900 | $FM_{58}$ | 0.411,0.531 | 0.850,0.075 | 0.028,0.897 |
| $FM_{11}$ | 0.084,0.837 | 0.460,0.514 | 0.150,0.750 | $FM_{35}$ | 0.535,0.392 | 0.668,0.225 | 0.411,0.531 | $FM_{59}$ | 0.535,0.392 | 0.900,0.025 | 0.157,0.743 |
| $FM_{12}$ | 0.076,0.849 | 0.596,0.300 | 0.137,0.767 | $FM_{36}$ | 0.185,0.715 | 0.821,0.095 | 0.781,0.125 | $FM_{60}$ | 0.070,0.855 | 0.900,0.025 | 0.481,0.507 |
| $FM_{13}$ | 0.025,0.900 | 0.523,0.427 | 0.481,0.507 | $FM_{37}$ | 0.084,0.837 | 0.850,0.075 | 0.523,0.427 | $FM_{61}$ | 0.084,0.837 | 0.869,0.052 | 0.514,0.453 |
| $FM_{14}$ | 0.083,0.840 | 0.550,0.350 | 0.481,0.507 | $FM_{38}$ | 0.076,0.849 | 0.821,0.095 | 0.816,0.099 | $FM_{62}$ | 0.157,0.743 | 0.900,0.025 | 0.367,0.545 |
| $FM_{15}$ | 0.084,0.837 | 0.523,0.427 | 0.514,0.453 | $FM_{39}$ | 0.025,0.900 | 0.596,0.300 | 0.821,0.095 | $FM_{63}$ | 0.137,0.767 | 0.850,0.075 | 0.076,0.849 |
| $FM_{16}$ | 0.138,0.766 | 0.500,0.500 | 0.826,0.092 | $FM_{40}$ | 0.443,0.520 | 0.771,0.114 | 0.788,0.120 | $FM_{64}$ | 0.460,0.514 | 0.850,0.075 | 0.507,0.475 |
| $FM_{17}$ | 0.523,0.427 | 0.411,0.531 | 0.291,0.608 | $FM_{41}$ | 0.025,0.900 | 0.198,0.701 | 0.460,0.514 | $FM_{65}$ | 0.070,0.855 | 0.900,0.025 | 0.148,0.753 |
| $FM_{18}$ | 0.507,0.475 | 0.367,0.545 | 0.826,0.092 | $FM_{42}$ | 0.076,0.849 | 0.301,0.598 | 0.514,0.453 | $FM_{66}$ | 0.083,0.840 | 0.884,0.037 | 0.781,0.125 |
| $FM_{19}$ | 0.668,0.225 | 0.460,0.514 | 0.867,0.054 | $FM_{43}$ | 0.541,0.374 | 0.523,0.427 | 0.083,0.840 | $FM_{67}$ | 0.137,0.767 | 0.900,0.025 | 0.460,0.514 |
| $FM_{20}$ | 0.523,0.427 | 0.367,0.545 | 0.596,0.300 | $FM_{44}$ | 0.332,0.568 | 0.185,0.715 | 0.795,0.115 | $FM_{68}$ | 0.076,0.849 | 0.884,0.037 | 0.596,0.300 |
| $FM_{21}$ | 0.076,0.849 | 0.693,0.202 | 0.693,0.202 | $FM_{45}$ | 0.535,0.392 | 0.523,0.427 | 0.076,0.849 | $FM_{69}$ | 0.083,0.840 | 0.869,0.052 | 0.185,0.715 |
| $FM_{22}$ | 0.137,0.767 | 0.795,0.115 | 0.795,0.115 | $FM_{46}$ | 0.391,0.537 | 0.652,0.241 | 0.138,0.766 | $FM_{70}$ | 0.433,0.520 | 0.900,0.025 | 0.535,0.392 |
| $FM_{23}$ | 0.367,0.545 | 0.821,0.095 | 0.693,0.202 | $FM_{47}$ | 0.596,0.300 | 0.668,0.225 | 0.028,0.897 | $FM_{71}$ | 0.083,0.840 | 0.900,0.025 | 0.781,0.125 |
| $FM_{24}$ | 0.084,0.837 | 0.885,0.036 | 0.750,0.150 | $FM_{48}$ | 0.025,0.900 | 0.749,0.153 | 0.148,0.753 | | | | |

Table D. Comparison of failure mode risk ranking

| Failure mode | Traditional FMEA | | | IF-MULTIMOORA | Failure mode | Traditional FMEA | | | IF-MULTIMOORA | Failure mode | Traditional FMEA | | | IF-MULTIMOORA |
|---|---|---|---|---|---|---|---|---|---|---|---|---|---|---|
| | (O,S,D) | RPN | Rank | Rank | | (O,S,D) | RPN | Rank | Rank | | (O,S,D) | RPN | Rank | Rank |
| $FM_1$ | (4,4,3) | 48 | 49 | 41 | $FM_{25}$ | (6,6,3) | 108 | 27 | 17 | $FM_{49}$ | (8,6,7) | 336 | 1 | 1 |
| $FM_2$ | (2,3,5) | 30 | 60 | 62 | $FM_{26}$ | (1,7,3) | 21 | 67 | 67 | $FM_{50}$ | (4,9,5) | 180 | 9 | 6 |
| $FM_3$ | (4,6,1) | 24 | 62 | 45 | $FM_{27}$ | (4,8,5) | 160 | 16 | 13 | $FM_{51}$ | (4,7,7) | 196 | 7 | 4 |
| $FM_4$ | (2,7,5) | 70 | 41 | 51 | $FM_{28}$ | (1,9,8) | 72 | 38 | 36 | $FM_{52}$ | (2,7,8) | 112 | 25 | 32 |
| $FM_5$ | (1,6,5) | 30 | 58 | 65 | $FM_{29}$ | (1,7,3) | 21 | 66 | 68 | $FM_{53}$ | (2,7,6) | 84 | 35 | 39 |
| $FM_6$ | (2,3,3) | 18 | 68 | 64 | $FM_{30}$ | (2,7,3) | 42 | 50 | 59 | $FM_{54}$ | (1,7,8) | 56 | 43 | 50 |
| $FM_7$ | (4,2,5) | 40 | 52 | 40 | $FM_{31}$ | (2,4,2) | 16 | 69 | 61 | $FM_{55}$ | (4,10,4) | 160 | 14 | 7 |
| $FM_8$ | (2,3,6) | 36 | 57 | 58 | $FM_{32}$ | (4,6,2) | 48 | 47 | 30 | $FM_{56}$ | (2,10,5) | 100 | 28 | 44 |
| $FM_9$ | (3,3,3) | 27 | 61 | 57 | $FM_{33}$ | (6,3,5) | 90 | 33 | 22 | $FM_{57}$ | (2,10,6) | 120 | 23 | 19 |
| $FM_{10}$ | (5,7,5) | 175 | 11 | 16 | $FM_{34}$ | (7,2,1) | 14 | 71 | 48 | $FM_{58}$ | (4,9,1) | 36 | 54 | 34 |
| $FM_{11}$ | (2,4,3) | 24 | 64 | 60 | $FM_{35}$ | (5,6,4) | 120 | 24 | 11 | $FM_{59}$ | (6,10,3) | 180 | 8 | 12 |
| $FM_{12}$ | (2,6,2) | 24 | 63 | 66 | $FM_{36}$ | (3,8,7) | 168 | 12 | 14 | $FM_{60}$ | (2,10,5) | 100 | 29 | 46 |
| $FM_{13}$ | (1,6,5) | 30 | 59 | 70 | $FM_{37}$ | (2,8,5) | 80 | 37 | 43 | $FM_{61}$ | (2,9,5) | 90 | 32 | 38 |
| $FM_{14}$ | (2,6,4) | 48 | 48 | 56 | $FM_{38}$ | (2,8,8) | 128 | 20 | 29 | $FM_{62}$ | (3,10,5) | 150 | 17 | 26 |
| $FM_{15}$ | (2,6,5) | 60 | 42 | 53 | $FM_{39}$ | (1,6,8) | 48 | 46 | 54 | $FM_{63}$ | (2,9,2) | 36 | 56 | 49 |
| $FM_{16}$ | (2,5,8) | 80 | 36 | 28 | $FM_{40}$ | (4,8,8) | 256 | 2 | 3 | $FM_{64}$ | (4,9,5) | 180 | 10 | 10 |
| $FM_{17}$ | (6,5,3) | 90 | 34 | 20 | $FM_{41}$ | (1,3,5) | 15 | 70 | 71 | $FM_{65}$ | (2,10,2) | 40 | 51 | 55 |
| $FM_{18}$ | (5,4,8) | 160 | 15 | 9 | $FM_{42}$ | (2,4,5) | 40 | 53 | 63 | $FM_{66}$ | (2,9,8) | 144 | 19 | 25 |
| $FM_{19}$ | (6,4,9) | 216 | 5 | 2 | $FM_{43}$ | (6,6,2) | 72 | 39 | 33 | $FM_{67}$ | (3,10,5) | 150 | 18 | 27 |
| $FM_{20}$ | (6,5,7) | 210 | 6 | 15 | $FM_{44}$ | (4,3,8) | 96 | 31 | 24 | $FM_{68}$ | (2,9,6) | 108 | 26 | 42 |
| $FM_{21}$ | (2,7,7) | 98 | 30 | 47 | $FM_{45}$ | (6,6,2) | 72 | 40 | 37 | $FM_{69}$ | (2,9,3) | 54 | 45 | 52 |
| $FM_{22}$ | (2,8,8) | 128 | 21 | 18 | $FM_{46}$ | (4,7,2) | 56 | 44 | 31 | $FM_{70}$ | (4,10,6) | 240 | 3 | 5 |
| $FM_{23}$ | (4,8,7) | 224 | 4 | 8 | $FM_{47}$ | (6,6,1) | 36 | 55 | 35 | $FM_{71}$ | (2,10,8) | 160 | 13 | 21 |
| $FM_{24}$ | (2,9,7) | 126 | 22 | 23 | $FM_{48}$ | (1,7,3) | 21 | 65 | 69 | | | | | |

*Fisheries & aquaculture*

*Trends in Maritime Technology and Engineering – Guedes Soares & Santos (eds)*


# Assessment of macroinvertebrates culture in an integrated multitrophic aquaculture system

J.P. Garcês
*IPMA - Portuguese Institute for the Ocean and Atmosphere, EPPO - Aquaculture Research Station, Olhão, Portugal*

N. Diogo
*UALG – University of Algarve, Faculty of Science and Technology, Faro, Portugal*

S. Gamito
*UALG and CCMAR – Center of Marine Sciences, University of Algarve, Faro, Portugal*

ABSTRACT: A continuously expanding world population along with less availability in natural resources requires an aquaculture sector that must be able to produce more with less. Integrated Multitrophic Aquaculture Systems (IMTA) are more efficient food production systems, allowing to grow together different organisms using an ecosystem-based approach. This work represents a preliminary assessment of the potential of production of invertebrates, including filter feeders (oysters), herbivorous/detritus feeders (sea urchins), and detritus feeders (sea cucumbers and peanut worms), using pumped nutrient enriched water from a fish pond. Good results were attained concerning oysters' growth, while for sea urchins the growth was temperature-dependent, with replacement by sea cucumbers when the temperature raised above a certain level. Regarding the peanut worms, several reasons may have contributed to the poor results, such as high temperature, or excessive water renewal in the experimental tanks and lack of appropriate food items. Further research and experimental work are necessary to define the best environmental conditions appropriate to the culture of some of the invertebrates tested, such as the peanut worms.

## 1 INTRODUCTION

Integrated multi-trophic aquaculture (IMTA) combines the production of fed species (finfish or shrimps) with extractive species which utilize the inorganic (e.g., seaweeds or other aquatic vegetation) and organic (e.g., suspension and deposit feeders) excess nutrients for their growth (Chopin et al. 2008). Through IMTA a bio-mitigation process occurs, because inorganic and organic wastes from fed-aquaculture organisms are assimilated by co-cultured autotrophic and heterotrophic species, respectively (Neori et al. 2004; Chopin 2013). According to Wilfart et al. (2020), IMTA systems are designed to increase system efficiency by optimizing the use of nutrients and energy in production, by decreasing nutrient loss, diversifying aquaculture products and generating and using different types and levels of ecosystem services.

Presently, aquaculture is still operating in mono-species systems, with consequent loss of nutrients and organic matter (Islam 2005; Franco-Nava et al. 2004; Herbeck et al. 2013). The diversification, as an alternative to monoculture, is one of the best paths to environmental sustainability, economic viability and social acceptance.

However, sustainable expansion of aquaculture worldwide requires technologies for the recycling of matter and energy (Blancheton et al. 2009; Jegatheesan et al. 2011; Martins et al. 2010). The IMTA concept represents a win-win solution (Lazaro and Sanchez-Jerez 2020), allowing to increase the yields of low-trophic level species through the extra food supply, while reducing the input of organic waste, limiting the environmental impact (Soto, 2009). Waller et al (2015) emphasize that a possible way to improve sustainability of aquaculture is by reproducing natural nutrient cycles on small scale. In practise, a major challenge for IMTA is a balanced system between waste production by fish and remediation by the extractive species (heterotrophic filter feeders and detritivores), as well as autotrophic seaweeds and a range of plant species (Reid et al. 2008).

Using only water from a fish pond, Garcês et al (2020) developed a multitrophic system, with five different species of invertebrates, in perfect balance, demonstrating the possibility to combine the production of these valuable species at a very low cost.

The primary aims of this study were to (a) evaluate the growth of peanut worm (*Sipunculus nudus*,

DOI: 10.1201/9781003320289-59

a detritivore) in a multitrophic culture, associated with a filter feeder or an herbivorous/ detritivore species.

## 2 METHODS

### 2.1 *Experimental design*

An experimental set of 9 tanks (50x50x50 cm) (Figure 1) was used, from April to September 2019, with four different species: oysters (*Crassostrea angulata*), sea urchins (*Paracentrotus lividus*), peanut worms (*Sipunculus nudus*) and sea cucumbers (*Holothuria arguinensis*). Peanut worms are subsurface detritus feeders relatively common on the low intertidal sandy muds in Ria Formosa and Ria de Alvor, (Algarve, South Portugal). All other invertebrate species used in the experiment naturally live above the sediment (oysters and sea cucumbers) or on hard substrate (sea urchins). Oysters are filter feeders, while sea urchins are herbivorous/detritus feeders and sea cucumbers are surface detritivores.

To evaluate the possibility of interaction between these species, belonging to different functional groups, three different treatments were compared and replicated three times. The combinations in each treatment are shown in Table 1.

Figure 1. Rearing tank used. EPPO, April 2019.

Each tank was filled with a 10 cm sand layer collected in Ria Formosa, previously screened to remove any associated macrofauna, sun dried and rinsed several times with freshwater and seawater alternately. The tanks positions and the organisms' combinations were randomly chosen to reduce any positions effects.

The water was pumped from a semi-intensive fish pond and circulated from the bottom to top, allowing a uniform water movement across the substrate and preventing the creation of stratified and anoxic zones

The food was provided by the nutrient enriched water and by the waste generated from the organisms present in the tanks, mainly oysters and sea urchins. No additional food was provided, except in tanks with sea urchins, which were daily supplied *ad libitum* with a macroalgae, *Ulva* spp.

The juvenile's oysters and sea urchins were placed inside the tanks in a suspended tray (Figure 1), avoiding close contact with the bottom and to allow a good dispersion of the faeces. The tanks were regularly monitored to determine mortality. The sea urchins were found to be negatively affected by water temperatures above 22°C, being thus replaced in June by sea cucumbers.

Water parameters: temperature, dissolved oxygen and turbidity were measured by a multiparameter water quality meter Hanna instruments model HI9829, salinity by a refractometer Hanna instruments HI83306, and Chlorophyll by a meter Hanna instruments HI7609829-4.

Table 1. Species combination and number in each tank.

| Tanks | Species Combination | Number |
|---|---|---|
| 1 | Peanut Worm (Pw) | 5 |
| 2 | Peanut Worm | 5 |
| | Sea urchin (Surch) | 17 |
| | Sea cucumber (Scu) (in July) | 3 |
| 3 | Peanut Worm | 5 |
| | Oyster (Oyst) | 20 |
| 4 | Peanut Worm | 5 |
| 5 | Peanut Worm | 5 |
| | Oyster | 20 |
| 6 | Peanut Worm | 5 |
| 7 | Peanut Worm | 5 |
| | Sea urchin | 17 |
| | Sea cucumber (in July) | 3 |
| 8 | Peanut Worm | 5 |
| | Oyster | 20 |
| 9 | Peanut Worm | 5 |
| | Sea urchin | 17 |
| | Sea cucumber (in July) | 3 |

### 2.2 *Indicators*

Growth data were monthly collected for each combination and tank, in a total of five samplings. Wet weight (g), for all the species, length (mm) and width (mm), only for oysters and sea urchins were adopted as biometric measures. To compare the growth between the different species used, the daily growth rate standardized (DGRs) by average body mass or the daily growth rate (DGR) were calculated as follows:

$$DGRs\left(gd^{-1}g^{-1}\right) = \left(\left(W_f - W_i\right)/\left(t_f - t_i\right)\right)/\bar{W} \quad (1)$$

$$DGR\left(gd^{-1}\right) = \left(W_{t2} - W_{t1}\right)/\left(t_2 - t_1\right) \quad (2)$$

Where: $\bar{W}$ – average biomass (g) of each experimental species; $W_f$ - $W_i$ = observed final and initial weight, over the total experimental time ($t_f$ - $t_i$); $W_{t2}$

and $W_{t1}$ = observed mean weight in two consecutive sampling dates ($t_2$ and $t_1$), in each treatment.

Growth was measured as the increase in weight, using an electronic 2 decimals point balance, the remaining parameters, with a digital calliper with a resolution of 0.1 mm and 2 mm accuracy.

### 2.3 *Statistical analysis*

Statistical analysis was performed using SPSS software (version 18; SPSS inc, Chicago, Il). The differences of weight, length, width and thickness between the different species combinations for each treatment were analysed via one-way Anova. The relationships between combinations and time on species growth was analysed by two-factor ANOVA, followed by a Tukey test when significant differences were found, considered at a 0.05 significance level. All data are presented as the mean ± confidence interval for the mean with a significance level of $\alpha=0.05$ (CI) of three replicates.

## 3 RESULTS AND DISCUSSION

### 3.1 *Water physical parameters*

The number of days in tank from introduction with measurement to the end, differed for the species in study, corresponding to 126 and 134 for peanut worms and oysters, respectively, while for sea urchins it was of 64 days and for sea cucumbers, 67. There were no significant differences between replicates ($p>=0.05$) for the different parameters under study. However, along the 4 months' duration of the study, gradual and slight variations in water temperature (18-26°C) and salinity (34-36 ppm) were observed (Figure 2).

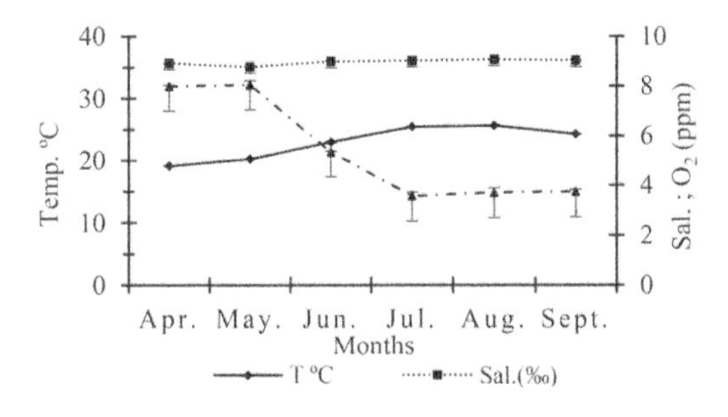


Figure 2. Water physical and chemical parameters: Temperature (°C), Salinity (Sal) and Dissolved oxygen concentrations (ppm) throughout the study.

The marked decrease in dissolved oxygen from June onwards, coincident with an increase in water temperature, did not affect the normal activity of the oysters, which could be confirmed by their growth. However, the sea urchins were found to be negatively affected by water temperatures above 22°C. At this point, they were at low vitality levels, although still alive (Figure 3), leading to their replacement by the more temperature-resistant sea cucumbers, in July.

Figure 3. Photo showing the poor status of sea urchins in June. EPPO, 2019.

Water turbidity and chlorophyll (Figure 4) remained relatively low and constant, with a slight drop in the summer months.

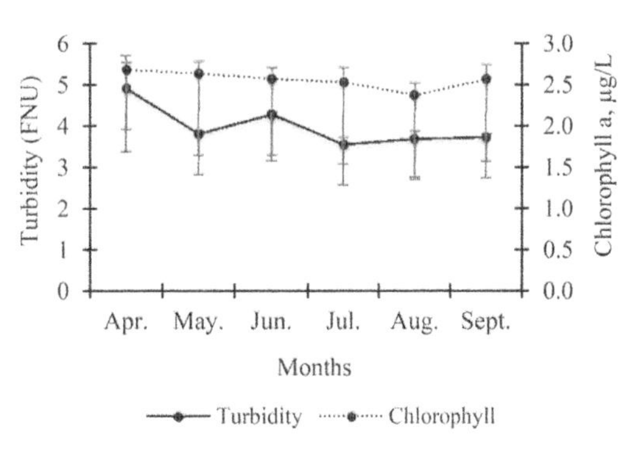


Figure 4. Turbidity (FNU) and chlorophyll a, µg/L concentrations throughout the study.

### 3.2 *Integrated-culture system*

The results obtained showed that the different species combinations used worked quite well both in terms of quality and efficiency of production. No significant differences ($p>0.05$) were found in growth between replicates of each treatment during the 4 months of the study.

However, significant differences were found among species. Only peanut worms did not present positive daily growth rates (Figure 5). Oysters, sea urchins and sea cucumbers grow at different rates, with oysters presenting the highest values.

Despite the negative result obtained in tank 9 for sea urchins, they grew well in the other tanks, till the rise of temperature above 22°C. The weak daily growth observed for sea cucumbers in tank 2 was attributed to their already considerable size at the time of the introduction in the tanks.

Mean weight variations for peanut worms in each treatment is represented in Figure 6. Comparing the three treatments significant differences were only observed in summer months (August and September) for all treatments, suggesting the existence of negative temperature effects.

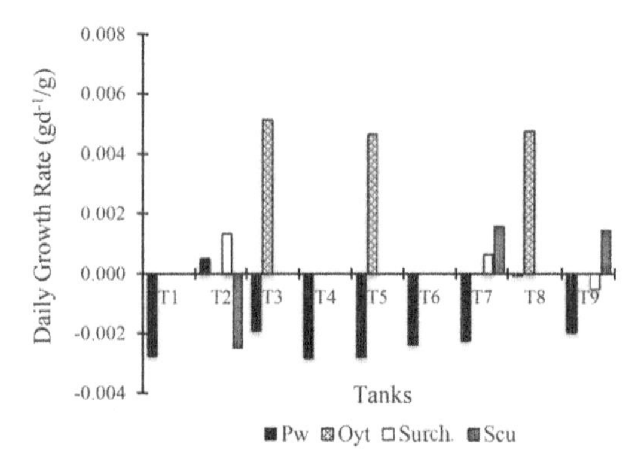


Figure 5. Daily growth rate standardized by average body mass (DGRs) in each tank. Tanks: T1 to T9. $\Delta t_{Pw}$=126 days; $\Delta t_{Oyst.}$=134 days; $\Delta t_{Surch}$=64 days; $\Delta t_{Scu}$=67 days.

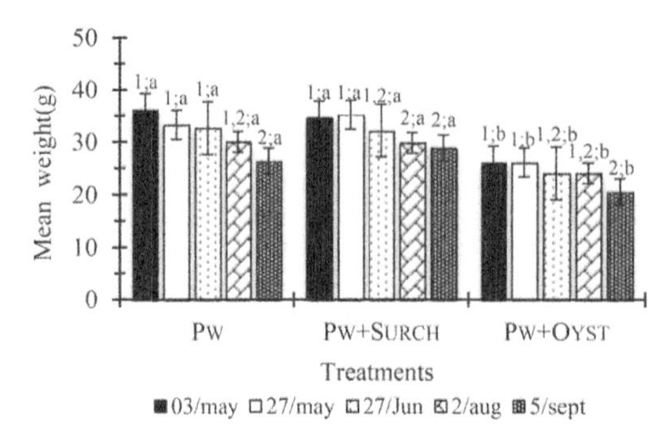


Figure 6. Mean weight variations for peanut worms in each treatment along the five sampling periods. Numbers correspond to significant differences (P<0.05) for the same treatment over time. Letters correspond to significant differences (P<0.05) between the three treatments for the same month. Means that have the same superscripts are not significantly different (P> 0.05).

In relation to the treatment "peanuts with oysters" (Pw+Oyst), despite the smaller initial weight of the former, which was significantly different from the others, the decrease in growth was only significant ($P$<0.05) in September, compared with initial weights suggesting a good relationship.

Nevertheless, and according to the daily growth rate in Table 2, only peanuts with sea urchins presented positive values in May, suggesting a positive effect in peanuts growth.

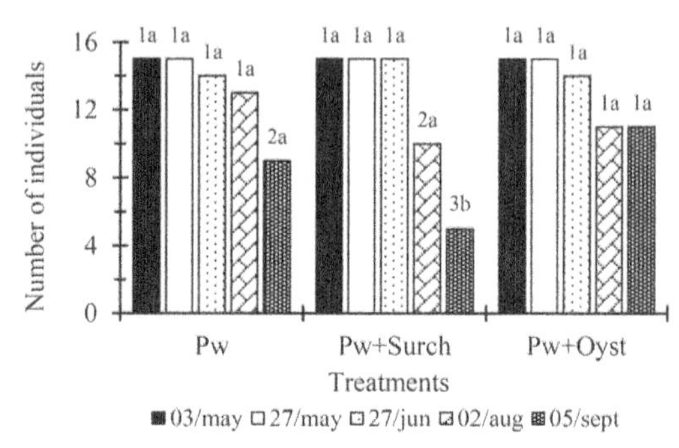


Figure 7. Peanut survival along the study. Numbers correspond to significant differences (P<0.05) for the same treatment over time Letters correspond to significant differences (P<0.05) between the 3 treatments for the same month. Means that have the same superscripts are not significantly different (P> 0.05).

Table 2. Peanut daily growth rate (DGR) for all treatments without mortality.

| | Daily growth rate (g/d$^{-1}$) | |
|---|---|---|
| **Treatments** | **27/may** | **27/jun** |
| **PwC** | -0,12 | -0,02 |
| **Pw+Surch** | 0,02 | -0,10 |
| **Pw+Oyst** | -0,07 | -0,07 |

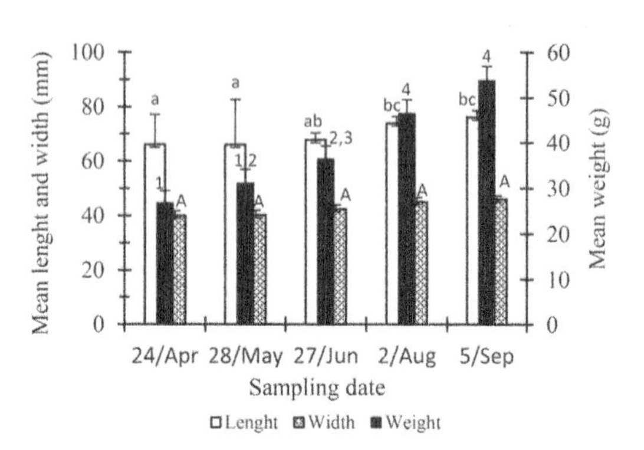


Figure 8. Mean length, width and weight variation throughout the experience. The numbers correspond to weight(g), lowercase letters for length(mm) and uppercase letters for width (mm). Means that have the same superscripts are not significantly different (P> 0.05).

The low density of oysters (see Table 1), combined with the high-water temperature, may have contributed to the negative growth occurred, in this treatment. Cai and Jiang (2011) reported a high survival rate of the worm, *S. nudus* when co-cultured with the clam, *Meretrix petechialis* in the intertidal zone.

Regarding the oysters, despite the low values of turbidity and Chlorophyll (see Figure 4), a high and continuous growth was observed, essentially in biomass (Figure 8).

However, significant differences were only observed in August and September. Survival rate was of near 100% along the entire study.

The sea urchins also grew during the study (Figure 9). The decrease in their weight from June onwards reflects the rise in water temperature showed a great deterioration of their condition as the water temperature increased above 22 °C in July and had to be replaced by sea cucumbers (see Figure 3).

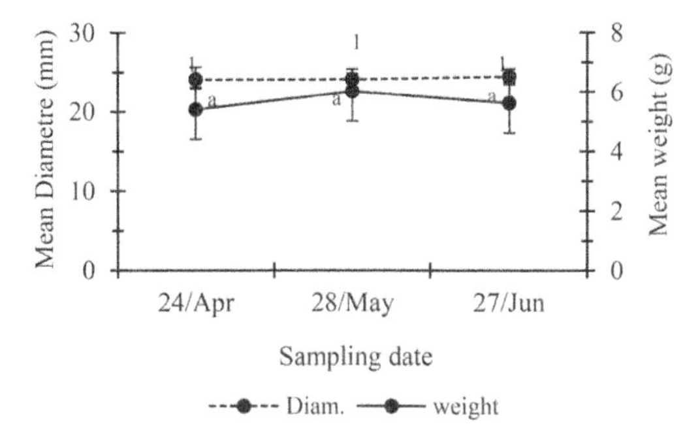


Figure 9. Sea urchins mean diameter and weight throughout the experience. The numbers correspond to weight(g), and lowercase letters for diameter (mm). Means that have the same superscripts are not significantly different (P> 0.05).

The poor growth shown by sea cucumbers may be related to the fact that the individuals used are of considerable size (Figure 10). However, there was no mortality.

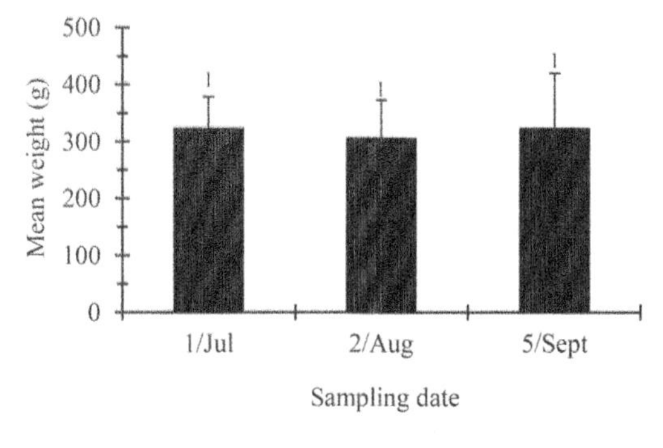


Figure 10. Sea cucumber mean weight variation throughout the experience. The numbers correspond to weight(g). Means that have the same superscripts are not significantly different (P> 0.05).

Although the negative growth obtained for *S. nudus* in this study, the positive growth rate obtained in May co-cultured with sea urchins (0.02g $d^{-1}$) (see Table 2) was higher than the growth obtained by Juwhei et al., (2015) (0.002 $gd^{-1}$), and slightly lower than the achieved by Wen et al. (2012) (0.07 $gd^{-1}$). Peanuts worms are deposit feeders that settle on various substrata and ingest particulate food (microalgae, protista, meiofauna, detritus, and fecal pellets) contained in the water column, sediment interface, and sediment (Murina, 1984; Kedra and Kowalczuk, 2008; Li et al., 2016 and references there in). According to Junwei et al. (2017), sands with a small grain size and organic materials are a key factor for the growth of *S. nudus*. The *S. nudus* weight loss occurred throughout the study may be due to the lack of enough appropriate food in the tanks that the food may not have been enough, as suggested by Junwei et al. (2015). These authors, in experiments of co-cultered sea peanuts with *Mugil cephalus*, concluded that the *S.nudus* weight loss observed in high densities were due to food limitation, because the waste produced by *Mugil cephalus* could only satisfy the requirements of a restricted number of *S. nudus*; therefore, *S.nudus growth was* density-dependent since the highest weight gain (0.1 g) was observed at the lowest stocking density.

In fact, the poor growth found in this study for peanut worms may be related with a low food availability which was limited to the waste produced by co cultured species along with water nutrients. However, other reasons may have contributed for this result, such the high temperatures and the water renewal from the bottom to the surface, creating an upward water current preventing the deposition of organic matter and detritus in the sediment.

On the contrary, the other species used in this study grew. The slight growth evidenced by sea cucumbers reveals that there was enough food on the sediment surface. The fact that peanut worms had lost weight when in association, suggests a greater nutrient requirement for this species.

The oysters showed a continuous growth in weight and shell length, revealing the presence of sufficient phytoplankton and dissolved organic matter in the water pumped from the fish ponds, and a good ability to adapt to the culture conditions associated with other species.

## 4 CONCLUSIONS

Despite the poor growth obtained for peanut worms, this study demonstrated the suitability of this species for culture with other animals. Their mortality rate increased with the water temperature rise, in all treatments, what indicates that the cultivation conditions need to be improved regarding this environmental parameter. The water circulation might also need to be improved in order to allow an accumulation of organic matter and increase food availability.

Oysters and sea-cucumbers showed to be appropriate species to live exclusively from fish pond effluents, in an IMTA, presenting positive growth and almost null mortality rates.

Further research and experimental work are necessary to define the best environmental conditions appropriate to the culture of some of the invertebrates tested.

## ACKNOWLEDGEMENTS

This study was carried out at the scope of the project: Integrate - *Integrate Aquaculture: an eco-innovative solution to foster sustainability in the Atlantic Area* EAPA_232/2016. This study received Portuguese national funds from FCT - Foundation for Science and Technology through project UIDB/04326/2020.

## REFERENCES

Blancheton, J.P., Bosc, P., Hussenot, J.M.E., d'Orbcastel, E.R., Romain, D. 2009. The 'new' European fish culture systems: recirculating systems, offshore cages, integrated systems. Cahiers d'Agriculture. 18: 227–234.

Cai, D.J., Jiang, Y., 2011. The polyculture of *Sipunculus nudus* with *Lioconcha castrensis* on the beaches. Sci. Fish Farming 11, 37–38 (Chinese with English abstract).

Chopin T., Robinson, S.M.C., Troell, M., Neori, A., Buschmann, A.H., Fang, J. 2008 Multitrophic integration for sustainable marine aquaculture. In: Jørgensen S. E., Fath, B.D. (eds) The Encyclopedia of Ecology, Ecological Engineering, Elsevier, Oxford, 3: 2463–2475.

Chopin, T. 2013. Aquaculture, Integrated Multi-Trophic (IMTA). Encyclopedia of Sustainability Science and Technology, 12:542–564.

Franco-Nava, M.A., Blancheton, J.P., Deviller, G., Charrier, A., Le-Gall, J.Y. 2004. Effect of fish size and hydraulic regime on particulate organic matter dynamics in a recirculating aquaculture system: elemental carbon and nitrogen approach. Aquaculture. 239:179–198.

Garcês, J.P., Quental-Ferreira, H., Theriaga, M., Neto, D., Pousão-Ferreira, P. 2021. A new approach to sustainable integrated cultures. In: Developments in Maritime Technology and Engineering, Guedes Soares, C., Santos, T. A. (Eds), Taylor & Francis, UK, 2: 699–704.

Herbeck, L.S., Unger, D., Wub, Ying., Jennerjahn, T.C. 2013. Effluent, nutrient and organic matter export from shrimp and fish ponds causing eutrophication in coastal and back-reefwaters of NE Hainan, tropical China. Cont. Shelf. Res. 57:92–104.

Islam, M.S. 2005 Nitrogen and phosphorus budget in coastal and marine cage aquaculture and impacts of effluent loading on ecosystem: review and analysis towards model development. Mar. Pollut. Bull. 50:48–61

Jegatheesan, V., Shu, L., Visvanathan, C. 2011 Aquaculture Effluent: impacts and remedies for protecting the environment and human health. reference module in earth systems and environmental sciences. Encycl. Environ. Health. 201, 123–135.

Junwei, L., Xiaoyong, X., Changbo Z., Yongjian, G., Suwen, C. 2017. Edible Peanut Worm (*Sipunculus nudus*) in the Beibu Gulf: Resource, Aquaculture, Ecological Impact and Counterplan. *J. Ocean Univ. China* (Oceanic and Coastal Sea Research) Review, 16 (5): 823–830.

Junwei, L., Changbo, Z., Yongjian, G., Xiaoyong, X., Guoqiang H., Suwen C. 2015. Experimental study of bioturbation by Sipunculus nudus in a polyculture system. Aquaculture 437:175–181.

Kedra, M., and Maria, W. K., 2008. Distribution and diversity of sipunculan fauna in high Arctic fjords (west Svalbard). Polar Biology, 31: 1181–1190.

Li, J. W., Zhu, C. B., Guo, Y. J., Xie, X. Y., Huang, G. Q., Chen, S. W. 2015. Experimental study of bioturbation by *Sipunculus nudus* in a polyculture system. *Aquaculture*, 437:175–181.

Martins, C.I.M., Eding, E.H., Verdegem, M.C.J., Heinsbroek, L.T.N., Schneider, O., Blancheton, J.P., d'Orbcastel, E.R., Verreth, J.A.J., 2010. New developments in recirculating aquaculture systems in Europe: a perspective on environmental sustainability. Aquacultural Engineering 43: 83–93.

Murina, G.V.V., 1984. Ecology of sipuncula. Mar. Ecol. Prog. Ser. 17, 1–7.

Neori, A., Chopin, T., Troell, M., Buschmann, A. H., Kraemer, G. P., Halling, C., Shpigel, M. and Yarish, C. 2004. Integrated aquaculture: rationale, evolution and state of the art emphasizing seaweed biofiltration in modern mariculture. Aquaculture, 231: 361–391.

Reid, G.K., Robinson, S.M., Chopin, T.R., Mullen, J., Lander, T., Powell, F. et al. 2008. Recent developments and challenges for open water, integrated multi-trophic aquaculture (IMTA) in the Bay of Fundy, Canada. Bull. Aquac. Assoc. Can. 12.

Sanz-Lazaro, C., Sanchez-Jerez, P. 2020. Regional Integrated Multi-Trophic Aquaculture (RIMTA): Spatially separated, ecologically linked. Journal of Environmental Management, 271

Soto, D. 2009. Integrated Mariculture: a Global Review. Food and Agriculture (No. 529). Food and Agriculture Organization of the United Nations (FAO).

Waller, U; Buhmann, A. K., Ernst, A., Hanke, V., Kulakowski, A., Wecker, B., Orellana, J., Papenbrock, J. 2015. Integrated multi-trophic aquaculture in a zero-exchange recirculation aquaculture system for marine fish and hydroponic halophyte production. Aquaculture International, 23(6).

Wen, X., Wang, Z.C., Liang, Z.H., Jiang, Y., Peng, H.J., Yang, J.L., 2012. The polyculture experiment of *Sipunculus nudus* with shrimp *Litopenaeus vannamei*. Fish. Sci. Technol. Inf. 39, 263–265.

Wilfart, A., Favalier, N., Metaxa, I., Platon, C., Pouil, S., et al. 2020. Integrated multi-Trophic Aquaculture in ponds: what environmental gain? An LCA point of view. 12th Inter- national Conference on Life Cycle Assessment of Food 2020 (LCA Food 2020), Towards Sustainable Agri-Food Systems: 206–208.

*Trends in Maritime Technology and Engineering – Guedes Soares & Santos (eds)*


# Experimental investigation of an array of vertical flexible net-type structures under regular waves

Y.C. Guo, Z.C. Liu, S.C. Mohapatra & C. Guedes Soares
*Centre for Marine Technology and Ocean Engineering (CENTEC), Instituto Superior Técnico, Universidade de Lisboa, Lisbon, Portugal*

ABSTRACT: The interaction between an array of vertical flexible net-type structures and regular waves is investigated by model tests. The present study focuses on the effect of the reflection, the transmission, and the dissipation coefficients through the system consisting of an array of nets made of nylon silky and supported by the vertical steel frame. Several tests are conducted for different wave periods, wave heights, and distances among vertical nets on the reflection, the transmission, and the dissipation coefficients. The performance of the net model is analysed by comparing the performance of different number of vertical nets. It is found that the arrangement of three nets reduces more wave heights than those of two- and one- vertical flexible net models and in this sense, it operates as a floating breakwater. The present model tests analysis will be helpful for better understanding and modelling the effect of several flexible fish cages using net-type structure in offshore aquaculture.

## 1 INTRODUCTION

Vertical flexible porous structures could be used to solve the problem associated with increasing wave disturbance which are common shore protection structures as breakwaters and application to fish cages. They can dissipate the incident wave energy and reduce the reflected wave height and are considered an effective replacement for traditional impermeable breakwater, especially for the reduction of wave energy in harbours and fishing ports. There has been a very small number of studies made by very few researchers on the wave interaction with a porous vertical net-type structure to analyse the different wave quantities by the various vertical porous structures (see Lee and Lo, 2002; Esmaeili et al., 2019) of multiple surface penetrating flexible wave barriers based on both analytically and experimentally the wave energy dissipation of a surface-piercing slotted barrier. On the other hand, recently, the dynamic analysis of an array of fish cages under the action of waves and currents based on numerical methodologies were studied (see Zhang et al., 2021) and the effect of mooring lines on the multiple fish cages (see Liu et al., 2022).

A series of experimental studies of different types of net panels were conducted to investigate the hydrodynamic forces on the net panels and the effect of net solidity (see Song et al., 2006; Lader et al., 2007; Esmaeili et al., 2019; Donga et al., 2019; Xu et al., 2021; Føre et al., 2021).

Another interesting aspect of the porous membrane and flexible net-type structures and applications in breakwaters and fish cage models is based on various methodologies (see Guo et al., 2020c). Different analytical models associated with submerged flexible porous membrane in finite water depth are developed to analyse various wave quantities such as reflection, transmission, and dissipation coefficients for application as an effective breakwater (see Guo et al. 2020a, 2020b). Further, Guo et al. (2021) studied analytically the effect of the submerged horizontal flexible porous membrane near a vertical rigid wall over flat bottom and the effect of the vertical flexible porous membrane in Guo et al. (2022).

Fish cage models associated with flexible net and frame system has been investigated by several researchers, providing rigorous analysis of net deformations and hydrodynamic of frame system by experimental studies (see Shen et al., 2021; Miao et al., 2021; Brizzi and Sabbagh, 2021) and numerically by Liu et al. (2021). Mohapatra et al. (2021) studied the wave-induced loads on the flexible net-type fish cage based on analytical and numerical approaches. However, experimental investigations concerned with vertical flexible net structure (VFNS) as breakwater is not available till-date to the literature.

Therefore, in the present study, a couple of model tests are conducted at LNEC (National Laboratory for Civil Engineering), Portugal, to investigate the wave reflection, the transmission, and the wave energy dissipation by an array of VFNSs under regular waves. The paper is organized as follows: Section 2 demonstrates the experimental setup, model description, and wave characteristics. In the present experiment, the

DOI: 10.1201/9781003320289-60

data process method is presented in Section 3. In Section 4, the effect of VFNSs on different design parameters are discussed and comparisons between one, two, and three vertical net-type models of the reflection, the transmission, and the dissipation coefficients are analysed. Finally, in Section 5, the findings from the present model tests are highlighted and the future scope of the work is discussed.

## 2 EXPERIMENTAL SETUP

### 2.1 *Experimental equipment*

The model tests are conducted in the National Laboratory for Civil Engineering (LNEC), Portugal (as shown in Figure 1). A piston-type wavemaker is installed at one end of the flume to generate regular waves whilst at the other end of the flume, a rock slope absorption area is provided to effectively absorb the incident waves energy. In this experimental study, the length, width, and height of the two-dimensional wave flume are 35m, 0.62m, and 1m, respectively. The free surface elevation inside the wave flume was measured by wave gauges (WGs) and the voltage signals were captured by a signal box, which can amplify or decrease the signal and sent to the computer. The sampling frequency is 40 Hz.

Figure 1. The wave flume in LNEC.

Further, 4 WGs are placed in front of and behind the tested net-type breakwater (see Figure 2) to measure the reflection and the transmission coefficients. The WG8 is placed 4.2 m before the front edge of the rock slope absorption area, whilst the WG4 is placed 3.65 m in front of the WG5. The distances between WG1 & WG2, WG2 & WG3, and WG3 & WG4 are 30 cm, 20 cm and 38 cm, respectively. The distances between WG5 & WG6, WG6 & WG7, WG7 & WG8 are the same as those between WG1-WG4.

### 2.2 *Details of the vertical net-type structure*

The vertical flexible net-type breakwater model is made based on the Froude Similarity with a ratio scale of 1:30. As shown in Figure 3, the vertical net is made of nylon material and the diameter of the mesh hole is $d$. The mass density of the mesh is $m_s$. The porosity of the net is defined as the ratio of the total area of holes and the whole net panel area. The porosity of the current nylon net is $\varepsilon$.

The vertical flexible net-type system model consists of two components, a nylon net panel, and a stainless-steel frame (see Figure 4). The steel frame has a steel plate and two steel rods (cylinder-type) of equal diameter.

The net is placed on the stainless-steel frame and stretched with light hand force to form a smooth panel, then the four edges of the net are fixed into the frame by thin wires. The dimensions and properties of the flexible net-type structure are provided in Table 1. Throughout the experiment, the stainless-steel frames were put on the top of the flume and fixed by placing heavy stones on top of them.

This experimental test mainly focuses on the efficiency of incident wave attenuation for a for a set of one, two, and three consecutive vertical nets with different distances among them, as plotted in Figure 2. The back edge of the vertical system of nets is placed 2.65 m behind the WG4 and 1m in front of the WG5. The distance between the first (the front edge) and second net is defined as $L_1$, whilst the distance between the second and third net (the back edge) is defined as $L$.

### 2.3 *Characteristics of waves*

In the present experimental study, all incident waves are considered to be regular waves and the wave characteristics used in the experiments are indicated in Table 2.

## 3 DATA ANALYSIS METHOD

A four-point method extended from the 3-points method of Mansard and Funke (1980) is used to separate the incident, the reflected, and the transmitted waves, and to calculate the reflection and the transmission coefficients $K_r$, $K_t$, from the records of composite waves by using four-wave gauges before and behind the VFNS. After the reflection coefficient $K_r$ and the transmission coefficient $K_t$ are known, the dissipation coefficient $K_d$ can be computed as:

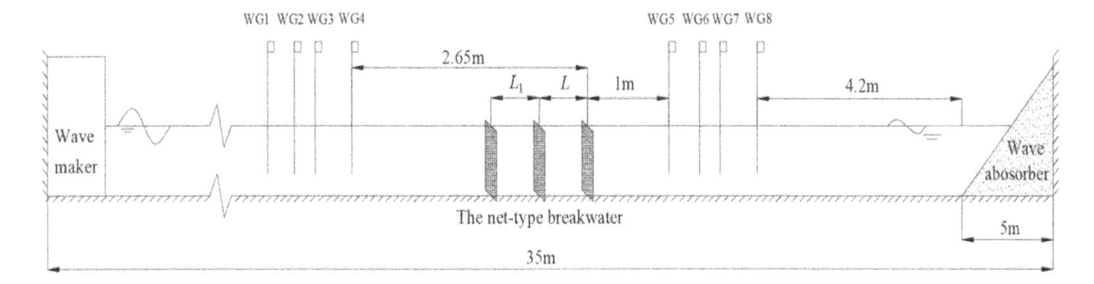


Figure 2. Sketch of an array of three-consecutive net-type structures and wave gauge arrangement.

Figure 3. Indicative mesh of the nylon net.

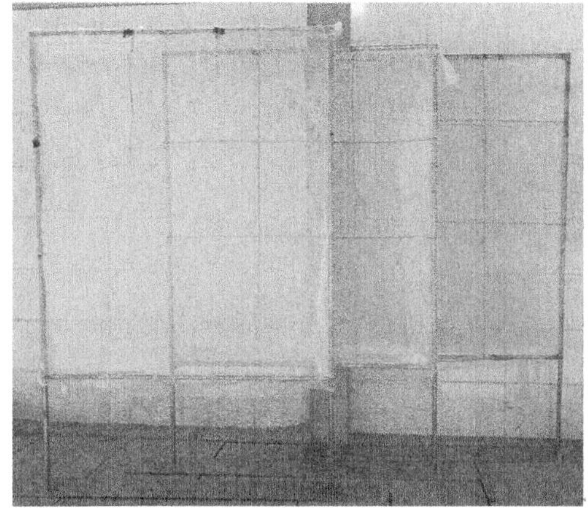

Figure 4. The tested net panels and the stainless-steel frames.

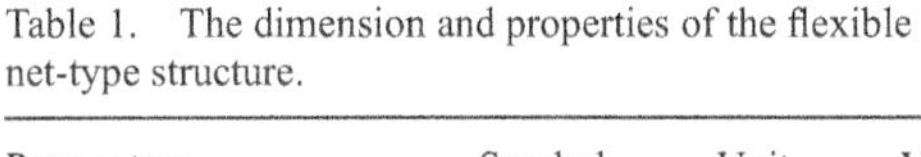

Table 1. The dimension and properties of the flexible net-type structure.

| Parameters | Symbol | Unit | Value |
|---|---|---|---|
| Diameter of mesh hole | $d$ | mm | 2 |
| Mass density | $m_s$ | kg/m$^2$ | 0.124 |
| Porosity | $\varepsilon$ | - | 0.32 |
| Net width | $b$ | mm | 600 |
| Net height | $l$ | mm | 750 |

Table 2. Wave parameters used to set up the experiment.

| Parameters | Range |
|---|---|
| Water depth (m) | 0.35 |
| Wave height (m) | 0.04, 0.06, 0.08 |
| Wave period (s) | 0.8-3.0 (increment 0.2) |

$$K_d = 1 - K_r^2 - K_t^2. \quad (1)$$

## 4 EXPERIMENTAL RESULTS AND DISCUSSIONS

### 4.1 *One vertical flexible net*

Figure 5 shows the effect of different wave heights $H$ on (a) the reflection and the transmission coefficients $K_r$, $K_t$, and (b) the dissipation coefficient $K_d$ of a single vertical net panel as a function of wave period $T$. In Figure 5(a), it is observed that for all three wave heights, the transmission coefficient $K_t$ is greater than 0.9 and the reflection coefficient $K_r$ is less than 0.2. This means that less than 10% of the incident waves are prevented and up to 20% of the incident wave is reflected by the vertical net with a single net panel. Further, the vertical net reflects more incident waves with a wave height $H$=0.08 m, when compared with the other two wave height cases. Nevertheless, the least wave transmission happens in waves with medium wave height $H$=0.06 m.

Figure 5(b) shows that very less incident wave energy can be dissipated by the vertical flexible net-type structure with a single net panel when the incident wave height is large or small ($H$=0.08 m,

0.04 m). For the waves with a medium wave height ($H$=0.06 m), this structure has better performance in dissipating the incident wave energy in the short-wave regime (1.0s≤$H$≤1.2s) and longwave regime (2.2s≤$H$≤3.0s). It can be concluded that a single vertical flexible net panel is inefficient for preventing incident waves.

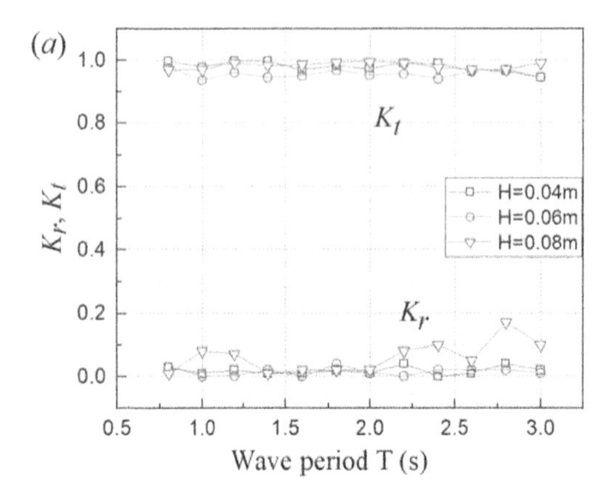


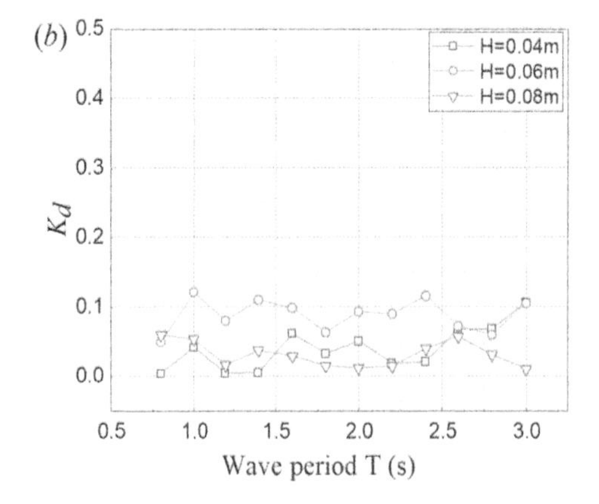


Figure 5. Effects of different wave height $H$ on (a) the $K_r$, $K_t$, and (b) the $K_d$ of the VFNS with one net.

## 4.2 *System of two vertical flexible nets*

Figure 6 reveals the effect of different wave height $H$ on (a) the reflection and the transmission coefficient $K_r$, $K_t$ and (b) the dissipation coefficient $K_d$ of the two consecutive VFNS with a distance $L$ =$h$ versus wave period $T$. It can be found that the newly added net panel in the two nets system does not change the incident wave heights where the minimum incident wave transmission and the maximum wave energy reflection and dissipation appear as that is observed from the one net model. Similarly, when the wave height is medium ($H$=0.06 m), the two VFNS have the minimum incident wave transmission and the maximum wave energy dissipation, whilst the greatest wave height ($H$=0.08 m) leads to the maximum wave reflection.

Figure 7 shows the effect of different distance $L$ between the two nets on (a) the reflection and the transmission coefficient $K_r$, $K_t$ and (b) the dissipation coefficient $K_d$ as a function of wave period $T$, under the action of an incident wave with wave height $H$=0.08 m.

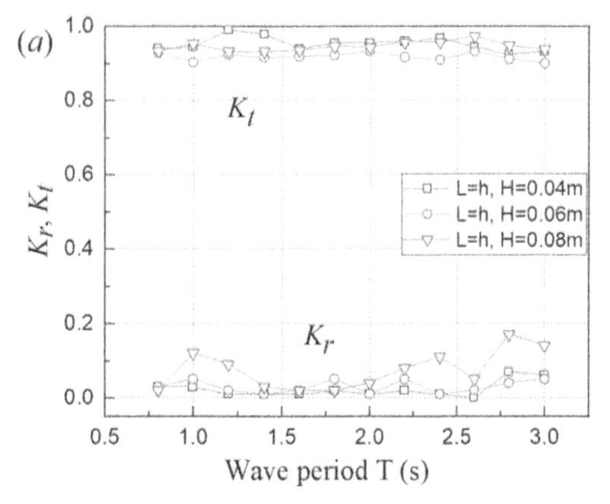


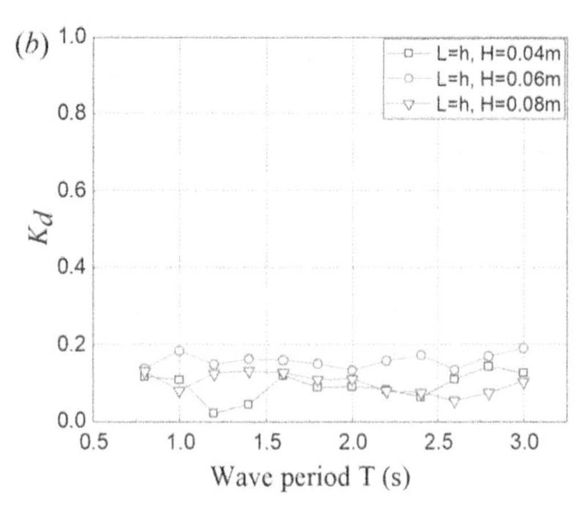


Figure 6. Effects of different wave height $H$ on (a) the $K_r$, $K_t$, and (b) $K_d$ of the two consecutive nets.

It is observed that the distance $L$ between the two nets has a negligible effect on the transmission coefficient $K_t$ and a small effect on the reflection and dissipation coefficients $K_r$, $K_d$. As the distance $L$ increases from $h$ to $2h$, the reflection coefficient $K_r$ almost has no change except at $T$=1.0 s and 2.6 s, whilst the reflection coefficient $K_r$ decreases slightly for $T$≥1.6 s when the distance $L$ increases to $3h$. When 1.0s≤$T$≤1.4s, the largest distance $L$=$3h$ leads to the most incident wave reflection while the trend is opposite as $T$≥1.6 s. Moreover, with an increase of wave period $T$, the reflection coefficient $K_r$ fluctuates periodically and peaks occur near $T$=1.0 s, 2.4 s, and 2.6 s. This phenomenon was also found by Karmakar et al. (2013) where the interaction of surface waves with multiple vertically moored surface-piercing membrane breakwaters was investigated.

It can be concluded that the distance $L$ between two vertical nets has a small effect on the system performance whilst the incident wave period has a greater effect on the reflection and the dissipation coefficients.

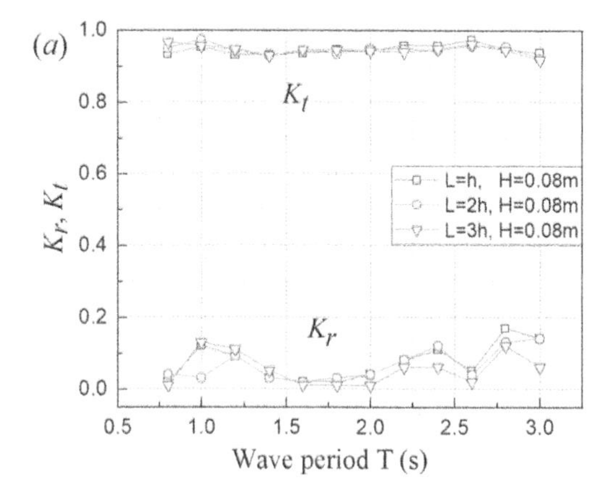


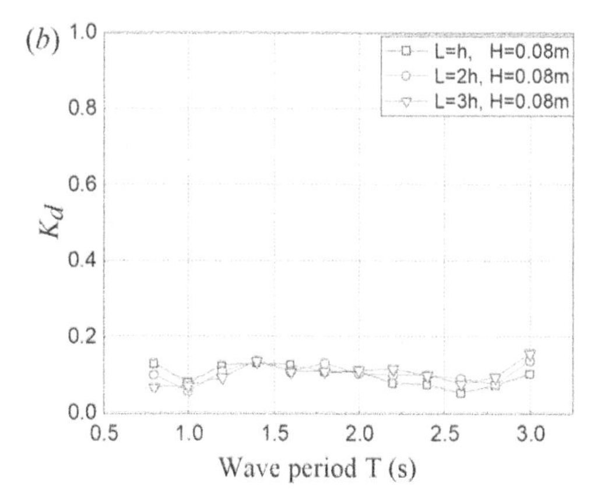


Figure 7. Effects of different distance $L$ between nets of the two consecutive nets on (a) the $K_r$, $K_t$, and (b) $K_d$.

## 4.3 *System of three consecutive vertical nets*

Figure 8 plots the effect of different wave height $H$ on (a) the reflection and the transmission coefficient $K_r$, $K_t$ and (b) the dissipation coefficient $K_d$ of the three consecutive nets as a function of wave period $T$. The distance among the three vertical nets is $L=L_1=h$.

In Figure 8, it is observed that the performance of the three-net system is more complicated than one or two-net models. With the third added net panel, when the wave height is medium ($H$=0.06 m), this system has the minimum incident wave transmission and the maximum wave energy dissipation only in the long-wave regime (1.4s≤$T$≤3.0s). The reflection coefficient $K_r$ still fluctuates in the period, but with a different pattern. When $T$=2.6 s, there is a trough for the reflection coefficient $K_r$ as $H$=0.08 m, whilst $K_r$ reaches a peak as $H$=0.04m and 0.06m.

Further, Figure 8(a) shows that the wave height $H$ has a small effect on the transmission coefficient $K_t$ when the three nets are under the action of long incident waves ($T$≥2.6 s). When $H$=0.06 m, more than 10% incident waves can be transmitted and more than 20% wave energy can be dissipated.

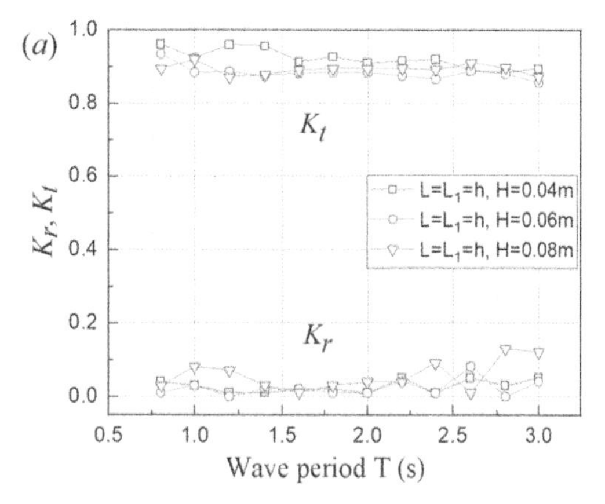


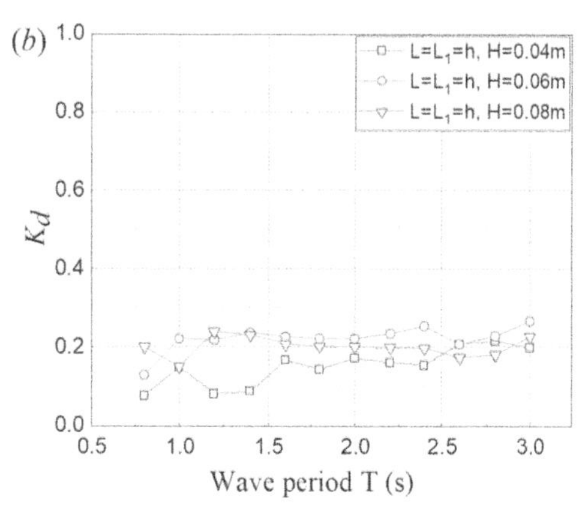


Figure 8. Effects of different wave heights $H$ on the $K_r$, $K_t$, and $K_d$ of the three-nets system.

Figure 9 shows the effect of different distances $L$, $L_1$ among nets of the three nets on (a) the reflection and the transmission coefficient $K_r$, $K_t$ and (b) the dissipation coefficient $K_d$ as a function of wave period $T$. It can be found that distances $L$, $L_1$ among the three vertical nets has a very small effect on the $K_r$, $K_t$, and $K_d$.

Moreover, it is observed from Figure 9(a) that for all three kinds of distances $L$, $L_1$, as the wave period $T$ increases, the transmission coefficient of the three nets decreases slightly, and there is a 10% reduction of $K_t$ between $T$=0.8 s and

$T$=3.0 s. In contrast, with the increase of the wave period $T$, the dissipation coefficient $K_d$ has a general trend of increasing and more than 20% of the wave energy can be dissipated by the net system when $T$=3.0 s.

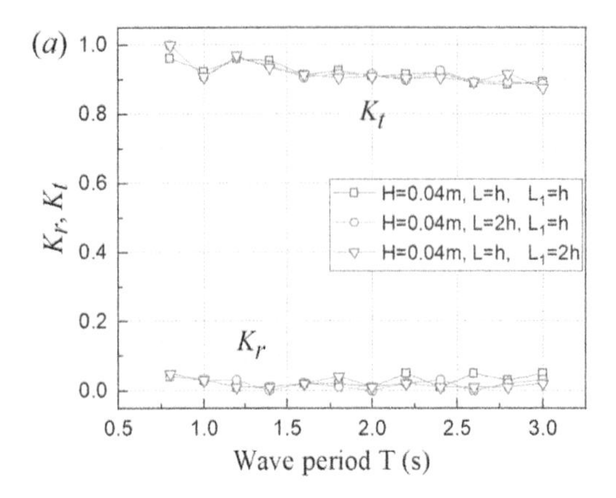


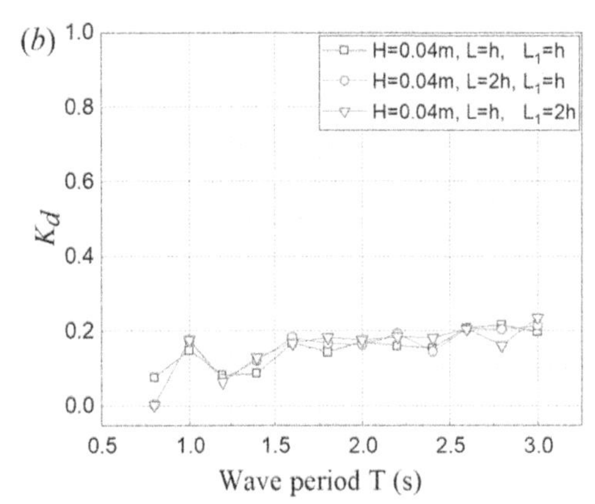


Figure 9. Effects of different distances $L$, $L_1$ among nets of the three nets on the $K_r$, $K_t$, and $K_d$.

### 4.4 *Comparison of results among the net-type structure of a single, two-, and three-conjugative vertical nets*

Figure 10 plots the effect of a different number of vertical net models on (a) the reflection and the transmission coefficient $K_r$, $K_t$ and (b) the dissipation coefficient $K_d$ versus wave period $T$, under the action of incident waves with $H$=0.04 m. The distance between adjacent net is $L=2h$ for the two-conjugative VFNS, whilst the distance is $L=L_1=h$ for the three nets.

It is observed from Figure 10(a, b) that VFNS with more nets makes less incident wave transmission and more wave energy dissipation, whilst the number of vertical nets has a very small effect on the reflection coefficient in the short-wave regime (0.8s≤$T$≤2.0s). This may be because the small mass density and thickness of the vertical net leads to high net flexibility and makes it very deformable. Therefore, the net-type structure is not rigid enough to reflect the incident wave and only dissipates the wave energy of the incident waves. Then the aforementioned description for the transmission, the reflection, and the dissipation coefficients is reasonable. Further, with the increase of the wave period $T$, the transmission coefficient $K_t$ has a slightly decreasing trend, while the dissipation coefficient $K_d$ fluctuates periodically and has a slightly increasing trend.

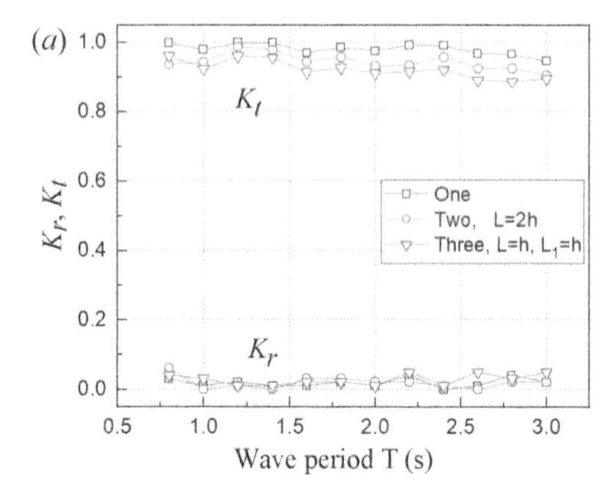


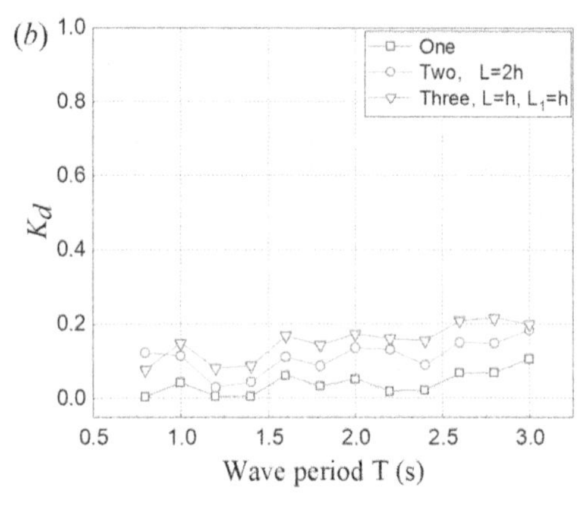


Figure 10. Comparison between one, two, and three net systems on the (a) $K_r$, $K_t$, and (b) $K_d$ with $H$=0.04 m.

From Figure 10, it can be concluded that the number of vertical nets has an obvious influence on the variation of the dissipation coefficient $K_d$ and the transmission coefficient $K_t$ when compared with the reflection coefficient $K_r$.

Figure 11 reveals the effect of a different number of vertical nets on (a) the reflection and the transmission coefficient $K_r$, $K_t$ and (b) the dissipation coefficient $K_d$ against wave period $T$, under the action of incident waves with $H$=0.08m. The distance between adjacent net is $L=3h$ for the two nets system, whilst the distance is $L$=2h and $L_1=h$ for the three nets system. The length of this net system is longer than that shown in Figure 10.

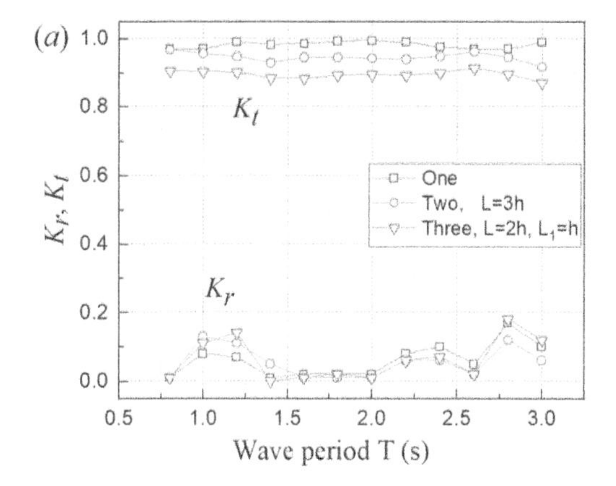


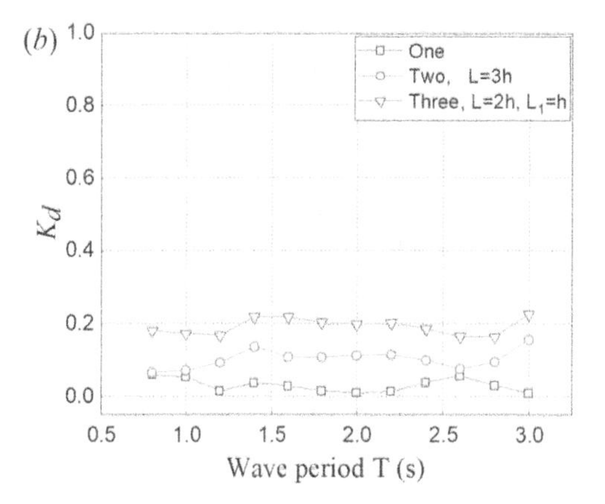


Figure 11. Comparison of the number of nets in the system on (a) the $K_r$, $K_t$, and (b) $K_d$ with $H$=0.08m.

It is observed from Figure 10(a) that with the increase in the number of nets, the transmission coefficient $K_r$ decreases significantly whilst the dissipation coefficient $K_d$ has a noticeable increase. This phenomenon is very similar to that observed in Figures 10(a, b). However, the reflection coefficient $K_r$ of all three kinds of nets with different numbers of vertical nets shows the same periodic fluctuation and peaks near $T$=1.0 s, 2.4 s, and 2.8 s.

Comparing Figure 10(a) and Figure 11(a), it can be found that the reflection coefficient $K_r$ of the net-type structure with greater length shows obvious periodic fluctuations when it is under the action of incident waves with large wave height $H$=0.08 m. This fluctuation has not been observed in the lower wave height case of $H$=0.04 m which is plotted in Figure 9(a).

## 5 CONCLUSIONS

The reflection, transmission, and dissipation of regular waves by systems of up to three consecutive vertical nets are studied by experimental tests in a two-dimensional wave flume. The vertical net-type structure is made of nylon silky and supported by a vertical steel frame. The influence of wave height and period, the distance among nets, and a number of nets on the reflection, transmission, and dissipation coefficients is examined. A four-point method based on Mansard and Funke (1980) is applied to predict the reflection coefficient, transmission coefficient, and dissipation coefficient for regular waves. The results of this study are summarized as follows:

1. A single vertical net is inefficient in preventing the incident waves.
2. The distance among vertical nets has a small effect on the performance of the net system
3. The performance of the three net system is more complicated than a single net or two nets system.
4. The net-type structure with a greater number of vertical nets makes less incident wave transmission and more wave energy dissipation.
5. It is found that wave action is efficiently reduced by adding more vertical net models.

The present results are helpful for better understanding the effect of different vertical nets arrangements through the transmission, the reflection, and the dissipation coefficients in the future. This can have applications in systems of breakwaters made of several porous membranes and in arrays of net-type fish cages used in offshore aquaculture.

## ACKNOWLEDGEMENTS


The paper is performed within the project Hydroelastic behaviour of horizontal flexible floating structures for applications to Floating Breakwaters and Wave Energy Converters (HYDROELASTWEB), which is co-funded by the European Regional Development Fund (Fundo Europeu de Desenvolvimento Regional - FEDER) and by the Portuguese Foundation for Science and Technology (Fundação para a Ciência e a Tecnologia - FCT) under contract 031488_770 (PTDC/ECI-EGC/31488/2017). The work of the first author has been supported in part by the scholarship from China Scholarship Council (CSC) under Grant No. 201707920004. The second author is funded by the Portuguese Foundation for Science and Technology (Fundação para a Ciência e a Tecnologia-FCT) through a doctoral fellowship under contract no. SFRH/BD/147178/2019. The third author has been contracted as a Researcher by the Portuguese Foundation for Science and Technology (Fundação para a Ciência e Tecnologia-FCT), through Scientific Employment Stimulus, Individual support under the Contract No. CEECIND/04879/2017. This work contributes to the Strategic Research Plan of the Centre for Marine Technology and Ocean Engineering (CENTEC), which is financed by the Portuguese Foundation for Science and Technology (Fundação para a Ciência e Tecnologia - FCT) under contract UIDB/UIDP/00134/2020.

## REFERENCES


Brizzi, G. & Sabbagh, M. 2021. A new criterion for multi-purpose platforms siting: Fish endurance to wave motion within offshore farming cages. *Ocean Engineering* 224: 108751.

Donga, G.H., Tanga, M.F., Xua, T.J., Bia, C.W. & Guo, W. J. 2019. Experimental analysis of the hydrodynamic force on the net panel in wave. *Applied Ocean Research* 87: 233–246.

Esmaeili, M., Rahbani, M., Khaniki. A.K. 2019. Experimental investigating on the reflected waves from the caisson-type vertical porous seawalls. Acta Oceanologica Sinica 38 (6): 117–123.

Føre, H. M., Endresen, P. C., Norvik, C., & Lader, P. 2021. Hydrodynamic Loads on Net Panels with Different Solidities. *Journal of Offshore Mechanics and Arctic Engineering, ASME*, October 2021, 143(5): 051901.

Guo, Y.C., Mohapatra, S.C. & Guedes Soares, C. 2022. Submerged breakwater of a flexible porous membrane with a vertical flexible porous wall over variable bottom topography. *Ocean Engineering* 243, 109989.

Guo, Y. C., Mohapatra, S.C. & Guedes Soares, C. 2021. Effect of vertical rigid wall on a moored submerged horizontal flexible porous membrane. In: Guedes Soares, C. & Santos, T. A. (Eds.), *Developments in Maritime Technology and Engineering*, Taylor & Francis Group, London, UK, Vol 2, pp. 757–766, DOI: 10.1201/9781003216599-81.

Guo, Y.C., Mohapatra, S.C. & Guedes Soares, C. 2020a. Wave energy dissipation of a submerged horizontal flexible porous membrane under oblique wave interaction. *Applied Ocean Research* 94: 101948.

Guo, Y.C., Mohapatra, S.C. & Guedes Soares, C. 2020b. Composite breakwater of a submerged horizontal flexible porous membrane with a lower rubble mound. *Applied Ocean Research* 104: 102371.

Guo, Y.C., Mohapatra, S.C. & Guedes Soares, C. 2020c. Review of developments in porous membranes and net-type structures for breakwaters and fish cages. *Ocean Engineering* 200: 107027.

Karmakar, D., Bhattacharjee, J., Guedes Soares, C. 2013. Scattering of gravity waves by multiple surface-piercing floating membrane. *Applied Ocean Research* 39: 40–52.

Lader, P., Jensen, A., Sveen, J.K., Fredheim, A., Enerhaug, B. & Fredriksson, D. 2007. Experimental investigation of wave forces on net structures. *Applied Ocean Research* 29: 112–127.

Lee, W.K. & Lo, E.Y.M. 2002. Surface-penetrating flexible membrane wave barriers of finite draft. *Ocean Engineering* 29 (14): 1781–1804.

Liu, Z., Mohapatra, S.C., Guedes Soares, C. 2021. Finite Element Analysis of the Effect of Currents on the Dynamics of a Moored Flexible Cylindrical Net Cage. *Journal of Marine Science and Engineering* 9(2): 159.

Liu, Z.C., Wang, S., & Guedes Soares, C. 2022. Numerical Study on the Mooring Force in an Offshore Fish Cage Array. *Journal of Marine Science and Engineering* 10 (3): 331.

Mansard, E.P.D., Funke, E.R., 1980. The measurement of incident and reflected spectra using a least squares method. *Proceedings of 17th Conference on Costal Engineering ASCE*, Sidney, Australia, pp. 154–172.

Miao, Y.J., Ding, J., Tian, C., Chen, X.J. & Fan, Y.I. 2021. Experimental and numerical study of a semi-submersible offshore fish farm under waves. *Ocean Engineering* 225: 108794.

Mohapatra, S.C., Bernardo, T.A. & Guedes Soares, C. 2021. Dynamic wave induced loads on a moored flexible cylindrical net cage with analytical and numerical model simulations. *Applied Ocean Research* 110: 102591.

Shen, Y., Firoozkoohi, R., Greco, M. & Faltinsen, O.M. 2021. Experimental investigation of a closed vertical cylinder-shaped fish cage in waves. *Ocean Engineering* 236: 109444.

Zhen-hua, H. 2007. Reflection and transmission of regular waves at a surface-pitching slotted barrier. *Applied Mathematics and Mechanics* 28(9): 1153–116.

Song, W., Liang, Z., Chi, H., Huang, L., Zhao, F., Zhu1, L. & Chen, B. 2006. Experimental study on the effect of horizontal waves on netting panels. *Fisheries Science* 72: 967–976.

Xu, T.J., Dong, G.H., Tang, M.F., Liu, J. & Guo, W.J. 2021. Experimental analysis of hydrodynamic forces on net panel in extreme waves. *Applied Ocean Research* 107: 102495.

Zhang, D.P., Bai, Y., & Guedes Soares, C. 2021. Dynamic Analysis of an Array of Semi-rigid "Sea Station" Fish Cages subjected to waves. *Aquaculture Engineering* 94: 102172.

*Trends in Maritime Technology and Engineering – Guedes Soares & Santos (eds)*


# Spatial characterization of pelagic fisheries in the Northeast Atlantic: The e-shape pilot "Monitoring Fishing Activity"

P. Gaspar
*IPMA, Portuguese Institute for Sea and Atmosphere, Algés, Portugal*

M. Chapela, R. Silva, G. Mendes, D. Cordeiro & N. Grosso
*DEIMOS Engenharia S.A., Lisboa, Portugal*

P. Ribeiro
*CoLAB +Atlantic, Cascais, Portugal*

V. Henriques & A. Campos
*IPMA, Portuguese Institute for Sea and Atmosphere, Algés, Portugal*
*CCMAR, Centre of Marine Sciences, Universidade do Algarve, Faro, Portugal*

ABSTRACT: The pilot application "Monitoring Fishing Activity", under the scope of the showcase Water Resources Management of the H2020 e-shape project, aims to develop a web-based tool for the exploration and visualization of spatio-temporal information on fishing fleets activities. The dynamics and landings of two Portuguese fisheries targeting migratory pelagic species were analysed: the tuna pole and line fishery and the swordfish drifting longline fishery taking place within the Portuguese Exclusive Economic Zone and surrounding high seas areas, in the Northeast Atlantic. This paper presents the first results on the activity of these two fishing fleets, from three main data sources: vessels sales, e-logbook records and spatial data from AIS, for the 2012-2018 period. The results are provided as maps on fishing intensity, landings and vessels fishing trajectories in multiple time scales that will be accessible to stakeholders on an interactive web platform.

## 1 INTRODUCTION

Fishing is considered the most important human activity directly impacting the marine ecosystems, and there is a special concern on its impacts over the oceanic high seas' areas. Many of these areas are under high fishing pressure, and we are still trying to answer some questions on what fleets are involved in these fisheries, the species exploited, and where and how do these fleets operate. This is often the case of fleets operating in the high seas and less monitored areas, targeting highly migratory species.

To control and understand the activity of these fleets on marine ecosystems, different monitoring tools are being used, allowing to collect fisheries dependent data (FDD), including sales declarations and, more recently, electronic logbooks and vessels' tracking systems, with an emphasis on Vessel Monitoring Systems (VMS). However, in many cases the information generated by these systems is still incomplete. In fact, e-logbooks with the identification of fishing events, including the gear used and their respective geographic position, despite being mandatory for vessels equal or above 12 meters, are often not readily available, or correctly filled out.

Vessels tracking systems have been implemented in the last decade in many countries. VMS was designed for fishing control, inspection and enforcement purposes and is currently mandatory for fishing vessels of 12 meters length overall and above fishing in the European Union waters (EC, 2009). Considering the difficulty to access VMS data for scientific purposes, mostly due to confidentiality and personal data protection policies (Natale et al., 2015), AIS (Automatic Identification System) has gradually come to use as an alternative source of vessel positioning data for scientific studies. The AIS was originally designed for navigational safety purposes, and is currently required on all vessels equal or above 15 meters, irrespectively of their activity. Nowadays, VMS or, to a lesser extent and as a complement, the AIS data provide valid means to assess the fishing activity in many fisheries. When coupled with e-logbooks data, the data from the fishing vessel tracking systems have allowed the analysis of the spatio-temporal footprint and patterns of fishing effort or intensity in many fisheries. (Lee et al.,

DOI: 10.1201/9781003320289-61

2010; Hintzen et al., 2012; Natale et al., 2015; Mccauley et al., 2016; Russo et al., 2016; Shepperson et al., 2018; Campos et al. 2018, Campos et al., 2021).

The e-shape is an European H2020 project (2019-2023) aiming to foster the development of operational Earth Observation services through co-designed pilot applications build with and for the users, delivering economic, social and policy value across the private and public sectors, to key stakeholders and to the public in general. (https://e-shape.eu/).

The aim of the pilot S5P5 "Monitoring Fishing Activity", within the e-shape project, is to develop a web-based tool providing a set of functionalities, including the exploration and visualization of spatio-temporal fishing information. The objective is to provide access to maps on fishing intensity, landings, catch rates and environmental characterization for the Northeast Atlantic high seas pelagic fisheries for the 2012-2018 period. In this study, preliminary results show the fishing intensity, landings and vessel fishing patterns of two Portuguese fisheries targeting highly migratory pelagic species: the pole and line fishery targeting tuna species and the drifting longline fishery targeting swordfish, taking place in high seas areas of the Northeast Atlantic. Data analysis combined AIS, vessels e-logbooks and port sales datasets.

## 2 METHODS

The study area corresponds to the Northeast Atlantic, comprising areas of Portuguese jurisdiction including the EEZs of the Azores and Madeira Islands and the extended continental shelf (Figure 1). For these areas, access to anonymized fisheries dependent data for the period 2012 to 2018 were provided by the Portuguese Directorate-General for Natural Resources, Safety and Maritime Services (DGRM), the Portuguese authority responsible for fisheries control. Vessel tracking data, namely satellite-based AIS data, were purchased from a commercial company for the same time period.

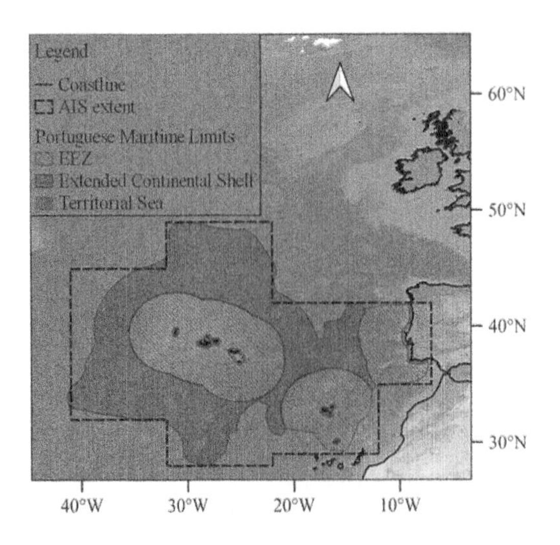


Figure 1. Pilot's coverage area, including the maritime areas under Portuguese jurisdiction the exclusive economic zones (EEZs) of the Azores and Madeira Islands, the Portuguese proposed extended continental shelf (ECS), and the extent of the AIS dataset.

Fisheries dependent data analysed included: a) e-logbooks, containing information on date, start and end coordinates for fishing events, as well as the species captured and respective catch in weight; b) sales records per vessel, with the date of sale, landing port and species weight and value; c) vessels' technical characteristics, specifying length overall, gross tonnage, engine power, year of construction and port of registration; and d) fishing licenses per vessel including the gear and year of licence.

The AIS data contains the geographic position, heading, speed, date and time and MMSI code (Maritime Mobile Service Identity) of individual vessels.

The datasets were initially pre-processed for detecting data inconsistencies, removing blunders, and formatting data tables for convenient database storage, access and optimized data analysis. The exploratory analysis allowed the selection of the fishing vessels that best identifies with the drifting longline and pole and line fleets, based on the amount of both target species and by-catch recorded in e-logbooks, as well as corresponding fishing trips occurring within the study area.

The spatial analysis of the AIS data mainly explored the segmentation of speed and course patterns occurring in the fishing trip trajectories, and later on validated from e-logbook corresponding records. The analysis resulted in maps showing the global, yearly, and seasonal fishing footprint aggregated per fleet, as well as the spatio-temporal dynamics of each vessel.

Further analysis of the e-logbook and sales records led to a global characterization of the captures of the target species and most relevant by-catch species per year, season, and landing port, for the 2012-2018 period.

These analyses resulted in the development of three mapping products: the Fishing Footprint, the Fishing Trips, and the Fishing Sales maps. These products will be available to stakeholders through the geoportal that is being developed under the e-shape pilot application S5P5 "Monitoring Fishing Activity".

## 3 RESULTS

From the analyses of fish captures reported in the e-logbooks from the Portuguese fleet, a total of 71 and 62 fishing vessels were identified as part of the pole and line and drifting longline fleets respectively, from which 34 and 40 vessels have fishing trips defined by AIS trajectories. The vessels of the pole and line fleet are based in ports of the Azores and Madeira islands while those of the drifting longline fleet are based in the Portuguese mainland ports.

Concerning the fishing footprint product, global, yearly, and seasonal fishing intensity were obtained as a series of maps of the activity for each fleet for the 2012-2018 period. In Figure 2, the global map estimated for the pole and line fleet is presented, showing that this fleet operated mostly in geographical areas around the Azores and Madeira islands. Similar maps were also created for the drifting longline fleet, with global, yearly, or seasonal time resolution.

The second product, the fishing trips maps, was also developed from the spatial analysis of the AIS data, and show individual fishing trips for vessels of both fleets, tentatively depicting their activity status along time as potentially fishing or non-fishing. Figure 3 is an example of this type of maps. The discrimination between fishing and non-fishing status within each fishing trip was based on the pattern analysis of both AIS velocity and route attributes.

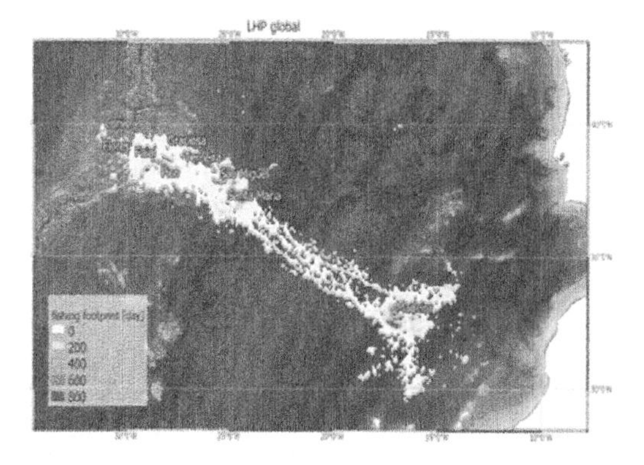


Figure 2. Map showing the fishing footprint of the pole and line fleet aggregated for the period between 2012 and 2018. The map was created from the AIS data for the vessels of the pole and line fleet. The spatial resolution of this product is 0.1 degree.

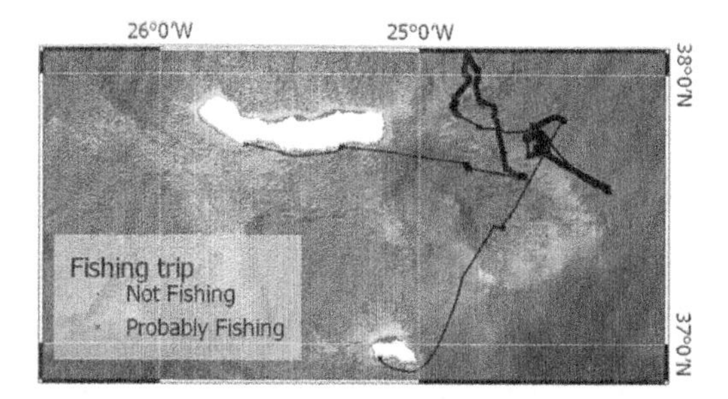


Figure 3. Example of the fishing trips map showing the track of a pole and line vessel fishing east of the island of S. Miguel, Azores. The thicker points tentatively correspond to the fishing activity, while the thinner points correspond to sailing.

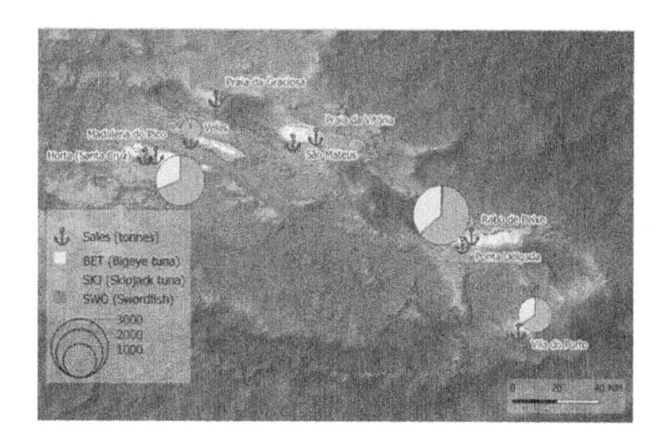


Figure 4. Map showing the aggregated sales of 3 main species (Bigeye Tuna, Skipjack Tuna and Swordfish) for the year 2018 at the central and eastern island groups of the Azores.

The third product, fishing sales maps, present the estimation of the annual sales by port of the main target species of each fleet, as a result from the analysis of sales declarations in Portuguese ports. Figure 4 is an example of such product depicting the total amount sold in 2018 of the three main target species: bigeye tuna, skipjack tuna and swordfish, in the central and eastern island groups of the Azores.

Complementary to these products, estimates of monthly captures and sales from both fleets were also obtained, displaying seasonal distributions of fishing yields. Figure 5 presents the total quantity sold of the three main target species of the pole and line fishery aggregated per month for the entire period covered by the pilot, clearly showing that the fishery is concentrated in the second and third quarters of the year, starting in April, with the catch of bigeye tuna (*Thunnus obesus*), and ending in October with skipjack (*Katsuwonus pelamis*). This chart evidences that the fishing period for bigeye is carried out mostly from April to August/September, while the season for skipjack happens mostly from June to September/October and for the albacore (*Thunnus alalunga*) is May and June.

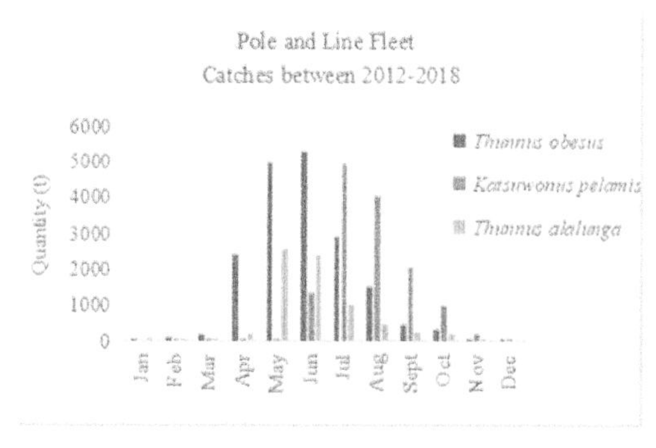


Figure 5. Total catch of the three main target species of the pole and line fleet, namely bigeye, skipjack, and albacore tunas, aggregated by month.

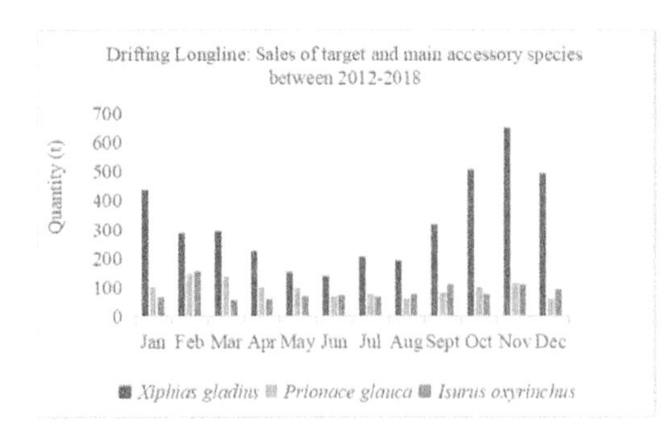


Figure 6. Swordfish captures in the drifting longline fleet along with the main by-catch species aggregated by month, for the 2012-2018 period.

Similarly, Figure 6 represents the total amount sold of the main target (swordfish) and by-catch species for the drifting longline fishery, aggregated per month, from 2012 to 2018.

The drifting longline fishery targeting swordfish (*Xiphias gladius*) develops mainly in the autumn and winter months, starting with higher catches in October progressively decreasing towards January/March. The main by-catch species of this fishery are the blue shark (*Prionace glauca*) and the shortfin mako shark (*Isurus oxyrinchus*). Figure 7 presents both the catches and sales in Portuguese ports of swordfish by this fleet, aggregated by month. A high discrepancy is observed between the e-logbooks and the sales datasets, which can partially be explained by the fact that a significant quantity of the catches are landed in Spanish ports, however, this cannot be confirmed with sales records as records of the sales made in Spanish ports have not been made available.

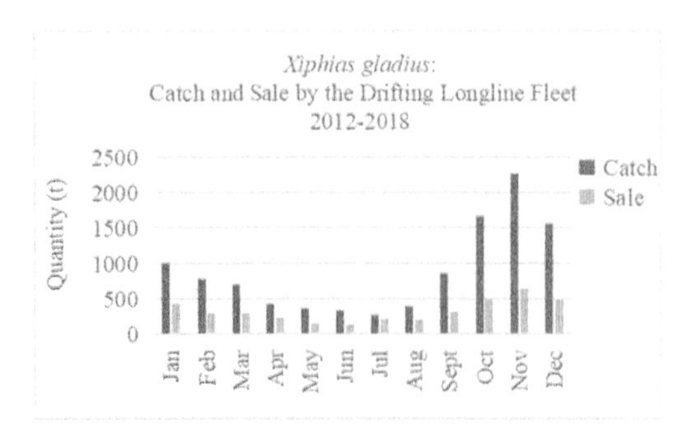


Figure 7. Captures and sales of swordfish (Xiphias gladius) from the drifting longline fleet, aggregated by month for the 2012-2018 period.

## 4 DISCUSSION AND CONCLUSIONS

The data used so far in the development of these products were daily sales in auction, vessel's technical characteristics and satellite AIS data. The e-logbooks were made available later and are being analysed. Its results are expected to improve the fishing footprint maps.

A new product relating fish distribution with environmental data is currently under development and represents a step further in this pilot exploring the potential of Earth Observation Data.

The results obtained from the correlated analyses of all datasets allowed the selection of vessels involved in the pole and line and in the drifting longline fisheries, the characterization of their fishing yields at different time scales, and the tentative discrimination of the spatio-temporal patterns of the fishing vessels.

The products involving the spatial extent and intensity of the fishing activity are important for the study of these oceanic fisheries with an emphasis on the lesser-known drifting longlining fishery due to its high seas features. However, along the different analysis steps we came across a number of inconsistencies and lack of information arising from poor data quality, posing limits to the exploration and interpretation of the data and the extent of the final results.

Although AIS was originally designed as a maritime safety tool, fisheries science has reported an interest in this data as a means to improve the knowledge on where many fishing activities occur. Despite its importance for the obtained spatial results, the positioning records of the fishing vessels have an irregular frequency and present coverage gaps. These issues with the AIS data are exacerbated for vessels operating mainly in the high seas, where these data are only available from satellite acquisition. In the present study, the combined analysis of AIS records with information from fishing e-logbooks allowed the identification on both datasets of data gaps, and poor and irregular spatial representation, leading to the underestimation of the fishing activity and footprint.

Unfortunately, VMS complementary data were not made available for this project, hindering the possibility of improving the spatial representation of the fishing activities as well as seeking for a better correlation between the spatial data and e-logbook data records.

Another factor hampering better results is the quality of the data provided in the e-logbooks. These electronic-based reporting data are mandatory and are partially filled out by fishermen during the fishing trips. However, significant misreporting on both datetime stamp and positioning information have been found in these datasets, preventing us to create maps of fishing effort for the targeted fish species for both fleets.

In this pilot, a co-design process is being undertaken involving key-users. For this purpose, online meetings with these users have been carried out, where expert information on the fisheries have been gathered, improving the interpretation and processing of the datasets. More recently an online survey was issued with the purpose to assess users' requirements, either in terms of key datasets or of characteristics of the products to be developed. Seven institutions were contacted, known to have an active voice in the Portuguese fishing sector, from which five answered the online survey. Their inputs were essential to allow us to produce mapping specifications based on their needs to better fit the outcome products to societal interests.

A further objective of this pilot is to raise awareness of key users for the human footprint extent in the marine ecosystems and motivate them towards more efficient and sustainable fishing strategies and practices, complying with international, European, and national regulations currently in place. Under development to fit this purpose, is a web-based tool that will include functionalities that allow the online exploration and visualization of spatial fishing information conveyed by the mapping products.

## ACKNOWLEDGEMENTS

This study was funded by H2020 e-shape (EuroGEOSS Showcases: Applications Powered by Europe) project. The Fisheries dependent data was supplied to IPMA and DEIMOS (Elecnor Group) by the Portuguese Directorate-General for Natural Resources, Safety and Maritime Services (DGRM).

## REFERENCES

Campos, A., Lopes, P., Fonseca, P., Araújo, G. & Figueiredo, I. 2018. Fishing patterns for a Portuguese longliner fishing at the Gorringe seamount—a first analysis based on AIS data and onboard Observations. Maritime Transportation and Harvesting of Sea Resources – Guedes Soares & Teixeira (Eds) © 2018 Taylor & Francis Group, London, ISBN 978-0-8153-7993-5.

Campos, A., Henriques, V, Erzini, K. & Castro, M. 2021. Deep-sea trawling off the Portuguese continental coast—Spatial patterns, target species and impact of a prospective EU-level ban. Marine Policy, Volume 128, 2021, 104466, ISSN 0308-597X, http://dx.doi.org/10.1016/j.marpol.2021.104466.

EC. 2009. Council Regulation (EC) No 1224/2009 of 20 November 2009, "establishing a Community control system for ensuring compliance with the rules of the Common Fisheries Policy". Official Journal of the European Union, 22- 12-2009.L 343/1-50.

Hintzen, N.T., Bastardie, F., Beare, D., Piet, G.J., Ulrich, C., Deporte, N., Egekvist J, & Degel, H. 2012. VMS tools: Open-source software for the processing, analysis and visualisation of fisheries logbook and VMS data. Fish Res 115–116:31–43.

Lee, J., South, a. B., & Jennings, S. 2010. Developing reliable, repeatable, and accessible methods to provide high-resolution estimates of fishing-effort distributions from vessel monitoring system (VMS) data. ICES Journal of Marine Science, 67: 1260–1271.

McCauley, D. J., Woods, P., Sullivan, B., Bergman, B., Jablonicky, C., Roan, A., Hirshfield, M., Boerder, K., & Worm B. 2016. Ending hide and seek at sea. New technologies could revolutionize ocean observation. Science, vol.351, Issue 6278: 1148–1150.

Natale F., Gibin, M., Alessandrini, A., Vespe, M., & Paulrud, A. 2015. Mapping Fishing Effort through AIS Data. PLoS ONE 10 (6):e0130746. doi:10.1371/journal.pone.0130746.

Russo, T., D'Andrea, L., Parisi, A., Martinelli, M., Belardinelli, A., Boccoli, F., Cignini, I. et al. 2016. Assessing the fishing footprint using data integrated from different tracking devices: Issues and opportunities. Ecological Indicators, 69: 818–827.

Shepperson, J. L., Hintzen, N. T., Szostek, C. L., Bell, E., Murray, L. G., & Kaiser, M. J. 2018. A comparison of VMS and AIS data: the effect of data coverage and vessel position recording frequency on estimates of fishing footprints. – ICES Journal of Marine Science, 75: 988–998.

*Trends in Maritime Technology and Engineering – Guedes Soares & Santos (eds)*


# Defining multi-gear fisheries through species association

P. Leitão
*IPMA, Portuguese Institute for the Sea and Atmosphere, Lisboa, Portugal*

A. Campos
*IPMA, Portuguese Institute for the Sea and Atmosphere, Lisboa, Portugal*
*CCMAR, Centre of Marine Sciences, Universidade do Algarve, Faro, Portugal*

ABSTRACT: The Portuguese coastal multi-gear fishing fleet operates along the entire continental coast (ICES Division IXa) where vessels use multiple gears to capture a great diversity of species. Fishing logbooks, allow the identification of the gear used and can assist in the identification of métiers; however, they are only available for a limited number of vessels, and thus most of the information on fishing gears used is unknown.
In this study, association rules were used to capture associations between species landed by the multi-gear fleet and assign these associations specific gear types according to previous knowledge. We relied on apriori algorithm and clustering to explore these species associations. A total of seven groups were set based on 44 species associations. From these, three well-known isolated groups were identified, and a four newly identified groupings were defined and attributed to trammel nets, drifting longlines and large and small meshed gillnets.

## 1 INTRODUCTION

The Portuguese coastal multi-gear fishing fleet, including vessels longer than 9 m and shorter than 33 m, operates all year round in the geographical area of the Northeast Atlantic waters of the Iberian Peninsula (ICES divisions VIIIc and IXa), excluding the Gulf of Cadiz, in an extension of 900 Km, over a great variety of ecosystems. The fishing activity is mainly concentrated in an area parallel to the coast, covering the continental shelf, until 200 m depth, and extending to vast areas of the slope, near the canyons. A fleet subsegment engages in deep-sea fisheries, operating in NE Atlantic ecosystems comprising seamount areas. The fleet is adapted to regional and seasonal availability in resources, and most vessels hold licenses for more than one gear, alternating between gears along fishing trips or even using more than one gear in the same trip, capturing a great diversity of benthic, demersal and pelagic species. Furthermore, they can change gears over the years, adding complexity to this analysis.

In this fleet, the identification of métiers (fishing operations targeting one species or a group of species, using similar gear, during the same period of the year and/or within the same area), is particularly important for management purposes(Biseau, 1998). Fishing logbooks, including the identification of the gear used at a haul level, can potentially assist the identification of métiers; however, despite they are mandatory for vessels equal or above 10 meters in length overall, often they are not readily available for analysis or correctly filled. With logbooks available only for a limited number of vessels, the information on fishing gears used is lost.

In this study, association rules were used to capture multiple relationships between species in daily landings composition by the coastal multi-gear fleet. These associations were then assigned specific gear types according to previous knowledge on species ecology and fisheries. Our main objective was to further contribute to a better understanding of the multi-gear fleet dynamics and the identification of potential métiers.

## 2 METHODS

### 2.1 *Data analysis*

Association rule learning is a rule-based machine learning method for discovering important relations between items in large databases. These techniques were first introduced by (Rakesh & Srikant, 1994) being one of the major techniques to detect and extract useful information from large scale transaction data, discovering regularities between products recorded by point-of-sale systems in supermarkets. Such information can be used as the basis for

DOI: 10.1201/9781003320289-62

decisions about marketing activities such as promotional pricing or product placements.

Apriori analysis is a common method for identifying these relationships. The algorithm uses a breadth-first search strategy to count the support of itemsets and a candidate generation function which exploits the downward closure property of support (Rakesh & Srikant, 1994). The same analysis was here extrapolated to fisheries, based on the principle that a given species is most often associated to other species present in the same fishing grounds.

Daily sales in weight (assumed to correspond to daily landings, i.e., the product of one fishing trip) comprising more than 250 000 sales for a total of 50 most relevant species (items) were analysed for the coastal multi-gear fleet. Three types of association rules were used to define associations between species:

a) Support, defined as the proportion of transactions (daily sales, corresponding to landings) in the data set where two items (species) are related in relation to the total number of daily sales:

$$Support\ (A \rightarrow B) = \frac{N^{\circ}\ of\ trips\ with\ A \rightarrow B}{Total\ number\ of\ trips}$$

b) Confidence, defining the probability of A being landed when B is landed:

$$Confidence\ (A \rightarrow B) = \frac{Support\ (A \rightarrow B)}{Support\ (B)}$$

c) Lift, defining the probability of B being landed when A is landed taking into account the proportion of landings of B. Lift > 1 means that the probability of B being landed increases if A is landed; Lift < 1 means that the probability of B being landed decreases if A is landed:

$$Lift\ (A \rightarrow B) = \frac{Support\ (A \rightarrow B)}{Support\ (A) \times Support\ (B)}$$

### 2.2 *Data visualization*

For the association rules visualization, the software Gephi 0.9.1 was used to apply force-directed graph drawing, using the force atlas algorithm, and modularity procedure for community detection for a total of 7 groups was conducted (Resolution=2.0) (Bastian et al., 2009) . These groups were based on the different and most common fishing licenses used by the multi-gear fleet: bottom longlines, traps and pots, gillnets, trammel nets, drifting longlines, pole and line and drifting gillnets. In the present study, a further step was explored when analysing these species associations: an attempt to assign fishing gears to the trips in analysis. In fact, distinct associations can be assigned specific gear types according to previous knowledge on the ecology and fisheries, including the types of fishing gears used, for the different species in the associations.

## 3 RESULTS

In the general diagram in Figure 1, a total of seven groups are shown, representing different fisheries; the species composition in the seven groupings identified can be seen in Table 1. The individualization of several species that interrelate, forming distinct associations corresponding to well individualized fisheries, can be observed. This is the case of the black scabbard fish *Aphanopus carbo*, captured with bottom longlines, associated to the tope shark *Galeorhinus galeus* and the lowfin gulper shark *Centrophorus lusitanicus*. It is also the case of the swordfish *Xiphias gladius*, captured with drifting longlines, associated to shortfin mako shark *Isurus oxyrinchus* and finally, the octopus *Octopus vulgaris*, captured in pots.

For the remaining species, a large number of association rules was defined (Figure 1), adding complexity to the analysis. However, four main species form what can be considered the centre of distinct groupings corresponding to species associations in the daily sales: the common sole, *Solea solea*, the thornback ray, *Raja clavata*, the axillary seabream, *Pagellus acarne* and the forkbeard, *Phycis phycis*, The associations formed define different types of fisheries: the common sole, closely associated with the soles *Dicologlossa cuneata* and *Pegusa lascaris;* with the cuttlefish, *Sepia officinalis*, and with the sea bass, *Dicentrarchus labrax*, all species being commonly captured in trammel nets. The thornback ray is closely associated to the spotted ray *Raja montagui*, the monkfishes *Lophius budegassa* and *Lophius piscatorius*, and two shark species, the smooth-hound *Mustelus mustelus* and the nursehound *Scyliorhinus stellaris*, commonly captured with large mesh size gillnets and trammel nets The axillary seabream is associated with the common seabream *Diplodus vulgaris*, the surmullet *Mullus surmuletus*, the Atlantic horse mackerel *Trachurus trachurus*, the Atlantic mackerel *Scomber scombrus* and flatfish (*Microchirus spp*), captured with small mesh size gillnets. Finally, the forkbeard is associated with the blackspot seabream *Pagellus bogaraveo*, the red porgy *Pagrus pagrus*, the blackbelly rosefish *Helicolenus dactylopterus*, the silver scabbardfish *Lepidopus caudatus* and the wreckfish *Polyprion americanus*, typically captured using longlines (Table 1).

Besides, two main species appear as the centre of multiple interrelations, the European hake *Merluccius merluccius* and the pouting *Trisopterus luscus*, two species frequently captured in more than one gear type. Association rules where hake is linked to the thornback ray and related species likely define landings from large mesh size gillnets or trammel nets, while rules where hake is linked to the Atlantic mackerel and related species define landings from small mesh size gillnets, and finally rules where hake is linked to forkbeard and related species define longline fisheries. Similarly, pouting is associated to soles and related species in trammel net fisheries, and to the Atlantic mackerel and the axillary seabream in small mesh size gillnets.

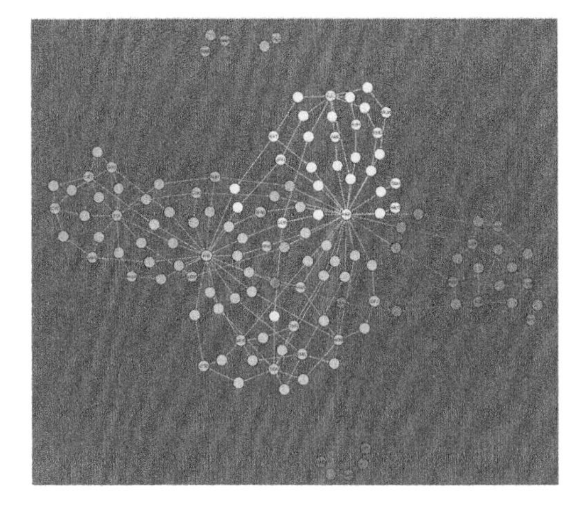

Figure 1. Diagram showing associations between 44 species in the daily sales of the coastal polyvalent fleet, along with 91 association rules. A total of seven groups are shown, representing different fisheries.

## 4 DISCUSSION AND CONCLUSIONS

The current analysis updates and enhances previous studies carried out by (Leitao et al., 2022) and (Campos et al., 2020) in the Iberian Peninsula waters, including the Portuguese continental coast, for the same period, 2012-2016.

Apriori algorithm allowed the identification of 7 groups based on 91 association rules. From these, three well-known isolated groups for the swordfish, black scabbardfish and octopus' fleets were identified, the first two associated with by-catch species. Each of these groups can be associated to a specific gear type. The swordfish fleet comprises a group of 25-30 drifting longline vessels with an average LOA of 23 meters, commonly engaged in fishing trips with 2-3 weeks duration, extending their activity to North Atlantic seamounts (NEAFC area), having as main by-catch species the shortfin mako and the blue shark. Similarly, the black scabbardfish fleet corresponds to a well-known group of 15-20 bottom longliners, with average LOA of 17 meters, operating along the continental slope around the 1 000 meters depth where fishing trips last around 24h (Bordalo-Machado & Figueiredo, 2009). The octopus fleet corresponds to a large group of 335 small opportunistic vessels, average LOA of 13 meters, where vessels use pots to specifically target octopus.

A total of four less known identified groupings were defined at the scope of this study: trammel nets for soles and cuttlefish; large mesh size gillnets and trammel nets for rays and monkfishes; longlines for forkbeards, congers, seabreams, silver scabbardfish, wreckfish and blackbelly rosefish, and small mesh

Table 1. Species composition in the seven groupings identified. In brackets, the number of fishing trips in each group. LLD – drifting longline; FPO – pots and traps; LLS* - black scabbard bottom longline; GTR** – small mesh size trammel nets; GNS/GTR*** - large mesh size gillnets/ trammel nets; LLS – bottom longline; GNS – small mesh size gill nets.

| Group 1 (520)LLD | Group 2 (54683)FPO | Group 3 (4959) LLS* | Group 4 (75066) GTR** | Group 5 (79044) GNS/GTR*** | Group 6 (48866) LLS | Group 7 (81655) GNS |
|---|---|---|---|---|---|---|
| Swordfish | Common octopus | Black scabardfish | Common sole | Thornback ray | Forbeard | Axillary seabream |
| Shortfin mako shark | | Lowfin gulper shark | Blonde ray | Angler | Atlantic wreckfish | Atlantic chub mackerel |
| | | Tope shark | Cuttlefish | Bastard sole | Blackbelly rosefish | Atlantic horse mackerel |
| | | | European bass | Blackbellied angler | Blackspot seabream | Atlantic mackerel |
| | | | Large-scaled gurnard | Blue shark | European Conger | Black seabream |
| | | | Meagre | Common smooth hound | Red porgy | Common two-banded sea bream |
| | | | Sand sole | Hake | Silver scabbardfish | *Micrchirus* spp |
| | | | Wedge sole | John Dory | | Neon flying squid |
| | | | | Nursehound | | Stripped red mullet |
| | | | | Red gurnard | | Whiting |
| | | | | Spotted ray | | Whiting-pout |

gillnets for seabreams, red mullet, axillary seabream and Atlantic mackerel.

Pouting and European hake, frequently captured in more than one gear type, appear as the center of multiple interrelations: pouting, associated to groups 4 and 7, and European hake, to groups 5, 6 and 7. These relationships with the remaining groupings constitute a valuable help in the distinction of the type of gear used in trips targeting these species.

Spatial visualization and clustering of the rules association facilitated the connection between species and fishing gears. However, association rule mining algorithms generated a large number of association rules, adding complexity to the analysis of these results. In this paper we presented rules summarily through an interactive visualization method which can be used to effectively explore large rule sets. Graph-based visualization offers a clear representation of rules but they tend to become easily cluttered and thus they were easily understandable only for small sets of associations such as the swordfish, the black scabbard and the octopus fisheries, with a relatively low number of well-defined by-catch species.

In this study, from the approximately 250 000 trips, over 90 000 were common between at least 2 groups since species from different groups may be captured in a single trip using a specific gear type. However, this can also be related to the use of different gears in the same trip, leading to false positive relationships between species.

Although apriori is a Bayesian algorithm, with some degree of certainty, it is able in the context of fisheries to identify which species are more likely to be landed when a certain target species is landed. This is of relevance within the context of small scale and coastal fisheries such as the one in study, for which electronic logbooks are not available.

Mapping of these distinct species associations/gear types, using georeferenced data associated to the fishing trips, can greatly contribute to the understand the fleet spatial dynamics and the definition of fishing grounds for the multi-gear fisheries off the Portuguese continental coast.

## ACKNOWLEDGEMENTS

This study was carried out at the scope of the project TECPESCAS – Project Mar 2020 16-01-04-FMP-0010 – IPMA. The fisheries data was supplied to IPMA by the Portuguese Directorate-General for Natural Resources, Safety and Maritime Services (DGRM).

## REFERENCES

Bastian M., Heymann S., & Jacomy M. (2009). *Gephi: an open source software for exploring and manipulating networks*. International AAAI Conference on Weblogs and Social Media.

Biseau, A. (1998). Definition of a directed fishing effort in a mixed-species trawl fishery, and its impact on stock assessments. *Aquatic Living Resources*, *11*(3), 119–136. https://dx.doi.org/10.1016/S0990-7440(98)80109-5

Bordalo-Machado, P., & Figueiredo, I. (2009). The fishery for black scabbardfish (Aphanopus carbo Lowe, 1839) in the Portuguese continental slope. *Reviews in Fish Biology and Fisheries*, *19*(1), 49–67. https://dx.doi.org/10.1007/s11160-008-9089-7

Campos, A., Henriques, V., & Fonseca, P. (2020). *Definition of landing profiles in the Portuguese coastal multi-gear fleet*.

Leitao, P., Sousa, L., Castro, M., & Campos, A. (2022). *Time and spatial trends in landing per unit of effort as support to fisheries management in a multi-gear coastal*. https://doi.org/https://doi.org/10.1101/2021.10.04.463092

Rakesh, A., & Srikant, R. (1994). Fast Algorithms for Mining Association Rules. *Proceedings of the 20th International Conference on Very Large Data Bases*, 487–499.

*Trends in Maritime Technology and Engineering – Guedes Soares & Santos (eds)*


# Numerical study on the mooring force of a gravity-type fish cage under currents and waves

Z.C. Liu & C. Guedes Soares
*Centre for Marine Technology and Ocean Engineering (CENTEC), Instituto Superior Técnico, Universidade de Lisboa, Lisbon, Portugal*

ABSTRACT: A numerical model of an entire fish cage under currents and waves is built using the finite element method. All the components in a fish cage are slender structures, thus, the hydrodynamic load is modelled by the Morison equation. The collar and the bottom ring are represented by pipe and beam elements. The net, ropes and mooring lines are modelled by link elements, which are uniaxial tension-compression elements. The initial shape of mooring lines is calculated based on mooring dynamics, using catenary equations. Because link elements are used, the weight and hydrodynamic load on the mooring lines are considered. The counterweight on the tapered bottom is modelled by a mass element. With this numerical model, the mooring forces and deformation of the cage are assessed for the cage in different sea states. The relationships between mooring forces, and current velocities, wave heights and periods are investigated. Both the mean value and the amplitude of the mooring forces increase with the current velocity, while the wave height and period mainly affect the amplitude rather than the mean value.

## 1 INTRODUCTION

The floating gravity fish cage is widely used in the offshore aquaculture industry, which is becoming an essential source of human food. A gravity cage is composed of four main parts. Collars float on the surface of the sea and provide buoyancy. For the cage with a conical bottom, sinkers usually include a bottom ring to stretch the net and a counterweight on the tip of the cone. The net connects the collar and sinkers, and provides a culturing volume. A mooring system is used to fix the cage. Advantages of a floating cage include large culturing volume, simple and visual operating process.

However, due to the large projected area of the gravity fish cage, the hydrodynamic load on it is also high and leads to large mooring force and rapid volume reduction. Thus, many publications studied the hydrodynamic characteristics of gravity fish cages and its mooring system. Moe et al. (2005) studied the stress distribution on the net structure and inclined the cross ropes to reduce the stress concentration. Lee et al. (2005) built a calculation model to simulate the dynamic behaviour of moored flexible structures. Berstad &Tronstad (2005) built a numerical model of a fish farm and studied the effect of regular and irregular waves on mooring forces.

Lader et al. (2008) did field tests at two Atlantic salmon farms and used pressure sensors to measure the instantaneous position of the net to estimate the cage volume. DeCew et al. (2013) measured the deformation of small gravity cages by acoustic sensors. However, due to the high signal interference under the water, this method cannot be applied for large cages. Fu et al. (2013) did a forced oscillation test to study the hydrodynamic coefficients of the floating collar. Endresen et al. (2013) considered the wake effect between twines nearby and the upstream twines to downstream twines in their numerical models. Bai et al. (2016) estimated the stress concentration on the junction of the collar to the mooring rope and calculated the fatigue life of the collar.

The solidity is an essential property for net structures which is equal to the projected area divided by the total area. Føre et al. (2016) built a numerical model of high solidity (>0.30) net cages and studied the volume reduction. They obtained a good agreement with their experimental results. Xu et al. (2020) studied the effects and properties of synthetic ropes in a mooring system. Detailed application of synthetic mooring cables for floating structures can be found in Wang et al (2019, 2020). Guo et al. (2020) provided a review of the theoretical, numerical, and experimental progress in the application of net-type structures to fish cages.

DOI: 10.1201/9781003320289-63

Bernardo & Guedes Soares (2021) used SIMA to calculate the motion of a fish cage subjected to currents and waves. Føre et al. (2021) measured the drag coefficients of net panels with different solidities. Mohapatra et al. (2021) compared the numerical and analytical methods to calculate the hydrodynamic load on a net cage subjected to waves and obtained a good agreement. Zhang et al. (2021) built a numerical model of a fish farm with 36 cages and studied the mooring forces.

In this work, a numerical model of an entire fish cage subjected to currents and waves is built using the finite element method by Ansys/APDL. In previous studies, the cage was simplified to one floating collar, the surrounding net and the bottom ring. The cage is fixed by four spring elements. A simplified numerical model was used in (Liu et al. 2021a) to compare with the theoretical model and the calculated the volume reduction of a fish cage subjected to uniform flow was determined in (Liu et al. 2021b). However, the complete numerical model in this work includes a two-ring collar with handles, the net with a tapered bottom, a bottom ring with the ropes connected to the collar and the net, and an entire mooring system with main, frame and bridle ropes.

The collar is modelled by pipe elements and the bottom ring is modelled by beam elements. The rest of the flexible structures are modelled by link elements. Thus, the hydrodynamic load and the weight of the mooring ropes are considered in this numerical model. The curved shape of the mooring ropes in the initial state can also be achieved with link elements.

From the calculation results, the mooring forces increase with current velocities. Wave heights and periods have little effect on the mean values of the mooring forces but a large effect on the amplitude. The detail of the numerical model is introduced in Section 2. The mooring forces and deformation of the cage are assessed for the cage in different sea states. The results are shown in Section 3. At last, a conclusion is drawn in Section 4.

## 2 NUMERICAL MODEL

The numerical model is built using the finite element method by Ansys/APDL. Figure 1 shows the fish cage model including a double-circle collar with handles, a net structure with a tapered bottom and a bottom ring with the connecting ropes, and a mooring system including main ropes, frame ropes and bridle ropes. The collar is modelled by pipe elements and the bottom ring is modelled by beam elements. The rest components including the net, mooring ropes and the connecting ropes are modelled by link elements, which are uniaxial tension-compression elements. More information about the numerical model can be found in Liu et al.(2021a).

The main dimensions and parameters of the components are listed in Table 1. In the numerical model of this work, the attack angle of current and waves is constant (positive direction of the x-axis) and the cage moves in the positive direction of the x-axis. The mooring forces of the two are very small and mainly from their own weight. The effect of the downstream main mooring cables on the upstream mooring forces is small. Thus, they are deleted, not to increase the computational efficiency.

Table 1. The main dimensions and material parameters of the numerical model.

| Component | Dimension/Parameter | Value |
|---|---|---|
| Handle | Density(kg/m$^3$) | 450 |
| | Height above the collar (m) | 1.2 |
| | Cross-section diameter (m) | 0.08 |
| | Thickness (mm) | 15 |
| Collar | Density(kg/m$^3$) | 450 |
| | Diameter(m) | 40 for the inner collar<br>42.4 for the outer collar |
| | Cross-section diameter (m) | 0.5 |
| | Thickness (mm) | 50 |
| Net | Material type | Nylon |
| | Young's modulus (GPa) | 1.4 |
| | Twine thickness (mm) | 2.7 |
| | Solidity | 0.225 |
| | Density (kg/m3) | 1140 |
| | Depth (m) | 12 for the cylinder part<br>3 for the tapered bottom |
| Bottom ring | Weight/length (kg/m) | 32 |
| | Diameter(m) | 42.4 |
| Counterweight at the top of the tapered bottom | Weight (kg) | 150 |
| Buoy | Buoyancy (kN) | 6 |
| Bridle/frame mooring rope | Material type | HMPE |
| | Stiffness, AE (N) | 1.25×108 |
| | Frame rope depth (m) | 3 |
| | Distance between cages (m) | 6 |
| | Submerged weight/ length (N/m) | 9.92 |
| | Breaking strength (kN) | 920 |
| Main mooring rope | Material type | Six strand wire rope (IWRC) HMPE |
| | Stiffness, AE (N) | 7.2×107 |
| | Anchor depth (m) | 50 |
| | Submerged weight/ length (N/m) | 54.4 |
| | Breaking strength (kN) | 840 |

In addition, the net has a large projected area subjected to currents and waves, which is the main source of the hydrodynamic loads on cages. If the link elements have the same diameter as the real net, which is between 1.5 mm to 2 mm, as usual, the model will include too many elements to calculate. Thus, the net structure is simplified and the parameters are adjusted to keep the same stiffness, weight and load on it. The total cross-section area of the adjusted structure, the density and the Young's modulus are kept the same as the original structure. Thus, to keep the hydrodynamic load consistent with the original structure, the drag and mass coefficients are adjusted. More information about the method to simplify the net structure can be found in Liu et al (2021a). The parameters of the original and adjusted net structure are shown in Table 2.

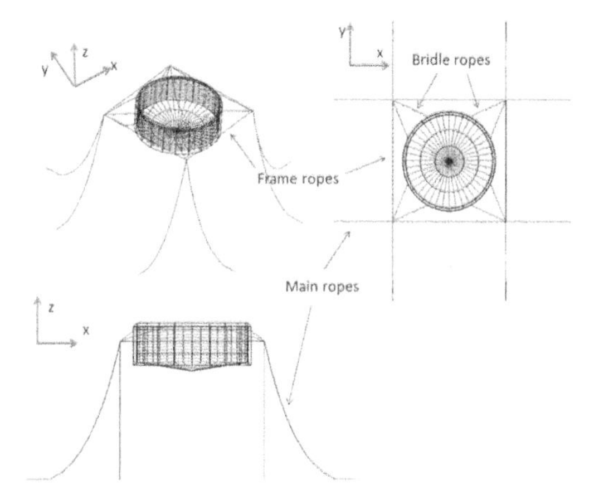


Figure 1. The model of the gravity fish cage with a mooring system.

The linear wave theory is used to model the waves and the Morison equation is used to calculate the hydrodynamic load. The velocity potential is expressed as:

$$\phi(x, z, t) = \frac{H_1 g}{2\omega} \frac{\cosh(k(h+z))}{\cosh(kh)} \sin(kx - \omega t) \quad (1)$$

where $k$ is wave number, $H_1$ is wave height, $h$ is the depth of the flow fluid, $\omega$ is angular frequency of the wave, x and z are coordinates of the point.

A modified Morison equation to take the relative motion between the structure and water particles into account is expressed as (Faltinsen, 1990):

$$f = \frac{1}{2}\rho C_D D(u - \dot{x})|u - \dot{x}| + \rho C_M \frac{\pi D^2}{4}\dot{u} - \rho(C_M - 1)\frac{\pi D^2}{4}\ddot{x} \quad (2)$$

where $f$ is the hydrodynamic force per length, $C_M$ is mass coefficient, $C_d$ is drag coefficient, $\rho$ is the density of sea water, $D$ is the diameter of the slender structure, $u$ and $\dot{u}$ are the velocity and acceleration of the water particle, $\dot{x}$ and $\ddot{x}$ are the velocity and acceleration of the structure element.

The pretension force is 1 kN. A part of the main mooring cable is assumed to be lying on the seabed. The initial configuration (x, z) of the main mooring cable can be expressed as (Faltinsen, 1990):

$$x(s) = \frac{H}{EA_0}s + \frac{H}{w}sinh^{-1}\left(\frac{ws}{H}\right) \quad (3)$$

$$z(s) = \frac{ws^2}{2EA_0} + \frac{H}{w}\left(\sqrt{1 + \left(\frac{ws}{H}\right)^2} - 1\right) \quad (4)$$

where $s$ is the length of the cable between the seabed and the point (x, z), $H$ is the horizontal component of the tension, $EA_0$ is the tensile stiffness of the mooring line, w is the wet weight per length of the mooring material.

Table 2. Parameters of the adjusted structure and original structure.

| | Original structure | Adjusted structure |
|---|---|---|
| Density (kg/m$^3$) | 1140 | 1140 |
| Young's modulus (GPa) | 1.4 | 1.4 |
| Twine diameter (mm) | 2.7 | N.A. |
| Length of the mesh bar(mm) | 24 | N.A. |
| Total Number of horizontal lines in the cylinder part | 500 | 12 |
| Total Number of horizontal lines in the tapered bottom | 125 | 2 |
| Total Number of vertical lines | 5236 | 40 |
| Cross section of each mesh bar (mm$^2$) | 5.73 | N.A. |
| Cross section of horizontal lines in the cylinder part (mm$^2$) | N.A. | 238.56 |
| Cross section of vertical lines in the cylinder part (mm$^2$) | N.A. | 749.47 |
| Cross section of the upper horizontal line in the tapered bottom (mm$^2$) | N.A. | 402.58 |
| Cross section of the lower horizontal line in the tapered bottom (mm$^2$) | N.A. | 268.38 |
| Cross section of vertical lines in the first layer of the tapered bottom (mm$^2$) | N.A. | 624.56 |
| Cross section of vertical lines in the second layer of the tapered bottom (mm$^2$) | N.A. | 374.73 |
| Cross section of vertical lines in the third layer of the tapered bottom (mm$^2$) | N.A. | 124.91 |

## 3 RESULTS

The calculation results are shown in this section. Section 3.1 includes the deformation and motion of the cage. The mooring forces of the cage subjected to different sea states are in the content of Section 3.2.

### 3.1 *Deformation and motion of the cage*

The deformation of the cage subjected to different sea states is shown in Figure 2. The upstream mooring lines have been straightened under the current from 0.5m/s to 1m/s. The upstream junctions of the main, frame and bridle mooring lines are also pulled to a deeper position. The upstream frame mooting line is lower than the main body of the cage. On the other hand, the depth of the downstream mooring lines has not changed much. It is due to the downstream mooring forces being relatively low, which is discussed in detail in Section 3.2

With the increase of the current velocity, the cage deforms more and the volume decreases., The ropes that connect the bottom rings in the previous study are removed and the bottom ring is added directly to the bottom circle of the net. The deformation of the bottom is small and the motion is close to being solid. However, in this numerical model, it can be observed that the bottom ring has a large deformation, especially the upstream part subjected to large current velocity, which is in agreement with the results from the experiments in (Vegard, 2016).

In addition, the vertical motion of the collar is also obtained from the numerical model. Except for the nodes connected to upstream bridle mooring lines and the nodes around them, the amplitudes of the vertical motion are close to the wave height. On the other hand, due to the pull of the mooring lines, the amplitudes of the vertical motion of the nodes connected to upstream bridle mooring lines are smaller. The vertical displacements of two nodes, subjected current velocity of 1m/s, wave height of 1 m and wave period of 10s, are plotted in Figure 3. The amplitude of Node179 is 99.99cm while the one of Node227 is only 91.53cm.

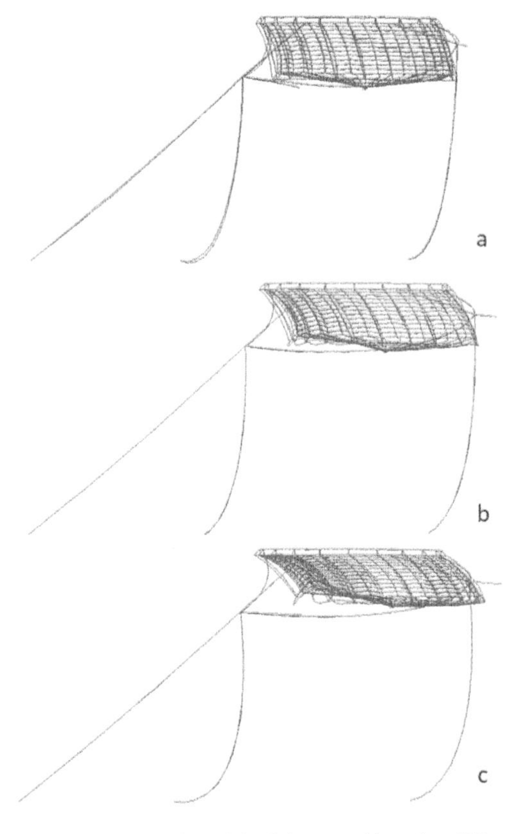


Figure 2. Deformation of the fish cage subjected to different current velocities (a. current velocity = 0.5m/s, b. current velocity = 0.75m/s, c. current velocity = 1m/s).

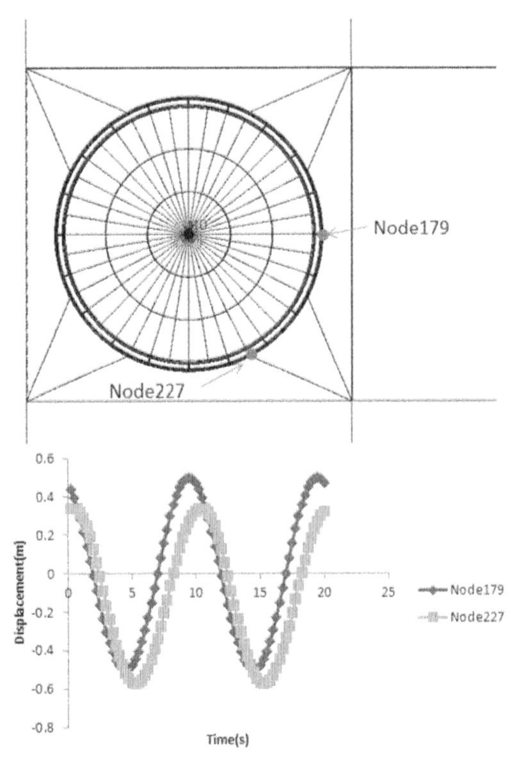


Figure 3. Vertical displacement of the collar subjected to a current velocity of 1m/s, a wave height of 1m and a wave period of 10s.

### 3.2 *Mooring forces subjected to different sea states*

The calculated mooring forces are shown in this Section. Figure 4 shows an instantaneous axial force distribution of the whole structure subjected to a current velocity of 1m/s, a wave period of 10s and a wave height of 1m. The distributions from other sea states are similar but the values of the mooring forces are different. The upstream main mooring ropes have the largest mooring force, while one of the bridle ropes on each side has a similar mooring force. The frame mooring ropes have relatively small forces which are mainly from the hydrodynamic load on themselves. The frame ropes do not take on the role of fixing the cage. Thus, the rest part mainly focuses on the mooring forces of the main ropes.

To study the effect of the current velocity, wave height, and wave period, two of them are constant while the rest one varies. Table 3 shows the maximum mooring forces under different current velocities while the wave period is 8s and the wave height is 1m. Obviously, both the mean values and the amplitudes of the mooring force increase with the current velocity. The mean value of the maximum mooring force is 53.52kN at a current velocity of 1m/s, which is 2.67 times that at a current velocity of 0.5m/s. On the other hand, the amplitude of the maximum mooring force is 11.26kN at a current velocity of 1m/s, which is 2.82 times that at a current velocity of 0.5m/s. The large current velocity increases the hydrodynamic load on the cage and leads to a large mooring force directly. In addition, the current velocity also changes the shape of the fish cage and the projected area, which indirectly affects the mooring force.

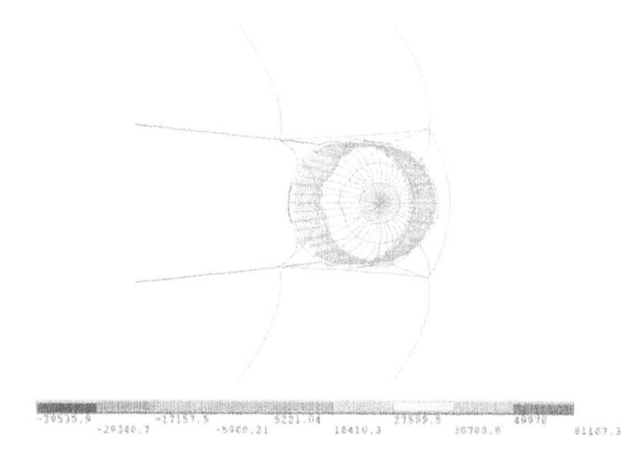

Figure 4. The instantaneous axial force distribution, unit: N. (current velocity = 1m/s, wave height =1m, wave period = 10s, time = 150s).

On the other hand, the effect of the wave period and wave height on the mean value of the mooring force is inconspicuous, compared with the current velocity. When the wave height increases from 0.75m to 1.25m, the mean value only increases by a factor of 0.04. Besides, the mean value decreases little with the wave period. However, at a wave height of 1.25m, the amplitude increases to 1.60 times the one at a wave height of 0.75m. When the wave period increase from 8s to 12s, the amplitude decreases by 16.87%. From the result, at the same current velocity, the mooring force is smaller subjected to long waves than short waves.

Table 3. Maximum mooring forces subjected a wave height of 1m and a wave period of 8s.

| Current velocity (m/s) | 0.5 | 0.75 | 1 |
|---|---|---|---|
| Mean value of the maximum mooring force (kN) | 20.02 | 36.73 | 53.52 |
| Amplitude of the maximum mooring force (kN) | 3.99 | 7.08 | 11.26 |

Table 4. Maximum mooring forces subjected a current velocity of 1m/s and a wave period of 10s.

| Wave Height (m) | 0.75 | 1 | 1.25 |
|---|---|---|---|
| Mean value of the maximum mooring force (kN) | 51.40 | 52.51 | 53.46 |
| Amplitude of the maximum mooring force (kN) | 7.29 | 9.44 | 11.68 |

Table 5. Maximum mooring forces subjected a current velocity of 1m/s and a wave height of 1m.

| Wave Period (s) | 8 | 10 | 12 |
|---|---|---|---|
| Mean value of the maximum mooring force (kN) | 53.52 | 52.51 | 52.08 |
| Amplitude of the maximum mooring force (kN) | 11.26 | 9.44 | 9.36 |

## 4 DISCUSSION

In this work, the layout of the mooring system is similar to Huang et al. (2008), Endresen et al. (2014) and Huo et al. (2020). Their models are fish arrays which include several cages. The frame ropes act to transmit mooring forces and connect other mooring ropes. However, the fish cage for offshore aquaculture in this work is large and a single one. From the instantaneous axial force distributions, the frame ropes have small mooring forces, which are mainly from the hydrodynamic load on themselves. Due to the horizontal hydrodynamic force from the current and the large distance between the junctions, the upstream frame rope has a large horizontal displacement. From Figure 2, the upstream frame rope has moved beyond the main body of the cage in the horizontal direction. In this numerical model, the frame ropes and the cage happen to have no contact because the upstream frame rope has moved under the cage. However, for the cage with a larger diameter or depth, the frame rope may rub against the net, which can lead to breakage of the net and escape of fish. Thus, for a large and single offshore fish cage, the necessity of the frame mooring ropes is questionable.

The purpose for applying the two adjacent bridle ropes on each side is to share the mooring forces and reduce the stress concentration, especially at the connection of the mooring to the collar. However, from the result, this design does not function as desired. The outer mooring rope has a much larger force than the inner one. For instance, under a current velocity of 0.5m/s, a wave period of 8s and a wave height of 1m, the average mooring force on the outer upstream bridle rope is 18.43kN, while the average mooring force on the inner one is less than 0.1kN. The inner mooring rope is almost slack. Whether it can be achieved by applying large pretension or adjusting the angle between the bridle ropes is a meaningful topic for future research.

## 5 CONCLUSIONS

This paper presents a numerical model of an entire fish cage subjected to both current and waves. Pipe and beam elements are used to model the collar and bottom ring. Link elements are used to model the flexible structure including the net, ropes and mooring ropes. The cage deformation and the mooring forces subjected to different sea states are calculated based on this numerical model. It has been concluded that:

1. The cage deforms more and the cage volume decreases with the increase of the current velocity.
2. The amplitude of vertical motion is close to the wave height for the most part of the collar. However, the motion is limited by the mooring ropes and the amplitudes of the nodes close to the junction of the upstream mooring ropes to the collar are smaller.
3. Current velocity is the main factor that influences both the mean value and the amplitude of the maximum mooring forces. Mooring forces increase with the current velocity.
4. Both wave height and period have little effect on the mean value of the maximum mooring force. The amplitude of the mooring force increases with the wave height. When for long waves, the amplitude is smaller than the one subjected to short waves.

## ACKNOWLEDGEMENTS

This work is performed within the project HYDROELASTWEB - Hydroelastic behaviour of horizontal flexible floating structures for applications to Floating Breakwaters and Wave Energy Converters, which is funded by the Portuguese Foundation for Science and Technology (Fundação para a Ciência e a Tecnologia - FCT) under contract 031488-770 (PTDC/ECI-EGC/31488/2017). The first author has been funded by the Portuguese Foundation for Science and Technology (Fundação para a Ciência e a Tecnologia-FCT), through a doctoral fellowship under the Contract no. SFRH/BD/147178/2019. This work contributes to the Strategic Research Plan of the Centre for Marine Technology and Ocean Engineering (CENTEC), which is financed by the Portuguese Foundation for Science and Technology (Fundação para a Ciência e Tecnologia - FCT) under contract UIDB/UIDP/ 00134/2020.

## REFERENCES

Bai, X., Xu, T., Zhao, Y., Dong, G. & Bi, C., 2016. Fatigue as-sessment for the floating collar of a fish cage using the de-terministic method in waves. *Aquacultural Engineering*, 74, 131–142.

Bernardo, T. A. and Guedes Soares, C. 2021. Validation of tools for the analysis of offshore aquaculture installations. Guedes Soares, C. & Santos T. A., (Eds.). *Developments in Maritime Technology and Engineering*. London, UK: Taylor and Francis; 2021; pp. Vol 2, pp. 685–692.

Berstad, A.J., & Tronstad, H., 2005. Response from current and regular/irregular waves on a typical polyethylene fish farm. In Guedes Soares, C. Garbatov Y. & Fonseca, N. (eds), *Maritime Transportation and Exploitation of Ocean and Coastal Resources*: Taylor & Francis Group, London. 1189–1196.

DeCew, J., Fredriksson, D.W., Lader, P.F., Chambers, M., Howell, W.H., Osienki, M., Celikkol, B., Frank, K., Høy, E., 2013. Field measurements of cage deformation using acoustic sensors. *Aquacultural. Eng*. 57, 114–125.

Endressen, C., Birkevold, J., Feae, M., Fredheim, A., Kristiansen, D., Lader, P., 2014. Simulation and validation of a numerical model of a full aquaculture net-cage system. *Proc. of 33rd International Conference on Ocean, Offshore and Arctic Engineering*, ASME Paper No: OMAE2014-23382, V007T05A006.

Endresen, P. C., Fore, M., Fredheim, A., Kristiansen, D. & Enerhaug, B., 2013, Numerical Modelling of Wake Effect on Aquaculture Nets, *Proc. of 32nd International Conference on Ocean, Offshore and Arctic Engineering*, ASME Paper No. OMAE2013-11446.

Faltinsen. O.M., 1990. Stationkeeping. In: *Sea Loads on Ships and Offshore Structures*. Cambridge University Press, pp.257–270.

Faltinsen. O.M., 1990. Viscous Wave Loads and Damping. In: *Sea Loads on Ships and Offshore Structures*. Cambridge University Press, pp.223–226.

Fu, S., Xu, Y., Hu, K. & Zhang, Y., 2013. Experimental investigation on hydrodynamics of floating cylinder in oscillatory and steady flows by forced oscillation test. *Marine Structures*, 34, 41–55.

Føre, H. M., Endresen, P. C., Norvik, C., and Lader, P. (February 3, 2021). "Hydrodynamic Loads on Net Panels With Different Solidities." *ASME. J. Offshore Mech. Arct. Eng*. October 2021; 143(5): 051901.

Føre, H. M., Lader, P., Lien, E. & Hopperstad, O., 2016. Structural response of high solidity net cage models in uni-form flow. *Journal of Fluids and Structures*, 65, 180–195.

Guo, Y.C., Mohapatra, S.C. & Guedes Soares, C. 2020. Review of developments in porous membranes and net-type structures for breakwaters and fish cages. *Ocean Engineering*, 200, 107027

Hou, H., Dong, G. and Xu, T., 2020. Analysis of probabilistic fatigue damage of mooring system for offshore fish cage considering long-term stochastic wave conditions. *Ships and Offshore Structures*, 1–12.

Huang, C., Tang, H. and Liu, J., 2008. Effects of waves and currents on gravity-type cages in the open sea. *Aquacultural Engineering*, 38(2),105–116.

Moe, H., Fredheim, A. & Heide, M.A., 2005. New net cage designs to prevent tearing during handling. In Guedes Soares, C. Garbatov Y. & Fonseca, N. (eds), *Maritime Transportation and Exploitation of Ocean and Coastal Resources*: Taylor & Francis Group, London. 1265–1272.

Mohapatra, S. C.; Bernardo, T. A., and Guedes Soares, C. 2021. Dynamic wave induced loads on a moored flexible cylindrical net cage with analytical and numerical model simulations. *Applied Ocean Research*. 2021; 110:102591.

Lader, P., Dempster, T., Fredheim, A., Jensen. F., 2008.Current induced net deformations in full-scale cages for Atlantic salmon (Salmo salar). *Aquacultural Eng*. 38, 52–65.

Lee, C.W., Kim, H.S., Lee, G.H., Koo, K.Y., Choe, M.Y., Cha, B.J. & Jeong, S.J., 2005. Computation modeling of the moored flexible structures. In Guedes Soares, C. Garbatov Y. & Fonseca, N. (eds), *Maritime Transportation and Exploitation of Ocean and Coastal Resources*: Taylor & Francis Group, London. 1239–1243.

Liu, ZC., Mohapatra, S. C., and Guedes Soares, C. (2021a), "Finite Element Analysis of the Effect of Currents on the Dynamics of a Moored Flexible Cylindrical Net Cage", *Journal of Marine Science and Engineering*, Vol. 9, 159.

Liu, ZC., Garbatov, Y. and Guedes Soares, C. (2021b), "Numerical modelling of full-scale aquaculture cages under uniform flow", *Developments in Maritime Technology and Engineering*, Guedes Soares, C. & Santos T. A., (Eds.), Taylor and Francis, London, UK, Vol 2, 705–712.

Vegard, A., 2016. Modelling of Aquaculture Net Cages in SIMA. Project Report. Marintek, Norway

Wang, S., Xu, S., Guedes Soares, C., Zhang, Y., Liu, H. and Li, L. (2020), "Experimental study of nonlinear behavior of a nylon mooring rope at different scales", Guedes Soares, C., (Ed.), *Developments in Renewable Energies Offshore*, Taylor and Francis Group, London, UK, pp. 690–697.

Wang, S., Xu, S., Xiang, G. and Guedes Soares, C. (2019), "An overview of synthetic mooring cables in marine application", Guedes Soares, C., (Ed.), *Advances in Renewable Energies Offshore*, Taylor and Francis Group, London, UK, pp. 853–863.

Xu, S., Wang, S. and Guedes Soares, C. (2020), "Experimental investigation on hybrid mooring systems for wave energy converters", *Energy*, Vol. 158, pp. 130–153.

Zhang, DP.; Bai, Y., and Guedes Soares, C. 2021. Dynamic Analysis of an Array of Semi-rigid "Sea Station" Fish Cages subjected to waves. *Aquacultural Engineering*. 94: 102172

*Trends in Maritime Technology and Engineering – Guedes Soares & Santos (eds)*


# Preliminary experiments of the behaviour of circular gravity cage in linear waves

Z.C. Liu & C. Guedes Soares
*Centre for Marine Technology and Ocean Engineering (CENTEC), Instituto Superior Técnico, Universidade de Lisboa, Lisbon, Portugal*

ABSTRACT: Preliminary experiments are conducted to study the dynamics of circular gravity cage in linear waves. Experiments are done in a flume 38.5 m long, with a simplified model of a cage, which includes a floating collar, a bottom ring and nets. The model is connected to supports above the flume by wires. Linear waves are applied in experiments. The upstream wave heights are around 4 cm and 8 cm, while wave periods are from 0.8 s to 3 s. For short waves, the wave height decreases significantly after passing the fish cage. For long waves, the fish cage model has little effect on the passing waves, not changing the wave height nor period. The preliminary tests show that for some ranges of periods the cages respond dynamically and they have a damping effect on the passing waves, while for longer periods they simply ride the waves. The limitations of the experiments and insights from the experiments for future research are discussed.

## 1 INTRODUCTION

Nowadays, due to overfishing and the increased demand for fish by human society, the traditional capture industry has been unable to meet the needs of society. Aquaculture is becoming an essential source of human food. From the State of World Fisheries and Aquaculture (FAO, 2020), the capture production in inland waters and marine waters barely increased since the 1990s. On the other hand, both aquaculture production in inland waters and at sea continues to grow. Gravity cages and cage arrays are widely used in aquaculture due to the sizeable culturing volume and simple operating process. A gravity cage usually includes four parts, a floating collar that provides buoyancy and supports the net that has bottom counterweights and a mooring system that keeps the cage in station.

The culturing volume of the gravity cage leads to a large projected area and hydrodynamic force under currents and waves. Thus, the mooring force of such a structure can be large but the volume of the flexible structure can reduce rapidly under currents. Thus, the hydrodynamic characteristics of gravity fish cages have been widely studied around the world. Lader et al. (2008) used pressure sensors to measure the deformation of the net and calculated the volumes of the real-scale cages at two different fish farms. The deformation of the flexible structure reduced the cage volume, while it lowers the total drag force than the stiff structure. DeCew et al. (2013) used acoustic sensors to measure the deformation of a gravity cage. Even though this method can directly measure the instant location of the net, due to the signal interference under the water, the acoustic sensors can be only applied for small-scale cages. Fu et al. (2013) did forced oscillation tests to measure the mass and drag coefficients of a floating cylinder structure under different Reynolds numbers (Re) and Keulegan-Carpenter numbers (KC).

Moe Føre et al. (2016) studied the deformation and hydrodynamic forces of the cages with various solidities from 0.19 to 0.43 and compared the results from experiment and numerical model. The deformation of the net was less dependent on solidity than on current velocity and bottom weights. Gansel et al. (2018) towed a full-scale fish cage in a fjord environment, where currents and waves were small and negligible. Both the drag force and volume reduction were measured under under different towing velocities. It was found that current calculation methods overestimate drag forces, especially at high flow velocities. Guo et al. (2020) presented a review of the theoretical and numerical methods in the application of net-type structures to fish cages.

Mohapatra et al. (2021) calculated the hydrodynamic load on a gravity cage by both numerical and analytical methods, and obtained a good agreement. Dong et al. (2021) built an experimental model to study the hydrodynamic characteristics of

DOI: 10.1201/9781003320289-64

a flexible net cage under uniform flow, including the drag force, remained cage volume and the flow field inside and outside the net cage. A strong negative correlation between the drag force and cage volumes was found from their results. It was also observed that the reduction of the current speed inside the net cage depends on the bottom weights and incoming current speed.

Moe Føre et al. (2021) measured the drag coefficients and velocity reduction of net material with different solidities by towing plat net panels in a flume. Dong et al. (2021) inserted vertical pipes inside a cage to represent the fish school and studied the flow field inside the cage and the drag force. Miao et al. (2021) built a model of semi-submersible offshore fish farm and studied the interaction between the fish farm (SOFF) and waves.

The study of Liu et al. (2021 & 2022) showed that waves have a big impact on mooring forces and usually, a fish farm includes arrays of gravity cages. Thus, in this paper, preliminary experiments are conducted to study the effect of the circular gravity cage on linear waves. The purpose of the preliminary experiments, which are mainly of qualitative nature, is to determine if experiments at the present scale allow observing different regimes of the floating cage under short waves and long waves, and secondly to check that whether it would be possible to measure the decay in wave height after the passage of the wave.

Section 2 includes the experimental environment, the fish cage model and the process to generate input files for the wave maker. The results are shown in Section 3. The flaws of the cage model and inspiration from experiments are discussed in Section 4. A conclusion is drawn in Section 5 at last.

## 2 EXPERIMENTAL ENVIRONMENT AND THE FISH CAGE MODEL

### 2.1 *Experimental environment*

The experiment was done in a flume at National Laboratory for Civil Engineering (LNEC), Lisbon, Portugal. The total length of the flume is 38.5 m. The width varies from 88 cm to 60 cm. Wave gauges and the fish cage model are located at the back part of the flume, where the width and water depth are constant: 60 cm and 35 cm respectively.

There are total 8 wave gauges: four in the front of the model to measure the upstream wave height and period, and four behind the model to measure the downstream wave height and period. The distances between the four gauges are 30 cm, 20 cm and 38 cm respectively.

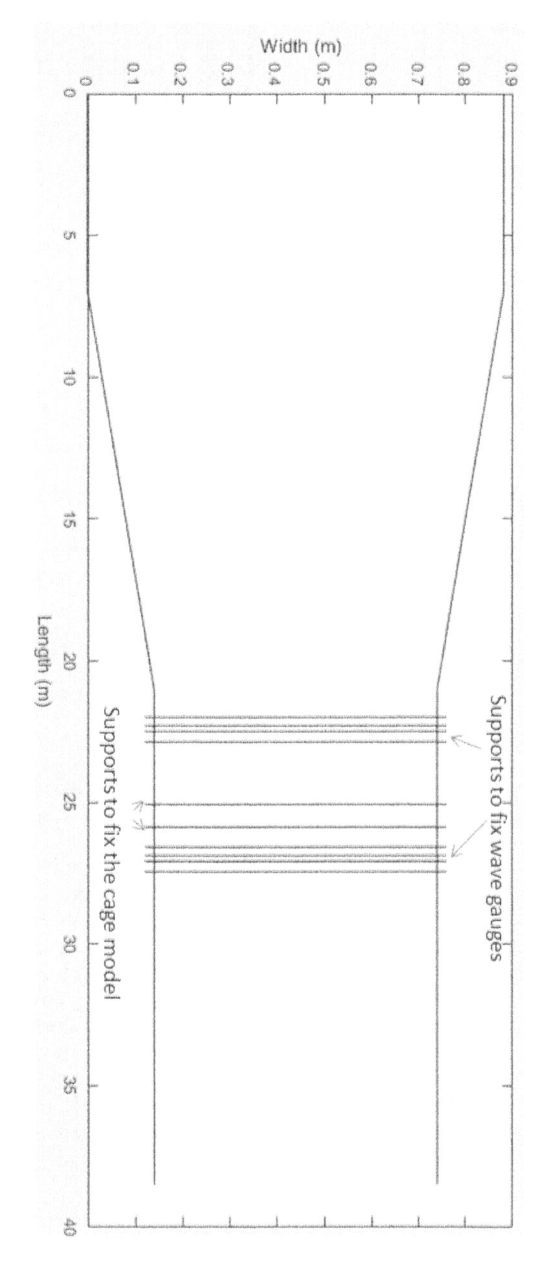


Figure 1. Top view of the flume and the location of supports to fix wave gauges and the cage model.

The distances between the model and the gauges are 2.6 m and 1.1 m. The wave gauges are fixed on slender supports and inserted into the water. The fish cage model is fixed by wires. The other ends of the wires are attached to supports at the top of the flume. The wires are slack to avoid the cage in the waves being lifted out of the water by the presence of the wires. Figure 1 shows the set-up of the flume and Figure 2 shows the wave gauges.

Figure 2. The wave gauge and the support.

## 2.2 *Gravity fish cage model*

The fish cage model includes three parts, the floating collar, the net and the bottom ring. The collar is made of plastic foam. The diameter of the collar is 46 cm and the diameter of the cross-section is 7 cm. The collar is solid and the weight is 0.06 kg, as shown in Figure 3.

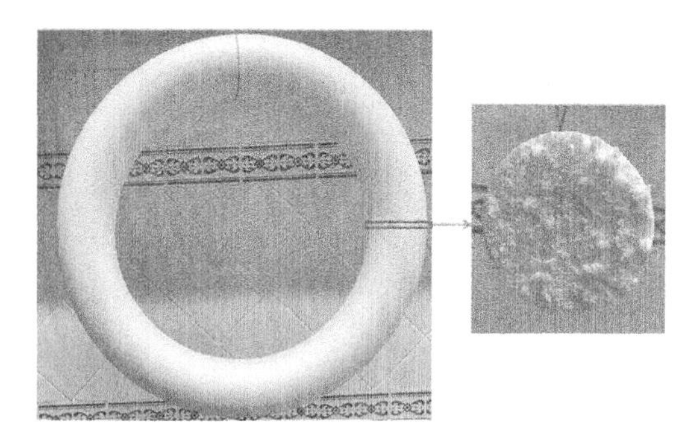

Figure 3. The floating collar and its cross-section.

To scale the parameter of the net as other parts of the cage, the thinnest net can be found is used in this experiment. The thickness of one twine is 0.3 mm and the length of the mesh bar is 2.6 mm. The net covers the perimeter and bottom of the fish cage. The bottom ring includes a leather cover and wires inside. The diameter of the bottom ring is 38.7 cm and the weight is 0.92 kg. The net is fixed on the collar and the bottom ring by plastic straps as shown in Figure 4. The diameters of different components of the fish cage are listed in Table 1.

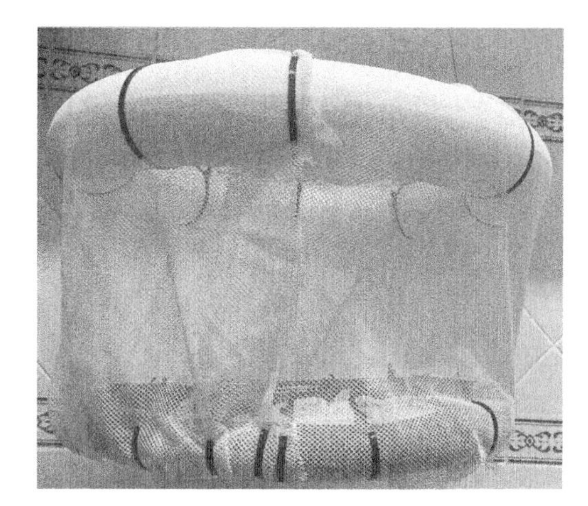

Figure 4. The fish cage model.

Table 1. The upstream and downstream wave heights with and without a fish cage model.

| Component | Dimension | Value |
|---|---|---|
| Collar | Diameter (cm) | 46 |
| | Cross-section diameter (cm) | 7 |
| | Weight (kg) | 0.06 |
| Net | Thickness of one twine (mm) | 0.3 |
| | Length of the mesh bar (mm) | 2.2 |
| Bottom ring | Diameter (cm) | 38.7 |
| | Cross-section diameter (cm) | 3.4 |
| | Weight (kg) | 0.92 |

## 2.3 *Wave generation*

Linear waves are applied in this experiment. The input files for wave generation are columns of voltage signals to the wavemakers to generate required waves. The required wave height in this experiment is 4 cm and 8 cm and the period varies from 0.8 s to 3 s. The first step is to input the required wave parameters and the dimensions of the flume to generate an initial input file. The initial file is input into the software to control the wavemaker and generate waves. If the wave height is between 90% and 110% of the required wave height, the file will be saved for that working condition. Otherwise, the voltages should be adjusted to generate a right file and the process above will be repeated.

# 3 RESULTS

The wave height will gradually decrease even there is no model in the flume. Thus, the upstream and downstream wave heights are also recorded when there is no model inside the fish cage. Table 2 shows the upstream and downstream wave heights with and without the cage model. From the results, the fish cage model does not change the wave period but reduces the wave height when the waves are short. The ratio of the downstream wave height with and without a model is plotted in Figure 5 When the wave period is 0.8 s, the model reduces the downstream wave height to 88% of the one without a cage model. However, when the wave period increases to 1.2 s or longer, the downstream wave height with a model is the same or at least larger than 98% of the one without a model. For long waves, the model has little effect on the wave height.

Table 2. The upstream and downstream wave heights with and without a fish cage model.

| Wave period(s) | Upstream wave height (m) | Downstream wave height without model (m) | Downstream wave height without a fish cage model (m) |
|---|---|---|---|
| 0.8 | 0.0375 | 0.0361 | 0.0318 |
| 1 | 0.0382 | 0.0325 | 0.0339 |
| 1.2 | 0.041 | 0.0332 | 0.0332 |
| 1.6 | 0.0417 | 0.0325 | 0.0332 |
| 2 | 0.0438 | 0.0368 | 0.0368 |
| 2.4 | 0.0424 | 0.0346 | 0.0346 |
| 3 | 0.0431 | 0.0368 | 0.0361 |
| 0.8 | 0.075 | 0.0728 | 0.0636 |
| 1 | 0.0806 | 0.0679 | 0.0721 |
| 1.2 | 0.0806 | 0.0672 | 0.0686 |
| 1.6 | 0.082 | 0.0664 | 0.0664 |
| 2 | 0.0785 | 0.0679 | 0.0672 |
| 2.4 | 0.0806 | 0.0672 | 0.0679 |
| 3 | 0.082 | 0.0735 | 0.0735 |

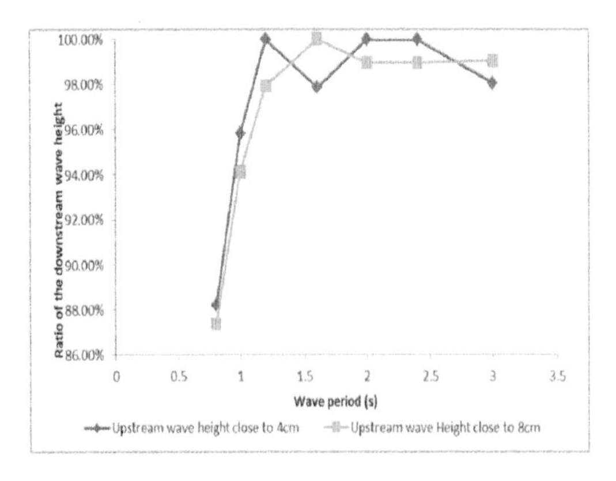


Figure 5. The ratio of the downstream wave height with/without a model.

In addition, the motion of the fish cage is also observed and photographed. Figures 6 and 7 show the cage model under a wave height of 8 cm and wave period of 0.8 s and 3 s, respectively.

For short waves, it can be observed clearly how the peak of the wave passes the collar from photos 1 to 6 in Figure 6. The relative motion between the floating collar and the water surface is heavy. From photos 5 and 6, after one wave the peak moves to the bottom of the collar, the front part of the collar is not in contact with the water surface. The next wave peak laps on the collar. On the other side, for long waves, the collar clings to the water surface, as shown in Figure 7. The relative motion between the floating collar and the

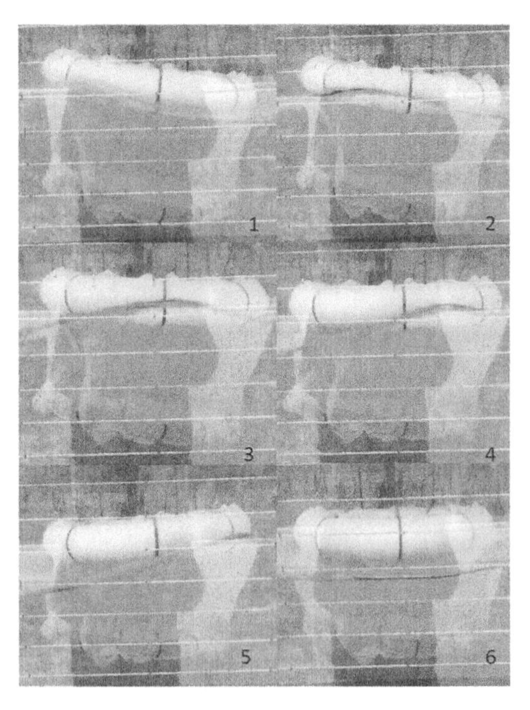


Figure 6. The motion of the cage model under wave with a period of 0.8s.

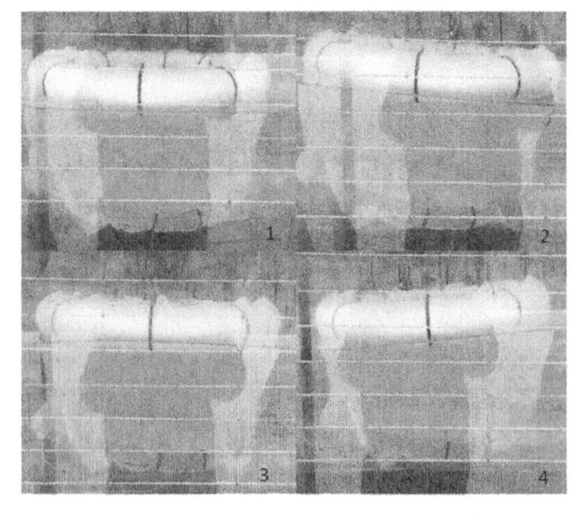


Figure 7. The motion of the cage model under wave with a period of 3s.

water surface, especially in the vertical direction is very small. That is one of the reasons why the cage does not affect the wave height for long waves.

## 4 DISCUSSION

As preliminary experiments to study qualitatively the behaviour of circular gravity cage in waves, the model used in this work is quite rough compared with a real gravity fish cage. The white foam ring in Figure 4 is a collar-like structure that can be bought in the market. However, the dimension of the ring is unique. The diameter of the cross-section is too large but the weight of the collar is too small as a downscale gravity cage model. The cross-section is solid which means it is impossible to add a counterweight in the collar. From Fig 8 and 9, it can be observed that most part of the collar is above the water surface.

In addition, there is no adequate proportion between the diameter of the collar and the length of the net. The total height of the cage is 23 cm while the diameter of the diameter of the collar is 46 cm. Their proportion do not match reality. Besides, the amount of the straps and the net around the collar and the bottom ring is far more than required to fix the net. The establishment of this model provides experiment for making more refined models in the future.

On the other hand, unlike the real mooring lines are fixed on the sea bed, the wires and supports are above the cage model. Even the wires are slack at the beginning of the experiment, the wire on one side is pulled taut as the cage moves to the other side with waves. It may make the collar have less contact with the water surface, which is contrary to the actual situation. When the mooring lines are pulled taut, more part of the collar will be pulled into the water.

Even though the model in this experiment is rough, the result represents a good inspiration for future experimental research and justifies further efforts to be invested. It is found that the model has a large effect on short waves. After a more refined physical model with different dimensions is built, the relationship of the diameters of the cage and the wave lengths of the waves which may be affected by the cages can be more properly studied. On the other hand, the effect of other dimensions like the net solidity, the weight of the bottom ring and the diameter of the cross-section can be studied experimentally.

## 5 CONCLUSIONS

This paper presents preliminary experiments to study qualitatively the behaviour of a circular gravity cage in linear waves. The experimental environment, the fish cage model and the manipulation of the wave-maker are introduced. The motion of the cage is observed. The relative motion between the floating collar and the water surface is heavy under short waves. However, the collar rides on long waves without apparent relative motion.

The upstream and downstream wave heights and periods are measured successfully. It can be concluded that the cage model does not change the wave period. For short waves, the wave height reduces significantly after the cage model. For long waves, the cage model has little effect and the downstream wave heights with and without a cage model are very close. Based on the result from the preliminary experiments, submerged flexible fish cage with mooring systems of different geometrical shapes will be developed to study the effect of dimensions of the fish cage including the diameter, the net solidity and the bottom weight in the future.

## ACKNOWLEDGEMENTS

The authors are grateful to Dr. Juana Fortes from LNEC for having allowed the use of the experimental facilities for this study. This work is performed within the project HYDROELASTWEB - Hydroelastic behaviour of horizontal flexible floating structures for applications to Floating Breakwaters and Wave Energy Converters, which is funded by the Portuguese Foundation for Science and Technology (Fundação para a Ciência e a Tecnologia - FCT) under contract 031488-770 (PTDC/ECI-EGC/31488/2017). The first author has been funded by the Portuguese Foundation for Science and Technology (Fundação para a Ciência e a Tecnologia-FCT), through a doctoral fellowship under Contract no. SFRH/BD/147178/2019. This work contributes to the Strategic Research Plan of the Centre for Marine Technology and Ocean Engineering (CENTEC), which is financed by the Portuguese Foundation for Science and Technology (Fundação para a Ciência e Tecnologia - FCT) under contract UIDB/UIDP/00134/2020.

## REFERENCES

DeCew, J., Fredriksson, D.W., Lader, P.F., Chambers, M., Howell, W.H., Osienki, M., Celikkol, B., Frank, K. & Høy, E., 2013. Field measurements of cage deformation using acoustic sensors. *Aquacultural Engineering*: 57, 114–125.

Dong, S, Park, S, Zhou, J, Li, Q, Yoshida, T, & Kitazawa, D. "Flow Field Inside and Around a Square Fish Cage Considering Fish School Swimming Pattern." *Proceedings of the ASME 2021 40th International Conference on Ocean, Offshore and Arctic Engineering. Volume 5: Ocean Space Utilization.* Virtual, Online. June 21–30, 2021. V005T05A003. ASME. https://doi.org/10.1115/OMAE2 021–63047

Dong, S., You, X. & Hu, F., 2021. Experimental investigation on the fluid–structure interaction of a flexible net cage used to farm Pacific bluefin tuna (Thunnus orientalis). *Ocean Engineering*: 226, p.108872.

Guo, Y.C., Mohapatra, S.C. & Guedes Soares, C. 2020. Review of developments in porous membranes and net-type structures for breakwaters and fish cages. *Ocean Engi-neering*: 200, 107027

Miao, Y., Ding, J., Tian, C., Chen, X. & Fan, Y., 2021. Experimental and numerical study of a semi-submersible offshore fish farm under waves. *Ocean Engineering*: 225, p.108794.

Moe Føre, H, Endresen, PC, & Bjelland, HV. "Load Coefficients and Dimensions of Raschel Knitted Netting Materials in Fish Farms." *Proceedings of the ASME 2021 40th International Conference on Ocean, Offshore and Arctic Engineering. Volume 6: Ocean Engineering*. Virtual, Online. June 21–30,2021. V006T06A044. ASME. https://doi.org/10.1115/OMAE2021-63401

Moe Føre, H., Lader, P. F., Lien, E., & Hopperstad, O. S. 2016. Structural response of high solidity net cage models in uniform flow. *Journal of Fluids and Structures*, 65, 180–195.

Mohapatra, S. C.; Bernardo, T. A., & Guedes Soares, C. 2021. Dynamic wave induced loads on a moored flexible cylin-drical net cage with analytical and numerical model simu-lations. *Applied Ocean Research*: 110,102591.

FAO. 2020. *The State of World Fisheries and Aquaculture 2020*. Fisheries and Aquaculture Department, Food and Agriculture Organization of the United Nations. Publishing Policy and Support Branch, Office of Knowledge Exchange, Research and Extension, Rome, Italy.

Fu, S., Xu, Y., Hu, K. & Zhang, Y., 2013. Experimental investigation on hydrodynamics of floating cylinder in oscillatory and steady flows by forced oscillation test. *Marine Structures*: 34, 41–55.

Gansel, L., Oppedal, F., Birkevold, J. & Tuene, S., 2018. Drag forces and deformation of aquaculture cages—Full-scale towing tests in the field. *Aquacultural Engineering*: 81, 46–56.

Lader, P., Dempster, T., Fredheim, A. & Jensen. F., 2008. Current induced net deformations in full-scale cages for Atlantic salmon (Salmo salar). *Aquacultural Engineering:* 38, 52–65.

Liu, ZC., Mohapatra, S. C., & Guedes Soares, C. (2021), Finite Element Analysis of the Effect of Currents on the Dynamics of a Moored Flexible Cylindrical Net Cage, *Journal of Marine Science and Engineering*: Vol. 9, 159.

Liu ZC, Wang S & Guedes Soares C. (2022) Numerical Study on the Mooring Force in an Offshore Fish Cage Array. *Journal of Marine Science and Engineering*: 10(3):331.

*Trends in Maritime Technology and Engineering – Guedes Soares & Santos (eds)*


# Variability in the structure of pelagic fish communities: A pitfall for management of herring in the Baltic Sea?

T. Raid & E. Sepp
*Estonian Marine Institute, University of Tartu, Estonia*

ABSTRACT: The Baltic herring (*Clupea harengus membras* L.) is traditionally one of the main targets of pelagic fisheries in the Baltic Sea, taken mostly in mixed fishery with sprat. The annual total landings amounted around 258 000 t on average for the most recent 20 years. The international management of the Baltic herring stocks rely on the Total Allowable Catch (TAC) agreements and on a few technical measures (gear restrictions in certain areas, closed areas and periods for fishery) as the operational management tools. There are three major agreed management units of herring in the Baltic: Central Baltic herring, Herring in the Gulf of Bothnia and Gulf of Riga herring. Despite of decades–long efforts in applying of regulatory measures, the fate of the stocks has been different: The Central Baltic herring has shown two major declines during its management history while the two other stocks have shown broadly opposite trends. The paper is discussing the possible reasons for the different outcome of management like compliance of fishery with the scientific advice and changes on pelagic communities of the Baltic, focusing on the dynamics in mean weight of herring as another factor potentially effecting on management success across the area.

## 1 INTRODUCTION

On the background of long-term low of cod stocks in the Baltic, the mixed pelagic fishery targeting the Baltic herring (*Clupea harengus membras* L and sprat (*Sprattus sprattus balticus* L.) has gained its economical importance in most of the countries around the Baltic sea.

The annual catches of herring have fluctuated from 187 000 to 367 000 tons and those of sprat from 37 000 to 529 000t since 1980. The total average annual catch of pelagic fishery has increased from just below 400 000 t in 1980s to 580 000 t in 1990 and 2000s, peaking at almost 800 000 t in 1997. (Figure 1). This most recent increase in total catch of pelagic fish has been driven mostly by the rapid recovery of sprat stock after the collapse of cod in mid-1980s (ICES 2019, Figure 1).

The Baltic pelagic fishery has historically sustained relatively high total catches despite of a number of regulatory measures taken to reduce the excess capacity in fishing fleets in 2000s, resulting in clear reduction in nominal effort in almost all principal fishing areas (Raid et al., 2015, Figure 2).

The management decisions follow the principles of the EU Common Fisheries Policy– CFP (EU 2013) in order to maintain the sustainable future for fish stocks, marine environment and the fisheries. The management of herring and sprat assessment units is implemented through imposing catch restrictions like Total Allowable Catch (TAC) aiming the Maximum Sustainable Yield (MSY) as the goal for each particular management unit. Additionally, a number of other technical measures like capacity restrictions, gear design and areal/temporal closures are implemented as the management tools. An EU Multi-annual Management Plan (MAP), serves as basis for taking management decisions since 2018. (EU 2016, 2019).

The fishery of the Baltic herring as well as of sprat is internationally managed since 1974. Currently the management is performed by the European Commission on the basis of the scientific catch advice from the International Council for the Exploration of the Sea (ICES). The scientific advice stems from the biological information on catch structure provided to ICES by all countries involved in the respective fisheries.

The Baltic herring is widely distributed all over the Baltic Sea forming around 10 local populations, which can be divided as gulf stocks and open sea stocks (Popiel 1958, Ojaveer 1981). The gulf herrings spend all year in the big gulfs, while open sea stocks perform annual migrations between spawning grounds, often located in the same gulfs, and feeding/wintering areas in the Baltic Proper. Such migrations have been well documented by tagging experiments, catch structure and the results of morphometric analyses (Aro 1989, Aro et al. 1990, Parmanne 1990, Parmanne et al. 1997).

DOI: 10.1201/9781003320289-65

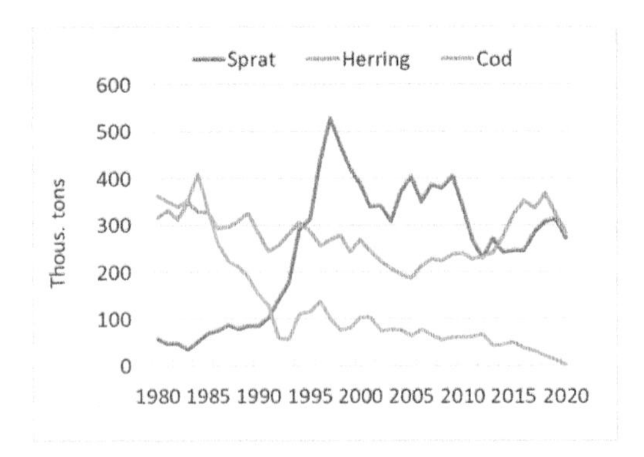


Figure 1. Catches of herring, sprat and Eastern Baltic cod in 1980-2020 (ICES, 2021).

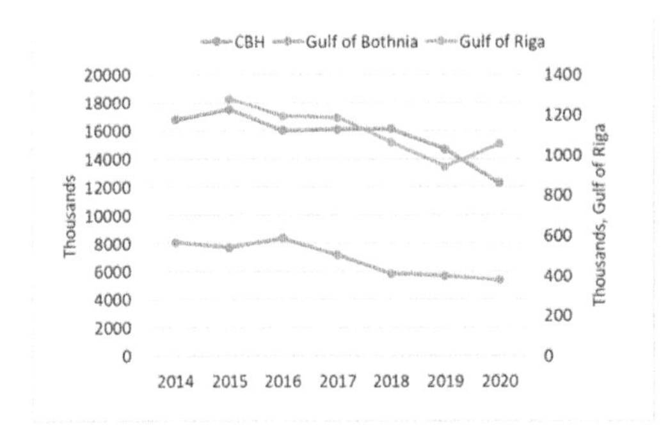


Figure 2. Trend in total effort (kW-days at sea) in EU fleet in the Baltic Sea, 2014-2020; (data: STECF 2021).

Ideally, the assessment and management units of the exploited fish stocks should follow the natural populations. This approach was used in the initial phase of the international assessment and management of Baltic herring in 1974-1990, when up to eight assessment units were considered (ICES 2001, 2002). However, the complex population structure makes the sustainable management of single herring stocks on routine basis challenging, both from scientific point of view and for costs. Therefore, in order to achieve the adequate management decisions, the internationally managed fish stocks are often treated by conventional management units during the advisory and management process (Zableckis *et al.*, 2009). In 1990, herring populations in southern, central and northern Baltic proper and in the Gulf of Finland were combined into one single Central Baltic Herring (CBH) assessment unit, as a compromise (ICES 2002).

At present the Baltic herring is assessed and managed by following units/stocks (Figure 3):

1. Herring in Sub-divisions 25-28.2, 29 &32; (Central Baltic herring-CBH)
2. Herring in Sub-divisions 30 and 31(Gulf of Bothnia herring);
3. Herring in Sub-division 28.1(Gulf of Riga herring).

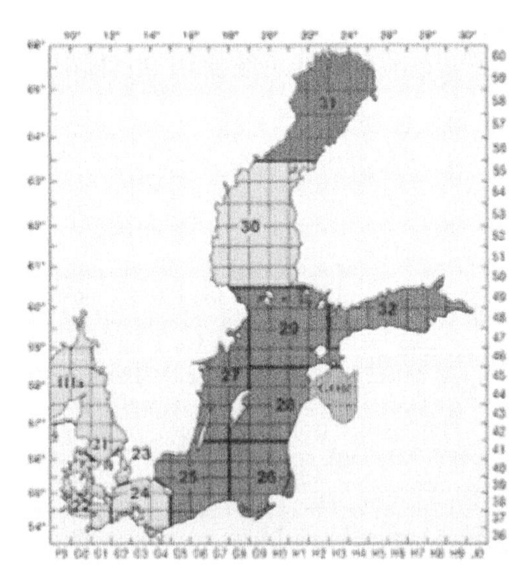


Figure 3. ICES Sub-divisions in the Baltic Sea and the herring assessment units. (GOR- Gulf of Riga herring).

The catch dynamics from the three management units reveals that the CBH as the major stock complex has contributed to the total herring catch the most, while the mean share of the Gulf of Bothnia and the Gulf of Riga herring has been around 24% and 10%, respectively (Figure 4).

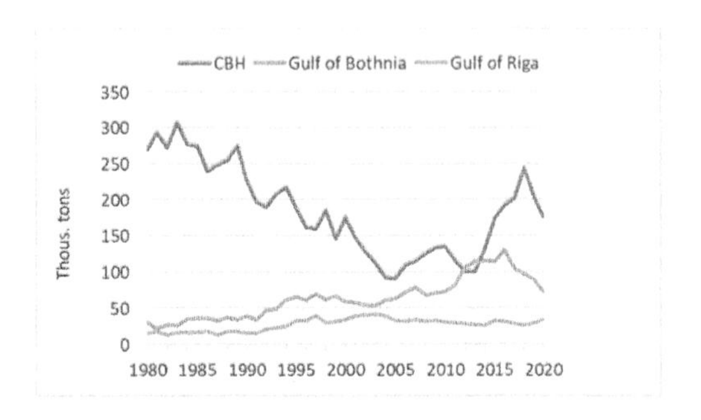


Figure 4. Catch dynamics in the Baltic herring management units in 1980-2020.

Despite of decades–long efforts in applying of regulatory measures, the fate of the stocks has been different: the Central Baltic herring has shown two major declines during its management history while the two other stocks have shown broadly the opposite trend like the Gulf of Riga herring or maintaining high stock level as the herring in the Gulf of Bothnia.

A few factors, potentially having effect on of management results of herring are discussed in the present study.

## 2 MATERIAL AND METHODS

### 2.1 *Material used*

The biological information used in preparation of the present study originates from the recent reports of the Baltic Fisheries Assessment working Group of the ICES (ICES 2021), as well as from the ICES Advice on the Baltic Ecoregion (ICES 2021c-f). Information on landings and stocks (SSB- Spawning Stock Biomass (SSB), Fishing Mortality (F), TAC, Spawning Stock Abundance (SSN) available in the reports of the ICES, what consolidate the fisheries and biological information reported by the ICES Baltic member countries annually. The data collection of EU Member States is performed as a part of National data Collection Programs for stock assessment purposes under the EU Data Collection Framework and the Multiannual Program for the collection and management of biological environmental, technical and socioeconomic data in the fisheries and aquaculture sectors (EU 2017, 2019). Additionally, the information on spatial distribution of fish stocks was derived from the results of the Baltic International Acoustic Survey (BIAS) carried out under the auspices of the ICES.

The information on effort used in the Baltic fisheries and fleet performance was taken from the reports of Scientific, Technical and Economic Committee for Fisheries (STECF) of the European Commission.

## 3 RESULTS AND DISCUSSION

### 3.1 *Trends in stock development and fisheries*

*Central Baltic herring*

The Central Baltic herring is the main herring management unit in the Baltic. The total reported catches in 2020 amounted to 177 079 t in 2020 (ICES, 2021). All countries around the Baltic sea are involved in fishery of that stock. The largest part of the catches in 2020 was taken by Sweden (26%), followed by Poland (20%) and by Finland (18%). Most of the catch is taken by the pelagic trawls and only in some areas like the Gulf of Finland the catches with passive gears play more significant role.

The spawning stock biomass (SSB) of the CBH declined from almost 1.9 million tons in 1974 to 0.33 million tons in 2002 (-83%). Since then, the SSB recovered to just below 0.7 million tons in 2014. The most recent years have shown the rapid decline again, reaching 0.36 million tons in 2020 (Figure 5). The SSB dynamics reveals that the SSB has been critically low in 2001- 2002, reaching the $B_{lim}$ (330 000t).

The fishing mortality has been particularly high in the period of 1989-2002, exceeding not only $F_{MSY}$ (0.21) but even $F_{PA}$ (0.32). Since 2016 the F has increased again above the $F_{PA}$.

The catches of CBH have been decreasing during the most of its management history, being restricted by the TAC constraints.

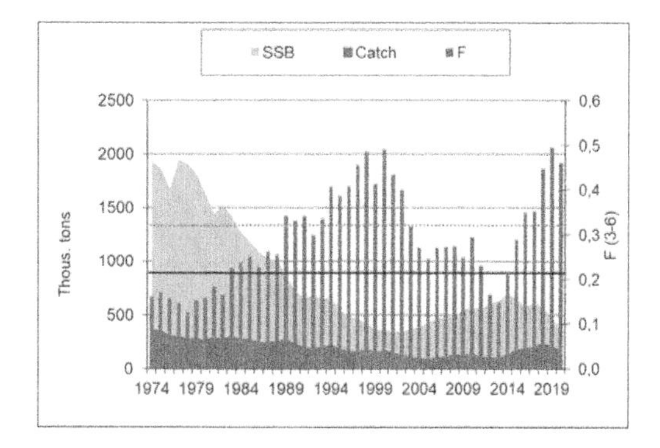


Figure 5. Herring in SD 25-29, 32: Spawning stock biomass (SSB), fishing mortality (F) in age groups 3-6 and catches in 1974-2020. Horizontal solid line indicates $F_{MSY}$ = 0.21 and the dotted line $F_{PA}$ = 0.32 (ICES, 2021).

*Herring in sub-division 28.1(Gulf of Riga)*

The herring fishery in the Gulf of Riga is performed by Estonia and Latvia, using both trawls and coastal trap-nets. The total catches of herring in the Gulf of Riga were 33 249 t in 2020; 78% of catches were taken by trawl fishery and the rest by coastal fishery with trap-nets (ICES, 2021).

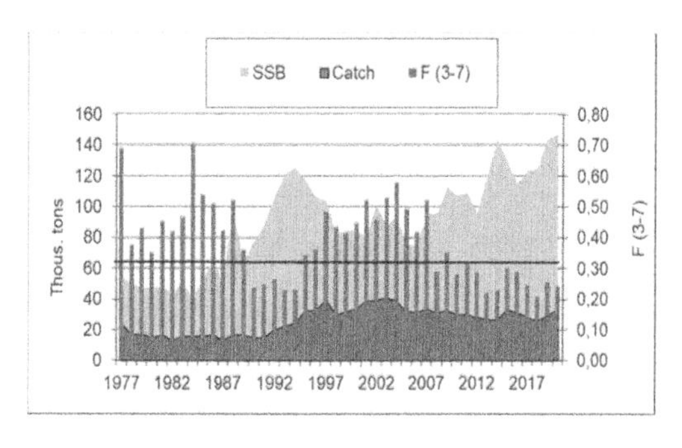


Figure 6. Gulf of Riga herring: Spawning stock biomass (SSB), fishing mortality (F) in age groups 3-7 and catches in 1977-2020. Horizontal solid line indicates $F_{MSY}$ = 0.32 (ICES, 2021).

The biomass of the Gulf of Riga herring has demonstrated rather different trends compared to the Central Baltic herring, proven both in analytical assessments and by the direct acoustic surveys (Raid et al., 2005, 2016). The spawning stock biomass of the Gulf of

Riga herring has been rather stable at the level of 40 000–50 000 t in the 1970s and 1980s. The SSB started to increase in the late 1980s, reaching peak values above 120 000 t in early 1990s. Afterwards SSB decreased and was fluctuating at the level below 100 000 t. until 2008, however never decreasing below the $B_{PA}$ (57 100 t). Since then the SSB increased again reaching close to 147 000 t in 2020 that is historical high. (Figure 6).

The mean fishing mortality in age groups 3–7 has been rather high in 1970s and 1980s fluctuating between 0.35 and 0.71until 1989 when it decreased below 0.4 and stayed at this level till 1996. Afterwards the fishing mortality increased above $B_{PA}$ (0.38). Since 2009, the fishing mortality has decreased below $F_{MSY}$ (0.32). In 2017-2020 the fishing mortality was below the $F_{MSY}$ (0.32) in the range of 0.21-0.26.

*Herring in sub-divisions 30 and 31 (Gulf of Bothnia)*
The fishery is performed by the Finnish and Swedish fishermen. In 2020, 94% of the Finnish landings came from trawl fishery, and the rest with trap-nets, and gill-nets. 96% of Swedish catches came from trawls and 4% with gill-nets. The total catches amounted to 72 956 tons in 2020 (ICES, 2021).

The history of SSB of the Gulf of Bothnia herring reveals two main trends: increase from around 600 000 t in 1970s to the highs of above 1.2 million tons in mid-1990s on the background of low fishing mortality. Since then, the decreasing trend in SSB has been dominating while the fishing mortality has quadrupled from historical lows in most recent decade. Yet, the Fishing mortality never reached the $F_{MSY}$ nor $F_{PA.}$ The catches have shown a general increasing trend except the probably TAC-driven decrease in most recent years (Figure 7).

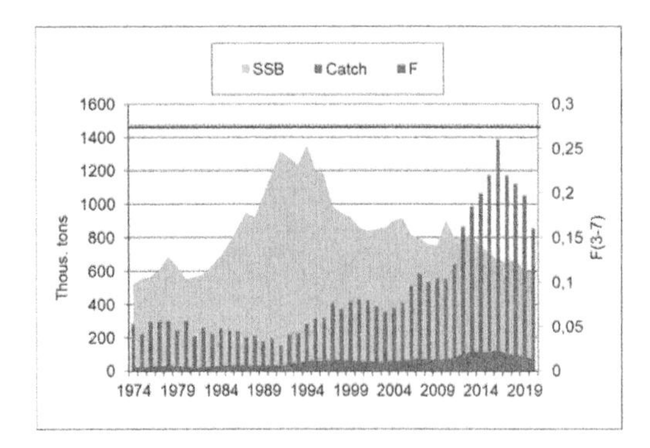


Figure 7. Herring in SD 30-31: Spawning stock biomass (SSB), catch and fishing mortality (F) in age groups 3-6 and catches in 1974-2020. Horizontal solid line indicates $F_{MSY}$ = 0.271 and the dotted line $F_{PA}$ = 0.272 (ICES, 2021).

## 3.2 *Factors potentially effecting on trends in herring stocks and fisheries.*

Despite of extensive biological knowledge on and TAC- based international management, the history of management has demonstrated different outcome in Baltic herring stocks: while stock size of the Central Baltic herring has prevailing decreasing trend on the background of periodically high fishing mortality, the herring in the Gulf of Riga has shown a rather stable or even increasing trend. The stock of the Gulf of Bothnia herring has sustained its high biomass so far, regardless to increasing fishing mortality. Below the following of the scientific advice, spatial variability of mean body weight and trends in sprat stock are discussed as factors, potentially effecting on the Central Baltic herring stock.

*Following the scientific advice*
Since fishery is the most eminent factor affecting the exploited stocks, the following of the scientific advice, produced on the best expertise available should be the key factor for a sustainable fishery.

The comparison of actual catches and the respective advice figures in 2005-2020 of the Central Baltic herring and in the Gulf of Riga herring show a rather good coherence in case of the Central Baltic herring: the catches exceeded the advised values in two periods only: 2010-2012 and in 2019 and 2020, in both cases up to 32%. (Figure 8). The respective situation in case of the Gulf of Riga herring has been even worse, however the range of deviations was lower here.

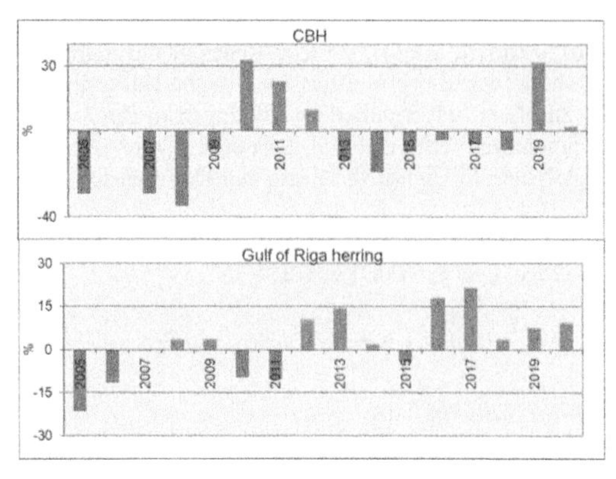


Figure 8. The coherence of catches and catch advice in 2005-2020 (%).

The comparison of this information with the assessment results suggests that high positive deviations from the advised catch limits were not accompanied by the elevated values of fishing mortality in 2010-2012 which, however, was observed in 2019-2020 (Figure 5). This allows concluding that deviations from the scientific advice probably have not had a negative impact on Central Baltic herring stock during the most pronounced period of deviations from the advice.

*Effect of the spatial and temporal diversity of mean weight*
Different environmental, particularly salinity conditions between the saline and milder South-eastern

Baltic and the North- eastern Baltic with relatively low salinity and severe winter conditions have resulted in different growth conditions for fishes inhabiting across the Baltic Sea. So, the mean weight at age of herring as well as sprat decreases from South-west to North-east (Ojaveer 1981).

Our data on herring mean weights at age clearly confirm the persistent existence of such a pattern: the overall average mean weight of herring in commercial catches observed in the Sub-divisions 25 and 26 exceeded the respective values in Sub-divisions 28.2, 29 and 32 by 40-50% in 1997 - 2010 (Raid et al., 2016), and also in 2011-2020 Figure 9).

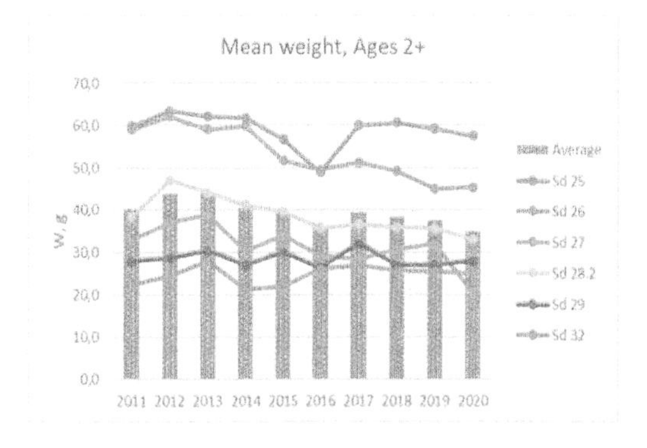


Figure. 9. Mean weight of herring at ages 2+ in 2011-2020.

The average mean weight of herring in SSB (ages 2+ years), observed in the Sub-divisions 25 and 26 was app 48% higher than average for Sub-divisions 29 and 32, (respectively 59.1 and 30.7 g) in 2011-2020.

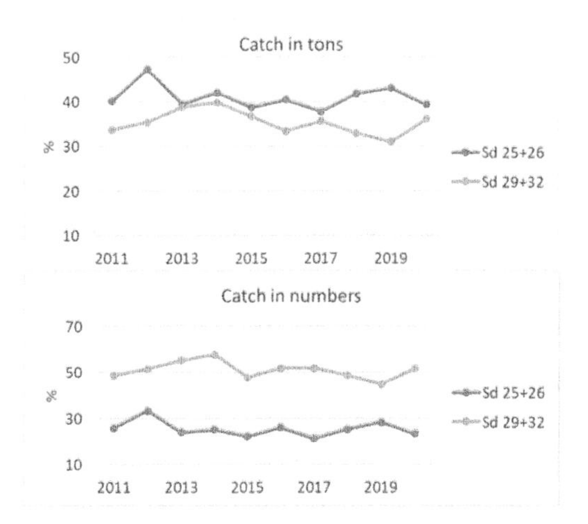


Figure 10. Proportion of catches in tons and in numbers of individuals, taken in Sd 25+26 and 29+32 in 2011-2020.

Consequently, in order to catch the given share of catch quota (TAC) in the North-eastern Baltic would result in bigger removal rate from stock in terms of individuals when compared to the respective value in the South-eastern Baltic. And indeed, the proportion of catch in tons and that of in numbers taken in Sub-divisions 25+26 and 29+32, show clearly different pattern in 2011-2020. The proportion of catch in tons from total catch of CBH has been quite similar, being around 40% in Sub-divisions 25 and 26 and between 30 and 40% in Sub-divisions 29 and 32. At the same time the catch proportion taken in the northern sub-divisions in terms of numbers fluctuated around 50% (51% on average) of total while the removals from the Sub-divisions 25 and 26 remained below 30% (26% on average, Figure 10).

Since the fishing mortality implies the stock losses in numbers of individuals as a result of fishing, the immediate conclusion would be that the same quantity of quota in tons taken in the South-eastern Baltic would mean lower fishing mortality compared to the North-eastern part of the sea. As we have demonstrated earlier (Raid et al., 2011, 2014, 2016), a rather strong correlation occurred between catch in numbers per 1000 t of landings and the fishing mortality estimated for the Sub-divisions of the North-eastern Baltic in 1997-2011. Applying the same approach in the present study, again a quite good correlation was found for the time period studied. (Figure 11).

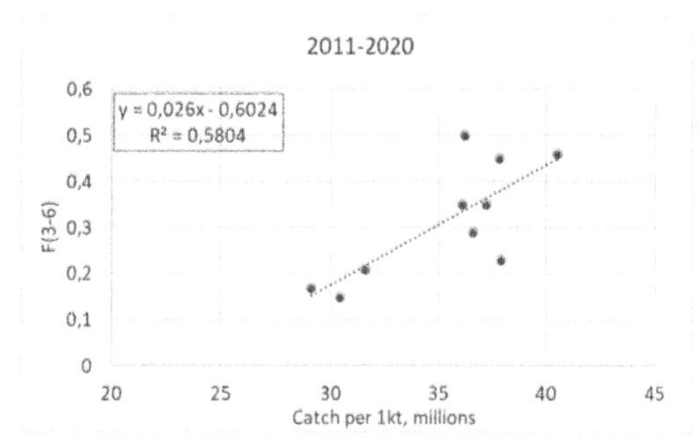


Figure 11. Central Baltic herring. Catch in millions per 1000 t of landings vs. fishing mortality in 2011-2020 (data from ICES, 2021).

Applying the relationship from the Figure 12 to the values of catch in numbers per 1000 t of landings observed in the Sub-divisions 29 and 32, suggest that considerably higher fishing mortality compared to the respective estimates for the whole Central Baltic stock occurred for in the Sub-divisions 29 and 32 in 2011-2020 – average values 0.32 and 0.73, respectively (Table 1).

Considering the relatively high proportion of herring catch taken in the Sub-divisions 29 and 32, the

Table 1. Estimates of fishing mortality for the Sub-divisions 29+32 and the values estimated for the Central Baltic herring stock (ICES, 2021).

| | CBH | | Sd 29+32 | |
|---|---|---|---|---|
| Year | CANUM | F(3-6) | CANUM | F(3-6) |
| 2011 | 37,9 | 0,23 | 55,0 | 0,83 |
| 2012 | 29,1 | 0,17 | 44,4 | 0,55 |
| 2013 | 30,4 | 0,15 | 46,8 | 0,61 |
| 2014 | 31,6 | 0,21 | 47,6 | 0,64 |
| 2015 | 36,6 | 0,29 | 50,1 | 0,70 |
| 2016 | 36,1 | 0,35 | 54,1 | 0,80 |
| 2017 | 37,2 | 0,35 | 54,8 | 0,82 |
| 2018 | 37,8 | 0,45 | 56,6 | 0,87 |
| 2019 | 36,2 | 0,5 | 52,4 | 0,76 |
| 2020 | 40,5 | 0,46 | 57,9 | 0,90 |
| Average | 35,3 | 0,32 | 52,0 | 0,75 |

overall effect of higher "local" fishing mortality of CBH stock might have been considerably higher than the estimated overall fishing mortality suggests.

*Effect of sprat stock*

Since herring trawl fishery in the Baltic Sea occurs as a mixed herring-sprat fishery, the credible information on species composition of pelagic catches is a key input for sustainable management of stocks in question. Therefore, high abundance of sprat stock, observed in the Baltic since the decline of cod in the mid-1980s and its spatial pattern, increase the risk of misinterpretation of catch statistics, rising concern as another potential factor affecting the results of assessment and management of the Central Baltic herring. So, the dynamics of spawning stock biomass of sprat and the fishing mortality of Central Baltic herring have demonstrated certain coherence throughout the period of assessment history (Figure 12).

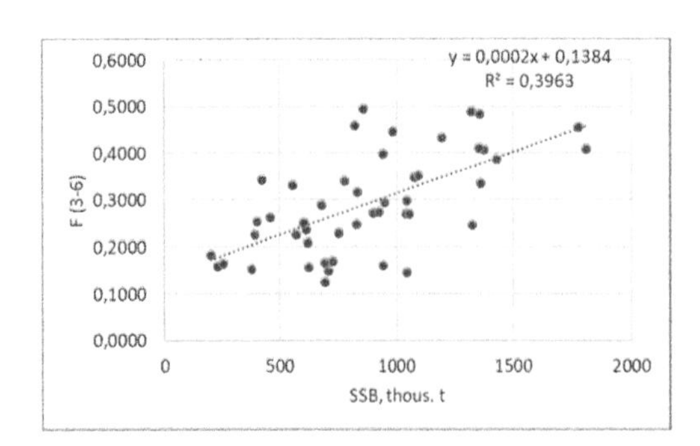


Figure 12. Spawning stock biomass of sprat vs. fishing mortality of the Central Baltic herring in 1974-2020 (data from ICES, 2021).

The comparison of herring and sprat catch dynamics indicate that the herring fishing mortality peaked in the period of highest sprat catches in late 1990s-early 2000s, and also in most recent years, when TAC has been restrictive for sprat catches (Figure 13).

The potential effect of sprat abundance in the mixed catches on herring fishing mortality suggest that the careful monitoring of species composition in mixed pelagic sprat and herring fishery is of critical importance. This is particularly essential in the light of increasing proportion of sprat in the pelagic fish communities, recently observed in the North-eastern Baltic (Figure 14), in order to reduce effect of possible species misreporting e.g. when the quota of one species has been exhausted. However, the effects of possible species misreporting have not been quantified nor included in the assessment notwithstanding they may affect the quality of the assessment (ICES 2021c).

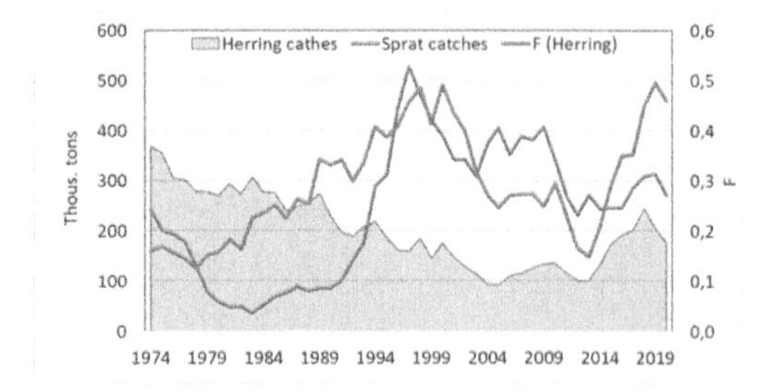


Figure 13. Catches of sprat and central Baltic herring and the estimates of herring fishing mortality in 1974-2020 (data from ICES 2021).

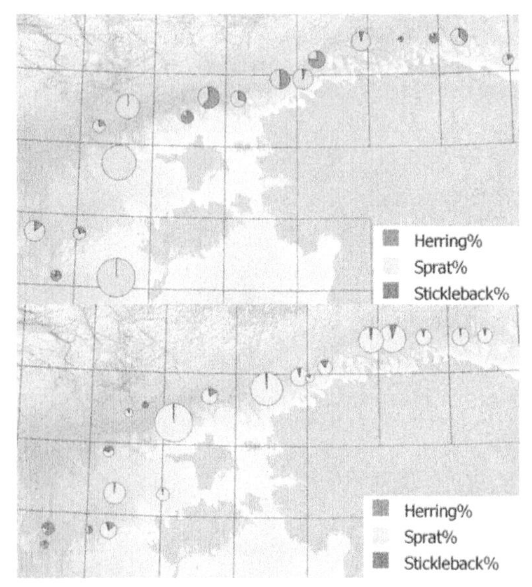


Figure 14. Proportion of herring and sprat in in control hauls during the BIAS surveys in 2018 (upper panel) and in 2021 (lower panel). (ICES, 2021b).

Low abundance (in Gulf of Riga) or missing of sprat (in Gulf of Bothnia) may have favoured to the sustainable fishing mortality in those herring stocks.

## 4 CONCLUSIONS

Our results indicate that probably several factors have been involved in causing periodically unsustainable fishing mortality in Central Baltic herring.

The spatial diversity observed in herring catch structure, particularly in mean weights at age, can potentially have a significant influence on the effect of fishery on Central Baltic herring stock, elevating locally fishing mortality in the areas with persistently lower mean weights at age.

There are also indications that the high abundance of sprat and its spatial distribution pattern may cause additional issues in providing high quality catch information for the stock assessment process. Consequently, the monitoring of the species composition of catches in the pelagic mixed fishery is vital for management process and deserves thorough attention.

## REFERENCES


Aro, E. 1989. A review of fish migration patterns in the Baltic. Rapp. P. v. Reun. Cons.Int. Explor. Mer., 190: 72–96.

Aro, E., Pushkin, S., Kotilainen, P., Mamylov, V., Flinkman, J. and Diogtev, A. 1990. Estimation of changes in abundance of Baltic herring and sprat stocks by combined hydroacoustic-trawl survey in the Gulf of Finland in autumn, winter and spring. ICES C.M. 1990/J: 25, 21 p.

EU 2013. Regulation (EU) No 1380/2013 of the European parliament and of the Council of 11 December 2013 on the Common Fisheries Policy, amending Council Regulations (EC) No 1954/2003 and (EC) No 1224/2009 and repealing Council Regulations (EC) No 2371/2002 and (EC) No 639/2004 and Council Decision 2004/585/EC. Official Journal of the European Union L 354(22), 40 p.

EU 2016. Regulation (EU) 2016/1139 of the European Parliament and of the Council of 6 July 2016 establishing a multiannual plan for the stocks of cod, herring and sprat in the Baltic Sea and the fisheries exploiting those stocks, amending Council Regulation (EC) No 2187/2005 and repealing Council Regulation (EC) No 1098/2007. Official Journal of the European Union, L 191, 15.7.2016. http://data.europa.eu/eli/reg/2016/1139/oj.

EU 2017. Regulation (EU) 2017/1004 of the European Parliament and of the Council of 17 May 2017 on the establishment of a Union framework for the collection, management and use of data in the fisheries sector and support for scientific advice regarding the common fisheries policy and repealing Council Regulation (EC) No 199/2008(recast). Official Journal of the European Union L 157, 21 p.

EU 2019. Regulation (EU) 2019/472 of the European Parliament and of the Council 19 March 2019 establishing a multiannual plan for stocks fished in the Western Waters and adjacent waters, and for fisheries exploiting those stocks, amending Regulations (EU) 2016/1139 and (EU) 2018/973, and repealing Council Regulations (EC) No 811/2004, (EC) No 2166/2005, (EC) No 388/2006, (EC). No 509/2007 and (EC) No 1300/2008. Official Journal of the European Union L 83 (22), 17 p.

ICES 2001. Report of the Study Group on the Herring Assessment Units in the Baltic Sea. ICES CM 2002/H: 10.

ICES 2002. Report of the Study Group on the Herring Assessment Units in the Baltic Sea. ICES CM 2002/H: 04.

ICES. 2019. Baltic Sea Ecosystem – Fisheries Overview. In Report of the ICES Advisory Committee, 2019. ICES Advice 2019, section 4.2. 28 pp. https://doi.org/10.17895/ices.advice.5566

ICES 2021. Baltic Fisheries Assessment Working Group (WGBFAS). ICES Scientific Reports. 3:53. 717 pp. https://doi.org/10.17895/ices.pub.8187

ICES. 2021a. ICES Working Group on Baltic International Fish Survey (WGBIFS; outputs from 2020 meeting).

ICES Scientific Reports. 3:02.539 p. https://doi.org/10.17895/ices.pub.7679.

ICES. 2021b. ICES Working Group on Baltic International Fish Survey (WGBIFS). ICES Scientific Reports. 3:80. 490 pp. https://doi.org/10.17895/ices.pub.8248.

ICES 2021c. Herring (*Clupea harengus*) in subdivisions 25-29 and 32, excluding the Gulf of Riga (central Baltic Sea). In Report of the ICES Advisory Committee, 2021. ICES Advice 2021, her.27.25-2932. https://doi.org/10.17895/ices.advice.7767.

ICES 2021d. Herring (*Clupea harengus*) in Subdivision 28.1 (Gulf of Riga). In Report of the ICES Advisory Committee, 2021. ICES Advice 2021, her.27.28. https://doi.org/10.17895/ices.advice.7768.

ICES 2021e. Herring (*Clupea harengus*) in Subdivisions 30 and 31 (Gulf of Bothnia). In Report of the ICES Advisory Committee, 2021. ICES Advice 2021, her.27.3031. https://doi.org/10.17895/ices.advice.7769.

ICES. 2021f. Sprat (*Sprattus sprattus*) in Subdivisions 22-32 (Baltic Sea). In Report of the ICES Advisory Committee, 2021. ICES Advice 2021, spr.27.22-32. https://doi.org/10.17895/ices.advice.7867.

Ojaveer, E. 1981. Marine pelagic fishes. In Aarno Voipio (ed) The Baltic Sea, Oceanography Series, 30: 276–292. Amsterdam: Elsevier Scientific Publishing Company.

Parmanne, R. 1990. Growth, morphological variation and migrations of herring (*Clupea harengus*) in the northern Baltic Sea. Finnish Fisheries Research, 10: 48 p.

Parmanne, R., Popov, A., Raid, T. 1997. Fishery and biology of herring (Clupea harengus L.) in the Gulf of Finland: A review. Boreal Environment Research, 2: 217–227.

Popiel. J. 1958. Differentiation of the biological groups of herring ion the southern Baltic. Rapp. P.-V. Reun. Cons. Int. Explor. Mer., 143(2): 114–121.

Raid, T., Kaljuste, O, Shvetsov, F., Strods, G. 2005. Acoustical estimations confirm the good health of herring stock in the Gulf of Riga assessed by analytical models. In: G.H. Kruse, V.F. Gallucci, D.E. Hay, R.J. Perry, R. M. Peterman, T.C. Shirley, P.D. Spencer, B. Wilson & B. Woodby (eds). Fisheres Assessment and Management in Data-limited Situations. Alaska Sea Grant College Program,University of Alaska Fairbanks. pp. 255–266.

Raid, T.; Shpilev, H.; Järv, L.; Järvik, A. 2011. Towards sustainable Baltic herring fishery: trawls vs. pound nets. Rizzutto, E.; Soares, C.G. (Eds). Sustainable Maritime Transportation and Exploitation of Sea Resources. Taylor & Francis Ltd: 1055–1059.

Raid, T., Järv, L., Hallang, A., Shpilev, H., Kaljuste, O., Järvik, A. 2014. Managing the spatial and temporal pattern of Baltic herring trawl fishery: a potential tool for sustainable resource management? In: Developments in Marine Transportation and Exploitation of Sea Resources: 15th International Congress of the International Maritime Association of the Mediterranean IMAM 2013 - Developments in Maritime Transportation and Exploitation of Sea Resources, 14.-17. October 2013, A Coruna, Spain. (Eds.) Soares, Guedes & Pena, Lopez. London: Taylor & Francis Ltd, 2014, 1091–109.

Raid, T., Arula, T., Kaljuste, O., Sepp, E., Järv, L., Hallang, A., Shpilev, H., Lankov, A. 2015. Dynamics of the commercial fishery in the Baltic Sea: what are the driving forces? Towards Green Marine Technology and Transport: Proceeding of the 16th International Congress of the International Maritime Assotiation of the Mediterranean (IMAM 2015), Pula, Croatia, 21-24 September. Ed. C. Guedes Soares; R. Dejhalla; D. Pavletic. Taylor & Francis, 897–906.

Raid, T., Järv, L., Pönni, J., Raitaniemi, J., Kornilovs, G. 2016. Managing the spatial and temporal pattern of Baltic herring trawl fishery: a potential tool for sustainable resource management? In: Guedes Soares, C. & Santos, T.A. (Ed.). Maritime Technology and Engineering (961–966). Boca Raton, London, New York, Leiden: Taylor & Francis Ltd.

STECF 2021. Scientific, Technical and Economic Committee for Fisheries (STECF) –Fisheries Dependent - Information – FDI (STECF-21-12). EUR 28359 EN, Publications Office of the European Union, Luxembourg, 2021, ISSN 1831-9424. 230 p. https://stecf.jrc.ec.europa.eu/reports/fdi.

Zableckis, S., Raid, T., Arnason, R., Murillas, A., Eliasen, Sverdrup-Jensen, S., Kuzebski, E. 2009. Cost of Management in Selected Fisheries. In: D. Wilson, K.H. Hauge (Editors). Comparative Evaluations of Innovative Fisheries Management – Global Experiences and European Prospectives, Springer, Heidelberg. 272 p.

*Trends in Maritime Technology and Engineering – Guedes Soares & Santos (eds)*


# Use of autonomous research vehicles in Baltic fisheries acoustic surveys: Potential benefits and pitfalls

E. Sepp, M. Vetemaa & T. Raid
*Estonian Marine Institute, University of Tartu, Tallinn, Estonia*

ABSTRACT: Direct abundance estimates obtained during the acoustic surveys are essential in stock assessment and management process of pelagic fish stocks. The conventional methodology of surveys includes direct hydroacoustic measurements of fish abundance along the pre-defined transects, combined with biological data, obtained from trawl hauls. Considerable amount of survey time and resource are currently allocated to steaming along transects. High operational costs of the research vessel are the major issue in performing such surveys. However, the advances in modern technologies potentially allow to overcome this problem by deploying the autonomous surface vehicles (ASV) for the routine acoustic measurements. A hybrid solution of reduced survey vessel usage by allocating suitable tasks to ASV, would be the most cost-effective solution, but currently not fully compatible with the requirements of maintaining the long -term time series. We argue that autonomous vessels should be considered as an economical alternative and tested through pilot studies.

## 1 INTRODUCTION

Autonomous and unmanned vessels have rapidly evolved during the last decade and are already used in various marine studies (Eriksen et al., 2001; Griffiths, 2002; Wynn et al., 2014; Verfuss et al., 2016, 2019; Barrera et al., 2021). Depending on the survey objectives and targets, Unmanned Aerial Systems (UAS), Autonomous Underwater Vehicles (AUV), Unmanned Surface Vehicles (USV) and Autonomous Surface Vehicles (ASV) have been developed. Despite the availability of several "off the shelf" vehicles, extensive offshore surveys still require tailored solutions (Verfuss et al., 2019).

Accuracy of scientific monitoring of fish stocks is of key importance in managing commercially exploited resources (Barange et al., 2009). Despite the advances in technology and modelling, the cost of marine surveys is still very high and therefore we are currently covering less than 0.01% (Sepp, 2021, unpubl. data) of the total water volume during single survey (Estonian acoustic-trawl survey on Baltic Sea). High operational costs of conventional research vessels drive the need for more economical solutions for conducting surveys in several marine research fields.

The main commercial pelagic fish species in Baltic Sea are European sprat Sprattus sprattus (Linnaeus 1758) and Baltic herring Clupea harengus membras (Linnaeus 1758), whose stocks are regularly monitored and assessed under the auspices of the International Council for the Exploration of the Sea (ICES) (ICES, 2021). Regular acoustic-trawl surveys covering the whole stock are a key source of independent input into their stock assessment (ICES, 2014, 2021). Information from acoustic sonars provide a cost-effective alternative to extensive trawl surveys, with higher coverage in limited time (Simmonds & MacLennan, 2008). These acoustic surveys on Baltic Sea have been conducted according to standardized procedures since 1998 and currently the acoustic sounding is conducted from the same vessel that is used for trawling. According to the methodology, the number of trawl hauls is fixed, but the exact location of hauls should be decided during the cruise in the locations with the strongest acoustic signals (ICES, 2014).

Fisheries research using autonomous or unmanned have been conducted or tested in several areas with variable success (Swart et al., 2016; Mordy et al., 2017; Chu et al., 2019; Verfuss et al., 2019). Current understanding from tests is that use of such vehicles provide cost-effective addition to the surveys, but cannot completely substitute the work of research vessels (Chu et al., 2019; De Robertis et al., 2021). Realtime data transfer/processing remains difficult in offshore areas and several surveys still require direct biological sampling (ICES, 2014; Verfuss et al., 2019; De Robertis et al., 2021). Several gaps and questions related to international and local maritime regulations are also currently preventing the full use of such vessels (Colet Maillo, 2021; MSC. 1/Circ. 1638, 2021; Parker, 2021).

The aim of the current study is to map the possibilities of implementing ASV/USV research vehicles to current acoustic-trawl surveys in Baltic Sea and provide recommendations for optimising the current survey methodology.

DOI: 10.1201/9781003320289-66

## 2 MATERIAL AND METHODS

In order to map the possibilities for implementing ASV/USV vessels in current survey methodology, we identified the parts of trawl-acoustic surveys that do not require large research vessels participation. Since the hauling locations are decided during the survey, they differ between years. For comparison we used actual acoustic track and hauling locations of Gulf of Riga herring surveys (GRAHS) during the years 2015-2021. The acoustic track covered during each survey calculated for comparison. The time of steaming the acoustic track and hauling were summarized according to the survey logs to get the networking time of the vessel (without the overnight stays in ports or at sea) (Figure 1,2).

Based on the log data of each survey, the shortest distance between hauling locations were calculated to get the minimum steaming distance needed for the trawling vessel. Average vessel speed of 7 kn were used to calculate the travelling time and additional 1 hour for completing each haul were combined to get the minimal survey time needed to complete only the trawl hauls and transit distance between hauls of each survey year.

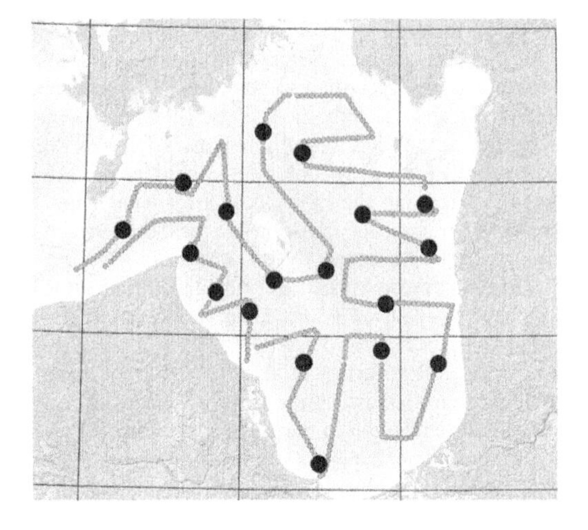

Figure 1. Gulf of Riga acoustic herring survey (GRAHS) track in 2020. Small dots represent each logged acoustic nautical mile and black dots represent locations of trawl hauls.

Based on this data we obtained the possible time that could have been saved in each survey, independent of the weather conditions factor. Storm-days and vessel repair time were not taken into consideration in the current comparison due to their sporadic nature. Relative difference in net realized survey time and calculated survey time was used to describe the possible savings of survey time from optimizing the trawl vessel survey routes.

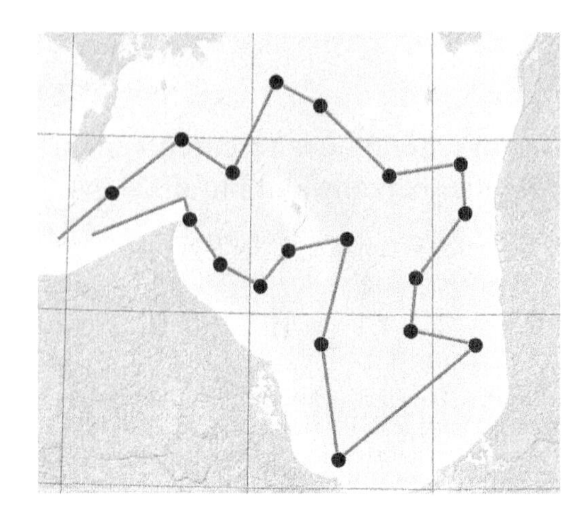

Figure 2. Gulf of Riga acoustic herring survey (GRAHS) calculated optimal track using actual trawl hauls of 2020. Green line represents the optimal track of trawler and black dots represent locations of trawl hauls.

## 3 RESULTS

The actual effective track covered (after removal of double tracks and faulty data) during the 7 years comparison period ranged from 411 to 480 nautical miles (average 434 nm, Figure 3) and number of control hauls completed was from 14 to 20 (average 17.9). The net time used for completion of the survey was between 66.5 and 87.5 hours (average 78.6h, Figure 4).

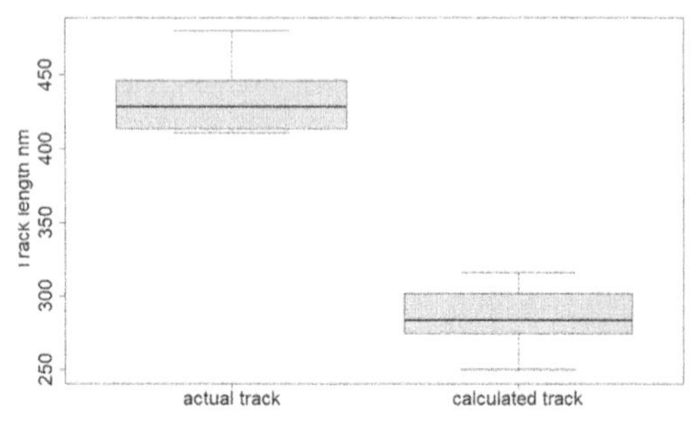


Figure 3. Boxplot of the median track length of actual track and calculated track based on GRAHS surveys in 2015-2021. The horizontal bold lines show the median, shaded boxes indicate the interquartile ranges with upper and lower whiskers (5% and 95%).

The calculated minimum track needed to complete the trawl hauls of each survey year was 250-316 nautical miles (average 286.1nm, Figure 3). Time needed to cover the calculated track with same number of hauls as conducted in respective survey, the net time needed for completion of the survey ranged between 49.7 and 65.1 hours (average 58.7h,

Figure 4). Relative time difference between net survey time of actual track and calculated track was on average 25.2% (19.1-29.7%)).

Difference between actual track length and length of calculated optimal track was statistically different (Mann-Whitney U test, n=14, W=49, p=0.0006). Same statistical results apply to differences in times.

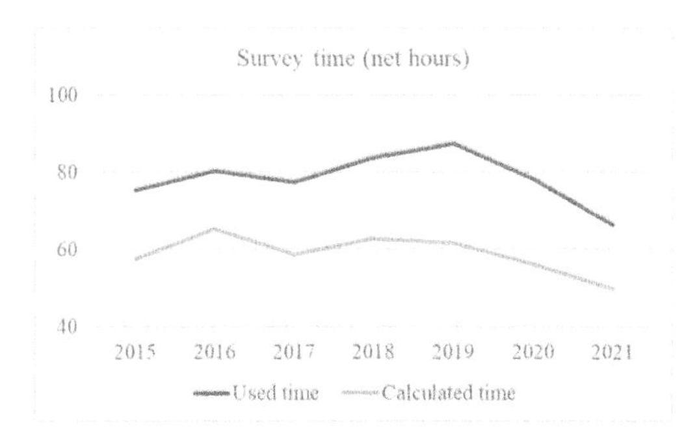


Figure 4. Net survey time in GRAHS surveys in 2015-2021 (black line) and calculated minimum survey times needed to complete only the trawl hauls (grey line).

## 4 DISCUSSION

Unmanned and autonomous vessels provide a cost-effective and safe alternative for research vessels. The research vessels are typically engaged in biological sampling for approximately 3-5 hours per day (of the 10-15 hours working day) and rest of the time is spent on covering the acoustic track and transit between hauling locations (based on log data from Estonian acoustic survey under the ICES GRAHS) (ICES, 2014). Some institutions have optimized the use of such vessels by implementing 24-hour work regime (Saari, 2015), which increases the daily survey coverage, but brings the need for additional crew. Still, large amount of the time, a large and expensive research vessel is used only to carry a small (typically less than 100kg) sonar. Any optimization in the use of research vessel would result in considerable economical saving.

Baltic herring and sprat assessments rely strongly on long time-series data, obtained on the basis of conventional methodology over the whole time period of surveys (ICES, 2014, 2021). Any alterations in survey methodology need to be thoroughly considered, and when necessary, possible conversion factors calculated. Unmanned vehicles are theoretically suitable for substituting parts of acoustic surveys (Griffiths, 2002; Fernandes et al., 2003; Verfuss et al., 2016, 2019), but their implementation and large-scale testing during surveys is still ongoing. Recent disruptions in regular surveys have led to accelerated experiments with only USV-conducted surveys (De Robertis et al., 2021). Such experiments provide us invaluable data of possibilities to minimise the use of expensive data acquisition methodologies.

Based on the GRAHS survey example in the current study, we argue that by using conventional research vessel for only biological sampling and acoustic measurements during transit, we could reduce the survey costs by 25%. Uncovered acoustic tracks could be covered simultaneously using USV or ASV with considerably lower operational costs. The saved costs could be allocated to USV to extend the survey area and collect more data. Therefore, even without modifying the current methodology, we could optimize the survey cost or increase the coverage to obtain more accurate information.

The main limitation of deploying unmanned vessels to acoustic-trawl surveys is the interpretation of sonar data. The methodology assumes that data from acoustic and biological sampling is obtained with minimal spatio-temporal variability (Simmonds & MacLennan, 2008; ICES, 2014). Unlinking the acoustic data acquisition from the biological sampling could increase the uncertainty of the data integration (De Robertis et al., 2021), since the acoustically observed fish schools are not biologically sampled at the same time/location. Meanwhile, the possible increase in survey coverage decreases the uncertainty in final estimates. By conducting the survey on several platforms simultaneously, USV could gather higher quality acoustic data while leaving more time to research vessel for biological sampling (Fernandes et al., 2003). Such possible trade-offs need extensive studies to find the optimal approach.

According to current methodology, the trawl hauls should be conducted in the areas with highest fish concentration (ICES, 2014). This approach has its limitations when using the same single vessel for acoustic measurements and trawling since the cruise leader has no information whether the acoustic signal will increase or decrease with propagation. Covering a large area with only sonar and then returning to highest fish school for trawling would be the most accurate approach, but have been seldom used due to increase in expensive survey time. USV could help overcome such limitation with providing real-time data while covering larger areas before the trawl vessel enters the area.

To obtain the most precise results, the time difference between acoustic measurements and trawling must be minimal. Real-time data transfer is one of the limitations in offshore areas. During acoustic surveys in GoR, approximately 5GB of data is recorded daily, which, in use of USV, should be transferred to acoustic expert before the trawling decisions can be made. Currently, the network coverage of mobile operators (4G) is limited in offshore areas (unpubl. data), which delays the data transfer. Use of expensive satellite data and delaying the transfer to inshore areas must be considered while using USV for offshore surveys.

The Gulf of Riga is a relatively small semi-enclosed part of Baltic Sea (Kotta et al., 2008), which would be suitable for testing new approaches to survey methodology. Estonian Marine Institute has started a program to design and test an autonomous survey vessel suitable for Baltic Sea acoustic surveys. Based on the current review, we suggest to use GRAHS survey as a pilot study to test the possibilities of implementing autonomous or unmanned vehicles in acoustic-trawl surveys. In the light of rapid technological developments, old survey methodologies should be constantly reviewed and tested to fully utilize the potential of optimizing the data acquisition.

## ACKNOWLEDGEMENTS

The study was partly supported by the European Maritime and Fisheries Fund (EMFF) through European Data Collection Framework.

## REFERENCES

Barange, M., Bernal, M., Cergole, MC., Cubillos, L., Daskalov, GM., and de Moor, CL. 2009. Current trends in the Assessment and Management of Small Pelagic Fish Stocks. In Climate change and small pelagic fish, pp. 191–255.

Barrera, C., Padron, I., Luis, F. S., and Llinas, O. 2021. Trends and challenges in unmanned surface vehicles (USV): From survey to shipping. TransNav: International Journal on Marine Navigation and Safety of Sea Transportation, 15.

Chu, D., Parker-Stetter, S., Hufnagle, L. C., Thomas, R., Getsiv-Clemons, J., Gauthier, S., and Stanley, C. 2019. 2018 Unmanned Surface Vehicle (Saildrone) acoustic survey off the west coasts of the United States and Canada. In OCEANS 2019 MTS/IEEE SEATTLE, pp. 1–7. IEEE.

Colet Maillo, J. 2021. Autonomous vessel surveying. Universitat Politècnica de Catalunya.

De Robertis, A., Levine, M., Lauffenburger, N., Honkalehto, T., Ianelli, J., Monnahan, C. C., Towler, R., et al. 2021. Uncrewed surface vehicle (USV) survey of walleye pollock, Gadus chalcogrammus, in response to the cancellation of ship-based surveys. ICES Journal of Marine Science, 78: 2797–2808. Oxford University Press.

Eriksen, C. C., Osse, T. J., Light, R. D., Wen, T., Lehman, T. W., Sabin, P. L., Ballard, J. W., et al. 2001. Seaglider: A long-range autonomous underwater vehicle for oceanographic research. IEEE Journal of oceanic Engineering, 26: 424–436. IEEE.

Fernandes, P. G., Stevenson, P., Brierley, A. S., Armstrong, F., and Simmonds, E. J. 2003. Autonomous underwater vehicles: future platforms for fisheries acoustics. ICES Journal of Marine Science, 60: 684–691. Oxford University Press.

Griffiths, G. 2002. Technology and applications of autonomous underwater vehicles. CRC Press.

ICES. 2014. Manual of International Baltic Acoustic Surveys (IBAS). Series of ICES Survey Protocols SISP 8 - IBAS. ICES.

ICES. 2021. Baltic Fisheries Assessment Working Group (WGBFAS). ICES Scientific Reports, 3:53. http://dx.doi.org/10.17895/ices.pub.8187.

Kotta, J., Lauringson, V., Martin, G., Simm, M., Kotta, I., Herkül, K., and Ojaveer, H. 2008. Gulf of Riga and Pärnu Bay. In Ecology of Baltic coastal waters, pp. 217–243. Springer.

Mordy, C. W., Cokelet, E. D., De Robertis, A., Jenkins, R., Kuhn, C. E., Lawrence-Slavas, N., Berchok, C. L., et al. 2017. Advances in ecosystem research: Saildrone surveys of oceanography, fish, and marine mammals in the Bering Sea. Oceanography, 30: 113–115. JSTOR.

MSC. 1/Circ. 1638. 2021. Outcome of the Regulatory Scoping Exercise for the Use of Maritime Autonomous Surface Ships (MASS). IMO London, UK.

Parker, J. 2021. The Challenges Posed by the Advent of Maritime Autonomous Surface Ships for International Maritime Law. Australian and New Zealand Maritime Law Journal, 35: 31–42.

Saari, T. 2015. Baltic international acoustic survey: MTA Arandan käyttö silakan ja kilohailin runsaustutkimuksissa. Turun ammattikorkeakoulu.

Simmonds, J., and MacLennan, D. N. 2008. Fisheries acoustics: theory and practice. John Wiley & Sons.

Swart, S., Zietsman, J. J., Coetzee, J. C., Goslett, D. G., Hoek, A., Needham, D., and Monteiro, P. M. 2016. Ocean robotics in support of fisheries research and management. African Journal of Marine Science, 38: 525–538. Taylor & Francis.

Verfuss, U. K., Aniceto, A. S., Biuw, M., Fielding, S., Gillespie, D., Harris, D., Jimenez, G., et al. 2016. Literature review: Understanding the current state of autonomous technologies to improve/expand observation and detection of marine species. SMRU Consulting.

Verfuss, U. K., Aniceto, A. S., Harris, D. V., Gillespie, D., Fielding, S., Jiménez, G., Johnston, P., et al. 2019. A review of unmanned vehicles for the detection and monitoring of marine fauna. Marine pollution bulletin, 140: 17–29. Elsevier.

Wynn, R. B., Huvenne, V. A., Le Bas, T. P., Murton, B. J., Connelly, D. P., Bett, B. J., Ruhl, H. A., et al. 2014. Autonomous Underwater Vehicles (AUVs): Their past, present and future contributions to the advancement of marine geoscience. Marine Geology, 352: 451–468. Elsevier.Berkeley, S. A., Hixon, M. A., Larson, R. J., and Love, M. S. 2004. Fisheries Sustainability via Protection of Age Structure and Spatial Distribution of Fish Populations. Fisheries, 29: 23–32.

# Author Index